Organic
Chemistry

Periodic Table of the Elements

Key:

atomic number
Symbol
name
standard atomic weight

1	2		13	14	15	16	17	18
1 **H** hydrogen [1.008]								2 **He** helium 4.003
3 **Li** lithium [6.941]	4 **Be** beryllium 9.012		5 **B** boron [10.81]	6 **C** carbon [12.01]	7 **N** nitrogen [14.01]	8 **O** oxygen [16.00]	9 **F** fluorine 19.00	10 **Ne** neon 20.18
11 **Na** sodium 22.99	12 **Mg** magnesium 24.31		13 **Al** aluminium 26.98	14 **Si** silicon [28.09]	15 **P** phosphorus 30.97	16 **S** sulfur [32.07]	17 **Cl** chlorine [35.45]	18 **Ar** argon 39.95

3	4	5	6	7	8	9	10	11	12
21 **Sc** scandium 44.96	22 **Ti** titanium 47.87	23 **V** vanadium 50.94	24 **Cr** chromium 52.00	25 **Mn** manganese 54.94	26 **Fe** iron 55.85	27 **Co** cobalt 58.93	28 **Ni** nickel 58.69	29 **Cu** copper 63.55	30 **Zn** zinc 65.38(2)
39 **Y** yttrium 88.91	40 **Zr** zirconium 91.22	41 **Nb** niobium 92.91	42 **Mo** molybdenum 95.96(2)	43 **Tc** technetium [99]	44 **Ru** ruthenium 101.1	45 **Rh** rhodium 102.9	46 **Pd** palladium 106.4	47 **Ag** silver 107.9	48 **Cd** cadmium 112.4
57-71 lanthanoids	72 **Hf** hafnium 178.5	73 **Ta** tantalum 180.9	74 **W** tungsten 183.8	75 **Re** rhenium 186.2	76 **Os** osmium 190.2	77 **Ir** iridium 192.2	78 **Pt** platinum 195.1	79 **Au** gold 197.0	80 **Hg** mercury 200.6
89-103 actinoids	104 **Rf** rutherfordium [267]	105 **Db** dubnium [268]	106 **Sg** seaborgium [271]	107 **Bh** bohrium [272]	108 **Hs** hassium [277]	109 **Mt** meitnerium [276]	110 **Ds** darmstadtium [281]	111 **Rg** roentgenium [280]	112 **Cn** copernicium [285]

Groups 1–2 (periods 4–7):
- 19 **K** potassium 39.10 / 20 **Ca** calcium 40.08
- 37 **Rb** rubidium 85.47 / 38 **Sr** strontium 87.62
- 55 **Cs** caesium 132.9 / 56 **Ba** barium 137.3
- 87 **Fr** francium [223] / 88 **Ra** radium [226]

Groups 13–18 (periods 4–7):

13	14	15	16	17	18
31 **Ga** gallium 69.72	32 **Ge** germanium 72.63	33 **As** arsenic 74.92	34 **Se** selenium 78.96(3)	35 **Br** bromine 79.90	36 **Kr** krypton 83.80
49 **In** indium 114.8	50 **Sn** tin 118.7	51 **Sb** antimony 121.8	52 **Te** tellurium 127.6	53 **I** iodine 126.9	54 **Xe** xenon 131.3
81 **Tl** thallium [204.4]	82 **Pb** lead 207.2	83 **Bi** bismuth 209.0	84 **Po** polonium [210]	85 **At** astatine [210]	86 **Rn** radon [222]
114 **Fl** flerovium	116 **Lv** livermorium				

Lanthanoids:

57 **La** lanthanum 138.9	58 **Ce** cerium 140.1	59 **Pr** praseodymium 140.9	60 **Nd** neodymium 144.2	61 **Pm** promethium [145]	62 **Sm** samarium 150.4	63 **Eu** europium 152.0	64 **Gd** gadolinium 157.3	65 **Tb** terbium 158.9	66 **Dy** dysprosium 162.5	67 **Ho** holmium 164.9	68 **Er** erbium 167.3	69 **Tm** thulium 168.9	70 **Yb** ytterbium 173.1	71 **Lu** lutetium 175.0

Actinoids:

89 **Ac** actinium [227]	90 **Th** thorium 232.0	91 **Pa** protactinium 231.0	92 **U** uranium 238.0	93 **Np** neptunium [237]	94 **Pu** plutonium [239]	95 **Am** americium [243]	96 **Cm** curium [247]	97 **Bk** berkelium [247]	98 **Cf** californium [252]	99 **Es** einsteinium [252]	100 **Fm** fermium [257]	101 **Md** mendelevium [258]	102 **No** nobelium [259]	103 **Lr** lawrencium [262]

Notes

The uncertainty of an atomic weight value is ±1 in the last digit unless a different value is given in parentheses. No values are given for elements which have no natural stable isotopes; for such elements, a mass number of a typical radioisotope is given in parentheses.

Organic Chemistry

a mechanistic approach

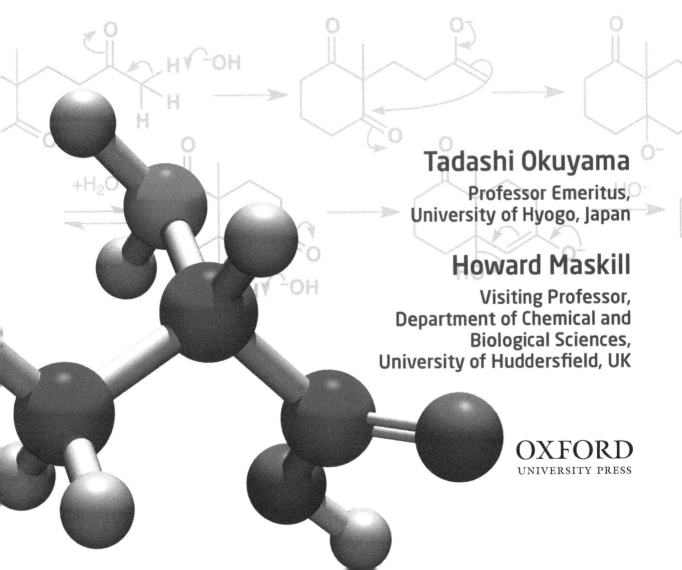

Tadashi Okuyama

Professor Emeritus,
University of Hyogo, Japan

Howard Maskill

Visiting Professor,
Department of Chemical and
Biological Sciences,
University of Huddersfield, UK

OXFORD
UNIVERSITY PRESS

UNIVERSITY PRESS

Great Clarendon Street, Oxford, OX2 6DP,
United Kingdom

Oxford University Press is a department of the University of Oxford.
It furthers the University's objective of excellence in research, scholarship,
and education by publishing worldwide. Oxford is a registered trade mark of
Oxford University Press in the UK and in certain other countries

© Tadashi Okuyama and Howard Maskill 2014

The moral rights of the authors have been asserted

Japanese version published by Maruzen Publishing Co., Ltd., Japan
© Tadashi Okuyama 2008.

British Library Cataloguing in Publication Data
Data available

ISBN 978–0–19–969327–6

Printed in Great Britain by
Ashford Colour Press Ltd

Foreword

Organic chemistry deals with the structures, synthesis, and functions of compounds whose molecules now include assemblies up to giant biomolecules such as nucleic acids, proteins, and polysaccharides. Because so many life processes are regulated by interactions between small organic molecules and gigantic biomolecules, James Watson, the 1962 Nobel Laureate in Physiology and Medicine, was able to say 'Life is simply a matter of chemistry'. It follows that organic chemistry applied to biological science is the basis of life science. Current organic chemistry is also central to burgeoning new areas of materials science whose applications extend to industrial products which support our daily lives. And, just as organic chemistry continues to develop, the way in which it is taught must adapt and, especially, use all the aids presently available to support the learning process.

By developing an appreciation of how organic reactions take place based on orbital interactions and electron flow, this book allows known reactions to be understood and new ones to be predicted. The book is organized in a manner which will facilitate the transition from high school chemistry to university level organic chemistry, and provides insights into some currently developing areas. In particular, the authors clearly present underlying principles and show how these bring order and logic to the subject.

Ryoji Noyori, 2013

Preface

Organic chemistry is a mature branch of science which continues to expand in the sense that new reactions and new compounds continue to be discovered. Some compounds newly isolated from natural sources support life; others, synthesized in the laboratory, are unknown in nature but have led to advances in medicine and other areas of science and technology. A consequence of the huge and increasing number of known organic compounds is that any chemist can have book-knowledge of only a tiny fraction and practical experience of an even smaller number. However, a molecule of an organic compound may generally be seen as a functional group bonded to a hydrocarbon residue and organic chemistry is essentially the chemistry of a relatively small number of functional groups. Consequently, comprehension of organic chemistry as a whole is achievable from knowledge of the characteristic reactions of functional groups and an understanding of how they occur, i.e. their mechanisms.

The Approach of this Book

There are different approaches to the teaching of organic chemistry at university level. In this book, we begin with a review of atomic and molecular structure and then look at factors which determine the shapes of molecules. Next, we cover acid–base (proton transfer) reactions since these are distinctive features of many reactions of organic compounds, especially ones of biological importance including reactions catalysed by enzymes. We then show that all overall reactions of organic compounds belong to one of a relatively small number of classes of reaction types. Moreover, when we introduce the concept of mechanism in organic chemistry, and look at how reactions take place, we see that only a small number of types of elementary steps are involved.

When features common to all organic reactions have been covered, we proceed to look at reactions of individual functional groups. Our approach, based upon a survey of teachers of organic chemistry in over 50 colleges and universities in Japan and guided by nine reviewers from different parts of Europe and North America, is to focus on underlying mechanistic principles as the unifying basis of organic chemistry. The outcome is a concise non-mathematical text which introduces molecular orbital considerations early on and uses 'curly arrows' (as appropriate) to describe mechanisms throughout. The book is not intended to be an encyclopaedic reference text of organic chemistry; it is a learning-and-teaching text and the coverage broadly corresponds to the organic chemistry syllabus of a typical honours degree in chemistry at a British university. However, we include connections to biological sciences wherever they are relevant to emphasize that organic chemistry is the basis of life science. To supplement the core chemistry, we have also included 'panels' containing material (sometimes topical) which relate the chemistry to current everyday life and biological phenomena. Consequently, depending on the level to which the subject is to be taught, the book could be appropriate for students of health sciences and technology, as well as premedical students.

Learning from this Book

To assist students, worked examples and exercises are embedded within each chapter; answers to in-chapter exercises are provided on the book's web site, which we describe further below. Each chapter also has a summary together with additional problems at the end. In addition, we include an early section on organic nomenclature, appendices which contain reference data, and flow charts encapsulating reactions and interconversions of functional groups, and a comprehensive index.

Online Support

Organic Chemistry: a mechanistic approach is accompanied by a website that features study and teaching aids.

For students:

- Answers to in-chapter exercises
- 3D-rotatable models of numerous compounds featured in the book
- Multiple-choice questions for each chapter to help you check your understanding of topics you have learned

For lecturers:

- Figures from the book in electronic format
- Answers to end-of-chapter problems
- Examples of organic synthesis reactions, related to topics covered in the book, for use in teaching
- Additional problems (with answers), to supplement those included in the book

 To find out more, go to **www.oxfordtextbooks.co.uk/orc/okuyama/**.

You can also explore organic reaction mechanisms at **www.chemtube3d.com**. This site provides a wide range of interactive 3D animations of some of the most important organic reactions you are likely to encounter during your studies.

Acknowledgements

This book is based on the Japanese text, *Organic Chemistry* (Maruzen Publishing Co., Ltd., Tokyo, 2008) by a group of authors including one of us. We are very grateful to the other coauthors of that book, especially Professors Mao Minoura and Hiroshi Yamataka (Rikkyo University), Akihiko Ishii (Saitama University), and Takashi Sugimura (University of Hyogo), for their help during our work on this book. We are also grateful to Dr Ryohei Kishi (Osaka University) for his assistance in the preparation of some of the molecular orbital diagrams, and to the editorial staff at OUP, especially Jonathan Crowe. In spite of all the help we have received, there will be residual errors in a book of this length; we welcome assistance in rooting out mistakes of any sort and will post corrections on the above mentioned website. Finally, we acknowledge with appreciation that this book could not have been completed without the forbearance and support of our wives.

A Note to Students

Some students occasionally find organic chemistry a formidable subject involving the memorization of an overwhelming number of compounds and their reactions. However, as we mention in the preface, organic compounds fall into a small number of classes characterized by the functional groups at which reactions take place; similarly, there is only a limited number of reaction types classified according to their mechanisms. Consequently, systematic learning of relatively few mechanisms brings order and logic to organic chemistry, and will allow you to appreciate the subject in all its glorious and fascinating diversity. This text, *Organic Chemistry: a mechanistic approach*, has been written to guide you along this path.

An organic chemical reaction—the transformation of one compound into another—is described in terms of the structures of compounds involved, and the *reactivity* of a compound (how it reacts and whether the reaction will be fast or slow) is determined by its structure (and the reaction conditions). How a reaction is believed to occur, i.e. its reaction mechanism, is nowadays represented by curly arrows describing the movement of electrons, and we use mechanistic schemes throughout this book. Usually, the schemes will show not just *how* the reaction occurs but *why* it occurs in the way shown, and why it is favourable. Our pictorial reaction schemes with structures of compounds and curly arrows showing how they react contain a lot of information. We have used several devices to assist their interpretation, including colour and annotations.

The following two schemes taken from the text illustrate some conventions in this book to describe reaction mechanisms. Some boxes contain text to indicate what facilitates a particular step, i.e. why it is favourable, and bonds newly formed in each step; text in other boxes identifies *types* of groups, e.g. nucleophile or electrophile. Coloured text under reaction arrows identifies the type of reaction which may be a single step (e.g. proton transfer) or an overall transformation (e.g. substitution). Text under a chemical species indicates its nature, e.g. an intermediate. Note that all steps in these two schemes are reversible in principle but, by including one arrow in the final step of the second scheme in parentheses, for example, we identify a step as being essentially unidirectional because of the reaction conditions and/ or the equilibrium constant.

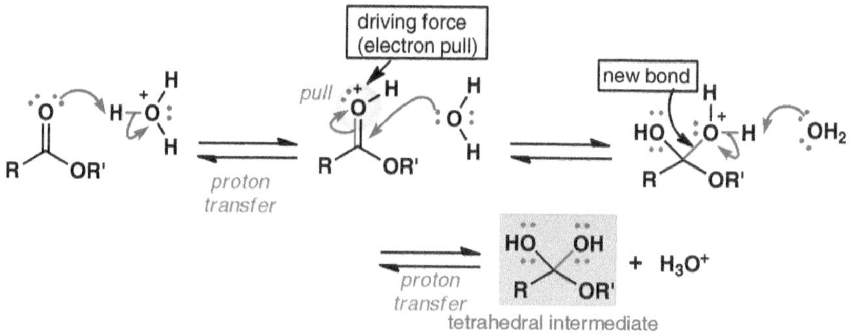

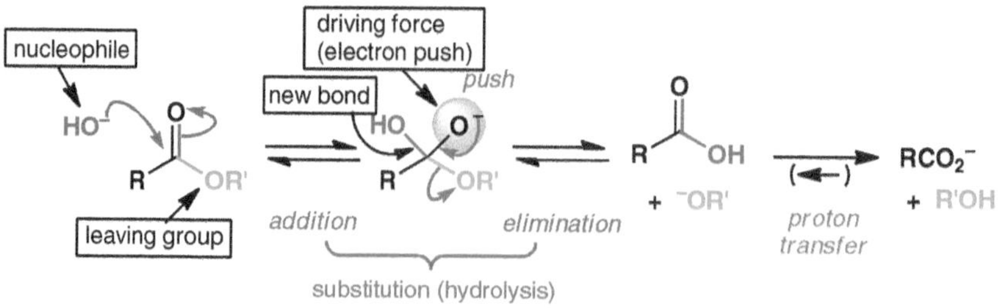

It is important that you can draw clearly in two dimensions organic structures which are generally three-dimensional. To do this, practice with pencil and paper is essential. In addition, you have to learn to use curly arrows to describe the movement of electrons corresponding to a reaction, i.e. bond breaking and bond making steps. Remember that organic chemistry can be communicated by drawing structures of molecules and curly arrow reaction mechanisms—it is as though we have a language with structures and mechanisms as the vocabulary and grammar; and, as with learning a language, fluency develops with practice.

Worked examples are embedded in the text to review what has just been covered and illustrate how to solve exercises and problems within and at the ends of chapters, respectively. In later chapters, we also have 'supplementary problems' which are a little more difficult and may relate to material in previous chapters. It will be most beneficial if you attempt exercises and problems without looking at the solutions first, even though they are available on the website associated with the book. If you find that you cannot do an exercise or problem, go back to the text to review the material upon which the exercise or problem is based, then try again. This iterative process is an important aspect of learning organic chemistry and will help you to learn how to solve problems generally (rather than just memorize facts). When you arrive at a reasonable answer, check it against the solution provided. However, note that there may be different ways of approaching some problems (and some may have more than a single correct answer); but when you are really stuck, always seek advice.

One final point: the names of chemists crop up from time to time throughout the book; they are usually eminent chemists who have made significant contributions to organic chemistry (which is, after all, an area of human endeavour) and their portraits are shown. Sometimes, reactions have been named after them. Although the use of chemists' names is a long-standing and often helpful short-hand way of referring to reactions and well-established empirical rules or general principles, knowing and understanding the chemistry involved is more important than remembering the names.

Abbreviations

Entry	Full name	Entry	Full name
Ac	acetyl	*i*-Pr	isopropyl
ADP	adenosine diphosphate	IR	infrared
AIBN	azobisisobutyronitrile	IUPAC	International Union of Pure and Applied Chemistry
AO	atomic orbital		
Ar	aryl	LDA	lithium diisopropylamide
ATP	adenosine triphosphate	LUMO	lowest unoccupied molecular orbital
BHA	butylated hydroxyanisole	MCPBA	*m*-chloroperoxybenzoic acid
BHT	butylated hydroxytoluene	Me	methyl
BINAP	2,2'-bis(diphenylphosphino)-1,1'-binaphthyl	MO	molecular orbital
		mp	melting point
Bn	benzyl	MS	mass spectrometry
Boc	*t*-butoxycarbonyl	NAD$^+$	nicotinamide adenine dinucleotide
bp	boiling point	NADH	reduced form of NAD
BPO	dibenzoyl peroxide	NBS	*N*-bromosuccinimide
Bu	butyl	n.g.p.	neighbouring group participation
Bz	benzoyl	NMF	*N*-methylformamide
Cbz	benzyloxycarbonyl	NMR	nuclear magnetic resonance
CIP	Cahn–Ingold–Prelog	NOE	nuclear Overhauser effect
CoA, CoASH	coenzyme A	O.P.	optical purity
DBN	1,5-diazabicyclo[4.3.0]non-5-ene	PCC	pyridinium chlorochromate
DBU	1,8-diazabicyclo[5.4.0]undec-7-ene	PET	poly(ethylene terephthalate)
DCC	*N,N*-dicyclohexylcarbodiimide	PG	prostaglandin
DEAD	diethyl azodicarboxylate	Ph	phenyl
DMAP	4-dimethylaminopyridine	Pr	propyl
DMF	*N,N*-dimethylformamide	PTC	phase transfer catalysis
DMSO	dimethyl sulfoxide	py	pyridine
DNA	deoxyribonucleic acid	rf	radio frequency
DOMO	doubly occupied molecular orbital	RNA	ribonucleic acid
E.A.	electron affinity	SAM	*S*-adenosylmethionine
E1	unimolecular elimination	SET	single electron transfer
E1cB	unimolecular elimination via conjugate base	S_N1	unimolecular nucleophilic substitution
E2	bimolecular elimination	S_N2	bimolecular nucleophilic substitution
EDG	electron-donating group	S_Ni	nucleophilic substitution, internal
ee	enantiomeric excess	SOMO	singly occupied molecular orbital
EPM	electrostatic potential map	THF	tetrahydrofuran
Et	ethyl	THP	tetrahydropyranyl
EWG	electron-withdrawing group	TBS	*t*-butyldimethylsilyl
FGI	functional group interconversion	*t*-Bu	*t*-butyl
Fmoc	fluorenylmethoxycarbonyl	TMS	tetramethylsilane
GC	gas chromatography	Tr	triphenylmethyl (trityl)
HOMO	highest occupied molecular orbital	TS	transition state, transition structure
HPLC	high performance liquid chromatography	Ts	*p*-toluenesulfonyl (tosyl)
		UMO	unoccupied molecular orbital
I.E.	ionization energy, ionization potential	UV	ultraviolet
		VSEPR	valence shell electron pair repulsion

Overview of Contents

Contents in Detail

Prologue
The History and Scope of Organic Chemistry

Chemistry of Organic Compounds

Compounds produced by living organisms

Indigo plant (*Polygonum tinctorium*).
Nancy Nehring/istockphoto

In the late eighteenth century, two broad kinds of substances began to be investigated systematically in what we now recognize as chemical science—organic and inorganic compounds. The early study of substances produced by living organisms became organic chemistry—essentially the chemistry of the compounds of carbon. Our bodies are composed of organic compounds, and numerous others are all around us, and always have been. Foods are organic and, until relatively recently, clothing was wholly derived from living sources (animal or vegetable). In contrast, the study of substances obtained from mineral sources in the first days of chemical science developed into what we now call inorganic chemistry and, at an early stage, the properties of organic and inorganic compounds were recognized as being characteristically different. Organic compounds were difficult for chemists to work with—they are often low-melting solids or liquids, their purification is difficult, they decompose readily, and they are generally combustible. In contrast, inorganic compounds are normally very stable, generally crystalline, and not usually combustible. The difference between organic and inorganic compounds was originally ascribed to an unexplainable, almost mystical, 'vital force' which was considered to be inherent in the compounds obtained from living sources, and this concept of 'vitalism' survived almost until the middle of the nineteenth century.

Pigments

In ancient times, clothing was made from fibres obtained from plants or animals and was dyed with pigments extracted from plants, e.g. blue indigo and red alizarin. Tyrian purple is the colour and the name of the prestigious dye used for liturgical vestments and the robes of royalty; it was produced from sea snails in eastern Mediterranean countries in extremely low yield (hence its costliness).

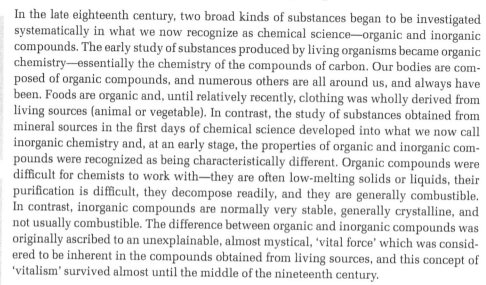

indigo (blue, from plants in the genus *Indigofera*)

alizarin (red,from the roots of madder)

the main ingradient of Tyrianpurple (from *Bolinus brandaris*, a type of sea snail)

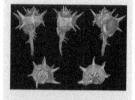

Medicines and toxins

Some medicines originally discovered in natural products long ago are still used today. Cinchona bark was used as a folk medicine by natives of South America to relieve pain and reduce fevers, and was brought to Europe by Jesuits in the sixteenth century. The main ingredient was later found to be quinine and is used to treat malaria (and as a bitter-tasting ingredient of some cocktails, e.g. gin and tonic).

Aspirin is an example of a compound developed from the use of plants for medical care. In ancient Greece, the bark of willow trees was used to relieve pain, and the active ingredient was later found to be salicin, which is a glucose derivative of salicyl alcohol. Testing modifications of this compound led chemists working for Bayer AG in Germany to acetylsalicylic acid in 1897. This first synthetic drug was marketed in 1899 as aspirin and is still commonly used as a pain-relieving, fever-reducing, and anti-inflammatory agent; it is also used in the treatment and prevention of heart attack, stroke, and blood clot formation owing to its anti-platelet forming properties.

quinine
(from cinchona)

salicin (from willow bark)

acetylsalicylic acid
(aspirin)

Chinchona tree.
(Courtesy of Forest and Kim Starr. This file is licensed under the Creative Commons Attribution-Share Alike 3.0 Unported license.)

Poisonous substances from plants are also known: coniine is found in the extract from hemlock (*Conium maculatum*), which was used to poison Socrates in 399 BC.

In suitably reduced amounts, some poisonous substances act as medical drugs. Digoxin, for example, may be extracted from foxgloves (*Digitalis*) and used for the treatment of various heart conditions.

Willow tree.
(Joe McDaniel/istockphoto)

coniine (from hemlock)

digoxin (from foxglove)

Hemlock.
(This file is licensed under the Creative Commons Attribution-Share Alike 3.0 Unported license.)

Antibiotics

Antibiotics are a relatively new type of drug produced by microorganisms. In 1928, the Scottish biologist Alexander Fleming working in London discovered a substance produced by a mould which killed staphylococcus bacteria. He named the substance *penicillin* and it is used for the treatment of bacterial infections; the term is now used generically for structurally related compounds. Streptomycin is another type of antibiotic which was the first effective drug for tuberculosis (TB). This was originally isolated in 1943 in the USA in the laboratory of S. Waksman who coined the term *antibiotics* (substances produced by microorganisms which kill other microorganisms, principally bacteria). Further antibiotics have subsequently been discovered and are widely used as antibacterial drugs.

Foxglove.
(Photograph by Varda-Elentari.)

Friedrich Wöhler (1800–1882).

Wöhler studied chemistry under the Swedish chemist, Jakob Berzelius in Stockholm, and taught chemistry in Berlin and later in Göttingen. He also contributed to chemistry by isolating a number of elements, including aluminium, beryllium, and titanium.

penicillin

streptomycin

A.W. Hermann Kolbe (1818–1884), then in Leipzig. He is known principally for his work on the electrolysis of salts of carboxylic acids (Kolbe electrolysis, Chapter 20) and the synthesis of salicylic acid (Kolbe–Schmitt reaction).

The Development of Organic Chemistry as a Science

The end of 'vitalism'

As mentioned above, chemists at the beginning of the nineteenth century believed that organic compounds found in nature could not be prepared in the laboratory—they could only be formed by living organisms. However, in 1828, Friedrich Wöhler discovered that urea, an organic compound found in urine, is formed when an aqueous solution of an inorganic salt, ammonium cyanate, was evaporated to dryness.

$$NH_4^+ \ NCO^- \xrightarrow{\text{heat}} H_2N-\overset{\overset{\displaystyle O}{\|}}{C}-NH_2$$

ammonium cyanate   urea

Subsequently in 1844, Hermann Kolbe showed that ethanoic (acetic) acid could be prepared from carbon disulfide, which was known to be obtainable from iron pyrites (FeS_2) and graphite (a form of carbon).

$$FeS_2 \ + \ C \longrightarrow CS_2 \Longrightarrow CH_3CO_2H$$

carbon disulfide   ethanoic acid (acetic acid)

These findings by Wöhler and Kolbe established that the concept of 'vitalism' in chemistry was no longer credible—organic compounds could be prepared in the laboratory—but its demise was not instantaneous and it lingered on in other areas of science. However, the development of modern organic chemistry had begun.

Concept of 'radical' and the school of organic chemistry in Giessen

It was recognized by 1830 that there were different compounds with the same composition; we know these as *isomers* and Wöhler's ammonium cyanate and urea are examples. This implied that the atoms of a pair of organic isomers were connected together differently. A collaboration in the 1830s between Wöhler and another important chemist of the time, Justus von Liebig, led to the concept of 'radicals'—groups of atoms joined together which occurred in different organic compounds. These 'radicals' were forerunners of what we now call *groups* (e.g. the ethyl group) in organic chemistry; it was thought that complex organic molecules were composed of radicals in the way that simple molecules are composed of atoms.

At the time, Liebig had been a Professor at the University of Giessen since 1824. He had improved analytical and experimental methods of organic chemistry in the 1820s and 1830s and built up a teaching and research school. His innovative methods of chemistry teaching and research in Giessen laid the foundation for the further development of organic chemistry, and provided a model for academic chemistry elsewhere. His school attracted many students who later became renowned in the field of organic chemistry: they include Hofmann, Kekulé, and Williamson, who are mentioned elsewhere in this book.

Justus von Liebig (1803–1873)

Reproduced from Duyckinick, Evert A. *Portrait Gallery of Eminent Men and Women in Europe and America.* New York: Johnson, Wilson & Company, 1873

Early modern history of organic chemistry

The concept of valence developed following the mid-decades of the nineteenth century, and August Kekulé and Archibald Couper independently proposed the tetravalency of carbon atoms in 1858. Kekulé also pointed out the possibility of cyclic structures of carbon compounds and proposed a six-membered ring structure for benzene in 1865 (Panel 5.1, p. 100). At that stage, only two-dimensional molecular structures were considered.

In 1874, Jacobus van't Hoff and Joseph Le Bel independently proposed tetrahedral bonding of carbon atoms (Chapter 2) to account for the optical activity of some carbon compounds and the enantiomers of tartaric acid which had been discovered by Louis Pasteur (Panel 11.4, p. 242) in 1848. The tetrahedral model of carbon bonding was supported by the carbohydrate studies of Emil Fischer (Panel 11.3, p. 238), and led towards an appreciation of three-dimensional (stereochemical) structures of organic compounds.

By the late nineteenth century, a comprehensive approach to organic chemistry based on structures of compounds had been established, and new organic reactions were increasingly being discovered; some of those reactions are still known by the names of their discoverers and examples will be discussed in this book. During this time and in the early years of the twentieth century, the accumulating knowledge of structures and reactions of organic compounds were being systematically organized to establish what we might call 'classical' organic chemistry.

Modern concepts and theories of chemical bonding and organic reactions

Following developments in physics on atomic structure at the beginning of the twentieth century, Gilbert Lewis (Chapter 1) and Irving Langmuir in the United States proposed the octet rule and the concept of covalent bonding in 1919. Quantum mechanics was established in the 1920s and provided the basis of modern theories of chemical bonding.

In the 1930s, Linus Pauling introduced concepts such as electronegativity, hybridization, and resonance which proved very useful in organic chemistry (Panel 2.1, p. 28) as will be discussed in Chapters 1 and 2. About this time, notions of how organic reactions occurred (mechanisms of organic reactions) were being developed based on contemporary electronic theories of chemical bonding. R. Robinson and C.K. Ingold, for example, described organic reactions in terms of the breaking and forming of chemical bonds involving the movement of electrons (Chapters 7 and 12). Although their representation of reactions was qualitative, it has developed into a simple and useful pencil-and-paper method of describing organic chemical reactions, and organic reactivity.

Quantum mechanics was applied to chemical bonding theory following its introduction in mathematical physics, but the application to organic compounds came later because of the huge computations which were required. In 1931, Erich Hückel introduced some simplifying approximations to molecular orbital theory which enabled the computation of energies of π electron systems of organic molecules (Chapter 5). These descriptions of ground and excited states of organic *molecules* were subsequently extended into the area of organic *reactions* when, in 1952, Kenichi Fukui proposed frontier orbital theory to describe organic reactivity (Chapter 7). Owing to developments in high-speed computers and ever-improving computational methodologies, we can now reliably analyse a wide range of chemical phenomena based, ultimately, on quantum mechanics.

William Henry Perkin (1838–1907).

Perkin discovered and started to manufacture aniline purple when he was only 18. His business was so successful that he was able to retire as a wealthy man at the age of 36. After that, he devoted the rest of his life to organic chemistry investigations in his private laboratory. The Perkin reaction for the preparation of cinnamic acid is named after him.

Chemical industry

Organic chemistry has always been applied by industry to improve the quality of life. The first major chemical industries developed for the production of synthetic dyestuffs in the UK and Germany. These followed the serendipitous discovery in 1856 of a dye in a reaction of impure aniline by the British chemist, W.H. Perkin; it gave a delicate shade of purple and Perkin marketed it as *aniline purple* or *mauve* (later, it was also called *mauveine*).

A component of mauve (a synthetic purple dye)

In the twentieth century, chemical industries developed rapidly using coal and later oil as starting materials. The American chemist, Wallace Carothers prepared nylon, the first synthetic fibre, by a polymerization reaction (Chapter 9) in 1935. Since then, various other polymerizations (Chapters 9 and 20) have led to other synthetic fibres and plastics which contribute enormously to the convenience and comfort of modern life.

Wallace H. Carothers (1896–1937).

Carothers worked for Du Pont and also contributed to the development of neoprene, a synthetic rubber.

Synthetic complexity as a measure of progress in organic chemistry

The increasing complexity of compounds synthesized can be seen in the work of one of the most celebrated synthetic chemists of the twentieth century, the American, R.B. Woodward. His syntheses of natural products including quinine (1944, see above), cholesterol (1952), chlorophyll (1960), and (jointly with the Swiss, A. Eschenmoser) vitamin B_{12} (1972) inspired many by their originality.

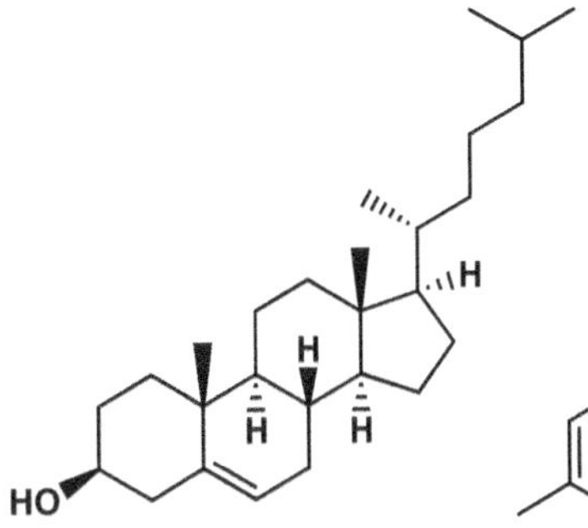

cholesterol

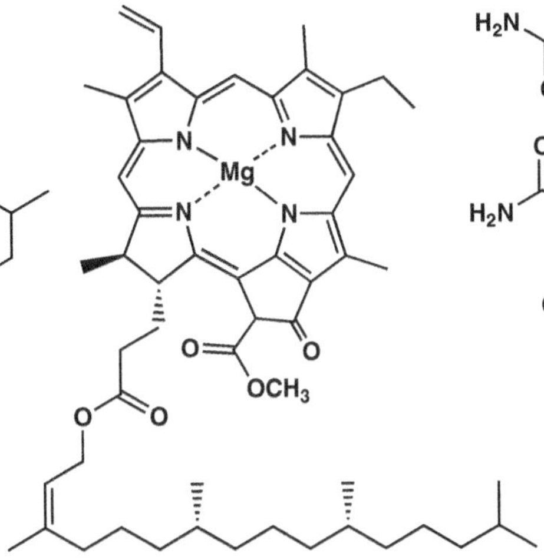

chlorophyll a

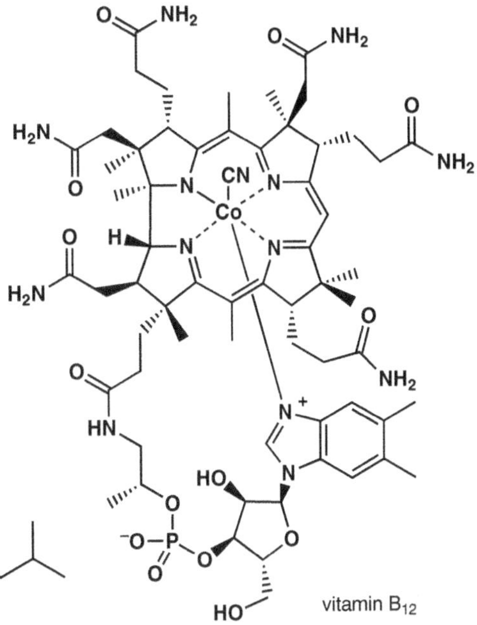

vitamin B$_{12}$

The increasing stereochemical complexity (as well as the size of molecules) has been a feature of compounds synthesized in the twenty-first century. Perhaps one of the most difficult so far is ciguatoxin (M. Hirama, 2001) with a molecular formula $C_{60}H_{86}O_{19}$ and 33 chirality centres, which is found in fish in tropical waters and causes a type of fatal food poisoning called ciguatera.

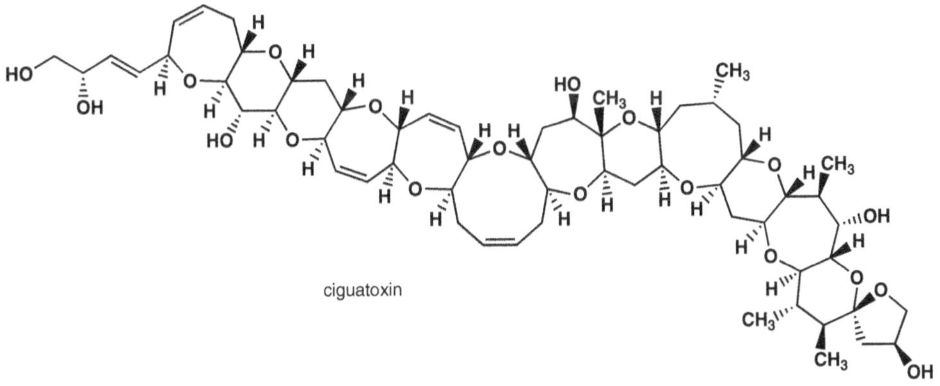

ciguatoxin

Robert Burns Woodward (1917–1979).

Woodward was a professor at Harvard University and was awarded the 1965 Nobel Prize in Chemistry for his outstanding achievements in organic synthesis. In addition to his achievements in synthesis, his collaboration with the theoretician, R. Hoffmann, led to the widespread appreciation of the importance of orbital symmetry considerations in concerted reactions (Chapter 21).

Organic Chemistry: Now and in the Future

Organic chemistry can trace its scientific origins back to the enlightenment in Europe, so is a relatively mature science, and continuous progress has led to developments which now support our daily lives in countless ways. Synthetic fibres and dyes for clothing, agricultural chemicals and fertilizers, plastics, paint, and adhesives, for example, are all industrial products based on conventional organic chemistry. Following the exploitation of the properties of liquid crystals for displays, for example, at the interface with physics, we are currently witnessing major developments in the formulation of further new organic materials (e.g. semiconductors, photovoltaic compounds, compounds which show electroluminescence, and single molecules which function as electronic components). Ongoing progress in the prevention and cure of diseases benefits from the contributions of organic chemistry to life sciences (e.g. the investigation of the molecular basis of diseases and the ongoing discovery of new pharmaceutical compounds).

Although the organic chemical industry unquestionably produces compounds and materials now deemed essential for civilized life, progress has occasionally been accompanied by adverse side-effects. By-products of unregulated industrial chemistry and irresponsible disposal of chemical wastes pollute our environment, and unanticipated medicinal hazards occasionally compromise well-intentioned use of pharmaceutical compounds. Some of these harmful aspects of industrial and medicinal applications of organic chemistry are because of our incomplete understanding of details of the chemistry involved, and addressing such matters is an ongoing task for chemists.

A more environmentally benign chemical industry, increased use of renewable energy, and more efficient use of carbon resources (including better recycling of materials) are major issues in the twenty-first century in which organic chemists will surely be engaged. Impossible to predict, however, are the applications of compounds yet to be discovered in the curiosity-led investigations of organic chemists.

Atoms, Molecules, and Chemical Bonding—a Review

1

Topics we shall cover in this chapter:

- Atomic structure
- Valence electrons
- Lewis representations of atoms
- Ionic and covalent bonds
- Electronegativity and bond polarity
- Lewis structures of simple molecules and ions
- Introduction to resonance

Organic chemistry deals with the compounds of carbon and is a major part of the wider subject. There are hugely more compounds of carbon than of all other elements combined. Why is this so? The answer must lie in the special properties of the element, and the characteristics of the carbon atom. Carbon is in the middle of the second period of the periodic table of elements; its atoms form strong bonds both to other carbon atoms and to atoms of other elements. As a result, more than ten million carbon compounds are known, and more remain to be discovered. Properties of chemical bonds between atoms within molecules, of individual molecules themselves, and of organic compounds and materials which we encounter as bulk material, are all ultimately dependent on the electronic structures of the atoms involved. We begin our study of organic chemistry by reviewing the nature of atoms and, in particular, their electronic structures. This will lead on naturally to a basic description of chemical bonds in simple molecules which, in subsequent chapters, we shall be able to develop to a level sufficient to account for the wide-ranging reactivities of organic compounds.

1.1 The Electronic Structure of Atoms

1.1.1 Atomic structure

At the beginning of any study of chemistry, we learn that compounds are built up from atoms, that a single atom consists of a nucleus and surrounding electrons, and that the nucleus consists of protons and neutrons.

An element is uniquely identified by its **atomic number** (Z), which is the number of protons in the nucleus (the magnitude of its positive charge) and equal to the number of electrons around the nucleus of a (neutral) atom of the element. If it is needed, Z is given as a lower prefix to the chemical symbol of an element; see Figure 1.1. As indicated

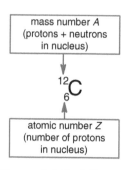

mass number A
(protons + neutrons
in nucleus)

$^{12}_{6}\text{C}$

atomic number Z
(number of protons
in nucleus)

Figure 1.1 Symbolic
representation of an element.

The *relative atomic mass* of
an element is sometimes
called its **atomic weight.**

Some elements, including
fluorine (^{19}F) and sodium
(^{23}Na), occur naturally as
single stable isotopes.

above, the nucleus of an atom of a specified element has neutrons in addition to protons, and the sum of the number of protons and neutrons in the nucleus is the **mass number** (A); this is indicated as an upper prefix to the chemical symbol of the element if it is needed.

The forms of an element with different numbers of neutrons in the nucleus are called **isotopes,** and they are chemically equivalent; they have the same value for Z but different values for A. Isotopes exist in proportions (natural abundances) which vary only slightly according to the distribution of the element in nature.

Masses of atoms are exceedingly small and commonly expressed in atomic mass units (1 amu $\approx 1.66 \times 10^{-27}$ kg). More conveniently, however, an atomic mass is usually expressed as its **relative atomic mass** (A_r), the standard being the mass of one atom of a specific isotope of a specific element; the modern standard is the ^{12}C isotope of carbon whose mass is defined as 12.0000 amu. Normally, however, we are not dealing with isotopically pure elements, but with the mixtures which occur in nature. Consequently, the relative atomic mass of an element (as opposed to that of just one of its isotopes) is the *weighted mean of the values of A_r of the naturally occurring isotopes*. For ^{13}C, the less common stable isotope of carbon whose natural abundance is about 1.11%, $A_r = 13.0034$, and the value for the element carbon is 12.011.

Exercise 1.1

How many protons and neutrons do nuclei of the following atoms have?

(a) $^{11}_{5}\text{B}$ (b) $^{23}_{11}\text{Na}$ (c) $^{14}_{7}\text{N}$ (d) $^{19}_{9}\text{F}$

Exercise 1.2

How many protons and neutrons does the nucleus of an atom of each of the following isotopes of carbon have: $^{12}_{6}\text{C}$, $^{13}_{6}\text{C}$, and $^{14}_{6}\text{C}$? ($^{14}_{6}\text{C}$ is a radioactive isotope, a radioisotope, of carbon used for carbon dating: see Panel 1.1.)

1.1.2 Electrons and atomic orbitals

According to quantum theory, the energy of an electron outside the nucleus of an atom cannot be continuously variable—it is *quantized*—and only certain energy levels, which are called **atomic orbitals** (AOs), are available to the electron. In addition to being an energy level, an AO has spatial character which is identified by letters of the Roman alphabet, s, p, d, and f (an s orbital, for example, is spherical). In other words, an AO restricts the space available to an electron in an atom in addition to limiting its energy.

An electron is characterized by *spin* as well as by its energy and spatial properties. This is a property which originates in quantum theory and can have only one of two possible values. It does not matter whether we call these values plus and minus, left and right, or up and down (we cannot attach a simple physical significance to *spin*). Any AO can accommodate a single electron of either spin, or two electrons if they are of opposite spin (when they are said to be *spin-paired*).

Atomic orbitals available to the electrons around the nucleus of an atom are grouped into *shells* of increasing energy according to their **principal quantum number**, n (1, 2, 3, ...); n also determines the types and number of orbitals within the shell. The shell of lowest energy with $n=1$ has only a single s orbital, labelled 1s, so it can contain only two electrons. The next shell ($n=2$) also contains an s orbital (labelled 2s) and, in addition, three p orbitals ($2p_x$, $2p_y$, and $2p_z$); these three are *degenerate*—they are of the

Atomic orbitals involved in
shells of $n = 1$–4:

Shell 1: 1s

2: 2s, 2p

3: 3s, 3p, 3d

4: 4s, 4p, 4d, 4f

Panel 1.1 Radiocarbon dating

The radioisotope ^{14}C is produced in the upper layers of the atmosphere by the nuclear reaction of thermal neutrons (produced by cosmic rays) with nitrogen ^{14}N.

$$^{14}N + {}^1n \text{ (neutron)} \rightarrow {}^{14}C + {}^1H$$

It then reacts rapidly with oxygen to form radioactive carbon dioxide which becomes distributed throughout the atmosphere mixed with ^{12}C carbon dioxide.

The ^{14}C radioisotope undergoes decay by emission of an electron to give the stable ^{14}N isotope of nitrogen with a half-life of about 5730 years (one half of the ^{14}C decays every 5730 years).

$$^{14}C \rightarrow {}^{14}N + e^-$$

The balance between its formation and decay leads to a stationary state natural abundance of ^{14}C in atmospheric carbon dioxide of about one part in one trillion (~1 : 10^{12}). Atmospheric carbon dioxide is absorbed in plants by photosynthesis (this process is called the *fixation* of CO_2) and the carbon is transferred to animals which consume plants as food. Consequently, as long as CO_2 from the atmosphere is being incorporated, the ^{14}C/^{12}C ratio within a living system will remain constant. Once the fixation stops, however, and the radioactive decay of ^{14}C continues, the ^{14}C/^{12}C ratio in the fixed carbon decreases with time. As we know the half-life of ^{14}C, analysis of the radioactivity of organic materials of bioorganic origin enables us to estimate the time since the carbon dioxide was fixed. This technique is called *radiocarbon dating*, or simply carbon dating, and was developed in 1949 by Willard Libby (University of Chicago) who was awarded the Nobel Prize in Chemistry in 1960 for the work. Times of up to about 60 000 years can be estimated and the method is widely applied in archaeology. Libby and his team first demonstrated the accuracy of the method by showing that the age of wood from an ancient Egyptian royal barge estimated by radiocarbon dating agreed with the age of the barge known from historical records.

---

same energy. The four AOs of this second shell ($n=2$) can accommodate a total of up to eight electrons. The third shell ($n = 3$) contains one 3s and three 3p orbitals ($3p_x$, $3p_y$, and $3p_z$) plus a set of five degenerate 3d orbitals—a total of 9 AOs which (together) can hold up to 18 electrons.

The relative energies of some of the atomic orbitals mentioned above for an unspecified atom are shown in Figure 1.2. The s orbitals of increasing energy with principal quantum numbers 1–5 are shown in the column on the left; in the centre column, the p orbitals are seen to increase in energy starting from $n = 2$; the five degenerate d orbitals only start with $n = 3$ (and no higher ones are shown). None of the seven-fold degenerate f orbitals are shown as they are higher in energy and do not start until $n = 4$; they are of minimal importance in organic chemistry.

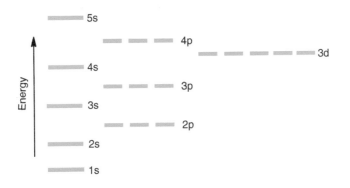

Figure 1.2 Energy levels of atomic orbitals.

Figure 1.3 The shapes of s and p atomic orbitals.

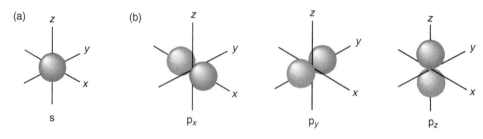

As mentioned above, the atomic orbital occupied by an electron indicates the space available to it as well as its energy. An s orbital is spherical, while each p orbital is elongated and circularly symmetrical about one of the three mutually perpendicular Cartesian axes (so they are labelled p_x, p_y, and p_z), as illustrated in Figure 1.3 (see also Sub-section 2.2.1).

1.1.3 Electronic configuration of an atom

The number of electrons around the nucleus of an isolated neutral atom is determined by its atomic number (Z, equal to the number of protons in its nucleus). In principle, these electrons can be distributed amongst the atomic orbitals in many ways, and any one distribution is referred to as an *electronic configuration* (or *electronic structure*). The different configurations correspond to different total electronic energies, and the most important is the one of lowest energy; this is called the **ground-state electronic configuration**. We can imagine a nucleus of an atom and a number of electrons equal to its atomic number being fed into the available orbitals; this is done according to the following three rules (sometimes known collectively as the *Aufbau Principle* from the German word meaning 'building up'):

(1) Electrons are added to orbitals in the order of their increasing energy (see Figure 1.2).

(2) Any orbital can hold one electron of either spin or two electrons of opposite spin.

That an orbital cannot contain two electrons of the same spin is called the *Pauli exclusion principle*.

(3) When the next available orbitals are *degenerate*, electrons with the same spin (i.e. unpaired) are added to them one at a time until they are all singly occupied (*Hund's rule*); a second electron of opposite (or paired) spin may then be added to each of them in turn.

To give a specific example, the result of following these rules for carbon ($Z=6$, so there are 6 electrons to be fed in) leads to the ground-state electronic configuration $1s^2 2s^2 2p_x^1 2p_y^1$ shown in Figure 1.4.

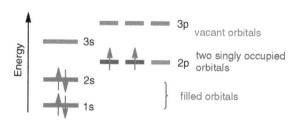

Figure 1.4 Ground-state electronic configuration of a carbon atom ($1s^2 2s^2 2p^2$). Electrons of opposite spin are represented by arrows pointing up and down.

Table 1.1 shows the ground-state electronic configurations of elements of the first three periods of the periodic table. Orbital occupancy is shown by a suffix (1 or 2) to the orbital designation. The first column corresponds to the first period with the addition of electrons to the 1s orbital to give hydrogen first then the noble gas element, helium. This completes the first shell ($1s^2$) which then becomes the inner shell, abbreviated by [He], for elements of the second period listed in the second column of Table 1.1 where electrons are added to the second shell ($n = 2$). Amongst these, for example, the electronic configuration of $_6$C is given as $[He]2s^22p_x^12p_y^1$ (see Figure 1.4); this shows that the ground state of a C atom contains the filled inner shell of He ([He]), 2 electrons in the 2s orbital, and 1 electron in each of $2p_x$ and $2p_y$ orbitals.

The second shell is complete (two electrons in each of the four orbitals available) with the electronic configuration of neon. The third column of Table 1.1 corresponds to elements of the third period where electrons are being added to the third shell ($n = 3$), the inner first and second shells being complete; this period ends with the third noble gas element, argon.

Any two of the three 2p orbitals could contain an electron since they are degenerate; but if the electrons were spin paired in just one of the three, i.e. contrary to Hund's rule, the configuration would not be the one of lowest energy. The ground-state electronic configuration of $_6$C can be represented by $[He]2s^22p^2$, it being understood that the two 2p electrons occupy different orbitals.

Exercise 1.3

Give the ground-state electronic configuration of each of the following elements.

(a) $_{35}$Br   (b) $_{38}$Sr   (c) $_{50}$Sn

We have already seen that an element is uniquely identified by its atomic number (Z) which, in the neutral atom, is equal to the *total* number of electrons around the nucleus. However, it is the electrons in the outermost shell (the **valence shell**) which characterize the *nature* of the element, and these are called the **valence electrons** of the atom. The electrons of the full inner shells are called **core electrons**; they have only a minor influence on the chemical properties of the element and are not involved in the formation of chemical bonds. For example, lithium has one valence electron ($2s^1$), fluorine has seven ($2s^22p^5$), and carbon has four ($2s^22p^2$) in the $n = 2$ valence shell. Atoms of these three elements have $1s^2$ inner core electrons and, as for all elements, *their chemical properties are determined principally by their valence electrons*.

Following completion of the third period, the next two electrons enter the 4s orbital (potassium $[Ne]3s^23p^64s^1$ and calcium $[Ne]3s^23p^64s^2$) in the normal way. As we saw in Figure 1.2, however, the next lowest orbitals are the five degenerate 3d orbitals, not the three 4p orbitals. Filling these orbitals corresponds to the first transition metals (d-block elements), Sc, Ti, V, etc.

Exercise 1.4

How many valence electrons does an atom of each of the following elements have?

(a) O   (b) Cl   (c) B   (d) N   (e) Mg

Table 1.1

Ground-state electronic configurations of elements[a]

Period 1		Period 2		Period 3	
$_1$H	$1s^1$	$_3$Li	$[He]2s^1$	$_{11}$Na	$[Ne]3s^1$
$_2$He	$1s^2$	$_4$Be	$[He]2s^2$	$_{12}$Mg	$[Ne]3s^2$
		$_5$B	$[He]2s^22p_x^1$	$_{13}$Al	$[Ne]3s^23p_x^1$
		$_6$C	$[He]2s^22p_x^12p_y^1$	$_{14}$Si	$[Ne]3s^23p_x^13p_y^1$
		$_7$N	$[He]2s^22p_x^12p_y^12p_z^1$	$_{15}$P	$[Ne]3s^23p_x^13p_y^13p_z^1$
		$_8$O	$[He]2s^22p_x^22p_y^12p_z^1$	$_{16}$S	$[Ne]3s^23p_x^23p_y^13p_z^1$
		$_9$F	$[He]2s^22p_x^22p_y^22p_z^1$	$_{17}$Cl	$[Ne]3s^23p_x^23p_y^23p_z^1$
		$_{10}$Ne	$[He]2s^22p_x^22p_y^22p_z^2$	$_{18}$Ar	$[Ne]3s^23p_x^23p_y^23p_z^2$

a. The symbol of each element is given with its atomic number. The filled inner shells of the second and third period elements are indicated by the bracketed symbol of the last noble gas before the element, [He] or [Ne]; these are called the *core electrons* (see above).

Table 1.2

Lewis representations of atoms								
Group no.	1	2	13	14	15	16	17	18
No. of valence electrons	1	2	3	4	5	6	7	8
Period 1	H·							He:
Period 2	Li·	Be:	B:	·C:	·N:	:O:	:F:	:Ne:
Period 3	Na·	Mg:	Al:	·Si:	·P:	:S:	:Cl:	:Ar:

1.1.4 Lewis representation of atoms

In 1902, the American G.N. Lewis proposed a method of representing atoms which gave prominence to their valence electrons and facilitated comparisons between different elements. The Lewis representation of an atom is the normal chemical symbol of the element with valence electrons shown by dots, i.e. the chemical symbol corresponds to the nucleus and the core electrons (those in the filled inner shells). Table 1.2 shows part of the periodic table with Lewis representations of atoms.

By comparing the Lewis representations with the ground-state electronic configurations in Table 1.1, we see that the four dots around the C for carbon correspond to two electrons in the 2s orbital and one electron in each of two 2p orbitals. For the oxygen atom, two electrons in the 2s orbital and four electrons in 2p orbitals are represented by six dots around the O. The maximum number of dots for the valence electrons of the main group elements shown here (periods 1–3) is eight—an **octet**.

Exercise 1.5

To which atomic orbitals do the electrons in the following Lewis representations of atoms belong?

(a) Mg   (b) :N·   (c) :P·   (d) :S·   (e) :Br·

1.2 Chemical Bonding

1.2.1 Formation of ions

Gilbert N. Lewis (1875–1946)
(Kindly supplied by Edward Lewis.)

Lewis proposed what we now call Lewis structures to represent the number of valence electrons of atoms and ions, as well as bonds in simple molecules (see Section 1.3), and the Lewis acid–base concept (see Sub-section 6.1.2).

Lewis realized that the chemical inertness of the noble gas elements of Group 18 is attributable to the stability of their electronic configurations. The outermost shell of an atom of these elements (except helium) contains eight electrons which fill the s and p orbitals (s^2p^6); the outermost shell for helium is the single 1s orbital, and its complement of electrons is just two. According to Lewis, atoms are most stable when their outermost shell is complete, and they have a tendency to achieve a complete outer shell by losing, gaining, or sharing electron(s). This concept is sometimes known as the **octet rule**. Atoms on the left side of the periodic table (metals) tend to lose one or more electrons to give positively charged ions (cations), e.g. eqn 1.1 for lithium. The outermost shell of an ion formed in this way was part of the core of the neutral atom, and the cation has the same electronic configuration as an atom of the noble gas which precedes the metal in the periodic table.

$$\text{Li} \cdot \longrightarrow \text{Li}^+ + e^- \tag{1.1}$$

$$:\!\text{F}\cdot + e^- \longrightarrow :\!\text{F}\!:^- \tag{1.2}$$

Scheme 1.1 Formation of a cation by electron loss and of an anion by electron gain.

On the other hand, those elements on the right side of the periodic table (non-metals) tend to achieve complete valence shell octets by gaining one or more electrons to give negatively charged ions (anions), e.g. eqn 1.2 for fluorine. The complete electronic configuration of the anion so formed is the same as that of an atom of the noble gas element which follows the non-metal in the periodic table.

Atoms can also achieve the noble gas electronic configuration by sharing valence electrons with other atoms, as will be discussed in the next section.

Example 1.1

Show the ground-state electronic configurations of Li^+ and F^-.

Solution
Loss of the one valence electron from Li gives Li^+ which has the same electronic configuration as He, while gain of one electron by F gives F^- with the same electronic configuration as Ne.

Li^+: $1s^2$ F^-: $1s^2 2s^2 2p^6$ (or $1s^2 2s^2 2p_x^2 2p_y^2 2p_z^2$)

Exercise 1.6

Show the ground-state electronic configurations of Na^+ and Cl^-.

a. Ionization energy and electron affinity

The energy *needed* to remove an electron from an atom is called its **ionization energy** (or ionization potential), *I.E.* This process always requires energy so ionization energies are always positive. The smaller the positive charge experienced by the valence electrons (the charge of the nucleus shielded to some extent by the core electrons), the easier it is for one to be removed and, therefore, the smaller the ionization energy. An atom of an element of low ionization energy can readily become a cation and is said to be **electropositive**.

On the other hand, the energy *released* when an atom gains an electron is called the **electron affinity**, *E.A.* An element of high electron affinity is said to be strongly **electronegative**; an atom of such an element can readily become an anion with the evolution of appreciable energy.

Table 1.3 lists values for the ionization energies (red) and electron affinities (blue) of atoms in electron volts (eV) in the form of the periodic table.

It follows from the definition of electron affinity that a *positive E.A.* converts into a *negative* enthalpy of reaction if the same process is represented as a conventional thermochemical equation, i.e. $E.A. = -\Delta_{EA}H$ (although enthalpies of reaction are normally given in kJ mol^{-1}).

Table 1.3

Ionization energies and electron affinities

H							He
13.60							24.59
0.75							~0
Li	Be	B	C	N	O	F	Ne
5.39	9.32	8.30	11.26	14.53	13.62	17.42	21.56
0.62	~0	0.24	1.27	~0	1.47	3.34	~0
Na	Mg	Al	Si	P	S	Cl	Ar
5.14	7.65	5.99	8.15	10.49	10.36	12.97	15.76
0.55	~0	0.46	1.24	0.77	2.08	3.61	~0

Values are given in eV (1 eV = 96.485 kJ mol^{-1}).

The thermochemical data associated with eqns 1.3 and 1.4 indicate that it costs more energy to remove an electron from Na than is gained by adding an electron to Cl. The two equations together, however, yield isolated Na^+ and Cl^- ions and do not take into account the very favourable electrostatic energy of formation of the ion pair or, on the molar scale, the huge lattice energy of crystalline NaCl.

We saw above that *electron affinity* is a measure of the tendency of *an isolated atom of an element* to accept an electron. It is related to, but distinct from, *electronegativity* which is a measure of *an atom within a molecule* to attract surrounding electron density.

The chemical processes corresponding to the formation of ions by electron loss (ionization energy, *I.E.*) and gain of an electron (electron affinity, *E.A.*) are exemplified in eqns 1.3 and 1.4 for sodium and chlorine, respectively.

$$\text{Na} \longrightarrow \text{Na}^+ + \text{e}^- \qquad I.E. = 5.14 \text{ eV} \tag{1.3}$$

$$\text{Cl} + \text{e}^- \longrightarrow \text{Cl}^- \qquad E.A. = 3.61 \text{ eV (or } \Delta_{EA}H = -348 \text{ kJ mol}^{-1}) \tag{1.4}$$

Scheme 1.2 Formation of a cation and an anion with associated energy changes.

b. Electronegativity

Several scales have been proposed to quantify **electronegativity**. The most commonly used is the one proposed by Pauling (in the 1930s), and some values are given in Table 1.4. The most electronegative element is F, and the values become smaller (the elements become less electronegative) upon going from the right to the left, or from the top to the bottom, of the periodic table; across the second period and down group 17, therefore, electronegativities decrease as follows:

F > O > N > C > B > Be > Li

F > Cl > Br > I

These trends conform to the influence of the positive charge of the nucleus, shielded to a degree by the core electrons, at the periphery of the atom. The *effective nuclear charge* experienced by the valence electrons becomes increasingly strong on going from the left to the right across the periodic table. On the other hand, the outermost (valence) shell of electrons becomes increasingly distant from the nucleus on going from the top to the bottom of the periodic table; hence the attractive force they experience becomes weaker.

Table 1.4

Electronegativity values on the Pauling scale[a]

H 2.20						
Li 0.98	Be 1.57	B 2.04	C 2.55	N 3.04	O 3.44	F 3.98
Na 0.93	Mg 1.31	Al 1.61	Si 1.90	P 2.19	S 2.58	Cl 3.16
K 0.82	Ca 1.00	Ga 1.81	Ge 2.01	As 2.18	Se 2.55	Br 2.96
						I 2.66

decreasing ←

decreasing ↓

a. Values revised by A.L. Allred in 1961.

1.2.2 Ionic and covalent bonds

a. Ionic bonds

We have seen above in our consideration of ionization energies and electron affinities that, in accord with Lewis's insight, atoms have a strong tendency to attain noble gas electronic configurations. This can be achieved by complete electron transfer between atoms of different elements if their electronegativities are sufficiently different. The more electropositive atom donates one or two electrons to the more electronegative atom and the two become a cation and an anion, both with the noble gas electronic configuration. There is then a strong electrostatic attractive force between the ions which constitutes an **ionic bond**. The classic example is sodium chloride, NaCl, eqn 1.5.

$$Na \cdot + \cdot \ddot{\underset{\cdot\cdot}{Cl}} : \longrightarrow Na^+ \; : \ddot{\underset{\cdot\cdot}{Cl}} :^- \tag{1.5}$$

Compounds with ionic bonds are usually crystalline solids with high melting points. In the crystal lattice, each cation is surrounded by several anions and interacts electrostatically with all of them. Likewise, each anion interacts electrostatically with several surrounding cations. Because of these multiple interactions, we cannot usually say that a particular cation of an ionic compound is bonded to a particular anion and NaCl, for example, does not form ion pairs in the crystalline state.

Depending upon the particular compound and the solvent, the cations and anions of an ionic compound $M^+ X^-$ may exist as ion pairs in solution, held in contact electrostatically, or M^+ and X^- may dissociate and diffuse through the solution independently. At extremely high temperatures in the gas phase, some very stable ionic compounds do exist as ion pairs.

> The real reaction between sodium and chlorine to give sodium chloride, which is hugely exothermic and would be explosive on anything except the smallest scale, does not take place according to our simplistic description above, but that does not invalidate our account of the nature of the bonding in NaCl (see also the margin note in Sub-section 1.2.1a, p. 8).

b. Covalent bonds

Electron transfer between atoms of two elements whose electronegativities are not very different will not be energetically favourable (the cost of removing an electron from one will not be provided by the energy of adding an electron to the other and, on the molar scale, the lattice energy). However, if atoms of elements need to gain electrons (rather than lose them) in order to attain the noble gas configuration, they can achieve this by sharing electrons. In eqn 1.6, for example, two fluorine atoms, each with seven valence electrons, can both achieve the noble gas configuration if they share a pair of electrons. This sharing of a pair of electrons constitutes a **covalent bond** and the two shared electrons are called a **shared (electron) pair** or a **bonding (electron) pair**. Most covalent bonds are formed between atoms of different elements, e.g. as shown for HCl in eqn 1.7.

> Shared electrons are counted as in the valence shells of both of the bonded atoms.

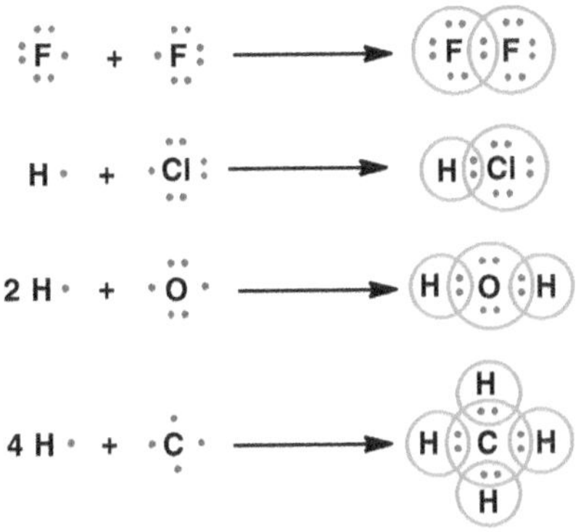

$$:\!\overset{\cdot\cdot}{\underset{\cdot\cdot}{F}}\!\cdot \; + \; \cdot\!\overset{\cdot\cdot}{\underset{\cdot\cdot}{F}}\!: \longrightarrow :\!\overset{\cdot\cdot}{\underset{\cdot\cdot}{F}}\!(\overset{\cdot\cdot}{\cdot\cdot})\!\overset{\cdot\cdot}{\underset{\cdot\cdot}{F}}\!: \tag{1.6}$$

$$H\cdot \; + \; \cdot\!\overset{\cdot\cdot}{\underset{\cdot\cdot}{Cl}}\!: \longrightarrow H(\overset{\cdot\cdot}{\cdot\cdot})\!\overset{\cdot\cdot}{\underset{\cdot\cdot}{Cl}}\!: \tag{1.7}$$

$$2\,H\cdot \; + \; \cdot\!\overset{\cdot\cdot}{\underset{}{O}}\!\cdot \longrightarrow H(\overset{}{\cdot\cdot})\!\overset{\cdot\cdot}{\underset{\cdot\cdot}{O}}\!(\overset{}{\cdot\cdot})H \tag{1.8}$$

$$4\,H\cdot \; + \; \cdot\!\overset{\cdot}{\underset{\cdot}{C}}\!\cdot \longrightarrow H(\overset{}{\cdot\cdot})\!\overset{H}{\underset{H}{C}}\!(\overset{}{\cdot\cdot})H \tag{1.9}$$

Scheme 1.3 Formation of covalent bonds.

One atom may share two or more pairs of electrons by making bonds with two or more other atoms in order to bring its valence shell occupancy up to eight electrons. An oxygen atom has six valence electrons, so needs two more, and forms covalent bonds with two H atoms, eqn 1.8, for example; its valence shell then comprises two bonding pairs and two unshared pairs of electrons. Just as a shared electron pair in the valence shell of an atom in a molecule is called a bonding pair, an *unshared electron pair* is often called a **nonbonding pair**, or a **lone pair**.

A carbon atom (4 valence electrons) achieves a filled valence shell (an octet) by making four bonds, e.g. with H atoms (eqn 1.9) so its filled valence shell comprises four bonding pairs. The four simple examples collected together in Scheme 1.3 show circled atoms which have achieved complete valence shells (octets for all except H whose 1s shell is full with just two electrons) by sharing electron pairs; only C and H do not include lone pairs in their complete valence shells.

1.2.3 Polar covalent bonds and dipoles

When a covalent bond is formed between atoms of different electronegativity, the bonding pair of electrons is not shared equally but is attracted towards the more electronegative atom. Such a bond is said to be *polarized*, or described as a **polar** bond, and the charge separation is represented by symbols $\delta-$ and $\delta+$ which represent *partial* negative and positive charges.

$$\overset{\delta+}{H}\!\!-\!\!\overset{\delta-}{Cl} \qquad \overset{\delta+}{H}\diagdown\!\!\overset{\delta-}{O}\!\!\diagup\overset{}{H}\,\delta+$$

If the difference in electronegativity is very large, an electron would be completely transferred from the less to the more electronegative atom to give a cation and an anion, and the bond becomes ionic (as discussed above). Bonds between atoms of non-metallic elements (e.g. C, N, O, and halogens) are typically covalent, and bonds between atoms of elements more electronegative than carbon (N, O, and halogens) and electropositive elements (metals) are usually ionic. However, the borderline between ionic and polar covalent bonding is not clear.

Example 1.2

Would a bond between each of the following pairs of atoms be covalent or ionic?

(a) O, H   (b) C, F   (c) Li, F   (d) C, Mg

Solution
The bond is ionic only when the electronegativity difference is appreciable (typically, >1.8 on the Pauling scale); otherwise, the bond is covalent.

(a) covalent   (b) covalent   (c) ionic   (d) covalent

Exercise 1.7

Would a bond between each of the following pairs of atoms be covalent or ionic?

(a) N, H   (b) C, Cl   (c) Li, C   (d) Mg, Cl

The charge separation (polarity) of a covalent bond may also be represented as a dipole with an arrow pointing from the positive end to the negative, with a plus sign (+) at the positive end of the arrow. This symbolism is used when we wish to emphasize that the polar bond has an electrical **dipole**. The magnitude of the dipole is expressed by the

dipole moment μ ($\mu = e \times d$ where e is the charge and d is the charge separation). The dipole moment of a bond is called a **bond moment**, and the dipole moment of a molecule is the vector sum of the bond moments involved.

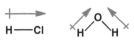

bond moments   dipole moment of a water molecule

Show the polarity of each of the following covalent bonds by partial charges and by a dipole arrow.

(a) O–H   (b) C–O   (c) C–Mg   (d) B–H

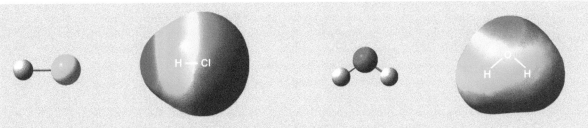

Molecular models and electrostatic potential maps of HCl and H$_2$O

The negative regions of the molecules are red; both Cl and O are electronegative, so the bonds to H are polar, and they also have unshared electron pairs.

Molecular models and electrostatic potential maps (EPMs). The diagrams in Scheme 1.3 are very simplistic and do not indicate the shapes of molecules, or the distribution of electrical charge around their surfaces. Molecular structures may be determined experimentally, or derived from quantum mechanical calculations, and represented graphically on the page by images of several types of molecular models, e.g. ball-and-stick or space-filling models (illustrated below) which show the correct relative sizes of atoms (see Panel 1.2 on sizes of atoms). An *electrostatic potential map* (EPM) can also be used if we want to show the distribution of electrical charge around the surface of a molecule. To construct an EPM, a total electron density surface (which is similar to a van der Waals surface) is first obtained computationally by joining together all the points in space around the molecule of a (specified) electron density *isovalue* (*isovalue* means 'equal value' from the Greek word for 'equal'). It is an *isosurface*—a surface joining together *isovalues*, in this case of electron density. Then, the electrostatic potential is calculated from the computed interactions of a hypothetical unit positive charge at each point on the isosurface with the adjacent electric charge (electron density) and nearby nuclei within the surface. The results are shown by colour, from red (most negative) to yellow or green (neutral) to blue (most positive).

Molecular models and an EPM of CH$_4$

Panel 1.2 Masses and sizes of atoms

Masses of atoms are so small that we express them using atomic mass units (1 amu = $1.6605402 \times 10^{-27}$ kg = 1/12 of the mass of ^{12}C). As far as chemistry is concerned, the nucleus of an atom of the carbon isotope $^{12}_{6}C$ may be regarded as comprising 6 protons and 6 neutrons although modern nuclear physics indicates that this is a simplification; surrounding the nucleus are 6 electrons. The currently accepted values of the masses of these particles at rest are:

proton: $1.6726231 \times 10^{-27}$ kg

neutron: $1.6749286 \times 10^{-27}$ kg

electron: $9.1093897 \times 10^{-31}$ kg

We see that the mass of an electron is nearly 1/2000 that of a proton or a neutron, i.e. the mass of an atom is concentrated in its nucleus and the mass of its electrons is effectively insignificant.

We also notice that the mass of an atom, e.g. a ^{12}C atom, is somewhat (nearly 1%) smaller than a simple sum of the masses of its constituent particles (1 amu is smaller than the mass of either a proton or a neutron). This is called the *mass defect* of the atom, which comes from the binding energy of the nucleus according to the Einstein equation, $E = mc^2$, where E is the binding energy, m is the mass defect, and c is the velocity of light. In other words, the strong binding energy of a nucleus comes from the conversion of mass into energy when the nucleus is formed from its constituent nuclear particles (nucleons).

The size of a *nucleus* defined by its diameter is 10^{-14} to 10^{-15} m (= 10^{-2} to 10^{-3} pm), but the size of an *atom* is more difficult to quantify. This is because, according to quantum mechanics, the location of an extranuclear electron can only be expressed by the probability of it being at a specified position outside the nucleus. However, if we think of the extra-nuclear electrons as a cloud of electron density, we can specify a surface around the nucleus which contains a selected proportion (e.g. 90, 95, or 99%) of the total electron density. The surface which is calculated to join together all points with the same (specified) electron density (the *isovalue*) is called an *isosurface*, as described in the note on *electrostatic potential maps* (EPMs) on p. 11. This makes the radius of an isolated atom approximately 10^{-10} m (= 100 pm).

In practice, the radius of an atom can be defined as half the distance between the nuclei when two identical atoms are as close as they can be without repelling each other or losing their spherical shape. According to the states of atoms, we can define different atomic radii for a particular element. One is the *covalent radius*, which is obtained when the two atoms are covalently bonded (e.g. the covalent radius of a chlorine atom is half the distance between the nuclei in a Cl_2 molecule). Another is the *van der Waals radius* which is determined from the closeness of approach without bonding. Values for both are given in the periodic table opposite. For a metallic element, the *metal*

1.3 Lewis Structures of Molecules and Ions

1.3.1 How to draw Lewis structures

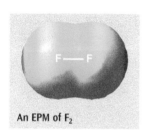

An EPM of F_2

We have already seen that valence electrons are represented by dots in the Lewis description of atoms; in the same way, lone pairs of electrons in molecules and the shared electron pairs of covalent bonds may also be represented by pairs of dots. Some molecules were presented in this way in Section 1.2. However, if all the molecular valence electrons are explicitly shown by dots, larger molecules are tedious to draw and too complicated to view. Consequently, from now on we shall follow the simpler convention that *the shared electron pair of a covalent bond is represented by a short line*, and that only lone pairs and unpaired valence electrons are shown as dots. The molecules given in Scheme 1.3 are shown below as simplified Lewis structures.

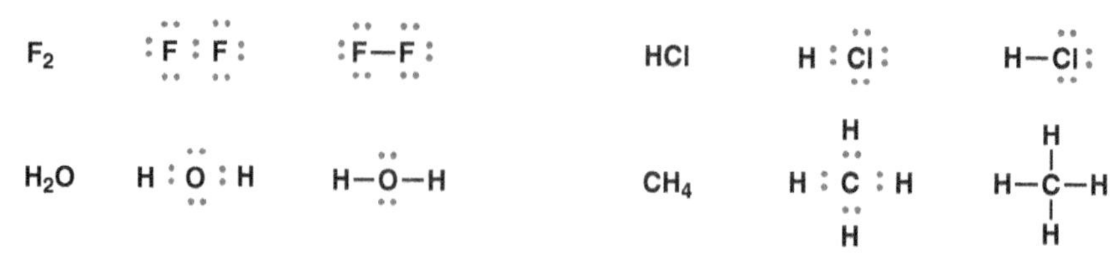

bond radius is determined from the atomic separations in the metal. The *ionic radius* of an element is determined from measurements on crystals of an ionic compound which contain the element as monoatomic ions (e.g. Na^+, Ca^{2+}, Cl^-, and O^{2-}).

Notice from the values in the table that the sizes (radii) of atoms tend to become smaller on going from left to the right across a period, and to become larger on going down a group. These tendencies result first from the increasing positive charge of the nucleus and its consequent hold on electrons on going from left to right across a period, and secondly from the expanding valence electron shell of the atom upon descending a group.

Covalent and van der Waals radii of atoms

1	2	13	14	15	16	17	18
H 32 120							He – 140
Li 134 182	Be 89 –	B 82 –	C 77 170	N 73 155	O 70 152	F 68 147	Ne – 154
Na 154 227	Mg 137 173	Al 126 –	Si 117 210	P 111 180	S 105 180	Cl 99 175	Ar – 188
K 196 275	Ca 174 –	Ga 126 187	Ge 122 210	As 119 185	Se 117 190	Br 114 185	Kr – 202
Rb 216 –	Sr 191 –	In 146 193	Sn 142 217	Sb 139 –	Te 136 206	I 133 198	Xe – 216

The upper values are covalent radii and the lower ones are van der Waals radii in pm.

It is very important to be able to draw correct (simplified) Lewis structures in organic chemistry, especially when we look at how organic reactions occur (since this involves the redistribution of valence electrons). Note that all the electrons in almost all stable molecules and ions are spin paired; a chemical species with an unpaired electron is called a radical and is usually very reactive, see Chapter 20.

For each of the following compounds, show the connectivity of the atoms and all valence electrons, then draw the (simplified) Lewis structure representing covalent bonds (if there are any) by lines.

Example 1.3

(a) NH_3 (b) BH_3 (c) NaCl

Solution
The neutral N and B atoms have five and three valence electrons, respectively. The N in ammonia has a full valence shell (an octet) which includes one lone pair. Boron with just three valence electrons can form no more than three simple covalent bonds, e.g. with three H atoms to give the molecule shown, borane. Even then, it still has only six valence electrons, i.e. an incomplete valence shell. NaCl is an ionic compound.

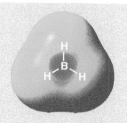

EPMs of NH₃ and BH₃

There is partial negative charge (red) on the N in NH₃ owing to the lone pair, but the B in BH₃ is slightly positive.

Exercise 1.9

Draw Lewis structures of the following compounds. In each case, first show the atom connectivity and all the valence electrons, then draw the Lewis structure representing covalent bonds (if there are any) by lines.

(a) CH_3OH (b) CH_3NH_2 (c) $B(OH)_3$ (d) LiF

The arrangement of atoms in a molecule (the molecular structure) needs to be determined experimentally, but we shall not deal with this here where we only use examples whose molecular structures are well established.

The following is a guide for drawing correct Lewis structures of more complex molecules, with bonding pairs represented by short lines.

(1) First, arrange the atoms of the molecule or ion according to how they are bonded together; add dots around each atom corresponding to the number of valence electrons of the neutral atom. Then, add extra dots according to the negative charge for an anion, or delete dots according to the positive charge for a cation.

As examples, consider Lewis structures of the series water (H_2O), hydroxide ion (HO^-), and oxonium ion (H_3O^+). First, put the atoms in place with their valence electrons. Then identify each bond between the O atom and an H with a (shared) pair of electrons—one from the O and one from an H. Thirdly, add an extra electron to the O of the anion, and remove one from the O of the cation.

Oxonium ion is the systematic name for H_3O^+ (also called hydronium ion).

$$H_2O \qquad HO^- \qquad H_3O^+$$

$$H:O:H \qquad H:O: \qquad \overset{\displaystyle H}{H:O:H}$$

(2) Replace pairs of dots representing shared electron pairs (covalent bonds) by short lines, and confirm that each non-hydrogen atom has a valence octet. Here, up to three bonds are possible, and the number of bonds plus lone pairs is four (i.e. eight valence electrons around the O in each case).

Exceptionally, the valence shells of atoms in and beyond the third period, e.g. S and P, can accommodate more than an octet (see Section 14.6).

$$H-\overset{..}{\underset{..}{O}}-H \qquad H-\overset{..}{\underset{..}{O}}: \qquad H-\overset{\displaystyle H}{\underset{..}{O}}-H$$

(3) Determine the **formal charge** for each atom and add the + or − value as a superscript to its symbol. The formal charge on an atom is the number of *valence electrons* in the neutral atom (before bonding), minus the number of *unshared electrons* around the bonded atom, minus the number of *bonding pairs*; in abbreviated form, this formula is given by eqn 1.10:

$$\text{Formal charge} = \text{(valence electrons)} - \text{(unshared electrons)} - \text{(bonding pairs)} \qquad (1.10)$$

Calculation of the formal charge on the oxygen atom in H_2O, HO^-, and H_3O^+:

	H–Ö–H	H–Ö:	H–Ö–H (with H above)
Valence electrons of neutral O	6	6	6
(–) Unshared electrons	–4	–6	–2
(–) Bonding electron pairs	–2	–1	–3
Formal charge	0	–1	+1

Consequently, the correct complete Lewis structures are:

$$\text{H–Ö–H} \qquad \text{H–Ö:}^- \qquad \text{H–Ö–H}^+$$

Note that the calculation of *formal* charges is really just an electron accounting exercise to determine the total charge of a chemical species; the *actual* charge (which must be numerically equal to the algebraic sum of the formal charges on all the atoms) is distributed throughout the molecule or ion, as will be discussed later. For example, the formal charge is +1 on the O of H_3O^+ and zero on each H, but the positive electrical charge is mainly on the H atoms of the ion. (Note also that the O still has one lone pair of electrons.)

An EPM of H_3O^+ showing partial positive charge on each H atom.

In neutral organic compounds, constituent atoms generally have characteristic numbers of bonds and lone pairs according to how many valence electrons they have in their unbonded states:

- H (1 valence electron) has 1 bond.

- C (4 valence electrons) has 4 bonds.

- N (5 valence electrons) has 3 bonds and 1 lone pair.

- O (6 valence electrons) has 2 bonds and 2 lone pairs.

- F, Cl, Br, and I (7 valence electrons) have 1 bond and 3 lone pairs each.

1.3.2 Further examples of drawing Lewis structures

- **Methanol CH_3OH**
  (i) Arrange all the constituent atoms and put in dots for valence electrons (for the sake of clarity, numbers of valence electrons are summarized on the left); (ii) identify pairs of electrons which correspond to covalent bonds and replace each by a line (this gives the Lewis structure); confirm that all non-hydrogen atoms have valence octets; (iii) calculate the formal charges (equal to zero).

In the following, as previously in this section, the drawings are to illustrate the electronic configurations of molecules and ions, not their shapes; we cover molecular structure in the next chapter.

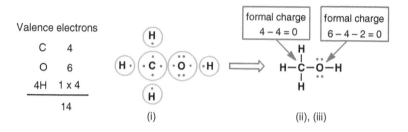

Valence electrons

C	4
O	6
4H	1 x 4

14

(i)

(ii), (iii)

• **Methanal** CH_2O

Proceeding as above, we find that making one bond between C and O leaves one unpaired electron on each of C and O. These electrons are paired to form a second bond between C and O to give the C=O double bond; both C and O now have full valence octets. Formal charges are all zero.

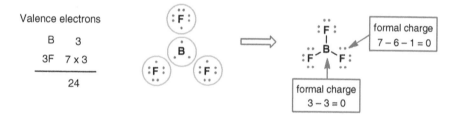

Valence electrons

C	4
O	6
2H	1 x 2

12

• **Boron trifluoride** BF_3

This molecule contains only three bonds since B has only three valence electrons. It follows that B in BF_3 has only six electrons whereas all three F atoms have full valence octets.

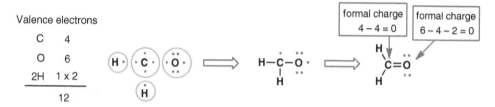

Valence electrons

B	3
3F	7 x 3

24

• **Ammonium ion** NH_4^+

This ion has a positive charge so, according to step (1) in the protocol for drawing Lewis structures described in Sub-section 1.3.1, one electron is deleted from the normal number of valence electrons for N. Calculation of the formal charge on the N shows that it is +1 even though the N has a complete valence octet.

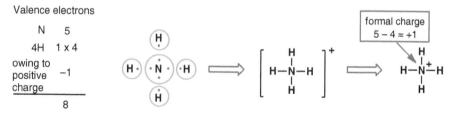

Valence electrons

N	5
4H	1 x 4
owing to positive charge	−1

8

• **Methyl anion** CH_3^-

This ion has a negative charge so one extra electron is put onto the C. Bond lines are drawn and the formal charge is calculated: with an unshared electron pair on the C, the formal charge is −1.

Valence electrons

$$\begin{array}{ll} \text{C} & 4 \\ \text{3H} & 1 \times 3 \\ \text{owing to} & \\ \text{negative} & +1 \\ \text{charge} & \\ \hline & 8 \end{array}$$

formal charge
$4 - 2 - 3 = -1$

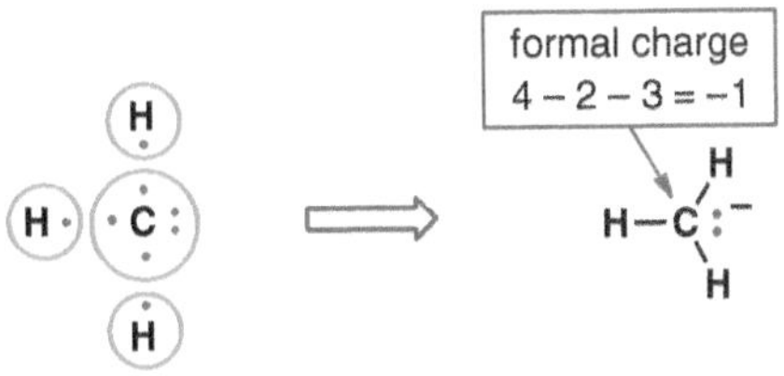

- **Ethanoate anion** $CH_3CO_2^-$
  One extra electron is added to one of the O atoms for this anion. In the structure with only single bonds, the O carrying the extra electron has an octet, but the other O has only seven electrons. The C to which it is bonded also has an unpaired electron, however, so a C=O double bond can be formed and both of these atoms then achieve complete octets. When the procedure for calculating formal charge is carried out on the singly bonded O atom with the extra electron, it turns out to be −1; all other formal charges are zero.

  The common name for ethanoate is acetate.

Valence electrons

$$\begin{array}{ll} \text{2C} & 4 \times 2 \\ \text{2O} & 6 \times 2 \\ \text{3H} & 1 \times 3 \\ \text{owing to} & \\ \text{negative} & +1 \\ \text{charge} & \\ \hline & 24 \end{array}$$

formal charge
$6 - 4 - 2 = 0$

formal charge
$6 - 6 - 1 = -1$

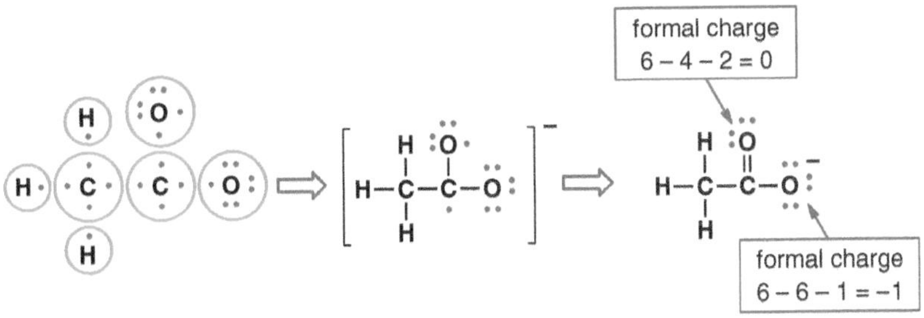

  In the above, the extra electron was arbitrarily put on one of the O atoms, and we obtained Lewis structure **1a** below. We could equally well have chosen the other O, in which case we would have finished up with Lewis structures **1b**; both **1a** and **1b** are legitimate Lewis structures for ethanoate anion. The significance of this will be discussed in the next section.

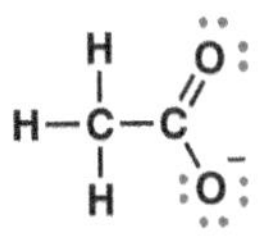

1a 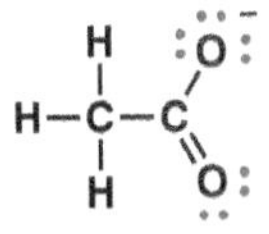   **1b**

- **Nitromethane** CH_3NO_2
  Finally, we construct the Lewis structure of nitromethane. The connection of atoms in the nitro group (NO_2) may not be familiar yet, but it is shown below. Each of the two O atoms in the singly bonded structure has seven electrons, so each has an unpaired electron (all other atoms have complete valence shells). A bond between the two O atoms would satisfy the valence requirements but would give a three-membered ring which would be very strained and hence unstable (see Chapter 4). Alternatively, if we tried to make two N=O (double) bonds using another two electrons on N, the result would be an impossible structure with 10 valence electrons on the N. Thirdly, if one electron is transferred from the N lone pair to one O, the unpaired electron left on the N can form a second bond to the other O. By this procedure, the valence requirements of all atoms are satisfied and the outcome is the reasonable Lewis structure shown; it simply needs the addition of formal charges.

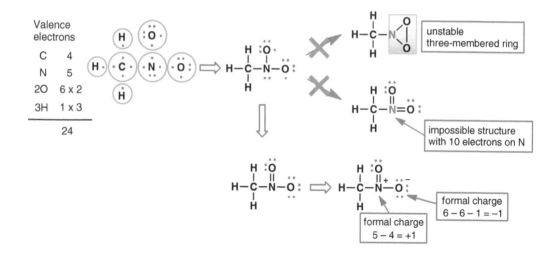

Note that replacement of the N^+ in **2a** and **2b** with C gives structures **1a** and **1b**, respectively, and both N^+ and C have 4 valence electrons. In other words, the nitromethane molecule and ethanoate anion have the same number of electrons; such structures are said to be **isoelectronic**.

The nitrogen has four bonds and no lone pairs, so its formal charge by eqn 1.10 is $(5-4)=+1$. The value for the singly-bonded O is $(6-6-1)=-1$ since it has 6 unshared electrons (three lone pairs) and one bond. The other (double-bonded) O has four unshared electrons and two bonds so the formal charge on this atom is $6-4-2=0$. This Lewis structure, therefore, involves charge separation even though the molecule is neutral overall.

The bonding shown for nitromethane is similar to that within the ethanoate ion described above. One of the two O atoms is double bonded and the singly bonded one bears a formal negative charge, as shown in structure **2a**. And just as in the case of ethanoate, the O atoms could be switched around to give an alternative equivalent Lewis structure, **2b**.

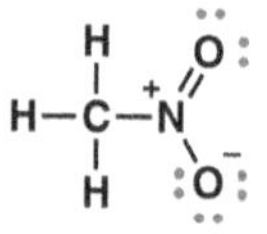

2a 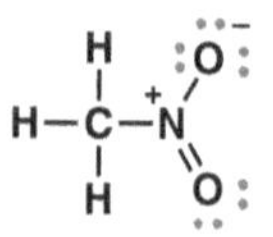　　　　　**2b**

Exercise 1.10

Add dots, to represent unshared electrons, to the following ionic species, and calculate formal charges to complete the Lewis structures.

(a) 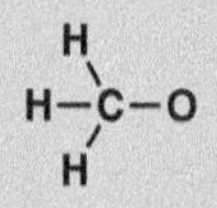 H—C—O (anion)　　(b) 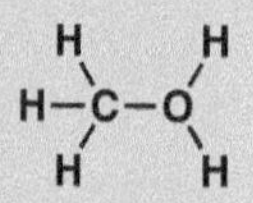 H—C—O (cation)　　(c) 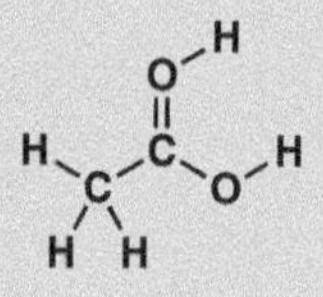 (cation)　　(d) 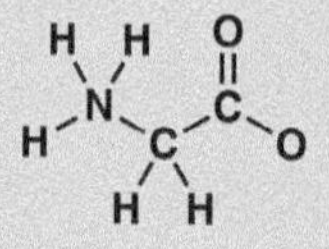 (charge separation)

Exercise 1.11

Draw Lewis structures for the following molecules and ions.

(a) HCN　　　(b) H_3C^+　　　(c) H_2N^-　　　(d) HCO_2H

1.4 Introduction to Resonance

We were able to represent both the ethanoate anion and the neutral nitromethane molecule by alternative Lewis structures. In both **1a** and **1b** for ethanoate, the two O atoms are different, one being single-bonded and the other double-bonded. However, the two oxygen atoms in the real ethanoate ion are known to be equivalent with equal C–O bond lengths. In other words, the *real* structure of the ethanoate ion cannot be described properly by a single Lewis structure. The actual structure is intermediate between the two structures **1a** and **1b** as indicated by drawings **1c** or **1d** with dotted lines to represent partial bonds. We invoke the concept of **resonance** and say that the real electronic structure of the ethanoate ion is better represented as a **resonance hybrid** of the **resonance contributors**, **1a** and **1b**. A widely used convention to represent this relationship between **1a** and **1b** is to connect the two Lewis structures with a double headed arrow (↔) as shown below.

Bond lengths in an ethanoate ion

(1 pm = 10^{-12} m)

1a **1b** **1c** **1d**

The concept of resonance was proposed by Linus Pauling (see Panel 2.1) in the 1930s, and will be discussed more fully in Section 5.4.

Resonance allows us to describe the electronic structure of a molecule or ion as a blend of two or more Lewis structures (resonance contributors) which individually are inadequate.

We saw above that nitromethane, like ethanoate, can be described by two Lewis structures and neither alone is adequate. The two O atoms are equivalent in the real molecule with equal N–O bond lengths; it too is best described as a resonance hybrid of resonance contributors, **2a** and **2b**, and can be represented by **2c** or **2d** with dotted lines representing partial bonds.

The lengths of the two N–O bonds in nitromethane are the same:

121.1 pm

2a **2b** **2c** **2d**

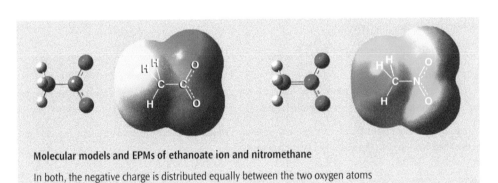

Molecular models and EPMs of ethanoate ion and nitromethane

In both, the negative charge is distributed equally between the two oxygen atoms

Draw two or more Lewis structures for the following and, in each case, show the relationship between them.

(a) CO_3^{2-} (carbonate ion)　　　(b) $HONO_2$ (i.e. HNO_3, nitric acid)

Example 1.4

Solution
The three oxygen atoms are equivalent in carbonate ion as are two of the three in nitric acid. Three and two Lewis structures (resonance contributors), respectively, are possible,

(continues …)

(… continued)

and the electronic structure of each species is better described as a resonance hybrid of the contributors.

Exercise 1.12

Draw two or more Lewis structures (resonance contributors) for each of the following ions, and show the relationship(s) between them.

(a) NO_3^- (b) $CH_3OCH_2^+$

Summary

- The **atomic number** (Z) of an element is the number of protons in the nucleus of an atom of the element (equal to the number of electrons outside the nucleus); the **mass number** (A) is the sum of the number of protons and neutrons in the nucleus.

- Different **isotopes** of an element have the same atomic number but different mass numbers, i.e. different numbers of neutrons in the nucleus.

- The chemical properties of an element are determined by the electrons accommodated in **atomic orbitals** around the nucleus of an atom of the element. Among these electrons, those in the outermost shell (the **valence electrons**) are the most important, and the chemistry of atoms is driven by the strong tendency for the valence shells to become complete (two electrons for H, and an **octet** for elements in the second and third periods of the periodic table).

- Atoms (other than H) attain octets of valence electrons by losing or gaining electrons to give ions, or by sharing electrons to form covalent bonds (**octet rule**).

- Covalent bonds are polar if the **electronegativities** of the bonded atoms are different.

- **Lewis structures**, in which valence electrons are represented by dots and covalent bonds by lines, are used to describe the electronic configurations of molecules. Valence electrons in molecules exist as **shared** (**bonding**) **pairs** or **unshared** (**nonbonding**) **pairs**, also called **lone pairs**.

- The electronic structures of molecules which cannot be described adequately by single Lewis structures may be described by **resonance hybrids** of contributing Lewis structures.

Problems

1.1 Which elements have the following ground-state electronic configurations?

(a) $1s^2 2s^2 2p^3$ (b) $1s^2 2s^2 2p^6 3s^1$

(c) $1s^2 2s^2 2p^6 3s^2 3p^2$ (d) $1s^2 2s^2 2p^6 3s^2 3p^5$

1.2 How many valence electrons does each of the following atoms have?

(a) Li   (b) O   (c) F

(d) S   (e) Al

1.3 Chlorine has two isotopes of mass numbers 35 and 37 in the natural abundance ratio of 75.8 : 24.2.

(a) How many nuclear protons and neutrons does each isotope have?

(b) Calculate the approximate relative atomic mass (atomic weight) of Cl from these data.

1.4 Will the bonding between each of the following pairs of atoms be covalent or ionic?

(a) C, O   **(b)** C, N   **(c)** B, H

(d) O, Mg   **(e)** C, P

1.5 Show the polarity of each of the following bonds by partial charges and by a dipole arrow.

(a) H–F   **(b)** N–H   **(c)** C–N

(d) C–Cl   **(e)** Li–C

1.6 Identify which of the following molecules have dipole moments and show the polarities by dipole arrows.

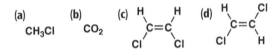

(a) CH_3Cl **(b)** CO_2

1.7 Draw a Lewis structure for each of the following molecules.

(a) H_2O_2 **(b)** CH_3Cl **(c)** NH_2OH

(d) CH_3CN **(e)** CS_2

1.8 Draw a Lewis structure for each of the following molecules.

(a) N_2 **(b)** CH_2NH **(c)** CH_2NOH

(d) CH_3COCl **(e)** H_2CO_3

1.9 Draw one or more Lewis structures for each of the following ions.

(a) $CH_3NH_3^+$ **(b)** CH_3NH^- **(c)** $H_2C=OH^+$

(d) HOO^- **(e)** $HC=CH_2$ with O^- substituent

1.10 Draw a Lewis structure for methyl nitrite (CH_3ONO) which has the same molecular formula as nitromethane (they are isomers, see Chapter 2).

1.11 Draw two or more Lewis structures for each of the following ions.

(a) NO_2^- (nitrite ion)   **(b)** CH_3CO^+ (acetylium ion)

1.12 Complete the Lewis structure for the adduct of BF_3 and Et_2O (diethyl ether) by adding lone pairs and formal charges.

F–B–O structure with F, F, Et, Et   (Et = C_2H_5)

2 Molecular Structure and Shapes of Organic Molecules

Related topics we have already covered:

- **Atomic orbitals** (Sub-section 1.1.2)
- **Covalent bonds** (Sub-section 1.2.2)
- **Lewis structures of molecules** (Section 1.3)

Topics we shall cover in this chapter:

- **Shapes of organic molecules**
- **Molecular orbital model of covalent bonding**
- **Hybridization of atomic orbitals**
- **Bonding in methane, ethene, and ethyne**
- **σ and π bonds**
- **Isomerism**
- **Drawing organic structures**

Molecules have three-dimensional shapes. What do they look like? Why does a molecule have a particular shape? The investigation of structures of small molecules by experimental and theoretical techniques progressed together in the early and middle parts of the last century, and advances in one led to advances in the other. Once the structure of a molecule was reliably known, it was necessary to be able to explain it theoretically. As confidence in bonding theory increased, novel structural types were predicted and subsequently confirmed experimentally. When a predicted structure was shown to be wrong, the theoretical method was refined until it could account for the experimental result. Towards the end of the last century, developments in high-speed computing led to further and ongoing advances in computational chemistry. This is now a widely used technique for the elucidation of molecular structure, and is applicable for compounds which are too unstable to be investigated experimentally.

We shall see in this chapter that three-dimensional shapes of simple molecules are the result of electrostatic repulsion between electron pairs and can be explained in terms of shapes of molecular orbitals involved in chemical bonding. The s and p atomic orbitals we encountered in Chapter 1 may be transformed into hybrid atomic orbitals with different shapes which lead to two types of covalent bonds, σ and π, and account for the various shapes of molecules.

2.1 Shapes of Molecules and the VSEPR Model

2.1.1 Tetrahedral shapes

Molecules are composed of atoms, and their shapes are defined by the different **bond lengths** and **angles** within the molecule. When a molecule comprises two or more atoms bonded to a single central atom, its shape can be predicted by considering the electrostatic

Figure 2.1 Molecular structure of methane, CH_4. (a) Lewis structure. (b) Three-dimensional structural formula.* (c) A molecular (ball-and-stick) model. (d) A space-filling model which shows the correct relative sizes of the atoms (see Panel 1.2 for sizes of atoms).

The basis of the VSEPR model was originally proposed in 1939 by R. Tsuchida in Japan, and independently by N. Sidgwick and H. Powell in England in 1940. The theory was further refined and popularized in the 1950s by R. Gillespie and R.S. Nyholm.

An electrostatic potential map (EPM, see Sub-section 1.2.3) of methane.

repulsion between electron pairs (bonding and nonbonding) around the central atom. These electron pairs originate in the valence shells of the atoms from which the molecule is formed, and the molecule will assume the shape in which the total electrostatic repulsion between electron pairs is minimized, i.e. the shape in which electron pairs are as far apart as possible. This is the **valence shell electron pair repulsion (VSEPR)** model or theory.

For example, the methane molecule (CH_4) has eight valence electrons distributed as four single bonds (four localized bonding electron pairs) from the central carbon atom; see Figure 2.1(a). According to the VSEPR model, the four electron pairs radiate from the carbon to be as far away from each other as possible. The unique result is that the methane molecule is tetrahedral, which requires the angle between any pair of C–H bonds to be 109.5°. This 'prediction' for methane is in accord with the known experimental structure; see Figures 2.1(b) and (c). The length of the C–H bond, which is not predictable from VSEPR theory (but see later), is 110 pm.**

The shape of the ammonia molecule (NH_3) can be predicted in a similar manner. The Lewis structure of NH_3 shows that the nitrogen has three bonding pairs of electrons (the N–H single bonds) and one nonbonding (lone) pair. The VSEPR theory predicts that electrostatic repulsion between the four valence electron pairs will cause them to be tetrahedrally distributed around the N atom, and the ammonia *molecule*, therefore, is pyramidal. The observed H–N–H bond angle is 107.3° (Figure 2.2), somewhat smaller than the regular tetrahedral angle of 109.5°. This small difference is reasonably attributed to the greater repulsion between the lone pair and an N–H bonding pair than that between two bonding pairs because the electron density of the lone pair is held more closely to the N.

As the Lewis structure of the water molecule (H_2O) shows in Figure 2.3(a), the oxygen atom is surrounded by four electron pairs: two O–H bonds and two lone pairs. Again, VSEPR theory predicts that electrostatic repulsion between the valence electron pairs of the H_2O molecule causes them to be tetrahedral around the O atom. The observed H–O–H bond angle is 104.5° (Figure 2.3), which is smaller still than the H–N–H bond angle of NH_3. This observation is also wholly compatible with the proposal that the lone pairs repel each other more strongly than they repel the O–H bonding pairs.

As we shall see in further examples, whenever a Lewis structure shows that there are four *separate* electron pairs (bonding or lone pairs) around a central atom, they radiate tetrahedrally, and this determines the shape of the molecule or ion.

*A bold wedge represents a bond coming out of the page, while a broken wedge represents a bond pointing behind the page. All other lines represent bonds in the plane of the page.

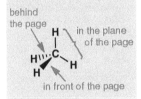

**Bond lengths are often given in Å (ångström, 1 Å = 10⁻¹⁰ m) but in this book we use the SI unit, pm (picometre, 10⁻¹² m). Bond lengths and angles are actually average values as bonds stretch and bend rapidly even at 0 K.

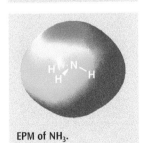

EPM of NH_3.

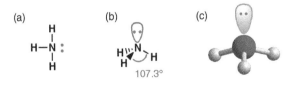

Figure 2.2 Molecular structure of ammonia, NH_3. (a) Lewis structure. (b) Three-dimensional structural formula. (c) A molecular (ball-and-stick) model showing the lone pair.

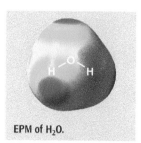

EPM of H₂O.

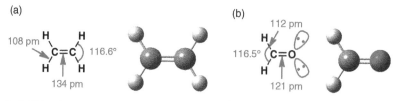

(a)   (b)   (c)

H—O—H   104.5°

Figure 2.3 Molecular structure of water, H₂O. (a) Lewis structure. (b) Three-dimensional structural formula. (c) A molecular model showing two lone pairs.

2.1.2 Trigonal planar shapes

What is the shape around a carbon atom from which the valence electrons radiate in three directions? The adverse electrostatic interactions are smallest when the three directions are as far apart as possible, i.e. at 120° to each other, so VSEPR theory predicts a trigonal planar arrangement. Ethene and methanal (formaldehyde) are molecules with trigonal planar C atoms, as shown in Figure 2.4 (although the H–C–H angles are somewhat smaller than 120°). In both cases, the double bond containing four valence electrons counts much the same as a single bonding pair or a lone pair in the VSEPR model.

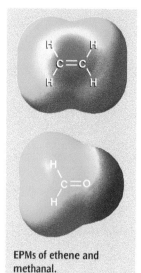

EPMs of ethene and methanal.

(a)

108 pm   H   H
     C=C   116.6°
     H   H
      134 pm

(b)

         112 pm
H
116.5°   C=O
H
    121 pm

Figure 2.4 Molecular structures of (a) ethene and (b) methanal.

2.1.3 Linear shapes

The coplanarity of the two CH₂ groups of ethene cannot be predicted by VSEPR theory but it follows naturally from orbital overlap considerations, which we discuss in Section 2.5.

In both ethyne (acetylene) and carbon dioxide, electron density in the form of bonds emanates from each C atom in only two directions. The electrostatic repulsion between these bonds causes them to be as far apart as possible, which is when they are collinear, i.e. with an angle of 180° between them; see Figure 2.5. According to VSEPR theory, therefore, the triple bond of ethyne counts as a single electron-rich region with six electrons, and is collinear with the C–H bonds. Correspondingly, carbon dioxide has two C=O double bonds from the one C, each of which counts as a single electron-rich region, so the molecule is linear.

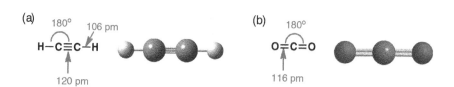

(a)   180°   106 pm

H—C≡C—H

120 pm

(b)   180°

O=C=O

116 pm

Figure 2.5 Molecular structures of (a) ethyne and (b) carbon dioxide.

Example 2.1

Predict approximate bond angles within the following species

(a) CH₂Cl₂   (b) CH₃⁺   (c) H₂C=C=O

Solution

(a) The Lewis structure of CH₂Cl₂ (dichloromethane) shows that four electron pairs (single bonds) radiate from the carbon so CH₂Cl₂ must be tetrahedral with bond angles about 109.5°. In fact, the Cl–C–Cl angle is widened slightly owing to electrostatic repulsion between the two Cl atoms.

(b) Three bonding pairs of electrons emanate from the C of CH_3^+, the methyl cation. Electrostatic repulsion between them leads to a planar trigonal structure and symmetry requires that the angles be equal at 120°.

(c) The methylene carbon of the $CH_2=C=O$ molecule (ketene) has two single bonds and one double bond, and the associated valence electrons radiate from it in three directions at angles of about 120°. The central carbon has two double bonds pointing in opposite directions, i.e. at 180°.

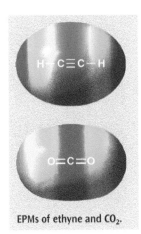

EPMs of ethyne and CO_2.

We have now seen that single and multiple bonds in trigonal and linear molecules can be treated in the same way within the VSEPR model for the prediction of molecular shapes.

Exercise 2.1

Predict approximate bond angles around the carbon atoms in the following species.

(a) CH_3OH (b) HCO_3^- (c) HCN

2.2 Orbital Description of Covalent Bonding

We have already seen that, according to the Lewis model, a covalent bond is formed by two atoms sharing an electron pair, and that bond angles can be predicted by VSEPR theory. However, these models alone cannot explain the difference between the two pairs of bonding electrons of a double bond, or the difference in chemical reactivity between single and double bonds, or why the two ends of an alkene are coplanar.

In this section, we shall see that interactions of *atomic orbitals* lead to *molecular orbitals*, and that a covalent bond is the accommodation of a pair of electrons in a bonding molecular orbital. We shall also see how optimal sideways overlap (interaction) of p orbitals requires coplanarity of the two ends of an alkene.

2.2.1 Atomic orbitals and their shapes

We saw in Chapter 1 that the electronic energy of an atom is quantized, and the electrons are accommodated in atomic orbitals (AOs), such as s and p orbitals. An orbital of specified energy can be thought of either as the space around the nucleus of an atom available to an electron, or as a mathematical formula (wave function) from which the probability of finding an electron at a particular location may be calculated.

In order to appreciate how a molecular orbital (MO) is derived from constituent AOs, we need to develop further our review of AOs in Chapter 1, in particular their spatial features. In order to visualize an AO, or represent it graphically, we need to think in terms of the spatial distribution of electron density, or of the probability of finding the electron in space.

One method is to draw an 'electron cloud', as shown in Figure 2.6(a); the probability of finding the electron (the electron density) is shown qualitatively by a gradation of colour

The s and p orbitals are by far the most important types of AOs in organic chemistry.

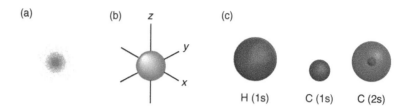

Figure 2.6 Representations of s orbitals. (a) An electron cloud. (b) An electron density isosurface. (c) Images calculated theoretically showing the relative sizes of the 1s AO of a hydrogen atom, and the 1s and 2s AOs of a carbon atom (each singly occupied).

which fades away as the probability of finding the electron decreases with the increasing distance from the nucleus. An alternative, which was introduced in Sub-section 1.2.3 in the context of electrostatic potential maps (EPMs), is to draw an electron density *isosurface*—a boundary surface which contains most of the electron density; see Figure 2.6(b). By this method, an s orbital appears as a sphere with the nucleus at its centre; there is only one s orbital for each principal quantum number, and it can accommodate up to two electrons. The size of the sphere increases with the principal quantum number: 1s<2s<3s, but it depends largely on the positive charge of the nucleus: the 1s AO on a hydrogen atom is relatively large because the nuclear charge is small; see Figure 2.6(c).

For principal quantum number 2, there are three 2p orbitals as well as the single 2s orbital. The three 2p AOs, which are mutually perpendicular and designated $2p_x$, $2p_y$, and $2p_z$ in Figure 2.7(c), are of the same energy (they are *degenerate*, see Section 1.1.2). A **2p orbital** has the approximate shape of a pair of distorted spheres (often called a dumbbell shape but actually the two spheres are almost touching), as illustrated in Figure 2.7(a) by a computer-generated image and in Figure 2.7(b) by cartoons.

The two **lobes** of each 2p AO in Figure 2.7 are shown in different colours to represent the + and the − signs of the mathematical equation (the wave function) of the orbital; the mathematical forms of the wave function for the two lobes are identical except for the plus and minus signs. Consequently, the probability of finding an electron (the electron density) associated with a 2p orbital at a specified distance from the nucleus is the same for the two lobes because it is equal to the *square* of the orbital wave function with the coordinates corresponding to the location. The probability of finding the electron in the plane which separates the two lobes of a 2p orbital (where the sign of the wave function changes from + to −) is zero: this plane is called a **node** or **nodal plane**.

Figure 2.7 Different representations of the shapes of 2p atomic orbitals. (a) A computer-generated image of a 2p AO. (b) Alternative cartoons of a single 2p AO. (c) Cartoons representing the three mutually orthogonal 2p AOs.

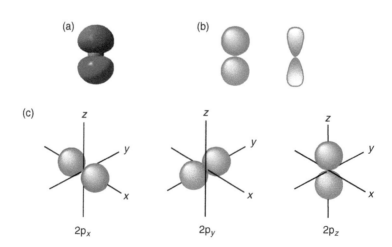

2.2.2 Overlap of atomic orbitals to give molecular orbitals

a. Bonding and antibonding molecular orbitals

According to molecular orbital (MO) theory, a covalent bond originates in the interaction (overlap) of an atomic orbital (AO) of one atom with an AO of another atom. When two AOs overlap, they combine to form two new orbitals called **molecular orbitals** (MOs), and the electrons from the original AOs are now accommodated in the new MO(s). Whereas AOs are localized on atoms, MOs are associated with molecules and sometimes extend over a whole molecule.

The simplest example is the formation of the H–H bond of a hydrogen molecule (H_2), which involves the interaction of the singly occupied 1s AOs of two H atoms. When the two AOs of equal energy overlap, two MOs are produced. One is lower in energy than the original AOs, while the other is higher (Figure 2.8), and the new MO of lower energy becomes occupied by the two electrons which were originally in the two AOs. *It is this pair of electrons which constitutes the covalent bond.* The H_2 molecule is lower in energy (more stable) than the two separate H atoms, and the energy difference, which is liberated as heat, is the bond energy of the H_2 molecule. In the reverse direction, the bond energy is the energy required to dissociate the molecule into two separate hydrogen atoms.

The lower **occupied** (or **filled**) MO in Figure 2.8 is called the **bonding** (σ) MO, and the one of higher energy is the **antibonding** (σ^*) MO which is **unoccupied** (or **vacant**). The unoccupied antibonding orbital does not contribute to the bond energy since no electrons are involved. The bonding MO corresponding to the covalent bond is formed by the constructive combination of two AOs, which is an *in-phase* interaction (expressed by 1s + 1s in Figure 2.8), and the two electrons accommodated in this MO are distributed mainly in the region between the two atoms, i.e. as the bonding (shared) electron pair of the covalent bond.

In contrast, the negative combination of the two AOs results in the antibonding MO, which is an *out-of-phase* interaction (expressed by 1s – 1s in Figure 2.8). There is a *nodal plane* between the two atoms which separates the two lobes of this orbital. If there were an electron in this antibonding orbital, there would be zero probability of finding it in this nodal plane. When this MO is occupied by such an additional electron to give H_2^-, the system is destabilized and the bond is weakened. A second additional electron would give the completely unstable H_2^{2-} anion which would dissociate to give two hydride ions ($2 \times H^-$).

The extent of the overlap between the 1s orbitals of the two H atoms (and hence the energy gap between the bonding and antibonding orbitals) increases as the internuclear distance decreases. However, the electrostatic repulsion between the two positively charged nuclei also increases as they become closer. A balance is achieved between these opposing effects at a certain distance, the equilibrium **bond length** of the hydrogen molecule, 74 pm.

Why is the He$_2$ molecule unstable? Two helium atoms do not form a stable He_2 molecule for the same reason that H_2^{2-} does not exist; He_2 and H_2^{2-} are *isoelectronic*. If two He atoms come close enough together, interactions between their 1s AOs lead to MOs similar to those of the H_2 molecule. However, two He atoms have four valence electrons which fill the bonding *and* the antibonding MOs of the hypothetical diatomic He_2 molecule, so it is not more stable than the two separate He atoms.

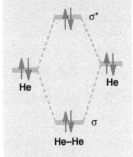

MOs of the unstable He$_2$ molecule

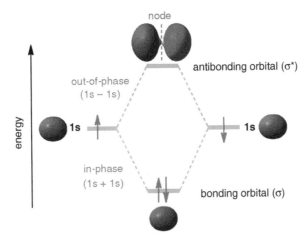

Figure 2.8 Formation of the H–H bond by the overlap of AOs.

Figure 2.9 σ Bond in H_2 and the associated MOs.

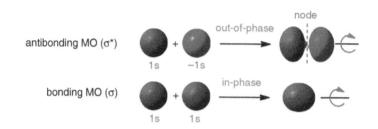

b. Symmetry of sigma molecular orbitals and sigma bonds

A covalent bond whose electron distribution is circularly symmetrical around the bond axis is called a **sigma (σ) bond**, and the appearance of the associated σ **MOs** does not change upon rotation about the bond axis. The H–H bond is an example of a σ bond, and the bonding (σ) and antibonding (σ*) MOs are illustrated in Figure 2.9.

Sigma MOs corresponding to σ bonds may also be generated by (i) the end-on overlap of one lobe of a p orbital with an s orbital, and (ii) the end-to-end overlap of lobes of a pair of p orbitals (or *hybrids* derived from p orbitals, see below). All single bonds are σ bonds, and the rotation about a single bond is usually easy owing to its circular symmetry.

2.3 Hybridization of Atomic Orbitals

Linus Carl Pauling (1901–1994)
The concept of *hybridization* was introduced by Linus Pauling around 1930 (see Panel 2.1).

We saw in Section 2.1 that carbon is tetra-covalent and forms compounds with tetrahedral coordination (e.g. methane) as well as trigonal planar (e.g. methanal) and linear coordination (e.g. carbon dioxide). However, the ground-state electronic configuration of carbon is $[\text{He}]2s^2 2p_x^1 2p_y^1$, so it needs four more electrons in its valence shell. If we were to proceed as we did in the case of the hydrogen molecule (Sub-section 2.2.2), we would invoke interaction (overlap) of the 2s or 2p valence shell orbitals of a C atom with orbitals on other atoms to construct MOs. However, it is not obvious how this would account for the four equivalent bonds in methane, for example. How can we reconcile the electronic configuration of a C atom with its tetra-covalency and the structural variety amongst its simple compounds? The way forward is to derive **hybrid** AOs on C from its 'pure' AOs by the 'mixing' (**hybridization**) of its 2s and up to three of its 2p AOs.

2.3.1 Three types of hybrid orbitals

Mixing the 2s AO and just *one* of the 2p AOs leads to a pair of **sp hybrid** AOs. These are equivalent in the sense that they are of the same energy and have the same shape; they

---

Panel 2.1 The work of Linus Pauling

Linus Pauling was born in Portland, Oregon, USA; he received the BSc degree in chemical engineering from the Oregon State College in 1922. In 1925, he was awarded the PhD degree in chemistry from the California Institute of Technology where he spent the rest of his working life. In the late 1920s and the 1930s, he carried out seminal research on chemical bonding based on quantum mechanics, and developed the orbital description of covalent bonds as well as the concepts of electronegativity, orbital hybridization, and resonance, which are amongst the foundations of modern chemistry. His early work was summarized in his celebrated book, *The Nature of the Chemical Bond*, in 1939 and he was awarded the Nobel Prize in Chemistry in 1954. His work also included X-ray crystallographic studies on structures of biochemically important molecules; this included identification of secondary structural features of proteins such as α-helices and β-pleated sheets. He was also active in peace movements and campaigned against nuclear weapons from the late 1950s; he was awarded the Nobel Peace Prize for 1962.

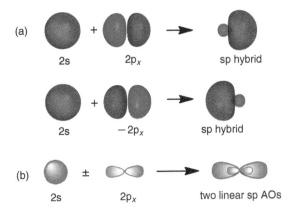

(a) 2s + 2p$_x$ → sp hybrid

2s + −2p$_x$ → sp hybrid

(b) 2s ± 2p$_x$ → two linear sp AOs

Figure 2.10 Shapes of sp hybrid AOs. (a) Computer-generated sp AOs. (b) Cartoon representations.

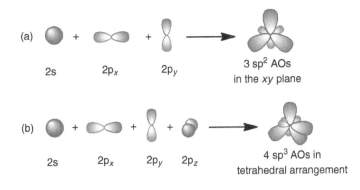

(a) 2s + 2p$_x$ + 2p$_y$ → 3 sp^2 AOs in the xy plane

(b) 2s + 2p$_x$ + 2p$_y$ + 2p$_z$ → 4 sp^3 AOs in tetrahedral arrangement

Figure 2.11 Cartoon representations of (a) sp^2 and (b) sp^3 hybrid AOs.

Computer-generated images of sp^2 and sp^3 hybrid AOs:

sp^2 hybrid   sp^3 hybrid

are linear but point in opposite directions (Figure 2.10)—like the lobes of the 2p orbital from which they are derived. The shape of a hybrid orbital can be generated computationally as a mix of 2s and 2p AOs; this is illustrated in Figure 2.10 for sp hybrids each of which consists of two lobes of different mathematical signs and different sizes. An sp-hybridized carbon atom still has two perpendicular 2p AOs remaining unhybridized.

Alternatively, mixing the 2s and *two* 2p AOs leads to a set of three **sp^2 hybrids**; these are also equivalent and spread at 120° to each other in the plane of the two 2p orbitals from which they are derived; see Figure 2.11(a). An sp^2-hybridized carbon has a residual 2p AO perpendicular to the plane of the sp^2 hybrid AOs.

The third possibility is to mix the 2s and *all three* 2p AOs, which leads to a set of four equivalent **sp^3 hybrids** radiating tetrahedrally at angles of 109.5°, as in Figure 2.11(b).

Which of these three hybridization schemes is invoked for a particular second-period element (not just carbon) depends on the shape of the molecule around the central atom. Note that the number of orbitals involved in hybridization is always conserved—hybridization cannot increase or decrease the number of orbitals.

2.3.2 Energies of hybrid orbitals

The energy of each of a set of hybrid AOs is the weighted mean of the energies of its constituent 2s and 2p AOs (Figure 2.12); in other words, the total energy of the orbitals involved is not changed by the hybridization. We have already seen that s and p AOs are *qualitatively* different, e.g. different symmetries, and a hybrid has properties derived from each of its constituent orbitals according to the proportions of the s and p AOs in its composition. We sometimes refer to the **s character** of a hybrid orbital: the sp, sp^2,

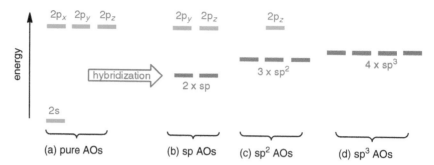

Figure 2.12 Energy levels of hybrid AOs.

and sp³ hybrids have 50%, 33%, and 25% s character, respectively. The greater the s character of a hybrid orbital, the lower its energy and the smaller its size, i.e. the closer the electrons are to the nucleus.

2.4 Bonding in Methane

Methane is a tetrahedral molecule. We can account for its four C–H bonds by invoking MOs derived from combinations of the four sp³ hybrid AOs of the C and 1s AOs of four H atoms; this leads to eight MOs in all—four bonding and four antibonding. When the eight valence electrons (four from C and one from each H atom) are then fed in, they occupy the four bonding (σ) MOs corresponding to the four C–H bonds of methane, and the associated four antibonding (σ*) MOs are vacant; the result is the tetrahedral molecule shown in Figure 2.13.

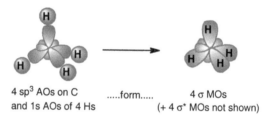

4 sp³ AOs on C   .....form.....   4 σ MOs
and 1s AOs of 4 Hs   (+ 4 σ* MOs not shown)

Figure 2.13 Molecular orbitals of methane.

A molecular model and EPM of methane.

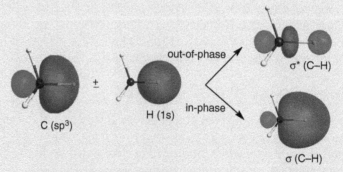

The two MOs corresponding to a single C–H bond in methane (generated theoretically by combination of an sp³ AO of the C and the 1s AO of one H) are shown below superimposed on a molecular model. Both the occupied σ MO of the C–H bond and the associated (but empty) σ* MO are circularly symmetrical about the bond axis; see Sub-section 2.2.2b.

Note that there is just one nodal plane (which contains the nucleus and is perpendicular to the C–H bond axis) in the bonding σ MO, and an additional one (part way along the bond) in the σ* MO.

What kinds of AOs (hybridized and unhybridized) are involved in forming bonding MOs in ethane and carbon tetrachloride CCl_4?

Explain why the bonding MO associated with the C–C bond in ethane is a σ MO.

2.5 Bonding in Ethene

2.5.1 Trigonal planar carbons

The bonds from each carbon in ethene are trigonal planar, and the bonding associated with this structure requires sp^2 hybrid AOs. In ethene, the MO corresponding to the C–C σ bond is formed by end-to-end overlap of one sp^2 AO on each C, and the MOs corresponding to the four C–H σ bonds are formed by the end-on overlap of the other two sp^2 AOs on each C with 1s AOs of four H atoms (Figure 2.14). Feeding in 10 of the 12 available valence electrons provides the σ bonded framework of the molecule; the associated five antibonding (σ*) MOs remain vacant.

*The sideways overlap of the two p orbitals is maximal when the two CH_2 groups are coplanar; this complements the discussion of the shape of an ethene molecule based on VSEPR theory in Section 2.1.2.

2.5.2 The pi (π) orbitals

Each sp^2 carbon involved in the σ bonding of ethene still has an out-of-plane unhybridized $2p_z$ orbital. When both C atoms and the four H atoms are coplanar, these two AOs are *parallel* and their sideways overlap allows formation of molecular orbitals of a different type.* The two new MOs both have a node in the (xy) plane of the molecule and the remaining two valence electrons are distributed above and below this nodal plane when they are fed into the bonding orbital, as represented in Figure 2.14.** This type of bonding orbital, formed from p AOs and with a nodal plane, is called a **pi (π) MO** and the occupying electron pair constitutes a **π bond**; the associated antibonding (π*) orbital which is not shown in Figure 2.14 remains vacant. So, ethene has a C–C σ bond formed by end-to-end sp^2–sp^2 overlap and a π bond formed by $2p_z$–$2p_z$ sideways overlap which, together, constitute the *double bond* between the two carbon atoms.

The bonding π MO of ethene shown in Figure 2.14 is of lower energy than the $2p_z$ orbitals from which it was derived, and the antibonding π* MO (not shown) is of higher energy***, as illustrated in Figure 2.15 with theoretically generated shapes of both MOs. Consequently, occupancy of the π MO by the two electrons from the two AOs, i.e. formation of the π bond, lowers the electronic energy of the system by an amount corresponding to the π bond energy of ethene (280 kJ mol⁻¹).

**This does not mean that one electron is above the nodal plane and the other below it. Two electrons are not differentiated and the *electron pair of the bond* is equally distributed above and below the plane of the molecule.

***The antibonding π* MO of ethene has a nodal plane between the two C atoms in addition to the one in the (xy) plane of the molecule.

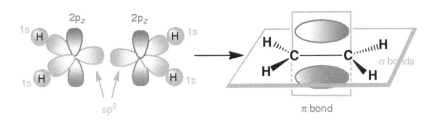

Figure 2.14 Bonding in ethene.

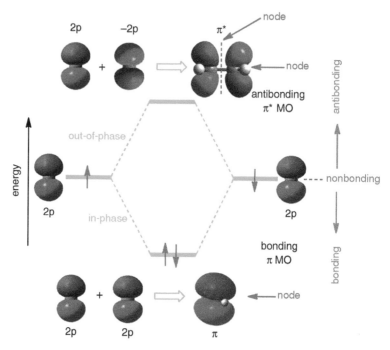

Figure 2.15 Energy diagram of the π orbitals of ethene.

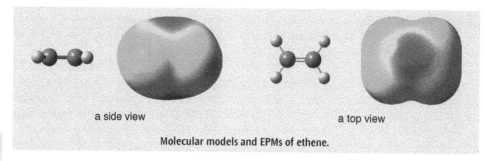

a side view   a top view

Molecular models and EPMs of ethene.

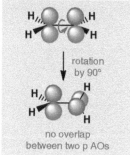

rotation by 90°

no overlap between two p AOs

The sideways overlap of the two unhybridized $2p_z$ orbitals to form the π bond of ethene is maximal when the AOs are parallel. If one CH_2 group is rotated about the C–C bond, the sideways overlap of the unhybridized p orbitals decreases and becomes zero at 90° when there can be no π bond (as illustrated in the margin). In other words, rotation about the double bond of an alkene is a high-energy process which does not take place at normal temperatures.

Example 2.2

Identify the hybridization of both carbons in a molecule of ethanal (acetaldehyde); from which AOs are the MOs of the C=O bond conceptually derived? What is the difference between the two types of C–H bonds in ethanal?

$$\underset{\text{ethanal}}{CH_3\overset{\overset{\displaystyle O}{\|}}{C}H}$$

Solution

First draw the Lewis structure to predict the bond angles (based on VSEPR theory, see Section 2.1) and hence the hybridizations of the C atoms, as shown below. End-to-end overlap of sp^2 hybrids on C and O leads to the MO corresponding to the σ bond of the carbonyl, and sideways overlap of the unhybridized 2p AOs on C and O gives the MO corresponding to the π bond; this leaves lone pairs in equivalent sp^2 orbitals on O. The aldehydic σ C–H bond is derived from overlap of the H 1s orbital and an sp^2 orbital on the C; the methyl σ C–H bonds are derived from 1s H AOs and three sp^3 hybrid AOs of the C.

Generally, the π–π^* MO energy gap is smaller than the σ–σ^* MO energy gap because the sideways overlap of two p AOs is poorer than the end-on overlap of orbitals of appropriate symmetry. Consequently, the order of MO energy levels is usually as follows:

High σ^* (antibonding σ MO)
π^* (antibonding π MO)
nonbonding MO (lone pair, if present)
π (bonding π MO)
Low σ (bonding σ MO)

Explain how the bonds in methanal are derived from atomic orbitals.

Exercise 2.4

2.6 Bonding in Ethyne

Ethyne is a linear molecule and the bonding is explained by invoking sp hybrid AOs of the two carbons. The triple bond consists of a σ bond and two orthogonal π bonds. The σ MO is formed by end-to-end overlap of one sp AO from each C atom, and the MOs corresponding to the two π bonds are formed by sideways overlap of two orthogonal pairs of 2p AOs ($2p_y$ and $2p_z$), as illustrated in Figure 2.16. Each of the π bonds has a nodal plane, the xz plane for the $2p_y$–$2p_y$ bond and the xy plane for the $2p_z$–$2p_z$ bond. These three bonding MOs accommodate six of the valence electrons. The MOs corresponding to the two σ C–H bonds are formed by end-on overlap of the other sp AO of each C with the 1s AO of an H, so the H–C–C–H framework of ethyne is linear, and these two bonding MOs accommodate the other four valance electrons (ten in all). As for methane and ethene, antibonding (σ^* and π^*) MOs generated by the out-of-phase combinations of AOs remain vacant.

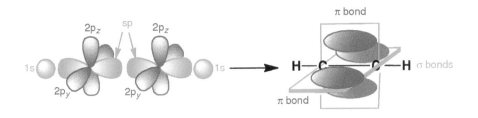

Figure 2.16 Bonding in ethyne showing the π MOs.

The two orthogonal π MOs of ethyne generated by theoretical calculations:

An electron pair in each of the two orthogonal π MOs leads to circularly symmetrical electron density about the axis of the ethyne molecule as shown by the EPM.

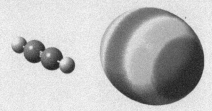

A molecular model and EPM of ethyne.

Example 2.3

Explain the bonding in the propadiene molecule (allene, $CH_2=C=CH_2$).

Solution

The two terminal C atoms are sp^2 hybridized and the central C is sp hybridized, as shown below. The central C is able, therefore, to form σ bonds to both terminal C atoms using its sp hybrid AOs. In addition, it can form *orthogonal* π bonds to both terminal C atoms. The C–H bonds on the terminal carbons are formed by end-on overlap of their sp^2 hybrid AOs with 1s AOs of four H atoms. Note that the two σ-bonded H atoms on C1 are in a plane at 90° to the plane which contains the two σ-bonded H atoms on C3.

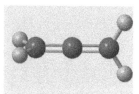

A molecular model of allene.

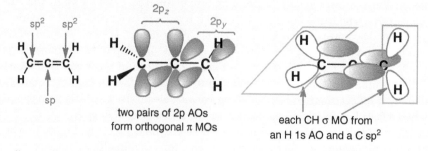

Bonding in allene and orthogonal π MOs.

Exercise 2.5

Describe how the MOs corresponding to the bonding in CO_2 are conceptually derived from AOs of the C and O atoms. See Problem 2.14.

Exercise 2.6

Identify the hybridization of the carbon and nitrogen atoms in the following molecules; identify the AOs involved in bonding, and describe the type of each bond.

(a) $CH_3CH=CH_2$ (b) CH_3NH_2 (c) HCN

2.7 Hybridization of Carbon and Bond Lengths

The 2s AO is lower in energy than the 2p AO, and electrons in the 2s AO are held closer to the nucleus than those in the 2p AO. These differences affect molecular properties so, for example, the lengths of bonds from a carbon atom depend upon its hybridization (sp^3, sp^2, or sp). As summarized for ethane, ethene, and ethyne in Table 2.1, the C–H bonds become shorter with the increasing s character of the hybridization of the C from 25%, to 33%, to 50%—they become shorter in the order: $C(sp^3)$–H > $C(sp^2)$–H > C(sp)–H.

The same is true for the lengths of bonds between C atoms, but the increasing *bond order* (the number of bonding electron pairs between the two C atoms) is more influential; the bond lengths shorten rapidly in the order: C–C single bond > double bond > triple bond.

Lengths of C–C and C–H bonds

Table 2.1

	$\begin{array}{c} \text{H H} \\ | \quad | \\ \text{H}-\text{C}-\text{C}-\text{H} \\ | \quad | \\ \text{H H} \end{array}$	$\begin{array}{c} \text{H} \qquad \text{H} \\ \diagdown \qquad \diagup \\ \text{C}=\text{C} \\ \diagup \qquad \diagdown \\ \text{H} \qquad \text{H} \end{array}$	$\text{H}-\text{C}\equiv\text{C}-\text{H}$
C–H bond	110 pm	108 pm	106 pm
C–C bond	154 pm	134 pm	120 pm

2.8 Drawing Organic Structures

We saw earlier that simple organic molecules may be drawn as Lewis structures in which dots represent electrons, and lines represent covalent bonds. It would be complicated and time-consuming to draw structures of large molecules in this way, and such structures would also take up a lot of space in books and documents. Consequently, we normally use **condensed formulas** or **line-angle** drawings (even for small molecules), either of which still show organic structures unambiguously. Some examples are given in Table 2.2.

Condensed formulas show all the atoms, with numbers as subscripts, but usually omit lines representing bonds (except for double and triple bonds).

Line-angle drawings, in which the carbon framework is represented by short lines with all C atoms and H atoms attached to them (normally) left out, are an alternative useful shorthand method of representing a structure. Each short line represents a C–C bond, and the points where lines meet, and the ends of chains, correspond to the carbons, each being understood to have the appropriate number of H atoms. For a long carbon chain, lines representing bonds are drawn in a zig-zag fashion, and each corner (angle) and end of a chain represents a carbon. Heteroatoms and H atoms attached to them are shown explicitly. In the following example, the structure of an unsaturated long chain

Table 2.2	Examples using different methods of representing molecular structures[a]			
Compound	Structural formula		Condensed formula	Line-angle drawing
Butane C_4H_{10}			$CH_3CH_2CH_2CH_3$ or $CH_3(CH_2)_2CH_3$	
Methylpropane C_4H_{10}			$CH_3CH(CH_3)CH_3$ or CH_3CHCH_3 $\quad$ CH_3	
But-1-ene C_4H_8			$CH_3CH_2CH=CH_2$	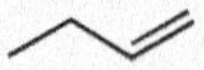
Ethanol C_2H_6O			CH_3CH_2OH or C_2H_5OH (EtOH)	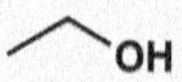
Methoxymethane (dimethyl ether) C_2H_6O			CH_3OCH_3 or $(CH_3)_2O$ (MeOMe)	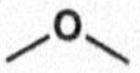
Methoxybenzene C_7H_8O			$C_6H_5OCH_3$ (PhOMe)	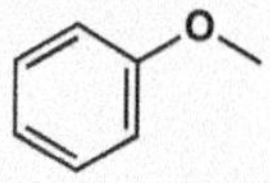

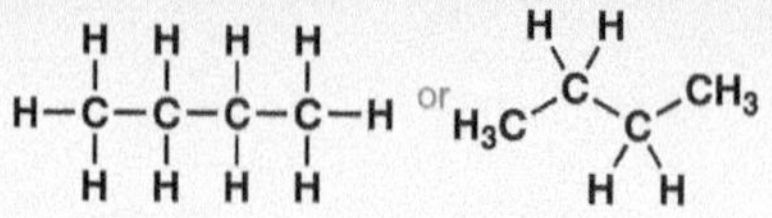

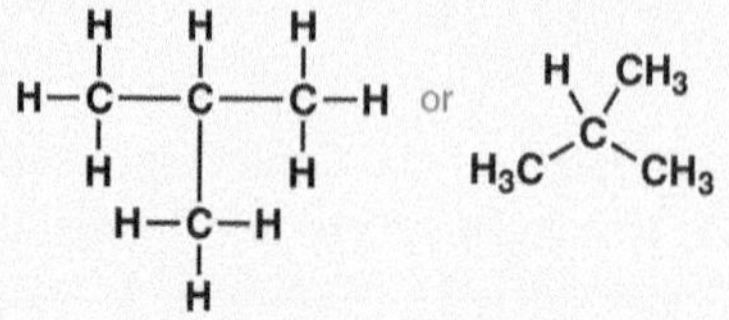

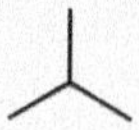

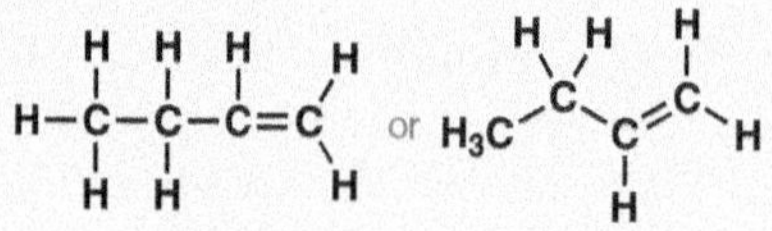

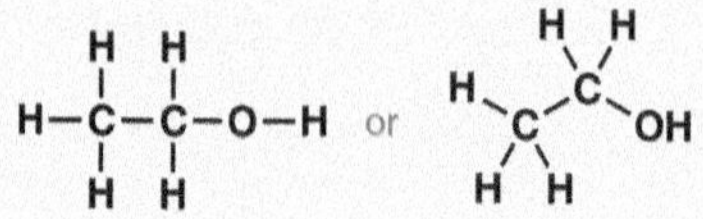

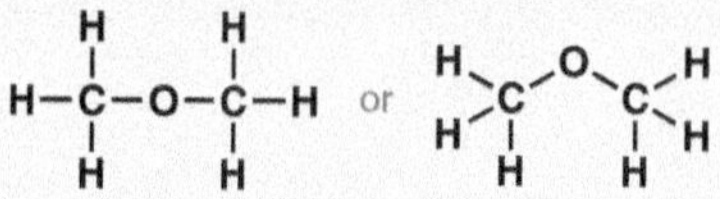

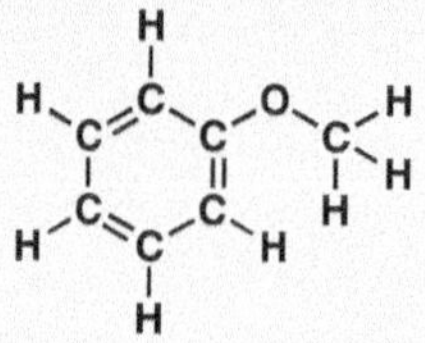

a. Abbreviations: Et=ethyl, Me=methyl, and Ph=phenyl.

carboxylic acid is represented by its condensed formula and as a line-angle drawing together with some notes for clarification.

oleic acid

$$CH_3(CH_2)_7CH=CH(CH_2)_7COOH$$

the *cis* double bond is difficult to represent in the condensed formula

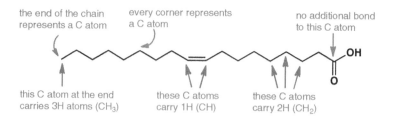

the end of the chain represents a C atom

every corner represents a C atom

no additional bond to this C atom

this C atom at the end carries 3H atoms (CH$_3$)

these C atoms carry 1H (CH)

these C atoms carry 2H (CH$_2$)

A line-angle structure is easy to draw, economical of space, and important features of the molecule (functional groups) can be shown clearly, so organic chemists generally favour this method of drawing organic structures. Three-dimensional structures can also be represented by line-angle drawings using wedged bonds as appropriate. These were used in Figure 2.1(b) to show the tetrahedral carbon of methane. The three-dimensional structure of an amino acid is illustrated below.

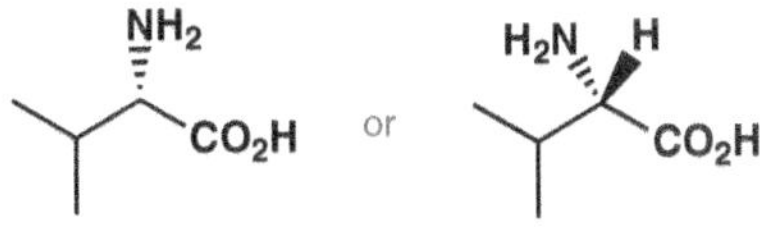

valine (an amino acid)

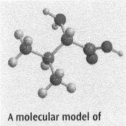

A molecular model of valine.

Write a condensed formula showing all the atoms for each of the compounds represented by the following line-angle drawings.

Exercise 2.7

(a) 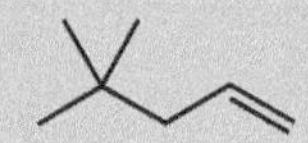 (b) 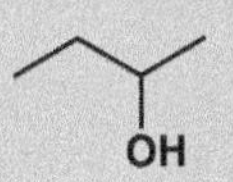 (c) 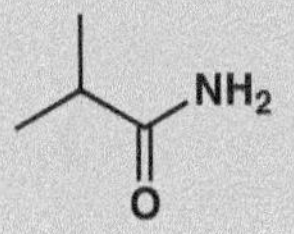 (d)

Give a line-angle drawing of each compound represented by the following condensed formulas.

Exercise 2.8

(a) $(CH_3)_2C=CHCHCH_2CCH_2CH_3$ with substituents CH_3 and CH_2CH_3 and CH_3

(b) $CH_3CH_2CH(CH_3)CH_2C\equiv CCH_3$

(c) $CH_3C(O)CH_2COOH$

(d) $CH_2=CHCH_2CHCHCH_2CH_3$ with CH_2CH_2OH and Cl substituents

---

2.9 Isomerism

2.9.1 Constitutional isomers

In Table 2.2, butane and methylpropane have the same molecular formula, C_4H_{10}, but the atoms are bonded together differently; such compounds are called **constitutional isomers** (or *structural isomers* in the earlier literature). Ethanol and dimethyl ether (both have the molecular formula C_2H_6O) are also constitutional isomers. The feature which identifies a set of compounds of the same molecular formula as constitutional isomers is that they have *different atom connectivities*.

2.9.2 *cis–trans* isomerism

Another kind of isomerism is known in which isomers have the same atom connectivity but the spatial arrangements of their atoms are different; such isomers are called **stereoisomers**. The simplest type of stereoisomerism occurs amongst compounds with an alkene group and is known as ***cis–trans* isomerism**. But-1-ene and but-2-ene (C_4H_8) are constitutional isomers—their atoms are bonded together differently. But there are two

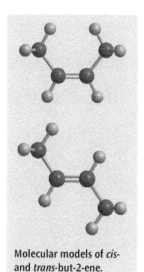

Molecular models of *cis*- and *trans*-but-2-ene.

structures of but-2-ene (below) which cannot be distinguished by the condensed formula on the left: one structure has the two methyl groups on the same side of the molecule and they are on opposite sides in the other isomer. These stereoisomers are called *cis* and *trans* isomers, respectively. This pair of *cis* and *trans* isomers exist as different compounds, each with its own properties, because there is no free rotation about the central C=C double bond (as there is about a C–C single bond); see Sub-section 2.5.2. Pairs of stereoisomers such as these, which cannot be interconverted without bond breaking, are also known as **configurational** isomers, and the property which distinguishes them is their **configuration**. (Other types of stereoisomerism will be discussed in Chapters 4 and 11.)

$CH_3CH{=}CHCH_3$

but-2-ene

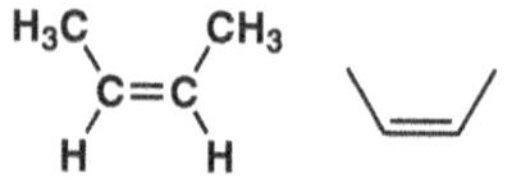

cis-but-2-ene

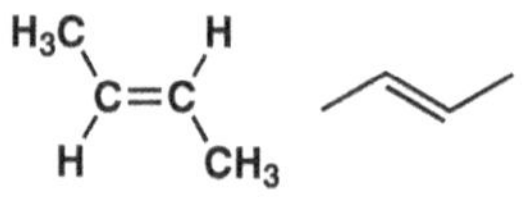

trans-but-2-ene

Exercise 2.9

Give a line-angle drawing of an alkene constitutionally isomeric with but-1-ene and but-2-ene.

Exercise 2.10

Give line-angle drawings of all isomeric alkenes with the molecular formula C_5H_{10}, and identify any stereoisomeric relationships.

2.9.3 *E,Z* nomenclature for *cis–trans* isomers of alkenes

Stereoisomers of a simple 1,2-disubstituted alkene such as but-2-ene are named *cis* or *trans* depending on whether the two substituents are on the same or opposite sides of the double bond. However, the terms *cis* and *trans* can be ambiguous if there are three or four substituents on the carbons of the double bond. For example, compounds **1** and **2** shown below are stereoisomers of 3-methylpent-2-ene, but it is not obvious from what we have seen so far which should be called *cis* and which should be called *trans*.

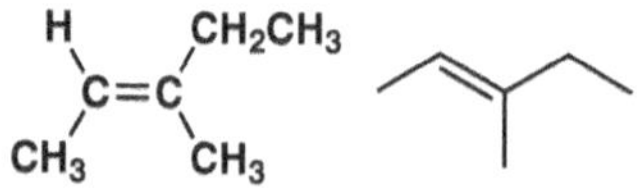

1

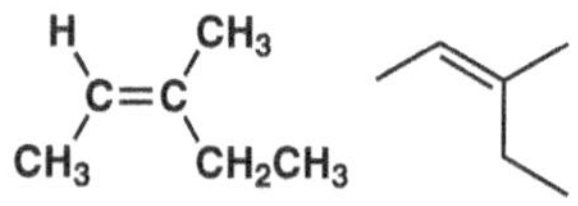

2

Stereoisomers of 3-methylpent-2-ene.

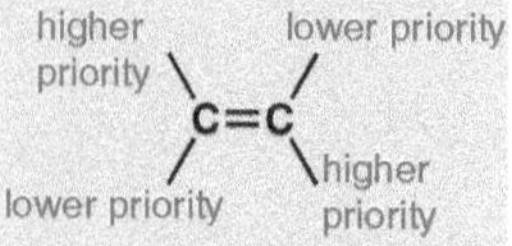

higher priority

lower priority

lower priority

higher priority

E

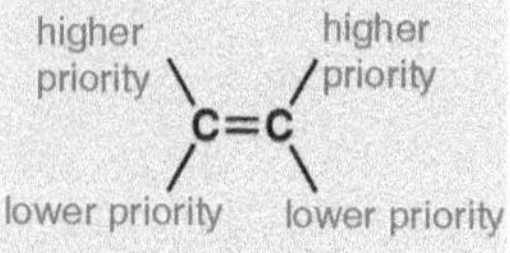

higher priority

higher priority

lower priority

lower priority

Z

To avoid this kind of ambiguity in the naming of stereoisomeric alkenes, the ***E,Z*** **nomenclature** for *cis–trans* isomers was devised.

First, we establish the **priority** of the two groups at one end of the double bond, and then of the two groups at the other end. The stereoisomer with the two higher priority groups on the same side of the double bond is called *Z* (from the German word, *zusammen* meaning 'together'). The stereoisomer with the two higher priority groups on opposite sides of the double bond is called *E* (from the German word *entgegen*, meaning 'opposite').

Priorities are assigned to atoms and groups according to the **Cahn–Ingold–Prelog (CIP) sequence rules**. The principal rules are as follows.

Rule 1: *Of two atoms, the one with the higher atomic number is of higher priority.*

$$\underset{1}{H} < \underset{6}{C} < \underset{7}{N} < \underset{8}{O} < \underset{9}{F} < \underset{16}{S} < \underset{17}{Cl} < \underset{35}{Br} < \underset{53}{I}$$

The *trans* and *cis* isomers of but-2-ene are assigned *E* and *Z* configurations, respectively, because the methyl group is of higher priority than H and the two methyl groups are on opposite sides in the *trans* isomer (so this is *E*) and on the same side in the *cis* isomer (so this is *Z*).

From left: Cahn, Ingold, and Prelog (1966).

(Kindly supplied by J. Seeman.)
R.S. Cahn (1899–1981, UK), C.K. Ingold (1893–1970, see Chapter 12), and V. Prelog (1906–1998; Prelog was born in Bosnia-Herzegovina to Croatian parents, but became a Swiss citizen in 1959; he shared the 1975 Nobel Prize in Chemistry with Australian, J.W. Cornforth).

higher H$_3$C   H lower

lower H   CH$_3$ higher

(*E*)-but-2-ene

higher H$_3$C   CH$_3$ higher

lower H   H lower

(*Z*)-but-2-ene

Rule 2: *If the first atoms of two groups are the same, look at the second, third, or fourth atoms, etc., until a difference is found.*

So, ethyl -CH$_2$CH$_3$ is of higher priority than methyl -CH$_3$, and -CH$_3$ is of higher priority than -H. This makes compound **1** above (*E*)-3-methylpent-2-ene, and its stereoisomer (**2**) with the opposite configuration is (*Z*)-3-methylpent-2-ene.

lower H   CH$_2$CH$_3$ higher

higher CH$_3$ CH$_3$ lower

1

(*E*)-3-methylpent-2-ene

lower H   CH$_3$ lower

higher CH$_3$ CH$_2$CH$_3$ higher

2

(*Z*)-3-methylpent-2-ene

Rule 3: *Multiply-bonded atoms are treated as though they are duplicated or triplicated.*

For example, ethenyl, ethynyl, and formyl groups are treated as illustrated:

An order of increasing priorities, therefore, is as follows:

$$-CH_2CH_3 \; < \; -CH(CH_3)_2 \; < \; -CH{=}CH_2 \; < \; -C{\equiv}CH \; < \; -CH_2OH \; < \; -CH{=}O$$

Exercise 2.11

Which of each pair of groups is of higher priority according to the CIP rules?

(a) -CH(CH$_3$)$_2$, -CH$_2$CH(CH$_3$)$_2$ (b) -F, -Cl   (c) -OCH$_3$, -N(CH$_3$)$_2$

(d) -Cl, -SCH$_3$ (e) -CH=CH$_2$, -C(CH$_3$)$_3$

Exercise 2.12

Assign *E* or *Z* configuration to the following compounds.

(a)   CH$_3$ CH$_2$CH$_3$ / Cl   OCH$_3$ (b)   H   H—C=O / CH$_3$ CH$_2$OH

Summary

○ Shapes of molecules are determined by electrostatic repulsion between electron pairs (bonding pairs and lone pairs) according to the **valence shell electron pair repulsion (VSEPR) model**.

○ **Molecular orbitals** (MOs) which correspond to covalent bonds are formed by the overlap of atomic orbitals (AOs).

○ AOs on the same atom can be hybridized prior to bonding which results in new AOs called **hybrid orbitals**; sp³ hybrid orbitals account for the tetrahedral structures of saturated carbon compounds, sp² hybrids account for the trigonal planar structures of carbon compounds containing double bonds, and sp hydrids account for the linear structures of carbon compounds with triple bonds and 1,2-dienes.

○ Covalent bonds may be classified as **σ bonds** (formed by end-on overlap of AOs with p character and from s AOs) or **π bonds** (formed by sideways overlap of p AOs); a σ bond and its associated MOs are circularly symmetrical about the axis of the bond whereas a π bond and its associated MOs have nodal planes which contain the axis of the bond.

○ Organic structures can be represented by shorthand condensed formulas or line-angle drawings.

○ Isomers are different compounds with the same molecular formulas. There are two principal kinds of isomers, constitutional isomers and stereoisomers. The former have different atom connectivities; stereoisomers have the same atom connectivities but different spatial arrangements.

○ *cis–trans* Isomerism of alkenes is one type of stereoisomerism; *cis* and *trans* isomers may be named unambiguously by *E,Z* nomenclature based on the Cahn–Ingold–Prelog sequence rules.

Problems

2.1 Predict approximate bond angles in the following molecules according to VSEPR theory.

(a) CCl₄   (b) H₃O⁺   (c) H₂C=CHCl   (d) HO—NH₂

2.2 Predict approximate bond angles around the coloured atoms in the following molecules according to VSEPR theory.

(a) ClCH₂—C(H)=CH₂

(b) H₃C—C(=O)—OH

(c) H₃C—C(H)=NH

(d) H₃C—C≡N

2.3 Indicate the hybridization of the atoms other than hydrogen in the following, and identify the atomic orbitals participating in bonding. Also, identify the types of bonds involved (σ or π).

(a) CH₃COOH   (b) NH₄⁺   (c) CH₃CN   (d) B(CH₃)₃

2.4 Indicate the hybridization of all atoms other than hydrogen in the following molecules.

(a)   (b) [pyridine ring with NH₂ substituent]

2.5 Give line-angle drawings of the following molecules.

(a) (CH₃)₃CCH₂CH(CH₃)₂   (b) CH₃CH₂OCH₂COOH

(c) CH₃CH(CH₂CH₃)CH₂CH₂CH(OH)CH₃

(d) CH₃CH₂CH(CH(CH₃)₂)CH₂COCH₃

2.6 Give line-angle drawings of the following molecules. If *cis–trans* isomers are possible, draw the structure of the all-*trans* isomer.

(a) HOCH(CH₃)C≡CC(CH₃)HOH

(b) (CH₃)₂C=CHCH=CHC(CH₃)=CHCH₂OH

(c) (CH₃)₂C=CHCOCH₃

(d) [cyclohexene ring] H₂C, C, C, CH=CHC(CH₃)=CHCH=CHC(CH₃)=CHCH₂OH, H₂C, C, C, CH₃, H₂

2.7 Give the molecular formulas of the following.

(a)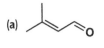

(b)

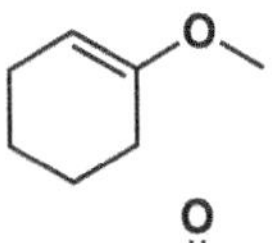

(c)

(d)

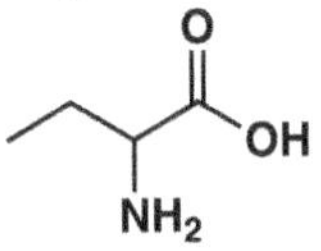

2.8 Give a condensed formula (without showing stereoisomerism) for each of the following line-angle structures.

(a)

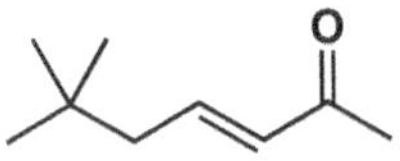

(b)

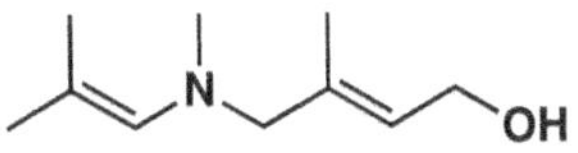

(c)

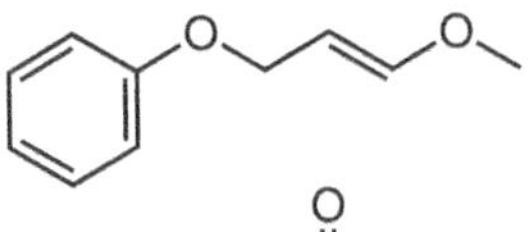

(d)

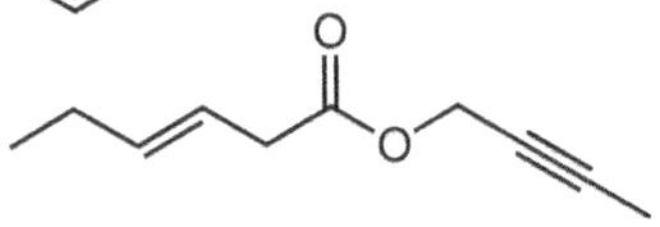

2.9 Indicate whether each of the following pairs of molecules are constitutional isomers, stereoisomers, the same compound, or none of these.

(a) 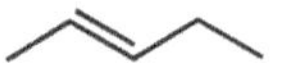   and

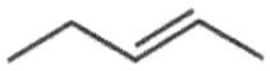

(b) 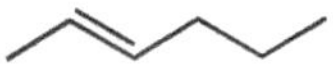   and

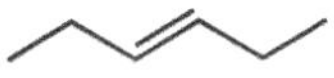

(c) 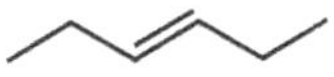   and

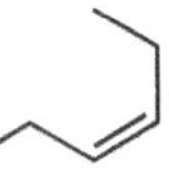

(d)    and

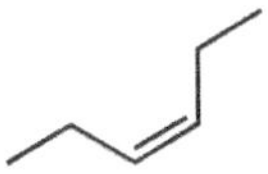

(e)    and

(f)    and

2.10 Give condensed formulas for all the possible constitutional isomers with the molecular formula $C_4H_{10}O$.

2.11 Which of each of the following pairs of groups is of higher priority according to the CIP rules?

(a) $-CH_3$ and $-NH_2$ (b) $-CH_2OH$ and $-CH(CH_3)_2$

(c) $-\overset{\overset{O}{\|}}{C}OCH_3$ and $-\overset{\overset{O}{\|}}{C}N(CH_3)_2$ (d) 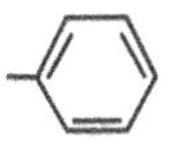 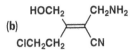 and $-C(CH_3)_3$

(e) $-C\equiv CH$ and $-C\equiv N$ (f) $-C\equiv N$ and $-CH=NCH_3$

2.12 Assign E or Z configuration to the following compounds (Ph stands for the phenyl group, C_6H_5).

(a)

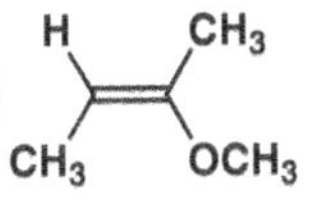

(b)

(c)

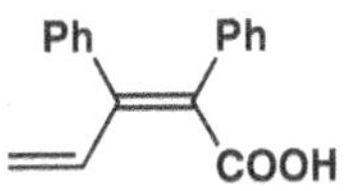

(d)

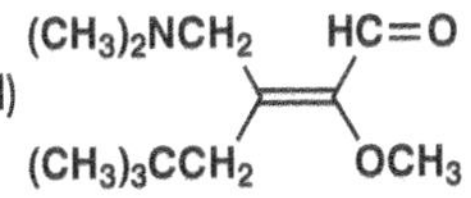

2.13 Give line-angle drawings for all the possible constitutional isomers with molecular formula C_5H_8 according to the following categories.

(a) Acyclic isomers.

(b) An isomer with a five-membered ring.

(c) Isomers with a four-membered ring.

(d) Isomers with one three-membered ring.

(e) An isomer with two three-membered rings.

2.14 Sketch lobes of the atomic orbitals used to form the σ and π bonds in carbon dioxide (cf. Figure 2.16). See Exercise 2.5.

3 Organic Compounds: their Functional Groups, Intermolecular Interactions, and Physical Properties

Related topics we have already covered:

- **Chemical bonding** (Sections 1.2 and 2.2)
- **Lewis structures of molecules** (Section 1.3)
- **Structural formulas** (Section 2.8)
- **Isomerism** (Section 2.9)

Topics we shall cover in this chapter:

- **Functional groups**
- **Classification of organic compounds**
- **Elements of organic nomenclature**
- **Intermolecular interactions and physical properties of organic compounds**

It was recognized at an early stage in the development of organic chemistry that a molecule of an organic compound generally consists of what we now call a functional group and a hydrocarbon residue, and that the distinctive properties of the compound are principally due to the functional group. This is very encouraging information for the student of organic chemistry because it means that knowledge of the properties of a relatively small number of functional groups allows reliable predictions of the chemical properties and reactions of an enormous number of different organic compounds. The functional group consists of a few atoms bonded together and they react as a unit. Consequently, organic compounds are classified and named according to their functional groups and, in this chapter, we shall identify the principal classes of organic compounds which are covered in this book, and how to name them. We shall also see how some properties of a compound can only be understood in terms of large assemblies of molecules and the physical interactions between them (as opposed to properties of individual isolated molecules).

3.1 Functional Groups

A **functional group** is an identifiable group of connected atoms in a molecule that is mainly responsible for the characteristic physical and chemical properties of the compound. The properties and reactions of a functional group are, to a large degree, independent of an attached hydrocarbon residue. The huge and increasing numbers of known organic compounds are classified according to a relatively small number of functional groups, which brings order to an otherwise bewildering range of compounds. Table 3.1 summarizes the main families of organic compounds and their functional groups.

Identify the functional groups within each of the following compounds.

(a) glyceraldehyde (the smallest carbohydrate)

(b) threonine (an amino acid)

(c) α-ionone (the scent of violets)

(d) atenolol (Tenormin®) (a cardioselective β-blocker)

3.2 Hydrocarbons

Compounds containing only carbon and hydrogen are called **hydrocarbons**. They are divided into saturated and unsaturated hydrocarbons according to the nature of the bonds between the C atoms (Figure 3.1). Compounds containing only C–H and C–C single bonds (alkanes) are said to be *saturated*, while those only containing at least one double or triple bond are *unsaturated* (alkenes, alkynes, and arenes).

3.2.1 Alkanes and cycloalkanes

Alkane is the preferred technical name for a saturated open-chain (or *acyclic*) hydrocarbon and they are collectively represented by the general formula C_nH_{2n+2}; they are also called *aliphatic hydrocarbons* (or sometimes *paraffins* in the older literature). Alkanes are the main component of petroleum as it occurs in nature (see Panel 3.1). Because alkanes have only C–C and C–H single bonds and no functional groups, they are unreactive (although they burn readily). Simple alkanes have a linear chain of the form $CH_3(CH_2)_{n-1}H$, and are named according to the number of carbon atoms, n, as listed in Table 3.2.

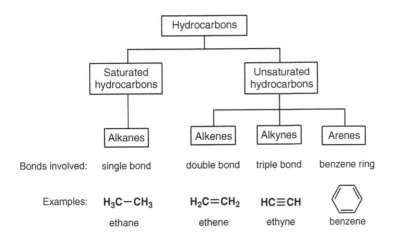

Figure 3.1 Classification of hydrocarbons.

Table 3.1

Classes of organic compounds and functional groups

Class	General form	Functional group		Example	
alkane	C_nH_{2n+2} (RH)	none	(C–C/C–H bonds)	CH_3CH_3	ethane
alkene	C_nH_{2n}	$C{=}C$	double bond	$CH_2{=}CH_2$	ethene
alkyne	C_nH_{2n-2}	$-C{\equiv}C-$	triple bond	$HC{\equiv}CH$	ethyne
arene	ArH	(benzene ring)	benzene ring	(benzene)	benzene
haloalkane	RX (X = F, Cl, Br, or I)	$-X$	halogeno	CH_3CH_2Cl	chloroethane (ethyl chloride)
alcohol	ROH	$-OH$	hydroxy	CH_3CH_2OH	ethanol (ethyl alcohol)
ether	ROR	$-OH$	alkoxy	$(C_2H_5)_2O$	ethoxyethane (diethyl ether)
amine	RNH_2, R_2NH, R_3N	$-NR_2$ (R = H or alkyl)	amino	$CH_3CH_2NH_2$	ethanamine (ethylamine)
aldehyde	RCHO	$C{=}O$ / $-\overset{O}{C}-H$	carbonyl / formyl	CH_3CHO	ethanal (acetaldehyde)
ketone	R_2CO	$C{=}O$ / ${=}O$	carbonyl / oxo	CH_3COCH_3	propanone (acetone)
carboxylic acid	RCO_2H	$-\overset{O}{C}-OH$	carboxy	CH_3CO_2H	ethanoic acid (acetic acid)
ester	RCO_2R	$-\overset{O}{C}-OR$	alkoxycarbonyl	$CH_3CO_2C_2H_5$	ethyl ethanoate (ethyl acetate)

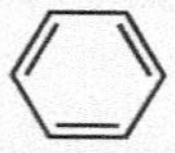

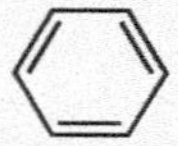

Classes of organic compounds and functional groups (Continued)

Class	General form	Functional group		Example	
acid anhydride	$(RCO)_2O$	$\overset{O}{\overset{\|}{C}}-O-\overset{O}{\overset{\|}{C}}$		$(CH_3CO)_2O$	acetic anhydride
acyl halide	RCOX	$\overset{O}{\overset{\|}{C}}-X$	haloformyl	CH_3COCl	acetyl chloride
amide	$RCONR_2$ (R=H or alkyl)	$\overset{O}{\overset{\|}{C}}-NR_2$	carbamoyl	CH_3CONH_2	ethanamide (acetamide)
nitrile	RCN	$-C\equiv N$	cyano	CH_3CN	ethanenitrile (acetonitrile)
nitro compound	RNO_2	$\overset{O}{\overset{\|}{\underset{}{N^+}}}-O^-$	nitro	CH_3NO_2	nitromethane
thiol	RSH	$-SH$	mercapto	CH_3CH_2SH	ethanethiol
sulfide	RSR	$-SR$	alkylthio	CH_3SCH_3	methylthiomethane (dimethyl sulfide)
sulfoxide	RS(O)R	$\overset{O}{\overset{\|}{S}}-R$	alkanesulfinyl	CH_3SOCH_3	dimethyl sulfoxide
sulfone	RSO_2R	$\overset{O}{\overset{\|}{\underset{\underset{O}{\|}}{S}}}-R$	alkanesulfonyl	$CH_3SO_2CH_3$	dimethyl sulfone
sulfonic acid	RSO_3H	$-SO_3H$	sulfo	$C_6H_5SO_3H$	benzenesulfonic acid

Panel 3.1 Organic resources: coal, oil, and natural gas

Organic compounds can be synthesized in the laboratory and manufactured on an industrial scale, but the raw materials for all such processes ultimately come from natural sources. Organic materials in nature are made, or have been made in the distant past, by living organisms. Natural gas, oil (petroleum), and coal (known collectively as *fossil fuels*) are believed to have been formed over millions of years from prehistoric organisms, although the origin of oil is still a matter of debate. These fossil fuels support our modern life as principal sources of energy, and as raw materials for chemical manufacturing industries. However, they are non-renewable and known reserves, especially of oil and gas, will be exhausted in the near future. Furthermore, burning fossil fuels produces enormous amounts of CO_2 which is implicated as a so-called greenhouse gas in the controversial phenomenon of global warming (climate change).

The primary sources of alkanes are natural gas and crude oil. Natural gas is mainly methane with smaller amounts of other compounds including ethane, propane, and butane. Crude oil is a complex mixture of compounds, principally alkanes and other hydrocarbons. Coal has a very complicated and variable structure from which a wide range of aromatic compounds (some containing heteroatoms) can be obtained.

Natural gas, like coal, is used directly as fuel whereas crude oil is first fractionally distilled to give fractions of different boiling points, as summarized below. The lower boiling fractions are in great demand as motor fuels, fuel oils, and aviation fuels. The petroleum industry has also developed methods for converting higher boiling alkanes into ones of lower bp for which there is a greater demand. *Catalytic cracking* is the name of the process by which long-chain alkanes are converted into hydrocarbons of low molecular weight; the *reforming* process converts unbranched alkanes into branched ones which have superior properties as motor fuel (higher 'octane rating', see below).

Typical fractions obtained by distillation of crude oil			
Fraction name	bp/°C	Carbon no.	Typical use
Petroleum gas	<20	C_1–C_4	Bottled gas (LPG)*
Light petroleum	20–120	C_5–C_7	Chemical feedstock
Gasoline	100–200	C_5–C_{12}	Automobile fuel
Kerosene	175–275	C_{12}–C_{16}	Jet fuel
Light fuel oil	250–350	C_{15}–C_{20}	Diesel fuel
Heavy oil	>350	C_{20}–C_{28}	Lubricants, heating oil
Residues	non-volatile		Asphalt (bitumen)

*LPG (liquefied petroleum gas) is principally either propane or butane.

Table 3.2

Linear alkanes and their boiling points							
No. of Cs (n)	Molecular formula	Name	bp/°C (1 atm)	No. of Cs (n)	Molecular formula	Name	bp/°C (1 atm)
1	CH_4	methane	−167.7	8	C_8H_{18}	octane	127.7
2	C_2H_6	ethane	−88.6	9	C_9H_{20}	nonane	150.8
3	C_3H_8	propane	−42.1	10	$C_{10}H_{22}$	decane	174.0
4	C_4H_{10}	butane	−0.5	11	$C_{11}H_{24}$	undecane	195.8
5	C_5H_{12}	pentane	36.1	12	$C_{12}H_{26}$	dodecane	216.3
6	C_6H_{14}	hexane	68.7	20	$C_{20}H_{42}$	icosane	343.0
7	C_7H_{16}	heptane	98.4	30	$C_{30}H_{62}$	triacontane	449.8

Table 3.2 includes only *unbranched* alkanes, but *branched* structures become possible (constitutional isomers, see Sub-section 2.9.1) when $n \geq 4$. The methyl in 2-methylpropane

Panel 3.1 Continued

Typical industrial fractionating towers.
(Michael Utech/istockphoto.)

Gasoline (petrol) is a complex mixture of hydrocarbons containing different proportions of C_5–C_{12} alkanes according to the source of the crude oil, and one aspect of its quality as a motor fuel is measured by its *octane rating* (or *number*). The highly branched alkane, iso-octane (2,2,4-trimethylpentane), ignites and burns very smoothly (without auto-ignition or *knocking*) in an internal combustion engine, while unbranched alkanes induce engine knocking. The tendency of a gasoline to cause knocking is quantified on a scale which includes isooctane with a rating of 100 and heptane (a poor fuel) with a rating of 0. The octane rating of a gasoline was originally the percentage of isooctane in a binary mixture of isooctane and heptane which has a performance equal to that of the gasoline, but it has been extended to accommodate fuels which are even better in this respect than isooctane. The octane rating of octane itself, being unbranched, is very poor (–20, i.e. worse than heptane) whereas ethanol, benzene, and toluene have ratings higher than isooctane (105, 106, and 120, respectively).

2,2,4-trimethylpentane
(isooctane, octane rating 100)

heptane
(octane rating 0)

Ethanol can be produced by fermentation of *biomass* and performs well as a motor fuel (high octane rating, but its heat of combustion is lower than that of gasoline); it is beginning to be used on a large scale in some countries (see Panel 14.1). Various additives can also be used to improve the octane rating of lower grade hydrocarbons. Tetraethyl-lead (Et₄Pb) was widely used in the past for this purpose, but its use in automobiles was abandoned because of fears about atmospheric lead pollution, and its adverse effects on catalytic converters in exhaust systems. It was replaced as a gasoline additive by methyl *t*-butyl ether (MTBE). However, this also leads to environmental problems: leaking underground gasoline tanks contaminate groundwater with MTBE which then diffuses widely owing to its solubility in water. Although the previously reported carcinogenicity of MTBE has been found to be low, it gives water an unpleasant taste at very low concentrations which is not easily removed. The above problems associated with fuels derived from oil have stimulated the search for more environmentally-friendly fuels for automobiles such as bioethanol and hydrogen, as well as the development of new automotive technologies, e.g. electric motor vehicles.

is the simplest **alkyl** group or substituent (an alkane lacking an H atom, i.e. with a *free valence*, is an *alkyl* group, usually represented generically by R, see later).

$CH_3CH_2CH_2CH_3$

butane (C_4H_{10})

CH_3
|
CH_3CHCH_3

2-methylpropane (C_4H_{10})

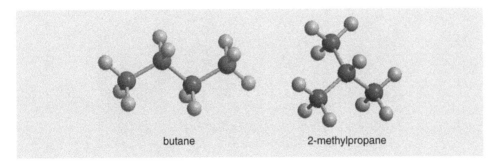

butane

2-methylpropane

Example 3.1

Give structural formulas of all the constitutional isomers of acyclic alkanes with five carbon atoms.

Solution

An acyclic C_5 alkane has the formula C_5H_{12}, and there are three constitutional isomers.

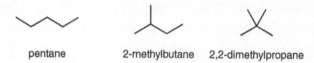

pentane   2-methylbutane   2,2-dimethylpropane

Exercise 3.2

Give condensed formulas for the constitutional isomers of C_5H_{12} given in Example 3.1.

In branched compounds, there are differently substituted carbons which may behave differently in chemical reactions. A carbon bonded to one, two, three, or four other carbons is called **primary** (often designated 1°), **secondary** (2°), **tertiary** (3°), or **quaternary** (4°), respectively, as shown below. Correspondingly, the hydrogen atoms bonded to each type of carbon are called primary, secondary, and tertiary, respectively.

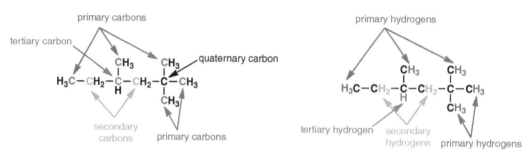

Exercise 3.3

Identify the primary, secondary, tertiary, and quaternary carbons in each constitutional isomer of C_5H_{12} given in Example 3.1.

Cyclic alkanes are called **cycloalkanes**, and have the molecular formula C_nH_{2n} or $(CH_2)_n$.

Examples of cycloalkanes:

cyclopropane   cyclobutane   cyclopentane   cyclohexane

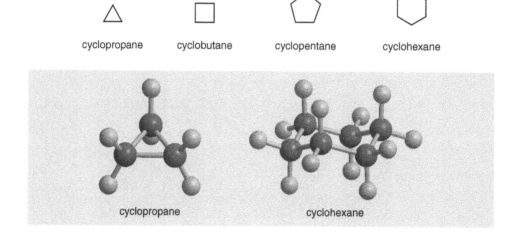

cyclopropane   cyclohexane

3.2.2 Alkenes and alkynes

An **alkene** has a C=C double bond and an **alkyne** has a C≡C triple bond. These unsaturated hydrocarbons show physical properties similar to those of alkanes, but their unsaturated bonds are sites of enhanced chemical reactivity (they undergo addition reactions) and count as functional groups. Rules for naming alkenes and alkynes, as well as other families of organic compounds, are given in Section 3.8.

Comparison of simple saturated and unsaturated hydrocarbons:

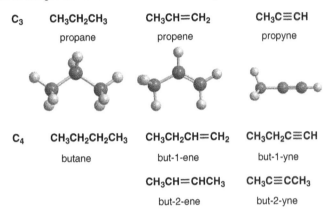

C_3	$CH_3CH_2CH_3$	$CH_3CH{=}CH_2$	$CH_3C{\equiv}CH$
	propane	propene	propyne

C_4	$CH_3CH_2CH_2CH_3$	$CH_3CH_2CH{=}CH_2$	$CH_3CH_2C{\equiv}CH$
	butane	but-1-ene	but-1-yne
		$CH_3CH{=}CHCH_3$	$CH_3C{\equiv}CCH_3$
		but-2-ene	but-2-yne

The pheromone by which a female housefly attracts a mate is the *cis* isomer of a C_{23} alkene:

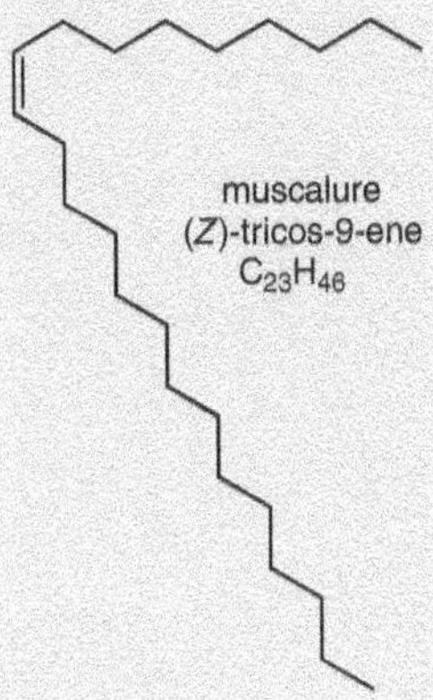

muscalure
(*Z*)-tricos-9-ene
$C_{23}H_{46}$

Note that but-2-ene has *cis* and *trans* isomers (see Sub-section 2.9.2).

Panel 3.2 Ethene as an industrial raw material

Ethene (ethylene) is produced industrially on a massive scale, principally by 'cracking' petroleum; world production in 2007 reached 1.21×10^8 tonnes. It is converted directly into polyethylene (polythene), and used to make starting materials for various other industrial chemicals used in manufacturing.

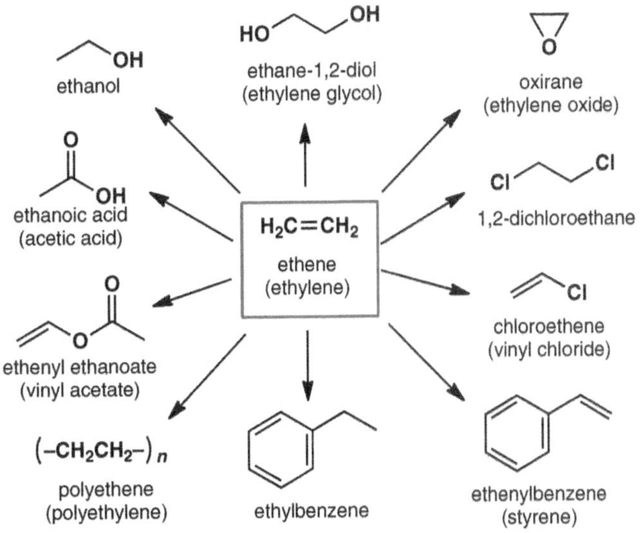

ethanol

ethane-1,2-diol
(ethylene glycol)

oxirane
(ethylene oxide)

ethanoic acid
(acetic acid)

1,2-dichloroethane

$H_2C{=}CH_2$
ethene
(ethylene)

chloroethene
(vinyl chloride)

ethenyl ethanoate
(vinyl acetate)

$(-CH_2CH_2-)_n$
polyethene
(polyethylene)

ethylbenzene

ethenylbenzene
(styrene)

Exercise 3.4

Cycloalkanes and acyclic alkenes are constitutional isomers. Draw structural formulas of cycloalkanes which are isomeric with butenes, C_4H_8.

Exercise 3.5

We considered constitutionally isomeric alkenes of molecular formula C_5H_{10} in Exercise 2.10 of Chapter 2. Draw structural formulas of cyclic constitutional isomers with the same molecular formula.

3.2.3 Arenes

Benzene and its derivatives are particular kinds of cyclic unsaturated compounds, and exhibit properties significantly different from those of cyclic alkenes and alkynes. They constitute a different and diverse family of compounds called **aromatic compounds**, and aromatic hydrocarbons are called **arenes**. The generic term for an arene with a free valence, e.g. $C_6H_5–$, phenyl (derived from benzene, C_6H_6), is **aryl**, its normal symbol being Ar (cf. R is used for alkyl). The characteristic feature of an arene is its planar ring structure with *delocalized* π electrons which are principally responsible for its distinctive physical and chemical properties. Arenes are usually represented with specific numbers of alternating double and single bonds (see Chapter 5).

benzene   toluene   naphthalene   anthracene

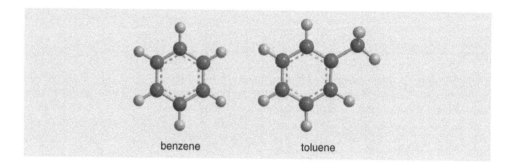

benzene   toluene

3.3 Alcohols, Ethers, and their Sulfur Analogues

The functional group of an **alcohol** is the **hydroxy** group, OH, and an alcohol may be regarded as a derivative of water in which one H atom has been replaced by an alkyl group; alternatively, it can be regarded as the product of replacing an H atom of an alkane with an OH group. If both H atoms of water are replaced by alkyl or aryl groups (the same or different) we have an **ether**. The sulfur analogues of alcohols and ethers are thiols and sulfides, respectively.

When R is an aryl, ROH is called a phenol.

H_2O	ROH	ROR'	RSH	RSR'	(R, R' = alkyl or aryl)
water	alcohol	ether	thiol	sulfide	

3.3.1 Alcohols

The simplest alcohol is methanol (methyl alcohol), and the next is ethanol (ethyl alcohol) which is the essential ingredient of intoxicating 'alcoholic beverages' and sometimes simply called *alcohol*.

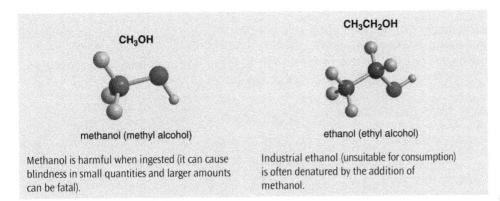

CH_3OH

methanol (methyl alcohol)

Methanol is harmful when ingested (it can cause blindness in small quantities and larger amounts can be fatal).

CH_3CH_2OH

ethanol (ethyl alcohol)

Industrial ethanol (unsuitable for consumption) is often denatured by the addition of methanol.

$HOCH_2CH_2OH$
ethane-1,2-diol
(ethylene glycol)
bp 197.3 °C, mp −12.9 °C
(used as an automotive antifreeze)

$HOCH_2CH(OH)CH_2OH$
propane-1,2,3-triol
(glycerol)
bp 290 °C, mp 17.8 °C
(Fats and oils are derived from glycerol; see Section 3.7.)

β-D-glucose
mp 150 °C

Alcohols of low molecular weight have some properties similar to those of water and are completely miscible with water (Table 3.3). This is principally a consequence of hydrogen bonds, an important type of intermolecular interaction which we shall discuss in Section 3.9.2.

Alcohols are classified as primary, secondary, or tertiary depending on the substitution pattern of the carbon bearing the hydroxy group. Some compounds have two or more hydroxy groups and are called diols, triols, ..., or polyols; glucose and other carbohydrates are polyols. In general, the more hydroxy groups there are in a molecule, the more soluble the compound is in water, and the higher its bp and mp.

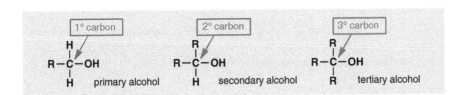

1° carbon	2° carbon	3° carbon
primary alcohol	secondary alcohol	tertiary alcohol

Names of some alcohols and their physical properties				
Chemical formula	Name	mp/°C	bp/°C (1 atm)	Solubility[a]
CH_3OH	methanol	−97	64.7	∞
CH_3CH_2OH	ethanol	−114	78.3	∞
$CH_3CH_2CH_2OH$	propan-1-ol	−126	97.2	∞
$CH_3CH(OH)CH_3$	propan-2-ol	−88	82	∞
$CH_3(CH_2)_3OH$	butan-1-ol	−85	118	7.9
$(CH_3)_3COH$	2-methylpropan-2-ol	25	82	∞
$CH_3(CH_2)_4OH$	pentan-1-ol	−78	138	2.3

Table 3.3

a. Solubility in g/100 g of water at 25 °C.

Exercise 3.6

Give structural formulas of the constitutionally isomeric alcohols with molecular formula $C_4H_{10}O$, and classify them as primary, secondary, or tertiary.

3.3.2 Ethers

An **ether** has two groups, alkyl or aryl, bonded to oxygen and diethyl ether (sometimes simply called *ether*) is representative. It is a volatile mobile liquid (bp 34.6 °C) whose systematic name is ethoxyethane (Section 3.8), and was once used as an anaesthetic. Polyethers and cyclic ethers are also well known.

$CH_3CH_2OCH_2CH_3$

ethoxyethane (diethyl ether)

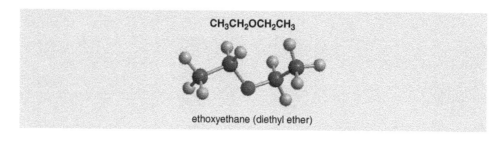

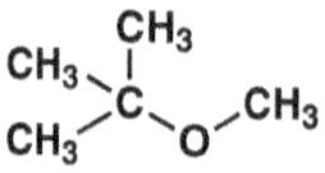

2-methoxy-2-methylpropane
(*t*-butyl methyl ether, TBME)

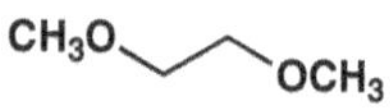

1,2-dimethoxyethane

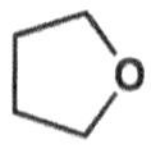

oxolane
(tetrahydrofuran, THF)

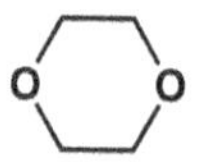

1,4-dioxane

As we shall see later, the physical properties of ethers are more like those of alkanes than of alcohols, but they have a few chemical similarities to alcohols (see Chapter 14).

3.3.3 Thiols

Thiols were also called mercaptans in the older literature.

A **thiol** is the sulfur analogue of an alcohol and has the **mercapto** functional group, SH. Thiols of low boiling point are notorious for their strong and often unpleasant odours.* The following are some simple examples. Thiol groups are often found in biomolecules too.

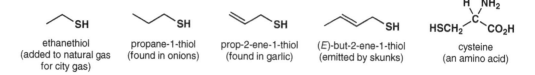

ethanethiol
(added to natural gas
for city gas)

propane-1-thiol
(found in onions)

prop-2-ene-1-thiol
(found in garlic)

(*E*)-but-2-ene-1-thiol
(emitted by skunks)

cysteine
(an amino acid)

*However, the distinctive flavour of grapefruit is attributable to a thiol:

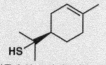

(*R*)-2-(4-methylcyclohex-
3-enyl)propane-2-thiol
(grapefruit mercaptan)

3.4 Haloalkanes

Compounds formed by replacing a hydrogen atom of an alkane with a halogen atom are called **haloalkanes** or **alkyl halides**, RX. These resemble alkanes in their physical properties but they are much more chemically reactive (see Chapters 12 and 13). Monohaloalkanes are classified as primary, secondary, or tertiary depending on the substitution pattern of the carbon bearing the halogen atom; polyhalogenated alkanes are also well known.

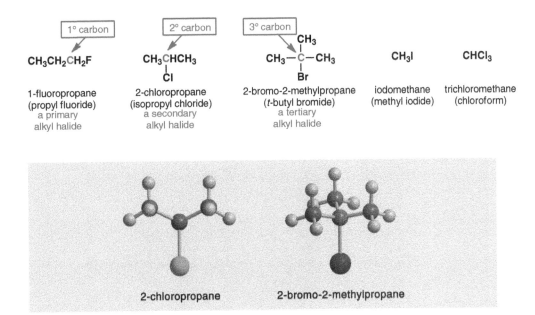

1° carbon	2° carbon	3° carbon		
$CH_3CH_2CH_2F$	$CH_3\overset{\underset{Cl}{\textstyle |}}{C}HCH_3$	$CH_3-\overset{\overset{\textstyle CH_3}{\textstyle |}}{\underset{\underset{\textstyle Br}{\textstyle |}}{C}}-CH_3$	CH_3I	$CHCl_3$
1-fluoropropane (propyl fluoride) *a primary alkyl halide*	2-chloropropane (isopropyl chloride) *a secondary alkyl halide*	2-bromo-2-methylpropane (*t*-butyl bromide) *a tertiary alkyl halide*	iodomethane (methyl iodide)	trichloromethane (chloroform)

2-chloropropane   2-bromo-2-methylpropane

3.5 Nitrogen Compounds

3.5.1 Amines

Amines are carbon derivatives of ammonia, NH_3, and are classified as **primary** (RNH_2), **secondary** (RR^1NH), or **tertiary** (RR^1R^2N) according to the number of carbon atoms attached to the nitrogen; R, R^1, and R^2 may be alkyl or aryl, and the same or different (and there are many cyclic amines as well). Correspondingly, NH_2 and its substituted variants, NHR and NRR^1, are referred to as primary, secondary, and tertiary amino groups. Note that the classification of amines is not the same as the one for alcohols and alkyl halides; it is based on the substitution pattern of the *nitrogen*, not of an alkyl group attached to the nitrogen. For example, butan-2-amine is a primary amine with a secondary alkyl group.

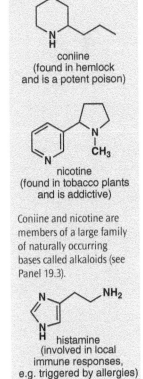

coniine
(found in hemlock
and is a potent poison)

nicotine
(found in tobacco plants
and is addictive)

Coniine and nicotine are
members of a large family
of naturally occurring
bases called alkaloids (see
Panel 19.3).

histamine
(involved in local
immune responses,
e.g. triggered by allergies)

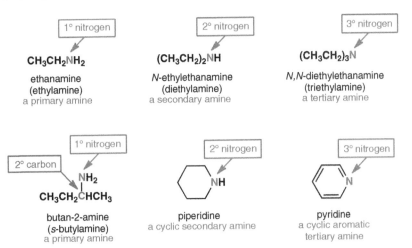

1° nitrogen	2° nitrogen	3° nitrogen
$CH_3CH_2NH_2$	$(CH_3CH_2)_2NH$	$(CH_3CH_2)_3N$
ethanamine (ethylamine) *a primary amine*	*N*-ethylethanamine (diethylamine) *a secondary amine*	*N,N*-diethylethanamine (triethylamine) *a tertiary amine*

2° carbon → 1° nitrogen

$CH_3CH_2\overset{\overset{\textstyle NH_2}{\textstyle |}}{C}HCH_3$

butan-2-amine
(*s*-butylamine)
a primary amine

2° nitrogen

NH

piperidine
a cyclic secondary amine

3° nitrogen

N

pyridine
a cyclic aromatic tertiary amine

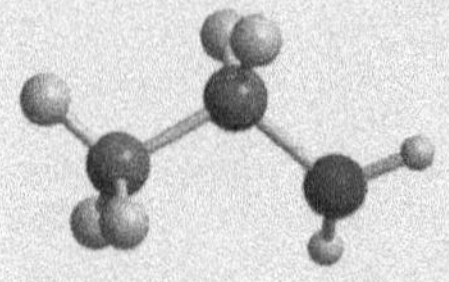

ethanamine    piperidine

Write structural formulas of amines with molecular formula C_3H_9N, and classify them as primary, secondary, or tertiary.

Solution

Primary amines: $CH_3CH_2CH_2-NH_2$ and $(CH_3)_2CH-NH_2$

Secondary amine: CH_3CH_2-NH
$\qquad\qquad\qquad\qquad |$
$\qquad\qquad\qquad\quad CH_3$

Tertiary amine: CH_3-N-CH_3
$\qquad\qquad\qquad\qquad\quad |$
$\qquad\qquad\qquad\qquad CH_3$

Write structural formulas of three secondary amines with molecular formula $C_4H_{11}N$.

Amines act as bases since the nitrogen has a lone pair of electrons which can accept a proton (Chapter 6), and are widely found in nature (e.g. we include amino acids, proteins, and nucleic acids in Chapter 24).

3.5.2 Nitro compounds

CH_3NO_2

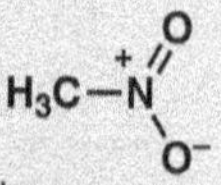

$H_3C-\overset{+}{N}\overset{\displaystyle O}{\underset{O^-}{\diagdown}}$

nitromethane

Nitroalkanes and nitroarenes are known and both are important intermediates for organic synthesis; nitromethane is a typical non-protic polar solvent (see Sub-section 12.3.2) and its electronic structure was discussed in Section 1.3.2. Polynitro compounds are unstable and may be explosive, e.g. trinitrotoluene (TNT).

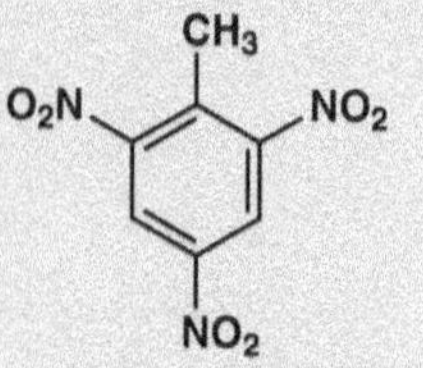

2,4,6-trinitrotoluene
(TNT)

Nitroglycerin (1,2,3-trinitroxypropane) is widely used as an explosive, e.g. as a main component of dynamite. Strictly, it is not a nitro compound but an ester of nitric acid as an alternative name, glyceryl trinitrate, indicates. It is also used medically as a vasodilator to treat heart conditions such as angina, and may be administered as tablets, patches, or sprays which are sometimes called 'nitro'. It releases nitric oxide, NO, which relaxes constricted blood vessels and lowers the blood pressure (nitric oxide is a natural vasodilator in the body).

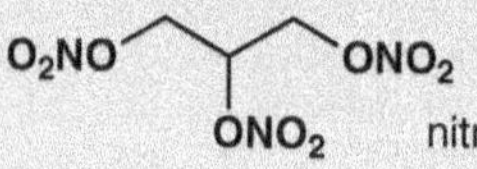

O_2NO $\qquad\qquad$ ONO_2
$\qquad\quad ONO_2$ $\quad$ nitroglycerin

3.6 Aldehydes and Ketones

Aldehydes and **ketones** are both **carbonyl compounds**. In an aldehyde, the C=O group bears one H atom and an alkyl or aryl group; in a ketone, the H of an aldehyde is replaced by another alkyl or aryl group.

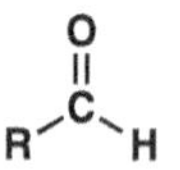

$$R-\overset{\displaystyle O}{\overset{||}{C}}-H \qquad\qquad R-\overset{\displaystyle O}{\overset{||}{C}}-R' \qquad (R,R' = \text{alkyl or aryl})$$

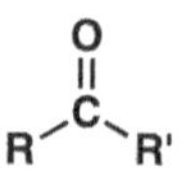

aldehyde $\qquad\qquad\qquad$ ketone

The bond of the carbonyl group is a typical polar unsaturated bond, and highly reactive (Chapter 8). Because aldehydes have an H atom on the carbonyl carbon, they are readily oxidized to carboxylic acids (Section 14.4). Some sugars show properties of being aldehydes (e.g. glucose) and others of being ketones (e.g. fructose), see Chapter 24.

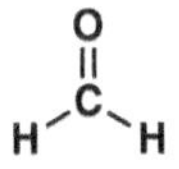

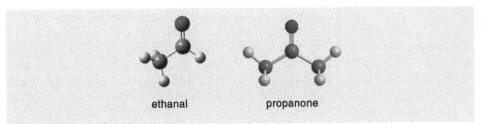

methanal
(formaldehyde)

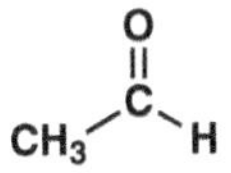

ethanal
(acetaldehyde)

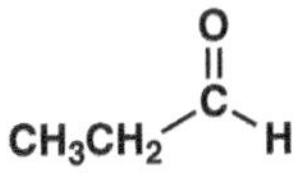

propanal
(propionaldehyde)

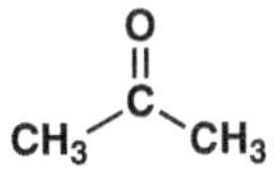

propanone
(acetone)

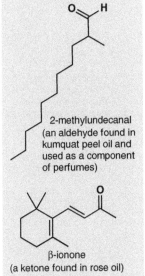

2-methylundecanal
(an aldehyde found in kumquat peel oil and used as a component of perfumes)

β-ionone
(a ketone found in rose oil)

ethanal

propanone

Write structural formulas of three ketones with molecular formula $C_5H_{10}O$.

Exercise 3.8

3.7 Carboxylic Acids and their Derivatives

The functional group of **carboxylic acids** is the **carboxy group**, COOH; carboxylic acids can be seen as aldehydes with the H on the carbonyl replaced by an OH, and they show acidic properties. Ethanoic acid (acetic acid) is the main ingredient of vinegar and is responsible for its distinctive smell and taste.

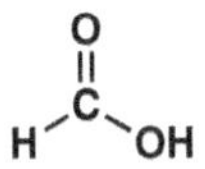

methanoic acid
(formic acid)

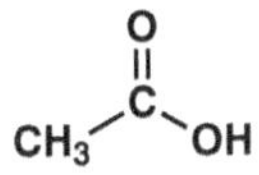

ethanoic acid
(acetic acid)

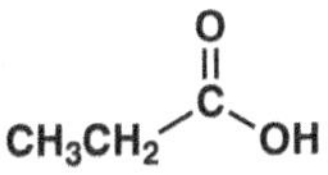

propanoic acid
(propionic acid)

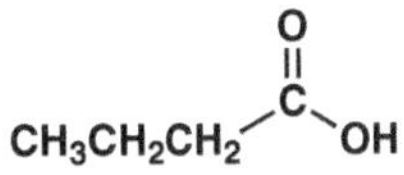

butanoic acid
(butyric acid)

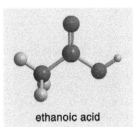

ethanoic acid

Compounds derived from carboxylic acids by replacement of the hydroxy group by a halogen, or another group bonded through a heteroatom (usually N or O), are called **carboxylic acid derivatives**. They include esters, acid anhydrides, amides, and acid halides, and all give the parent carboxylic acids when hydrolysed (see Chapter 9).

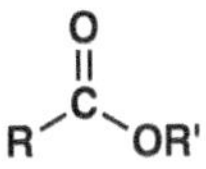

ester

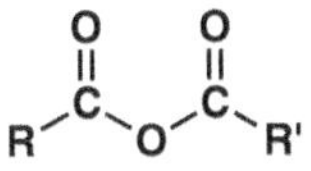

acid anhydride

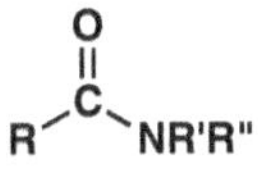

amide
(R',R" = H, alkyl, or aryl)

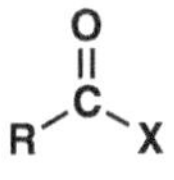

acyl halide (acid halide)
(X = halogen)

Write structural formulas of two carboxylic acids with molecular formula $C_4H_8O_2$.

Exercise 3.9

Fats and vegetable oils are glyceryl esters of long-chain carboxylic acids (fatty acids), see Chapter 24; glycerol is propane-1,2,3-triol. Amino acids, from which proteins are formed, are carboxylic acids bearing an amino substituent.

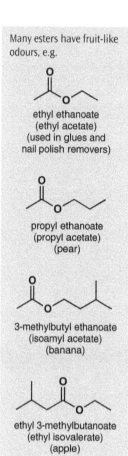

Many esters have fruit-like odours, e.g.

ethyl ethanoate
(ethyl acetate)
(used in glues and
nail polish removers)

propyl ethanoate
(propyl acetate)
(pear)

3-methylbutyl ethanoate
(isoamyl acetate)
(banana)

ethyl 3-methylbutanoate
(ethyl isovalerate)
(apple)

*If the suffix begins with a consonant, the final 'e' of the name of the alkane is left on, thus *ethanol* but *ethanenitrile* (acetonitrile).

**Alternatively, the locant for an infix or suffix can come *before* the parent name, so 3-methylbut-*2*-en-1-ol could also be called 3-methyl-2-buten-1-ol. As another simple example, butan-*1*-ol could be called *1*-butanol. However, according to a 1993 IUPAC recommendation, the locant is better placed just before the infix or suffix.

$CH_3CH_2CH_2CH_2OH$

butan-1-ol (1-butanol)

$$H_2C-OC(O)CH_2(CH_2)_nCH_3$$
$$HC-OC(O)CH_2(CH_2)_nCH_3$$
$$H_2C-OC(O)CH_2(CH_2)_nCH_3$$

typical components of fats and oils (*n* = 9–19)

amino acid

3.8 Elements of Organic Nomenclature

3.8.1 IUPAC nomenclature

Organic compounds were once named fairly arbitrarily on the basis of their source or, perhaps, some particular property. Some natural products, and other compounds with complicated structures, are still known by their common names, but it was recognized early in the last century that a systematic protocol was essential for naming the ever increasing number of new compounds being discovered or synthesized. We now name organic compounds according to the rules (or system of **nomenclature**) recommended by the International Union of Pure and Applied Chemistry (IUPAC). In this book, we generally use systematic IUPAC names along with some common names allowed by IUPAC. When both are used, the common names are shown in parentheses.

According to the IUPAC system, the name of an organic compound is based on the name of the parent alkane (or arene—see later) of which it is a derivative, if this is possible; the names of alkanes are based on the number of carbons as listed in Table 3.2 (p. 46). The main functional group is generally indicated by a suffix appended to the name of the alkane after the final 'e' has been deleted;* other functional groups and substituents are indicated by prefixes with the position of each shown by its *locant* (the C-number along the backbone to which it is bonded)—see Figure 3.2. So, the form of a typical IUPAC name runs *locant–prefix–parent–locant–suffix*. If the compound is an alkene or an alkyne, i.e. if the C-backbone includes double or triple bonds, these must be indicated by the *infixes* -en- and/or -yn- as appropriate, each with its locant.

As an example, 3-methylbut-2-en-1-ol is an alcohol derived from the methyl-substituted C_4 alkene.**

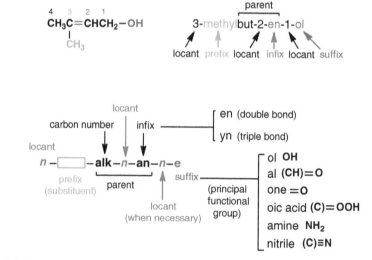

Figure 3.2 The essence of the IUPAC nomenclature of organic compounds.

3.8.2 Naming a non-aromatic hydrocarbon

The following four paragraphs show (with examples) how to name a non-aromatic hydro-carbon based on the IUPAC rules outlined above.

(1) Name an **alkane** according to its longest carbon chain (see Table 3.2 in Sub-section 3.2.1, p. 46). A branch is given as an alkyl substituent—see paragraph (4) below. The following are different representations of the same alkane, 3-methylhexane; regardless of how it is drawn, we need to recognize that the longest chain (red) is C$_6$, so any compound (not just an alkane) based on this carbon backbone is a derivative of hexane. The position (locant) of a substituent is numbered from the end of the carbon backbone which gives the smallest number possible.

CH$_3$
CH$_3$CH$_2$CHCH$_2$CH$_2$CH$_3$
1 2 3 4 5 6

4 5 6
CH$_2$CH$_2$CH$_3$
CH$_3$CH$_2$CHCH$_3$
1 2 3

4 5 6
CH$_2$CH$_2$CH$_3$
CH$_3$CHCH$_2$CH$_3$
3 2 1

3-methylhexane (not 4-methylhexane)

(2) Name **alkenes** and **alkynes** according to the longest carbon chain *which includes the unsaturated bond(s)**, and number the chain from the end which gives the smallest possible locants for the C=C and C≡C groups; if there is an alkyl substituent, add its name as a prefix (with its locant). When there are two or more substituents, put their prefixes in alphabetical order, and show multiple substitution of a group by an additional prefix (di-, tri-, tetra-, etc.).

*We choose the carbon chain which includes the greatest number of unsaturated bonds if there are two or more.

CH$_3$
CH$_2$=CHCHCH$_2$CH$_3$
1   2 3 4 5

3-methylpent-1-ene

CH$_3$
CH$_3$CH$_2$C=CHCH$_3$
5   4 3 2 1

3-methylpent-2-ene
(not 3-methylpent-3-ene)

CH$_2$CH$_3$
CH$_3$CH$_2$C=CH$_2$
4   3 2 1

2-ethylbut-1-ene

CH$_3$
CH$_3$C≡CCCH$_3$
1   2 3 |
CH$_3$

4,4-dimethylpent-2-yne
(not 2,2-dimethylpent-3-yne)

(3) For **cyclic structures**, the prefix 'cyclo' is added to the parent name corresponding to the number of C atoms in the ring.

cyclopentene

$_1$ CH$_3$
4
CH$_3$CH$_2$
2
3

4-ethyl-1-methylcyclohexene
(not 1-methyl-4-ethylcyclohexene)

6   1
5   2
4   3

cycloocta-1,5-diene

(4) To name a hydrocarbon *substituent*, identify its longest carbon chain starting with the free valence at C1. If the substituent is saturated, replace the final 'ane' of the alkane name corresponding to its C chain length with 'yl'; common (unsystematic) names of some alkyl groups which are allowed by the IUPAC rules are given in Table 3.4. If the substituent is unsaturated, replace the 'e' of the alkene/alkyne name corresponding to its C chain length with 'yl'. If the substituent *itself* is branched, this is described by substituent(s) within the main substituent (e.g. 1-methylpropyl below).

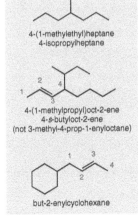

4-(1-methylethyl)heptane
4-isopropylheptane

2   4
1
3

4-(1-methylpropyl)oct-2-ene
4-s-butyloct-2-ene
(not 3-methyl-4-prop-1-enyloctane)

1   3
2   4

but-2-enylcyclohexane

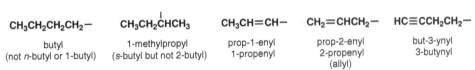

CH$_3$CH$_2$CH$_2$CH$_2$—

butyl
(not *n*-butyl or 1-butyl)

CH$_3$CH$_2$CHCH$_3$

1-methylpropyl
(*s*-butyl but not 2-butyl)

CH$_3$CH=CH—

prop-1-enyl
1-propenyl

CH$_2$=CHCH$_2$—

prop-2-enyl
2-propenyl
(allyl)

HC≡CCH$_2$CH$_2$—

but-3-ynyl
3-butynyl

Table 3.4

Common names of some alkyl groups allowed by IUPAC

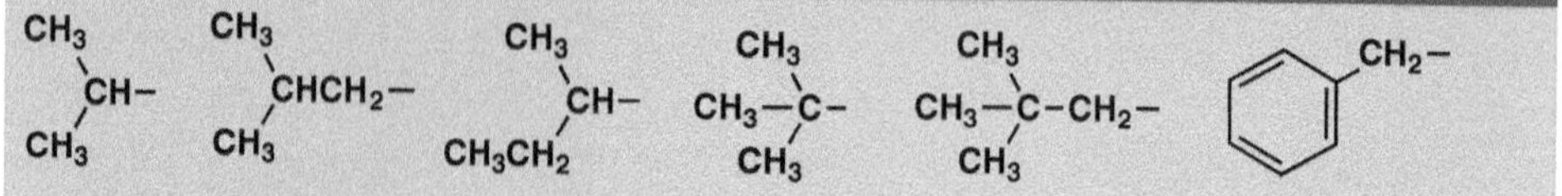

| isopropyl | isobutyl | s-butyl | t-butyl | neopentyl | benzyl |

Prefixes *s*- and *t*- (meaning secondary and tertiary) may also be spelled out as *sec*- and *tert*-.

Example 3.3

Give the IUPAC name for the alkane below.

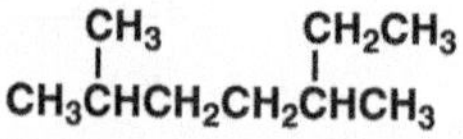

$$CH_3CHCH_2CH_2CHCH_3$$
(with CH_3 and CH_2CH_3 substituents)

Solution

First find the longest carbon chain, and number the carbons from the end which gives the smallest locants for substituents. In this alkane, the longest chain is C_7 and the carbons bearing the two methyl groups are 2 and 5 (lower than 3 and 6 which would be obtained if we numbered from the other end of the C_7 chain). So, the IUPAC name is 2,5-dimethylheptane.

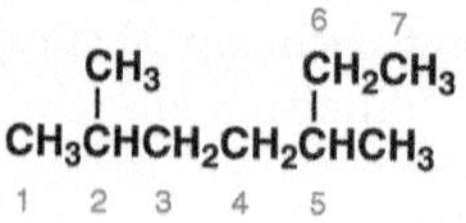

$$\underset{1\ 2\ 3\ 4\ 5}{CH_3CHCH_2CH_2CHCH_3}\quad{}^{6\ 7}$$

Exercise 3.10

Give the systematic IUPAC name for each alkyl group listed in Table 3.4.

3.8.3 Naming a compound with one or more functional groups

Commonly used abbreviations for some substituents

Me: methyl
Et: ethyl
Pr: propyl
i-Pr: isopropyl
Bu: butyl
t-Bu: *t*-butyl
Ph: phenyl
Bn: benzyl (PhCH₂)
Ac: acetyl (CH₃CO)
Bz: benzoyl (PhCO)

In the examples, the corresponding parts of structures and names are shaded in the same colours.

A compound which has one of the functional groups listed in Table 3.5 must be named with the suffix indicated. If the compound has more than one of the listed groups, we use the one of highest priority (the *principal group*) as the suffix and others are included as prefixed substituents. Other common functional groups listed in Table 3.6 are included in the name of a compound only as prefixed substituents.

Names of some representative compounds are given below.

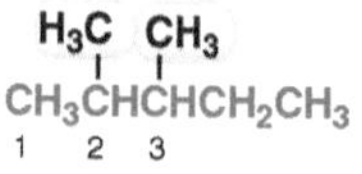

$CH_3CHCHCH_2CH_3$ — 2,3-dimethylpentane
$CH_3OCH_2CH_2CH_3$ — 1-methoxypropane (methyl propyl ether) 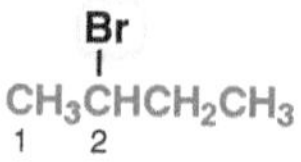
$CH_3CHCH_2CH_3$ (Br) — 2-bromobutane (1-methylpropyl bromide) (*s*-butyl bromide)
CH_2Cl_2 — dichloromethane (methylene chloride)

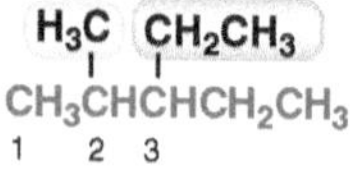

$CH_3CHCHCH_2CH_3$ — 3-ethyl-2-methylpentane (not 2-methyl-3-ethylpentane) 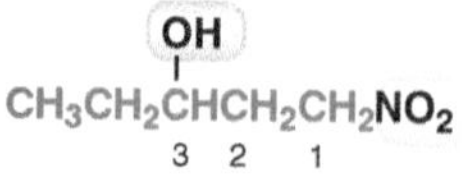
$CH_3CH_2CHCH_2CH_2NO_2$ (OH) — 1-nitropentan-3-ol (1-nitro-3-pentanol) 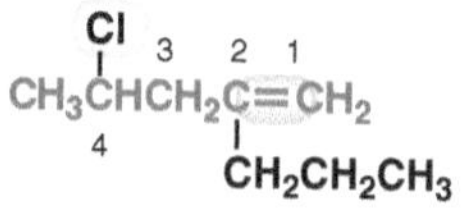
$CH_3CHCH_2C{=}CH_2$ (Cl, CH₂CH₂CH₃) — 4-chloro-2-propylpent-1-ene

Some compounds with multiple functional groups are given below to show how principal groups are chosen according to the priorities given in Table 3.5.

Table 3.5

Priorities and nomenclature of common functional groups

Priority	Class name	Structure[a]	as a suffix	as a prefix
1	anion		-ate	ato-
	cation		-onium	onio-
2	carboxylic acid	-(C)OOH	-oic acid	
		-COOH	-carboxylic acid	carboxy-
	sulfonic acid	-SO$_3$H	-sulfonic acid	sulfo-
3	acid anhydride	-COOCO-	-oic anhydride	
	ester	-(C)OOR	R -oate	
		-COOR	R -carboxylate	alkoxycarbonyl-
	acid halides	-(C)OX	-oyl halide	
		-COX	-carbonyl halide	haloformyl-
	amide	-(C)ONH$_2$	-amide	
		-CONH$_2$	-carboxamide	carbamoyl-
4	nitrile	(C)≡N	-nitrile	
		C≡N	-carbonitrile	cyano-
5	aldehyde	(C)H=O	-al	oxo-
		CH=O	-carbaldehyde	formyl-
6	ketone	(C)=O	-one	oxo-
7	alcohol	OH	-ol	hydroxy-
	thiol	SH	-thiol	mercapto-
8	amine	NH$_2$	-amine	amino-
	imine	=NH	-imine	imino-

a. A C in parentheses (C) indicates that it is part of the parent structure/name.

Table 3.6

Common functional groups included only as prefixed substituents in the names of compounds

Substituent	Prefix	Substituent	Prefix
RO	alkoxy	F	fluoro
RS	alkylthio	Cl	chloro
NO$_2$	nitro	Br	bromo
=N$_2$	diazo	I	iodo

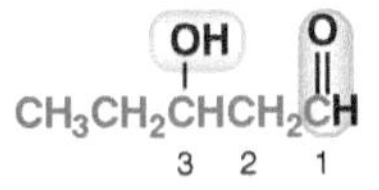

CH$_3$CH$_2$CHCH$_2$CH (OH, O)
    3 2 1

3-hydroxypentanal
(not 1-oxo-3-pentanol
or 1-formyl-2-butanol)

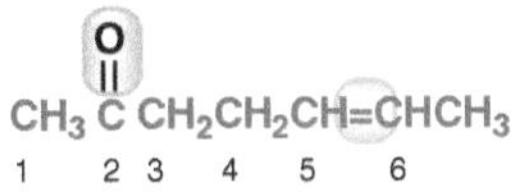

CH$_3$ C CH$_2$CH$_2$CH=CHCH$_3$ (O)
1 2 3 4 5 6

hept-5-en-2-one
(not hept-2-en-6-one)

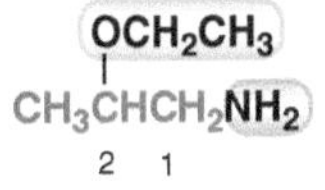

CH$_3$CHCH$_2$NH$_2$ (OCH$_2$CH$_3$)
    2 1

2-ethoxypropan-1-amine
(2-ethoxypropylamine)

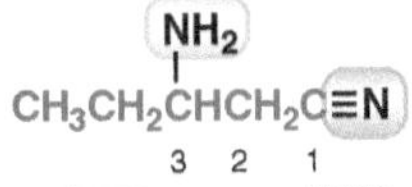

CH$_3$CH$_2$CHCH$_2$C≡N (NH$_2$)
    3 2 1

3-aminopentanenitrile

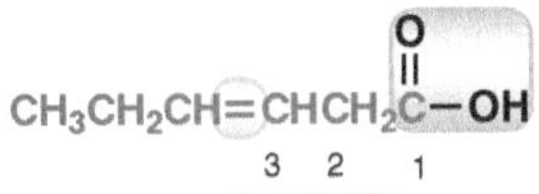

CH$_3$CH$_2$CH=CHCH$_2$C—OH (O)
    3 2 1

hex-3-enoic acid

HO—⬡—COOH

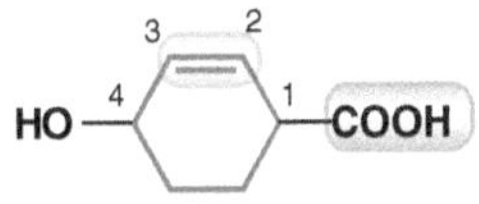

4-hydroxycyclohex-
2-enecarboxylic acid

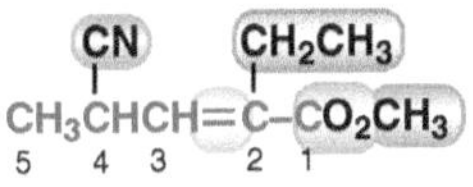

CH$_3$CHCH=C-CO$_2$CH$_3$ (CN, CH$_2$CH$_3$)
5 4 3 2 1

methyl 4-cyano-2-ethylpent-2-enoate

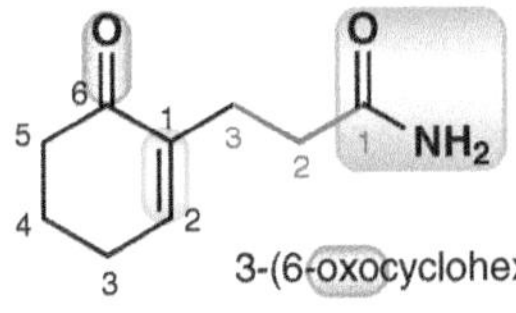

3-(6-oxocyclohex-1-enyl)propanamide

Example 3.4

Show the structure of 4-chlorohexan-2-ol.

Solution

First, draw the C-backbone of the parent alkane (hexane), and attach the OH functional group on C2 and the Cl substituent on C4. Complete the structure by adding the hydrogen atoms.

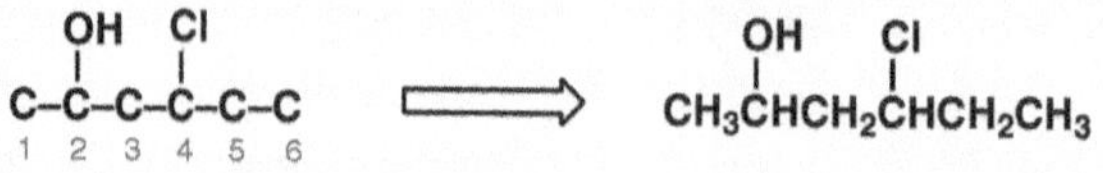

Exercise 3.11

Show structures of the following.

(a) 3-methylhexan-2-ol (b) 2-chlorobutanal (c) 2-aminobutanoic acid

(d) ethyl cyclopent-2-enecarboxylate (e) oct-4-en-6-yn-3-one

3.8.4 Naming aromatic compounds

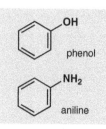

phenol

aniline

All parent aromatic compounds have IUPAC names (e.g. benzene, naphthalene, and pyridine) and numbering systems, which makes the naming of substituted derivatives easy. Additionally, many familiar derivatives have common names approved by IUPAC, e.g. phenol, aniline, and toluene.

For names and numbering of polycyclic and heterocyclic aromatic compounds, see Chapter 19.

Constitutionally isomeric 1,2-, 1,3- and 1,4-disubstituted benzenes are commonly named using o- (*ortho*), m- (*meta*), and p- (*para*) nomenclature, respectively.

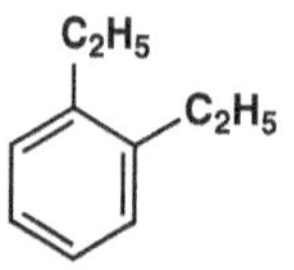

1,2-diethylbenzene
(*o*-diethylbenzene)

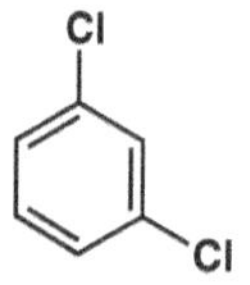

1,3-dichlorobenzene
(*m*-dichlorobenzene)

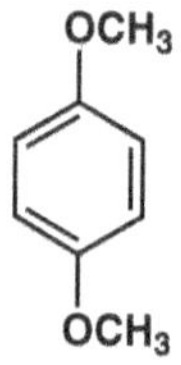

1,4-dimethoxybenzene
(*p*-dimethoxybenzene)

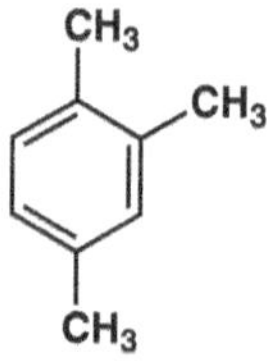

1,2,4-trimethylbenzene

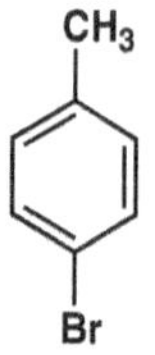

4-bromotoluene
(*p*-bromotoluene)

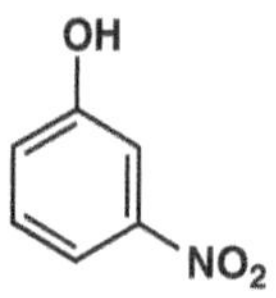

3-nitrophenol
(*m*-nitrophenol)

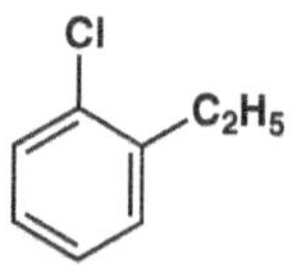

1-chloro-2-ethylbenzene

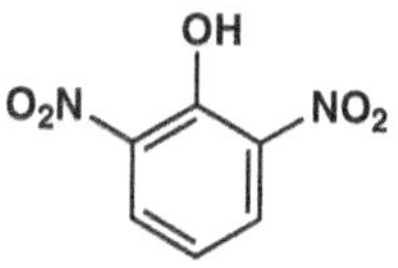

2,6-dinitrophenol

Exercise 3.12

Give the names of the following benzene derivatives.

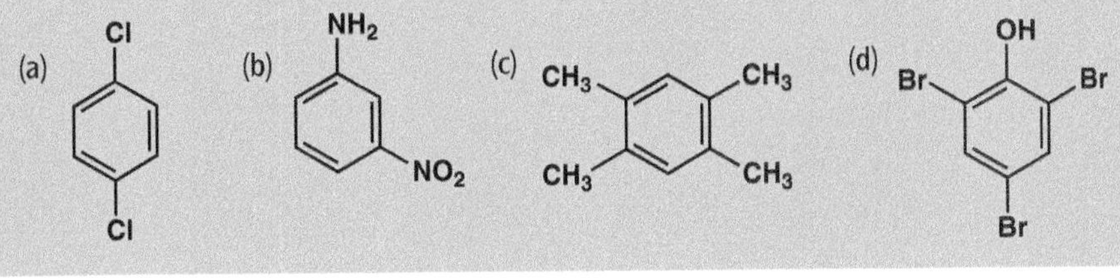

(a) [structure: 1,4-dichlorobenzene with Cl groups]

(b) [structure: benzene with NH₂ and NO₂ groups]

(c) CH₃ ... CH₃ / CH₃ ... CH₃ [tetramethylbenzene structure]

(d) OH with Br, Br, Br [tribromophenol structure]

Scrutiny of the names of compounds used in the text for structures shown will further illustrate the IUPAC system, but only practice at naming compounds will lead to mastery of nomenclature. Generally, however, it is sufficient that organic chemists be able to give a name to a compound which unambiguously specifies its structure, and be able to draw the correct structure from a proper name. If a strictly correct IUPAC name is required, IUPAC literature may be consulted, and some computer software for drawing chemical structures very usefully includes tools for naming compounds.

3.9 Intermolecular Interactions and Physical Properties of Organic Compounds

We can distinguish between properties of individual molecules (e.g. bond lengths and angles, and charge distribution) and properties of large assemblies of identical molecules, or of substances in general (e.g. boiling point, viscosity). The physical properties of compounds and other bulk materials are affected by molecular properties, but they are also strongly affected by **intermolecular interactions**.

Intermolecular relates to an interaction or effect *between* two (or more) molecules in contrast to *intramolecular* which relates to an interaction or effect *within* a (single) molecule. *Intermolecular interactions* are types of *non-bonded interactions* (interactions between atoms or groups which are not *directly* covalently bonded together). As we shall see, they can be attractive or repulsive depending upon the electrical charges (if any) on the interacting non-bonded groups, and the distances between them.

All intermolecular interactions are ultimately electrostatic. The strongest are the **coulombic interactions** between ions of opposite charge (cation–anion interactions). The interaction between an ion and a polar (but overall neutral) molecule (*ion–dipole* interaction) is also strong. These interactions influence solubility and the properties of ionic substances in solution (see below). Intermolecular interactions between electrically neutral molecules include **van der Waals forces** and **hydrogen bonds**; van der Waals forces include dipole–dipole interactions, dipole–induced dipole interactions, and (London) dispersion forces.*

J.D. van der Waals (1837–1923) was a Dutch physicist, famous for his work on an equation of state for gases and liquids. He received the 1910 Nobel Prize in Physics.

3.9.1 van der Waals forces

A molecule of a so-called polar compound, by definition, has a dipole, so there is an attractive **dipole–dipole** (electrostatic) interaction between the positive end of one such molecule and the negative end of another (Figure 3.3).

*London refers to the German-born physicist, F. London (1900–1954), not the English city.

Figure 3.3 Dipole–dipole interaction between chloromethane molecules shown by EPM diagrams of the molecules.

The dipole of a polar molecule will also deform the electron cloud of any nonpolar molecule nearby, and induce a dipole in it. Consequently, there will be an attractive **dipole-induced dipole** interaction between a polar molecule and a nonpolar molecule.

> The polarizability of an atom or ion depends upon how loosely its electrons are held, so the larger the atoms within a molecule, the greater the molecular polarizability and the greater the van der Waals forces it will experience.

Nonpolar molecules also experience mutually attractive **dispersion forces**. The average polarity of a nonpolar molecule over time is zero (or close to zero), but the instantaneous electron distribution is constantly fluctuating. Even a nonpolar molecule, therefore, can have a small temporary dipole, and this momentary dipole in one molecule can induce opposite (attractive) dipoles in surrounding molecules. These attractive dispersion forces are only weak (0.1–8 kJ mol^{-1}) but are stronger with molecules whose electron clouds are more easily distorted (i.e. high **polarizability**), and with large molecules of high surface (or contact) areas.

> *It is possible to calculate a surface around a molecule at which steric interactions by approaching molecules begin to become repulsive. Such a surface is called a *van der Waals surface* which looks similar to the isosurface of an EPM.

In the liquid phase, all the above mentioned attractive interactions drop off rapidly with increasing intermolecular distances, and are negligible in the gas phase. However, when two molecules approach each other too closely, repulsive forces intrude. These are due to the electrostatic repulsion between the electron clouds of the two molecules, and are called van der Waals repulsion or **steric repulsion**.*

3.9.2 Hydrogen bonds

A hydrogen atom bonded to an electronegative atom such as oxygen (as in the OH of a water molecule, or an alcohol) is electrically positive due to polarization of the bond between the H and the O, so there is a relatively strong electrostatic interaction between the hydrogen of an OH of one molecule and a lone pair on a heteroatom of another. This attractive force is called a **hydrogen bond** (Figure 3.4). The strength of a hydrogen bond is typically in the range 5–40 kJ mol^{-1}, which is appreciably greater than van der Waals forces, but it is still much weaker than a covalent single bond (200–400 kJ mol^{-1}). Hydrogen bonds are generally observed for compounds containing O–H or N–H bonds, and are especially important in the context of solute–solvent phenomena, e.g. when a solvent has OH groups and solutes have lone pairs.

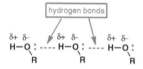

Figure 3.4 Hydrogen bonding of an alcohol.

> Hydrogen bonding between molecules, illustrated in Figure 3.4, is a favourable *intermolecular* interaction. However, as we shall see in Chapter 6 (Section 6.6), there can sometimes be a hydrogen bond between two OH groups within the same molecule. This can still be called a *non-bonded interaction* since the two OH groups are not *directly* covalently bonded to each other, but it is *intramolecular* rather than *intermolecular* since the interacting groups are within the same molecule. Correspondingly, there can be *adverse* intramolecular non-bonded interactions, e.g. steric repulsion (see above) when two bulky alkyl groups in the same molecule become too close together (see Chapter 4).

Clearly, formation of an intermolecular H-bond requires a close encounter of the OH of one molecule with the heteroatom bearing the lone pair in the other molecule. If close proximity is inhibited, e.g. by other parts of either the H-bond donor molecule or the acceptor, H-bond formation may be prevented. For example, the OH group of tri(*t*-butyl)-methanol is so congested that pairs of molecules cannot form intermolecular H-bonds in nonpolar solvents. This is an example of the general phenomenon known as **steric hindrance** caused by steric repulsion, and we shall see examples in later chapters of steric hindrance to various kinds of organic chemical reactions.

3.9.3 States of matter and phase changes

A compound exists as a gas, liquid, or solid depending on the balance between the strengths of interactions between molecules and the collective kinetic energy of individual molecules. Strengths of intermolecular interactions and (especially) molecular kinetic energy depend upon the temperature; consequently, whether a compound is a gas, liquid, or solid depends upon the temperature.

In the **gas phase**, individual molecules move independently with only very weak intermolecular interactions (none at all for the hypothetical ideal gas), and the thermal molecular kinetic energy is more than sufficient to overcome these very weak attractive forces. In a **crystalline solid**, molecules are packed regularly in a lattice at specific average distances apart because the molecules do not have sufficient kinetic energy to overcome the attractive intermolecular interactions, and each molecule undergoes only small oscillations about its average position.

There are attractive intermolecular interactions between molecules in a **liquid** which prevent it from expanding to occupy all the space available (as a gas does). However, they are not strong enough to enforce an orderly close-packed crystalline structure, and there is considerable freedom for molecules to diffuse about within the liquid. In some liquids, there can be short-range order or loose structure involving small clusters of molecules, but individual members of these groupings exchange rapidly with those of the rest of the material, and the bulk liquid has little or no time-averaged large-scale order.

When crystals are heated, they melt (fuse) at the **melting point** (mp) when both the solid and liquid are stable (i.e. they are in equilibrium at the mp—if *latent heat* is supplied, crystals melt; if it is extracted, the liquid solidifies). When a liquid open to the atmosphere is heated, it boils when its vapour pressure becomes equal to the ambient pressure. The temperature at which this happens (the **boiling point**, bp) depends upon the ambient pressure; it is when both gas and liquid forms are stable and in equilibrium (if *latent heat* is supplied, liquid becomes gas; if it is extracted, gas condenses). The melting and boiling phase changes commonly occur over specific and narrow temperature ranges (often less than a degree) and can be taken as an indication of the identity and purity of the substance.

Most liquids and some solids (e.g. solids which have odours) evaporate below their boiling or melting points. When this occurs, the phase transfer is not thermodynamically reversible. Some solids become gases before they reach a melting point, e.g. CO_2 at −78 °C and 1 atm pressure. This process is called *sublimation* and occurs when the vapour pressure of the solid becomes equal to the ambient pressure, and the molecules have sufficient thermal energy to overcome the intermolecular forces which hold them in the crystal lattice.

3.9.4 Boiling points of organic compounds

Boiling points of alkanes become higher with increasing molecular weight (Table 3.2 in Sub-section 3.2.1) because of increasing dispersion forces. Alcohols have much higher bps than alkanes of similar molecular weight (Table 3.7) because alcohol molecules

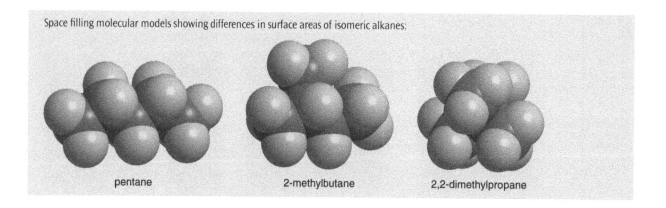

Figure 3.5 The hydrogen-bonded dimer of a carboxylic acid.

associate strongly through hydrogen bonds in the liquid phase (Figure 3.4). Still higher bps are observed for carboxylic acids (Table 3.7) due to their stronger hydrogen bonds; they form cyclic dimers with two H-bonds (Figure 3.5) in addition to more random H-bondings. The high boiling point of water for a compound of such a low molecular weight reflects the strong intermolecular hydrogen bonding in liquid water.

Table 3.7

Boiling points of organic compounds of similar molecular weight		
Compound	m.w.	bp/°C (1 atm)
$CH_3CH_2CH_2CH_2CH_3$ (pentane)	72	36.1
$CH_3CH_2OCH_2CH_3$ (diethyl ether)	74	34.5
$CH_3CH_2CH_2CHO$ (butanal)	72	75
$CH_3CH_2CH_2CH_2OH$ (butan-1-ol)	74	118
CH_3CH_2COOH (propanoic acid)	72	141

Example 3.5

Explain the different boiling points of the constitutional isomers of C_5 alkanes given below.

36.1 °C   27.9 °C   9.5 °C

Solution

Dispersion forces are greater for molecules with greater surface areas. Branched alkane molecules, being more compact, have smaller surfaces than unbranched ones, so their dispersion forces are smaller. Consequently, the bps decrease along the three compounds shown as the degree of branching increases.

Space filling molecular models showing differences in surface areas of isomeric alkanes:

pentane   2-methylbutane   2,2-dimethylpropane

Exercise 3.13

Explain why the boiling point of butanal is intermediate between those of pentane and butan-1-ol, as listed in Table 3.7.

Fluoroalkanes (sometimes loosely called fluorocarbons) have extraordinarily low bps. For example, perfluoropentane C_5F_{12} (m.w. 288) has a slightly lower bp than pentane C_5H_{12} (m.w. 72). This is mainly due to the very low polarizability of fluorine atoms resulting in very small van der Waals forces. For the same reasons, the fluorocarbon polymer known as Teflon, $-(CF_2)_n-$, has lubricating properties, and is used as a surface coating of non-stick frying pans.

Melting points are also dependent on intermolecular interactions, but they are also affected by the shapes of molecules—close packing in the crystal lattice is easier when molecules are more compact and/or more symmetrical. This contributes to the unexpectedly high mp of 2-methylpropan-2-ol (*t*-butyl alcohol) shown in Table 3.3.

3.9.5 Solubility

a. Solubility of covalent compounds

It is often said that 'like dissolves like', which suggests that polar compounds dissolve well in polar solvents and nonpolar compounds dissolve in nonpolar solvents. In a pure liquid or solid, identical molecules interact only with each other. When a compound is dissolved in a solvent, intermolecular interactions develop between molecules of the *solute* (the material being dissolved) and the solvent. Interactions between identical solute molecules and between identical solvent molecules, therefore, are replaced by interactions between solute and solvent molecules (Figure 3.6).

In order to achieve good solubility, intermolecular interactions between the solute and the solvent should be comparable with, or more favourable than, those of the pure solute and solvent separately. Consequently, different compounds with similar properties can mix without large changes to the strengths of interactions, and they dissolve each other well.

Now consider the case of a nonpolar compound being dissolved in a polar solvent. In the pure solvent, polar solvent molecules experience dipole–dipole interactions and perhaps hydrogen bonding, while the nonpolar solute experiences only weak dispersion forces. When the nonpolar molecules dissolve in the polar solvent, weak dispersion forces between polar and nonpolar molecules replace stronger interactions between polar solvent molecules; see Figure 3.7. Although the new dispersion forces may be somewhat stronger than those in the pure nonpolar solute, the net energy change in intermolecular interactions on dissolution is unfavourable owing to the loss of the stronger dipole–dipole interactions of the solvent. Consequently, a nonpolar compound such as an alkane is not very soluble in a polar solvent such as water.

An alcohol, like water, has a hydroxy group that can form hydrogen bonds, but the alkyl residue has nonpolar properties similar to those of an alkane. This explains why low molecular-weight alcohols are soluble in water, but the aqueous solubility decreases with increasing size of the alkyl portion of the alcohol (Table 3.3).

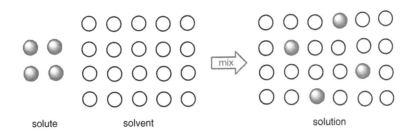

solute   solvent   solution

Figure 3.6 Schematic representation of dissolution.

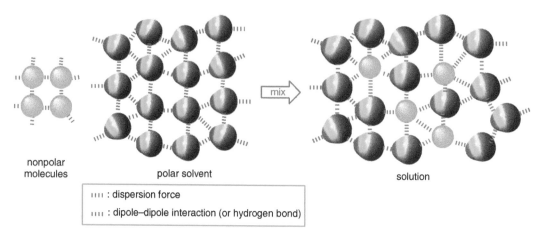

nonpolar molecules

polar solvent

solution

⊩⊩ : dispersion force

⊩⊩ : dipole–dipole interaction (or hydrogen bond)

Figure 3.7 Schematic representation of dissolution of a nonpolar solute in a polar solvent.

Panel 3.3 Chromatography

Methods for the separation of a mixture of compounds and the purification of a single compound depend on the partitioning of a compound between two phases which, in turn, depends upon intermolecular interactions:

Distillation – partitioning between liquid and vapour phases (boiling point)
Recrystallization – partitioning between solid and liquid phases (solubility)
Extraction – partitioning between two phases, usually aqueous and organic liquid phases, but sometimes between a solid and a liquid phase (solubility)

Chromatography, a particularly effective technique for separation and purification, is based on the multiple partitioning between a **mobile phase** (a flowing liquid or gas) and a **stationary phase** (an immobile solid or liquid surface). There are several forms of chromatography and methods have been developed for analytical purposes as well as for the isolation of samples obtained from natural sources or by synthesis. The physical principles are the same, however, in analytical and preparative scale chromatography.

The most general type of chromatography is **liquid chromatography** (LC), in which the mobile phase is a liquid and the stationary phase is a finely powdered insoluble solid such as silica gel or alumina.* In a typical form of LC called *column chromatography* (Figure 1), the mixture of compounds to be separated is added to the top of a vertical column of the stationary phase just covered by the mobile phase (an organic solvent) in a glass tube with a stop-cock at the bottom. More of the mobile phase** is then added to the top and run slowly through the column as illustrated below. As the compounds travel slowly down the column in solution, their passage is impeded by attractive intermolecular interactions with the stationary phase (including any water adsorbed on its surface). The stronger these interactions, the more slowly the compounds move down the column, i.e. the separation of the mixture is achieved by repeated different partitioning of the components between the liquid mobile phase and the solid stationary phase.

*The granules of the solid stationary phase invariably have a surface layer of adsorbed water molecules (the material is still a dry-looking powder).

**The mobile phase is sometimes called an *eluent*, and can be a mixture of solvents.

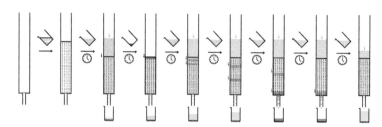

Figure 1 A schematic representation of column chromatography.

Acetanilide is a crystalline compound which can be recrystallized easily from water—it is moderately soluble in hot water, but less so in cold water. Which features of this molecule contribute to its solubility in water, and which adversely affect it?

Exercise 3.14

$$\underset{\text{acetanilide}}{\text{—NH—}\overset{\overset{\displaystyle O}{\|}}{\text{C}}\text{—CH}_3}$$

b. Solubility of ionic compounds

In solution, ionic compounds exist as ion pairs or dissociated ions. An ion pair is formed by electrostatic (Coulombic) attractive forces between ions of opposite charges. Dissociation of ion pairs is facilitated by weakening the electrostatic attractive forces between oppositely charged ions by using solvents of high *dielectric constant* (one of several measures of solvent polarity); the higher the dielectric constant of the medium, the weaker the electrostatic forces between dissolved oppositely charged ions. This

---

Panel 3.3 Continued

The efficiency of a separation in LC depends on the surface area of the stationary phase; consequently, the smaller the particles, the more efficient the column. However, the resistance to the flow of the liquid phase increases as the particle size decreases. To solve this problem, high pressure is applied in *high performance liquid chromatography* (HPLC: sometimes called high pressure liquid chromatography). For HPLC, a very finely divided stationary phase is packed in a metal column and a pump drives the mobile phase through at up to 40 MPa (about 400 atm). The column and the pump are assembled together with a detector (and a recorder) into an instrument called an (HPLC) chromatograph which may be analytical or preparative.

An ion-exchange resin can be used as a stationary phase for the separation of ionizable compounds such as amino acids, and the technique is then called *ion-exchange chromatography.*

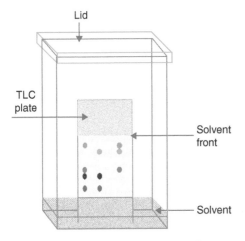

Figure 2 Thin layer chromatography.
(This file is licensed under the Creative Commons Attribution-Share Alike 3.0 Unported license.)

Thin layer chromatography (TLC) is another form of chromatography based on the same principles. The powdered stationary phase is coated as a thin layer of uniform thickness on a plastic (or metal or glass) plate whose dimensions depend on whether the application is analytical or preparative. The sample to be separated is applied a short distance from one end of the plate, and the plate is put into the shallow mobile phase in a vessel. A lid minimizes evaporation of the mobile phase which creeps up the stationary phase by capillary action, as illustrated in Figure 2. Analytical *paper chromatography* to separate polar compounds is similar but uses a sheet of special paper as the stationary phase, and the mobile phase is aqueous.

In **gas chromatography** (GC, also known as vapour-phase or gas–liquid chromatography, GLC), the mobile phase is a gas (helium or nitrogen and called the carrier gas). The stationary phase is usually a non-volatile liquid coated on an inert solid support packed in glass or metal tubing (called a column) or (in analytical applications) on the inside surface of a long capillary column. The sample is vaporized into the carrier gas flow in a heated sample injector which, with the column (thermostat usually set well above room temperature) and the detector, are part of the instrument called a *gas chromatograph.*

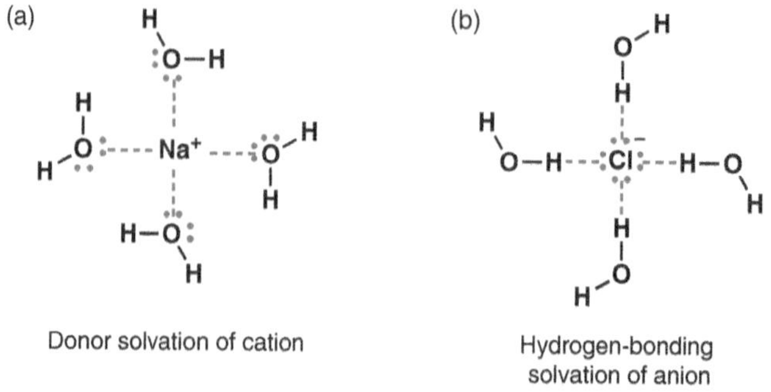

Figure 3.8 Schematic representation of the stabilizing solvation of a cation (Na^+) and an anion (Cl^-) by water. In both cases, the arrangement around the ion is tetrahedral.

property of a solvent is a *medium effect* and not directly attributable to specific intermolecular phenomena.

An ion, positive or negative, interacts favourably with polar solvent molecules by loose ion–dipole interactions; this enhances the dipoles of the solvent molecules involved which, in turn, leads to them having stronger dipole–dipole interactions with other solvent molecules. In addition, a cation may have more specific favourable electrostatic interactions with a limited number of solvent molecules, each with at least one lone pair. The cation acts as a Lewis acid and the solvent molecules as Lewis bases (see Chapter 6), as illustrated in Figure 3.8(a) for water. In a complementary manner, a dissolved anion can be stabilized by the formation of a specific number of hydrogen bonds with water or an alcoholic solvent (i.e. a *protic solvent*—an H-bond donor), Figure 3.8(b).

> A layer of solvent molecules surrounding a solute molecule or ion is called a *solvation shell* or *solvent cage. Solvation* involves favourable (attractive) interactions between a solute molecule or ion and (usually) a specific number of solvent molecules.

Exercise 3.15 Explain why propanone (acetone) is completely miscible with water.

Exercise 3.16 Explain why formamide ($HCONH_2$) is a good solvent.

Summary

- Principal characteristics of an organic compound are determined by its **functional group**, which is also a main centre of chemical reactivity, and the classification of organic compounds is based on their functional groups.

- Organic compounds are named according to the **IUPAC rules**. An IUPAC name is generally based on the parent hydrocarbon structure with a suffix indicating the main functional group and prefixes indicating substituents (Figure 3.2).

- Molecules of a pure substance associate through **intermolecular interactions** to form the bulk material which may be a liquid (which has a freezing point and a boiling point) or a solid (which has a melting point). When the kinetic energy of its molecules is sufficient to overcome intermolecular interactions, the material exists as a gas. A substance in any of the three states may dissolve in another (liquid) material to give a solution.

- Intermolecular interactions are essentially electrostatic and those of neutral molecules include van der Waals forces (dipole–dipole, dipole–induced dipole, and dispersion interactions) and hydrogen bonding.

- Solvation of ions in solution occurs by ion–dipole interactions as well as by lone pair donor interaction of the solvent to a cation, and hydrogen bonding of an OH of the solvent with an anion.

Problems

3.1 Identify the functional groups of each compound.

(a) MeO
vanillin
(vanilla flavour)

(b)
paracetamol
(pain reliever)

(c) CN OMe
a component
of super glue

(d) CO₂H
aspartame (NutraSweet®)

(e) Me₂N
ranitidine
(Zantac®, anti-ulcer drug)

3.2 Draw structural formulas of constitutional isomers of alkanes with molecular formula C_6H_{14}, and give the IUPAC name of each.

3.3 Give structural formulas of constitutional isomers of alkenes with molecular formula C_6H_{12}, and give the IUPAC name of each.

3.4 Draw structural formulas of compounds described as follows:

(a) Alcohols with molecular formula C_3H_8O.
(b) An ether with molecular formula C_3H_8O.
(c) An aldehyde with molecular formula C_3H_6O.
(d) A ketone with molecular formula C_3H_6O.
(e) A carboxylic acid with molecular formula $C_3H_6O_2$.
(f) Esters with molecular formula $C_3H_6O_2$.

3.5 Give structural formulas of compounds described as follows:

(a) Aldehydes with molecular formula C_4H_8O.
(b) A ketone with molecular formula C_4H_8O.
(c) Alkenyl ethers with molecular formula C_4H_8O.
(d) A cycloalkyl ether with molecular formula C_4H_8O.
(e) Cyclic ethers with molecular formula C_4H_8O.

3.6 Constitutional isomeric amines of molecular formula $C_4H_{11}N$ include primary and tertiary amines in

addition to the secondary amines featured in Exercise 3.6. Give structural formulas of primary and tertiary amines $C_4H_{11}N$.

3.7 Draw structural formulas of all the alcohols with the molecular formula $C_5H_{12}O$, classify them as primary, secondary, or tertiary, and give their IUPAC names.

3.8 Give structural formulas of all the chloroalkanes with the molecular formula C_4H_9Cl, classify them as primary, secondary, or tertiary, and give their IUPAC names.

3.9 Give the correct IUPAC name for each of the following compounds identified by its common name.

(a) neopentane (b) isobutene
(c) isooctane (d) isobutyl alcohol
(e) methyl ethyl ketone (f) valeric acid
(g) ethanolamine (h) carbon tetrachloride

3.10 Give the correct IUPAC name for each of the following compounds.

(a) (b) (c) (d) (e) (f)

3.11 Give the structural formula of the compound represented by each of the following names (some of which are not correct IUPAC names). If a name does not follow the IUPAC rules, point out why and give the correct IUPAC name.

(a) 2-ethyl-4-methylhexane

(b) 3-chloro-4-methylpentane

(c) 2-propyl alcohol

(d) 3-methoxy-5-ethylheptanal

(e) 2-penten-1-ol

(f) butenedioic acid

3.12 Explain why 2-methylpropan-2-ol (*t*-butyl alcohol) is completely miscible with water whereas butan-1-ol (butyl alcohol) is only slightly soluble, as indicated in Table 3.3.

3.13 Although ethanoic acid and methyl methanoate are constitutional isomers with the molecular formula $C_2H_4O_2$, their boiling points (32 and 118°C) are very different. Which of the two isomers has the higher boiling point, and why?

3.14 Explain the order of the boiling points of the isomeric alcohols, butan-1-ol (bp 118 °C), butan-2-ol (bp 99 °C), and 2-methylpropan-2-ol (bp 82 °C).

3.15 The following amines are constitutional isomers and have seemingly similar structures.

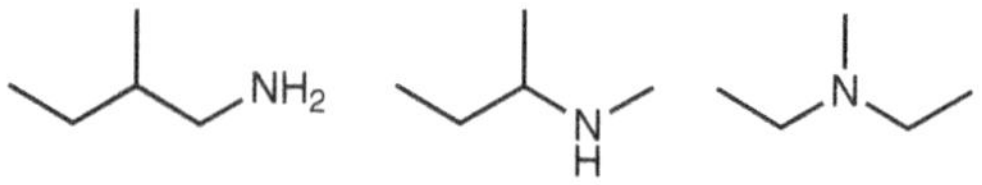

| bp/°C | 87 | 84 | 65 |

(a) Give the IUPAC name of each compound.

(b) Comment on their different boiling points.

3.16 Explain why tetrabutylammonium perchlorate dissolves in water and in some organic solvents.

Conformation and Strain in Molecules

4

Related topics we have already covered:

- **Covalent bonds** (Sections 1.2 and 2.2)
- **Shapes of molecules** (Section 2.1)
- **Drawing organic structures** (Section 2.8)
- **Alkanes** (Sub-section 3.2.1)
- **Non-bonded interactions** (Section 3.9)

Topics we shall cover in this chapter:

- **Molecular vibrations and internal rotation**
- **Configuration and conformation**
- **Strain in molecules**
- **Conformations of alkanes and cycloalkanes**
- **Chair and flexible forms of cyclohexane**
- *cis–trans* **Isomerism in cycloalkanes**

We saw in Chapter 2 that four single bonds radiate in tetrahedral directions from a saturated (sp^3-hybridized) carbon atom, which leads to organic molecules having three-dimensional structures. However, molecules are not rigid because molecular vibrations cause bond lengths and bond angles to be oscillating constantly about their average values. In this chapter, we move on to see how internal rotations about single bonds cause some molecules to be flexible rather than to have single fixed shapes. Also, we shall see what causes strain in a molecule, and how molecular strain affects the three-dimensional shape of a molecule. We shall use several methods to represent the three-dimensional structures of organic molecules in two dimensions; these will help us to generate mental images of three-dimensional shapes of organic molecules and hence understand better stereochemical aspects of organic chemistry.

4.1 Molecular Vibrations and Internal Rotation

A molecule is not a rigidly bonded assembly of atoms. It has bond stretching and bending vibrations and, depending upon its structure, one or more internal rotations in which one part of the molecule rotates with respect to the rest of the molecule.

4.1.1 Bond stretching and bending vibrations

The stretching vibration of the R–Y bond about an equilibrium bond length r_e and the bending of the Y–C–Z angle about its equilibrium value of α_e are represented in Figure 4.1. In general, more energy is required to stretch a bond than to bend it (stretching force constants are greater than those for bending vibrations).

The pyramidal inversion shown in Scheme 4.1 is a special kind of molecular vibration which involves deformation of several bond angles; it is available to some amines and their phosphorus analogues (molecules with an atom which has three σ bonds and

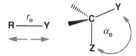

Figure 4.1 Simple bond stretching and bending vibrations.

Scheme 4.1 Pyramidal inversion of an amine.

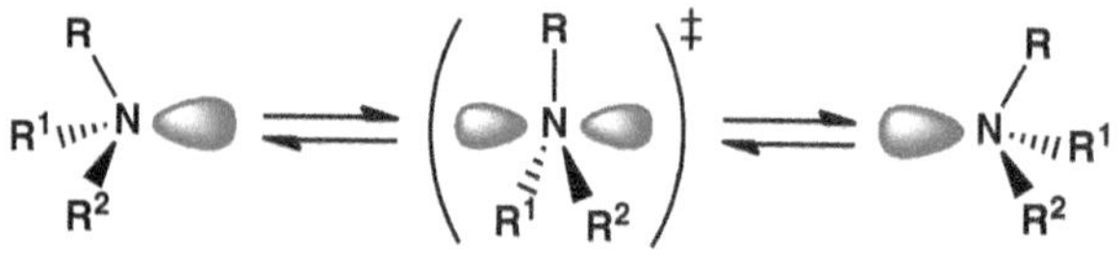

a lone pair). The planar molecular configuration corresponding to the energy maximum for the inversion is indicated by ‡. Ammonia and simple amines invert rapidly at room temperature (the energy barrier for the inversion is low) but phosphorus analogues, and amines where the N atom is incorporated into a three-membered ring, have appreciably higher barriers to inversion.

4.1.2 Internal rotation

Rotation of one part of an open chain organic molecule with respect to the rest of the molecule about a single bond is usually relatively free and called an *internal rotation* (not to be confused with rotation of the molecule as a whole); this is illustrated in Scheme 4.2, where Y is a polyatomic group.

Scheme 4.2 Internal rotation.

If some structural feature inhibits the internal rotation without making it impossible, it becomes a *hindered* or *restricted* rotation. Different shapes of a molecule which can be interconverted by rotations with low energy barriers about single bonds, or of amines which interconvert easily by inversion through the nitrogen, are called **conformations**.

4.2 Conformations of Alkanes

Remember that a σ bond is circularly symmetrical about the axis of the bond.

Two parts of an organic molecule joined simply by a C–C σ bond can rotate with respect to each other relatively easily because this kind of internal rotation does not affect the overlap of the orbitals from which the MO corresponding to the σ bond was formed. The different stereochemical structures which can be interconverted by rotations about single bonds (conformations) must not be confused with *configurationally different compounds* whose interconversions require the breaking of bonds (Sub-section 2.9.2). A stable conformation corresponding to an energy minimum is also called a **conformer** or **conformational isomer** (or sometimes a *rotational isomer*).

4.2.1 Ethane and torsional strain

Ethane, with just two sp³-hybridized carbon atoms, has only one internal rotation. When the two methyl groups rotate relative to each other, there is, in principle, an infinite number of possible conformations. However, only two are significant. These are the *staggered* and *eclipsed* conformations shown in Figures 4.2 and 4.3. In these figures, we illustrate alternative ways of representing in two dimensions the three-dimensional shape of a molecule of ethane, including *Newman projections*. In Figure 4.2, we also include guidance for constructing a Newman projection.

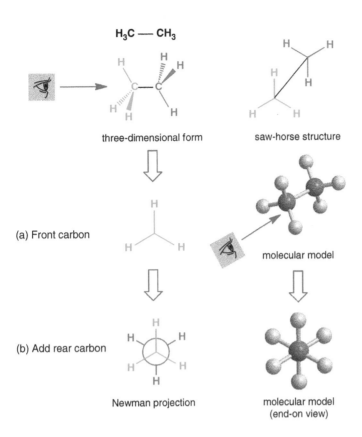

Figure 4.2 Staggered conformation of ethane and a method for constructing a Newman projection.

H₃C ── CH₃

three-dimensional form

saw-horse structure

(a) Front carbon

molecular model

(b) Add rear carbon

Newman projection

molecular model (end-on view)

A **Newman projection** provides an end-on view along a C–C bond, and the one of the **staggered conformation** of ethane is shown in Figure 4.2. The front carbon is represented by a dot and the three C–H bonds which radiate tetrahedrally from it appear at angles of 120° in the projection in Figure 4.2(a). In the projection in Figure 4.2(b), the rear carbon is represented by a circle centred at the dot of the front carbon, and its three C–H bonds also appear at angles of 120°. The relationship between the C–H bonds of the two carbons in this staggered conformation of ethane is described by the **dihedral angle** which is 60°. The saw-horse structure and ball-and-stick molecular model shown on the right in Figure 4.2 provide three-dimensional views of the staggered conformer of ethane. The electron pairs of the C–H bonds on the adjacent carbons are as far apart as they can be, and this is the most stable conformation of ethane.

The dihedral angle (ϕ, sometimes called the *torsion* angle) is the angle seen between bonds on the front and back carbons in a Newman projection. It is the angle between the two planes shown in the saw-horse projection below (60°).

dihedral angle
$\phi = 60°$

a staggered form

dihedral angle
$\phi = 60°$

Figure 4.3 Eclipsed conformation of ethane.

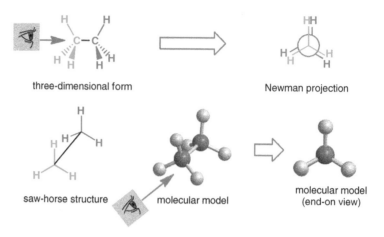

three-dimensional form   Newman projection

saw-horse structure   molecular model   molecular model (end-on view)

A 60° rotation of one CH_3 relative to the other about the C–C bond of the staggered conformer of ethane gives the **eclipsed conformation**, which is represented by a Newman projection together with a saw-horse structure and molecular model in Figure 4.3. The C–H bonds of the two carbons in this conformation are aligned (dihedral angle=0°), and their electron pairs are closer together than in any other conformation; this is the most unstable conformation of ethane.

The interconversion of eclipsed and staggered conformations of ethane (shown as Newman projections) by internal rotation about the C–C bond is illustrated in Scheme 4.3, and the periodic relationship between energy and the angle of rotation is shown in

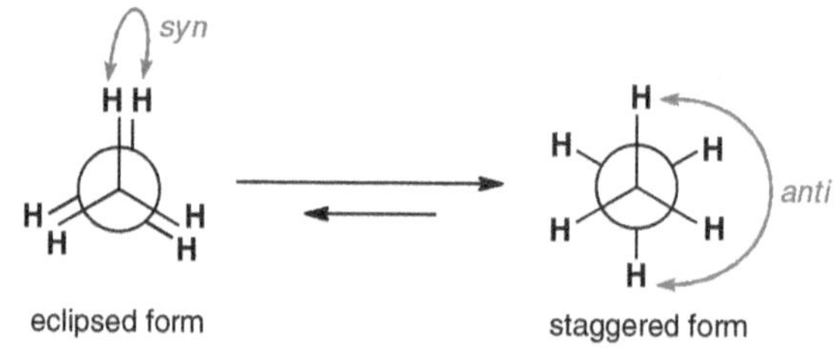

Scheme 4.3 Interconversion of conformations of ethane showing *syn* pairs of C–H bonds in the eclipsed form and *anti* pairs in the staggered form.

eclipsed form   staggered form

Figure 4.4. We see that there is only one stable conformation of ethane, the staggered form with a dihedral angle (ϕ) of 60°, corresponding to the minima in the rotational profile. The unstable eclipsed form at the top of each energy barrier in the rotational interconversion

Figure 4.4 Energy profile for internal rotation about the C–C bond of ethane.

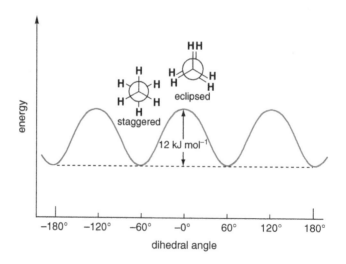

energy

staggered

eclipsed

12 kJ mol⁻¹

−180°   −120°   −60°   −0°   60°   120°   180°

dihedral angle

of equivalent staggered conformers has no appreciable lifetime so it cannot be described as a conformational 'isomer'.

The energy difference between the minima and the maxima in Figure 4.4 (the rotational or *torsional* barrier) has been measured experimentally and is very low (about 12 kJ mol^{-1}). Consequently, the thermal energy of molecules is sufficient for C–C bond rotation in ethane to be very rapid (within about 10^{-10} s) at room temperature.

> The (translational) kinetic energy of ethane is only about 2.5 kJ mol^{-1} at room temperature, so how do the molecules achieve this rapid internal rotation when the barrier is 12 kJ mol^{-1}? The kinetic energy is not distributed equally amongst all molecules. A proportion have more energy than the average (according to the Boltzmann distribution law)—sufficient to overcome the rotational barrier. And because the kinetic energy of molecules is constantly being exchanged by extremely frequent intermolecular collisions, all molecules in turn undergo the rapid internal rotation.

There are believed to be two principal causes of the energy difference between staggered and eclipsed conformations of ethane, i.e. the torsional energy (or **torsional strain**) of the eclipsed conformations. The first is the *destabilizing* electrostatic repulsion between the electron pairs of the C–H σ bonds on the two carbons, as implied above. It is smallest when the pairs of bonds are staggered, and greatest when they are eclipsed, as shown in Figure 4.5(a). Additionally, there is a *stabilizing* feature which is greater for the staggered form than for any other conformation. In Figure 4.5(b), we see that the occupied σ MO of one C–H bond and the vacant σ* MO of the C–H bond in the *anti* position on the adjacent carbon are aligned for maximal interaction; see Sub-section 2.2.2 for σ* (antibonding) orbitals. This favourable type of orbital interaction involving σ and σ* orbitals on adjacent atoms is known as *hyperconjugation* (see also Section 12.4.3).

> Since the torsional barrier is about 12 kJ mol^{-1} and there are three pairs of C–H bonds which are concurrently *syn* in the eclipsed conformation, the contribution of a *single* pair of eclipsed C–H bonds to the torsional strain is about 4 kJ mol^{-1}.

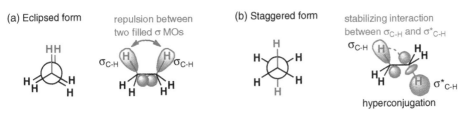

(a) Eclipsed form   repulsion between two filled σ MOs   (b) Staggered form   stabilizing interaction between σ$_{C-H}$ and σ*$_{C-H}$

hyperconjugation

Figure 4.5 Interactions of MOs associated with C–H bonds of ethane. (a) Repulsive interaction between electron pairs in MOs in the eclipsed conformation. (b) Hyperconjugation in the staggered conformation.

Draw Newman projections for the staggered and eclipsed conformations of propane.

Example 4.1

Solution

staggered form   eclipsed form

Draw saw-horse structures of the staggered and eclipsed conformations of propane.

Exercise 4.1

Draw Newman projections of the staggered and eclipsed conformations of chloroethane.

4.2.2 Butane and steric strain

Steric strain (also known as *steric repulsion* or *van der Waals strain*) occurs when non-bonded groups, i.e. groups which are not directly bonded, approach each other too closely; the more closely the groups come together, the greater the steric strain (see Section 3.9).

If one hydrogen on each of the two carbons of ethane is replaced by a methyl group, we generate the butane molecule which is a somewhat more complicated conformational system. Rotation about the central C2–C3 bond of butane gives three staggered conformations corresponding to energy minima and three eclipsed forms corresponding to energy maxima, as shown in the rotational energy profile in Figure 4.6. The single staggered form with the *anti* methyl groups ($\phi=180°$) is the most stable conformation of butane. The other two staggered forms, which are equivalent with dihedral angles of +60° and –60° between the two methyl groups, are called *gauche* conformations and are about 3.8 kJ mol^{-1} less stable than the *anti* form owing to *steric strain* between the two methyl groups.

The eclipsed forms at the energy maxima in Figure 4.6 are all unstable. Methyl groups eclipse hydrogens in the two equivalent forms at the 15 kJ mol^{-1} barriers ($\phi=\pm120°$), and the most unstable form at the 20 kJ mol^{-1} barrier has the methyl groups eclipsed ($\phi=0°$). One cause of the instability of the eclipsed forms is torsional strain (more or less as in ethane). However, the steric strain when a methyl group eclipses an H ($\phi=\pm120°$) or the other methyl ($\phi=0°$) in the *syn* form is absent in ethane, and this is likely to be the principal cause of the torsional energy barriers for butane *steric hindrance* to rotation being appreciably greater than the one for ethane (compare Figures 4.6 and 4.4).

Figure 4.6 Conformational energy profile for rotation about the C2–C3 bond of butane.

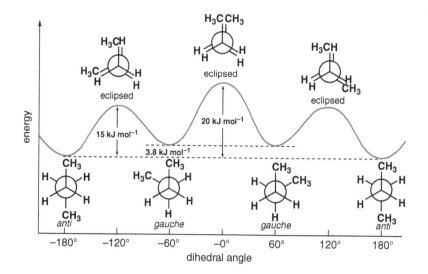

Give three-dimensional representations (using wedged bonds) of the *anti* and *gauche* conformations of butane.

Draw Newman projections of the *anti*, *gauche*, and *syn* conformations of 1,2-dichloroethane.

Conformations of butane: Newman projections and space-filling models showing how steric congestion between methyl groups increases in the order *anti < gauch < syn*.

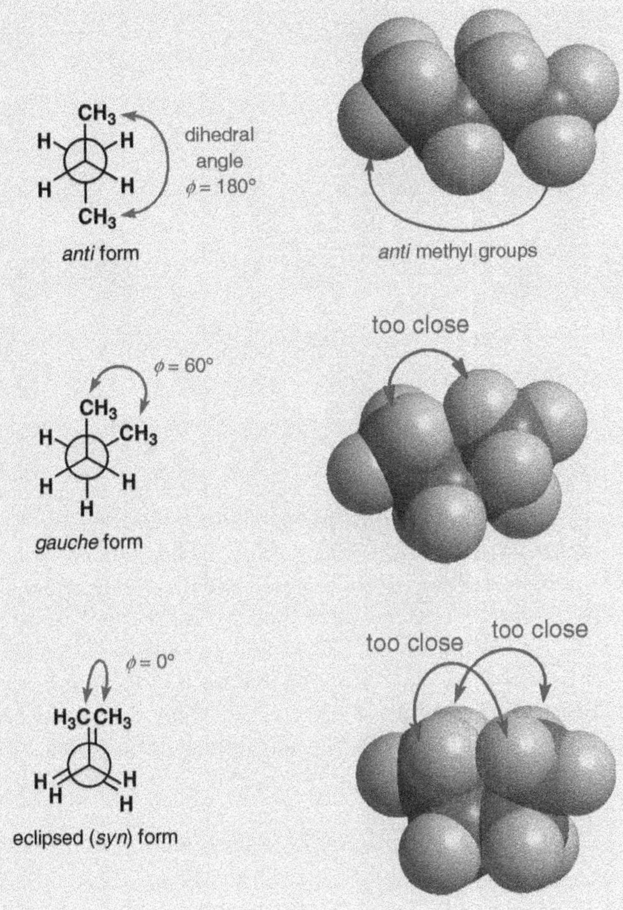

anti form

anti methyl groups

dihedral angle $\phi = 180°$

$\phi = 60°$

gauche form

too close

$\phi = 0°$

eclipsed (*syn*) form

too close   too close

4.3 Cycloalkanes

In open chain organic molecules, internal rotation about a C–C σ bond through 360° is possible except where substituents or branching in the carbon backbone lead to excessive adverse steric interactions. In cyclic systems, however, the angle through which there can be rotation about a ring C–C σ bond is limited to a degree determined by the ring size.

4.3.1 Cyclopropane and angle strain

The smallest cycloalkane is the three-membered cyclopropane. Three points are necessarily in a single plane, so the carbons of cyclopropane form an equilateral triangle (Figure 4.7) which allows no rotation about a ring C–C σ bond at all. Also, each C–C–C angle must be 60°, which represents substantial distortion from the normal tetrahedral angle of 109.5° for sp³-hybridized carbons. This compression of the three C–C–C angles leads to **angle strain** which destabilizes the cyclopropane molecule.

A molecular orbital description of cyclopropane shows that the angle between the orbitals corresponding to the bonds at each of the three C atoms (about 104°) is much larger than the C–C–C angle. In contrast to a normal σ bond made by end-on overlap of the constituent AOs along the line between the atoms being bonded, the AOs on two

Figure 4.7 Different representations of cyclopropane and the corresponding views of a molecular model.

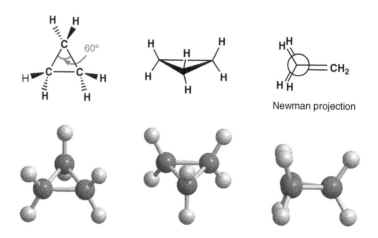

Newman projection

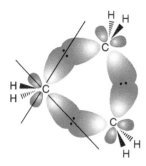

Figure 4.8 Orbital overlap for the C–C bonds in cyclopropane.

carbon atoms in cyclopropane overlap at an angle and thereby form a *bent* (or *banana-shaped*) *bond* (Figure 4.8). In other words, some of the electron density of each C–C bond in cyclopropane lies outside the straight line joining the bonded C atoms.

Cyclopropane experiences not only angle strain but also torsional strain because any pair of the three C–H bonds on each face of the ring is inevitably eclipsed, as shown in the Newman projection in Figure 4.7. The total **ring strain** of cyclopropane can be estimated from the heat of the C–C bond-breaking reaction (which opens the ring and thereby releases the strain). The heat of reaction for C–C bond cleavage of cyclopropane is 270 kJ mol^{-1}, while that of ethane is 380 kJ mol^{-1}. It follows that cyclopropane is destabilized by about 110 kJ mol^{-1} and this corresponds to the total ring strain energy of the molecule.

Example 4.2

From the torsional barrier in ethane, the contribution of a single pair of eclipsed C–H bonds has been calculated to be about 4 kJ mol^{-1} (Sub-section 4.2.1). Use this result to calculate the desta-bilization of cyclopropane attributable to the eclipsed C–H bonds.

Solution

Since cyclopropane has six pairs of eclipsed C–H bonds, the destabilization attributable to them (torsional strain) is estimated to be about $4 \times 6 = 24$ kJ mol^{-1}.

4.3.2 Cyclobutane and cyclopentane

If cyclobutane took the form of a square, the C–C–C angle would be 90°, and there would be appreciable angle strain. In addition, the eight pairs of eclipsed C–H bonds would involve considerable torsional strain (Figure 4.9). However, a four-membered ring does not need to be planar. The ring may distort from planarity which would modify both angle and torsional strain. Experiments show that cyclobutane undergoes a ring-pucker-ing vibration and is nonplanar by about 28° in its most stable conformation with C–C–C angles of about 88° (Figure 4.9). This deformation appreciably reduces the torsional strain though the angle strain is slightly increased.

Cyclopentane would have very little angle strain if the carbon framework were planar since the internal angle of a regular pentagon is 108°. However, this planar form would be appreciably strained because of the 10 pairs of eclipsed C–H bonds. This torsional strain is largely relieved in a nonplanar form which still has only little bond angle strain. The most stable conformation of cyclopentane is the **envelope form** (Figure 4.10) and the

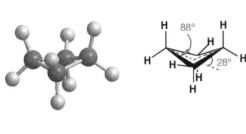

Newman projection

Figure 4.9 Nonplanar structure and views of a molecular model of cyclobutane.

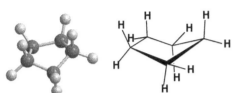
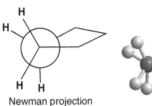

Newman projection

Figure 4.10 The conformationally mobile envelope form of cyclopentane.

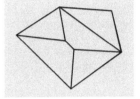

considerable flexibility of the molecule allows a kind of rapid rippling motion in which each carbon in turn becomes either above or below the approximate plane of the other four. This ripple moving around the ring is called a *pseudo*-rotation since the molecule as whole does not rotate. Five-membered ring systems are usually stable and found widely in naturally occurring compounds.

The internal angle, α, of a regular planar polygon with n corners can be calculated by the equation, $\alpha = 180° - (360°/n)$. Calculate the internal angles of (a) a regular hexagon, (b) a regular heptagon, and (c) a regular octagon.

Exercise 4.5

4.3.3 Cyclohexane: chair conformations

The most stable form of cyclohexane is the **chair conformation** which, compared with cyclopentane, is relatively inflexible (but see below). In this conformation, all the neighbouring C–H bonds are staggered and all bond angles are equal to the normal angle of tetrahedral carbon, so the molecule is essentially free of torsional strain and bond angle strain. Such ring structures are very common amongst naturally occurring compounds, especially steroids (see Sub-section 24.4.4), and the hardness of diamond has been attributed to its structure being a matrix of chair-form six-membered carbocyclic rings rigidly locked in three dimensions.

Structure of the steroidal framework:

Structure of diamond:

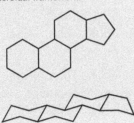
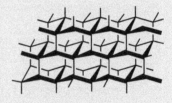

The chair form of cyclohexane has two kinds of C–H bonds on the ring, **axial** (red) and **equatorial** (blue) in Figure 4.11, and each carbon atom has one of each. The axial bonds are parallel to the axis of the ring and alternately point upwards and downwards. The

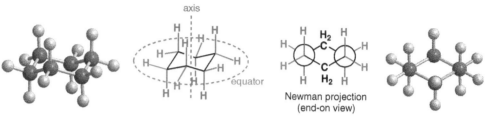

Figure 4.11 The chair conformation of cyclohexane. Axial bonds are red and equatorial bonds are blue.

The equatorial C–H bonds can be seen as three sets of pairs in Figure 4.12, each pair being parallel to two C–C bonds within the ring, as shown by the thick blue bonds. This feature helps us to draw realistic chair cyclohexanes with the correct perspective.

axial bonds   equatorial bonds

Figure 4.12 Axial and equatorial bonds of the chair conformation of cyclohexane.

equatorial bonds project outwards from the ring around the 'equator', again alternately upwards and downwards. Axial and equatorial bonds are also illustrated in Figure 4.12 to show clearly the stereochemical relationships.

It might be thought that a monosubstituted cyclohexane such as methylcyclohexane should have two isomeric forms since the methyl could be either axial or equatorial. However, the two forms are conformational isomers which interconvert rapidly by *ring inversion* (or *ring flip*); see Scheme 4.4. This interconversion is so rapid that we cannot separate the two forms at room temperature.

Scheme 4.4 Ring flip of methylcyclohexane.

equatorial methylcyclohexane
95%

ring flip

axial methylcyclohexane
5%

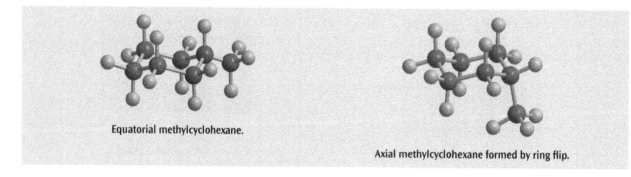

Equatorial methylcyclohexane.

Axial methylcyclohexane formed by ring flip.

A similar rapid ring flip occurs with unsubstituted chair cyclohexane, so we cannot differentiate its axial and equatorial hydrogens by conventional methods. The axial hydrogens become equatorial by ring flip, and the equatorial hydrogens become axial (Scheme 4.5), but note that ring flip does not change whether a C–H (or a substituent) is pointing up or down.

About 95% of methylcyclohexane molecules have the methyl group equatorial, the remainder having the methyl axial, and the two forms are rapidly interconverting

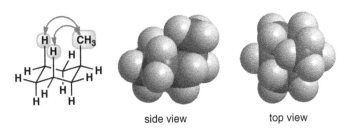

Scheme 4.5 Ring flip of the chair conformation of cyclohexane. Only one hydrogen atom is shown on each carbon of the ring.

(Scheme 4.4). These proportions indicate that the conformer with the equatorial methyl is more stable than the one with the methyl axial by about 7.6 kJ mol^{-1}. The instability of the axial methylcyclohexane is caused by unfavourable steric interactions between the axial methyl and the two axial hydrogens at C3 and C5 (Figure 4.13). This destabilizing feature of a chair cyclohexane conformer with an axial substituent is called a **1,3-diaxial interaction**.

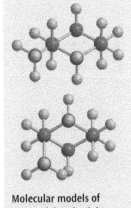

side view   top view

Figure 4.13 Two 1,3-diaxial interactions in axial methylcyclohexane and views of space-filling molecular models.

Example 4.3

Draw Newman projections along the C1–C2 bond of both chair conformers of methylcyclohexane, and explain the 1,3-diaxial interaction in terms of *anti* and *gauche* interactions.

Solution
If we look at Newman projections along the C1–C2 bond (shown below) and C1–C6 bond (not shown), we can identify different *gauche* and *staggered* interactions in the two conformers. On the one hand, the equatorial methyl group is *anti* to the C2–C3 (red) bond (and to the C6–C5 bond), which involves no appreciable strain. In contrast, the axial methyl is *gauche* to the C2–C3 (red) bond (and to the C6–C5 bond), which does involve steric strain. The axial methyl is close to the axial hydrogen atom on C3 (and the one on C5) so this conformer is destabilized by 1,3-diaxial interactions. This sort of analysis shows that the greater stability of the equatorial methyl (compared with an axial one) in a chair cyclohexane is analogous to the greater stability of the *anti* conformer of butane (compared with the *gauche*) as discussed in Section 4.2.2.

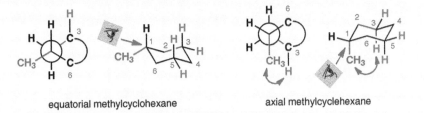

equatorial methylcyclohexane   axial methylcyclehexane

Molecular models of equatorial and axial methylcyclohexane viewed along the C1–C2 bond.

Exercise 4.6

Draw a structure of the most stable conformation of chlorocyclohexane.

Panel 4.1 Heterocyclic chair compounds: tetrodotoxin

Stable chair forms of six-membered rings are ubiquitous in nature, and the rings often contain heteroatoms. Six-membered rings in sugar contain one oxygen atom and coniine, a potent toxin encountered in the Prologue, has a nitrogen-containing ring.

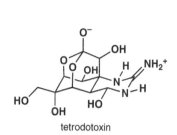

β-D-glucopyranose   coniine

Tetrodotoxin is a particularly interesting molecule consisting of five chairs including oxygen- and nitrogen-containing heterocycles and a carbocycle all joined together. It is a powerful neurotoxin which is isolated from the ovaries and liver of the pufferfish (*Tetraodontidae*), so named because it inflates itself when frightened. Although this toxin is found in the pufferfish and some other fish, it is actually produced by symbiotic bacteria, and pufferfish which do not contain the toxin can be cultivated. Some species of pufferfish (called *fugu* in Japanese) are well known Japanese delicacies, but only specially licensed restaurants and chefs are allowed to serve them.

tetrodotoxin   (Holger Feroudj/istockphoto.)

A trace amount of this toxin causes weakness, paralysis, and eventually death. The $=NH_2^+$ group of tetrodotoxin binds to the sodium ion channel in nerve cell membranes and interferes with the transportation of Na^+.

The most stable conformer of any monosubstituted cyclohexane is invariably the one with the substituent equatorial, and the difference in steric strain between this and the axial conformer depends on the bulk of the substituent.

Exercise 4.7

Draw the two chair conformations of *t*-butylcyclohexane, and explain which is more stable.

4.3.4 Cyclohexane: ring inversion of chair conformations

The lowest energy route between chair cyclohexane conformers is a sequence involving, first, conversion of one chair form into a family of more flexible conformations. The energy barrier for this step is about 45 kJ mol⁻¹, and the molecule at the energy maximum has a *half-chair* conformation (Scheme 4.6 and Figure 4.14), which suffers severe angle and torsional strain in the planar part of the molecule. The flexible forms themselves interconvert very rapidly by a *pseudo*-rotation analogous to that which interconverts the flexible cyclopentane conformations. The six local energy minima on this *pseudo*-rotation are called *twist-boat* conformations, and the six maxima are the characteristic

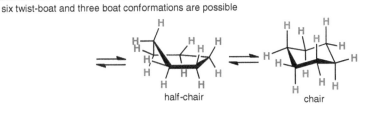

Scheme 4.6
Interconversion of cyclohexane
conformations.

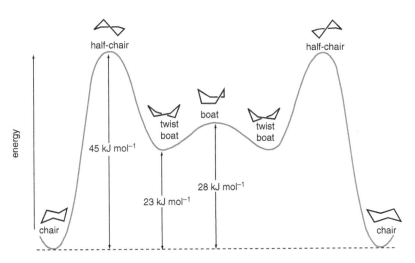

six twist-boat and three boat conformations are possible

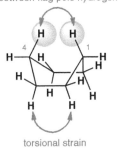

Figure 4.14 Energy profile for interconversion of cyclohexane conformations.

boat conformations (Figure 4.14). The inversion is completed by the flexible conformation reverting to a (relatively inflexible) chair form via another half-chair form.

Although angle strain is minimal in the unstable boat form, there is appreciable torsional strain owing to the four pairs of C–H bonds eclipsed along the C2–C3 and C5–C6 bonds. Furthermore, there is high steric strain due to the closeness of one of the hydrogen atoms on C1, the so-called flag-pole hydrogen, and the other flag-pole hydrogen on C4 (Figure 4.15). Both are partially relieved in the twist-boat forms. In all, the distinctive boat form is about 28 kJ mol^{-1} more strained, i.e. less stable, than the chair form, and about 5 kJ mol^{-1} less stable than the twist-boat form.

Figure 4.15 Strain in the unstable boat conformation of cyclohexane.

Draw a Newman projection along the C2–C3 bond of the boat cyclohexane in Figure 4.15.

Exercise 4.8

4.4 Disubstituted Cycloalkanes: *cis–trans* Isomerism

If a cycloalkane has a substituent on each of two ring carbons, **stereoisomers** are possible: the two substituents can be on the same face of the cycloalkane ring or on opposite faces. These stereoisomers, which are *configurational* isomers like the *cis–trans* isomers of an alkene (Sub-section 2.9.2), cannot interconvert without breaking a C–C bond. Structures of *cis-* and *trans*-1,3-dimethylcyclobutane are shown as examples.

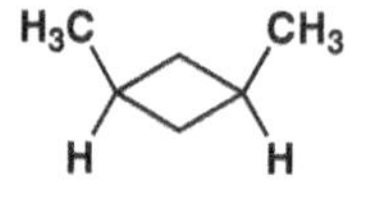

cis-1,3-dimethylcyclobutane 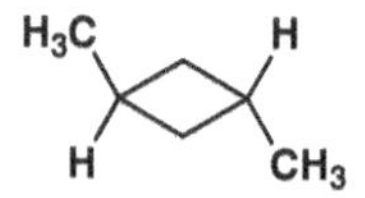   trans-1,3-dimethylcyclobutane

<table>
<tr><td>

Exercise 4.9

</td><td>

Draw the isomers of 1,2,3-trimethylcyclopropane and discuss their relative stabilities.

</td></tr>
</table>

Conformational complexities of cycloalkanes do not complicate their *cis–trans* isomerism. It is possible to identify *cis–trans* isomers by considering wholly unrealistic planar conformations!

Although the *conformational analysis* of a disubstituted cyclohexane is not straightforward because of alternative chair conformations, this does not complicate the *cis–trans* isomerism. As shown in Scheme 4.7, for example, *trans*-1,4-dimethylcyclohexane (**1**) may be depicted simply (but unrealistically) as a planar ring structure (**1a**), or as the diaxial chair form (**1b**), or as the alternative chair conformer (**1c**) with both substituents equatorial. In both **1a** and **1b**, the substituents are unmistakably *trans* to each other, but both are unstable conformations (the former much more so than the latter). The most stable conformation of this compound, by far, is **1c** but the *trans* relationship of the substituents may not be quite so obvious.

Scheme 4.7
Conformations of *trans*-1,4-dimethylcyclohexane.

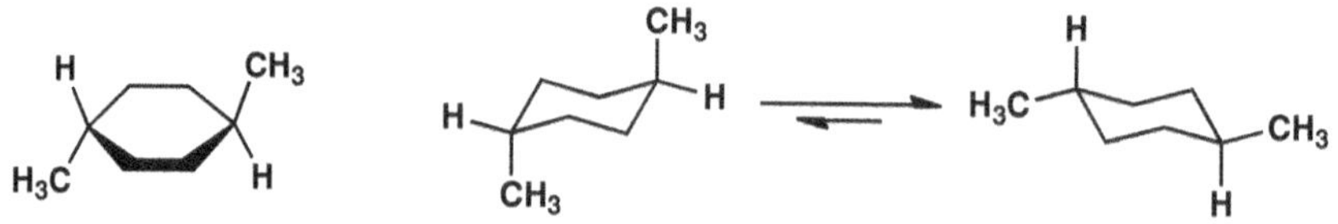

1a   **1b** (*ax-ax*)   **1c** (*eq-eq*)

<table>
<tr><td>

Exercise 4.10

</td><td>

Draw a chair form of *cis*-1,4-dimethylcyclohexane.

</td></tr>
</table>

<table>
<tr><td>

Example 4.4

</td><td>

Draw possible chair conformations of *cis-* and *trans*-1-*t*-butyl-4-methylcyclohexane; explain which conformation is the more stable in each case, and indicate which of the two isomers is the more stable.

</td></tr>
</table>

Solution
The highly unfavourable 1,3-diaxial interactions of the very bulky *t*-butyl group when it is axial mean it has an extremely strong tendency to be equatorial. Consequently, the *eq-eq* conformer of the *trans* isomer will be hugely more stable than the diaxial form, and this isomer will be virtually entirely the single conformer (*eq*=equatorial and *ax*=axial).

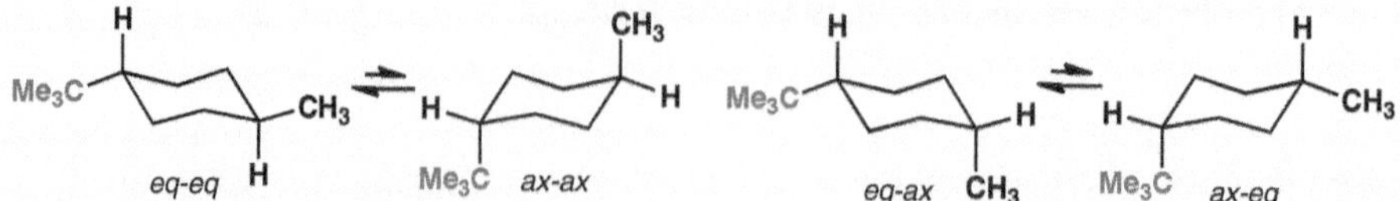

eq-eq   Me₃C ax-ax   eq-ax CH₃   Me₃C ax-eq
trans isomer   *cis* isomer

Both conformers of the *cis* isomer have one equatorial and one axial substituent. However, the much stronger tendency of the *t*-butyl group to be equatorial means that the *eq-ax* conformer will be more stable than the *ax-eq* form, so this isomer will also be predominantly a single conformer. The *eq-eq* conformer of the *trans* isomer must be more stable than the *eq-ax* conformer of the *cis* isomer so we can safely predict that the *trans* isomer is the more stable of the two compounds. But the *cis* isomer cannot isomerize into the *trans* without C–C bond cleavage.

Indicate whether the substituents in each of the following structures are *cis* or *trans*; how many different compounds are represented by these four structures?

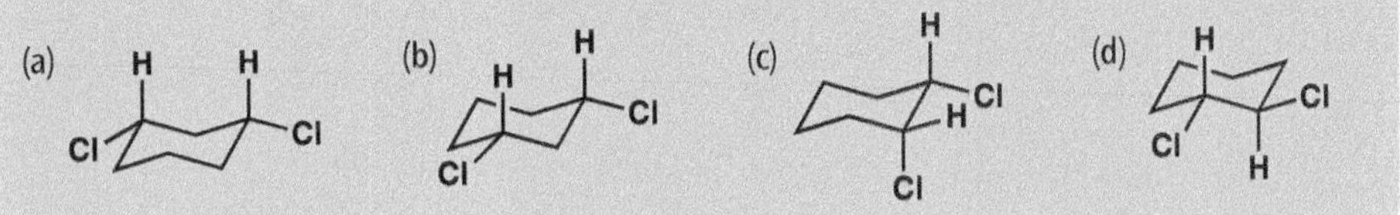

Exercise 4.11

(a) (b) (c) (d)

Draw the most stable conformation of each of *cis*- and *trans*-1-*t*-butyl-3-methylcyclohexane, and predict which isomer is the more stable.

Exercise 4.12

4.5 Strain in Cycloalkanes: Heat of Combustion

Small-ring cycloalkanes are strained as discussed in Sub-sections 4.3.1 and 4.3.2, and the strain energy can be determined from the heat of combustion. Alkanes are generally unreactive and stable in air but they burn readily when ignited; burning (combustion) is, of course, vigorous oxidation by atmospheric oxygen. The heat of this reaction when carried out under controlled conditions is known as the **heat of combustion**. Since the general formula of cycloalkanes is $(CH_2)_n$, their oxidation is described generically by eqn 4.1.

$$(CH_2)_n + 1.5n\,O_2 \longrightarrow n\,CO_2 + n\,H_2O \qquad \Delta H = \text{heat of combustion} \qquad (4.1)$$

Experimentally determined values of molar heats of combustion of cycloalkanes with $n=3-8$ are summarized in Table 4.1 (all values are large and negative because combustion is strongly exothermic); we observe that the values per CH_2 are not the same, and the value for cyclohexane is the lowest. Since the products of combustion, CO_2+H_2O, are identical for all cycloalkanes as seen in eqn 4.1, the differences must come from the release of ring strain energy (see Section 4.2) not present in cyclohexane. In other words, the difference between the heat of combustion per CH_2 for one cycloalkane and

n	cycloalkane	$-\Delta H$	$-\Delta H/CH_2$	strain/CH_2
Heats of combustion for cycloalkanes $(CH_2)_n$ in kJ mol^{-1}				
3	cyclopropane	2091.3	697.1	37.1
4	cyclobutane	2745.0	686.3	26.3
5	cyclopentane	3319.6	663.9	3.9
6	cyclohexane	3959.9	660.0	0
7	cycloheptane	4636.7	662.4	2.4
8	cyclooctane	5310.3	663.8	3.8

Table 4.1

Figure 4.16 Determination of the strain energy of a cycloalkane from its heat of combustion per CH_2.

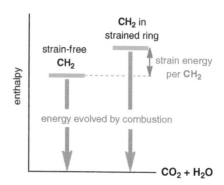

the value for cyclohexane (the putative strain-free reference) is attributed to the strain energy of the cycloalkane. This interpretation of the thermochemical results is illustrated in Figure 4.16.

Cyclopropane and cyclobutane have high strain energies as expected. The total strain energy of cyclopropane determined in this way is $37.1 \times 3 = 111.3$ kJ mol^{-1}, which agrees well with the value, 110 kJ mol^{-1}, calculated from the C–C bond dissociation energies of ethane and cyclopropane (Sub-section 4.3.1). The strain energy of cyclopentane is small, as anticipated above, and cycloheptane and cyclooctane also have only little strain. However, as the ring size increases beyond cyclooctane, there is a tendency for the strain energy to increase again.

Panel 4.2 Bicycloalkanes

Some organic compounds have two or more rings and these can be joined together in different ways. **Bicyclic compounds** have two rings sharing one or more atoms, and there are three types: **fused** bicyclic compounds, **bridged** bicyclic compounds, and **spiro** compounds. The atoms at which the two rings of a bicyclic compound are joined are called **bridgehead** atoms. When these two bridgehead atoms are *adjacent*, we have a *fused* bicyclic compound, e.g. **1**. If a compound has two bridgehead atoms but they are *not* adjacent, it is a *bridged* bicyclic compound, e.g. **2**. If the two rings of a compound have only a *single* common (tetrahedral) atom, we have a *spiro* compound, e.g. **3**.

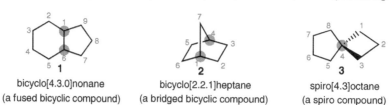

| bicyclo[4.3.0]nonane | bicyclo[2.2.1]heptane | spiro[4.3]octane |
| (a fused bicyclic compound) | (a bridged bicyclic compound) | (a spiro compound) |

The naming of these compounds needs some explanation: a parent alkane is named by the number of carbons, so compounds **1**, **2**, and **3** are a nonane, a heptane, and an octane as they have 9, 7, and 8 carbons, respectively, and the prefix 'bicyclo' or 'spiro' is added as appropriate. The name of a parent hydrocarbon is then uniquely completed by specifying the numbers of carbons in the bridges. These numbers in descending order are separated by dots within square brackets (no hyphens or spaces), as illustrated in the above examples: three numbers for a fused bicyclic compound, the third of which is 0; three for bridged bicyclic compounds; two for spiro compounds.

If we need to indicate the position of a substituent on a bicycloalkane, we must also number the carbon framework. For bridged and fused compounds, we start with a bridgehead as C1 and work to the other bridgehead by the longest bridge, then back to the first bridgehead by the second longest route, and finally back again by the remaining (shortest) route. In this way, compound **2** above is numbered as shown, and the derivative with a hydroxy group at C7, for example, is 7-hydroxybicyclo[2.2.1]heptane. The system is different for spiro compounds; for these, the numbering goes round the small ring starting at the carbon next to the spiro atom, it then passes through the spiro atom and round the larger ring.

Summary

- Organic molecules undergo bond stretching and bending vibrations. If there is a nitrogen (or a phosphorus) in the molecule, there may also be a pyramidal inversion vibration.

- Three-dimensional structures generated by internal rotations about single bonds in a molecule are called **conformations**. In the most stable conformation of ethane which can be viewed as a **Newman projection**, the C–H bonds of the two methyl groups are **staggered**; this minimizes torsional strain and maximizes hyperconjugative stabilization. They are **eclipsed** in the most unstable conformation of ethane.

- The stability of a particular conformation of an open chain compound depends not only on the **torsional strain** owing to the repulsive interactions between bonds from adjacent carbon atoms, but also on **steric strain** attributable to adverse interactions between non-bonded groups.

- Small-ring cycloalkanes also have **angle strain** owing to the distortion of bond angles from the normal tetrahedral value. Cyclopropane has no conformational flexibility; four-membered and larger ring cycloalkanes become nonplanar to relieve torsional strain, and conformational interconversions are rapid at normal temperatures.

- The **chair conformation** of cyclohexane is the most stable and taken as the standard against which strain in other cycloalkanes is compared. The total ring strain energy of any cycloalkane can be determined by comparison of its heat of combustion with that of cyclohexane.

- Each carbon in a chair cyclohexane has an **axial** bond and an **equatorial** bond. The lower stability of an axial substituent is attributable mainly to adverse steric **1,3-diaxial interactions**—the larger a substituent, the greater its tendency to be equatorial.

- Cycloalkanes bearing a substituent on each of two ring carbons show *cis–trans* **isomerism**.

Problems

4.1 Does each of the following pairs of structures show the same molecule or different molecules?

(a)
(b)
(c)
(d)

4.2 Draw three-dimensional structures using wedged bonds for the staggered and eclipsed conformations of propane.

4.3 Draw saw-horse structures for the *anti* and *gauche* conformations of butane.

4.4 Draw Newman projections of all the staggered and eclipsed conformations obtained by rotation about the C1–C2 bond of 1-chloropropane.

4.5 Draw Newman projections along the C2–C3 bond for two staggered forms of different energy of 2-methylbutane, and explain which is the more stable.

4.6 Represent the most stable conformation of each of the following by a Newman projection along the bond indicated:
(a) 3-methylpentane along the C2–C3 bond;
(b) 2,3-dimethylpentane along the C2–C3 bond;
(c) 2,3-dimethylpentane along the C3–C4 bond;
(d) 3,3-dimethylhexane along the C3–C4 bond.

4.7 Draw the *anti* conformation of 1,2-dibromoethane and all the conformations obtained by sequential 60° internal rotations about its C–C bond; further, sketch an approximate energy profile to show the energy change upon rotation about the C–C bond through 360°.

4.8 Draw structures of *cis* and *trans* isomers of (a) 1,2-dimethylcyclobutane, and (b) 1,3-dichlorocyclopentane.

4.9 Draw two chair conformations of bromocyclohexane.

4.10 Draw possible chair conformations of *cis*- and *trans*-1,2-dimethylcyclohexane.

4.11 Draw a chair form of *trans*-1,3-dimethylcyclohexane and also Newman projections along the C1–C6 and C3–C4 bonds (cf. Figure 4.11); further, show any 1,3-diaxial interactions with double-headed curved arrows (⤺⤻).

4.12 Draw two chair forms of *cis*-1,3-dimethylcyclohexane and also the Newman projections along the C1–C6 and C3–C4 bonds for each conformation (cf. Figure 4.11); further, show any 1,3-diaxial interactions with double-headed curved arrows (⤺⤻).

4.13 Draw possible chair conformations of *cis*- and *trans*-1-*t*-butyl-2-methylcyclohexane, and compare their stabilities (cf. Example 4.4).

4.14 Draw all possible chair conformations of each of the following; if different chair forms are possible, compare their stabilities.

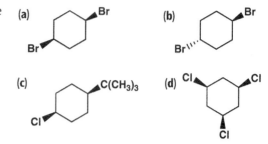

4.15 Draw the most stable conformation of each of *cis*-1,4-dihydroxycyclohexane and *cis*-1,3-dihydroxycyclohexane, and compare stabilities of the two isomers. (Hint: an intramolecular hydrogen bond may be possible.)

4.16 The most stable form of glucose has the chair conformation of a six-membered ring with all the five substituents in equatorial positions. The first planar structure below shows glucose with no stereochemical information. Show the stereochemistry of glucose (a) by adding substituents to the chair structure in the centre, and (b) by adding substituents with wedged bonds to the structure on the right.

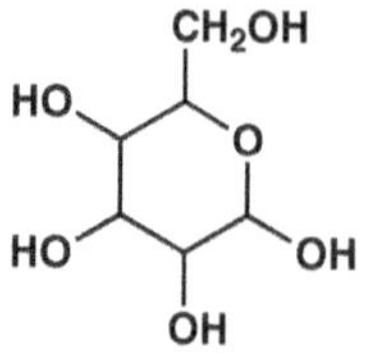

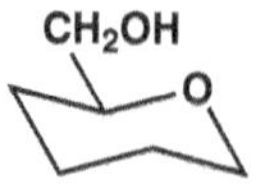

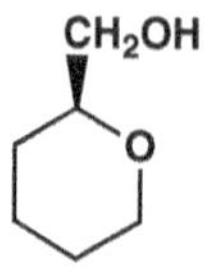

Conjugation, π-Electron Delocalization, and Aromaticity

5

Related topics we have covered:

- **Lewis structures** (Section 1.3)
- **Introduction to resonance** (Section 1.4)
- **Orbital description of covalent bonding** (Section 2.2)
- **Hybridization of atomic orbitals** (Section 2.3)
- **Bonding in ethene** (Section 2.5)

Topics we shall cover in this chapter:

- **Extended π-bond systems and the concept of conjugation**
- **Electron delocalization**
- **MOs of butadiene and allylic systems**
- **Resonance**
- **Electronic structure of benzene**
- **Aromaticity**
- **Photoexcited states**

When two or more C=C double bonds alternate with C–C single bonds, the double bonds are said to be *conjugated*, and the chemical properties of a conjugated system are different from those of a system with a single C=C double bond, or with multiple but non-conjugated double bonds. The double bonds of a conjugated system do not react independently. Physical properties of a conjugated system are also characteristically different from those of isolated or non-conjugated multiple C=C double bonds. Colour, for example, is a familiar physical property of a substance which can be related to its molecular properties (as we shall see in Section 25.2 of Chapter 25). Simple alkenes are colourless, as are compounds with several non-conjugated C=C double bonds. However, extended conjugated polyenes are coloured and the colour can be related to the extent of the conjugation (see Panel 5.2). Lycopene contains 11 alternating double bonds and is a natural red pigment found in tomatoes and other red fruit and vegetables. β-Carotene has a similar structure and is responsible for the colour of carrots.

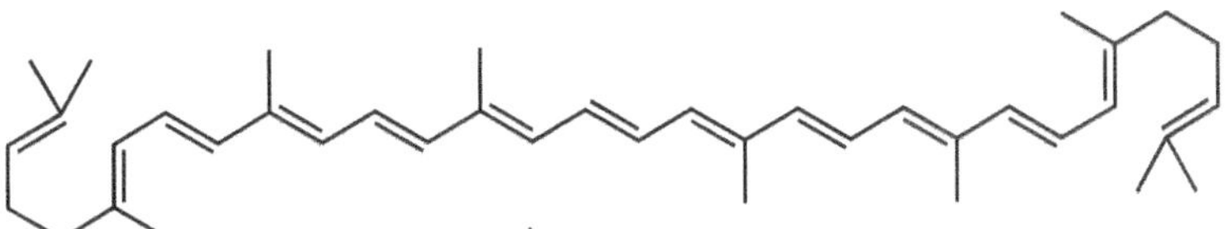

lycopene

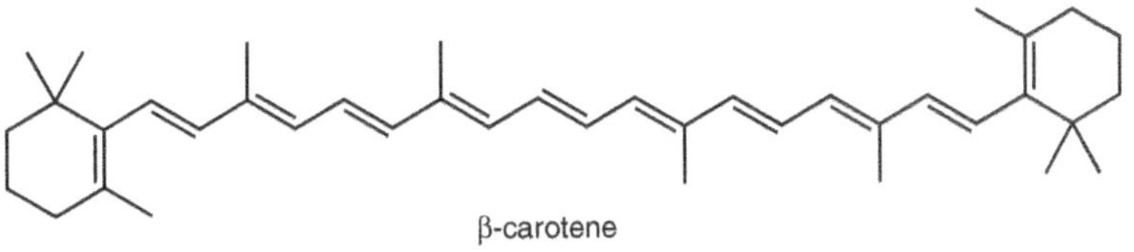

β-carotene

Benzene and other aromatic compounds contain cyclic alternating double bonds—they are a special kind of conjugated system with distinctive properties which are characteristically different from those of linear conjugated polyenes. In this chapter, we shall see how constituent π and p orbitals of appropriate energy and symmetry within a molecule (or ion) interact to give the new MOs of a conjugated system which are responsible for its properties, and how the conjugation can be described by resonance.

5.1 Extended π Bonds and the Concept of Conjugation

The C=C double bond is made up of a σ bond and a π bond, as discussed for ethene in Section 2.5, and the characteristic chemical properties of ethene come mainly from the π bond. In contrast to the end-on overlap of the sp² atomic orbitals (AOs) of the carbons which leads to the σ and σ* molecular orbitals (MOs) of ethene, the MOs for π bonding are formed by the sideways overlap of 2p AOs of the C atoms. In much the same way, *two C=C double bonds separated by a single σ bond are also able to interact by sideways overlap of their π MOs* (called a conjugative interaction) in a **conjugated system**.

Conjugated C=C double bonds are not the only type of conjugative interaction—a C=C can also be conjugated with atomic orbitals of the correct symmetry, e.g. a vacant 2p AO on an adjacent C or a 2p AO on a heteroatom containing a lone pair. The overlap of orbitals arising from conjugation allows increased **delocalization** of π and p electrons, which stabilizes the system. The difference in energy between the real conjugated system and the energy of the hypothetical system of non-interacting independent π bonds and occupied (or vacant) 2p orbitals is called the **stabilization** or **delocalization energy**.

5.2 Bonding in Butadiene

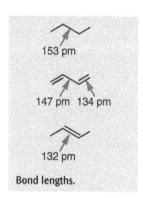

153 pm

147 pm 134 pm

132 pm

Bond lengths.

Buta-1,3-diene, with its two C=C groups joined by a single bond, is the simplest example of a **conjugated diene**. Experimentally, both C=C bonds are longer than in ethene, and the C–C bond is a little shorter than in ethane; these measurements indicate that the molecule is not exactly like two independent ethene units simply joined together.

The π bonding in buta-1,3-diene originates in the four sp²-hybridized carbon atoms, each with an out-of-plane 2p AO containing a single electron. In the stable conformation with all sp² orbitals in the same plane, the four singly occupied 2p AOs are parallel (Figure 5.1) and can combine by sideways overlap to give the four π MOs shown in Figure 5.2. These four new MOs (like the orbitals from which they were constructed) all have the nodal plane characteristic of π orbitals, and extend over the whole four-carbon skeleton (but not equally or evenly); in other words, the π electrons are no longer localized between two pairs of carbon atoms.

The π MO of butadiene of lowest energy, π_1, is formed by the in-phase combination of all four 2p AOs and has only a single node in the plane of the molecule; the highest, π_4, is formed by their fully out-of-phase combination and has three additional nodes (shown by red vertical broken lines). Between these two, as indicated, π_2 and π_3 have one and two additional nodal planes, respectively. There are four electrons from the four original 2p orbitals to feed in, so these occupy the bonding MOs, π_1 and π_2, (two in each) in the

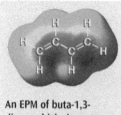

An EPM of buta-1,3-diene, which shows delocalization of electron density over the whole carbon framework (in red).

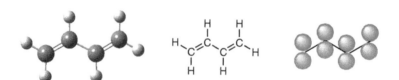

Figure 5.1 Structure of buta-1,3-diene and component 2p AOs.

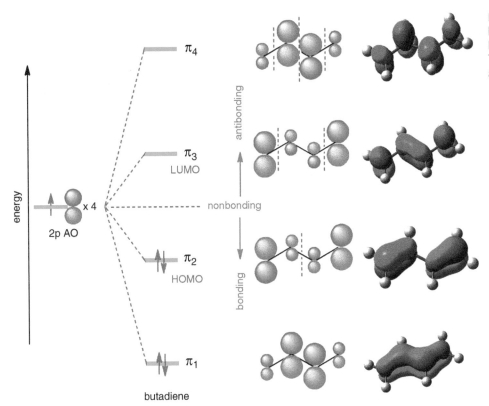

Figure 5.2 π MOs of butadiene. Red vertical broken lines show nodal planes. Theoretically generated MOs are shown on the right.

electronic ground state of butadiene. Both the higher energy MOs, π_3 and π_4, are anti-bonding, and unoccupied.

Note that the MOs of butadiene in Figure 5.2 illustrate a general principle: as the energy of MOs formed from the same constituent orbitals increases, so does the number of nodal planes.

Discuss whether the double bonds in the following molecules are conjugated.

(a) $H_2C{=}C{=}CH_2$ (allene)   (b) $H_2C{=}CHCH_2CH{=}CH_2$ (penta-1,4-diene)

Example 5.1

Solution

(a) The central carbon of allene is sp-hybridized and π MOs of the two double bonds are perpendicular to each other, as shown in the diagram. Consequently, the π orbitals cannot interact (they are orthogonal), and the double bonds are not conjugated.

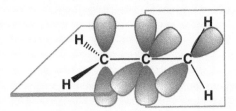

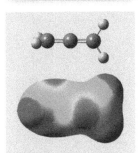

A molecular model and EPM of allene.

The red parts of the EPM show that π electrons are above and below the double bond on the left, while they are in front of and behind the double bond on the right.

(b) Between the two double bonds of penta-1,4-diene, there is an sp³-hybridized C atom which has no p AO; consequently, the π orbitals cannot interact with each other (they can be thought of as being 'insulated' from each other by the sp³-hybridized C) so, here also, the double bonds are not conjugated.

Identify conjugated double bonds in the following molecules.

(a) $H_2C=CH-CH=CH-CH_3$ (b) $H_2C=CH-CHO$ (c) CO_2 (d) 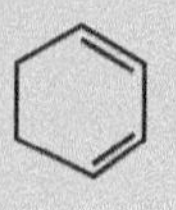 ⬡   (e) 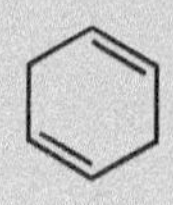 ⬡

5.3 Allylic Systems

The allyl cation, radical, and anion (illustrated below) can be seen as an ethenyl (vinyl) group ($CH_2=CH-$) bonded to an sp² hybridized carbon which has two other bonds and an unhybridized 2p orbital which is unoccupied in the cation, singly occupied in the radical, and filled in the anion. The parent **allylic systems** shown, or derivatives obtained by replacing the hydrogens by alkyl groups, are important intermediates in organic reactions that we shall encounter later.

$$CH_2=CH\overset{+}{C}H_2 \qquad CH_2=CH\overset{\bullet}{C}H_2 \qquad CH_2=CH\overset{-}{C}H_2$$

allyl cation   allyl radical   allyl anion

We shall see next how the molecular orbitals of the parent allylic systems can be derived from 2p AOs of the three constituent carbon atoms. Then, we shall use allylic systems to demonstrate the use of resonance (introduced in Section 1.4) as a shorthand method of describing the bonding in molecules and ions which cannot be described accurately by single Lewis structures.

5.3.1 Molecular orbitals of allylic systems

To construct a generic allylic system, we start with three connected sp²-hybridized carbon atoms in a conformation with their 2p AOs aligned parallel with each other, as shown in Figure 5.3.

Combination of the three 2p AOs gives three π MOs, $\pi_1-\pi_3$ (Figure 5.4); the MOs are distributed over the three C atoms, and electrons which are fed into these MOs are delocalized. As for butadiene, the plane containing the three atoms is the characteristic nodal plane of these π MOs. The lowest MO, π_1, has no additional node; π_2 and π_3 have one and two additional nodes, respectively.

The allyl cation has only two π electrons, so they are accommodated in π_1. In the allyl anion, each of π_1 and π_2 is occupied by a pair of electrons. The allyl radical has three π electrons: two in π_1 and just one in π_2. The MO π_2 in the allyl radical, therefore, is a *singly occupied MO (SOMO)*, as shown in Figure 5.4.

> In general, reactions of a molecule or ion are most strongly influenced either by its highest occupied MO (**HOMO**) or its lowest unoccupied MO (**LUMO**) (see Chapter 7 for further discussion), so the π_2 MO is very important in controlling the reactions of allylic intermediates (cations, anions, and radicals). A distinctive feature of this MO (when occupied) is that its π electron density at the central carbon is zero because it lies in a nodal plane. This is one reason why reactions of allylic systems start at one of the terminal carbons.

Draw energy level diagrams for the π systems of the allyl cation and anion, with small arrows representing occupying electrons, and identify the LUMOs and HOMOs.

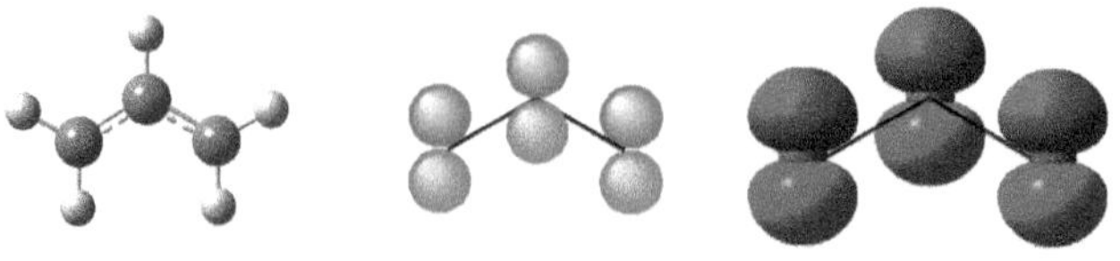

Figure 5.3 Constituent 2p AOs of an allylic system.

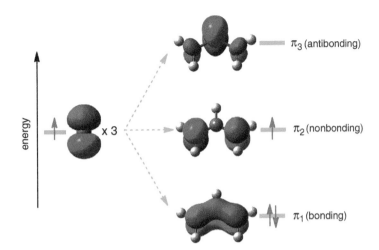

Figure 5.4 π MOs of the allyl system with electron occupancy for the allyl radical.

π_3 (antibonding)

π_2 (nonbonding)

π_1 (bonding)

5.3.2 Resonance description of allylic systems

Lewis structure **1a** represents the ion obtained if an H of the planar methyl cation (CH_3^+) is replaced by an ethenyl group ($CH_2=CH-$). In this ion, the localized π MO of the ethenyl substituent is derived from 2p AOs of its constituent C atoms. However, the 2p AO on the central C in **1a** could equally well form a localized π MO with the 2p AO of the C to its right and, if the two electrons are in this MO, we have Lewis structure **1b**. These two equivalent structures are both legitimate Lewis structures of the allyl cation, but neither adequately describes the real ion. For example, the two terminal carbon atoms are equivalent, with the positive charge shared equally between them; the two C–C bonds are also equivalent, neither being a single bond or a double bond. The actual structure is better represented as a *resonance hybrid*, **1c**, of the two Lewis structures (**1a** and **1b**) which are examples of *resonance contributors* (or *resonance forms*, Section 1.4).

1a	**1b**	**1c**

> The different locations of the electron pairs in resonance contributors **1a** and **1b** are correlated with curly arrows, as shown.

If two more electrons are added to the allyl cation, we generate the allyl anion which can be represented by the two Lewis structures **2a** and **2b**. However, as for **1a** and **1b** in the case of the cation, neither of these equivalent Lewis structures is an adequate description of the real anion. This time, the real ion is best described as a resonance hybrid of **2a** and **2b** with the negative charge distributed equally between the terminal carbons as shown in **2c**.

2a	**2b**	**2c**

> Note that pairs of curly arrows are needed to correlate resonance contributors **2a** and **2b**.

According to *valence shell electron pair repulsion (VSEPR)* theory as encountered in Chapter 2, the methyl anion (CH_3^-) is pyramidal with a tetrahedral arrangement of valence electrons. When one H is replaced by an ethenyl group ($CH_2=CH-$), the delocalization energy associated with the rebonding leading to the allyl anion as described above is more than sufficient to compensate for the necessary configurational change from sp^3 to sp^2 for the one carbon.

Each of the two ions described above contains three carbons and, in both cases, the π electrons (two in the cation and four in the anion) are delocalized along all three. Thus, the cation is a *three-centre two-π-electron system* and the anion is a *three-centre four-π-electron system*.

If just one electron is added to the allyl cation (or one is removed from the anion), we obtain the uncharged **allyl radical** which is represented by a resonance hybrid of **3a** and **3b**, and can be depicted as **3c**. In this system, three π electrons (including one which is unpaired) are delocalized along the three carbon atoms. Just as for the cation and anion, the two terminal carbons of the allyl radical are equivalent, and the two C–C bond lengths are equal.

Note that (three) singly barbed curly arrows are used to relocate single electrons and thereby correlate resonance contributors **3a** and **3b** (we show just one set in **3a**).

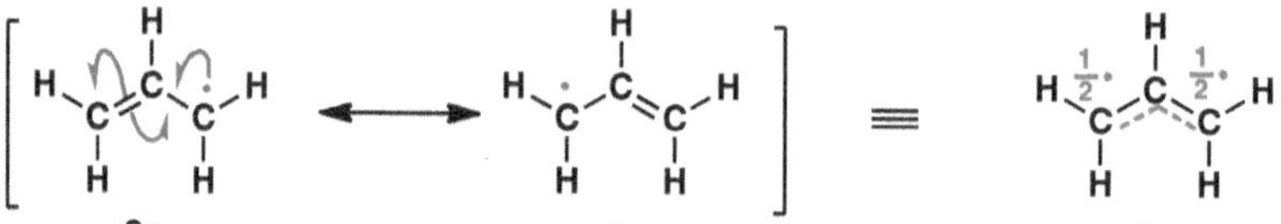

3a   **3b**   **3c**

5.3.3 Allyl anion analogues

The conjugated systems in the ethanoate (acetate) ion and nitromethane molecule, encountered in Sub-section 1.3.2 and described by resonance in Section 1.4, also have three-centre four-π-electron systems, *isoelectronic* with the allyl anion. As with the allyl anion, the known symmetry of both ethanoate and nitromethane is explained by equal contributions of two resonance forms in the overall hybrids (Section 1.4).

ethanoate ion   nitromethane

In general, any system containing a double (or triple) bond from an atom which is singly bonded directly to a heteroatom such as O, N, and halogen with a lone pair in a 2p AO will have a delocalized π-electron system analogous to that of the allyl anion. These molecules and ions are collectively known as **allyl anion analogues**; further examples with their resonance representations are given below.

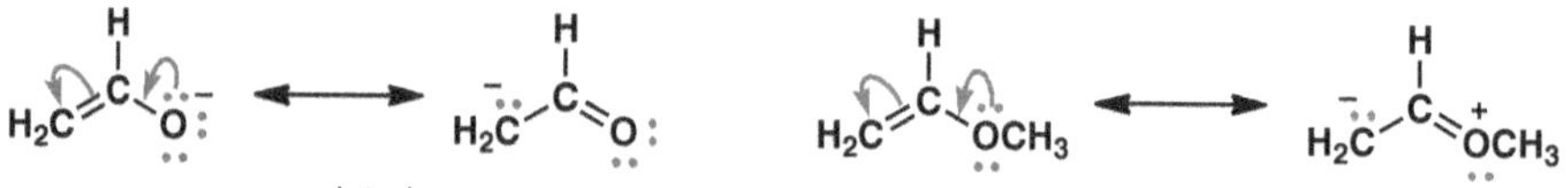

an enolate ion   methyl vinyl ether

an enamine   an amide

Note that the two Lewis structures for each of these four examples (unlike the pairs for ethanoate, nitromethane, and the allyl cation, radical, and anion) do not contribute equally to the resonance hybrid; this matter will be discussed in Sub-section 5.4.2.

Exercise 5.3

Draw Lewis structures of the two principal resonance contributors of each of the following allyl anion analogues.

(a) $H_2C=CHOH$ (enol)   (b) $CH_3CO_2CH_3$ (an ester)

5.4 Resonance Revisited

We first encountered resonance in Section 1.4 as a method of describing the bonding in molecules and ions which cannot be described accurately by single Lewis structures, and used it in the preceding section. In this section, we shall explore the concepts further, and identify the features which affect the relative weightings of individual **resonance contributors** (also called *resonance structures, resonance forms* or, sometimes, *canonical structures*) to the overall hybrid.

   The starting point for identifying a set of resonance contributors is a single legitimate Lewis structure expressed as lines representing bonds, pairs of dots representing lone pairs, and (if present) single dots representing unpaired electrons. This will describe the atom connectivity, the number of valence electrons, and one possible distribution of them. It may be inadequate in the sense that the properties expected of the molecule or ion on the basis of the single Lewis structure do not correspond with those observed experimentally. If so, additional Lewis structures need to be identified. For each of ethanoate and nitromethane illustrated above, only a second Lewis structure was invoked, but later we shall encounter compounds for which more are required. In all cases, the real electronic structure of the molecule is a weighted blend (the **resonance hybrid**) of all resonance contributors. The hybrid cannot be accurately represented by a single structure but it can be approximated by the set of individual contributors connected by double-headed arrows.

5.4.1 The nature of resonance

As seen above for the allyl cation and anion, the conceptual process by which one resonance form is converted into another, i.e. the relocation of 2p or π electron pairs within the unchanging σ-bonded connection of atoms, is described by curly arrows, and it follows that the total number of valence electrons and the overall charge do not change. This use of a curly arrow is quite different from its use to describe the motion of an electron pair in a mechanistic equation (see Chapter 7). In that application, the curly arrow describes the movement of an electron pair in a chemical transformation—a dynamic process. In the context of resonance, the curly arrow is just an aid in relating one resonance form to another—a relationship indicated by *the double-headed arrow* ($\leftrightarrow$): it does not describe a chemical process—no atoms relocate, nothing changes—the species has just a single hybrid structure. Consequently, it is important that the double-headed arrow for resonance be not confused with an equilibrium double arrow ($\rightleftharpoons$) or a reaction arrow ($\rightarrow$). In the allyl cation example given above, the resonance symbol relating **1a** and **1b** does not indicate that there are two ions, and that single and double bonds interconvert. There is a single ion with two carbon–carbon bonds which are identical, and intermediate between single and double bonds.

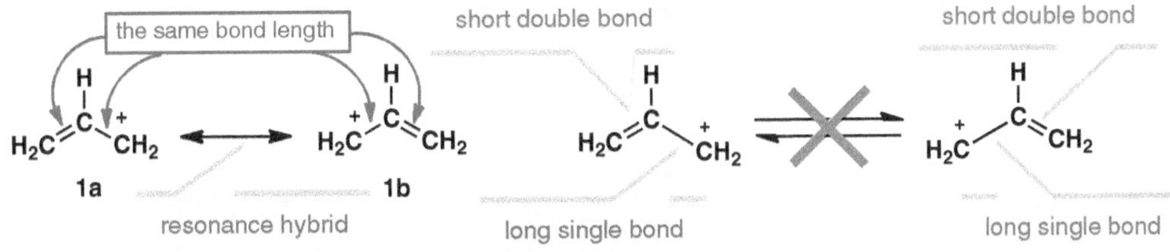

5.4.2 Resonance forms and their relative contributions

Based upon the precepts given above, the protocol for writing resonance contributors for a molecule or ion is as follows.

1. Write a legitimate Lewis structure.

2. Move electron pair(s) using curly arrow(s) to give another Lewis structure; curly arrows start from a π bonding pair or a 2p lone pair of electrons

3. Repeat process (2) to draw as many Lewis structures as possible.

In ethanoate, for example, there are only two credible resonance contributors and, by symmetry, they are equivalent; in such cases, the two forms contribute equally to the resonance hybrid. When they are not equivalent, we have to rank the individual contributors in order of stabilities of their electronic configurations. This process is informed by knowledge of relative stabilities of ions and molecules which can be represented by single Lewis structures. We then take it as self-evident that the more stable one of a set of resonance contributors is, the greater its contribution will be to the resonance hybrid, and (therefore) the nearer it is to the electronic structure of the real molecule or ion.

The following is a guide for establishing the relative stabilities (and hence relative contributions) of resonance forms.

1. The more bonds (and hence the more atoms with complete valence shells) a contributor has, the more stable it is. For example, resonance contributing structures **4b** and **4c** can be drawn for buta-1,3-diene, but these structures both have fewer bonds than **4a** so contribute much less to the resonance hybrid (they also fail by point 2 below).

Another example is the methoxymethyl cation which is represented as a resonance hybrid of **5a** and **5b**. Structure **5b** has one more covalent bond than **5a** (and all its atoms have full valence octets), so **5b** is more important even though there is a positive charge on the more electronegative O atom (contrary to point 4 below).

2. Charge separation is unfavourable. Methanoic (formic) acid is another allyl anion analogue, and could be described as a resonance hybrid of **6a** and **6b**. However, **6b** is a very minor contributor because of its charge separation, and the bonding of methanoic acid is reasonably described by just **6a**.

3. Unpairing of electrons is unfavourable. It produces two atoms with incomplete valence shells and a structure with fewer bonds.

4. A contributor in which the electron distribution conforms to the electronegativities of the constituent atoms will be more important than ones where this is not the case: negative charge is better located on a more electronegative atom, and positive charge on a less electronegative atom. The enolate ion obtained by deprotonation of ethanal (acetaldehyde), the allyl anion analogue introduced in Sub-section 5.3.3, is described by the resonance hybrid of **7a** and **7b**. Since O is more electronegative than C, the negative charge is better located on the O. Consequently, **7a** contributes to the hybrid to a greater extent than **7b**, and corresponds better to the electron density on the O being higher than on the terminal carbon.

7a ⟷ **7b**

The carbonyl (C=O) bond is polar owing to the electronegativity difference between C and O. When methanal (formaldehyde) is described by resonance, two charge-separated structures, **8b** and **8c**, seem possible. In **8b**, the charge separation conforms to the different electronegativities of C and O, so **8b** is a significant contributor to the resonance hybrid, and accounts for the bond polarity. In contrast, the charge separation in **8c** involves a negative charge on the less electronegative C, and a positive on the O; this is very unfavourable and contributor **8c** may be neglected. Consequently, the dipolar methanal molecule with a partial positive charge on the carbonyl carbon (**8d**), is best described as a resonance hybrid of **8a** and **8b**.

> The second most important resonance contributor often provides insight into the chemical reactivity of a compound, e.g. the tendency of carbonyl compounds to undergo nucleophilic addition at the C.

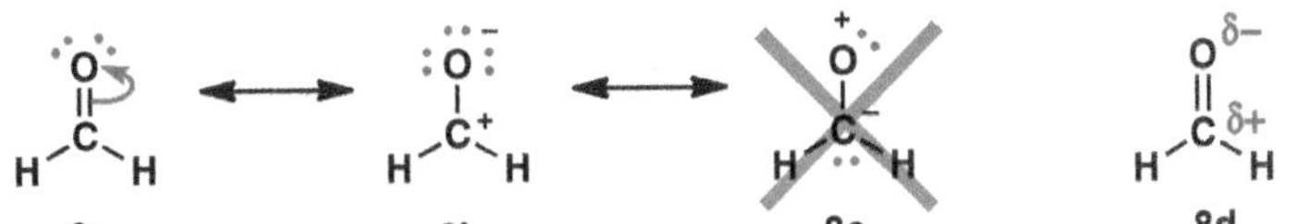

8a ⟷ **8b** ⟷ **8c** ~~⟷~~ **8d**

> **Example 5.2**

Represent the following molecules or ions by resonance, and explain which resonance contributor is the most important in each case.

(a) cyanide ion (CN⁻) (b) diazomethane ($H_2C=N_2$) (c) acetylium ion ($CH_3-C^+=O$)
(d) chloroethene ($H_2C=CHCl$)

Solution

(a) 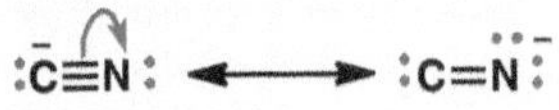 :C≡N: ⟷ :C=N:

Both C and N have octets in the first contributor, whereas the C in the second has only six valence electrons. Consequently, the first is more important.

(b)  $H_2C=N=N$: ⟷ $H_2C-N≡N$:

All the C and N atoms have octets in both contributors. The first one is more important, however, since it has the negative charge on the more electronegative N rather than on the C. Note that it is not possible to write a sensible resonance form for the neutral diazomethane molecule which does not involve charge separation.

(c)  $CH_3-C=O$: ⟷ $CH_3-C≡O$:

All atoms have octets in the second contributor, whereas the positive C has only six valence electrons in the first. The second contributor is more important.

(d)

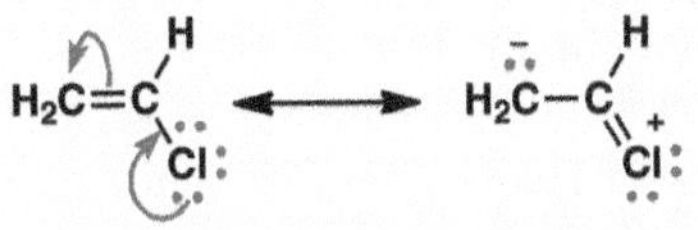

The second contributor has charge separation and a positive charge on the electronegative Cl atom. The first contributor is more important.

> **Exercise 5.4**

Nitrate ion NO_3^- is often represented by the Lewis structure shown below. However, experiments indicate that lengths of all the N–O bonds are the same. Explain.

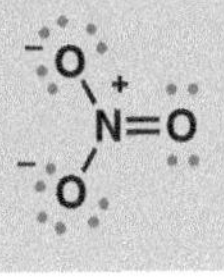

nitrate ion

Exercise 5.5

Draw additional resonance contributors for the following, and explain which is the most important in each case.

(a) H_3CCNH_2 (with O double-bonded to C)

(b) $H_2C{=}CHOCH_3$

(c) $H_2\overset{+}{C}{-}NH_2$

(d) $H_3C{-}\overset{-}{C}H{-}\overset{O}{\overset{||}{C}}{-}CH_3$

5.5 Benzene

Benzene is a very important compound in the history of chemistry and the development of theories of bonding. For many years after its isolation in 1825, its structure was a puzzle as it was far less reactive than its highly unsaturated molecular formula (C_6H_6) suggested.

5.5.1 Structure of benzene

Benzene is usually represented by structure **9** with a six-membered ring and alternating C=C double and C–C single bonds, or by one of the line-angle drawings, **9a** or **9b**. Any of these three drawings alone is known as a **Kekulé structure** after the chemist who proposed it as the structure of benzene (see Panel 5.1); it appears to imply that there are two types of bonds between C atoms in benzene, which is not the case. We know from physical methods of structure determination that the benzene molecule is a regular planar hexagon, and all six bonds between C atoms are equivalent and 139.5 pm long, which is between the normal C–C single bond (154.1 pm) and a normal C=C double bond (133.7 pm).

Structure **9c** is convenient for representing benzene and its simple derivatives, but not for complicated derivatives, or if we need to monitor the flow of electrons in a mechanism, for example. We shall follow the normal convention of using a single Kekulé structure (**9**, **9a**, or **9b**) with no implication that it accurately represents the molecular and electronic structure of benzene.

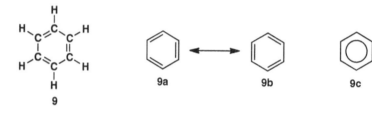

By analogy with our description of the allyl systems, we now describe benzene as a resonance hybrid of **9a** and **9b**, which is often represented by **9c** to emphasize the delocalization of π electrons (see below) and the six-fold symmetry.

5.5.2 Molecular orbitals of benzene

Benzene has a planar circular structure, and all six carbons are sp² hybridized with their 2p orbitals perpendicular to the plane of the ring (Figure 5.5).

Combination of the six 2p AOs leads to the six π MOs of benzene shown to the right in Figure 5.6. All have the nodal plane characteristic of a π orbital in the plane of the six-membered ring. The MO of lowest energy, π_1 with no additional node, is circularly distributed above and below the plane of the molecule. The next two MOs, π_2 and π_3, are

Figure 5.5 Structure of benzene and its constituent 2p orbitals.

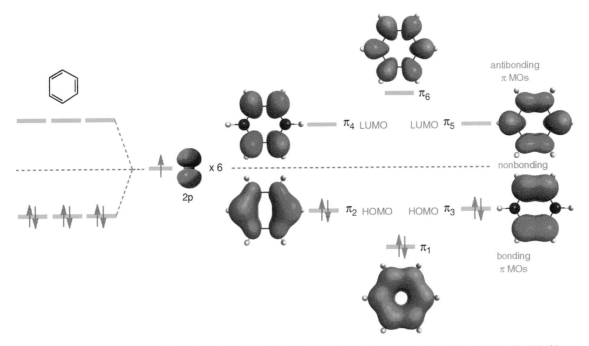

Figure 5.6 π MOs of benzene (viewed from above) derived from six 2p AOs, and the energy levels of MOs of the isolated double bonds of the hypothetical cyclohexa-1,3,5-triene to the left.

of equal energy (they are *degenerate*) and each has one additional nodal plane. The three MOs π_1–π_3 are bonding orbitals, and each is occupied by two electrons; the two degenerate MOs π_2 and π_3 are the HOMOs. The MOs next higher in energy are the degenerate pair, π_4 and π_5; these antibonding MOs have a total of three nodal planes and, as they are unoccupied, are the LUMOs. The single unoccupied MO highest in energy, π_6, is formed by the out-of-phase combination of all six constituent AOs, and has a total of four nodal planes.

The six electrons originally in the 2p AOs occupy the three π bonding MOs in the ground-state electronic configuration of benzene. This provides benzene with the special stability known as **aromaticity** (Section 5.6).

5.5.3 Stabilization energy of benzene

We see in Figure 5.6 that the electronic energy of benzene is appreciably lower with the π electrons being delocalized around the whole molecule, i.e. by occupying $\pi_1 - \pi_3$, than if there were no interactions between the six 2p AOs. Benzene is also lower in energy, i.e. more stable, than the hypothetical 'cyclohexatriene' molecule constructed with just pairwise interactions between the six 2p AOs. The energy difference between the molecule with the electronic structure described to the right of Figure 5.6 and the hypothetical cyclohexatriene shown to the left can be calculated by modern computational methods; this is known as the **delocalization energy** of benzene. It may also be taken as the energy difference between the resonance hybrid and just one of the two contributing forms shown, in which case it is usually called the **resonance energy** of benzene. We are also able to determine the stabilization of benzene experimentally.

Both cyclohexene and benzene react with H_2 in the presence of a catalyst to give the same product, cyclohexane (eqns 5.1 and 5.2). The reactions are exothermic, and the enthalpy of reaction (ΔH) is called the **heat of hydrogenation**. Experimental values of ΔH are given in eqns 5.1 and 5.2.

Panel 5.1 The structure of benzene and Kekulé's dreams

In the early 19th century, London streets were illuminated by lamps burning a gas manufactured from whale oil. In 1825, Michael Faraday (1791–1867) isolated a liquid containing only carbon and hydrogen in equal proportions from the residues of the gas. That liquid isolated by Faraday, a sample of which is still kept in the Royal Institution (London), is now called benzene.

Although benzene is highly unsaturated, it is quite stable and was known to be relatively unreactive in marked contrast to known alkenes. It was deduced that the molecular formula of benzene is C_6H_6, but its structure remained a puzzle until the German chemist, August Kekulé, proposed a cyclic hexagonal structure in 1865.

Kekulé had developed a structural theory of organic chemistry during 1857–1858 based on the idea that carbon is tetravalent, and the structural proposal for benzene was a consequence of his valence theory. During his speech at the *Benzolfest* in 1890 (the 25th anniversary of the benzene theory) he explained that the idea of the tetravalency of carbon occurred whilst dreaming on a horse-drawn London bus one summer evening. The cyclic structure for benzene occurred in a similar way. The story goes that, one evening whilst working on a textbook in Ghent in Belgium (where he served as a chemistry professor), his thoughts drifted away. He turned his chair around and, gazing towards the fire, imagined atoms in the flames, joined in long rows and all twisting and turning in snake-like motion. He described how one of the snakes had seized hold of its own tail, and the image whirled before his eyes; then, as if by a flash of lightning, he woke up and spent the rest of the night working out the consequences of his ideas.

August Kekulé (1829–1896)

Kekulé's theories about organic compounds and his proposal for the structure of benzene were the basis for major developments in organic chemistry in the late 19th century and the foundation of the chemical dyestuffs industry.

In the 1930s, the sixfold symmetrical cyclic structure of benzene was strongly supported by molecular orbital theory based on quantum mechanics developed in the 1920s. However, it took nearly 100 years from Kekulé's proposal until the molecular structure of benzene was confirmed experimentally by X-ray and neutron diffraction methods.

$$+ \text{H}_2 \xrightarrow{\text{catalyst}} \qquad \Delta H = -120 \text{ kJ mol}^{-1} \qquad (5.1)$$

$$+ 3\text{H}_2 \xrightarrow{\text{catalyst}} \qquad \Delta H = -208 \text{ kJ mol}^{-1} \qquad (5.2)$$

The heat of hydrogenation of the hypothetical cyclohexa-1,3,5-triene (with three alternating non-interacting double bonds) should be three times that of cyclohexene, i.e. $\Delta H = 3 \times (-120) = -360$ kJ mol^{-1}. The difference between this and the observed result for benzene ($\Delta H = -208$ kJ mol^{-1}) is the experimental **stabilization energy** of benzene (152 kJ mol^{-1}) shown in Figure 5.7, and corresponds to its (theoretical) delocalization or resonance energy.

5.6 Aromaticity in General

5.6.1 Hückel's rule

The sideways interactions of a set of parallel unhybridized 2p orbitals of a planar cyclic system of sp^2-hybridized Cs lead to a set of delocalized MOs, and the distinctive arrangement of their energy levels depends upon the number of Cs. We have already seen the arrangement for benzene with six Cs in Fig. 5.6, and that benzene gains appreciable stabilization by the occupancy of its three MOs of lowest energy by the six electrons originally in the 2p orbitals. Other cyclic systems of sp^2-hybridized Cs are known but the

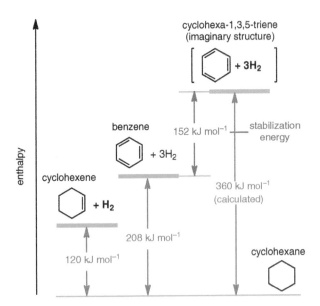

Figure 5.7 Determination of the stabilization energy of benzene from heats of hydrogenation.

The relationship between the number of π electrons and aromatic stabilization was predicted by Hückel in 1931 from quantum theory, and is called **Hückel's 4n + 2 rule**. Note that it is the number of electrons, not the number of atoms, which characterizes the system as aromatic or not.

E.A.A.J. Hückel (1896–1980)

stabilization is observed only when the total number of delocalized π electrons is 2, 6, 10, ..., i.e. $4n + 2$ where $n = 0, 1, 2, ...$ Such compounds, like benzene, are said to be **aromatic** and the consequences of **aromaticity** are observed in their physical properties and chemical reactions. We shall cover reactions of benzene and its derivatives in Chapters 16 and 18, polycyclic and heterocyclic aromatic compounds are discussed in Chapter 19, and the highly characteristic magnetic (NMR) properties of aromatic compounds are included in Chapter 25 and, particularly Panel 25.1.

In the 1930s, the German physical chemist, Erich Hückel, developed an approximation method for quantum mechanics which he applied to π electron systems and derived the $4n + 2$ rule of aromaticity; his method became known as Hückel Molecular Orbital (HMO) theory and has been the basis of many subsequent developments. In 1923, Hückel and the Dutch physicist, P.J.W. Debye introduced what we now call the Debye–Hückel theory of electrolytic solutions.

The capacity of an array of MOs derived from a set of parallel unhybridized 2p orbitals to accommodate $4n + 2$ electrons depends upon the number of sp²-hybridized Cs in the planar cyclic system; so far we have only considered the six-carbon system of benzene whose MOs are shown in Figure 5.6. We can generalize Figure 5.6 for rings containing 3–8 atoms, and leave out the diagrams of the orbitals, to obtain Figure 5.8(a) for cyclic systems with 4, 6, and 8 atoms, i.e. even numbers of atoms, and Figure 5.8(b) for cyclic systems with 3, 5, and 7 atoms, i.e. odd numbers of atoms.

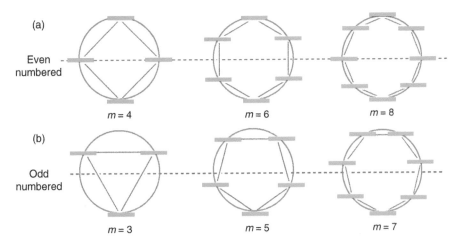

The red broken lines indicate the energy levels of constituent atomic orbitals. Notice that MO energy levels are at the vertices of a regular polygon of m sides for $(CH)_m$ inscribed within the circle with one vertex central at the bottom.

Figure 5.8 Molecular orbital energy diagrams for planar cyclic π systems, $(CH)_m$: (a) MOs for systems with an even number of sp²-hybridized C atoms in the ring (4, 6, and 8); (b) MOs for systems with an odd number (3, 5, and 7).

In Figure 5.8(a), we see a single lowest energy MO, then degenerate pairs of MOs of increasing energy, and a single highest energy MO; this is the pattern for all systems with an *even* number of atoms in the ring. In Figure 5.8(b) there is again a single lowest energy MO then simply degenerate pairs of MOs; this is the pattern for all systems with an *odd* number of atoms in the ring. If we add six electrons to the second system in Figure 5.8(a), we have the familiar MO energy level pattern of benzene (see Figure 5.6).

Some other examples of aromatic compounds and ions are given in Figure 5.9. The smallest aromatic system ($n = 0$ in the Hückel formula) is the cyclopropenium ion with just two π electrons, and they occupy the single lowest MO in the first system of Figure 5.8(b), the three-membered ring. Systems with $n = 1$ are the most common and have six π electrons (sometimes called *aromatic sextets*) which fill the lowest and the first degenerate pair of π MOs. Benzene is prototypical, but a ring CH can be replaced by one of a few other atoms (principally N to give pyridine, for example) with retention of aromaticity (but the symmetry of the energy level diagram in Figure 5.6 will be reduced); see Chapter 19 for aromatic heterocyclic compounds. The next higher aromatic compounds contain 10 π electrons ($n = 2$), and examples include naphthalene (Figure 5.9).

pyridine

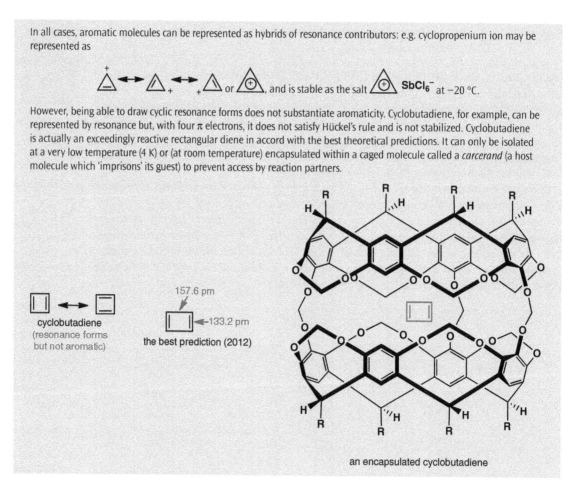

In all cases, aromatic molecules can be represented as hybrids of resonance contributors: e.g. cyclopropenium ion may be represented as

, and is stable as the salt $SbCl_6^-$ at −20 °C.

However, being able to draw cyclic resonance forms does not substantiate aromaticity. Cyclobutadiene, for example, can be represented by resonance but, with four π electrons, it does not satisfy Hückel's rule and is not stabilized. Cyclobutadiene is actually an exceedingly reactive rectangular diene in accord with the best theoretical predictions. It can only be isolated at a very low temperature (4 K) or (at room temperature) encapsulated within a caged molecule called a *carcerand* (a host molecule which 'imprisons' its guest) to prevent access by reaction partners.

cyclobutadiene
(resonance forms
but not aromatic)

157.6 pm
←133.2 pm
the best prediction (2012)

an encapsulated cyclobutadiene

Figure 5.9 Hückel's $4n+2$ rule and some aromatic hydrocarbons and ions.

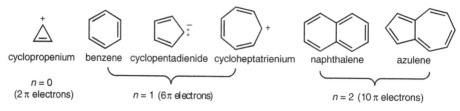

cyclopropenium   benzene cyclopentadienide cycloheptatrienium   naphthalene   azulene

$n = 0$
(2 π electrons)

$n = 1$ (6π electrons)

$n = 2$ (10 π electrons)

The cyclopentadienide ion (obtained by proton abstraction from cyclopentadiene using a strong base) has a cyclic arrangement of five constituent 2p AOs, and there are seven in the cycloheptatrienium ion (obtained by hydride abstraction from cycloheptatriene); both ions are planar, contain six π electrons, and are aromatic. Also, both have an odd number of sp²-hybridized C atoms in their rings so have degenerate pairs of highest energy MOs corresponding to the second and third entries in Figure 5.8(b). Just as for benzene, there are aromatic analogues of the cyclopentadienide ion in which an sp²-hybridized carbon (or more) is replaced by a heteroatom with a lone pair (but the degeneracy of their orbitals is reduced), and these are neutral molecules.

cyclopentadiene

cycloheptatriene

Example 5.3

Explain why pyrrole is aromatic but the pyrrolium ion is not.

 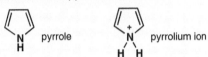

Solution
The N atom of pyrrole is sp² hybridized and has a lone pair of electrons in the unhybridized 2p orbital. This orbital interacts by sideways overlap with the constituent 2p orbitals of the diene system to give a total of five cyclic delocalized π MOs; the three of lowest energy accommodate the six available electrons. The N becomes sp³ hybridized when it is protonated to give the pyrrolium ion, and the lone pair becomes an N–H bond, so cyclic interaction with the diene system is prevented. (As we shall see in Section 19.3, protonation of pyrrole at the C next to the N is more favourable than on the N, but the cation so produced is highly reactive.)

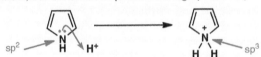

furan   thiophene

Each of pyrrole and its analogues with the NH replaced by O and S (furan and thiophene, respectively, see Chapter 19) can be represented as a hybrid of resonance contributors (which do not contribute equally in the way that the five equivalent ones for cyclopentadienide do).

Exercise 5.6

Describe the electronic configurations of pyridine and the pyridinium ion, and discuss whether they are aromatic.

 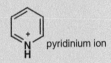

pyridine   pyridinium ion

Exercise 5.7

Are any of the following compounds aromatic? Explain.

 (a)   (b)   (c)   (d)   (e)

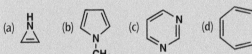

Exercise 5.8

Using curly arrows, show how additional resonance contributors of the following can be identified.

(a)   (b)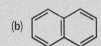

5.6.2 Annulenes

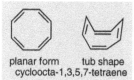

planar form   tub shape
cycloocta-1,3,5,7-tetraene

Hückel predicted that planar delocalized cyclic $4n$ π electron systems, in contrast to $(4n + 2)$ ones, would have no aromatic stabilization (and have subsequently been described as *antiaromatic*). Cycloocta-1,3,5,7-tetraene is a cyclic molecule, (CH)$_8$, with four conjugated double bonds, so a planar conformation with eight π electrons delocalized around the whole molecule should have no aromatic stability. In fact, cycloocta-1,3,5,7-tetraene is a flexible non-planar tub-shaped molecule and its properties are those of a typical polyene.

Benzene and other aromatic compounds comprising six-membered rings built up from sp^2-hybridized C atoms are largely free of the types of molecular strain we encountered in Chapter 4. However, molecular strain needs to be taken into account when considering possible aromaticity of compounds with larger rings. For example, aromatic stabilization of the planar [10]annulene in Figure 5.10 is not sufficient to overcome its adverse non-bonded interactions, so the compound is neither planar nor aromatic—it is a reactive polyene.

The strain is relieved in the bridged analogue 1,6-methano[10]annulene which, though not completely planar, is stable and aromatic. Similarly, the planar all-*cis* [14]annulene with all H atoms pointing outwards would be highly strained and is not stable, but the structure shown with four H atoms inside the ring is planar and aromatic.

5.7 Photoexcited Organic Molecules

As we have seen, the electronic ground state of a typical organic molecule has all its electrons paired; this type of electronic configuration is called a **singlet state**. Bonding MOs are usually fully occupied, and antibonding MOs are unoccupied. The examples of ethene, buta-1,3-diene, and benzene are shown in Figures 2.15, 5.2, and 5.6 where the HOMOs and the LUMOs are identified.

5.7.1 Interactions of organic molecules with electromagnetic radiation

When an unsaturated compound is irradiated, individual molecules are energized by the absorption of photons. If the radiation is of the correct frequency, an electron may be promoted from the HOMO (π) to the LUMO (π*), a so-called ππ* (π→π*) transition, to give an electronically **excited state**. The frequency (v) of the radiation needed to effect such a transition is given by the familiar equation,

$$\Delta E = hv,$$

where ΔE is the energy difference between the HOMO and the LUMO and h is the Planck constant.

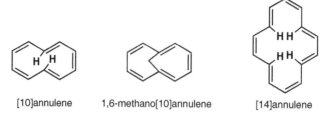

[10]annulene   1,6-methano[10]annulene   [14]annulene

Figure 5.10 Annulenes.

Radiation corresponding to energy differences between the ground states and the first electronically excited states of unsaturated organic compounds is typically in the ultraviolet (UV) or visible range (200–800 nm) which converts (by the above equation) to 600–150 kJ mol^{-1}. Radiation in a different range of the electromagnetic spectrum will lead to different types of transitions (see Chapter 25). For example, radiation in the infrared range (4000–250 cm^{-1} or 48–3 kJ mol^{-1}) broadly corresponds to differences between vibrational energy levels of organic compounds. Electronic transitions from the HOMO to high energy σ^* orbitals are seldom observed because they require very short wavelength radiation, which presents experimental difficulties.

The electrons are still spin-paired immediately following interaction of an organic molecule with a photon, which is an extremely fast process, so the photoexcited molecule is still a singlet. However, although a pair of electrons in the same orbital must be spin paired, they can have the same (*parallel*) spin when they are in different orbitals. A molecule in which two electrons have parallel spin is called a **triplet state**, and a photoexcited triplet is normally lower in energy than its corresponding singlet. Consequently, the initially formed singlet photoexcited state, whose lifetime is exceedingly short, may convert into the triplet state by spin inversion; this takes place in competition with (1) return to the ground state with emission of a photon (fluorescence), (2) return to the ground state with dissipation of energy to the surroundings as heat, and (3) a chemical reaction of the photoexcited singlet molecule (photochemistry); see Figure 5.11.

In competition with reacting chemically, the photoexcited *triplet* state may also return to the singlet ground state (which involves spin inversion) either by emission of a photon (phosphorescence) or dissipation of energy to the surroundings as heat.

The dioxygen molecule is unusual as it has a triplet ground state with one electron in each of a pair of degenerate π^* orbitals in accord with Hund's rule (Sub-section 1.1.3); it is a diradical (see Chapter 20). The singlet state, with the two electrons spin-paired in the same antibonding π^* orbital, is 92 kJ mol^{-1} higher in energy and may be formed photochemically or otherwise.

5.7.2 Properties of photoexcited states

Ground and photoexcited states of molecules often have different shapes and physical properties because they have different electronic configurations; the principal difference, of course, is their stability (lifetime). Methanal, $H_2C=O$, is a trigonal planar molecule in its ground state (Sub-section 2.1.2) and its dipole moment (μ, see Sub-section 1.2.3) is 2.33 D. Irradiation at 280 nm excites an electron from the localized lone pair on the O into the delocalized antibonding π^* orbital (a so-called $n\pi^*$ ($n{\rightarrow}\pi^*$) transition). The singlet photoexcited molecule is pyramidal and less polar ($\mu = 1.56$ D); the triplet is also pyramidal and even less polar ($\mu = 1.29$ D).

naphthalen-2-ol

$pK_a(S_0) = 9.5$
$pK_a(S_1) = 3$
$pK_a(T_1) = 8$

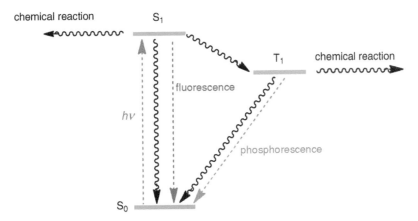

Figure 5.11 Processes following photoexcitation of singlet ground state S_0 to give singlet excited state S_1 and triplet excited state T_1.

The symbol $h\nu$ indicates irradiation; processes involving photons are in colour, thermal processes are in black.

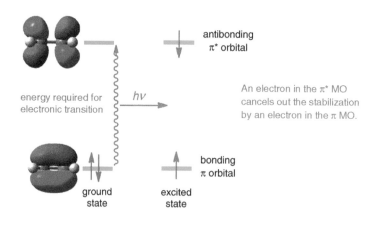

Panel 5.2 The perception of colours

When we see transmitted light, e.g. through a solution of a coloured substance, we see that part of the light which has not been absorbed by the solute. In other words, we see the colour *complementary* to that of the absorbed part of the visible light (which corresponds to the energy gap between the LUMO and HOMO of the substance). For example, if the absorbed light is blue ($\lambda \sim 450$ nm), the transmitted light which we see is orange coloured, and if red light is absorbed ($\lambda \sim 640$ nm), we see greenish-blue transmitted light.

The more extensive the conjugated system of a compound and its associated molecular orbitals are, the smaller the energy gap between the HOMO and the LUMO. It follows that the wavelength of the absorbed light corresponding to an electronic excitation becomes longer as the conjugated system of a compound becomes more extensive. For example, ethene ($\lambda_{max} = 174$ nm) and buta-1,3-diene ($\lambda_{max} = 217$ nm) absorb only ultraviolet radiation so have no colour whereas β-carotene (which we saw in the introduction to this chapter) has 11 conjugated double bonds and absorbs greenish-blue light (455 nm), so it looks orange. Structures of some coloured compounds are given below (the structural feature which gives rise to the absorption of light, and hence colour, is called a *chromophore*).

flavonol (yellow of plants) — chlorophyll (green of plants) — crystal violet (a dye used to stain bacteria for classification)

Chemical properties can also be different. Naphthalen-2-ol (2-naphthol, also called β-naphthol) is a typical weakly acidic phenol ($pK_a = 9.5$, see Chapter 6) in its ground state (S_0), but its photoexcited singlet state (S_1) is more acidic ($pK_a = 3$) than a carboxylic acid. Its photoexcited triplet state (T_1) is similar in acidity to its ground state.

5.7.3 Photochemical reactions

When an alkene is irradiated at the correct wavelength, an electron may be promoted from the HOMO (π) to the LUMO (π^*) to give an electronically excited state (Figure 5.12).

Figure 5.12
Photochemical formation of an electronically excited alkene. The red arrows in the π and π^* orbitals indicate the singlet state; this or the corresponding triplet may undergo bond rotation.

energy required for electronic transition — hv — antibonding π* orbital — An electron in the π* MO cancels out the stabilization by an electron in the π MO. — bonding π orbital — ground state — excited state

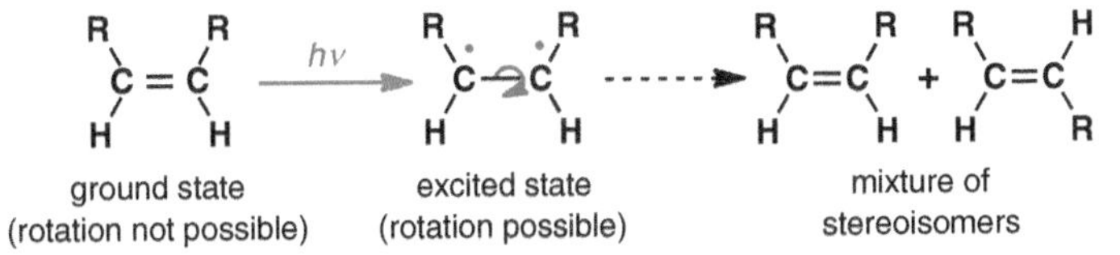

Scheme 5.1
Photochemical isomerization
of a 1,2-disubstituted alkene.

As a result of the $\pi\pi^*$ transition, the π bond is effectively broken and internal rotation about the C–C bond becomes possible. Re-formation of the double bond following internal rotation and return to the electronic ground state leads to *cis–trans* isomerization (Scheme 5.1).

Sometimes, photochemical *cis–trans* isomerizations lead to non-equilibrium proportions of isomers. For example, cyclohept-2-enone may be isomerized by irradiation at 321 nm at −160 °C in an alcohol matrix to give 80% of the unstable *trans* isomer, eqn 5.3. As the temperature rises, it reverts thermally to the more stable *cis* isomer.

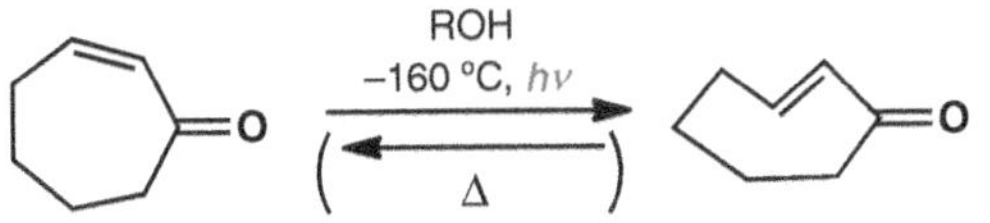

(5.3)

Some transformations can be achieved photochemically which do not take place thermally, e.g. [2+2] cycloaddition reactions of alkenes (see Sub-section 21.2.5 in Chapter 21).

Panel 5.3 The chemistry of vision

Our eyes sense light by a photochemical reaction involving an imine, rhodopsin, formed from a polyunsaturated aldehyde, 11-*cis*-retinal, and an amino group of the protein, opsin.

When rhodopsin absorbs light, the double bond between C11 and C12 undergoes *cis–trans* isomerization. This causes a change in the tertiary structure of the protein and triggers a signal to the brain that the eye has detected light. However, the all-*trans* imine is unstable within the protein complex and undergoes enzyme-catalysed hydrolytic dissociation back to opsin and all-*trans*-retinal (also known as vitamin A aldehyde). To complete the cycle, all-*trans*-retinal is converted by enzymes back to 11-*cis*-retinal, which is used again to make rhodopsin.

All-*trans*-retinal is formed in the body from dietary vitamin A (retinol), a deficiency of which leads to defective vision.

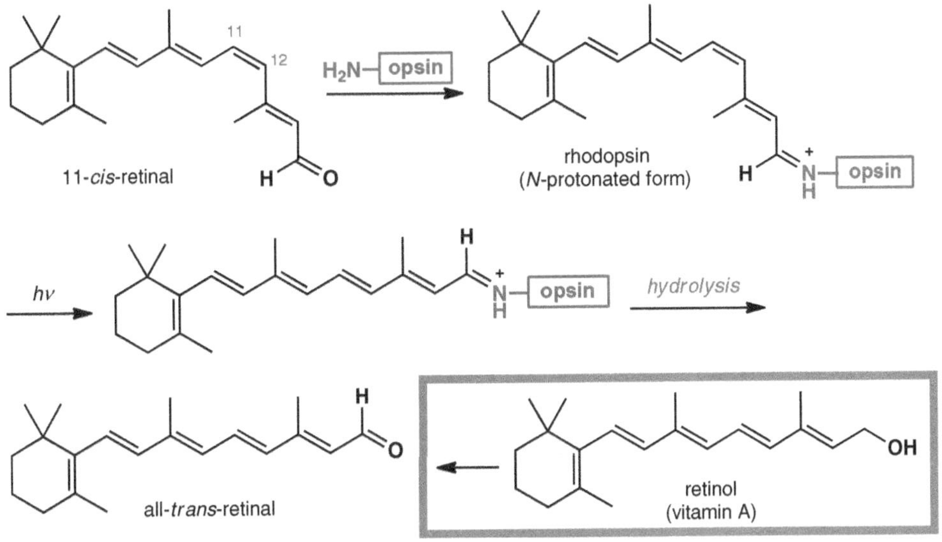

The example of eqn 5.4 is an intramolecular [2+2] photo-cycloaddition which yields a highly strained compound; non-photochemical synthesis of the product would be a major challenge.

(5.4)

Summary

○ When a double (or triple) bond has an adjacent double bond or a heteroatom bearing a 2p AO, the orbitals with π/p symmetry interact and their electrons are *delocalized* in extended π MOs. This phenomenon is called **conjugation**, and the system is stabilized.

○ Conjugation can be described by the sideways overlap (interaction) of π and p orbitals, but the electronic structure of the delocalized (conjugated) system cannot be represented accurately by a single Lewis structure. Most comprehensively, a set of delocalized MOs can be derived from the constituent π and p orbitals. A **resonance hybrid** of all contributing Lewis structures (resonance contributors) is a more convenient short-hand description.

○ Benzene, a planar cyclic molecule with six π electrons, is the prototype aromatic compound; delocalization of the $(4n + 2)$ π electrons of such compounds imparts a special stability which is one of the characteristics of their **aromaticity**.

○ Photochemical excitation of an electron from the HOMO of a molecule into a vacant higher orbital gives a short-lived photoexcited state which has its own distinctive physical and chemical properties.

Problems

5.1 Which of the following have conjugated systems?

(a) $H_2C=CHCO_2H$ (b) $H_2C=CHC≡CH$

(c) $H_2C=C=CHCH_3$

(d)   (e)

(f)   (g)

(h)   (i)

(j) ⬡   (k)

5.2 Are the following pairs of structures related by resonance? If not, give reasons.

(a) $H_2C=C\overset{\ddot{O}H}{\underset{H}{}}$ $H_3C-C\overset{\ddot{O}}{\underset{H}{}}$

(b) $H_3C-\overset{\ddot{O}}{\underset{NH_2}{C}}$ $H_3C-C\overset{\ddot{O}H}{\underset{NH}{}}$

(c) $H_3C-\overset{\ddot{O}}{\underset{NH_2}{C}}$ $H_3C-C\overset{\ddot{O}^-}{\underset{NH_2^+}{}}$

(d) $O=\overset{CH_3}{\underset{}{N^+}}-O^-$ $O=\overset{CH_3}{\underset{}{N}}=O$

(e) ⬡   ⬡

5.3 Draw additional resonance forms for each of the following.

(a) CH₃CH=ȮH (b)

(c) CH₃ĊH−O−CH₃ (d) CH₃CH=CH | Ȯ⁻ (e) O=C O⁻ / O⁻

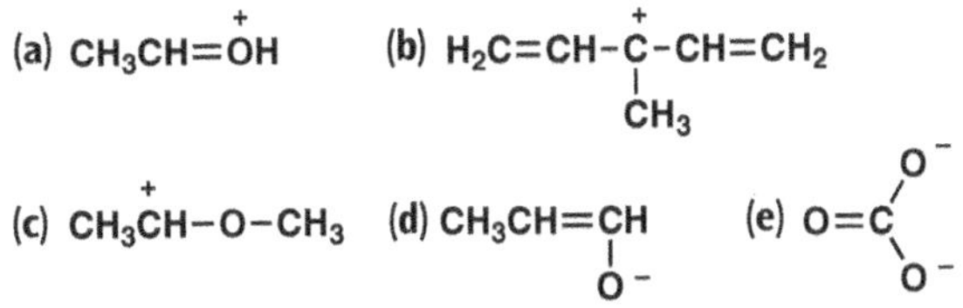

5.4 Draw additional resonance forms for each of the following.

(a)   (b)   (c)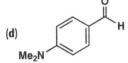

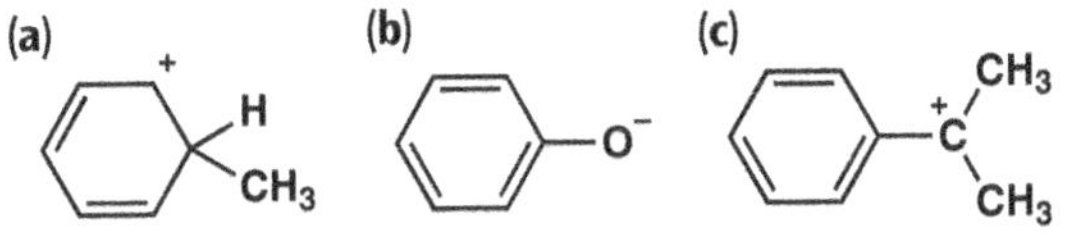

5.5 Draw additional resonance forms for each of the following, and explain which is the most important contributor to the resonance hybrid.

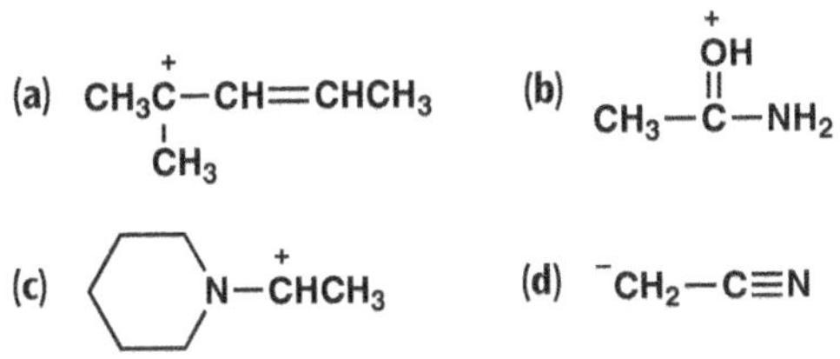

(a) CH₃Ċ−CH=CHCH₃ | CH₃ (b)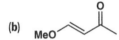

(c)   (d) ⁻CH₂−C≡N

5.6 For each of the following, draw the resonance contributor with the largest charge separation.

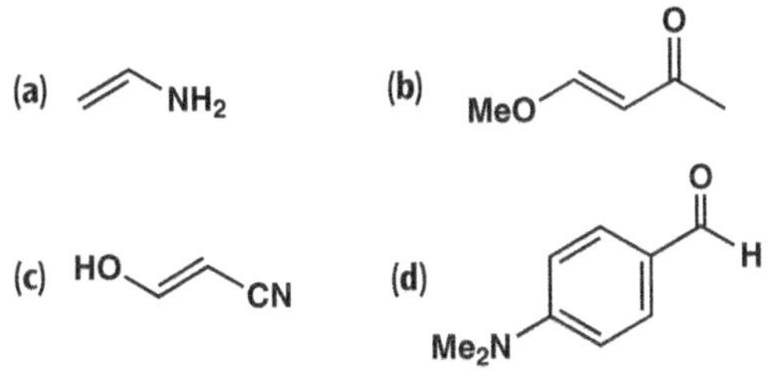

(a) CH₂=CH−NH₂ (b) MeO...

(c) HO...CN (d) Me₂N...

5.7 Discuss possible aromaticity in each of the following.

(a) 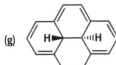 (b) 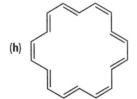 ... CH₃ (c) ... CH₃

(d) (e) ...CH (f)

(g) H...ᴵᴵH (h)

5.8 Discuss possible aromaticity in each of the following.

(a)   (b)   (c)   (d)

(e)   (f)   (g)

(h)   (i) (j)

5.9 Draw additional resonance forms of the following; in each case, discuss the possible aromaticity of the ion/molecule.

(a)   (b)

(c) 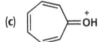 =ȮH (d)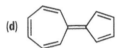

5.10 Azulene is a dark blue solid (the name comes from *azul*, the Spanish word for blue) whereas its isomer, naphthalene (see Figure 5.9), is colourless. Draw additional resonance contributors for azulene, including ones with charge separation, and comment upon its aromaticity.

 azulene

6 Acids and Bases

Related topics we have already covered:	Topics we shall cover in this chapter:
• **Electronegativity and polar bonds** (Sub-section 1.2.3)	• **Brønsted acids and bases**
	• **Lewis acids and bases**
• **Functional groups of organic compounds** (Chapter 3)	• **Acid dissociation constants and** pK_a
	• **Buffer solutions**
• **Electron delocalization and resonance** (Chapter 5)	• **Factors which affect strengths of acids and bases**
• **Aromaticity** (Section 5.6)	• **Carbon acids and carbanions**
	• **Basicity of organic compounds**
	• **Solvent effects on acid–base reactions**

Substances that taste sour have long been known as acids. Ethanoic (acetic) acid is the essential ingredient of vinegar, and citric acid contributes to the sour taste of lemons, limes, and other citrus fruit (hence its name). Citric acid is produced industrially by fermentation of sugar and used on a massive scale as a flavouring agent and preservative in soft drinks. Bases are compounds which counteract or neutralize acids and an acid–base reaction is most widely understood simply to involve transfer of a proton from the acid to the base. The base sodium hydrogencarbonate (bicarbonate) is a familiar domestic item which is taken orally to relieve the discomfort of 'acid indigestion'. With tartaric acid (which occurs naturally in various fruit, especially grapes), sodium bicarbonate is a component of baking powder; the tartaric acid transfers a proton to the bicarbonate anion, which leads to the liberation of carbon dioxide as tiny bubbles within the dough and makes it lighter.

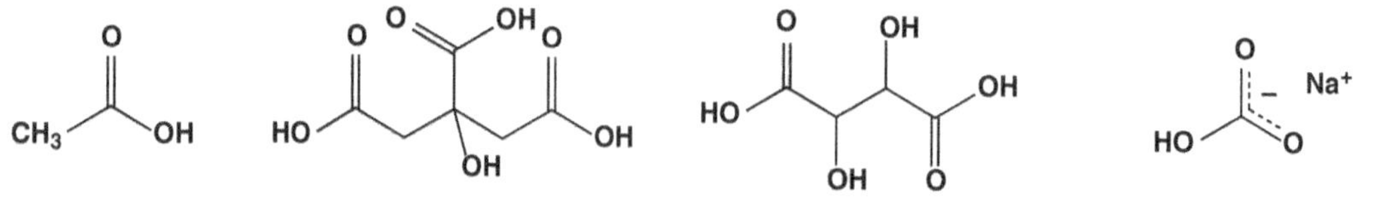

ethanoic acid (acetic acid)   citric acid   tartaric acid   sodium hydrogencarbonate (sodium bicarbonate)

Many organic compounds show acidic or basic properties (some show both), and acid–base reactions are involved as intermediate steps in many multistep organic reactions, as will be discussed in Chapter 7. The strength of an acid is measured by the equilibrium constant for the reversible proton transfer to the solvent (usually water) and, since the deprotonated acid is a base, the same equilibrium constant is a measure of base strength. Factors which affect acid and base strengths are largely the same as those which affect equilibrium and rate constants of other organic reactions, and these will be discussed in this chapter.

6.1 Definitions of Acids and Bases

6.1.1 Brønsted acids and bases

As defined by Brønsted in Denmark in the early part of the twentieth[1] century, *an acid is a compound which donates a proton (H^+) and a base is a compound which can accept a proton*; such compounds are now often called **Brønsted acids** (or sometimes proton acids) and bases.

Gaseous hydrogen chloride, HCl, dissolves in water and immediately reacts by **proton transfer** in a typical Brønsted acid–base reaction, to give hydrochloric acid. In this *acid dissociation reaction*, HCl donates a proton to a water molecule so HCl is an acid and H_2O is a base (eqn 6.1). In principle, this reaction is reversible and, in the reverse direction, the chloride accepts a proton from the oxonium ion H_3O^+ (also called hydronium ion); Cl^- is now a base (the **conjugate base** of HCl) and H_3O^+ is an acid (the **conjugate acid** of H_2O). In practice, however, no covalent HCl molecules can be detected at equilibrium in dilute aqueous solution (it is an example of a **strong acid**, see later).

Johannes N. Brønsted (1879–1947)

The Danish chemist J.N. Brønsted introduced the proton-transfer model of acid–base reactions in 1923 and also contributed to our understanding of the catalysis of chemical reactions by acids and bases. In particular, he established the relationship between the strength of a base and its effectiveness as a catalyst for base-catalysed reactions.

$$HCl \ + \ H_2O \ \rightleftharpoons \ Cl^- \ + \ H_3O^+ \qquad (6.1)$$

$\quad$ acid $\qquad$ base $\qquad\qquad$ conjugate base $\quad$ conjugate acid
$\qquad\qquad\qquad\qquad\qquad$ of HCl $\qquad\qquad$ of H_2O

Gaseous hydrogen cyanide, HCN, also dissolves in water and reacts as an acid, but the reaction does not proceed to completion; HCN (the solute) is a **weak acid** (see later) and covalent HCN molecules coexist in aqueous solution with the dissociated acid (eqn 6.2).

$$HCN \ + \ H_2O \ \rightleftharpoons \ CN^- \ + \ H_3O^+ \qquad (6.2)$$

$\quad$ acid $\qquad$ base $\qquad\qquad$ conjugate base $\quad$ conjugate acid
$\qquad\qquad\qquad\qquad\qquad$ of HCN $\qquad\qquad$ of H_2O

In another example (eqn 6.3), ammonia dissolves in water with proton transfer from H_2O to NH_3, i.e. the solute ammonia acts as a base with water acting as solvent and as an acid.

$$NH_3 \ + \ H_2O \ \rightleftharpoons \ NH_4^+ \ + \ OH^- \qquad (6.3)$$

$\quad$ base $\qquad$ acid $\qquad\qquad$ conjugate acid $\quad$ conjugate base
$\qquad\qquad\qquad\qquad\qquad$ of NH_3 $\qquad\qquad$ of H_2O

The reaction of eqn 6.3, however, like that of eqn 6.2, does not proceed to completion; in fact, in dilute aqueous solution, most of the ammonia remains unprotonated because it is only a **weak base**.

In contrast, there is an immediate and virtually complete proton transfer reaction when sodium amide is added to water (even though it is still written as an equilibrium in eqn 6.4): NH_2^- is a **strong base** in water.

$$Na^+ NH_2^- \ + \ H_2O \ \rightleftharpoons \ NH_3 \ + \ Na^+ OH^- \qquad (6.4)$$

$\quad$ base $\qquad\qquad$ acid $\qquad\qquad$ conjugate acid $\qquad$ conjugate base
$\qquad\qquad\qquad\qquad\qquad\qquad$ of NH_2^- $\qquad\qquad\qquad$ of H_2O

The word *conjugation* was introduced in Chapter 5 to describe alternating double and single bonds; we say the double bonds in buta-1,3-diene, for example, are *conjugated*, and *conjugative* effects are sometimes synonymous with *resonance* effects. We now use the word *conjugate* to describe the relationship between an acid and the species obtained by proton removal. This is a reciprocal relationship: an acid has a *conjugate* base, and a base (by proton addition) has a *conjugate* acid. These different meanings of the same words should not be confused since they are used in quite different contexts.

---

[1] A publication about the same time by the British chemist T.M. Lowry included ideas relating Brønsted's proton transfer model of acids and bases, but without explicit definitions.

Example 6.1

Correlate acid with its conjugate base and base with its conjugate acid in the following acid–base reaction.

$$CH_3CO_2H + NH_3 \rightleftharpoons CH_3CO_2^- + NH_4^+$$

Solution

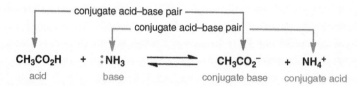

conjugate acid–base pair

conjugate acid–base pair

$$CH_3CO_2H \quad + \quad :NH_3 \rightleftharpoons CH_3CO_2^- \quad + \quad NH_4^+$$

acid   base   conjugate base   conjugate acid

Exercise 6.1

Give the conjugate base of each of the following.

(a) HBr   (b) HCO_2H (c) CH_3SH (d) H_2SO_4 (e) CH_3NH_2

Exercise 6.2

Give the conjugate acid of each of the following.

(a) CH_3OH (b) CH_3NH_2 (c) Cl^- (d) CH_3O^- (e) $(CH_3)_2C=NH$

Exercise 6.3

Complete the following two acid–base reactions, and correlate the conjugate acid–base pairs.

(a) $CH_3-\overset{+}{\underset{|}{O}}-H + CH_3NH_2 \rightleftharpoons$ (b) $CH_3OH + NH_2^- \rightleftharpoons$

The movement of an electron pair is represented by a curly arrow when the lone pair becomes a new bonding pair as in the Lewis acid–base association (red arrow, eqn 6.5), and when a bonding pair becomes a lone pair as in the dissociative reverse reaction (blue arrow, eqn 6.5). Further aspects of curly arrow representations of reaction mechanisms are discussed in Chapter 7.

6.1.2 Lewis acids and bases

A more general definition of acids and bases was proposed by Lewis in 1923 with the focus on the electron pair of the base rather than the proton of the acid; a *Lewis base is an* **electron pair donor** *and a Lewis acid is an* **electron pair acceptor**. The central atom in a Lewis acid is able to accept the lone pair of a base into its valence shell and thereby form a covalent bond between them. Typical Lewis acids include $AlCl_3$, $FeBr_3$, $ZnCl_2$, and BF_3. For example, BF_3 with only six valence electrons on the boron reacts with ammonia, which has a lone pair on the nitrogen, to form an adduct, as shown in eqn 6.5. In this way, the valence needs of both B and N are satisfied when a covalent bond is formed between them. The shared electron pair of the bond, however, comes wholly from the N, so there is a transfer of charge and the adduct is dipolar.

Note that a Brønsted base has a lone pair of electrons to accept a proton, so is necessarily also a Lewis base. A Brønsted acid, however, *donates* a proton and is itself not necessarily a Lewis acid (but H⁺, if it had an independent existence, would be a Lewis acid).

Lewis acid–base association

$$F-\underset{\underset{F}{|}}{\overset{\overset{F}{|}}{B}} \quad + \quad :\underset{\underset{H}{|}}{\overset{\overset{H}{|}}{N}}-H \rightleftharpoons F-\underset{\underset{F}{|}}{\overset{\overset{F}{|}}{B}}-\overset{+}{\underset{\underset{H}{|}}{\overset{\overset{H}{|}}{N}}}-H$$ (6.5)

Lewis acid   Lewis base   adduct

Lewis acid–base dissociation

Brønsted and Lewis published their proposals in the same year, 1923.

Reaction 6.5 represented by molecular models and EPMs:

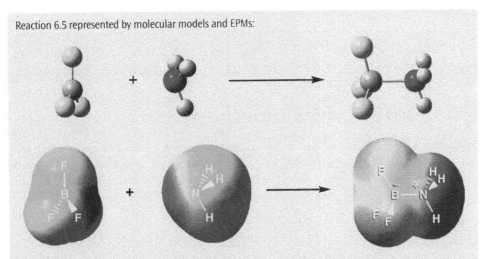

The EPMs show a transfer of charge to give the positive NH_3 moiety (blue) and the negative BF_3 (yellow to orange). This movement of electrons is described by the red curly arrow in eqn 6.5.

Example 6.2

Complete the following Lewis acid–base reaction by showing the electron pairs involved.

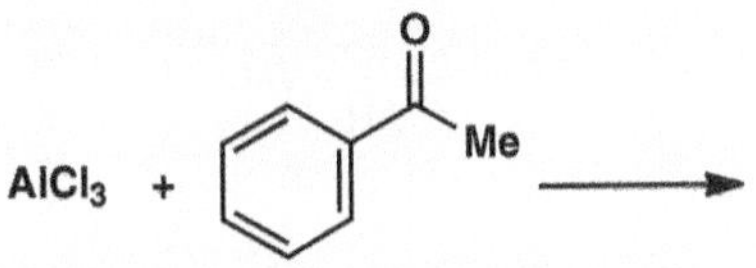

$$AlCl_3 \ + \ \text{(benzophenone-type structure with Me)} \longrightarrow$$

Solution

The ketone is a Lewis base and $AlCl_3$ is a Lewis acid in this reaction. Movement of the electron pair to give the dipolar adduct is represented by a curly arrow.

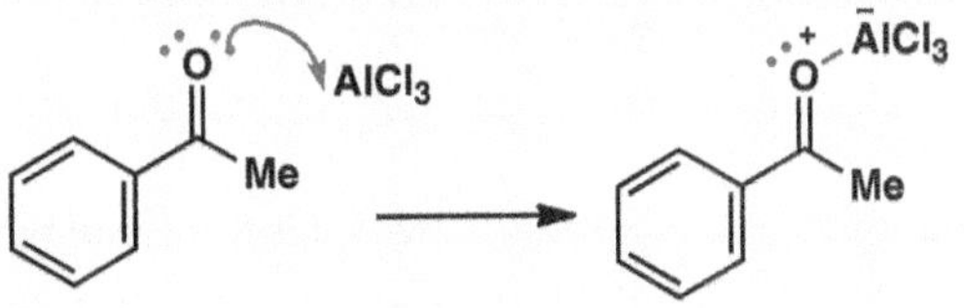

Exercise 6.4

Classify each of the following as a Lewis acid or base.

(a) $(CH_3)_2O$ (b) $FeCl_3$ (c) $(CH_3)_2NH$ (d) $B(CH_3)_3$ (e) CH_3^+

Exercise 6.5

Complete the following two Lewis acid–base reactions by showing the structures of the products.

(a) $BF_3 \ + \ (CH_3)_2O \longrightarrow$ (b) $CH_3CH_2Cl \ + \ AlCl_3 \longrightarrow$

6.2 Equilibrium in Brønsted Acid–Base Reactions

As indicated above, a Brønsted acid–base reaction (proton transfer) is, in principle, reversible and very many are reversible in practice. When the reaction

$$\underset{\text{acid}}{AH} \ + \ \underset{\text{base}}{B} \ \rightleftharpoons \ \underset{\text{conjugate base}}{A^-} \ + \ \underset{\text{conjugate acid}}{BH^+}$$

comes to equilibrium in aqueous solution, the proportions of the various species depend upon the relative *acidities* of AH and BH⁺ (or, equivalently, the relative *basicities* of A⁻ and B) under the conditions of the reaction.[2] The strength of a Brønsted acid in aqueous solution is expressed by its acid dissociation constant, K_a, which we will consider next.

6.2.1 Acid dissociation constants and pK_a

Dissociation of a Brønsted acid in aqueous solution is an acid–base reaction with a solvent water molecule acting as the base as shown in eqn 6.6 for a generic acid, AH. A proton transfers from acid AH to H_2O to give the conjugate base A⁻ and H_3O^+ (oxonium ion).

$$AH + H_2O \underset{}{\overset{K_a}{\rightleftharpoons}} A^- + H_3O^+ \tag{6.6}$$

The extent of proton transfer when the system achieves equilibrium is expressed by an equilibrium constant, K_a, defined in the conventional way by eqn 6.7, where [AH], [A⁻], and [H_3O^+] stand for the equilibrium concentrations of AH, A⁻, and H_3O^+.* The equilibrium constant, K_a, is called the **acid dissociation constant** (or **acidity constant**), and provides a measure of the acid strength (or acidity) of a compound, i.e. its effectiveness as a proton donor.

$$K_a = \frac{[A^-][H_3O^+]}{[AH]} \tag{6.7}$$

*More properly, the terms in eqn 6.7 should be the *activities* of the various species, and there should be another term in the denominator for the activity of the solvent on the left-hand side of eqn 6.6. The molarity (mol dm⁻³) scale is convenient for the activities of solutes and the mole fraction (m.f.) scale for the activity of the solvent. For dilute solutes, the actual molar concentration is a good approximation for the activity. For the solvent of a dilute solution, the mole fraction activity is essentially the same as for the pure solvent, i.e. m.f.=1, which is why the term is generally left out of eqn 6.7.

Organic acids are generally weak and K_a is correspondingly small. In order to deal conveniently with a wide range of very small numbers, a logarithmic scale is used and **pKa** is defined as the *negative* logarithm of the numerical value of K_a (eqn 6.8). It follows that *the stronger an acid, the smaller its pK_a* (acids as strong as HCl or H_2SO_4 have negative pK_a values). For the dissociation of ethanoic (acetic) acid, a typical weak organic acid, $K_a=1.74\times10^{-5}$ mol dm⁻³ (eqn 6.9).

$$pK_a = -\log K_a \tag{6.8}$$

$$CH_3CO_2H + H_2O \rightleftharpoons CH_3CO_2^- + H_3O^+ \tag{6.9}$$

ethanoic acid ⟶ ethanoate ion

$$K_a = 1.74\times10^{-5}\text{mol dm}^{-3} \quad \text{and} \quad pK_a = 4.76$$

[2] *Basicity* was originally used by chemists to mean the number of dissociable hydrogens of a Brønsted acid, thus H_2SO_4 was *dibasic* (we prefer *diprotic*, see later). Correspondingly, but less commonly, *acidity* was used to mean the number of sites at which a base could accept a proton, so $H_2NCH_2CH_2NH_2$ was *diacidic*. These usages led to confusion since *acidity* and *acidic* were already commonly used English words referring to the property of being an acid, and could be qualified (e.g. low acidity, strongly acidic, etc.). In those early days of chemistry, there was no abstract noun to describe the property which characterizes a base in the way that *acidity* is the characteristic property of an acid. Since there is obviously a need for such a word, *basicity* began to be used in this way, and we now follow this usage.

Panel 6.1 pK_a values for water and the oxonium ion

If we want to compare water *as a weak acid* with other weak acids in aqueous solution, we need to think of it in two ways: as a weak acid (the proton donor which is being compared with other weak acids using eqn 6.7) and as the solvent (which becomes protonated in just the same way as it does by any acidic solute). The activity of water as the proton donor is again approximated by its concentration using the molarity scale, as explained in Sub-section 6.2.1. The more convenient mole fraction scale is used for water *as the solvent* exactly as it is for aqueous solutions of all acids, and its value is again unity so it does not appear in eqn 6.7.

$$H_2O \quad + \quad H_2O \rightleftharpoons HO^- \quad + \quad H_3O^+$$

$$\text{acid} \qquad \text{solvent} \qquad \qquad \substack{\text{conjugate base} \\ \text{of acid}} \quad \substack{\text{conjugate acid} \\ \text{of solvent}}$$

We now apply eqn 6.7 to the chemical equation above,

$$K_a = \frac{[HO^-][H_3O^+]}{[H_2O]}$$

$[HO^-]$ and $[H_3O^+]$ in pure water are both equal to 10^{-7} mol dm^{-3} and $[H_2O]$, *as the acid in the above chemical equation*, is equal to $1000/18 = 55.5$ mol dm^{-3}. These values lead to $K_a = 1.80 \times 10^{-16}$ so $pK_a = 15.7$, which compares with values of about 16 for simple alcohols, ROH, as expected.

In a similar way, the acidity of the oxonium ion, H_3O^+, can be quantified from eqn 6.7 and the chemical equation:

$$H_3O^+ \quad + \quad H_2O \rightleftharpoons H_2O \quad + \quad H_3O^+$$

$$\text{acid} \qquad \text{solvent} \qquad \qquad \substack{\text{conjugate base} \\ \text{of acid}} \quad \substack{\text{conjugate acid} \\ \text{of solvent}}$$

In pure water, $[H_3O^+]$ on both sides of the chemical equation is 10^{-7} mol dm^{-3} and the concentration of $[H_2O]$ *as the conjugate base of* H_3O^+ on the right of the equation is 55.5 mol dm^{-3}. Again, water *as the solvent* (mole fraction $= 1$) does not come into the arithmetic so $K_a = 55.5 \times [H_3O^+]/[H_3O^+]$, i.e. 55.5, and the pK_a of H_3O^+ is -1.74.

The strengths of a range of acids are given in Figure 6.1, and pK_a values of more are summarized in Appendix 1 at the end of the book.

We also see from eqn 6.6 that a weak acid AH corresponds to a relatively strong conjugate base, A$^-$, and vice versa—a strong acid has a relatively weak conjugate base. In other words, the base strength of a compound may be expressed by the pK_a of its conjugate acid (the larger the pK_a of its conjugate acid, the stronger the base) so acid strengths and base strengths can be expressed using the single (pK_a) scale (see Section 6.5).

> **Exercise 6.6**
>
> Calculate K_a values from the pK_as given below; which is the stronger acid?
>
> (a) $(CH_3)_3CCO_2H$ ($pK_a = 5.0$)   (b) C_6H_5OH ($pK_a = 10.0$)

6.2.2 Equilibrium in acid–base reactions

Hydrogen chloride dissociates completely in aqueous solution so, expressed rather loosely, equilibrium in the acid–base reaction of HCl with H_2O lies far to the right hand side of eqn 6.10. In other words, HCl is a stronger acid than H_3O^+, $pK_a(HCl) < pK_a(H_3O^+)$, and Cl$^-$ is a weaker base than H_2O.

$$HCl + H_2O \rightleftharpoons Cl^- + H_3O^+ \qquad (6.10)$$

$$pK_a \qquad -7 \qquad\qquad\qquad -1.7$$

$$\text{strong acid}$$

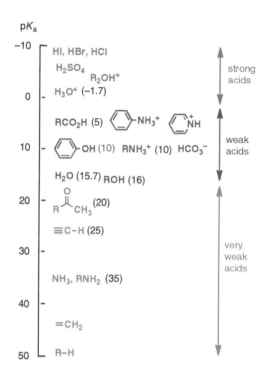

Acids which dissociate
completely in water are
known as **strong** acids; this
must not be confused with
a *concentrated* acid. An acid
which dissociates to only a
limited extent in aqueous
solution is a **weak** acid
(not to be confused with a
dilute acid).

On the other hand, ethanoic acid is a weak acid; $pK_a(CH_3CO_2H) > pK_a(H_3O^+)$ and ethanoate (acetate) is a stronger base than H_2O. It dissociates in aqueous solution to only a very small extent; equilibrium in eqn 6.11 lies well over to the left hand side, and most of the ethanoic acid molecules in aqueous solution are undissociated.

$$CH_3CO_2H \quad + \quad H_2O \quad \rightleftharpoons \quad CH_3CO_2^- \quad + \quad H_3O^+ \qquad (6.11)$$

pK_a 4.76   −1.7

weak acid

These relationships between the equilibrium position and acid strength can be applied to acid–base reactions between solutes in aqueous solution. In such reactions, acid–base equilibria favour the side of the weaker acid + weaker base. For example, the acid–base equilibrium between ethanoic acid and ammonia (eqn 6.12) is towards the side of ammonium ethanoate because $pK_a(NH_4^+) > pK_a(CH_3CO_2H)$.

$$CH_3CO_2H \quad + \quad NH_3 \quad \rightleftharpoons \quad CH_3CO_2^- \quad + \quad NH_4^+ \qquad (6.12)$$

pK_a 4.76   9.24

stronger acid   stronger base   weaker base   weaker acid

Example 6.3

Calculate the value of the equilibrium constant for the acid–base reaction between ethanoic acid and ammonia in aqueous solution (eqn 6.12).

Solution

The equilibrium constant, K, for the reaction of eqn 6.12 is given in the conventional manner (where Ac stands for the acetyl group, CH_3CO) in terms of the equilibrium concentrations of reactants and products:

$K = [AcO^-][NH_4^+]/[AcOH][NH_3]$

$\quad = ([AcO^-][H_3O^+]/[AcOH])([NH_4^+]/[NH_3][H_3O^+])$

$\quad = K_a(AcOH)/K_a(NH_4^+) = 10^{-4.76}/10^{-9.24} = 10^{4.48} = 3.0 \times 10^4$

We see that $K \gg 1$ so equilibrium lies far towards the right hand side.

Exercise 6.7

Calculate the equilibrium constants for the following acid–base reactions, and indicate which side the equilibrium favours.

(a) $HCO_2H + CH_3NH_2 \rightleftharpoons HCO_2^- + CH_3NH_3^+$

$\quad$ pK_a 3.75 $\qquad\qquad\qquad\qquad\qquad$ 10.64

(b) $C_6H_5CO_2H + C_6H_5NH_2 \rightleftharpoons C_6H_5CO_2^- + C_6H_5NH_3^+$

$\quad$ pK_a 4.20 $\qquad\qquad\qquad\qquad\qquad$ 4.60

Exercise 6.8

For the following generic proton transfer reaction in aqueous solution,

$$AH + B \xrightarrow{\quad K \quad} A^- + BH^+$$

show that the equilibrium constant, K, is represented by the equation, pK = pK_a(AH) − pK_a(BH$^+$).

6.2.3 Acidity of aqueous solutions and ratios of conjugate acid–base pairs

By taking logarithms of eqn 6.7, we can derive eqn 6.13 which relates the acidity of an aqueous solution (its pH) to the pK_a of a dissolved weak acid, AH. In this equation, the origins of AH and A$^-$ do not matter. For example, A$^-$ could be simply by dissociation of AH, or it could be from addition of the sodium salt, Na$^+$A$^-$; and AH could be added as the covalent compound or could be generated by addition of a strong acid, such as hydrochloric, to a solution of Na$^+$A$^-$.

> Equation 6.13 is sometimes called the Henderson–Hasselbalch equation.

$$pK_a = -\log[H_3O^+] - \log([A^-]/[AH])$$
$$= pH + \log([AH]/A^-) \tag{6.13}$$

Equation 6.13 indicates that the value of log ([AH]/[A$^-$]) increases as the pH decreases; i.e. the higher the acidity of the solution, the greater the proportion of the acid form of the conjugate acid–base pair.

When concentrations of an undissociated acid and its conjugate base in an aqueous solution are equal ([AH]=[A$^-$]), pK_a=pH according to eqn 6.13. In other words, the pK_a of an acidic solute is equal to the pH of the solution when [AH]=[A$^-$]. The so-called titration curve of Figure 6.2 is obtained by plotting the fraction of the conjugate acid against pH.

From Figure 6.2 (or eqn 6.13), we can find the proportions of acid and conjugate base in aqueous solution from the pK_a and the pH. For example, benzoic acid (pK_a=4.2) is almost completely ionized and exists as its conjugate base, the benzoate anion, in a neutral solution (pH=7). In contrast, phenol (pK_a=10) is virtually wholly un-ionized in a neutral aqueous solution.

> The pK_a of an acidic solute indicates its protonating power, whereas the pH of a solution indicates the protonating power of the solution as a medium towards a (basic) solute.

> The ionized forms of acids are usually soluble in aqueous solution as sodium salts, for example, but not in organic solvents, so un-ionized phenol can be extracted into ether from a neutral aqueous solution while benzoic acid in its ionized form cannot.

Figure 6.2 The graphical relationship between the proportion of a weak acid AH and the pH of a solution of AH and its conjugate base.

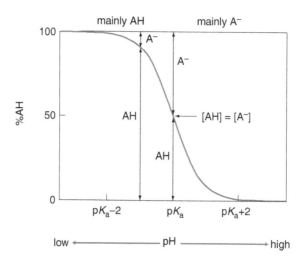

In which form does each of the following acids exist in aqueous solutions of controlled pH 2, pH 7, and pH 12. Numbers in parentheses are pK_a values.

(a) HCO_2H (3.75)   (b) NH_4^+ (9.24)   (c) $C_6H_5NH_3^+$ (4.6)   (d) HCN (9.1)

6.2.4 Buffer solutions

Buffer solutions have two important properties. First, they allow us to prepare an aqueous solution of a desired pH using a weak acid and its conjugate base (a salt of the acid); the pH of the solution is given by eqn 6.13 rewritten as eqn 6.14.

$$pH = pK_a - \log([AH]/[A^-]) \tag{6.14}$$

The *buffer ratio* ([AH]/[A⁻]) is usually close to unity and almost invariably within the range 0.1–10, in which case the pH of the solution will be equal to the pK_a of the weak acid ±1; a weak acid is usually selected whose pK_a is similar to the pH required. For example, ethanoic acid (pK_a = 4.76) and sodium ethanoate would be used if a buffered solution of pH about 5 is required and the precise pH is given by eqn 6.14.

As alternatives to mixing conjugate acid and base, equivalent practical methods of making a buffer solution are either to add a strong base (e.g. NaOH) to an aqueous solution of a weak acid, or to add a strong acid (e.g. hydrochloric) to an aqueous solution of a weak base. The method will be chosen, at least in part, according to the pH required.

The second important property of a buffer solution is that its pH will remain approximately constant if relatively small amounts of a further acid or base are added. This is very useful if a constant pH is required for a reaction which generates or consumes either H_3O^+ or HO^-. For example, if 0.001 mol of HCl is added to 1 dm³ of an ethanoic acid buffer made with [AcOH] = 0.1 and [AcONa] = 0.2 mol dm⁻³, AcO⁻ is protonated and the [AcOH]/[AcO⁻] ratio changes from 0.1/0.2 to 0.101/0.199. The pH, however, is only reduced from 5.06 by <0.01. If the same amount of HCl were added to 1 dm³ of unbuffered water, the pH would change from 7.0 to 3.0.

In addition to being used for chemical reactions in aqueous solution at constant pH, buffers are also used for handling and storing biochemicals, e.g. enzymes and other proteins, whose stability depends upon the pH of the solution.

Calculate the pH of the following solutions (pK_a of AcOH is 4.76).

(a) A mixture of 0.2 dm³ of 0.1 mol dm⁻³ aqueous AcONa and 0.1 dm³ of 0.1 mol dm⁻³ aqueous HCl.

(b) A mixture of 0.2 dm³ of 0.1 mol dm⁻³ aqueous AcONa and 0.1 dm³ of 0.1 mol dm⁻³ aqueous AcOH.

(c) A mixture of 0.3 dm³ of 0.1 mol dm⁻³ aqueous AcOH and 0.2 dm³ of 0.1 mol dm⁻³ aqueous NaOH.

(d) A mixture of 0.3 dm³ of 0.1 mol dm⁻³ aqueous AcOH and 0.1 dm³ of 0.1 mol dm⁻³ aqueous NaOH.

Panel 6.2 pH indicators and colours of flowers

A pH indicator (sometimes called a halochromic compound) is a compound which changes its colour in solution as the pH of the solution changes. It has acid–base properties itself and dissociates according to eqn 6.13,

$$pH = pK_a - \log([\text{H-Ind}]/[\text{Ind}^-]),$$

where H-Ind and Ind⁻ are the acid and base forms; these are equally stable when the pH of the solution is numerically the same as the pK_a value of H-Ind.

One of the most widely used indicators is phenolphthalein and its dissociation, hence the colour change, in a dilute aqueous solution occurs around pH 9; it is colourless below pH ~ 8 and pink above pH ~ 9 (both phenolic hydrogens dissociate between pH 8 and 9). However, at very low pH (<0), phenolphthalein becomes orange, and at very high pH (>13), it becomes colourless.

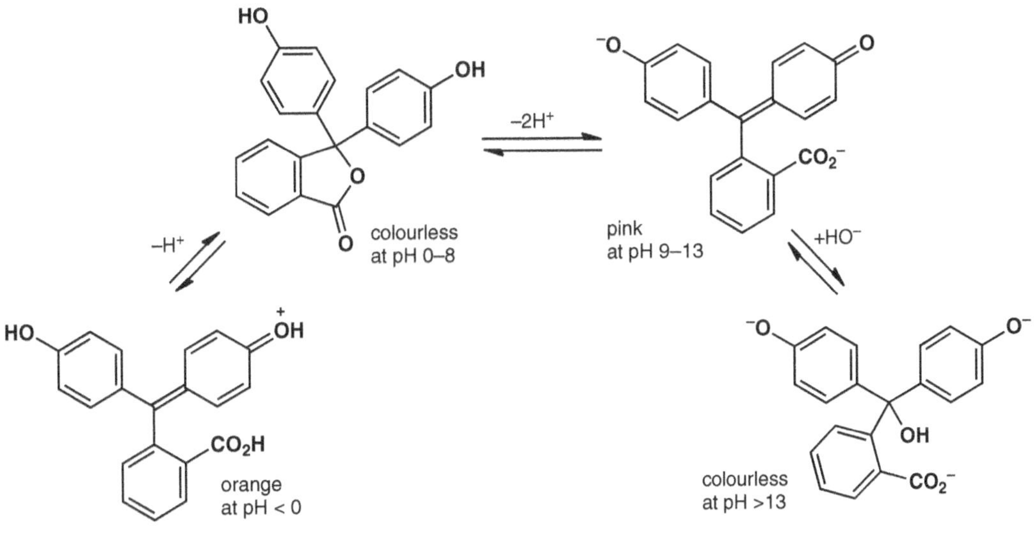

Colour changes of other pH indicators occur in a similar manner but at different pHs depending on their pK_a values.

Anthocyanins are pigments of flowers and other parts of plants, and their colours change with pH: in general, they change from red to purple to blue as their aqueous solution changes from strongly acidic, to weakly acidic, to weakly basic. The major structures of one anthocyanin are given below. However, colours of flowers depend on the coordination of metal ions and the stacking structure of the pigment molecules as well as on the pH.

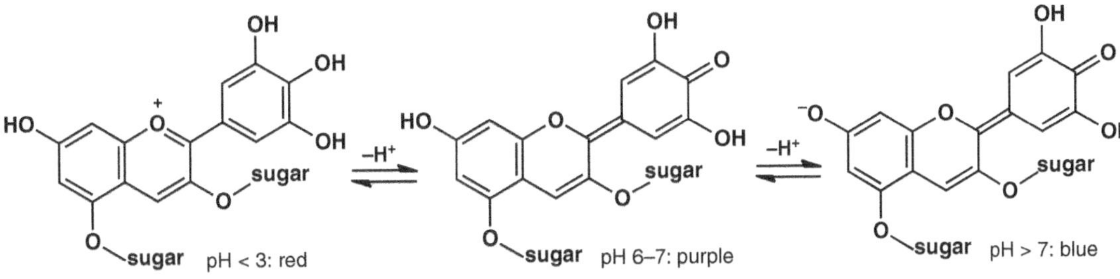

6.3 Factors which affect the Strength of an Acid

6.3.1 The element bearing the acidic hydrogen

HF is an anomalously weak acid. This is because the ion pair $H_3O^+F^-$ formed upon initial proton transfer does not completely dissociate. We need to distinguish between *ionization* of an acid (proton transfer to a solvent molecule) and the subsequent *dissociation*.

Dissociation of an acid in aqueous solution involves transfer of a proton from the acidic site of the acid to a water molecule to give the conjugate base of the acid and H_3O^+. However, since H_3O^+ is generated in all acid dissociations in aqueous solution, we can ignore its formation when considering the relative strengths of different acids. The strength of the bond to the dissociable H in the acidic site, therefore, is a principal factor when considering relative acidities. The pK_a values of hydrogen halides HX decrease (the acidity increases) in the order shown:

Acid:	HF	HCl	HBr	HI
Conjugate base:	F⁻	Cl⁻	Br⁻	I⁻
pK_a:	3.2	−7	−9	−10

The bond strength of H–X decreases as the size of the atom X increases owing to the increasingly poor overlap between the orbital on X which is involved and the smaller 1s orbital of hydrogen. This trend is also seen in the relative acidities of H_2O ($pK_a = 15.7$) and H_2S ($pK_a = 7.0$).

The conjugate base of any neutral acid XH is an anion, and the stability of the anion, X⁻, is also a major factor in affecting the acidity of XH. The more *electronegative* the atom to which the proton is bonded, the better able it will be to accommodate a negative charge following dissociation. This trend is illustrated by the increasing acidities of the hydrides of the elements as we proceed across the second period from left to right.

Acid:	CH_4	NH_3	H_2O	HF
Conjugate base:	CH_3^-	NH_2^-	OH⁻	F⁻
pK_a:	~49	~35	15.7	3.2

Exercise 6.11

Which of ethanol (EtOH) and ethanethiol (EtSH) is the more acidic and why?

In the particular case of carbon, the acidities of ethane, ethene, and ethyne increase in the order shown:

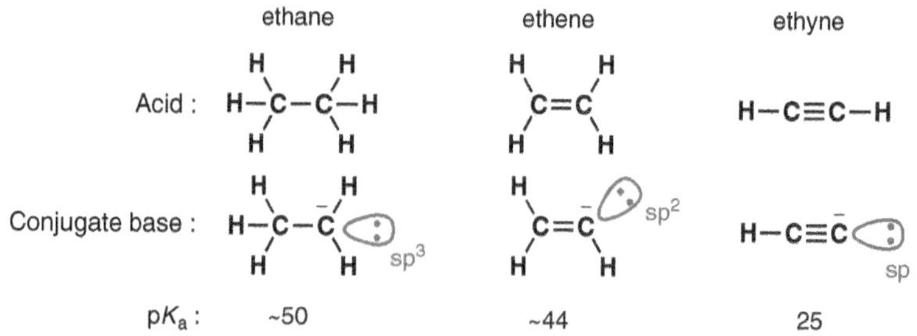

The negative charge of the conjugate bases resides as a lone pair in sp^3, sp^2, and sp hybrid orbitals of the carbon, respectively. As the s character of the orbital increases, the energy level of the hybrid orbital becomes lower (Section 2.3) and the anion becomes more stable (electrons are held closer to the nucleus). In other words, the electronegativity of the carbon increases in the hybridization order: sp^3-C $<$ sp^2-C $<$ sp-C as the s character increases from 25 to 33 to 50% along the series, and this leads to increasing acidity.

Explain the difference in acidity of the conjugate acids of piperidine and pyridine, which have pK_a values of 11.1 and 5.25, respectively.

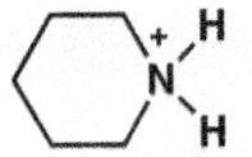

Example 6.4

piperidinium ion
$(pK_a = 11.1)$

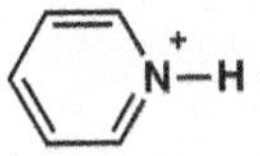

pyridinium ion
$(pK_a = 5.25)$

Solution

The N atoms in the conjugate bases, piperidine and pyridine, are sp^3- and sp^2-hybridized, respectively; the latter has more s character so the N here is more electronegative. As a result, the lone pair of pyridine is more tightly held and less available for protonation, so pyridine is the weaker base and its conjugate acid is more acidic than that of piperidine.

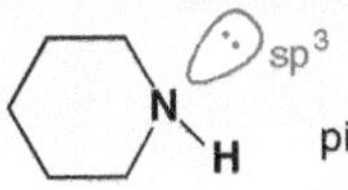

sp^3

piperidine

sp^2

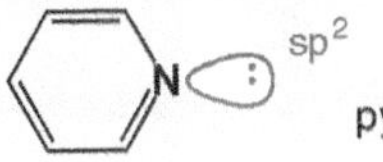

pyridine

6.3.2 Charge delocalization in anions

Anions are stabilized by dispersion (delocalization) of negative charge (or electrons). The relative acidities of the oxy-acids of chlorines are typical:

Acid:	HClO	$HClO_2$	$HClO_3$	$HClO_4$
pK_a:	7.5	2	−1	$\sim$ −10

The acidity of the oxy-acids increases with the increasing number of oxygen atoms bonded to the central Cl as these are able to delocalize the negative charge in the anionic conjugate bases. This is illustrated below by the four equivalent resonance forms of the highly stabilized tetrahedral perchlorate anion, ClO_4^-: perchloric acid is a very strong acid.

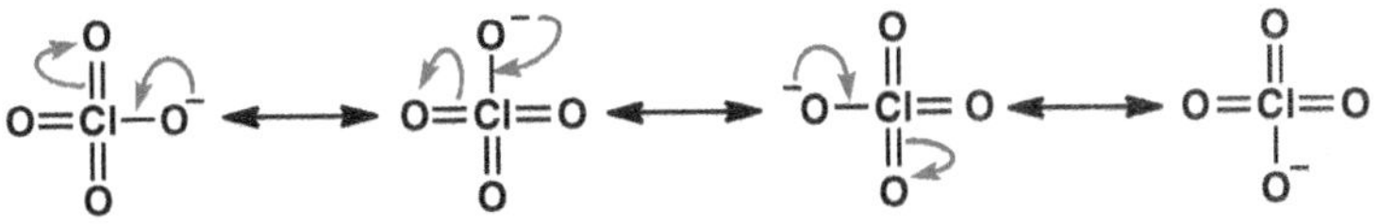

Resonance stablization of the perchlorate ion

The higher acidity of carboxylic acids (pK_a of AcOH $= 4.76$) compared with alcohols (pK_a of ethanol $= 15.9$) is also explained by electron delocalization in the conjugate base of the acid. The negative charge of ethoxide, EtO^- is localized on a single O atom whereas the charge of the carboxylate anion is delocalized symmetrically, and the two oxygen atoms become equivalent (eqn 6.15).

The electron-withdrawing effect of the carbonyl group also contributes to the acidity of a carboxylic acid being greater than that of an alcohol.

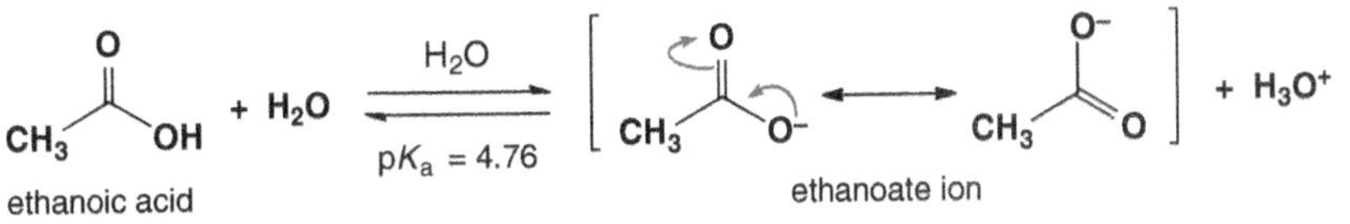

$$CH_3COOH + H_2O \rightleftharpoons \left[\begin{array}{c} \text{ethanoate ion} \end{array} \right] + H_3O^+ \quad (6.15)$$

ethanoic acid
$pK_a = 4.76$

ethanoate ion

Phenol ($pK_a \sim 10$) is considerably more acidic than cyclohexanol ($pK_a \sim 16$, i.e. about the same as ethanol); see eqns 6.16 and 6.17. This is principally because the phenoxide anion is stabilized by resonance delocalization of the negative charge into the benzene ring without charge separation (as shown below eqn 6.17); the corresponding lone pair delocalization of the un-ionized phenol is weaker because it involves charge separation.

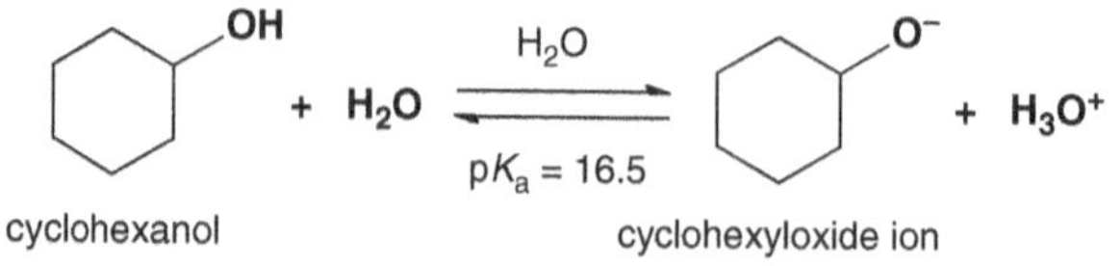

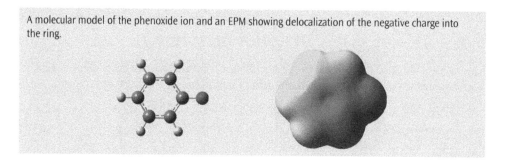

$$\text{cyclohexanol} + H_2O \underset{pK_a = 16.5}{\overset{H_2O}{\rightleftharpoons}} \text{cyclohexyloxide ion} + H_3O^+ \quad (6.16)$$

$$\text{phenol} + H_2O \underset{pK_a = 10}{\overset{H_2O}{\rightleftharpoons}} \text{phenoxide ion} + H_3O^+ \quad (6.17)$$

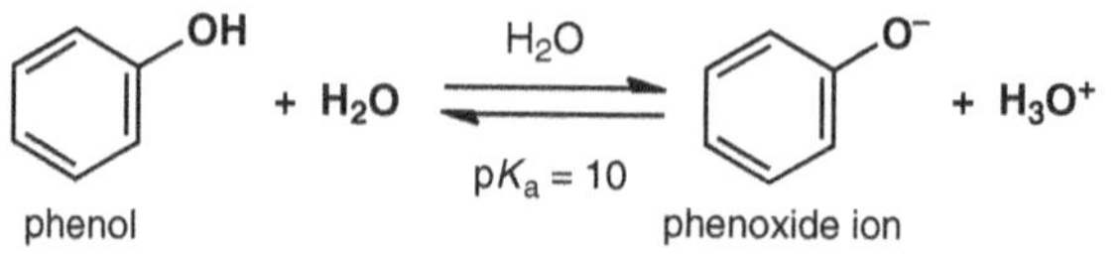

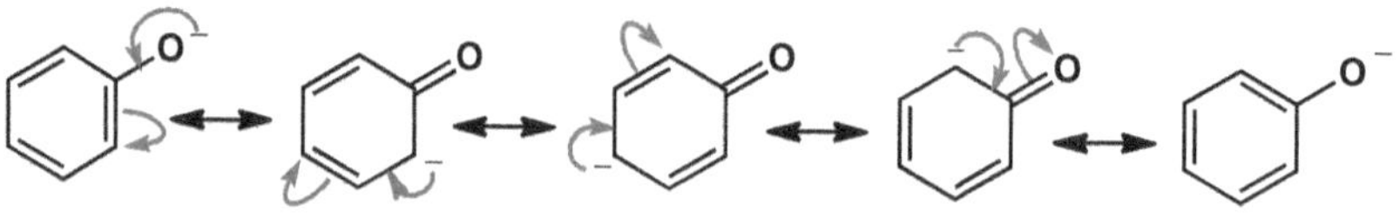

resonance of phenoxide ion

A molecular model of the phenoxide ion and an EPM showing delocalization of the negative charge into the ring.

6.3.3 Substituent effects

When a hydrogen atom on a carbon in a molecule is replaced by another atom or group, the atom or group introduced is called a *substituent* and its effects on properties of other parts of the molecule, especially a functional group, are called **substituent effects**. We saw in Chapter 3 that organic compounds are classified according to the functional group, which is that distinctive part of the molecule responsible for the characteristic chemical properties of the compound. Such a property is often modulated by a substituent elsewhere in the molecule without qualitatively changing the *nature* of the property. Two kinds of substituent effects have been recognized which affect not only the acidity of a compound but other chemical and physical properties as well. This ability to attribute a quantitative change in a property of a functional group to a systematic change elsewhere in the molecule is a major unifying feature of mechanistic organic chemistry.

a. Inductive effects

When a proton is abstracted from an acidic functional group of an uncharged molecule, the conjugate base becomes an anion and, in the absence of any special effect, the charge resides as a lone pair localized on the functional group. An **electron-withdrawing (electron-attracting)** group elsewhere in the molecule will stabilize the anion by dispersing the negative charge and thereby enhance the acidity of the molecule. The highly electronegative chlorine, for example, is a substituent which stabilizes an anion by attracting electron density.

The pK_a values of chloroethanoic acids given below illustrate this effect:

	CH_3CO_2H	$ClCH_2CO_2H$	Cl_2CHCO_2H	Cl_3CCO_2H
pK_a :	4.76	2.86	1.35	−0.5

When the methyl hydrogen atoms of ethanoic acid are replaced by chlorine atoms one by one, the acid becomes increasingly stronger. The origin of this **inductive effect** is the strong polarization of the C–Cl σ bond.

Inductive effects are transmitted through the σ bonds of a molecule and weaken rapidly as the number of σ bonds between a substituent and the reaction site increases.[3] This tendency can be seen in a series of chlorobutanoic acids:

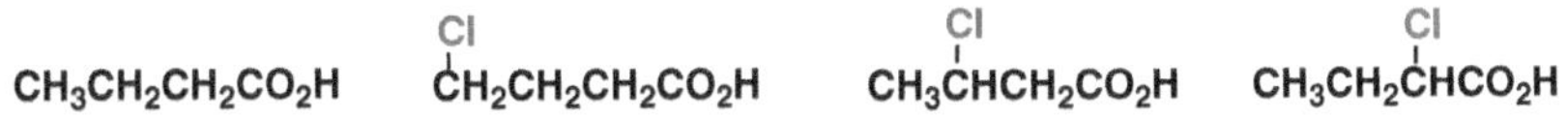

	CH₃CH₂CH₂CO₂H	$\overset{Cl}{\underset{|}{C}}$H₂CH₂CH₂CO₂H	CH₃$\overset{Cl}{\underset{|}{C}}$HCH₂CO₂H	CH₃CH₂$\overset{Cl}{\underset{|}{C}}$HCO₂H
pK_a	4.8	4.5	4.1	2.8

> Predict the relative acidities of the following carboxylic acids.
>
> CH_3CO_2H FCH_2CO_2H $HOCH_2CO_2H$

Exercise 6.12

Methyl and other alkyl groups are **electron donating (electron releasing)**, although the effect is small, so ethanoic acid is less acidic than methanoic (formic) acid.

	HCO_2H	CH_3CO_2H
pK_a :	3.75	4.76

b. Conjugative effects

Substituents can also affect the electron distribution within a molecule, and hence its properties, by conjugation. Such effects of substituents are known as **conjugative (or resonance) effects** and are found in molecules (or ions) containing π electron systems; they can be represented by drawing the contributing resonance forms.

The effect is exemplified by the acidity of nitrophenols. A *m*-nitro group enhances the acidity of phenol through its inductive effect but a *p*-nitro group enhances the acidity even more strongly even though it is further away from the reaction site.

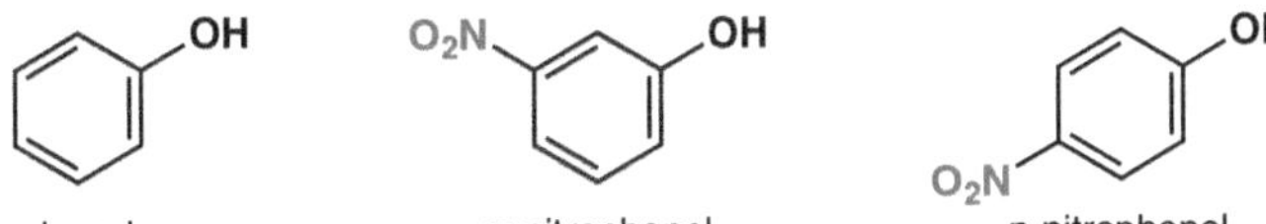

	phenol	*m*-nitrophenol	*p*-nitrophenol
pK_a	9.99	8.35	7.14

The negative charge of *p*-nitrophenoxide ion is delocalized over the molecule as illustrated by a resonance contributor with direct conjugation between the anionic oxygen

---

[3] Although the acidities of carboxylic acids can be correlated with the electronic properties of substituents elsewhere in the molecule, as illustrated here, the effect is expressed through the entropy of dissociation, not the enthalpy, see Chapter 7. For example, trifluoroethanoic acid is much more acidic than ethanoic acid. The pK_a values at 25 °C are −0.6 and 4.76, respectively, corresponding to ΔG° values of −3.4 and +27.2 kJ mol⁻¹ for dissociation (see Sub-section 7.4.5). But the contribution of ΔH° to both of these ΔG° values (remember $\Delta G = \Delta H - T\Delta S$) is virtually zero, so ΔS° values are +11 and −91 J K⁻¹ mol⁻¹ for trifluoroethanoic and ethanoic acids, respectively. In other words, the entropy change for the dissociation of trifluoroethanoic acid in water is slightly favourable, but hugely unfavourable for ethanoic acid. How this happens is understood in terms of the solvation of undissociated and dissociated acids, and the effects of the solvated species upon the structure of water, but discussion is beyond the scope of this book.

and the nitro group through the π system of the benzene ring. Although a lone pair on the OH of un-ionized *p*-nitrophenol can also be delocalized by resonance into the NO$_2$, this leads to charge separation which is absent in the corresponding delocalization in the *p*-nitrophenoxide. Consequently, resonance stabilization is greater in *p*-nitrophenoxide than in the *p*-nitrophenol, and *p*-nitrophenol is a stronger acid than phenol. For the *m*-nitro derivative, such direct conjugation is impossible, and only the inductive electron-withdrawing effect of the nitro group stabilizes the *m*-nitrophenoxide ion relative to the un-ionized *m*-nitrophenol.

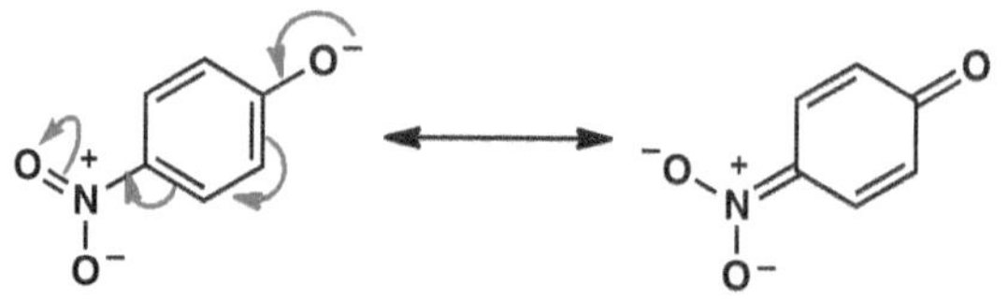

p-nitrophenoxide ion

Exercise 6.13

Represent the electronic configuration of *p*-nitrophenoxide ion by showing as many resonance contributors as possible.

We use abbreviations EWG and EDG for electron-withdrawing and electron-donating groups, respectively.

A *p*-methoxy substituent is an example of an electron-donating group (EDG) by resonance, in contrast to the electron-withdrawing inductive effect of the methoxy group (oxygen is an electronegative element). These opposing properties of the methoxy group are illustrated by the acidities of the methoxybenzoic acids: *p*-methoxybenzoic acid is weaker than the unsubstituted benzoic acid, whereas the *m*-methoxy analogue is stronger than both.

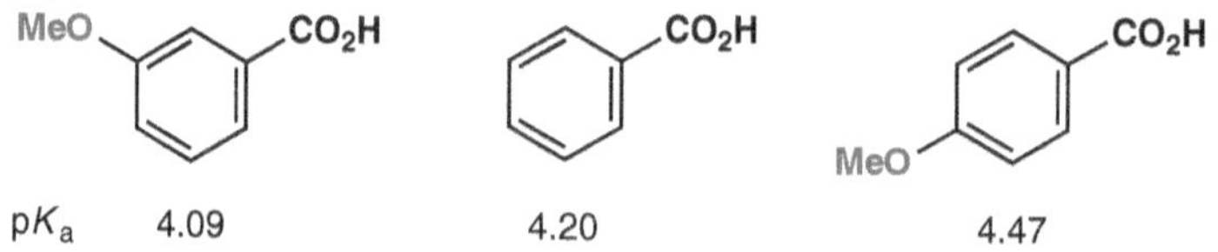

| pK$_a$ | 4.09 | 4.20 | 4.47 |

The *m*-methoxy group can only attract electrons through σ bonds, and this effect is stronger in the anionic conjugate base than in the carboxylic acid. In contrast, a *p*-methoxy group can conjugate with the carboxy group of benzoic acid through the π system of the benzene ring (a lone pair of the methoxy is delocalized to some degree into the carbonyl of the carboxy group). And because this resonance effect is greater in the carboxylic acid than in the carboxylate anion, the acid is stabilized with respect to its conjugate base so the acidity is decreased by the *p*-methoxy substituent.

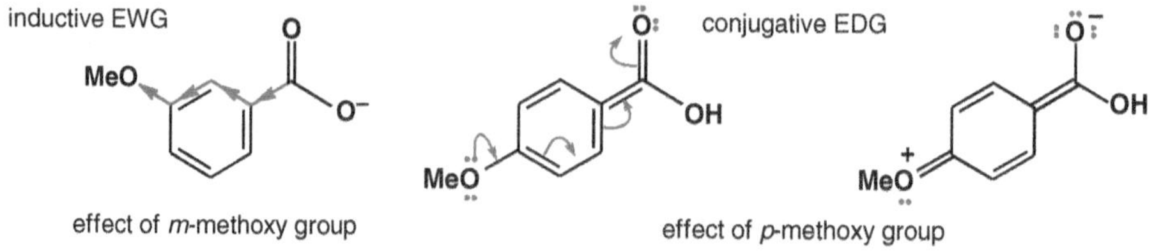

effect of *m*-methoxy group

effect of *p*-methoxy group

6.4 Carbon Acids and Carbanions

6.4.1 Hydrocarbons

Hydrocarbons in general are extremely weak acids because of the strength and low polarity of the C–H bond. Nonetheless, an acetylenic hydrogen bonded to an sp-hybridized carbon has a pK$_a$ of about 25: ethyne is a stronger acid than ammonia (pK$_a$ = 35), and hugely more acidic than its saturated analogue (ethane, $pK_a \sim 50$, eqn 6.18), although still very weak. Acids which release a proton by heterolysis of the C–H bond are called **carbon acids**, and the conjugate bases are known as **carbanions** which carry a formal negative charge on the carbon. Carbon acids become stronger as the conjugate carbanion bases are stabilized by electron-withdrawing substituents or by resonance.

We encountered the allyl anion as a representative delocalized carbanion in Subsection 5.3.2. The conjugate acid of the allyl anion is propene, the pK_a of which is about 43 (eqn 6.19). Propene, therefore, is considerably more acidic than a simple alkane ($pK_a \sim 50$); this appreciably enhanced acidity is attributed to the resonance stabilization of the allyl anion.

> The large pK_a values (very low acidities) quoted here for carbon acids relate to aqueous solution, so they may be compared with values quoted for other acids in water, but they were determined by indirect methods or estimated from data in other solvents. In general, there is greater uncertainty in such large pK_a values compared with results obtained by direct measurements in aqueous solution.

$$CH_3\text{–}CH_3 \rightleftharpoons CH_3\text{–}\overset{-}{C}H_2 \ + \ H^+ \qquad (6.18)$$
$$\text{ethane} \qquad pK_a \sim 50$$

$$CH_2{=}CH\text{–}CH_3 \rightleftharpoons \left[CH_2{=}CH\text{–}\overset{-}{C}H_2 \longleftrightarrow \overset{-}{C}H_2\text{–}CH{=}CH_2 \right] + H^+ \quad (6.19)$$
$$\text{propene} \qquad pK_a \sim 43 \qquad\qquad \text{allyl anion}$$

Toluene is somewhat more acidic than propene (eqn 6.20). The conjugate base of toluene is the benzyl anion, which is stabilized by extensive delocalization of its negative charge.

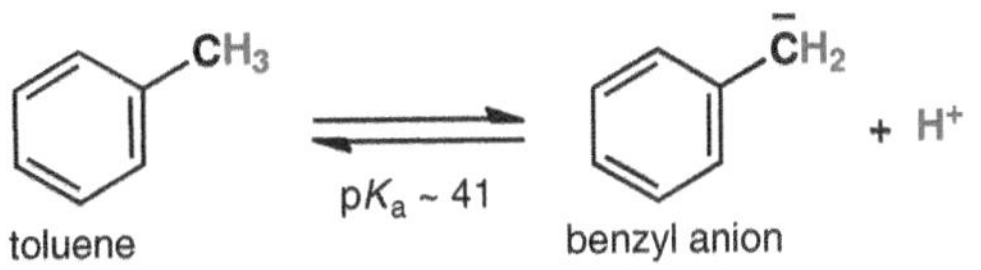

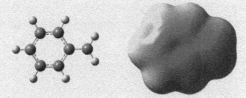

$$(6.20)$$

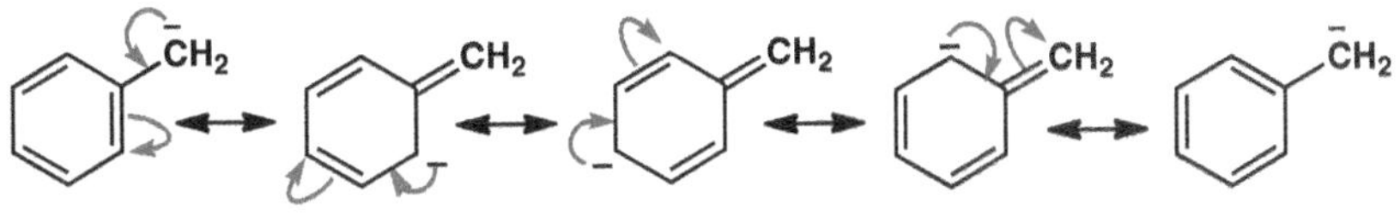

toluene $pK_a \sim 41$ benzyl anion

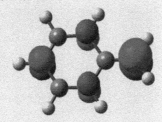

Resonance stabilization of the benzyl anion

A molecular model of the benzyl anion and an EPM showing delocalization of the negative charge:

The HOMO of the benzyl anion showing how delocalized electron density from the CH_2 group is accommodated at the *ortho* and *para* positions.

The π system of the benzyl anion includes the six electrons of the benzene ring and an unshared electron pair originating from the methylene (CH_2) side-chain. This electronic system is the same as that of phenoxide ion (eqn 6.17). Two such systems are said to be **isoelectronic.** The resonance contributors are similar but the relative importance of each contributor to the resonance hybrid is different because the atoms are different.

The stabilizing conjugative effect of a phenyl attached to a carbanionic centre is cumulative so, if there is an additional phenyl group, it will further enhance the acidity of the C–H. However, the third phenyl group in the triphenylmethyl (trityl) anion shows only a small additional effect when the pK_a values of diphenylmethane and triphenylmethane are compared.

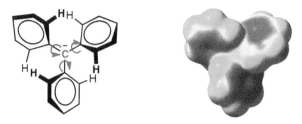

benzyl anion	diphenylmethyl anion	triphenylmethyl anion (trityl anion)
pK_a of conjugate acid 41	33.4	31.5

For greatest stabilization of a π system by conjugation, all the p orbitals involved need to be parallel for maximal sideways overlap. Consequently, the phenyl group should be coplanar with the sp²-hybridized carbon of the carbanionic centre in order to achieve the highest possible conjugative stabilization (delocalization of electrons). However, a perfectly planar trityl anion would involve very close proximity of pairs of the six *ortho* hydrogen atoms of the three phenyl groups; this severe steric strain is avoided by the three phenyl groups being twisted to give a propeller shape (Figure 6.3). Because of this non-planarity, however, sideways orbital overlap is less than maximal and the additional resonance stabilization arising from the third phenyl group is weak; triphenylmethane is not as acidic as expected for a planar structure.

The carbanion formed from triptycene, like the trityl anion, has three benzene rings attached to the anionic carbon, but their stabilities are very different as the pK_a values of their conjugate acids show. In the triptycyl anion, the three benzene rings cannot conjugate at all with the lone pair of the carbanion.

The alignments of p orbitals for benzyl and triptycyl anions are illustrated in Figure 6.4. In the former system, all the p orbitals are parallel, including that of the lone pair of the anionic centre, whereas the lone pair and the p orbitals of the benzene rings are orthogonal in the latter.

The cyclopentadienide anion consists of a five-membered ring with six π electrons. A cyclic six π electron system is aromatic as discussed in Section 5.6, and this *aromatic*

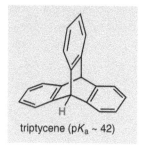

triptycene (pK$_a$ ~ 42)

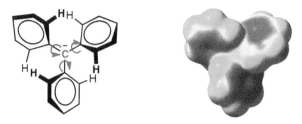

Figure 6.3 The propeller-shaped molecular structure of triphenylmethyl anion and an EPM which also shows the electron delocalization.

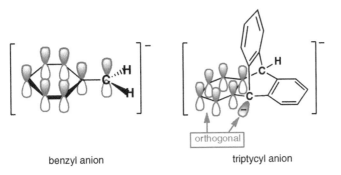

benzyl anion   triptycyl anion

A molecular model of the triptycyl anion.

Figure 6.4 Comparison of p orbitals of benzyl and triptycyl anions.

Exercise 6.14

The pK_a of 9-phenylfluorene is 18.5. Account for the difference between this and the value for triphenylmethane (pK_a 31.5); both compounds have three benzene rings attached to the potentially acidic C–H.

9-phenylfluorene (pK_a = 18.5)

stabilization is responsible for the low pK_a (16) of cyclopenta-1,3-diene (eqn 6.22) compared with the corresponding linear system, penta-1,4-diene (pK_a = 35, eqn 6.21).

$$C-H \quad \rightleftharpoons \quad C-H \ + \ H^+ \tag{6.21}$$

penta-1,4-diene   pK_a ~ 35

$$C-H \quad \rightleftharpoons \quad + \ H^+ \tag{6.22}$$

cyclopenta-1,3-diene   pK_a = 16   cyclopentadienide ion

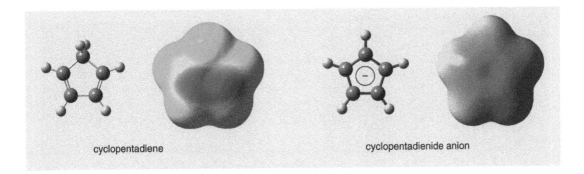

cyclopentadiene   cyclopentadienide anion

6.4.2 Effects of electron-withdrawing groups on C–H acidity

When a conjugatively electron-withdrawing group (EWG) is attached to a hydrogen-bearing carbon, the acidity increases tremendously; examples of such groups include nitro, carbonyl, and cyano. The nitro group is especially effective, and the pK_a of nitromethane (10.2) is similar to that of phenol. The conjugate base of propanone (acetone, eqn 6.23) is an enolate anion, previously encountered as an allyl anion analogue in Sub-section 5.3.3.

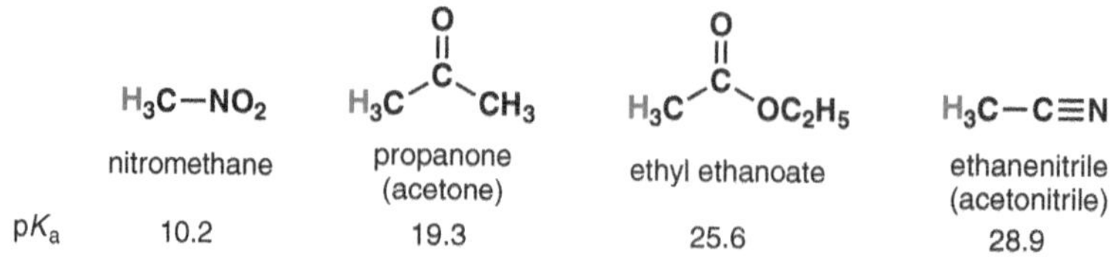

H$_3$C–NO$_2$ nitromethane   propanone (acetone)   ethyl ethanoate   ethanenitrile (acetonitrile)

	nitromethane	propanone (acetone)	ethyl ethanoate	ethanenitrile (acetonitrile)
pK_a	10.2	19.3	25.6	28.9

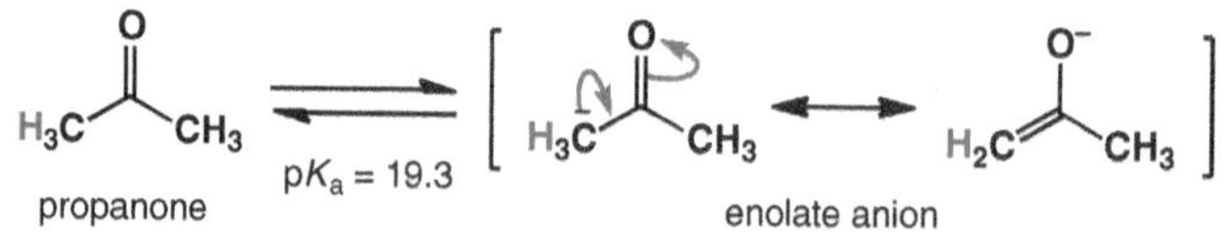

propanone   pK_a = 19.3   enolate anion   + H$^+$ (6.23)

Exercise 6.15

Show how the conjugate base of nitromethane is stabilized by resonance.

Exercise 6.16

Why is ethyl ethanoate (pK_a 25.6) less acidic than propanone (pK_a 19.3)?

Carbon acids become stronger (i.e. their conjugate bases become more stable) if two or even three EWGs are bonded to the C bearing the acidic H. The conjugate bases of carbon acids are important as reactive intermediates, and reactions involving them will be covered in Chapter 17.

Example 6.5

The pK_a of pentane-2,4-dione (acetylacetone) is 8.84. Of the two kinds of hydrogen, which is the more acidic, and why is this compound much more acidic than propanone?

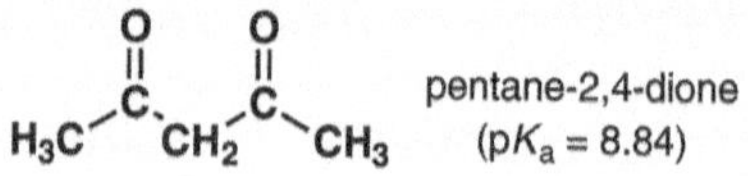

pentane-2,4-dione (pK_a = 8.84)

Solution

The hydrogens on C3 of pentane-2,4-dione are the more acidic, and deprotonation from this position gives a resonance stabilized conjugate base as illustrated below with the negative charge delocalized mainly onto the two oxygen atoms (as opposed to onto just one as in the enolate from propanone).

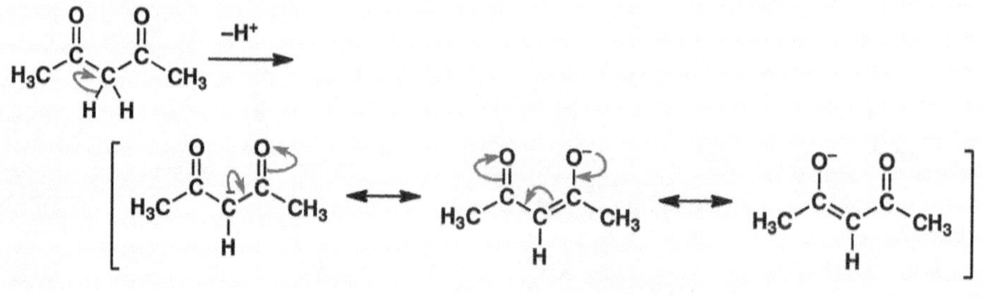

A molecular model and EPM of pentane-2,4-dione:

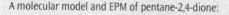

A molecular model and EPM of the enolate ion of pentane-2,4-dione:

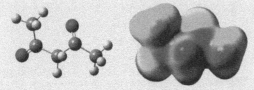

Exercise 6.17

The pK_a of ethyl 3-oxobutanoate (ethyl acetoacetate) is 10.7. Which are the most acidic hydrogens, and why is this compound much more acidic than ethyl ethanoate (pK_a 25.6)?

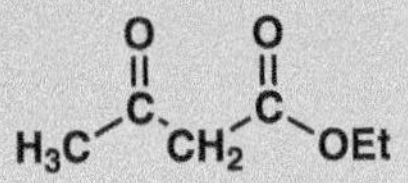

$$H_3C \overset{\overset{O}{\parallel}}{C} CH_2 \overset{\overset{O}{\parallel}}{C} OEt$$

ethyl 3-oxobutanoate
($pK_a = 10.7$)

6.5 Basicity of Organic Compounds

6.5.1 Definition of base strengths

A Brønsted base is always related to its conjugate acid, so the pK_a of the conjugate acid can be used as a quantitative measure of the basicity of a base (see Sub-section 6.2.1). However, when a compound (B) is uncharged and the focus is on its base strength with the reaction written as eqn 6.24, we usually refer to the pK_{BH^+} of B rather than the pK_a of BH^+ (eqn 6.25), but the numerical value is, of course, the same. We see, therefore, that the larger the value of the pK_{BH^+}, the more basic compound B is.

$$\underset{\text{base}}{B} + H_3O^+ \rightleftharpoons \underset{\text{conjugate acid}}{BH^+} + H_2O \qquad (6.24)$$

$$BH^+ + H_2O \overset{K_{BH^+}}{\rightleftharpoons} B + H_3O^+$$

$$K_{BH^+} = \frac{[B][H_3O^+]}{[BH^+]} \qquad (6.25)$$

> We shall see in Sub-section 17.4.2 that the inductively electron-withdrawing halogens can enhance the acidity of Hs attached to the same C; the pK_a of chloroform ($HCCl_3$) is about 24 and values for $CHBr_3$ and CHI_3 are even lower.

> Base strengths of various organic compounds are listed in Appendix 1 at the end of this book as pK_a values of their conjugate acids.

An alternative definition of base strength used in older books is based on the reaction of the base with water:

$$B + H_2O \overset{K_b}{\rightleftharpoons} BH^+ + HO^- \qquad K_b = \frac{[BH^+][HO^-]}{[B]} \text{ and } pK_b = -\log K_b$$

Since the autoprotolysis constant (ionic product) of water is defined as $K_w = [H_3O^+][HO^-] = 10^{-14}$, we can work out that $pK_b + pK_{BH^+} = 14$.

6.5.2 Nitrogen bases

Amines are the most commonly encountered organic bases. For most typical alkyl-amines, pK_{BH^+} is about 10 (e.g. eqn 6.26), but arylamines are much weaker ($pK_{BH^+} = 4.6$ for aniline, eqn 6.27). This relationship is similar to the one between an alcohol and a phenol (see eqns 6.16 and 6.17 in Sub-section 6.3.2) except that some degree of delocalization is possible for the phenol as well as phenoxide.

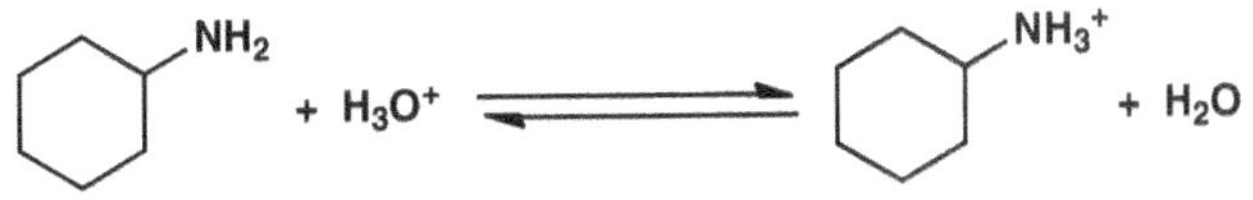

cyclohexylamine ($pK_{BH^+} = 10$)

$$\qquad (6.26)$$

$$\text{aniline } (pK_{BH^+} = 4.6) \quad + H_3O^+ \rightleftharpoons \text{anilinium ion} \quad + H_2O \qquad (6.27)$$

Aniline is isoelectronic with phenoxide and the benzyl anion. The lone pair on the nitrogen of aniline is delocalized into the benzene ring as the resonance forms below show. But this is not possible for the protonated form, the anilinium ion, so appreciable resonance stabilization is lost upon protonation.

Resonance stabilization of aniline

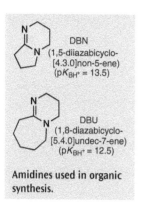

DBN
(1,5-diazabicyclo-
[4.3.0]non-5-ene)
$(pK_{BH^+} = 13.5)$

DBU
(1,8-diazabicyclo-
[5.4.0]undec-7-ene)
$(pK_{BH^+} = 12.5)$

Amidines used in organic synthesis.

Amines are often used as bases in synthetic reactions and, when a simple alkylamine is not sufficiently basic, modified versions are available. The pK_{BH^+} values of acetamidine and guanidine are 12.4 and 13.6, respectively, so they are quite strong as neutral bases in spite of the hybridization of the nitrogen (sp^2 compared with sp^3 of the N in simple alkylamines). The high basicity of these amines compared with the analogous imines is attributable to the extra resonance stabilization of their conjugate acids.

	an imine	an amidine	guanidine
pK_{BH^+}	~8	12.4	13.6

resonance of the conjugate acid of an amidine

resonance of the conjugate acid of guanidine

The conjugate acid of an
α-amino acid containing a
guanidine group:

arginine
$(pK_a\ 2.17,\ 9.04,\ 12.48)$

If still stronger bases are needed for organic syntheses, anionic bases such as alkoxides RO^- $(pK_{BH^+} \sim 16)$ or amides R_2N^- $(pK_{BH^+} \sim 35)$ have to be used.

6.5.3 Weakly basic organic compounds

Neutral organic compounds which contain heteroatoms such as nitrogen or oxygen have lone pairs capable of being protonated, and pK_{BH^+} values of representative compounds are given below:

Weak organic base	alcohol ROH	ether R_2O	ketone $R_2C=O$	ester RCOOR'	amide $RCONH_2$
Conjugate acid					
pK_{BH^+}	−2	−2.5	−3	−4	−0.2

Methanol, for example, is weakly basic and strongly acidic compounds dissociate in it (see later).

$$HCl \; + \; MeOH \rightleftharpoons MeOH_2^+ \; + \; Cl^-$$

Carbonyl compounds are less basic than alcohols but (very usefully) protonation makes them more electrophilic, i.e. susceptible to nucleophilic attack; this is the basis of *acid catalysis* in many reactions of carbonyl compounds with nucleophiles (see Chapter 8).

The lower basicity of carbonyl oxygens in ketones and esters compared with alcohols and ethers is attributable to the lone pair on the carbonyl oxygen being in an sp²- rather than sp³-hybridized orbital—the orbital with the greater s-character holds the lone pair closer to the nucleus (see Sub-section 6.3.1 for hybridization effects). However, the carbonyl and ether oxygens in an ester are not independent; protonation of the carbonyl oxygen gives a structure which is resonance stabilized to a degree not possible following protonation of the ether oxygen. Also, the electron-withdrawing effect of the carbonyl group reduces the basicity of the ether oxygen of the ester. Consequently, the carbonyl oxygen of an ester is the more basic site.

Poorer stabilization of the ether oxygen-protonated form:

Resonance stabilization of the conjugate acid of an ester

Example 6.6

Explain why the carbonyl oxygen of an amide is its most basic site even though an amine is generally much more basic than a ketone or aldehyde.

Solution
There are two basic sites in an amide, O and N, so two isomers are possible upon protonation. The *O*-protonated isomer has three resonance contributors none of which involves charge separation. The *N*-protonated isomer has only two resonance contributors and one of them involves charge separation and two formal positive charges on adjacent atoms, which is very unfavourable, so the second form hardly contributes. In addition, the basicity of the NH₂ group will be reduced by the electron-withdrawing effect of the carbonyl. Consequently, the *O*-protonated isomer is the more stable.

This result could not be predicted from the normal properties of the elements: in an organic compound, an isolated N is usually more basic than an isolated O. But note that protonation of the carbonyl here is similar to protonation of the imino group discussed above for amidine basicity.

O-protonated form   *N*-protonated form

6.6 Polyfunctional Acids and Bases

Some acids have two or more acidic hydrogens—sulfuric and phosphoric acid are familiar examples. Sulfuric acid with two acidic hydrogens is said to be *diprotic* (or *dibasic* in the older literature) while phosphoric acid is a triprotic (tribasic) acid. The dissociations of a poly-protic acid are stepwise, and the acid has multiple pK_a values.

(H_2SO_4) $pK_{a1} = -3$ $pK_{a2} = 1.99$

Both hydrogensulfate (HSO_4^-) and sulfate (SO_4^{2-}) are resonance stabilized. Draw resonance contributors for each ion.

Dicarboxylic acids are typical diprotic organic acids; some examples and their pK_a values are shown below:

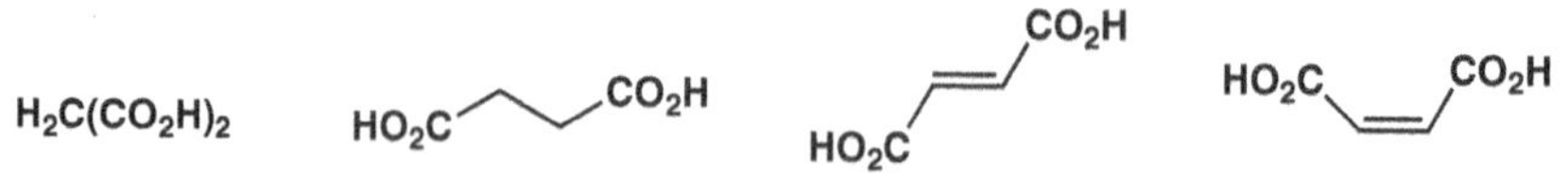

$H_2C(CO_2H)_2$

propanedioic acid
(malonic acid)
pK_a 2.83, 5.69

butanedioic acid
(succinic acid)
4.19, 5.48

(E)-butenedioic acid
(fumaric acid)
3.02, 4.38

(Z)-butenedioic acid
(maleic acid)
1.92, 6.34

A statistical factor also makes pK_{a1} smaller than pK_{a2}. In the first dissociation, there are two acidic (carboxy) groups, and there are two basic (carboxylate) groups to accept a proton in the reverse of the second dissociation. The mono-dissociated form, however, can only lose or gain a single proton.

When one of the two acidic groups dissociates, the other group acts as a substituent and modifies the acidity of the first. The carboxy group –CO_2H is electron withdrawing, which causes the dissociation of the other to be more favourable than that of an isolated carboxy group. Dissociation of the second carboxy group is affected by the (slightly) electron-donating carboxylate. More important, however, this second dissociation involves removal of a proton from an anion, which is more difficult than proton abstraction from a neutral molecule. Consequently, $pK_{a1} < pK_{a2}$ and the difference between the two pK_a values is greater when the two acidic groups are closer together. However, proximate carboxy groups can sometimes form an intramolecular hydrogen bond, especially in the mono-dissociated state. In this event, the first dissociation is further enhanced and the second is further inhibited because of the strongly stabilized cyclic hydrogen-bonded structure of the mono-dissociated ion. Scheme 6.1 shows the dissociations of (E)- and (Z)-butenedioic acid.

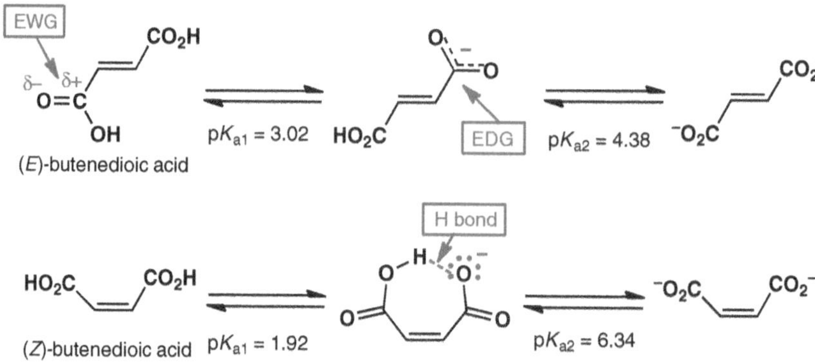

Scheme 6.1 Acid dissociations of isomeric butenedioic acids.

Explain why the pK_{a1} of phthalic acid (benzene-1,2-dicarboxylic acid) is smaller than that of terephthalic acid (benzene-1,4-dicarboxylic acid) but pK_{a2} for phthalic acid is larger than that for terephthalic acid.

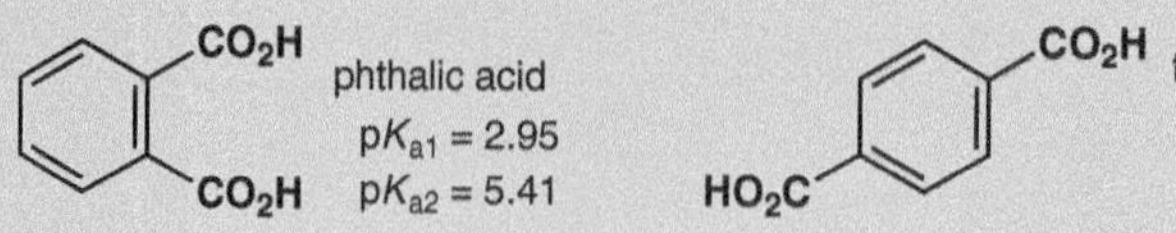

phthalic acid
pK_{a1} = 2.95
pK_{a2} = 5.41

terephthalic acid
pK_{a1} = 3.54
pK_{a2} = 4.46

Compounds with two basic groups can be considered in the same way starting with their diprotonated conjugate acids. The pK_a values for conjugate acids of some diamines are shown:

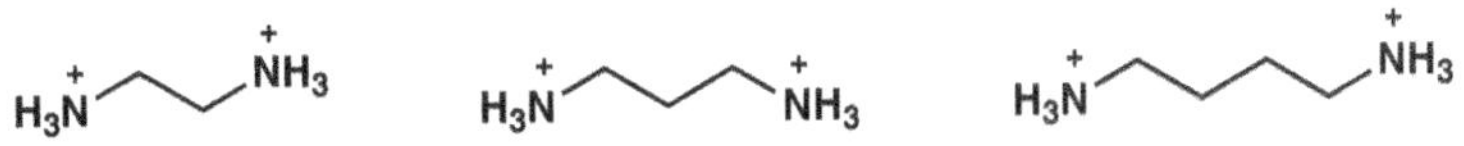

ethane-1,2-diamine
pK$_a$ 6.85, 9.93

propane-1,3-diamine
8.5, 10.4

butane-1,4-diamine
9.2, 10.7

The ammonio group with a positive charge is strongly electron withdrawing, while the amino group is only weakly so; this accounts for the large difference between the pK$_{a1}$ and pK$_{a2}$ values of the conjugate acids of ethanediamine. The difference diminishes with the increasing number of intervening CH$_2$ groups.

Some organic compounds have both acidic and basic groups, and are said to be *amphoteric*. Amino acids, which are important as constituents of proteins, are typical (Section 24.3) and glycine is the simplest example (Scheme 6.2).

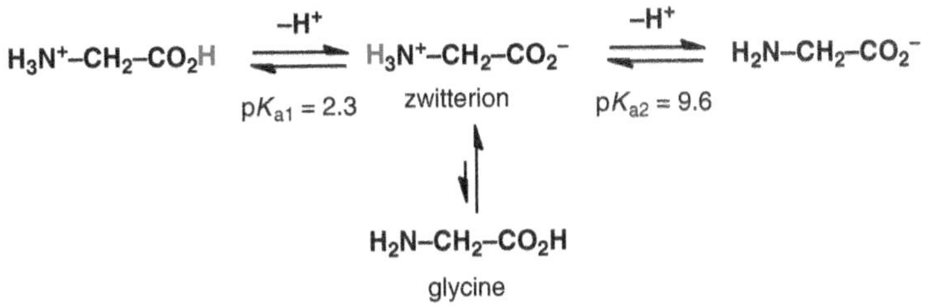

Scheme 6.2 Dissociation of glycine, an amino acid.

In acidic aqueous solution, glycine exists in its protonated form: this is a diprotic acid, and the first dissociation (pK$_{a1}$ = 2.3) is of the carboxy group whose acidity is enhanced by the ammonio group. This dissociation gives a *zwitterion* (or dipolar ion). The second dissociation (pK$_{a2}$ = 9.6) is of the ammonio group, which is hardly affected by the carboxylate group ($-$CO$_2^-$). An amino acid as crystals, or in aqueous solution in the absence of additional acids or bases, is in the zwitterion form.

Amphoteric usually means a compound which shows both acidic and basic properties in aqueous solution, as described above. However, it can also be used for other solvent systems, in which case a solute may be amphoteric in one solvent but not in another. For example, CH$_3$CO$_2$H, is a commonly used acidic solvent (it is also very weakly basic—but sufficiently so that HCl dissociates in it to a limited extent, i.e. HCl is a weak acid in CH$_3$CO$_2$H). In ethanoic acid, 4-aminophenol (which is amphoteric in water) acts only as a strong base and is not measurably acidic. It follows that a solvent must be specified (or assumed to be water) when a compound is described as amphoteric.

Exercise 6.20

Show the structures of the conjugate base and conjugate acid of each of the following amphoteric compounds.

(a) H$_2$O   (b) NH$_2$CN   (c) 4-aminobenzoic acid   (d) 2-hydroxypyridine   (e) 2-aminophenol

6.7 Solvent Effects on Acid–Base Reactions

6.7.1 The levelling effect of water

We saw above that water is amphoteric. For example, it can accept a proton from HCl to give H$_3$O$^+$ or, in a different solution, it can donate a proton to NH$_3$ to give HO$^-$. Only a limited range of acids and bases can be used in an amphoteric solvent; in water, the range is from acids with pK$_a$s > ~ −2 to bases with pK$_{BH^+}$ < ~16). This is because an acid

Panel 6.3 Extraction of morphine from opium

Opium (sometimes called poppy tears) is the dried latex obtained from opium poppies (whose Latin name, *Papaver somniferum*, means 'sleep-bringing poppy'); it is used as a narcotic drug and its trade has affected world history in the past and still influences international politics. The physiological effects are due to opium alkaloids, basic heterocyclic amines exemplified below. Morphine is the most abundant constituent of opium (12–15%); it is a powerful analgesic which is used medicinally for the relief of severe pain.

Some opium alkaloids:

$pK_a \sim 10$

$pK_{BH^+} \sim 8$

morphine   codeine   noscapine

Alkaloids (see Panel 19.3) can be isolated using their basic properties. They form water-soluble ammonium salts, so they can be extracted with an aqueous acid. Morphine has a weakly acidic phenolic group ($pK_a \sim 10$) in addition to an amino group ($pK_{BH^+} \sim 8$), so is amphoteric. Since the other alkaloids have only basic groups, morphine can be separated from them using its acidic properties.

$pK_a \sim 10$

$pK_a \sim 8$ $-H^+$ $-H^+$

An opium poppy.
(This file is licensed under the Creative Commons Attribution-Share Alike 3.0 Unported license.)

When an acidic aqueous solution containing protonated opium alkaloids is brought to pH~9, the liberated unprotonated forms of all the alkaloids become soluble in an organic solvent such as dichloromethane, and can be extracted from the aqueous solution. The organic phase is then shaken with aqueous NaOH whereupon morphine alone is extracted into the aqueous layer as its sodium phenolate; all the other alkaloids remain in the organic layer. By bringing the separated aqueous solution containing the conjugate base of morphine back to pH~9 (the pH at which morphine is neither protonated nor deprotonated to any appreciable extent), morphine can be isolated by extraction with an organic solvent.

which is more acidic than the conjugate acid of the solvent is completely deprotonated by the solvent, while a base which is more basic than the conjugate base of the solvent becomes completely protonated by the solvent. In other words, the strongest acid and base which can exist in a solvent are the conjugate acid and conjugate base of the solvent, respectively. It follows that H_2SO_4 ($pK_a = -3$) and HCl ($pK_a = -7$) cannot be distinguished as acids in water ($pK_{BH^+} = -1.7$); both dissociate completely in dilute aqueous solution to give H_3O^+. This property of a solvent is called its *levelling effect*. So, if we want to exploit the greater acidity of HCl compared with H_3O^+, for example, we need to use a solvent less basic than water. Ethanoic acid is a commonly used weakly basic solvent and we can even use H_2SO_4 as an extremely weakly basic (and strongly acidic) medium.

Correspondingly, in order to use bases stronger than OH^-, e.g. NH_2^- or H^- as their sodium salts which would simply deprotonate water to give NaOH, we need to use a solvent less acidic than water. Liquid ammonia ($pK_a = 35$, bp $= -33$ °C), which is also appreciably basic as we have already seen ($pK_{BH^+} = 9.25$), or dimethyl sulfoxide (DMSO, $pK_a = 33$), which is only weakly basic and a good solvent for organic compounds, are such solvents used in organic synthesis.

6.7.2 Acid–base reactions in non-aqueous solvents

Previously, we have been concerned mainly with acids and bases in aqueous solution even though some *aqueous* pK_a values have been estimated from results in other solvents (because they cannot be measured in aqueous solution when water is either too acidic or too basic). However, acids and bases are widely used, e.g. as catalysts, for organic reactions in non-aqueous solvents, especially so-called *polar aprotic solvents,* e.g. dimethyl sulfoxide, so measures of relative acidity (and basicity) in such solvents are desirable.

Use of such solvents in substitution reactions and their effects on nucleophilicity are be covered in Chapter 12, Sub-section 12.3.2.

Some pK_a values in water and DMSO are compared in Table 6.1 where $K_{a(S)}$ of acid AH in solvent (S) is defined as $K_{a(S)} = [SH^+][A^-]/[AH]$ (eqn 6.28):

$$\text{AH} + \text{S} \underset{}{\overset{K_{a(S)}}{\rightleftharpoons}} \text{SH}^+ + \text{A}^- \tag{6.28}$$

We expect the $K_{a(S)}$ value of an acid to depend on the basicity of the solvent S because the acid donates a proton to the solvent in its dissociation (see Sub-section 6.7.1). However, water and DMSO are similarly basic (pK_{BH^+} values -1.7 and -1.5, respectively, in water) and yet we find interesting differences. The most distinctive feature in the data of Table 6.1 is that the neutral acids (HCl and $MeCO_2H$) are much *less* dissociated (larger pK_a values) in DMSO than in water whereas cationic acids (NH_4^+ and Et_3NH^+) are similarly dissociated (similar pK_a values) in these two solvents. This is because dissociation of the neutral acids results in charge separation (eqn 6.28), while dissociation of the cations does not (eqn 6.29), and solvents exert a strong effect upon acid–base behaviour only when proton transfer changes the ionic states of the compounds involved. The charge-separated state is more stable in a more polar solvent, so neutral acids dissociate more readily in water than in a less polar solvent.

$$\text{BH}^+ + \text{S} \underset{}{\overset{K_{a(S)}}{\rightleftharpoons}} \text{SH}^+ + \text{B} \tag{6.29}$$

In addition to a general medium effect, specific solvation of H_3O^+ and of anionic conjugate bases, e.g. Cl^-, by hydrogen bonding is much more effective in water than in DMSO (Figure 6.5).

pK$_a$ values of some acids in H$_2$O and DMSO		
Acid	H$_2$O	DMSO
HCl	−7	2.0
MeCO$_2$H	4.76	12.6
NH$_4^+$	9.24	10.5
Et$_3$NH$^+$	10.75	9.0

Table 6.1

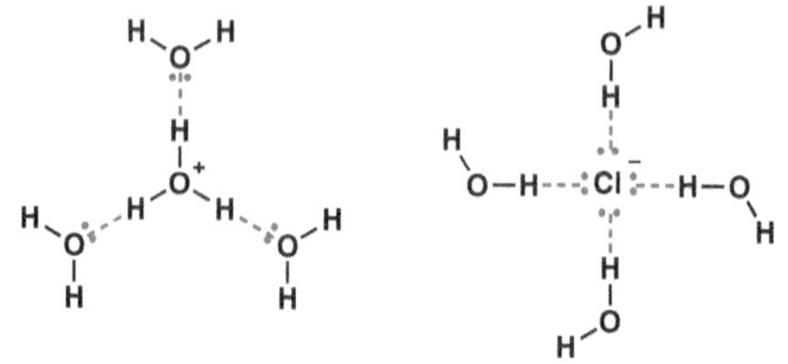

Figure 6.5 Solvation of the oxonium ion and Cl⁻ formed upon dissociation of HCl in water.

Summary

- Brønsted acids and bases are proton donors and acceptors. The more general definition of Lewis acids and bases is in terms of electron-pair acceptors and donors.

- The acidity of a Brønsted acid AH is quantified by the acid dissociation constant K_a or pK_a ($= -\log K_a$). The basicity of a Brønsted base B can be expressed by its pK_{BH^+} (equal to the pK_a of its conjugate acid, BH^+).

- The principal molecular factors which determine acid or base strength include the strength of the bond to the hydrogen, the electronegativity of the atom to which the H is bonded, electronic effects (inductive and conjugative) of substituents elsewhere in the molecule, and electron delocalization if the system is unsaturated.

- Medium effects and solvation, as well as the acid–base properties of the solvent itself, account for very appreciable differences in acid–base properties in different solvents.

Problems

6.1 Classify each of the following as a Lewis acid or base.

(a) $ZnBr_2$ (b) CH_3Cl (c) $H_2C=CH_2$

(d) $AlEt_3$ (e) $B(OCH_3)_3$

6.2 The following can act as both Brønsted acids and bases. Show the structures of the conjugate base and conjugate acid of each.

(a) 4-aminophenol (b) HCO_3^- (c) CH_3OH

(d) CH_3NH_2 (e) CH_3CO_2H

6.3 Explain whether phenol ($pK_a = 10$) is soluble or essentially insoluble in aqueous solutions of each of the following (refer to Appendix 1 for pK_a values).

(a) HCl (b) $NaHCO_3$ (c) Na_2CO_3 (d) NaOH

6.4 Explain whether aniline ($pK_{BH^+} = 4.6$) is soluble or essentially insoluble in aqueous solutions of each of the following.

(a) HCl (b) $NaHCO_3$ (c) Na_2CO_3 (d) NaOH

6.5 Draw the structure(s) of the major form(s) of a small sample of glutamic acid in each of the following aqueous solutions:

Glutamic acid: $H_3\overset{+}{N}-\underset{CH_2CH_2CO_2H}{\overset{CO_2H}{\underset{|}{\overset{|}{C}}}}-H$ (pK_a 2.19, 4.25, 9.67)

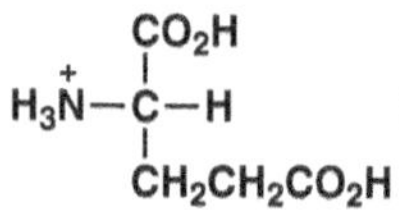

(a) 1 mol dm⁻³ HCl (b) pH 4 buffer

(c) pH 7 buffer (d) 1 mol dm⁻³ NaOH

6.6 Complete the following two acid–base reactions in aqueous solution and calculate their equilibrium constants from the pK_a and pK_{BH^+} values given.

(a) HCO_2H + $C_6H_5NH_2$ ⇌

   pK_a 3.75 pK_{BH^+} 4.60

(b) HCN + $HOCH_2CH_2NH_2$ ⇌

   pK_a 9.1 pK_{BH^+} 9.5

6.7 Show structures of the adducts produced by the following Lewis acid–base pairs.

(a) $BF_3 + H_2O$ (b) $BF_3 + F^-$ (c) $EtMgBr + Et_2O$

(d) $Br_2 + FeBr_3$ (e) $Br_2 + Br^-$ (f) $(CH_3)_3C^+ + H_2O$

6.8 Explain the relative base strengths of the following.

$CH_3OH, CH_3O^-, CH_3CO_2^-$

6.9 Explain the relative acidities of the following carboxylic acids.

FCH_2CO_2H (pK_a 2.59), F_2CHCO_2H (pK_a 1.34), F_3CCO_2H (pK_a −0.6)

6.10 Explain the relative acidities of the following carboxylic acids.

FCH_2CO_2H (pK_a 2.59), $ClCH_2CO_2H$ (pK_a 2.86), $BrCH_2CO_2H$ (pK_a 2.90), ICH_2CO_2H (pK_a 3.18)

6.11 Explain the relative acidities of the following alcohols.

OH (pK_a 16.1)   OH (pK_a 15.5)

OH (pK_a 13.6)

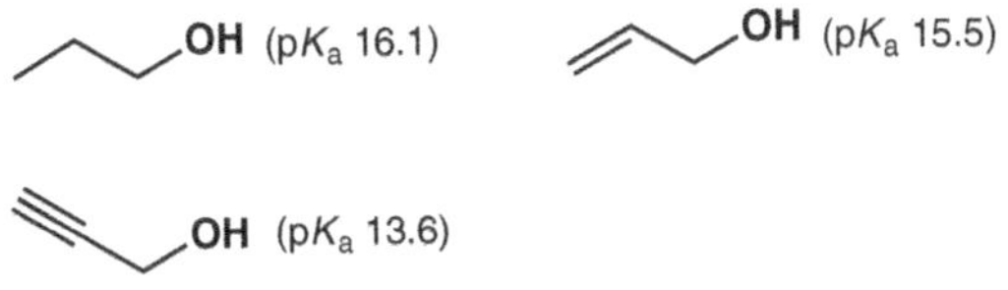

6.12 Explain the relative acidities of the following alcohols.

OH (pK_a 15.5)   OH (pK_a 11)

6.13 Explain why *m*-fluorobenzoic acid (pK_a=3.86) is stronger than *p*-fluorobenzoic acid (pK_a=4.14).

6.14 Explain the relative acidities of the following phenols.

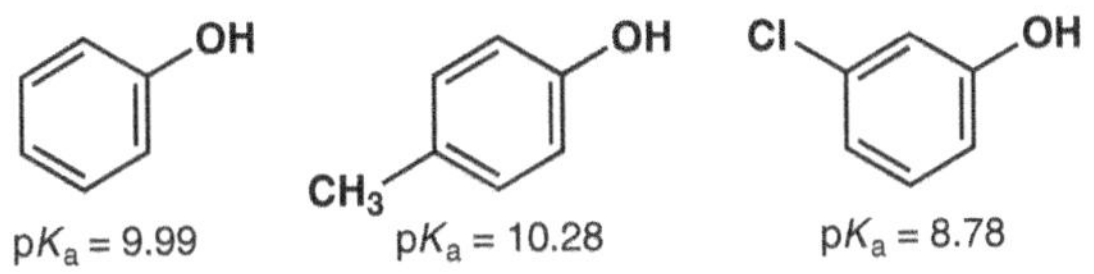

OH   OH   Cl OH

$pK_a = 9.99$ CH_3 $pK_a = 10.28$ $pK_a = 8.78$

6.15 Explain why 4-nitrotoluene is a stronger carbon acid than toluene itself.

6.16 Explain the relative base strengths of each pair of substituted anilines:

(a)
pK_{BH^+} 2.80 pK_{BH^+} 1.75

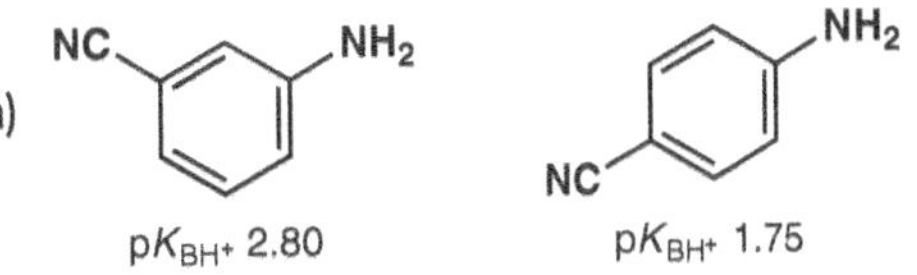

(b)
pK_{BH^+} 4.23 pK_{BH^+} 5.34

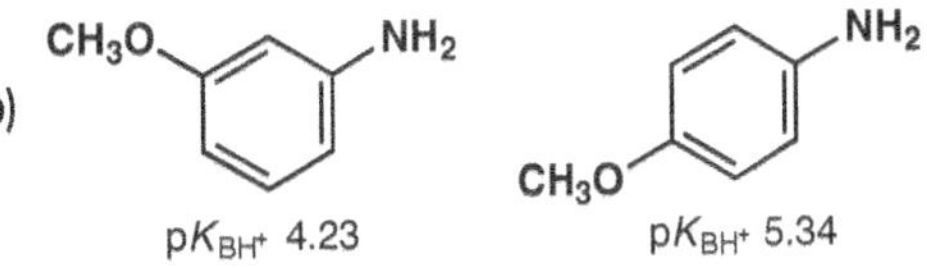

6.17 Answer the questions below regarding the pK_a values of benzoic acid and its three isomeric hydroxy-substituted derivatives.

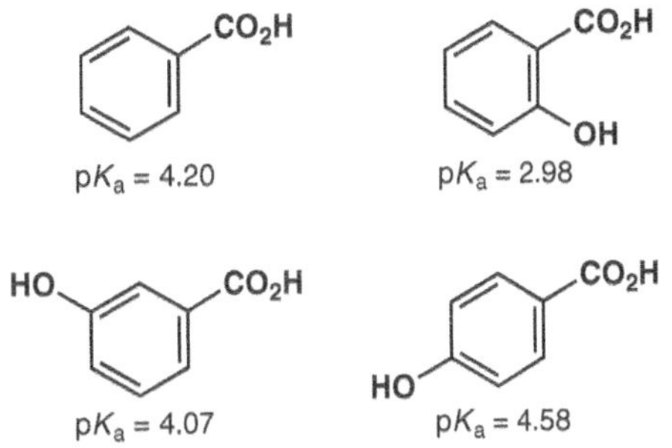

$pK_a = 4.20$ $pK_a = 2.98$

$pK_a = 4.07$ $pK_a = 4.58$

(a) Why is the pK_a of *m*-hydroxybenzoic acid smaller than that of the parent?

(b) Why is the pK_a of *p*-hydroxybenzoic acid larger than that of the parent?

(c) Why is the pK_a of *o*-hydroxybenzoic acid significantly smaller than the other two?

6.18 Why is the pK_{a2} value for the second dissociation of *o*-hydroxybenzoic acid (12.62) appreciably larger than that (9.09) of the *para* isomer?

6.19 Explain the relative base strengths of the following two amines.

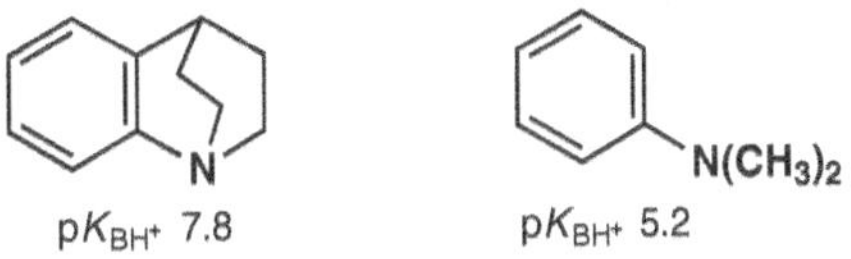

pK_{BH^+} 7.8 pK_{BH^+} 5.2

6.20 4-Aminopyridine (pK_{BH^+}=9.11) is a much stronger base than either pyridine (pK_{BH^+}=5.2) or aniline (pK_{BH^+}=4.60). Show its major conjugate acid and account for its base strength.

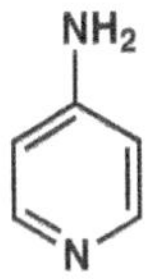

4-aminopyridine
(pK_{BH^+} = 9.11)

7 Organic Reactions and the Concept of Mechanism

Related topics we have already covered:

- **Covalent bonds** (Sections 1.2 and 2.2)
- **Lewis structures** (Sections 1.3)
- **Intermolecular interactions** (Sections 3.9)
- **Steric hindrance** (Sections 4.2)
- **Molecular orbitals** (Chapter 5)
- **Acid–base reactions** (Chapter 6)

Topics we shall cover in this chapter:

- **Classes of organic reactions**
- **Types of elementary reactions**
- **Curly arrow representation of reaction mechanisms**
- **Orbital interactions in rebonding**
- **Energetics of reaction**
- **Reaction profiles**
- **Reversibility of reactions and equilibrium**
- **Investigation of organic reaction mechanisms**

Chemical transformations involve the redistribution of valence electrons of the reactant molecules (lone pair electrons and those of chemical bonds), and how this happens is a major aspect of what is called the reaction **mechanism**. Electrons in molecules exist in molecular orbitals and we shall see in this chapter how the reorganization of chemical bonds as an organic molecule reacts can be related to transformations of the molecular orbitals involved. We shall also illustrate how the redistribution of valence electrons associated with a chemical transformation can be represented by *curly arrows*. The reorganization of bonds associated with an organic chemical reaction is not instantaneous, however: some organic reactions are very fast and become complete as quickly as the reactants can be mixed whereas some reactions are imperceptibly slow. Consequently, it is important to have an appreciation of the dynamics of organic reactions and associated energy changes. These may be described qualitatively in a *reaction profile* which includes a representation of the rate of the reaction (the Gibbs energy of activation) and the feasibility of the reaction (the Gibbs energy of reaction). Concepts and principles introduced in this chapter will be illustrated by just a few reactions taken from later chapters where they are discussed in context and in more detail.

7.1 Classes of Organic Chemical Reactions

Each of many types of organic compounds may be transformed into one or more other compounds, so the number of possible organic reactions is huge, and some of them appear complicated. However, just as we saw in Chapter 3 that organic *compounds* belong to classes according to their functional groups, every known organic *reaction* either belongs to one of only four fundamental classes of reaction, or is a combination of two or more of these four. Consequently, if we understand these four classes of reaction, we shall be able to understand the whole range of organic reactions and, therefore, all general features of organic chemistry.

The four fundamental classes of reaction are **substitution, addition, elimination,** and **rearrangement**, and even the most complicated overall organic transformations can be represented as combinations of these four classes (along with *acid—base* reactions which facilitate some of them). As we shall see, there is a degree of mechanistic diversity within each class of reaction; consequently, to identify a reaction as an addition, for example, says nothing about mechanism. Examples of each reaction class are given below.

Substitution:

$$CH_3{-}H \ + \ Br_2 \longrightarrow CH_3{-}Br \ + \ H{-}Br \qquad (7.1)$$

$$CH_3{-}Br \ + \ {}^-OH \longrightarrow CH_3{-}OH \ + \ Br^- \qquad (7.2)$$

$$(CH_3)_3C{-}Cl \ + \ H{-}OH \longrightarrow (CH_3)_3C{-}OH \ + \ H{-}Cl \qquad (7.3)$$

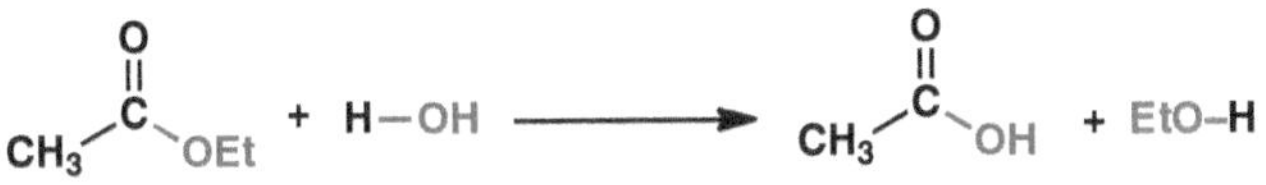

(7.4)

Addition:

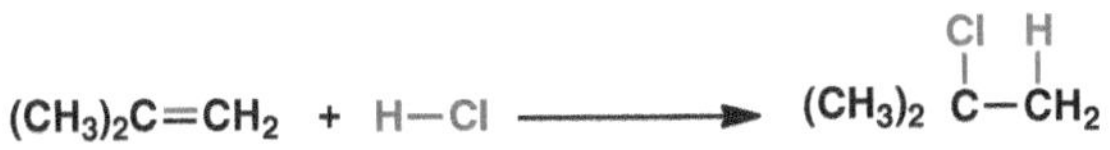

$$(CH_3)_2C{=}CH_2 \ + \ H{-}Cl \longrightarrow (CH_3)_2 \ \overset{Cl}{\underset{}{C}}{-}\overset{H}{\underset{}{C}}H_2 \qquad (7.5)$$

(7.6)

Elimination:

$$(CH_3)_2 \ \overset{Cl}{\underset{}{C}}{-}\overset{H}{\underset{}{C}}H_2 \longrightarrow (CH_3)_2C{=}CH_2 \ + \ H{-}Cl \qquad (7.7)$$

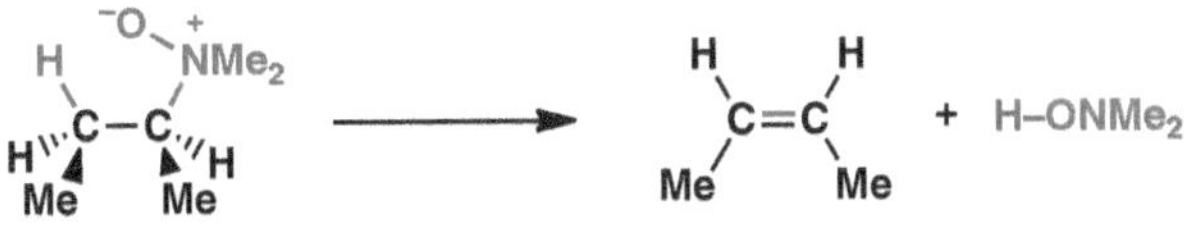

(7.8)

Rearrangement:

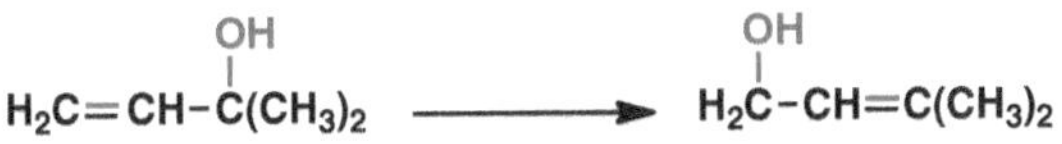

$$H_2C{=}CH{-}\overset{OH}{\underset{}{C}}(CH_3)_2 \longrightarrow H_2\overset{OH}{\underset{}{C}}{-}CH{=}C(CH_3)_2 \qquad (7.9)$$

(7.10)

In a substitution reaction, an atom or group within a molecule is replaced by another atom or group. In reaction 7.1, a hydrogen of methane is replaced by a bromine atom whilst, in reactions 7.2–7.4, the halogen atoms of two haloalkanes and the ethoxy group of an ester are replaced by the OH group. The *mechanisms* of these four substitution reactions are different—the term *substitution* alone implies nothing about mechanism, as indicated above; similarly, the terms *addition, elimination,* and *rearrangement* imply nothing about *how* the reactions take place.

In an addition reaction, one molecule (or its constituent atoms) adds to a second to give a single product molecule; such reactions are characteristic of unsaturated compounds. Examples are the additions of HCl to an alkene (eqn 7.5) to give a chloroalkane, and of HCN to the carbonyl group of an aldehyde to give a cyanohydrin (eqn 7.6).

In an elimination reaction, one molecule (or its constituent atoms) is eliminated from the reactant molecule; conceptually, it is the reverse of an addition reaction. The elimination of HCl from a chloroalkane to give an alkene in eqn 7.7 is the reverse of reaction 7.5, and (in this case) the experimental conditions will determine the direction in which the reaction proceeds.

Reactions 7.9 and 7.10 are examples of rearrangements in which bond reorganization simply leads to an isomeric product. In reaction 7.9, one C–O bond is broken and a new C–O bond forms with migration of the double bond; in reaction 7.10, one C–O bond is broken and the other becomes a C=O bond, two C=C bonds migrate, and a new C–C bond is formed.

Acid–base reactions are not usually classified as organic reactions—they are important throughout the whole of chemistry, as discussed in Chapter 6. They often play a crucial role in organic reactions as intermediate steps which facilitate the overall transformation.

Exercise 7.1

To which fundamental class of reaction (substitution, addition, elimination, or rearrangement) does each of the following transformations belong? Note that you do not need to be familiar with a reaction to be able to identify the class to which it belongs.

(a)

(b)

(c)

(d)

(e)

(f)

7.2 Elementary Steps in a Chemical Reaction

In the above consideration of classes of organic reactions, we did not concern ourselves with *how* a reaction takes place, i.e. the mechanism of a reaction. We shall now begin to look at this topic and, just as there are only a few classes of compounds and of reactions, there are also relatively few types of mechanisms. First, we need to recognize that, although some overall transformations occur in a single step (A → C), others involve two or more steps, i.e. some reactions are **stepwise** (e.g. A → B → C). The individual steps in a multistep (or stepwise) sequence are sometimes called *elementary* reactions, and some may be reversible (e.g. A ⇌ B → C). In stepwise reactions, of course, the product of the first step is the reactant in the second, and so on if there are more than two steps. A molecule or ion which intervenes between initial reactant and final product (e.g. B in the above examples) is an **intermediate**, and these range in stability between ones which are very reactive and (consequently) short-lived, and ones which are stable enough to be isolated and characterized.

There is relatively little diversity amongst types of elementary steps. The simplest is when a single covalent bond between two atoms (or groups) is broken or made, and this can occur in one of two ways.

7.2.1 Homolysis

In homolysis, the two electrons of a σ bond are unpaired and the two parts of the molecule separate, each with an unpaired electron; either fragment (or both) may be single atoms or polyatomic groups. This process is illustrated for a bromine molecule in eqn 7.11; each of the products of homolysis with a single unpaired electron (bromine atoms in eqn 7.11) is called a **radical** (formerly, they were called *free* radicals). Radicals are generally reactive intermediates, and one of their characteristic reactions is combination with another to form a new covalent bond (for example, the reverse of eqn 7.11). Mechanisms of reactions involving radicals are described by curly arrows with a singly barbed head (also called fish-hook arrows, ⌒) which indicates movement of a single electron; the tails of the two singly barbed curly arrows describing a homolysis begin in the centre of the bond.

> Radicals are not usually charged although we shall encounter radical anions in Section 20.9 and radical cations in Section 25.6.

homolysis

$$Br-Br \longrightarrow Br\cdot + \cdot Br \qquad (7.11)$$

radical   radical

The substitution reaction of methane with Br_2 to give bromomethane (eqn 7.1) involves a Br atom (radical) as an intermediate, which is formed in reaction 7.11 and then reacts with a methane molecule in a mechanism we shall discuss in Chapter 20.

7.2.2 Heterolysis

Cleavage of a simple σ bond can also occur without unpairing the bonding electrons in the process known as **heterolysis**; the electron pair ends up on the more electronegative of the two atoms originally bonded. If the reactant is a neutral molecule, one fragment is a cation and the other an anion. Heterolysis of the chloroalkane in eqn 7.12 is an example in which the cation (a **carbenium ion**) is a reactive intermediate which reacts further.

> A *carbenium ion* has a central trigonal trivalent carbon with only six valence electrons; it is a type of *carbocation*.

heterolysis

$$(CH_3)_3C-Cl: \longrightarrow (CH_3)_3C^+ + :Cl:^- \qquad (7.12)$$

2-chloro-2-methylpropane   cation   anion
(*t*-butyl chloride)

> The heterolysis in eqn 7.12 is represented by a single curly arrow with double barbs which indicates movement of a pair of electrons. Curly arrows were used differently in Sub-section 5.4.2 to correlate hypothetical contributors to a resonance hybrid; in the context of resonance, a curly arrow does not represent movement of an electron pair in a real dynamic process.

One possible reaction of the cation formed in eqn 7.12 is recombination with the chloride anion, i.e. eqn 7.13 (the reverse of eqn 7.12). The electron pair of the new covalent bond of the product comes wholly from just one of the two reactants—the **nucleophile** or nucleophilic reagent. A nucleophile is a type of Lewis base (Chapter 6). Its partner, the species which accepts the electron pair to form the new bond, is called an **electrophile** or electrophilic reagent, and is a type of Lewis acid.

> In heterolysis, the group which departs with the bonding electron pair, e.g. Cl^- in eqn 7.12, is sometimes called a *nucleofuge*.

$$(CH_3)_3C^+ + :Cl:^- \longrightarrow (CH_3)_3C-Cl: \qquad (7.13)$$

electrophile   nucleophile
(Lewis acid)   (Lewis base)

Example 7.1

The reaction between a Lewis base and a Lewis acid generates a new covalent bond. Write a curly arrow description of the reaction between diethyl ether (Et_2O) and BF_3.

Solution

The ether is a Lewis base (nucleophile) and the boron trifluoride is a Lewis acid (electrophile) in this reaction. Note that both reactants are neutral so, although charge separation (polarity) is developed in the *adduct*, it must be neutral overall.

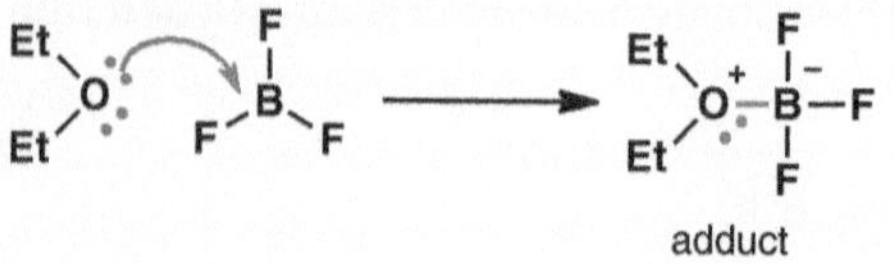

adduct

As noted above, a curly arrow indicates movement of an electron pair; it does *not* represent mutual approach of reactants. Consequently, protonation of ethanal (acetaldehyde) on the oxygen, for example, should be represented by eqn 7.14 and *not* by eqn 7.15!

Correct representation:

(7.14)

Wrong representation:

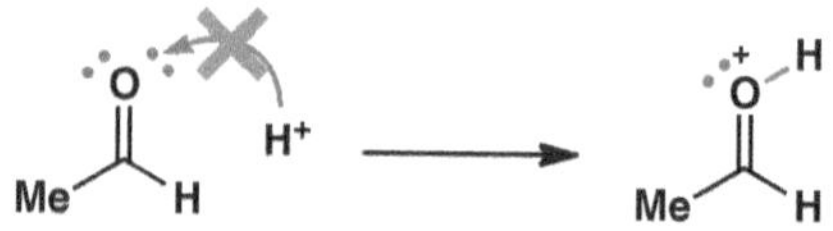

(7.15)

Reactions involving heterolysis and its reverse are called **polar** (or sometimes **ionic**) **reactions**; most organic reactions in solution, but very few in the gas phase, are polar. Reactions not involving heterolysis or its reverse, e.g. radical reactions, are sometimes called **nonpolar reactions**; most gas-phase reactions are nonpolar.

Exercise 7.2

Mark all the unshared electron pairs in reactants and products in the following reactions, and represent movements of electron pairs by curly arrows.

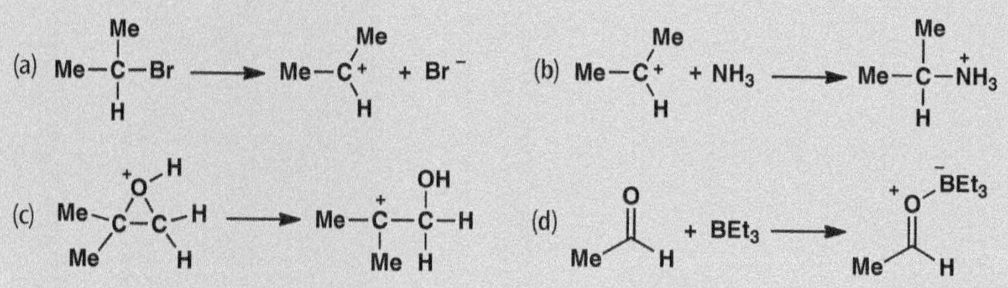

Exercise 7.3

Identify nucleophiles, electrophiles, and nucleofuges in the following reactions.

(a) $CH_3Cl + C_2H_5O^- \longrightarrow CH_3OC_2H_5 + Cl^-$ (b) $(CH_3)_3CBr \longrightarrow (CH_3)_3C^+ + Br^-$

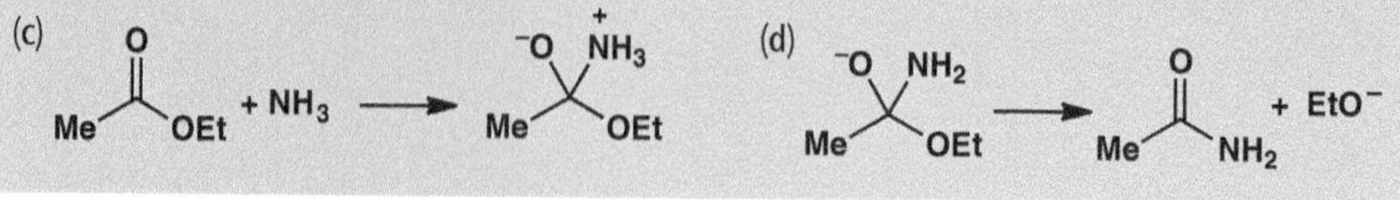

7.2.3 Concerted bond formation and cleavage in an elementary reaction

Homolysis or heterolysis of a single σ bond is the simplest type of elementary step in a chemical reaction but either type alone seldom constitutes a whole chemical reaction. The next level of mechanistic complexity is when several bonds are made and/or broken together in a **concerted** single step. Many overall chemical reactions in solution occur by concerted single-step mechanisms in which the individual bond-making and/or bond-breaking components are heterolytic. The proton transfer reactions we encountered in the previous chapter are simple examples, e.g. eqn 7.16.

$$H_2O \ + \ H\text{--}Cl \ \longrightarrow \ H_3O^+ \ + \ Cl^- \tag{7.16}$$

In this reaction, an unshared electron pair on the base makes a new bond to the H of the acid, and *at the same time*, the old bond between the H and the Cl breaks heterolytically to leave a new unshared electron pair on the conjugate base. Consequently, two curly arrows are needed to show bond-making and bond-breaking and, as a result, H⁺ transfers from the acid to the base. The electronic reorganization associated with this overall transformation is a single concerted process and, as both aspects are heterolytic, this is a polar reaction. The *mechanism* is shown using curly arrows in Scheme 7.1.

When a mechanism requires multiple curly arrows to represent the concerted movement of electron pairs as in Schemes 7.1 and 7.3 (below), they always point in the same direction. This is because the electron flow they describe is always in a single direction starting from a site of high electron density. Unlike singly barbed curly arrows, doubly barbed ones never appear head to head (⤳⤶) or tail to tail (⤳⤳)—compare eqn 7.11 and Scheme 7.1.

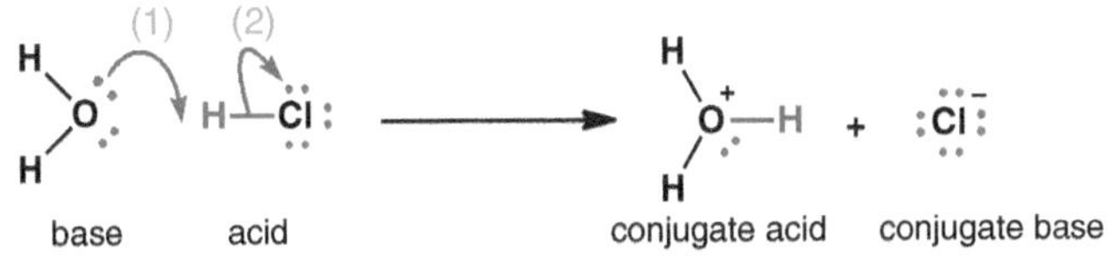

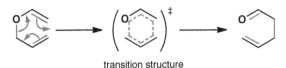

Scheme 7.1 Dissociation of HCl in water, an example of how to use curly arrows to represent a concerted, single-step, polar reaction. Curly arrow 1 starts at a lone pair on the O and ends between the O and the H; curly arrow 2 starts at the bond between H and Cl, and ends above the Cl.

A second type of elementary reaction comprising concerted bond-breaking and bond-making involves neither homolysis nor heterolysis as described above. These are called **pericyclic reactions** and the rearrangement of eqn 7.10 given earlier is an example. Its mechanism is now shown in Scheme 7.2; because this mechanism does not involve heterolysis, it is a *nonpolar* concerted reaction. Further examples will be discussed in Section 15.9 (cycloaddition reactions) and Chapter 21 (sigmatropic and electrocyclic reactions).

Scheme 7.2 Concerted pericyclic rearrangement of allyl vinyl ether.

transition structure

Note that the curly arrows in the concerted mechanism in Scheme 7.2 do not have the same significance as, for example, in Scheme 7.1. In Scheme 7.1, each curly arrow represents the movement of a specific pair of electrons. In Scheme 7.2, they simply identify the electrons involved in the transformation and help us to keep track of them; we shall use curly arrows in this way to describe mechanisms of other concerted pericyclic reactions in Chapter 21. We could write the curly arrows in Scheme 7.2 pointing anticlockwise if we want—it would make no difference; we could not point the two curly arrows describing the proton transfer in Scheme 7.1 in the opposite direction! Curly arrows are used differently again to correlate hypothetical contributors to a resonance hybrid (see also Sub-sections 7.2.2 and 5.4.2).

7.2.4 The transition structure in a concerted elementary reaction

Note that the transition structure (TS) in the pericyclic reaction in Scheme 7.2 has the aromatic character of a cyclic 6-electron system (see Chapter 21).

In the concerted proton transfer shown above in Scheme 7.1, there will be a stage at which the H$^+$ is partially bonded to the H_2O molecule and partially to the Cl$^-$ ion. The highly unstable species at this stage is called a **transition structure** and it is usually represented within round brackets and identified by the double dagger superscript (Figure 7.1).

The mechanism of the substitution reaction of bromomethane with hydroxide ion to give methanol (eqn 7.2) is shown in Scheme 7.3; it is a typical one-step, **bimolecular**, polar reaction. This so-called S$_N$2 mechanism will be discussed further in Chapter 12.

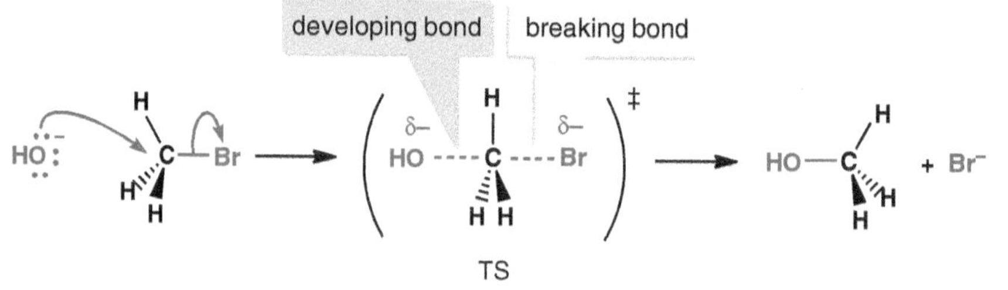

Scheme 7.3 The concerted S$_N$2 reaction of bromomethane with hydroxide anion.

The elementary steps in eqn 7.12 and Scheme 7.2 are described as **unimolecular** since each involves a single reacting molecule. In contrast, the elementary step in Scheme 7.3 in which the TS is derived from two chemical species (in this case a molecule and an ion) is **bimolecular**. The rebonding which leads to the formation of the transition structure in a bimolecular step takes place within an **encounter complex** formed by a collision between the reactants and transiently held together by weak intermolecular interactions. See Sub-section 7.4.1 for a discussion of transition structures in unimolecular bond-cleavage elementary reactions. Because *trimolecular* collisions are so infrequent, they contribute insignificantly to chemical reactions.

The carbon of the bromomethane is electrically slightly positive owing to the polarity of the C–Br bond. The electrostatic interaction between this and the nucleophile is important in bringing the reactants together in the initial stage of the reaction to give the *encounter complex*. Rebonding then takes place with electron flow from the HO$^-$ anion to the partially positive carbon to form the O–C bond as the C–Br bond is broken heterolytically. Here, the transition structure has *partial* bonds between O and C, and between C and Br.

Exercise 7.4

The elimination reaction given in eqn 7.8 is an example of a unimolecular single-step reaction. Give a curly arrow mechanism and draw a transition structure which shows that it involves the cyclic reorganization of 6 electrons.

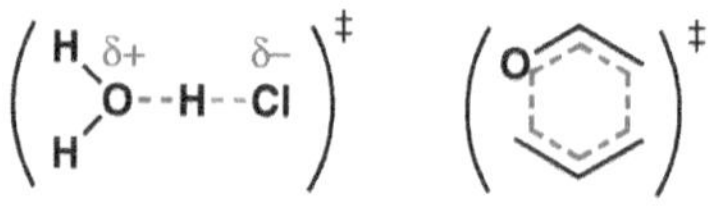

Figure 7.1 Polar and nonpolar transition structures in the reactions in Scheme 7.1 and 7.2.

7.2.5 Site of nucleophilic attack at a cationic electrophile

Since electrons flow from an electron-rich site in the nucleophile to an electron-poor site in the electrophile, it is natural to think that nucleophilic attack at a cationic electrophile should always be at the atom formally bearing the positive charge. This can only happen, however, if the atom at the cationic centre can accommodate the additional electron pair in its valence shell. This happens in the reaction of eqn 7.13 where the cationic centre of the carbenium ion has only six valence electrons.

However, if a nucleophile were to add to the positive nitrogen of an ammonium ion, as in eqn 7.17, an impossible electronic configuration would be obtained: *a second row element cannot have more than an octet of valence electrons* (see Sub-section 1.3.1).

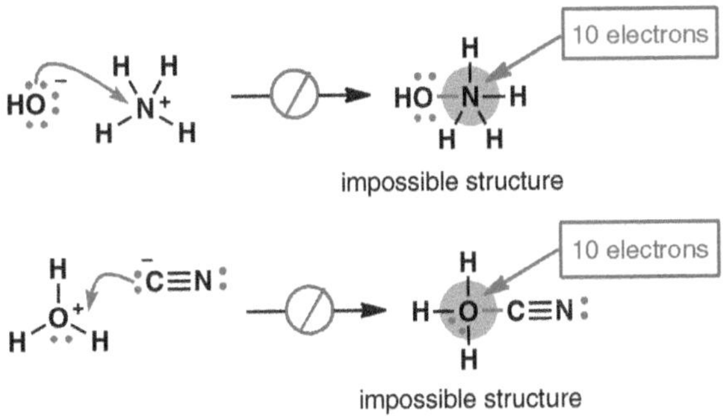

impossible structure

(7.17)

impossible structure

(7.18)

Scheme 7.4 Impossible reactions which generate species containing atoms with more valence electrons than they can accommodate.

Similarly, the nucleophilic addition reaction of eqn 7.18 is forbidden since it requires the oxygen to accommodate 10 valence electrons. This kind of mistake may be avoided by showing all unshared electron pairs on *heteroatoms*.

The reactions which actually take place between the reactants in Scheme 7.4 are shown in Scheme 7.5. These reactions are further examples of acid–base reactions where two curly arrows are needed to show concerted bond-breaking and bond-making and, as a result, H+ transfers from the acid to the base.

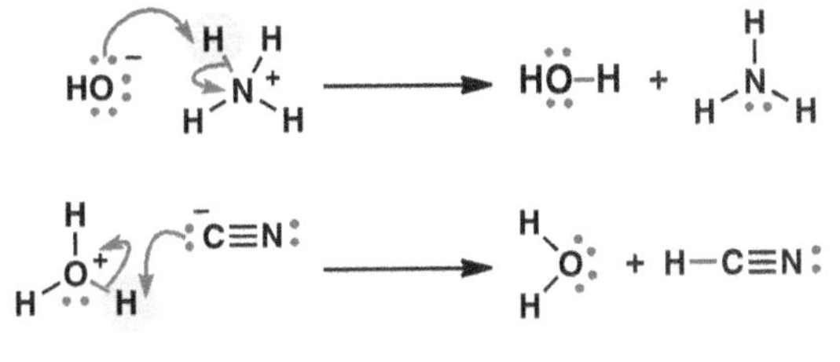

Heteroatoms are atoms other than C and H in organic molecules, e.g. O, N, and halogens.

Scheme 7.5 Proton transfer from positively charged acids.

The reactions of Scheme 7.5 are the usual ones between a nucleophile/base and a molecule protonated on a heteroatom with a complete valence shell. In general, nucleophilic attack occurs at the atom *next to a* positive heteroatom with a complete valence shell.

Correspondingly, a nucleophile reacts with a protonated carbonyl group not at the positive oxygen atom but at the carbon next to it: a new σ bond to the C is made as the π electrons become a lone pair on the O (eqn 7.19).

(7.19)

Example 7.2

In the substitution reaction of Scheme 7.3 shown earlier, the two curly arrows indicate that bond formation and cleavage are concerted; in other words, the substitution proceeds in one step via a single TS and not in two steps via an anionic intermediate. Why is the two-step mechanism shown below not possible?

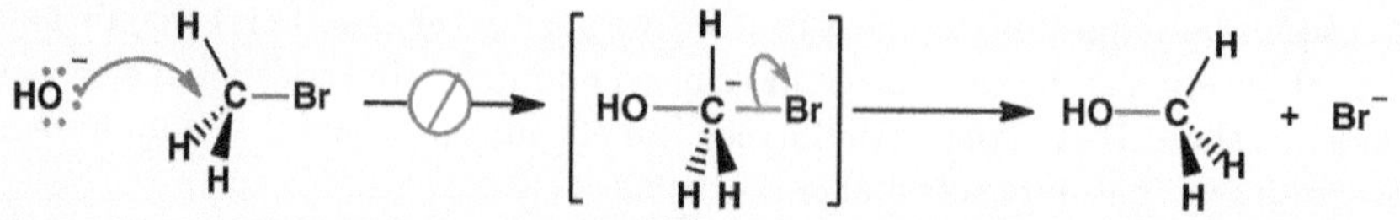

Solution
The carbon of the putative anionic intermediate is penta-coordinated and consequently has more than an octet of valence electrons; this is an impossible structure. If a nucleophile bonds to an atom which already has a complete valence shell, another bond to it must be broken heterolytically.

Exercise 7.5

An alternative reaction to the one in eqn 7.19 between the protonated carbonyl and EtOH may occur at the H on the positive O atom. Represent the reaction with curly arrows.

Exercise 7.6

Write curly arrow mechanisms for two possible reactions of the following combination of a cation and an anion.

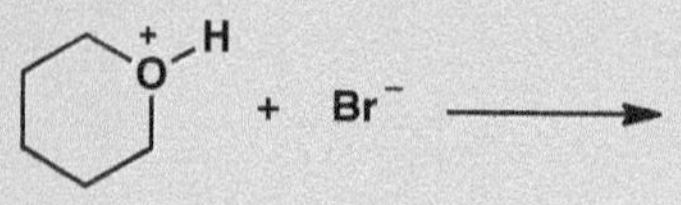

7.2.6 Sigma and pi bonds as nucleophilic centres

The nucleophilic centres we have encountered so far in neutral molecules and anions have been unshared electron pairs, and the tail of a curly arrow representing nucleophilic attack begins at the lone pair. This need not always be so. The reduction of carbonyl groups by anions such as BH_4^- and AlH_4^-, exemplified in eqn 7.20, is a well-known reaction (see Chapter 10). In eqn 7.20, the electron pair of a B–H σ bond is the nucleophilic centre, and the hydrogen transfers from B to C with its bonding electron pair. The reaction is formally a hydride (H^-) transfer and the electron pair of the old B–H bond becomes the shared electron pair of the new C–H bond; consequently, the curly arrow in the mechanism starts at the B–H bond.

Note that the central B in BH_4^- in eqn 7.20 carries a formal negative charge but no atom in the anion has a lone pair, and the BH_3 goes on to react further (see Chapter 10).

(7.20)

In other reactions, π bonds may be nucleophilic. The addition in eqn 7.5 at the beginning of this chapter is shown as a two-step mechanism in Scheme 7.6 below. In the first step of this reaction, the alkene acts as a π nucleophile—it uses a π electron pair to make a new σ bond to the H of the electrophile (eqn 7.21). The second step where the electrophilic carbenium ion is captured by the nucleophilic Cl⁻ is the reaction given earlier in eqn 7.13.

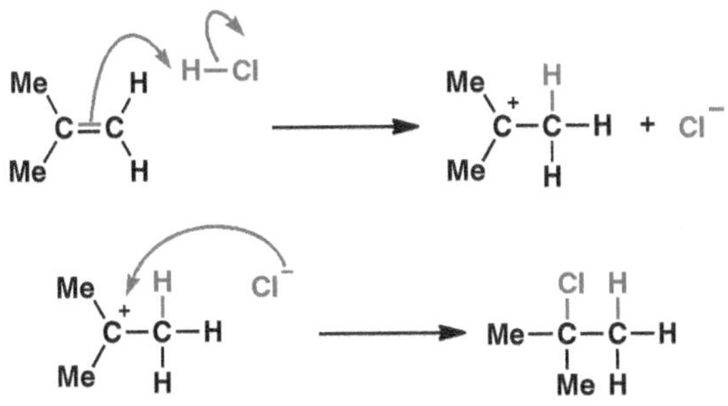

$$(7.21)$$

Scheme 7.6 Two-step mechanism for addition of HCl to an alkene.

$$(7.13)$$

Note that the alkene in Scheme 7.6 is unsymmetrical so it may appear that the addition of HCl could have been the other way round to give $Me_2CH–CH_2Cl$. For reasons to be discussed in Chapter 15, the initial electrophilic attack is exclusively as indicated in eqn 7.21, and this controls which product is formed. The curly arrow in eqn 7.21 also illustrates the convention we shall follow when, in principle, an electrophile could bond to either end of a nucleophilic double bond: *bond formation is to the end of the double bond on the **concave** side of the curly arrow.*

Exercise 7.7

Reaction 7.5 given earlier may be carried out using concentrated hydrochloric acid in which the HCl is almost wholly dissociated. Use curly arrows to describe the initial electrophilic reaction of the H_3O^+ with the alkene.

Exercise 7.8

The elimination reaction illustrated in eqn 7.7 at the beginning of the chapter is the reverse of reaction 7.5. Give a two-step mechanism for this reaction in aqueous solution. Note that the first step is the same as the first step in the mechanism of the reaction in eqn 7.3.

In Scheme 7.6, lone pairs are omitted, and the curly arrow starts at the minus sign of the charge of the chloride anion in eqn 7.13. Since the negative charge here corresponds to an unshared electron pair, the arrow can start from the minus sign and there is no need to include pairs of dots for electron pairs. However, remember that some nucleophilic anions such as BH_4^- in eqn 7.20 do not have a lone pair.

7.3 A Molecular Orbital Description of Polar Elementary Reactions

7.3.1 Orbital interactions in bimolecular elementary reactions

We saw in Chapter 2 that a covalent bond between two atoms is formed by the sharing of a pair of electrons in a molecular orbital (MO) formed by the overlap of atomic orbitals (AOs); in the case of two H atoms, it is by the overlap (interaction) of singly occupied 1s AOs (Sub-section 2.2.2). Bond-making between two polyatomic groups (radicals, ions, or neutral molecules) occurs in the same way, but by interactions between more complicated molecular orbitals.

Each interaction between two orbitals of compatible symmetry results in two new orbitals; one (the bonding orbital) is lower in energy than either of the originals and the other (the antibonding orbital) is higher in energy. The interaction between two singly occupied MOs (SOMOs) is illustrated in Figure 7.2(a), and corresponds to bonding between two radicals. Interaction between a doubly occupied MO (DOMO) and an unoccupied MO (UMO) is illustrated in Figure 7.2(b), and corresponds to a bimolecular

Figure 7.2 Interactions between two orbitals: (a) SOMO–SOMO, (b) DOMO–UMO, (c) DOMO–DOMO, and (d) UMO–UMO.

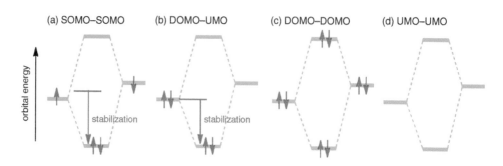

reaction between a nucleophile and an electrophile. In both of these reactions, nonpolar (Figure 7.2(a)) and polar (Figure 7.2(b)), there are just two electrons from the original interacting orbitals and these are accommodated in the new (bonding) orbital of lower energy. Consequently, the overall electronic energy following bond formation is lower in both cases: the bonded states are more stable than the unbonded states.

However, if there is an interaction between two DOMOs (Figure 7.2(c)) with four electrons to be accommodated, the new bonding and antibonding orbitals would both be filled; this, however, would not lead to an overall lowering of energy. Similarly, interaction between two UMOs (Figure 7.2(d)) would also have no effect on the electronic energy of the system because, this time, no electrons are involved. Consequently, interaction between two DOMOs, or between two UMOs, does not lead to bond formation.

In a bimolecular polar reaction, therefore, the important interaction is between a DOMO of the nucleophile (Nu or Nu⁻) and a UMO of the electrophile (E or E⁺) to give a pair of new orbitals, and the original unshared electron pair in the MO of the nucleophile becomes shared with the electrophile and accommodated in the new bonding orbital.

7.3.2 HOMO–LUMO interactions

There are constraints on which orbitals can interact favourably and lead to bond formation as shown in Figure 7.2 above. First, the interacting orbitals must be of compatible symmetries. It is possible to have end-to-end overlap between orbitals with s character (e.g. sp, sp^2, and sp^3 orbitals), and side-to-side overlap between orbitals with p character (e.g. p and π orbitals); but there cannot be favourable end-to-side overlap between, for example, an sp^3 orbital and a p orbital. Also, the closer two orbitals of compatible symmetries are in energy, the better the interaction between them. This is illustrated in Figure 7.3 where we see that the system with the smaller energy gap between the interacting orbitals on Nu⁻ and E⁺ has the greater stabilization upon bonding.

We also indicate in Figure 7.3 that the occupied orbital of Nu⁻ is lower in energy than the unoccupied orbital of E⁺. Consequently, any feature which raises the energy of the former, or lowers the energy of the latter, will reduce the energy gap between them. This will enhance the orbital interaction and lead to a stronger bond.

Figure 7.3 includes interactions between only two orbitals and shows the two new orbitals which are formed; however, both the nucleophile and electrophile will have other orbitals (occupied and unoccupied). In Figure 7.4, the bonding interaction is between the **highest occupied molecular orbital (HOMO)** of the nucleophile and the **lowest unoccupied molecular orbital (LUMO)** of the electrophile as these are the closest in energy. In general, the new orbitals corresponding to the formation of bonds in polar reactions arise from **HOMO–LUMO interactions** (Figure 7.4) or, at least, they are the most significant. The HOMO and LUMO orbitals which principally determine organic reactivity are usually called *frontier orbitals*.

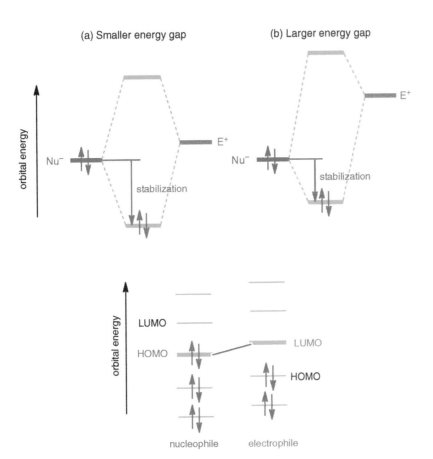

Figure 7.3 Orbital interactions between a nucleophile and an electrophile.

Figure 7.4 HOMO–LUMO interaction between a nucleophile and an electrophile.

In order to avoid an overcomplicated diagram, the bonding and antibonding orbitals derived from the HOMO–LUMO interaction in Figure 7.4 are not shown as they are in Figure 7.3.

7.3.3 Orbital overlap and orientation

A principal factor which determines the quality of an interaction between orbitals when a bond is being formed is the extent of their overlap. Except for s AOs (which are spherical), orbitals are directional. It follows that the proper relative orientation of the two reactants with alignment of relevant orbitals of compatible symmetries is crucial for effective orbital overlap. For example, in reaction 7.13, the 3p orbital (HOMO) of Cl^- containing an unshared electron pair must be aligned for end-to-end overlap with one lobe of the unoccupied 2p orbital (LUMO) of the carbenium ion (Figure 7.5). Bond formation involves transformation of these two orbitals into the (occupied σ) bonding and (unoccupied σ^*) antibonding orbitals between the Cl and the (newly sp^3-hybridized) C.

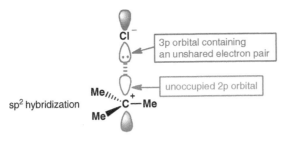

Figure 7.5 Orbital alignment for optimal interaction leading to bond formation between a chloride ion and a carbenium ion.

In a concerted reaction, bonds are broken as others are being formed. Although the orbital implications for such reactions are more involved, the principles are just the same. Reaction 7.2 occurs in one step as illustrated in Scheme 7.3 above: the nucleophile attacks the bromomethane from the side opposite the bromine and collinear with it. This direction of the nucleophilic approach allows interaction between the lone pair orbital on the O (the HOMO of the HO⁻) and the vacant antibonding σ* orbital of the C–Br bond (the LUMO of the CH_3Br), as will be discussed in Section 12.2.3.

7.4 Reaction Energetics, Reaction Profiles, and Equilibria

7.4.1 Energy change for a one-step reaction of a single molecule

In a one-step reaction such as Scheme 7.2, some bonds break (which costs energy) and others form (which releases energy) as a reactant molecule is transformed into a product molecule. If the sum of the bond energies is greater for the product than for the reactant, the overall reaction releases energy (usually as heat). However, there is a barrier which needs to be overcome for the reaction to proceed. This is illustrated in the **reaction profile** shown in Figure 7.6. The transient molecular species at the energy maximum, which is neither a reactant nor a product molecule but has some features of both, is the transition structure (TS) which we encountered in Section 7.2.4. Its lifetime is on the femtosecond time-scale (1 fs $=10^{-15}$ s)—the time-scale of molecular vibrations.

There are two ways in which the energy required to overcome the barrier may be provided. One is by collisions with other molecules and, since the energy transferred in intermolecular collisions is greater at higher molecular velocities, the barrier will be more easily overcome at higher temperatures. Such reactions are sometimes called *thermal reactions.* The other is by irradiation when the reaction takes place through a photoexcited state in a so-called *photochemical reaction* (see Section 5.7).

7.4.2 From reaction of a single molecule to reaction on a molar scale

Figure 7.6 represents a single molecule reacting via the transition structure (TS) to give a product molecule. The horizontal axis, the **reaction coordinate**, is a composite of the molecular structural parameters which change as the reactant molecule is transformed into the product molecule. The vertical axis is the energy of the molecular species defined

Figure 7.6 Molecular energy profile for a one-step reaction.

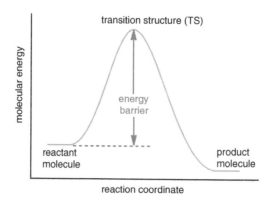

Panel 7.1 Reaction profiles for unimolecular bond-cleavage elementary reactions

---

Cleavage of a single covalent bond can be a step in different kinds of multi-step reactions, e.g. homolysis of Br_2 (eqn 7.11) in the gas-phase bromination of alkanes, and the heterolysis of *t*-butyl chloride (eqn 7.12) in its solvolysis. They involve separation of atoms, radicals, or ions following a stretching vibration of increasing amplitude when the molecule is sufficiently energized.

The gas-phase homolysis of a diatomic molecule, e.g. Br_2, is best described by a so-called *Morse curve*, a vibrational potential energy diagram with increasing anharmonicity as the amplitude of the stretching vibration increases, and the molecule finally dissociates as shown in (a) below. There is no transition structure in this process, and no minimum in the potential energy diagram beyond the one for the bonded molecule; the dissociated atoms simply recombine with no barrier to overcome, or otherwise react if they encounter different reaction partners. In principle, a heteronuclear diatomic molecule, A–B, could undergo homolytic or heterolytic dissociation and, superficially, the molecular potential energy curves would look similar. However, the energy required for heterolysis in the gas phase is much greater than for homolysis, and heterolysis in the gas phase does not occur.

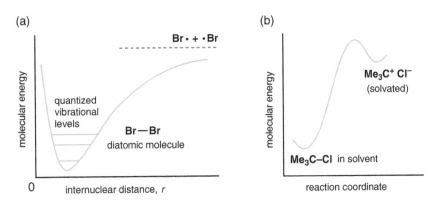

Reaction profiles for (a) the gas-phase homolysis of Br_2, and (b) the solvolytic heterolysis of *t*-Bu–Cl.

The dissociation of a more complex molecule involves more than just the vibrational stretching of a σ bond followed by its cleavage because other bond lengths (and bond angles) within the reactant molecule may change as it dissociates. For example, the bond angles around the quaternary C atoms in $Me_3C–CMe_3$ are approximately tetrahedral whereas they become trigonal in the *t*-butyl radicals, $Me_3C\cdot$, which are obtained upon homolysis of the central bond.

For reactions in solution where heterolysis becomes feasible, solvation of the reactant and of the fragments will be different and the enhanced solvation of ions is strongly stabilizing. Consequently, although energy is required initially to stretch the bond in the formation of the TS, solvent reorganization stabilizes the developing polarity which leads to the formation of a solvated intermediate ion pair. For example, the TS for the heterolysis of $Me_3C–Cl$ (eqn 7.12) in solution has a partial bond between the C and the Cl, and developing solvation of the partial charges on the C and the Cl (see Scheme 7.10 later in the chapter). The energy of the system then decreases as the bond cleavage becomes complete and the ions become fully solvated. In the molecular reaction profile describing this solvolytic heterolysis, there is a single barrier corresponding to the TS between the $Me_3C–Cl$ molecule and the intermediate ion pair, as shown in (b).

---

by its position along the horizontal axis; this is sometimes called the molecular *potential* energy since it includes the vibrational and rotational energy of the molecule, but not its *translational* kinetic energy. But in practice (and organic chemists are practical people), we deal with matter on the macroscopic scale—amounts of material which we measure out in grammes or moles. Consequently, it is sometimes more appropriate to represent our organic reaction by a *molar enthalpy* profile rather than a *molecular energy* profile.

The molar enthalpy of a compound is directly related to the *total* molecular energy using the Avogadro constant. Consequently, the enthalpies of the initial and final states* present no conceptual difficulties, and are properties of real compounds under specified

*These are the thermodynamic states corresponding to one mole of reactant molecules and one mole of product molecules, respectively, under the conditions of the experiment.

Figure 7.7 Enthalpy profiles for (a) exothermic and (b) endothermic reactions.

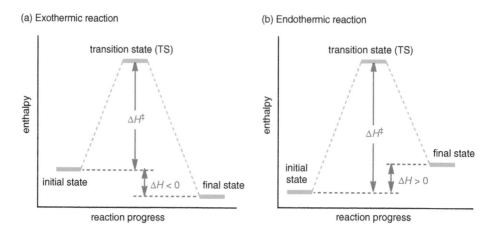

experimental conditions. The difference between the enthalpies of the initial and final states is the enthalpy (or heat) of reaction (ΔH); see Figure 7.7. When ΔH is negative, the reaction is **exothermic** (heat is given out); when ΔH is positive, the reaction is **endothermic** (heat is absorbed from the surroundings).

Correspondingly, the difference in enthalpy between initial state and **transition state** is the enthalpy of activation ($\Delta H^{\ddagger}$) which is often called the *activation energy* (and can be determined experimentally).

Note that we use TS as an abbreviation for transition state and transition structure—the context will indicate whether we mean the one or the other—and there is no inconsistency in referring to transition structures, i.e. the molecular species, within the transition state.

The *transition state* in the molar enthalpy profile corresponds to the *transition structure* in the molecular energy profile. However, it exists only in our imagination—it is a hypothetical construct which is very useful as we shall see later. It is not real because we can never have one mole of transition structures—a reaction does not take place with all molecules reacting together in unison! In reality, when the reaction is, say, 50% complete, we have essentially 50% starting material and 50% product with an infinitesimally small proportion of molecules in the process of actually reacting. It follows that there is no reaction coordinate for a reaction on the molar scale. Consequently, the meaningful continuous line we see in the molecular energy profile of a single-step reaction is replaced in the enthalpy profile by broken lines connecting the initial state, transition state, and final state. However, the broken lines represent nothing more than the hypothetical progression of one state into the next.

7.4.3 Gibbs energy reaction profiles

As will be familiar from basic physical chemistry, the Gibbs energy change (ΔG) for a process is related to the changes in enthalpy (ΔH) and entropy (ΔS) by $\Delta G = \Delta H - T\Delta S$. Gibbs energy is sometimes called Gibbs *free* energy or simply *free energy*. When ΔG is negative (an **exergonic** reaction), the reaction is thermodynamically favourable; when ΔG is positive (an **endergonic** reaction), it is thermodynamically unfavourable.

We saw above how a single-step mechanism can be described either by a molecular energy profile involving a transition structure or by a molar enthalpy profile. The latter describes conversion of the (real) initial state to the (real) final state via a single (imagined) transition state. Since molar quantities of material allow us to introduce the concept of entropy, we can extend our imagination and create a *Gibbs energy reaction profile* for a reaction.

As an example, Figure 7.8 shows the Gibbs energy profile corresponding to the one-step molecular energy profile shown in Figure 7.6. This could represent the concerted unimolecular mechanism shown in Scheme 7.2 for the isomerization of eqn. 7.10. The Gibbs energies of the initial and final states in Figure 7.8 are properties of real compounds (one mole of an unsaturated ether and one mole of an unsaturated aldehyde) under the conditions of the experiment, and the difference between them is the Gibbs energy of reaction (ΔG). If a reaction is reversible, the Gibbs energy of reaction is related to the *equilibrium constant*, K, which can be determined experimentally (see Sub-section 7.4.5).

In contrast, the Gibbs energy of the transition state (TS) in Figure 7.8 is a property of an imaginary one mole of the transition structures depicted in Scheme 7.2. Nevertheless,

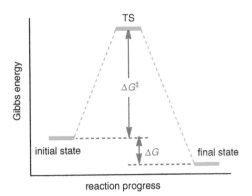

Figure 7.8 Gibbs energy profile of a single-step exergonic reaction.

*The relationship between the rate constant and the Gibbs energy of activation is expressed in the Eyring equation: $k = (k_B T/h) \exp(-\Delta G^{\ddagger}/RT)$, where T is the absolute temperature (K), and k_B, h, and R are the Boltzmann constant, Planck constant, and the gas constant. If k has been measured experimentally, the value of $\Delta G^{\ddagger}$ can be easily calculated using this equation.

the Gibbs energy difference between the initial state and the hypothetical transition state (the Gibbs energy of activation, $\Delta G^{\ddagger}$) is a very useful concept because it is directly related to the rate constant (k) of the reaction.*

7.4.4 Profiles of multistep reactions

The replacement of the Cl in 2-chloro-2-methylpropane (*t*-butyl chloride, see eqn 7.3 and Sub-section 7.2.2) by OH takes place by the multistep mechanism in aqueous solution illustrated in Scheme 7.7; this is the S_N1 mechanism which will be discussed further in Chapter 12. In the first step, the C–Cl bond undergoes heterolysis to give a carbenium ion, the *t*-butyl cation, and chloride anion (eqn 7.12 seen earlier). The short-lived carbenium ion intermediate reacts rapidly with a water molecule to give the protonated alcohol (eqn 7.22). The reaction is completed by a third step, an acid-base reaction between the protonated alcohol and another water molecule to form the neutral alcohol and H_3O^+ (eqn 7.23). For an overall reaction of this complexity, we do not normally attempt to give a single *molecular* reaction profile, but we can give a composite *molar* reaction profile for the overall transformation, as shown in Figure 7.9.

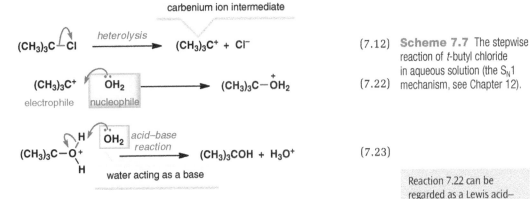

(7.12)

(7.22)

(7.23)

Scheme 7.7 The stepwise reaction of *t*-butyl chloride in aqueous solution (the S_N1 mechanism, see Chapter 12).

Reaction 7.22 can be regarded as a Lewis acid–base reaction.

The Gibbs energy reaction profile in Figure 7.9 includes sequential maxima (TS_1, TS_2, and TS_3) and minima. The molecular species corresponding to the maxima are transition structures, and those corresponding to the minima are intermediates. The distinction between an intermediate and a transition structure is crucial. The lifetimes of intermediates cover a wide range; some are extremely short whereas others are sufficiently long

Figure 7.9 Molar reaction profile for the three-step mechanism of Scheme 7.7.

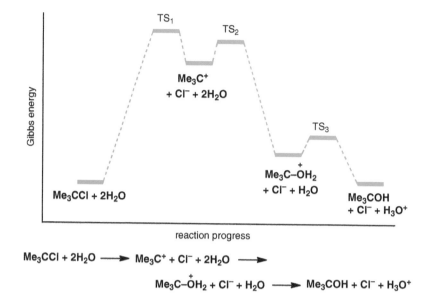

$$Me_3CCl + 2H_2O \longrightarrow Me_3C^+ + Cl^- + 2H_2O \longrightarrow$$

$$Me_3C\overset{+}{-}OH_2 + Cl^- + H_2O \longrightarrow Me_3COH + Cl^- + H_3O^+$$

that the intermediate can be detected spectroscopically or even isolated. For example, the *t*-butyl carbocation in the reaction above is very reactive and can only be observed under extremely non-nucleophilic conditions at low temperatures whereas the protonated alcohol corresponding to the second minimum in Figure 7.9 is not so unstable. Transition structures, however, are always extremely short-lived (see Sub-sections 7.2.4 and 7.4.1).

In a multistep reaction, all intermediate steps generally have an effect on the rate of the overall process. However, when the Gibbs energy barrier corresponding to one step is appreciably higher than the others, that step is dominant and often called the **rate-determining** (or rate-limiting) **step**. In the profile describing the mechanism of the hydrolysis of *t*-butyl chloride (Figure 7.9), we see that the initial step (the heterolysis) is rate determining. When this is the case, the elementary rate constant of the initial step is equal to the rate constant of the overall reaction.

Example 7.3

Draw the reaction profile for an endergonic two-step reaction with a rate-determining second step.

Solution

The final state in the Gibbs energy profile is higher than the initial state, and TS$_2$ is higher than TS$_1$.

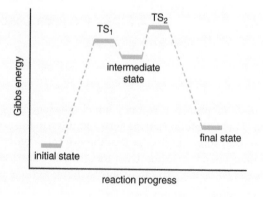

Note that the forward barrier from the intermediate over TS_2 is much smaller than the barrier from the initial state over TS_1 to the intermediate. This means that the elementary rate constant for the second step (k_2) is *much larger* than the one for the initial step (k_1) even though the second step is rate determining. This paradox is understood when we realize that, during most of the reaction, the *rate* of the second step $(= k_2[\text{intermediate}])$ is smaller than the *rate* of the first $(= k_1[\text{reactant}])$ because [intermediate] is very much smaller than [reactant]. Note also that the intermediate returns to reactant faster than it proceeds to product (the barrier back over TS_1 is lower than the barrier forward over TS_2, i.e. $k_{-1} > k_2$).

7.4.5 Equilibrium constant

We saw in Chapter 6 that proton transfer in acid-base reactions is usually reversible and, in Section 7.1 of this chapter, that addition is the reverse of elimination. In fact, very many organic reactions are reversible and, when a reversible system achieves equilibrium, the proportions of reactants and products are described by an experimental *equilibrium constant*, K. For a generic reaction between reactants A and B (left hand side of the chemical equation) which give products C and D (right hand side),

$$A + B \xrightleftharpoons{K} C + D$$

the relationship between the molar concentrations of the reactants and products at equilibrium is described by

$$K = \frac{[C]\,[D]}{[A]\,[B]}$$

The value of K for a particular reaction will depend upon the experimental conditions, e.g. the temperature and the solvent for a reaction in solution.

We also saw in Sub-section 7.4.3 that the value of K for a reversible reaction is related to its Gibbs energy, and the relationship for a reaction at temperature T is given by

$$K = \exp\left(-\Delta G^{\circ}/RT\right) \quad \text{or} \quad \Delta G^{\circ} = -RT \ln K,$$

where ΔG° is the *standard* Gibbs energy of the reaction (the Gibbs energy of one mole of reaction, reactants and products being in their *standard states*—normally 1 mol dm^{-3} for solution reactions in a specified solvent under 1 standard atmosphere pressure).

It follows from the above equation that K is greater than unity $(K>1)$ when ΔG° is negative (an exergonic reaction), and K is smaller than unity $(K<1)$ when ΔG° is positive (an endergonic reaction), as illustrated in Figure 7.10. In other words, when the Gibbs energy of the final state (the products) is lower than that of the initial state (starting materials), equilibrium favours the product side. On the other hand, if ΔG° is positive, the products are less stable than the starting materials, the equilibrium favours starting materials, and the reaction proceeds to only a limited extent. Sometimes, a reversible organic reaction at equilibrium appears to be entirely the product(s) $(K \gg 1)$ and looks like an irreversible reaction.

An exergonic reaction (negative ΔG) may be described as *thermodynamically favourable*, but this information alone gives no indication about whether it has a high or a low rate constant, which is determined by the reaction's Gibbs energy of activation, $\Delta G^{\ddagger}$ (see Sub-section 7.4.3). If an element or compound can undergo a strongly exergonic reaction, but does not react because $\Delta G^{\ddagger}$ is very large, it is said to be *thermodynamically unstable* but *kinetically stable*. Most organic compounds in the presence of atmospheric oxygen are thermodynamically unstable with respect to oxidation (combustion) but are kinetically stable in the absence of an ignition. Carbon itself (either impure in the form of coal or pure, e.g. as diamonds) is thermodynamically unstable in the presence of oxygen, but kinetically stable.

Figure 7.10 Gibbs energy changes for equilibria with single step mechanisms.

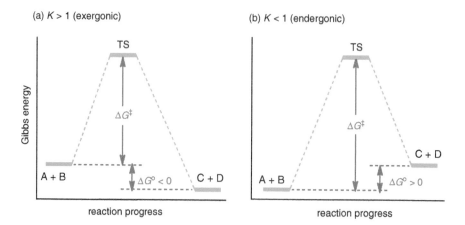

(a) $K > 1$ (exergonic)

(b) $K < 1$ (endergonic)

Formation of the cyanohydrin from HCN and the carbonyl compound shown earlier in eqn 7.6 is a typical reversible reaction, details of which will be discussed in Chapter 8. The equilibrium constant for this type of reaction depends principally upon the structure of the carbonyl compound. Generally, $K > 1$ for aldehydes (e.g. reaction 7.24) and $K < 1$ for ketones (e.g. reaction 7.25). This indicates that good yields of aldehyde cyanohydrins may be obtained at equilibrium whereas only modest or poor conversions are obtained from ketones. However, this information tells us nothing about how long the reactions take to reach equilibrium; see Chapter 8 for further discussion.

$$K = 200 \qquad (7.24)$$

an aldehyde   a cyanohydrin

$$K = 0.8 \qquad (7.25)$$

a ketone   a cyanohydrin

Note that in practical matters relating to equilibria, e.g. the determination of K by measuring yields at equilibrium, the mechanism of the reaction is seldom relevant—Figure 7.10 could have included multistep mechanisms. However, the value of the equilibrium constant for a reaction in solution may be affected by the nature of the solvent.

7.5 Characterization of Organic Reactions and Investigation of their Mechanisms

Characterization of a reaction of an organic compound is seldom achieved by just a couple of experiments, nor is it ever complete in the sense that everything about a reaction becomes reliably known. Knowledge about a chemical reaction is usually accumulated bit by bit, and new evidence could lead to a refinement or perhaps even a major revision of our mechanistic understanding of it.

7.5.1 Product studies and mechanistic proposals

The first steps in the investigation of a reaction usually include identification of the product(s), and then quantitative product analysis. For these purposes, we use

Panel 7.2 The Hammond postulate

In the 1950s, the relationship between structure and reactivity of organic compounds was a major concern of organic chemists. Central to the debate was the nature of transition structures: how could one probe the structure and properties of a species which, according to Transition State Theory, has such a fleeting existence?

In a two-step reaction, the reactive intermediate is flanked by two transition structures (see figure below). If we focus on the highly endothermic first step, we see that the *energy* of TS_1 is fairly close to that of the intermediate, i.e. the maximum corresponding to TS_1 and the subsequent minimum in the reaction coordinate corresponding to the intermediate are fairly close together in the reaction coordinate. In 1955, the American chemist G.S. Hammond postulated that when this energy difference between TS_1 and the subsequent intermediate is *very* small, the transformation of TS_1 into the intermediate involves only a very small *structural* change. It follows that most of the structural reorganization in the endothermic conversion of reactant molecule(s) into reactive intermediate occurs in the formation of the TS. This is sometimes expressed by saying that a *strongly endothermic* elementary step has a *late* transition structure.

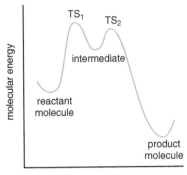

In the highly exothermic second step of the reaction in the figure, the energy of TS_2 is also similar to that of the intermediate (although it now *follows* the intermediate in the reaction coordinate). Consequently, according to what is now known as the Hammond postulate, the *structure* of TS_2 will be similar to that of the intermediate to a degree depending upon their closeness in the energy axis. It follows that most of the structural reorganization in a *highly exothermic* elementary step occurs between the TS and the product molecule—such a step has an *early* TS.

The Hammond postulate is very helpful in the interpretation of reaction profiles derived from kinetics results. It can provide credible estimates of bond lengths and angles of a transition structure if the TS can be related to an intermediate close to it in the reaction coordinate. Since it was first proposed, there have been major developments in computational chemistry which support the validity of the Hammond postulate.

G.S. Hammond (1921–2005).

(Kindly supplied by C. Wamser.)

chromatographic (see Panel 3.3) and spectroscopic methods (see Chapter 25). Once the identities and yields of products are known, alternative possible 'paper mechanisms' can be devised. These are normally based upon reactions of related compounds whose mechanisms are believed to be known, and usually expressed using curly arrows (but the ability to write a credible curly arrow mechanism is not proof that it is correct!). Ratios of multiple products (product selectivity) or the stereochemical course of a reaction sometimes serve as mechanistic evidence, especially regarding the involvement of reactive intermediates (see below).

Mechanistic experiments should be designed to exclude wrong possibilities, or test probable ones, rather than to confirm what we believe is the correct one.

The next stage is to carry out experiments which will allow us to exclude wrong possibilities. We should then be left with a single mechanism which accommodates all the evidence, and we take this to be the true one. We need to bear in mind, however, that we may have overlooked an alternative possible mechanism which also fits all the evidence so far. Consequently, in principle, we can never be 100% certain that we know the detailed mechanism of a reaction. In practice, of course, we can be *reasonably* confident about the mechanisms of the majority of organic chemical reactions which have been investigated, including the ones covered in this text.

7.5.2 Detection of intermediates in stepwise mechanisms

Scheme 7.8 includes stepwise and concerted possibilities for the irreversible conversion of one compound into another.

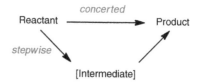

Scheme 7.8 Alternative concerted and stepwise mechanisms for the irreversible conversion of a reactant into a product.

Detection of a putative intermediate is solid evidence that the reaction is stepwise. The detection could be spectroscopic or the intermediate could be trapped by an added reagent which is known to react very rapidly with it. Failure to detect a possible intermediate, however, does not prove the reaction is concerted; the intermediate may be just too short lived to be detected or intercepted under the normal conditions of the experiment. In such cases, it is sometimes possible to carry out the reaction under different conditions designed to facilitate the detection, e.g. at a lower temperature, or in a different solvent. But it remains generally true that it is more difficult to prove a reaction is concerted than that it is stepwise.

The reaction of bromine with alkenes illustrates how an intermediate was proposed to account for the product stereochemistry, and how the reaction was modified to allow confirmation of the stepwise nature of the reaction. Cyclohexene reacts with bromine to give *trans*-1,2-dibromocyclohexane in CCl_4, and *trans*-2-bromocyclohexanol in aqueous alkaline solution (Scheme 7.9). These results are best accommodated by invoking a three-membered cyclic bromonium ion intermediate which can be trapped by different

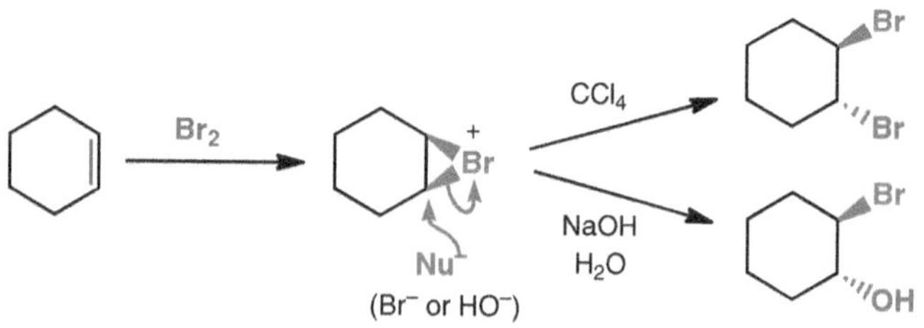

Scheme 7.9 Reaction of bromine with cyclohexene involving nucleophilic capture of an intermediate implicated by product stereochemistry.

nucleophiles to give the different products. It was known that S_N2 reactions proceed with complete inversion of configuration (see Scheme 7.3 above and Chapter 12 for details). By analogy, the cyclic bromonium ion in Scheme 7.9 is expected to suffer rear-side nucleophilic attack by Br^- or HO^- to give the *trans* products shown.

The intermediate in Scheme 7.9 is extremely short-lived in the presence of powerful nucleophiles. However, when the reactant was modified such that it led to a sterically hindered intermediate, nucleophilic capture was inhibited which allowed the cyclic bromonium ion to be isolated as its tribromide salt, **1**.

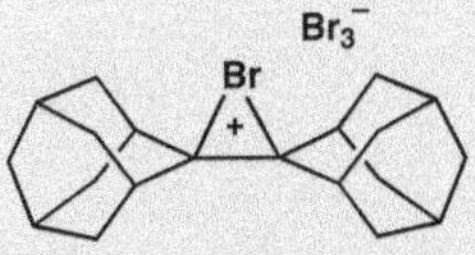

1
a bromonium ion salt

The intermediate *t*-butyl cation in Scheme 7.7 has been isolated as a crystalline salt with a caged carborane counterion; the salt is is stable under extremely non-nucleophilic conditions and its structure was determined by X-ray crystallography in 2005.

The bromonium ion in Scheme 7.9 could, in principle, be in equilibrium with a 2-bromocarbenium ion as shown below. If the open carbenium ion were involved as an intermediate in the bromination in CCl_4, what would you expect the stereochemistry of the product(s) to be?

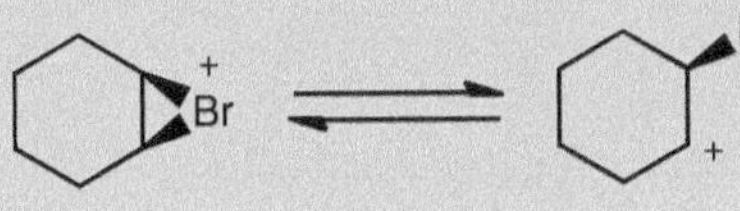

7.5.3 The rate law as an indicator of mechanism

The rate of a chemical reaction is directly proportional to the rate at which reactant molecules are converted into transition structures, and then on to product molecules. In Section 7.4, we saw that a transition structure in a simple unimolecular reaction is derived from a single reactant molecule. It follows that the rate of such a reaction is directly proportional to the concentration of reactant molecules. A unimolecular reaction of compound A, therefore, necessarily has a first-order rate law which is written, rate $= k$ [A], with [A] expressed in units mol dm^{-3}, and k is the first-order rate constant. Since the rate of a reaction has units mol dm^{-3} s^{-1}, a first-order rate constant has units s^{-1}.

Correspondingly, the rate of a simple bimolecular reaction between different reactants A and B is proportional to the rate at which a TS is formed which, in turn, is proportional to the product of the concentrations of the two reactants, rate $= k$ [A] [B]. If a bimolecular reaction involves a single reactant (A), a TS is formed from two molecules of A, so rate $= k$ [A]^2. In both, the second-order rate constant has units dm^3 mol^{-1} s^{-1}.

We see above how the mechanism determines what the experimental rate law will be; it follows that experimental determination of the rate law is powerful evidence in the elucidation of the mechanism of a reaction. A simple unimolecular reaction necessarily has a first-order rate law, and a simple bimolecular reaction must have a second-order rate law. However, the converse is not true. It does not follow that a first-order rate law proves a unimolecular mechanism, or that a second-order rate law proves a bimolecular mechanism; some quite complicated mechanisms can lead to deceptively simple rate laws. Nevertheless, if we observe a second order rate law, it cannot be of a reaction with a simple unimolecular mechanism; and if a reaction between two solutes is first order in one of them and zero-order in the other, it cannot have a simple bimolecular mechanism.

The concentrations of reactants decrease as a reaction proceeds so, for a first- or second-order reaction, the *rate* of the reaction decreases. However, the *rate constant* does not change with time; it is a parameter which characterizes the dynamics of the reaction under specified constant physical conditions.

If the rate-determining step in a multistep reaction is unimolecular, the overall rate law may not be first order if there are prior steps. We can safely say, however, that the concentration terms in the rate law indicate the composition (though not necessarily the structure) of the TS of the rate-determining step.

Example 7.4

Reaction of *t*-butyl chloride (Scheme 7.7) proceeds by unimolecular heterolysis of the C–Cl bond. The reverse reaction to give *t*-butyl chloride from *t*-butyl alcohol in concentrated hydrochloric acid occurs by a unimolecular heterolysis of the C–O bond (but needs the help of an acid catalyst). Explain why the former reaction is kinetically first order whereas the latter is second order.

Solution

The initial unimolecular step (eqn 7.12) of the former reaction is rate determining; consequently, the reaction is simply first order in the substrate concentration. The reaction of *t*-butyl alcohol in concentrated hydrochloric acid takes place by initial reversible protonation *followed by* the rate-determining step (r.d.s.)—departure of the H_2O molecule with heterolysis of the C–O bond (see Section 14.1).

$$Me_3C-OH \; + \; H_3O^+ \; \rightleftharpoons \; Me_3C\overset{+}{-}OH_2 \; + \; H_2O$$

$$\xrightarrow{\text{r.d.s.}} \; Me_3C^+ \; + \; 2H_2O \; \xrightarrow{Cl^-} \; Me_3C-Cl$$

The rate of this reaction depends upon the concentration of the protonated substrate, which is proportional to the concentrations of *both* the substrate *and* the acid: rate = $k[t\text{-BuOH}][H_3O^+]$. Note that the protonated alcohol returns to reactants faster than it undergoes the rate-determining forward step.

Exercise 7.10

The rearrangement reaction of the allylic alcohol given in eqn 7.9 occurs with acid catalysis. The reaction proceeds by reversible protonation followed by heterolysis to give an allylic cation which is then trapped by water. (a) Write an equation, including curly arrows, for each step of the mechanism. (b) Draw a Gibbs energy profile for this reaction.

7.5.4 Effect of substrate structure and reaction conditions on rate constants

As indicated in Sub-section 7.4.3, the rate constant of a reaction (k) is directly related to $\Delta G^{\ddagger}$, the Gibbs energy of activation (the Gibbs energy change between initial state and transition state). Gibbs energies of compounds depend upon the physical conditions (principally, the temperature and the nature of the solvent for compounds in solution). Consequently, magnitudes of rate constants depend upon the reaction conditions as well as on the structure(s) of the reactant(s). The absolute value of k or $\Delta G^{\ddagger}$ at a particular temperature seldom provides much mechanistic information. In contrast, we obtain considerable insight by investigating how these parameters change at constant temperature when *small* systematic changes are made to the structure(s) of the reactants, or to the nature of the solvent for reactions in solution (and most organic reactions are carried out in solution).

If a *large* structural change is made to a reactant, or if the reaction conditions are changed drastically, we may cause a different reaction to take place rather than simply induce a change in the rate constant of the original reaction. A minimal structural change to a reactant can be made by substitution of one isotope of an element for another, e.g. replacement of a particular protium (^1H) by a deuterium (^2H). Comparison of the rate constants of the two forms of the compound under identical conditions (i.e. measurement of the *kinetic isotope effect*) can provide detailed information about the mechanism of the reaction; for example, it can show whether a particular C–H bond is broken during the reaction.

We can illustrate the effect of structure upon reactivity by regarding Scheme 7.7 (Figure 7.9) as a mechanistic proposal for the hydrolysis of *t*-butyl chloride with rate-determining ionization to give the intermediate ion pair, $Me_3C^+ \, Cl^-$. Any modification to the substrate which will lead to a more stable carbenium ion will stabilize the preceding TS (an example of the Hammond postulate; see Panel 7.2). According to this mechanism, therefore, any such change will facilitate the reaction, i.e. lead to a higher rate constant under the same reaction conditions. Consequently, if a Me group in Me_3CCl is replaced by a Ph group, which is much better than Me at stabilizing a carbenium ion, the rate constant for hydrolysis should become greater. Indeed, k values increase by more than 10^6 along the series Me_3CCl, $PhC(Me)_2Cl$, $Ph_2C(Me)Cl$, and Ph_3CCl. The mechanism also requires Me_3CBr to be more reactive than Me_3CCl because the C–Br bond is weaker than the C–Cl bond; this is also observed. These findings, therefore, are in accord with the mechanism in Scheme 7.7 for the hydrolysis of t-butyl chloride.

Explain why the carbenium ion intermediate in the hydrolysis of $PhC(Me)_2Cl$ is more stable than the one in the hydrolysis of Me_3CCl.	**Exercise 7.11**

We can also use the solvolysis (reaction with a solvent) of t-butyl chloride to illustrate the investigation of solvent effects as a mechanistic tool. The mechanistic proposal illustrated in Scheme 7.7 and Figure 7.9 includes an initial rate-determining ionization of the nonpolar reactant (Scheme 7.10). In contrast, the mechanism shown in Scheme 7.2 for the isomerization in eqn 7.10 involves no appreciable change in polarity when a reactant molecule becomes a transition structure. On the basis of these two mechanisms, we can expect the rate constant for the solvolysis to be much more dependent upon the nature of the solvent than is the one for the isomerization. Indeed, first-order rate constants for isomerizations like eqn 7.10 change relatively little over a wide range of solvents whereas t-butyl chloride does not react at all in hydrocarbon solvents, and is 3×10^5 times more reactive in H_2O than in EtOH! These findings are again in accord with the abbreviated mechanism shown in Scheme 7.10 for the solvolysis of t-butyl chloride.

$$Me_3C\text{—}Cl \longrightarrow \left(Me_3C^{\delta+} \text{- - -} Cl^{\delta-} \right)^{\ddagger} \longrightarrow Me_3C^+ + Cl^- \xrightarrow{\text{solvent}} Products$$

Scheme 7.10 Solvolysis of t-butyl chloride.

A more detailed discussion of solvent effects on rate constants is given in Section 12.3, but we can generalize as follows regarding their investigation as a technique for mechanistic studies. Insensitivity of the rate constant of a reaction to the nature of the solvent is evidence that the reaction has a nonpolar mechanism; on the other hand, if the rate constant changes appreciably as the nature of the solvent is changed, the mechanism is likely to be polar.

Summary

- Organic reactions fall into one of four fundamental classes (*substitution*, *addition*, *elimination*, and *rearrangement*), or combinations of them.

- Every overall reaction (e.g. a substitution) is either a single step (in which all changes are concerted) or stepwise, and every *elementary step* is either **polar**, **radical**, or **pericyclic**. Additionally, acid–base reactions play an important role in facilitating many organic reactions.

- Molecules require activation to react and this may be brought about thermally or photochemically.

- Cleavage of a single bond occurs either by **homolysis** or **heterolysis**.

- Homolysis of a σ bond leads to the formation of two **radicals**, each with an *unpaired electron* in its incomplete valence shell. Reactions involving radicals are represented using *singly barbed curly arrows*, each corresponding to movement of a single electron.

- Polar reactions involve heterolysis and/or bond formation between a **nucleophile** and an **electrophile**, the former donating an electron pair to the latter in bond-making. They are described by representing the flow of electron pairs by *doubly barbed curly arrows*.

- The critical interaction leading to bond formation between a nucleophile and an electrophile corresponds to the overlap of the **HOMO** of the nucleophile with the **LUMO** of the electrophile, the so-called frontier molecular orbitals.

- The relationship between the molecular energy, molar enthalpy, or Gibbs energy of a reaction and a coordinate which corresponds to the transformation of reactant(s) into product(s) may be represented graphically in a **reaction profile**.

- The maximum in the profile of a single-step reaction corresponds to the **transition structure** or **transition state** (**TS**) depending upon whether the energy axis is molecular or molar.

- The enthalpy difference between the initial state and the TS is often called the **activation energy**, and the Gibbs energy difference between the initial state and the TS is the **Gibbs energy of activation** which is related to the rate constant.

- Multistep reactions involve sequential TSs and intermediates; if the barrier corresponding to one step is appreciably higher than all others, that step is called the **rate-determining step**.

- The standard Gibbs energy of a reversible reaction is directly related to the **equilibrium constant** for the reaction.

- Characterization of an organic reaction starts with product identification and analysis from which possible alternative mechanisms may be proposed. *Experiments are then designed to exclude wrong mechanisms*. Methods of investigation include attempted detection of possible reactive intermediates, stereochemical studies, rate measurements, and the effects of structural modifications to the substrate(s) and of solvent changes upon rate constants.

Problems

7.1 Identify each of the following as an electrophile or a nucleophile.

(a) H_3O^+ (b) OH^- (c) NH_3 (d) I^-

(e) Cl_2 (f) $(CH_3)_3C^+$ (g) CN^-

7.2 Classify each of the following reactions as substitution or addition, and identify each reactant as an electrophile or a nucleophile.

(a) 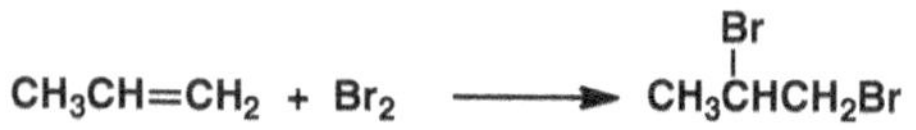 $CH_3CH=CH_2 + Br_2 \longrightarrow CH_3\overset{Br}{\underset{|}{C}}HCH_2Br$

(b)    Me—C(=O)—OEt $+ H_2O \longrightarrow$ Me—C(=O)—OH $+ EtOH$

(c)   benzene $+ SO_3 \longrightarrow$ C₆H₅—SO_3^- $+ H^+$

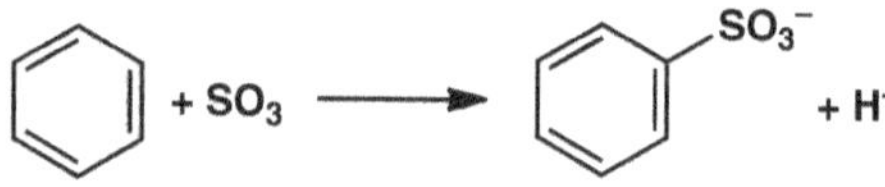

(d) 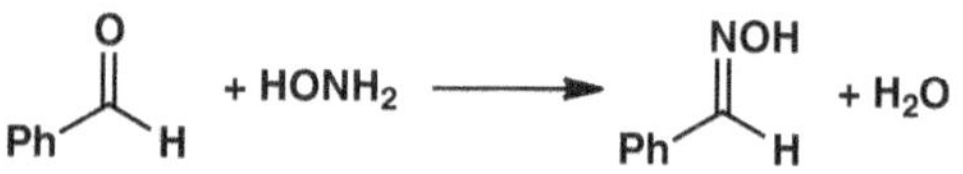   Ph—C(=O)—H $+ HONH_2 \longrightarrow$ Ph—C(=NOH)—H $+ H_2O$

(e)   Me—C(=O)—Me $+ \bar{B}H_4 \longrightarrow Me_2CHO-\bar{B}H_3$

7.3 Classify each of the following reactions as substitution, addition, elimination, or an acid-base reaction, and identify each reactant as an electrophile, an acid, a nucleophile, a base, or none of these.

(a) $CH_3CH_2I + (CH_3CH_2)_3N \longrightarrow (CH_3CH_2)_4N^+ \ I^-$

(b)  [structure: allylic bromide with Br] $+ \ Et_2NH \longrightarrow Et_2N^+$ [allyl chain] $+ \ Br^-$

(c) $Me_3COH + H_3O^+ \longrightarrow Me_3COH_2^+ + H_2O$

(d) $Me_3COH_2^+ \longrightarrow Me_3C^+ + H_2O$

(e) $Me_2\overset{+}{C}CH_3 + H_2O \longrightarrow Me_2C{=}CH_2 + H_3O^+$

7.4 Explain the difference between a transition structure and a reactive intermediate.

7.5 Draw the reaction profile for an exergonic two-step reaction with the second step rate determining.

7.6 Draw the reaction profile for a three-step reaction (two reactive intermediates) with the following features: the overall reaction is exergonic, the first intermediate is less stable than the second, and the second step is rate determining.

7.7 Explain why each of the following statements is true.

(a) The higher the energy of the HOMO of a nucleophile, the more reactive it is.

(b) The lower the energy of the LUMO of an electrophile, the more reactive it is.

(c) Methylamine is more nucleophilic than methanol.

7.8 Show all lone pair electrons of reactants in the following acid-base reactions, and depict the flow of electrons with curly arrows. Formal charges should be added where necessary.

(a) 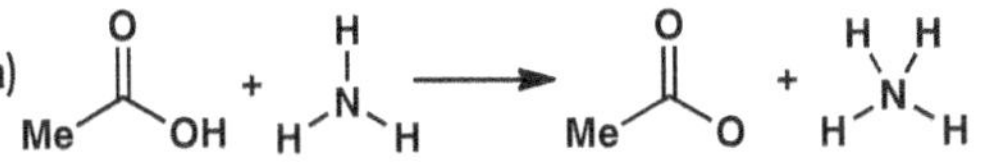 [structure: acetic acid Me-C(=O)-OH + H-N(H)-H → Me-C(=O)-O + H-N⁺(H)(H)-H]

(b) $CH_3CH_2O^- + H{-}C{\equiv}N \longrightarrow CH_3CH_2OH + \ ^-C{\equiv}N$

(c) 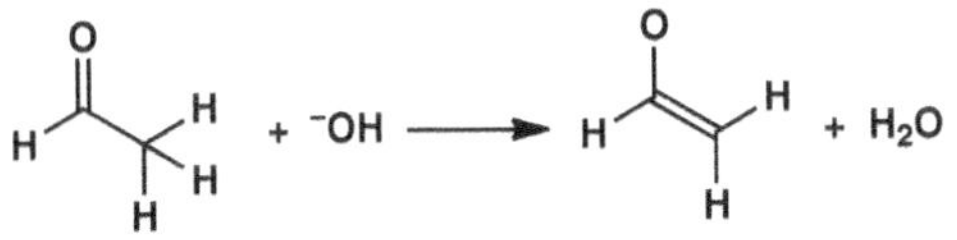 [structure: H-C(=O)-CH₂-H + ⁻OH → H-C(=O)-CH=CH H + H₂O]

(d)  [structure: H-C(-O⁻)=CH₂ + O(H)(H) → H-C(-OH)=CH₂ + ⁻OH]

7.9 Indicate the flow of electron pairs in the following reactions by curly arrows.

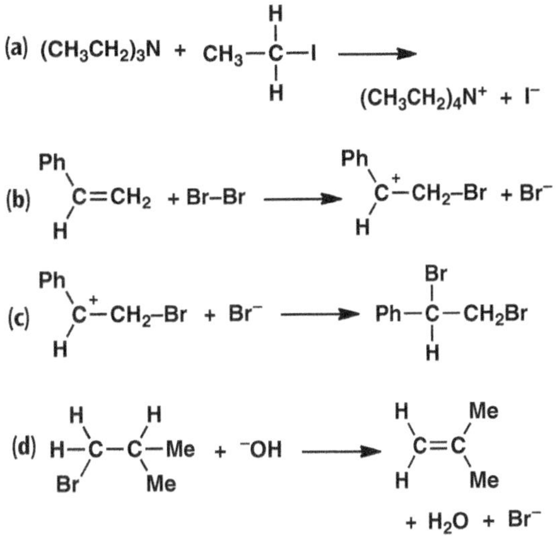

(a) $(CH_3CH_2)_3N + CH_3{-}\overset{H}{\underset{H}{\overset{|}{C}}}{-}I \longrightarrow$ $(CH_3CH_2)_4N^+ + I^-$

(b) [structure: Ph(H)C=CH₂ + Br–Br → Ph(H)C⁺–CH₂–Br + Br⁻]

(c) [structure: Ph(H)C⁺–CH₂–Br + Br⁻ → Ph–C(Br)(H)–CH₂Br]

(d) [structure: H–C(Br)(H)–C(Me)(H)–Me + ⁻OH → (H)(H)C=C(Me)(Me) + H₂O + Br⁻]

7.10 Addition of HCN to ethanal in eqn 7.6 (at the beginning of the chapter) is catalysed by base B which generates CN^- from HCN in a preliminary step,

$$HCN + \ :B \rightleftharpoons CN^- + HB^+$$

The stepwise reaction then proceeds as follows.

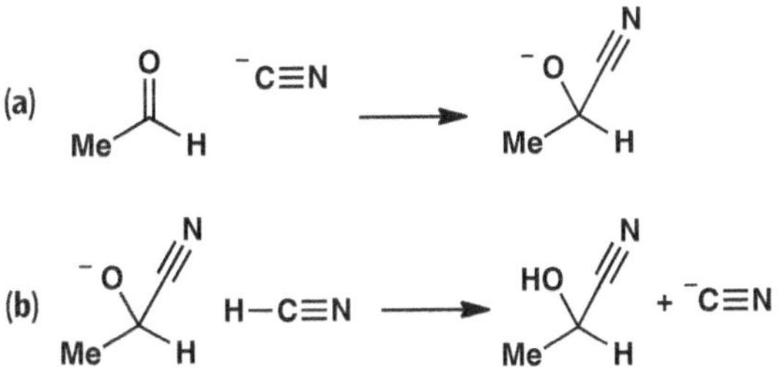

(a) [structure: Me-C(=O)-H + ⁻C≡N → Me-C(-O⁻)(H)-C≡N]

(b) [structure: Me-C(-O⁻)(H)-C≡N + H-C≡N → Me-C(-OH)(H)-C≡N + ⁻C≡N]

Mark in all the lone pairs in molecules and ions involved, and represent movements of electrons by curly arrows.

7.11 The following reaction is the cleavage of an ether with hydrogen iodide. Show how each step takes place using curly arrows; include all unshared electron pairs and formal charges.

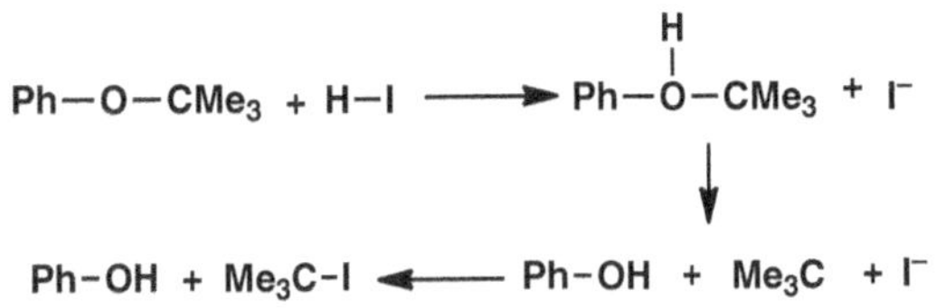

$$Ph{-}O{-}CMe_3 + H{-}I \longrightarrow Ph{-}\overset{H}{\overset{|}{O}}{-}CMe_3 + \ I^-$$

$$\downarrow$$

$$Ph{-}OH + Me_3C{-}I \longleftarrow Ph{-}OH + Me_3C^+ + \ I^-$$

7.12 Hydrolysis of the simple ester in eqn 7.4 (at the beginning of the chapter) is an example of a type of substitution reaction which occurs in alkaline or acidic solution. Alkaline hydrolysis of the ester proceeds as shown below. Represent the flow of electron pairs in each step by means of curly arrows, and include all lone pair electrons.

(a)

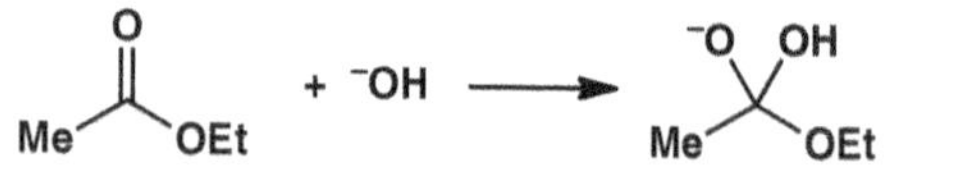

(b)

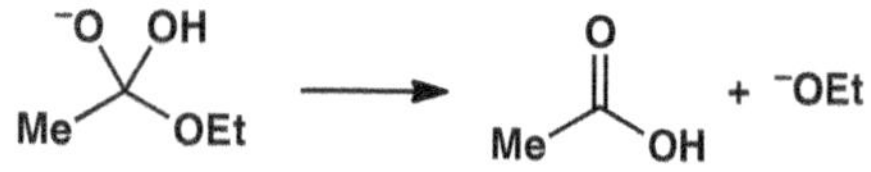

(c)

7.13 Acid-catalysed hydrolysis of an ester starts with protonation at the carbonyl oxygen as shown below. Represent the flow of electron pairs in each step by curly arrows, and include all lone pair electrons.

(a)

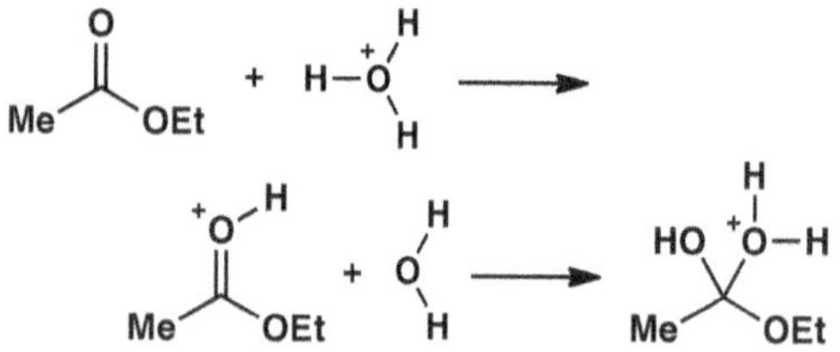

(b)

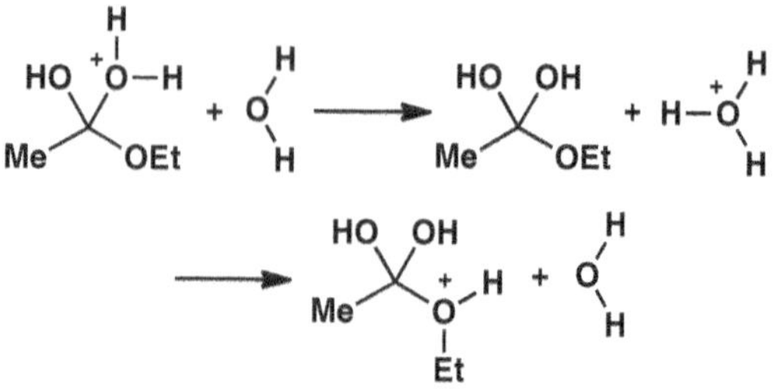

(c)

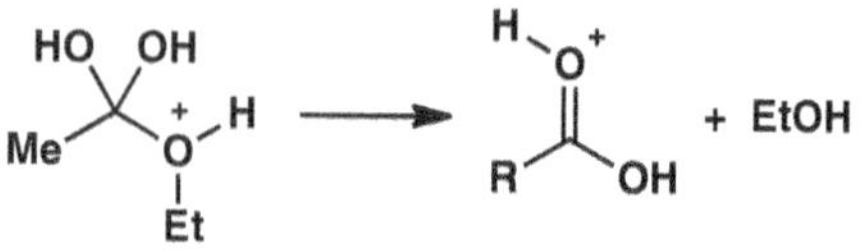

(d)

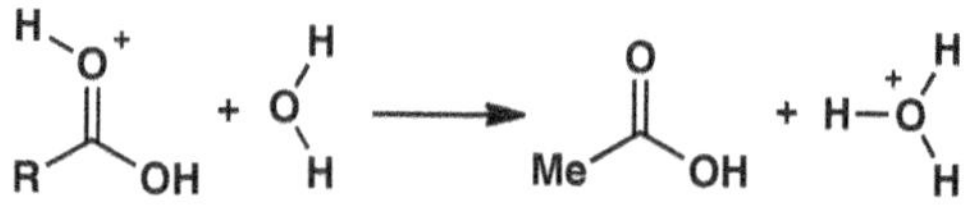

7.14 Some relatively stable carbenium ions have extended lifetimes in strongly acidic solution, and can sometimes be observed spectroscopically; explain why strongly acidic conditions make this possible.

7.15 Addition of HCl to an alkene occurs by the stepwise mechanism exemplified in Scheme 7.6. Give the rate law for this reaction and predict (giving your reasoning) the order of increasing rate constants for the reactions of the following alkenes with HCl.

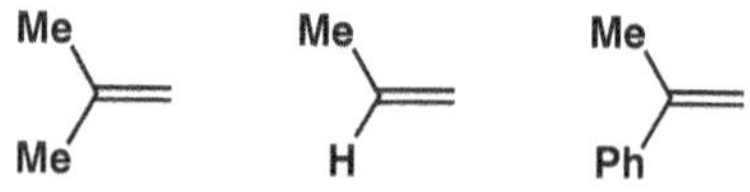

Nucleophilic Addition to the Carbonyl Group in Aldehydes and Ketones

8

Related topics we have already covered:

- **Polarity of a covalent bond** (Sub-section 1.2.3)
- **π Bond of ethene** (Sub-section 2.5.2)
- **Acids and bases** (Chapter 6)
- **Classes of organic reactions** (Section 7.1)
- **Bond cleavage and formation** (Section 7.2)
- **Curly arrow representation of reaction mechanisms** (Section 7.2)

Topics we shall cover in this chapter:

- **Polarity of the carbonyl bond**
- **Nucleophilic addition of CN⁻, H_2O, ROH, and HSO_3^- to the carbonyl group**
- **Catalysis of nucleophilic addition to carbonyl groups by acids and bases**
- **Acetal and dithioacetal formation**
- **Formation of imines and enamines**
- **Wittig reaction**

The carbonyl (>C=O) group is a functional group found in aldehydes and ketones as well as carboxylic acids and their derivatives, and is one of the most important in organic and biological chemistry. The C=O bond is polarized and unsaturated, so undergoes addition reactions initiated by nucleophilic attack at the partially positive carbon atom.

In this chapter, the first in which we consider in detail the reactions of a functional group, we shall look at nucleophilic additions to the carbonyl group of aldehydes and ketones. This is partly because these reactions provide clear examples of several fundamental principles of organic reaction mechanisms, but also because they include some of the most useful reactions for organic synthesis, and they relate to important biochemical processes.

Some fragrant aldehydes and ketones in nature:

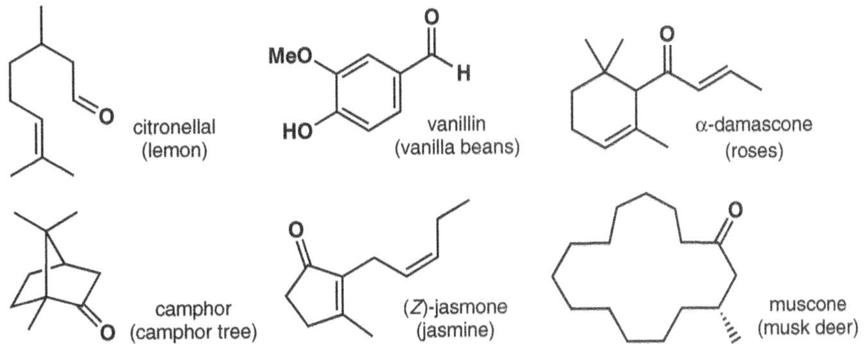

citronellal (lemon)

vanillin (vanilla beans)

α-damascone (roses)

camphor (camphor tree)

(Z)-jasmone (jasmine)

muscone (musk deer)

8.1 Polarity of the Carbonyl Bond

Because oxygen is more electronegative than carbon, the shared electrons of the C=O double bond are attracted more towards the oxygen; consequently, there is a partial positive charge on the carbon and a partial negative on the oxygen in **1**. This polarization can be represented by resonance hybrid **1c** (see Sections 1.4 and 5.4) with (unequal) contributions of the two forms shown below (**1a** and **1b**). One resonance contributor (**1a**) has a double bond between the oxygen and carbon atoms, and the other (**1b**) has singly bonded C and O atoms with full charges on both. Note that the oxygen in the doubly bonded contributor (**1a**) has two pairs of unshared electrons (lone pairs), and three in the singly bonded form (**1b**).

Carbonyl bond:

The polarized C=O bond may also be represented by an electron density diagram (electrostatic potential map, EPM) calculated theoretically; one for methanal (formaldehyde) is shown in Figure 8.1.

The polarization corresponding to the electron distribution in the π molecular orbital (MO) of the C=O bond can be deduced theoretically from the energy levels of the interacting atomic orbitals (AOs) of the carbon and oxygen atoms. The π bond is formed by the combination of 2p AOs of the two atoms in the same way that the π bond of ethene is formed, as described in Section 5.1, although the energy levels are different. The energy level of the 2p AO of oxygen is lower than that of carbon owing to the higher electronegativity of oxygen. The 'sideways' overlap (interaction) of these AOs is illustrated in Figure 8.2.

The new bonding π MO has a greater contribution from the AO of lower energy (the one on the oxygen), so the two electrons in this π orbital are distributed more towards the O atom and, therefore, the C atom is partially positive. In contrast, the corresponding antibonding MO (π*) has a greater contribution from the AO of higher energy (the one on the carbon), but this MO is unoccupied (it is the LUMO).* The (polarized) σ bond between the C and the O of the carbonyl group is also derived from AOs on the C and the O. The low energy occupied bonding σ orbital and the high energy unoccupied antibonding (σ*) orbital are derived from end-to-end overlap of AOs of appropriate symmetry. These AOs and MOs are not included in Figure 8.2 because they are not appreciably affected by addition reactions.

When an anionic nucleophile (Nu⁻) reacts with the carbonyl group, the HOMO of Nu⁻ containing a lone pair of electrons interacts with the LUMO (π*) of the C=O group. Since the main component of the LUMO is the AO on the C atom, the nucleophile interacts principally with this, and the new σ bond is from Nu to the C atom (eqn 8.1). The plane of the π* MO of the carbonyl group is perpendicular to the trigonal plane of the

Although the π MO is the LUMO, its corresponding π MO is not the HOMO; the oxygen of the carbonyl group has lone pairs in nonbonding MOs (not shown in Fig. 8.2) and these are the HOMOs of simple aldehydes and ketones.

Figure 8.1 Molecular model and an electrostatic potential map (EPM) of methanal. See Sub-section 1.2.3 for EPMs in which the calculated electron density is shown with colours ranging from red (high) to blue (low).

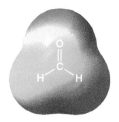

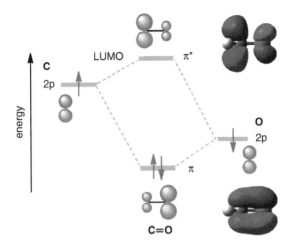

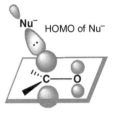

Figure 8.2 π MOs of the C=O bond. Shapes of theoretically calculated π and π* MOs of methanal are shown on the right.

For clarity, the two occupied 2p AOs on O containing the lone pairs (the HOMOs) are not included in Fig. 8.2.

LUMO of C=O (π*)

Figure 8.3 Molecular orbital interaction between the HOMO of a nucleophile (Nu⁻) and the LUMO (π*) of a carbonyl group in the addition reaction of Nu⁻ to the C=O.

sp²-hybridized carbon; consequently, to achieve maximal overlap between the interacting orbitals, nucleophilic attack begins from above the C of the carbonyl group as shown in eqn 8.1 and Figure 8.3. As the new σ-bond develops more fully, the hybridization of the C atom changes from sp² to sp³ in the addition product.

$$Nu^- \ \overset{sp^2}{C=O} \longrightarrow \overset{Nu}{\underset{sp^3}{C-O^-}}$$

(8.1)

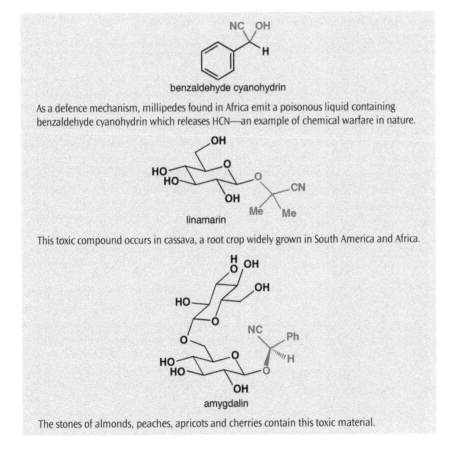

benzaldehyde cyanohydrin

As a defence mechanism, millipedes found in Africa emit a poisonous liquid containing benzaldehyde cyanohydrin which releases HCN—an example of chemical warfare in nature.

linamarin

This toxic compound occurs in cassava, a root crop widely grown in South America and Africa.

amygdalin

The stones of almonds, peaches, apricots and cherries contain this toxic material.

Panel 8.1 Common carbonyl compounds: methanal, ethanal, and propanone

Methanal (formaldehyde), the simplest aldehyde, is a gas at room temperature (bp −20 °C) but is too reactive to be stored in pure form; it readily polymerizes to give paraformaldehyde and also gives trioxane, a cyclic trimer. An aqueous solution, which may be up to 37% at saturation and in which methanal exists principally as its hydrate, is called formalin. It is used as a disinfectant and to preserve biological samples, for example, but only with great care owing to its toxicity and carcinogenicity.

(CH_2O)	$(CH_2O)_3$	$(CH_2O)_n$	
methanal (formaldehdye)	trioxane	paraformaldehyde	methanal hydrate (in formalin)

Methanal is produced by the oxidation of methanol on a large scale (annual world production: almost 29 million tons, 2010), and is used to make plastics (e.g. melamine, phenol–formaldehyde resin, and urea–formaldehyde resin), adhesives, sealants, varnishes, and paints, which are used in the manufacture of modern construction and furnishing materials such as plywood, chipboard, laminates, and wall coverings. When newly manufactured, these materials release low concentrations of methanal which, like other volatile organic compounds, can lead to 'sick building syndrome' and 'sick car syndrome'. The notoriety of methanal owing to its harmful effect on human health is closely related to its high reactivity with nucleophiles. Biomolecules of our bodies and other organisms contain many hydroxy and amino groups which react as potent nucleophiles with methanal.

Ethanal (acetaldehyde), which is a volatile liquid (bp 20 °C), is a more representative aldehyde and widely used as a starting material in organic synthesis. It is produced industrially by the Wacker process, a selective oxidation of ethene, and used for the production of ethanoic acid and other chemicals.

$$H_2C=CH_2 \ + \ O_2 \ \xrightarrow[\textit{Wacker oxidation}]{PdCl_2-CuCl_2} \ CH_3CH=O$$

ethene — ethanal (acetaldehyde)

Ethanal is formed during the metabolism of ethanol (see Panel 10.3) but is toxic and is the cause of hangovers. The efficiency of the enzymes involved in this metabolism is known to depend upon ethnicity. Many East Asian people have effective ethanol dehydrogenase, which produces the ethanal, but inefficient acetaldehyde dehydrogenase which converts ethanal to ethanoic acid. Consequently, such people do not metabolize alcohol in beverages very effectively.

Propanone (acetone) is the simplest ketone, and is a useful solvent since it dissolves organic materials well and is miscible with water. It is employed not only as a solvent in laboratories and chemical manufacturing industries, but also as a component of consumer products such as nail polish remover. Industrial methods of production include the Wacker oxidation of propene, oxidation of propan-2-ol, and the so-called cumene process for the production of phenol from isopropylbenzene (cumene).

isopropylbenzene (cumene) + O_2 → rearrangement (Chap. 22) → propanone (acetone) + phenol

Propanone is also used as a raw material for the production of bisphenol A which is used to make a polycarbonate plastic and epoxy resins. Another industrial product from propanone is methyl methacrylate which polymerizes to give a transparent thermoplastic sometimes known as acrylic glass (Plexiglass, Perspex, etc.): see Section 18.3 for anionic polymerization.

bisphenol A

8.2 Formation of Cyanohydrins

One of the most typical nucleophilic addition reactions of carbonyl compounds is the addition of cyanide ion, CN^-, to give a cyanohydrin following protonation (Scheme 8.1); specific examples of cyanohydrins which are found in nature (some as sugar derivatives) are shown in Section 8.1 on p. 167. This mechanistically simple reaction is significant because cyanohydrins, with a new C–C bond and the CN group capable of further transformation, are useful in synthesis.

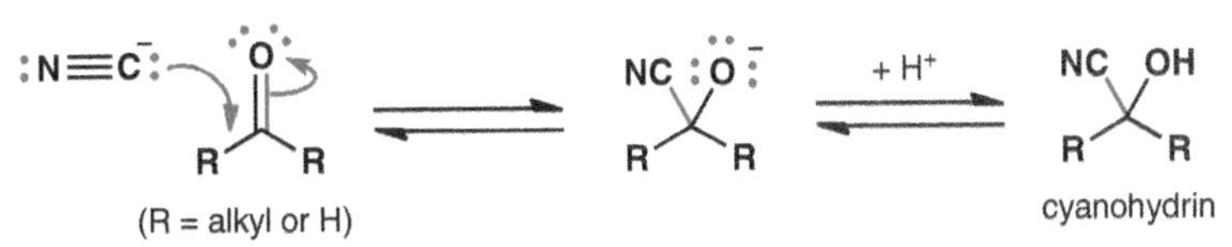

(R = alkyl or H)

cyanohydrin

Scheme 8.1 Formation of a cyanohydrin.

The reaction starts with electrostatic attraction between the negatively charged nucleophile, CN^-, and the partially positive carbonyl carbon. The cyanide ion has unshared electron pairs on both the C and the N. However, the nucleophilic centre is the carbon because the HOMO of CN^- is a lone pair located mainly on the carbon. Consequently, the lone pair on the carbon atom of CN^- forms a covalent σ bond with the carbonyl carbon via the HOMO–LUMO interaction, and the π electron pair of the C=O bond is concurrently pushed onto the oxygen atom. The resulting tetrahedral intermediate has a full negative charge on the oxygen, so is an alkoxide. These electron pair movements are represented by curly arrows in the first step of Scheme 8.1. Upon protonation of the alkoxide in the second step of Scheme 8.1, the final product is the neutral cyanohydrin.

Example 8.1

In buffered aqueous reaction mixtures initially containing R_2CO, HCN, and CN^-, the rate of cyanohydrin formation was found to be proportional to the concentrations of both the carbonyl compound and cyanide ion, but independent of the acid concentration:

$$Rate = k[R_2C=O][CN^-]$$

(a) What does this rate law indicate regarding the rate-determining step of the reaction?
(b) The yield of cyanohydrin, though not its rate of formation, depends upon the acidity. What mechanistic information does this result provide?

Solution

(a) The first-order rate dependence upon each of $[R_2C=O]$ and $[CN^-]$ and zero-order rate dependence upon [HCN] and $[H_3O^+]$ indicate that only $R_2C=O$ and CN^- are involved in the rate-determining step of the reaction. Of the two steps in the mechanism of Scheme 8.1, therefore, the first is rate determining.

(b) Although the acidity does not affect the rate of the reaction, it does affect the proportion of cyanohydrin in the product mixture. This is evidence that the second step of the mechanism of Scheme 8.1 is an equilibrium, and the yield of cyanohydrin in the product mixture is determined by the acidity.

Exercise 8.1

Possible proton sources in the second step of Scheme 8.1 include HCN, H_3O^+, and H_2O depending upon the pH of the reaction mixture. Write a curly arrow mechanism for each of these acids in the proton transfer step.

The overall reaction shown in Scheme 8.1 is, in fact, reversible and may be described more succinctly as an equilibrium (eqn 8.2) with the equilibrium constant defined in the usual way ($K=$[cyanohydrin]/[R_2CO][HCN] where the square brackets represent concentrations of reactants and products at equilibrium). However, hydrogen cyanide (HCN) itself cannot react directly as a nucleophile because it has no lone pair of electrons on the carbon. The reaction must, therefore, actually proceed in two steps: addition of cyanide anion followed by protonation of the intermediate, as shown in Scheme 8.1. So, although eqn 8.2 describes the overall transformation, it does not indicate *how* the reaction takes place.

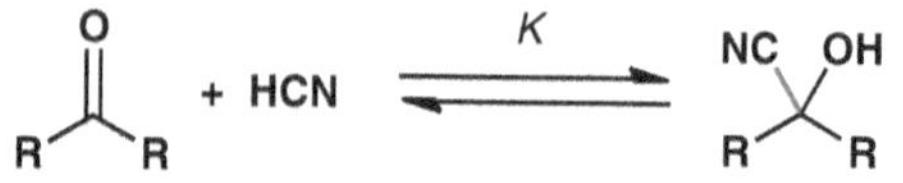

$$\text{(structure: } R_2C{=}O + HCN \overset{K}{\rightleftharpoons} R_2C(CN)(OH)\text{)} \tag{8.2}$$

Equilibrium constants (~25 °C) for formation of cyanohydrins (eqn 8.2)			
Aldehyde	K/mol^{-1} dm^3	Ketone	K/mol^{-1} dm^3
CH_3CHO	7100	CH_3COCH_3	28
C_6H_5CHO	200	$C_6H_5COCH_3$	0.67

Table 8.1

*This operation must be done carefully in a fume-hood in case any of the highly poisonous HCN gas escapes from the reaction mixture.

In practice,* the reaction is carried out by addition of an appropriate amount of a strong acid (e.g. H_2SO_4) to an aqueous solution of sodium cyanide (NaCN) and the carbonyl compound (eqn 8.3). At equilibrium, $R_2C{=}O$, CN^-, HCN, cyanohydrin, cyanohydrin anion, and H_3O^+ will all be present at concentrations determined by the amounts of starting materials used and the magnitudes of various equilibrium constants.

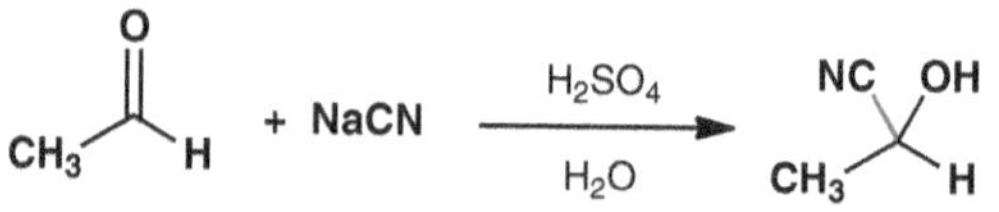

$$\text{(structure: } CH_3CHO + NaCN \xrightarrow[H_2O]{H_2SO_4} CH_3CH(CN)(OH)\text{)} \tag{8.3}$$

Without added acid, the reverse of the first step of Scheme 8.1 takes place easily, so an equilibrium between reactants and the intermediate is established, and the reaction will not proceed to completion. Correspondingly, if an aqueous solution of a cyanohydrin is made basic, the deprotonated cyanohydrin will equilibrate with the original carbonyl compound and CN^- (Scheme 8.2).

Scheme 8.2 Reversible reaction between a cyanohydrin and a carbonyl compound plus cyanide under basic conditions.

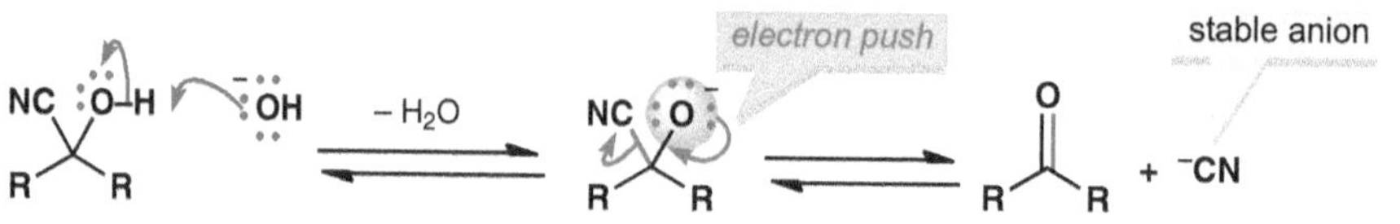

Equilibrium constants, K, for the formation of cyanohydrins (eqn 8.2) depend on the structure of the carbonyl compound. Aldehydes generally have larger K values than ketones, as exemplified in Table 8.1; this difference is mainly due to steric effects on the reaction. With the change in hybridization from sp^2 of the reactant carbonyl carbon to sp^3 of the product, the bond angle between the two groups attached to the carbonyl contracts from 120° to 109.5° (eqn 8.4), and the groups become closer together. If the H atom of an aldehyde is replaced by a larger alkyl group in a ketone, the increase in steric strain during the reaction is greater (see space-filling models next), so the equilibrium constant is smaller. Similar effects of the structure of the carbonyl compound are also observed for other addition reactions, including hydration* as will be discussed in the next section.

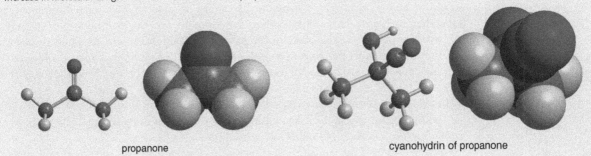

$$(8.4)$$

Hydration, the addition reaction of water to a compound, needs to be distinguished from **hydrolysis**, which is any other reaction brought about by water. Elimination of H_2O from a compound is, of course, **dehydration**.

Increase in molecular congestion with the conversion of propanone into its cyanohydrin, shown with ball-and-stick and space-filling models:

propanone   cyanohydrin of propanone

Explain the order of equilibrium constants for formation of cyanohydrins:

$$CH_3CHO > CH_3COCH_3 > CH_3COCH(CH_3)_2$$

Exercise 8.2

Equilibrium constants K for formation of cyanohydrins from cyclohexanone and butan-2-one are 1700 and 28 mol^{-1} dm^3, respectively; explain the difference.

Example 8.2

Solution

Cyclohexanone has appreciable strain due to the trigonal planar carbonyl carbon whereas its cyanohydrin has only little strain (from 1,3-diaxial interactions, as discussed in Chapter 4) in its chair form. Consequently, formation of the cyanohydrin from cyclohexanone relieves strain and the equilibrium constant is large. In contrast, there is an increase in steric strain in the formation of a cyanohydrin from a simple acyclic ketone like butan-2-one as discussed in the text, and K is much smaller.

8.3 Addition of Water to Aldehydes and Ketones

8.3.1 Hydration equilibrium

Aldehydes and ketones react reversibly with water to give 1,1-diols (also called *gem*-diols or hydrates); see eqn 8.5. Hydration of these carbonyl compounds is very easily reversible, which usually prevents isolation of the hydrate.

$$(8.5)$$

hydrate (1,1-diol)

The prefix '*gem*' is an abbreviation of 'geminal' (from the Latin word *gemini* meaning twins), which indicates two identical groups are attached to the same carbon.

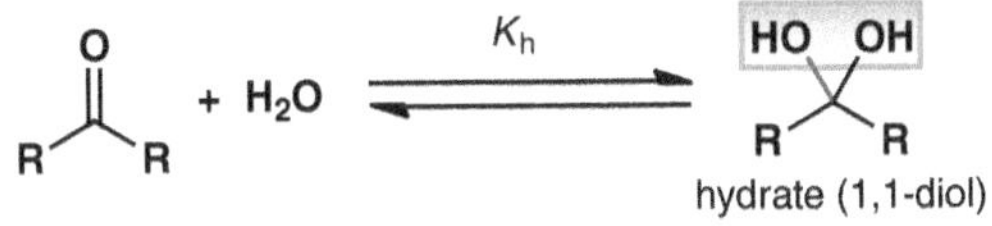

Table 8.2

Equilibrium constants (25 °C) for hydration of carbonyl compounds (eqn 8.5)			
Carbonyl compd	K_h	Carbonyl compd	K_h
H_2CO	2000	C_6H_5CHO	0.008
CH_3CHO	1.06	$p\text{-}NO_2C_6H_4CHO$	0.17
$ClCH_2CHO$	37.0	CH_3COCH_3	0.0014
Cl_3CCHO	$\sim 10^4$	$ClCH_2COCH_3$	0.11
$(CH_3)_2CHCHO$	0.43	$Cl_2CHCOCH_3$	10

Equilibrium constants for hydration, like those for cyanohydrin formation, are very dependent on the structure of the carbonyl compound, and the features which affect the one reaction also affect the other in the same way. Table 8.2 contains some values for equilibrium constants for hydration, K_h (= $[R_2C(OH)_2]/[R_2C{=}O]$)* in aqueous solution.

In aqueous solution, methanal exists almost completely in its hydrate form whereas only about 0.1% of propanone (acetone) exists as its hydrate. Ethanal (acetaldehyde) is hydrated to the extent of about 50% in aqueous solution, and replacement of one of its methyl hydrogens by a chlorine atom increases the extent of hydration. The hydrate of trichloroethanal (chloral) is stable enough for it to be isolated.

The magnitude of the equilibrium constant is affected by steric and electronic effects. The **steric effect** is related to the change in hybridization of the carbonyl carbon from sp² to sp³ (eqn 8.4), as discussed for cyanohydrin formation. If small R groups are replaced by larger ones, the increase in steric strain when the carbonyl group is hydrated will be greater. Consequently, we expect that the equilibrium constants for hydration will be smaller for aldehydes and ketones with bulkier R groups. This is observed in Table 8.2. Both R groups in methanal are hydrogen; in other aldehydes, one is alkyl and the other hydrogen; in ketones, both are alkyl. Consequently, K_h for hydration of methanal is large, but it becomes smaller for other aldehydes, and smaller still for ketones.

Figures 8.4(a) and (b) illustrate these relationships in energy terms. The value of K_h for ethanal is close to 1, that is to say, the energies of ethanal and its hydrate are about the same, as illustrated in (a). Hydration of propanone leads to an increase in energy due to the increase in steric strain, as shown in (b), and K_h is much less than unity.

An **electronic effect** is observed when an electron-withdrawing group replaces a hydrogen on the α-carbon of the aldehyde, as in chloroethanal. In this case, the resonance form with the positive charge on the carbonyl carbon is destabilized by the partial positive charge on the adjacent carbon induced by the electron-withdrawing Cl atom. Consequently, its contribution to the resonance hybrid decreases and the modified aldehyde is less stable. In other words, the carbonyl compound is destabilized by the electron-withdrawing group.

Figure 8.4 Energy changes in the hydration of carbonyl compounds.

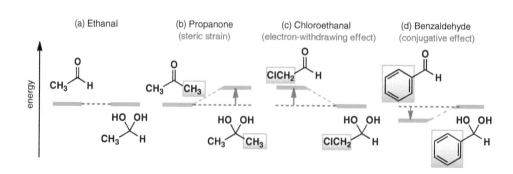

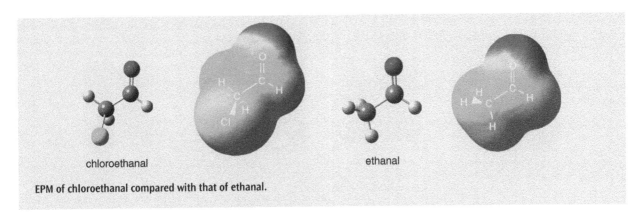

EPM of chloroethanal compared with that of ethanal.

The destabilizing electrostatic effect is weaker in the hydrate which, therefore, becomes more stable relative to the carbonyl compound (K_h becomes larger). This effect is illustrated in Figure 8.4(c): chloroethanal is electronically destabilized, so its energy level is raised; the electrostatic effect is smaller in the hydrate, so K_h is now greater than unity.

Comparison of the K_h values for ethanal (CH_3CHO) and benzaldehyde (PhCHO) in Table 8.2 illustrates another type of electronic effect. PhCHO is stabilized by a conjugative (resonance) interaction between the Ph group and the C=O bond, an effect absent in CH_3CHO. This conjugative stabilization in PhCHO is lost upon formation of its hydrate (Figure 8.4(d)), so the K_h value for hydration of PhCHO is much smaller than that for CH_3CHO.

> **Exercise 8.3**
>
> Draw the resonance forms which contribute towards the resonance hybrid of benzaldehyde (PhCHO).

There are similar steric and electronic effects upon equilibrium constants for other nucleophilic addition reactions of carbonyl compounds. For the same reasons, similar trends are also observed for the rate constants of all these reactions, and the general order of reactivity of carbonyl compounds toward nucleophilic addition is:

$$H_2C=O \quad > \quad RCHO \quad > \quad R_2C=O$$

methanal   aldehydes   ketones

> **Exercise 8.4**
>
> Explain the order of rate constants for addition of CN^-.
>
> (a) $CH_3CHO > (CH_3)_3CCHO$ (b) $C_6H_5CHO < p\text{-}NO_2C_6H_4CHO$

8.3.2 The mechanism of hydration of carbonyl compounds and catalysis

In the hydration of aldehydes and ketones, eqn 8.5 above, water adds as a nucleophile to the carbonyl group. However, water is not a very powerful nucleophile and the carbonyl is only a modest electrophile.

Hydroxide ion is a much more powerful nucleophile than water so, under basic aqueous conditions, the first step of the hydration of a carbonyl compound is the addition of hydroxide ion (HO⁻) to the C=O double bond. The anionic tetrahedral intermediate then abstracts a proton from a water molecule of the solvent to produce the hydrate and HO⁻ (Scheme 8.3). This two-step reaction is similar to that for cyanohydrin formation (Scheme 8.1) but, in this reaction, the hydroxide is used in the first step then regenerated in the second. Such a reaction is said to be **base-catalysed** since the HO⁻ facilitates the reaction but is not consumed in the overall process.

Scheme 8.3 Base-catalysed hydration of a carbonyl group.

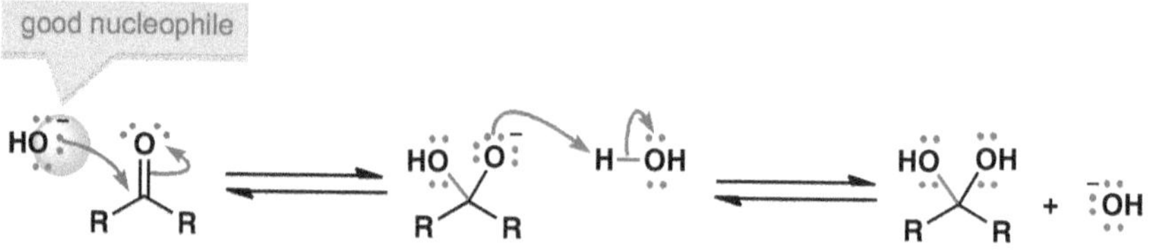

Under acidic aqueous conditions, the carbonyl oxygen is first reversibly protonated on a lone pair by H_3O^+ (Scheme 8.4). In this *pre-equilibrium*, the carbonyl group becomes activated towards nucleophilic attack by a (relatively weakly nucleophilic) solvent water molecule. The product of this nucleophilic attack is a cationic tetrahedral intermediate. In the final step, a proton is released from the intermediate to give the hydrate and regenerate H_3O^+, so the overall reaction is **acid-catalysed**.

EPM of protonated methanal: the shades of blue show the distribution of the positive charge.

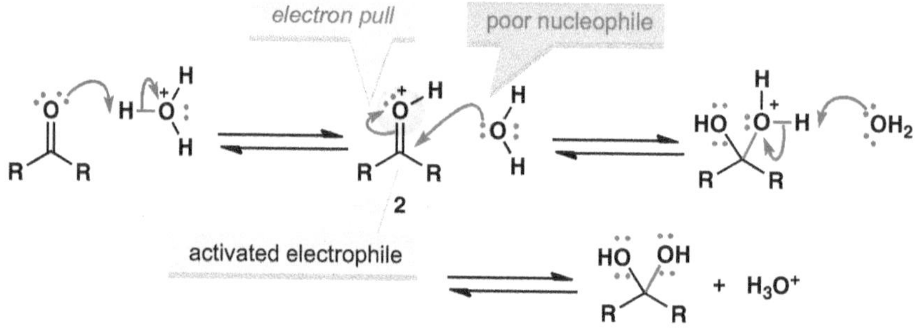

Scheme 8.4 Acid-catalysed hydration of a carbonyl group.

In the above, we see that a more powerful nucleophile is generated by base catalysis than in the uncatalysed reaction whereas the substrate is rendered more susceptible to nucleophilic attack (more electrophilic) by acid catalysis.

Example 8.3

Show how resonance explains the enhanced electrophilicity of the carbonyl group upon protonation.

Solution

A protonated ketone (**2**) may be represented as a hybrid (**2c**) of two resonance contributors, one (**2a**) with a formal positive charge on the oxygen (an oxonium ion form) and the other (**2b**) with a formal positive charge on the carbon (a carbenium ion form).

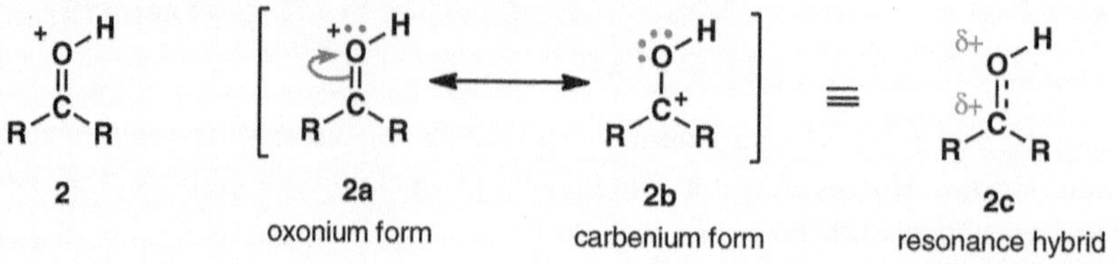

2	**2a** oxonium form	**2b** carbenium form	**2c** resonance hybrid

(continues . . .)

(... continued)

The carbenium ion form here (**2b**) does not include a separation of opposing formal charges as the corresponding contributor (**1b**) does to the unprotonated carbonyl **1** in Section 8.1. p.166 Consequently, **2b** will contribute to the resonance hybrid **2c** to a greater extent than **1b** does to the hybrid **1c**.

8.3.3 Reversibility of hydration and oxygen isotope exchange

We have already seen that hydration of the carbonyl group is reversible. The mechanism of the back reaction is exactly the reverse of the forward reaction.* Consequently, acid- and base-catalysed dehydrations of a 1,1-diol (hydrate) proceed as shown in Scheme 8.5. In the acid-catalysed reaction (eqn 8.6), a very poor potential nucleofuge (HO$^-$) is turned into a good one (H$_2$O) by protonation; in the base-catalysed reaction (eqn 8.7), the same very poor (anionic) nucleofuge is expelled by a push of electron density from the adjacent O$^-$ group.

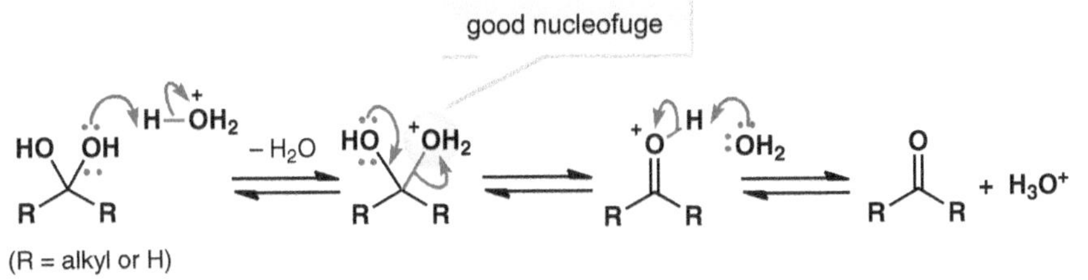

good nucleofuge

$$+ H_3O^+ \qquad (8.6)$$

(R = alkyl or H)

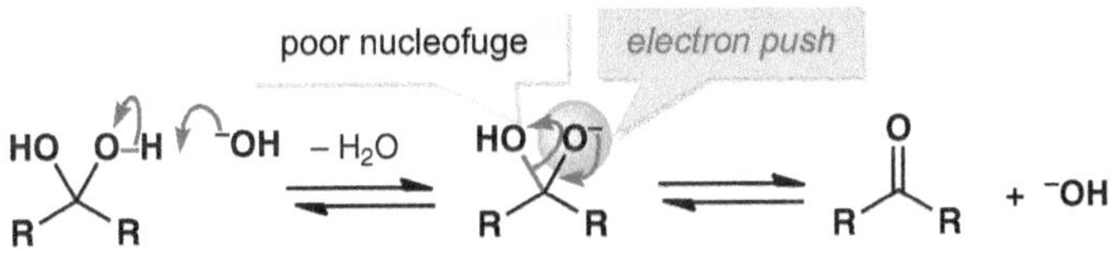

poor nucleofuge   *electron push*

$$+ \ ^-OH \qquad (8.7)$$

Scheme 8.5 Acid- and base-catalysed dehydrations of a 1,1-diol.

*The transition structure in a single-step, thermally induced, reversible reaction (see Chapter 7) may be formed from either the reactant or the product. It follows that there is a common lowest energy pathway for a reversible thermal chemical reaction which may be followed in the forward direction or the reverse. This is sometimes called the *principle of microscopic reversibility*. It is not generally true of reactions which involve photoexcited states.

The two hydroxy groups of a hydrate molecule formed from a single water molecule and a single molecule of a carbonyl compound are equivalent. Consequently, in the reverse reaction, there is a 50% probability that the oxygen of the original carbonyl group will be lost in the water molecule. It follows that, if a carbonyl compound undergoes reversible hydration in water which is labelled to some extent with the ^{18}O isotope (eqn 8.8), the oxygen of the carbonyl will become isotopically labelled. In this **isotope exchange** reaction, which may be catalysed by either H$_3$O$^+$ or HO$^-$, the label is said to have become *scrambled* between the carbonyl compound and the solvent.

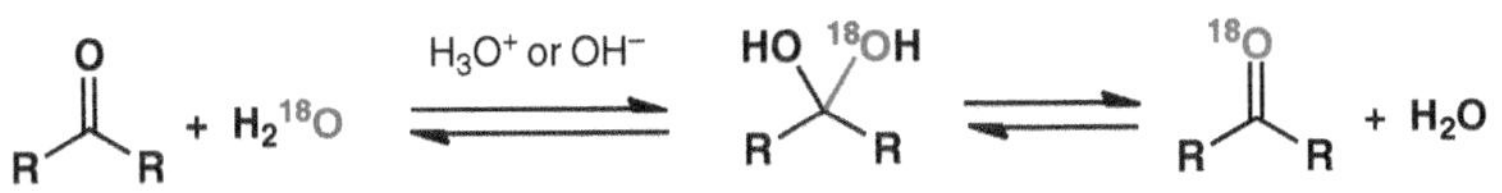

$$+ H_2O \qquad (8.8)$$

Example 8.4

Write a curly arrow mechanism for acid-catalysed isotope exchange of an ^{18}O-labelled ketone in aqueous solution.

Solution

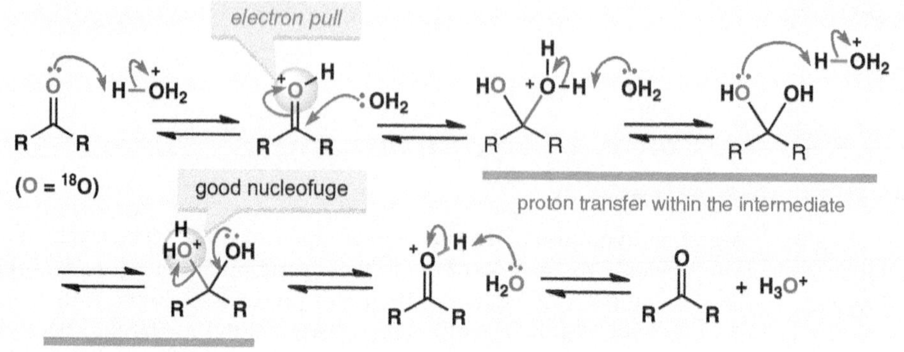

Exercise 8.5

Propose a curly arrow mechanism for base-catalysed isotope exchange of an ^{18}O-labelled ketone in aqueous solution.

8.4 Addition of Alcohols to Aldehydes and Ketones

8.4.1 Formation of hemiacetals

Addition of an alcohol to an aldehyde or ketone (using the alcohol as the solvent) proceeds in the same way as the addition of water, which was discussed in the preceding section. The similarity is apparent if the mechanisms for base- and acid-catalysed addition of an alcohol ROH (Schemes 8.6 and 8.7) are compared with those for hydration (Schemes 8.3 and 8.4). The product of alcohol addition to an aldehyde or ketone is called a **hemiacetal**.

Scheme 8.6 Base-catalysed addition of an alcohol to a carbonyl group.

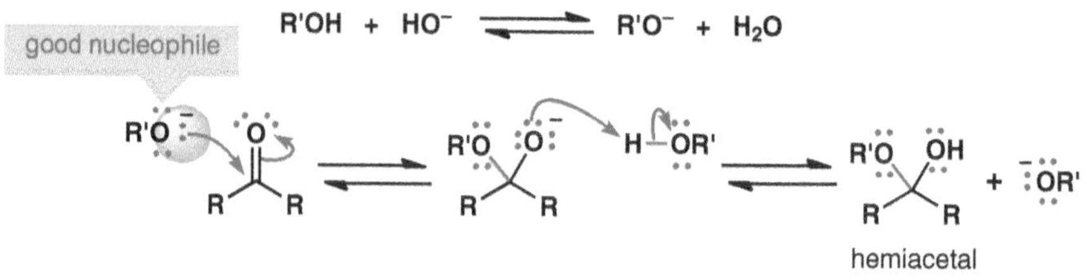

Scheme 8.7 Acid-catalysed addition of an alcohol to a carbonyl group.

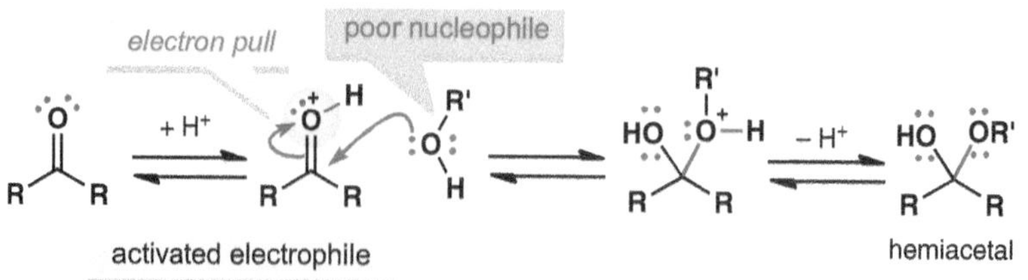

These reactions, like hydration, are reversible and the hemiacetal cannot usually be isolated due to the ease of the reverse reaction. The stability of hemiacetals in solution, like that of hydrates, is dependent on steric and electronic effects. However, a cyclic hemiacetal formed by intramolecular reaction of a hydroxy group elsewhere in the carbonyl compound is more stable than one formed in an intermolecular reaction. Five- and six-membered cyclic hemiacetals are quite often observed (eqn 8.9).

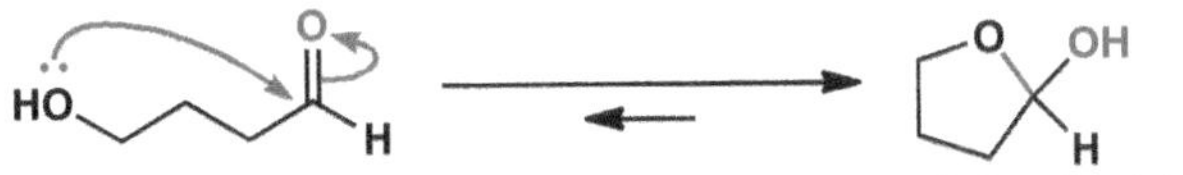

(8.9)

a cyclic hemiacetal

Glucose, a carbohydrate, exists essentially as an equilibrium mixture of two stereoisomeric six-membered cyclic hemiacetals in solution (eqn 8.10), and both isomers can be isolated as stable crystalline compounds. The chair forms are similar to cyclohexane with most of the hydroxy groups in equatorial positions (see Sub-section 24.1.1).

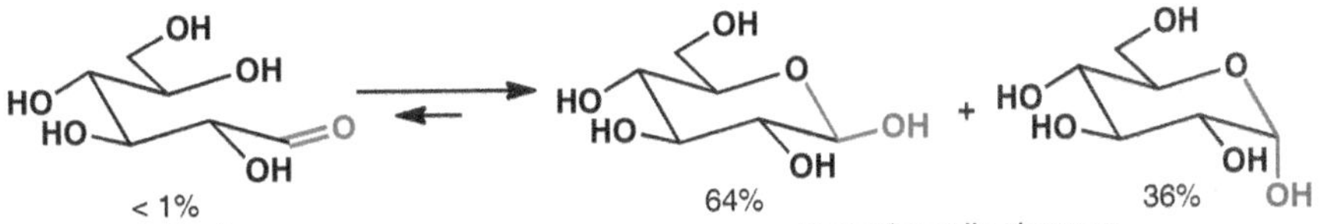

(8.10)

< 1%
open-chain glucose

64%   36%
isomeric cyclic glucoses

The more favourable intramolecular formation of a cyclic hemiacetal compared with the intermolecular reaction is due to an entropy effect. In the intermolecular reaction, two molecules become one, which appreciably decreases the entropy of the system and contributes unfavourably to the free energy of hemiacetal formation. In the intramolecular reaction, however, one molecule of hydroxy aldehyde (or hydroxy ketone) forms one molecule of the hemiacetal with only a small adverse change in entropy.

Exercise 8.6

Draw a structure of the hemiacetal formed from each of the following combinations of carbonyl compound and alcohol.

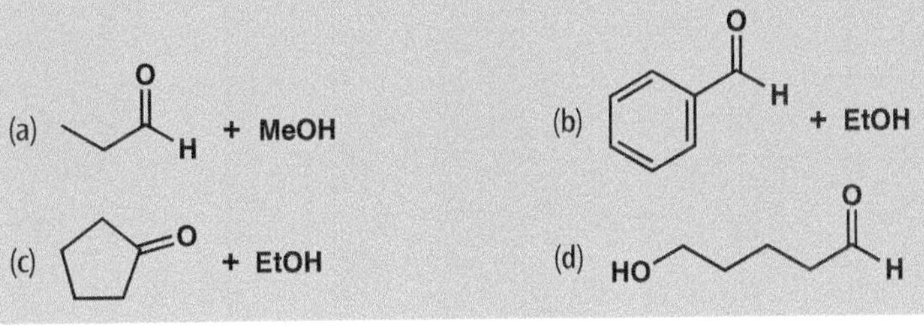

(a) [structure] + MeOH   (b) [structure] + EtOH

(c) [structure] + EtOH   (d) [structure]

Exercise 8.7

Propose a reaction mechanism for the base-catalysed hydrolysis of a hemiacetal.

8.4.2 Formation of acetals

Consider the reverse of the acid-catalysed formation of a hemiacetal from a carbonyl compound dissolved in an alcohol, R'OH (Scheme 8.7). The protonated hemiacetal intermediate is shown again as **3** in Scheme 8.8 where departure of the alcohol and formation of the (protonated) carbonyl compound (**2**) occur by bond fission at the protonated oxygen of **3**. An isomeric structure (**4**) is also possible for the protonated hemiacetal and is present in equilibrium with **3**. This form is protonated on the OH oxygen and, because OH and OR' are similarly (weakly) basic, **3** and **4** will be present at similar (low) concentrations in solution.

Just as intermediate **3** releases R'OH to give the protonated carbonyl compound **2**, intermediate **4** loses H_2O to give cationic intermediate **5** which has a structure similar to that of **2**. The electrophilic carbon of **5**, therefore, suffers nucleophilic attack by another

Scheme 8.8 Reversible acid-catalysed reaction of a carbonyl compound with an alcohol.

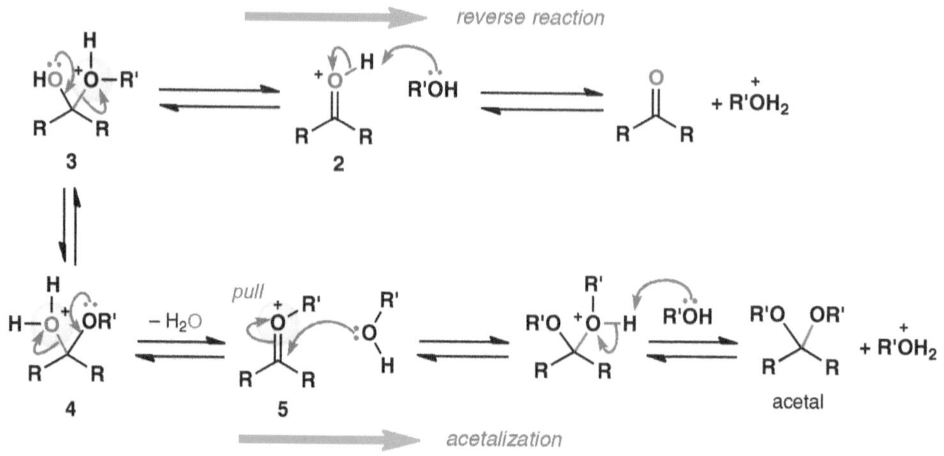

molecule of R'OH to give an adduct, deprotonation of which yields a new product called an **acetal**. The acetal results from addition of two molecules of R'OH to the carbonyl compound, with the result that the carbonyl oxygen is replaced by two alkoxy groups and a water molecule is eliminated (eqn 8.11).

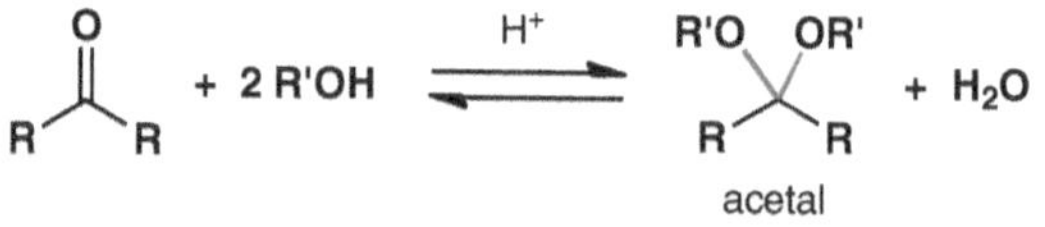

(8.11)

Exercise 8.8

Write the expression for the equilibrium constant for the reversible acid-catalysed formation of the acetal from $R_2C{=}O$ and R'OH, and explain how the reaction can be driven in either direction according to the choice of reaction conditions.

In practice, **acetalization** (the forward direction of eqn 8.11) is usually carried out by dissolving the carbonyl compound in the alcohol containing a catalytic amount of anhydrous acid, such as H_2SO_4 or HCl. Water is liberated in the reaction and the equilibrium is driven towards acetal formation by the large excess of alcohol, e.g. eqn 8.12. The reaction, however, is readily reversible so an acetal may be hydrolysed in good yield by the large excess of water in dilute aqueous strong acid, e.g. eqn 8.13.

Acid-catalysed acetalization of an aldehyde:

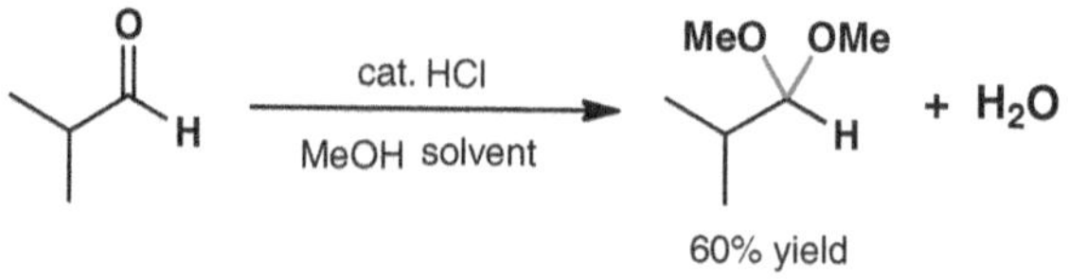

(8.12)

Acid-catalysed hydrolysis of an acetal:

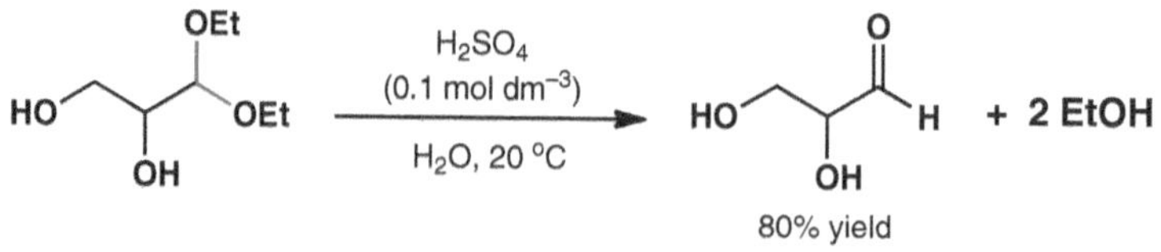

(8.13)

Give a curly arrows reaction mechanism for acid-catalysed hydrolysis of an acetal.

Example 8.5

Solution

R'O OR'   H—OH₂⁺ ⇌ R'O O—R' − R'OH *pull* O⁺—R' :OH₂ ⇌
R   R   H   R   R   R   R

R'O O—H *H⁺ transfer* R'—O⁺ OH − R'OH O⁺ H OH₂ ⇌ O + H₃O⁺
R   R   R   R   R   R   R   R

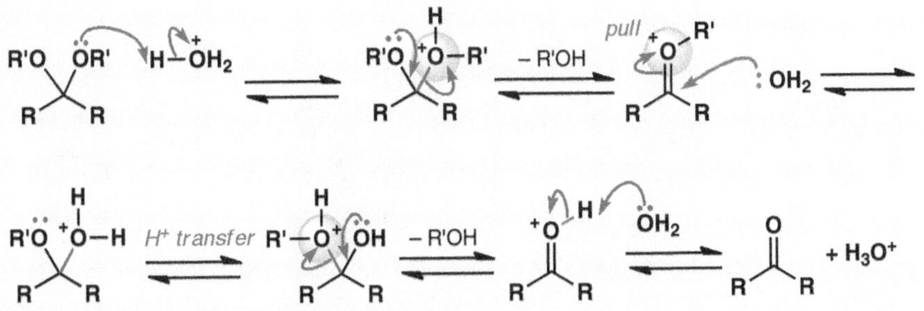

The interconversion of the intermediate protonated hemiacetals (labelled 'H⁺ transfer') occurs mainly by stepwise bimolecular (intermolecular) proton transfer reactions involving the solvent, as shown in the solution to Example 8.4.

Give the stepwise mechanism for the H⁺ transfer in the reaction scheme in the solution to Example 8.5.

Exercise 8.9

Reaction of aldehyde RCHO with ethane-1,2-diol and an anhydrous acid gives a five-membered cyclic acetal. Write a curly arrows reaction mechanism for this transformation.

Exercise 8.10

Draw structures of the hemiacetal and acetal formed from each combination of carbonyl compound and alcohol.

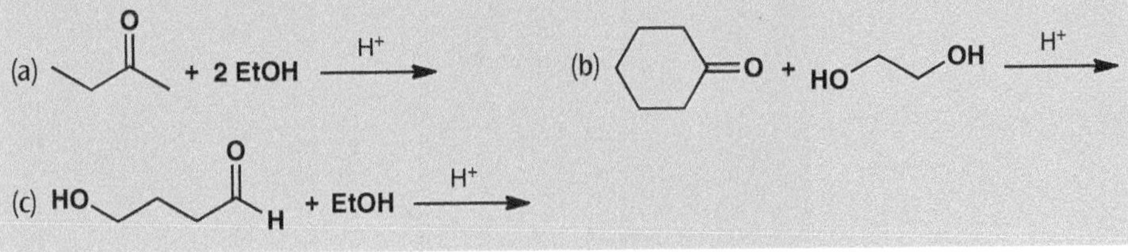

Exercise 8.11

(a) [structure] + 2 EtOH —H⁺→

(b) [cyclohexanone]=O + HO⌒OH —H⁺→

(c) HO⌒⌒CHO + EtOH —H⁺→

What are the products of acid-catalysed hydrolysis of each acetal?

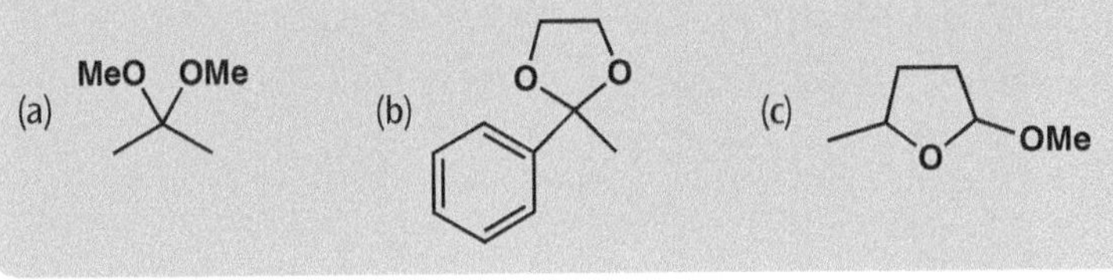

Exercise 8.12

(a) MeO OMe [structure]   (b) [structure]   (c) [structure] OMe

Hemiacetals are formed either by acid or base catalysis from carbonyl compounds and alcohols. Why are acetals not formed under basic conditions?

Example 8.6

Solution
A hemiacetal is a kind of alcohol and an acetal is a kind of ether, so the reaction to form an acetal from a hemiacetal is the replacement of OH by OR'. Although such a reaction may be thermodynamically favourable, it does not take place directly because there is no viable mechanism. In particular, it is because OH⁻ is such a poor leaving group (nucleofuge) in nucleophilic

(continues ...)

(... continued)

substitution reactions; such matters will be discussed further when reactions of alcohols are covered in Chapter 14.

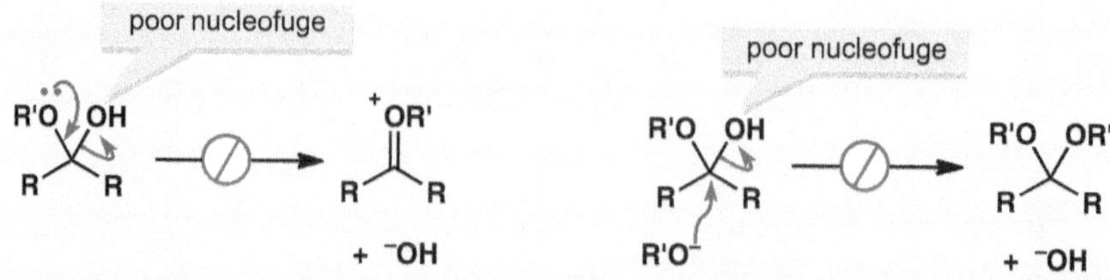

Under basic conditions, the only reaction of a simple hemiacetal is deprotonation from the OH group, which would simply drive the reaction back toward the carbonyl compound and the alcohol. In acidic solution, however, protonation of the OH group gives $-(OH_2)^+$, a very good leaving group, so the mechanism of Scheme 8.8 which drives the reaction toward formation of the acetal becomes possible.

The above illustrates a general principle. A thermodynamically favourable transformation must have an accessible reaction pathway in order to take place.

Exercise 8.13

Why cannot acetals be hydrolysed under basic conditions?

8.4.3 Addition of thiols

Thiols, RSH, are the sulfur analogues of alcohols, and they react with an aldehyde or ketone to form dithioacetals (also called thioacetals); the mechanism of the reaction is analogous to that for the formation of an acetal from an alcohol. The reaction is normally carried out in an aprotic solvent such as $CHCl_3$ with either a protic or a Lewis acid as catalyst (eqn 8.14). A thiol is generally a better nucleophile than the corresponding alcohol.

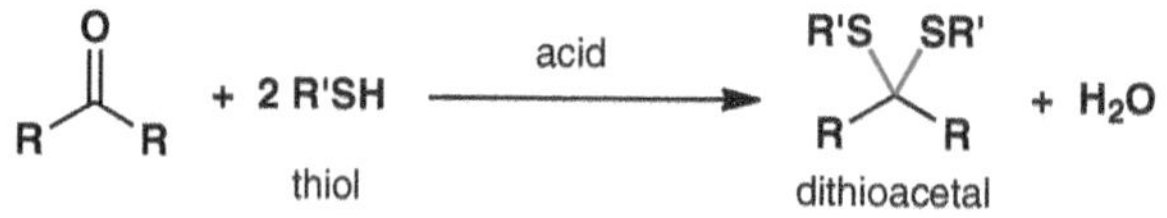

$$(8.14)$$

Exercise 8.14

Propose a mechanism for the dithioacetal formation of eqn 8.14 with BF_3 as a Lewis acid catalyst.

Acetals and thioacetals are used as protecting groups for the carbonyl group during the transformation of other functionality in complex aldehydes and ketones. This protection–deprotection strategy is important in planning organic syntheses, and will be discussed in Chapter 10. Dithioacetals can also be used for the reduction of >C=O to >CH₂ (see Sub-section 10.2.2).

8.5 Addition of Bisulfite to Aldehydes and Ketones

Sodium hydrogensulfite (also called sodium bisulfite), $NaHSO_3$, reacts with aldehydes and ketones to give crystalline products called bisulfite adducts. The bisulfite ion is usually written with a formal negative charge on the oxygen as in eqn 8.15. However, it reacts with the carbonyl carbon through the sulfur atom which, like the oxygens, carries a lone pair of electrons, and the HOMO of the nucleophile corresponds to the lone pair on the S.

Sulfur, being an element in the third period of the periodic table, can accommodate more than an octet of electrons in its valence shell; see also Sub-section 14.6.4.

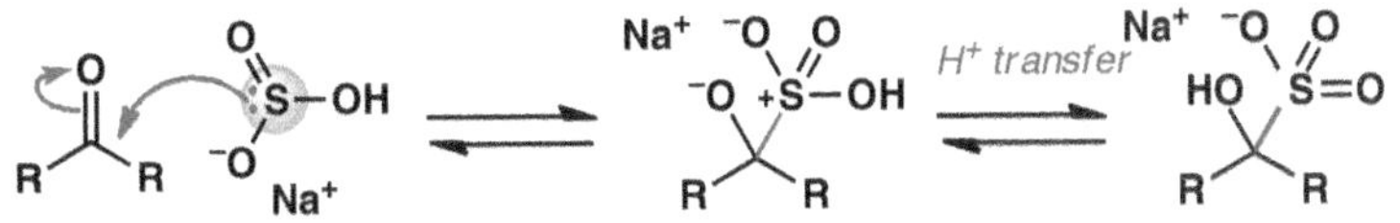

$$\text{(8.15)}$$

This addition reaction is reversible, and the adduct can be converted back to the aldehyde or ketone easily by hydrolysis in acidic or basic aqueous solution. As the adducts are usually easily isolated and purified as crystalline solids, they were previously used for the isolation and purification of aldehydes and ketones.

8.6 Imines and Enamines

8.6.1 Reactions of primary amines with aldehydes and ketones

Primary amines react with aldehydes and ketones (often without acid catalysis) to give, initially, tetrahedral intermediate adducts (Scheme 8.9). The neutral tetrahedral intermediate in this reaction is usually even less stable than a 1,1-diol, and undergoes acid-catalysed dehydration to give an **iminium ion** containing a C=N double bond. Subsequent loss of a proton gives an **imine** (also known as a Schiff's base after the discoverer), which may be isolated.

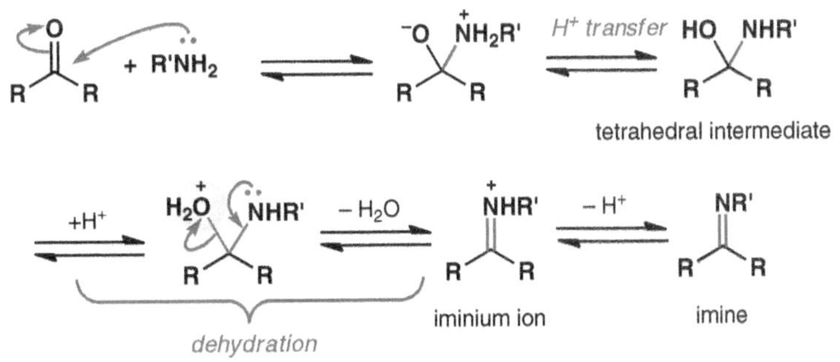

tetrahedral intermediate

iminium ion   imine

dehydration

Scheme 8.9 Reaction mechanism for acid-catalysed imine formation.

The overall reaction of imine formation is described in two steps by eqn 8.16.

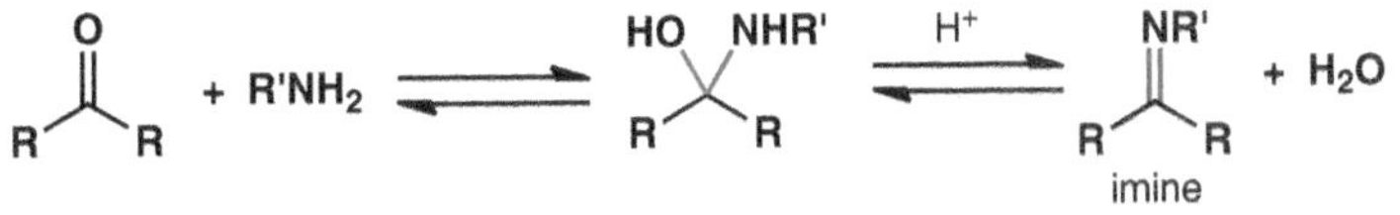

$$\text{(8.16)}$$

imine

Amine derivatives such as hydroxylamine $HONH_2$ or (substituted) hydrazines $R'NHNH_2$, which have an O or N group directly attached to the N, give more stable imine derivatives due to conjugation of the lone pair electrons of the heteroatom with the C=N double bond; see eqns 8.17 and 8.18. The products (sometimes called condensation products) are usually crystalline and easily isolated; they were once used as derivatives of aldehydes and ketones for purposes of purification and identification.

A reaction in which two molecules combine with an accompanying loss of a small molecule such as water or an alcohol is sometimes called a *condensation* reaction.

$$RCHO + H_2NOH \xrightarrow{H^+} RCH{=}NOH + H_2O \qquad (8.17)$$

hydroxylamine   oxime

$$RCHO + H_2NNH_2 \xrightarrow{H^+} RCH{=}NNH_2 + H_2O \qquad (8.18)$$

hydrazine   hydrazone

Exercise 8.15

Draw resonance forms of an oxime and a hydrazone.

Exercise 8.16

Semicarbazide forms a semicarbazone with an aldehyde. Why does reaction take place at only the one amino group of semicarbazide as indicated?

$$RCHO + H_2NNHCNH_2 \xrightarrow{H^+} RCH{=}NNHCNH_2 + H_2O$$

semicarbazide   semicarbazone

Example 8.7

Propose a reaction mechanism for the acid-catalysed hydrolysis of an oxime.

Solution

The mechanism for the acid-catalysed hydrolysis of an oxime is simply the reverse of the mechanism for its formation (principle of microscopic reversibility, see above). In practice, whether the reversible reaction yields oxime or hydroxylamine plus carbonyl compound depends upon the reaction conditions. Hydroxylamine is usually nucleophilic enough to add to the carbonyl group of an aldehyde or ketone without catalysis. The last step of the hydrolysis (departure of hydroxylamine) is just the reverse of addition and takes place from the zwitterionic tetrahedral intermediate.

In the last step, the equilibrium concentration of the zwitterionic intermediate is small, but the electron push from the O⁻ facilitates the departure of hydroxylamine as leaving group. Its conjugate base (HONH⁻) is a much poorer nucleofuge, and cannot leave from the more abundant neutral intermediate.

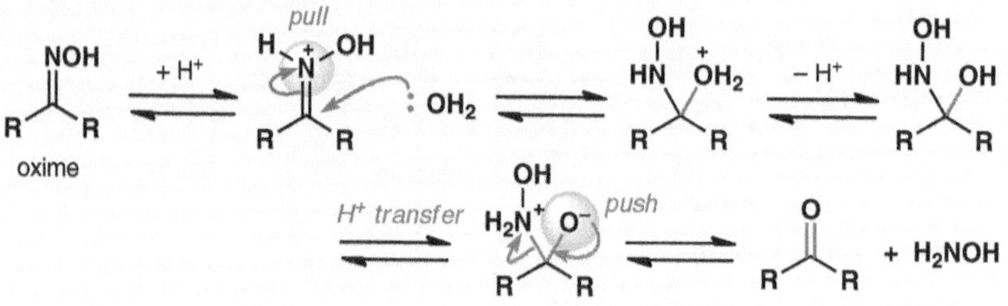

Exercise 8.17

Draw a structural formula of the product in each of the following reactions.

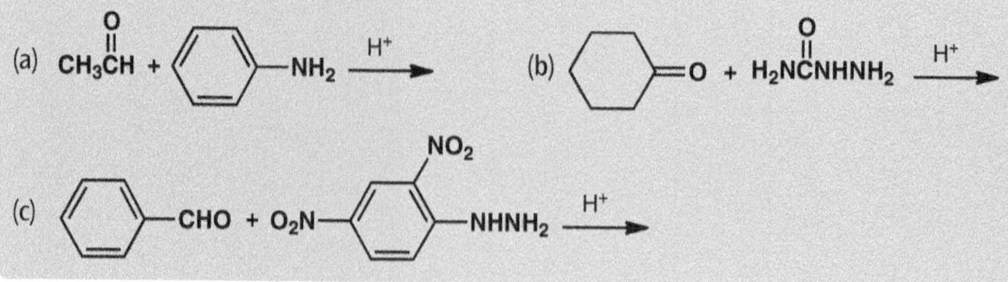

(a) $CH_3CH{=}O$ + [benzene ring]—NH_2 $\xrightarrow{H^+}$

(b) [cyclohexane ring]$=O$ + $H_2NCNHNH_2$ $\xrightarrow{H^+}$

(c) [benzene ring]—CHO + O_2N—[benzene ring with NO_2]—$NHNH_2$ $\xrightarrow{H^+}$

8.6.2 Reactions of secondary amines with aldehydes and ketones

Secondary amines, like primary amines, react with carbonyl groups of aldehydes and ketones to form iminium ions. However, the iminium ion formed from a secondary amine has no proton on the nitrogen which may be abstracted. But if there is a hydrogen on an adjacent carbon, it can be lost to give a C=C double bond, as shown in the last step of Scheme 8.10 for an aldehyde. The product is an **enamine** (en+amine), a highly nucleophilic alkene which is also readily hydrolysed in aqueous acid.

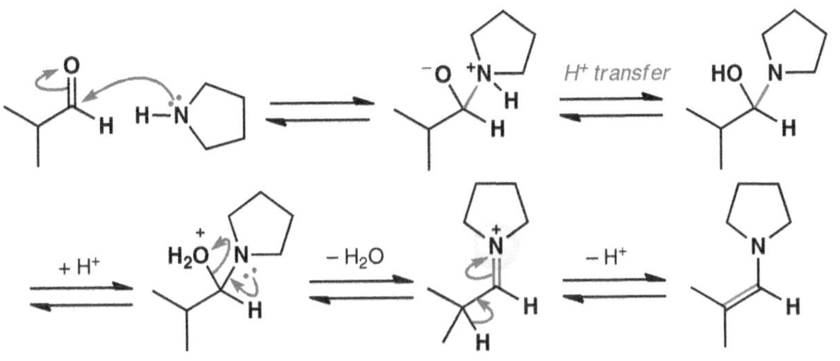

Scheme 8.10 Reaction mechanism for the formation of an enamine.

an iminium ion   an enamine

Write a curly arrow reaction mechanism for the acid-catalysed hydrolysis of the following enamine.

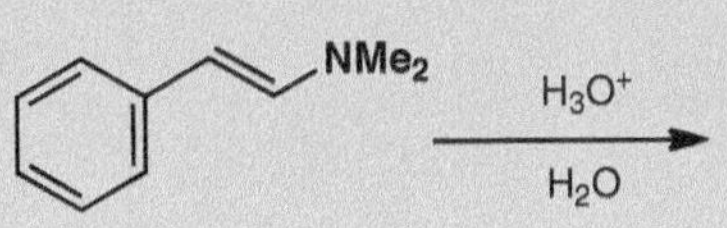

Exercise 8.18

8.7 The Wittig Reaction

The **Wittig reaction** converts the C=O bond of an aldehyde or ketone into a C=C double bond using a charge-separated reagent called a *phosphonium ylide*. The ylide has a P^+–C^- bond and a minor resonance contributor with a P=C double bond since P can accommodate more than an octet of valence electrons. It is generated by treating a phosphonium salt with a strong base such as sodium hydride (eqn 8.20). The phosphonium salt is the phosphorus analogue of an ammonium salt, and is prepared from a phosphine R_3P (usually Ph_3P) and an alkyl halide (eqn 8.19).

This reaction was discovered by the German chemist, G. Wittig (1897–1987), who shared the 1979 Nobel Prize in Chemistry with the American, H.C. Brown (1912–2004).

$$(C_6H_5)_3P : \quad CH_3-Br \longrightarrow (C_6H_5)_3\overset{+}{P}-CH_3 \quad Br^- \qquad (8.19)$$

triphenylphosphine   a phosphonium salt

$$(C_6H_5)_3\overset{+}{P}-CH_3 \ Br^- \xrightarrow[-H_2]{NaH} \left[(C_6H_5)_3\overset{+}{P}-\overset{-}{C}H_2 \longleftrightarrow (C_6H_5)_3P{=}CH_2 \right] \qquad (8.20)$$

a phosphonium ylide

A typical example of the Wittig reaction is given in Scheme 8.11. The nucleophilic carbanion residue of the phosphonium ylide will be attracted electrostatically to the

Panel 8.2 Imines in biochemical reactions

Imines are important intermediates in biochemical transformations. Typical reactions involve pyridoxal phosphate, a coenzyme derived from vitamin B_6, which catalyses transformations of amino acids. The reaction scheme below illustrates the oxidative deamination of alanine to give pyruvic acid. Pyridoxal first bonds to the amino group of an enzyme through an imine linkage with the loss of a molecule of water, and becomes protonated. Imine exchange with alanine then gives the protonated alanine–pyridoxal adduct. Deprotonation from the alanine residue relocates the imine bond, and hydrolysis of the isomeric imine leads to pyridoxamine and pyruvic acid. Other important physiological processes of amino acids such as transamination, racemization, and decarboxylation also take place through their imine adducts with pyridoxal.

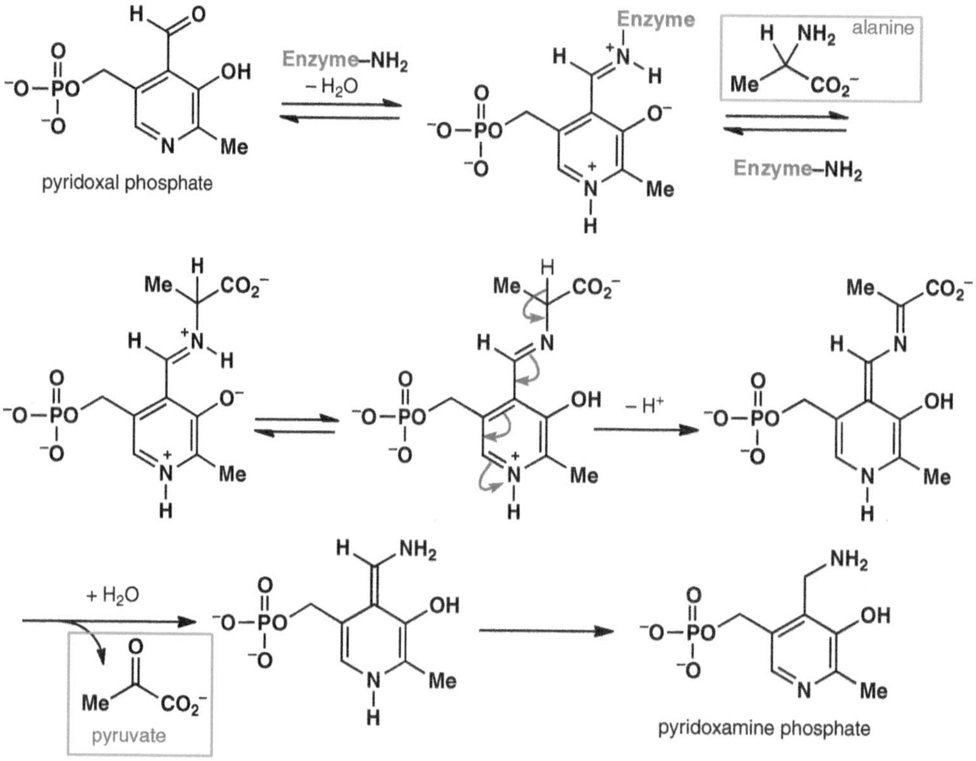

Initial formation of a zwitterionic intermediate (called a betaine) by nucleophilic addition of the phosphonium ylide to the carbonyl was once postulated, but this is not supported by any evidence. Presumably, the transition structure of the addition is appreciably dipolar.

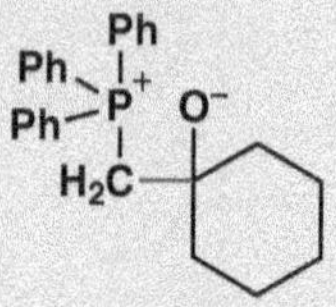

carbonyl of an aldehyde or ketone, but the rebonding appears to be concerted. The product, a four-membered cyclic intermediate called an oxaphosphetane, undergoes elimination of $R_3P=O$ (a very stable molecule) to form a C=C double bond. When *cis–trans* isomeric alkenes are possible, their ratio depends on the structures of the reagents.

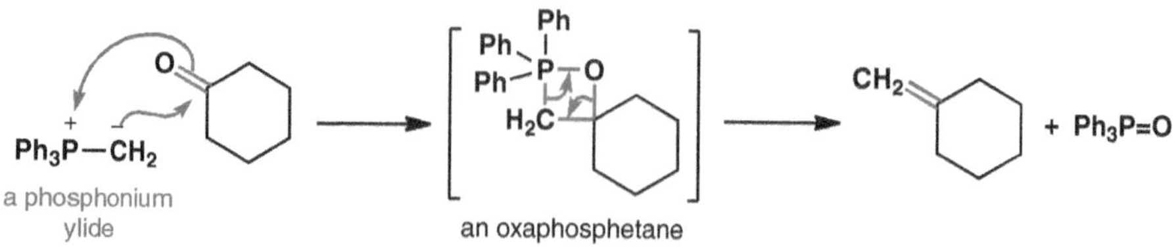

Scheme 8.11 An example of the Wittig reaction.

The Wittig reaction has been a useful synthetic method for preparing alkenes without any ambiguity about the location of the double bond. A classic example from the field of natural product synthesis is shown in eqn 8.21.

vitamin A aldehyde
(retinal)

a phosphonium ylide

(8.21)

β-carotene
(orange pigment in carrots and other vegetables)

However, the reaction is not very atom-economical, especially for alkenes of relatively low molecular weight, in the sense that there is wastage of a large part (PPh$_3$) of the phosphonium ylide reagent. In subsequent chapters, we shall discuss alternative methods for alkene synthesis, each with its own advantages and disadvantages.

Other well-known reactions of aldehydes and ketones are covered in later chapters. Hydride reduction and organometallic additions are synthetically important further nucleophilic additions to carbonyl compounds, and will be discussed in Chapter 10; related reactions of carboxylic acid derivatives are covered in Chapter 9. Enols (isomers of carbonyl compounds) and their conjugate bases (enolates) are nucleophilic forms of carbonyl compounds; these important compounds and their reactions are included in Chapter 17.

Summary

- The carbonyl (>C=O) group in aldehydes and ketones has a polar double bond which reacts with nucleophiles at the electrophilic carbon. The initial nucleophilic attack gives a tetrahedral intermediate which, in protic solvents, becomes protonated; the overall stepwise reaction is often reversible.

- The reactions are adversely affected by steric hindrance and stabilizing resonance interactions in the carbonyl compound, and facilitated by electron-withdrawing groups.

- Nucleophilic additions of water and alcohols are subject to **acid and base catalysis**. The nucleophile in the base-catalysed reactions is the strongly nucleophilic HO$^-$ or R'O$^-$ (the conjugate base of the solvent). In the acid-catalysed reactions, a weakly nucleophilic neutral solvent molecule, H$_2$O or R'OH, adds to the protonated carbonyl group (>C=OH$^+$), which is much more electrophilic than the unprotonated carbonyl in the uncatalysed reaction.

- These reactions illustrate a general principle: the nucleophile is made more reactive by base catalysis and the electrophile is made more reactive by acid catalysis.

- Addition of amines to aldehydes and ketones is usually followed by the elimination of water from the tetrahedral intermediate to give **imines** with primary amines, or **enamines** with secondary amines.

- Cyanide and bisulfite anions, and thiols, are other nucleophiles which add to the carbonyl of aldehydes and ketones.

- The synthetically important Wittig reaction with phosphonium ylides converts C=O to C=C.

Problems

8.1 Give the structures of the carbonyl compounds and alcohols needed to form each of the following unstable hemiacetals in solution.

(a) 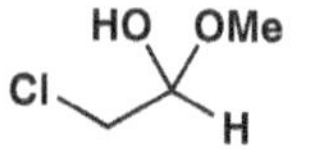 [structure: HO OMe group on carbon with Cl, H]

(b) 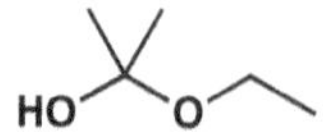 [structure: HO O with ethyl]

(c) 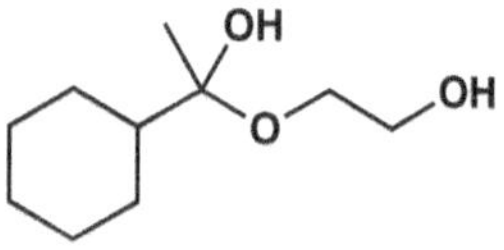 [structure: cyclohexane with OH and O-CH₂CH₂-OH]

(d) 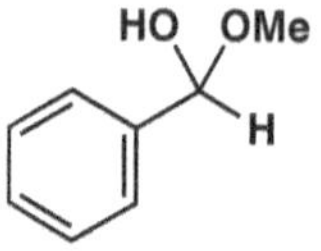 [structure: benzene with HO OMe on carbon with H]

8.2 Draw structures of the carbonyl compound and alcohol needed to give each of the following acetals.

(a) 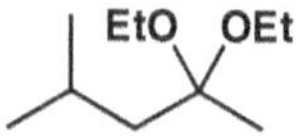 [structure: EtO OEt]

(b) 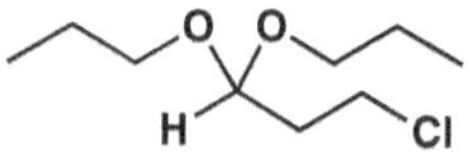 [structure: O O with H and Cl]

(c) 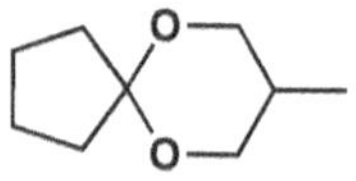 [structure: spiro bicyclic acetal]

(d)  [structure: dioxolane on cyclohexane]

(e) 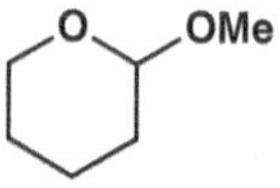 [structure: tetrahydropyran with OMe]

8.3 Show structures of the hemiacetal and acetal formed from each combination of carbonyl compound and alcohol below.

(a) 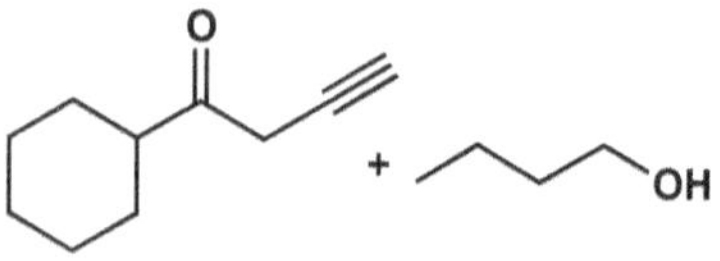 [structure: cyclohexyl ketone with alkyne] + [structure with OH]

(b) 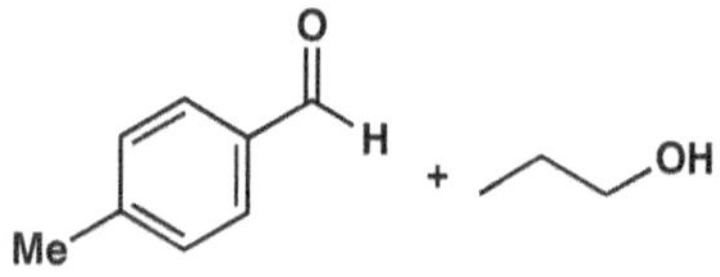 [structure: Me-benzaldehyde] + [structure with OH]

(c) 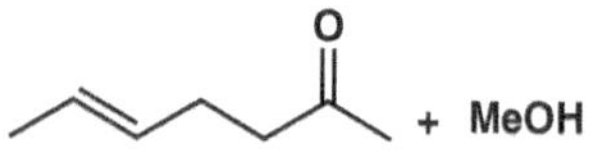 [structure: enone] + MeOH

(d) 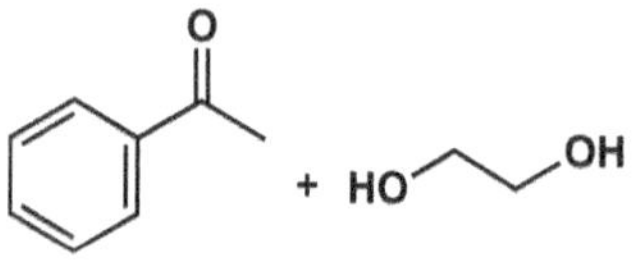 [structure: acetophenone] + HO-CH₂CH₂-OH

8.4 What is the main product of each of the following reactions?

(a) 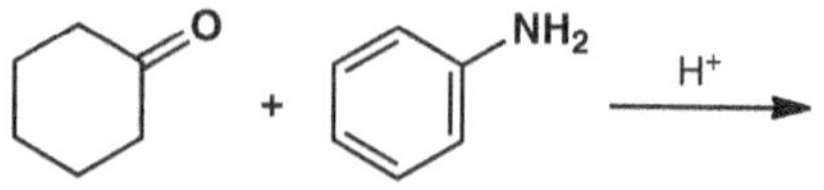 [structure: cyclohexanone] + [structure: aniline NH₂] $\xrightarrow{H^+}$

(b) 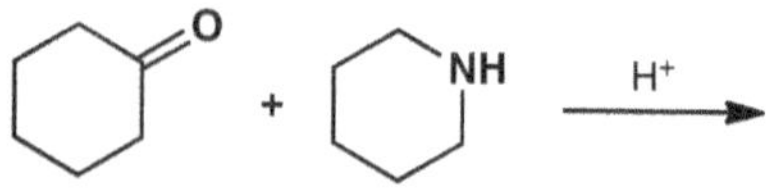 [structure: cyclohexanone] + [structure: piperidine NH] $\xrightarrow{H^+}$

(c) 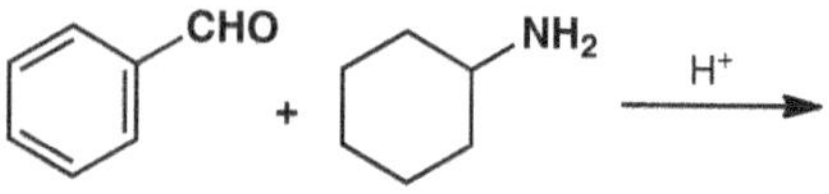 [structure: benzaldehyde CHO] + [structure: cyclohexylamine NH₂] $\xrightarrow{H^+}$

(d) 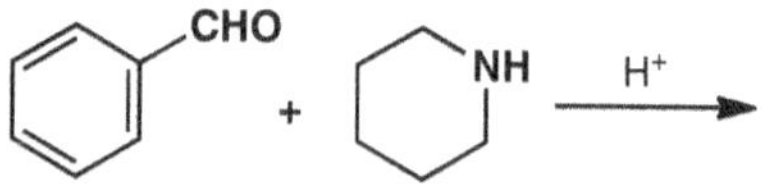 [structure: benzaldehyde CHO] + [structure: piperidine NH] $\xrightarrow{H^+}$

8.5 How would you prepare each of the following alkenes by the Wittig reaction?

(a) 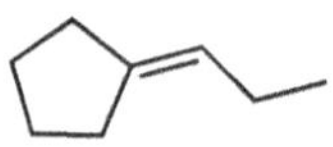 [structure: cyclopentylidene alkene]

(b) 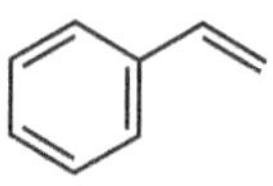 [structure: styrene derivative]

8.6 Explain the relative reactivities of ethanal and chloroethanal in cyanohydrin-forming reactions.

8.7 Explain the following relative magnitudes of equilibrium constants for hydration.

(a) $CH_3CHO > (CH_3)_2CHCHO$ (b) $ClCH_2COCH_3 < Cl_2CHCOCH_3$

8.8 Explain the following relative magnitudes of equilibrium constants for hydration.

[structures: CHO with OMe para] < [CHO] < [CHO with CN para]

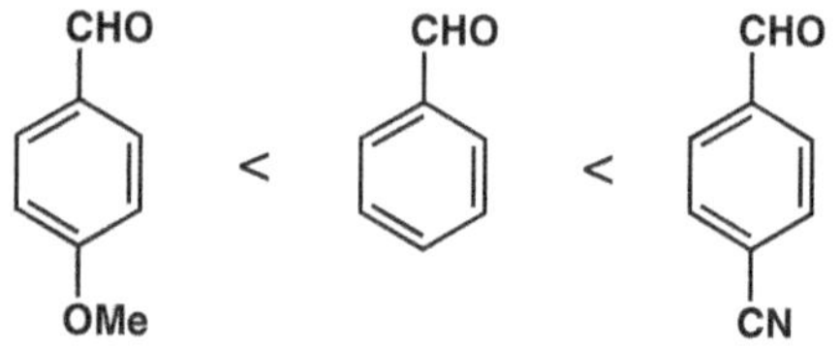

8.9 Why is very little cyanohydrin formed when an aldehyde in aqueous solution is treated with (a) just hydrogen cyanide (HCN) or (b) just sodium cyanide (NaCN)?

8.10 Propose a mechanism for the formation of a hemiacetal from 4-hydroxybutanal by acid catalysis.

8.11 Propose a mechanism for the acid-catalysed hydrolysis of a hemiacetal.

8.12 When an aldehyde RCH=O is dissolved in water containing $H_2^{18}O$, the ^{18}O isotope is incorporated into the carbonyl group. This incorporation is accelerated by a trace of acid or base. Write reaction mechanisms for the acid- and base-catalysed incorporation of the heavy isotope.

$$RCH{=}O \ + \ H_2^{18}O \ \rightleftharpoons \ RCH{=}^{18}O \ + \ H_2O$$

8.13 Cyclohexanone yields different products in methanol solution under acidic and basic conditions. Give reaction mechanisms and products for both reactions.

8.14 What is the product of the following reaction? Give a mechanism for the reaction using curly arrows.

8.15 Identify the product and any intermediates in the following reaction by proposing a curly arrow mechanism.

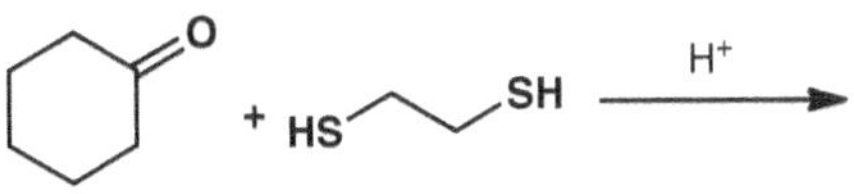

8.16 Ninhydrin, which is used for colorimetric identification of amino acids, is the monohydrate of

the triketone whose structure is given below. Show the structure of ninhydrin and explain why it, rather than isomers, is most readily formed.

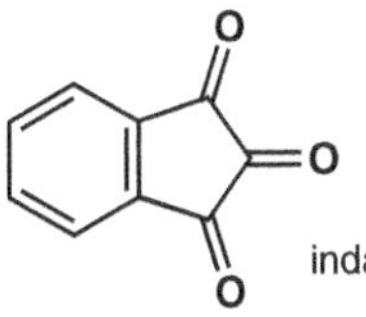

indane-1,2,3-trione

8.17 Explain why cyclopropanone exists mainly as a monohydrate in aqueous solution.

8.18 Give a mechanism for the formation of a hydrazone from aldehyde RCHO and the substituted hydrazine R'NHNH$_2$.

8.19 Propose a curly arrow mechanism for the following reaction.

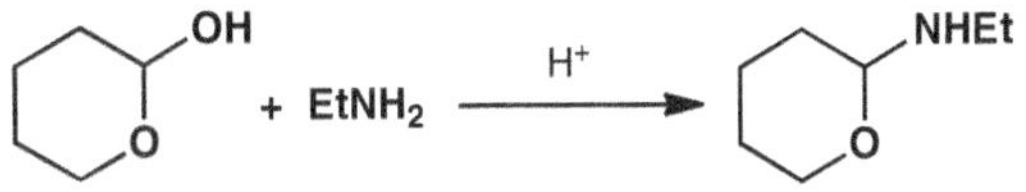

8.20 Propose a mechanism for the acid-catalysed hydrolysis of the bisulfite adduct of an aldehyde.

9 Nucleophilic Substitution Reactions of Carboxylic Acid Derivatives

Related topics we have already covered:

- Acids and bases (Chapter 6)
- Organic reactions and the concept of mechanism (Chapter 7)
- Curly arrow representation of reaction mechanisms (Section 7.2)
- Nucleophilic additions to carbonyl groups (Chapter 8)

Topics we shall cover in this chapter:

- Addition–elimination mechanism of nucleophilic substitution
- Tetrahedral intermediates
- Ester hydrolysis
- Acid-base catalysis
- Relative reactivities of carboxylic acid derivatives

Carboxylic acids contain a **carboxy** functional group (a carbonyl with a hydroxy group attached) bonded to an alkyl residue, and are represented by the general formula RCO_2H or RCOOH; analogues with an aryl group (Ar) instead of an alkyl are also well known. **Carboxylic acid derivatives** are obtained by replacing the hydroxy with another group bonded through O, N, S, or halogen (general formula, RCOY). This class of compounds includes esters (Y=OR'), anhydrides (Y=OCOR'), amides (Y=NR'$_2$), and acyl halides (Y=halogen); the RCO residue is called **acyl** so carboxylic acid derivatives are also known as **acyl compounds**. All carboxylic acid derivatives can be converted back to carboxylic acids by hydrolysis.

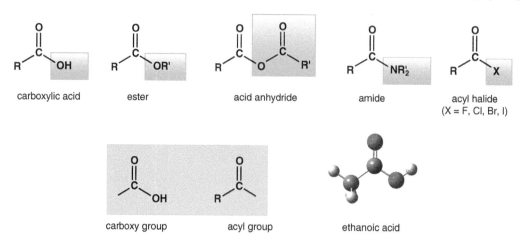

carboxylic acid   ester   acid anhydride   amide   acyl halide (X = F, Cl, Br, I)

carboxy group   acyl group   ethanoic acid

In this chapter, we shall cover the interconversions of carboxylic acid derivatives which generally take place by nucleophilic substitution via addition–elimination mechanisms.

9.1 Reactions of Carboxylic Acid Derivatives

Aldehydes and ketones, RCOR' in which R and R' are H, alkyl, or aryl, characteristically undergo nucleophilic addition to the CO group (Chapter 8), and the stable bonds to R and R' remain intact. The carbonyl group of carboxylic acid derivatives, RCOY, also undergoes nucleophilic addition, but with different consequences due to the possible elimination of the heteroatom group, Y. In these reactions, the initial addition is followed by elimination, so the overall reaction is substitution, as illustrated in eqn 9.1 for an anionic nucleophile, Nu⁻.

$$\text{(9.1)}$$

Nucleophilic substitution by addition–elimination is the main reaction of carboxylic acid derivatives, RCOY; it is the mechanism by which they interconvert, and by which they may be converted back to the parent carboxylic acid by hydrolysis.

Panel 9.1 Common names of carboxylic acids

Many carboxylic acids are known by their common names rather than systematic IUPAC names. Some representative examples are given below with the common names in red.

HCO₂H	CH₃CO₂H	CH₃CH₂CO₂H	CH₃(CH₂)₂CO₂H	CH₃(CH₂)₃CO₂H	CH₃(CH₂)₄CO₂H
methanoic acid	ethanoic acid	propanoic acid	butanoic acid	pentanoic acid	hexanoic acid
formic acid	acetic acid	propionic acid	butyric acid	valeric acid	caproic acid

HO₂C–CO₂H	HO₂CCH₂CO₂H	HO₂CCH₂CH₂CO₂H	HO₂C(CH₂)₃CO₂H	HO₂C(CH₂)₄CO₂H
ethanedioic acid	propanedioic acid	butanedioic acid	pentanedioic acid	hexanedioic acid
oxalic acid	malonic acid	succinic acid	glutaric acid	adipic acid

CH₂=CHCO₂H — propenoic acid / acrylic acid

trans-but-2-enoic acid / crotonic acid

2-methylpropenoic acid / methacrylic acid

cis-butenedioic acid / maleic acid

trans-butenedioic acid / fumaric acid

trans-3-phenyl-propenoic acid / cinnamic acid

benzenecarboxylic acid / benzoic acid

benzene-1,2-dicarboxylic acid / phthalic acid

(S)-2-hydroxy-propanoic acid / (S)-lactic acid

2-hydroxybenzene-carboxylic acid / salicylic acid

As we saw in Section 3.8, the numbering of the carbon chain of carboxylic acids (alkanoic acids) and their derivatives begins with the carbonyl carbon as C1 in the IUPAC nomenclature system. However, the Greek symbols α, β, γ, δ, etc., may also be used (but only with common names) in an alternative system starting at the C *next to* the carbonyl.

CICH=CHCO₂H — 3-chloropropenoic acid / β-chloroacrylic acid

H₂NCH₂CH₂CH₂CO₂H — 4-aminobutanoic acid / γ-aminobutyric acid (GABA) (not γ-aminobutanoic acid) (inhibitory neurotransmitter)

9.2 Hydrolysis of Esters

An aqueous solution is sometimes described as alkaline when its pH is greater than 7.0, regardless of the source of the HO⁻. For example, an aqueous solution could become alkaline by the direct addition of hydroxide ions in the form of KOH or NaOH, or by addition of a compound which yields HO⁻ (perhaps to just a limited extent) upon reaction with water, e.g. Na_2CO_3.

*Under alkaline conditions, the carboxylic acid is converted into its conjugate base.

Esters undergo hydrolysis in aqueous solution, but the reaction is normally slow unless the solution is acidic or alkaline (eqn 9.2). The reaction is a typical nucleophilic substitution reaction at an unsaturated carbon atom, and its mechanism has been comprehensively investigated. We shall consider this reaction in detail and then extend the concepts involved to analogous reactions.

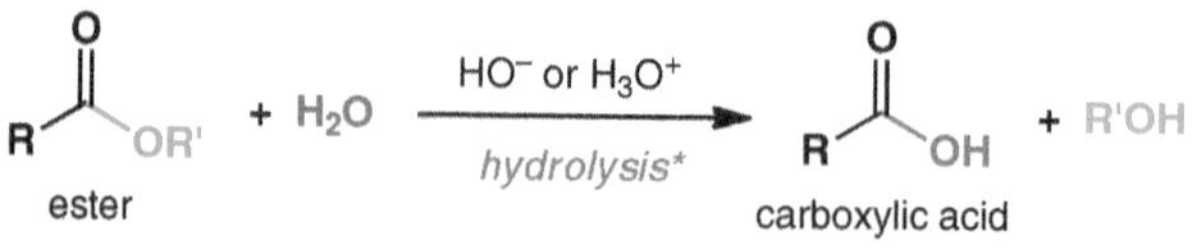

$$\underset{\text{ester}}{R-C(=O)-OR'} + H_2O \xrightarrow[\text{hydrolysis*}]{HO^- \text{ or } H_3O^+} \underset{\text{carboxylic acid}}{R-C(=O)-OH} + R'OH \qquad (9.2)$$

9.2.1 Hydration of the carbonyl group

We saw that hydration of an aldehyde or ketone is promoted by acid and base catalysis (Sub-section 8.3.2). Esters also undergo hydration in the same way, as shown in Schemes 9.1 and 9.2. These reactions are reversible and the hydrate is a **tetrahedral intermediate** since the initially trigonal planar central carbon becomes tetrahedral upon nucleophilic attack, and it is unstable (present at very low concentrations at equilibrium).

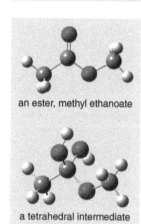

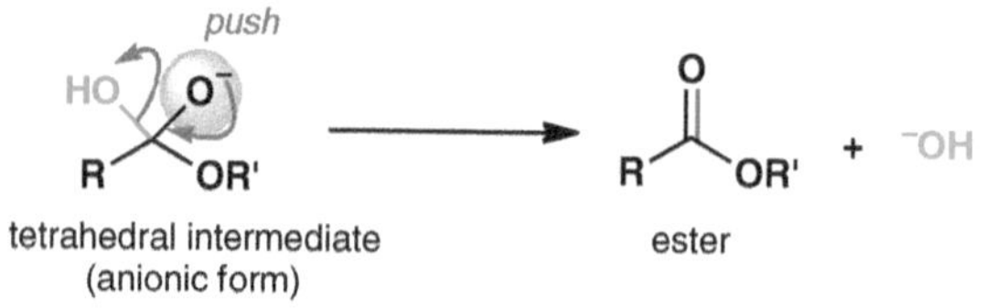

an ester, methyl ethanoate

a tetrahedral intermediate

$$\underset{\substack{\text{tetrahedral intermediate}\\ \text{(anionic form)}}}{\overset{push}{HO-C(O^-)(R)(OR')}} \longrightarrow \underset{\text{ester}}{R-C(=O)-OR'} + \ ^-OH \qquad (9.3)$$

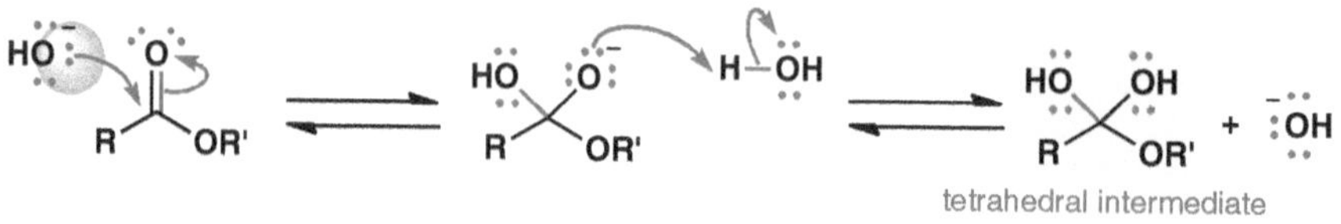

Scheme 9.1 Base-catalysed hydration of an ester.

tetrahedral intermediate

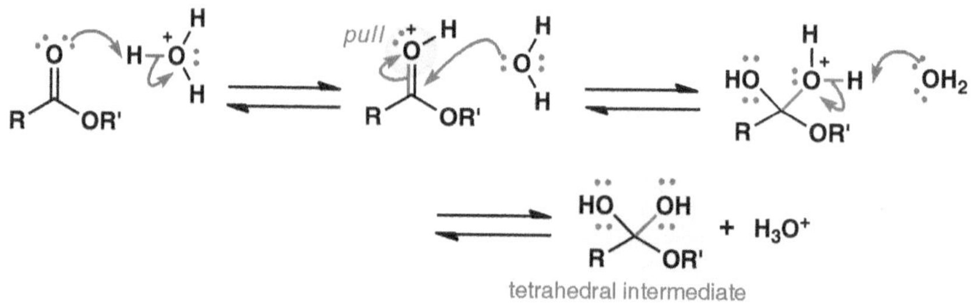

Scheme 9.2 Acid-catalysed hydration of an ester.

9.2.2 Reaction under alkaline conditions

The reverse of the first step in the base-catalysed hydration of an ester is described by curly arrows in eqn 9.3. An electron pair on the negatively charged oxygen of the anionic tetrahedral intermediate feeds in (*electron push*) to expel the hydroxide ion and re-form the carbonyl group of the original ester.

Alternatively, the anionic tetrahedral intermediate may expel the alkoxide ion (RO⁻), instead of the HO⁻ (eqn 9.4), and leave the carboxylic acid.

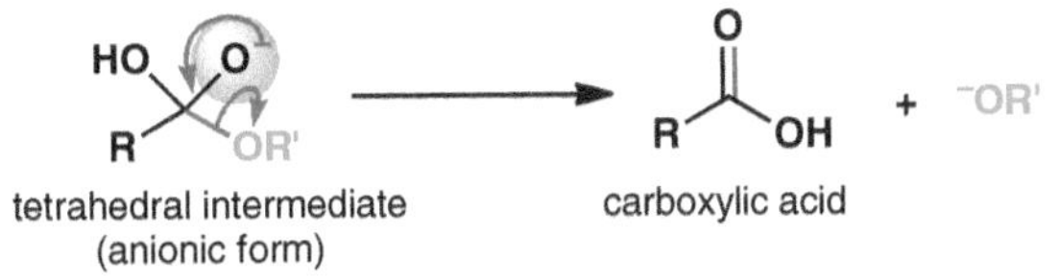

tetrahedral intermediate   carboxylic acid
(anionic form)

(9.4)

The final step of this reaction overall (Scheme 9.3) is proton transfer from the carboxylic acid to the alkoxide ion to give the carboxylate anion and alcohol. This acid–base reaction is very favourable owing to the large difference in acidity between a carboxylic acid ($pK_a \sim 5$) and an alcohol ($pK_a \sim 16$), i.e. RO⁻ is a much stronger base than RCO_2^-, and the large equilibrium constant for this final step effectively prevents the overall reaction being reversible.

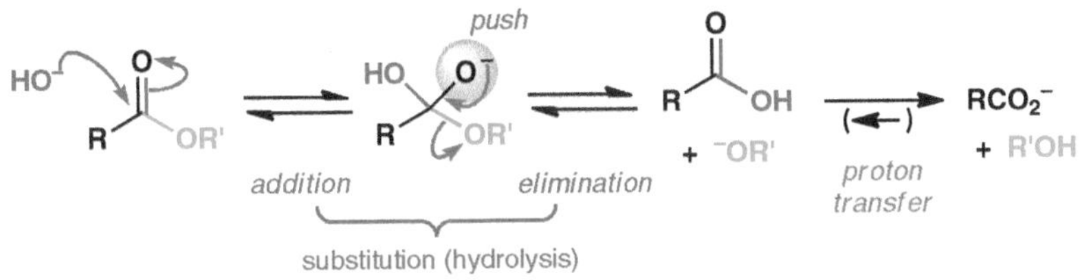

Scheme 9.3 Mechanism of the alkaline hydrolysis of a simple ester.

In the reaction in Scheme 9.3, the hydroxide is consumed, so the reaction is not catalytic. For this reason, the irreversible hydrolysis of esters under aqueous alkaline conditions is better simply called **alkaline hydrolysis**, and *not* base-catalysed hydrolysis.

This reaction is used to make soap from fats (eqn 9.5 below, see also Sub-section 24.4.1), which is why the alkaline hydrolysis of esters is sometimes called *saponification*.

Although the neutral tetrahedral intermediate in Scheme 9.1 is not stable, *direct* heterolysis to an ion pair is extremely unfavourable, e.g. for hydrolysis:

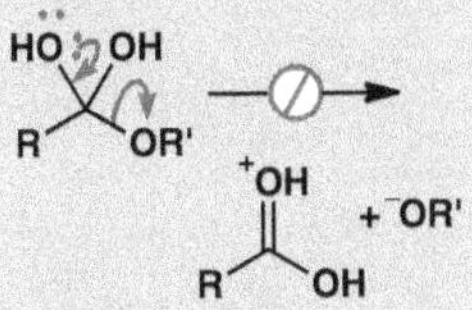

However, the 'electron push' from the anionic oxygen within the tetrahedral intermediate in the base catalysis mechanism facilitates expulsion of RO⁻ or HO⁻ which are both poor nucleofuges ($pK_{BH^+} \sim 16$).

Example 9.1

Draw the structural formulas of the two esters below, and give the structures and names of the carboxylic acid and alcohol products obtained by alkaline hydrolysis of each followed by acidification.

(a) pentyl ethanoate   (b) isopropyl propanoate

Solution
The structures and names are given in the following equations.

(a)

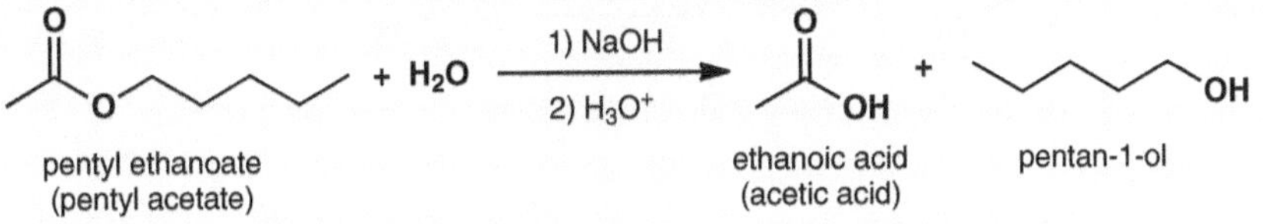

pentyl ethanoate
(pentyl acetate)

ethanoic acid
(acetic acid)

pentan-1-ol

(b)

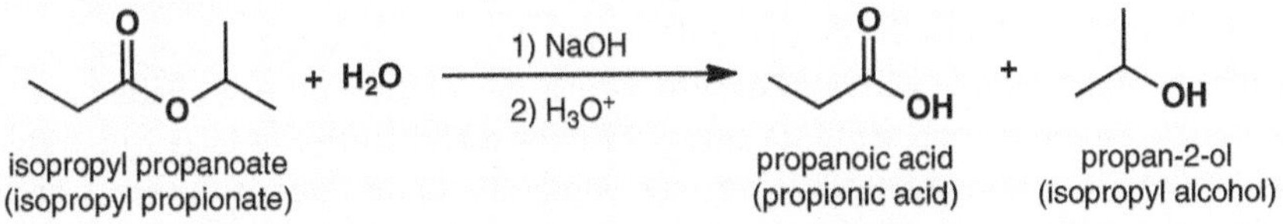

isopropyl propanoate
(isopropyl propionate)

propanoic acid
(propionic acid)

propan-2-ol
(isopropyl alcohol)

Exercise 9.1

Draw the structural formulas of the two esters below, and give the structures and names of the carboxylic acid and alcohol products obtained by alkaline hydrolysis of each followed by acidification.

(a) propyl 3-methylbutanoate   (b) 1-methylpropyl ethanoate

Saponification:

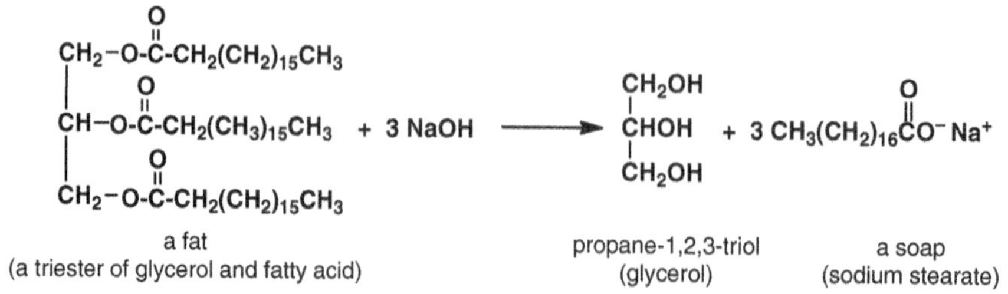
CH2-O-C-CH2(CH2)15CH3
CH-O-C-CH2(CH3)15CH3
CH2-O-C-CH2(CH2)15CH3
+ 3 NaOH
CH2OH
CHOH
CH2OH
+ 3 CH3(CH2)16CO⁻ Na⁺
a fat
(a triester of glycerol and fatty acid)
propane-1,2,3-triol
(glycerol)
a soap
(sodium stearate)

(9.5)

9.2.3 Acid-catalysed hydrolysis

Note that 'electron pull' provided by protonation in the acid catalysis mechanism is important in both the addition and elimination steps: it helps attack at the carbonyl by a weak nucleophile (H_2O), and departure of the nucleofuge (ROH) from the tetrahedral intermediate. This mechanism exemplifies another general principle: a catalyst in a reversible reaction necessarily catalyses the reaction in both directions.

The reverse of acid-catalysed hydration of a simple ester (Scheme 9.2) occurs by elimination of the protonated OH group as H_2O from the cationic tetrahedral intermediate. The basicity of the OR' group in the neutral tetrahedral intermediate is similar to that of the OH, so protonated OR' and OH groups should exist to similar extents in acidic solution, and departure of R'OH (which leads to ester hydrolysis) will compete with departure of H_2O. The complete overall reaction shown in Scheme 9.4, therefore, is fully reversible, but use of a large excess of water (as solvent) will drive the equilibrium in the direction of hydrolysis. Since H_3O^+ is consumed in the first step but regenerated in the final step, it is a catalyst in the overall reaction.

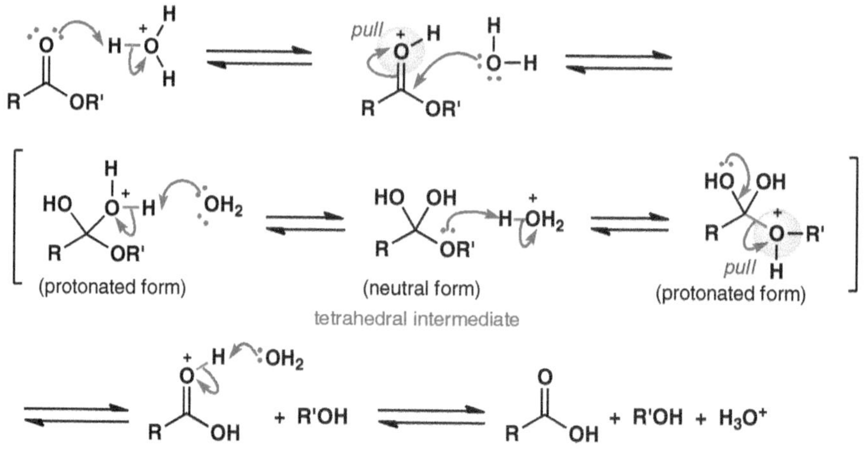
pull
(protonated form)
(neutral form)
tetrahedral intermediate
pull
(protonated form)
+ R'OH
+ R'OH + H3O⁺

Scheme 9.4 Mechanism of the reversible acid-catalysed hydrolysis of a simple ester.

The reaction of Scheme 9.4 can be driven in the reverse direction by using the carboxylic acid in the alcohol as solvent, i.e. with the alcohol now in large excess, and a small amount of a strong acid as catalyst. This procedure, one of the well-established methods for preparing esters in high yield, is called **Fischer esterification** and is illustrated in eqn 9.6. If the water is removed, e.g. by azeotropic distillation out of the reaction mixture, the process becomes unidirectional and a virtually complete conversion of the carboxylic acid can be achieved.

The symbol Δ as used in eqn 9.6 means that the reaction is heated.

$$\underset{\text{ethanoic acid}}{CH_3CO_2H} + EtOH \xrightarrow[(\leftarrow)\ \ EtOH,\ \Delta]{H_2SO_4\ \text{catalyst}} \underset{\text{ethyl ethanoate}}{CH_3CO_2Et} + H_2O \qquad (9.6)$$

The following esters smell like apples (a) and pears (b). Name the esters and show how each can be made by Fischer esterification giving the names of all organic reactants.

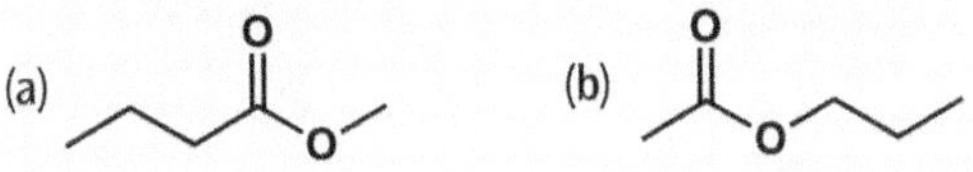

(a) [structure]   (b) [structure]

Solution

Names of compounds are given in the equations.

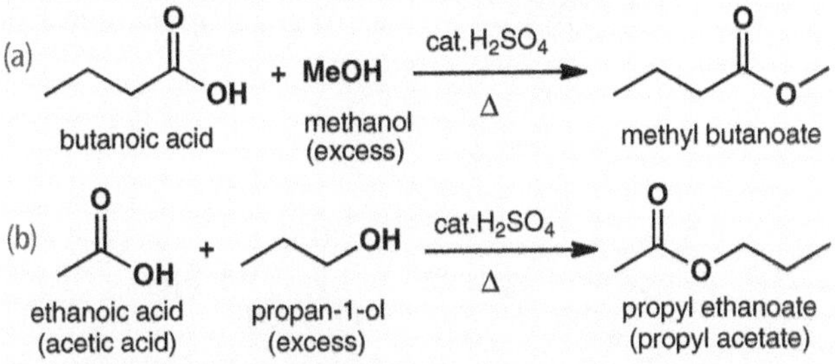

(a)

[reaction scheme]

butanoic acid + **MeOH** → methyl butanoate
                   methanol
                   (excess)
                                 cat.H₂SO₄ / Δ

(b)

[reaction scheme]

ethanoic acid + propan-1-ol → propyl ethanoate
(acetic acid)   (excess)   (propyl acetate)
                                cat.H₂SO₄ / Δ

Write a curly arrow mechanism for the acid-catalysed ester formation from RCO₂H and R'OH.

9.2.4 Evidence for a tetrahedral intermediate

Ester hydrolysis is a nucleophilic substitution reaction which proceeds by a two-step addition–elimination mechanism via a tetrahedral intermediate. Part of the evidence for the existence of this intermediate is the oxygen isotope exchange observed during the hydrolysis of an ^{18}O-labelled ester.

In 1951, the American chemist Myron Bender showed that the ^{18}O isotope of carbonyl-labelled ethyl benzoate was partially replaced by the normal ^{16}O isotope during alkaline hydrolysis; this was established by isotopic analysis of unreacted ethyl benzoate recovered before completion of the reaction (Scheme 9.5). Rapid proton transfer between the two oxygens of the tetrahedral intermediate in the two-step mechanism allows either of the oxygens to depart in the reverse process which regenerates the ester. This mechanism accounts for the observed loss of ^{18}O in unreacted ester; if the substitution had been by a direct one-step mechanism, or if the first step of the two-step mechanism were not reversible, the unreacted ester could not have lost labelled oxygen.

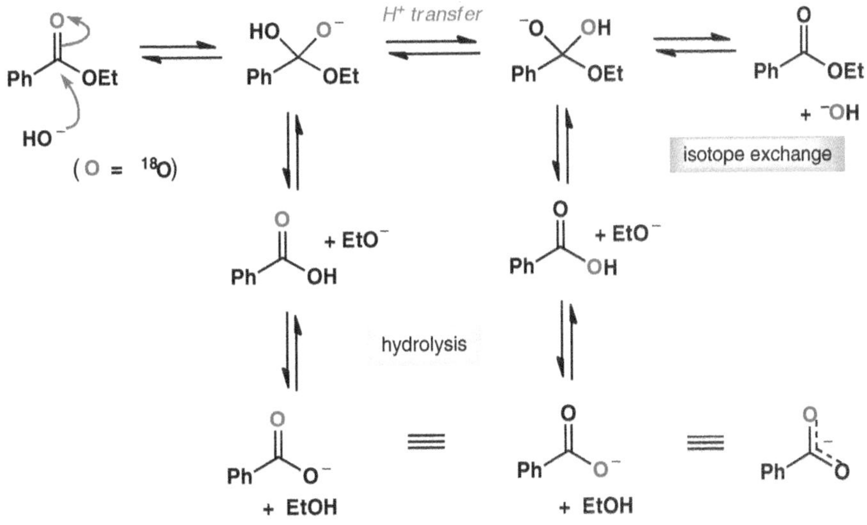

Scheme 9.5 Hydrolysis and oxygen isotope exchange of a labelled ester.

Exercise 9.3

Give curly arrow mechanisms for the reactions of the tetrahedral intermediates formed in the alkaline hydrolysis of labelled ethyl benzoate (Scheme 9.5) leading to (a) hydrolysis, (b) reverse reaction, and (c) isotope exchange.

When ethyl benzoate ^{18}O-labelled in the alkoxy oxygen was hydrolysed under the same alkaline conditions, the label was found only in the ethanol and not in the benzoate anion (Scheme 9.6). This observation indicates that acyl–O (not alkyl–O) bond cleavage occurs during the hydrolysis, a result required by the two-step mechanism involving the tetrahedral intermediate.

Scheme 9.6 The fate of the ^{18}O-labelled alkoxy oxygen in the alkaline hydrolysis of ethyl benzoate.

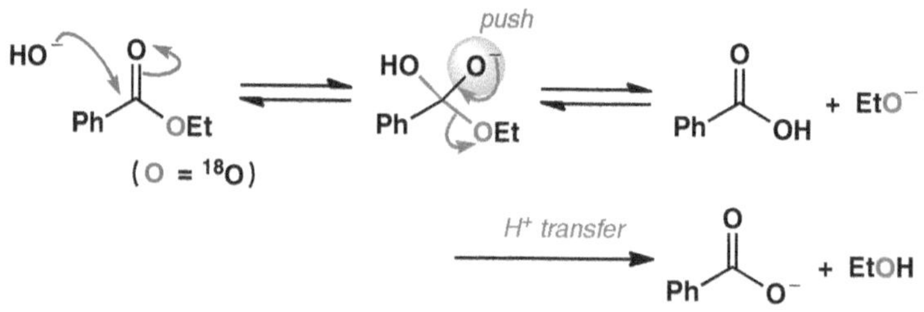

Example 9.3

Isotope exchange is also observed in unreacted starting material during the acid-catalysed hydrolysis of the carbonyl-labelled ester, PhC(^{18}O)OEt. Give a curly arrow mechanism which accounts for how this occurs.

Solution

Proton transfers occur within the tetrahedral intermediate and the isomeric forms lead to the reverse reaction, isotope exchange, and hydrolysis.

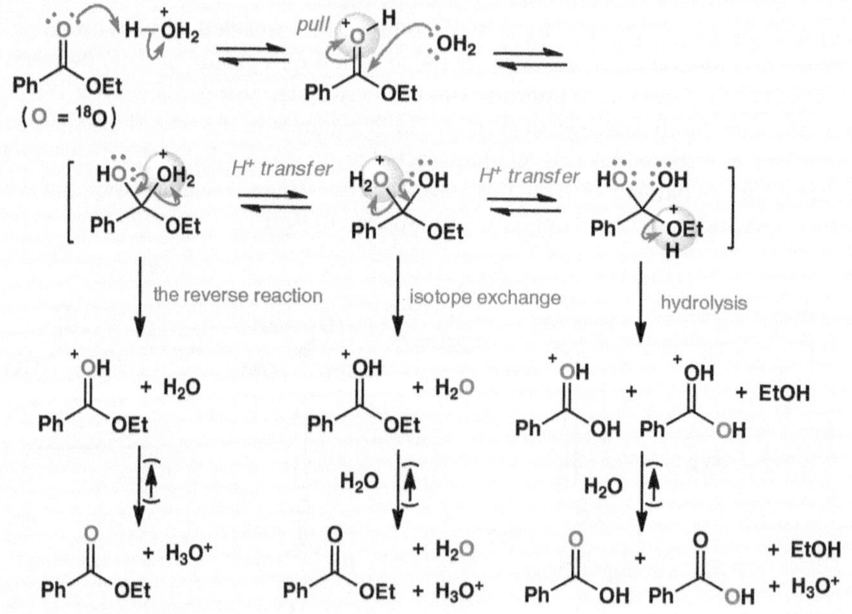

Exercise 9.4

Show by a curly arrow mechanism how the ^{18}O isotope is distributed in the products when ethyl ethanoate is hydrolysed under alkaline conditions in water labelled with ^{18}O.

9.3 Other Reactions of Esters

9.3.1 Ester exchange reactions

When acid or base is added to a solution of an ester in an alcohol, the alkoxy group of the ester and the alcohol undergo exchange (**ester exchange** or **transesterification**, eqn 9.7).

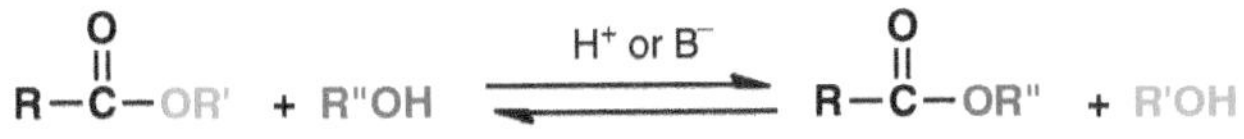

$$\text{(9.7)}$$

This reaction proceeds by a mechanism closely similar to that of hydrolysis in aqueous solution. However, the reversible ester exchange is catalysed by both acids (Scheme 9.7) and bases (Scheme 9.8) in contrast to hydrolysis which is catalysed by H_3O^+ (Sub-section 9.2.3) but consumes HO^- (Sub-section 9.2.2).

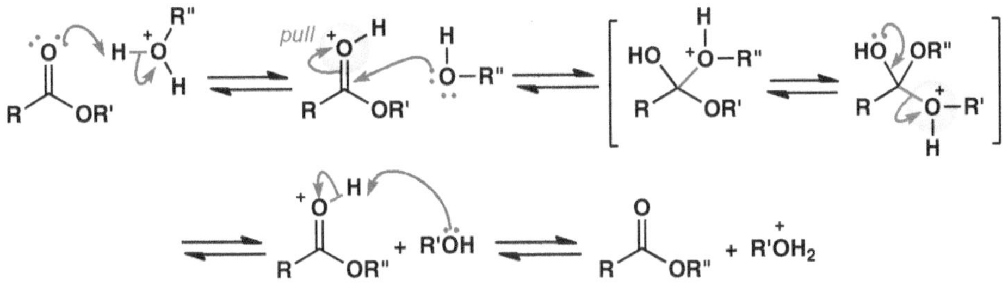

Scheme 9.7 Acid-catalysed ester exchange.

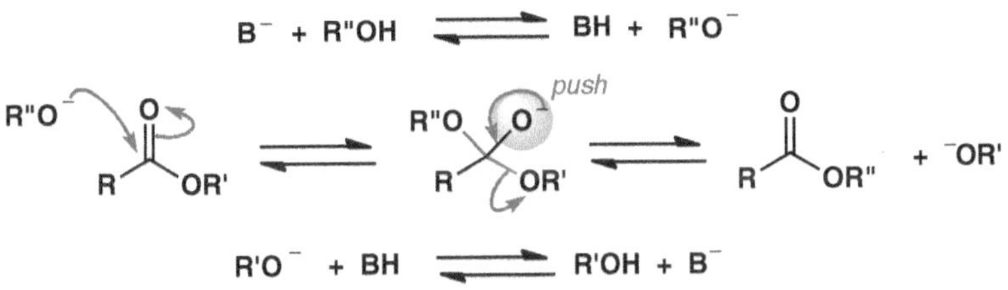

Scheme 9.8 Base-catalysed ester exchange.

| Exercise 9.5 |
Explain why base-catalysed ester exchange is reversible, whereas ester hydrolysis is irreversible under alkaline conditions and not catalytic.

| Exercise 9.6 |
What are the products of the following ester exchange reactions? How can we maximize the yield of the exchanged ester product?

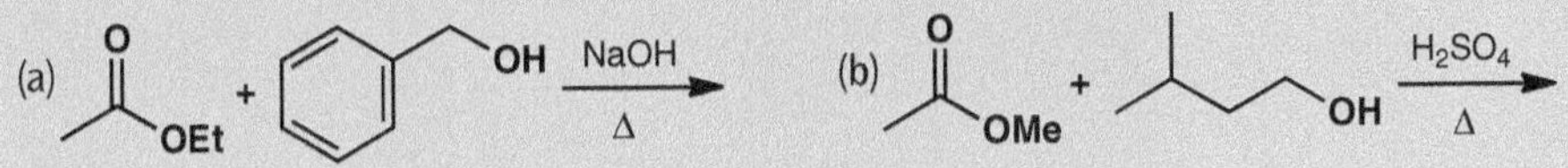

9.3.2 Reactions of esters with amines

Esters react with ammonia, and primary and secondary amines, to give amides, as illustrated in Scheme 9.9. Amines are usually sufficiently nucleophilic to add to an ester without the need for catalysis and, under the nonaqueous basic conditions, alkoxide is sufficiently good as a nucleofuge in the elimination step.

In principle, other nucleophiles can react with esters, but the relative reactivities of the nucleophile and nucleofuge limit the effective formation of some products in practice, as will be discussed in the next section.

Scheme 9.9 Formation of
an amide from an ester.

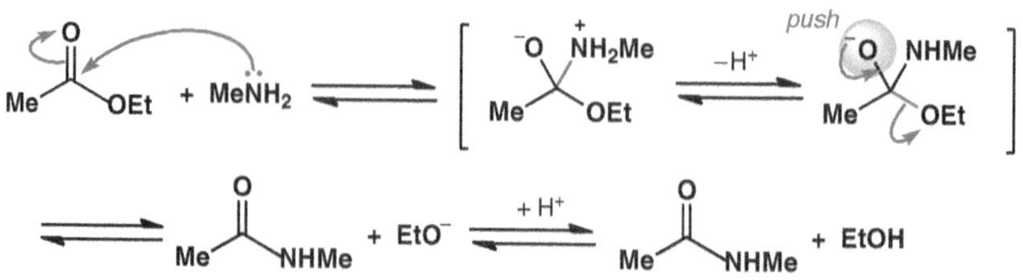

Exercise 9.7

Complete the following reactions.

(a) PhCH₂COOMe + NH₃ ⟶   (b) + PhCH₂NH₂ ⟶

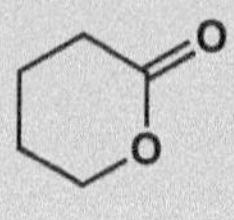

9.4 Generalized Nucleophilic Addition–Elimination Reactions

9.4.1 Reaction mechanism

So far in this chapter, we have seen some reactions of esters which are representative carboxylic acid derivatives, RCOY. They all involved addition of nucleophiles to give tetrahedral intermediates, followed by elimination of the nucleofuge Y, which results in overall **nucleophilic substitution** at the carbonyl group. In the reactions of an ester (Y=OR), good nucleophiles Nu⁻ or Nu (e.g. HO⁻, RO⁻, and RNH₂) add to the carbonyl double bond without catalysis. The reaction is represented generically in Scheme 9.10, where expulsion of the nucleofuge in the elimination step is facilitated by electron push from the anionic oxygen of the negatively charged tetrahedral intermediate. However, poor nucleophiles Nu–H (e.g. H₂O and ROH) require catalysis.

With base catalysis, a base B⁻ first generates the more powerfully nucleophilic conjugate base of the nucleophile Nu–H (for example: B⁻+ROH⇌BH+RO⁻) which adds directly to the C=O bond, as exemplified for transesterification in Scheme 9.8 above. The nucleofuge in the anionic tetrahedral intermediate is then expelled as Y⁻, just as in Scheme 9.10. In the final step, Y⁻+BH ⇌ YH+B⁻, the catalytic base is regenerated.

Scheme 9.10
Nucleophilic substitution
reactions of carboxylic
acid derivatives with good
nucleophiles, usually
conducted under basic
conditions.

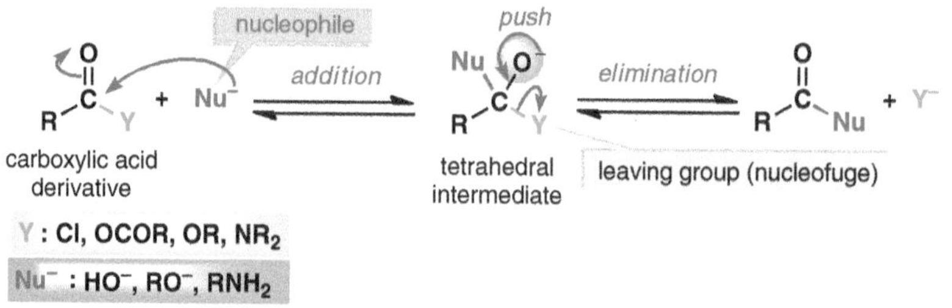

Y : Cl, OCOR, OR, NR₂

Nu⁻ : HO⁻, RO⁻, RNH₂

Reactions of acid derivatives with poor nucleophiles, Nu–H, can also be assisted by acid catalysis (Scheme 9.11). Protonation of the carbonyl group enhances its electrophilicity (*electron pull*, see Sub-section 8.3.2), which facilitates attack by a poor nucleophile. Proton transfer within the positively charged tetrahedral intermediate then converts Y into a much more effective nucleofuge—the uncharged molecule, HY. Departure of HY is followed by proton transfer to the solvent, which regenerates the catalytic acid and yields the neutral transformed acid derivative. There are, therefore, two

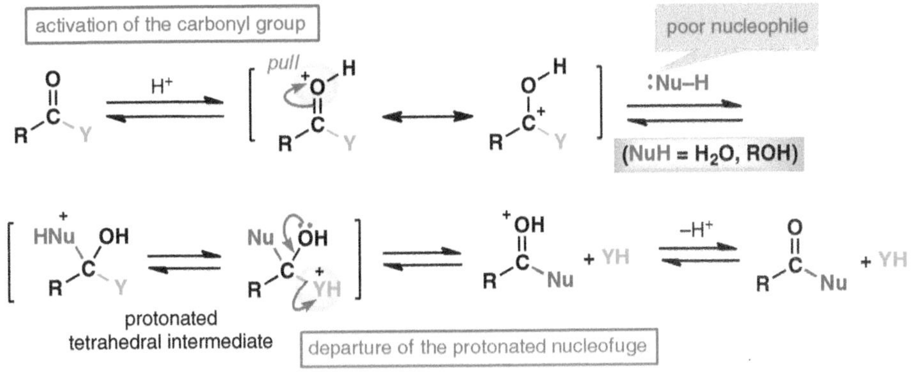

Scheme 9.11 Acid-catalysed nucleophilic substitution reaction of carboxylic acid derivatives.

aspects of acid catalysis in a nucleophilic substitution reaction such as ester hydrolysis or ester exchange: facilitation of initial nucleophilic attack *and* of subsequent departure of the nucleofuge.

9.4.2 Relative reactivities of carboxylic acid derivatives

The characteristic feature of reactions of carboxylic acid derivatives RCOY which distinguishes them from reactions of aldehydes and ketones RCOR' is that the heteroatom group Y can depart as a nucleofuge, Y$^-$, from the tetrahedral addition intermediate.

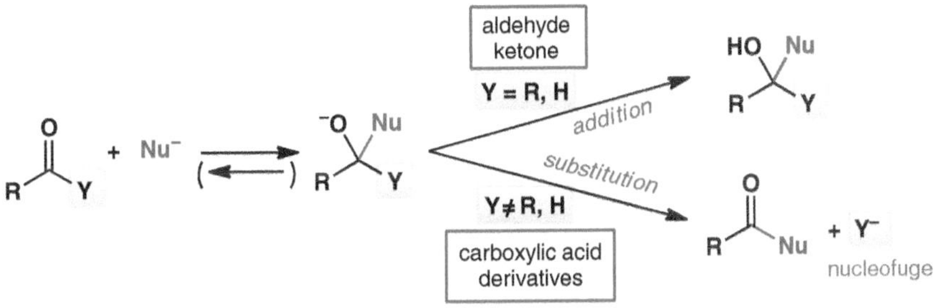

Scheme 9.12 Comparison of reactions of nucleophiles with carboxylic acid derivatives and with aldehydes and ketones.

The **leaving ability** (nucleofugality) of Y$^-$ is one of the factors which determines the relative reactivities of carboxylic acid derivatives. It is closely related to the acidity of the conjugate acid of Y$^-$ (HY). The acid strength of HY indicates the ease of formation of the anion Y$^-$ from HY, and the leaving ability of Y$^-$ represents the ease of formation of Y$^-$ from a tetrahedral intermediate, which corresponds to the relative reactivities of carboxylic acid derivatives in the elimination step.

Remember that acid strength measured by pK_a relates to an equilibrium (the dissociation of an acid in water, see Chapter 6), i.e. it is a *thermodynamic* parameter. Leaving group ability, however, relates to the rate of a step in a mechanism, i.e. it is a *kinetic* aspect.

Acidity of HY:	HCl	>	HOCR (O)	>	HOR	>	H$_2$NR
Approximate pK_a of HY:	-7		5		16		35
Leaving ability of Y$^-$:	Cl$^-$	>	RCOO$^-$	>	RO$^-$	>	RNH$^-$

| acyl chloride | acid anhydride | ester | amide |

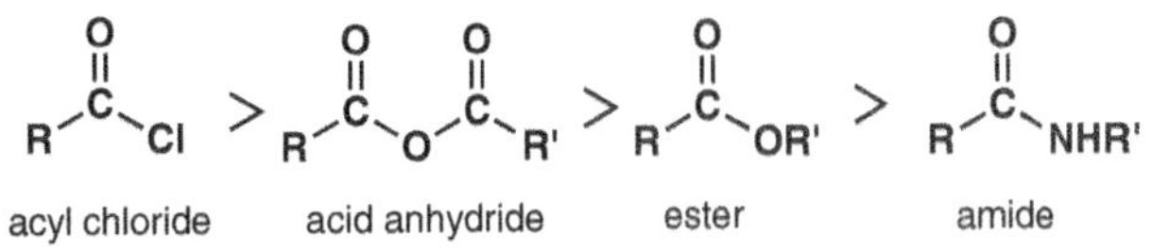

Reactivity in the initial nucleophilic addition step, i.e. the *electrophilicity* of the carbonyl group, is also important, and is affected by the nature of the attached group (Y)

by both induction and resonance. The heteroatom through which Y is bonded will be more electronegative than C so there is an intrinsic polarization of the C–Y σ bond. This effect is principally responsible for the high electrophilicity of acyl halides. In addition, the heteroatom bonded to the carbonyl carbon has an unshared electron pair which can conjugate with the carbonyl bond. A carboxylic acid derivative can be represented as a resonance hybrid of **1a–1c**, and the greater the contribution from the (minor) resonance form **1c**, the greater the stabilization of the reactant which is lost upon nucleophilic attack to give the tetrahedral intermediate.

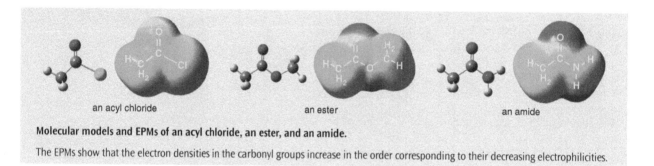

The electron-donating ability of the group Y increases in the order Cl<OCOR<OR<NHR. It follows that the contribution of **1c** to the hybrid, and hence the resonance stabilization of the reactant RCOY, increases in the same order. So, although this resonance effect opposes the inductive effect of Y and reduces the reactivity of RCOY (since it stabilizes the reactant), *it reinforces the trend in the electrophilicities.*

Reactivity towards nucleophilic addition: (electrophilicity)

acyl chloride > acid anhydride > ester > amide

We see, therefore, that the order of **electrophilicity** of carboxylic acid derivatives is the same as the order of reactivity in the elimination step discussed above.

an acyl chloride   an ester   an amide

Molecular models and EPMs of an acyl chloride, an ester, and an amide.

The EPMs show that the electron densities in the carbonyl groups increase in the order corresponding to their decreasing electrophilicities.

9.4.3 Comparison of reactions of nucleophiles with carboxylic acid derivatives and with aldehydes and ketones

In aliphatic compounds RCOY, aldehydes and ketones have groups Y=H and alkyl, respectively, strongly bonded to the carbonyl group. In contrast, Y in carboxylic acid derivatives RCOY is a nucleofuge bonded through a heteroatom which is invariably more electronegative than C or H. This polarization of the C–Y bond in acid derivatives (compared with aldehydes and ketones) suggests that the acid derivatives will be the more electrophilic. However, acid derivatives, unlike aldehydes and ketones, also have a lone pair on the group Y, as discussed in the previous section; this leads to the stabilizing resonance contributor **1c** above which reduces their electrophilicity. The result of these two opposing electronic effects is that acyl halides and acid anhydrides (dominant inductive effect) are more reactive than aldehydes and ketones, but esters and amides (dominant resonance effect) are less reactive.

We also saw in Chapter 8 that ketones are normally less reactive than aldehydes due to steric hindrance to nucleophilic attack, and steric factors must also be taken into account when considering reactivities of acyl compounds. The outcome of these various effects is that, experimentally, electrophilicity is usually in the order:

Electrophilicity: acyl chloride > acid anhydride > aldehyde > ketone > ester > amide

As discussed above, aldehydes and ketones do not undergo *substitution* at the carbonyl because neither H nor alkyl acts as a leaving group in the tetrahedral intermediate. However, in a special case, there is an exception to this generalization. In the final step of the *haloform reaction* (Scheme 9.13, see also Sub-section 17.4.2), the carbanion CX_3^- departs from a tetrahedral intermediate carrying a trihalomethyl group, CX_3. This carbanion is stabilized by three electron-withdrawing halogen groups, so is a viable leaving group (pK_a of $CHCl_3$ is about 24). Ketones $RCOCX_3$ can be seen as a link, therefore, between other ketones and aldehydes on the one hand, and carboxylic acid derivatives on the other.

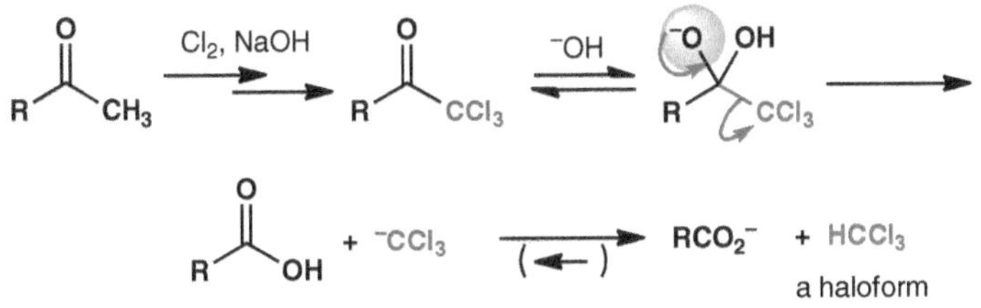

Scheme 9.13 The haloform reaction involving a trihalomethyl ketone.

Exercise 9.8

Discuss the relative electrophilicities of $PhCO_2Et$ and $MeCO_2Et$.

9.5 Interconversion of Carboxylic Acid Derivatives

9.5.1 Acyl chlorides

Acyl halides (sometimes called acid halides) are the most reactive carboxylic acid derivatives, and acyl chlorides are the most common of them by far. An acyl chloride reacts spontaneously with water and alcohols in hydrolysis and esterification reactions (eqns 9.8 and 9.9), respectively. However, the reactions are usually carried out in the presence of a base (often a tertiary amine such as pyridine or triethylamine) which neutralizes the HCl byproduct.

$$\underset{\text{acyl chloride}}{R-C(=O)-Cl} + H_2O \longrightarrow \underset{\text{carboxylic acid}}{R-C(=O)-OH} + HCl \quad (9.8)$$

$$R-C(=O)-Cl + R'OH \longrightarrow \underset{\text{ester}}{R-C(=O)-OR'} + HCl \quad (9.9)$$

Acyl chlorides react even more readily with ammonia or primary and secondary amines, but two equivalents of NH_3 or the amine are required for complete conversion of the acyl chloride to the amide and neutralization of the HCl, eqn 9.10.

$$R-C(=O)-Cl + 2\,R'NH_2 \longrightarrow \underset{\text{amide}}{R-C(=O)-NHR'} + R'\overset{+}{N}H_3\,Cl^- \quad (9.10)$$

As the name implies, a *simple* anhydride is obtained by dehydration of a carboxylic acid: 2 RCO₂H → (RCO)₂O + H₂O, a reaction easily reversed by hydrolysis.

Reaction of an acyl chloride with a carboxylate salt gives an acid anhydride (eqn 9.11), and this reaction is used to prepare *mixed* anhydrides as illustrated.

$$R-\overset{O}{\underset{\|}{C}}-Cl \;+\; R'-\overset{O}{\underset{\|}{C}}-O^-\,Na^+ \;\longrightarrow\; R-\overset{O}{\underset{\|}{C}}-O-\overset{O}{\underset{\|}{C}}-R' \;+\; NaCl \qquad (9.11)$$

mixed acid anhydride

We see from the above that acyl chlorides are important because they can be used to prepare esters, amides, and anhydrides. How, then, can the acyl chloride itself be prepared? It is normally prepared by reaction of the carboxylic acid with thionyl chloride, $SOCl_2$, as shown in eqn 9.12, or phosphorus pentachloride, PCl_5.

SOCl₂ is usually preferred because it is volatile (bp 75 °C) and both the byproducts are gaseous, so purification of the acyl chloride is easy.

$$R-\overset{O}{\underset{\|}{C}}-OH \;+\; SOCl_2 \;\longrightarrow\; R-\overset{O}{\underset{\|}{C}}-Cl \;+\; SO_2 \;+\; HCl \qquad (9.12)$$

thionyl chloride

Exercise 9.9

When an acyl chloride is added to an aqueous solution of a tertiary amine, the amine reacts first to give an intermediate which is then hydrolysed to give the final products. Write equations and mechanisms for this sequence of reactions.

9.5.2 Acid anhydrides

Acid anhydrides react similarly to acyl halides, but are less reactive. They are hydrolysed in aqueous solution, and give esters and amides in reactions with alcohols and amines, as shown in eqns 9.13 and 9.14, respectively. As in the analogous reactions of acyl chlorides, at least two equivalents of the amine are required to ensure a high conversion of the anhydride in the reaction of eqn 9.14.

$$R-\overset{O}{\underset{\|}{C}}-O-\overset{O}{\underset{\|}{C}}-R \;+\; R'OH \;\longrightarrow\; R-\overset{O}{\underset{\|}{C}}-OR' \;+\; R-\overset{O}{\underset{\|}{C}}-OH \qquad (9.13)$$

acid anhydride   ester

$$R-\overset{O}{\underset{\|}{C}}-O-\overset{O}{\underset{\|}{C}}-R \;+\; 2\,R'NH_2 \;\longrightarrow\; R-\overset{O}{\underset{\|}{C}}-NHR' \;+\; R'NH_3\,^-O_2CR \qquad (9.14)$$

amide

Example 9.4

Give complete balanced equations for the reactions between the following pairs of compounds.

Solution

(a) acetic anhydride + phenol → phenyl ethanoate + MeCO₂H

(b) propanoyl chloride + 2 pyrrolidine → N-propanoylpyrrolidine + pyrrolidinium chloride

Give complete balanced equations for the reactions between the following pairs of compounds.

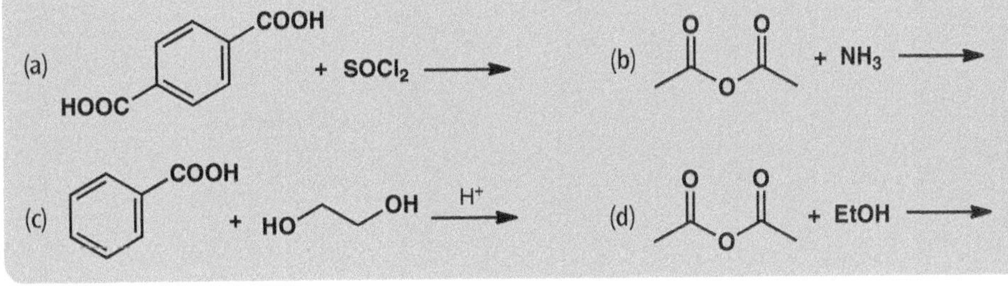

Exercise 9.10

(a) [structure: benzene ring with COOH and HOOC groups] + SOCl$_2$ ⟶

(b) [structure: anhydride] + NH$_3$ ⟶

(c) [structure: benzene ring with COOH] + HO⌒OH $\xrightarrow{H^+}$

(d) [structure: anhydride] + EtOH ⟶

9.5.3 Amides

Amides are the least reactive of the carboxylic acid derivatives; they do not react with ammonia or amines (in exchange reactions), or with alcohols under normal conditions. However, amides can be hydrolysed in hot concentrated acidic or basic aqueous solution.

$$R-\overset{\displaystyle O}{\overset{\|}{C}}-NH_2 + \boxed{H_2O} \xrightarrow[H_2O, \ \Delta]{NaOH \ or \ HCl} R-\overset{\displaystyle O}{\overset{\|}{C}}-\boxed{OH} + NH_3$$

amide

RCO$_2$Na is formed with NaOH, and NH$_4$Cl is formed with hydrochloric acid.

Give chemical equations for the hydrolysis of each of the following amides in aqueous solutions of (i) a strong acid, and (ii) a strong base. Ensure that all products are in forms appropriate to the reaction conditions.

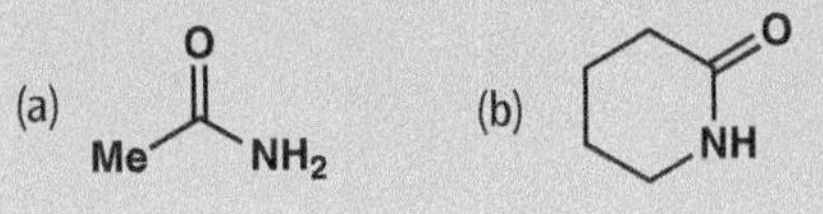

Exercise 9.11

(a) [structure: Me—C(=O)—NH$_2$]

(b) [structure: cyclic amide, piperidinone with NH]

Give a curly arrow mechanism for the hydrolysis of an amide under acidic conditions.

Exercise 9.12

9.5.4 Carboxylic acids

Some reactions of a carboxylic acid are similar to those of esters since OH and OR groups are similar in some respects. However, with nucleophiles which are also basic, acid–base reactions predominate (eqn 9.15) because proton transfer is hugely faster than nucleophilic addition. The carboxylate anion then has little electrophilicity, and does not react with nucleophiles.

$$R-\overset{\displaystyle O}{\overset{\|}{C}}-OH + Nu^- \rightleftharpoons R-\overset{\displaystyle O}{\overset{\|}{C}}-O^- + NuH \qquad (9.15)$$

poor electrophile

carboxylate

In principle, a carboxylic acid can react with a weakly basic nucleophile, such as an alcohol to give an ester, but the reaction is exceedingly slow in practice and acid catalysis is normally required; the so-called Fischer esterification method uses an anhydrous acid such as HCl or H$_2$SO$_4$ (eqn 9.16).

$$R-\overset{\displaystyle O}{\overset{\|}{C}}-OH + \boxed{R'OH} \xrightarrow{H^+} R-\overset{\displaystyle O}{\overset{\|}{C}}-\boxed{OR'} + H_2O \qquad (9.16)$$

acid catalysis

ester

Panel 9.2 Lactones and lactams

Cyclic esters and cyclic amides are called lactones and lactams, respectively. In principle, they may be formed by intramolecular reactions of carboxylic acids carrying a hydroxy or an amino group within the molecule. The five- and six-membered ring compounds are particularly stable; they are usually called γ-lactones/γ-lactams and δ-lactones/δ-lactams, respectively, using the common names of the parent carboxylic acids. However, the IUPAC names for lactones have the suffix –olide (e.g. butan-4-olide and pentan-5-olide) and the suffix –lactam is used for lactams (e.g. butane-4-lactam and pentane-5-lactam). (See Panel 9.1, p. 189, for an explanation of Greek prefixes in organic chemistry nomenclature.)

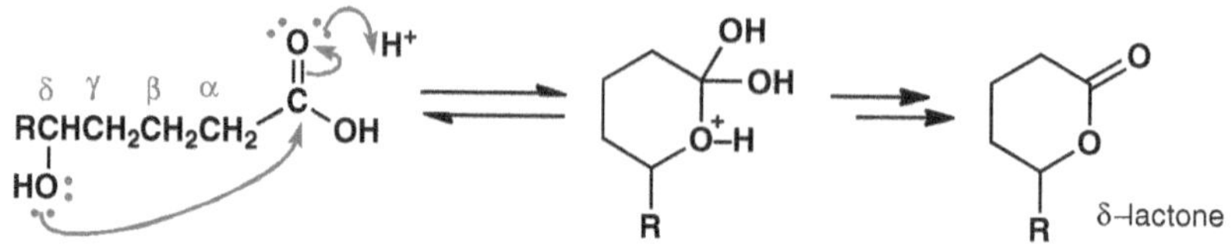

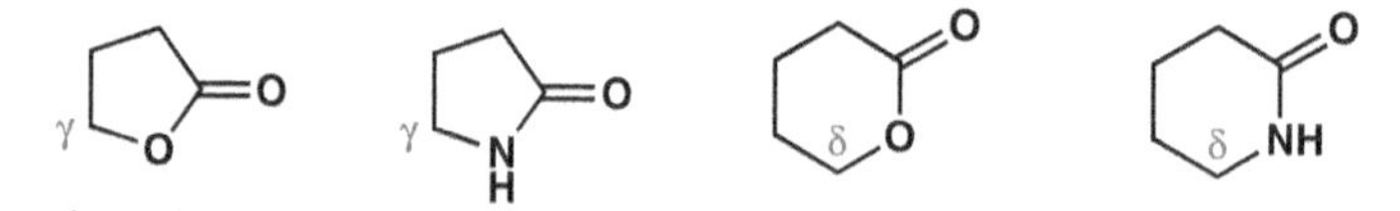

γ–butyrolactone	γ–butyrolactam	δ–valerolactone	δ–valerolactam
butan-4-olide	butane-4-lactam	pentan-5-olide	pentane-5-lactam

Like acyclic esters, lactones are easily hydrolysed in aqueous alkali, but the 4- and 5-hydroxycarboxylates often spontaneously cyclize back to the γ- and δ-lactones, respectively, upon addition of excess acid.

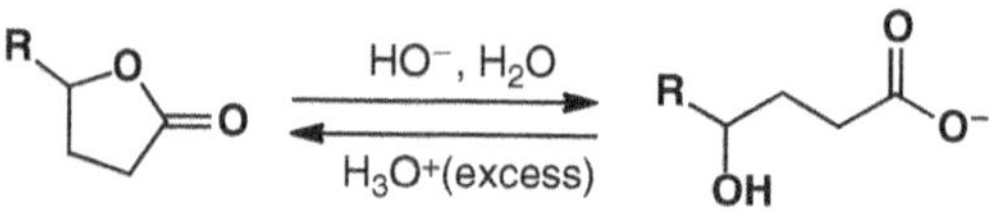

However, just as amides are more difficult to hydrolyse than esters, lactams are more difficult to hydrolyse than lactones. Furthermore, formation of γ- and δ-lactams from 4- and 5-amino acids is more complicated than formation of the corresponding lactones. If reaction conditions are too acidic, the amino group is protonated and loses its nucleophilicity; if conditions are too basic, the carboxy group is deprotonated and becomes non-electrophilic.

Lactones and lactams are abundant in nature. For example, vitamin C (ascorbic acid) is a γ-lactone, and some antibiotics are large ring lactones (sometimes called macrolide antibiotics).

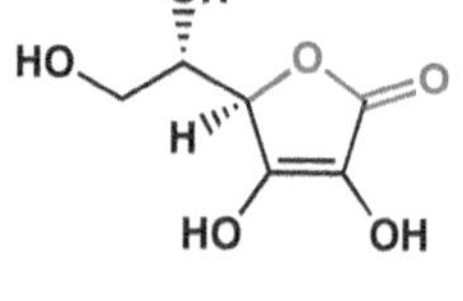

vitamin C (ascorbic acid)	erythromycin A

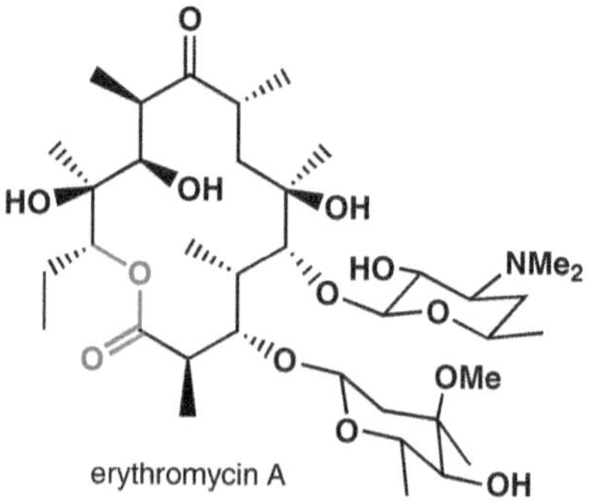

Penicillins and cephalosporins are two families of widely used antibiotics which contain four-membered cyclic amide groups and, hence, are known collectively as β-lactam antibiotics.

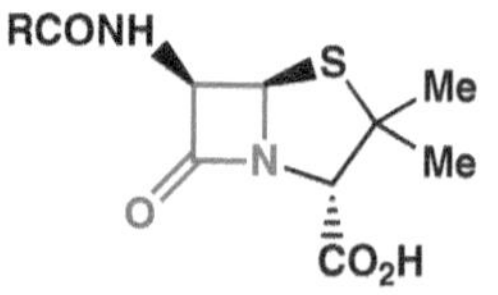

penicillin	cephalosporin

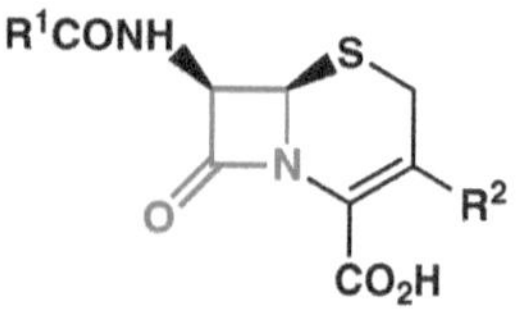

The seven-membered cyclic amide, ε-caprolactam, is the raw material used in the manufacture of nylon-6, as described in the text.

The mechanism of Fischer esterification (eqn 9.16) is just the reverse of the acid-cata-lysed hydrolysis of the ester (Sub-section 9.2.3), which was considered in Exercise 9.2.

9.5.5 Summary of relative reactivities

The relative reactivities of the common classes of carboxylic acid derivatives are summa-rized in Figure 9.1 where we see that acyl chlorides are the most reactive and carboxylate anions are the least. This chart shows that less reactive derivatives can be prepared from more reactive derivatives, but the opposite is not possible by simple nucleophilic substi-tution reactions. For example, an acyl chloride cannot be obtained by the reaction of an ester and chloride ion.

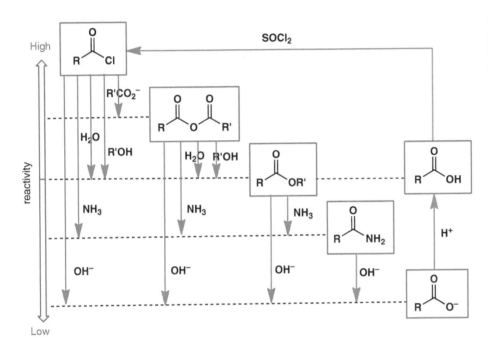

Figure 9.1 Relative reactivities and interconversions of carboxylic acid derivatives.

What are the products, if any, of each of the following reactions?

(a) CH_3CCl + CH_3CONa ⟶

(b) CH_3COCCH_3 + Cl^- ⟶

(c) $CH_3COC_2H_5$ + $(CH_3)_2NH$ ⟶

(d) CH_3CNHCH_3 + CH_3OH ⟶

Exercise 9.13

9.6 Polycondensation

Sequential reactions of bifunctional compounds can provide high molecular weight com-pounds called **polymers** which have repeating units. For example, repeated esterification reactions between a dicarboxylic acid and a diol under appropriate reaction conditions will yield a polyester. One of the most important polyesters is poly(ethylene terephthalate) or PET.

$$n \; HOC{-}\bigcirc{-}COH \; + \; n \; HOCH_2CH_2OH \xrightarrow[-2n\,H_2O]{\text{acid, } \Delta} \left(\!\!\begin{array}{c} O \\ \| \\ C \end{array}\!{-}\bigcirc{-}\begin{array}{c} O \\ \| \\ C \end{array}{-}OCH_2CH_2O\!\!\right)_n$$

a dicarboxylic acid   a diol   a polyester (PET)
(terephthalic acid)   (ethane-1,2-diol)

Base-catalysed transesterification is employed to produce PET in an important industrial process.

$$n\ \text{MeOC}-\text{C}_6\text{H}_4-\text{COMe} + n\ \text{HOCH}_2\text{CH}_2\text{OH} \xrightarrow[-2n\ \text{MeOH}]{\text{base, } \Delta} \text{poly(ethylene terephthalate)}$$

dimethyl terephthalate   ethane-1,2-diol   (PET)
(ethylene glycol)

PET is widely used to manufacture synthetic fibres and films, as well as for bottles and food containers.

Panel 9.3 Recycling of PET

Plastic materials are abundant and contribute enormously to the convenience and comfort of modern life. However, the huge and increasing accumulation world-wide of plastic waste, and concern about the depletion of natural resources, are driving a realization that we should be doing much more about the recycling of plastics. There are three types of plastic recycling: (1) mechanical recycling of the material; (2) recycling recovered feedstock (the chemicals from which plastics were made); and (3) recycling some of the energy used in the initial production of the plastics.

Mechanical recycling involves reusing the plastic materials (polymers) without chemical transformation. The plastics are simply melted down and often extruded in the form of pellets which are then used to manufacture other products. In the second option, the polymer is transformed chemically (depolymerization) into smaller units (oligomers) or even monomers from which new polymers can be made. The third option is just to burn the plastics as fuel. However, whilst this may be better than using the material as landfill, much of its value is lost.

It is important to separate plastics into their different chemical types before they can be mechanically recycled or chemically reprocessed. A numerical code (1–7) surrounded by a triangle of arrows is used to help consumers separate the different types of plastics. The most common, and the ones easiest to recycle, are made of poly(ethylene terephthalate) (PET) and assigned the number 1.

01	02	03	04	05	06	07
PET	PE-HD	PVC	PE-LD	PP	PS	O
poly(ethylene terephthalate)	high-density polyethylene	poly(vinyl chloride)	low-density polyethylene	polypropylene	polystyrene	other plastics

Rates of recovery are very different in different countries: in 2011, 79.6% of PET bottles in Japan were recovered, compared with 51.0% in Europe and 29.3% in the USA, and about 90% of the PET was reused in the manufacture of textiles, sheets, and other products. However, PET bottles are only a small percentage of total plastic waste. In Japan, 78% of *all* plastic waste was recovered, but about 67% of this was used as fuel in 2011.

Condensation polymers such as PET can be depolymerized to give their constituent monomers; this is simply taking advantage of the reversible nature of the transformations of carboxylic acid derivatives. One industrial process is the trans-esterification of PET with methanol to recover ethane-1,2-diol and dimethyl terephthalate which are then used again.

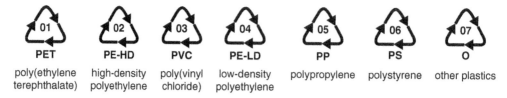

This type of polymer-forming reaction is called *polycondensation*. Polyamides, including the nylon family, are further important industrial polymers produced by polycondensation. Nylon 66 is obtained from hexanedioic acid and hexane-1,6-diamine.

See the margin note in Sub-section 8.6.1, p. 181, about simple condensation reactions.

$$n\ \text{HOC(CH}_2)_4\text{COH} \ + \ n\ \text{H}_2\text{N(CH}_2)_6\text{NH}_2 \ \xrightarrow[-2n\,H_2O]{\Delta} \ \left(\!\text{NH(CH}_2)_6\text{NHC(CH}_2)_4\text{C}\!\right)_{\!n}$$

hexanedioic acid   hexane-1,6-diamine   a polyamide (nylon 66)
(adipic acid)

Nylon 6 is manufactured by a ring-opening polymerization of the cyclic amide, hexane-6-lactam (ε-caprolactam). In this polymerization, a catalytic amount of water converts some of the lactam into the C_6 amino acid which then initiates polymerization of the rest of the lactam to produce the polyamide.

$$n\ \text{(hexane-6-lactam)} \ \xrightarrow[\Delta]{H_2O} \ \left[\text{H}_2\text{N(CH}_2)_5\text{COH}\right] + (n-1)\ \text{(lactam)} \ \xrightarrow[\Delta]{-H_2O} \ \left(\!\text{NH(CH}_2)_5\text{C}\!\right)_{\!n}$$

hexane-6-lactam   6-aminohexanoic acid   nylon 6
(ε-caprolactam)

Proteins are naturally occurring polyamides produced by enzyme-catalysed mechanisms in living organisms from twenty different α-amino acids, as we shall see in more detail Section 24.3. The following shows how three amino acids link together to form part of a protein chain.

$$\text{H}_2\text{N}\overset{R^1}{\underset{}{\diagup}}\text{CO}_2\text{H} \ + \ \text{H}_2\text{N}\overset{R^2}{\underset{}{\diagup}}\text{CO}_2\text{H} \ + \ \text{H}_2\text{N}\overset{R^3}{\underset{}{\diagup}}\text{CO}_2\text{H} \ + \ \cdots \ \xrightarrow{-3H_2O}$$

α-amino acids   a partial structure of a protein

Exercise 9.14

Kevlar® is an aromatic polyamide which is stronger than steel and used in the manufacture of racing car and bicycle tyres, and even in the material used for combat helmets and body armour. It is synthesized from terephthalic acid and 1,4-diaminobenzene. Draw the partial structure (repeat unit) of Kevlar®.

Summary

- Carboxylic acid derivatives (RCOY) contain a heteroatom group attached to the carbonyl carbon.

- Compounds RCOY undergo substitution reactions with nucleophiles Nu⁻ (or H–Nu) by nucleophilic addition–elimination mechanisms via tetrahedral intermediates; the heteroatom group Y⁻ (or Y–H) is the nucleofuge in the elimination step.

- Some interconversions of acid derivatives show acid–base catalysis.

- Reactivities in nucleophilic substitution reactions decrease in the following order.

Reactivity :   acid chloride > acid anhydride > ester > amide

          R—C(=O)—Cl   R—C(=O)—O—C(=O)—R'   R—C(=O)—OR'   R—C(=O)—NR'$_2$

- Acyl chlorides are made from carboxylic acids with thionyl chloride, and most other acid derivatives can be made from acyl chlorides.
- All acid derivatives can be converted back to carboxylic acids by hydrolysis.
- Polyesters (e.g. PET) and polyamides (e.g. nylon) are important materials manufactured industrially to make a wide range of consumer products.

Problems

9.1 What will be the products, if any, from ethyl ethanoate and the following reagent(s) and/or reaction conditions?

(a) H_2O, H_3O^+

(b) H_2O, HO^-

(c) CH_3NH_2 (excess)

(d) CH_3CO_2Na

(e) phenol

(f) CH_3OH, HCl

9.2 Predict the products when ethanoyl (acetyl) chloride, CH_3COCl, replaces ethyl ethanoate in Problem 9.1.

9.3 Predict the products when ethanamide (acetamide), CH_3CONH_2, replaces ethyl ethanoate in Problem 9.1.

9.4 Indicate the products of the following reactions.

(a) 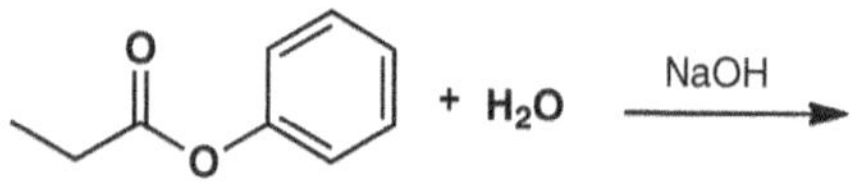 [structure] + H_2O —NaOH→

(b) 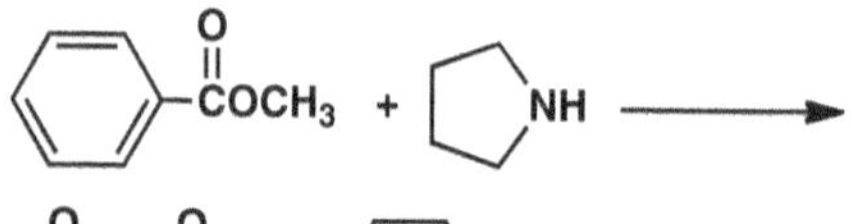 [structure] $COCH_3$ + [structure] NH —→

(c) 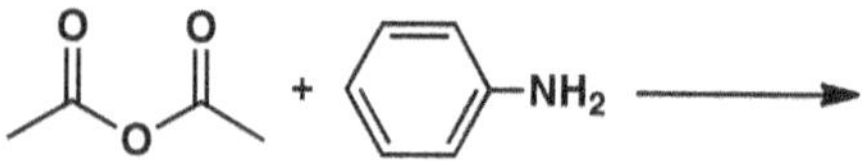 [structure] + [structure] —NH_2→

(d) 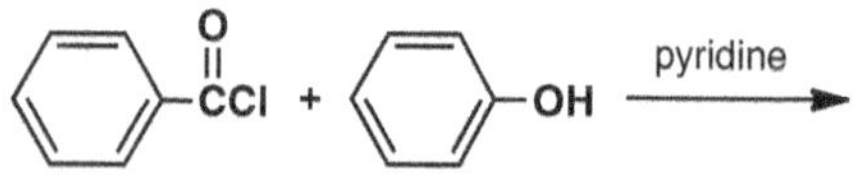 [structure] CCl + [structure] OH —pyridine→

9.5 Predict the order of reactivity in nucleophilic substitution reactions of each of the following sets of carboxylic acid derivatives RCOY with R or Y as specified.

(a) 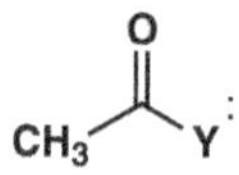 CH_3—(C=O)—Y: Y = OC_2H_5, Cl, $NHCH_3$, $OCCH_3$

(b) 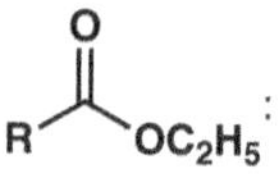 R—(C=O)—OC_2H_5: R = H, CH_3, $(CH_3)_2CH$

(c) 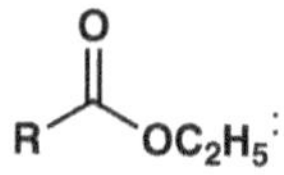 R—(C=O)—OC_2H_5: R = CH_3CH_2, $ClCH_2CH_2$, Cl_2CHCH_2

9.6 Give a mechanism for the formation of an ester from an acyl chloride (RCOCl) and an alcohol (R'OH).

9.7 Give a mechanism for the reaction of an acid anhydride, $(RCO)_2O$, with water (hydrolysis).

9.8 Nucleophilic substitution of carboxylic acid derivatives proceeds via tetrahedral intermediates. What are the main products from the decay of each of the following intermediates when they are formed in methanol solution?

(a) 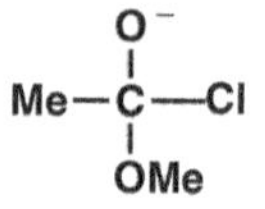 Me—C(O⁻)(Cl)—OMe
(b) Me—C(O⁻)(NHMe)—OMe
(c) Me—C(O⁻)(Cl)—NHMe
(d) 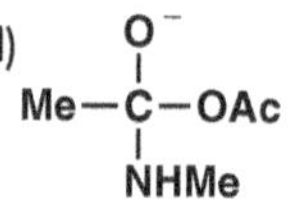 Me—C(O⁻)(OAc)—NHMe
(e) 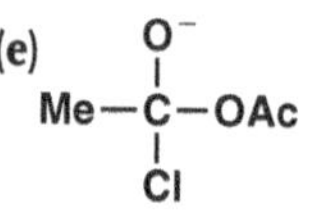 Me—C(O⁻)(OAc)—Cl

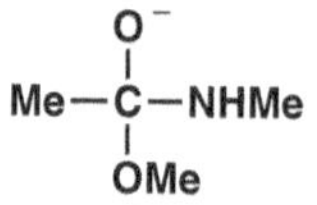

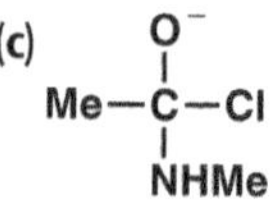

9.9 In order to prepare phenyl ethanoate, pyridine is added as a catalyst to a mixture of acetic anhydride and phenol to facilitate the reaction. In this reaction, pyridine reacts with the anhydride first and the intermediate which is formed then undergoes further substitution to give the required product. Give a curly arrow mechanism for the overall reaction.

9.10 Explain the following contrasting observations.

(a) When an ester is prepared from a carboxylic acid and an alcohol, strong acids catalyse the reaction whereas bases retard or even prevent the esterification.

(b) The esterification of an alcohol with an acyl chloride or acid anhydride is facilitated by addition of a tertiary amine such as pyridine or triethylamine.

9.11 Explain why the following reactions do not proceed in aqueous solution.

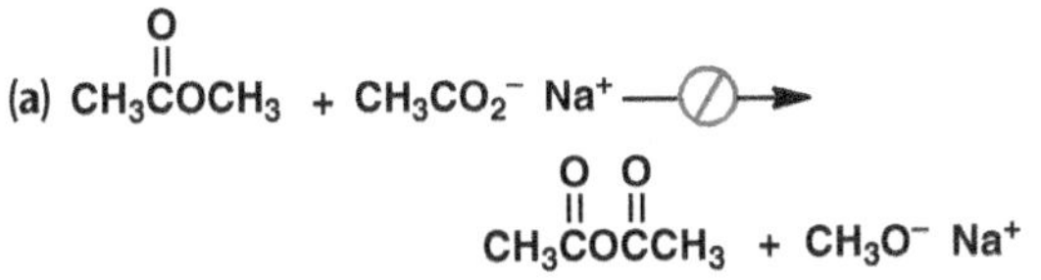

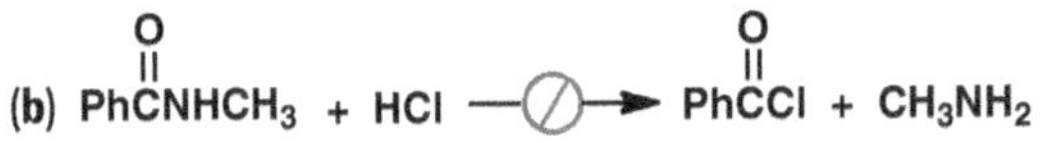

(a) $CH_3\overset{O}{\overset{||}{C}}OCH_3 + CH_3CO_2^- \ Na^+ \longrightarrow\!\!\!\!\oslash$

$CH_3\overset{O}{\overset{||}{C}}O\overset{O}{\overset{||}{C}}CH_3 + CH_3O^- \ Na^+$

(b) $Ph\overset{O}{\overset{||}{C}}NHCH_3 + HCl \longrightarrow\!\!\!\!\oslash Ph\overset{O}{\overset{||}{C}}Cl + CH_3NH_2$

9.12 What are the products of each of the following reactions?

(a) $\xrightarrow[\text{MeOH}]{\text{MeONa}}$

(b) $\xrightarrow[\text{MeOH}]{\text{HCl}}$

(c) $+ \ MeNH_2 \longrightarrow$

(d) $\xrightarrow{\Delta}$

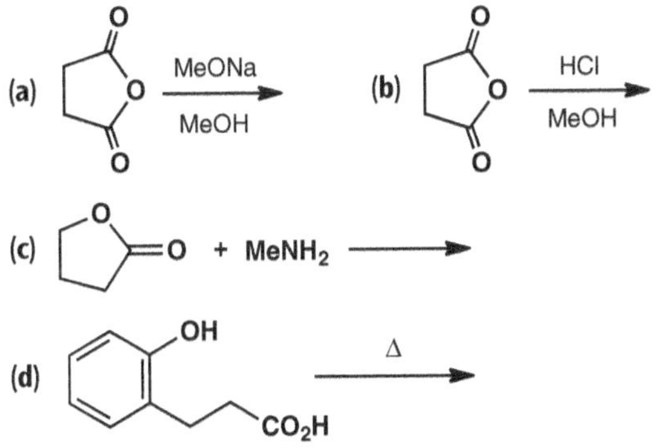

9.13 Propose a reaction for the transformation of o-hydroxybenzoic acid (salicylic acid) into each of the following compounds.

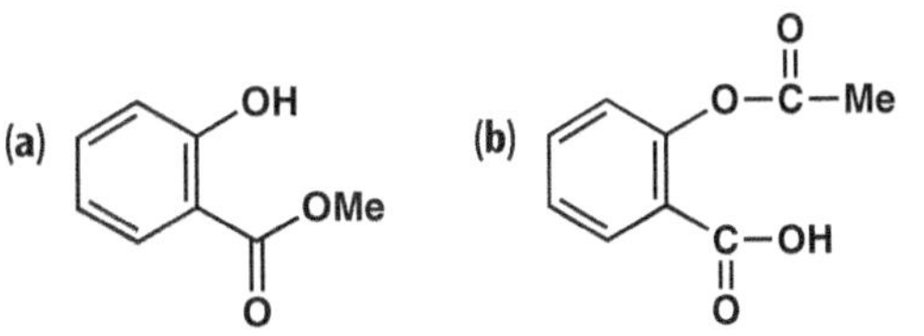

(a)

methyl salicylate (oil of wintergreen)

(b)

acetylsalicylic acid (aspirin): a drug which releaves pain, fever, and inflammation

9.14 When ethanoic acid is dissolved in water containing $H_2{}^{18}O$, the isotopic label gradually exchanges with both the oxygen atoms of the ethanoic acid. Propose a reaction mechanism for this process.

$\underset{Me}{\overset{O}{\underset{}{\diagup}}}OH + H_2{}^{18}O \ \rightleftharpoons \ \underset{Me}{\overset{{}^{18}O}{\underset{}{\diagup}}}{}^{18}OH + H_2O$

9.15 When a solution of methyl benzoate in methanol was treated with sodium methoxide, some dimethyl ether was formed. When ester labelled with the carbon isotope ^{14}C at the methyl carbon was employed for the reaction under the same conditions, the label was lost from the ester more rapidly than the dimethyl ether was formed. Explain this observation.

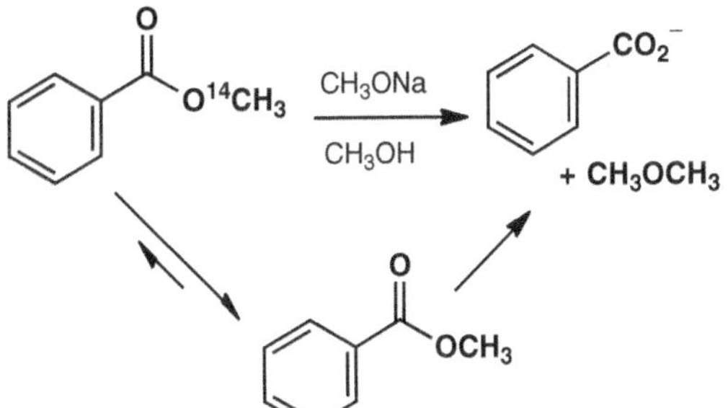

10 Reactions of Carbonyl Compounds with Hydride Donors and Organometallic Reagents

Related topics we have already covered:

- **Polarity of covalent bonds** (Sub-section 1.2.3)
- **Acids and bases** (Chapter 6)
- **Curly arrow representation of reaction mechanisms** (Section 7.2)
- **Nucleophilic addition to the carbonyl group** (Chapter 8)
- **Nucleophilic substitution of carboxylic acid derivatives** (Chapter 9)

Topics we shall cover in this chapter:

- **Nucleophilicity of metal–hydrogen and metal–carbon reagents**
- **Hydride reduction of C=O**
- **C–C bond formation by Grignard reactions**
- **Carbonyl group protection/deprotection**
- **Planning in organic synthesis**

We have already seen in Chapters 8 and 9 that carbonyl compounds are electrophilic and react with nucleophiles. The nucleophiles involved in the reactions discussed earlier all had lone pairs and, in each case, the lone pair was the nucleophilic centre. **Metal hydrides** and **organometallic compounds** are a different type of nucleophile in which the nucleophilic site is the σ bonding pair of electrons of the M–H or M–C bond (M = metal). They are important reagents for (i) reducing carbonyl compounds to alcohols, and (ii) synthesizing alcohols with new C–C bonds from carbonyl compounds. This chapter is concerned with these reactions of the carbonyl group. We also begin to consider aspects of designing an organic synthesis, and functional group protection.

10.1 Hydride Reduction of Carbonyl Groups

10.1.1 Reduction of aldehydes and ketones

Addition of a hydride ion H$^-$ to the carbonyl of an aldehyde or ketone gives an alkoxide ion, protonation of which gives an alcohol; the overall reaction is a reduction. Simple metal hydrides such as NaH behave like salts of the hydride ion, but cannot be used as reducing agents since the hydride reacts preferentially as a strong base (and abstracts a proton) rather than as a nucleophile. **Sodium borohydride** (NaBH$_4$) is a very convenient hydride donor which is easy to use in the laboratory in alcoholic or aqueous solvents. It is a mild reducing agent and the solvent protonates the alkoxide to give the alcohol directly (eqn 10.1).

$$\text{(10.1)}$$

Although the tetrahedral anion of $NaBH_4$ is the nucleophile, the central atom bearing a formal negative charge has no lone pair electrons; its valence octet comprises four M–H bonds. The HOMO of BH_4^- involved in the reaction is a B–H bonding orbital, and the bonding electron pair moves to the carbonyl group as the nucleophilic addition occurs (Scheme 10.1).

$$\text{(10.2)}$$

$$\text{(10.3)}$$

Gaseous hydrogen with a metal catalyst, and sodium in an alcohol (see Sub-section 20.9.2), are other methods for reducing aldehydes and ketones to alcohols, but the mechanisms of these reactions are quite different.

See Panel 10.1 for the molecular orbitals in BH_4^-.

Scheme 10.1 Reduction of an aldehyde to an alcohol by hydride transfer.

Consequently, the curly arrow starts from the B–H bond (not from the minus symbol on the B) when this mechanism is described by eqn 10.2 of Scheme 10.1. The alkoxide ion then immediately bonds to the other product, BH_3, to give the alkoxyborohydride which, in turn, also acts as a reductant, and so on. One BH_4^- anion, therefore, can reduce up to four carbonyl molecules (eqn 10.3, Scheme 10.1).

Panel 10.1 Bonding in BH_4^-

The molecular orbitals corresponding to a B–H bond in BH_4^- (and its derivatives with at least one B–H bond) are derived from an sp^3 hybridized orbital on B and the 1s AO of a hydrogen atom (see figure). Boron has metal-like tendencies, for example low electronegativity and high-energy valence electrons (low ionization energy). Since the energy gap between the interacting AOs is large, the stabilization of the bonding (σ) orbital is small. Consequently, this bonding MO is more like the original 1s AO of the H than the sp^3 AO of the B, and the occupying (σ) electron pair is polarized towards the more electronegative H atom. As a result, the bonding electron pair in this HOMO is a site of nucleophilic reactivity and the polarity facilitates hydride transfer from borohydride.

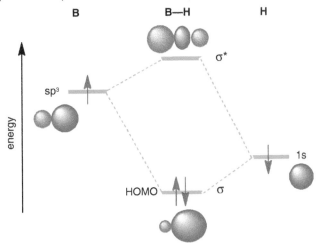

Molecular orbitals in the formation of one of the four B–H bonds in BH_4^-.

The bonding in other metal hydrides is similar, as is the M–C bonding of organometallic reagents, which makes them effective as carbon nucleophiles.

10.1.2 Reduction of carboxylic acid derivatives

The mild reagent, $NaBH_4$, is not sufficiently reactive to reduce less electrophilic carbonyl compounds such as esters. A more powerful hydride donor is required for these and other carboxylic acid derivatives, and the usual reagent of choice is lithium aluminium hydride, $LiAlH_4$; this compound will, of course, also reduce aldehydes and ketones. Lithium aluminium hydride also reacts vigorously with protic compounds such as alcohols and water with the evolution of H_2, so the reduction is usually carried out in anhydrous ethers, diethyl ether or tetrahydrofuran (THF) for example.

Also, the protonation of the alkoxide formed in the hydride transfer must be carried out in a subsequent step, e.g. by addition of water, with acidification to dissolve the hydrated Al(III) compounds (eqn 10.4).

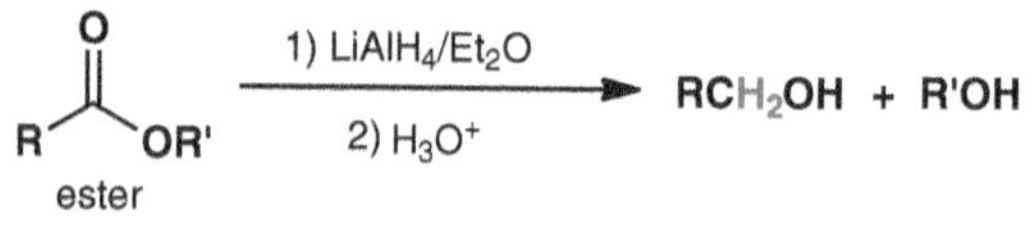

> Protic compounds are those bearing a hydrogen which can be readily released as a proton—they are Brønsted acids. The hydrogen is usually bonded to a heteroatom such as O, N, or S.

$$\text{(10.4)}$$

The tetrahedral intermediate formed by the addition of hydride ion to the carbonyl group of an ester decomposes with the release of an alkoxide ion, as in the alkaline hydrolysis of an ester (Sub-section 9.2.2). The aldehyde so formed is then itself further reduced to the primary alcohol since it is more electrophilic than the original ester (Scheme 10.2). Besides esters, amides and even carboxylic acids themselves can be reduced by $LiAlH_4$.

> Note that the alkoxy group of RCO_2R' in Scheme 10.2 is liberated as R'OH in the workup following reduction of the ester with $LiAlH_4$.

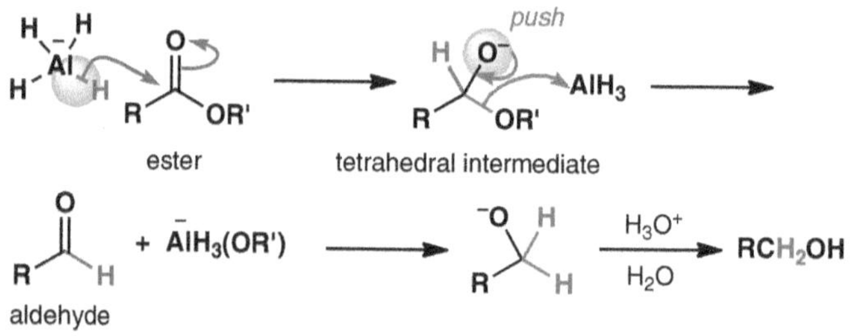

Scheme 10.2 Reduction of an ester to a primary alcohol.

When $LiAlH_4$ is added to a carboxylic acid in an ether, an immediate reaction gives the carboxylate anion (an acid–base reaction); see eqn 10.5 in Scheme 10.3. Although the carboxylate is much less electrophilic than the carboxylic acid, $LiAlH_4$ is sufficiently nucleophilic to reduce it to the alcohol (eqn 10.6).

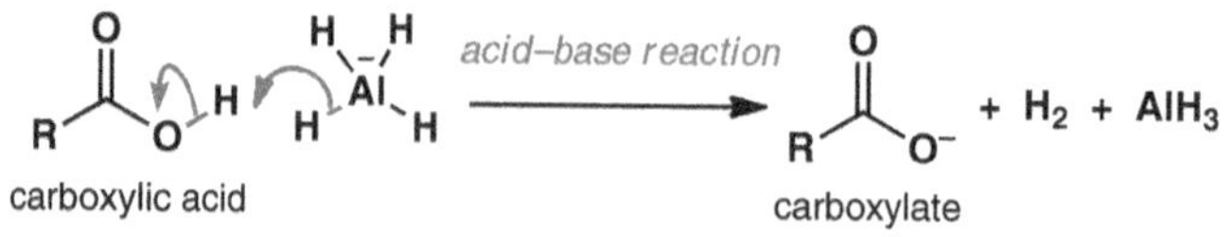

$$\text{(10.5)}$$

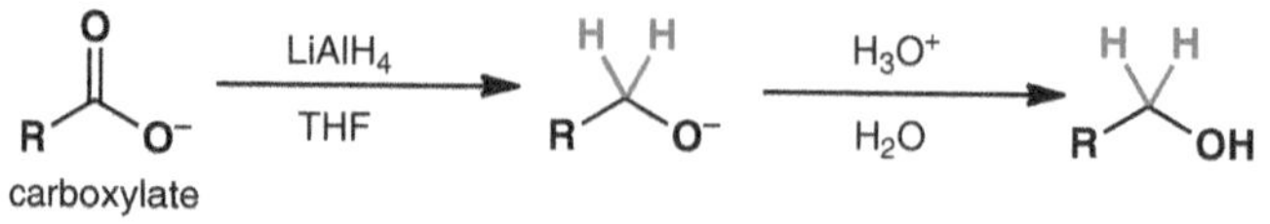

$$\text{(10.6)}$$

Scheme 10.3 Reduction of a carboxylic acid by $LiAlH_4$ via the carboxylate anion.

Reaction of an amide with $LiAlH_4$ proceeds via an iminium ion to give an amine as the final product after an aqueous workup; see eqn 10.7 in Scheme 10.4. However, it is possible to obtain the aldehyde by a controlled reaction under milder conditions, and hydrolysis of the reaction intermediate (eqn 10.8).

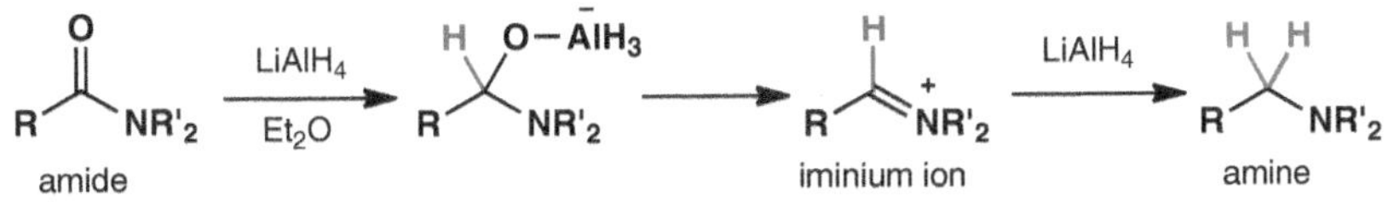

$$(10.7)$$

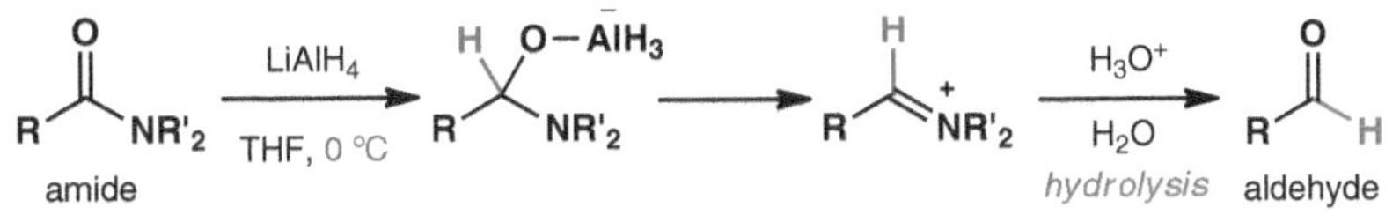

Scheme 10.4
Controlled LiAlH$_4$ reduction of an amide to give an amine or an aldehyde.

$$(10.8)$$

The ranges of application of NaBH$_4$ and LiAlH$_4$ as reducing agents, which follow the electrophilicity of carbonyl compounds, are summarized below:

Ease of reduction:

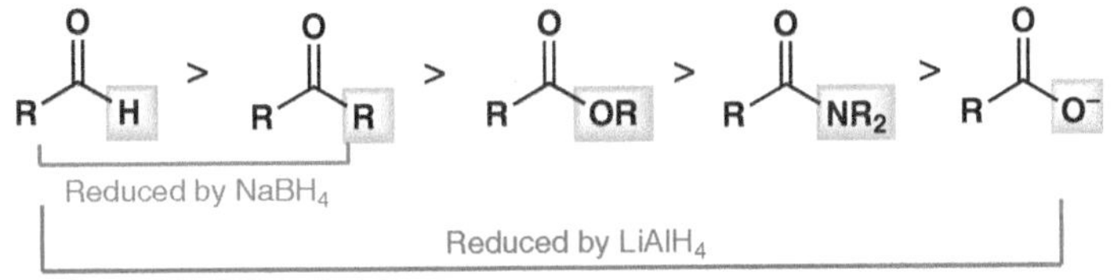

Bifunctional compounds having both ketone and ester groups can be reduced selectively by the choice of reducing agent (Scheme 10.5).

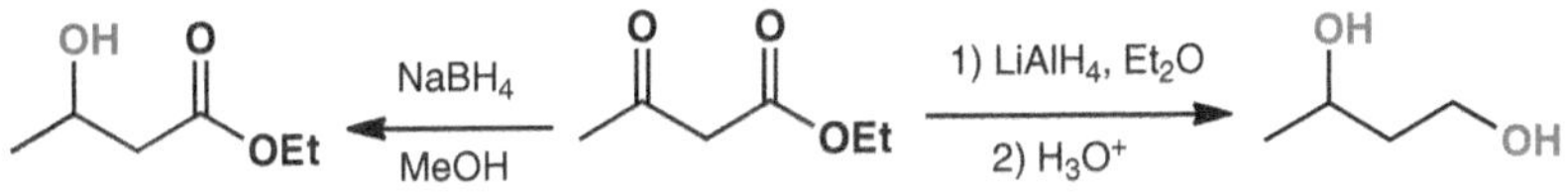

Scheme 10.5
Selective reduction of a ketoester.

The controlled transformation of the keto ester into a hydroxy ketone is also possible (see Sub-section 10.5.3).

Exercise 10.1

What is the main product when each of the following carbonyl compounds is reduced with NaBH$_4$ in an alcoholic solution?

(a) 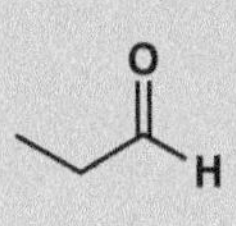 (b) 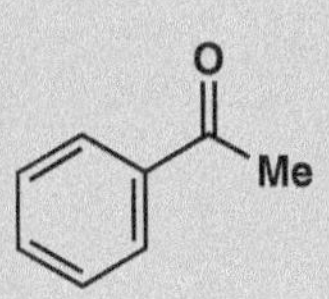 (c)

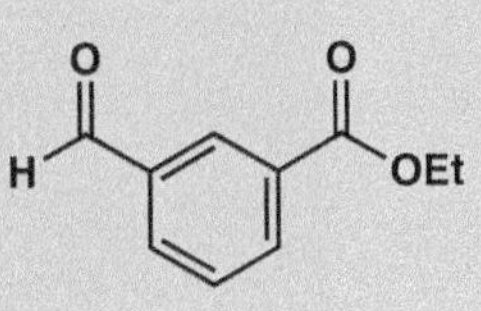

Exercise 10.2

What is the main product when each carbonyl compound in Exercise 10.1 is reduced with LiAlH$_4$ in an ether, and the reaction is worked up with aqueous acid?

10.2 Indirect Reduction of Aldehydes and Ketones

In Section 10.1 we discussed the direct reduction of carbonyl groups by hydride transfer to give alcohols. Alternative reduction methods are available which do not give alcohols

See Sub-section 8.6.1 for the formation of imines.

and, in the case of aldehydes and ketones, involve the intermediacy of carbonyl derivatives with a C=N bond, i.e. imines. The C=N bond is a polarized double bond and undergoes some reactions similar to ones of the C=O bond.

10.2.1 Reductive amination

Complete reduction of the C=N bond of an imine, e.g. by hydrogenation, gives an amine. In practice, an aldehyde or ketone may be treated with ammonia or an amine in the presence of a reducing agent to give the amine reduction product without isolation of the imine (Scheme 10.6). The reducing agent is often H_2 in the presence of a metal catalyst, or sodium cyanoborohydride $NaBH_3CN$. This very mild hydride donor will not reduce a carbonyl group or a neutral imine, but it does reduce a protonated imine, i.e. the iminium ion, $>C=N^+(R)H$, which is generated under mildly acidic conditions. This overall reaction is known as **reductive amination**, and is one of several good methods for the preparation of amines.

Electrophilicity:

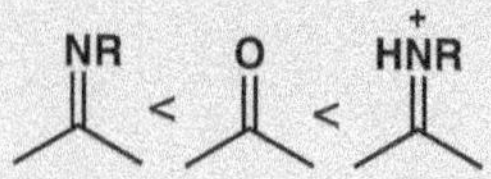

NR   O   HNR
(with < and + marks)

Under the acidic conditions of the reaction, the amine product will be protonated.

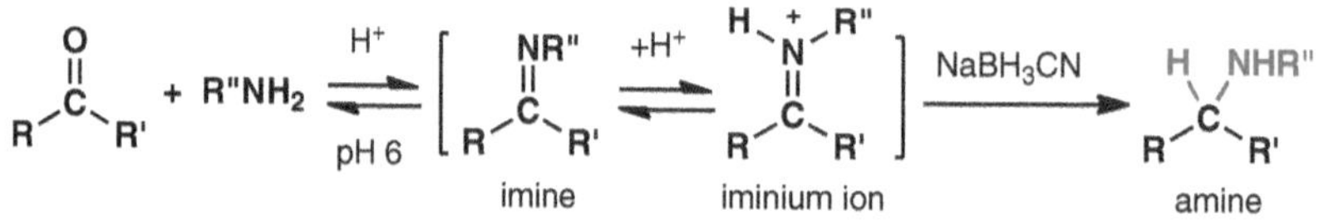

Scheme 10.6 Reductive amination of an aldehyde or ketone.

Example 10.1

Show the structure of the amine formed by each reductive amination.

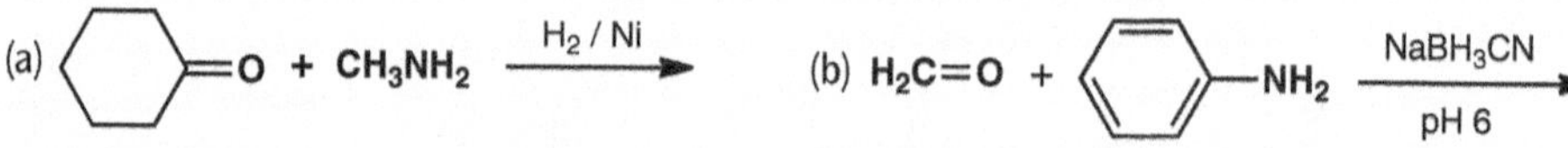

(a) cyclohexanone =O + CH₃NH₂ →(H₂ / Ni)   (b) H₂C=O + [phenyl]–NH₂ →(NaBH₃CN, pH 6)

Solution
First draw the structure of the intermediate imine, and then deduce the structure of the product obtained by hydrogenation/reduction of the C=N bond.

(a) [cyclohexane with NHCH₃ and H]   (b) [phenyl]–NHCH₃

Exercise 10.3

Show how to prepare each of the following amines by reductive amination. For (c), give reactants other than those in Example 10.1(a).

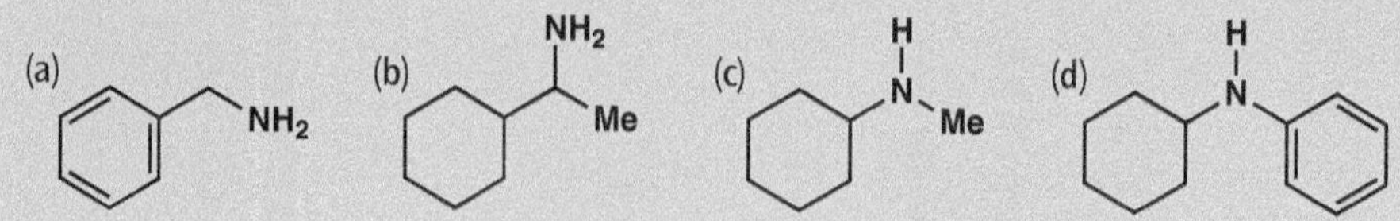

(a) [benzyl CH₂NH₂]   (b) [cyclohexyl CH(Me)NH₂]   (c) [cyclohexyl-NH-Me]   (d) [cyclohexyl-NH-phenyl]

10.2.2 Reduction of the C=O of aldehydes and ketones to give CH₂

When hydrazones of aldehydes and ketones (see Chapter 8) are heated in a concentrated aqueous solution of NaOH, the C=NNH₂ groups are converted to CH₂ with the evolution of N₂. It is not usually necessary to isolate the hydrazone so, starting with an aldehyde or ketone and hydrazine, the overall reaction becomes the reduction of the >C=O to >CH₂ (eqn 10.9). This transformation is sometimes called the *Wolff–Kishner reduction*, and the mechanism in Scheme 10.7 accounts for the experimental observations.

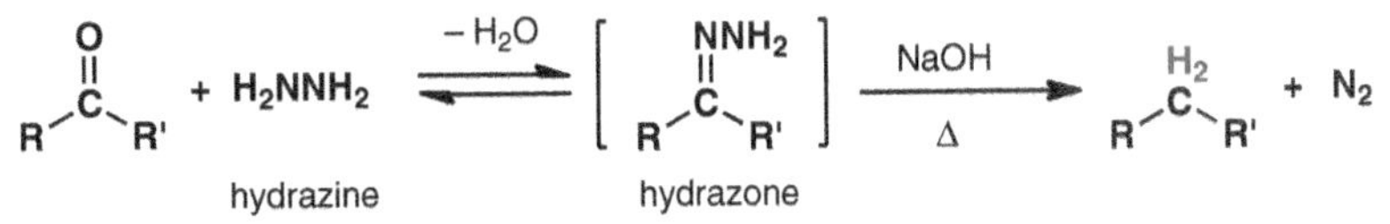

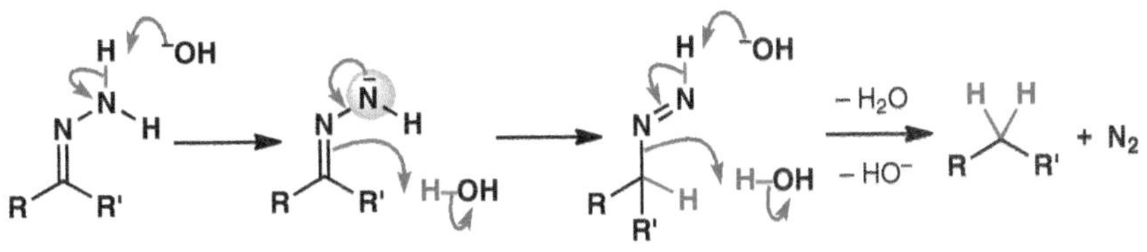

Scheme 10.7 The Wolff–Kishner reduction and its mechanism.

A milder method often used for the same purpose, which avoids strongly acidic and basic conditions, is desulfurization of a dithioacetal (see Sub-section 8.4.3) by hydrogenolysis with *Raney nickel*,[1] a black pyrophoric form of the metal saturated with hydrogen (eqn 10.10).

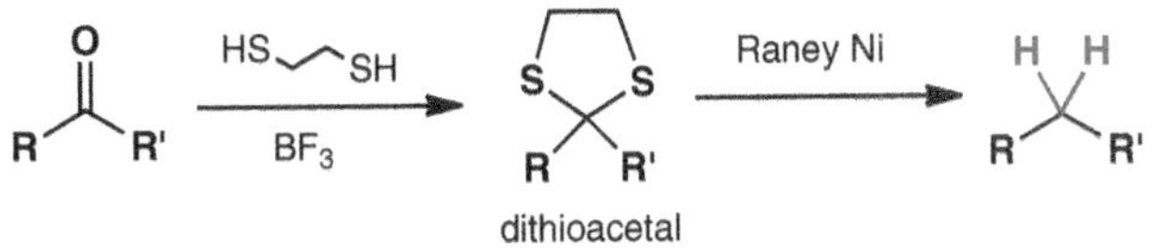

$$(10.10)$$

10.3 Hydride Transfer from Carbon

A hydrogen on the C next to a carbonyl group (a so-called α hydrogen) is weakly acidic, so when an aldehyde or ketone bearing such a hydrogen dissolves in an alkaline solution, an equilibrium concentration of the enolate anion is formed. We shall discuss reactions involving enolates more generally in Chapter 17.

When a carbonyl compound having no α hydrogen dissolves in concentrated aqueous NaOH, the only possible initial reaction is reversible nucleophilic addition (see Chapter 8). If the carbonyl compound without the α hydrogen is an aldehyde, equal amounts of the corresponding alcohol and sodium salt of the carboxylic acid are formed, as exemplified in eqn 10.11, in what is known as the **Cannizzaro reaction**.

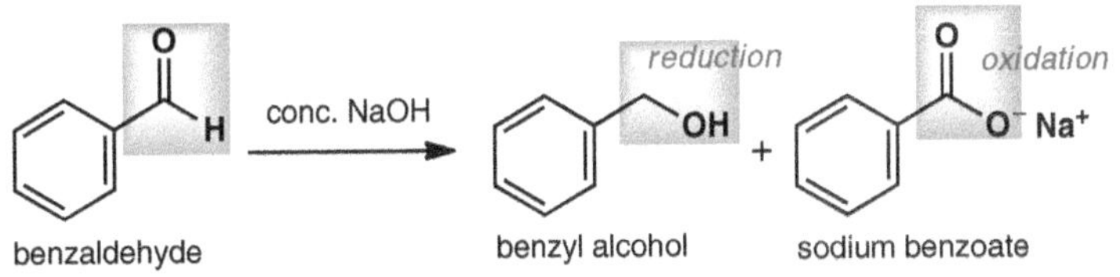

$$(10.11)$$

One half of the aldehyde in eqn 10.11 is reduced and the other half is oxidized, and the mechanism is given in Scheme 10.8. Initial nucleophilic addition of HO⁻ gives the anion A⁻ and the –O⁻ within A⁻ has a strong tendency to push electrons in and regenerate the stable C=O bond by expelling another group. We saw this previously when an anionic tetrahedral intermediate in ester hydrolysis regenerates a carbonyl compound (Sub-section 9.2.2) by expelling an alkoxide. In the Cannizzaro reaction, however, departure of the hydroxide is simply the reverse of the initial nucleophilic attack. But if another aldehyde

An alternative method for replacing the >C=O of an aldehyde or ketone by >CH$_2$, which is complementary to the Wolff–Kishner reduction, is worth mentioning here. In contrast to the Wolff–Kishner method, which employs strongly basic reaction conditions, the *Clemmensen reduction* uses zinc metal (or amalgamated zinc) in fairly concentrated hydrochloric acid.

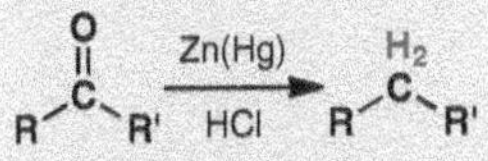

Thus, one uses the Wolff–Kishner method for compounds with other functionality which can tolerate basic conditions, and the Clemmensen method if other functionality is stable to acidic conditions.

---

[1]   Beware confusion between the following terms.
Hydration is the addition of H$_2$O to a double bond;
Hydrolysis is the decomposition of a compound in/by H$_2$O;
Hydrogenation is the addition of H$_2$ to a multiple bond;
Hydrogenolysis is the cleavage of a C–Z bond (Z = an atom or group other than H) and replacement of Z by H.

Stanislao Cannizzaro (1826–1910).

The Italian chemist Cannizzaro reported the reaction named after him in 1853. He also contributed to the recognition of the periodic table of elements.

molecule is in the vicinity of A⁻, the second aldehyde can accept a hydride anion (H⁻) from A⁻ as an electron pair from –O⁻ pushes out H⁻ (instead of OH⁻) to reform the C=O bond. Consequently, the aldehyde which receives the H⁻ is reduced to the corresponding alcohol, while the one which supplies the H⁻ is oxidized to the corresponding carboxylate ion after equilibration involving proton transfer.

Scheme 10.8 Mechanism of the Cannizzaro reaction.

Methanal (formaldehyde) is especially reactive as an aldehyde lacking an α hydrogen, and is a good hydride donor under strongly alkaline conditions, as in the 'crossed' Cannizzaro reaction of eqn 10.12.

$$PhCHO + H_2C=O \xrightarrow{\text{conc. NaOH}} PhCH_2OH + HCO_2^- \qquad (10.12)$$

Exercise 10.4

Propose a mechanism using curly arrows for reaction 10.12.

Panel 10.2 The Meerwein–Ponndorf–Verley–Oppenauer reaction

The equation below shows an example of the extravagantly named Meerwein–Ponndorf–Verley–Oppenauer reaction which is mechanistically similar to the Cannizzaro reaction. The forward reaction is the reduction of an aldehyde or ketone using a secondary alcohol in the form of an aluminium alkoxide (usually isopropoxide) as the hydride donor.

$$R^1R^2C=O + Me_2CHOH \underset{}{\overset{Al(OCHMe_2)_3}{\rightleftharpoons}} R^1R^2CH\text{-}OH + Me_2C=O$$

First, the carbonyl compound, a Lewis base, coordinates the Al(III), a Lewis acid, which activates the carbonyl group towards nucleophilic attack, and brings it into the vicinity of the hydride donor. Within the coordination complex, hydride transfer reduces the carbonyl compound to an alkoxide bonded to the aluminium and, at the same time, an isopropoxide is oxidized to propanone which is released. To complete the reaction, propan-2-ol (isopropyl alcohol) reacts with the alkoxy-aluminium di-isopropoxide, which liberates the reduction product itself and regenerates aluminium tri-isopropoxide. The equilibrium shown above is driven to the right by using an excess of propan-2-ol and selectively distilling out the propanone (which is invariably the most volatile component in the reaction mixture).

$$R_2CHOAl(OCHMe_2)_2 + Me_2CHOH \rightleftharpoons R_2CHOH + Al(OCHMe_2)_3$$

The reaction is reversible, and can be used to oxidize secondary alcohols to ketones. In this direction, the secondary alcohol and an excess of propanone are heated with aluminium isopropoxide in toluene. The secondary alcohol exchanges with an isopropoxide on the Al(III) complex, and hydride is transferred to a propanone which is Lewis acid-base complexed to the Al(III).

Panel 10.3 Nature's hydride donor: NADH

In our bodies, the oxidation of ethanol to ethanal is catalysed by an enzyme called alcohol dehydrogenase, and the hydride acceptor is an oxidized form of a coenzyme, nicotinamide adenine dinucleotide, NAD$^+$ (see Chapter 24). Its reduced form, NADH, can work as a hydride donor to reduce carbonyl compounds.

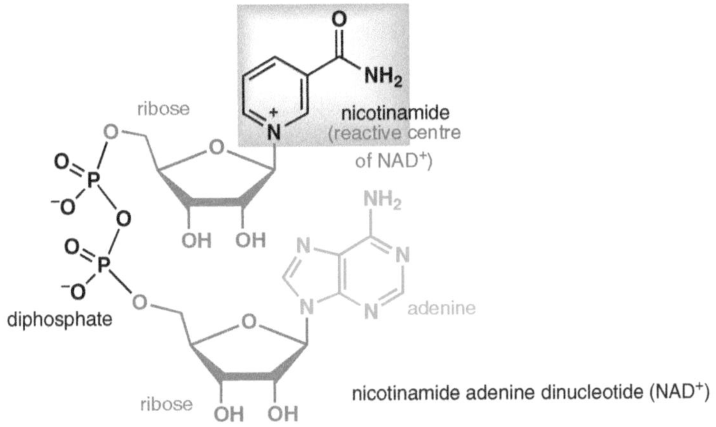

nicotinamide adenine dinucleotide (NAD$^+$)

Oxidation of ethanol occurs by hydride transfer to NAD$^+$, which is facilitated by electron-pull by the positive nitrogen of the pyridinium residue at the reactive site, and gives NADH and the aldehyde. The driving force for the conversion of NADH back to NAD$^+$ by reaction with another hydride acceptor is the recovery of the aromatic stability of the pyridine ring. The toxic ethanal is quickly converted into acetate to be used by the cell.

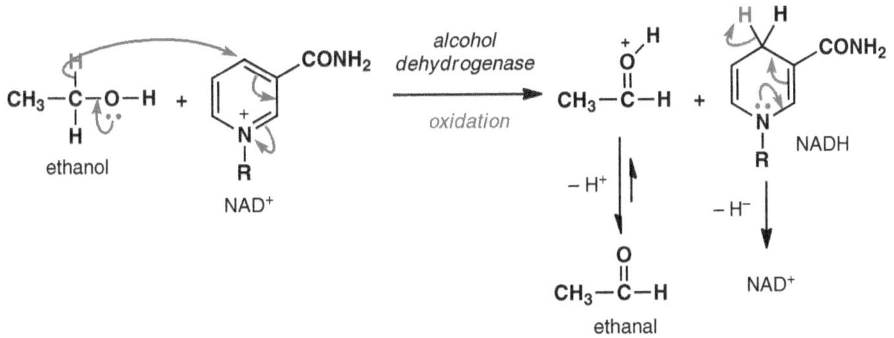

10.4 Reactions with Organometallic Reagents: C–C Bond Formation

10.4.1 Organometallic compounds

A compound containing a metal–carbon (M–C) bond is called an organometallic compound. Since the metal is electropositive, the M–C bond is polarized and the carbon bears a partial negative charge. The MO corresponding to the M–C bond is the HOMO of the organometallic compound in the same way that the MO corresponding to the M–H bond is the HOMO of a metal hydride.

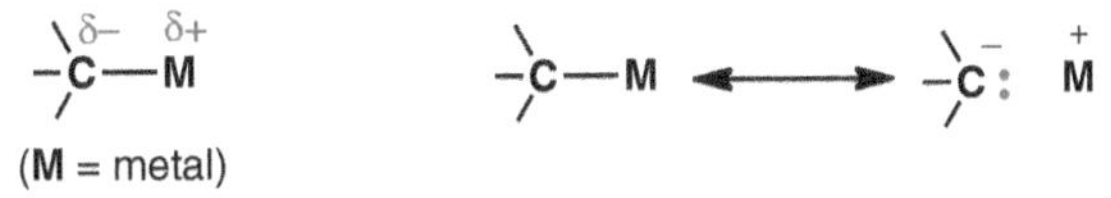

(M = metal)

Organolithium and organomagnesium compounds are important organometallic reagents and they react as carbon nucleophiles with carbonyl groups to form new C–C bonds

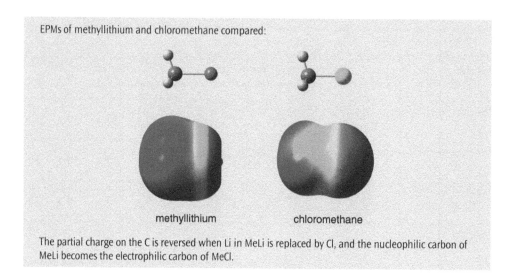

EPMs of methyllithium and chloromethane compared:

methyllithium   chloromethane

The partial charge on the C is reversed when Li in MeLi is replaced by Cl, and the nucleophilic carbon of MeLi becomes the electrophilic carbon of MeCl.

(eqn 10.13). Protonation of the initially formed adduct by addition of water gives the product, which is an alcohol with an extended carbon framework. Such reactions are widely used in organic synthesis.

$$\text{(10.13)}$$

Organolithium and organomagnesium compounds are prepared from haloalkanes or haloarenes by metal–halogen exchange reactions, as shown in eqns 10.14 and 10.15.

$$\text{RX} + 2\,\text{Li} \xrightarrow{\text{Et}_2\text{O}} \text{RLi} + \text{LiX} \qquad (10.14)$$
$$(\text{X} = \text{Cl, Br, I})$$

$$\text{RX} + \text{Mg} \xrightarrow{\text{Et}_2\text{O}} \text{RMgX} \qquad (10.15)$$
$$(\text{X} = \text{Cl, Br, I})$$

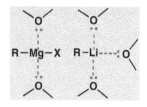

Because of the polarity of M–C bonds, the metal atoms of RLi and RMgX bear partial positive charges so are Lewis acid sites. Consequently, solvents with Lewis base properties (lone pairs of electrons), such as ethers (typically diethyl ether or tetrahydrofuran), facilitate the metal exchange reactions by coordinating (solvating) the metal component of the product.

Compounds RLi and RMgX are also strong Brønsted bases since they are essentially the conjugate bases of alkanes, RH (pK_a ~ 50). They react vigorously with protic compounds such as water and alcohols, therefore, to form alkanes RH (eqn 10.16). Reaction with D_2O gives a deuterated product (eqn 10.17).

$$\text{RMgX} + \text{H}_2\text{O} \longrightarrow \text{RH} + \text{MgX(OH)} \qquad (10.16)$$

$$\text{RMgX} + \text{D}_2\text{O} \longrightarrow \text{RD} + \text{MgX(OD)} \qquad (10.17)$$

Terminal alkynes (pK_a ~ 25) are considerably more acidic than alkanes, so they react readily with RLi or RMgX to form alkynyl–metal compounds (eqn 10.18).

$$\text{R}-\text{C}{\equiv}\text{C}-\text{H} + \text{R'MgX} \xrightarrow{\text{Et}_2\text{O}} \text{R}-\text{C}{\equiv}\text{C}-\text{MgX} + \text{R'H} \qquad (10.18)$$

10.4.2 The Grignard reaction

Although both organolithium and organomagnesium compounds react similarly, the former are more reactive and difficult to handle, so the latter (which were discovered earlier) have been more widely used in organic synthesis. Organomagnesium reagents are called **Grignard reagents** after their discoverer.

Reactions of Grignard reagents with carbonyl compounds provide alcohols after aqueous workup. The class of alcohol which is obtained (primary, secondary, or tertiary) depends on the nature of the carbonyl compound employed (eqns 10.19–10.22).

$$RMgX + H_2C=O \xrightarrow{\text{1) Et}_2\text{O, 2) H}_3\text{O}^+} RCH_2OH \qquad (10.19)$$

methanal — 1° alcohol

$$RMgX + \underset{H}{\overset{R'}{\diagdown}}C=O \xrightarrow{\text{1) Et}_2\text{O, 2) H}_3\text{O}^+} \underset{R'}{\overset{R}{\diagdown}}CH\text{-}OH \qquad (10.20)$$

higher aldehyde — 2° alcohol

$$RMgX + \underset{R''}{\overset{R'}{\diagdown}}C=O \xrightarrow{\text{1) Et}_2\text{O, 2) H}_3\text{O}^+} R-\underset{R''}{\overset{R'}{\underset{|}{\overset{|}{C}}}}-OH \qquad (10.21)$$

ketone — 3° alcohol

$$2\,RMgX + R'-\overset{O}{\overset{\|}{C}}-OR'' \xrightarrow{\text{1) Et}_2\text{O, 2) H}_3\text{O}^+} R-\underset{R}{\overset{R'}{\underset{|}{\overset{|}{C}}}}-OH + R''OH \qquad (10.22)$$

ester — 3° alcohol

F.A.V. Grignard (1871–1935).

The French chemist Victor Grignard discovered the reaction which now bears his name in 1900, and shared the 1912 Nobel Prize in Chemistry with his fellow-countryman, Paul Sabatier.

In eqns 10.19–10.22, the designation '1) Et₂O, 2) H₃O⁺' above the arrows indicates that the reactions are carried out in two stages: the first is carried out in diethyl ether (Et₂O), and the reaction mixture is treated (quenched) with aqueous acid only after the first stage is complete. A description such as 'Et₂O, H₃O⁺' without numbering implies that reactants are mixed in acidic water and ether together. If this had happened for eqns 10.19–10.22, the RMgX would have been destroyed immediately by its much faster reaction with H₂O (eqn 10.16). It follows that, up to the aqueous quench, Grignard reactions must be carried out with strict exclusion of moisture, and all solvents, apparatus, and reagents must be anhydrous.

Write a reaction sequence to show how the transformation of eqn 10.20 takes place. | **Exercise 10.5**

The reactions of Grignard reagents with aldehydes and ketones give primary, secondary, or tertiary alcohols depending on the pattern of alkyl/aryl substitution on the carbonyl carbon. The reaction with esters is slightly different; an ester reacts with *two* molar equivalents of the Grignard reagent to give (ultimately) a tertiary alcohol bearing two identical alkyl groups on the α carbon. This is due to the nature of the tetrahedral intermediate (alkoxide) initially formed by addition of RMgX to the ester (Scheme 10.9). This intermediate is similar to the one formed during ester hydrolysis under alkaline conditions (Sub-section 9.2.2), and spontaneously releases R''O⁻ to give a ketone which reacts further with another RMgX.

In practice, a Grignard reaction is carried out by starting with the preparation of the organomagnesium reagent, RMgX, from metallic Mg and the haloalkane (RX) in ether (eqn 10.15), and a solution of the carbonyl compound in ether is then added, all under

Scheme 10.9
Mechanism of the Grignard
reaction with an ester.

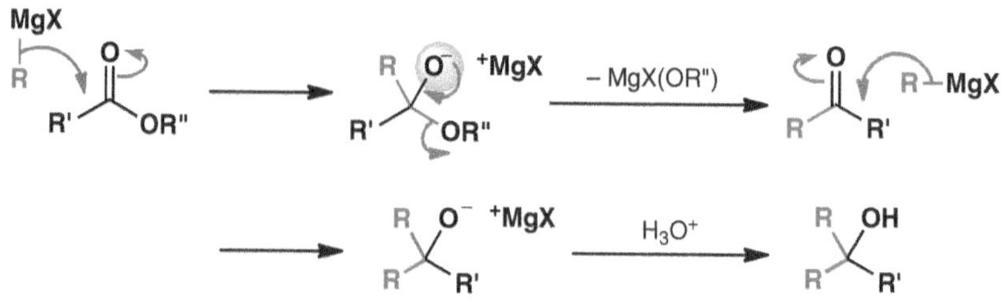

Occasionally, the Grignard
reagent is added to the
solution of the carbonyl
compound; this is called
a Grignard reaction with
inverse addition.

rigorously anhydrous conditions. The overall reaction is completed by the aqueous acidic quench and can be described as in eqn 10.23.

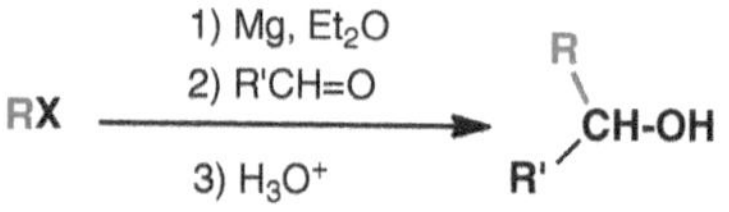

$$RX \xrightarrow[\substack{\text{1) Mg, Et}_2\text{O} \\ \text{2) R'CH=O} \\ \text{3) H}_3\text{O}^+}]{} \quad \substack{R \\ R'}CH\text{-}OH \qquad (10.23)$$

Exercise 10.6

What is the product of each of the following Grignard reactions?

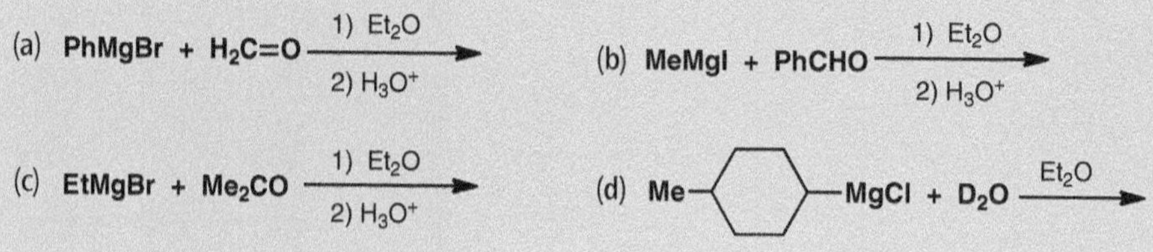

(a) PhMgBr + H₂C=O $\xrightarrow[\text{2) H}_3\text{O}^+]{\text{1) Et}_2\text{O}}$

(b) MeMgI + PhCHO $\xrightarrow[\text{2) H}_3\text{O}^+]{\text{1) Et}_2\text{O}}$

(c) EtMgBr + Me₂CO $\xrightarrow[\text{2) H}_3\text{O}^+]{\text{1) Et}_2\text{O}}$

(d) Me—⟨ ⟩—MgCl + D₂O $\xrightarrow{\text{Et}_2\text{O}}$

Reaction of an *N,N*-dimethylamide with RMgX provides a ketone (eqn 10.24). This occurs because the initial adduct cannot readily decompose owing to the poor leaving ability of the amide ion. Hydrolysis of the adduct gives the ketone during aqueous acidic workup (Scheme 10.10), when protonation of the NMe₂ group provides a much better nucleofuge.

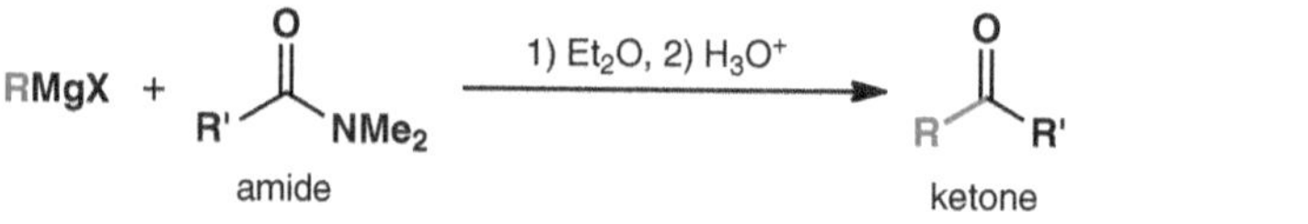

$$(10.24)$$

amide   ketone

Scheme 10.10
Mechanism of the Grignard
reaction of a tertiary amide.

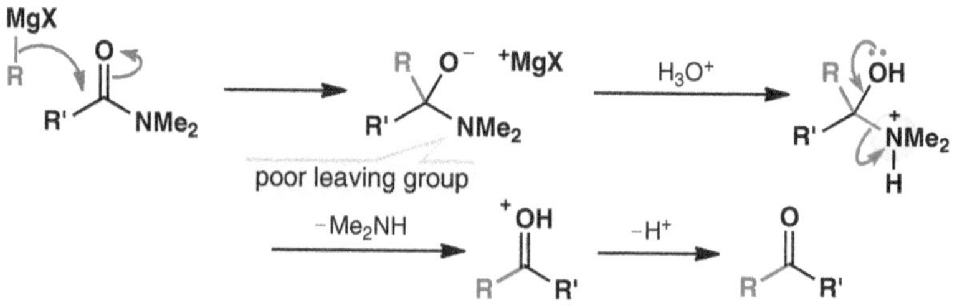

poor leaving group

Carboxylic acids can be prepared by the reaction of RMgX with carbon dioxide (eqn 10.25); note that the carbon chain has been increased by one C unit in this reaction.

$$RMgX + CO_2 \xrightarrow{\text{1) Et}_2\text{O, 2) H}_3\text{O}^+} RCO_2H \qquad (10.25)$$

carbon dioxide   carboxylic acid

Grignard reagents also react with oxiranes (epoxides) the reactivity of which is attributable to the strain of the three-membered ring (see Section 14.5); ethylene oxide gives a primary alcohol with two additional carbon units (eqn 10.26).

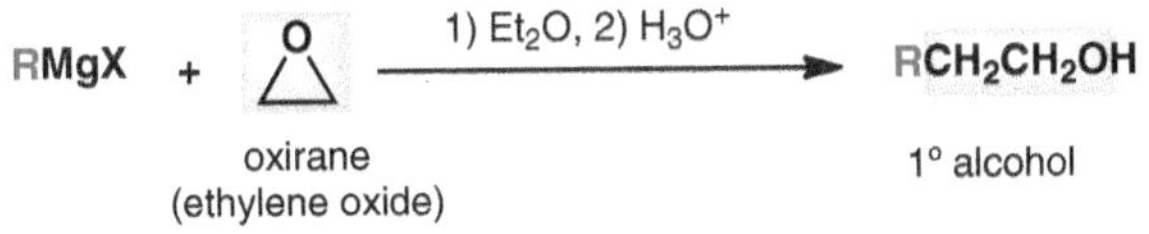

$$\text{RMgX} + \triangle\!\!\!\overset{O}{} \xrightarrow{\text{1) Et}_2\text{O, 2) H}_3\text{O}^+} \text{RCH}_2\text{CH}_2\text{OH} \qquad (10.26)$$

oxirane
(ethylene oxide)

1° alcohol

What is the main product of each of the following Grignard reactions?

(a) $PhCO_2Et$ + 2 MeMgI $\xrightarrow[\text{2) H}_3\text{O}^+]{\text{1) Et}_2\text{O}}$

(b) HCO_2Me + 2 EtMgBr $\xrightarrow[\text{2) H}_3\text{O}^+]{\text{1) Et}_2\text{O}}$

(c) 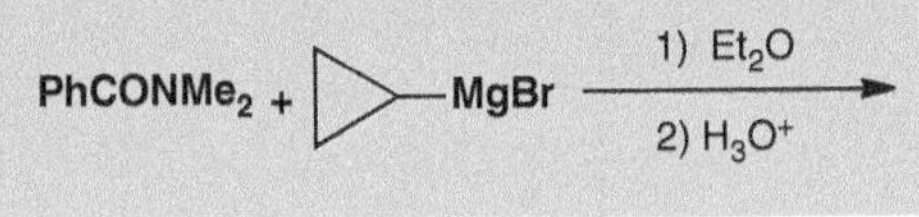 $PhCONMe_2$ + —MgBr $\xrightarrow[\text{2) H}_3\text{O}^+]{\text{1) Et}_2\text{O}}$

(d) PhMgBr + CO_2 $\xrightarrow[\text{2) H}_3\text{O}^+]{\text{1) Et}_2\text{O}}$

10.4.3 Side reactions with Grignard reagents

Two types of side reactions are possible in Grignard addition reactions to carbonyl groups. One is possible only if the α C of the carbonyl compound bears a hydrogen; such hydrogens are weakly acidic and may be abstracted by a strong base such as a Grignard reagent. Proton transfer from the α C of the carbonyl compound to RMgX gives an enolate, as shown in Scheme 10.11 (see Chapter 17 for a general discussion of enols and enolates). The enolate ion is reprotonated by water during the aqueous workup to give the original carbonyl compound.

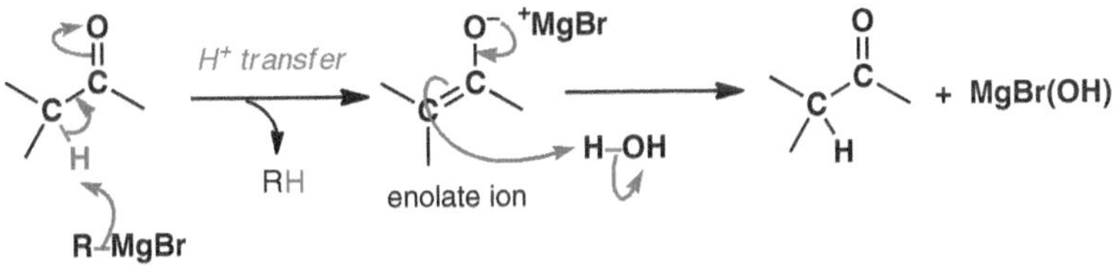

Scheme 10.11
Enolate formation during a Grignard reaction.

The apparent result is the recovery of unreacted carbonyl compound and a low yield of the addition product. This side reaction involving an α C–H occurs when the nucleophilic addition is inhibited by steric effects.

Hydride transfer from RMgX to the carbonyl compound, i.e. reduction (Scheme 10.12), is another reaction which intrudes when nucleophilic addition is sterically inhibited.

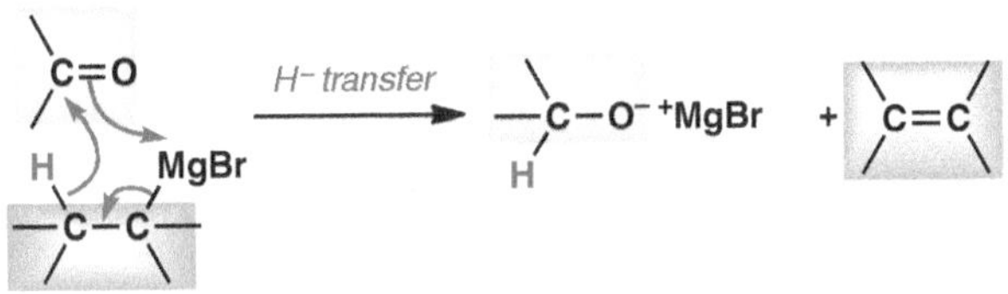

Scheme 10.12
Reduction during a Grignard reaction.

This hydride transfer reaction can only occur when there is a hydrogen on the carbon next to the nucleophilic C bonded to the Mg; the reaction is similar to the hydride transfer from aluminium isopropoxide (see Panel 10.2). The products are the alcohol formed by reduction of the carbonyl compound and the alkene from RMgX. An example of this complication is observed with the sterically congested ketone shown in eqn 10.27.

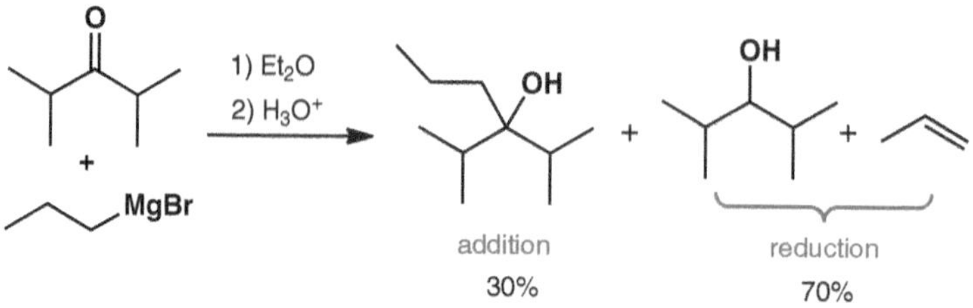

(10.27)

10.5 Planning Organic Syntheses: Synthesis of Alcohols

10.5.1 An introduction to organic synthesis

Practice in devising an efficient synthesis involves drawing on our total knowledge of organic chemistry and is an excellent vehicle for revising and consolidating such knowledge—it helps us to see the subject as a whole rather than as a collection of unrelated facts.

Organic synthesis is essentially the preparation of more complex organic compounds from readily available simpler precursors, and should be distinguished from functional group transformations (interconversions) which usually involve no extension of the carbon framework of a molecule. However, as new target molecules become more complex, the planning of an efficient route becomes increasingly challenging, and an appreciation of *how* reactions take place is critical in this exercise.

To find a reasonable route for the synthesis (and there will invariably be more than one), we work backwards from the target molecule to immediate precursors. From these, we work further backwards, one step at a time from each, until we arrive at suitable starting materials. This process is called **retrosynthetic analysis**, and each step is represented by a special arrow ($\Rightarrow$).

In this section, we shall discuss only the synthesis of alcohols by hydride reduction and Grignard reactions. As we shall see, alternative routes are usually possible; in such cases, the availabilities and costs of different starting materials, and the efficiencies of alternative routes, should be taken into account. These matters and further aspects of organic synthesis will be covered in Chapter 23.

10.5.2 Examples of alcohol synthesis

We shall consider two examples here: the first is the synthesis of a representative tertiary alcohol with three different groups on the carbon bearing the OH group. Three alternative *disconnections* of C–C bonds, shown in Scheme 10.13 by wavy lines, indicate three alternative combinations of Grignard reagent plus ketone.

Scheme 10.13
Retrosynthetic analysis of 2-phenylbutan-2-ol.

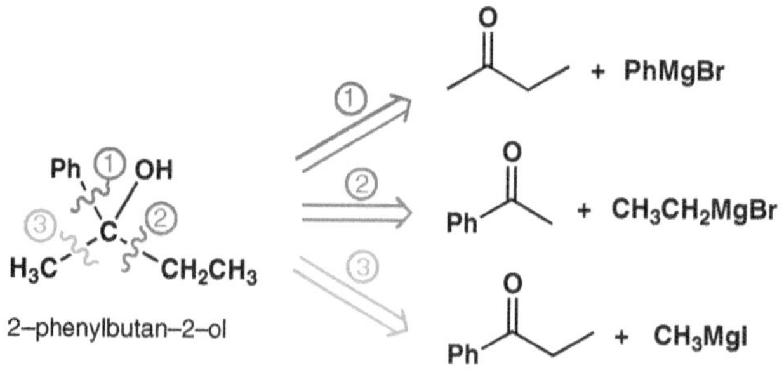

A possible synthesis corresponding to disconnection 1 is given in eqn 10.28.

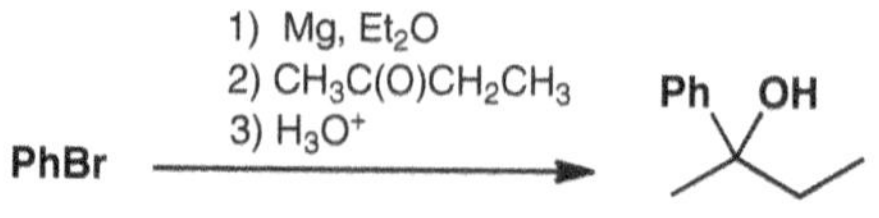

(10.28)

Our second example is a secondary alcohol which may be obtained by reduction of the corresponding ketone (corresponding to disconnection 1), or addition of a Grignard reagent to an aldehyde (disconnection 2 *or* 3). However, when an alcohol has two identical groups on the α carbon, the addition of two equivalents of a Grignard reagent to an ester is also a synthetic possibility, as in eqn 10.22 above. In the present case, disconnections 2 *and* 3 of Scheme 10.14, the ester must be a methanoate.

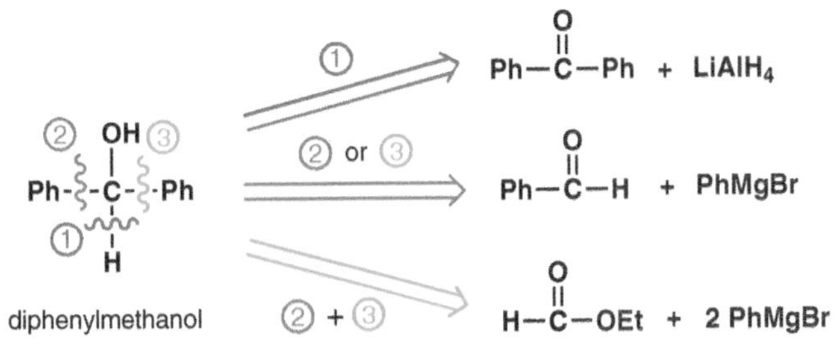

Scheme 10.14 Retrosynthetic analyses of diphenylmethanol.

Exercise 10.8

Give syntheses corresponding to the retrosyntheses given in Scheme 10.14.

Exercise 10.9

After retrosynthetic analysis of the following three alcohols, show how a single Grignard reagent can be used in a preparation of each.

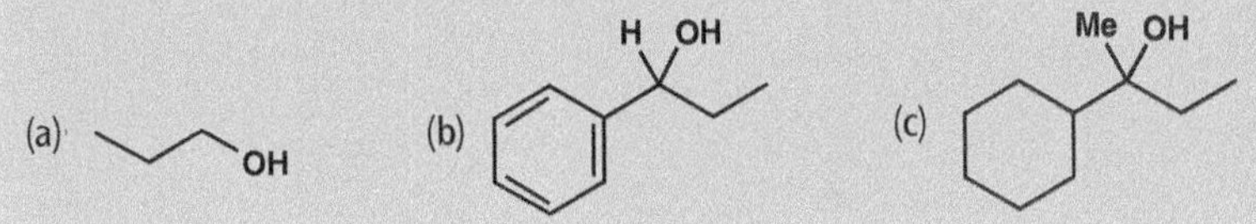

10.5.3 Protection of carbonyl groups and deprotection

Carbonyl compounds are reactive under wide-ranging experimental conditions. Consequently, when a carbonyl compound has a second functional group, and some transformation of the second functional group is planned, it is often necessary to *protect* the carbonyl group during the transformation. Conversion of aldehyde and ketone carbonyl groups into acetals is one strategy because acetals are particularly stable except under acidic conditions. Acetal formation is usually easy with a suitable alcohol in the presence of an acid catalyst, and its hydrolysis readily regenerates the aldehyde or ketone, as discussed in Sub-section 8.4.2.

For example, if the C–Br group in a bromo ketone needs to be converted into the Grignard reagent for a reaction with another carbonyl compound (eqn 10.29), the carbonyl group of the bromo ketone is a complication.

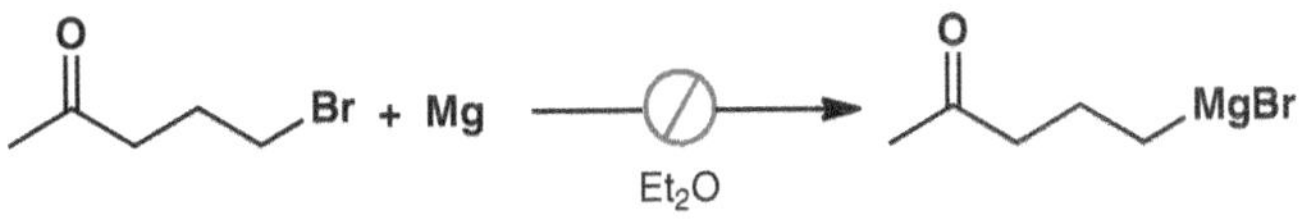

(10.29)

In this example, the bromo ketone can be transformed to its acetal (**protection**) then, after the Grignard reaction, the acetal can be hydrolysed to regenerate the carbonyl group (**deprotection**), as shown in Scheme 10.15.

Scheme 10.15
Protection of a bromo ketone
prior to a Grignard reaction,
and subsequent deprotection.

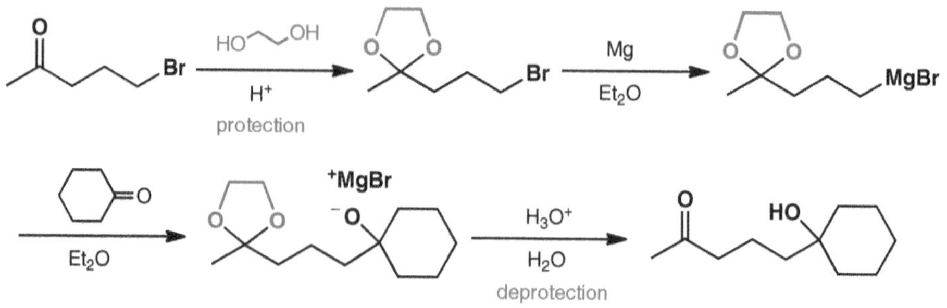

Protection and deprotection are important general aspects of synthetic strategy; another example is the selective reduction of the ester group of a keto ester (Scheme 10.16). Selective reduction of the ketone group of a keto ester is a relatively easy task with a choice of reducing agents, since the ketone carbonyl is more reactive than the ester (Subsection 10.1.2). However, the more reactive ketone group would not survive the conditions necessary for the reduction of the ester. We can accomplish the task if we protect the ketone group as an acetal according to the reaction sequence given in Scheme 10.16.

Scheme 10.16
Protection of a ketoester prior
to LiAlH$_4$ reduction of the ester
function, and subsequent
deprotection.

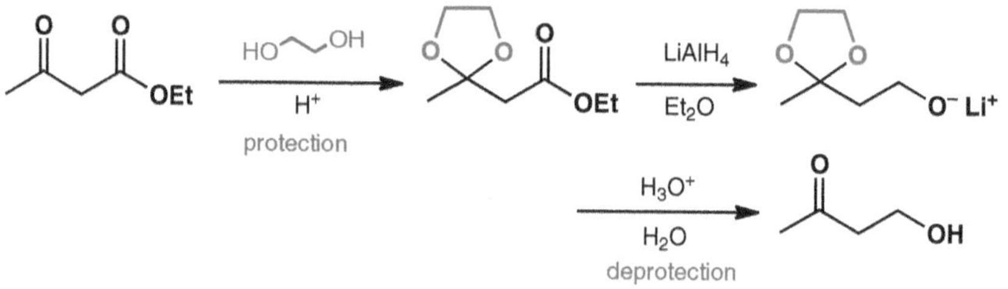

Exercise 10.10   Explain why only one of the two carbonyl groups of the ketoester in Scheme 10.16 forms an acetal in the first step.

Dithioacetals can also be used as protecting groups for aldehydes and ketones. They are considerably more stable in acidic solution than acetals, but may be hydrolysed with catalysis by Hg(II) salts. Partial oxidation of sulfur (for example, using *m*-chloroperbenzoic acid, MCPBA) also facilitates their hydrolysis back to the carbonyl compounds.

Protection of a ketone as a dithioacetal and deprotection:

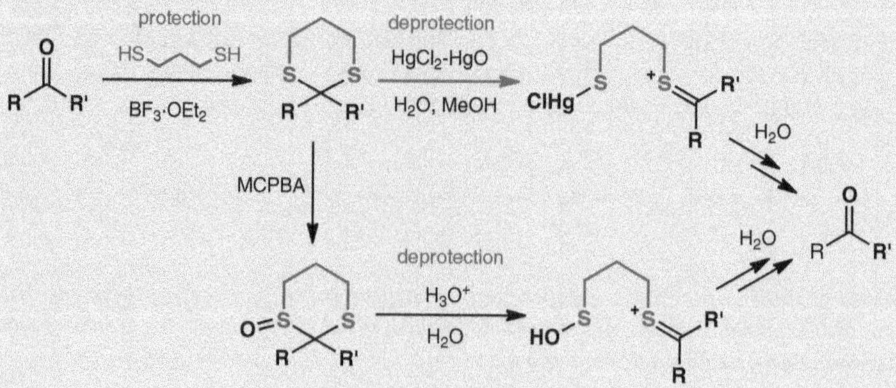

Summary

- Metal hydrides (M–H) and organometallic compounds M–C (M = metal) react with carbonyl groups by transfer of H⁻ and C⁻ nucleophiles, respectively. Alcohols are formed in these additions to the C=O bonds by **hydride reduction** or **C–C bond formation**.

- The hydride donors NaBH₄ and LiAlH₄ reduce aldehydes and ketones, but only the latter is sufficiently reactive to reduce carboxylic acids and their derivatives.

- Aldehydes and ketones under reductive conditions in the presence of ammonia or a simple amine (reductive amination) give amines.

- Two molecules of an aldehyde without α hydrogens under strongly basic conditions yield one molecule of the corresponding alcohol (reduction product) and one of the carboxylic acid (as its conjugate base, the oxidation product) by intermolecular hydride transfer.

- Other methods of reduction of aldehydes and ketones (C=O → CH₂) include the use of hydrazine in aqueous NaOH, Zn in hydrochloric acid, and hydrogenolysis of dithioacetals.

- **Grignard reagents**, RMgX, are particularly important organometallic reagents for the formation of new C–C bonds; their reactions with carbonyl compounds generally give alcohols.

- An organic synthesis can be planned by **retrosynthetic analysis** which involves disconnection of strategic bonds in the target molecule to identify an appropriate sequence of reactions and suitable starting materials.

Problems

10.1 Give the structural formulas of carbonyl compounds which would yield the following alcohols upon hydride reduction. Also, give the IUPAC names of the carbonyl compounds.

(a) propan-1-ol (b) propan-2-ol
(c) cyclohexanol (d) 1-phenylpropan-2-ol
(e) but-3-en-1-ol

10.2 What are the main products when the following carbonyl compounds react with NaBH₄ in methanol?

(a)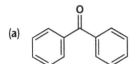

(b)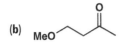

(c)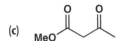

(d)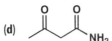

(e)

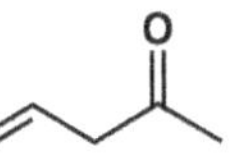

10.3 What are the main products when the carbonyl compounds in Problem 10.2 react with LiAlH₄ in ether followed by treatment with aqueous acid?

10.4 Show how to prepare each of the following amines by reductive amination.

(a) 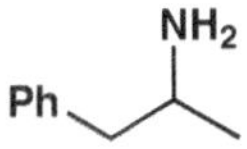 Ph, NH₂ (b) 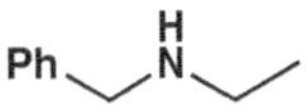 Ph, H N (c) 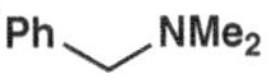 Ph, NMe₂

10.5 What are the main product(s) of each of the following reactions?

(a) 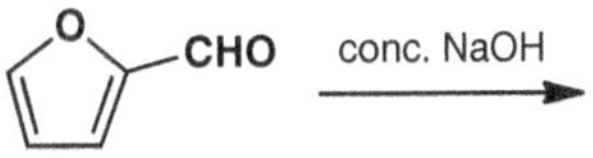 CHO, conc. NaOH
(b) 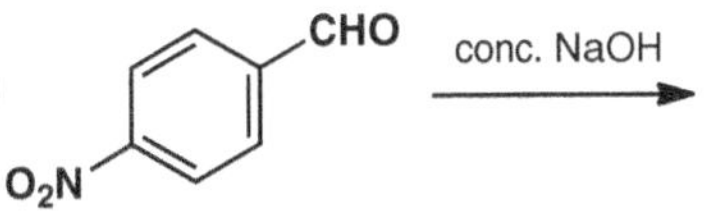 CHO, conc. NaOH, O₂N
(c) 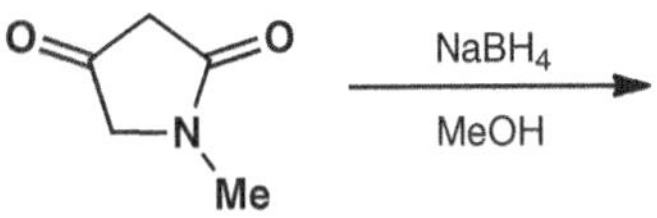 NaBH₄, MeOH
(d) 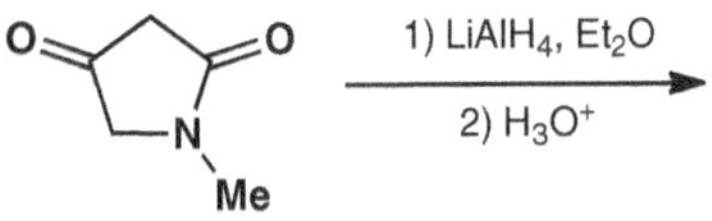 1) LiAlH₄, Et₂O 2) H₃O⁺
(e) 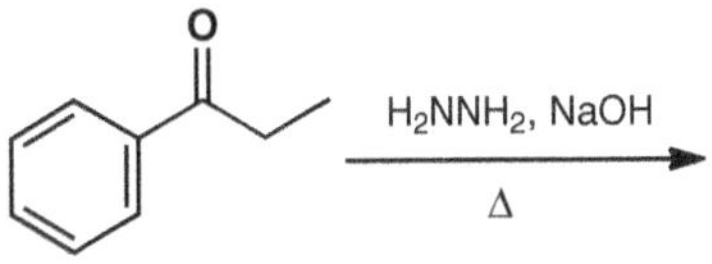 H₂NNH₂, NaOH, Δ

10.6 What is the main product of each of the following reactions when up to two equivalents of the Grignard reagent are added?

(a) EtMgBr + [cyclohexanone]   1) Et₂O 2) H₃O⁺

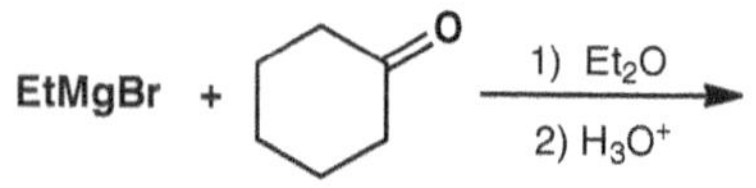

(b) ═══─MgBr + CO₂   1) Et₂O 2) H₃O⁺

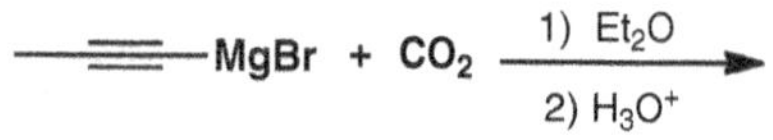

(c) PhMgBr + PhCO₂Me   1) Et₂O 2) H₃O⁺

(d) PhMgBr + PhCONMe₂   1) Et₂O 2) H₃O⁺

10.7 Dimethyl carbonate gives 3-ethylpentan-3-ol upon reaction with three equivalents of ethylmagnesium bromide in ether followed by an aqueous acidic workup. Show the reaction sequence of this overall transformation.

MeOCOMe + 3 EtMgBr   1) Et₂O 2) H₃O⁺   Et₃COH

10.8 Propose at least two methods of preparation for each of the following alcohols by hydride reduction and/or Grignard reactions.

(a) PhCH₂OH   (b) PhC(CH₃)₂OH   (c) Ph₂CHOH

(d) Ph₃COH   (e) PhCHCH₂CH₃ (OH)

10.9 Propose at least two methods of preparation for each alcohol below by using Grignard reactions.

(a) hexan-2-ol   (b) 2-methylbutan-2-ol
(c) 2-phenylethanol   (d) hept-2-yn-4-ol
(e) 2–cyclobutylpentan-2-ol

10.10 How can we prepare the ketoalcohol from the keto ester shown below? Propose a reaction sequence for this transformation.

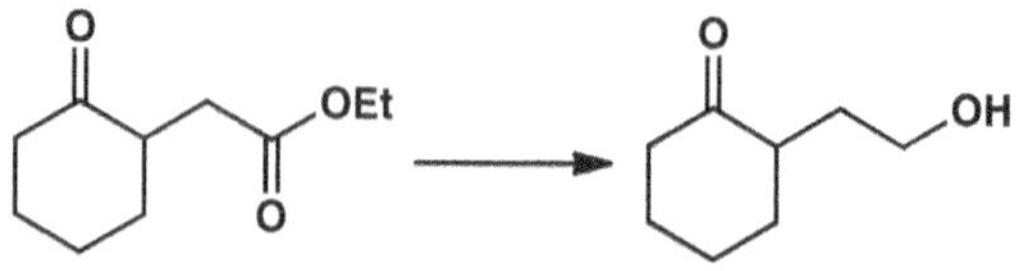

[keto ester with OEt] → [ketoalcohol with OH]

10.11 What is the main product from reduction of the following lactone, with LiAlH₄ followed by an aqueous workup? Propose a mechanism for the formation of the product. (See Panel 9.2 for lactones.)

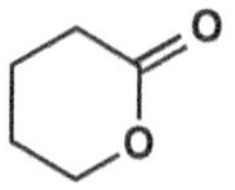

[lactone structure]   pentan-5-olide

10.12 Reaction of a Grignard reagent with a nitrile provides a synthetic method for ketones. Write a reaction sequence for the following transformation.

RMgX + R'─C≡N   1) Et₂O 2) H₃O⁺   [ketone R-CO-R']

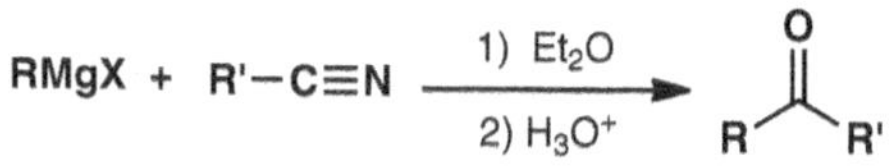

10.13 Explain why one of the two carbonyl groups of succinimide can be reduced with NaBH₄, but not both. What would you predict the product to be?

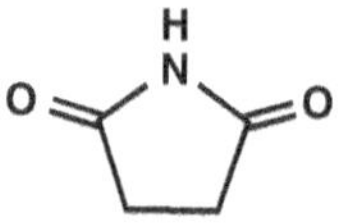

[succinimide structure]   succinimide

10.14 A C=C double bond conjugated with a carbonyl group can sometimes undergo reduction with a hydride donor. Write a reaction sequence to show how the following transformation proceeds.

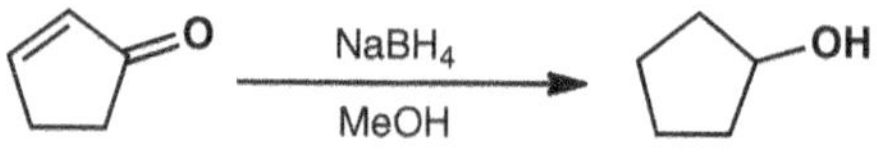

[cyclopent-2-enone]   NaBH₄ / MeOH   [cyclopentanol with OH]

cyclopent-2-enone   cyclopentanol

Stereochemistry and Molecular Chirality

11

Related topics we have already covered:

- **Tetrahedral carbon** (Chapter 2)
- **Drawing organic structures** (Section 2.8)
- *E,Z* **stereoisomers and the Cahn–Ingold– Prelog sequence rules** (Section 2.9)
- **Conformations of alkanes** (Chapter 4)

Topics we shall cover in this chapter:

- **Three-dimensional structures of molecules and their mirror images**
- **Chirality and chirality centres**
- **Enantiomers and diastereoisomers**
- *R,S* **nomenclature**
- **Optical activity**
- **Reactions which produce enantiomers**
- **Prochirality**

We have already discussed three-dimensional structures and stereochemical features of some organic molecules, e.g. *cis–trans* stereoisomerism in alkenes and cycloalkanes, and conformations of alkanes. This chapter introduces further types of stereoisomerism which occur when a molecule and its mirror image are not identical. Some molecules have the property of handedness or **chirality**, which relates to the non-superposability of mirror image forms. Actually, most naturally occurring organic compounds have this property and the biochemical world in general is chiral at the molecular level; consequently, chirality is tremendously important in the chemistry of life. In this chapter, we shall see how chirality occurs in molecules, and introduce some of its consequences in chemical reactions.

11.1 Chirality

Images of some objects are superposable on their mirror images,[1] while others are not. Objects whose images cannot be superposed on their mirror images are said to be **chiral**, while those whose images are superposable are **achiral**. Our hands, which are obviously not identical, are typical chiral objects; the image of one cannot be superposed on its mirror image. In fact, one hand looks very much like the mirror image of the other: they bear a mirror image relationship. This property of handedness, the non-superposability of the image of an object upon its mirror image, is called **chirality**, from the Greek word meaning 'hand'.

---

[1] We use the verb *superpose* to mean to put one object or image on top of another such that all like parts coincide exactly; two items which are exactly superposable are identical. To *superimpose* is simply to put one object or image on top of another with no implication of any other relationship between them. In principle, one can superimpose, but not superpose, anything on anything else. In this book, when we refer to the superposability (or non-superposability) of *three-dimensional objects*, we mean *of their images*.

Exercise 11.1

Which of the following are chiral?

(a) an ear   (b) a horseshoe   (c) a pair of spectacles   (d) a screw

(e) a screwdriver   (f) a ship's propeller

Objects with and without a mirror plane:

Cups and saucers have mirror planes, so they are *achiral*.

(This file is licensed under the Creative Commons Attribution-Share Alike 3.0 Unported license.)

A hand has no mirror plane—it is *chiral*.

An α-amino acid usually exists as a zwitterion (see Section 24.3).

Let us consider possible chirality in some every-day objects. Gloves, like hands, are chiral but socks, before they are put on feet, are not; golf clubs are chiral but tennis rackets are not. Cups are achiral but a corkscrew is chiral. Spirals and helices are other familiar chiral structures (e.g. spiral staircases and drill bits); they are either right- or left-handed, and the one is not superposable upon the other—they have mirror image relationships (see Panel 11.1). What is the *significant* difference between chiral and achiral objects? The overwhelming majority of achiral objects have (at least) one *mirror plane* (also called a *symmetry plane* or *plane of symmetry*), i.e. a plane which divides the object into two halves, each half being the mirror image of the other.

Now consider some consequences of chirality in every day life. You would feel awkward trying to shake hands if your guest offers his/her left hand to your right hand, and you cannot comfortably put your right shoe on your left foot. A bolt will only screw into a nut of the same size if they both have either left handed or right handed threads. Without analysing too precisely the problem in each of the above examples, we can say that chiral objects do not fit well with other chiral objects unless their chiralities match.

11.1.1 Chiral molecules

Molecules, being three-dimensional objects, are either chiral or achiral. Naturally occurring amino acids, carbohydrates, and nucleic acids are almost all chiral; mirror image structures of the amino acid alanine are shown in Figure 11.1 as a simple example of a chiral biological molecule. Clearly, the atom connectivities of the two forms of alanine shown are the same yet the two structures are spatially different; according to the definition given earlier (Chapter 4), therefore, the two forms of alanine are stereoisomers.

Since chiral molecules are the principal building blocks of biological materials, we can say that biology is essentially chiral at the molecular level. Consequently, if some biological process involves one form of a chiral compound, the mirror image form cannot normally be substituted—the same biological process will not usually take place. For example, many pharmaceutical compounds are chiral, and their mirror image forms will not generally have the desired beneficial effects; indeed, they may have very serious adverse effects.

Figure 11.1 An α-amino acid as an example of a chiral molecule; the two mirror image forms are not superposable.

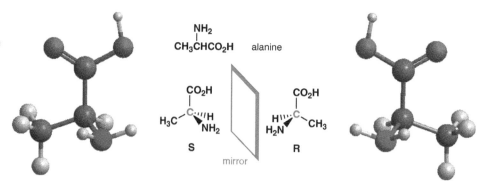

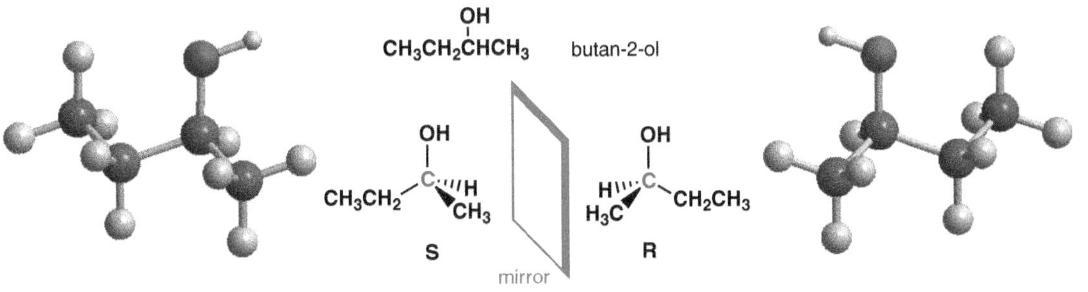

Figure 11.2 Enantiomers of butan-2-ol.

11.1.2 The basis of chirality in molecules

A tetrahedral carbon atom with four different substituents is the main cause of chirality in organic molecules, and butan-2-ol is a simple example. Either by drawing three-dimensional structures (with wedged bonds), or by using molecular models, we can show that there are two, and only two, different structures for butan-2-ol, as shown in Figure 11.2. Each of these is the mirror image of the other, and the two cannot be superposed; such pairs of stereoisomers are called **enantiomers**, and **enantiomerism** is a subdivision of stereoisomerism (see Panel 11.2). The two enantiomers of alanine were shown in Figure 11.1.

In the two pairs of enantiomers shown above, the characteristic feature which imparts chirality is a tetrahedral carbon bearing four different substituents; such a carbon is an example of a **chirality centre** (also called a *chiral centre* and the less satisfactory term 'asymmetric carbon atom' was used in older books). In butan-2-ol, the chirality centre is C2 which has H, CH_3, C_2H_5, and OH groups bonded to it; the carbon at the chirality centre of alanine in Figure 11.1 is bonded to H, CH_3, CO_2H, and NH_2. We can now generalize: any molecule with a single chirality centre, e.g. a tetrahedral carbon bearing four different substituents, exists either as a single enantiomer, or as a pair of enantiomers (in equal or unequal amounts).

Let us now consider a molecule in which two of the four groups attached to a tetrahedral carbon are the same, e.g. propan-2-ol. This molecule has a mirror plane which contains the OH, C2, and the H (Figure 11.3). We can also see that the mirror image (**B**) of structure **A** can be superposed on **A** when **B** is rotated by 60° around the C–O bond. In other words, a molecule which contains a mirror plane cannot be distinguished from its mirror image—such a molecule is **achiral**. In general, any molecule with a mirror plane is achiral.[2]

The enantiomers of butan-2-ol can also be represented by zigzag drawings (see Chapter 2) with wedges:

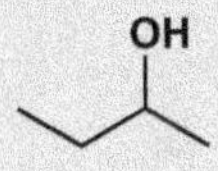

Unspecified configuration at the chirality centre.

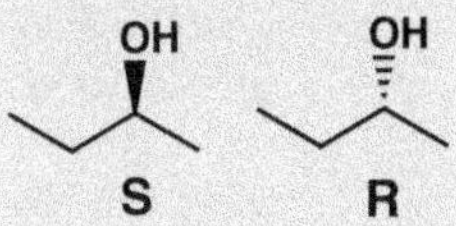

*This type of chirality centre, a tetrahedral carbon bearing four different substituents, is an example of a *stereogenic centre* (or *stereocentre*) which is one type of *stereogenic element*. We saw earlier that a C=C double bond can also give rise to stereoisomerism, but of a different type, i.e. *cis* and *trans* isomers. So, depending on its substituents, a double bond is also a stereogenic element, i.e. a feature of a molecule which gives rise to stereoisomerism, but this time, a stereogenic *axis* rather than a stereogenic *centre*.

---

[2] The converse of this statement is not true—molecules of a small number of compounds do not have a mirror plane but are still achiral. The 'ultimate' test is whether a molecule has an *alternating axis of symmetry*: if it does, it is achiral; if it does not, it is chiral. *trans*-2,5-Dimethylcyclohexane-1,4-dione (**1**), for example, does not have a mirror plane but is achiral because it has a (twofold) alternating axis of symmetry.

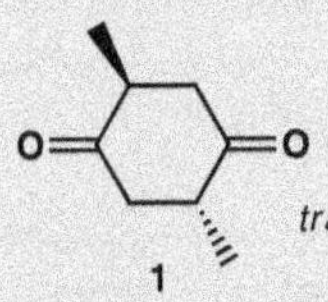

trans-2,5-dimethylcyclohexane-1,4-dione

(When a molecule has an *n*-fold alternating axis of symmetry, rotation through an angle of 360/*n* degrees about this axis followed by reflection in a plane perpendicular to the axis gives a molecule which is indistinguishable from the original. A mirror plane is equivalent to a 1-fold alternating axis, and a centre of symmetry is equivalent to a 2-fold alternating axis as seen for 1.)

Panel 11.1 Right- and left-handed helices

Right-handed and left-handed helices are non-superposable mirror images of each other—they are chiral and, along with spiral shapes, are commonly found in nature and in man-made structures. At the molecular level also, we occasionally encounter helical structural features.

A helix is right-handed if the turns of the helix appear to slope from lower left to the upper right when viewed from the side; if they appear to slope from lower right to upper left, it is a left-handed helix. Perhaps it is easier to relate the helix to a normal screw, a common man-made object with a spiral thread which is said to be right-handed when the screwdriver is turned clockwise to drive the screw in. If the 'thread' of a helix is the same as that of a normal screw, it is a right-handed helix.

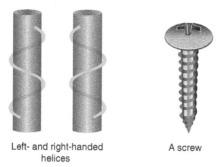

Left- and right-handed helices

A screw

The helical twines of the morning glory climbing vine are right-handed like those of many similar plants, but the honeysuckle twines are left-handed. Springs are often helical and architecture occasionally incorporates spiral and helical features. Spiral staircases in mediaeval European castles are normally right-handed, but left-handed ones can be found and the helical external stairway on the Bòbila Almirall Chimney (Spain) is a spectacular example. Many spiral structures are found in old churches: some pillars in the cloisters of the Basilica di San Paolo fuori le Mura (Rome) are right- and left-handed double helices.

Morning glory
(Svetlana Prikhodko/istockphoto.)

Honeysuckle
(This file is licensed under the Creative Commons Attribution-Share Alike 3.0 Unported, 2.5 Generic, 2.0 Generic and 1.0 Generic license.)

Figure 11.3
Representations of the structure of propan-2-ol, an achiral molecule, showing (a) a mirror plane, and (b) the identity relationship between the mirror image forms.

OH
|
CH_3CHCH_3 propan-2-ol

(a)   mirror plane

(b)   rotate around C–O
                     bond by 60°

mirror

The Bòbila Almirall chimney
(Terrassa, Spain)

(This file is licensed under the Creative Commons
Attribution-Share Alike 3.0 Unported license.)

Pillars in the Basilica di San Paolo
fuori le Mura (Rome)

(This file is licensed under the Creative Commons
Attribution-Share Alike 3.0 Unported license.)

Nonbonded interactions between the six fused benzene rings in hexahelicene prevent the molecules from being planar, and the compound exists as helical enantiomers as shown below. On the larger molecular scale, nucleic acids have
helical structures, and protein structures often include helical sections; the double helix of the most common DNA is
right-handed, as are the α helix sections of proteins (see Chapter 24).

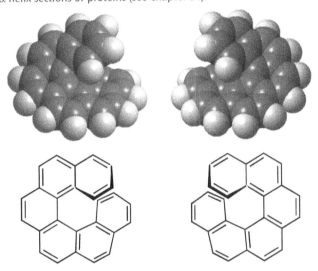

Enantiomers of hexahelicene

Draw structures which indicate the stereochemistry of both enantiomers for each of the following chiral molecules.

Example 11.1

(a) 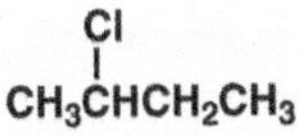 CH₃CHCH₂CH₃ (with Cl substituent)

$$CH_3\overset{\displaystyle Cl}{\underset{|}{C}}HCH_2CH_3$$

(b) 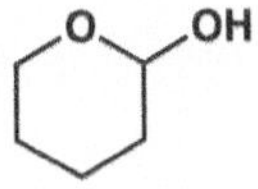 [tetrahydropyran ring with O and OH]

(continues . . .)

(...continued)

Solution

In each case, find the chirality centre, i.e. the C bearing four different groups, and draw both enantiomers using wedged bonds at this carbon.

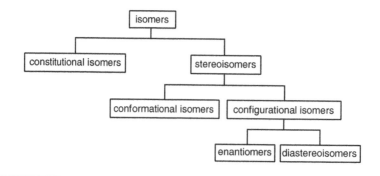

Exercise 11.2

Classify the following molecules as chiral or achiral, and draw stereochemical structures of both enantiomers for those which are chiral.

(a) CH₃CHCO₂H (with OH substituent) (b) CH₃CHCO₂H (with CH₃ substituent) (c) (d)

Panel 11.2 Summary of isomeric hierarchy

Isomers are different compounds with the same molecular formula. Those with different atom connectivities are **constitutional isomers** (Sub-section 2.9.1), while those with different three-dimensional arrangements of atoms, but with the same atom connectivity, are **stereoisomers** (Sub-section 2.9.2). The *cis–trans* **isomers** in alkenes and cycloalkanes (Section 4.4) exemplify one type of stereoisomerism. In Chapter 4, we also encountered conformations of alkanes (stable conformations are also called conformers or **conformational isomers**); these are different three-dimensional structures which may be interconverted simply by rotation about single bonds. Conformational equilibrations are usually rapid at normal temperatures so individual conformers cannot usually be isolated, and they are not normally regarded as different compounds. In contrast to conformational isomers, stereoisomers which cannot interconvert except by bond cleavage are known as **configurational isomers** (Sub-section 2.9.2 and Section 4.4), and the property which differentiates one of these stereoisomers from another is the **configuration** attributable to a *stereogenic element*. **Enantiomers** are non-superposable mirror images of each other and arise if the species has a *chirality centre*, a *chirality axis*, or a *chirality plane* (so these *chirality elements* are the stereogenic elements of enantiomerism). All other kinds of configurational isomers (including *cis–trans* isomers, i.e. non-enantiomeric stereoisomers) are called **diastereoisomers** (diastereomers).

```
                    isomers
                       |
        +--------------+--------------+
        |                             |
constitutional isomers          stereoisomers
                                      |
                         +------------+------------+
                         |                         |
              conformational isomers      configurational isomers
                                                   |
                                      +------------+------------+
                                      |                         |
                                 enantiomers          diastereoisomers
```

11.2 *R,S* nomenclature for Chirality Centres

A pair of enantiomers, each with a single chirality centre, are identical in all respects except their configuration at the chirality centre. However, we need to be able to refer to them individually because they are different compounds. This is achieved by labelling the chirality centres of the two stereochemical forms as ***R*** and ***S***.

The assignment of the ***R*** or ***S*** configuration to an enantiomer with a single chirality centre starts by determining the priority of the four different groups attached to the chirality centre according to the Cahn–Ingold–Prelog sequence rules which have previously been introduced for designation of the *E,Z* configuration of alkenes (Sub-section 2.9.3). The arrangement of these four groups in space around the chirality centre determines whether a single enantiomer is of ***R*** or ***S*** configuration. In the example of butan-2-ol above, the priority order of the four groups is: $H < CH_3 < CH_2CH_3 < OH$, and in alanine it is: $H < CH_3 < CO_2H < NH_2$.

An enantiomer is then viewed along the bond from the central C to the lowest priority group pointing backwards behind the plane of the paper. We now see the remaining three groups arranged around the central C atom. The path of decreasing priority from the highest to the lowest will appear either clockwise or anticlockwise. If it is clockwise, the chirality centre is assigned the ***R* configuration**, while if it is anticlockwise, the chirality centre is of ***S* configuration**. Since a molecule with one chirality centre has only two enantiomers, the *R,S* convention allows their unambiguous differentiation, and adding the prefix (*R*)- or (*S*)- to the name of the compound specifies which of the two we are dealing with. We now see that butan-2-ol (***R***) and alanine (***S***) shown above are, in fact, (*R*)-butan-2-ol and (*S*)-alanine.

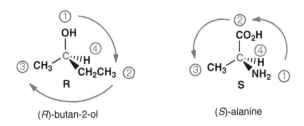

(*R*)-butan-2-ol (*S*)-alanine

Decreasing priority: 1 > 2 > 3.

Remembering *R* and *S* You can easily remember the *R,S* convention by thinking of how you turn the steering wheel of a car. Imagine that the column holding the steering wheel corresponds to the group of lowest priority, and you turn the wheel in the direction of decreasing priority from the highest to the lowest. Turning the steering wheel clockwise (a) leads to a right turn, while turning it anticlockwise (b) leads to a left turn. The former corresponds to the *R* configuration from the Latin word *rectus* meaning 'right', and the latter corresponds to the *S* configuration from the Latin word *sinister* meaning 'left'.

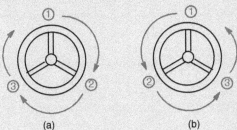

(a) (b)

(a) Right turn corresponds to (*R*). (b) Left turn corresponds to (*S*). (Priorities: 1 = highest, 3 = lowest.)

Example 11.2

Assign *R* or *S* configuration to the following molecule of butan-2-ol.

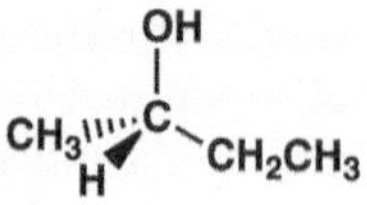

$$CH_3\cdots C(OH)(H)-CH_2CH_3$$

Solution

The structure shown should be re-drawn after rotating it through 120° about the C–O bond to make the lowest priority group (H) point backwards; this gives the structure below. The priorities of the three remaining groups are then established, and we note that the path of decreasing priorities when we look along the C–H bond is anticlockwise, so this enantiomer is (*S*)-butan-2-ol.

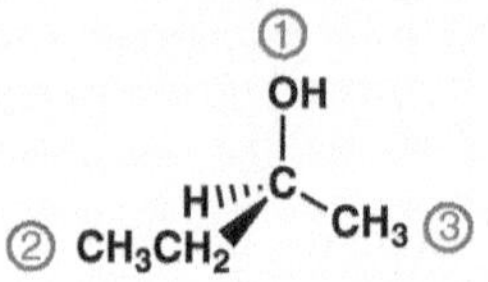

It follows that, if we view the molecule with the lowest priority group pointing *towards us*, an apparent clockwise direction for the order of decreasing priorities of the other groups corresponds to the (*S*) configuration. Note also that interchanging any two groups on the chirality centre corresponds to an inversion of configuration. Thus, if we interchange the CH₃ and H groups attached to the chirality centre of the butan-2-ol molecule shown in Example 11.2, we obtain (*R*)-butan-2-ol.

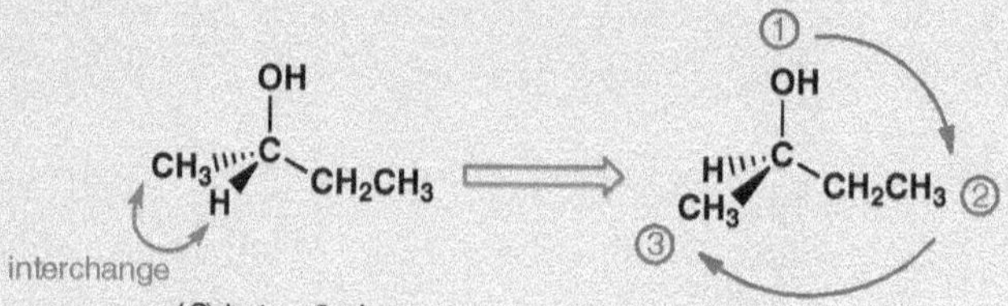

interchange
(*S*)-butan-2-ol

(*R*)-butan-2-ol

Exercise 11.3

Assign the *R* or *S* configuration to each of the following structures of butan-2-ol.

(a) 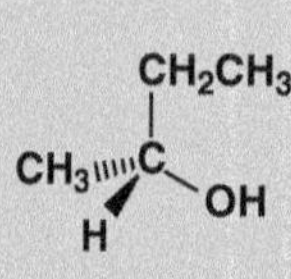 (b) 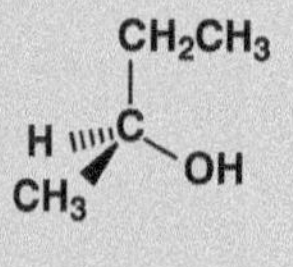 (c)

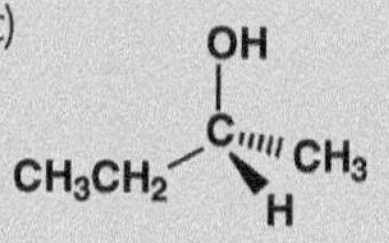

Exercise 11.4

Assign the *R* or *S* configuration to the following.

(a) 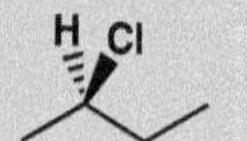 (b) 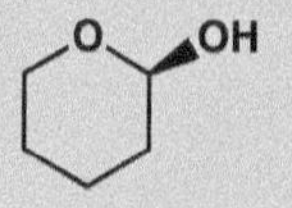 (c)

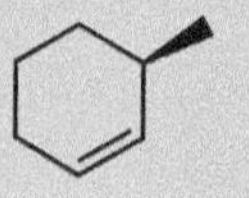

11.3 The Fischer Convention for representing the Configuration of Chirality Centres

We have already used wedged bonds to represent configurations of chirality centres as shown above for butan-2-ol, for example. An earlier method of representing stereochemistry around a tetrahedral C was introduced by the German chemist Emil Fischer who was awarded the Nobel Prize in chemistry in 1902 for his work on carbohydrates (see Panel 11.3). According to the Fischer convention, a single tetrahedral C is represented by the centre of a cross, and the two vertical bonds are understood to point backwards (away from us) while the horizontal bonds are understood to point forwards from the plane of the paper (towards us). When several carbon atoms are joined together, the longest carbon chain is usually drawn vertically with the carbon of highest oxidation state at the top; the structure of (R)-2,3-dihydroxypropanal (glyceraldehyde) is represented according to this convention as follows.

> Note that rotating the Fischer projection of a molecule with a single chirality centre through 90° *in the plane of the paper* leads to a projection which represents the molecule with *inverted* configuration; rotation through a further 90° leads to a different projection of the original molecule.

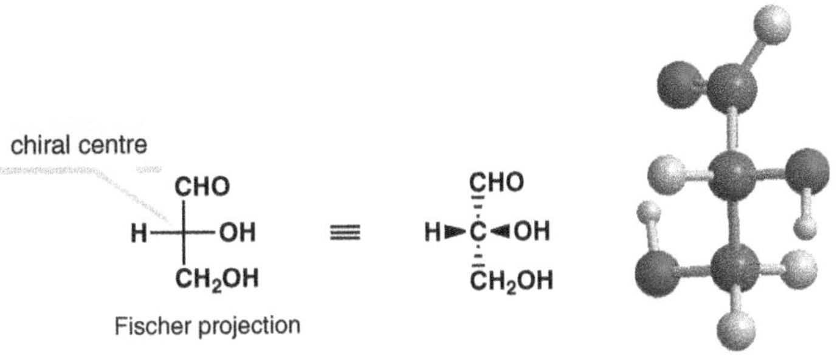

chiral centre

CHO
H——OH
CH₂OH

Fischer projection

$\equiv$

CHO
H►C◄OH
CH₂OH

(R)-2,3-dihydroxypropanal

Draw Fischer projection formulas of (S)- and (R)-alanine. See Figure 11.1 for the structure of alanine.

Exercise 11.5

Fischer projections are convenient for representing linear molecules containing more than one chirality centre, and are still widely used for representing open chain forms of carbohydrates. However, extended C chains in other types of molecules are now usually represented as zigzag chains with wedges to indicate the configuration at chirality centres as exemplified below.

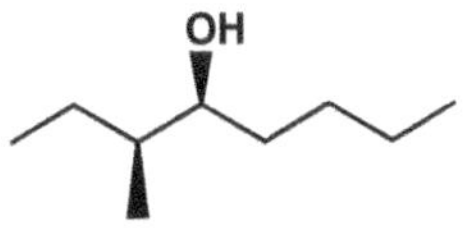

(3S,4S)-3-methyloctan-4-ol
(a pheromone of a palm weevil)

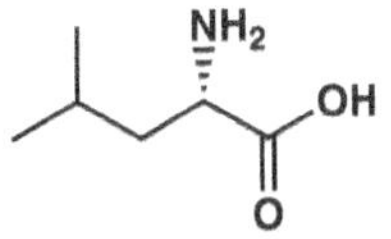

(S)-leucine
(an α-amino acid)

11.4 Compounds with two Chirality Centres

When a molecule has two chirality centres, and each can be R or S, four stereoisomers are *normally* possible: RR, SS, RS, and SR. We say *normally* because, in particular circumstances which we shall explore below, the RS and SR pair are identical, in which case there are only three stereoisomers.

11.4.1 Enantiomers and diastereoisomers

Let us examine stereoisomers of 2,3,4-trihydroxybutanal as a typical example of a compound with two chirality centres. Carbons C2 and C3 are the chirality centres and we can represent the four stereoisomers, (2R,3R), (2S,3S), (2R,3S), and (2S,3R), by Fischer projections.

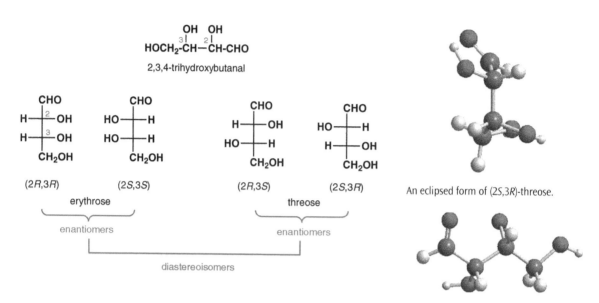

An eclipsed form of (2S,3R)-threose.

A zigzag, staggered form of (2S,3R)-threose.

In the Fischer projection of an open-chain molecule with four or more C atoms, *both* vertical lines *from any* C *within the vertical chain* are always drawn in the plane of the paper, but they are still understood to be pointing backwards *from that C.* And although a Fischer projection unambiguously shows the configuration at each chirality centre within the vertical chain, it gives little information about the *shape* of the molecule, and none about the relative stabilities of its different conformations. For these purposes, the Fischer projection of an open-chain molecule needs to be converted into another type of stereochemical representation.

These Fischer projections correspond to the following molecules in which the stereo-chemistry is now represented by wedged bonds.

When assigning configuration to a (non-terminal) chirality centre of a Fischer projection, note that the bond to the group of lowest priority (invariably H) is horizontal and, hence, points towards us. The decreasing order of priorities on C2 of the left-most isomer of 2,3,4-trihydroxybutanal above is OH → CHO → C3, and on C3 it is OH → C2 → CH_2OH; looking along the H–C bond *from the front* in each case, the decreasing order of priorities at *both* chirality centres *appears* anticlockwise, in accord with the compound having (2R,3R) configuration.

The eclipsed conformations above are redrawn below in a format which allows a better appreciation of possible conformational interconversions, e.g. formation of the staggered forms.

These compounds are carbohydrates: the (2R,3R) and (2S,3S) pair are the enantiomers of erythrose, and the (2R,3S) and (2S,3R) pair are the enantiomers of threose. Erythrose and threose are *stereoisomers which are not enantiomers*; such stereoisomers are called **diastereoisomers**, and they cannot be interconverted without bond-breaking.

Draw saw-horse structures of the stable staggered forms of the enantiomers of erythrose, and give their zigzag representations.

Example 11.3

Solution

Internal rotation about the C2–C3 bond of the three-dimensional structures of the eclipsed forms above gives the staggered forms below with *anti* relationships of the CHO and CH$_2$OH groups, and the zigzag representations are obtained by drawing the carbon chain horizontally in the plane of the paper.

(2*R*,3*R*) isomer   (2*S*,3*S*) isomer

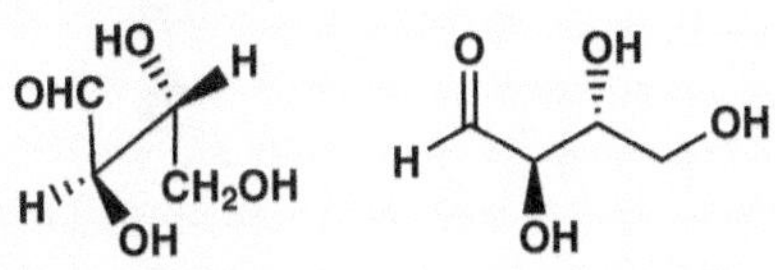

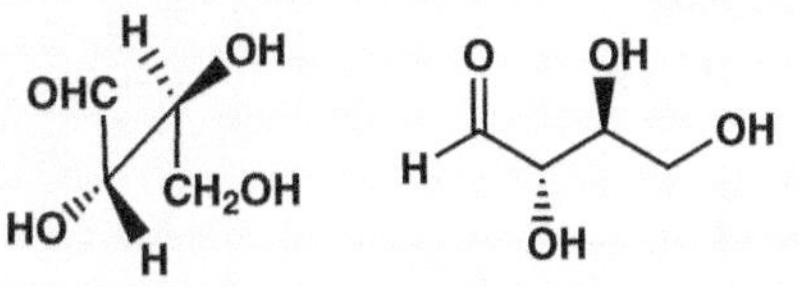

Note that a zigzag structure can be drawn directly from a Fischer projection by extending the carbon chain horizontally, and joining substituents to chiral carbons by solid or broken wedges according to the *R* or *S* configuration in each case.

Draw Newman projections of stable staggered forms of the enantiomers of erythrose.

Exercise 11.6

Draw saw-horse and zigzag structures of the stable staggered conformers of the enantiomers of threose.

Exercise 11.7

Draw Fischer projections of the stereoisomers of butane-1,2,3-triol, and assign the configuration (*R* or *S*) of every chirality centre. Furthermore, identify enantiomeric and diastereoisomeric relationships amongst the stereoisomers.

Exercise 11.8

11.4.2 Meso compounds

We shall now consider a compound with two chirality centres but which, because of a particular feature, has only *three* stereoisomers: 2,3-dihydroxybutanedioic acid (tartaric acid). This dicarboxylic acid is a crystalline compound found in plants, and is especially abundant in grapes. Fischer projections corresponding to (2*R*,3*R*), (2*S*,3*S*), (2*R*,3*S*), and (2*S*,3*R*) configurations are drawn below. The first two, (2*R*,3*R*) and (2*S*,3*S*) are clearly a pair of enantiomers, i.e. non-superposable mirror images of each other. The second pair as drawn are mirror images of each other but, when one of them is rotated through 180° *within the plane of paper*, we see that they can be superposed. In other words, this second pair are identical and achiral—each is superposable on its mirror image. We come to the same conclusion by noting that each of these two Fischer projections has a mirror plane between the C2 and C3 atoms, which divides each molecule into identical halves. There are, therefore, only three stereoisomers of tartaric acid even though it has two chirality centres.

As noted in Section 11.3, rotation of a Fischer projection through 180° within the plane of the paper does not alter the identity of the molecule represented. In contrast, turning the projection over inverts the configuration at all the chirality centres and gives the enantiomer.

OH OH
  3| 2|
HO₂C-CH—CH-CO₂H

2,3-dihydroxybutanedioic acid
(tartaric acid)

CO₂H	CO₂H	CO₂H	CO₂H
H ——— OH	HO ——— H	H ——— OH	HO ——— H
HO ——— H	H ——— OH	H ——— OH	HO ——— H
CO₂H	CO₂H	CO₂H	CO₂H
(2R,3R)	(2S,3S)	(2R,3S)	(2S,3R)

≡   mirror plane

enantiomers

a meso compound

diastereoisomers

meso-tartaric acid

<table>
<tr><td>Note that *meso* is italicized when it is part of the name of a compound.</td><td>Compounds such as the (2R,3S) stereoisomer of tartaric acid, whose molecules have two (or more) chirality centres but have a mirror plane so are achiral as whole, are called **meso compounds** (or **meso isomers**). The meso isomer of tartaric acid is a diastereoisomer of each of the enantiomers.</td></tr>
</table>

Exercise 11.9

Draw saw-horse and zigzag structures of *meso*-tartaric acid, and a Newman projection.

Example 11.4

Show all the stereoisomers of (a) cyclobutane-1,2-diol and (b) cyclobutane-1,3-diol, and assign R or S configuration to every chirality centre. Also, explain the stereochemical relationships between the isomers.

Solution

(a) There are two enantiomers of *trans*-cyclobutane-1,2-diol, and *cis*-cyclobutane-1,2-diol is a meso compound, i.e. a molecule of the latter has two chirality centres and a mirror plane.

Note that the rapidly puckering conformational vibration of cyclobutanes does not complicate the configurational considerations; in fact, the time-averaged structure of cyclobutane itself has coplanar C atoms (see also the margin note in Section 11.6).

mirror plane

H	OH	HO	H		HO	OH
R	R	S	S		S	R
HO	H	H	OH		H	H

trans-cyclobutane-1,2-diol   *cis*-cyclobutane-1,2-diol

enantiomers   meso compound

(b) Cyclobutane-1,3-diol has *cis* and *trans* isomers. Molecules of both have a mirror plane so are achiral, and neither has a chirality centre.

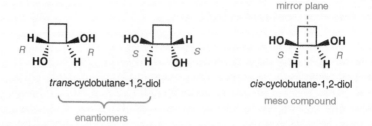

---HO ◁▷ ᴴᴵOH --- mirror plane   -- HO ◁---▷ OH -- mirror plane

trans-cyclobutane-1,3-diol   *cis*-cyclobutane-1,3-diol

The following are Fischer projection formulas representing stereoisomers of butane-2,3-diol.

(a) Assign *R* or *S* configuration to each chirality centre.
(b) Identify any meso compounds.
(c) Identify enantiomeric pairs and diastereoisomeric relationships.

11.5 Properties of Stereoisomers

11.5.1 Properties in achiral environments

In an achiral environment, enantiomers have identical physical and chemical properties, and cannot be differentiated. Table 11.1 shows some properties of tartaric acid: the two enantiomers have the same mp, density, solubility, and pK_a values. The only difference listed is in the *specific rotation*, which will be explained and discussed in the next sub-section. However, the meso isomer, which is a diastereoisomer of the enantiomers, has properties which are all significantly different from those of the enantiomers. In general, diastereoisomers show different physical and chemical properties whenever the properties of enantiomers are identical.

Some properties of stereoisomers of tartaric acid			
	(*R,R*)-tartaric acid	(*S,S*)-tartaric acid	*meso*-tartaric acid
specific rotation	+12.7	−12.7	0
mp/°C	171–174	171–174	146–148
density/g cm^{-3} (20 °C)	1.7598	1.7598	1.660
solubility[a]	139	139	125
pK_{a1} (25 °C)	2.98	2.98	3.23
pK_{a2} (25 °C)	4.34	4.34	4.82

a. In g/100 ml H_2O at 20 °C.

Table 11.1

The French physicist J.B. Biot discovered that light can be plane-polarized early in the nineteenth century. He also found that the plane of polarization is rotated when the light passes through a solution of certain organic compounds such as sugar, and other naturally occurring compounds (including some non-organic examples). Compounds which are able to rotate the plane of polarized light are said to be *optically active*. The relationship between optical activity and enantiomerism was established by Louis Pasteur when he discovered the enantiomers of tartaric acid in 1848 (see Panel 11.4).

11.5.2 Optical activity

Enantiomers have a very special property: *they rotate the plane of plane-polarized light*. A beam of ordinary light consists of electromagnetic waves oscillating in an infinite number of planes which contain the line of the direction of the light beam. However, when a beam of ordinary light passes through a *polarizer*, only the light wave oscillating in a single plane passes through, and the transmitted light is called **plane-polarized light**.

A solution of a pure enantiomer rotates the plane of polarized light in one direction and its enantiomer, at the same concentration, rotates the plane in the opposite direction by the same amount.

Panel 11.3 Configurations of sugars and amino acids

The stereochemical course of a chemical reaction is frequently known; for example, an S_N2 reaction always proceeds with inversion of configuration at the reaction centre. However, although we may know the configuration of the product *relative to* that of the reactant in such a reaction, this is not the same as knowing its *absolute* configuration (whether it is *R* or *S*) unless we know the absolute configuration of the reactant. It was necessary, therefore, that the absolute configuration of one compound had to be known with complete certainty in order that the absolute configurations of other compounds could be established by reference to it. The first determination of absolute configuration was accomplished in 1951 by X-ray crystallographic analysis when the rubidium sodium salt of (+)-tartaric acid was shown to have the *R,R* configuration.

The phenomenon we now know as enantiomerism (see Panel 11.2) was firmly established by the late 19th century after it had been shown that some stereoisomers exhibited equal but opposite optical activities. To differentiate them, the dextrorotatory isomer was labelled by addition of the prefix (+)- or *d*- to the name of the compound, while the levorotatory isomer was labelled by the prefix (–)- or *l*-. However, chemists at the time could not assign what we now call the *R* or *S* configuration based on the (+) or (–) sign of the optical activity of an enantiomer.

It was proposed that (+)-glyceraldehyde (2,3-dihydroxypropanal) had what we now call the (*R*) configuration, and there was obviously a 50% chance of being correct. At the time (before the introduction of the *R,S* descriptors), its chirality centre was assigned the prefix D- (small capital letter), and any compound that could be related to D-glyceraldehyde without disturbing this chirality centre was also labelled a D-compound. Correspondingly, the (–)-enantiomer of glyceraldehyde was assumed to be of what we now call the (*S*) configuration, and the prefix L- was added to its name. Then, if a C had the same configuration as that of the central C of L-glyceraldehyde, it too was labelled an L-compound. It subsequently turned out that the chirality centre of (+)-glyceraldehyde does indeed have the (*R*) configuration and, through this good fortune, structural assignments based on the above-mentioned assumption are correct.

Although D and L have been redundant as structural descriptors since the introduction of the *R,S* system, they are still used, especially in carbohydrate chemistry and, to a lesser degree, in amino acid chemistry.

D-glyceraldehyde [(*R*)-glyceraldehyde]
($[\alpha]_D$ +8.7)

L-glyceraldehyde
[(*S*)-glyceraldehyde]
($[\alpha]_D$ –8.7)

The fluctuating electron density in a molecule generates a minute fluctuating electric field, which interacts with electromagnetic radiation, i.e. light. The total electron distribution of an enantiomer, like the enantiomer itself, is chiral and its interaction with the plane-polarized light induces a rotation of the plane of polarization. A similar interaction between the other enantiomer and the plane polarized light results in an equal rotation of the plane of the polarized light in the opposite sense.

The rotation of the plane of the polarized light can be measured with an instrument called a **polarimeter**. Light passing through a polarizer becomes plane-polarized, and will then pass through another polarizer only when the two are exactly aligned; when the polarizers are at 90° to optimal alignment, no light passes through the second polarizer. Consequently, a second polarizer can be used as an analyser in a polarimeter, as illustrated in Figure 11.4. The light passes through the first polarizer and becomes polarized; it then passes through a sample tube containing a solution of an optically active compound whereupon its plane of polarization becomes rotated. The magnitude of this rotation of the plane of the polarized light is measured by rotating the analyser to find the new plane of polarization. (Because it is experimentally easier, full blackout of light passing through the analyser is detected, first in the

Panel 11.3 Continued

Emil Fischer studied the stereochemistry of sugars in the 1860s, and found that naturally occurring sugars all have the same configuration as D-glyceraldehyde at the chiral carbon farthest from their terminal carbonyl group. This is illustrated below in the Fischer projection of the open-chain form of the dextrorotatory enantiomer of glucose; the OH group on the second carbon from the bottom points to the right, i.e. this C has the same configuration (*R*) as D-(+)-glyceraldehyde, so this isomer of glucose is D-(+)-glucose.

D-(+)-glucose

L-α-amino acid

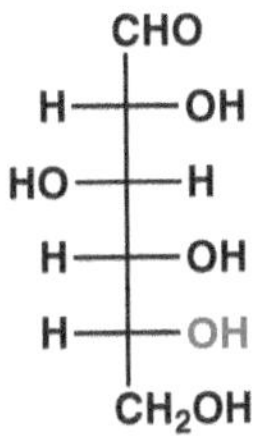

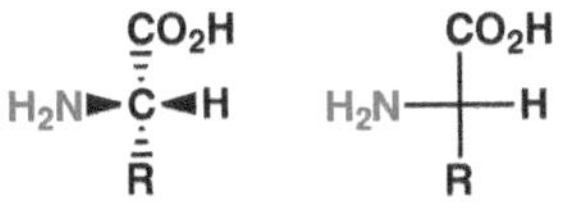

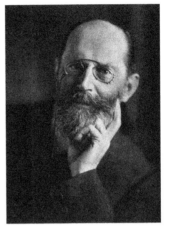

Hermann Emil Fischer (1852–1919)

In contrast, naturally occurring α-amino acids (see below) relate to L-(–)-glyceraldehyde since the NH_2 group is on the left in their Fischer projections. The alkyl (R) groups of most α-amino acids are of lower priority than CO_2H according to the Cahn–Ingold–Prelog sequence rules; consequently, with the exception of L-cysteine (for which R = CH_2SH), L-α-amino acids are of *S* configuration.

L-α-amino acids:

(*S*)-serine

(*R*)-cysteine

absence of the sample solution, then with the sample tube charged with the solution of known concentration.) This gives the **optical rotation**, the magnitude of which depends on the nature of the sample, its concentration in the solution, and the length of the tube; the **specific rotation** $[\alpha]_D$, which is a characteristic property of the sample, is defined by

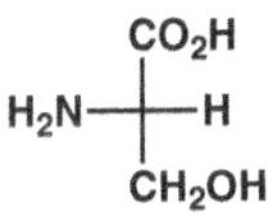

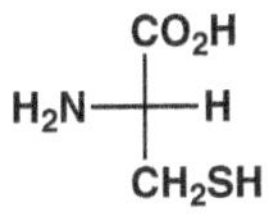

$$[\alpha]_D = \text{observed angle of rotation in degrees}/lc$$

where l is the light path length through the sample tube in dm, and c is the concentration of the solution of the sample in g ml^{-1}. The suffix D indicates the light source used in the measurement—almost invariably the so-called D line in the emission spectrum of sodium (it is actually a doublet centred at 589.3 nm); the temperature (in °C) is often indicated by an upper suffix.

An enantiomer which rotates the plane of polarization clockwise (when viewed facing the oncoming light beam) is said to be **dextrorotatory** (from the Latin *dexter*, meaning 'right'); in this event, the specific rotation is given a positive value and the prefix (+)- is added to its chemical name or formula. The other enantiomer, which rotates the plane of polarization in the opposite sense, is said to be **levorotatory** (from Latin

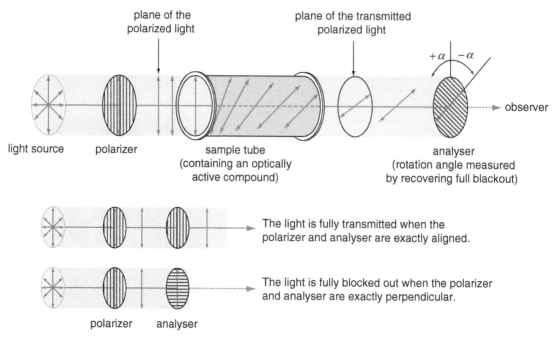

plane of the
polarized light

plane of the transmitted
polarized light

$+\alpha$ $-\alpha$

→ observer

light source   polarizer   sample tube
(containing an optically
active compound)

analyser
(rotation angle measured
by recovering full blackout)

The light is fully transmitted when the
polarizer and analyser are exactly aligned.

The light is fully blocked out when the polarizer
and analyser are exactly perpendicular.

polarizer   analyser

Figure 11.4 Schematic representation of a polarimeter.

Strictly, the units of specific
rotation are 10^{-1} deg cm^2
g^{-1} but, in accord with an
IUPAC recommendation,
values of $[\alpha]_D$ are now
usually given without units.

lævus, meaning 'left'); it has a negative $[\alpha]_D$ and the prefix (–)- is attached to its name
or chemical symbol.

(*R*)-(–)-butan-2-ol

$[\alpha]_D^{25} = -13.52$

(*S*)-(+)-butan-2-ol

$[\alpha]_D^{25} = +13.52$

Note that there is no simple correlation between the *R,S* configuration of a compound
and the sign of its specific rotation. Occasionally, very closely related compounds of the
same *R,S* configuration have specific rotations of opposite sign; for example, levorotatory
(*R*)-(–)-2-hydroxypropanoic acid (lactic acid) becomes dextrorotatory upon ionization
but, obviously, remains *R*.

NaOH

(*R*)-(–)-2-hydroxypropanoic
acid (lactic acid)

$[\alpha]_D^{25} = -3.8$

sodium (*R*)-(+)-2-hydroxy-
propanoate (lactate)

$[\alpha]_D^{25} = +13.5$

11.5.3 Resolution of enantiomers

Pure enantiomers have specific rotations of the same absolute value but with opposite
signs, so a mixture of equal amounts of two enantiomers is not optically active. This
mixture is called a **racemate** (or a racemic mixture), and the material is said to be **racemic**; its chemical name or formula is prefixed by (±)- to distinguish it from either of the

enantiomers. For a mixture of unequal amounts of enantiomers, the **optical purity** (OP) is defined (usually) as a percentage as follows,

OP = 100×(the observed specific rotation)/(specific rotation of the pure enantiomer),

which is numerically equal to the **enantiomeric excess** ($ee = |R\% - S\%|$).

> **Racemic conglomerate and racemic compound** In the solid state, crystals of a racemate exist in one of two ways depending upon the nature of the compound, the solvent from which it crystallized, and the temperature of the crystallization. Each of a pair of enantiomers may independently form as separate homogeneous enantiomeric crystals, and an equimolar mixture of the different crystals is called a *racemic conglomerate*. Its formation is sometimes called *spontaneous resolution*, since pure or nearly pure enantiomers can be obtained by separating the enantiomeric crystals from the conglomerate. Alternatively, equal amounts of the two enantiomers may crystallize together in a well-defined arrangement within homogeneous achiral crystals; such a racemate is called a *racemic compound*. Formation of a racemic compound is more common, and its crystal lattice is different from that of the pure enantiomer, with different intermolecular interactions. Consequently, any properties which depend upon intermolecular interactions within the crystal lattice (e.g. the melting point) will be different for the two crystal forms.

Chemical reactions of achiral or racemic starting materials carried out under normal laboratory conditions (an achiral environment) always give racemic products if they are chiral. However, most organic compounds found in living organisms are chiral and usually occur in the form of single enantiomers. Consequently, biological organisms provide chiral environments, so the biochemical properties of a chiral compound will depend on the identity (configuration) of the enantiomer. Pharmaceutical compounds and other substances used for biological purposes, therefore, need to be available as pure enantiomers if they are chiral. It follows that synthetic racemates need to be separated into pure enantiomers. This process is called **resolution** (or *enantiomeric resolution*). Alternatively (and preferably), the synthesis itself must lead to stereochemically pure products *via stereoselective* reactions (see Sections 11.8 and 23.5)

Since the properties of enantiomers are identical in an achiral environment, their direct resolution requires use of a chiral environment which can differentiate the enantiomers. Chromatography, a very common method of separation (see Panel 3.3), can be used with a chiral stationary phase. As the racemate dissolved in a mobile phase passes through the chiral environment of the adsorbent, intermolecular interactions between enantiomers and the stationary phase will be different, so one enantiomer will be eluted more quickly than the other.

Alternatively, the mixture of enantiomers may be converted into diastereoisomeric derivatives using an enantiomerically pure (*enantiopure*) reagent (often a naturally occurring compound); these are then separated by conventional methods. Clearly, formation of the derivatives must be readily reversible so that the diastereoisomers can be converted back to the enantiomers after separation. If the racemate is acidic, it will form diastereoisomeric salts with a chiral base. For example, with a racemic carboxylic acid, (R)-AH+(S)-AH, an optically active amine (R)-B will form diastereoisomeric salts, (R)-BH⁺ (R)-A⁻ and (R)-BH⁺ (S)-A⁻. The diastereoisomeric salts will generally have different solubility properties and be separable by recrystallization. Correspondingly, a racemic base (R)-B+(S)-B will react with a chiral acid, e.g. (S)-AH, to give diastereoisomeric salts, which can be separated. In another reaction type, a racemic alcohol could react with an enantiopure chiral carboxylic acid to give separable diastereoisomeric esters.

Scheme 11.1 shows a procedure for the resolution of naproxen, the S form of which is an anti-inflammatory pharmaceutical compound. It is a carboxylic acid, and an enantiopure amino sugar is used as the chiral base to give separable diastereoisomeric salts.

Scheme 11.1 Resolution of naproxen, an anti-inflammatory drug.

Panel 11.4 Pasteur's resolution of a salt of (±)-tartaric acid

Louis Pasteur (1822–1895).

The first resolution of what we now call a racemate was achieved in 1848 by the French scientist Louis Pasteur with a salt of tartaric acid (2,3-dihydroxybutanedioic acid). In the early 1840s, two forms of tartaric acid were known. One was dextrorotatory and the other, then called 'racemic acid', was optically inactive; both were available as byproducts of wine production ('racemic' comes from the Latin word *racemus* meaning a cluster of grapes). The young Pasteur, shortly after completing his studies in Paris, started studying 'racemic acid'.

He was actually working on the ammonium sodium salt of 'racemic acid', and he noticed that there were two crystalline forms of the salt which were mirror images of each other. He separated them carefully with tweezers under a microscope and obtained samples of the right-handed and the left-handed crystals. He dissolved crystals of each kind in water and measured their optical rotations: one solution was dextrorotatory and the other was levorotatory. Furthermore, a solution of equal amounts of the two kinds of crystals was optically inactive. From these observations, he deduced that not only the crystals but also the molecules themselves were 'handed', and that he had succeeded in separating the mirror image forms (what we now call enantiomers). Pasteur had discovered that his 'racemic acid' was a 1 : 1 mixture of enantiomers, only one of which (the dextrorotatory) was readily available in pure form from grapes. 'Racemic acid' was renamed *racemic tartaric acid*, and the term 'racemic' was adopted to describe any 1 : 1 mixture of enantiomers. This discovery of enantiomerism subsequently led to the proposal of tetrahedral bonding around saturated carbon atoms in organic compounds by van't Hoff and Le Bel independently in 1874 (see the Prologue).

11.6 Chirality of Conformationally Mobile Molecules

In principle, a tertiary amine RR¹R²N, which has a tricoordinate sp³-hybridized nitrogen atom and a lone pair, is chiral. However, inversion of configuration at N converts the molecule into its mirror image, i.e. the molecule has enantiomerized which (in turn) leads to racemization of the compound. This process is usually very rapid at normal temperatures for tertiary amines, which prevents isolation of pure stable enantiomers. Phosphines RR¹R²P, which are also tricoordinate with a lone pair on the central P atom, are also chiral. Compared with tertiary amines, however, they have high energy barriers to molecular inversion, so their rates of racemization are generally slow at room temperature, and enantiomers are configurationally stable.

> The energy barrier to inversion is appreciably greater for the nitrogen of an *N*-alkylaziridine when the three-membered ring prevents the C–N–C angle opening up to 120°.
>
> a chiral *N*-alkylaziridine

Tricoordinate sulfur(IV) compounds, which also have lone pairs, are nonplanar but, like phosphines, their rates of inversion are sufficiently low that individual 'invertomers' are stable and isolable; the unsymmetrical sulfoxide shown in the margin is an example of a chiral sulfur compound.

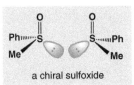

a chiral sulfoxide

If structures which are potentially enantiomeric can be interconverted rapidly simply by conformational changes, i.e. by rotation about C–C single bonds, the compound will actually be achiral (or racemic, see below). This is exemplified by *cis*-1,2-dimethylcyclohexane. Chair form **A** in Figure 11.5 is not superposable on its mirror image (**B**), and chair–chair ring flip of **A** gives **C**; we can also see that rotation of **C** through 120° about

---

Panel 11.4 Continued

(a) Crystal of the dextrorotatory salt

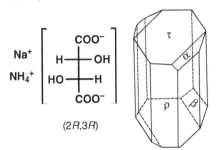

(2R,3R)

(b) Crystal of the levorotatory salt

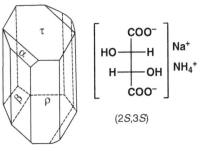

(2S,3S)

The enantiomeric crystalline forms of ammonium sodium tartrate.

Pasteur was lucky to have used ammonium sodium tartrate during cool weather in Paris. It is now known that only this salt of tartaric acid forms an enantiomeric crystalline conglomerate (see Sub-section 11.5.3), and that this happens only below 26 °C.

Although Pasteur started his career as a chemist, he is better known as a microbiologist. He discovered that fermentation in the production of wine, the spoiling of food, and some infections and diseases are all caused by what we now call *bacteria*. The process known as 'pasteurization', the heating of milk, for example, to destroy harmful bacteria is derived from his name.

If we pretend that the ring of any substituted conformationally mobile cycloalkane is planar, we shall not make mistakes in identifying stereoisomers because, as we observed in Chapter 4 in the context of *cis–trans* isomerism, *conformational changes do not affect configuration*; in other words, we can consider possible enantiomerism of conformationally mobile cyclic molecules using unrealistic planar conformations.

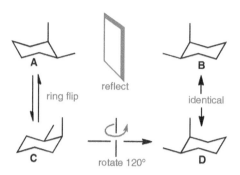

Figure 11.5 Non-enantiomerism of *cis*-1,2-dimethylcyclohexane when ring-flip is rapid.

an axis through the centre of the molecule perpendicular to the notional plane of the molecule gives **D**, which is identical with **B**. Consequently, at normal temperatures when ring flip is exceedingly rapid, *cis*-1,2-dimethylcycohexane is achiral. At very low temperatures when the rate of ring flip is much lower, the compound is better described as racemic since the enantiomers do not now rapidly interconvert. In the case of *cis*-1,2-dimethylcyclopropane whose ring has no conformational complications, the molecule has a mirror plane so is achiral under all conditions.

Exercise 11.11

Show all the stereoisomers of (a) cyclohexane-1,2-diol and (b) cyclohexane-1,4-diol, and assign the R,S configuration to each chirality centre. Also, explain the relationships between these isomers.

Exercise 11.12

Identify all stereoisomers of (a) 1,2-dichlorocyclopentane and (b) 1,3-dichlorocyclopentane.

Molecular models of 2,2′-diphosphino-1,1′-binaphthyl.

In sterically congested molecules, however, enantiomers related by *hindered* rotation about a C–C bond at normal temperatures can become separable, as in the case of the bidentate ligand BINAP (below). Conformational isomers whose interconversion is inhibited by a high internal rotation barrier are known as *atropisomers*.

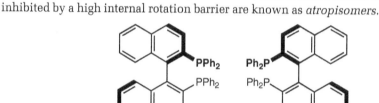

BINAP
2,2′-bis(diphenylphosphino)-1,1′-binaphthyl

11.7 Enantiomers with a Chirality Axis

The phenyl groups on both P atoms of BINAP have been changed to hydrogens for clarity.

With the exception of atropisomers, we have so far only seen molecules whose chirality is attributable to a single identifiable stereogenic centre: either a C atom bearing four different groups, or an atom (N, P, or S) with three different groups and a lone pair. Other molecular features, however, may also lead to chirality and hence enantiomerism. The stereogenic element in the *allene*, penta-2,3-diene, illustrated below is a *chirality axis*. The two structures shown are non-superposable mirror images of each other, and are said to show **axial chirality**.

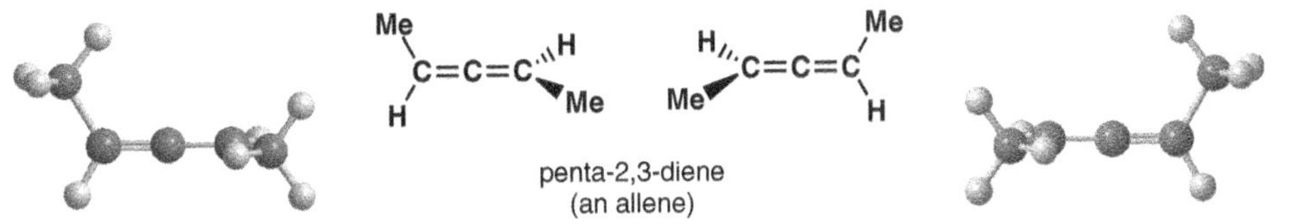

penta-2,3-diene
(an allene)

Other examples include *spiro compounds* in which one sp³-hybridized carbon atom is shared by two rings (see Panel 4.2 for the nomenclature of spiro compounds).

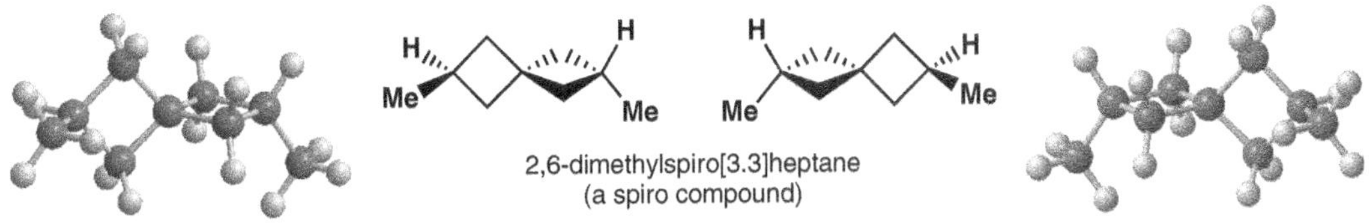

2,6-dimethylspiro[3.3]heptane
(a spiro compound)

Atropisomers mentioned above are also examples of axial chirality, but (unlike allenes and spiro compounds) they can interconvert thermally (in principle) without bond cleavage.

Which of the following are chiral?

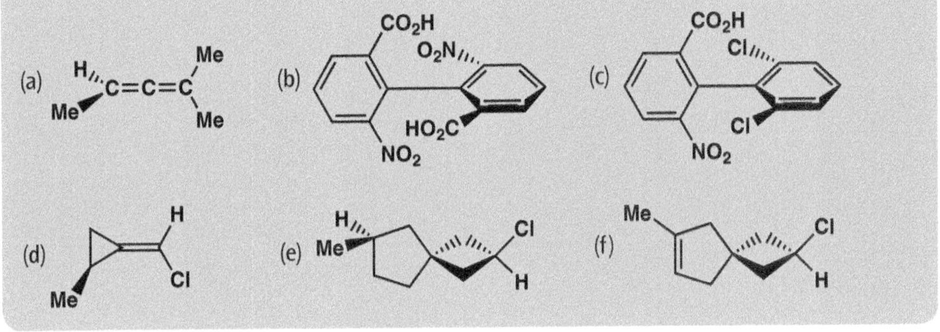

11.8 Reactions which give Enantiomeric Products

In preceding chapters, we have encountered reactions in which an achiral molecule gives a chiral product. Nucleophilic addition to the planar sp² carbon of a carbonyl group gives a product with a tetrahedral carbon which would be a chirality centre if the four bonds to it were all different. For example, a cyanohydrin formed from an aldehyde other than methanal is chiral. However, the products are always racemic under normal (achiral) reaction conditions.

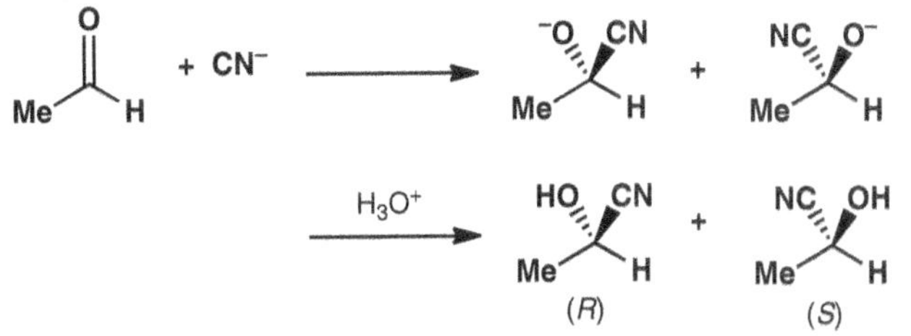

Catalytic hydrogenation of an unsymmetrical ketone gives a chiral secondary alcohol. Under normal conditions, only a racemic alcohol is obtained. However, selective formation of just one of the enantiomers is possible by using a chiral catalyst, and this kind of selective transformation is important in organic synthesis, as will be discussed in Section 23.5.

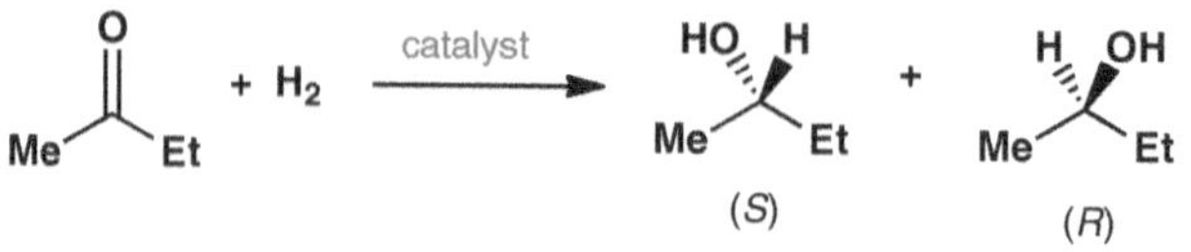

If there is any *enantioselectivity* in the nucleophilic addition to an unsymmetrical ketone, it must occur through selectivity by the nucleophile for one of the faces of the plane containing the trigonal C of the carbonyl. In Scheme 11.2, nucleophilic attack at the upper face gives one enantiomer, and attack at the lower face gives the other.

Scheme 11.2 Formation of enantiomers in the nucleophilic addition to the carbonyl of an unsymmetrical ketone.

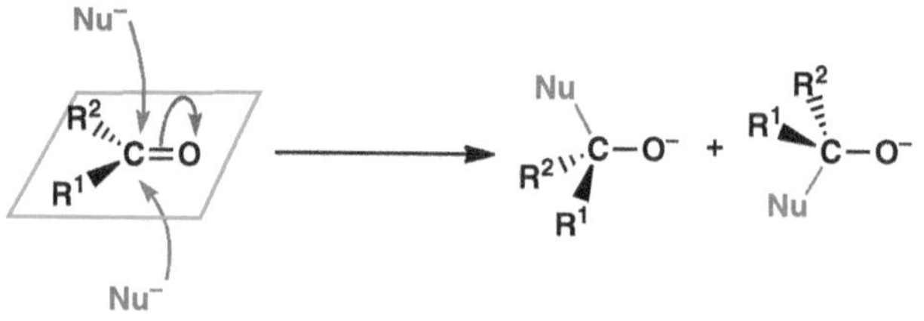

Prochirality The C of the carbonyl of the unsymmetrical ketone $R^1R^2C=O$ in Scheme 11.2 is an example of an achiral centre which can become chiral in a single step—such centres (or molecules) are said to be *prochiral*. In Scheme 11.2, addition of a group other than R^1 or R^2 (or H_2O) to the alternative faces of the carbonyl compound leads to different enantiomers—the *faces* are said to be *enantiotopic*. The *enantiotopicity* of the faces of an unsymmetrical ketone (or an aldehyde) arising from its prochiral carbonyl group is an aspect of the wider subject of *topicity* which is considered in Panel 25.2 in the context of NMR spectroscopy.

Note the essential difference between a *stereospecific* reaction (stereoisomeric *reactants* give different stereochemical outcomes), and a *stereoselective* reaction (an achiral or racemic compound gives stereoisomeric *products* in unequal yields).

Earlier in this chapter, we hinted that stereoselectivity (the preferential formation of one stereoisomer from an achiral or racemic reactant) is a very desirable feature of a synthetic reaction. In this section, we have indicated how catalytic hydrogenation of an unsymmetrical ketone can be made stereoselective by using a chiral catalyst. In later chapters, we shall discuss further stereoselective reactions.

Stereoselectivity, however, is just one aspect of the *stereochemical course of a reaction*. In later chapters, we shall encounter examples of how stereoisomeric reactants give different stereochemical outcomes under the same experimental conditions. Such reactions are said to be *stereospecific*. We shall see that the outcomes of reactions which involve stereoisomers, either as reactants or products, provide information about the mechanisms involved. We shall also see that an understanding of the mechanisms of such reactions is important in planning an organic synthesis.

Exercise 11.14

Which of the following reactions give racemic product?

(a) PhCHO + NaBH₄ $\xrightarrow[\text{MeOH}]{}$

(b) PhCHO + CH₃MgBr $\xrightarrow[\text{2) H}_3\text{O}^+]{\text{1) Et}_2\text{O}}$

(c) 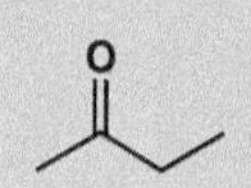 + LiAlH₄ $\xrightarrow[\text{2) H}_3\text{O}^+]{\text{1) Et}_2\text{O}}$

(d) 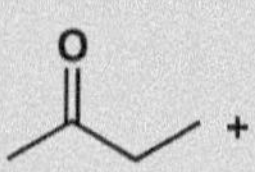 + CH₃MgBr $\xrightarrow[\text{2) H}_3\text{O}^+]{\text{1) Et}_2\text{O}}$

(e) 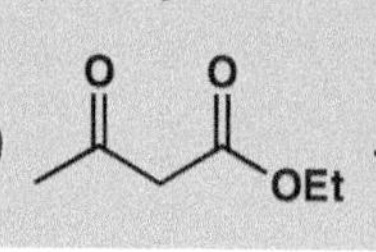 + NaBH₄ $\xrightarrow[\text{MeOH}]{}$

Summary

- Stereoisomers have the same molecular formula and the same atom connectivities, but atoms and groups are arranged differently in space; the origin of stereoisomerism in a particular molecule is called a **stereogenic element**.

- If two stereoisomers are non-superposable mirror images of each other, they are **enantiomers**; all other types of stereoisomers are **diastereoisomers**. An enantiomer, being non-superposable on its mirror image, is said to be **chiral** and, if the chirality is attributable to an identifiable atom of the molecule, that atom is called the **chirality centre**; an **achiral** molecule is superposable on (indistinguishable from) its mirror image and, hence, cannot be enantiomeric.

- Sometimes, a molecule with two or more chirality centres is superposable on its mirror image, i.e. the molecule as a whole is achiral. A compound comprising such molecules is called a **meso compound**.

- Enantiomers have identical physical and chemical properties in an achiral environment, but rotate the plane of polarized light in opposite directions, i.e. they are **optically active**. Diastereoisomers, in principle, have different chemical and physical properties.

- An equimolar mixture of enantiomers is called a **racemate**, and its separation is known as **resolution**.

- Achiral compounds and racemates normally give racemic products in reactions which generate new chirality centres.

Problems

11.1 Identify chiral objects amongst the following.

(a) scissors (b) tweezers (c) a clock (d) a bugle
(e) a spoon (f) a shoe

11.2 Classify the following as chiral or achiral compounds.

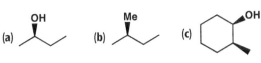

11.3 Classify the following as chiral or achiral compounds.

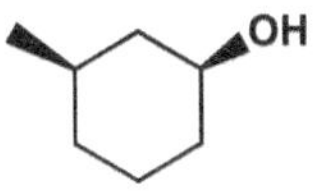

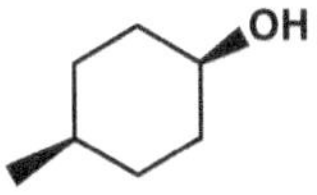

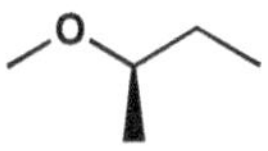

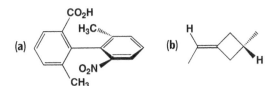

(c), 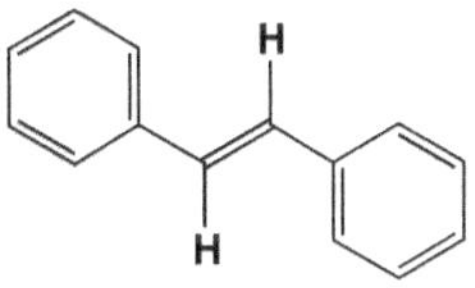 (d), 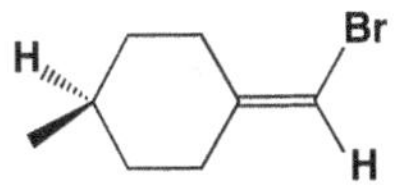 (e), 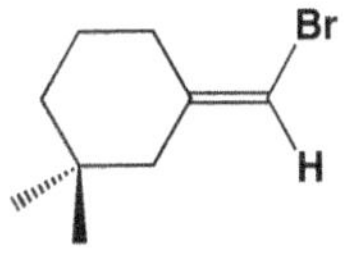 (f) 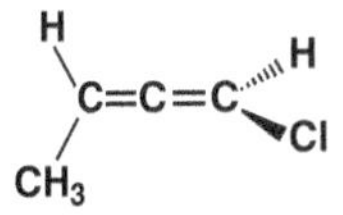 structures shown to the right.

11.4 For each of (a)–(d) below, is the compound identical with the one given above them, or the enantiomer?

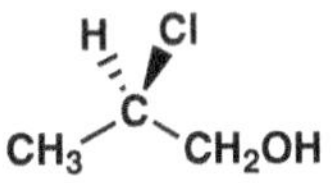

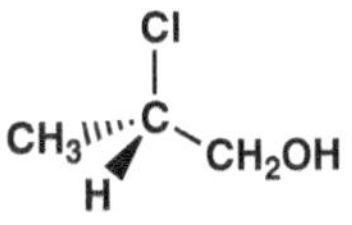

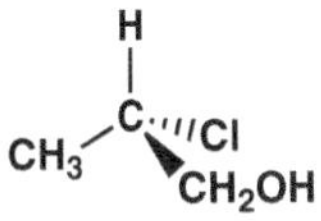

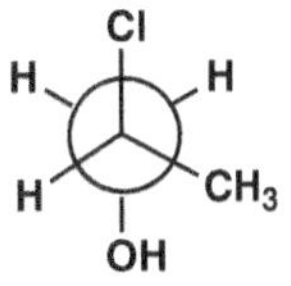

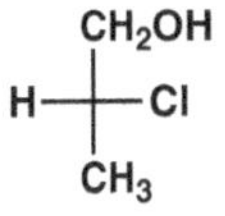

11.5 Name the compounds given in Problem 11.2 with assignment of *R* or *S* configuration to the chirality centres.

11.6 Assign *R* or *S* configuration to all the chirality centres in the following compounds.

(a) 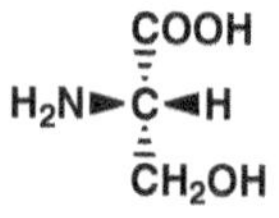
COOH
H₂N►C◄H
CH₂OH
serine

(b) 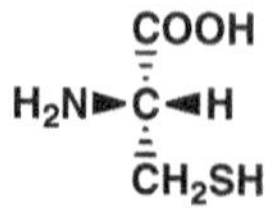
COOH
H₂N►C◄H
CH₂SH
cysteine

(c) 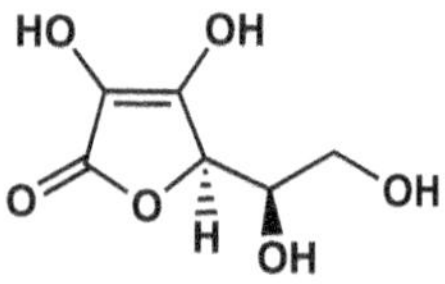
ascorbic acid
(vitamin C)

(d) 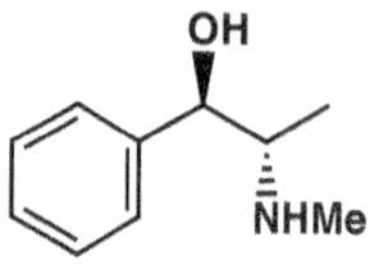
ephedrine
(stimulant)

(e) 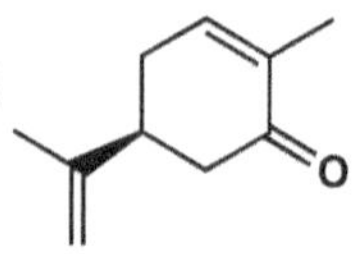
(+)-carvone
(caraway oil)

(f) 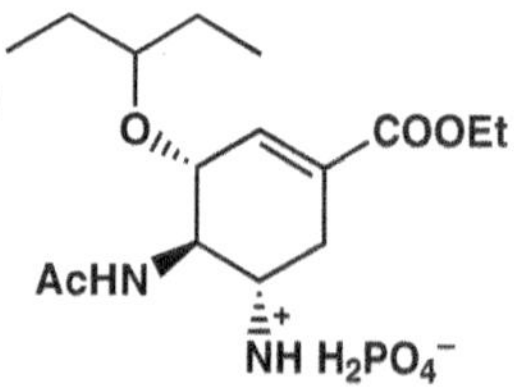
Tamiflu (oseltamivir)
(antiviral drug for influenza)

11.7 Draw zigzag structures of the following.
(a) (*R*)-pentan-2-ol   (b) (*S*)-3-hydroxybutanoic acid
(c) (*R*)-but-3-en-2-ol
(d) (*S*)-2,6-diaminohexanoic acid (lysine)
(e) (2*R*,3*R*)-2,3-dichloropentane

11.8 Identify pairs of compounds with different *R*,*S* configurations.

(a) 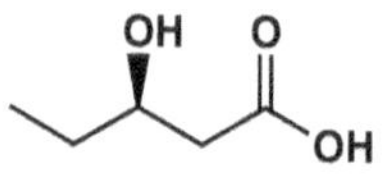 ... and ... 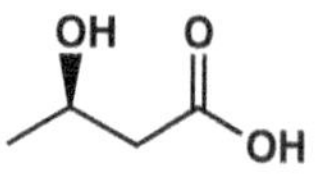
(b) 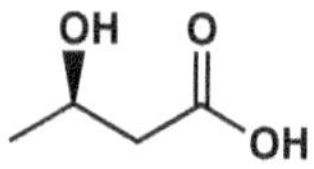 ... and ... 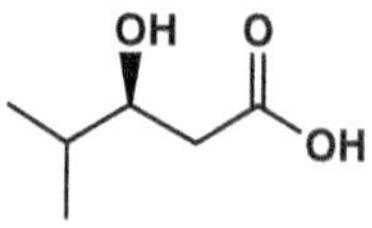
(c) 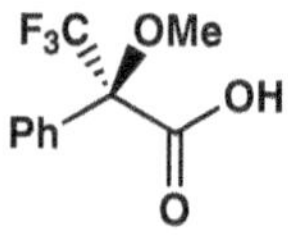 F₃C OMe Ph ... OH and F₃C OMe Ph ... Cl
(d) 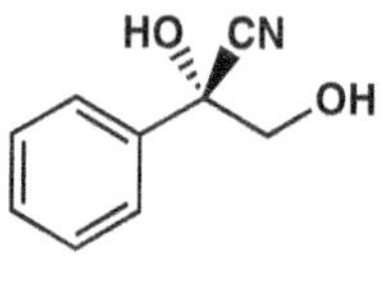 HO CN ... OH and HO COOH ... OH

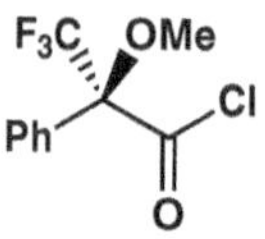

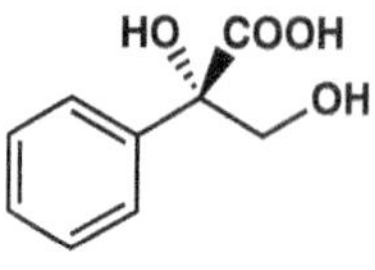

(e) 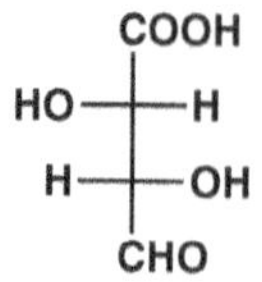
COOH
HO──H
H──OH
CHO
and
COOH
HO──H
H──OH
CH₂OH

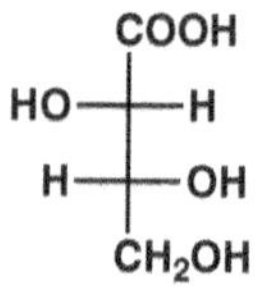

11.9 Which constitutionally isomeric alcohols with the molecular formula C₅H₁₂O allow stereoisomerism? If enantiomers are possible, show only the *R* isomer with wedged bonds.

11.10 Draw a stereochemical structure of the *S* enantiomer of a chiral ether with the molecular formula C₅H₁₂O.

11.11 For each of the following pairs of compounds, is the relationship enantiomeric, diastereoisomeric, or are they constitutional isomers?

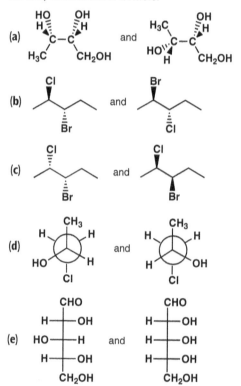

11.12 Write structural formulas of all the stereoisomers of cyclohexane-1,3-diol, and identify the stereochemical relationships.

11.13 Enantiomers have identical physical properties except for optical activity. However, crystals of a pure enantiomer and of the racemic compound usually have different melting points. Explain why this is so and predict any other differences in the properties of these crystals?

11.14 Draw structures of the main products of the following reactions of carbonyl compounds. When stereoisomers are possible, show them all and assign the *R* or *S* configuration to chirality centres.

(a) PhCHO + NaCN $\xrightarrow{H_3O^+}$

(b) 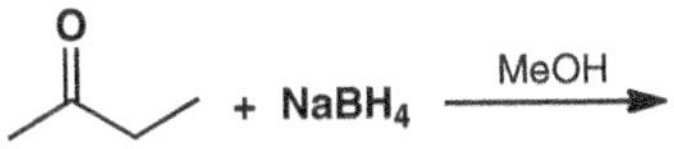 [structure: CH₃C(=O)CH₂CH₃] + NaBH₄ $\xrightarrow{MeOH}$

(c)  [structure: ketone] + MeMgBr $\xrightarrow[2) H_3O^+]{1) Et_2O}$

(d)  [structure: ester OEt] + 2PhMgBr $\xrightarrow[2) H_3O^+]{1) Et_2O}$

(e) 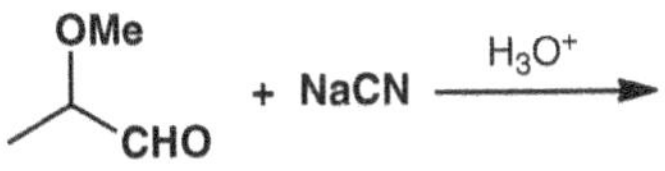 [structure: OMe, CHO] + NaCN $\xrightarrow{H_3O^+}$

(f) 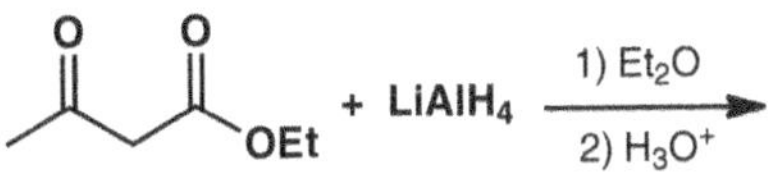 [structure: diketone ester OEt] + LiAlH₄ $\xrightarrow[2) H_3O^+]{1) Et_2O}$

11.15 Discuss possible stereoisomerism of the adamantane derivative shown below.

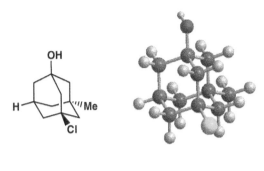

12 Nucleophilic Substitution Reactions of Haloalkanes and Related Compounds

Related topics we have already covered:

- Molecular strain and steric hindrance (Chapter 4)
- Curly arrow representation of reaction mechanisms (Section 7.2)
- Energetics of reaction (Section 7.4)
- Characterization of organic reactions and reaction mechanisms (Section 7.5)
- Nucleophilic addition and substitution at carbonyl groups (Chapters 8–10)
- Stereochemistry (Chapter 11)

Topics we shall cover in this chapter:

- Substitution reactions of haloalkanes and related compounds under nucleophilic/basic conditions
- Mechanisms of nucleophilic substitution: S_N1 and S_N2
- Stability of carbenium ions
- Stereochemistry of substitution at saturated carbon
- Solvent effects upon substitution reactions
- Neighbouring group participation

Haloalkanes and related compounds containing a single bond from a saturated (sp^3) carbon to a heteroatom can be represented generically as RY, where R is alkyl and Y is a halogen, a group such as a sulfonate ($-OSO_2R'$), or an ammonio residue ($-NR_3^+$). In all such compounds, the R–Y σ-bond is polarized because halogens and the other common heteroatoms (O, N, and S) are more electronegative than C, so the electrophilic C is susceptible to nucleophilic attack and Y is a potential nucleofuge.[1]

Polarity of the C–Y bond:

$$\overset{\displaystyle\ \ \ \delta+\ \ \delta-}{\underset{\overset{\displaystyle\uparrow}{\underset{\substack{\text{electrophilic}\\ \alpha\,C}}{}}}{{-}\overset{\displaystyle|}{\underset{\displaystyle|}{C}}{-}Y}}\quad\nwarrow\ \begin{array}{l}\text{a heteroatom}\\ \text{(or a heteroatom group)}\end{array}$$

The two most characteristic reactions of RY (substitution and elimination) under nucleophilic/basic conditions both involve heterolysis of the bond from the electrophilic carbon to the heteroatom and departure of the nucleofuge. In substitutions which we cover in this chapter, the nucleofuge is replaced by a nucleophile at the α C. The reaction of bromomethane with hydroxide ion was introduced in Chapter 7 as a typical nucleophilic substitution with a bimolecular mechanism (S_N2).

$$HO^- + CH_3–Br \longrightarrow HO–CH_3 + Br^-$$

---

[1] Atoms other than C and H are known as *heteroatoms* and include halogens, O, N, and S. To a major extent, Chapters 12–14 cover related material: the chemistry of carbon–heteroatom (C–Y) single bonds, where Y represents halogens and groups bonded through O, N, and S.

However, as we shall see, another substitution mechanism (S_N1) is also possible depending upon the structure of RY and the reaction conditions. In this chapter, we cover both mechanisms, and explore the relationships between them. This will be principally with haloalkanes as substrates but we shall anticipate corresponding reactions of some other compounds RY, i.e. those with single C–O, C–S, and C–N bonds at the α C, before they are discussed more fully in Chapter 14. Before that, however, we shall cover elimination reactions to give alkenes from RY in the next chapter. In these, departure of the nucleofuge is associated with proton abstraction from the β C rather than with a nucleophile bonding to the α C. The mechanisms of substitution and elimination from RY are related, as we shall see, so there are strong connections between these two chapters.

We shall see that simple oxygen groups like OH and OR in alcohols and ethers cannot readily depart as nucleofuges, as was first discussed in Chapter 9. However, substitution (and elimination) reactions of these compounds can be induced by a catalytic mechanism, as we shall see in Chapter 14.

12.1 Reactivity of Haloalkanes with Nucleophiles

The rates of reactions of haloalkanes (alkyl halides, RY where Y = I, Br, or Cl) with **nucleophiles**, Nu⁻, in solution (eqn 12.1) are invariably first order in [RY].

$$Nu^- + R-Y \longrightarrow Nu-R + Y^- \qquad (12.1)$$

Such reactions of simple primary or secondary alkyl halides are also first order in the concentration of the nucleophile:

rate $= k[RY][Nu^-]$ where k is the second-order rate constant.

The reaction is said to be **second order** overall.

We can illustrate these generalizations by the reactions of a series of bromoalkanes with aqueous hydroxide.

$$HO^- + R-Br \longrightarrow HO-R + Br^- \qquad (12.2)$$

For the reaction of bromomethane with hydroxide to give methanol (eqn 12.2, R = CH_3), the rate law may be expressed as follows.

Rate $= k[RBr][HO^-]$ (R = CH_3, $MeCH_2$, and Me_2CH)

Bromoethane and 2-bromopropane (which correspond to replacing first one then two hydrogens of bromomethane by methyl groups) also react with HO⁻, and obey the same rate law, but the rate constants decrease steeply.

CH_3–Br : $MeCH_2$–Br : Me_2CH–Br : Me_3C–Br

Relative k values: 1.0   0.08   0.014   see text

The C–F bond in fluoroalkanes is much stronger than the corresponding bonds in the other haloalkanes so fluoroalkanes do not normally undergo nucleophilic substitution reactions.

By the end of this chapter, we shall be able to compare reactions of nucleophiles at the electrophilic singly-bonded C in compounds RY with nucleophilic additions to the electrophilic carbons of C=O and C=N double bonds, which were covered in Chapters 8–10.

A rate law is the outcome of experimental observations and is a major component of the evidence leading to a mechanistic proposal. A second-order rate law *suggests*, but does not *require*, a bimolecular rate-determining step, see Chapter 7. For reactions which are kinetically second order but have *unimolecular* rate-determining steps, see Sections 13.3 (E1cB elimination reactions) and 14.1 (acid-catalysed reactions of alcohols).

The above rate law suggests that the reactions of CH_3Br, $MeCH_2Br$, and Me_2CHBr in aqueous hydroxide are all **bimolecular**, and that the transition structure in each case is derived from one molecule of each reactant, RBr and HO^-.

In contrast, the kinetics results for the reaction in aqueous hydroxide of the compound obtained by replacing all three hydrogens of bromomethane by methyl groups, 2-bromo-2-methylpropane (Me_3CBr, *t*-butyl bromide), are *qualitatively* different from those for the other three compounds above. At low concentrations of hydroxide, Me_3CBr reacts much faster than CH_3Br, $MeCH_2Br$, or Me_2CHBr, but the rate is independent of $[HO^-]$: it is simply first order in $[Me_3CBr]$. However, at higher concentrations of $[HO^-]$, the reaction becomes first order in $[HO^-]$ as well as in $[Me_3CBr]$. The different rate laws for Me_3CBr, first at low concentrations of $[HO^-]$ and then at higher values, indicate different mechanisms under the different sets of conditions.

12.2 The S_N2 Mechanism

Since the C–Y bond of a haloalkane RY is polar, we can expect that the nucleophilic attack in the bimolecular reaction of simple primary and secondary alkyl halides in eqn 12.1 will be at the partially positive electrophilic carbon. And, in order to avoid an adverse

---

Panel 12.1 Biological alkylation

Non-naturally occurring compounds such as haloalkanes which react as electrophiles in S_N2 reactions may be toxic owing to their ability to alkylate biological compounds. Alkylation can occur at nucleophilic sites such as NH_2 or SH groups of enzymes, which deactivates their catalytic properties, or at the bases of DNA in which case the electrophilic haloalkanes may be mutagenic. Other types of electrophiles such as aldehydes and epoxides are also capable of modifying these nucleophilic sites and, consequently, may also be mutagens.

One of the most notorious alkylating agents is mustard gas, bis(2-chloroethyl) sulfide, which was used as a chemical warfare agent during World War I. The sulfur atom acts as an intramolecular nucleophile to generate a cyclic sulfonium ion which is a very reactive alkylating agent. Mustard gas is a powerful vesicant which causes prolonged suffering, blindness, and (ultimately) death in the event of heavy doses.

mustard gas

In living organisms, *S*-adenosylmethionine (SAM, which is also a sulfonium ion) is a natural methylating agent. For example, SAM methylates norepinephrine in the biosynthesis of epinephrine (adrenaline) in our bodies.

norepinephrine

S-adenosylmethionine (SAM)

epinephrine (adrenaline)

electrostatic interaction with the nucleofuge as it approaches, this can reasonably be expected to take place as shown in Scheme 12.1. The nucleophile approaches the side of the carbon diametrically opposite the nucleofuge in 'rear-side nucleophilic attack'. This simple one-step mechanism, deduced from experimental kinetics evidence, is supported theoretically and by stereochemical evidence (see below).

Sir Christopher Ingold (1893–1970).
(Courtesy of the Royal Society of Chemistry.)

C.K. Ingold was born in London and served as a professor in University College, London. In the 1930s, he contributed to the establishment of modern concepts of reaction mechanisms including nucleophiles and electrophiles, inductive and resonance effects, and introduced terms such as S_N1, S_N2, E1, and E2.

Scheme 12.1 The S_N2 mechanism.

12.2.1 Steric hindrance in S_N2 reactions

We can now see why the rate constant decreases along the series CH_3–Br, $MeCH_2$–Br, Me_2CH–Br; increasing steric congestion at the α C inhibits close nucleophilic approach, as illustrated in eqn 12.3, and reactivity decreases in the order: methyl > ethyl > isopropyl. In other words, the more substituents there are at the electrophilic carbon, and the bulkier the substituents, the slower the S_N2 reaction will be.

$$(12.3)$$

In general, therefore, the S_N2 reactivity of RY is in the order shown below, with tertiary alkyl substrates essentially undergoing no S_N2 reaction at all (they do react, but by different mechanisms, as we shall see in Section 12.4 and the next chapter).

The EPM of CH_3Br shows the polarity of the C–Br bond.

S_N2 reactivity:

methyl > 1° alkyl > 2° alkyl >> 3° alkyl (no reaction)

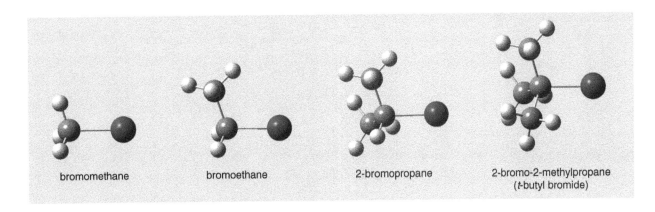

bromomethane   bromoethane   2-bromopropane   2-bromo-2-methylpropane (t-butyl bromide)

Example 12.1

Give structures of all the isomers of bromoalkane C_4H_9Br in order of decreasing reactivity in S_N2 reactions.

Solution

Two isomers are primary alkyl bromides; the one with the branched alkyl is the less reactive because of greater steric hindrance to nucleophilic approach.

$$CH_3CH_2CH_2CH_2Br \quad > \quad (CH_3)_2CHCH_2Br \quad > \quad CH_3CH_2\overset{\overset{\displaystyle CH_3}{|}}{C}HBr \quad >> \quad (CH_3)_3CBr$$

1-bromobutane	1-bromo-2-methylpropane	2-bromobutane	2-bromo-2-methylpropane
(butyl bromide)	(isobutyl bromide)	(s-butyl bromide)	(t-butyl bromide)
1°	1°	2°	3°

Exercise 12.1

Which of each of the following pairs of haloalkanes is more reactive by the S_N2 mechanism?

(a) $CH_3CH_2CH_2Cl$ and $(CH_3)_2CHCl$

(b) $(CH_3)_2CHCH_2Br$ and $(CH_3)_3CCH_2Br$

12.2.2 Stereochemistry of the S_N2 mechanism

The most distinctive characteristic of the S_N2 mechanism is the **inversion of configuration** of the electrophilic carbon in the transition structure (Scheme 12.1). The reaction is **stereospecific**. This was established by experiments with optically active substrates which were chiral at the reaction site. For example, reaction of (R)-2-bromobutane with hydroxide yields (S)-butan-2-ol (eqn 12.4), while the (S) substrate gives the (R) product.

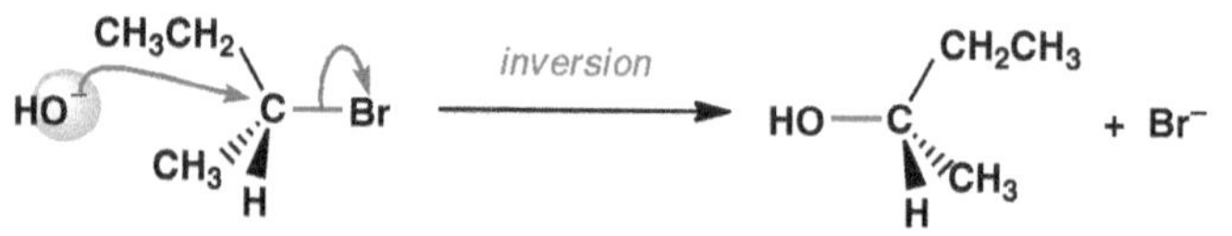

A reaction is stereospecific if stereoisomeric starting materials give different stereoisomeric outcomes (see Chapter 11).

$$(12.4)$$

(R)-2-bromobutane → (S)-butan-2-ol

Note that chirality is not essential for the observation of inversion of configuration by the S_N2 mechanism; S_N2 reactions of cis- and trans-1-bromo-4-methylcyclohexanes, for example, are stereospecific.

If (R)-2-bromobutane reacts first with iodide (which is a good nucleophile), and the product then reacts with hydroxide (iodide is also an excellent nucleofuge), the final product is (R)-butan-2-ol (Scheme 12.2). This is the result of two sequential inversions of configuration, leading to overall retention.

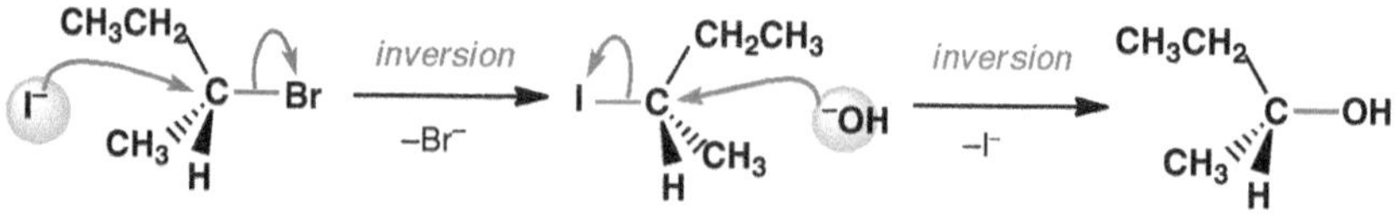

Scheme 12.2 Two consecutive inversions of configuration giving overall retention.

(R)-2-bromobutane → (S)-2-iodobutane → (R)-butan-2-ol

overall retention of configuration

Exercise 12.2

What is the substitution product upon reaction of each of cis- and trans-1-bromo-4-methylcyclohexanes in aqueous hydroxide?

12.2.3 Stereoelectronic description of the S$_N$2 mechanism

The S$_N$2 reaction of RY is represented generically by Scheme 12.1 above; the reactive site of the nucleophile Nu$^-$ is usually a lone pair and its direction of attack at the electrophilic C is collinear with the departure path of Y$^-$ which is not limited to being a halide (see below).

Reactions between a nucleophile and an electrophile occur through interactions of the HOMO of the former with the LUMO of the latter (Section 7.3). In the S$_N$2 reaction of Scheme 12.1, the HOMO (lone pair) of Nu$^-$ interacts with the LUMO of RY. The latter is the (vacant) antibonding σ* orbital of the C–Y bond, and has a large lobe on the opposite side of the C–Y bond, as illustrated in Figure 12.1 for the reaction of CH$_3$Br with HO$^-$. To achieve the best possible overlap of the HOMO and the LUMO, the HO$^-$ approaches the C from the side opposite to the C–Br bond. As the new MOs for the HO–C bond develop, the electrons from a lone pair on HO$^-$ enter the new bonding orbital. *This mechanism requires inversion of the configuration of the C.*

A computer-generated illustration of the shape of the LUMO of CH$_3$Br with a molecular model embedded within it; only the two small grey spheres of the model corresponding to two of the three H atoms are immediately obvious.

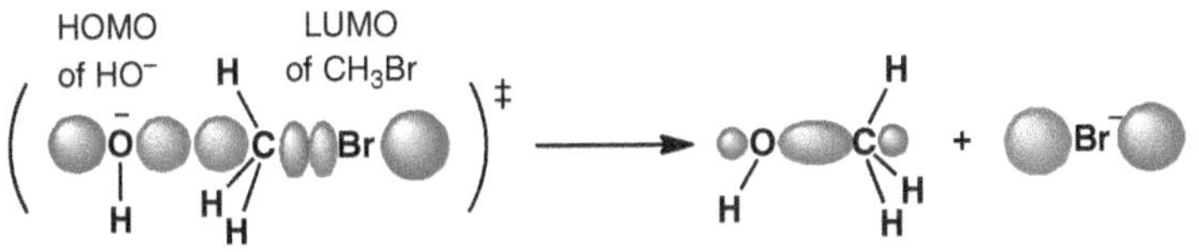

Figure 12.1 A schematic representation of the HOMO–LUMO interactions in the transition structure of the S$_N$2 reaction of HO$^-$ and CH$_3$Br.

At the same time, the electron pair of the C–Br bond becomes a lone pair on Br as it departs as Br$^-$. The breaking of the one bond and the formation of the other occur together: the whole rebonding process within the *encounter complex* is *concerted* once the reactants have come together in the right way.

> Draw a reaction profile for an S$_N$2 mechanism (see Section 7.4).

Exercise 12.3

12.2.4 Nucleophiles and nucleofuges

The following is a list of nucleophiles (including hydroxide) for S$_N$2 reactions in order of decreasing reactivity (**nucleophilicity**) in a protic solvent.

Nucleophiles:

$$RS^-, CN^-, I^- > RO^-, HO^- > Br^-, NH_3, RNH_2, N_3^- > Cl^- > RCO_2^- > F^- > H_2O, ROH$$

good ⟷ poor

If the reaction sites of a series of nucleophiles are elements of the same *period* of the periodic table, the more basic reagents are generally more nucleophilic. Basicity is a measure of affinity toward the proton (which is obviously an electrophile) while nucleophilicity is reactivity toward an electrophilic carbon, and the two trends are broadly parallel. For example,

nucleophilicity and basicity : $RNH^- > RO^-, HO^- > RNH_2 > RCO_2^- > H_2O, ROH.$

> Note that basicity is an equilibrium (thermodynamic) property (see Chapter 6), while nucleophilicity is a kinetic (reactivity) property.

In contrast, on going down a *group* of the periodic table of elements (e.g. halide ions), the nucleophilicity in protic solvents *increases* in spite of their *decreasing* basicity.

Lewis bases are stabilized by hydrogen-bonding solvation in protic solvents (Section 3.9); however, in order to react as a nucleophile, there must be some degree of desolvation.

The larger ions have more diffuse electrons (higher polarizability) and are less strongly solvated by hydrogen bonding. They also have HOMOs of higher energy which can over-lap more effectively at greater distances with the LUMO of the electrophilic carbon leading to stronger orbital interactions in the transition structures.

Nucleophilicity:	$I^- > Br^- > Cl^- > F^-$	$RS^- > RO^-$
Basicity:	$F^- > Cl^- > Br^- > I^-$	$RO^- > RS^-$

Nucleophilicity is much more affected by steric hindrance than is basicity (affinity for the small proton). For example, the t-butoxide ion is one of the most basic alkoxides, but is a very poor nucleophile due to its steric bulk.

Basicity:	$CH_3CH_2O^- < (CH_3)_3CO^-$
Nucleophilicity:	$CH_3CH_2O^- \gg (CH_3)_3CO^-$
Steric bulk:	small → large

ethoxide ion

t-butoxide ion

Halide ions other than F^- are typical good nucleofuges but others include sulfonates RSO_3^- and some neutral groups (e.g. H_2O from $R-OH_2^+$ following protonation of $R-OH$). In general, a weak base Y^- (strong conjugate acid, HY) will be a good nucleofuge. This is reasonable because the tendency of HY to release Y^- should be parallel with the tendency of RY to release Y^-.

Good nucleofuges (Y^-):	I^-,	Br^-,	Cl^-,	RSO_3^-,	H_2O,	ROH
Approximate pK_{BH^+} values:	−10	−9	−7	−3	−2	−2

However, although the leaving ability of halide ions (Y^-) decreases with decreasing acidity of HY,

leaving ability:	$I^- > Br^- > Cl^- \gg F^-$,
acidity of HY:	→ decreases →,

arenesulfonates ($ArSO_3^-$) are better nucleofuges than expected from their pK_{BH^+} values (they are generally better than iodide).

Example 12.2

Complete the following nucleophilic substitution reactions.

(a) ∕∖∕CI + NH₃ ⟶   (b) ⎯⎯Br + EtO⁻ Na⁺ ⟶

The S$_N$2 reaction of a haloalkane with an alkoxide is a general method for the synthesis of ethers and is sometimes known as the Williamson synthesis. Its susceptibility to steric inhibition and competing elimination is a disadvantage, however.

Solution

(a) NH$_3$ acts as a nucleophile in an S$_N$2 reaction with a primary alkyl chloride to give an alkylammonium salt.

∕∖∕CI + NH₃ ⟶ ∕∖∕NH₃⁺ Cl⁻

| 1-chloropropane (propyl chloride) | ammonia | propylammonium chloride |

(b) Here, ethoxide ion acts as a nucleophile in an S_N2 reaction with a primary alkyl bromide.

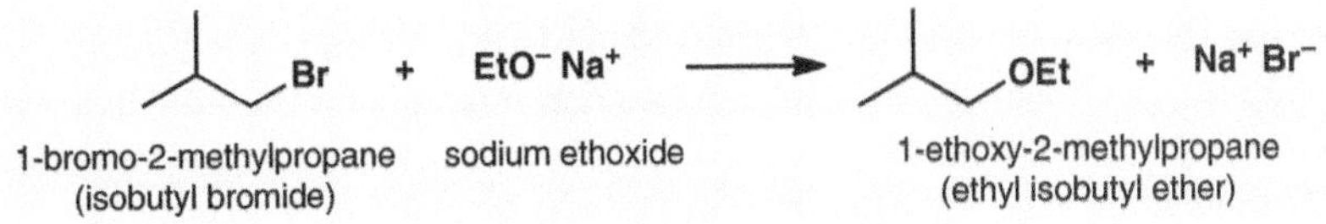

| 1-bromo-2-methylpropane (isobutyl bromide) | sodium ethoxide | 1-ethoxy-2-methylpropane (ethyl isobutyl ether) | |

Exercise 12.4

Complete the following nucleophilic substitution reactions.

(a) + Na⁺ CN⁻ ⟶

(b) + ⟶

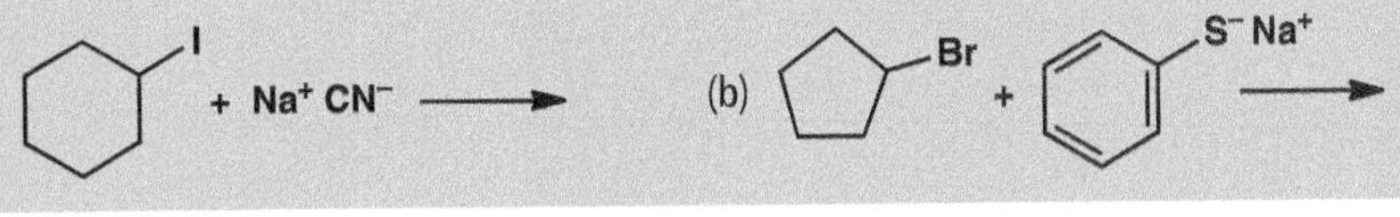

Exercise 12.5

Which reaction in each of (a) and (b) occurs more readily?

(a) (1) $CH_3CH_2CH_2Br$ + EtO^- $\xrightarrow{EtOH}$ $CH_3CH_2CH_2OEt$ + Br^-

(2) $CH_3CH_2CH_2Br$ + EtS^- $\xrightarrow{EtOH}$ $CH_3CH_2CH_2SEt$ + Br^-

(b) (1) $CH_3CH_2CH_2Br$ + CN^- $\xrightarrow{propanone}$ $CH_3CH_2CH_2CN$ + Br^-

(2) $CH_3CH_2CH_2I$ + CN^- $\xrightarrow{propanone}$ $CH_3CH_2CH_2CN$ + I^-

12.3 Solvent Effects

12.3.1 Polarity of the transition structure

Homolytic reactions are not greatly influenced by the nature of the solvent and often occur without a solvent or in the gas phase. Most organic reactions, however, are carried out in solution and the majority involve heterolytic bond cleavages. Such reactions of neutral molecules rarely occur in the gas phase because of the high energies required for charge separation. It follows that the choice of solvent is very important for any reaction in which the conversion of reactant molecule(s) into transition structure and then into product molecule(s) is accompanied by a redistribution of charge.

A compound dissolves in a solvent when intermolecular interactions between the solute and solvent molecules in solution (solvation) are more stabilizing than solute–solute interactions in the pure solute (see Sub-section 3.9.5). In any reaction in solution, reactant–solvent interactions are replaced by product–solvent interactions, so the choice of solvent can have an effect upon the equilibrium constant.

When dealing with *reactivity*, however, we need to consider the effect of the solvent upon the initial state and upon the transition state (TS). If stabilization by a solvent is greater in the TS than in the initial state, the activation energy is smaller in the solvent, and the rate constant is larger (see Chapter 7). On the other hand, if stabilization by the solvent is greater in the initial state than in the TS, the activation energy is greater and the rate constant is smaller.

We can use the same ideas to predict the effect of a *change* in solvent upon a rate constant. In Sub-section 3.9.5, we saw that a polar compound is more soluble in a polar than a nonpolar solvent. For the same reasons, a transition structure which is more polar than the reactant molecule(s) will be stabilized to a greater degree. Consequently, the activation energy of such a reaction will be decreased (greater rate constant) by increasing the polarity of the solvent, as shown in Figure 12.2(a). Conversely, when the polarity *decreases* between the initial state and the TS, the rate constant will be smaller (greater activation energy) in a more polar solvent which preferentially stabilizes the initial state (Figure 12.2(b)).

We can illustrate the above by the S_N2 reaction of a neutral substrate RY when the nucleophile is an anion (eqn 12.5 in Scheme 12.3). The negative charge is dispersed in the TS (a decrease in polarity), so the initial state is more strongly stabilized by an increase in solvent polarity (Figure 12.2(b)) and the rate constant becomes smaller. In contrast, the rate constant of the reaction of RY with a neutral nucleophile (eqn 12.6) becomes larger as solvent polarity increases because the reaction involves charge separation (TS more polar than reactants, Figure 12.2(a)).

Figure 12.2 Effects of higher solvent polarity on a rate constant. (a) TS more polar than the reactant(s), greater rate constant. (b) TS less polar than the reactant(s), smaller rate constant.

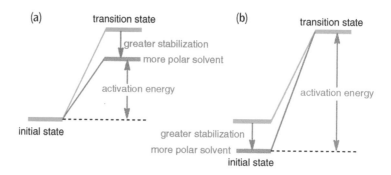

Scheme 12.3 S_N2 reactions of a neutral substrate (RY) with an anionic nucleophile (eqn 12.5) and with a neutral nucleophile (eqn 12.6).

$$\text{HO}^- + \text{RY} \longrightarrow \left(\overset{\delta-}{\text{HO}}\text{- -R- -}\overset{\delta-}{\text{Y}} \right)^{\ddagger} \longrightarrow \text{HOR} + \text{Y}^- \qquad (12.5)$$

an anionic nucleophile   charge dispersal

$$\text{H}_3\text{N} + \text{RY} \longrightarrow \left(\overset{\delta+}{\text{H}_3\text{N}}\text{- - - -R- -}\overset{\delta-}{\text{Y}} \right)^{\ddagger} \longrightarrow \overset{+}{\text{H}_3\text{N}}\text{R} + \text{Y}^- \qquad (12.6)$$

a neutral nucleophile   charge separation

Exercise 12.6

Explain the solvent effects on the following substitution reactions.

(a) The rate constant for the reaction of trimethylsulfonium ion with ethoxide in ethanol is about 20 000 times greater than the one for its reaction with hydroxide in water.

$$\overset{+}{\text{CH}_3\text{S}(\text{CH}_3)_2} + \text{RO}^- \longrightarrow \text{CH}_3\text{OR} + (\text{CH}_3)_2\text{S}$$

(b) The rate constant for the analogous reaction of iodomethane with ethoxide in ethanol is also greater than the one for its reaction with hydroxide in water, but only slightly.

$$\text{CH}_3\text{I} + \text{RO}^- \longrightarrow \text{CH}_3\text{OR} + \text{I}^-$$

12.3.2 Classes of solvents

Solvents are often roughly classified as **polar** or **nonpolar,** but this is a simplification—there is a broad range from nonpolar alkanes to strongly polar solvents such as water. Additionally, solvents may be classified as **protic** or **aprotic** depending on the presence or absence of a weakly acidic hydrogen capable of hydrogen bonding. This is an important distinction because solvation of anions by hydrogen bonding can have a major influence on solvent effects, as discussed in Sub-section 3.9.5. Usually, protic solvents are also polar (higher alcohols and carboxylic acids other than methanoic acid are exceptions) so, taking this into account, we have an (approximate) composite three-way classification of solvents: we can label them as (a) weakly polar to polar protic, (b) polar aprotic, and (c) nonpolar to weakly polar aprotic solvents. Common solvents are broadly classified below with dielectric constants in parentheses as approximate measures of polarity.

(a) Protic solvents (polar to weakly polar)

H_2O (80) CH_3OH (32) C_2H_5OH (25) HCO_2H (58) CH_3CO_2H (6.2) $H\text{-}\overset{O}{\overset{\|}{C}}\text{-}NHCH_3$ (182)

water   methanol   ethanol   methanoic acid (formic acid)   ethanoic acid (acetic acid)   N-methyl formamide (NMF)

> The systematic IUPAC name for formamide is methanamide.

(b) Polar aprotic solvents

$CH_3\text{-}\overset{O}{\overset{\|}{C}}\text{-}CH_3$ (21) CH_3CN (36) $H\text{-}\overset{O}{\overset{\|}{C}}\text{-}N(CH_3)_2$ (37) $CH_3\text{-}\overset{O}{\overset{\|}{S}}\text{-}CH_3$ (47)

propanone (acetone)   ethanenitrile (acetonitrile)   N,N-dimethyl formamide (DMF)   diimethyl sulfoxide (DMSO)

(c) Aprotic solvents (nonpolar to weakly polar)

$CH_3(CH_2)_4CH_3$ (1.9) C_6H_6 (2.4) $CHCl_3$ (4.9) $(C_2H_5)_2O$ (4.4)   O (7.5)

hexane   benzene   trichloromethane (chloroform)   diethyl ether   tetrahydrofuran (THF)

Explain the dissolving of ionic substances (e.g. NaI) by illustrating how a cation (Na^+) and an anion (I^-) are solvated by a polar aprotic solvent (DMSO) and a protic solvent (ROH).

Example 12.3

Solution
Consult Sub-section 3.9.5b for the effects of the dielectric constant of a solvent and of dipoles of individual solvent molecules. The cation is solvated by lone pairs of molecules of both solvents by Lewis acid–base interactions (DMSO better than ROH). The anion will be solvated by dipole–dipole interactions in DMSO and by hydrogen bonding in the protic solvent.

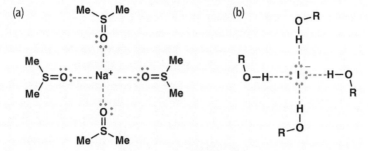

Solvation of (a) Na^+ in DMSO and (b) I^- in an alcohol (the solvating molecules are distributed in three dimensions in both cases).

Cyclic (crown) ethers and bicyclic (cryptand) ethers and amines are also able to complex metal cations and consequently solubilize anionic nucleophiles in aprotic solvents. (See Panel 14.5 for more on crown ethers).

Nucleophiles are often anionic and generally have lone pair(s), so they are effectively stabilized by hydrogen-bonding solvation in protic solvents, which greatly reduces their nucleophilic reactivity. It follows that nucleophiles will be much more reactive when they are unsolvated. A practical problem is that a solvent which does not solvate a solute will not dissolve it! However, although ionic substances are not soluble in nonpolar solvents, *polar aprotic solvents* often solvate a cation effectively and thereby solubilize the ionic substance in spite of only very weak solvation of the counter anion. The anion in such solvents is sometimes described as 'naked' and often very reactive as a nucleophile (and base); consequently, polar aprotic solvents are generally excellent for reactions of nucleophiles such as S_N2 reactions. Many of the S_N2 reactions encountered in this book are carried out in propanone (acetone), which is one of the most common solvents because it is inexpensive and easy to use.

Exercise 12.7

Which will be the better solvent for each of the following reactions? Assume that the counterion of the nucleophile will be chosen so that the salt is soluble in the solvents specified.

(a) $CH_3I + Cl^-$ ⟶ $CH_3Cl + I^-$

Solvent : CH_3OH or CH_3CN

(b) 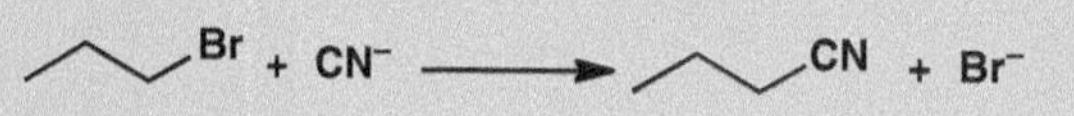 [structure] Br + CN^- ⟶ [structure] CN + Br^-

Solvent : $HCONMe_2$ (DMF) or $HCONHMe$ (NMF)

12.4 The S_N1 Mechanism

It was suggested in Section 12.1 that the substitution reaction of t-butyl bromide, Me_3CBr, with aqueous hydroxide is not by the S_N2 mechanism which is followed by simple primary and secondary alkyl bromides. As indicated earlier, the rate of reaction of Me_3CBr is independent of hydroxide concentration at low values of $[HO^-]$ (rate $= k\,[Me_3CBr]$).

Panel 12.2 Phase-transfer catalysis ───────

Most organic reactions are carried out in single-phase solutions but, when a water-soluble ionic reactant which does not dissolve well in organic solvents is needed to react with an organic reactant which is only slightly soluble in water, finding a suitable solvent may be problematical. A typical difficulty is getting anions of salts to serve as nucleophiles in S_N2 reactions in organic solvents. One approach is to use an organic counterion of the anion, and quaternary ammonium cations with lipophilic alkyl or aryl substituents have been used. Tetrabutylammonium azide ($Bu_4N^+N_3^-$), for example, is much more soluble in organic solvents than, say, sodium azide if N_3^- is required as a nucleophile.

Another strategy is to use a *phase-transfer catalyst* in a vigorously stirred two-phase system starting with the organic substrate in an organic solvent and the ionic reagent in water. For example, tetrabutylammonium hydrogensulfate helps nucleophilic anions such as cyanide to migrate from the aqueous phase into the organic phase and facilitates an S_N2 reaction, as illustrated in Figure 1. The rate-limiting step in such a reaction is sometimes the transfer for the reagent across the phase boundary. Phosphonium salts can also be used in **phase-transfer catalysis**.

In the following two-phase system, benzyltrimethylammonium hydroxide (sometimes called Triton B) facilitates formation of a carbanion from the ketone using sodium hydroxide as the source of the hydroxide in the aqueous phase (see Chapter 17 for this type of reaction). In such reactions, the deprotonation takes place predominantly at the interface.

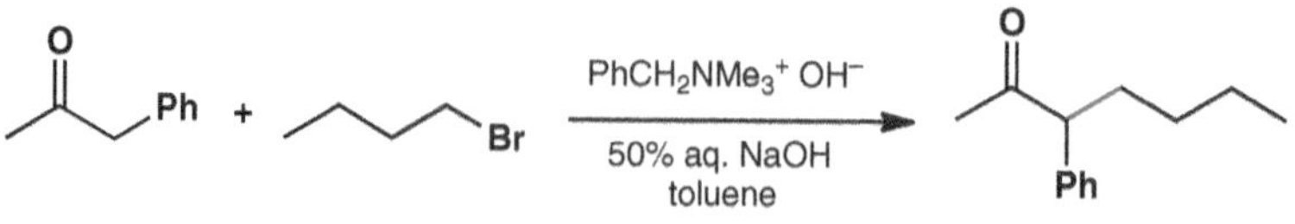

[reaction scheme]

O
‖
[structure] Ph + [structure] Br → $PhCH_2NMe_3^+$ OH^- / 50% aq. NaOH / toluene → O‖[structure] Ph [structure]

t-Butyl bromide and similar tertiary alkyl substrates (RY) do undergo nucleophilic substitution reactions, as indicated earlier in eqn 12.1,

$$Nu^- + R-Y \longrightarrow Nu-R + Y^- \qquad (12.1)$$

and the reactions are **first order** in [RY]; however, they are independent of the concentration of the nucleophile:

$$\text{reaction rate} = k[RY]$$

It follows that *the nucleophile is not involved in the rate-determining transition structure* which contains only one molecule of RY. The mechanism, therefore, is a **unimolecular** nucleophilic substitution (**S$_N$1**). However, we know from eqn 12.1 that the nucleophile participates in the overall transformation, so it must become involved *after* the rate-determining step.

12.4.1 Carbenium ion intermediates

A reasonable two-step mechanism for an S$_N$1 reaction (Scheme 12.4) involves initial heterolysis of the C–Y bond to give a **carbenium ion intermediate** R$^+$ (eqn 12.7), followed by rapid trapping of R$^+$ by the nucleophile Nu$^-$ to give the product (eqn 12.8).

$$RY \xrightarrow{\text{r.d.s.}} [R^+] + Y^- \qquad (12.7)$$
$$\text{carbenium ion intermediate}$$

$$[R^+] + Nu^- \xrightarrow{\text{fast}} RNu \qquad (12.8)$$

Scheme 12.4 The two-step S$_N$1 mechanism.

This mechanism accommodates the observed first-order kinetics and, as will be discussed in the following sections, the dependence of the rate constant upon the stability

The rate becomes first order in [HO$^-$] at concentrations of HO$^-$ > 0.05 mol dm^{-3}, but product analysis shows that the reaction is elimination to give alkene (see Chapter 13), not substitution.

A polar solvent will facilitate the S$_N$1 reaction of a neutral substrate because the rate-determining step is ionization. However, common polar solvents such as water and the lower alcohols are also nucleophilic to a degree so will compete with a nucleophilic solute for the intermediate carbenium ion. Consequently, mixtures of products are a common feature of S$_N$1 reactions, which reduces their synthetic utility.

─────── **Panel 12.2** Continued ───────

Cyclic polyethers (crown ethers) also act as phase-transfer catalysts by solubilizing ionic reagents M$^+$X$^-$ by complexation of the cation and allowing X$^-$ to react with the substrate in organic solvents (see Panel 14.5).

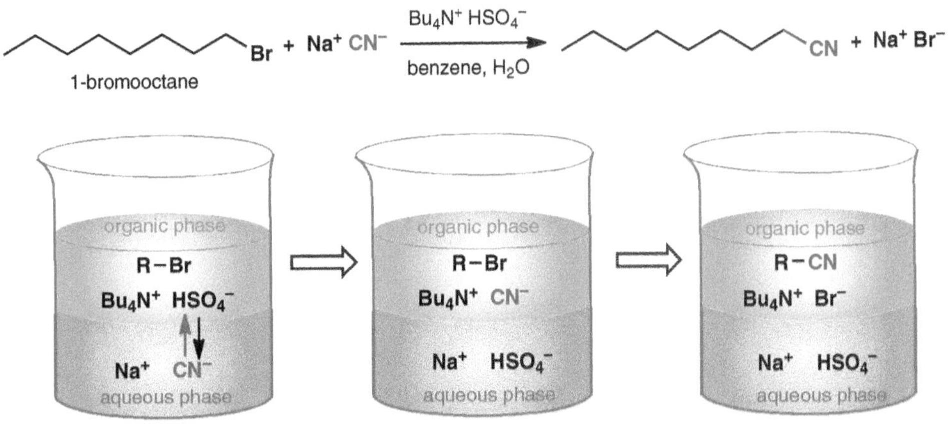

Figure 1 Schematic representation of phase-transfer catalysis of the S$_N$2 reaction, RBr + CN$^-$ → RCN + Br$^-$.

of the intermediate carbenium ion and the polarity of the solvent; it also accounts for the stereochemical evidence (see next sub-section).

Exercise 12.8

Draw a reaction profile for an S_N1 mechanism (cf. Section 7.4).

12.4.2 Stereochemistry of the S_N1 mechanism

If the S_N1 reaction proceeds *via* a carbenium ion, the stereochemical outcome must be quite different from that of the S_N2 reaction because the central carbon of the intermediate is sp² hybridized and planar (Scheme 12.5). The lobes of the vacant 2p orbital (LUMO) are perpendicular to the plane of the carbenium ion and the HOMO of the nucleophile may interact with either of the lobes. If the reaction site of substrate RY is a chiral centre, and nucleophilic attack at the carbenium ion is equally possible from both sides, the outcome must be the formation of racemic product regardless of the enantiomeric purity of the substrate RY.

Experimentally, however, racemization is usually only partial because the nucleofuge is still present as a counter-ion on the side of the carbenium ion from which it departed. In other words, the carbenium ion initially exists as an **ion pair** (Scheme 12.5) and its counter-ion Y⁻ inhibits nucleophilic capture from the side leading to product with retention of configuration. The result is **partial inversion** of configuration, the degree of which depends on the stability (lifetime) of the ion-pair and the nature of the solvent.

Scheme 12.5
Stereochemistry of the S_N1 reaction of a chiral substrate.

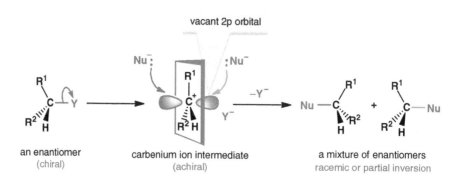

vacant 2p orbital

an enantiomer (chiral)   carbenium ion intermediate (achiral)   a mixture of enantiomers racemic or partial inversion

Example 12.4

Complete the following substitution reactions.

(a) [structure] Br + H₂O $\xrightarrow{H_2O}$ (b) [structure] I + MeOH $\xrightarrow{MeOH}$

Solution

(a) 2-Bromo-2-methylbutane is a tertiary alkyl bromide, and reacts by the S_N1 mechanism in a weakly nucleophilic solvent.

[structure] Br + 2 H₂O $\xrightarrow[H_2O]{S_N1}$ [structure] OH + H_3O^+ + Br⁻

2-bromo-2-methylbutane   2-methylbutan-2-ol

(b) Iodocyclohexane is a secondary halide with a good nucleofuge, and also reacts by the S$_N$1 mechanism in a weakly nucleophilic solvent.

$$\text{iodocyclohexane} + \text{MeOH} \xrightarrow[\text{MeOH}]{S_N1} \text{methoxycyclohexane} + \text{HI}$$

Note that HI, like HBr and HCl, dissociates extensively in MeOH:

$$\text{HI} + \text{MeOH} \rightleftharpoons \text{MeOH}_2^+ + \text{I}^-$$

Explain why the rate of hydrolysis of Me$_3$CBr in water can be measured by monitoring the acidity (pH) of the solution.

Exercise 12.9

Complete the following equations for solvolytic substitution reactions.

Exercise 12.10

(a) [structure with Cl] $+ \text{AcOH} \xrightarrow{\text{AcOH}}$ (b) [structure with Br] $+ \text{EtOH} \xrightarrow{\text{EtOH}}$

12.4.3 Stability of carbenium ions

Any feature which stabilizes the intermediate carbenium ion in an S$_N$1 reaction will also stabilize the transition structure for its formation (see the Hammond postulate in Panel 7.2). Stabilities of alkyl cations decrease in the order 3° > 2° >> 1°, so the S$_N$1 reactivity of alkyl halides, for example, decreases in the same order.

Stability of carbenium ions:

$$\underset{\text{R}}{\overset{\text{R}}{\underset{|}{\text{R}-\overset{+}{\text{C}}-\text{R}}}} > \underset{\text{H}}{\overset{\text{R}}{\underset{|}{\text{R}-\overset{+}{\text{C}}-\text{H}}}} > \underset{\text{H}}{\overset{\text{R}}{\underset{|}{\text{H}-\overset{+}{\text{C}}-\text{H}}}} > \underset{\text{H}}{\overset{\text{H}}{\underset{|}{\text{H}-\overset{+}{\text{C}}-\text{H}}}}$$

S$_N$1 reactivity of RY:

3°   >   2°   >>   1°   methyl

How is a carbenium ion centre stabilized by an alkyl substituent? Delocalization of the positive charge is the most likely explanation, and we can identify two possible mechanisms. On purely electrostatic grounds, we can expect a localized positive charge to polarize adjacent bonds and thereby attract electron density through the σ bonds. This is the so-called **inductive effect** we have encountered previously (Sub-section 6.3.3). Since the electrons around a carbon atom are more polarizable than those around a hydrogen atom, a C–C σ bond is more polarizable than a C–H bond, and better able to supply electron density to the positive centre. Alkyl substituents, therefore, stabilize a carbenium ion centre by acting as **electron-donating groups** (but note that this electron-donation is purely a response to the strong electron-attracting effect of the carbenium ion centre). Consequently, as the H atoms of a methyl cation are successively replaced by alkyl groups, the positive charge is increasingly dispersed into the alkyl groups, and the carbenium ion is stabilized.

The second mechanism by which alkyl substituents stabilize carbenium ions is best described by the molecular orbital theory of bonding. The vacant 2p AO of the carbenium

Simple primary alkyl and methyl cations are too unstable to exist in common solvents, so their derivatives do not react by the S$_N$1 mechanism.

A carbenium ion intermediate sometimes rearranges to a more stable ion which leads to rearrangement products. This complication will be discussed in Chapter 14 in the context of acid-catalysed reactions of alcohols.

Electron-donating inductive effect:

Figure 12.3 Stabilization of a carbenium ion by hyperconjugation.

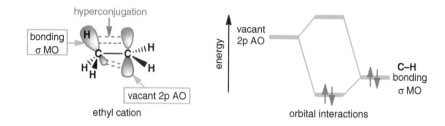

ethyl cation   orbital interactions

We encountered hyperconjugation as a contributor to the stabilization of the staggered conformation of ethane in Sub-section 4.2.1.

centre is able to interact with the bonding σ orbital of an adjacent C–H bond of the substituent alkyl group when the orbitals are coplanar, as illustrated in Figure 12.3. Although the interaction between the relatively low energy σ bonding MO and the vacant 2p AO is not large, there is a transfer of some C–H σ bonding electron density into the vacant 2p orbital of the carbenium ion: the alkyl group acts as an electron-donating group. In this way, the carbenium ion is stabilized by delocalization of positive charge from the central sp² carbon. This type of interaction, delocalization involving a σ orbital, is called **hyperconjugation**.

Carbenium ions are more strongly stabilized by **conjugation** of the vacant p orbital with adjacent π bonds or unshared electron pairs (see Chapter 5), so a compound which yields a conjugatively stabilized carbenium ion will be a reactive S_N1 substrate. For example, rates of S_N1 reactions increase in the order:

Exercise 12.11

Write resonance forms for allylic, benzylic, and 1-methoxyethyl cations formed during the reactions of the substrates given above.

Exercise 12.12

What are the organic products in the reaction of 1-chloro-1-methoxyethane in aqueous solution? Write a mechanism for this reaction.

Electronic stabilization of a carbenium ion is not the only reason for the relatively easy formation of tertiary carbenium ions by heterolysis: there is a steric reason as well. The hybridization of the central carbon changes from sp³ to sp² upon formation of a carbenium ion by heterolysis of its precursor. Consequently, the bond angles around the central carbon widen from about 109.5° to 120° so there is a relief of steric strain as substituent alkyl groups are able to move further apart (Figure 12.4). This **relief of steric strain**, which facilitates bond cleavage in the first step of the S_N1 mechanism (*steric acceleration*), is largest for a tertiary alkyl substrate.

This is the opposite of what happens in nucleophilic addition to a carbonyl group (Section 8.2).

The highly congested tri(*t*-butyl)methyl chloride, (Me₃C)₃CCl, is a particularly reactive S_N1 substrate—it is much more reactive than *t*-butyl chloride.

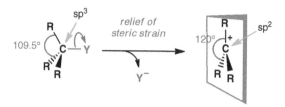

Figure 12.4 Relief of steric strain upon heterolysis in an S_N1 reaction.

Panel 12.3 The S_N1 mechanism in biological substitution reactions

The S_N1 mechanism occurs when a substrate has a good nucleofuge and gives a relatively stable carbenium ion intermediate. The most commonly found good nucleofuges in nature are phosphate and diphosphate (also called pyrophosphate), which are conjugate bases of strong acids (phosphoric and diphosphoric acid, respectively). The usual abbreviation for diphosphate (pyrophosphate) as a substituent is PPO (or OPP): this polyprotic acid residue will dissociate and bear a charge (which is not usually indicated) according to the pH of its aqueous environment. Then, following departure, diphosphate is represented as PPO⁻ regardless of its actual charge which, again, will depend upon the pH.

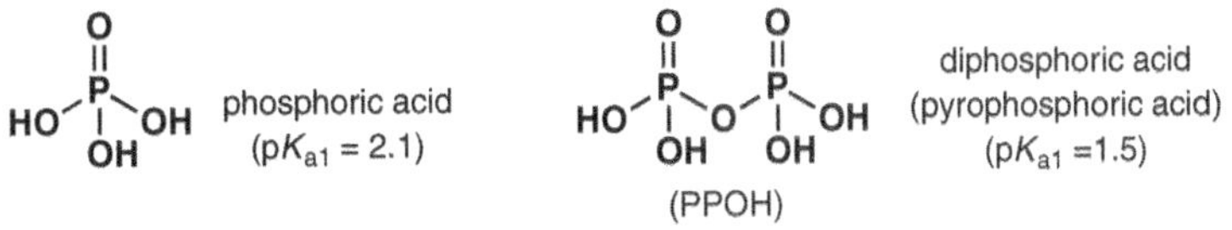

3-Methylbut-2-enyl (dimethylallyl) diphosphate is a precursor for the terpene class of natural products (also called isoprenoids) which are built up from branched C_5 (isoprene) subunits (see Chapter 24). In addition to having a good nucleofuge, dimethylallyl diphosphate is a good substrate for the S_N1 mechanism because the intermediate dimethylallyl cation is resonance-stabilized. Nucleophilic centres which capture this intermediate in biological S_N1 reactions include the OH groups of alcohols, and electron-rich carbon residues.

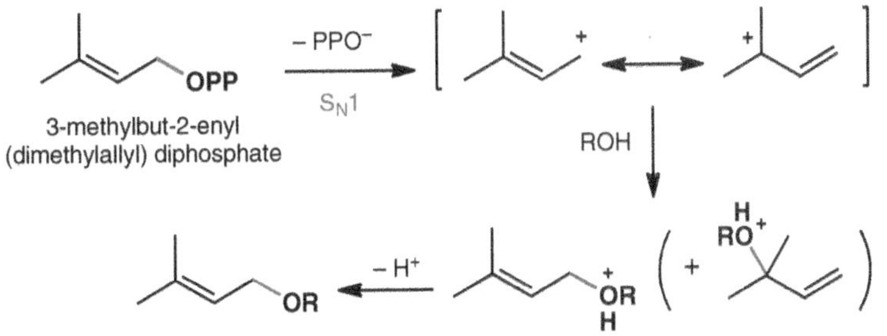

When the nucleophile is 3-methylbut-3-enyl diphosphate (an isomer of dimethylallyl diphosphate), the product immediately following deprotonation is geranyl diphosphate, the hydrolysis product of which is geraniol—a component of rose oil.

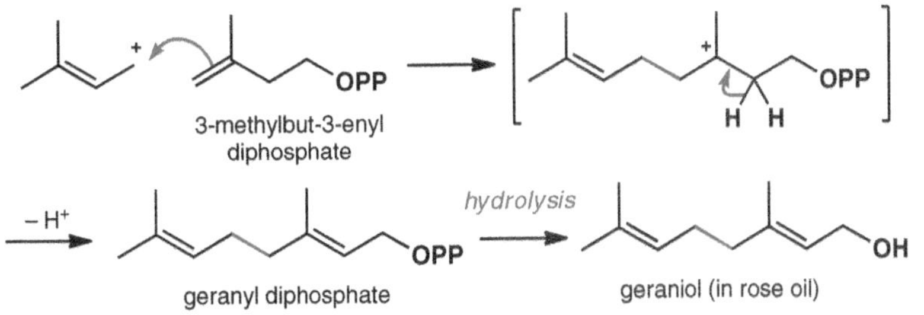

Geranyl diphosphate is similar in reactivity to its dimethylallyl analogue as an S_N1 substrate and reacts in the same way with a further molecule of the carbon nucleophile, 3-methylbut-3-enyl diphosphate. Deprotonation of the intermediate gives farnesyl diphosphate with the additional isoprene unit, and hydrolysis yields farnesol, which is found in lily of the valley. Still higher members of this family of natural products are produced in a similar manner.

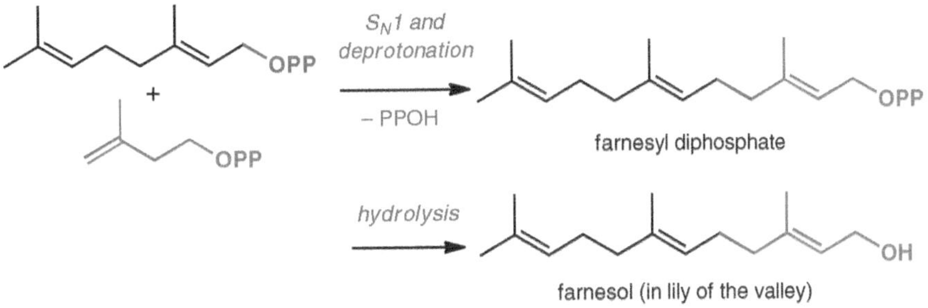

Exercise 12.13

Which of each pair of haloalkanes is more reactive in an S_N1 reaction?

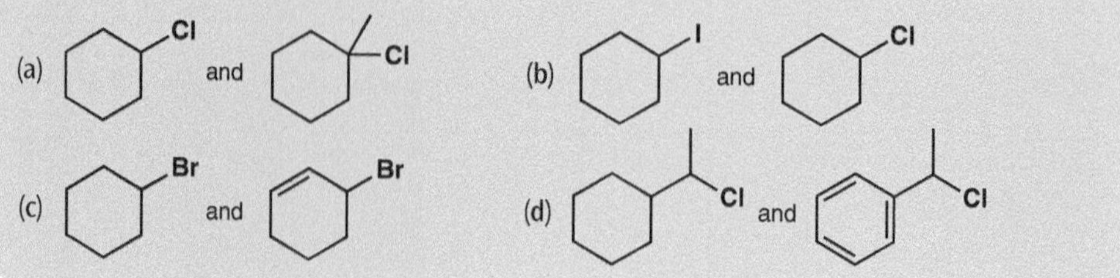

(a) [structure] and [structure] (b) [structure] and [structure]

(c) [structure] and [structure] (d) [structure] and [structure]

12.5 Intramolecular Nucleophilic Displacement: Neighbouring Group Participation

See Section 14.5 for the chemistry of oxiranes (epoxides).

The normal S_N2 mechanism is intermolecular with a bimolecular transition structure. But suppose the molecule with the nucleofuge also contains a nucleophilic group—could the internal nucleophile compete with an intermolecular nucleophile?

Following extended reaction in alkaline solution, *trans*-1-chloro-2-hydroxycyclohex-ane yields *trans*-1,2-dihydroxycyclohexane, i.e. replacement of Cl by OH with *reten-tion* of configuration. The most credible explanation of this stereochemical result is the mechanism shown in Scheme 12.6 involving a bicyclic intermediate containing a three-membered cyclic (oxirane) function. Isolation of the intermediate (cyclohexene oxide) under less forcing conditions supports this mechanism.

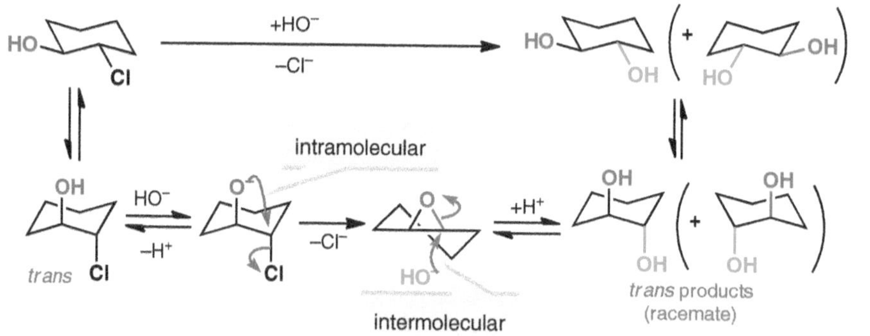

Scheme 12.6 Nucleophilic displacement of chloride by an internal nucleophile and subsequent reaction.

*A single half-chair form of the oxirane which has appreciable conformational flexibility is shown in Scheme 12.6 and the time-averaged structure has a mirror plane.

The alkoxide, generated in a pre-equilibrium, displaces Cl⁻ to give an oxirane in an intramolecular nucleophilic displacement reaction. This is an example of **neighbouring group participation** (n.g.p.), and the oxirane then suffers ring opening by intermolecular nucleophilic attack by HO⁻. The double substitution, each with inversion, accounts for the overall retention of configuration. Note, however, that the HO⁻ can attack either of the two electrophilic carbon atoms;* consequently, regardless of the enantiomeric purity of the reactant, the product must be racemic.

In contrast, the diastereoisomeric *cis*-1-chloro-2-hydroxycyclohexane (in which the internal nucleophile cannot reach the rear side of the C bearing the nucleofuge) yields the *same* product, *trans*-1,2-dihydroxycyclohexane, in a simple S_N2 reaction (Scheme 12.7). The rate constant for this reaction, however, is about 100 times smaller than the one for the *trans* diastereoisomer, and the product from enantiomerically enriched starting material would be optically active.

The rate enhancement associated with neighbouring group participation, as found for *trans*-1-chloro-2-hydroxycyclohexane in alkaline solution, is sometimes called *anchimeric assistance.*

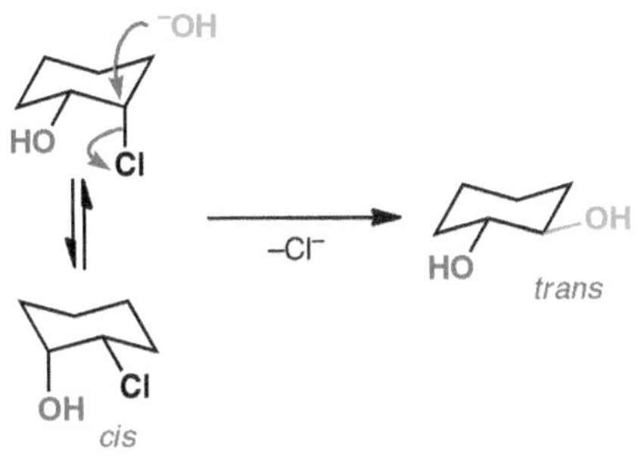

Scheme 12.7 Intermolecular nucleophilic displacement of chloride by hydroxide.

Neighbouring group participation may also be observed in S_N1 reactions *via* carbenium ion intermediates. The solvolysis of *trans*-1-acetoxy-2-bromocyclohexane in ethanoic acid containing silver ethanoate is appreciably faster than expected, and the product is *trans*-1,2-diacetoxycyclohexane, i.e. Br^- has been replaced by ethanoate with retention of configuration. The stereochemical outcome and the enhanced rate are classic evidence of n.g.p. in a substitution reaction, and a reasonable mechanism is shown in Scheme 12.8. The departure of the nucleofuge, Br^-, is assisted by nucleophilic participation by the acetoxy group, then the bicyclic cationic intermediate is captured at either of the equivalent electrophilic ring carbons to give the product.

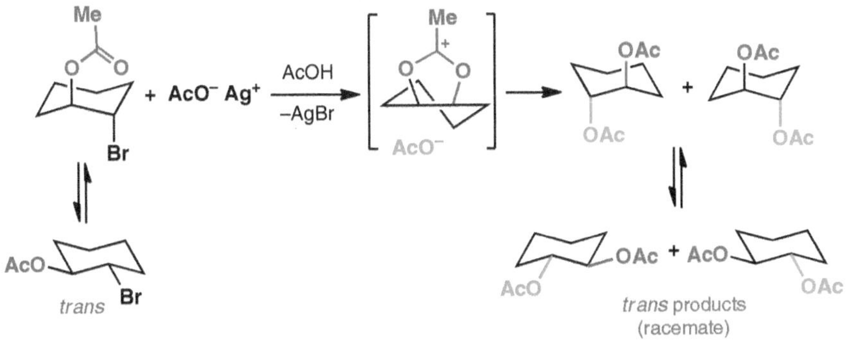

Ag^+ ions complex the Br to form a more effective nucleofuge in the solvolysis of an alkyl bromide. Note also that the products are an equimolar mixture of enantiomers (racemate) due to the symmetry of the intermediate even if the starting material is the pure stereoisomer.

Scheme 12.8 Neighbouring group participation in the solvolysis of *trans*-1-acetoxy-2-bromocyclohexane in ethanoic acid containing AcOAg.

Note that n.g.p. always involves formation of a cyclic intermediate, e.g. the oxirane in Scheme 12.6 and the resonance-stabilized acetoxonium ion in Scheme 12.8.

The neighbouring acetoxy group in *cis*-1-acetoxy-2-bromocyclohexane is unable to approach the rear side of the C which bears the Br in either chair conformation, so n.g.p. is not possible. Consequently, its substitution reaction under the same solvolytic conditions is much slower and gives predominantly inversion of configuration (Scheme 12.9).

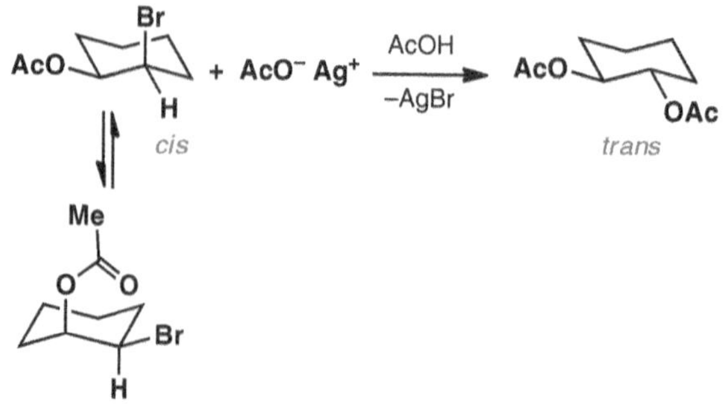

Scheme 12.9
Solvolysis of *cis*-1-acetoxy-2-bromocyclohexane in ethanoic acid containing AcAg.

Exercise 12.14

If the solvolyses of enantiomerically enriched *cis*- and *trans*-1-acetoxy-2-bromocyclohexanes were monitored in polarimeters, explain what you would expect to observe.

Internal nucleophilic groups able to participate as neighbouring groups are not limited to those with lone pairs, as we shall see in Chapter 22. We shall also explore in more detail in that chapter the formation of rearrangement products in reactions involving neighbouring group participation.

12.6 Competition between S_N1 and S_N2 Mechanisms

Characteristic features of S_N1 and S_N2 mechanisms are compared in Table 12.1 which we use below to summarize and review what we have covered so far about nucleophilic substitution reactions at saturated carbon.

Simple primary alkyl substrates do not react by the S_N1 mechanism because primary alkyl cations are too unstable to form in solution, and tertiary alkyl substrates are too sterically hindered to react by the S_N2 mechanism. Between these, secondary alkyl substrates may react by S_N1 and/or S_N2 mechanisms, depending on the reaction conditions.

One defining difference between the two mechanisms is that the nature of the nucleophile and its concentration affect the rate of an S_N2 reaction, but not of an S_N1. So, good nucleophiles favour the S_N2 mechanism, whereas the S_N1 mechanism is more likely in weakly nucleophilic reaction conditions. Polar aprotic solvents are good for S_N2 reactions and, as they are usually only weakly nucleophilic, they will not compete effectively with a nucleophilic solute. Polar protic solvents, such as water and lower alcohols and

Table 12.1

Comparison of S_N1 and S_N2 mechanisms		
	S_N1	S_N2
Rate-determining step	unimolecular	bimolecular
Rate law	1st order	2nd order
Mechanistic steps	two steps	one step
Intermediate	carbenium ion	none
Reactivity	3° > 2° >> 1°, methyl (carbenium ion stability)	methyl > 1° > 2° >> 3° (steric hindrance)
Effect of nucleophile	none	faster with better
Good solvent	polar protic	polar aprotic
Stereochemistry	racemization/partial inversion	inversion

carboxylic acids, are good for S_N1 reactions because their polarity facilitates formation of carbenium ion intermediates which are stabilized by solvation (electron-pair donation). S_N1 reactions in which the solvent is an active participant (a solvent molecule acts as a nucleophile in the product-determining step) are called **solvolytic reactions**, or **solvolyses** (e.g. hydrolysis in water, methanolysis in methanol, etc.).

A second defining difference is that an S_N2 mechanism involves complete inversion of configuration at the reaction site, and often gives a high yield of a single product. In contrast, the S_N1 mechanism involves an achiral carbenium ion intermediate (initially within an ion pair), even if the starting material is chiral at the reaction site. Consequently, the product is always racemic to some degree (often a high degree) but with an excess of inversion of configuration due to the departing nucleofuge inhibiting nucleophilic capture with retention. S_N1 reactions may give a multiplicity of products, including alkenes by elimination (see next chapter), which reduces their synthetic value.

Example 12.5

Predict a mechanism for each of the following reactions and give structures of substitution products with stereochemistry where applicable.

(a) 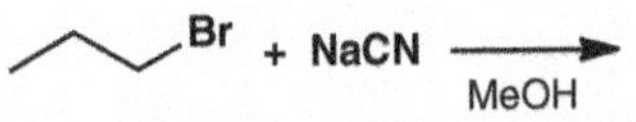 + NaCN $\xrightarrow{\text{MeOH}}$

(b) 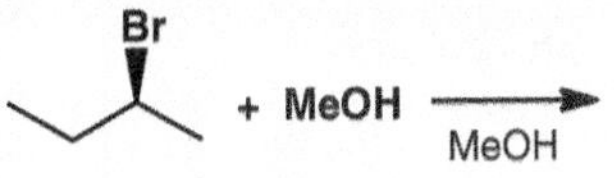 + MeOH $\xrightarrow{\text{MeOH}}$

(c) 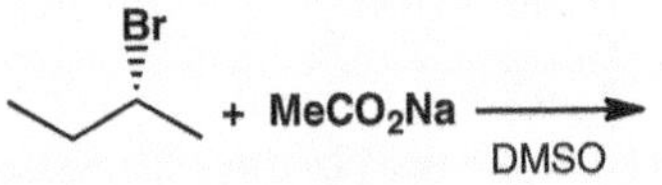 + MeCO$_2$Na $\xrightarrow{\text{DMSO}}$

(d) 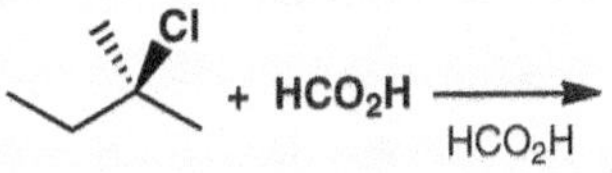 + HCO$_2$H $\xrightarrow{\text{HCO}_2\text{H}}$

Solution

(a) Reaction of a primary alkyl bromide with a good nucleophile proceeds by the S_N2 mechanism. Products are:

+ NaBr

butanenitrile

(b) Solvolysis of a secondary alkyl bromide in methanol (no nucleophilic solute) proceeds by the S_N1 mechanism. The product is the largely racemic ether.

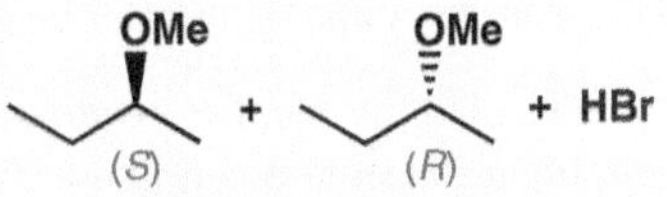

+ + HBr

(S)   (R)

2-methoxybutane

(c) Ethanoate (acetate) ion is usually a poor nucleophile in hydrogen-bonding solvents, but is sufficiently good in the aprotic DMSO to give an S_N2 reaction with a secondary alkyl bromide with inversion of configuration.

+ NaBr

(S)-1-methylpropyl ethanoate

(d) Solvolysis of a tertiary (achiral) alkyl chloride in the highly polar and only moderately nucleophilic methanoic acid (formic acid) occurs by the S_N1 mechanism.

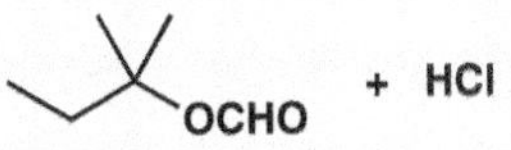

+ HCl

1,1-dimethylpropyl methanoate

Exercise 12.15

Predict a mechanism for each of the following reactions and give structure(s) of substitution product(s) with stereochemistry where applicable.

(a) 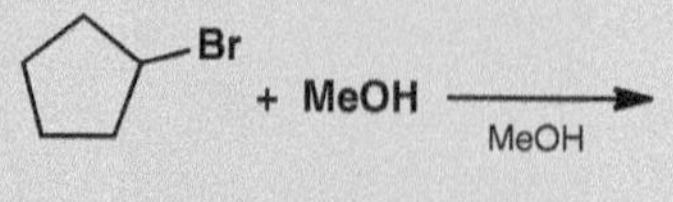 (cyclopentyl–Br) + MeOH $\xrightarrow{\text{MeOH}}$

(b)  (cyclopentyl–Br) + NaSH $\xrightarrow{\text{propanone}}$

(c) 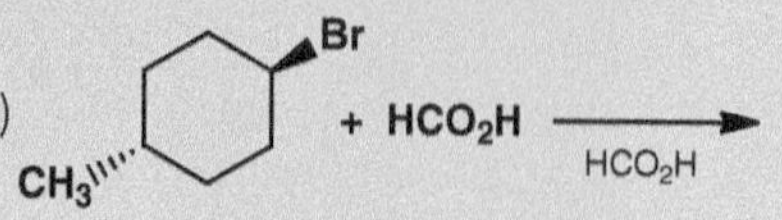 (CH_3-cyclohexyl–Br) + HCO_2H $\xrightarrow{HCO_2H}$

(d) 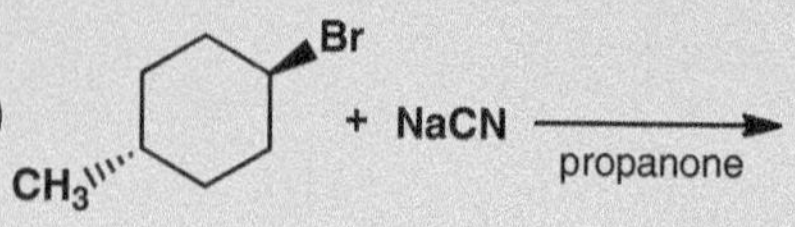 (CH_3-cyclohexyl–Br) + NaCN $\xrightarrow{\text{propanone}}$

Summary

- The C–Y bond of haloalkanes and related compounds RY is polar, and the partially positive (electrophilic) carbon is susceptible to direct **nucleophilic** attack.

- Nucleophilic attack occurs at the C from the side opposite to the C–Y bond by the S_N2 mechanism, whereupon the **nucleofuge** (leaving group, Y⁻) is displaced to give a substitution product with stereochemical **inversion**.

- Being bimolecular, the **single-step** S_N2 mechanism is subject to **steric hindrance** and the reactivity generally decreases in the order methyl > 1° > 2° alkyl derivatives; 3° alkyl derivatives do not react by the S_N2 mechanism.

- Solvents have major effects upon the rate constant of a reaction when there is an appreciable difference in polarity between reactant molecule(s) and the (rate-determining) transition structure. Polar aprotic solvents are generally good for S_N2 reactions with anionic nucleophiles.

- The alternative S_N1 mechanism occurs in polar but weakly nucleophilic/basic solution by initial **unimolecular** heterolysis of the C–Y bond to give a **carbenium ion intermediate** which is trapped by a nucleophile (usually the solvent) in the second step to complete the substitution; the more stable the intermediate carbenium ion, the faster the reaction.

- Reactivity by the S_N1 mechanism decreases in the order, 3° > 2° for simple alkyl derivatives; allylic and benzylic derivatives are particularly reactive whereas simple primary alkyl substrates do not react by the S_N1 mechanism.

- The rate constant of a reaction is appreciably enhanced in the event of **neighbouring group participation** by an internal nucleophilic group; this involves intramolecular rate-determining formation of a cyclic intermediate which is then trapped by external nucleophiles to give the final products.

Problems

12.1 Complete the following equations.

(a) 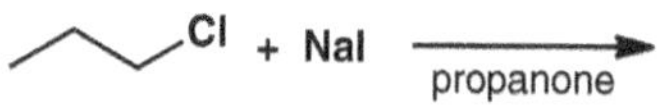 (propyl)Cl + NaI $\xrightarrow{\text{propanone}}$

(b) 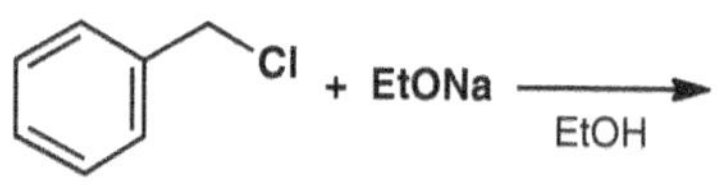 (benzyl)Cl + EtONa $\xrightarrow{\text{EtOH}}$

(c) 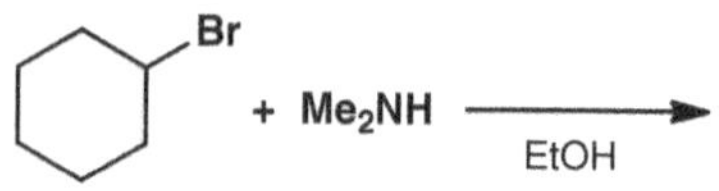 (cyclohexyl–Br) + Me_2NH $\xrightarrow{\text{EtOH}}$

(d)  (sec-butyl–Br) + EtSNa $\xrightarrow{\text{EtOH}}$

12.2 What is the main product in each of the following substitution reactions?

(a) 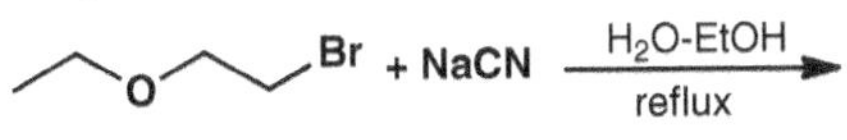 [structure: CH₃CH₂-O-CH₂CH₂-Br] + NaCN →(H₂O-EtOH, reflux)

(b) 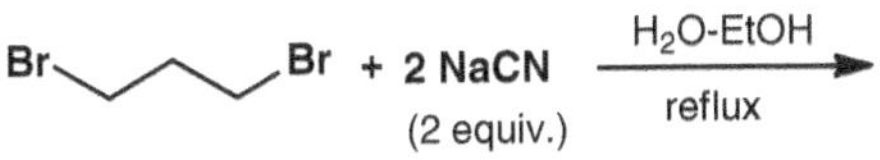 Br[structure] Br + 2 NaCN (2 equiv.) →(H₂O-EtOH, reflux)

(c) PhCH₂Cl + PhNH₂ →(NaHCO₃, H₂O, 90–95 °C)

(d) 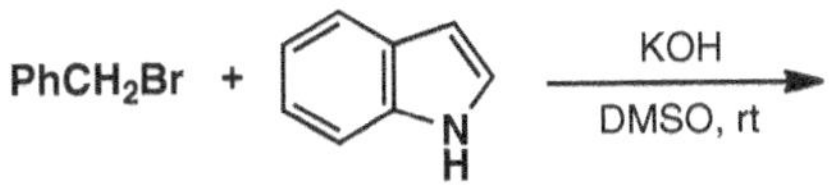 PhCH₂Br + [indole structure] →(KOH, DMSO, rt)

(e) 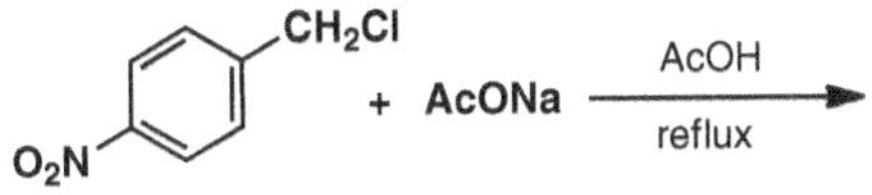 [structure: O₂N-C₆H₄-CH₂Cl] + AcONa →(AcOH, reflux)

(f) 2 PhCH₂Cl + Na₂S (2 equiv.) →(95% ethanol, reflux)

12.3 Draw the structure of the main product in each of the following reactions, and give its IUPAC name including its R,S stereochemical designation.

(a) 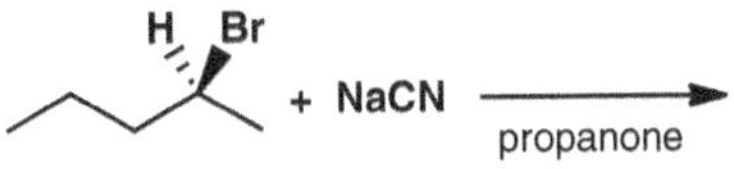 [structure with H, Br] + NaCN →(propanone)

(b) 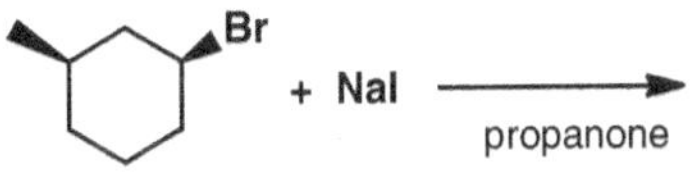 [cyclohexane with Br] + NaI →(propanone)

(c) 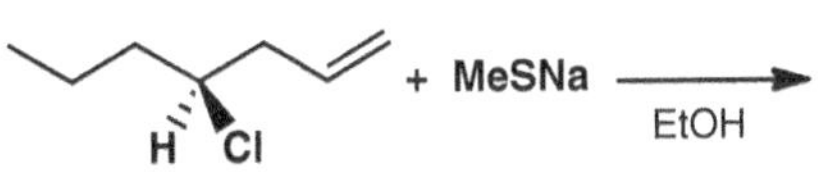 [structure with H, Cl] + MeSNa →(EtOH)

(d) 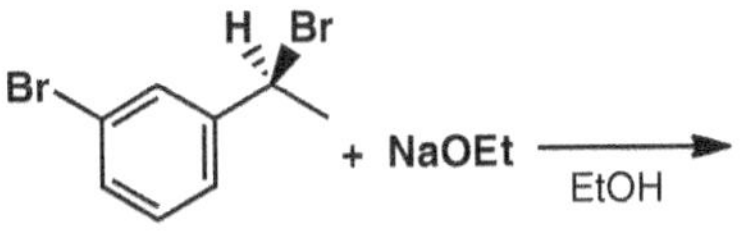 [structure Br-C₆H₄ with H, Br] + NaOEt →(EtOH)

12.4 Which of each of the following will be the faster S$_N$2 reaction?

(a) 1-Bromobutane or 2-bromobutane with sodium iodide in propanone.

(b) 2-Bromopentane or 2-iodopentane with sodium cyanide in ethanol.

(c) Ethanamine or 2-methylpropan-2-amine (*t*-butylamine) with 1-bromobutane.

(d) Sodium ethoxide or sodium ethanoate (acetate) with iodomethane.

12.5 Which of each of the following pairs of haloalkanes will be the more reactive S$_N$1 reactant, and explain why?

(a) 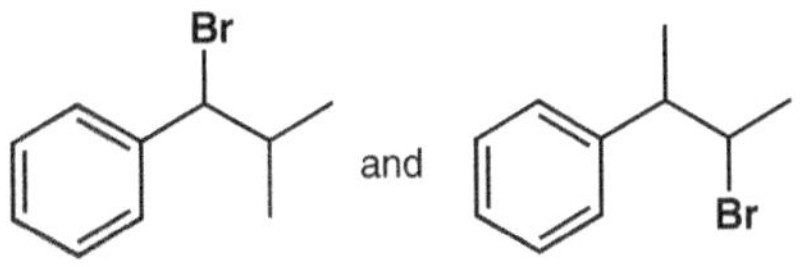 [structure with Br] and [structure with Br]

(b) 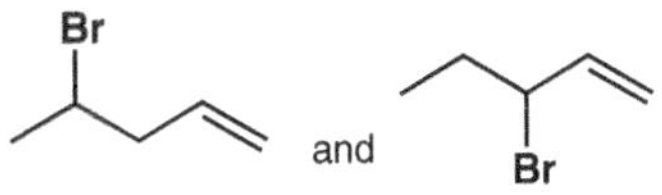 [structure with Br] and [structure with Br]

(c) 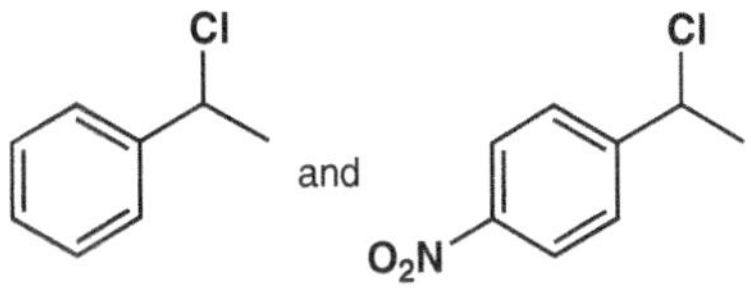 [structure with Cl] and [O₂N structure with Cl]

(d) 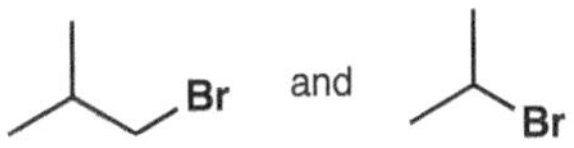 [structure with Br] and [structure with Br]

12.6 When benzyl and *p*-methoxybenzyl bromides are heated in ethanol, the corresponding ethyl ethers are formed. Which of the two bromides reacts faster, and explain why?

12.7 What is the main product from each of the following dihalides in the reactions indicated?

(a)  Cl[structure]Br + KCN (1 equiv.) →(EtOH-H₂O, reflux)

(b) 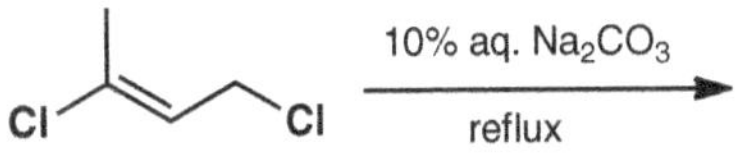 [structure: Cl-C=C-CH₂Cl] →(10% aq. Na₂CO₃, reflux)

(c) Br-[C₆H₄]-CH₂Br + Me₂NH →(propanone, reflux)

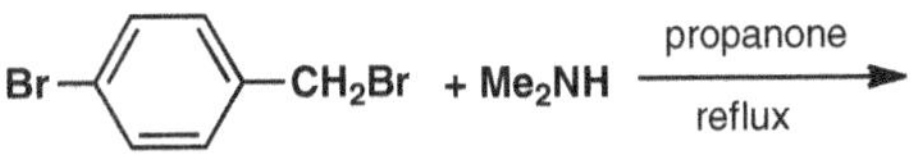

12.8 Draw a free energy profile for the reaction in ethanol given in Exercise 12.6(a), and show how the profile would be different for the reaction in water.

12.9 Draw a free energy profile for the reaction of iodomethane with dimethylamine, and use it to explain how a change from a less to a more polar solvent would affect the rate constant of the reaction.

12.10 When 1,4-dibromobutane reacts with Na₂S in ethanol, a product with the molecular formula C₄H₈S is obtained. Give the structure of the product by showing how it is formed.

12.11 Explain the following reaction and comment on the outcome if the reactant had been a single enantiomer.

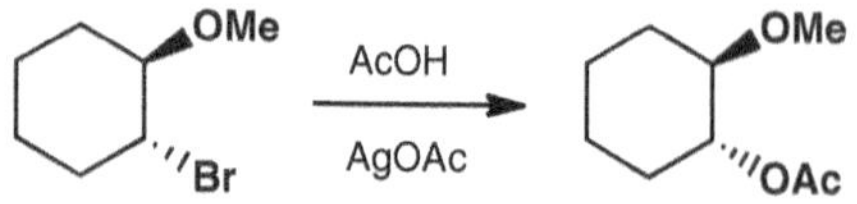

12.12 Give a mechanism for the following reaction and indicate the stereochemistry of the main product.

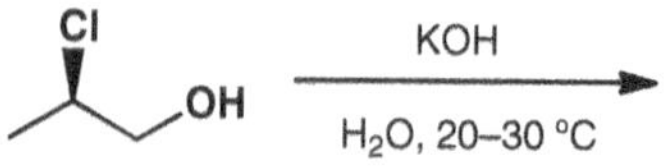

Elimination Reactions of Haloalkanes and Related Compounds

13

Related topics we have already covered:

- Molecular strain and steric hindrance (Section 4.2)
- Curly arrow representation of reaction mechanisms (Section 7.2)
- Energetics of reaction (Section 7.4)
- Characterization of organic reactions and reaction mechanisms (Section 7.5)
- Stereochemistry (Chapter 11)
- S_N1 and S_N2 nucleophilic substitution at saturated carbon atoms (Chapter 12)

Topics we shall cover in this chapter:

- Mechanisms of elimination reactions: E1, E2, and E1cB
- Regioselectivity in elimination reactions
- Competition between substitution and elimination

Under nucleophilic/basic conditions, haloalkanes and related compounds, RY, frequently undergo elimination reactions to give alkenes along with nucleophilic substitution reactions; both involve heterolytic departure of the nucleofuge Y^- from the electrophilic α C. In elimination, however, departure of the nucleofuge is associated with proton abstraction from the β C rather than with the bonding of a nucleophile to the α C which occurs in nucleophilic substitution. The elimination of HCl from 2-chloro-2-methylpropane is an example given in Chapter 7:

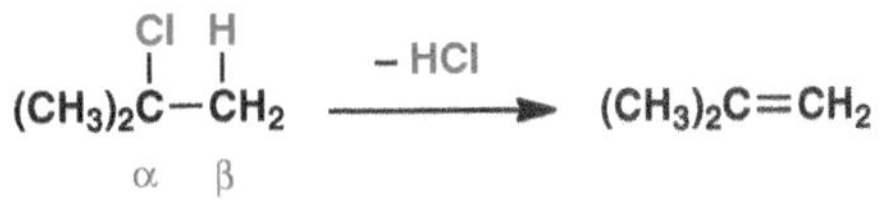

$$(CH_3)_2\overset{\underset{|}{Cl}}{\underset{\alpha}{C}}-\overset{\underset{|}{H}}{\underset{\beta}{CH_2}} \quad \xrightarrow{-HCl} \quad (CH_3)_2C=CH_2$$

Just as in substitution, there is some mechanistic diversity amongst elimination reactions and, because nucleophiles often also act as bases, substitution and elimination reactions of compounds RY are generally in competition with each other. In this chapter, we cover the mechanisms of elimination reactions of RY (principally haloalkanes) to give alkenes, and explore the relationships between these and substitution reaction mechanisms covered in the previous chapter.

13.1 The E1 Elimination Mechanism

We saw in the previous chapter that tertiary alkyl compounds such as *t*-butyl bromide undergo substitution reactions by the S_N1 mechanism; however, solvolysis products in alcohols and water were also found to include appreciable amounts of an alkene elimination product, as exemplified by the reaction in methanol (eqn 13.1).

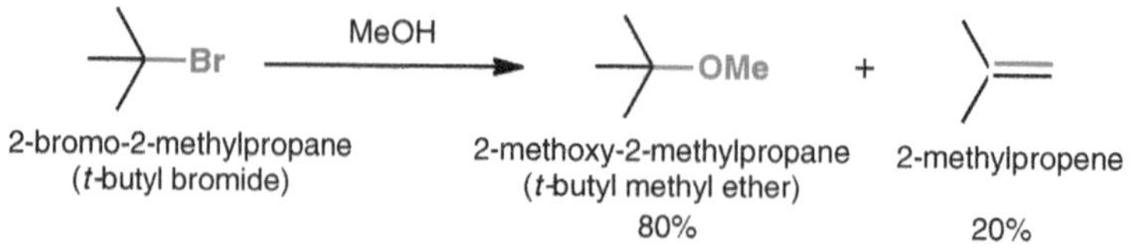

$$(13.1)$$

2-bromo-2-methylpropane
(*t*-butyl bromide)

2-methoxy-2-methylpropane
(*t*-butyl methyl ether)
80%

2-methylpropene

20%

With only low concentrations of methoxide, the reaction remains first order in just *t*-butyl bromide, so the elimination reaction accompanying the substitution must also be unimolecular. The most economical explanation is that elimination and substitution are parallel routes *after* a common rate-determining unimolecular ionization of the neutral substrate. When the focus is on the alkene-forming path, and sometimes this is predominant, the mechanism is called **unimolecular elimination (E1)**. In other words, there is a single initial rate-determining ionization to give an *intermediate carbenium ion which then undergoes either nucleophilic capture (S$_N$1 substitution), or proton loss (E1 elimination)*, as illustrated in Scheme 13.1.

Scheme 13.1 Composite S$_N$1 and E1 mechanisms via a common carbenium ion intermediate.

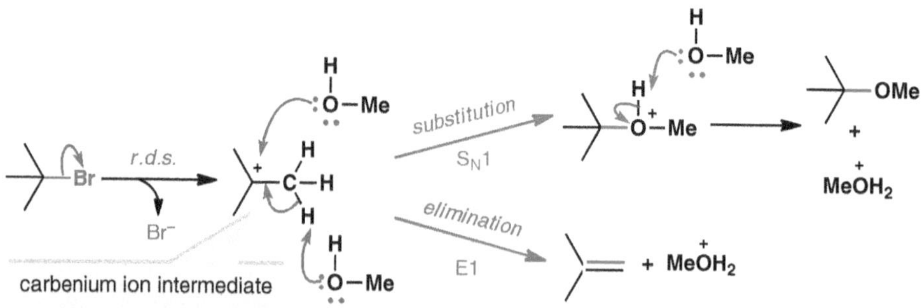

The partitioning of the carbenium ion intermediate between S$_N$1 and E1 paths occurs in the *product-determining* (or *product-forming*) *steps* following the rate-determining step. A methanol molecule either attacks the carbenium ion centre to give the substitution product (S$_N$1), or it abstracts a proton from the adjacent methyl group to provide the alkene elimination product (E1). In these alternative parallel paths, the solvent molecule acts *either* as a nucleophile (Lewis base) to form a new bond, or as a (Brønsted) base to deprotonate the carbenium ion. In most cases, therefore, S$_N$1 substitution and E1 elimination occur competitively.

Example 13.1

Give the elimination and substitution products in the following reactions.

(a)

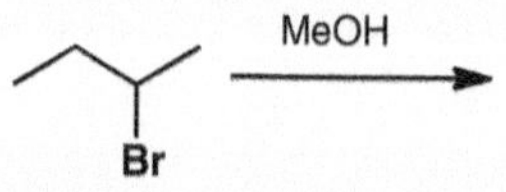

(b)

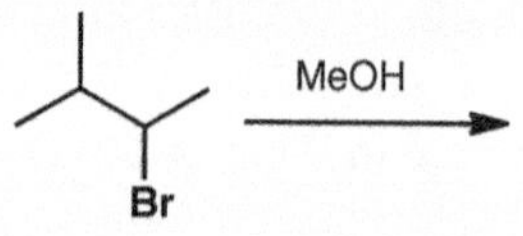

Solution

When there are different alkyl groups bonded to the C bearing the nucleofuge, constitutionally isomeric alkenes are usually obtained; this issue of **regioselectivity** is discussed below in Section 13.5.

(a) In addition to 2-methoxybutane and but-1-ene, *E* and *Z* isomers of but-2-ene are formed.

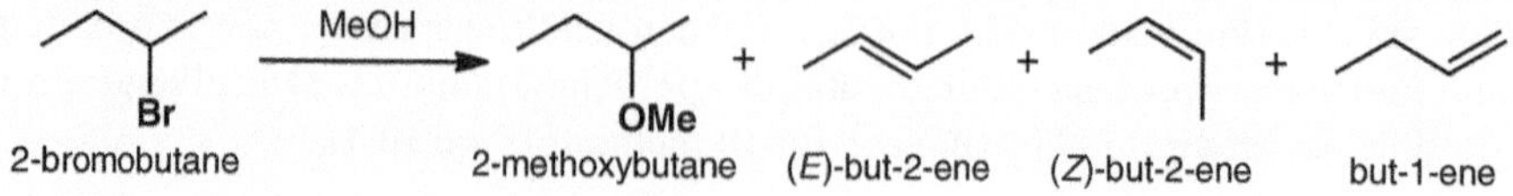

2-bromobutane

2-methoxybutane   (*E*)-but-2-ene   (*Z*)-but-2-ene   but-1-ene

(b) 2-Methoxy-3-methylbutane is accompanied by constitutionally isomeric alkenes.

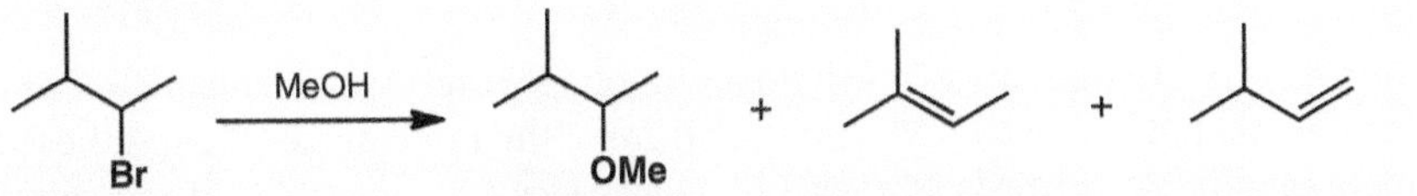

| 2-bromo-3-methylbutane | 2-methoxy-3-methylbutane | 2-methylbut-2-ene | 3-methylbut-1-ene |

Exercise 13.1

Give the elimination and substitution products in the following reactions.

(a)

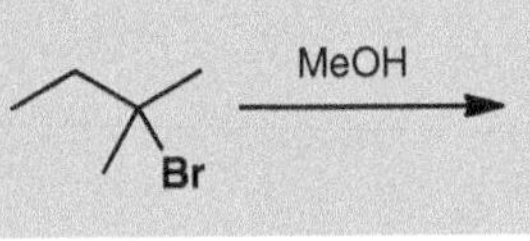

(b)

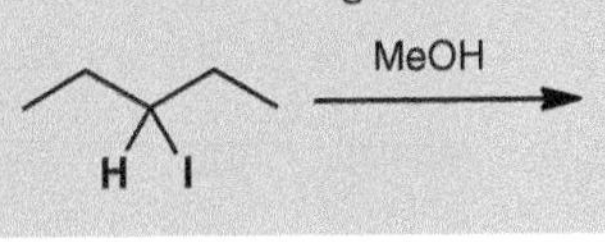

13.2 The E2 Elimination Mechanism

In the E1 elimination reaction of *t*-butyl bromide described above, the alkene is formed by deprotonation of the intermediate carbenium ion by a weak base, usually the solvent. Under different conditions, the same haloalkane can give the alkene *directly*, but this requires the involvement of a strong base (eqn 13.2). In general, such reactions are first order in haloalkane and in base, so are **second order** overall (eqn 13.3).

$$(CH_3)_3CBr + HO^- \xrightarrow{H_2O} (CH_3)_2C{=}CH_2 + H_2O + Br^- \qquad (13.2)$$

$$\text{Reaction rate} = k[\text{RY}]\,[\text{base}] \qquad (13.3)$$

The rate law of eqn 13.3 indicates that the transition structure in the rate-determining step involves one molecule of haloalkane and one of the base, and the simplest possibility (which also accommodates other evidence—see later) is a single-step **bimolecular elimination (E2)**.

A tertiary alkyl halide alone in aqueous solution will invariably undergo a first-order $S_N1/E1$ solvolytic reaction (Scheme 13.1) but this will be dominated by the second-order reaction at high concentrations of the base, $[HO^-]$ (eqn 13.3). This reaction is essentially wholly E2 elimination (eqn 13.2) because the alternative second-order reaction, S_N2 substitution, is too slow owing to steric hindrance. However, under these reaction conditions (high concentrations of base/nucleophile), secondary and primary alkyl halides may undergo both second-order reactions, E2 and S_N2 (they cannot be distinguished simply from the rate law). The proportions, which can be determined from the product analysis, will depend on the substrate structure and the relative basicity/nucleophilicity of the reagent. Contrasting examples are given in eqns 13.4 and 13.5.

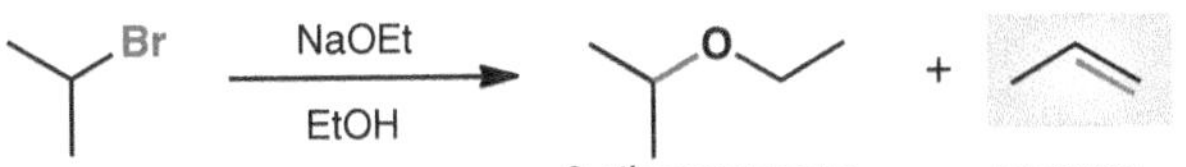

(13.4)

| 2-bromopropane 2° | | 2-ethoxypropane (ethyl isopropyl ether) 13% | propene 87% |

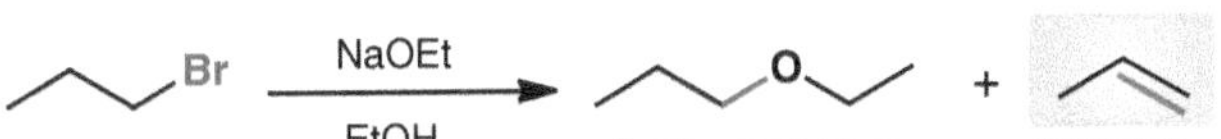

(13.5)

| 1-bromopropane 1° | | 1-ethoxypropane (ethyl propyl ether) 91% | propene 9% |

13.2.1 Stereoelectronic description of the E2 mechanism

The bimolecular E2 elimination mechanism involves three events: abstraction of a proton, development of a new π bond, and departure of the nucleofuge, Y⁻, and these occur in a single step. In other words, the two bond cleavages are *concerted* with the formation of the new π bond and the new σ bond (from O to H), as shown in Scheme 13.2.

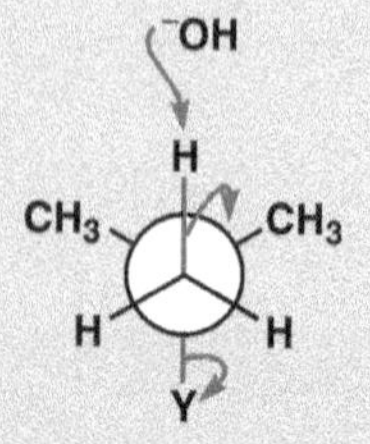

A Newman projection representation of the E2 mechanism in Scheme 13.2.

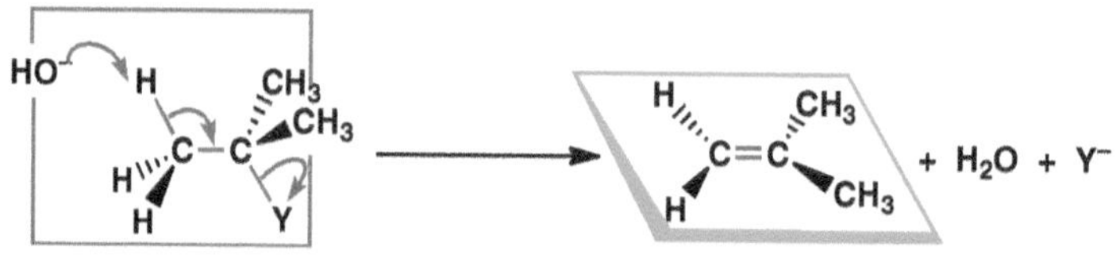

Scheme 13.2 Concerted bond cleavages and bond formations in the E2 mechanism of reaction 13.2.

For optimal orbital interactions in the formation of the new π bond and the accompanying concerted C–H and C–Y bond cleavages, the molecular orbitals involved must be aligned to achieve maximum overlap. This requires that the reactants come together in a very particular way with the H–C–C–Y bonds of the haloalkane coplanar in an **antiperiplanar** conformation, as represented in Scheme 13.2 and the margin note. As a result, H⁺ is abstracted and Y⁻ departs from coplanar opposite ends of the bimolecular encounter complex, and the reaction is said to be an *anti* elimination. The relevant interacting orbitals are represented in Figure 13.1.

Abstraction of H⁺ by HO⁻ involves interaction of the HOMO (lone pair) of HO⁻ with the σ* orbital of the C–H bond (Figure 13.1(a)). This leads to the formation of two new orbitals: the occupied σ bonding orbital of the new O–H bond of the H_2O molecule, and its associated vacant σ* orbital. In the centre of the system, interaction of the σ orbital of the C–H bond with the σ* orbital of the C–Cl bond leads to the π and π* orbitals of the alkene product (Figure 13.1(b)). In the associated redistribution of electrons, the C–C π bond is formed and, at the same time, the C–Cl bond undergoes heterolysis with the bonding electron pair of the C–Cl bond becoming a lone pair of the Cl⁻.

Chronologically, it was the observed stereospecificity which led to our mechanistic understanding of E2 reactions and the subsequent stereoelectronic description.

There are specific stereochemical consequences of the E2 mechanism through the *anti* arrangement shown in Scheme 13.2 and Figure 13.1 which we can elucidate by looking at the formation of (*E*)- and (*Z*)-3-methylpent-2-enes from stereoisomers of 3-methyl-2-bromopentane. The reactions are illustrated in Scheme 13.3 in which reactants are drawn in conformations appropriate for *anti* elimination. *Which reactant gives which product is determined by the mechanism.* As shown, the *R,R* and *S,S* pair of enantiomers give the (*E*)-alkene, whereas the diastereoisomeric *R,S* and *S,R* pair of enantiomers give the (*Z*)-alkene: the E2 elimination mechanism is **stereospecific**.

Figure 13.1 Orbital interactions in the concerted elimination of HCl from a chloroalkane by HO⁻.

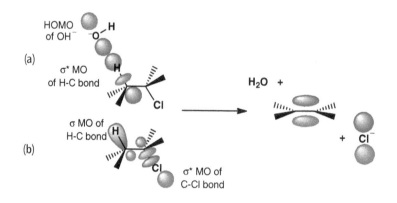

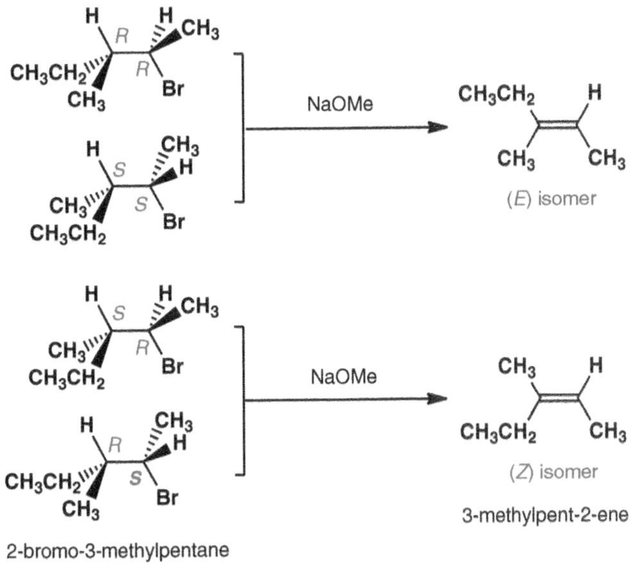

Scheme 13.3
Stereospecific E2 elimination of HBr from four stereoisomers of 3-methyl-2-bromopentane.

Give the structure of the elimination product when (2S,3R)-2-chloro-3-phenylbutane is treated with sodium ethoxide in ethanol.

Example 13.2

Solution
The mechanism is an *anti* E2 elimination. First, draw a stereochemical structure of the substrate with the antiperiplanar relationship of the H and Cl to be eliminated.

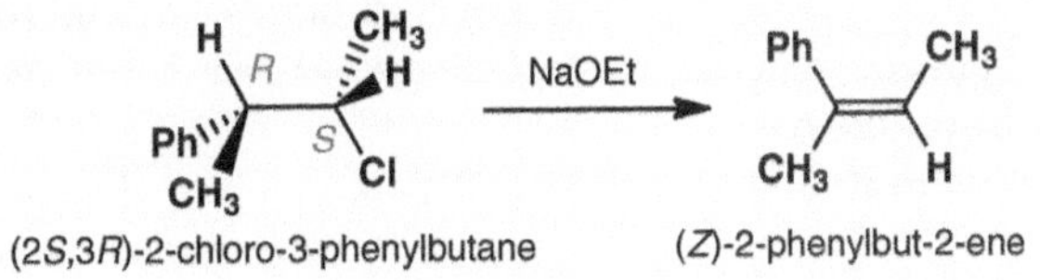

Show the structures of the other three stereoisomers of 2-chloro-3-phenylbutane and give the products when they are treated with sodium ethoxide in ethanol.

Exercise 13.2

13.3 The E1cB Elimination Mechanism and Graded Transition Structures in the E2 Mechanism

In the E2 elimination, abstraction of the proton and departure of the nucleofuge are concerted: they occur together in a single-step mechanism. In contrast, we saw in our earlier discussion of the E1 mechanism that heterolysis of the C–Y bond is complete *before* the proton is abstracted from the carbenium ion intermediate in the second step (eqn 13.6 in Scheme 13.4).

But what about another two-step reaction at the other mechanistic extreme? We can imagine abstraction of the proton occurring *first* to give an intermediate **carbanion** which *then* gives an alkene upon departure of the nucleofuge Y^- (eqn 13.7). This alternative stepwise elimination mechanism has been detected and is known as the **E1cB**. Here, 'cB' stands for 'conjugate base' since the carbanion intermediate is the conjugate base of the substrate, and the departure of Y^- from the conjugate base is unimolecular.

In Chapter 8, we encountered the reverse of cyanohydrin formation and the dehydration of carbonyl hydrates under basic conditions (Scheme 8.2 and eqn 8.7). The mechanisms of these reactions which form C=O bonds are very similar to the E1cB elimination mechanism which forms the C=C bond.

E1:

carbenium ion intermediate

(13.6)

E1cB:

carbanion intermediate

(13.7)

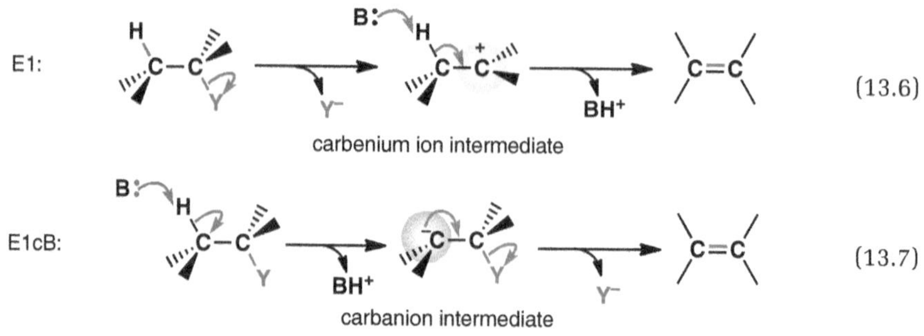

Scheme 13.4 Mechanistic stepwise extremes: the E1 and E1cB elimination mechanisms.

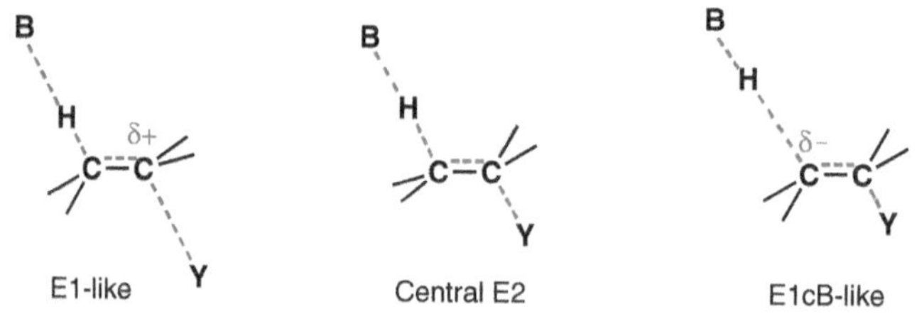

Figure 13.2 Different transition structures within the range for E2 eliminations.

E1-like   Central E2   E1cB-like

The E1cB mechanism is not very common, but the base-catalysed dehydration step in the aldol reaction (Chapter 17) is a typical and important example.

The two mechanistic extremes shown in Scheme 13.4 are both stepwise, each involving an intermediate, and to be contrasted with the concerted (single step) E2 alternative. However, although heterolyses of the C–H and C–Y bonds proceed *together* in the one step of the E2 mechanism, the *degrees of cleavage* of these bonds in the E2 transition structure (TS) may be different.

As illustrated in Figure 13.2, there can be an E1-like TS in which cleavage of the bond to the nucleofuge is well advanced but the proton is still largely bonded to the β C. This TS will be favoured by features which stabilize developing positive charge at the C bearing the nucleofuge. On the other hand, there can be an E1cB-like TS in which proton abstraction is well advanced whilst the bond to the nucleofuge is still largely intact. This TS will be favoured by features which stabilize developing negative charge at the C bearing the hydrogen.

When there are no features stabilizing either the E1-like TS or the E1cB-like TS, the E2 mechanism will follow the 'central' path with no appreciable build-up of charge at one C or the other in the TS.

13.4 Reaction Maps

Two-dimensional *reaction profiles* were introduced in Chapter 7 to describe the energetics of reactions. They may be seen as graphical representations of mechanisms, i.e. our views of how reactions proceed; as well as initial and final states, they include intermediates if reactions are stepwise. We saw, for example, how to use them to describe (and predict) solvent effects upon rate constants of reactions (Section 12.3). The y-axis of a reaction profile corresponds to the energy of the system and the x-axis is the reaction coordinate which is a composite *one-dimensional variable* corresponding to the progress of the reaction as reactants are transformed into products.

If we want to include alternative possible graded transition structures for an E2 mechanism (Figure 13.2) in a single graphical description of an elimination, however, we need *two* configurational coordinates—one corresponding to cleavage of the C–H bond and one corresponding to cleavage of the C–Y bond. These are the *x*- and *y*-axes of the three-dimensional 'chicken wire' energy surface diagram in Figure 13.3(a) for the elimination reaction of eqn 13.8 with the *z*-axis representing the energy of the system.

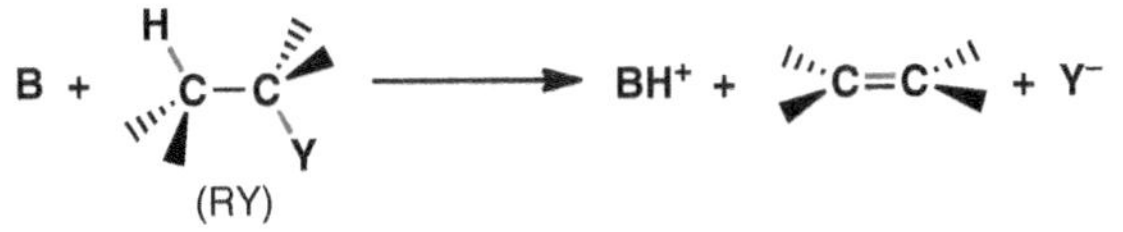

$$B + \overset{H}{\underset{Y}{\text{''''C}-\text{C}}} \longrightarrow BH^+ + \text{''''C}=\text{C'''} + Y^-$$

(13.8)

(RY)

The energy of the reaction system can also be represented by contour lines as in Figure 13.3(b) which is an example of a *reaction map*. The lower right corner corresponds to the carbenium ion intermediate formed by heterolysis of only the C–Y bond, while the upper left corresponds to the carbanion intermediate formed by cleavage of only the C–H bond. The path from the bottom left of the reaction map up the least steep energy gradient (cutting across the most widely spaced contours) corresponds to the trough diagonally across the energy surface of Figure 13.3(a); this path representing the E2 mechanism is of lower energy than alternatives corresponding to stepwise (E1 or E1cB) mechanisms via intermediates.

Different possible reaction paths for the elimination reaction involving unspecified compounds R–Y and B are shown in a simplified *single* reaction map (Figure 13.4), where the energy contours are implicit. Reaction path (1) represents the E1 mechanism *via* the

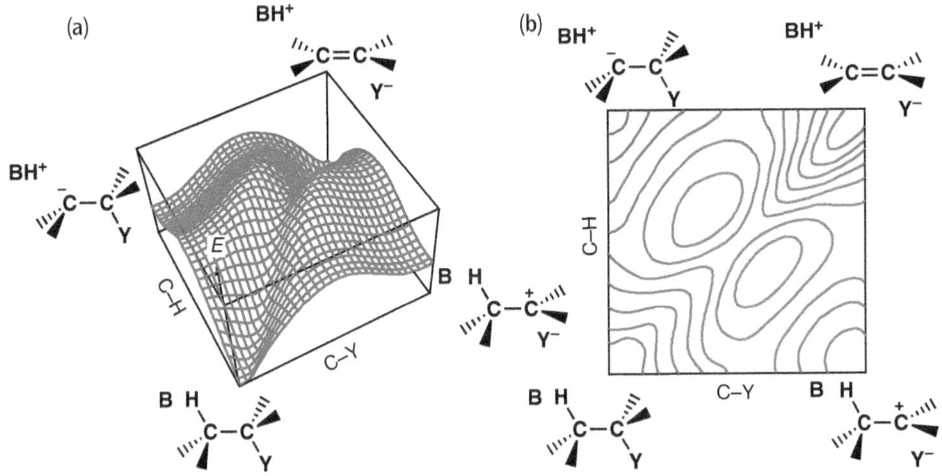

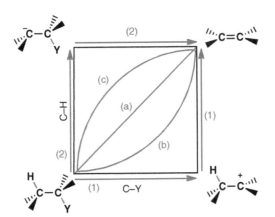

Figure 13.3 (a) Three-dimensional potential energy diagram for an elimination reaction, and (b) the corresponding reaction map.
Reproduced from Noyori, R. et al Ed. (1999) *An Advanced Study of Organic Chemistry*, Vol. 1. With kind permission from Tokyo Kagaku Dozin.

Figure 13.4 Simplified reaction map showing paths for (1) E1, (2) E1cB, and (a) central E2, (b) E1-like E2, and (c) E1cB-like E2 mechanisms.

Different cross-sections through the three-dimensional energy diagram of Figure 13.3(a) from the lower left corner to the upper right (reactants to products) correspond to different two-dimensional reaction profiles for elimination. The lowest-energy path (up the valley, over the pass, and down the other side) for the case shown corresponds to a concerted E2 elimination with the 'central' TS in Figure 13.2.

Note that a reaction map does not include the coming-together of the reactants if the mechanism is bimolecular, or the separation of products (if there are more than one) after the rebonding is complete.

Diagrams like Figure 13.3(b) were introduced in the context of mechanisms of elimination reactions by R.A. More O'Ferrall (1937–2012) of University College, Dublin.

carbenium ion; this would be followed if Y⁻ were a good nucleofuge and R⁺ a relatively stable carbenium ion. Path (2) corresponds to the E1cB mechanism *via* the carbanion intermediate; this would be the lowest energy path if, for example, there were carbanion-stabilizing substituents in R. The three paths inside the diagram represent the subtly different graded E2 mechanisms when the C–H and C–Y bond cleavages are concerted: path (a) corresponds to the central E2 mechanism; path (b), which curves towards the lower right corner, represents an E1-like E2 mechanism; path (c), which curves towards the upper left corner, corresponds to an E1cB-like E2 mechanism. The three transition structures represented in Figure 13.2 above will be located on the lines (b), (a), and (c), respectively.

> Other reactions can also be represented by reaction maps; it is necessary only to be able to identify the *x*- and *y*-axes of the map with two bonds which break or form during the reaction, e.g. the C–Y and C–Nu bonds for the nucleophilic substitution reaction of RY with Nu⁻: R—Y + Nu⁻ ⇌ R—Nu + Y⁻.

Exercise 13.3

> What type of substituents will favour an E2 mechanism described by path (b) in the reaction map of Figure 13.4, and what type will favour a mechanism described by path (c)?

13.5 Regioselectivity in Elimination

13.5.1 Regioselectivity in E1 eliminations

Solvolysis of the tertiary alkyl substrate, 2-bromo-2-methylbutane, in ethanol gives two constitutionally isomeric alkenes by E1 elimination, in addition to the S_N1 substitution product (eqn 13.9).

| 2-bromo-2-methylbutane | 2-ethoxy-2-methylbutane 64% | 2-methylbut-2-ene 30% | 2-methylbut-1-ene 6% | (13.9) |

S_N1 product   E1 products

One alkene is obtained by deprotonation from the ethyl and the other by deprotonation from one of the two (equivalent) methyls, and the yield of the trisubstituted alkene is about five times greater than that of the disubstituted constitutional isomer; this is an example of a **regioselective** reaction (from the Latin word *regio* for *direction*). It arises because the substrate in the elimination reaction has different alkyl groups on the C carrying the leaving group, and different alkenes will be obtained according to which of non-equivalent hydrogens is abstracted in the elimination.

Amongst constitutionally isomeric alkenes, the more alkyl substituents there are on the double bond, the more stable the alkene is likely to be, and a carbenium ion will generally give isomeric alkenes preferentially according to their stabilities. Consequently, *the alkene isomer with the most alkyl substituents is usually the main product in an E1 elimination*. This empirical generalization of the **regioselectivity** in E1 eliminations was based on early experimental observations, and is sometimes known as the **Zaitsev rule**.* It has a sound mechanistic basis, however, because the transition structures in the alternative competing product-determining proton abstractions from the carbenium ion intermediate have partial double bond character (Hammond postulate, see Chapter 7), so the one with most alkyl substituents will be the most stable.

*Different attempts at transliterating this Russian chemist's name from Cyrillic into Roman script have led to other spellings, e.g. Saytzeff is used in some older books. A.M. Zaitsev (1841–1910) published his results on elimination reactions in 1875, i.e. some years after Hofmann's findings on other elimination reactions reported in 1851; see next sub-section.)

Exercise 13.4

Give the structure of the carbenium ion intermediate in the reaction of eqn 13.9 and, using curly arrows, show how each of the three products is derived from it.

Exercise 13.5

Predict the main alkene amongst the elimination products in each of the reactions in Example 13.1.

Give the structure of the carbenium ion intermediate in the reaction of eqn 13.9 and, using curly arrows, show how each of the three products is derived from it.

Predict the main alkene amongst the elimination products in each of the reactions in Example 13.1.

13.5.2 Regioselectivity in E2 eliminations

When the TS in an E2 reaction is E1-like, or towards the central region of the range indicated in Figure 13.2, formation of the more substituted alkene isomer is the preferred path (as for the E1 mechanism, and for the same reason). The elimination reaction of 2-bromopentane with methoxide, for example, gives 72% of pent-2-ene (Scheme 13.5).

This preference for the most stable alkene can be subverted, however, if a particularly bulky base is used. The same substrate with *t*-butoxide gives only 20% of the internal (most substituted) alkene (Scheme 13.5). This tendency for some elimination reactions to give the *least* alkyl-substituted of the possible isomeric alkenes is also well established experimentally, and is known as the **Hofmann rule**. Although it contradicts the Zaitsev rule, there is a mechanistic explanation. Steric hindrance is a known complication in bimolecular reactions, and when a bulky base such as *t*-butoxide is used in an E2 reaction, the TS leading to the more substituted alkene becomes less stable than the one leading to the least substituted isomer. Abstraction of a proton from the C1 methyl group of 2-bromopentane by a bulky base in eqn 13.10 in Scheme 13.6 leads preferentially to a TS less hindered than the one for proton abstraction from C3 (eqn 13.11).

The tendency to give the least alkyl-substituted alkene in a bimolecular elimination reaction is also observed when the steric congestion is caused by a bulky nucleofuge (i.e. in reactions of substrates other than alkyl halides), or by a bulky alkyl residue within the substrate.

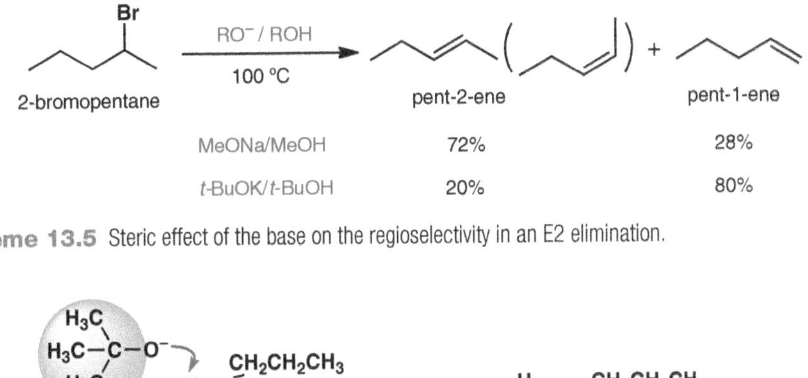

August W. von Hofmann (1818–1892).

Hofmann was a student of Liebig's. The Hofmann rule was based upon his investigations of elimination reactions of the type given in Example 13.3(b) and described in Sub-section 14.7.1. In addition to the elimination reaction, a rearrangement (Chapter 22) is also named after him.

	pent-2-ene	pent-1-ene
MeONa/MeOH	72%	28%
t-BuOK/*t*-BuOH	20%	80%

Scheme 13.5 Steric effect of the base on the regioselectivity in an E2 elimination.

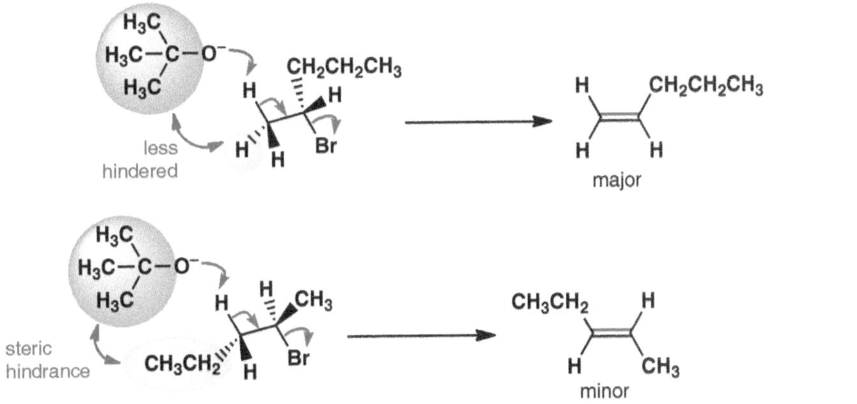

$$(13.10)$$

$$(13.11)$$

Scheme 13.6 Schematic representations of steric effects in the reaction using *t*-BuOK/*t*-BuOH in Scheme 13.5.

We have already seen that the C bearing the functionality—in the current context the nucleofuge—is often call the α C, and an adjacent C atom is a β C; if two C atoms next to the α C are different, as in 2-bromopentane, they may be called the β C and the β′ C.

There is also an electronic reason why Hofmann regioselectivity may be observed in an E2 reaction. When the TS is E1cB-like (Fig. 13.2), there is a build-up of negative charge at the β C from which a proton is being abstracted, and substituents which stabilize an electron-rich TS are not the same as those which stabilize an E1-like TS. Alkyl substituents (which stabilize electron-deficient centres—see Section 12.4.3) generally have a *destabilizing* effect upon carbanionic centres. Consequently, *less* alkyl-substituted alkene isomers are formed preferentially when the TS is E1cB-like, as illustrated in Scheme 13.7. The internal alkene (Zaitsev rule) is formed preferentially when the leaving halogen is I, Br, or Cl. However, the proportion of the more alkyl-substituted alkene decreases as the halogen is changed from I to Br to Cl, i.e. as it becomes a worse leaving group but more electron-attracting. When the nucleofuge is F, the main product is the terminal isomer (Hofmann rule) because F is both a very poor leaving group *and* strongly electron-attracting: the TS in this E2 reaction is E1cB-like.

Scheme 13.7 Effect of nucleofuge on the regioselectivity in E2 eliminations.

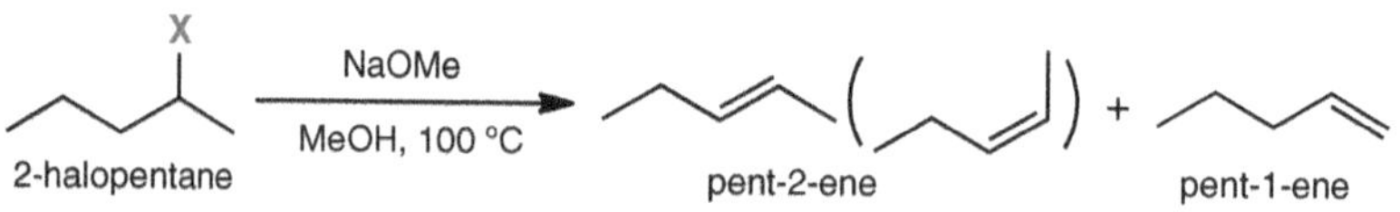

	pent-2-ene	pent-1-ene
X = I	81%	19%
X = Br	72%	28%
X = Cl	67%	33%
X = F	30%	70%

Example 13.3

Show all possible alkenes formed in the following elimination reactions, and predict which is the main product in each case.

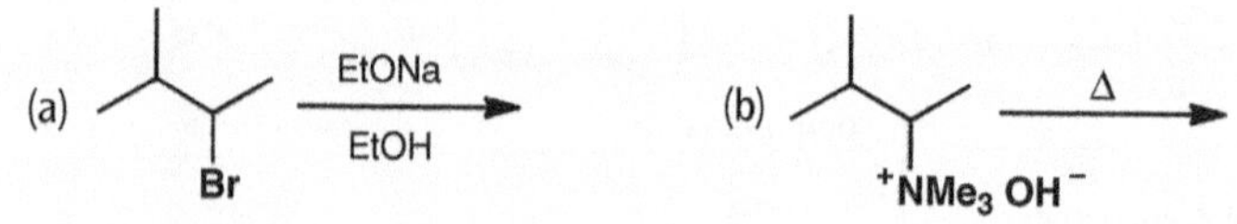

Solution

(a) This is a normal E2 reaction and the main product is 2-methylbut-2-ene.

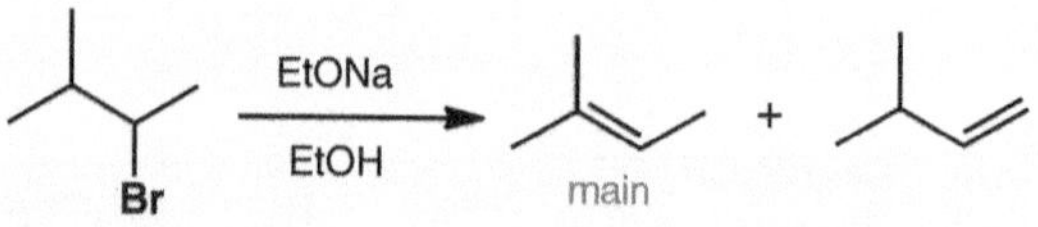

(b) Thermolysis of an *N*-alkyltrimethylammonium hydroxide gives mainly a terminal alkene as the elimination product (this reaction is sometimes known as the Hofmann elimination; see Sub-section 14.7.1). Trimethylamine is a poor leaving group and the positively charged ammonio group is strongly electron withdrawing which stabilizes a carbanionic TS. The trimethylammonio group is also sterically bulky which additionally directs attack of the HO⁻ to the less hindered β C–H of the methyl group. These factors combine to favour predominant formation of the less substituted alkene isomer.

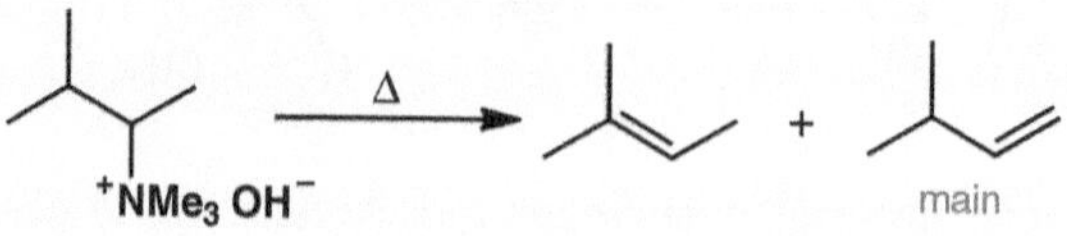

Give all the possible alkenes formed in each of the following elimination reactions, and predict which is the main product in each case.

(a) [structure: bromocyclohexane with methyl] → EtONa / EtOH

(b) [structure] → EtONa / EtOH

(c) [structure with Cl] → t-BuONa / t-BuOH

(d) [structure with +NMe₃ OH⁻] → Δ

Panel 13.1 Hofmann and Zaitsev regioselectivity, and Bredt's rule

---

We saw in Section 13.5 that regioselectivity in elimination reactions when the C–Hs β and β' to a C bearing a nucleofuge are different is generally covered by the empirical Zaitsev and Hofmann 'rules'. In the absence of steric effects, we normally observe formation of the more substituted (stable) alkene (Zaitsev regioselectivity). If the base, the nucleofuge, or the alkyl residue is bulky and steric congestion is an issue in E2 eliminations, however, the least substituted alkene (the one *via* the least congested TS) is generally obtained (Hofmann regioselectivity). However, exceptions to these generalizations *as originally proposed* are known, but they can be understood if we consider the mechanistic basis of the 'rules'.

On the left below, we have a typical Hofmann bimolecular elimination in which ethene is formed preferentially, as anticipated. On the right, however, a similar reaction of a sterically hindered substrate does *not* give the 'Hofmann' product. Instead, we see elimination to give the *more substituted* product in which the new C=C bond is conjugated with the phenyl group: the developing resonance stabilization of the TS for the formation of the phenylethene (styrene) more than compensates for the adverse steric effect involved.

[reaction schemes]

93% / < 1%

The only alkene in the elimination of HY from the bridged 2-substituted bicyclo[2.2.1]heptane (see Panel 4.2), regardless of the nature of Y, the reaction conditions, or mechanism, is the one shown in blue. The more substituted isomeric alkene shown in red is far too strained to exist under normal reaction conditions (although it has been implicated as a transient intermediate in the dehalogenation of 1,2-dihalobicyclo[2.2.1]heptanes at low temperatures). The four bonds on the two sp²-hybridized carbons of this structure cannot be coplanar as is required for proper sideways overlap of the 2p orbitals to give a stable π bond.

[reaction scheme]

bicyclo[2.2.1]hept-1(2)-ene   – HY   bicyclo[2.2.1]hept-2-ene

The above illustrates another empirical generalization, first reported in 1924 and originally known as Bredt's rule (J. Bredt, 1855–1937): an elimination reaction cannot give a bridgehead alkene if the double bond would be in a ring with seven or fewer carbons. As with the 'rules' of Zaitsev and Hofmann, however, we now understand its mechanistic basis.

13.6 Competition between Elimination and Substitution

In principle, a nucleophile (Lewis base) can induce substitution and/or elimination with an alkyl halide because it can also act as a Brønsted base; nucleophilic attack at the electrophilic carbon results in substitution, whereas proton abstraction from a β C leads to elimination.

In a weakly nucleophilic protic solvent with no added nucleophile/base, secondary and tertiary alkyl halides (except fluorides) form carbenium ion intermediates which give substitution products (S_N1) by nucleophilic capture by the solvent, or alkenes (E1) by proton transfer to the solvent. Simple primary alkyl halides undergo neither substitution nor elimination at measurable rates under these conditions.

When a weakly nucleophilic solvent contains a nucleophilic solute which can also act as a base, primary alkyl halides usually undergo an S_N2 reaction, but a *strong* base can also induce elimination. A sterically hindered strong base like *t*-butoxide invariably leads to the E2 reaction exclusively. With tertiary alkyl derivatives, or other substrates branched at the β position, the S_N2 reaction is sterically hindered, so the E2 reaction will occur to give the more substituted, more stable alkene.

Reactivities of secondary alkyl derivatives are between those of the primary and tertiary substrates, and the partitioning between substitution and elimination depends on the nucleofuge, nucleophile/base, solvent, and temperature. When nucleophilicity and basicity of the solute are both high, as for hydroxide and ethoxide ions, the E2 reaction is generally preferred. When the reagent is both weakly nucleophilic and weakly basic, *e.g.* ethanoate (acetate) ion, substitution is the usual reaction. At higher temperatures, however, the elimination/substitution ratio is usually greater. This is because $\Delta H^{\ddagger}$ is generally larger for an E2 than for an S_N2; consequently, as the temperature is raised, the increase in the rate constant is steeper for the elimination than for the substitution (see Chapter 7). Table 13.1 summarizes the above generalizations.

Panel 13.2 Polyhalogenated compounds and the environment

Polyhalogenated compounds have been used in agriculture and industry but they have caused a number of environmental problems. An example is the case of DDT (the abbreviation of its common name, dichlorodiphenyltrichloroethane). It was introduced in 1939 as the first synthetic insecticide and was very effective in combatting malaria (by killing mosquitoes) and typhus (by killing lice) in many parts of the world during and immediately following World War II.

After the war, DDT was used extensively in agriculture: it is effective against a wide variety of insects but not very toxic to mammals, and it is cheap and persistent. However, the properties which made it valuable as an insecticide caused it to be an environmental hazard. DDT is very stable and lipophilic (soluble in nonpolar compounds but insoluble in aqueous media) so it accumulates in the environment, particularly in the fatty tissues of animals through the food chain. Although the long-term effect on humans is not known, it was connected to the defective eggshell formation of some predator birds such as eagles and hawks. The problems associated with persistent organochlorine compounds in the biosphere were first publicized by the American biologist Rachel Carlson in her famous book, *Silent Spring* (1962). DDT is now banned in most countries, but is still used in developing countries where malaria remains a serious problem.

DDT   2,4-D   2,4,5-T

2,4-Dichlorophenoxyethanoic acid (2,4-D) is a widely used systemic herbicide. It was used on a massive scale in a mixture with 2,4,5-T (2,4,5-trichlorophenoxyethanoic acid which is not now widely used) as the notorious 'Agent

Table 13.1

Substitution and elimination reactions of alkyl halides and related compounds

	Weakly nucleophilic solvents (H_2O, ROH, RCO_2H)	Weakly basic nucleophiles (I^-, Br^-, RS^-)	Strongly basic nucleophiles	
			Not bulky (EtO^-)	Bulky (t-BuO$^-$)
RCH_2Y (1°)	no reaction	S_N2	S_N2	E2
R_2CHCH_2Y (β branch)	no reaction	S_N2	E2	E2
R_2CHY (2°)	$S_N1/E1$ (slow)	S_N2	E2	E2
R_3CY (3°)	$E1/S_N1$ (fast)	$S_N1/E1$	E2	E2

Y = Cl, Br, I, or another nucleofuge such as $R'SO_3$ or R'_3N^+.

Example 13.4

Explain whether substitution or elimination occurs preferentially in the following reactions, and give the structure of the main product in each case.

(a)
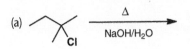

(b) + NaBr →
propanone

Solution

(a) With a tertiary alkyl substrate under strongly basic conditions, the more substituted alkene is formed as the main product by the E2 mechanism.

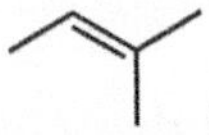

2-methylbut-2-ene

--- **Panel 13.2** Continued ---

Orange' to defoliate trees during in the Vietnam War. The highly carcinogenic tetrachlorodioxin shown below (which causes DNA damage) was a contaminant of the 2,4,5-T, and led to major problems including severe birth defects. Dioxins and dioxin-like compounds have been associated other medical problems including oxidative stress, damage to the immune system, and endocrine disruption. Strictly, *dioxin* is the systematic name for an aromatic compound in which two CH groups have been replaced by Os. However, it is widely used as the common name for the particular compound shown, 2,3,7,8-tetrachlorodibenzo-1,4-dioxin, arguably the most hazardous synthetic compound known. Structurally related compounds such as the polychlorinated dibenzofuran shown are also often included in the collective term *dioxins*. All are generally formed as industrial by-products and have no known use but are persistent, highly toxic environmental pollutants.

A dioxin and dioxin-like compounds:

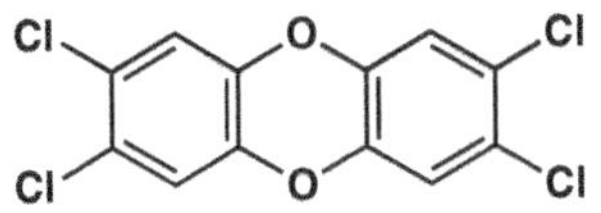

2,3,7,8-tetrachlorodibenzo-1,4-dioxin 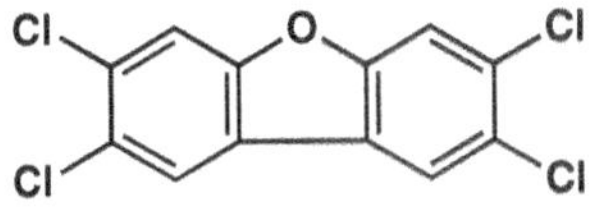   2,3,7,8-tetrachlorodibenzofuran 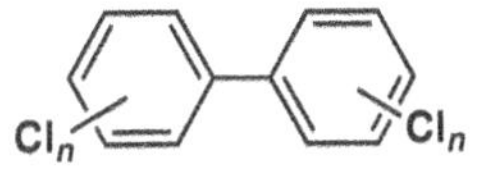   polychlorobiphenyls (PCBs)

Polychlorinated biphenyls (PCBs) were once widely used as dielectric fluids in electrical transformers and capacitors as well as heat transfer agents, but they are now also classified as persistent organic pollutants because they are similar to dioxins in toxicity. Also, disposal of PCBs by industrial incineration produces dioxins.

(b) Reaction of a secondary alkyl iodide (good nucleofuge) with a good nucleophile (Br⁻, which is only weakly basic) leads mainly to S_N2 substitution.

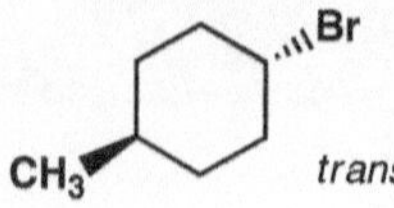

trans-1-bromo-4-methylcyclohexane

Exercise 13.7

Explain which of substitution or elimination dominates in the following reactions, and give the structure of the main product in each case.

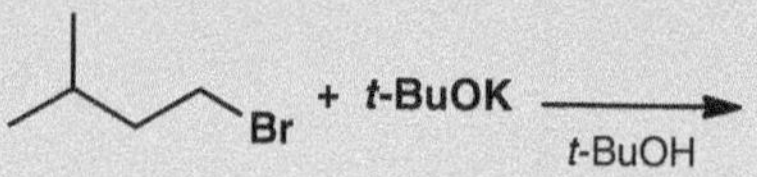

Exercise 13.8

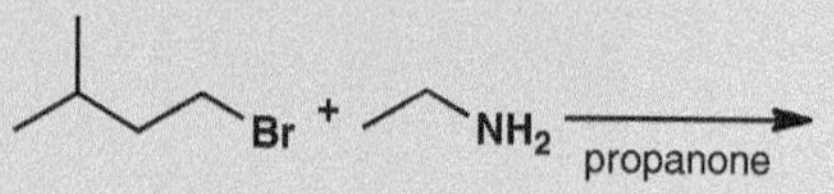

In the preparation of unsymmetrical ethers, ROR', by the S_N2 reaction, there is a choice of which pair of haloalkane plus alkoxide to use: RY + R'O⁻ or RO⁻ + R'Y. Propose suitable pairings of haloalkane and alkoxide which will minimize alkene formation (by competing elimination reactions) in preparations of the following ethers.

(a) $CH_3CH_2CH_2OCH(CH_3)_2$

(b) $CH_3CH_2OCH_2CH(CH_3)_2$

(c) 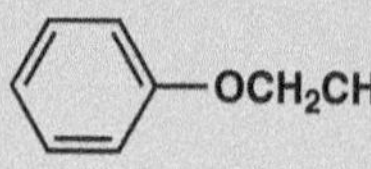 —OCH₂CH₃

(d) 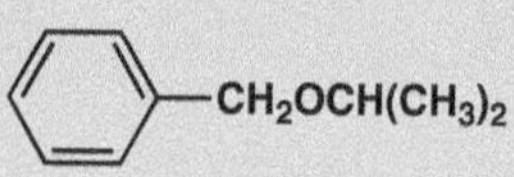 —CH₂OCH(CH₃)₂

Summary

- Abstraction of a proton from a β C together with departure of the nucleofuge Y⁻ from RY leads to elimination to form an alkene.

- With a strong base, deprotonation is **concerted** with departure of Y⁻ in an *anti* arrangement by the bimolecular **E2** mechanism.

- In weakly basic solutions, tertiary alkyl derivatives form carbenium ion intermediates which undergo **E1** elimination in competition with S_N1 substitution.

- From substrates with different β carbons, elimination usually leads to the more substituted, more stable alkene (Zaitsev rule). However, the alternative **regioselectivity** leading to the less substituted alkene (Hofmann rule) is possible when (i) formation of the TS leading to the more substituted alkene is sterically hindered, or (ii) the TS in an E2 reaction is E1cB-like.

- The less common stepwise E1cB mechanism of elimination, which is only encountered with poor leaving groups and strong bases, involves initial deprotonation of a β C–H followed by departure of the nucleofuge from the α C of the carbanion intermediate in the second step.

Problems

13.1 Show all possible alkenes formed in the elimination reactions below, and indicate the main product in each.

(a) [structure with Br, EtOK/EtOH] (b) [structure with Br, t-BuOK/t-BuOH]

13.2 Show all possible alkenes formed in the elimination reactions below, and indicate the main product in each.

(a) [Br structure, EtONa, EtOH, 80 °C]

(b) [I⁻ NMe₃ structure, EtONa, EtOH, 130 °C]

13.3 Show all possible alkenes formed in the elimination reactions below, and indicate the main product in each.

(a) [Br structure, EtOK/EtOH] (b) [Br structure, t-BuOK/t-BuOH]

13.4 Show all possible alkenes formed in the elimination reactions below, and indicate the main product in each.

(a) [Br, EtOK/EtOH] (b) [Br, t-BuOK/t-BuOH]

(c) [Br, t-BuOK/t-BuOH] (d) [Br, Δ, EtOH]

13.5 Give the main product in each of the following elimination reactions, and show how it is formed.

(a) BrCH₂CH(OEt)₂ $\xrightarrow{t\text{-BuOK}, t\text{-BuOH}, 120-130\,°C}$

(b) R—CF₂Br $\xrightarrow{DBU, \text{propanone}, \Delta}$

(DBU = [structure])

13.6 What is the main product obtained from each of the diastereoisomeric (1R,2S)- and (1R,2R)-1-bromo-1,2-diphenylpropanes by treatment with sodium ethoxide in ethanol? Write curly arrow reaction mechanisms.

13.7 Explain why the main product of the elimination reaction of *trans*-1,2-dibromocyclohexane is cyclohexa-1,3-diene, and not 1-bromocyclohexene.

[cyclohexane structure with Br, H, Br, H] $\xrightarrow[\substack{CH_3O(CH_2CH_2O)_3CH_3 \\ 100-110\,°C}]{NaOCH(CH_3)_2}$ [cyclohexadiene]

13.8 In the reactions of isomeric 1-bromo-4-*t*-butylcyclohexanes with potassium *t*-butoxide to give the same product, 4-*t*-butylcyclohexene, the *cis* isomer reacts about 500 times faster than the *trans* isomer. Explain the reason for this difference in reactivity.

13.9 Menthyl chloride and neomenthyl chloride are stereoisomers. Predict their relative reactivity upon treatment with sodium ethoxide in ethanol, and the main elimination product from each of them.

(CH₃)₂CH [cyclohexane] CH₃, Cl — menthyl chloride   (CH₃)₂CH [cyclohexane] CH₃, Cl — neomenthyl chloride

13.10 Aldol is obtained by the base-catalysed dimerization of ethanal (acetaldehyde, see Chapter 17). If aldol is heated in a basic solution, it undergoes dehydration (elimination of H₂O) by the E1cB elimination mechanism to give an α,β-unsaturated aldehyde. Write a curly arrow mechanism for this reaction and identify the feature of aldol which facilitates the E1cB mechanism.

[aldol structure OH, O, H] $\xrightarrow[\Delta]{NaOH}$ [unsaturated aldehyde O, H]

Supplementary Problems

13.11 Relative rate constants for solvolysis of three bromoalkanes in 60% EtOH-H$_2$O and in water are as follows:

	Br	Br	Br
60% EtOH-H$_2$O (55 °C)	1.00	1.78	2.41 x 10^4
H$_2$O (50 °C)	1.00	11.6	1.2 x 10^6

(a) Explain why 2-bromo-2-methylpropane undergoes solvolysis more than 10^4 times faster than bromoethane and 2-bromopropane in both solvents.

(b) Explain why the relative solvolytic reactivity of 2-bromopropane and bromoethane is significantly larger in H$_2$O than in 60% EtOH–H$_2$O.

13.12 The product ratio of substitution to elimination is not very different for the solvolysis of t-butyl chloride and a t-butyldimethylsulfonium salt in 80% aqueous ethanol. What does this observation imply regarding the reaction mechanisms?

Me–C(Me)(Me)–Y $\xrightarrow[65\,°C]{80\%\ EtOH}$ Me–C(Me)(Me)–OH + (Me)(Me)C=CH$_2$

| Y = Cl | 63.7% | 36.3% |
| Y = $\overset{+}{S}Me_2$ | 64.3% | 35.7% |

13.13 Rate constants for the solvolysis of t-butyl chloride depend tremendously on the solvent; the one for reaction in water is about 3 x 10^5 times larger than the one in ethanol. In contrast, the solvolytic reactivity of t-butyldimethylsulfonium chloride is not very dependent on the solvent.

$(CH_3)_3C–Cl$

2-chloro-2-methylpropane (t-butyl chloride)

$(CH_3)_3C–\overset{+}{S}(CH_3)_2\ Cl^-$

t-butyldimethylsulfonium chloride

(a) Explain the different solvent effects on the solvolysis of the two substrates.

(b) What would be the effect upon the reactivity of each substrate if chloride was replaced by iodide?

13.14 When benzenethiol is first treated with an excess of sodium ethoxide, and then with 1,2-dibromoethane, the main product is phenylthioethene (phenyl vinyl sulfide). Explain this reaction by giving a curly arrow reaction mechanism.

PhSH $\xrightarrow[EtOH]{NaOEt}$ PhSNa $\xrightarrow{Br\frown Br}$ =SPh

13.15 Using the C–Y and C–Nu bonds as coordinates, construct a simplified reaction map for the S$_N$2 reaction,

R–Y + Nu$^-$ ⇌ R–Nu + Y$^-$

Include the reaction path for a mechanism with an S$_N$1-like transition structure; what type of substituents in R will favour such a mechanism?

13.16 1,2-Dihalides undergo nucleophile-induced dehalogenation to give alkenes in reactions whose mechanism is closely similar to the E2:

Nu$^-$ + –C(X)–C(X)– ⟶ C=C + Nu–X + X$^-$

Give a curly arrow mechanism and identify the product of the reaction of (2S,3S)-2,3-dibromobutane with NaI in acetone. What is the product starting from meso-2,3-dibromobutane.

Reactions of Alcohols, Ethers, Thiols, Sulfides, and Amines

14

Related topics we have already covered:

- Hydrogen bonding in alcohols (Section 3.9)
- Curly arrow representation of reaction mechanisms (Section 7.2)
- Acids and bases (Chapter 6)
- Acid–base catalysis (Sub-section 8.3.2)
- Preparation of alcohols (Chapter 10)
- Nucleophilic substitution and elimination reactions of haloalkanes and related compounds (Chapters 12 and 13)

Topics we shall cover in this chapter:

- Acid catalysis to make the OH/OR group a good leaving group
- Acid-catalysed reactions of alcohols and ethers: substitution and elimination
- Ring-opening reactions of epoxides (oxiranes)
- Reactions of thiols and sulfides
- Reactions of amines

Alcohols and ethers may be seen as derivatives of H_2O in which first one then two H atoms are replaced by alkyl (R) or aryl (Ar) groups. They are classes of compounds with general structures ROH and ROR (or ROR'), respectively, with hydroxy (OH) and alkoxy (OR) as their functional groups. Their physical properties are largely different but they have some common features regarding chemical reactivity.

The principal methods of preparation of alcohols are hydride reduction and reactions of Grignard reagents with carbonyl compounds, which were covered in Chapter 10. The characteristic properties of alcohols originate in the hydroxy group. Their Brønsted acidity was discussed in Chapter 6, and physical properties attributable to hydrogen bonding of the OH were reviewed in Section 3.9. Protonation of the O in alcohols and ethers converts poor leaving groups, HO^- and RO^-, into good ones, H_2O and ROH. We shall see in this chapter that this enhancement of reactivity by protonation is another example of acid catalysis; examples we shall cover include acid-catalysed nucleophilic substitution and elimination reactions. In these reactions, $R–OH_2^+$ and $R–O(H)R^+$ are comparable with haloalkanes, e.g. RBr, in reactivity and we can represent them all generically as RY. Ring strain (see Section 4.3) causes three-membered cyclic ethers (epoxides, also called oxiranes) to be much more reactive than their acyclic analogues, and they undergo both acid- and base-catalysed ring-opening reactions.

In this chapter, we shall also cover reactions of the sulfur analogues of alcohols (thiols, RSH) and of ethers (sulfides, RSR'), which may be regarded as derivatives of hydrogen sulfide, H_2S. Correspondingly, amines (RNH_2, R_2NH, and R_3N) may be regarded as derivatives of NH_3 in which H atoms are successively replaced by alkyl (or aryl) groups. Thiols, sulfides, and amines can all be made from haloalkanes by S_N2 substitution reactions (Chapter 12). A major difference between amines on the one hand, and O and S analogues on the other, is the much higher basicity of amines.[1] Amines also differ from O and S analogues by undergoing characteristic reactions with nitrous acid (HNO_2).

---

[1] The amines included in this chapter are invariably *alkylamines*; there are some differences between their properties and those of *arylamines* which will be discussed when we cover aromatic compounds later in the book (Section 16.6).

Many of the reactions we shall introduce in this chapter involve heterolysis of the C–O, C–S, and C–N bonds. These reactions can be seen as extensions of the chemistry of the carbon–halogen bonds discussed in Chapters 12 and 13, but we shall see that there are some distinctive features of groups bonded through O, S, and N atoms. In particular, the -OH, -SH, and -NH$_2$ groups can be transformed without breaking the bonds to attached alkyl groups.

14.1 Acid-Catalysed Reactions of Alcohols and Ethers

14.1.1 Leaving ability of hydroxide and alkoxide

Alkoxide (RO⁻) and hydroxide ions (HO⁻) are very poor nucleofuges even in S$_N$2 reactions (eqn 14.1) although, as discussed in Sub-section 9.2.2, alkoxide ion can be expelled from the tetrahedral intermediate in ester hydrolysis under alkaline conditions by *electron push of the anionic oxygen* (eqn 14.2).

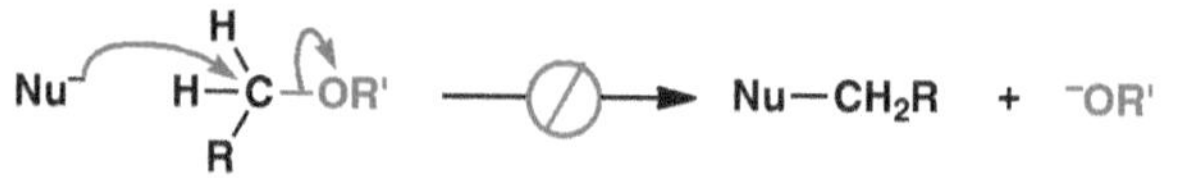

$$\text{Nu}^- \quad \text{H}-\overset{\text{H}}{\underset{\text{R}}{\text{C}}}-\text{OR}' \quad \longrightarrow \oslash \longrightarrow \quad \text{Nu}-\text{CH}_2\text{R} \quad + \quad {}^-\text{OR}' \tag{14.1}$$

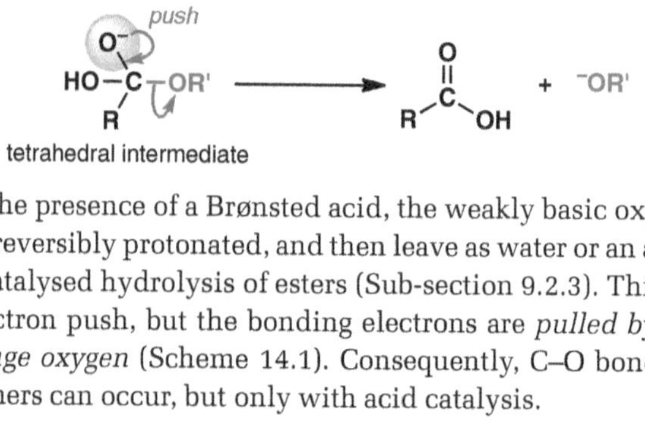

$$\underset{\text{tetrahedral intermediate}}{\overset{push}{\text{HO}-\overset{\text{O}^-}{\underset{\text{R}}{\text{C}}}-\text{OR}'}} \quad \longrightarrow \quad \underset{\text{R}}{\overset{\text{O}}{\underset{}{\text{C}}}}{-}\text{OH} \quad + \quad {}^-\text{OR}' \tag{14.2}$$

However, in the presence of a Brønsted acid, the weakly basic oxygen of an alcohol or an ether can be reversibly protonated, and then leave as water or an alcohol, respectively, as in the acid-catalysed hydrolysis of esters (Sub-section 9.2.3). This occurs without the help of any electron push, but the bonding electrons are *pulled by the positive charge on the nucleofuge oxygen* (Scheme 14.1). Consequently, C–O bond cleavage of simple alcohols and ethers can occur, but only with acid catalysis.

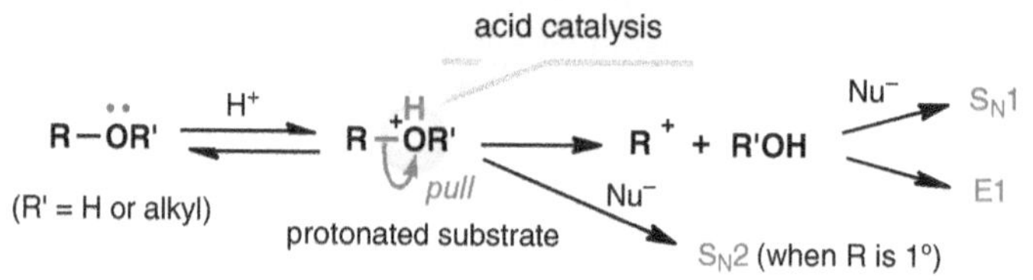

Scheme 14.1 Acid-catalysed cleavage of an alcohol or ether.

acid catalysis

$$\underset{(\text{R}' = \text{H or alkyl})}{\overset{\cdot\cdot}{\text{R}}-\text{OR}'} \quad \underset{\text{H}^+}{\overset{}{\rightleftharpoons}} \quad \underset{\text{protonated substrate}}{\overset{+\text{H}}{\underset{pull}{\text{R}-\text{OR}'}}} \quad \longrightarrow \quad \text{R}^+ + \text{R}'\text{OH} \quad \overset{\text{Nu}^-}{\underset{\text{Nu}^-}{<}} \quad \begin{matrix} \text{S}_N1 \\ \text{E1} \end{matrix}$$
$$\text{S}_N2 \text{ (when R is 1°)}$$

Reactions of a protonated alcohol or ether (the conjugate acid of the substrate) with loss of H$_2$O or R'OH as the nucleofuge very much resemble the reactions of haloalkanes discussed in the preceding two chapters. They undergo substitution and elimination reactions and the mechanisms (E1/S$_N$1 or S$_N$2 according to the nature of R) are closely similar to those of alkyl halides (RX, X = Cl, Br, I). However, the range of nucleophiles that can function under acidic conditions is limited. Those which are basic lose their nucleophilicity upon protonation in acidic media (e.g. CN⁻), and those which are not protonated in acidic solution are almost invariably poor nucleophiles (halide ions are exceptions—see next sub-section). For example, sulfate and phosphate, being conjugate bases of strong acids, are not protonated in dilute acidic solution, and are only weakly nucleophilic. In an acid-catalysed reaction of an alcohol *without* a solvent, the unprotonated alcohol will participate as a nucleophile to give an ether by an S_N1 or S_N2 mechanism depending upon the alkyl group (e.g. eqn 14.3), or as a weak base to give an alkene (see Sub-section 14.1.3).

$$2\ CH_3CH_2OH \xrightarrow[130\ °C]{cat.\ H_2SO_4} CH_3CH_2OCH_2CH_3\ +\ H_2O \qquad (14.3)$$

14.1.2 Reactions with hydrogen halides

The hydrogen halides HX (X = Cl, Br, and I) are strong acids in water ($pK_a \ll 1$) and dissociate to give hydrochloric, hydrobromic, and hydriodic acids,* three of the four so-called hydrohalic acids:

$$HX\ +\ H_2O \rightleftharpoons H_3O^+\ +\ X^-$$

These halide ions, being so weakly basic, remain unprotonated even in moderately concentrated acidic solution, and are exceptional in being strongly nucleophilic (see previous sub-section). Consequently, these three acids react with alcohols and ethers to give alkyl halides in acid-catalysed nucleophilic substitution reactions.

The reaction is initiated by equilibrium protonation by H_3O^+ and leads to the substitution product by the S_N1 or S_N2 mechanism. Tertiary alcohols react by the S_N1 mechanism in hydrohalic acids, as illustrated in Scheme 14.2 for the formation of *t*-butyl bromide from *t*-butyl alcohol and hydrobromic acid.

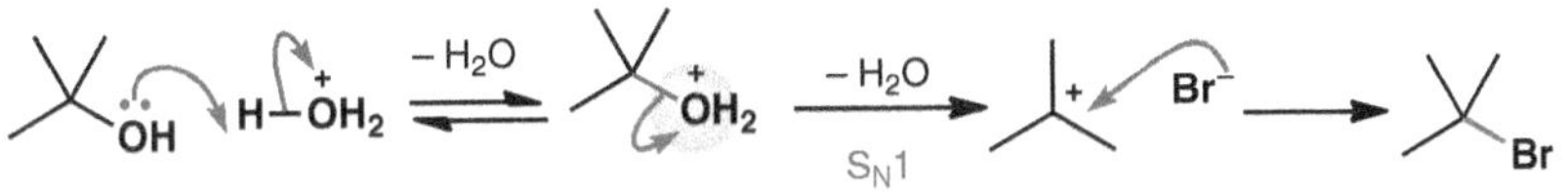

Scheme 14.2 Acid-catalysed S_N1 reaction of a tertiary alcohol.

In contrast, primary alcohols usually require heating with a concentrated hydrohalic acid to drive the less favourable S_N2 reaction (e.g. Scheme 14.3). Secondary alcohols are between their primary and tertiary analogues in reactivity, and the rate and mechanism depend on the particular compound and the reaction conditions.

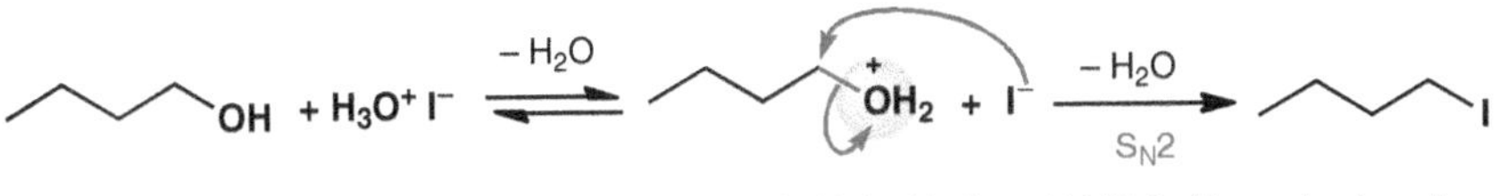

Scheme 14.3 S_N2 reaction of a primary alcohol with hydriodic acid initiated by proton transfer.

Ethers react with hydrohalic acids in the same way as alcohols, but selectivity is an issue with unsymmetrical ethers, R–O–R': which of the two C–O bonds undergoes heterolysis? Under acidic conditions in the absence of strong nucleophiles or bases, reactions occur by the S_N1/E1 mechanism if a viable carbenium ion intermediate can be generated. If alternatives are possible, C–O bond cleavage is normally the one which forms the *more stable carbenium ion* (e.g. Scheme 14.4). With secondary alkyl ethers, the mechanism will be S_N1 or S_N2 reaction depending on the reaction conditions.

The acid-catalysed reaction of an alcohol to give an ether is not a general method for the synthesis of ethers. They are usually prepared by S_N2 reactions of alkoxides with haloalkanes (see Example 12.2(b) in Chapter 12).

*These three acids are often represented in chemical equations as HCl, HBr, and HI even though virtually none of the un-ionized acids exist in even moderately concentrated aqueous solution. The fourth, hydrogen fluoride, undergoes the same reaction to give hydrofluoric acid; however, since HF is only a weak acid (pK_a = 3.2), it is dissociated to only a limited extent; see Chapter 6.

The rate of the S_N1 reaction of Scheme 14.2 in aqueous acid is proportional to the concentration of the protonated alcohol which, in turn, is proportional to [H_3O^+] and [*t*-BuOH]. Consequently, although the mechanism involves a *unimolecular* rate-determining step, the reaction has a *second-order* rate law:

rate = k[ROH][H_3O^+].

A reagent made from concentrated hydrochloric acid and $ZnCl_2$ (sometimes called Lucas reagent) is often used for the preparation of primary alkyl chlorides from the corresponding alcohols. Zn(II) is a Lewis acid and the reagent contains the complex anion, $ZnCl_4^{2-}$.

Low equilibrium concentrations of tertiary carbenium ions can be generated from an ether in dilute acid, but ether cleavage reactions are usually carried out using concentrated aqueous acid or dry HX in a nonaqueous solvent.

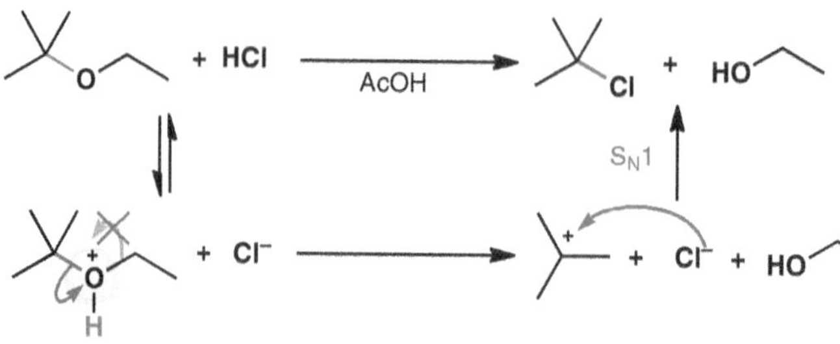

Scheme 14.4 Cleavage of an ether by hydrogen chloride in ethanoic acid.

Example 14.1

When 2-ethoxypropane is treated with hydriodic acid, propan-2-ol and iodoethane are obtained. Explain this selectivity.

Solution

Since iodide ion is a powerful nucleophile, the protonated ether undergoes the S_N2 reaction which occurs more easily at the primary carbon than at the secondary carbon.

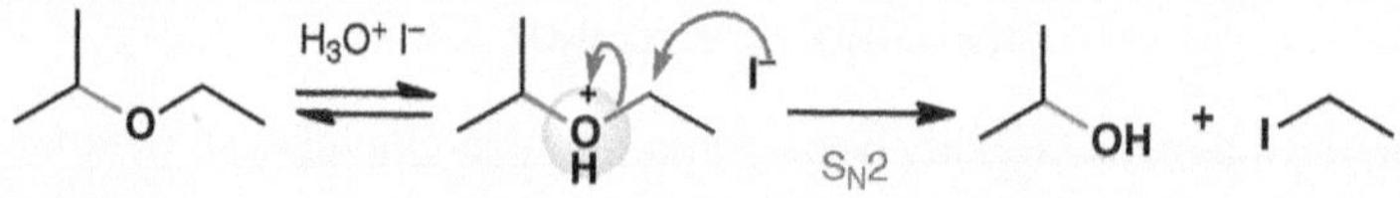

Exercise 14.1

Predict the products when each of the following isomeric compounds is treated with hydriodic acid, and propose reaction mechanisms.

(a)

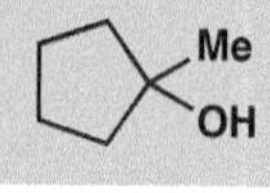

(b)

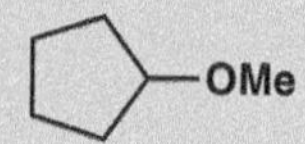

Exercise 14.2

Upon heating ethanol containing a small amount of H_2SO_4, diethyl ether is formed (eqn 14.3 given earlier). Propose a mechanism for this reaction.

Exercise 14.3

When 2-ethoxypropane is treated with a large excess of hydriodic acid, propan-2-ol is initially formed (see Example 14.1). However, this product is further converted to another compound after a prolonged reaction time. What is the final product and how is it formed?

14.1.3 Dehydration of alcohols

Elimination of H_2O from an alcohol to give an alkene is an example of a **dehydration**. This reaction is promoted by acid catalysis; sulfuric or phosphoric acids are commonly used since their conjugate bases are only weak nucleophiles (see above) and do not lead to competing substitution reactions (Scheme 14.5). Under these conditions, dehydrations of secondary and tertiary alcohols usually occur by the E1 mechanism; owing to the instability of primary alkyl cations, primary alcohols are rather unreactive and any elimination will be by a very slow E2 mechanism.

When two different alkenes can be formed, as in the example of eqn 14.4, the more stable, more alkyl-substituted isomer is preferred (Zaitsev selectivity, Section 13.5).

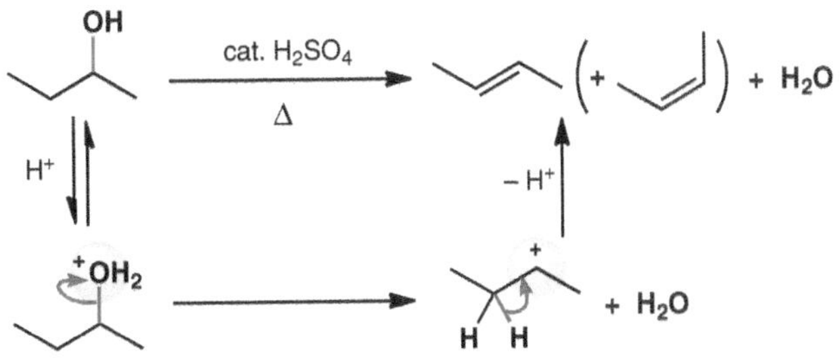

Scheme 14.5 Mechanism of the acid-catalysed dehydration of a secondary alcohol.

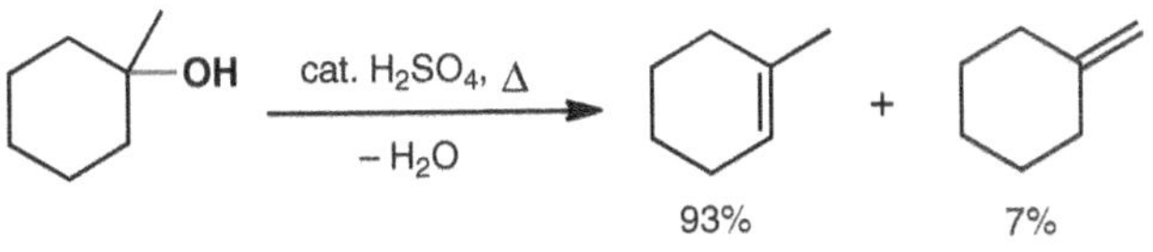

(14.4)

93%   7%

Give the structure of another alkene possibly formed in the reaction in Scheme 14.5 in addition to the but-2-enes.

Exercise 14.4

Propose a mechanism for the dehydration of 1-methylcyclohexanol with sulfuric acid as a catalyst (eqn 14.4), and explain the observed regioselectivity.

Exercise 14.5

Show structures of the alkenes formed by acid-catalysed dehydration of each of the following alcohols, and predict the main product in each reaction.

Exercise 14.6

(a) 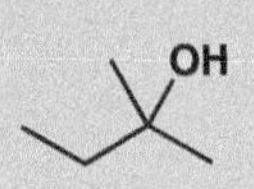   (b)

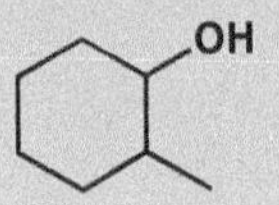

14.2 Rearrangements involving Carbenium Ions

The formation of rearrangement products is a characteristic feature of reactions involving carbenium ion intermediates and will be discussed in more detail in Chapter 22. During the acid-catalysed reaction of a secondary alcohol with a branched adjacent carbon, the intermediate secondary carbenium ion often rearranges by a 1,2-shift of hydrogen or an alkyl group from the carbon adjacent to the positive centre, e.g. Scheme 14.6. This type of **rearrangement** is also observed during the $S_N1/E1$ reactions of haloalkanes discussed in the preceding chapters, and the rearranged carbenium ion is usually more stable than the one initially formed.

Primary alcohols under acidic conditions, like primary alkyl halides, do not react *via* highly unstable primary carbenium ions, but rearrangement products are still sometimes observed. How can this be when the reactions are expected to be direct S_N2 substitutions

A *rearrangement* has occurred when the connectivity of atoms/groups (taking into account that one atom or group may have replaced another, e.g. a Cl may have replaced an OH) is different between reactants and products. If one compound or ion gives another with the same molecular formula but different atom connectivity, the process is usually called an *isomerization*.

Panel 14.1 Industrial production of alcohols

Alcohols with four carbons or fewer are produced industrially on a large scale and used as solvents and starting materials for other chemical products.

Methanol is manufactured by catalytic hydrogenation of carbon monoxide.[1] A mixture of H_2 and CO called synthesis gas (or syngas) is obtained principally from natural gas these days by *steam reforming* ($CH_4 + H_2O \rightarrow CO + 3H_2$). This mixture is then passed over a metal/metal oxide catalyst at high temperature and pressure.

$$CO + 2H_2 \xrightarrow[250\ °C,\ 50–100\ atm]{Cu–ZnO–Cr_2O_3} CH_3OH$$

Ethanol is produced by fermentation of sugars and this method should always be used for ethanol in alcoholic beverages and other foods. Ethanol for other purposes is now mainly produced industrially by acid-catalysed hydration of ethene.

$$H_2C{=}CH_2 + H_2O \xrightarrow[300\ °C,\ 70\ atm]{H_3PO_4} CH_3CH_2OH$$

In recent years, fermentation methods to produce 'bioethanol' from cellulose-containing materials (agricultural products and other forms of biomass) have become increasingly important for the production of 'green' fuels.

Distillation of a mixture of ethanol and water gives an azeotropic mixture of 95 : 5 ethanol : water boiling at 78.15 °C, which is slightly lower than the bp of pure ethanol (78.3 °C); consequently, pure ethanol cannot be obtained by distillation of a water–ethanol binary mixture. If just sufficient benzene is added to this 95% ethanol/water mixture, all the water can be distilled out as a ternary azeotrope (water/ethanol/benzene) boiling at 64.9 °C, and anhydrous ethanol (sometimes called absolute ethanol) can then be obtained from the residue. However, anhydrous ethanol produced by this method contains a low concentration (several p.p.m.) of benzene which is carcinogenic; consequently, this material should not be added to beverages or used in the preparation of food.

Ethane-1,2-diol (ethylene glycol) can be produced by oxidation of ethene, but is usually produced directly from syngas using a metal catalyst. Its unique properties (high boiling point, stability, low toxicity, and miscibility with water) make it a good antifreeze additive in automobile cooling systems.

$$2\,CO + 3\,H_2 \xrightarrow{Rh\ or\ Ru\ cat.} \text{HO}\diagup\!\diagdown\!\diagup\text{OH}$$

Propan-2-ol (which is used as an industrial solvent and in embrocations/liniments) is synthesized by acid-catalysed hydration of propene, and propan-1-ol is produced by hydrogenation of propanal, which is obtained by hydroformylation (addition of an H and a CHO group) of ethene.

$$H_2C{=}CH_2 + CO + H_2 \xrightarrow[\text{hydroformylation}]{Rh\ cat.} \!\!\!\diagup\!\diagdown\!\diagup\!\!\!{=}O \xrightarrow[Ni]{H_2} \!\!\!\diagup\!\diagdown\!\diagup\text{OH}$$

2-Methylpropan-2-ol (*t*-butyl alcohol, which is used industrially as a solvent, a cosolvent, and as an automotive fuel additive) can be obtained industrially by acid-catalysed hydration of 2-methylpropene (isobutene). A mixture of its isomers, butan-1-ol and 2-methylpropan-1-ol (isobutyl alcohol), is produced by hydroformylation of propene followed by hydrogenation.

$$\diagup\!\diagdown\!\diagup + CO + H_2 \xrightarrow{Rh\ cat.} \quad + \quad$$

$$\xrightarrow[Ni]{H_2} \!\!\!\diagup\!\diagdown\!\diagup\text{OH} \quad + \quad \text{OH}$$

---

[1] Methanol is oxidized to methanal in our bodies, and this reactive aldehyde interferes with the formation of the biological imine rhodopsin (the visual pigment in eyes) from the polyunsaturated aldehyde retinal and the protein, opsin (see Panel 5.3); this can lead to blindness. In larger quantities, ingested methanol is poisonous.

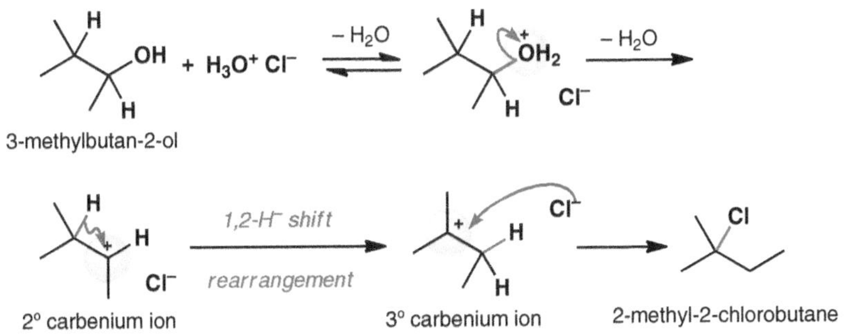

3-methylbutan-2-ol

2° carbenium ion   3° carbenium ion   2-methyl-2-chlorobutane

Scheme 14.6 Rearrangement of a carbenium ion by a 1,2-hydride shift in the acid-catalysed conversion of an alcohol into an alkyl chloride.

Take care drawing the curly arrow representing the 1,2-shift (e.g. in Schemes 14.6 and 14.7). The curvature of the tail of the curly arrow should initially be concave on the side of the H (or alkyl group) which moves with its bonding pair: the curly arrow is S-shaped. This 1,2-shift is often called a 1,2-*hydride* shift because the migrating hydrogen moves with its bonding pair of electrons.

In carbenium ion rearrangements, the migrating group may be methyl or another alkyl group. Give mechanisms for the following product distribution in the dehydration of 3,3-dimethylbutan-2-ol induced by phosphoric acid.

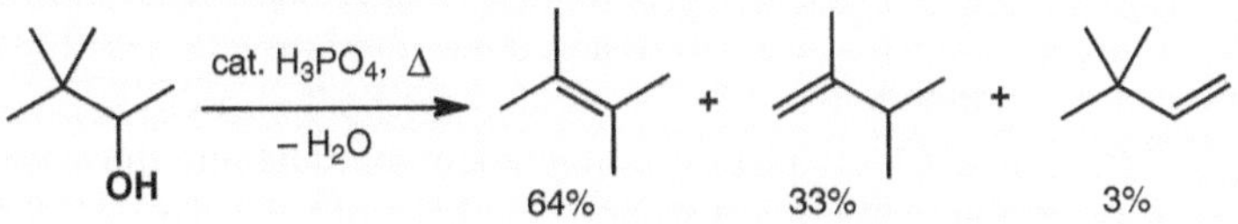

Example 14.2

64%   33%   3%

Solution
The initially formed secondary carbenium ion suffers only a small extent of deprotonation to give the low yield of unrearranged alkene. Predominantly, it rearranges to a tertiary cation by the 1,2-shift of a methyl group with its bonding electron pair. Proton loss from the tertiary cation gives the rearranged alkenes.

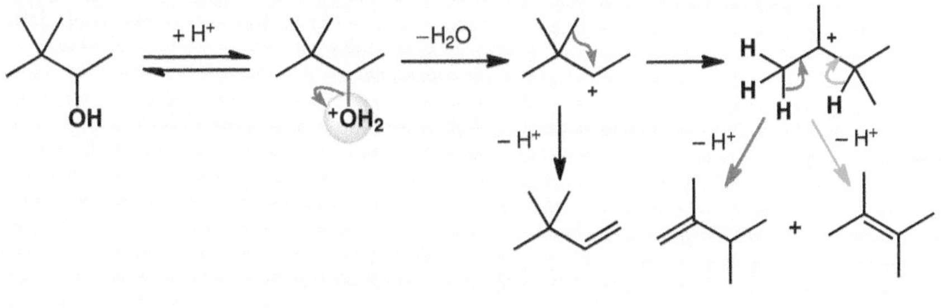

Upon treatment of 3,3-dimethylbutan-2-ol with hydrobromic acid 2-bromo-2,3-dimethylbutane is formed as a major product. Propose a mechanism for its formation.

Exercise 14.7

as in Scheme 14.3 given earlier? Such reactions are observed when the S_N2 mechanism for substitution is sterically hindered, as in the case of a branched primary alcohol under acidic conditions, and an alternative reaction with rearrangement becomes faster. Protonation of 2,2-dimethylpropan-1-ol (neopentyl alcohol) in concentrated hydrobromic acid generates a good nucleofuge (Scheme 14.7) but the S_N2 reaction is sterically hindered, and the primary alkyl carbenium ion is too unstable to form. However, a 1,2-methyl shift *concerted with departure of the nucleofuge* gives a relatively stable (tertiary) carbenium ion intermediate, capture of which by bromide gives the rearranged bromoalkane.

Note that the 1,2-methyl shift in Scheme 14.7 is *concerted with* the departure of the nucleofuge, i.e. there is no initial formation of a very unstable primary carbenium ion. In Scheme 14.6, however, the 1,2-shift *follows* departure of the nucleofuge, i.e. the first-formed secondary carbenium ion rearranges (isomerizes) to the more stable tertiary isomer. It is often difficult to distinguish between stepwise and concerted rearrangements (see Chapter 22).

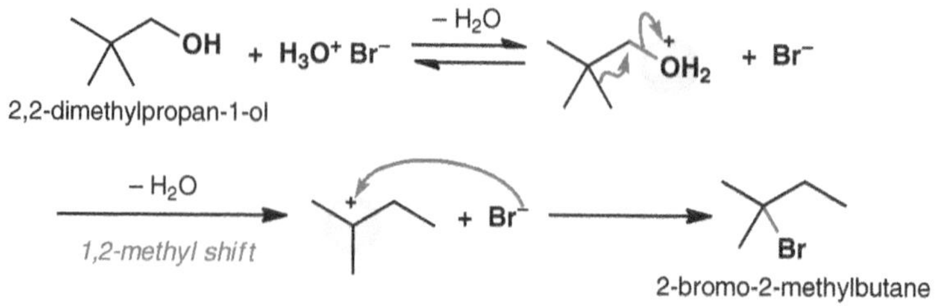

2,2-dimethylpropan-1-ol

1,2-methyl shift

2-bromo-2-methylbutane

Scheme 14.7 Rearrangement involving a 1,2-methyl shift in the acid-catalysed conversion of a primary alcohol into an alkyl bromide.

14.3 Conversion of OH into a Better Nucleofuge

14.3.1 Sulfonate esters

Acid catalysis is required in order to enhance the leaving ability of OH and OR groups in substitution and elimination reactions of alcohols and ethers as discussed above. In such reactions, the conjugate acids of the alcohols and ethers are formed reversibly at low concentration as reactive intermediates. Another strategy for alcohols ROH is to transform the OH into a better leaving group; for example, conversion of ROH into the ester of a strong acid provides a nucleofuge which is the conjugate base of the strong acid. Sulfonates, especially *p*-toluenesulfonates (tosylates, Sub-section 12.2.4), are widely used in this way and can often be isolated. Alkyl sulfonates are good substrates for S_N1 and S_N2 reactions which can be carried out under neutral or basic conditions (Scheme 14.8).

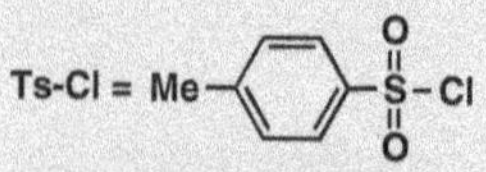

p-toluenesulfonyl chloride (tosyl chloride); the configuration is tetrahedral at the S.

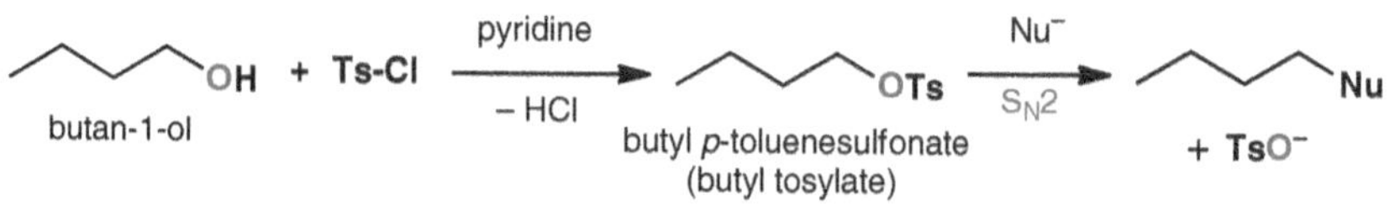

butan-1-ol   pyridine   butyl *p*-toluenesulfonate (butyl tosylate)

Scheme 14.8 Substitution reaction of an alcohol via its tosylate: conversion of a poor leaving group into a good one.

As discussed in the preceding section, rearrangement of the alkyl group of alcohols ROH sometimes becomes problematical in acid-catalysed reactions. In such cases, conversion to the alkyl tosylate and reaction under non-acidic S_N2 conditions is often an effective method to avoid this complication for primary and secondary alcohols.

Exercise 14.8

Explain why butan-2-ol cannot be converted into 2-methylbutanenitrile by direct reaction with NaCN. Write a reaction sequence for the preparation of 2-methylbutanenitrile starting with NaCN and (*R*)-butan-2-ol via the *p*-toluenesulfonate; show the stereochemistry of the intermediate and the product.

14.3.2 Sulfur and phosphorus reagents

Thionyl chloride ($SOCl_2$) can be used to convert alcohols directly into alkyl chlorides without rearrangement. In the presence of pyridine or another base, this reaction proceeds through an intermediate chlorosulfite ester which undergoes an S$_N$2 reaction with Cl$^-$ (Scheme 14.9; note the fragmentation of the nucleofuge).

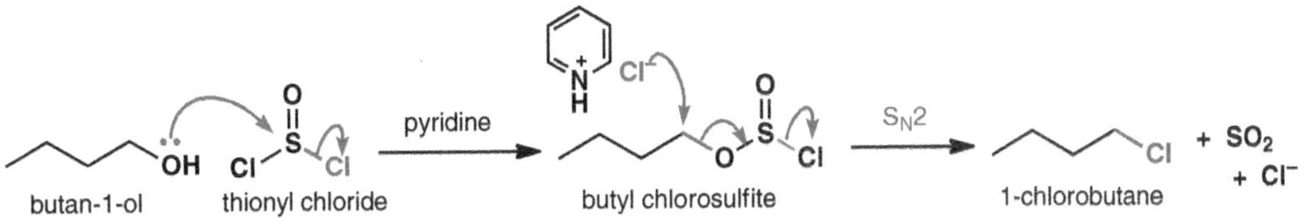

Scheme 14.9 Conversion of an alcohol into the corresponding chloride using thionyl chloride in the presence of pyridine.

Note that gaseous HCl is evolved in the reaction of an alcohol with thionyl chloride in diethyl ether *without* an added base; the intermediate chlorosulfite undergoes heterolysis to give an ion pair which yields the alkyl chloride and SO$_2$. If the reactant is a single enantiomer of a chiral secondary alcohol, the product is of retained configuration (Scheme 14.9a; this is sometimes called the S$_N$i mechanism, the 'i' standing for *internal*).

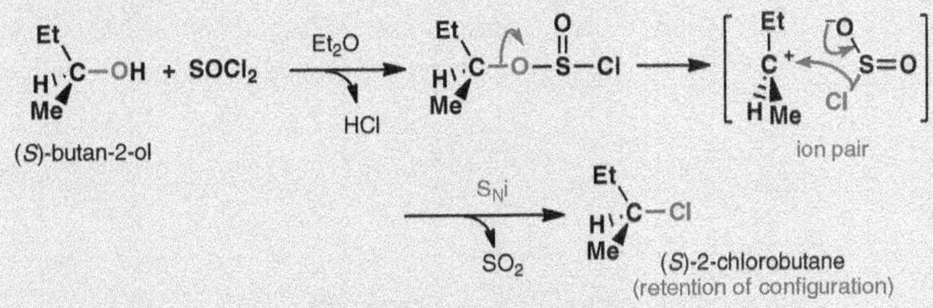

Scheme 14.9a Conversion of an alcohol into an alkyl chloride with retention of configuration using thionyl chloride in ether: the S$_N$i mechanism.

Phosphorus halides such as PCl$_5$ and PBr$_3$ can also be used (Scheme 14.10) but the reaction using SOCl$_2$ is preferred for alkyl chlorides as the by-products are gaseous so the workup is easy. The reaction using PBr$_3$ is generally used for the preparation of bromoalkanes.

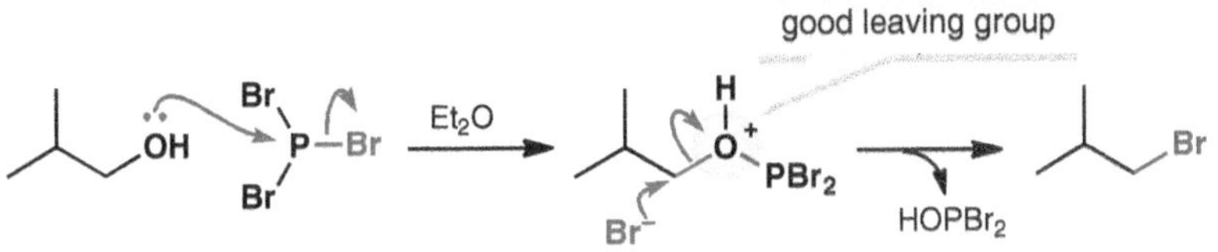

Scheme 14.10 Use of phosphorus tribromide to convert an alcohol into an alkyl bromide.

Exercise 14.9

Propose a method for the preparation of each of the following starting from propan-1-ol.
(a) CH$_3$CH$_2$CH$_2$Cl   (b) CH$_3$CH$_2$CH$_2$Br   (c) CH$_3$CH$_2$CH$_2$I   (d) CH$_3$CH$_2$CH$_2$CN

Panel 14.2 The Mitsunobu reaction

The Mitsunobu reaction is a mild method of reacting a chiral secondary alcohol with a carboxylic acid to obtain an ester with inversion of configuration. Experimentally, the alcohol is treated with diethyl azodicarboxylate (DEAD) and triphenylphosphine followed by the carboxylic acid.

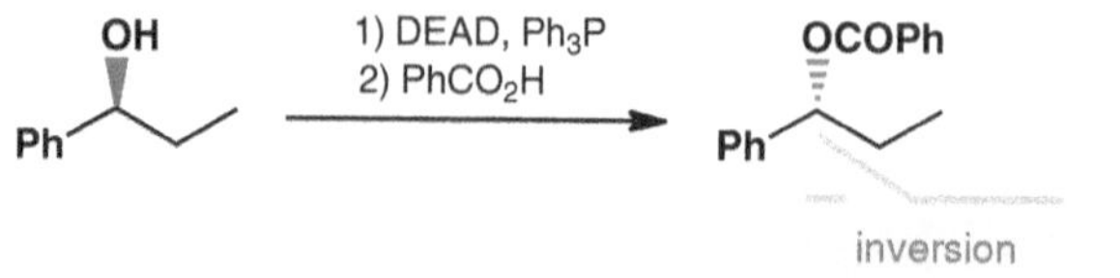

The triphenylphosphine first adds to the DEAD to give a dipolar species which deprotonates the alcohol, then the triphenylphosphine residue transfers to the alkoxide to give an electrophilic cation, as shown in the mechanism below. This suffers nucleophilic attack with inversion of configuration in an S_N2 reaction with the carboxylate anion obtained by deprotonation of the carboxylic acid. The ester may be hydrolysed to give the alcohol whose configuration is opposite to that of the starting material, i.e. its enantiomer.

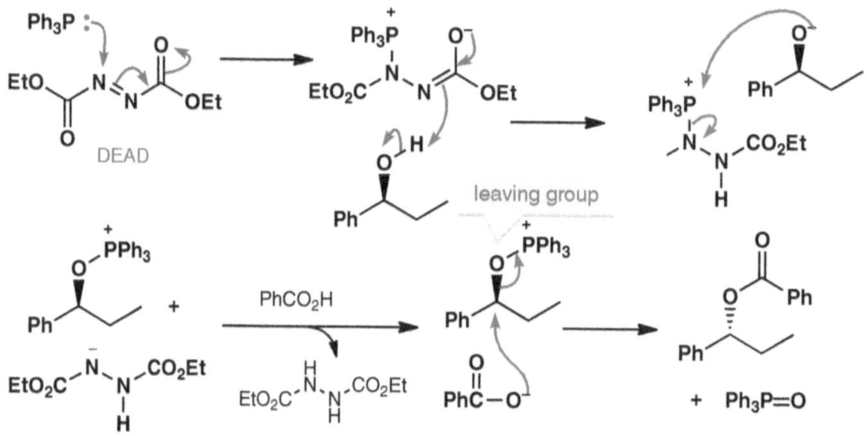

A proposed mechanism of the Mitsunobu reaction.

14.4 Oxidation of Alcohols

Chromium(VI) compounds are known to be carcinogens, and should be handled with great care.

Oxidation of a primary alcohol first provides an aldehyde and further reaction gives the carboxylic acid; oxidation of a secondary alcohol leads to a ketone. Chromium(VI) compounds such as chromium trioxide, CrO_3, or acidified sodium dichromate, $Na_2Cr_2O_7$, are widely used for these oxidations.

In an experimental method sometimes known as *Jones oxidation*, $Na_2Cr_2O_7$ in dilute aqueous H_2SO_4 (often called *chromic acid*) is added to a solution of the alcohol in propanone. The aldehyde formed from a primary alcohol is in rapid equilibrium with its hydrate in the acidic aqueous solution, and the hydrate is readily oxidized to the carboxylic acid (Scheme 14.11). It is seldom possible to stop the oxidation at the aldehyde stage by this method.

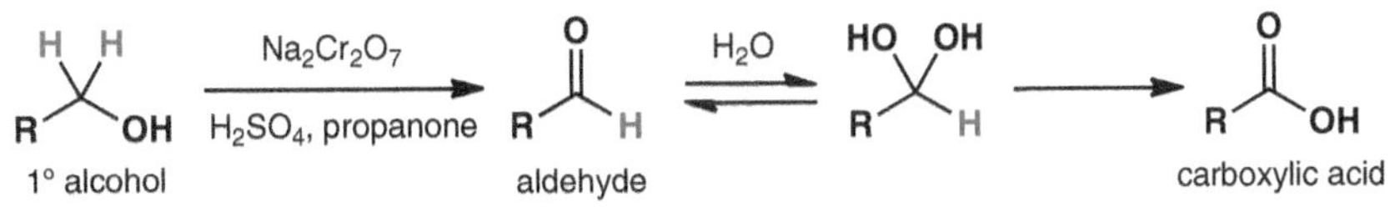

Scheme 14.11 Oxidation of a primary alcohol.

The mechanism of this oxidation involves an elimination step to give the C=O bond from an intermediate chromate ester, as shown in Scheme 14.12 for the oxidation of a secondary alcohol. The clear orange solution containing $Cr_2O_7^{2-}$ ions of the oxidant becomes green owing to the Cr(III) species, so it is easy to monitor the progress of the reaction. Chromate esters of tertiary alcohols have no hydrogen to participate in the elimination step so are not oxidizable; the same is true for hydrates of ketones.

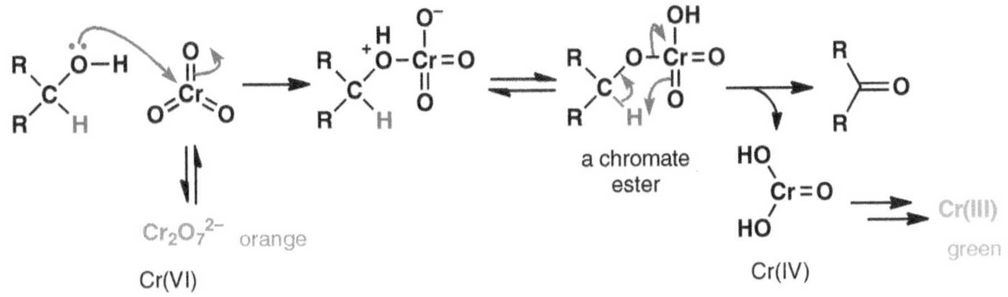

Scheme 14.12 Mechanism of chromic acid oxidation of an alcohol.

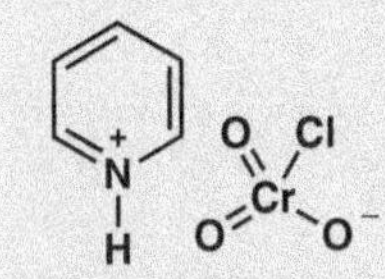

To avoid the oxidation of aldehydes, a non-aqueous solvent is used (which prevents the formation of hydrates) and an oxidant which is soluble in it. Pyridinium chlorochromate (PCC, in which Cr is also in the +6 oxidation state) in CH_2Cl_2 is a widely used oxidation method to obtain aldehydes from primary alcohols in good yield.

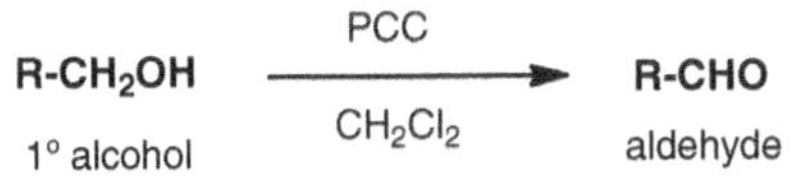

pyridinium chlorochromate (PCC)

$$R\text{-}CH_2OH \xrightarrow[CH_2Cl_2]{PCC} R\text{-}CHO$$
1° alcohol · · · · · · · · · · · aldehyde

Exercise 14.10

What is the main oxidation product of each of the following reactions?

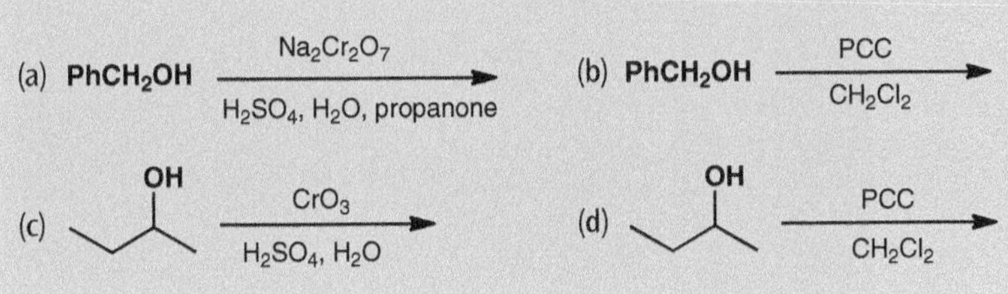

(a) PhCH₂OH $\xrightarrow[H_2SO_4,\ H_2O,\ propanone]{Na_2Cr_2O_7}$

(b) PhCH₂OH $\xrightarrow[CH_2Cl_2]{PCC}$

(c) $\xrightarrow[H_2SO_4,\ H_2O]{CrO_3}$

(d) $\xrightarrow[CH_2Cl_2]{PCC}$

Panel 14.3 Breath test for alcohol

When ethanol is oxidized by dichromate, Cr(VI) is reduced to Cr(III), and this is accompanied by a clear colour change from orange to green. This colour change is the basis of a preliminary test for whether someone has more than the legally permitted concentration of alcohol (the common word for ethanol in this context) in their blood (which varies from country to country). Ethanol in blood is in equilibrium with ethanol in the air in the lungs, which can be estimated by exhalation through a device which contains acidified potassium dichromate: the amount of ethanol in the exhaled breath is related to the amount of Cr(VI) reduced. Since the concentration of ethanol in the blood can be deduced from its concentration in the breath, this simple road-side test indicates whether a more accurate analysis is warranted. The more accurate analysis may be a further breath test under more controlled conditions, or by a direct analysis of ethanol in the blood or urine.

Panel 14.4 Swern oxidation

Several methods involving mild reagents have been developed for the oxidation of primary alcohols to aldehydes. One of them, which uses dimethyl sulfoxide (DMSO), oxalyl chloride, and a base, is known as *Swern oxidation*.

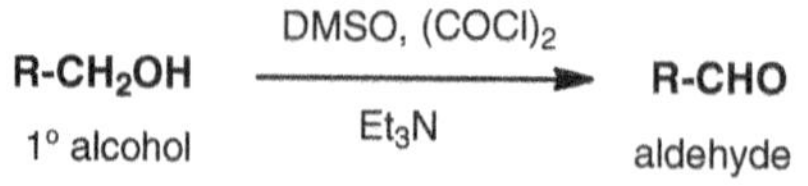

$$\text{R-CH}_2\text{OH} \xrightarrow[\text{Et}_3\text{N}]{\text{DMSO, (COCl)}_2} \text{R-CHO}$$

1° alcohol   aldehyde

A complex stepwise process shown below is initiated by nucleophilic addition of DMSO at a reactive carbonyl group of oxalyl chloride and loss of Cl⁻, CO, and CO_2 to give an electrophilic cationic sulfur species; this then suffers nucleophilic capture by a primary alcohol to give an alkoxysulfonium ion. The reaction is completed by deprotonation of the alkoxysulfonium ion and an intramolecular elimination of dimethyl sulfide to give the aldehyde.

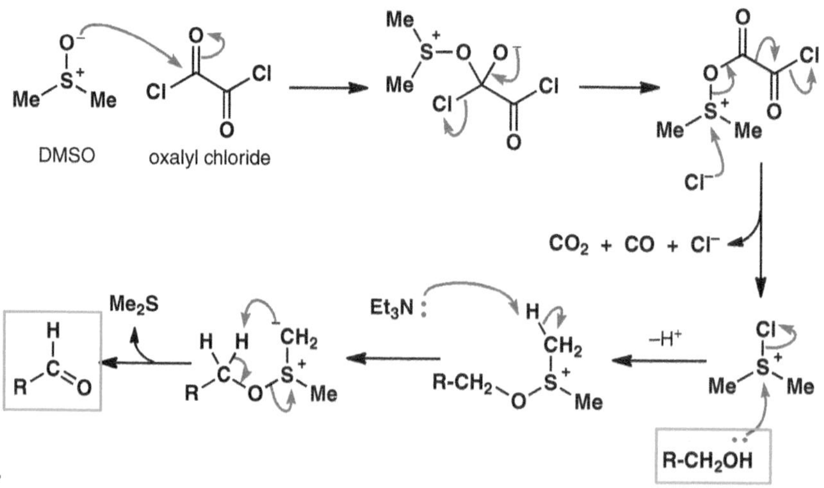

A proposed mechanism of Swern oxidation

14.5 Ring Opening of Epoxides

Epoxides, also called oxiranes, are three-membered cyclic ethers,[2] *i.e.* the oxygen atom is in the ring. They are more reactive than normal ethers owing to the angle strain of the small ring.

14.5.1 Acid-catalysed ring opening

The ring opening of alkyl-substituted oxiranes occurs readily with acid catalysis by the **S$_N$1 mechanism** and bond cleavage at the more substituted carbon. However, the

---

[2]   Epoxide is the common generic name for three-membered cyclic ethers, and oxirane is the IUPAC name. However, naming even relatively simple epoxides by the IUPAC oxirane system can be cumbersome. Consequently, trivial names as derivatives of alkenes are widely used, e.g. ethylene oxide for oxirane (the simplest epoxide and the parent for this class of compounds), and propylene oxide (methyloxirane). An alternative IUPAC system has also been approved using the prefix *n,m-epoxy-* where *n* and *m* specify the two carbons which, with the oxygen, form the three-membered ring, e.g. 1,2-epoxycyclohexane (cyclohexene oxide).

oxirane
(ethylene oxide)

methyloxirane
(propylene oxide)

1,2-epoxycyclohexane
(cyclohexene oxide)

intermediate is stabilized by a weak Lewis acid–base interaction between the cationic centre and the lone pair of the internal oxygen atom (Scheme 14.13).

Ethylene oxide is a hazardous inflammable gas produced industrially on a large scale by the catalytic oxidation of ethene. It is used for the manufacture of a range of organic compounds and as a disinfectant, e.g. in hospitals for sterilizing medical equipment.

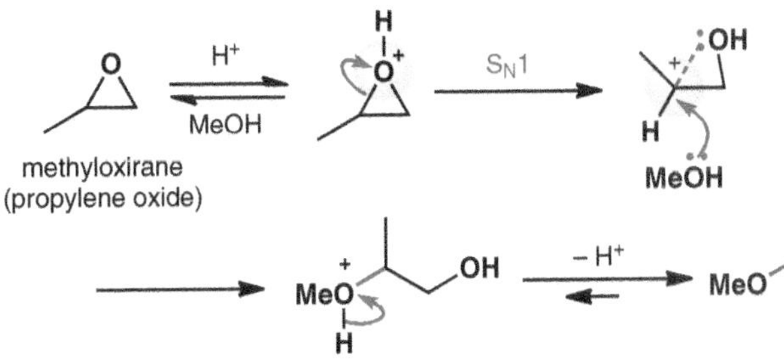

methyloxirane
(propylene oxide)

Scheme 14.13 Acid-catalysed ring opening of an epoxide.

If a protonated epoxide is not capable of yielding a viable carbenium ion, ring-opening will be by an S_N2 reaction.

14.5.2 Base-catalysed ring opening

In contrast to normal ethers, epoxides are susceptible to ring-opening attack by powerful nucleophiles; these reactions occur readily without catalysis owing to the release of ring strain. Grignard reagents, for example, open epoxides in a reaction which is useful for the synthesis of alcohols with an additional two-carbon unit, as mentioned in Sub-section 10.4.2.

Weak nucleophiles, however, require base catalysis. Consequently, solvolytic reactions with water or an alcohol containing a base proceed via HO^- or RO^- formed from the solvent, e.g. Scheme 14.14.

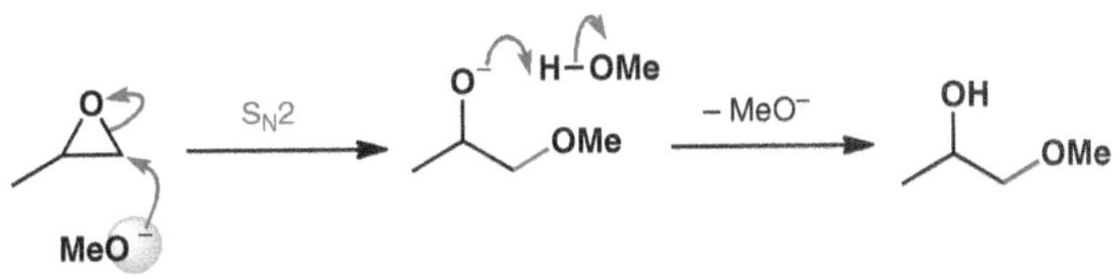

Scheme 14.14 Base-catalysed ring opening of an epoxide with NaOMe in MeOH.

With or without base catalysis, nucleophile-induced ring opening of an epoxide may be regarded as an S_N2 substitution in which the nucleofuge is tethered to the β C of the electrophilic alkyl group. For steric reasons, therefore, nucleophilic attack and consequent C–O bond cleavage occur at the less substituted carbon, as illustrated in Scheme 14.14.

We saw in Chapters 8 and 9 that acid catalysis enhances the reactivity of the electrophile whereas base catalysis enhances the reactivity of the nucleophile in nucleophilic addition and substitution reactions.

Exercise 14.11

Show the structure of the main product when 2,2-dimethyloxirane reacts in methanol (a) in the presence of an acid, and (b) with added sodium methoxide, and explain the results.

2,2-dimethyloxirane

Panel 14.5 Crown ethers and cryptands

The lone pairs on the oxygen atoms of ethers are responsible for their Lewis basicity and solvent properties. A series of cyclic polyethers known as crown ethers (this name comes from their shapes) exemplify dramatically the cooperative effect of multiple ether groups upon the ability of the molecules to solvate metal ions. Each crown ether is identified as n-crown-m, where n is the size of the ring and m the number of oxygen atoms in the ring.

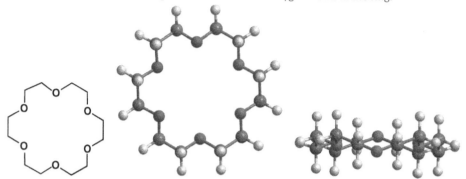

Structure of 18-crown-6 and a molecular model (top view and side view).

Although single oxygen–metal ion interactions are weak, they are cumulative; consequently, coordination of specific cations by crown ethers can be strong (depending on the sizes of the cation and the ring). The affinity between a cation and a crown ether is greatest when the cation just fits inside the ring with all interactions optimal; 12-crown-4 and 18-crown-6 are selective for Li^+ and K^+, respectively, and the flexibility of the cyclic ligands allows the coordination to be tetrahedral around the Li^+ and octahedral around the K^+.

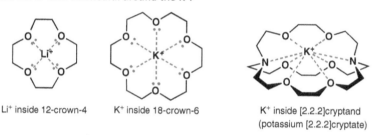

Li^+ inside 12-crown-4 K^+ inside 18-crown-6 K^+ inside [2.2.2]cryptand
(potassium [2.2.2]cryptate)

Cations complexed by crown ethers and a cryptand.

A three-dimensional extension of crown ethers was achieved using bridgehead nitrogen atoms to give **cryptands** which can encapsulate cations of the right size within their cage structures very effectively; [2.2.2]cryptand complexes K^+ very effectively to give a **cryptate**, potassium [2.2.2]cryptate.

Panel 14.6 Fluorodeoxyglucose in cancer diagnosis: rapid synthesis by an S_N2 reaction using a cryptand

One technique used widely in cancer diagnosis is positron emission tomography (PET) in which a positron-emitting radioisotope, typically ¹⁸F, is used. The nucleus of this unstable isotope of fluorine (9 protons and 9 neutrons) emits a positron (same mass as an electron, but with a positive charge). Upon collision with a nearby electron, both particles are annihilated, which generates two gamma rays 180° apart. These are detected by an imaging scanner which enables location of the radioisotope in the body.

The ¹⁸F is incorporated in a bioactive compound such as glucose for uptake by humans. Rapidly growing tumours require more glucose and take up greater amounts than normal tissue. 2-Fluoro-2-deoxy-D-glucose (FDG) labelled with ¹⁸F is used since it is taken up together with normal glucose by the cancerous tissue.

2-fluoro-2-deoxy-D-glucose-¹⁸F
(FDG)

D-glucose

The crown ether or cryptand is sometimes called a host and the complexed cation a guest, and we refer to guest–host interactions as being weak or strong according to the magnitude of the equilibrium constant for the formation of the complex. C.J. Pedersen, D.J. Cram, and J.-M. Lehn were awarded the 1987 Nobel Prize in Chemistry for their work on crown ethers and related compounds with structure-specific interactions of high selectivity. Their work allowed the development of a new field of chemistry and contributed to our understanding of what has become known as **molecular recognition**; the phenomenon is closely related to many biological processes such as enzyme–substrate and antibody–antigen interactions, neural transmission, and the transport of ions through cell membranes.

Selective crown ether–metal ion complexation is exploited in **phase-transfer catalysis** (see Panel 12.2) in organic synthesis. Complexation of metal ions by crown ethers causes their salts to become soluble in nonpolar solvents; for example, potassium salts such as KF, KCN, $KMnO_4$, and CH_3CO_2K can be solubilized in an organic solvent by 18-crown-6. When a solution of an electrophilic organic substrate in an organic solvent which is immiscible with water (e.g. toluene or chloroform) is mixed with an aqueous solution of the potassium salt of a nucleophilic anion, no S_N2 reaction will take place because the reactants are in different phases and do not come into contact. However, if 18-crown-6 is added, it will transport the complexed K^+, *accompanied by the anion to ensure electrical neutrality*, across the phase boundary into the organic phase. In the organic solvent, the anion will not be complexed by H-bonding as it was in the water, so is 'naked' and very reactive, and the reaction proceeds efficiently.

$$RCH_2Cl\ +\ KCN\ \xrightarrow[\text{C}_6\text{H}_5\text{CH}_3\text{–H}_2\text{O}]{\text{18-crown-6}}\ RCH_2CN\ +\ KCl$$

The solution obtained by solubilization of $KMnO_4$ in benzene with 18-crown-6 is known as 'purple benzene' (with the colour of MnO_4^-) and is very convenient for oxidation reactions in an organic phase.

Some antibiotics called **ionophores** act by transferring metal ions across biological phase boundaries. Nonactin, for example, is a cyclic polyether (with lactone functionality also) which selectively coordinates K^+, and can transport it through (nonpolar) lipid cell membranes. As a result, the concentration of the K^+ within a cell is upset which leads to cell death.

nonactin

The radioisotope ^{18}F has a half-life of about 110 min (it decays by one half in 110 min) to give ^{18}O, a stable isotope of oxygen of low natural abundance. This short half-life is good for patients as it minimizes their exposure to radiation, but the rapid synthesis of the compound and its quick use are crucial.

The ^{18}F isotope is generated as $K^+\ {}^{18}F^-$ by cyclotron irradiation of $H_2{}^{18}O$ in aqueous solution (with KOH to provide the K^+). However, the synthesis of FDG cannot be carried out effectively in aqueous solution since hydrated fluoride ion is virtually unreactive. The anionic nucleophile is only weakly solvated, and consequently much more reactive, in a polar aprotic solvent if it can be made soluble by complexation of its K^+ counter cation. This has been achieved using a crown ether, but a cryptand which binds K^+, [2.2.2]cryptand, is more effective. The aqueous solution of $K^+\ {}^{18}F^-$ is concentrated on an anion-exchange resin (see Panel 3.3), then eluted with a solution of [2.2.2]cryptand and K_2CO_3 in ethanenitrile (acetonitrile). The acetonitrile solution of $^{18}F^-$ and K^+ complexed by the cryptand obtained this way is then used for the preparation. The organic starting material, D-mannose, must also be available in a form which is soluble in ethanenitrile, and reactive towards $^{18}F^-$.

D-mannose is modified by conversion of the C2–OH into trifluoromethanesulfonate, an excellent leaving group, with the other four OH groups protected by acetylation. The modified mannose is then treated with the acetonitrile solution containing the $^{18}F^-$ and the K^+ [2.2.2]cryptate; this leads to a rapid S_N2 reaction with inversion of

(continues...)

14.6 Thiols and Other Sulfur Compounds

Sulfur is a third-period element in the same group as oxygen. Differences between the properties of sulfur and oxygen compounds are mainly because the valence electrons of S are shielded from the nuclear charge by an extra shell of core electrons. Consequently, compared with oxygen, the atomic radius of sulfur is larger, the valence electrons are more diffuse and polarizable, and the electronegativity of S is low (2.58 on the Pauling scale, i.e. comparable with 2.55 for carbon; see Section 1.2).

14.6.1 Thiols and their derivatives

For RSH, $pK_a \sim 10$ compared with ~ 16 for ROH.

Thiols (RSH, also known as mercaptans) are the sulfur analogues of alcohols but they are appreciably stronger Brønsted acids for reasons discussed in Section 6.3.1. However, they form only weak hydrogen bonds, which causes them to be much more volatile than alcohols; they are frequently detectable at very low concentrations by their strongly unpleasant odours (some examples were given in Sub-section 3.3.3).

Thiols are amongst the most nucleophilic neutral compounds known and, for example, react with aldehydes and ketones to give dithioacetals (Sub-section 8.4.3) which are useful intermediates for organic syntheses (Sub-sections 10.2.2 and 10.5.3).

Being anionic, thiolates are even more nucleophilic (and more so than alkoxides, as discussed in Sub-section 12.2.4); they participate in S_N2 reactions and react with acylating reagents to give thioesters.

Thioesters are generally more reactive than normal esters because conjugation between the carbonyl group and sulfur is less effective than with oxygen. This is principally owing to the disparity in size between the 2p orbital of the C of the carbonyl and the larger, more diffuse orbital of the correct symmetry on S, i.e. orbital overlap is poorer. Consequently, the acyl-S bond is weaker than the acyl-O bond, and thiolate (RS⁻) is a better nucleofuge than alkoxide.

a thioester

a dithioacetal

Thiolates are like iodide in being good nucleophiles *and* good nucleofuges.

thioester (Y = S)
ester (Y = O)

contribution of this form: S < O

Panel 14.6 Continued

configuration at C2 (a mannose derivative becomes a glucose). Subsequent acid hydrolysis of the acetate groups gives FDG.

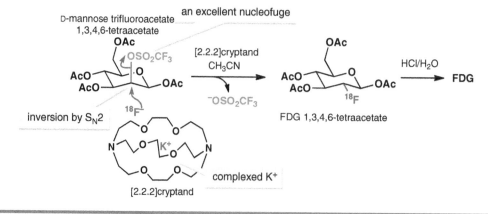

D-mannose trifluoroacetate 1,3,4,6-tetraacetate

an excellent nucleofuge

inversion by S_N2

[2.2.2]cryptand
CH_3CN

FDG 1,3,4,6-tetraacetate

HCl/H_2O

FDG

complexed K⁺

[2.2.2]cryptand

Complete the following reactions.

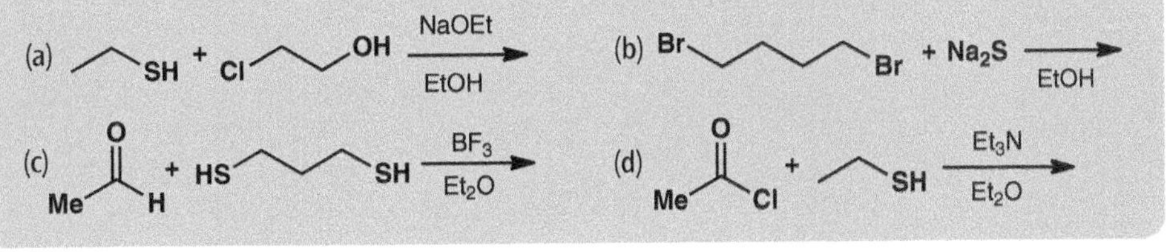

(a) $\xrightarrow[\text{EtOH}]{\text{NaOEt}}$

(b) $+ Na_2S \xrightarrow{\text{EtOH}}$

(c) $\xrightarrow[\text{Et}_2\text{O}]{\text{BF}_3}$

(d) $\xrightarrow[\text{Et}_2\text{O}]{\text{Et}_3\text{N}}$

Sulfides (RSR', sometimes called thioethers) are sufficiently nucleophilic to give sulfonium salts upon reaction with haloalkanes, e.g. eqn 14.5. The sulfonium salts formed in these reactions are stable and act as alkyl donors, i.e. as the electrophilic component in S_N2 reactions. Such reactions will be like the reverse of reaction 14.5 but with another nucleophile instead of iodide (see Panel 12.1 for a biological example).

Electrophilic sulfonium ions are involved in the Swern oxidation method in Panel 14.4.

$$ \underset{\text{dimethyl sulfide}}{(CH_3)_2S} + CH_3-I \xrightarrow{S_N2} \underset{\text{trimethylsulfonium iodide}}{(CH_3)_2\overset{+}{S}-CH_3 \; I^-} \qquad (14.5) $$

nucleophile

14.6.2 Biological thiols: their functions and derivatives

Biological thiols play important roles in living organisms, and we shall illustrate this by focussing on just a few. Cysteine is an amino acid with a thiol group, and is a unit of proteins and other peptides. Glutathione is a tripeptide biomolecule which contains a cysteine residue and is widely distributed throughout the body.

Glutathione is regenerated in the reverse of eqn 14.6 with NADH (Panel 10.3) as reductant and catalysis by the enzyme glutathione reductase.

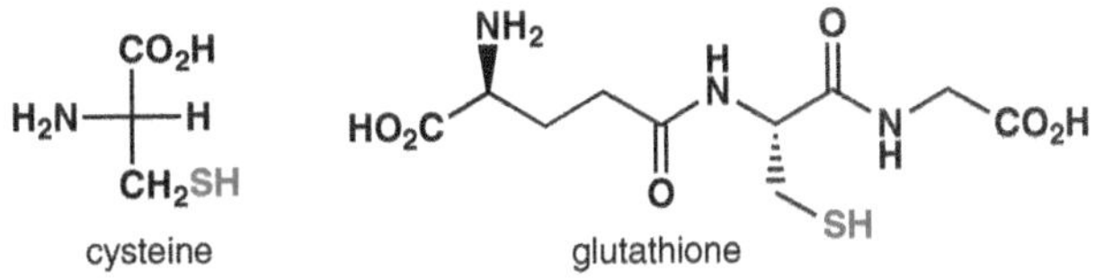

cysteine

glutathione

The biochemical functions of these two compounds originate in two distinctive properties of thiols. The first is the ease of the reversible oxidative coupling of two thiols to give a disulfide (eqn 14.6). Two cysteine residues at different locations in a protein may form a disulfide bridge, in which case it will be a major feature in determining the tertiary structure of the protein (see Chapter 24).

$$ \underset{\text{thiol}}{2\,R\text{-SH}} \underset{\xrightarrow{\;\;\text{reduction}\;\;}}{\xleftarrow{\;\;\text{oxidation}\;\;}} \underset{\text{disulfide}}{R\text{-S-S-R}} \qquad (14.6) $$

Perms

The reaction of eqn 14.6 is applied to the (permanent) waving of hair and also to hair straightening. Whether a hairstyle is straight or wavy depends on the shapes of individual hairs, and these are maintained by disulfide bonds and non-covalent interactions of proteins. If the hairstyle is to be changed, the disulfide bonds within individual hairs are first cleaved by exchange reactions (reduction) with proprietary thiol reagents (R'SH),

$$ RS\text{-}SR + R'SH \rightleftharpoons RSH + RS\text{-}SR' $$

which 'loosens' the shapes of the hairs. The hair is then constrained to the desired new style (straight or wavy) and 'set' by forming new S–S bonds by oxidation using H_2O_2 or another mild oxidizing agent.

Glutathione quenches harmful oxidants in our body by the same reaction, i.e. by acting as a mild reducing agent.

Exercise 14.13

Lipoic acid is an important disulfide involved in some biological oxidations. What is the reduced form of lipoic acid?

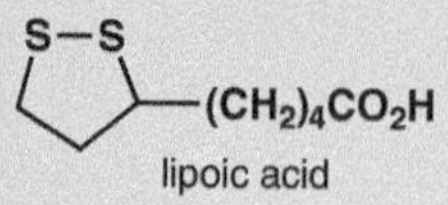

lipoic acid

The second distinctive property of thiols which is expressed biochemically is their high nucleophilicity. Glutathione reacts as a nucleophile with potentially toxic foreign compounds (electrophilic xenobiotic compounds) such as haloalkanes and epoxides, and makes them water-soluble for excretion (Scheme 14.15). See Panel 19.2 about the detoxification of carcinogenic aromatic epoxides for a specific example.

Scheme 14.15 The S_N2 reactions of glutathione with xenobiotic electrophiles.

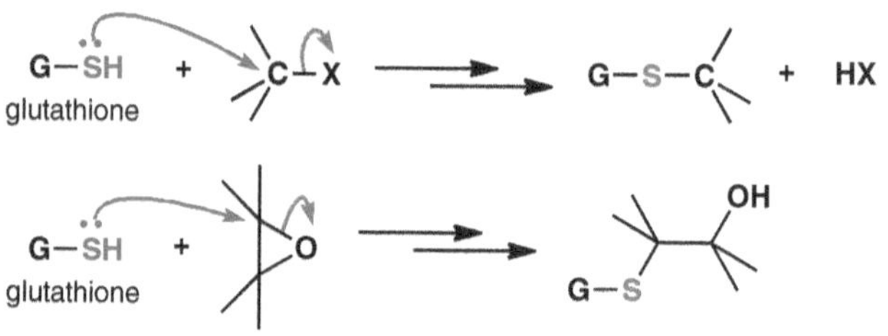

Coenzyme A (CoASH or simply CoA) is another thiol whose biological function depends on its high nucleophilicity; it forms various thioesters which are involved in metabolism, e.g. acetyl CoA.

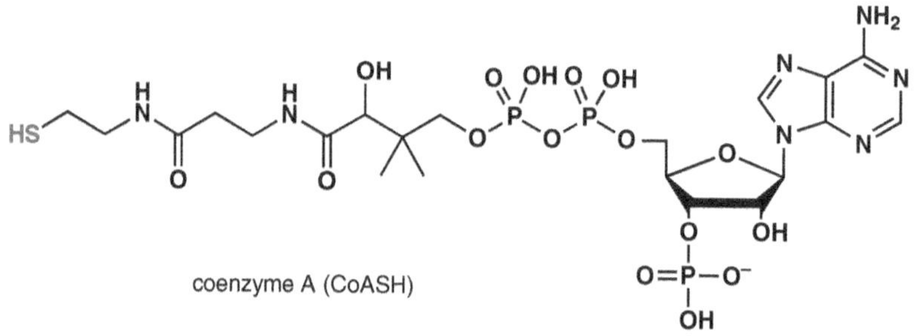

coenzyme A (CoASH)

Because the thiolate residue of these biological thioesters is a good nucleofuge (see above), acetyl CoA is an effective acetylating agent for biochemical nucleophiles (Nu⁻ in eqn 14.7; see Chapter 9); it is involved in biological condensation reactions, for example in the biosynthesis of fatty acids (see Panel 17.3).

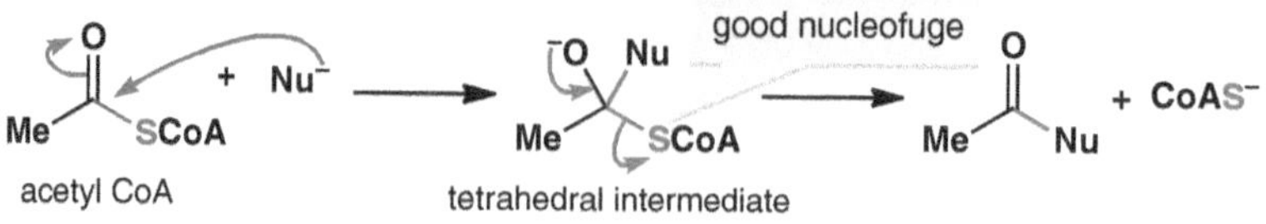

acetyl CoA   tetrahedral intermediate   (14.7)

14.6.3 Dual electronic effects of alkylthio groups

An alkylthio group (RS-) stabilizes carbenium ions by donating its sulfur lone pair, an effect which is analogous to that of the alkoxy group. Although the electron-donating conjugation from S to an electron-deficient sp^2 C is poorer than from O (see Sub-section 14.6.1), S is less electronegative so has a weaker electron-withdrawing inductive effect than O. The result of these opposing effects is that the stabilization of a carbenium ion by an RS group is broadly comparable with that by an RO group.

Interestingly, sulfur can also stabilize carbanions. The pK_a of 1,3-dithiane (a di-thioacetal) is 31.1 compared with about 50 for cyclohexane. The enhanced stability of the conjugate base of dithiane was once thought to be attributable to the vacant 3d orbitals on sulfur which can accept electrons. However, the hyperconjugative effect is now believed to be a more important factor in the stabilization of an adjacent negative charge (Figure 14.1). The occupied orbital of the carbanion lone pair interacts with the (vacant) antibonding orbitals of the two adjacent S–C bonds (interaction with only one is shown in Figure 14.1). In this way, negative charge is (partially) delocalized onto the S atoms. Note that in the hyperconjugative stabilization of a carbenium ion, an adjacent C–H bonding pair is partially delocalized into the vacant orbital of an electron-deficient centre (Sub-section 12.4.3), so the electron flow is in the opposite direction.

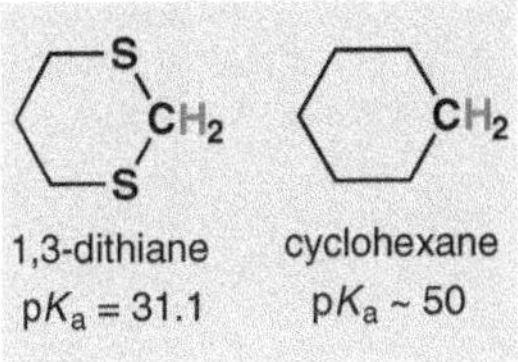

1,3-dithiane
$pK_a = 31.1$

cyclohexane
$pK_a \sim 50$

Figure 14.1 (a) Structure of the conjugate base of 1,3-dithiane, and (b) hyperconjugative stabilization of the carbanion.

By converting an *electrophilic* carbonyl group first into a dithiane then into a carban-ion by deprotonation, we generate a useful *nucleophilic* reagent; this conversion inverts the electrical nature of the reagent (an electrophile becomes a nucleophile) and is called 'Umpolung' (a German word meaning reversal of polarity). The carbanion can then react as a nucleophile with haloalkanes (e.g. R'Y in Scheme 14.16) by the S_N2 mechanism, or with other electrophiles. When the dithiane group is hydrolysed back to the carbonyl, we see that we have been able to replace the H of the aldehyde by an *electrophile* by exploita-tion of the Umpolung strategy.

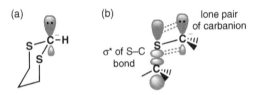

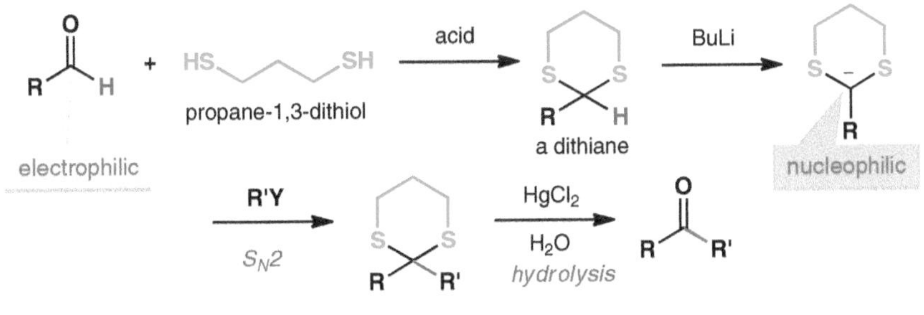

Scheme 14.16
Umpolung via a 1,3-dithiane.

14.6.4 Compounds of S(IV) and S(VI)

Because sulfur is in the third period, it can accommodate more than an octet of electrons in its valence shell. This means that S can form more than two covalent bonds:

$$
\begin{array}{cccc}
\underset{\text{sulfoxide}}{R-\overset{\overset{\displaystyle O}{\|}}{S}-R'} &
\underset{\text{sulfone}}{R-\overset{\overset{\displaystyle O}{\|}}{\underset{\underset{\displaystyle O}{\|}}{S}}-R'} &
\underset{\text{sulfinic acid}}{R-\overset{\overset{\displaystyle O}{\|}}{S}-OH} &
\underset{\text{sulfonic acid}}{R-\overset{\overset{\displaystyle O}{\|}}{\underset{\underset{\displaystyle O}{\|}}{S}}-OH}
\end{array}
$$

Sulfoxides and sulfones are often represented by structures with S=O double bonds, but they are not planar and are more properly drawn with charge-separated bonds or, more completely but less conveniently, as resonance hybrids:

$$
\underset{\text{sulfoxide}}{R-\overset{\overset{\displaystyle O}{\|}}{S}-R' \; \longleftrightarrow \; R-\overset{\overset{\displaystyle O^-}{|}}{\overset{+}{S}}-R'}
\qquad
\underset{\text{sulfone}}{R-\overset{\overset{\displaystyle O}{\|}}{\underset{\underset{\displaystyle O}{\|}}{S}}-R' \; \longleftrightarrow \; R-\overset{\overset{\displaystyle O^-}{|}}{\underset{\underset{\displaystyle O^-}{|}}{\overset{2+}{S}}}-R'}
$$

better representation

We have already encountered dimethyl sulfoxide (DMSO) as an excellent polar aprotic solvent (Sections 6.7 and 12.3), and as an oxidizing agent in the so-called Swern oxidation method (Panel 14.4).

14.7 Reactions of Amines

Simple alkylamines such as $EtNH_2$, Et_2NH, and Et_3N (pK_{BH^+} values 10.7, 11.0, and 10.7, respectively; see Sub-section 6.5.2) are slightly stronger Brønsted bases than ammonia (NH_3, $pK_{BH^+} = 9.25$). As bases, they are widely used in the preparation of aqueous buffer solutions, and as basic reagents, but are otherwise unexceptional. Substituents in their alkyl groups, or replacement of an alkyl by an aryl group, have the effects predicted from knowledge of the electronic effects of the substituents, e.g. $PhNH_2$ and $CF_3CH_2NH_2$ ($pK_{BH^+}=4.6$ and 5.7, respectively) are weaker bases than $EtNH_2$ whereas guanidine $HN=N(NH_2)_2$ is appreciably stronger ($pK_{BH^+} = 13.6$).

14.7.1 Amines as nucleophiles and nucleofuges

Ammonia itself reacts as a nucleophile with alkyl halides RX to form C–N bonds by the S_N2 mechanism.

$$
H_3N : \overset{\frown}{\;} R-X \xrightarrow{S_N2} R-NH_3^+ \; X^- \xrightarrow{NH_3} \underset{\text{a primary amine}}{R-NH_2 + NH_4^+ \; X^-}
$$

A primary amine formed in this way can then undergo further consecutive S_N2 reactions to give secondary and tertiary amines, and ultimately a quaternary ammonium salt (Scheme 14.17). Unless the ammonia is present in very large excess, its reaction with an alkyl halide usually provides a mixture of all of these compounds.

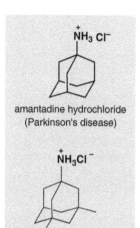

Scheme 14.17 Reactions of amines with an alkyl halide.

$$
RNH_2 \xrightarrow{RX} R_2NH_2^+ \; X^- \xrightarrow{NH_3} \underset{\text{a secondary amine}}{R_2NH} \xrightarrow{RX} R_3NH^+ \; X^-
$$

$$
\xrightarrow{NH_3} \underset{\text{a tertiary amine}}{R_3N} \xrightarrow{RX} \underset{\text{a quaternary ammonium salt}}{R_4N^+ \; X^-}
$$

Reactions of amines with carbonyl compounds to give imines (Section 8.6), enamines (Section 8.6), and amides (Sections 9.3 and 9.4) have already been discussed.

Exercise 14.14

What is the main product in each of the following reactions when the amine acts as a nucleophile?

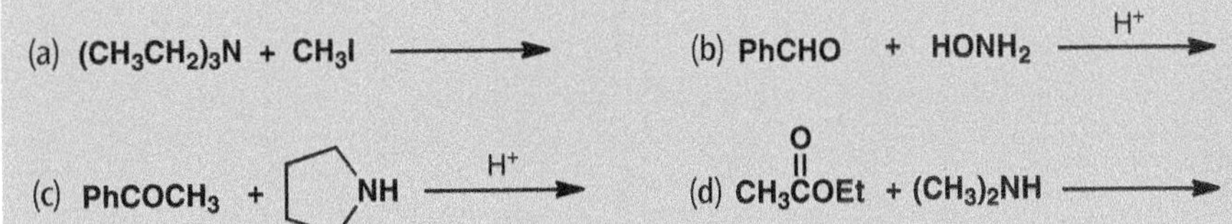

(a) $(CH_3CH_2)_3N + CH_3I \longrightarrow$

(b) $PhCHO + HONH_2 \xrightarrow{H^+}$

(c) $PhCOCH_3 + $ [cyclopentane ring]$NH \xrightarrow{H^+}$

(d) $CH_3\overset{\overset{\displaystyle O}{\|}}{C}OEt + (CH_3)_2NH \longrightarrow$

The amide anion (NH_2^-) is even worse as a nucleofuge than hydroxide (pK_{BH^+} of $NH_2^- = 33$ and pK_{BH^+} of $HO^- = 15.7$). However, an amino group can be converted into an ammonio residue in which case, as a leaving group, it is an amine which is much better (though still not very good). Reaction of a primary alkylamine with an excess of iodomethane (permethylation) followed by moist silver oxide gives the alkyltrimethylammonium hydroxide. Heating this compound leads to a regioselective (Hofmann) E2 elimination with NMe_3 as the nucleofuge to give the less substituted alkene as the main product (Scheme 14.18). This reaction is known as the Hofmann elimination (see Sub-section 13.5.2 and Example 13.3(b)).

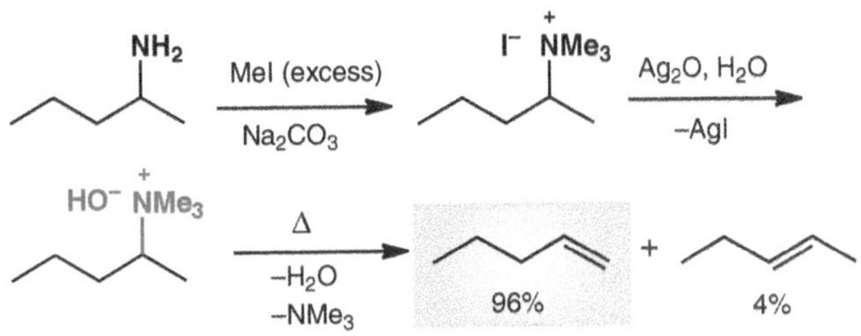

Scheme 14.18 The Hofmann elimination to give a terminal alkene.

14.7.2 Reactions of alkylamines with nitrous acid

Nitrous acid (HNO_2) is an unstable weak acid ($pK_a = 3.3$) which is easily generated *in situ* by adding sodium nitrite to an ice-cold dilute aqueous strong acid, usually hydrochloric acid (eqn 14.8).

$$H_3O^+ + NO_2^- \rightleftharpoons \underset{\text{nitrous acid}}{O=N-OH} + H_2O \qquad (14.8)$$

Treatment of a primary or secondary alkylamine with nitrous acid under these acidic conditions leads initially to nitrosation, i.e. the formation of an *N*-nitrosamine. In the case of a secondary amine, this is usually an isolable yellow oil, $R_2N–NO$ (see eqn 14.10).

A nitrosated primary amine, however, immediately reacts further in a process leading to **diazotization** to give an unstable alkanediazonium ion ($R–N_2^+$), which reacts rapidly (see next sub-section) with the evolution of N_2 (eqn 14.9). Tertiary amines do not *appear* to react with nitrous acid (eqn 14.11), but in fact usually undergo a complicated reaction of little use which involves cleavage of C–N bonds. The three obviously different reactions with nitrous acid in Scheme 14.19 were once used to distinguish between primary, secondary, and tertiary aliphatic amines.

Although nitrous acid is generated under acidic conditions, and amines are basic, the equilibrium concentration of free amine is sufficient for reaction to proceed.

Scheme 14.19 Reactions of nitrous acid with primary, secondary, and tertiary aliphatic amines.

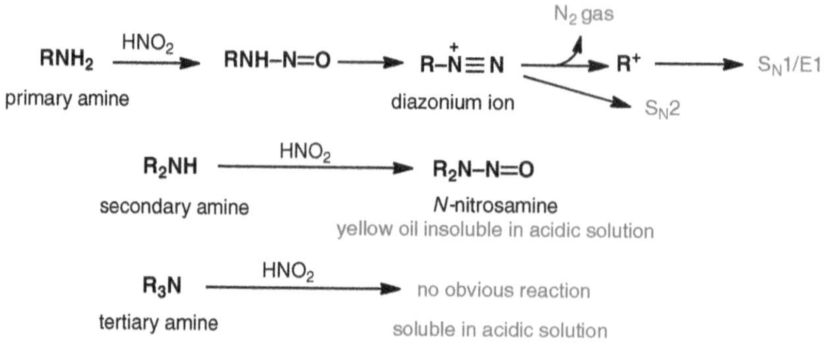

$$RNH_2 \xrightarrow{HNO_2} RNH-N=O \longrightarrow R-\overset{+}{N}\equiv N \xrightarrow{N_2 \text{ gas}} R^+ \longrightarrow S_N1/E1 \qquad (14.9)$$

primary amine   diazonium ion $\searrow S_N2$

$$R_2NH \xrightarrow{HNO_2} R_2N-N=O \qquad (14.10)$$

secondary amine   *N*-nitrosamine
                   yellow oil insoluble in acidic solution

$$R_3N \xrightarrow{HNO_2} \text{no obvious reaction} \qquad (14.11)$$

tertiary amine   soluble in acidic solution

14.7.3 Alkanediazonium ions

Beware possible confusion between the meanings of the terms *primary*, *secondary*, and *tertiary* according to whether they apply to amines or alkyl groups (and alcohols).

The nitrogen molecule is an extremely good leaving group; consequently, the alkanediazonium ion ($R-N_2^+$) formed from nitrous acid and a primary amine (eqn 14.9) decomposes immediately with the evolution of N_2 either by an $S_N1/E1$ mechanism *via* a carbenium ion (secondary and tertiary alkyl groups), or by an S_N2 mechanism (simple primary alkyl groups). When the reaction is carried out in aqueous solution, the substitution product is an alcohol, so the overall reaction (sometimes called *deamination*) corresponds to $R-NH_2 \rightarrow R-OH$. However, because of associated elimination and possible rearrangements, yields are seldom high.

Acidification of a solution of sodium nitrite (eqn 14.8 above) leads to a complicated mixture of several nitrosating species including the nitrosonium ion, NO^+, and dinitrogen trioxide (the anhydride of nitrous acid):

$$O=N-OH + H_3O^+ \rightleftharpoons NO^+ + 2 H_2O$$

$$O=N-OH + NO^+ + H_2O \rightleftharpoons O=N-O-N=O + H_3O^+$$

dinitrogen trioxide

A credible mechanism for the diazotization of a primary amine by dinitrogen trioxide (N_2O_3) is as follows:

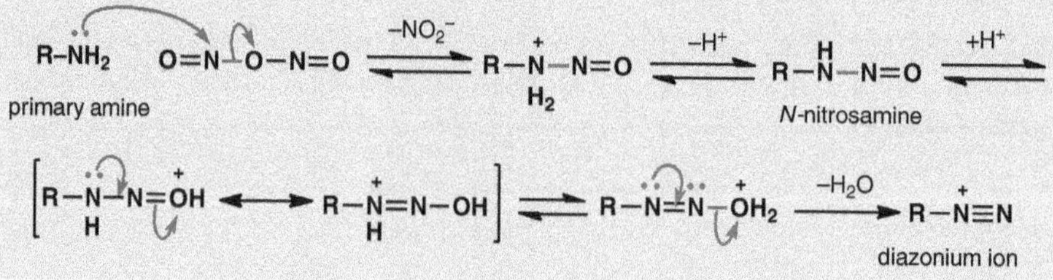

$$R-\ddot{N}H_2 \quad O=N-O-N=O \rightleftharpoons R-\overset{+}{\underset{H_2}{N}}-N=O \xrightarrow{-H^+} R-\overset{H}{\underset{}{N}}-N=O \xrightarrow{+H^+}$$

primary amine $-NO_2^-$ *N*-nitrosamine

$$\left[R-\overset{+}{\underset{H}{N}}=\overset{+}{N}=\overset{..}{O}H \leftrightarrow R-\overset{+}{\underset{H}{N}}=N-OH \right] \rightleftharpoons R-\ddot{N}\equiv\overset{+}{N}-\overset{+}{O}H_2 \xrightarrow{-H_2O} R-\overset{+}{N}\equiv N$$

diazonium ion

Exercise 14.15

Give the main product(s) of each reaction?

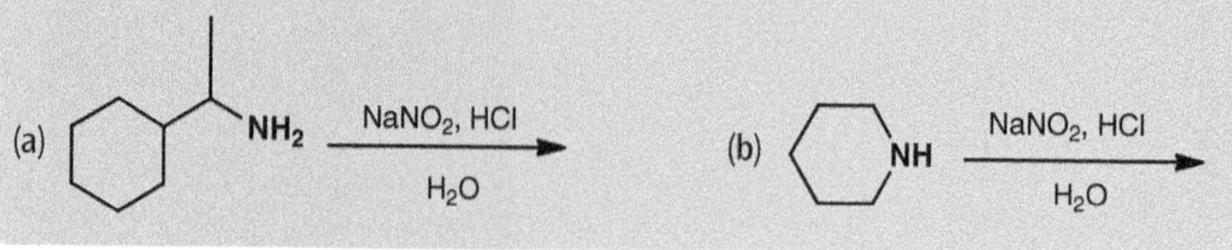

(a) [structure: cyclohexyl-CH(CH_3)-NH_2] $\xrightarrow[\text{H}_2\text{O}]{\text{NaNO}_2, \text{HCl}}$

(b) [structure: piperidine, NH ring] $\xrightarrow[\text{H}_2\text{O}]{\text{NaNO}_2, \text{HCl}}$

Summary

- Alcohols are important starting materials for the preparation of alkenes, haloalkanes, ethers, and esters.

- The C–O bonds of **alcohols** ROH and **ethers** ROR' undergo heterolysis in substitution and elimination reactions, just as the C–Y bonds of haloalkanes do, but only in the presence of an **acid catalyst**. The catalysed reaction is initiated by reversible protonation of the oxygen and proceeds with departure of H_2O or R'OH as the leaving group.

- The mechanisms of reactions of alcohols and ethers carried out under weakly nucleophilic acidic conditions are usually $S_N1/E1$ if a viable **carbenium ion** is possible (in which case it may rearrange by a 1,2-shift of hydrogen or an alkyl group). If not, the protonated substrate reacts by an S_N2 mechanism.

- **Oxidation** of primary alcohols gives aldehydes which are very easily oxidized further to give carboxylic acids. To obtain the aldehyde, mild methods under non-aqueous conditions are required. Secondary alcohols give ketones, while tertiary alcohols are resistant to oxidation.

- **Epoxides** (oxiranes) are reactive three-membered cyclic ethers which undergo ring-opening nucleophilic substitution under acidic or basic conditions.

- **Sulfur** is directly below oxygen in the periodic table, and the distinctive chemical properties of its compounds (high nucleophilicity and ease of oxidation) are mainly attributable to their highly polarizable, loosely held valence electrons. The thiol group is important in biological processes.

- **Amines** are typical organic bases, and react as nucleophiles; they undergo characteristic reactions with nitrous acid.

Problems

14.1 What are initial product(s) in each of the following reactions with hydrohalic acids?

(a)

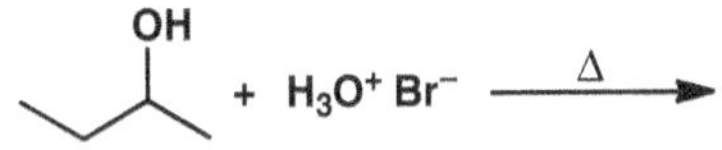

OH

$\diagdown\diagup\!\diagdown\diagup$ + H_3O^+ Br^- $\xrightarrow{\Delta}$

(b) $(CH_3)_3C-O-CH_3$ + H_3O^+ I^- $\xrightarrow{\Delta}$

(c) 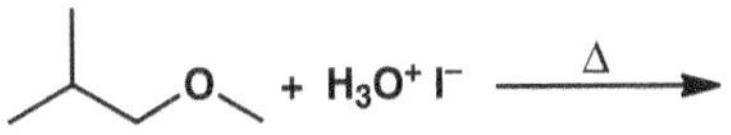 $\diagdown\!\diagup\!O\!\diagup$ + H_3O^+ I^- $\xrightarrow{\Delta}$

(d) 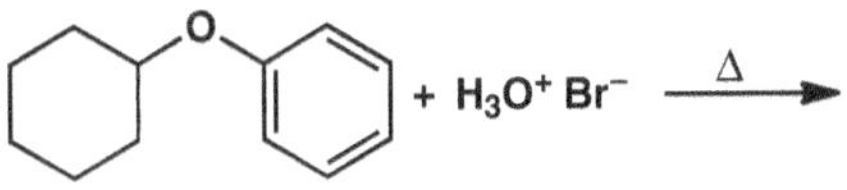 [cyclohexyl–O–phenyl] + H_3O^+ Br^- $\xrightarrow{\Delta}$

14.2 Propose a reaction mechanism for cleavage of each of the following ethers in hydriodic acid.

(a) 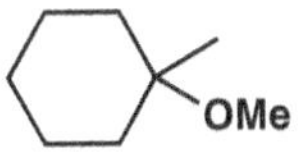 [cyclohexane with CH₃ and OMe]

(b) 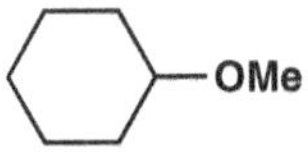 [cyclohexane with OMe]

14.3 What is the main product in the acid-catalysed dehydration of each of the following?

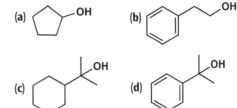

(a) [cyclopentane]—OH

(b) [benzene ring with CH₂CH₂]—OH

(c) [cyclohexane with C(CH₃)₂]—OH

(d) [benzene with C(CH₃)₂]—OH

14.4 Show the structures of all alkenes formed in the acid-catalysed dehydration of the following alcohols; indicate in each case which alkene is the main product, and explain why.

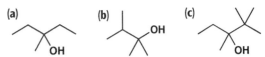

(a) [structure with OH]

(b) [structure with OH]

(c) [structure with OH]

14.5 Explain the regioselectivity from the rearranged carbenium ion in the reaction of Example 14.2.

14.6 When a mixture of methanol and t-butyl alcohol is treated with a catalytic amount of sulfuric acid, only *t*-butyl methyl ether is obtained in a high yield. However, a similar reaction of methanol and butan-1-ol gives three ethers upon heating. Explain the difference.

14.7 2-Ethoxy-2-methylpropane (*t*-butyl ethyl ether) is converted into 2-methylpropan-2-ol (*t*-butyl alcohol) and ethanol upon treatment by aqueous H_2SO_4.

(a) Write a reaction mechanism.

(b) Which product contains the ^{18}O when the oxygen of the starting ether is labelled with the ^{18}O isotope?

14.8 But-3-en-2-ol gives 3-bromobut-1-ene and 1-bromobut-2-ene on treatment with hydrobromic acid. Propose a reaction mechanism.

14.9 Show the reactivity order of the following alcohols in acid-catalysed dehydration.

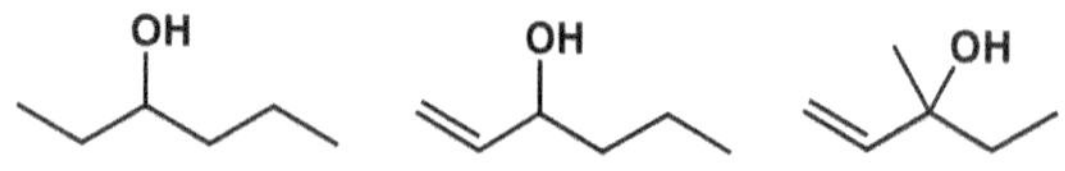

1   2   3

14.10 3-Methylbutan-2-ol gives 2-bromo-2-methylbutane as a main product on treatment with hydrobromic acid. Propose a reaction mechanism.

14.11 Dehydration of 2,2-dimethylpropan-1-ol with a catalytic amount of sulfuric acid gives 2-methylbut-2-ene as a main product. Propose a reaction mechanism.

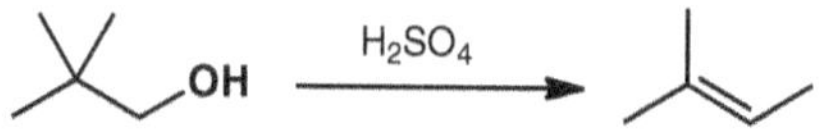

14.12 Propose a method for the preparation of each compound below from an appropriate alcohol.

(a) 1-chloropentane   (b) 2-iodopentane
(c) bromocyclopentane   (d) pentanenitrile (C_4H_9CN)
(e) pentan-3-amine

14.13 What is the main product in each of the following reactions?

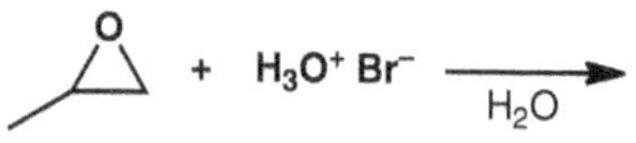

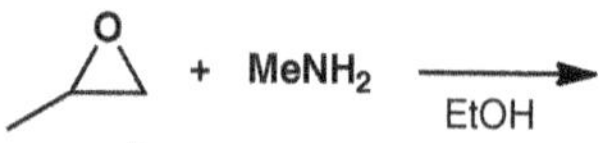

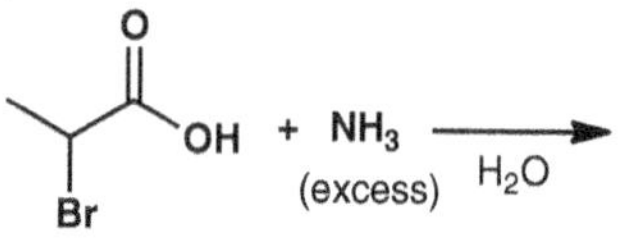

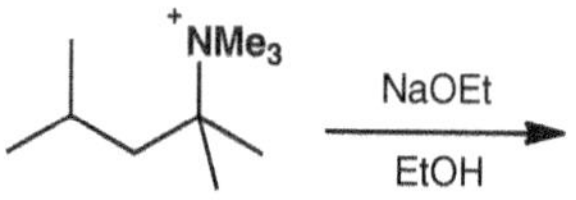

14.14 Give reagent(s) necessary for each conversion.

(a) $CH_3CH_2NH_2 \longrightarrow (CH_3CH_2)_4N^+ Br^-$

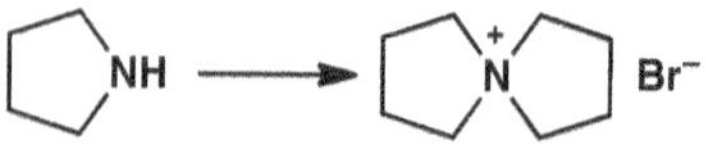

14.15 Propose a method for the preparation of each of the following alcohols using oxirane (ethylene oxide) and any other appropriate reagent(s).

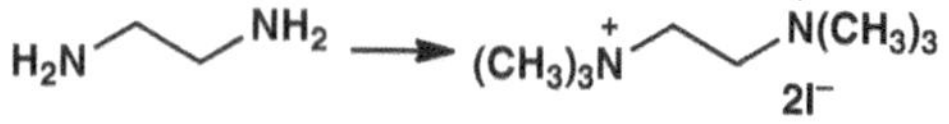

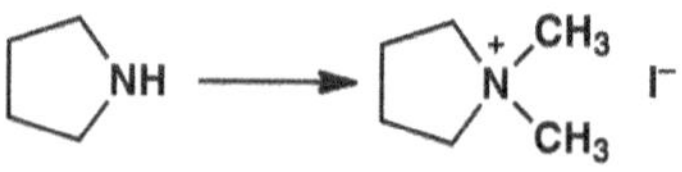

(a) $CH_3CH_2CH_2OCH_2CH_2OH$

(b) $CH_3CH_2CH_2CH_2CH_2OH$

(c) $CH_3C{\equiv}CCH_2CH_2OH$

(d) $CH_3OCH_2CH_2OCH_2CH_2OH$

(e) $PhOCH_2CH_2OH$

14.16 Propose a mechanism for the acid-catalysed hydrolysis of cyclohexene oxide showing stereochemistry as appropriate.

cyclohexene oxide

14.17 Predict the main product obtained from phenyloxirane (styrene oxide) upon treatment with (a) sodium methoxide in methanol, and (b) dilute sulfuric acid in methanol.

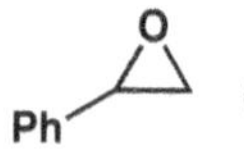

phenyloxirane
(styrene oxide)

14.18 Show how all the following five alcohols give 2-methylbut-2-ene as a main product in acid-catalysed dehydration.

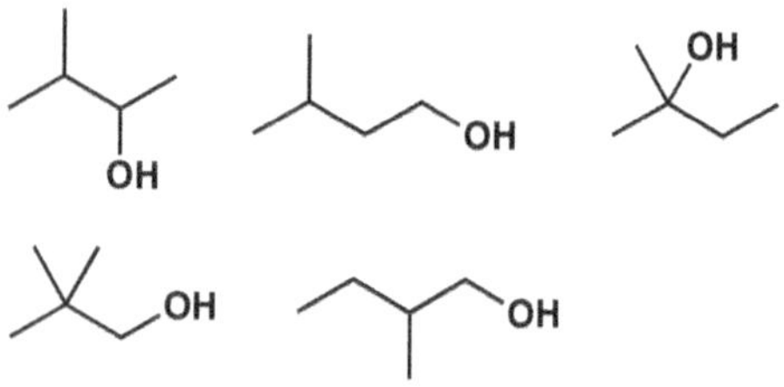

14.19 Explain why the rate constants for reaction (a) in aqueous solution increase according to the nature of HX in the order HCl < HBr < HI. Give the mechanism of the similar reaction (b) and predict how the rate constants depend upon HX for the same three acids.

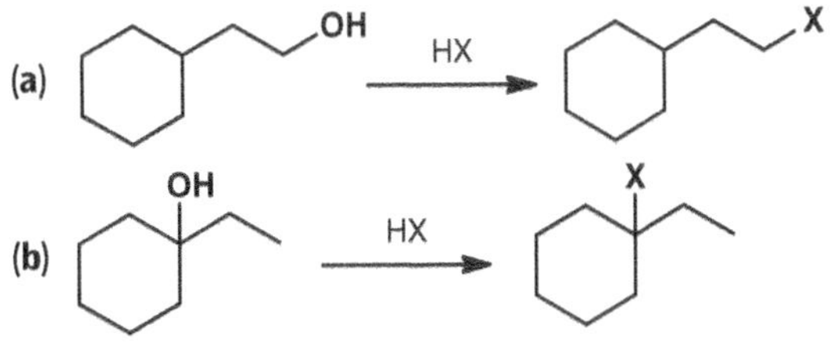
(a)
OH
HX
X
(b)
OH
HX
X

Supplementary Problems

14.20 Show the structure of the main initial product from each of the following reactions. Give the mechanism and main final product when each of the initial products is treated with sodium ethoxide in ethanol.

(a) benzyl alcohol + $SOCl_2$

(b) 2-methylbutan-1-ol + PBr_3

(c) cyclohexanol + *p*-toluenesulfonyl chloride + pyridine

(d) 2-chloroethanol + NaOH

14.21 An aldehyde can be transformed into a ketone via a 1,3-dithiane as illustrated in Scheme 14.16. Propose an alternative reaction sequence for the transformation of an aldehyde (RCHO) into a ketone (RCOR') using a Grignard reagent.

14.22 A single enantiomer of octan-2-ol was converted into an ester of ethanoic acid by two methods, and the optical rotations of the products were of opposite sign. Predict the stereochemical structures of the esters and explain the results.

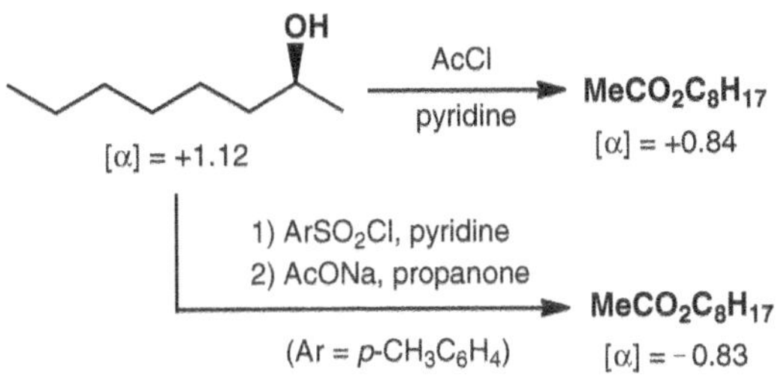
OH
[α] = +1.12
AcCl
pyridine
MeCO2C8H17
[α] = +0.84
1) ArSO2Cl, pyridine
2) AcONa, propanone
(Ar = p-CH3C6H4)
MeCO2C8H17
[α] = −0.83

14.23 To promote hydrolysis of a dithioacetal to the carbonyl compound, a mercuric salt is normally used, but mercuric salts are very toxic and introduce a disposal problem. The same overall conversion can be carried out quite effectively via an intermediate sulfoxide without the need for mercuric ion catalysis. Propose a mechanism for the acid-catalysed hydrolysis of the intermediate sulfoxide.

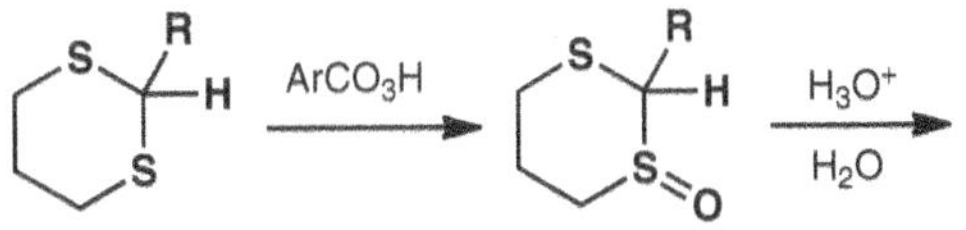
S
R
H
S
ArCO3H
S
R
H
S
O
H3O+
H2O

15 Addition Reactions of Alkenes and Alkynes

Related topics we have already covered:

- **Molecular orbitals of π electron systems** (Chapter 5)
- **Curly arrow representation of reaction mechanisms** (Section 7.2)
- **Stability of carbenium ions** (Sub-section 12.4.3)
- **Elimination reactions to form alkenes** (Chapters 13 and 14)

Topics we shall cover in this chapter:

- **Electrophilic attack at π bonds**
- **Regioselectivity of electrophilic addition to alkenes**
- **Stereochemistry of electrophilic addition to alkenes**
- **Electrophilic addition to alkynes**
- **Addition reactions to 1,3-dienes**
- **Kinetic and thermodynamic control**
- **Diels–Alder reactions**
- **Epoxidation of alkenes**
- **Cyclopropanation of alkenes with carbenes and carbenoids**
- **Hydrogenation of alkenes and alkynes**

Carbon–carbon double and triple bonds are the functional groups of alkenes and alkynes, respectively. The simplest alkene is ethene (ethylene) which acts as a plant hormone controlling the maturation of flowers and fruit (Panel 15.1); it is also an important source of industrial chemicals (Panel 3.2). Characteristic flavours and fragrances of plants and trees are often due to terpenes, which are a class of naturally occurring compounds, very many of which are alkenes (Sub-section 24.4.3). Compounds used by insects for communication are called pheromones and many of these are also alkenes. Alkynes are rather rare in nature, but one (ichthyothereol) is the active constituent of the leaves of a family of flowering plants (ichthyothere) found in South and Central America which are toxic to animals, especially to fish. Ethyne itself is used industrially to produce a high-temperature flame in oxyacetylene welding.

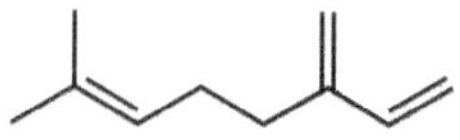

myrcene
(a terpene: a fragrant
component of bay leaf
(*Laurus nobilis*))

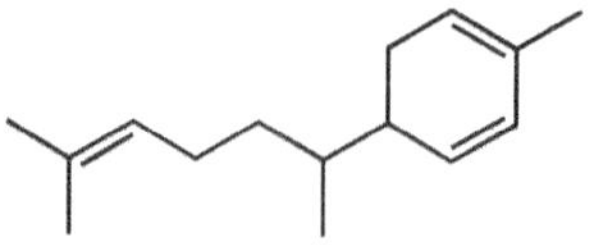

zingiberene
(a terpene: a main component
of the essential oil of ginger)

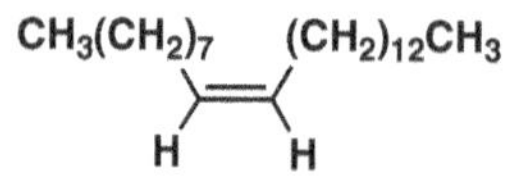

$CH_3(CH_2)_7$ $(CH_2)_{12}CH_3$

muscalure
(a sex attractant of the housefly)

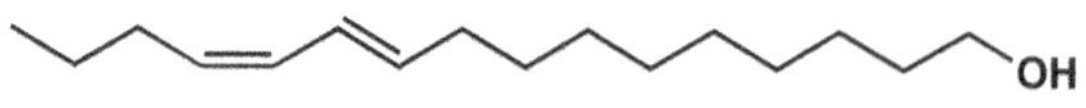

a sex attractant of the silkworm moth
(even 10^{-18} g is effective)

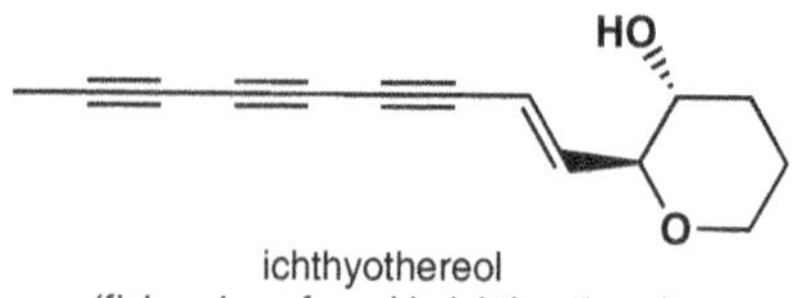

ichthyothereol
(fish poison found in ichthyothere)

The characteristic reactions of alkenes and alkynes are addition reactions which may be initiated in various ways depending on the nature of the alkene (or alkyne) and the reagent. The most common reaction of simple alkenes is electrophilic addition, which is usually a stepwise process via an intermediate carbenium ion. Typical electrophiles include hydrogen halides and other protic acids (in acid-catalysed additions of water and alcohols), halogens, and carbenium ions (in cationic polymerization). Electrophilic attack at 1,3-dienes leads to both 1,2- and 1,4-addition products, the partitioning between the two being controlled either kinetically or thermodynamically. Diels–Alder (cycloaddition) reactions between 1,3-dienes and alkenes to form cyclohexene derivatives are introduced in this chapter but will be covered in more detail in Chapter 21. Epoxidation and addition of carbenes to alkenes give three-membered cyclic compounds and are also cycloadditions; they are treated in this chapter, however, because they are similar to electrophilic addition reactions. Additions to alkenes initiated by nucleophiles and radicals will be covered in Chapters 18 and 20, respectively. Oxidation and reduction reactions generally are reviewed together in Panel 15.4 at the end of this chapter.

15.1 Electrophilic Addition to Alkenes

The π electrons associated with multiple bonds are less tightly bound and more diffuse than σ electrons, so are more susceptible to attack by an electrophile. In other words, simple alkenes are nucleophilic (although only weakly so compared with most anions and molecules with lone pairs) and **electrophilic addition** to the double bond is their most general and characteristic reaction. An electrophile first forms a new σ bond to one of the sp^2 carbons of the alkene; this gives a **carbenium ion intermediate** which is then captured by a nucleophile to complete the addition (Scheme 15.1).

The overall addition reaction involves an electrophile and a nucleophile but, *mechanistically*, it is called *electrophilic addition* because it is *initiated* by the rate-determining attack of the electrophile.

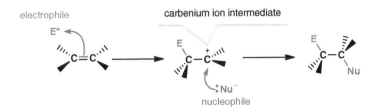

When bond rotation of the carbenium intermediate can occur, the stereochemistry of the addition process may be obscured.

Scheme 15.1 Electrophilic addition to an alkene.

Among the occupied molecular orbitals of an alkene, π MOs are of higher energy than the σ MOs, and the HOMO is a π MO which provides the electrons when the alkene acts as a nucleophile. The LUMO of an attacking electrophile interacts with the HOMO of the alkene to form a bond in the first step of an electrophilic addition reaction. Because the π electrons of an alkene are above and below its molecular plane, the electrophile attacks from above (or below). When the electrophile is HCl, for example, its LUMO is the σ* antibonding orbital, which is linear along the axis of the molecule and circularly symmetrical (Figure 15.1). Consequently, the positive end of the polar HCl molecule approaches from above the plane of the alkene with the vacant lobe of its antibonding orbital directed towards one lobe of the occupied π orbital of the alkene, as shown in Figure 15.1.

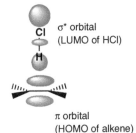

σ* orbital
(LUMO of HCl)

π orbital
(HOMO of alkene)

Figure 15.1 Interaction between the HOMO of an alkene (nucleophile) and the LUMO of HCl (electrophile).

Panel 15.1 Ethene as a plant hormone

Ethene acts at extremely low concentrations as a plant growth regulator. It regulates the germination of plants, and the growth, flowering, and ripening of various fruits. Fruit farmers can harvest unripe fruits, which are less easily spoiled, and then ripen them with ethene at the time of sending them out to consumers. For example, bananas are harvested whilst still unripe, and their ripening is controlled during shipping and storage.

Plants produce ethene naturally from an amino acid, methionine, via 1-aminocyclopropanecarboxylic acid.

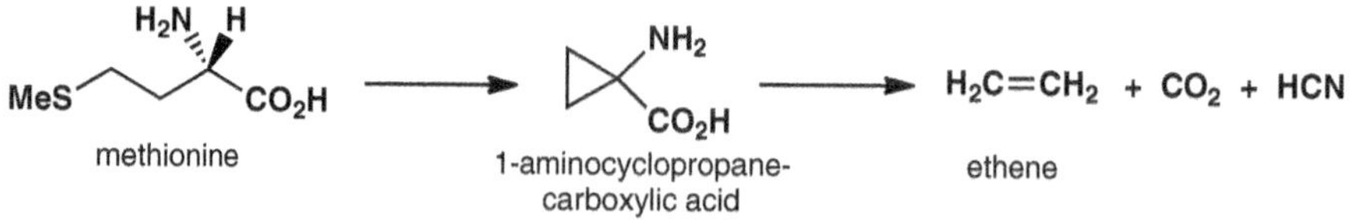

The synthetic material ethephon (2-chloroethanephosphonic acid) is used as an artificial precursor of ethene as plants are able to metabolize it into ethene. It is used to speed up the ripening of crops as wide ranging as soft fruit, tomatoes, sugar beet, and coffee.

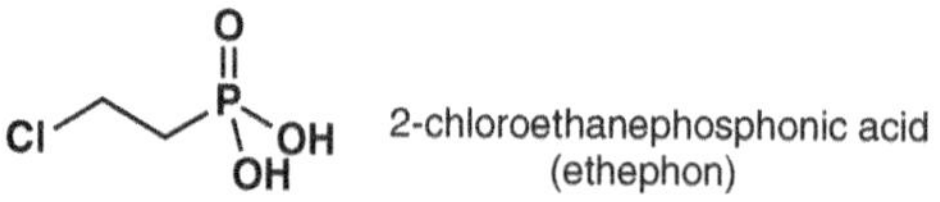

15.2 Addition of Hydrogen Halides: Hydrohalogenation

Solvents for hydrohalogenation include ethanoic acid, diethyl ether, dichloromethane, and nitromethane, but the reaction can also be carried out using concentrated aqueous hydrohalic acid or sometimes even without a solvent. In aqueous solution, HCl, HBr, and HI are fully dissociated and the reaction is initiated by proton transfer from H_3O^+.

Hydrohalogenation is the electrophilic addition of a hydrogen halide (HX, X=Cl, Br, I) to an alkene often in a non-aqueous solvent to give a haloalkane (eqn 15.1).

$$RCH{=}CH_2 + HX \longrightarrow RCH(X){-}CH_3 \qquad (15.1)$$

15.2.1 Reaction mechanism

An undissociated hydrogen halide (or H_3O^+ in aqueous solution) reacts by approaching the alkene as illustrated in Figure 15.1, and a new C–H bond is formed by proton transfer using the π electrons of the alkene to give a carbenium ion intermediate. A typical example is the addition of HCl to 2-methylpropene in ethanoic acid to give 2-chloro-2-methylpropane via the intermediate carbenium ion shown in Scheme 15.2. In this reaction, the intermediate carbenium ion is tertiary, and its relative stability strongly influences the ease of the reaction. Nevertheless, it is rapidly captured by the nucleophilic chloride ion

Scheme 15.2 The stepwise mechanism for hydrochlorination of 2-methylpropene to give 2-chloro-2-methylpropane.

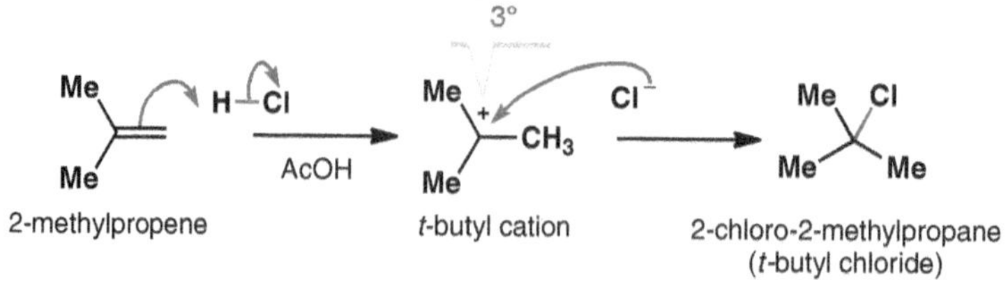

to give the product. The overall addition reaction is the reverse of the E1 elimination of haloalkanes described in Section 13.1.

15.2.2 Regioselectivity in addition to unsymmetrical alkenes

The two ends of the double bond in 2-methylpropene in Scheme 15.2 are different, so why does the HCl not also add the other way round to give 1-chloro-2-methylpropane along with the 2-chloro-2-methylpropane? The answer is shown in Scheme 15.3: it would require the intermediacy of a simple primary carbenium ion which would be too unstable to exist in solution—this reaction does not take place.

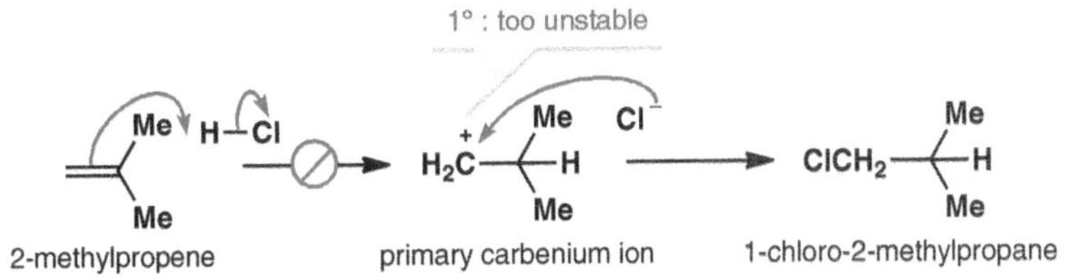

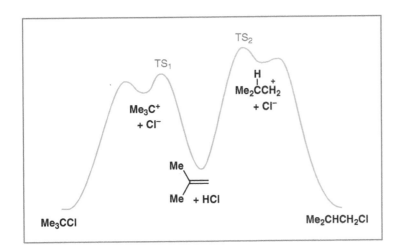

Scheme 15.3 The non-viable mechanism for hydrochlorination of 2-methylpropene to give 1-chloro-2-methylpropane.

Let us now consider the general case in which alternative isomeric carbenium ions appear possible by protonation of alternative ends of the double bond in the initial electrophilic attack at an unsymmetrical alkene; the preference for one over the other is a question of **regioselectivity** (see Example 13.1 and Section 13.5). The partitioning between the alternative outcomes is determined by the difference in the activation energies of the competing initial rate-determining steps, i.e. the relative heights of the transition structures of the competing reactions (Section 7.4). In Figure 15.2, the reactants are in the centre of the profile and the alternative competing reactions are to the left and the right. We see that each transition structure (TS) is close in energy to the intermediate which follows it in the reaction coordinate. To a good approximation, therefore, we may say that the partitioning between the competing reactions depends upon the relative stabilities of the intermediate carbenium ions in the competing paths. This is an example of the **Hammond postulate** (Panel 7.2); in the present context, *the regioselectivity in electrophilic addition to unsymmetrical alkenes is determined by the relative stabilities of the alternative intermediate carbenium ion intermediates in the competing pathways.*

Figure 15.2 A composite reaction profile for the hydrochlorination of 2-methylpropene. TS_1 is for the reaction shown in Scheme 15.2; TS_2 is for the non-viable reaction shown in Scheme 15.3.

In the addition of HCl to 2-methylpropene, the regioselectivity is overwhelmingly in favour of reaction via TS$_1$ leading to the tertiary alkyl cation which is much more stable than isomeric primary alkyl cation formed in the alternative path via TS$_2$. Consequently, the addition of HCl occurs as shown in Scheme 15.2 to give 2-chloro-2-methylpropane exclusively.

The regioselectivity in the addition of HX to an unsymmetrical alkene is now readily understood on the mechanistic basis given here. Previously, the outcome of such reactions was predicted using the empirical *Markovnikov rule* (named after the Russian chemist whose work it is based on): in the addition of HX to an unsymmetrical alkene, the hydrogen adds to the less-substituted carbon (the one which bears the greater number of H atoms). In a nonpolar solvent such as CH_2Cl_2 or Et_2O, fully formed ionic intermediates may not be formed, but transition structures in concerted additions will be highly polarized, so regioselectivities can still be predicted as if carbenium ions were involved.

Vladimir V. Markovnikov (1838–1904)

Markovnikov reported his findings in 1869. He also synthesized four- and seven-membered carbocyclic compounds when only the five- and six-membered analogues were known.

Example 15.1

What is the main product in each of the following reactions?

(a) $CH_3CH{=}CH_2$ + HBr $\xrightarrow{\text{AcOH}}$
propene

(b) ⬡=CH$_2$ + $H_3O^+\ I^-$ $\xrightarrow{H_2O}$
methylenecyclohexane

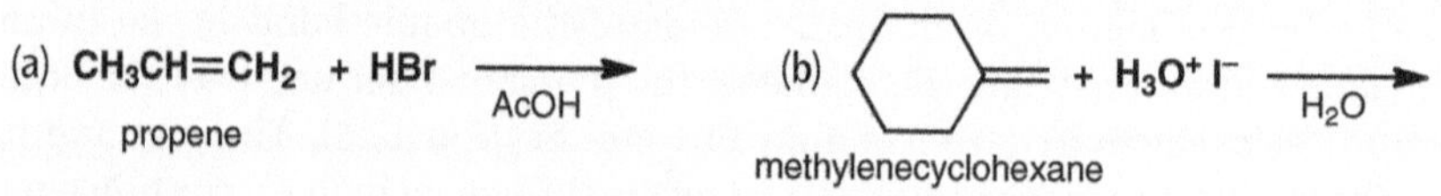

Solution

In each case, protonation of the double bond preferentially gives the more stable of the alternative possible carbenium ions which then gives the haloalkane shown below as the main product.[1] Stabilities of the alternative possible intermediates are compared in parentheses.

(a) $\overset{Br}{\underset{|}{CH_3CH}}{-}CH_3$ ($CH_3\overset{+}{CH}{-}CH_3$ > $CH_3CH_2{-}\overset{+}{CH_2}$)
2-bromopropane

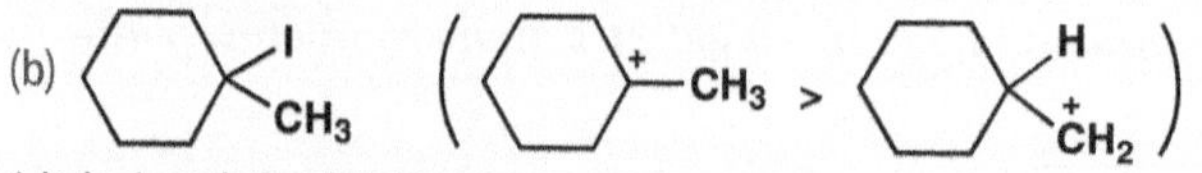

(b) 1-iodo-1-methylcyclohexane

Exercise 15.1

What is the main product in each of the following reactions?

(a) + HCl $\xrightarrow{\text{AcOH}}$

(b) + $H_3O^+\ Br^-$ $\xrightarrow{H_2O}$

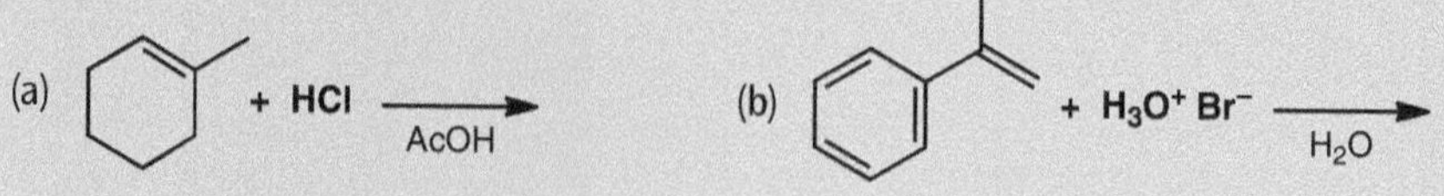

[1] In a nucleophilic solvent such as H_2O or AcOH, a solvent molecule may also act as a nucleophile in a parallel product-determining second step of the mechanism. The proportions of the alternative products will then be determined by the concentration of the HX (or X⁻) in the solvent, and by the relative nucleophilicities of HX (or X⁻) and the solvent.

Some compounds HX, including just HBr amongst the hydrogen halides, give the so-called *anti-Markovnikov* product (e.g. 1-bromo-2-methylpropane from 2-methylpropene+HBr) in nonpolar solvents in the presence of a trace of a peroxide (for example). This apparently anomalous regioselectivity is observed because the mechanism is not the polar one described above. The peroxide (or any other radical initiator) induces a radical addition mechanism which leads to *anti-Markovnikov* regioselectivity, as will be discussed in Section 20.5.

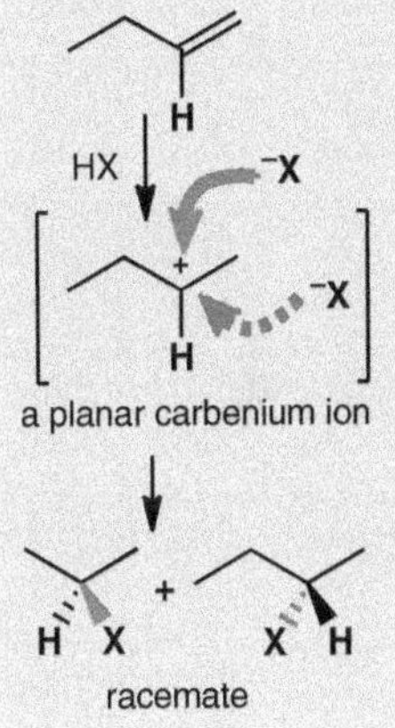

15.2.3 Stereochemistry of addition

Addition of a hydrogen halide, HX, to 1,2-dimethylcyclopentene gives *cis* and *trans* isomers of 1-halo-1,2-dimethylcyclopentane, **1** and **2**, by *syn* and *anti* additions (Scheme 15.4). Similar results have been found for other alkenes with the proportions of *syn* and *anti* additions depending on the reaction conditions (principally, the solvent and temperature). Consequently, we cannot generally be confident about predicting the stereoselectivity of HX addition to an alkene.

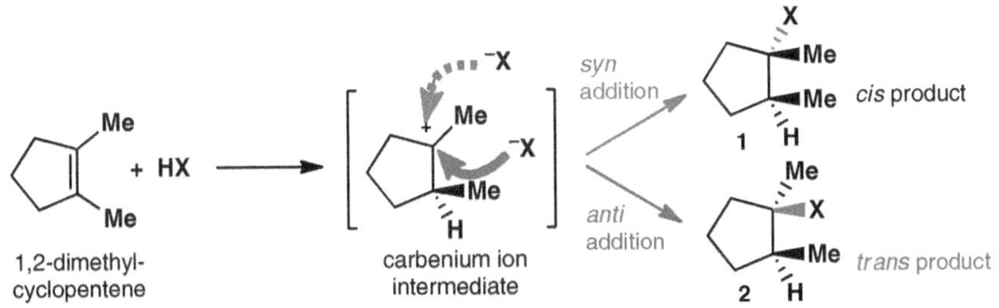

Scheme 15.4 Addition of hydrogen halide (HX) to 1,2-dimethylcyclopentene. The enantiomers of 1-halo-*cis*- and 1-halo-*trans*-1,2-dimethylcyclopentanes are also formed.

> **Exercise 15.2**
>
> (a) Draw the structures of the enantiomers of products **1** and **2** in Scheme 15.4, and assign the *R,S* configuration to each chiral centre. (b) Of the four products of the reaction in Scheme 15.4, identify pairs which are necessarily formed in equal yields and pairs which, in principle, will be formed in unequal yields.

15.2.4 Electrophilic addition to alkynes

Alkynes react with electrophiles in the same way as alkenes although they are less reactive. This lower reactivity is due at least in part to the different hybridizations of the unsaturated carbons: an sp carbon is more electronegative than an sp^2 carbon (see Subsection 6.3.1). Consequently, the alkyne π electrons are less diffuse and, therefore, less susceptible to electrophilic attack than the π electrons of an alkene (remember that an sp orbital is lower in energy than an sp^2 orbital, Section 2.3).

Addition of one equivalent of hydrogen halide (HX) to an alkyne provides a haloalkene, and (if the HX is in excess) the haloalkene further reacts with one more equivalent of HX to give a dihaloalkane, e.g. Scheme 15.5.

When an alkyne is not symmetrical (e.g. propyne in Scheme 15.5), the regioselectivity of the addition can be predicted by applying the principles we identified for electrophilic addition to an alkene: the reaction will be via the more stable of the alternative possible carbenium ions formed by protonation. The vinylic cation formed in the first step of Scheme 15.5 is methyl-substituted at the electron-deficient carbon, whereas protonation of the central C would have given the much less stable MeCH=CH$^+$ (not shown). The second HBr addition is slower than the first due to the electron-withdrawing inductive effect of the Br in the first-formed product (which means the initial alkene product can often be

---

When an achiral alkene undergoes electrophilic addition, both the initial electrophilic attack and the subsequent nucleophilic capture of the planar carbenium ion will be from above or below a planar species. Consequently, racemic material will inevitably be obtained if the product is chiral (see Section 11.8).

a planar carbenium ion

racemate

Note that the intermediate carbenium ion formed by protonation of an alkyne is vinylic and its positive carbon is sp-hybridized in contrast to the sp^2 carbon of an alkyl cation formed by protonation of an alkene; the sp carbon is less capable of accommodating a positive charge.

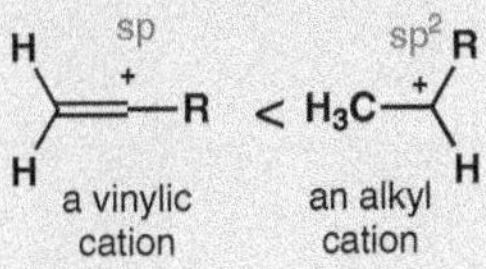

a vinylic cation $<$ an alkyl cation

Stabilities.

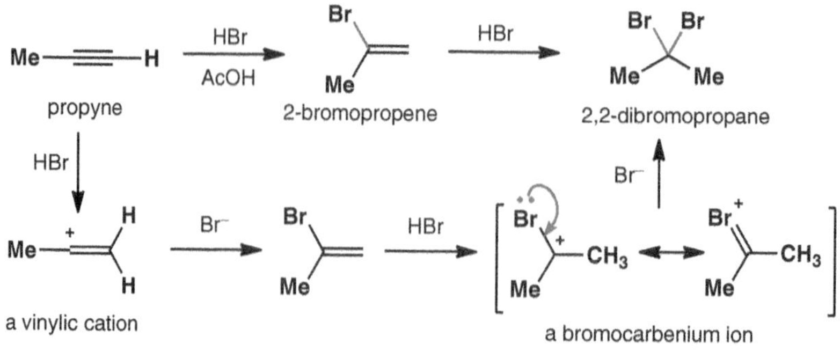

Scheme 15.5 Addition of HBr to propyne.

isolated). However, the electron-donating resonance effect of the Br directs the regioselectivity in the protonation of the bromoalkene. In other words, the carbenium ion formed in the second hydrobromination is stabilized by conjugation of a lone pair on the bromine with the carbenium ion centre (even though the *inductively* electron-withdrawing Br is rate-retarding), and the final product is the 2,2-dibromoalkane.

Panel 15.2 Cyclic enediyne antitumour antibiotics

Some antibiotics containing a cyclic enediyne structure have antitumour activity; examples are calicheamicin and kedarcidin (which consists of the molecule shown below and a carrier protein). They are very reactive and attack DNA to prevent cancer cells from proliferating.

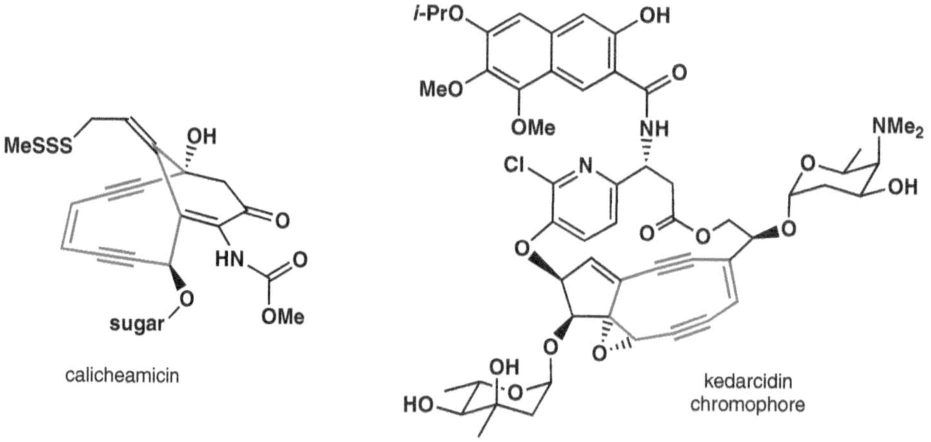

calicheamicin

kedarcidin
chromophore

The reaction of the cyclic enediyne part of the molecule in the drug action is interesting. It is a thermal rearrangement (following changes elsewhere in the molecule to allow proximity of the two triple bonds) to give a diradical (sometimes called a *p*-benzyne) which abstracts two hydrogen atoms from neighbouring DNA chains; this disrupts the DNA double helix structure and hence prevents its replication.

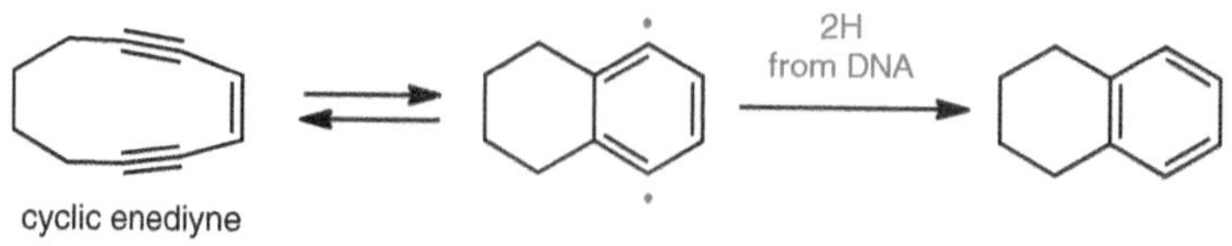

cyclic enediyne

15.3 Addition of Water

15.3.1 Acid-catalysed hydration

Addition of H_2O to the double bond of a molecule is called **hydration** and, in the case of an alkene, gives an alcohol; we have already encountered the reverse reaction (**dehydration**—a type of elimination, Section 14.1.3). The forward and reverse directions are illustrated for a typical example in eqn 15.2 and, since the dehydration is acid-catalysed (see Chapter 14), it necessarily follows that the reaction in the hydration direction must also be acid-catalysed.

(15.2)

> An acid-catalysed hydration/dehydration system involving an alcohol and an alkene, e.g. eqn 15.2, can often be driven in either direction, and the equilibrium concentrations will be determined by the initial conditions and the equilibrium constant. Typically, if conversion of an alkene to an alcohol is required, the reaction will be carried out with a large excess of water, e.g. in aqueous solution.

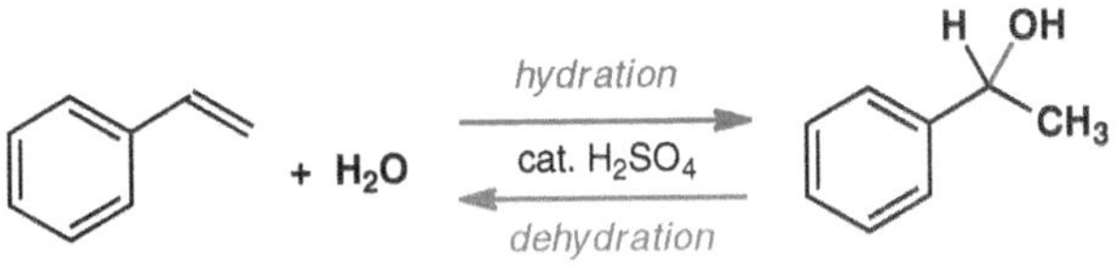

The hydration of styrene (phenylethene) in aqueous solution (eqn 15.2) is initiated by protonation of the alkene by the acid catalyst, H_3O^+, and the intermediate carbenium ion is then trapped by a water molecule; the final step gives the alcohol product and regenerates the catalytic H_3O^+ (Scheme 15.6). In the mechanism of Scheme 15.6, the intermediate carbenium ion is stabilized by conjugation with the phenyl group; the opposite regioselectivity involving protonation of the other end of the double bond would give $PhCH_2CH_2OH$ via a very unstable primary carbenium ion, and is not observed.

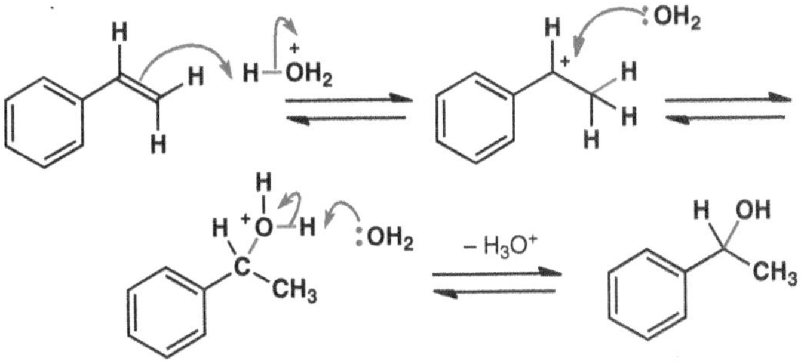

Scheme 15.6 Mechanism of the acid-catalysed hydration of styrene.

Exercise 15.3

Write a curly-arrow mechanism for the acid-catalysed hydration of 1-methylcyclohexene.

Exercise 15.4

What is the hydration product in each of the following reactions.

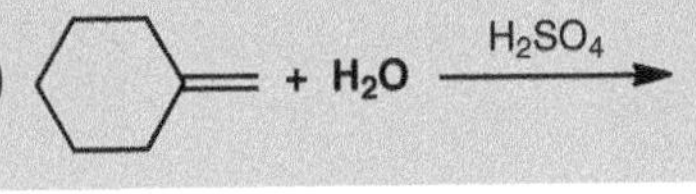

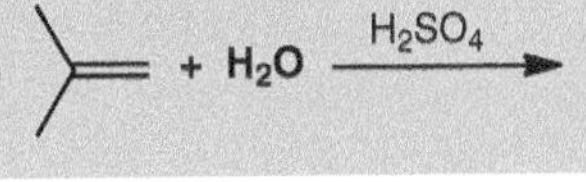

Example 15.2

Ethoxyethene (ethyl vinyl ether) is very reactive under acidic aqueous conditions, and its hydration product is a hemiacetal which reacts further. Identify the final products by giving the mechanism of the reaction.

Solution

Initial protonation of the $=CH_2$ gives the first intermediate which is stabilized by conjugation between the alkoxy group and the carbenium ion centre. Nucleophilic capture by water then gives

(continues ...)

(... continued)

the protonated hemiacetal which reacts further to give protonated acetaldehyde and ethanol. The former reacts with another water molecule to regenerate the catalyst and yield an equilibrium mixture of the aldehyde (not shown) and its hydrate (see Section 8.4).

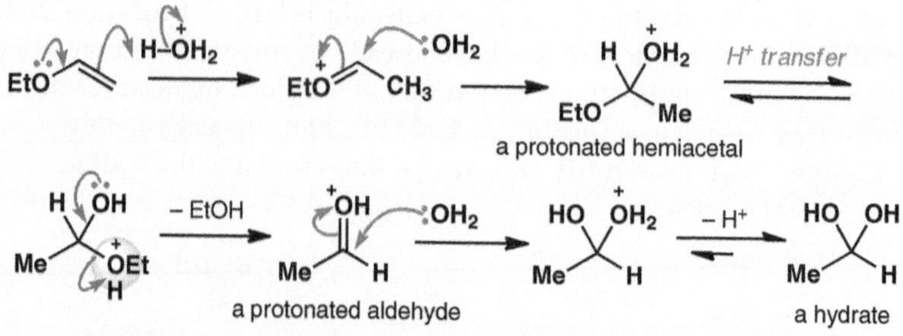

a protonated hemiacetal

a protonated aldehyde   a hydrate

Note that initial protonation at the other end of the alkene group would lead to a non-viable primary carbenium ion, so the observed regioselectivity is completely dominant.

15.3.2 Oxymercuration–demercuration

The acid-catalysed hydration of an alkene does not proceed well unless the intermediate carbenium ion is sufficiently stable, e.g. a tertiary carbenium ion or one stabilized by resonance. Furthermore, the acid-catalysed hydration is reversible. An alternative method which leads to the same product but avoids these disadvantages involves **oxymercuration** followed by reductive demercuration (Scheme 15.7).

Scheme 15.7
Oxymercuration–demercuration of an alkene to give an alcohol.

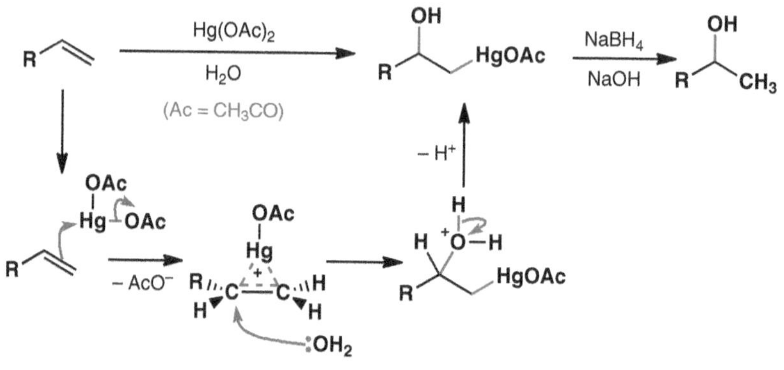

The alkene acting as a nucleophile displaces one ethanoate (acetate) group from the mercuric ethanoate to give an intermediate which then reacts with H_2O to give the hydroxy-organomercury compound shown. The Hg residue is readily displaced by reaction with $NaBH_4$ (reductive demercuration) to provide the alcohol. This reaction follows Markovnikov regioselectivity because the more alkyl-substituted C bonded to the Hg in the intermediate carries a greater partial positive charge than the less-substituted C; consequently, it more effectively attracts nucleophilic attack. Another merit of this procedure is that the first step is carried out in a neutral solution. Consequently, the reaction is not accompanied by rearrangements which are sometimes problematical in acid-catalysed hydrations via carbenium ion intermediates (see Section 14.2 and Chapter 22 for carbenium ion rearrangements). The toxicity of mercury, however, is a major disadvantage so the reaction must be carried out with great care.

If an alcohol is used as the solvent rather than water, the product will be an ether.

Toxicity of mercury

Mercury and its compounds are toxic and may be converted by microorganisms in the environment into methylmercury compounds CH_3HgX or $(CH_3)_2Hg$ which accumulate in the fatty tissues of fish. Ingested methylmercury compounds are neurotoxic and large-scale industrial releases of mercury in the past led to tragic cases of mercury poisoning, e.g. in Minamata (Japan) and Ontario (Canada). Reagents as toxic as mercury compounds must be used with the greatest of care with safe collection of residues for recycling or responsible disposal; their use also requires sound knowledge of how to deal with spillages or other mishaps. In oxymercuration followed by reductive demercuration, the mercury ends up as the metallic element, which can be collected and recycled.

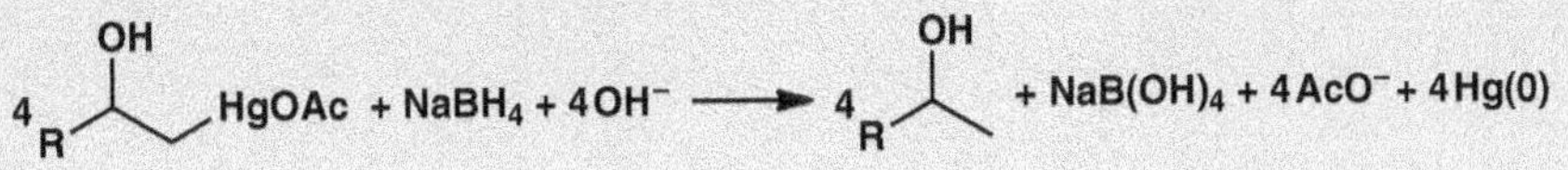

$$4 \underset{R}{\overset{OH}{\diagup}}\!\!\diagdown\!\diagdown HgOAc + NaBH_4 + 4\,OH^- \longrightarrow 4\underset{R}{\overset{OH}{\diagup}}\!\!\diagdown + NaB(OH)_4 + 4\,AcO^- + 4\,Hg(0)$$

15.3.3 Hydroboration–oxidation

Another useful method for the preparation of alcohols from alkenes is based on the **hydroboration** reaction developed by H.C. Brown (1912–2004) who was awarded the 1979 Nobel Prize in Chemistry. Hydroboration is the addition of borane, BH_3, to an alkene to form an alkylborane which can then be oxidized by alkaline hydrogen peroxide to give an alcohol by replacing the C–B bond with C–OH (Scheme 15.8).

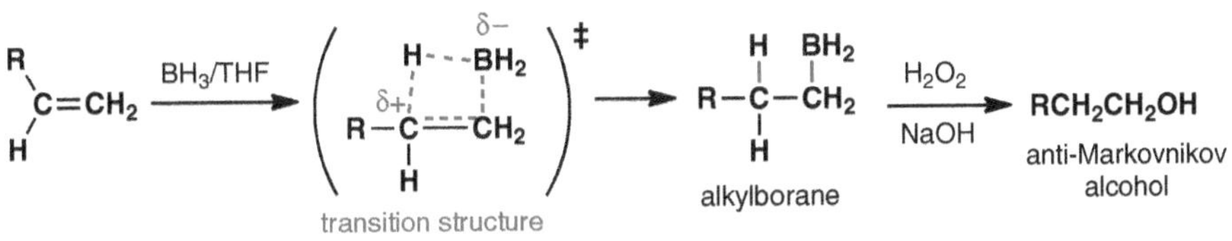

Scheme 15.8
Hydroboration–oxidation of an alkene to give an alcohol with anti-Markovnikov regioselectivity.

Borane, BH_3, is a very reactive compound which has only six valence electrons around the B, so is a Lewis acid. It is dimeric in the gas phase and in nonpolar solvents; as such, it is usually called diborane and has two H-bridged, three-centre, two-electron bonds (the detailed description of which is outside the scope of this book). Solvents with Lewis base properties react with diborane to form adducts of the monomer which is then called borane and acts as an electrophile.

diborane (B_2H_6)   borane-THF

Since the B–H bond of the first-formed alkylborane can add to another alkene molecule, one equivalent of BH_3 can in principle react with up to three equivalents of an alkene to give a trialkylborane, R_3B. However, depending upon how sterically hindered the alkene is, the addition may stop at either RBH_2 or R_2BH.

In borane addition, the boron acts as the electrophilic centre which bonds to the less alkyl-substituted carbon of the alkene as in other electrophilic additions, and the H adds to the more substituted C. Another feature of this addition is that it is concerted (a one-step reaction) via a four-membered cyclic transition structure, as shown in Scheme 15.8, which requires it to be a *syn* addition and susceptible to strong steric hindrance. Consequently, the regioselectivity governed by an electronic effect is enhanced by a steric effect. The product following oxidation is an alcohol which corresponds to hydration of the alkene with anti-Markovnikov regioselectivity.[2]

Example 15.3

Give structural formulas of the alcohols formed from but-1-ene by (a) oxymercuration–demercuration, and (b) hydroboration–oxidation.

Solution

The alcohol with Markovnikov regioselectivity is obtained in (a) but note that a chiral centre is generated (see Section 11.8 for similar reactions) so the product will be racemic; the isomeric anti-Markovnikov product is obtained in (b).

(a)
1) Hg(OAc)₂ / H₂O
2) NaBH₄ / NaOH
but-1-ene → butan-2-ol (OH)

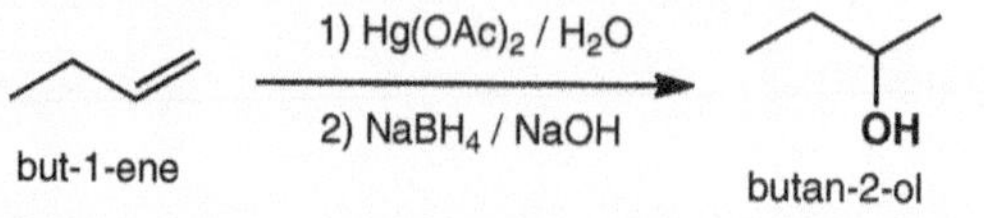

(b)
1) BH₃ / THF
2) H₂O₂ / NaOH
but-1-ene → butan-1-ol (OH)

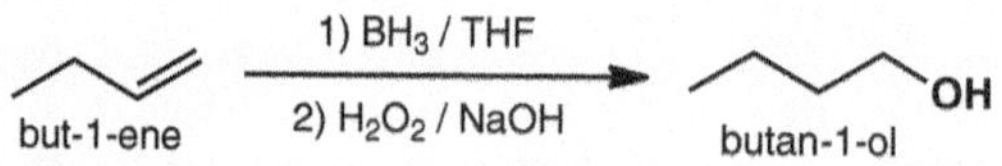

Exercise 15.5

Propose a method starting from an alkene for the preparation of each of the following alcohols.

(a) 3-methylpentan-1-ol   (b) 3-methylpentan-2-ol   (c) 3-methylpentan-3-ol

15.3.4 Hydration of alkynes

Although alkynes are not sufficiently reactive to undergo acid-catalysed hydration, they do undergo oxymercuration quite readily. A typical reaction is carried out under acidic conditions. Consequently, the first-formed enol-mercuric ethanoate is protonated, and this happens at the C which bears the Hg residue (Scheme 15.9) to give a protonated carbonyl compound (ketonization; see Section 17.1 for keto–enol tautomerism).

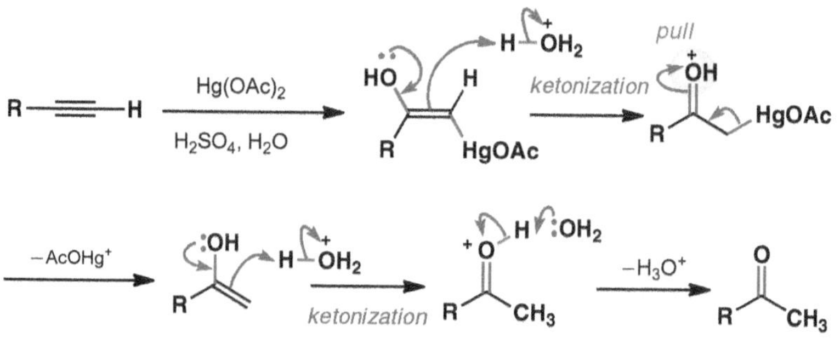

Scheme 15.9
Oxymercuration–demercuration of a terminal alkyne to give a ketone.

[2] Oxidation of an alkylborane under alkaline conditions occurs by nucleophilic attack of HOO⁻ at the boron followed by a 1,2-shift of the alkyl group to the oxygen facilitated by departure of the HO⁻, and hydrolysis of the alkoxyborane. Rearrangement reactions by 1,2-shifts will be discussed in Chapter 22.

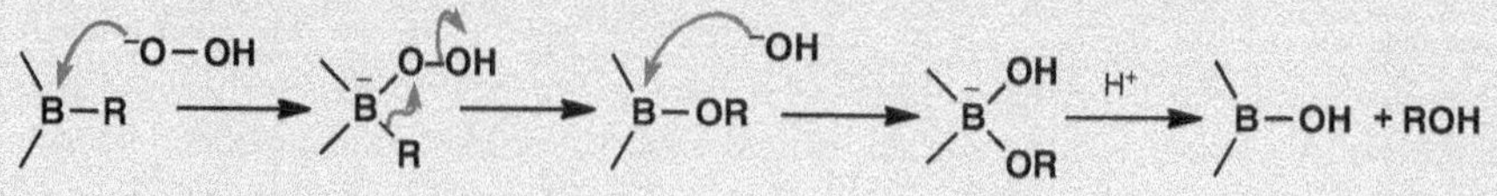

Spontaneous demercuration is then facilitated by the electron-withdrawing protonated carbonyl group to give an enol which, in the final step, rapidly isomerizes to the ketone.

The hydration of the alkyne above can also be carried out with catalytic $HgSO_4$ instead of $Hg(OAc)_2$ in dilute sulfuric acid, as shown in eqn 15.3.

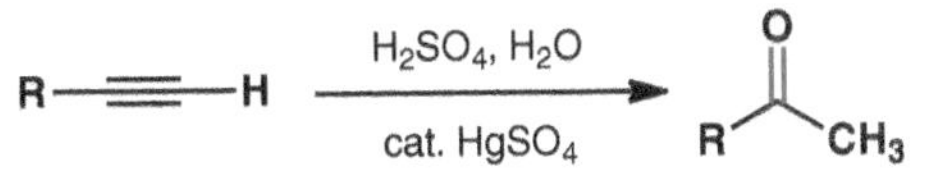

$$R{-}{\equiv}{-}H \xrightarrow[\text{cat. HgSO}_4]{\text{H}_2\text{SO}_4,\ \text{H}_2\text{O}} R{-}\overset{O}{\underset{}{C}}{-}CH_3 \qquad (15.3)$$

Hydroboration of an alkyne is also possible and the oxidative workup gives an enol with anti-Markovnikov regioselectivity in the case of a terminal alkyne which, under the alkaline conditions of the reaction, isomerizes to an aldehyde (Scheme 15.10).

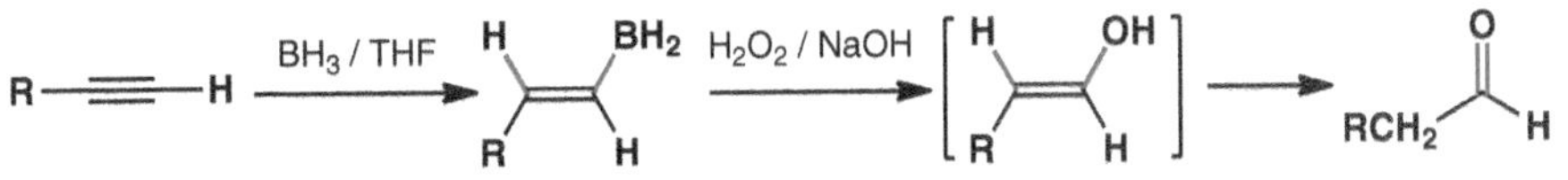

Scheme 15.10 Hydroboration–oxidation of a terminal alkyne.

Note that a terminal alkyne gives a methyl ketone by oxymercuration–demercuration, but an aldehyde by hydroboration–oxidation. A simple internal alkyne, $R^1C{\equiv}CR^2$, generally gives isomeric ketones with poor regioselectivity by both methods.

Give the structural formula of the main product of oxymercuration–demercuration under acidic conditions of each of the following alkynes.

(a) 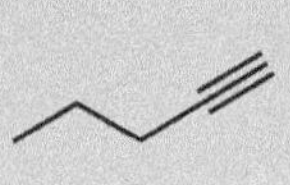   (b)

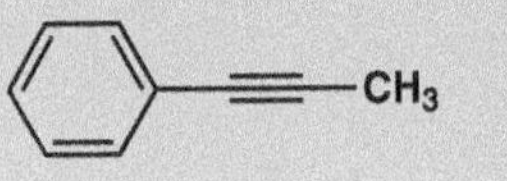

Exercise 15.6

15.4 Addition of Halogens

A halogen molecule is a potent electrophile because its bond is weak, the energy level of its antibonding orbital (σ^*) is low, and the element is strongly electronegative. The electrophilic addition of bromine to an alkene typically proceeds via an intermediate three-membered cyclic **bromonium ion**. This intermediate is captured by a nucleophile, usually bromide ion, to give the addition product; a nucleophilic solvent such as water or an alcohol, however, will compete with the Br⁻ in capturing the bromonium ion (Scheme 15.11).

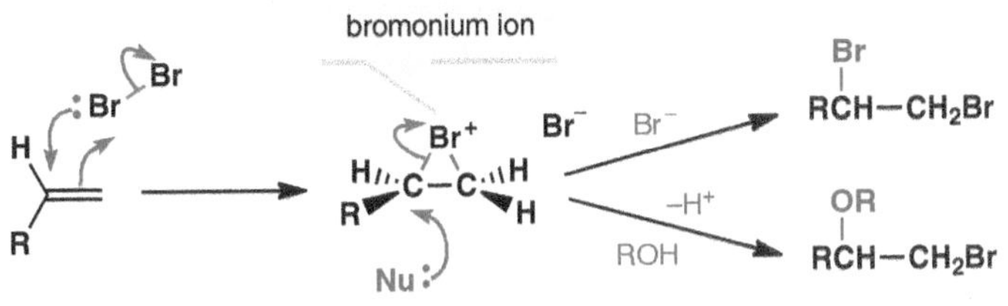

Scheme 15.11
Bromination of an alkene via a bromonium ion intermediate.

The second step of the bromination is a ring-opening reaction of the cyclic bromonium ion, and the nucleophile attacks the more alkyl-substituted carbon in spite of possible steric hindrance. (This is like the nucleophilic capture of the mercuric intermediate in Scheme 15.7 and in contrast to the ring opening of an epoxide discussed in Section 14.5). Nucleophilic attack is at the C bearing the greater (partial) positive charge rather than the less substituted C: if we represent the bromonium ion by resonance, the contribution of **1b** is greater than that of **1c**.

1a   **1b**   **1c**   **1d**

Furthermore, nucleophilic attack is from the side of the C–C bond opposite the Br of the three-membered ring (Scheme 15.11); this second step is like an S_N2 reaction, so the overall reaction is *anti* addition of Br and Nu. This can be illustrated most easily by the addition of bromine to a cyclic alkene, e.g. eqn 15.4.

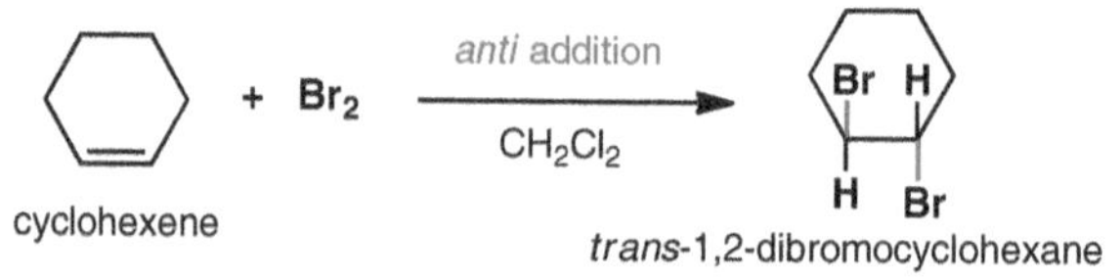

cyclohexene   *trans*-1,2-dibromocyclohexane   (15.4)

Chlorine and iodine react in much the same way, and reactivity increases in the order, $I_2 < Br_2 < Cl_2 \ll F_2$; the reaction with F_2 is normally extremely vigorous and difficult to control.

Example 15.4

In the addition of Br_2 to (*Z*)- and (*E*)-but-2-ene, products are either the racemate or the *meso* isomer of 2,3-dibromobutane. Show how these different stereoisomeric products are formed from the diastereoisomeric alkenes.

Solution
Draw the structures of the products of *anti* addition of Br_2 using wedged bonds and assign the R,S configuration to each chiral centre. When we consider all the possibilities, we observe that the *Z* isomer can give enantiomers with equal probability, i.e. the product is racemic, while the *E* isomer can give only a single product, the *meso* isomer.

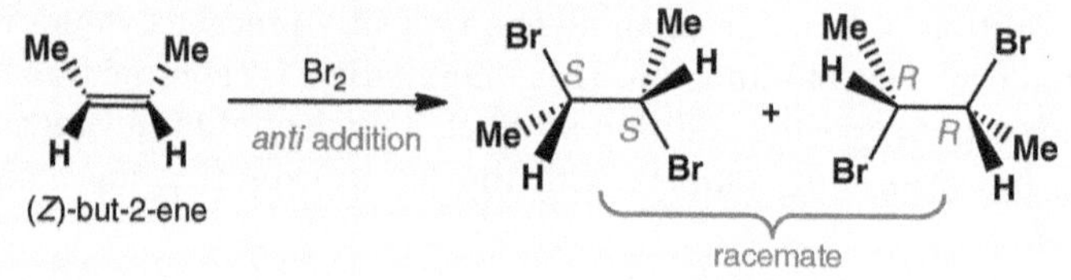

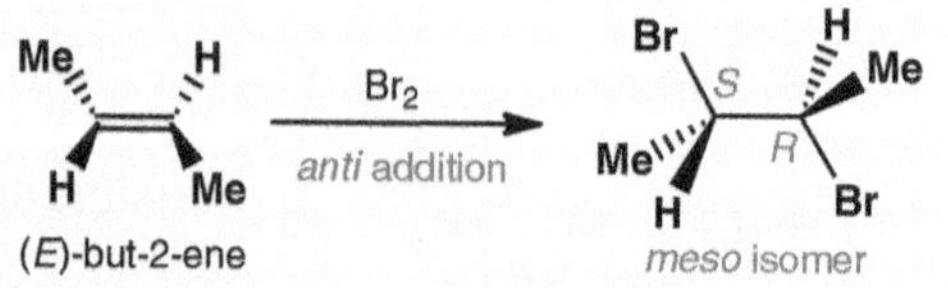

(*Z*)-but-2-ene   racemate

The addition of bromine to diastereoisomeric alkenes is a classic case of a stereospecific reaction (see Chapter 11); (*Z*)- and (*E*)-but-2-ene give different products.

(*E*)-but-2-ene   *meso* isomer

Exercise 15.7

What are the products of the following reactions in the solvents shown?

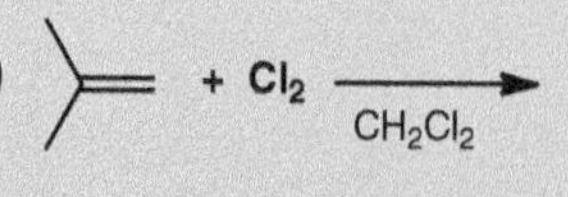

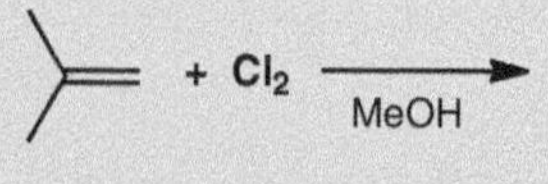

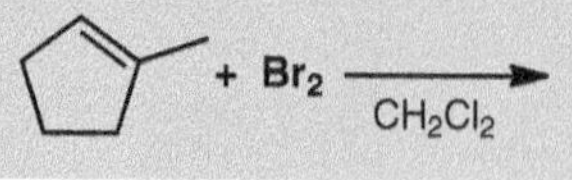

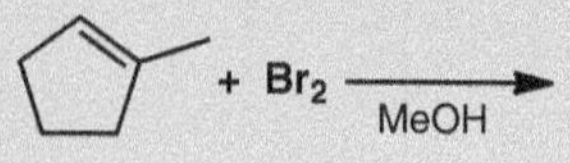

Being similar to alkenes, alkynes react with one equivalent of bromine to give *trans*-dibromoalkenes by *anti* addition (eqn 15.5); since a single stereoisomer is formed from a non-stereoisomeric reactant, this is a stereoselective reaction (see Chapter 11).

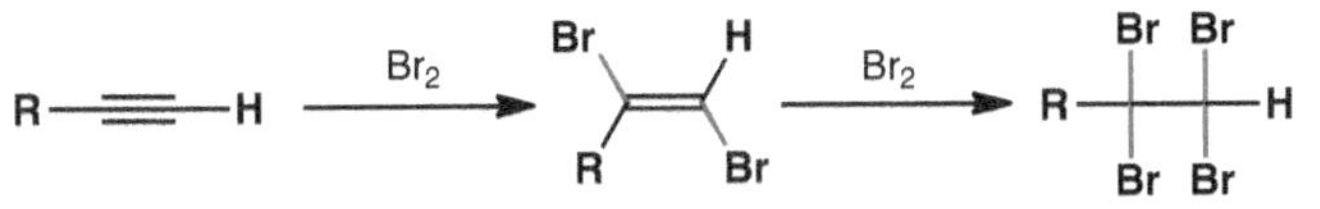

$$ (15.5) $$

Addition of one further equivalent of Br_2 leads to a tetrabromoalkane. Because the initially formed dibromoalkene is less susceptible to electrophilic attack than the original alkyne, it is usually possible to stop the reaction after the first addition.

15.5 Epoxidation

Peroxy acids, RCO_3H, which are also called peracids, have a carbonyl with a hydroperoxy group (–OOH) attached. Because the O–O bond is weak and carboxylate is a good nucleofuge, the HO oxygen is electrophilic (eqn 15.6).

$$ (15.6) $$

Reaction of a peroxy acid with an alkene gives a three-membered cyclic ether, an oxirane (epoxide; for reactions of oxiranes, see Section 14.5). This reaction is similar to the reaction of Br_2 to form a bromonium ion, but the oxirane product is a relatively stable compound. The most common peroxy acid used in laboratories is *m*-chloroperoxybenzoic acid (also called *m*-chloroperbenzoic acid, MCPBA), and the reaction with (Z)-but-2-ene is shown in eqn 15.7. The stereochemistry of the alkene is maintained, i.e. the reaction is stereospecific, which indicates that the two C–O bond formations are concerted.

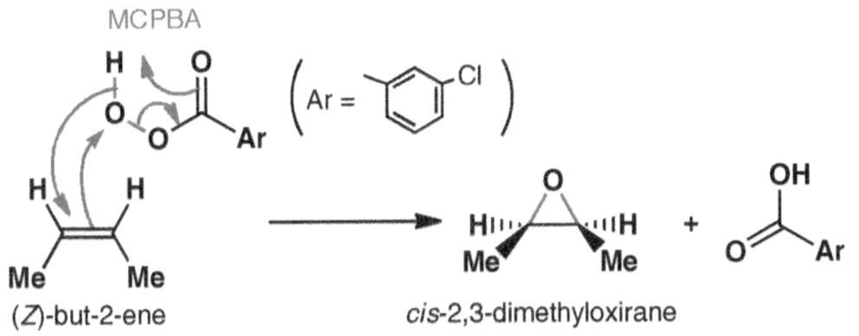

$$ (15.7) $$

(Z)-but-2-ene   cis-2,3-dimethyloxirane

Oxidation reactions of alkenes with ozone, osmium tetroxide, and permanganate are covered in Chapter 21.

Draw the structure of the oxirane formed by reaction of (E)-but-2-ene with MCPBA.

Exercise 15.8

15.6 Addition of Carbenes

Carbenes are unstable neutral compounds with a divalent carbon. The valence electrons around the C of the angular parent carbene (usually called methylene, $H_2C:$) are trigonal planar with the lone pair in one of the carbon's sp²-hybridized orbitals, and the 2p orbital is unoccupied (Figure 15.3). Consequently, the vacant 2p orbital provides an electrophilic centre while the lone pair contributes nucleophilicity.

Figure 15.3 Electronic
structure of the parent singlet
carbene.

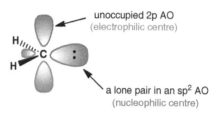

unoccupied 2p AO
(electrophilic centre)

a lone pair in an sp² AO
(nucleophilic centre)

This type of carbene in which the two nonbonding valence electrons are paired is called a *singlet*
carbene. Another electronic configuration (known as a *triplet* state) is possible in which the two electrons
are un-paired and accommodated in different orbitals.

Whether the electrophilicity or nucleophilicity of a carbene is dominant depends upon its substituents
and the nature of its reaction partner. In fact, most carbenes R_2C: (including those with $R,R=H,H$; F,F;
Cl,Cl; H,CO_2Et; Me,Cl) are electrophilic, and only those with strongly electron-donating groups (e.g.
$R,R=Me,MeO$ or MeO,MeO) are nucleophilic.

In the absence of an
unsaturated group in the
reaction partner, a carbene
may insert into a C–H, O–H,
or N–H bond:

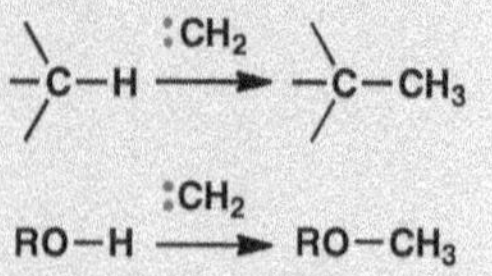

$$\overset{:CH_2}{-\overset{|}{\underset{|}{C}}-H} \longrightarrow -\overset{|}{\underset{|}{C}}-CH_3$$

$$RO-H \overset{:CH_2}{\longrightarrow} RO-CH_3$$

Alkylcarbenes
often rearrange by
intramolecular C–H
insertion, so they are
not generally used for
cyclopropanation.

Carbenes are very reactive and add to alkenes to give three-membered cyclic products,
i.e. cyclopropanes (e.g. Scheme 15.12). The reaction may be represented as concerted electrophilic and nucleophilic C–C bond-forming processes, as depicted in Scheme 15.12,
and the stereochemistry of the alkene is conserved, i.e. the addition is stereospecific.

$$R^1 \quad R^2 \qquad \overset{C}{\underset{H \quad H}{}} \text{ carbene} \longrightarrow R^1 \quad R^2 \quad \overset{C}{\underset{H \quad H}{}} \text{ cyclopropane}$$

Scheme 15.12 Stereospecific *syn* addition of singlet methylene to an alkene.

The simplest carbene, CH_2 (methylene, which is extremely reactive), can be generated by
the thermal or photolytic decomposition of diazomethane, a poisonous yellow gas (eqn 15.8);
however, diazomethane has a tendency to explode so these are seldom useful reactions.

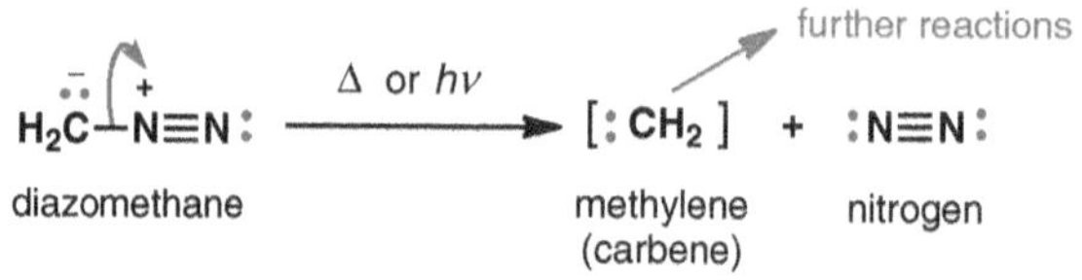

$$H_2\overset{-}{C}-\overset{+}{N}\equiv N: \xrightarrow{\Delta \text{ or } h\nu} [:CH_2] + :N\equiv N: \qquad (15.8)$$

further reactions

diazomethane   methylene   nitrogen
(carbene)

Exercise 15.9

Draw resonance forms of diazomethane.

In an α-elimination, the
two groups (or atoms)
are lost from the same
atom—it is also called a
1,1-elimination. The much
more common elimination
is a β- or 1,2-elimination to
form an alkene (Chapter 13).

Halo-substituted carbenes can be formed by an α-elimination reaction of a haloalkane;
for example, dichlorocarbene is generated by the reaction of trichloromethane (chloroform) with a strong base (Scheme 15.13 and eqn 15.9).

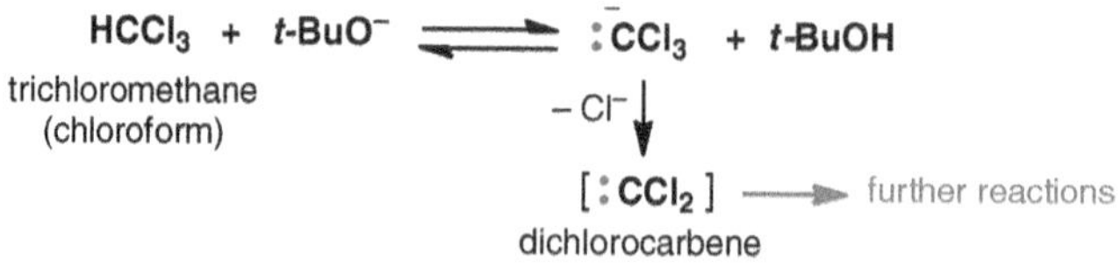

$$HCCl_3 + t\text{-BuO}^- \rightleftharpoons :\overset{-}{C}Cl_3 + t\text{-BuOH}$$

trichloromethane
(chloroform)

$$-Cl^- \downarrow$$

$$[:CCl_2] \longrightarrow \text{further reactions}$$

dichlorocarbene

Scheme 15.13 Formation of the reactive intermediate dichlorocarbene by α-elimination of HCl from $CHCl_3$.

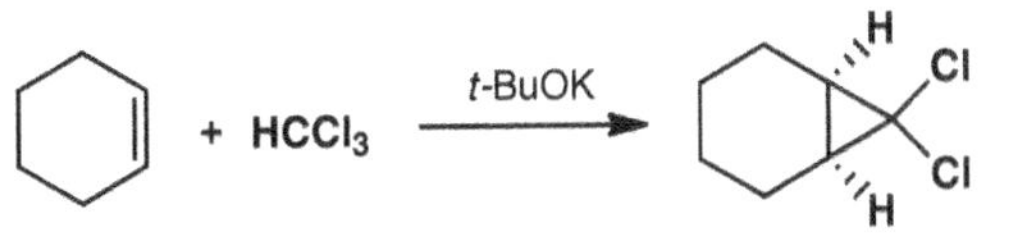

t-BuOK
+ HCCl3
H
Cl
Cl
H

cyclohexene — 7,7-dichlorobicyclo[4.1.0]heptane (15.9)

Note that $HCCl_3$ is a weak acid: $pK_a \sim 24$.

The most convenient practical method of generating a dihalocarbene CX_2 from HCX_3 for addition to an alkene exploits phase-transfer catalysis (see Panel 12.2) using a tetra-alkylammonium salt: for example,

Ph
CHCl3, PhCH2NEt3 Cl−
NaOH, H2O
Ph
CCl2

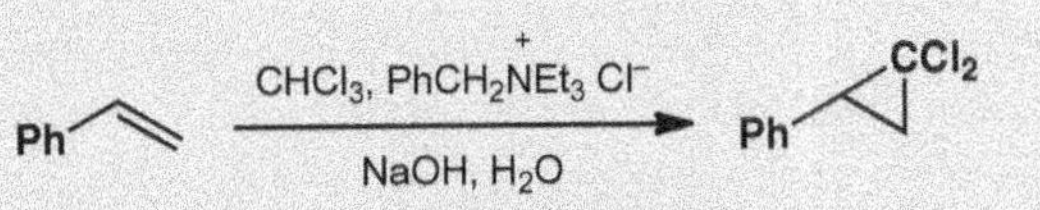

In this example, the two-phase mixture of styrene in chloroform plus benzyltriethylammonium chloride and sodium hydroxide in water is stirred vigorously to maximize the interfacial surface area. The lipophilic tetra-alkylammonium cation facilitates transfer of hydroxide into the organic phase where it reacts with the $CHCl_3$ to generate a low equilibrium concentration of Cl_3C^- which expels Cl^- to give CCl_2. This then adds to the styrene to give the substituted cyclopropane.

The addition of CH_2 itself to an alkene is not easy experimentally because of the associated inconvenience (and hazards). It is both safer and less trouble to generate a species called a **carbenoid** which acts as though it were CH_2. When diiodomethane is treated with a *zinc–copper couple* (finely divided zinc activated by copper), an organometallic carbenoid (usually represented as I-CH_2-ZnI) is formed which delivers CH_2 to an alkene to give a cyclopropane, e.g. eqn 15.10.*

*This reaction is sometimes called the Simmons–Smith reaction after the two DuPont chemists who discovered it. The Furukawa method provides a more effective zinc carbenoid for the addition of CH_2 by using diethylzinc rather than the zinc–copper couple:

$$CH_2I_2 + ZnEt_2 \longrightarrow I-CH_2-ZnI + C_4H_{10}$$

Me
Me
H
H
+ CH2I2
1) Zn-Cu, Et2O
2) H3O+
H2
C
H
H
Me
Me

(*Z*)-but-2-ene — diiodomethane — *cis*-1,2-dimethylcyclopropane (15.10)

Note that because carbenes and carbenoids are unstable, they are invariably generated *in situ*, i.e. in the presence of the reactant with which they are to react.

Exercise 15.10

What are the main products of the following reactions?

(a) (*Z*)-but-2-ene + $CHCl_3$ $\xrightarrow{t\text{-BuOK}}$

(b) (*E*)-but-2-ene + CH_2I_2 $\xrightarrow[\text{benzene}]{ZnEt_2}$

(c)
+ PhCH2Br
PhCH2NEt3 Cl−
NaOH, H2O, CH2Cl2

(d) (*Z*)-but-2-ene + CH_3CHI_2 $\xrightarrow[\text{Et}_2\text{O}]{\text{Zn-Cu}}$ (two stereoisomers)

15.7 Addition of Carbenium Ions to Alkenes and Cationic Polymerization

The intermediate in an electrophilic addition to an alkene is a carbenium ion, which can react as an electrophile itself. Consequently, in the absence of a nucleophile other than the alkene, the intermediate carbenium ion may react further with another alkene molecule. When an alkene bears electron-donating substituents, its nucleophilicity may be sufficient to allow many repeat reactions and the formation of a **polymer**. This mechanism of polymerization in which the *monomer* is a nucleophilic alkene, and propagation is by cations, is called *cationic polymerization*.

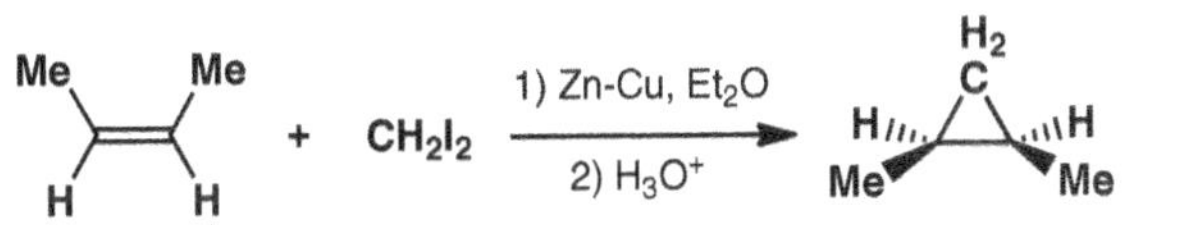

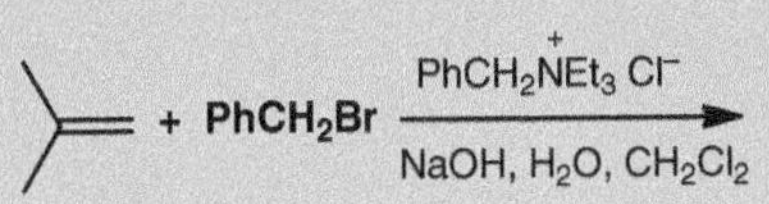

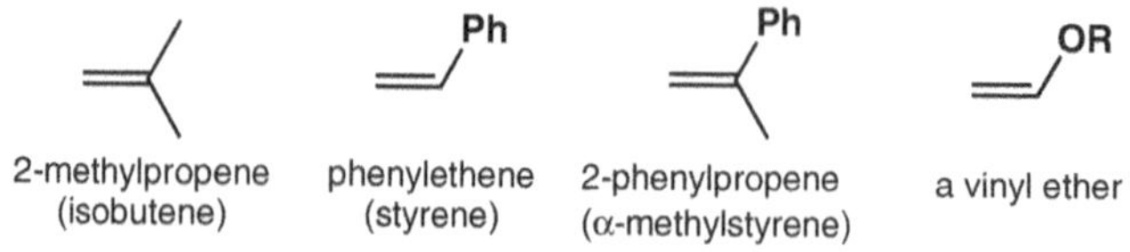

Figure 15.4 Some alkenes which act as monomers in cationic polymerization.

2-methylpropene (isobutene)   phenylethene (styrene)   2-phenylpropene (α-methylstyrene)   a vinyl ether

Vinyl ethers are particularly good monomers for cationic polymerization. Normally, a Lewis acid is used to start the process, its reaction with adventitious H$_2$O providing the initiating proton (Scheme 15.14). The polymeric cation is finally deprotonated by a base, or captured by a nucleophile, to terminate the reaction.

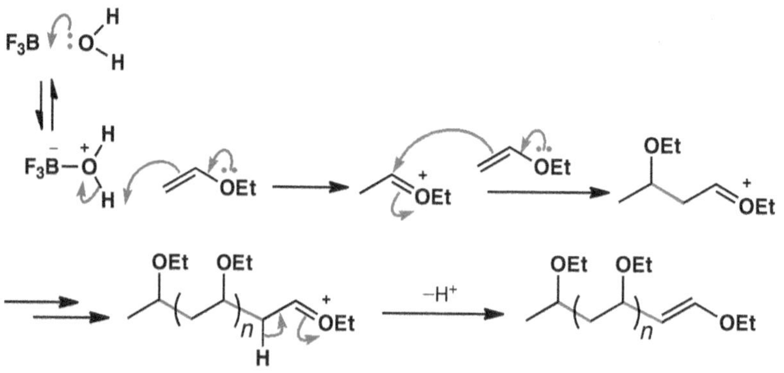

Scheme 15.14 Cationic polymerization of a vinyl ether.

Exercise 15.11

Draw the structure of the main chain of a polymer obtained by cationic polymerization of each of the monomers given in Figure 15.4.

15.8 Electrophilic Additions to Butadiene

15.8.1 1,2-Addition and 1,4-addition

Buta-1,3-diene has conjugated double bonds, and undergoes a distinctive reaction with electrophilic reagents. For example, HCl yields an addition product with a migrated double bond besides the expected product (eqn 15.11).

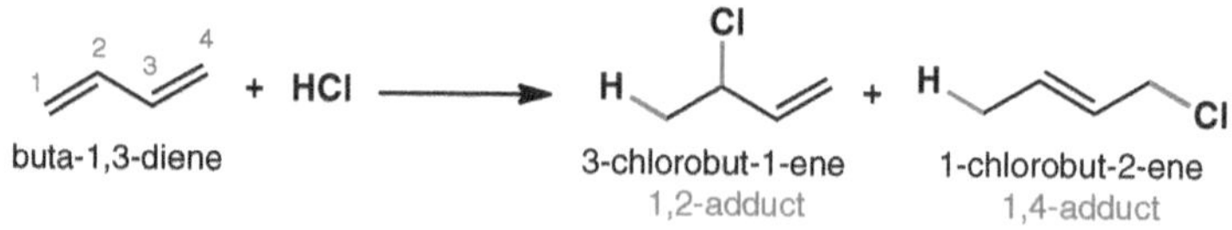

buta-1,3-diene   3-chlorobut-1-ene   1-chlorobut-2-ene   (15.11)
                    1,2-adduct   1,4-adduct

The numbers in the terms '1,2-' and '1,4-addition' are not necessarily the same as in the systematic names of the compounds; they simply describe the relative positions within the 1,3-diene part of the reactant where addition takes place.

Although the hydrogen is bonded to the terminal carbon (C1) of the original diene in both products, the chloride is bonded to C2 of the original diene in one product and to C4 in the other. The two products are called the 1,2-adduct and 1,4-adduct, and the two reactions, **1,2-addition** and **1,4-addition** (the latter also being called **conjugate addition**), occur together.

The outcome of the above reaction of buta-1,3-diene with HCl arises from the nature of the intermediate carbenium ion, the allylic cation (Scheme 15.15). Protonation at C1 of butadiene gives the allylic cation in which the positive charge is delocalized as described by resonance (see Section 5.4). Nucleophilic capture by Cl⁻ may then occur at C2 or C4 (addition of Cl⁻ to C3 cannot lead to a legitimately bonded molecule). The 1,2-adduct has a terminal (mono-substituted) double bond whereas the 1,4-adduct has an internal (di-substituted) double bond; the latter (in particular the *trans* (*E*) diastereoisomer) is generally more stable than the 1,2-adduct.

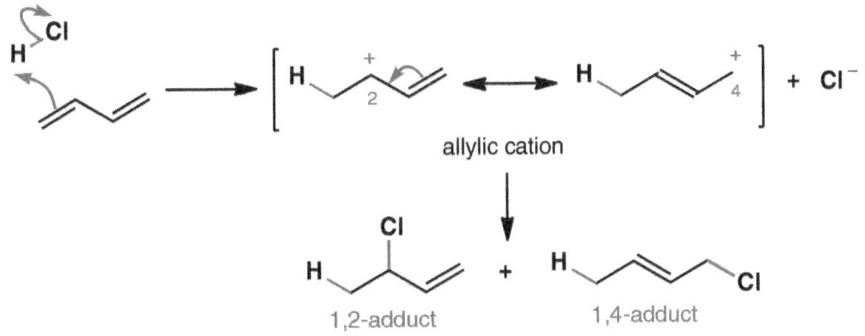

allylic cation

1,2-adduct   1,4-adduct

Scheme 15.15 1,2- and 1,4-addition of HCl to butadiene.

Note that the 1,2- and 1,4-addition products from buta-1,3-diene arise through a common intermediate; they are not the products of independent parallel reactions.

Why is a product not formed by initial addition of H⁺ at C2 in reaction 15.11?

Example 15.5

Solution
Protonation at C2 would give a primary carbenium ion which is not stabilized by conjugation with the double bond; reaction via such an unstable carbenium ion could not compete with the reaction initiated by protonation at C1.

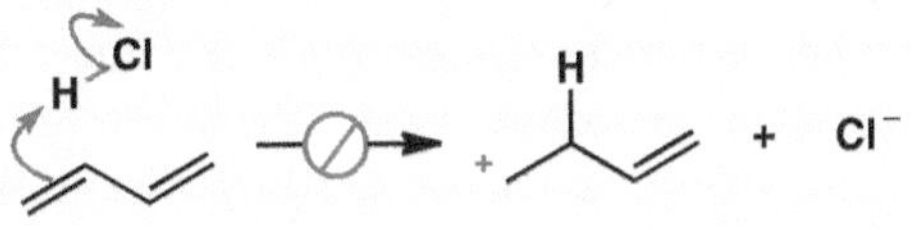

Show the structures of the main products when (3*E*)-penta-1,3-diene reacts with each of the following.

(a) 1 equivalent of HBr   (b) 1 equivalent of Br₂   (c) 2 equivalents of Br₂

Exercise 15.12

15.8.2 Kinetic and thermodynamic control

The proportions of the two products in addition reactions to buta-1,3-diene depend on the reaction conditions. For example, addition of HBr yields the 1,2-adduct preferentially at lower temperatures, but the 1,4-adduct becomes the main product at higher temperatures (Scheme 15.16).

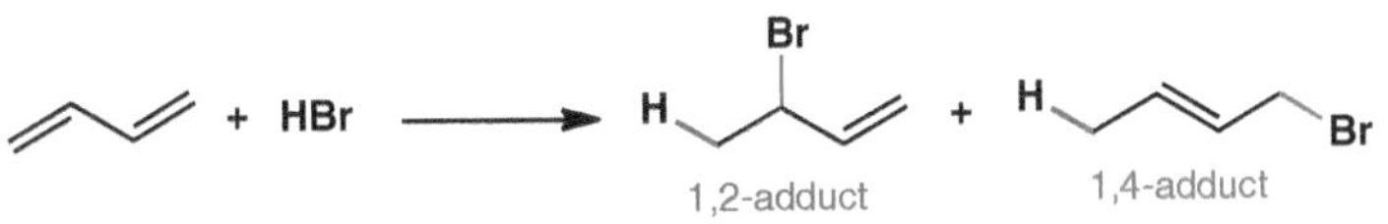

	1,2-adduct	1,4-adduct
−80 °C	90%	10%
45 °C	15%	85%

Scheme 15.16 Kinetic and thermodynamic control in the addition of HBr to butadiene.

Why should the proportions be different at the two temperatures? We have a clue in the observation that the relative amounts of the two products obtained at the lower temperature change when the product mixture is brought to the higher temperature: the

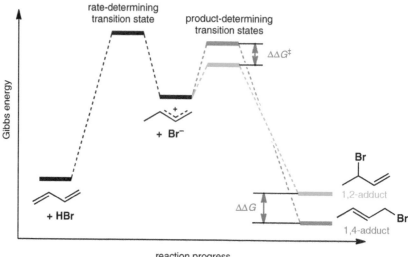

Figure 15.5 A Gibbs reaction profile for the 1,2-addition and 1,4-addition of HBr to buta-1,3-diene. The parallel second steps are product determining: under kinetically controlled (non-reversible) conditions the product ratio is determined by the difference in Gibbs energies of activation ($\Delta\Delta G^{\ddagger}$) in the competing second steps, while under thermodynamically controlled (reversible) conditions, the product ratio is determined by the difference in the Gibbs energies of the products ($\Delta\Delta G$).

new ratio of the products is then the same as that obtained in the reaction at the higher temperature. These results suggest that the addition reaction is reversible (or that the products simply interconvert), and that the 1,4-adduct predominates when the system achieves equilibrium. At the lower temperature, the less stable 1,2-adduct predominates because it forms *faster* than the 1,4-adduct.

This interpretation of the experimental observations can be illustrated by the reaction profile of Figure 15.5. Under conditions which allow re-formation of the intermediate ions from the (less stable) 1,2-adduct and the (more stable) 1,4-adduct, the products eventually become an equilibrium mixture. The proportions of the two products are then described by the equilibrium constant for their interconversion under the conditions of the reaction (principally, the temperature and the solvent).

The reaction at lower temperatures, when the ratio of products is determined by the rates of their formation from the common intermediate, is said to be **kinetically controlled**. In contrast, the reaction at higher temperatures when the ratio of the products is determined by their relative stabilities is said to be **thermodynamically controlled**. In the addition reactions of buta-1,3-diene, the 1,2-adduct is usually the kinetic product while the 1,4-adduct is the thermodynamic product.

Exercise 15.13

Draw the structural formulas of the products of 1,2- and 1,4-addition of HBr to 4-methylpenta-1,3-diene, and predict which of the two is more stable?

A Diels–Alder reaction is an example of a [4+2] **cycloaddition** because it involves the 4π electron system of a diene and the 2π electron system of an alkene (the **dienophile**) to form a cyclic product via a 6π electron TS. These reactions are covered more fully in Chapter 21.

15.9 Diels–Alder Reactions

We have seen above how alkenes and 1,3-dienes characteristically undergo addition reactions and, in all cases, the addition was of an electrophile, i.e. the alkene or diene acted as a nucleophile in a polar reaction. Another distinctive reaction of these unsaturated compounds is when a conjugated diene such as buta-1,3-diene reacts at positions 1 and 4 with an alkene to form a new cyclohexene ring. This is called a **Diels–Alder reaction** and the prototype is the reaction of buta-1,3-diene and ethene to form cyclohexene, as shown in Scheme 15.17.

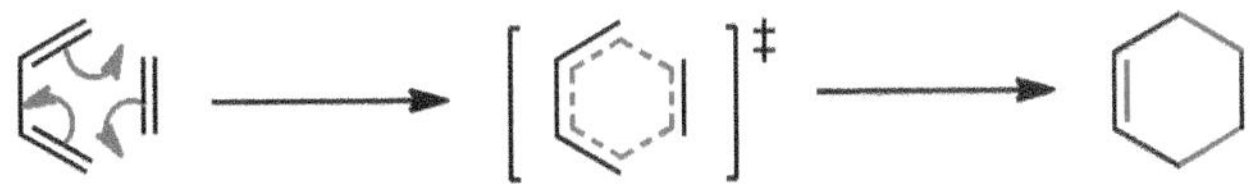

Scheme 15.17 The prototype Diels–Alder reaction between buta-1,3-diene and ethene.

We can estimate the heat of reaction for the Diels–Alder reaction in Scheme 15.17 from the bond energies of bonds broken and formed. The consequence of the overall reaction shown is the loss of two π bonds and the gain of two σ bonds. If we take the C–C π bond and σ bond energies to be about 280 and 370 kJ mol^{-1}, respectively, this simple Diels–Alder reaction is predicted to be exothermic by about 180 kJ mol^{-1} (i.e. $\Delta H \sim -180$ kJ mol^{-1}), which shows that the reaction is thermochemically strongly favourable. However, this particular reaction is very slow, and only occurs by heating the reactants under high pressure.

If the dienophile has at least one *electron-withdrawing* substituent and the diene has at least one *electron-donating* substituent, the reaction occurs more readily (see Chapter 21). For example, the reaction of butadiene and maleic anhydride proceeds quantitatively at 100 °C in a sealed vessel (eqn 15.12).

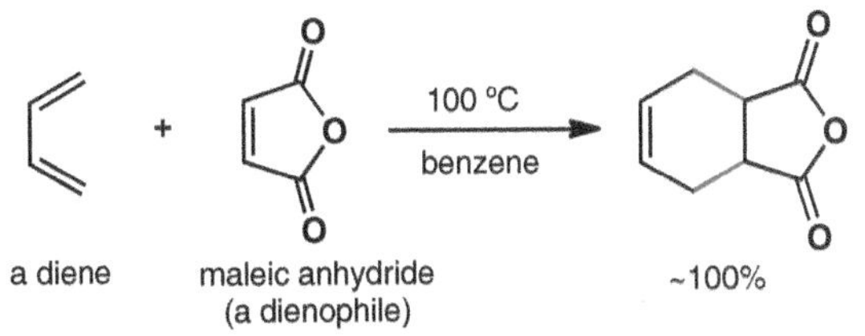

a diene   maleic anhydride   ~100%
(a dienophile)

(15.12)

15.9.1 Stereospecificity in Diels–Alder reactions

The reaction in eqn 15.12, like the overwhelming majority of [4+2] cycloadditions, is concerted and involves *syn additions at the diene and the dienophile*. Consequently, when reactants are stereoisomeric, the reaction is *stereospecific* and stereochemical relationships are retained, as illustrated in Scheme 15.18.

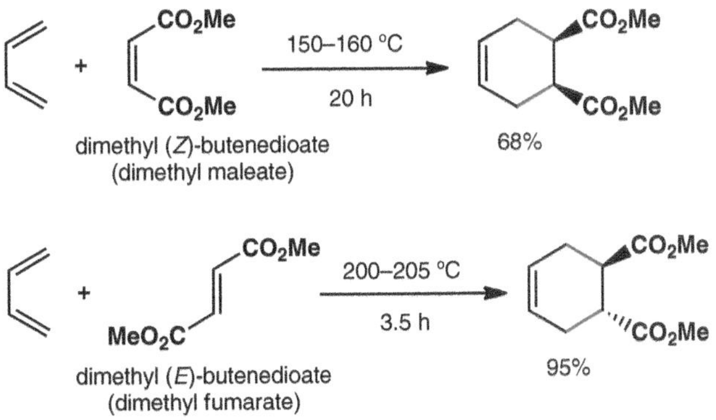

dimethyl (Z)-butenedioate
(dimethyl maleate)   68%

dimethyl (E)-butenedioate
(dimethyl fumarate)   95%

Scheme 15.18 Diels–Alder reactions of buta-1,3-diene with (Z)- and (E)-butenedioates: concerted *syn* additions with retention of stereochemical relationships.

Buta-1,3-diene exists as rapidly interconverting *s-cis* and *s-trans* conformers where *s* refers to the *single* bond connecting the two alkene groups; the *s-trans* conformer is the more stable but concerted Diels–Alder reactions can only proceed through the *s-cis* form.

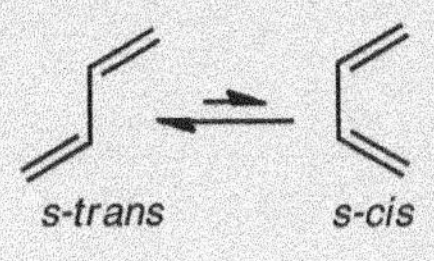

s-trans   *s-cis*

As we observed in Subsection 7.2.3, the curly arrows in a concerted mechanism like the one in Scheme 15.17 do not have the same significance as in, for example, the representation of an S_N2 mechanism. See also Section 21.2.

Otto Diels (1876–1954).

Diels studied under Emil Fischer and taught at the University of Kiel in Germany. He shared the Nobel Prize in Chemistry with his student Kurt Alder in 1950 for their discovery and development of the cycloaddition reactions now associated with their names.

Kurt Alder (1902–1958).

Alder later became a professor at the University of Cologne.

Cyclic conjugated dienes such as cyclopentadiene and furan are held in the *s-cis* form so are particularly good substrates for Diels–Alder reactions.

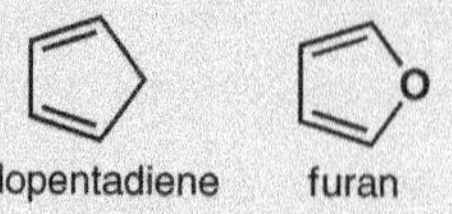

cyclopentadiene   furan

Alkynes can also react as dienophiles in concerted [4+2] cycloadditions; as expected, these are also stereospecific but this can only be observed in the diene (eqn 15.13).

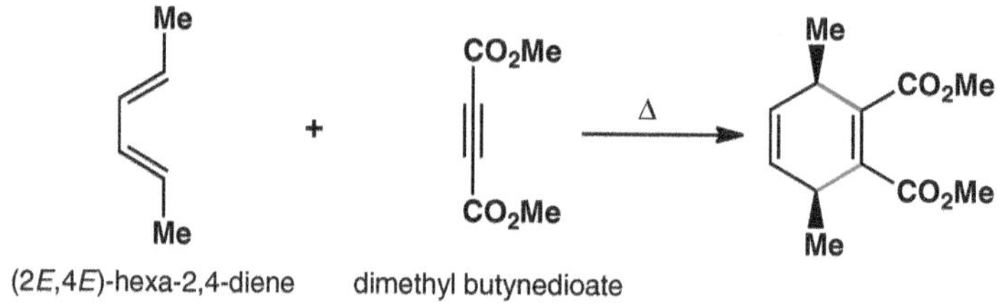

(15.13)

(2E,4E)-hexa-2,4-diene   dimethyl butynedioate

Exercise 15.14

What are the main products of the following Diels–Alder reactions?

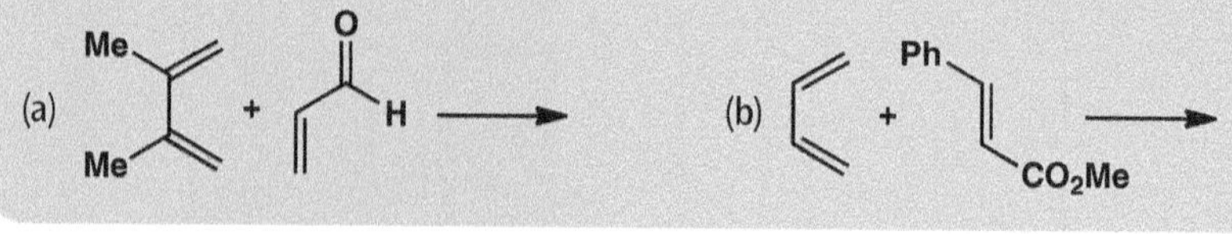

(a)   (b)

15.10 Addition of Hydrogen

Hydrogen adds to alkenes with catalysis by metals such as Pt, Pd, or Ni to form alkanes in the process known as **hydrogenation**, e.g. eqn 15.14; it is a type of reduction. Molecular hydrogen is adsorbed onto the surface of the metal and the two H atoms add to the double bond of an alkene from the same side in one step; it is another *syn* addition.

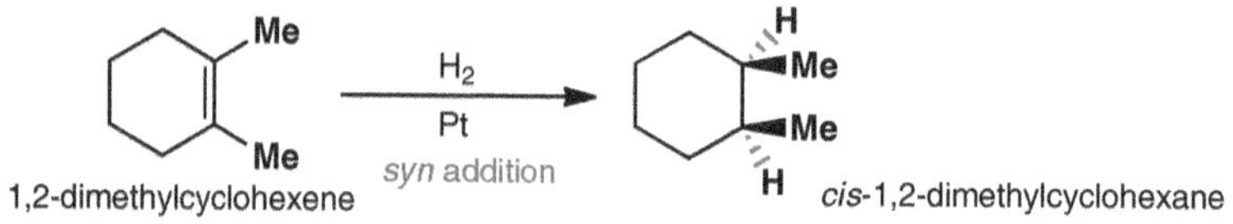

(15.14)

1,2-dimethylcyclohexene   *syn* addition   *cis*-1,2-dimethylcyclohexane

Under the same conditions, an alkyne will also be hydrogenated; however, the alkene initially formed is seldom isolable and the product obtained is almost invariably an alkane after addition of a second equivalent of H_2.

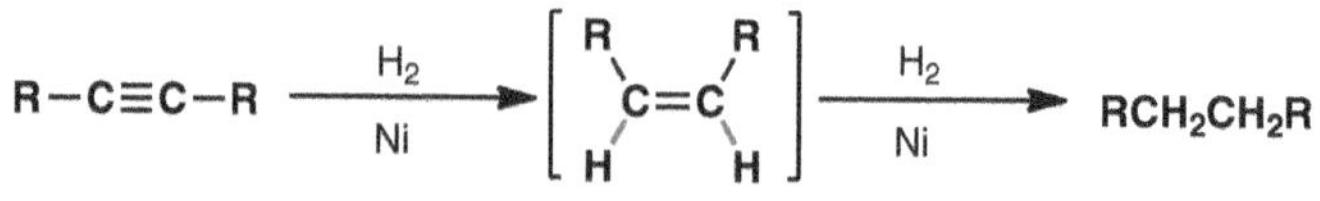

$$R-C\equiv C-R \xrightarrow[\text{Ni}]{H_2} \left[\begin{array}{c} R \quad R \\ C=C \\ H \quad H \end{array} \right] \xrightarrow[\text{Ni}]{H_2} RCH_2CH_2R$$

In order to obtain the alkene from an alkyne, a *deactivated* catalyst is required. The well-known Lindlar catalyst which is used for this purpose is Pd supported on calcium carbonate and deactivated with lead; an internal alkyne gives a (Z)-alkene by *syn* addition.

(E)-Alkenes may be obtained from internal alkynes by dissolving metal reduction (see Sub-section 20.9.1).

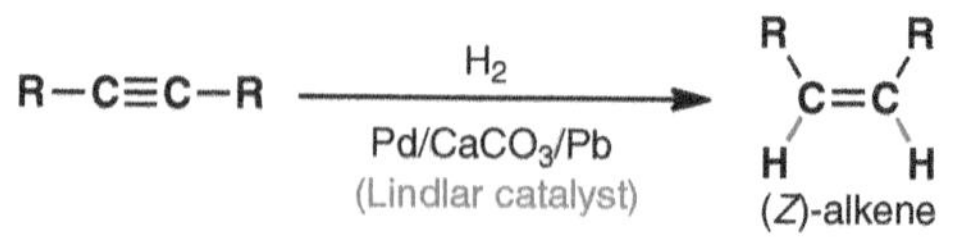

$$R-C\equiv C-R \xrightarrow[\substack{\text{Pd/CaCO}_3/\text{Pb} \\ \text{(Lindlar catalyst)}}]{H_2} \begin{array}{c} R \quad R \\ C=C \\ H \quad H \end{array}$$

(Z)-alkene

Panel 15.3 Relative stabilities and heats of hydrogenation of alkenes

---

The hydrogenation of alkenes is exothermic, i.e. values for the enthalpy of the reaction (ΔH) are negative, and the heat evolved is called the *heat of hydrogenation*. When isomeric alkenes give the same alkane upon hydrogenation, we can evaluate their relative (thermochemical) stabilities from their heats of hydrogenation; the methodology is the same as when we calculated the delocalization energy of benzene in Section 5.5.3.

No.	alkene	$-\Delta H/\text{kJ mol}^{-1}$
1	$CH_2{=}CH_2$	137
2	$CH_2{=}CHCH_3$	126
3	$CH_2{=}CHCH_2CH_3$	127
4	*cis*-$CH_3CH{=}CHCH_3$	120
5	*trans*-$CH_3CH{=}CHCH_3$	115
6	$CH_2{=}CHCH(CH_3)_2$	127
7	$CH_2{=}C(CH_3)CH_2CH_3$	119
8	$(CH_3)_2C{=}CHCH_3$	113
9	$(CH_3)_2C{=}C(CH_3)_2$	111

Heats of hydrogenation of some alkenes

The table lists values of ΔH for the hydrogenation of some representative alkenes. Among them, the isomeric butenes (Nos. 3–5) all give butane upon hydrogenation.

$$C_4H_8 + H_2 \rightarrow C_4H_{10}; \quad \Delta H = \text{heat of hydrogenation}$$

Values of ΔH become increasingly *less* negative (i.e. increasingly *less* energy is evolved) along the series but-1-ene → *cis*-but-2-ene → *trans*-but-2-ene, which indicates that the isomeric butenes become increasingly stable in this order. This is illustrated in the figure below: the more stable the isomeric alkene, the smaller the heat of hydrogenation to give a common product.

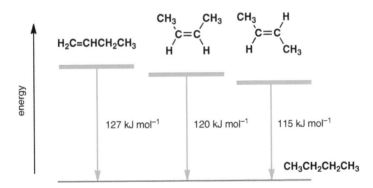

Heats of hydrogenation of isomeric butenes and their relative stabilities.

The alkenes of entries Nos. 6–8 in the table all give 2-methylbutane upon hydrogenation. Again, by comparing heats of hydrogenation for this set of isomeric alkenes, we see that the more substituted the double bond is, the more stable the alkene. Comparison of entries Nos. 1, 2, and 9 leads to the same conclusion.

Panel 15.4 Oxidation and reduction in organic chemistry

From the earliest days of organic chemistry, reactions involving bond formation from carbon to oxygen (or loss of hydrogen from carbon) have been known as **oxidations**, while processes involving bond formation to hydrogen (or loss of oxygen) have been defined as **reductions**. Epoxidation and dihydroxylation of alkenes described in this chapter, for example, are obviously oxidations; conversions of alcohols to carbonyl compounds (Section 14.4) are also oxidations. Conversely, the hydrogenation of alkenes is a reduction as is the conversion of a carbonyl group (>CO) to >CHOH or >CH$_2$, as covered in Chapter 10.

Other definitions of oxidation and reduction (probably encountered in courses on inorganic or physical chemistry) are in terms of the transfer of electrons (see also Section 20.9): *the loss of one or more electrons is an **oxidation**, and their gain is a **reduction**.* It follows that a reagent which donates electrons is a reducing agent and one which removes electrons is an oxidizing agent. Depending on the number of electrons lost or gained, a transformation may be called a two-electron oxidation, for example, or a four-electron reduction. In an electron transfer reaction, one molecule or ion gains an electron as another loses one, so oxidation and reduction by electron transfer necessarily occur together (this is analogous to a proton transfer reaction). But how do we reconcile these definitions of oxidation and reduction in terms of electron transfer with descriptions based upon gain or loss of oxygen, and loss or gain of hydrogen? In order to do this, we have to introduce the concept of oxidation numbers.

The **oxidation number** (ON) of the atom in a mono-atomic ion is its charge, e.g. +1 for sodium in Na$^+$ and −1 for chlorine in Cl$^-$. In determining the ON of an atom bonded to other atoms within a covalent organic molecule, we assume (i) that bonds to it are ionic rather than covalent, and (ii) that *both* electrons of each covalent bond 'belong' to the more electronegative of the two bonded atoms; we then proceed as above. In this way, for example, the ON of hydrogen in HBr is +1 and that of the bromine is −1, and the ON of carbon in CO$_2$ is +4 and that of both oxygen atoms is −2. In other words, for the purpose of calculating oxidation numbers, all polar bonds are imagined to be ionic, and the charge of a hypothetical ion gives the oxidation number of that atom in the molecule.[3] In these calculations, bonds to an identical atom are ignored and the algebraic sum of the oxidation numbers of all atoms in a (neutral) molecule must be zero, or equal to the overall charge on an ion. The ONs of carbon in some organic compounds are shown below.

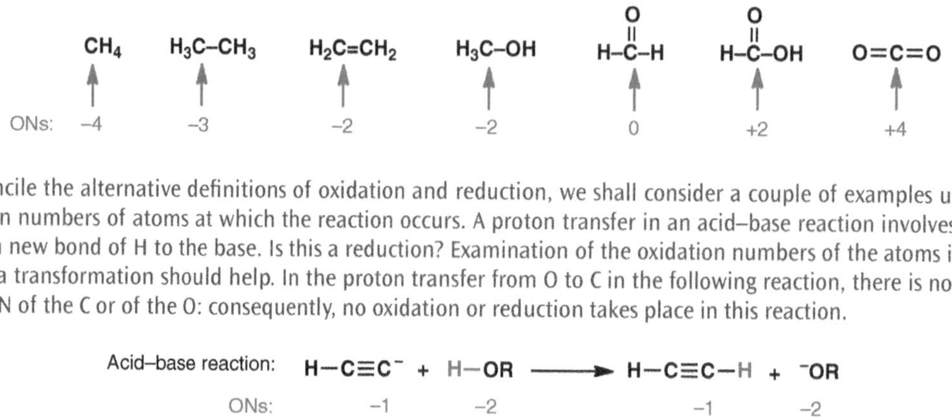

CH$_4$	H$_3$C–CH$_3$	H$_2$C=CH$_2$	H$_3$C–OH	H–C(=O)–H	H–C(=O)–OH	O=C=O
↑	↑	↑	↑	↑	↑	↑
ONs: −4	−3	−2	−2	0	+2	+4

To reconcile the alternative definitions of oxidation and reduction, we shall consider a couple of examples using the oxidation numbers of atoms at which the reaction occurs. A proton transfer in an acid–base reaction involves formation of a new bond of H to the base. Is this a reduction? Examination of the oxidation numbers of the atoms involved in such a transformation should help. In the proton transfer from O to C in the following reaction, there is no change in the ON of the C or of the O: consequently, no oxidation or reduction takes place in this reaction.

Acid–base reaction: H—C≡C$^-$ + H—OR ⟶ H—C≡C—H + $^-$OR

ONs: −1   −2   −1   −2

---

[3] Note that the *oxidation number* is not the same as the *formal charge* in a Lewis structure; the latter is calculated by assigning each of a bonding pair of electrons to each of the bonded atoms (regardless of the polarity of the bond, see Sub-section 1.3.1).

Now look at the transfer of a hydride from BH_4^- to a carbonyl group, a reaction covered in Chapter 10 and regarded by organic chemists as a reduction at carbon.

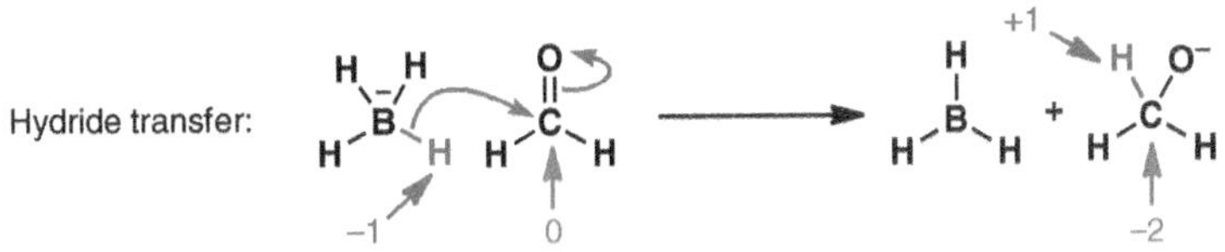

Hydride transfer:

The ON of the carbon changes from 0 to –2, meaning that a two-electron reduction takes place. Correspondingly, the ON of a hydrogen changes from -1 in BH_4^- (remember that B is less electronegative than H) to +1 in CH_3O^-, a *change* of +2; this corresponds to a two-electron oxidation. Note that the ON of boron is the same in BH_4^- as in BH_3 so it undergoes neither oxidation nor reduction. All is in order: C has been reduced and H has been oxidized, and the oxidation–reduction (we cannot have one without the other) in terms of hydride transfer to the organic compound is wholly compatible with notions based on electron transfer (oxidation numbers).

We can generalize the above as follows: when the oxidation number *increases* at a carbon, the reaction at the C is an *oxidation*; when the ON of the C *decreases*, the reaction is a *reduction*.

Our third example involves chromic acid, a representative oxidizing agent which was introduced for the oxidation of alcohols to carbonyl compounds in Section 14.4, as illustrated below. The oxidation number of chromium in chromic acid is +6 and it *decreases* to +4 in the elimination step which gives the carbonyl compound, as shown below; this corresponds to a *two-electron reduction* of the chromium. The associated transformation of $R_2CH–OH$ into $R_2C=O$ involves an *increase* in the ON of the central C from 0 to +2, i.e. a two-electron *oxidation*.

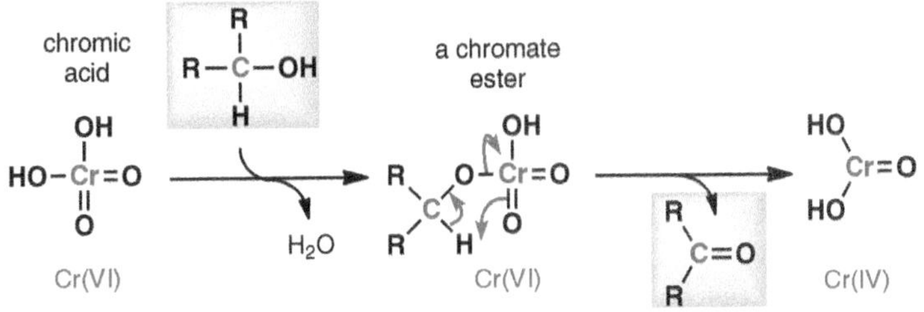

Oxidation numbers of chromium in the chromic acid oxidation of an alcohol.

In this example, we have been able to relate the familiar oxidation of an organic compound (loss of hydrogen) by a transition metal complex to an electron transfer process; we have again shown that the alternative ways of describing oxidation-reduction reactions are wholly consistent.

Finally, consider whether the following two addition reactions involve oxidation-reduction.

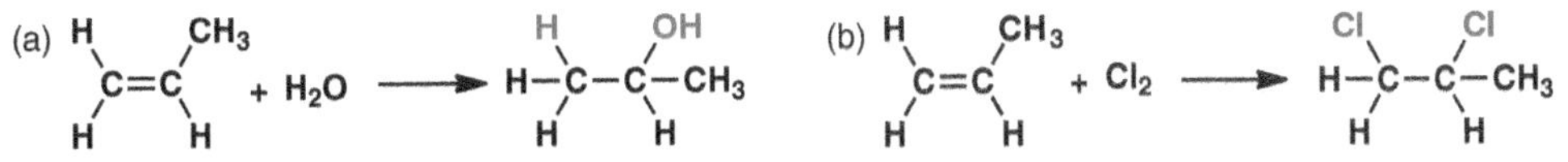

By proceeding as described above, we see that the *total* ONs of carbon atoms do not change in the hydration reaction (a), but the *total* CONs increase in the addition reaction of Cl_2 in (b). So, although both reactions are electrophilic additions to an alkene, the former does not involve oxidation–reduction but the latter is an oxidation of the organic compound (and the chlorine is correspondingly reduced).

Summary

- A typical alkene undergoes addition reactions initiated by electrophilic attack at the π bond to give a **carbenium ion intermediate**. The addition is completed by capture of the intermediate by a nucleophile. Alkynes also undergo electrophilic addition reactions but they are generally less reactive than alkenes.

- If an alkene is unsymmetrical, initial electrophilic attack gives the more stable of possible alternative carbenium ions, so the overall reaction is regioselective.

- Oxymercuration–demercuration and hydroboration–oxidation are useful methods of achieving alternative regioselectivities in the hydration of unsymmetrical alkenes.

- Halogenation of alkenes proceeds by *anti* addition and the reactions of diastereoisomeric alkenes with halogens are stereospecific.

- Carbenes (X_2C:) are reactive electrophilic divalent compounds which add stereospecifically to alkenes to give cyclopropanes. Alkenes act as nucleophiles in their reactions with peroxy acids to give oxiranes (epoxides).

- Electrophilic addition to conjugated dienes gives 1,2- and 1,4-adducts via intermediate **allylic cations**. 1,2-addition is usually favoured under conditions of **kinetic control**, whereas the 1,4-addition is usually the major reaction under **thermodynamic control**.

- Conjugated dienes combine with alkenes acting as **dienophiles** in stereospecific **Diels–Alder** (cycloaddition) reactions to form cyclohexene derivatives via six-membered cyclic 6π electron transition structures.

- Hydrogenation (addition of hydrogen to a multiple bond) is a *syn* addition which requires metal catalysis.

- Oxidation and reduction of organic compounds are reviewed in Panel 15.4.

Problems

15.1 What is the main product of each of the following reactions in a non-nucleophilic solvent?

(a) $CH_3CH{=}CHCH_3$ + HCl ⟶

(b) $CH_3CH{=}CH_2$ + HI ⟶

(c) 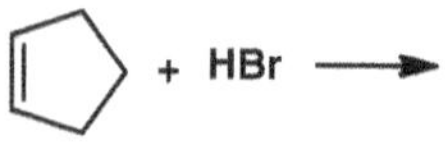 [structure] + HBr ⟶

(d) 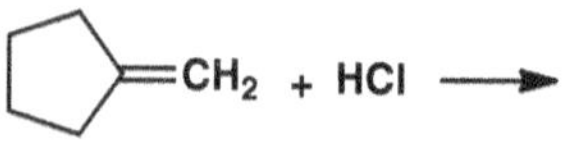 [structure]$=CH_2$ + HCl ⟶

15.2 Draw the structure of the main product in the reaction of HCl with each of the following alkenes.

(a) 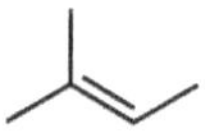 [structure] (b) 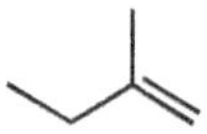 [structure] (c) 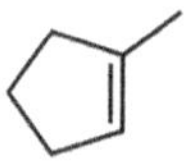 [structure]

(d) 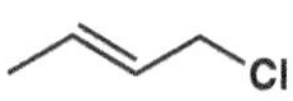 [structure] Cl (e) 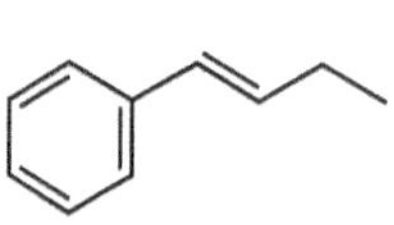 [structure]

15.3 Give a mechanism and indicate the final product for the addition of HCl to each of the following in a non-nucleophilic solvent.

(a) 1-methylcyclohexene (b) 1-phenylpropyne

15.4 Show the increasing order of second-order rate constants for the reaction of HBr with the following alkenes under the same conditions, and give reasons for your predictions.

(a) $CH_2{=}CH_2$ [structure] [structure]

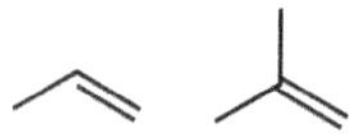

(b) 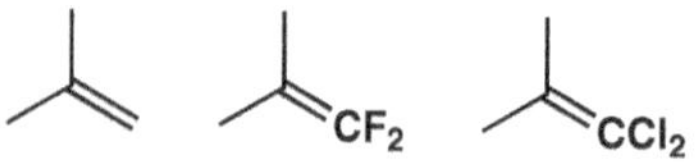 [structure] [structure] CF_2 [structure] CCl_2

15.5 What is the product of the acid-catalysed hydration of each of the following alkenes?

(a) propene (b) 2-methylbut-2-ene
(c) 2-methylpent-1-ene

15.6 What is the product of hydroboration followed by treatment with alkaline hydrogen peroxide for each alkene in Problem 15.5.

15.7 Give a mechanism and indicate the final product for the reaction of each of the following in an aqueous acid.

(a) 1-methylcyclohexene   (b) 1-methoxycyclohexene

15.8 Draw the structure (including stereochemical information) of the main product of the reaction of one equivalent of Br_2 with (a) (E)-pent-2-ene, (b) (Z)-pent-2-ene, and (c) pent-2-yne.

15.9 Give a mechanism for the addition of Br_2 to cyclopentene.

15.10 Show the structures (including stereochemical information) of products obtained by the reaction of Br_2 with 3-methylcyclohexene.

15.11 When a solution of propene in methanol is treated with a catalytic amount of sulfuric acid, a methyl ether is formed. Deduce the structure of the product by giving a mechanism for the reaction.

15.12 Methylenecyclohexane isomerizes to 1-methylcyclohexene in solution upon addition of a catalytic amount of an acid. Give a mechanism for this transformation.

15.13 Propose reactions to convert pent-1-yne into (a) pentanal and (b) pentan-2-one.

15.14 Propose a reaction to prepare each of the following isomeric alcohols from an appropriate alkene with the same number of carbons.

(a) 2-methylpentan-1-ol   (b) 2-methylpentan-2-ol
(c) 2-methylpentan-3-ol   (d) 4-methylpentan-2-ol
(e) 4-methylpentan-1-ol

15.15 In the reaction of Br_2 with buta-1,3-diene in CCl_4, the ratio of the products (3,4-dibromobut-1-

ene and 1,4-dibromobut-2-ene) was 67 : 33 at −15 °C whereas it was 10 : 90 at 60 °C. Explain these results.

15.16 Show the structure of the main Diels–Alder adduct from each of the following combinations of diene and dienophile; indicate the stereochemistry if stereoisomers are possible.

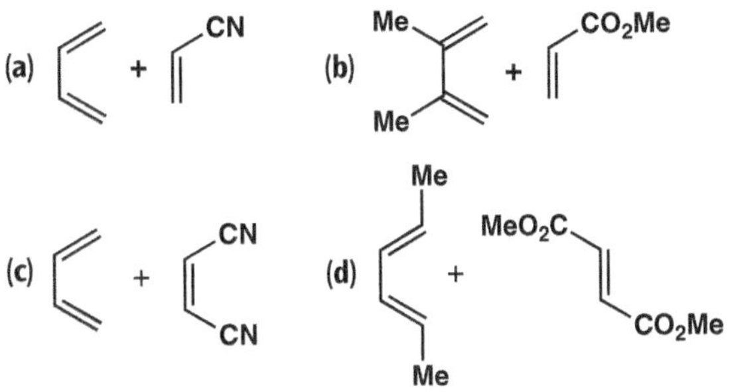

15.17 Explain why 3-methylenecyclohexene does not undergo a Diels–Alder reaction with maleic anhydride even though it is a conjugated diene.

15.18 Show the structure of the main product in each of the following reactions; indicate the stereochemistry if stereoisomers are possible.

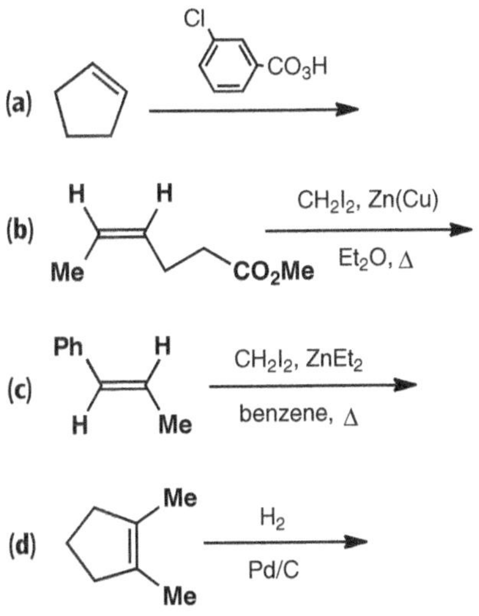

Supplementary Problems

15.19 Show the structures of compounds (A)–(D) in the following reactions.

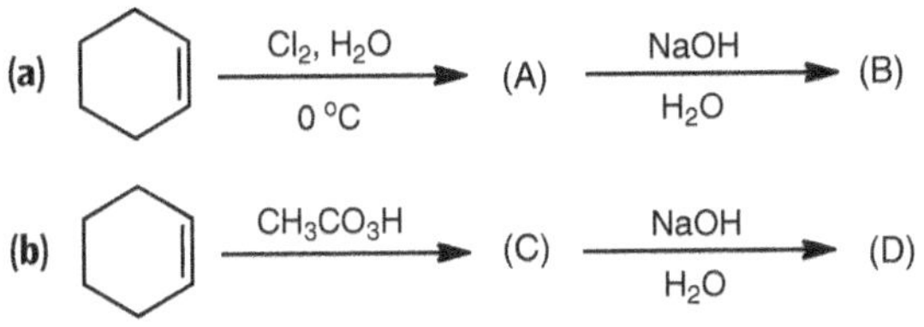

15.20 Give a mechanism for the conversion of propyne into propanone in aqueous sulfuric acid in the presence of mercuric sulfate.

15.21 Propose a mechanism for the formation of the two products shown below in the reaction of HCl with 3,3-dimethylbut-1-ene in ethanoic acid.

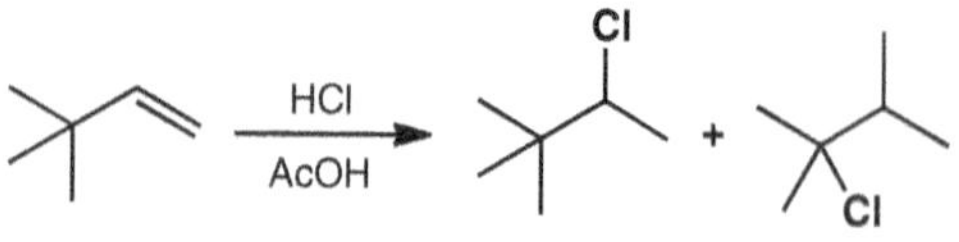

15.22 Acid-catalysed hydration of 3-methylbut-1-ene takes place with rearrangement.

(a) Give a mechanism for this reaction and identify the product.

(b) Propose alternative reactions of the same alkene to provide two un-rearranged isomeric alcohols.

15.23 Relative rate constants for the acid-catalysed hydration and bromination of alkenes are compared below. In the former reaction, the reactivities of propene and but-2-ene are very similar but 2-methylpropene is hugely more reactive. In contrast, bromination increases moderately along the series propene < but-2-ene < 2-methylpropene. Explain these results.

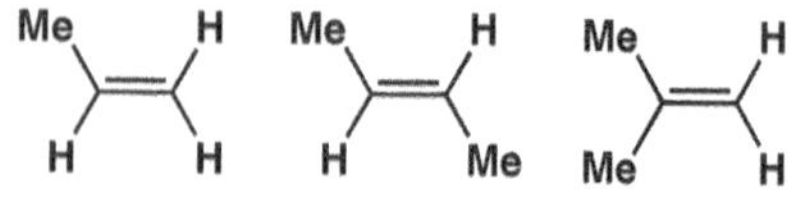

Hydration	1.0	0.8	1.6×10^5
Bromination	1.0	28	89

15.24 When phenylpropadiene (phenylallene) reacts with one equivalent of Br_2, contrasting regioselectivities are observed in CS_2 and in MeOH at 0 °C as shown below. Explain these results.

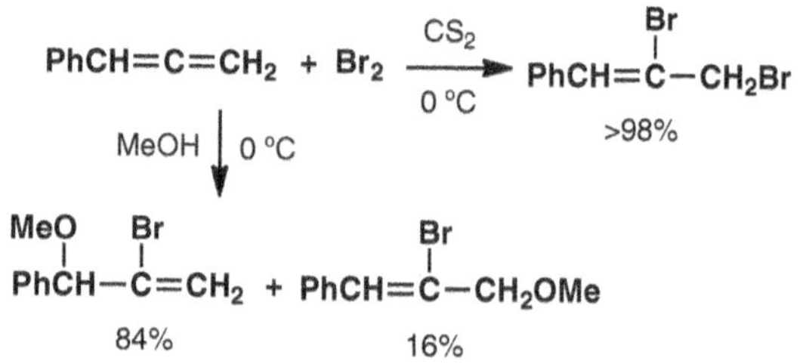

Electrophilic Aromatic Substitution 16

Related topics we have already covered:

- **Resonance** (Sections 1.4 and 5.4)
- **Structure and molecular orbitals of benzene** (Section 5.5)
- **Aromaticity** (Section 5.6)
- **Substituent effects on acidity** (Sub-section 6.3.3)
- **Curly arrow representation of reaction mechanisms** (Section 7.2)
- **Stability of carbenium ions** (Sub-section 12.4.3)
- **Electrophilic additions to alkenes** (Chapter 15)

Topics we shall cover in this chapter:

- Differences between alkenes and benzene
- Substitution reactions of benzene by electrophilic addition–elimination
- Halogenation
- Nitration
- Sulfonation
- Friedel–Crafts alkylation and acylation
- Substituent effects upon reactivity of benzene in electrophilic addition
- Regioselectivity of electrophilic substitution directed by substituents
- Reactivity of phenol and aniline
- Preparation of substituted benzenes

We saw in Chapter 5 that benzene is a conjugated unsaturated compound which has special stability attributed to **aromaticity**; it is the parent member of a large class of so-called *aromatic* compounds. However, an organic compound does not have to be either aromatic, or not, as a whole: molecules of many compounds, including natural products and biologically important compounds, contain aromatic and non-aromatic residues linked together (Section 5.6 and Panels 16.2–16.4).

The main industrial source of aromatic compounds used to be coal. Among them, benzene, toluene, and the isomeric xylenes, sometimes collectively called BTX, are important starting materials for industrial organic chemistry.

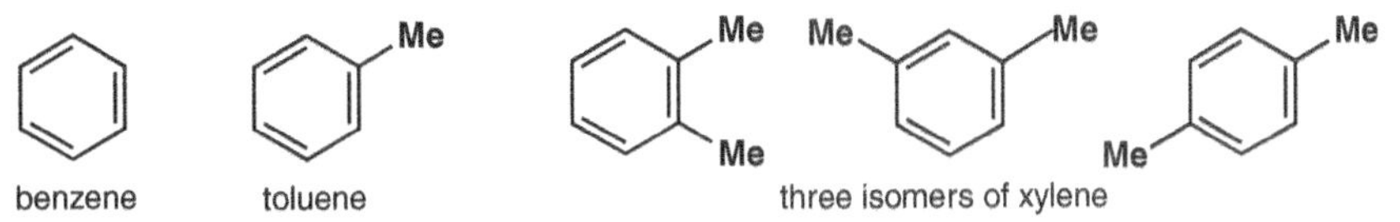

benzene   toluene   three isomers of xylene

However, they are now largely obtained from the naphtha fraction in the distillation of crude oil (Panel 3.1).

The distinguishing feature of an aromatic compound is its cyclic system of delocalized π electrons (as described in Chapter 5); this can be a centre of nucleophilicity like π bonds in alkenes (discussed in Chapter 15). Benzene, like alkenes, undergoes electrophilic addition, but a proton is then eliminated from the first-formed intermediate. Consequently, aromaticity is restored and the overall reaction is substitution: **electrophilic aromatic substitution**. These reactions include halogenation, nitration, sulfonation, and

Friedel–Crafts alkylation and acylation. This chapter is concerned mainly with these reactions of benzene and some of its simple derivatives.

We shall see that mono-substituted derivatives of benzene, C_6H_5-Y, give isomeric *ortho, meta*, and *para* disubstituted products, Z-C_6H_4-Y with electrophile Z^+, in proportions which depend on the nature of the initial substituent Y; Y also affects reactivity and may be *activating* or *deactivating*. Multiply substituted benzenes can be synthesized by judicious combinations of sequential electrophilic aromatic substitution reactions.

16.1 Structures of Substituted Benzenes

Before we study reactions of benzene, let us review some aspects of its electronic structure (see Chapter 5). We saw that the carbon framework of a benzene molecule is a planar, regular hexagon, and six π electrons are delocalized over the whole ring rather than localized as bonding pairs of three double (alkene) bonds (Section 5.5). This has been confirmed experimentally by structural studies (especially X-ray crystallography and NMR spectroscopy) and rationalized by molecular orbital theory.

A 1,2-disubstituted benzene is only ever a single compound, a fact which requires that all the C–C bonds of benzene are equivalent. Nonetheless, we still usually represent benzene by a **Kekulé structure** with alternate double and single bonds.

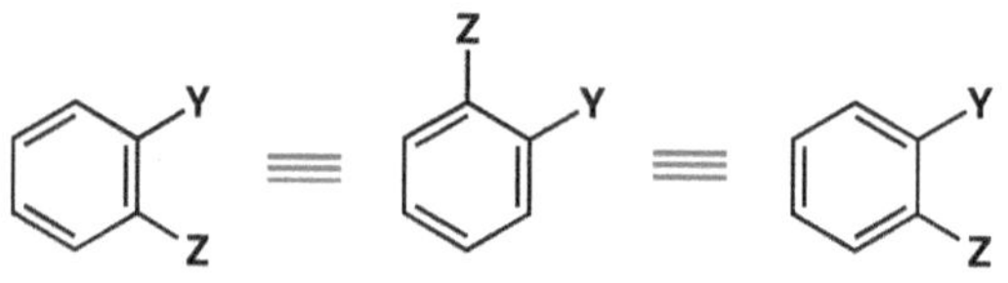

A 1,2-disubstituted benzene.

A single Kekulé structure is used simply for convenience, it being understood that benzene has a delocalized electronic structure more accurately represented by a resonance hybrid of the contributing forms. The electron-paired Lewis/Kekulé structure is especially useful when we describe reaction mechanisms by curly arrows which show movements of electron pairs.

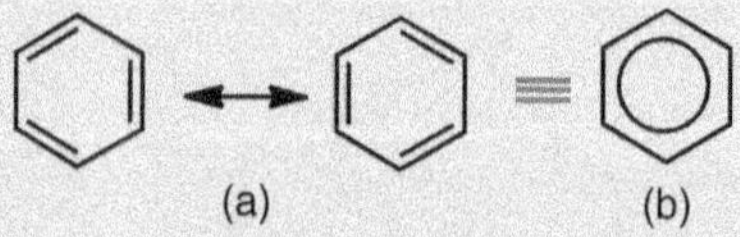

Kekulé structures (a) of benzene, which is also represented by (b).

16.2 Electrophilic Aromatic Substitution by an Addition–Elimination Mechanism

As mentioned above, benzene is an unsaturated compound rich in π electrons and, like an alkene, is susceptible to electrophilic attack. When cyclohexene is mixed with bromine in solution, the reddish colour of bromine fades rapidly as an immediate reaction gives an addition product (eqn 16.1; see Chapter 15). In contrast, benzene does not readily react with bromine (eqn 16.2).

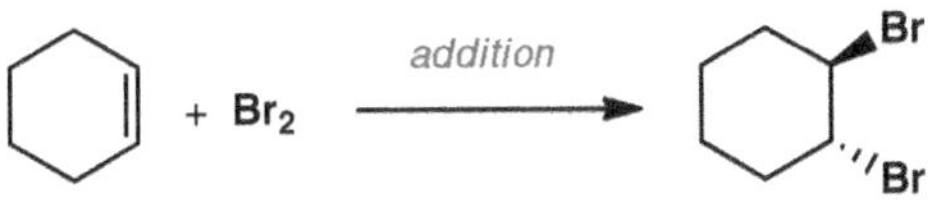

$$\text{cyclohexene} + Br_2 \xrightarrow{\text{addition}} \text{trans-1,2-dibromocyclohexane} \quad (16.1)$$

$$\text{benzene} + Br_2 \xrightarrow{\;\;\varnothing\;\;} \text{no immediate reaction} \quad (16.2)$$

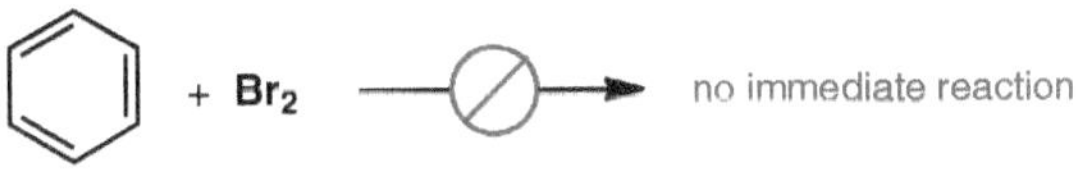

Upon addition of a *Lewis acid* such as $AlBr_3$ to benzene plus Br_2 in a nonpolar solvent, however, there is a reaction, but it is substitution to give bromobenzene (eqn 16.3) not addition.

$$\text{benzene} + Br_2 \underset{\text{substitution}}{\overset{\substack{\text{Lewis acid}\\ AlBr_3}}{\longrightarrow}} \text{bromobenzene (Br)} + HBr \quad (16.3)$$

bromobenzene

Bromine alone is not sufficiently electrophilic to react with the weakly nucleophilic benzene but the catalytic Lewis acid activates it by forming a highly electrophilic Lewis acid–base complex (eqn 16.4). The first step in the reaction with benzene (Scheme 16.1) is an addition but it is followed by the elimination of a proton, the *electrofuge*, to give a substitution product (eqn 16.5).

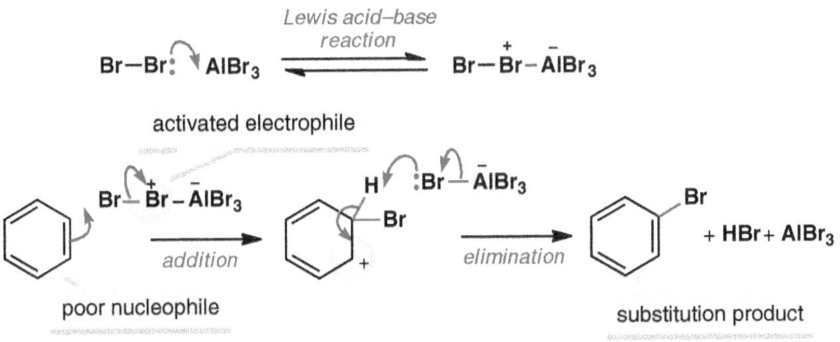

Just as a group which departs *with* its bonding pair of electrons in an E1, E2, or a nucleophilic substitution reaction is called a *nucleofuge*, a group which departs *without* its bonding pair, e.g. the H^+ in an electrophilic substitution reaction of benzene, is called an *electrofuge*.

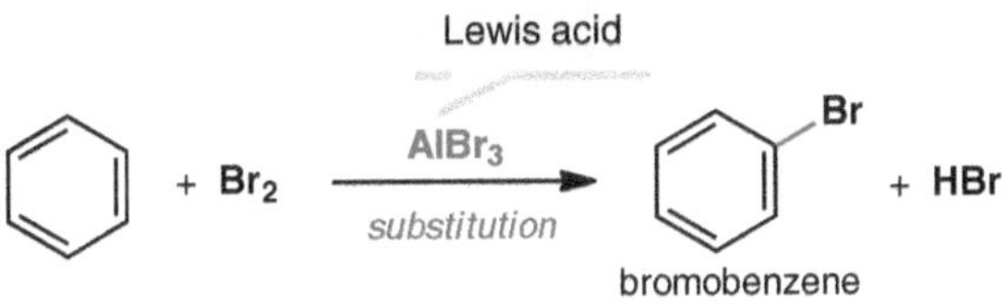

$$Br{-}Br{:} \; \; AlBr_3 \;\; \underset{\text{reaction}}{\overset{\text{Lewis acid–base}}{\rightleftharpoons}} \;\; Br{-}\overset{+}{Br}{-}\overset{-}{AlBr_3} \quad (16.4)$$

activated electrophile

$$\text{benzene} + Br{-}\overset{+}{Br}{-}\overset{-}{AlBr_3} \xrightarrow{\text{addition}} \text{benzenium ion} \xrightarrow{\text{elimination}} \text{bromobenzene} + HBr + AlBr_3 \quad (16.5)$$

poor nucleophile   substitution product

Scheme 16.1 Lewis acid-catalysed bromination of benzene.

Reactions of cyclohexene and benzene with an **electrophile** E^+ are compared in Schemes 16.2 and 16.3. The electrophile E^+ adds to unsaturated systems in both reactions to give **carbocation intermediates**, **1** and **2**. Although they look very similar, **1** is captured by a nucleophile (Nu^-) to give an addition product (Scheme 16.2) whereas **2** is deprotonated by Nu^- acting as a base in an elimination step to give a substitution product—overall, an H of benzene has been replaced by E (Scheme 16.3).

The alternative (competing) paths from the intermediate benzenium ion **2** formed by electrophilic addition to benzene are both included in Figure 16.1 which shows that one product (the arene) is much more stable than the other (a cyclohexadiene). As some of the aromaticity of the substituted arene is developing in the preceding transition structure,

The carbocation formed by electrophilic addition to benzene (an arene) is called a **benzenium ion** (an arenium ion). The ones formed in $E1/S_N1$ reactions are *carbenium ions*, as encountered in previous chapters.

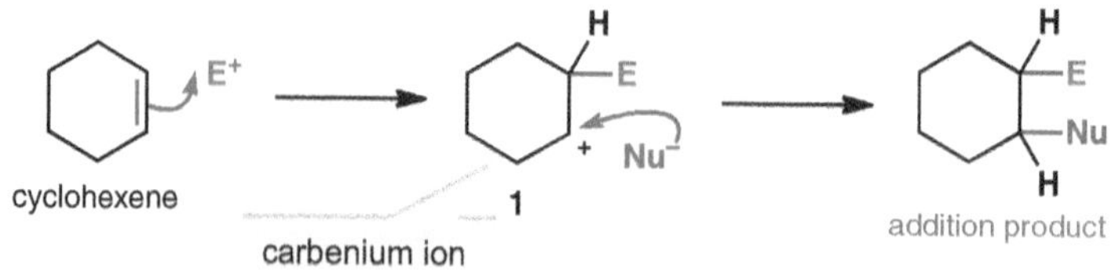

Scheme 16.2 Electrophilic addition to cyclohexene followed by nucleophilic capture.

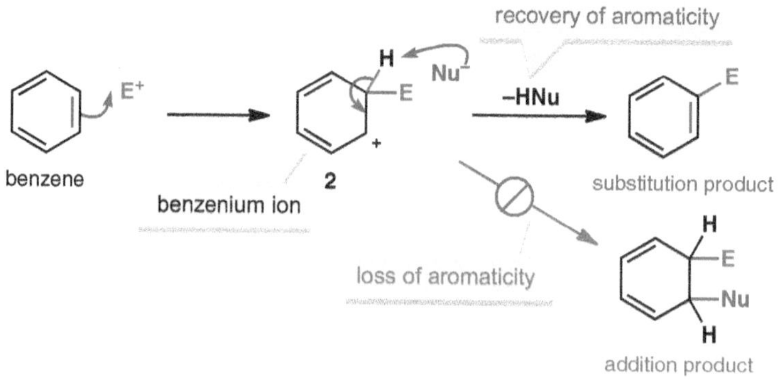

Scheme 16.3 Electrophilic addition to benzene followed by proton loss rather than nucleophilic capture (overall substitution rather than addition).

Reminder: arene is the generic term for an aromatic hydrocarbon.

the barrier to its formation (red line) is lower than the one leading to the cyclohexadiene. In other words, the preference for deprotonation leading to overall substitution over nucleophilic capture leading to addition is associated with the *recovery of the aromaticity* of the benzene ring in the **electrophilic addition-elimination** sequence.

The intermediate benzenium ion **2** in Scheme 16.3 can be represented by a resonance hybrid. It has four π electrons delocalized over five sp^2 hybridized C atoms with a partial positive charge at positions *ortho* and *para* to the sp^3 C which completes the

If we were to construct a reaction profile for the addition of E$^+$ to cyclohexene, the initial barrier would be lower than the one in Figure 16.1 because the reaction of cyclohexene does not involve disruption of an aromatic system.

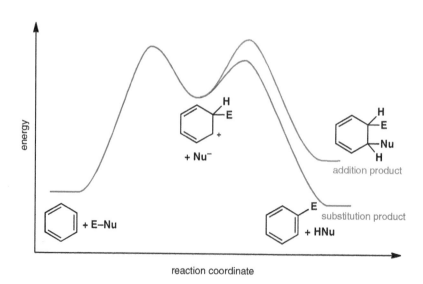

Figure 16.1 Superimposed reaction profiles for electrophilic substitution and addition reactions of benzene.

six-membered ring. The electronic structure of **2** can be represented more concisely by **2a** or the simplified form, **2b**.

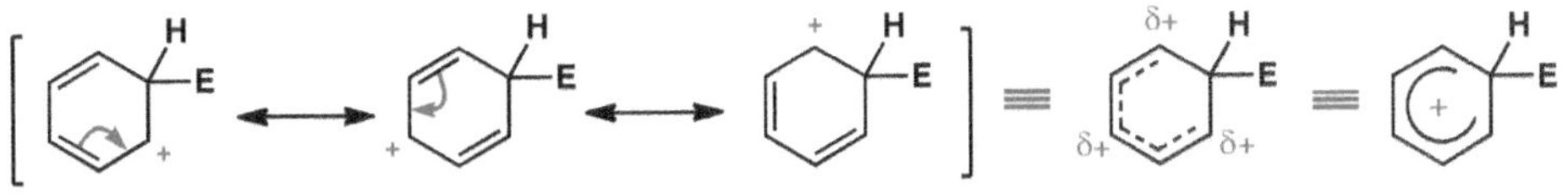

resonance forms of benzenium ion (**2**)   **2a**   **2b**

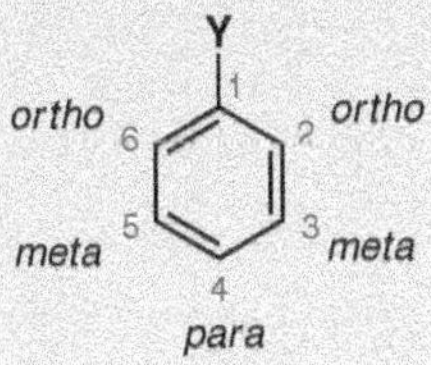

Although the benzenium ion is stabilized by electron delocalization, its stabilization energy is not as high as that of the fully aromatic cyclic 6π system of benzene and its simple derivatives. *The high aromatic stability of benzene is responsible for both its low reactivity and strong tendency to undergo substitution rather than addition.*

Positions 2 and 6 in **2**, or in any aromatic compound C_6H_5-Y, are usually said to be *ortho* to the substituent at C1; positions 3 and 5 are *meta* to it, and position 4 is *para*.

16.3 Main Classes of Electrophilic Aromatic Substitution

We can classify electrophilic aromatic substitution reactions according to the electrophile as summarized in Table 16.1 for the main classes. Note that we include separate columns for (i) the electrophiles which actually add, and (ii) the reagents which deliver the electrophiles; cations may be in the form of ion pairs or Lewis acid–base adducts.

Electrophiles and main classes of electrophilic aromatic substitutions				
Reaction	Electrophile	Reagent	Product	
Halogenation	X^+ (X=Cl, Br)	X_2–LA[a]	Ar-X	
Nitration	NO_2^+	HNO_3–H_2SO_4	Ar-NO_2	
Sulfonation	SO_3	H_2SO_4 (+ SO_3)	Ar-SO_3H	
Alkylation	R^+	RX–LA[a]	Ar-R	
Acylation	$RC{\equiv}O^+$	RCOCl–LA[a]	Ar-COR	

Table 16.1

a. LA=Lewis acid, e.g. AlX_3 or FeX_3.

16.3.1 Halogenation

Chlorination and **bromination** of benzene take place with the halogen (X_2; X=Cl or Br) and a Lewis acid catalyst such as AlX_3 or FeX_3, e.g. Scheme 16.4.

Iodination of benzene with iodine can be promoted by addition of an oxidizing agent such as HNO_3, $CuCl_2$, or H_2O_2. Iodine is first oxidized to a highly electrophilic I^+ species or IOH which brings about the substitution reaction (eqn 16.6).

Sometimes, finely divided metallic iron is used; the Fe is oxidized in the reaction mixture by the halogen to give FeX_3 which is the real catalyst. Some particularly reactive aromatic compounds do not require catalysis, e.g. phenol and aniline (see later).

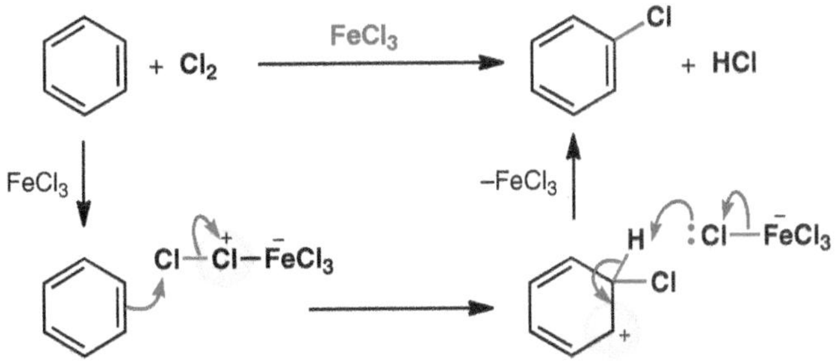

Scheme 16.4
Chlorination of benzene catalysed by $FeCl_3$.

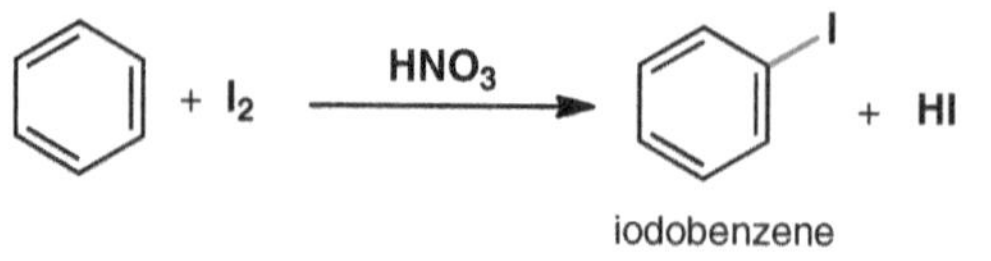

$$\text{benzene} + I_2 \xrightarrow{\text{HNO}_3} \text{iodobenzene} + \text{HI} \qquad (16.6)$$

16.3.2 Nitration

A low equilibrium concentration of **nitronium ion** (NO_2^+, a very reactive electrophile) is generated when concentrated sulfuric acid is added to concentrated nitric acid (eqn 16.7). This then reacts by the two-step mechanism shown in eqn 16.8 of Scheme 16.5 for the **nitration** of benzene.

$$\text{nitric acid} + \text{sulfuric acid} \xrightleftharpoons{-HSO_4^-} \xrightleftharpoons{-H_2O} \text{nitronium ion} \qquad (16.7)$$

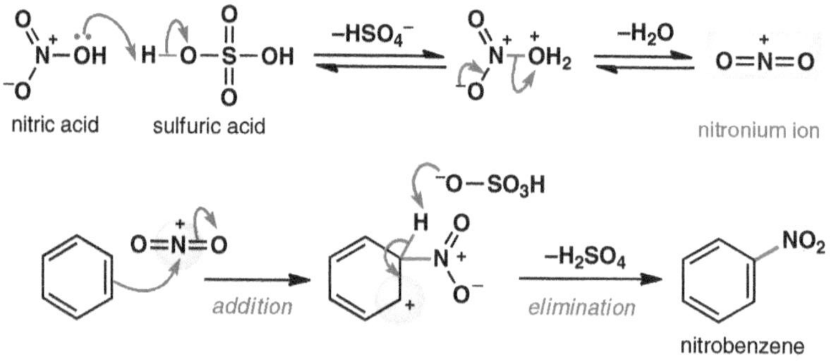

Scheme 16.5 Nitration of benzene.

Although NO_2^+ is very reactive, it is not so unstable under highly non-nucleophilic conditions; $NO_2^+ \ BF_4^-$ is commercially available as a crystalline salt, and is a potent nitrating agent.

$$\qquad (16.8)$$

Example 16.1

Why is concentrated nitric acid alone, which contains nitronium ion, as described in the following equation, not a very good reagent for the nitration of benzene?

$$2 \ HNO_3 \rightleftharpoons NO_2^+ + NO_3^- + H_2O$$

Solution
The reaction which generates nitronium ion using just nitric acid involves one molecule of HNO_3 acting as an acid and another acting as a base (autoprotolysis), see Chapter 6, then dissociation of the protonated molecule, $O_2N{-}OH_2^+$, to give NO_2^+.

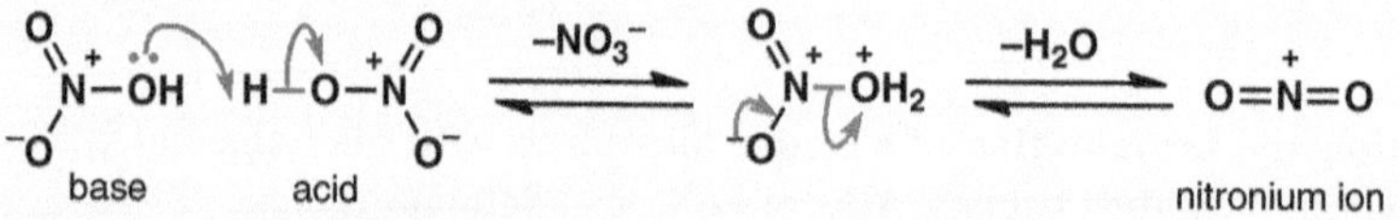

This reaction corresponds to eqn 16.7 above in which H_2SO_4 is the proton donor in the first step. Nitric acid ($pK_a \sim -1.4$) is a weaker Brønsted acid than sulfuric acid ($pK_a \sim -3$) so generates a much lower equilibrium concentration of $O_2N{-}OH_2^+$ which, in turn, does not provide a concentration of NO_2^+ sufficient for an effective reaction. Note that in this reaction and the one in eqn 16.7, the H_2O which is liberated will be extensively protonated by the strongly acidic reaction conditions.

16.3.3 Sulfonation

Benzene reacts slowly with concentrated sulfuric acid to form benzenesulfonic acid. The electrophile for **sulfonation** is sulfur trioxide SO_3 (the anhydride of sulfuric acid) or its protonated form HSO_3^+ (eqn 16.9). Fuming sulfuric acid, which is a solution of SO_3

in H_2SO_4 (and sometimes called oleum), is a more effective sulfonating agent. Note that the electrophilic centre of SO_3 is the sulfur and that all the steps in Scheme 16.6 are reversible. In concentrated sulfuric acid, equilibrium is towards the sulfonic acid, but the reverse reaction occurs when the sulfonic acid is heated with dilute aqueous acid.

$$\text{(16.9)}$$

sulfur trioxide

$$\text{(16.10)}$$

benzenesulfonic acid

Scheme 16.6 Reversible sulfonation of benzene.

Charles Friedel (1832–1899)

Friedel, a French chemist and mineralogist, had been a student of Pasteur's and became a professor at the Sorbonne in 1876; he developed what we call the Friedel–Crafts reactions in a collaboration with J.M. Crafts in 1877.

The product of eqn 16.10 in Scheme 16.6, benzenesulfonic acid, is the parent *arenesulfonic* acid; these are comparable in acidity with sulfuric acid and often used as acid catalysts for organic reactions.

16.3.4 Friedel–Crafts alkylation

An alkyl halide can form a carbenium ion (or its equivalent) upon addition of a Lewis acid. Electrophilic substitution by a carbenium ion intermediate to give an alkylarene (e.g. eqn 16.11 for the reaction of benzene) is called **Friedel–Crafts alkylation** after the two chemists who discovered the reaction. The reaction proceeds as shown in Scheme 16.7.

$$\text{(16.11)}$$

alkylbenzene

Friedel–Crafts alkylation can also involve carbenium ions generated from alcohols and alkenes with catalysis by Brønsted acids, e.g. H_2SO_4 or H_3PO_4.

James M. Crafts (1839–1917)

Crafts was an American who studied under Bunsen in Germany and Wurtz in France after graduation from Harvard University; he served as a professor and president of the Massachusetts Institute of Technology.

Scheme 16.7 Friedel–Crafts alkylation of benzene.

Scheme 16.7 includes heterolysis of an alkyl halide with Lewis acid catalysis, and benzene acting as the nucleophile. If the carbenium ion R⁺ is too unstable to be formed under the conditions of the reaction (e.g. if R is a simple primary group), the RX–Lewis acid adduct is usually sufficiently electrophilic to react with benzene.

Example 16.2

Reaction of benzene with propene in the presence of phosphoric acid gives isopropylbenzene (also called cumene, an industrial starting material for the synthesis of phenol, Section 16.5). Propose a mechanism for this reaction.

Solution

The carbenium ion involved in this alkylation is generated by protonation of the alkene.

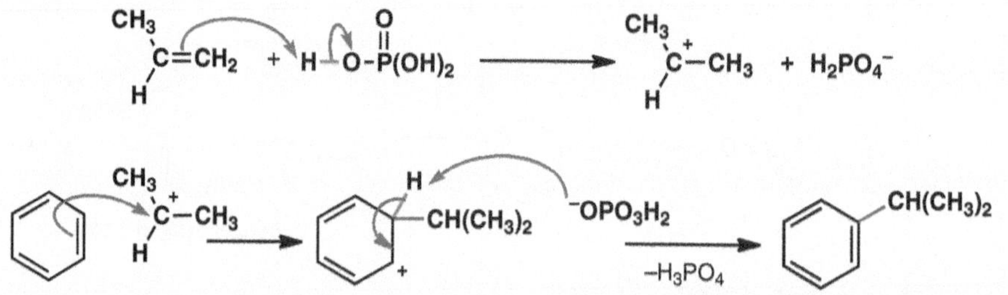

Exercise 16.1

Show using curly arrows how a carbenium ion is generated in the alkylation of benzene with propan-2-ol catalysed by sulfuric acid.

16.3.5 Friedel–Crafts acylation

Friedel–Crafts acylation (eqn 16.12) is a two-stage procedure which gives an aryl ketone: the first step involves the arene, an acyl chloride (acid chloride), and a Lewis acid; the second step is an aqueous quench.

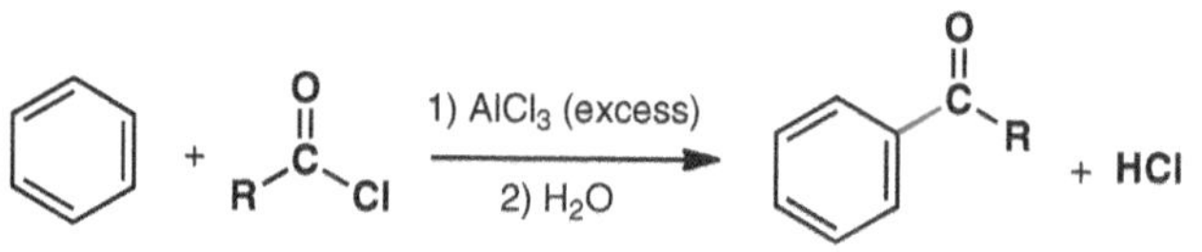

a phenyl ketone

$$(16.12)$$

Mechanistically, acyl chlorides act as Lewis bases with $AlCl_3$ to form adducts which dissociate to give acylium ions; these act as electrophiles in the acylation of benzene to give phenyl ketones, Scheme 16.8. However, the ketone product is also a Lewis base so it forms an adduct with the $AlCl_3$ which, therefore, is consumed in the reaction so is not truly catalytic. Consequently, more than one equivalent of $AlCl_3$ is required for a Friedel–Crafts acylation, and the product is not liberated until the hydrolysis of the ketone–$AlCl_3$ adduct in the aqueous quench.

Exercise 16.2

Friedel–Crafts acylations can also be carried out with acid anhydrides but more than two equivalents of a Lewis acid are needed. Give a mechanism for the acylation of benzene with anhydride $(RCO)_2O$ and $AlCl_3$, and explain why more than two equivalents of $AlCl_3$ are needed.

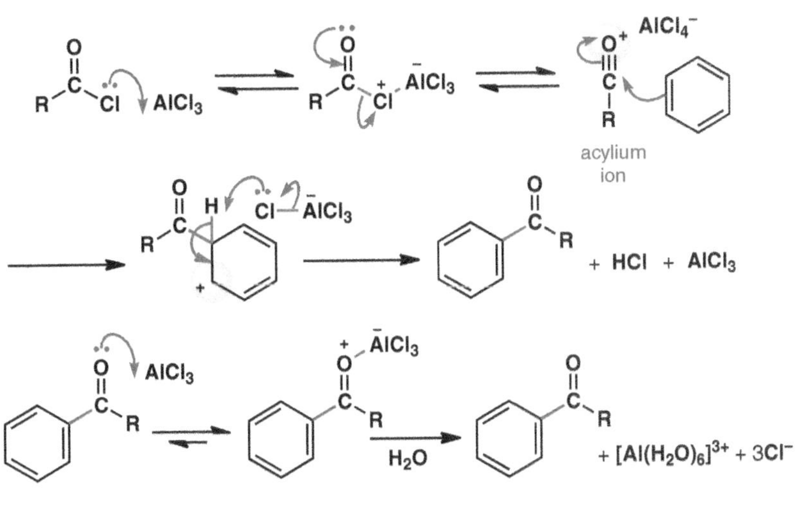

Scheme 16.8 Friedel–Crofts acylation of benzene.

What is the main product of the Friedel–Crafts reaction of benzene with each of the following and $AlCl_3$?

(a) $(CH_3)_3CCl$ (b) $C_6H_5CH_2Cl$ (c) C_6H_5COCl (d) CH_3CH_2COCl

Exercise 16.3

16.4 Reactivity of Substituted Benzenes and Regioselectivity

So far, we have presented electrophilic substitution reactions of only benzene itself; in the absence of isotopic labelling, the six C–H groups of benzene are indistinguishable to an electrophile so there is never a question of where the reaction will be—there is only one possibility. This is not the case for a mono-substituted benzene, C_6H_5-Y, which has two equivalent *ortho* positions, two equivalent *meta* positions, and one *para* position. Consequently, there can be three parallel independent reactions with electrophile Z^+ which, in principle, give three constitutionally isomeric products, Y-C_6H_4-Z. In other words, there is a question of **regioselectivity** (or orientation)—at which position of the mono-substituted benzene C_6H_5-Y does the electrophile bond: *ortho*, *meta*, or *para*?

If the reactions at the different positions had identical rate constants, we would expect the three products to be formed in the ratios *ortho* : *meta* : *para* = 2 : 2 : 1; if the yields are not in these ratios, the reaction is *regioselective* to a greater or lesser degree. Most electrophilic aromatic substitutions are not reversible and the products do not equilibrate. In other words, the reactions are *kinetically controlled* (see Sub-section 15.8.2) and the relative yields of the isomeric products are determined by the relative rates at which they are formed, i.e. by the rate constants of the parallel competing reactions or, equivalently, by the relative stabilities of the transition structures in the rate-determining steps of the respective routes.

16.4.1 Activating and deactivating substituents in electrophilic aromatic substitution

A second issue which arises in the electrophilic substitution of a mono-substituted benzene besides regioselectivity is relative **reactivity**: is the compound more or less reactive than benzene itself? However, as we shall see, regioselectivity and reactivity in electrophilic aromatic substitution are closely related.

If k_o, k_m, and k_p are the rate constants for individual reactions at each of the o, m, and p positions of PhY, the overall experimental second-order rate law for parallel bimolecular reactions with electrophile Z^+ will be:
rate $= k_{exp}$ [PhY] [Z^+], where
$k_{exp} = (2 k_o + 2 k_m + k_p)$.

In principle, reaction of C_6H_5-Y (PhY) comprises three concurrent parallel reactions, and the overall process (with a single experimental rate constant) is the weighted mean of the three competing component reactions. The (overall) rate constant for PhY in an electrophilic substitution reactions may be greater than the rate constant for benzene itself in the same reaction under the same conditions. In this event, the substituent Y is said to be *activating*, or to have an activating effect. For example, anisole (methoxybenzene) is very reactive, and its nitration occurs rapidly under mild conditions to give mainly *o*- and *p*-nitroanisoles (eqn 16.13).

anisole   *o*-nitroanisole   *p*-nitroanisole   *m*-nitroanisole
             71%   28%   <0.5%

In reaction 16.13, the main nitrating agent will be AcONO$_2$, which is readily formed by reaction of HNO$_3$ with the solvent, Ac$_2$O. AcONO$_2$ is a very effective NO$_2^+$ donor because AcO$^-$ is a good nucleofuge under these conditions.

In contrast, nitrobenzene is very much less reactive than benzene under the same reaction conditions, so NO$_2$ is *deactivating*; significantly, its main product in nitration is predominantly *m*-dinitrobenzene (eqn 16.14).

nitrobenzene   *m*-dinitrobenzene   *o*-dinitrobenzene   *p*-dinitrobenzene
               92%   6%   2%

The above two reactions illustrate that, broadly, there are two kinds of substituents: one, like methoxy, is activating and leads mainly to *ortho* and *para* substitution; the other, like nitro, is deactivating and mainly leads to *meta* substitution. It is not a coincidence that reactivity and regioselectivity are closely related—they are both governed by the relative stabilities of the alternative isomeric benzenium ion intermediates in the initial electrophilic attack (which is usually rate limiting—see Figure 16.1). In our discussion of the issues below, we shall invoke the Hammond postulate (see Panel 7.2) and begin by examining the different effects of MeO and NO$_2$ on the stability of the parent benzenium ion.

16.4.2 Effects of substituents on the stability of the benzenium ion

Upon reaction of anisole with an electrophile, E$^+$, three isomeric benzenium ions could, in principle, be formed depending on the position of the electrophilic attack. They are represented by resonance contributors as follows:

ortho attack:

relatively stable contributor

para attack:

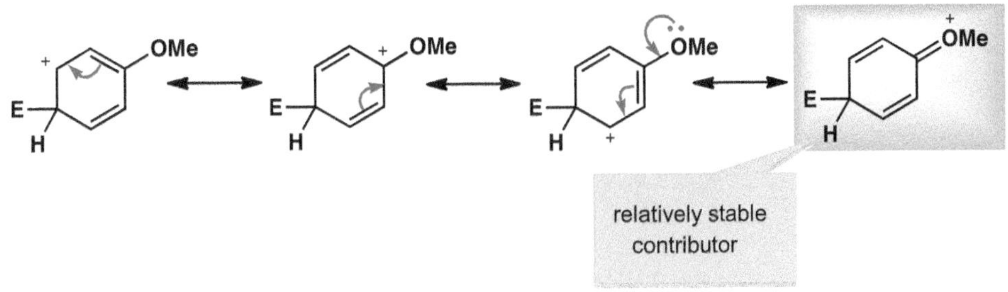

relatively stable contributor

meta attack:

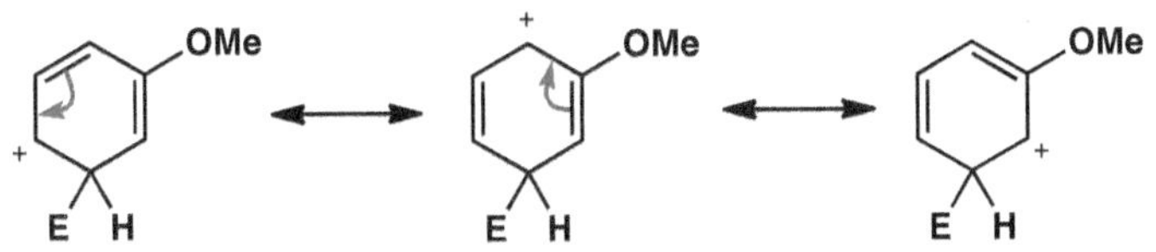

Note that we refer to electrophilic attack by E+ at positions *ortho, meta,* or *para* to Y in PhY. We then refer to the original substituent Y as being *ortho, meta,* or *para* to the C bearing E in the newly formed benzenium ion, i.e. the point of reference changes upon formation of the benzenium ion. Note also that *ortho, meta,* and *para* are abbreviated to *o-, m-,* and *p-* in the names of compounds, e.g. 4-nitrophenol can be written *p*-nitrophenol.

Each of the three isomeric benzenium ions has three resonance contributors representing delocalization of the positive charge within the six-membered ring. In addition, a fourth contributor is possible for the *ortho*- and *para*-substituted ions corresponding to donation of a lone pair on the oxygen of the MeO into the ring. In this contributor, the oxygen bears a formal positive charge and no atom has less than its full valence complement of electrons. In contrast, electrophilic attack *meta* to the MeO gives a benzenium ion which cannot be stabilized by a donating resonance contribution by MeO. Instead, the *m*-MeO substituent exerts a destabilizing electron-withdrawing inductive effect owing to the high electronegativity of the oxygen. We see, therefore, that electrophilic attack *ortho* or *para* to the MeO group gives benzenium ions which are more stabilized than the parent (unsubstituted) benzenium ion whereas electrophilic attack *meta* to the MeO gives a MeO-substituted benzenium ion which is less stable than the parent. This mechanistic analysis of the electrophilic substitution reaction of anisole accounts for its high reactivity and the high regioselectivity exemplified in eqn 16.13 above.

Now let us consider the three possible isomeric benzenium ions formed by electrophilic attack at nitrobenzene knowing that the nitro group is powerfully electron-withdrawing.

Overall electronic effects at C1 in benzene of MeO at other positions:

- *o*- and *p*-MeO: electron-donating (dominant resonance effect);
- *m*-MeO: electron-withdrawing (inductive effect).

ortho attack:

highly unstable contributor

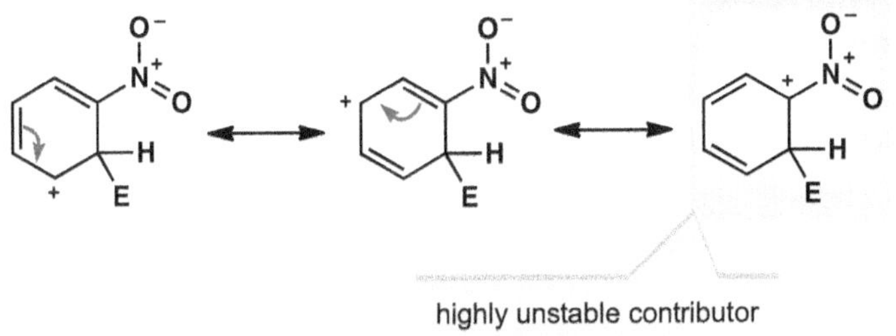

para attack:

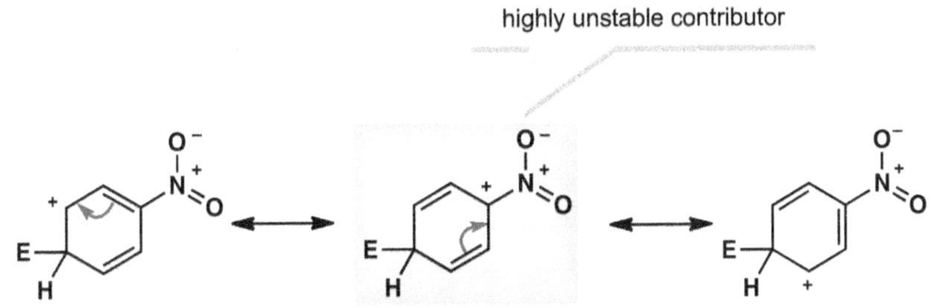

highly unstable contributor

meta attack:

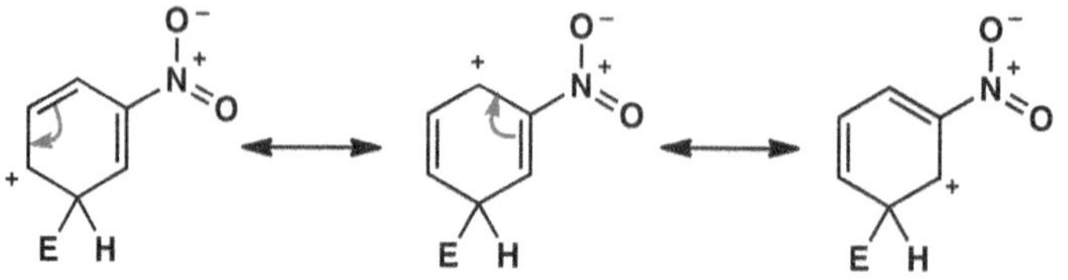

For electrostatic reasons, we can expect an electron-withdrawing nitro group to desta-bilize the benzenium cation and, consequently, inhibit electrophilic attack at any posi-tion. We can be more precise, however: each of the three isomeric benzenium ions has three resonance contributors, but the positive charge in one contributor (highlighted) for the cations formed in *ortho* and *para* attack is located on the ring C bearing the NO_2. Consequently, these particularly high energy contributors are not important components of the resonance hybrids of the benzenium ions following *ortho* and *para* electrophilic attack. The *o*- and *p*-nitrobenzenium intermediates, therefore, are even less stable than the *m*-nitro analogue which does not have such a contributor. In other words, the degree of destabilization of the benzenium ion intermediate (and, according to the Hammond postulate, the TSs before and after it in the reaction coordinate) by the NO_2 group is small-est for the *meta* isomer. We now see why nitrobenzene is not only much less reactive than benzene in electrophilic substitution, but also why the reaction is highly regioselective—a high yield of the *meta*-substituted product is obtained in a very slow reaction.

In summary, the methoxy group is activating and *ortho,para*-directing in electro-philic aromatic substitution reactions, whereas the nitro group is deactivating and *meta*-directing.

16.4.3 Classification of substituents

We have seen above how their contrasting electronic properties lead MeO and NO_2 to affect both reactivity and regioselectivity differently in electrophilic substitution reac-tions of anisole and nitrobenzene. We can now generalize the principles involved and, in the process, we shall identify a sub-group of *ortho* and *para* directing substituents—ones which are *deactivating* in electrophilic aromatic substitution.

(1a) Activating and *ortho,para*-directing groups: NH_2, NR_2, OH, OR, Ph, R (alkyl)

Nitrogen and oxygen groups (like MeO) have lone pairs capable of conjugation with the benzene ring; the π system of Ph is also conjugatively stabilizing; alkyl groups have electron-donating inductive (or hyperconjugative) effects.

(1b) Deactivating and *ortho,para*-directing groups: F, Cl, Br, I

Halogens have lone pairs capable of conjugation as illustrated below and, like MeO, are *ortho,para*-directing because of the influence of the fourth resonance form.

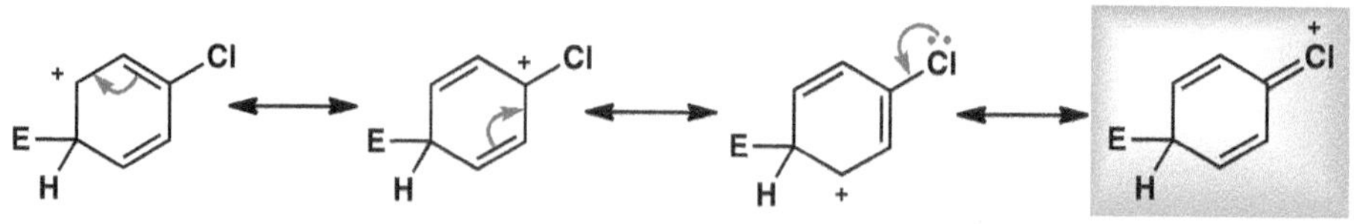

However, there are two complicating features. First, the electronegative halogens have electron-attracting inductive effects which weaken down the group from F to I. Secondly, the size of the p orbital which conjugates with the carbocyclic π system increases as we go down the group. Consequently, the quality of the overlap and, therefore, the electron-donating resonance effect also weakens down the group from F to I. The balance of these electronic effects for the different halogens is not well understood but the outcome is that they are all *ortho,para*-directing but deactivating.

(2) Deactivating and *meta*-directing groups: NO_2, C=O, CN, SO_3H, CF_3, NR_3^+

These groups all have electron-withdrawing resonance or inductive effects.

Substituents can be arranged from electron-donating to electron-withdrawing on a continuous scale, as shown in Figure 16.2 which correlates with the reactivity and the regioselectivity in electrophilic substitution as indicated.

The following reactions illustrate the effects of representative substituents in the classification above.

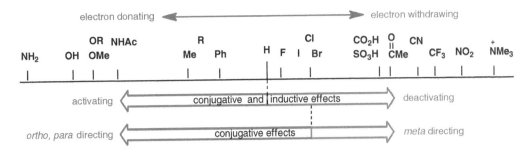

Figure 16.2 Effects of substituents Y on electrophilic substitution reactions of Ph-Y.

Exercise 16.5

What are the main mono-substitution products in the nitration of the following?

(a) [structure: benzene ring with CH₂CH₃]

(b) [structure: benzene ring with CF₃]

(c) [structure: biphenyl]

(d) [structure: benzene ring with C(=O)OCH₃]

(e) [structure: benzene ring with O-C(=O)-CH₃]

16.4.4 Reactions of disubstituted benzenes

In electrophilic substitution reactions of disubstituted benzenes, both substituents affect the reactivity and regioselectivity. When the effects are cooperative, the results are readily predictable. However, when the two substituents compete in their effects on regioselectivity, we need to think more carefully and consider each case individually.

In reaction 16.15, both substituents direct the electrophilic attack *ortho* to the methyl group, so we can safely predict the product.

Note that there are two pairs of equivalent positions for 1,4-disubstituted derivatives.

[reaction scheme 16.15]

$$p\text{-nitrotoluene} \xrightarrow[\text{}]{\text{HNO}_3,\ \text{H}_2\text{SO}_4} 2,4\text{-dinitrotoluene}$$

o to CH_3
m to NO_2

(16.15)

However, if two substituents direct the substitution to different positions, what will be the outcome? As we saw earlier, most electrophilic substitution reactions are kinetically controlled, so the proportions of regioisomers reflect the proportions of independent competing parallel reactions. In general, therefore, *a more strongly activating substituent will compete more effectively than a less strongly activating (or deactivating) substituent*, and dominate the regioselectivity.

In reaction 16.16, for example, the methoxy group and the methyl are both activating and *o,p*-directing, but in competition. Because the (conjugative) activating effect of MeO is much greater than the activating effect of Me, the reaction facilitated by the MeO leading to the product shown (2-bromo-4-methylanisole) is much faster than the alternative reaction (only weakly facilitated by Me) to give 3-bromo-4-methylanisole (not shown).

[reaction scheme 16.16]

o to CH_3

$$p\text{-methylanisole}\ (p\text{-methoxytoluene}) \xrightarrow{\text{Br}_2} 2\text{-bromo-4-methylanisole}$$

o to CH_3O

(16.16)

The outcome is not always so readily predicted. The situation in eqn 16.17 appears similar to that in eqn 16.16, i.e. competition between an *o,p*-directing substituent (Cl) and the *o,p*-directing Me. Now, however, the Cl is *deactivating*, so the path in which the regiochemistry is controlled by the *activating* Me is somewhat faster than the one in which the regioselectivity is directed by the Cl. The regioselectivity in eqn 16.17, however, is very modest and, in a different reaction, the outcome from the same substrate could be different.

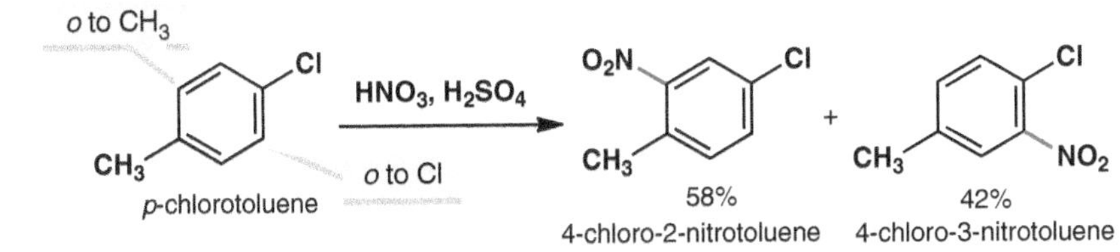
o to CH3
Cl
HNO3, H2SO4
CH3
o to Cl
p-chlorotoluene
O2N
Cl
CH3
58%
4-chloro-2-nitrotoluene
+
Cl
CH3
NO2
42%
4-chloro-3-nitrotoluene

(16.17)

The directing effects of the conjugatively deactivating, *m*-directing CN and the inductively activating, *o,p*-directing Me are in opposition in eqn 16.18, and the methyl group prevails. In other words, the reactions facilitated by the Me substituent which lead to the products shown are the preponderant components of the overall process.

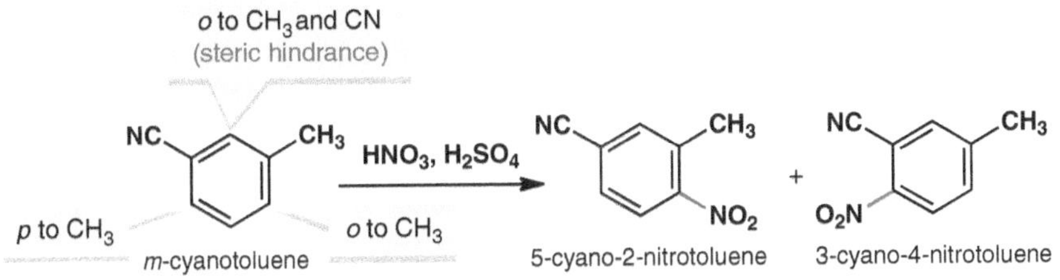
o to CH3 and CN
(steric hindrance)
NC
CH3
HNO3, H2SO4
p to CH3
o to CH3
m-cyanotoluene
NC
CH3
NO2
5-cyano-2-nitrotoluene
+
NC
CH3
O2N
3-cyano-4-nitrotoluene

(16.18)

Note, however, that the position between the substituents *meta* to each other in *m*-cyanotoluene in eqn 16.18 is not substituted even though it is *ortho* to the methyl. This is an example of steric hindrance influencing regioselectivity. Positions *ortho* to bulky substituents are particularly unlikely to be substituted as illustrated in eqn 16.19.

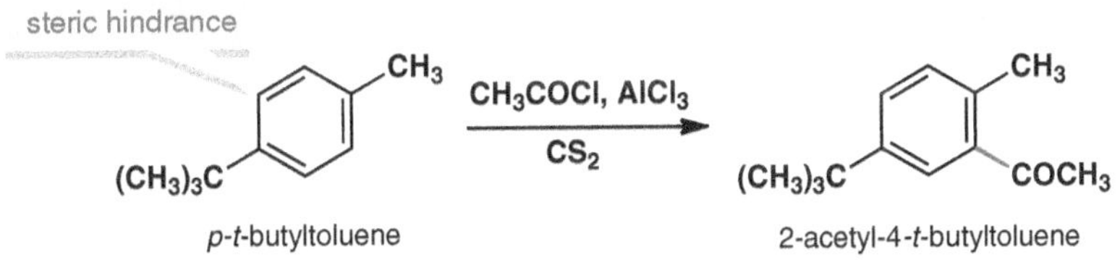
steric hindrance
CH3
(CH3)3C
p-t-butyltoluene
CH3COCl, AlCl3
CS2
CH3
(CH3)3C
COCH3
2-acetyl-4-t-butyltoluene

(16.19)

Exercise 16.6

What are the main mono-substitution products upon nitration of the following?

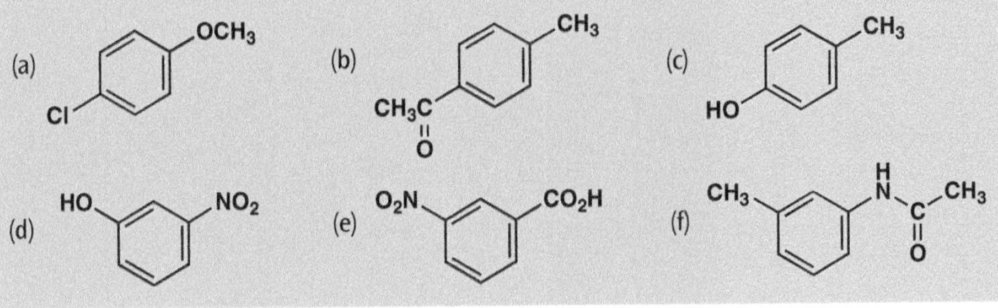
(a)
OCH3
Cl
(b)
CH3
CH3C
O
(c)
CH3
HO
(d)
HO
NO2
(e)
O2N
CO2H
(f)
CH3
H
N
C
CH3
O

16.5 Reactivity of Phenol

Phenol, an important industrial chemical, is produced industrially on a large scale by acid-catalysed rearrangement of cumene hydroperoxide (see Sub-section 22.4.1). It is very reactive in electrophilic aromatic substitution and easily brominated, for example,

Panel 16.1 Biological electrophilic aromatic substitution: thyroxine biosynthesis

Thyroxine is an iodine-containing thyroid hormone which increases the metabolic rate not only to effect synthesis of proteins, long bone growth, and neuronal maturation (essential for the proper development of cells and growth of children) but also to regulate the metabolism of proteins, carbohydrates, and fats (to provide energy used by the body). It is an iodinated derivative of an amino acid, tyrosine, and the biochemical iodination occurs by a kind of electrophilic aromatic substitution.

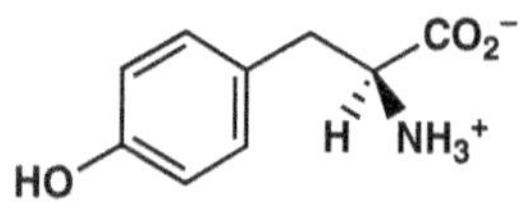

tyrosine 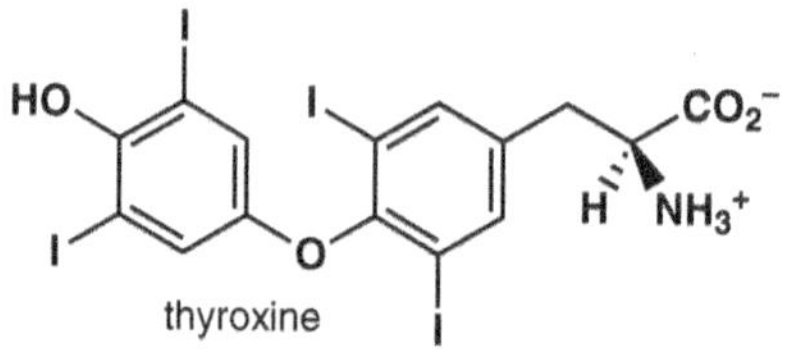   thyroxine

Tyrosine and iodine (as iodide ion in our diet) are stored in the thyroid gland. First, the iodide is oxidized to an electrophilic I(0) species by hydrogen peroxide in a reaction catalysed by the enzyme, iodoperoxidase. Two tyrosine units of a protein called thyroglobulin then undergo electrophilic iodination at positions *ortho* to their OH groups according to the directing effect of OH, and the OH of one of them is converted to OI in a further oxidative step. After the hydroxy-diiodophenyl group of one tyrosine unit has been transferred to the other tyrosine in another biological electrophilic substitution, the thyroxine is liberated and transported by the blood to almost every cell in the body to carry out its biological functions.

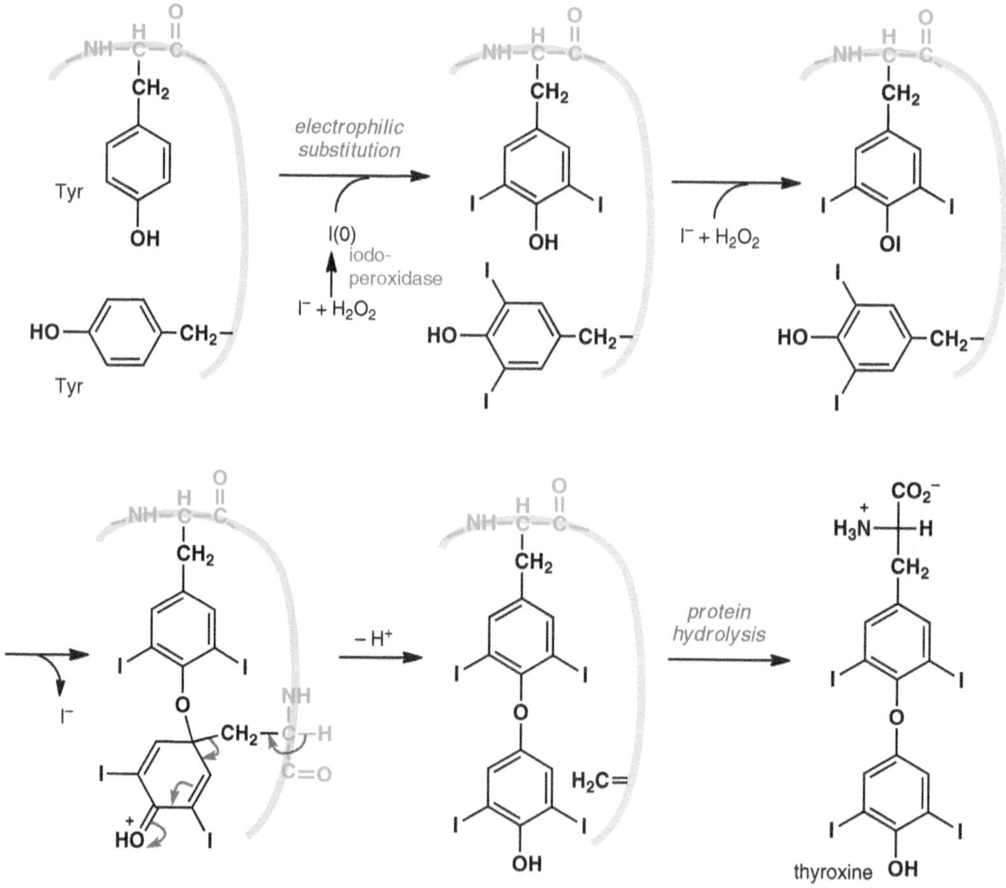

(The configuration of the thyroxine is given here as a Fischer projection)

An overactive thyroid, which may be recognized by a swelling in the neck (a goitre), causes excessive levels of thyroxine in the bloodstream and is known as thyrotoxicosis. An underactive thyroid gland is known as hypothyroidism and leads to decreased metabolic rates in adults.

under mild conditions without Lewis acid catalysis, as illustrated in the reaction of eqn 16.20.

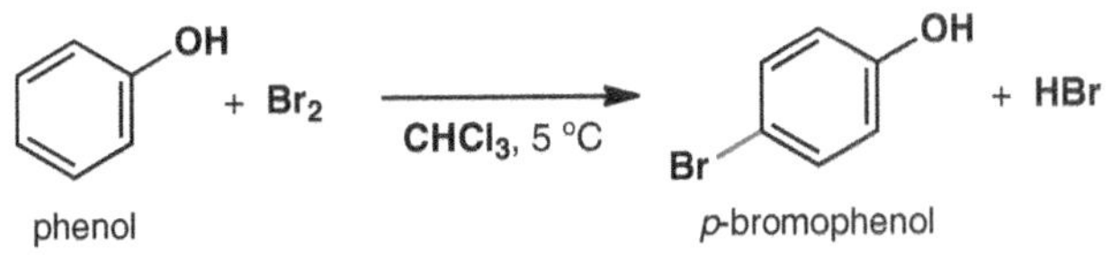

(16.20)

Before its toxicity was fully appreciated, phenol (previously known as carbolic acid) was used as one of the very earliest antiseptics (germicides). The vapour, aqueous solutions, and crystalline phenol all cause chemical burns, which can be fatal.

The reaction proceeds even more easily under more polar aqueous conditions and gives the trisubstituted product directly in nearly quantitative yield (eqn 16.21).

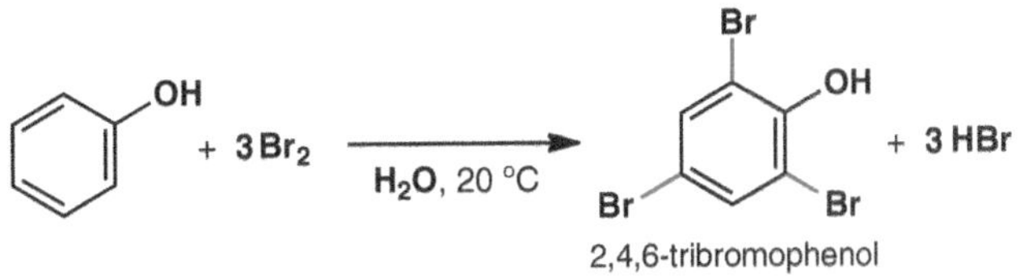

(16.21)

The acidity of the hydroxy group facilitates two routes for the bromination of phenol in aqueous solution, and their relative contributions depend upon the extent of the dissociation of the phenol. Deprotonation of the phenol concerted with electrophilic addition of bromine leads to the bromodienone in Scheme 16.9 which has been detected as a short-lived intermediate. Isomerization by protonation–deprotonation leads to the product.

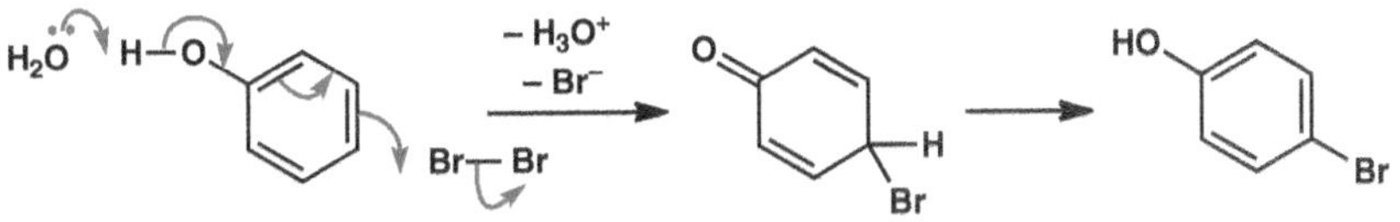

Scheme 16.9 Mechanism for bromination of undissociated phenol in aqueous solution.

Write a curly arrow mechanism for the isomerization of the bromodienone intermediate in the reaction of Scheme 16.9.

Exercise 16.7

Additionally, phenol ($pK_a=10$) dissociates to a low extent in water and the phenoxide ion is extremely reactive in electrophilic substitution in spite of its low concentration under non-basic conditions (the proportion of bromination by this route increases as the pH of the solution increases). Then, both the p-bromophenol and its conjugate base (the bromophenol product is more acidic than phenol itself) undergo further bromination, and so on until tribromophenol is the only bromination product.

Write a mechanism for the reaction of phenoxide ion with Br_2 to form the p-bromo product in aqueous solution.

Exercise 16.8

The phenoxide ion can undergo substitution with even very poor electrophiles such as carbon dioxide or carbonyl compounds. Sodium phenoxide reacts with CO_2 under

The reaction of eqn 16.22, sometimes called the Kolbe–Schmitt reaction, is important in the production of aspirin (acetylsalicylic acid), the widely used analgesic and antifebrile medication.

pressure to give mainly the *ortho* substitution product, salicylate (*o*-hydroxybenzoate), as shown in eqn 16.22.

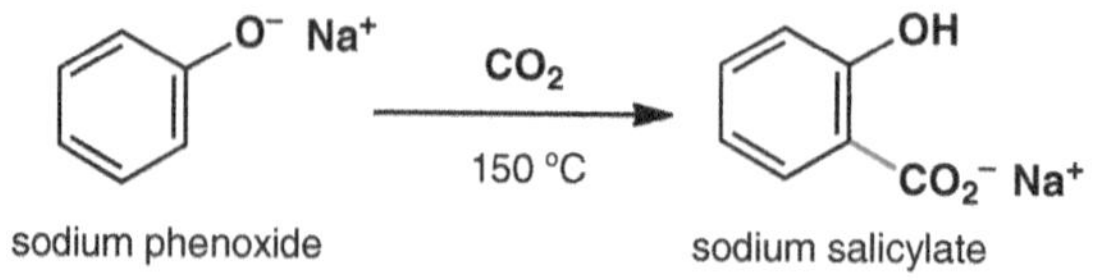

$$\text{sodium phenoxide} \xrightarrow[150\ ^\circ C]{CO_2} \text{sodium salicylate}$$

(16.22)

sodium phenoxide   sodium salicylate

Exercise 16.9

Write a mechanism for reaction 16.22.

Bakelite is widely regarded as the first synthetic 'plastic'; its preparation was announced in 1909 in the USA by the Belgian-born chemist, L.H. Baekeland.

Heating sodium phenoxide with methanal (formaldehyde) gives a polymeric phenol–formaldehyde resin which, upon cooling, sets hard to give the product known as Bakelite® (Scheme 16.10).

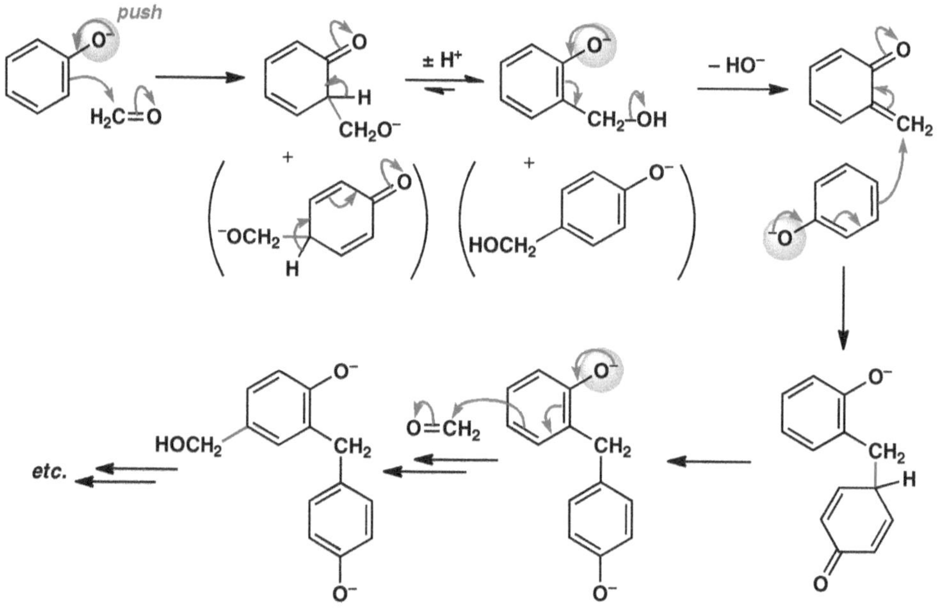

Reactions may occur both *ortho* and *para* to the O⁻ function of the phenolate and, ultimately, each methanal carbon connects two benzene rings, which produces a complex polymeric matrix.

Scheme 16.10 Reactions leading to phenol–formaldehyde resin

Example 16.3

Reaction of phenol with chloroform in strongly basic aqueous solution followed by acidification gives *o*-hydroxybenzaldehyde (salicylaldehyde). This reaction (sometimes called the Reimer–Tiemann reaction) proceeds by an electrophilic substitution with dichlorocarbene (Section 15.6) formed as a reactive intermediate from chloroform. Show how this reaction proceeds by a stepwise reaction sequence.

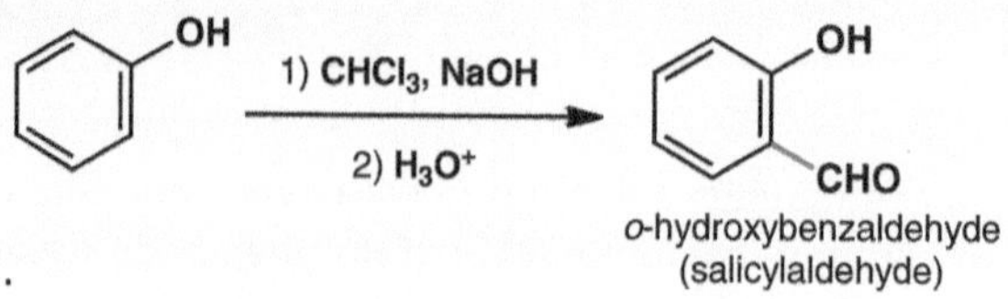

$$\text{phenol} \xrightarrow[2)\ H_3O^+]{1)\ CHCl_3,\ NaOH} \text{o-hydroxybenzaldehyde (salicylaldehyde)}$$

o-hydroxybenzaldehyde
(salicylaldehyde)

Solution

Dichlorocarbene is formed in low concentration by base-induced α-elimination of HCl from chloroform ($pK_a \approx 24$).

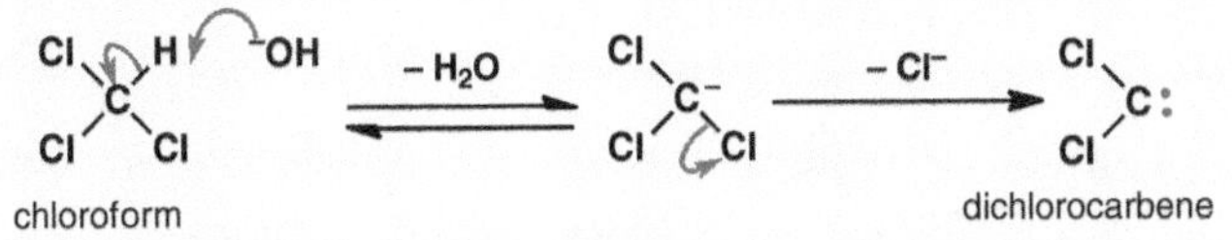

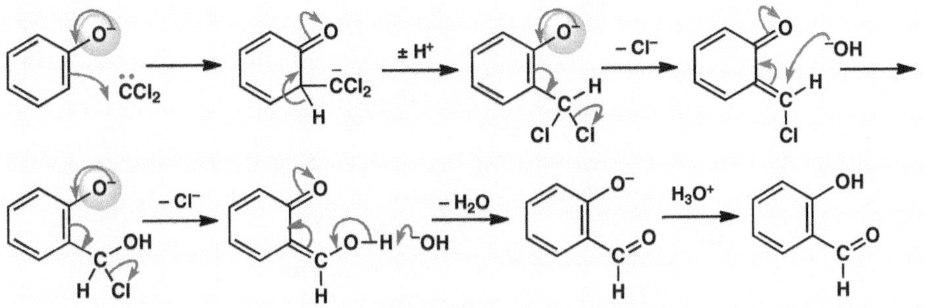

16.6 Reactivity of Aniline

Aniline is produced industrially on a large scale and used, for example, as a starting material in the manufacture of polyurethanes. It is easily obtained by reduction of the readily available nitrobenzene using iron or tin in moderately concentrated hydrochloric acid in the laboratory, or by catalytic hydrogenation on an industrial scale.

16.6.1 Electrophilic substitution

Aniline is even more reactive than phenol (see Section 16.5) due to the powerful electron-donating activating resonance effect of the NH_2 group, so controlling the reaction to give a mono-substitution product is often difficult (eqn 16.23).

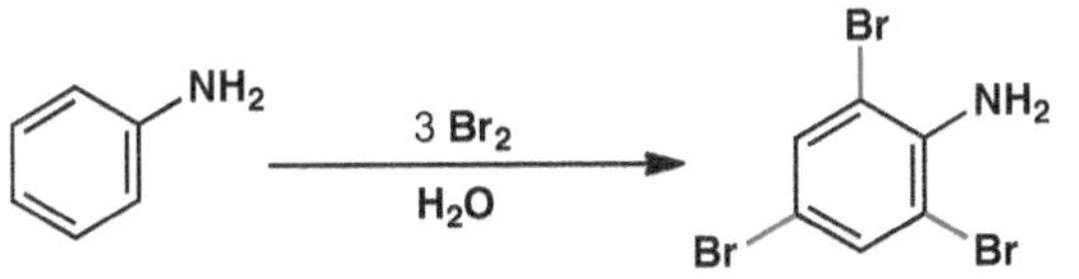

(16.23)

Write a reaction mechanism for mono-bromination of aniline in aqueous solution.

Exercise 16.10

   Furthermore, complications may occur because the NH_2 group is basic so the ortho,para-directing NH_2 is converted into the meta-directing NH_3^+ group under acidic conditions. Although aniline is almost wholly protonated in strongly acidic solutions, the rate constant for reaction of the low concentration of the free base is very much greater than that of the more abundant protonated form. Consequently, the proportions of regioisomeric substitution products depend on the acidity of the reaction conditions.

Panel 16.2 Quinones

Quinones are cyclohexadienediones and may be obtained by oxidation of substituted phenols or anilines. The simplest example is the oxidation of benzene-1,4-diol (hydroquinone) to 1,4-benzoquinone (*p*-benzoquinone), and suitable oxidizing agents include $H_2Cr_2O_7$, $(KSO_3)_2NO$ (Fremy's salt), Fe(III) salts, and $NaClO_3$–V_2O_5).

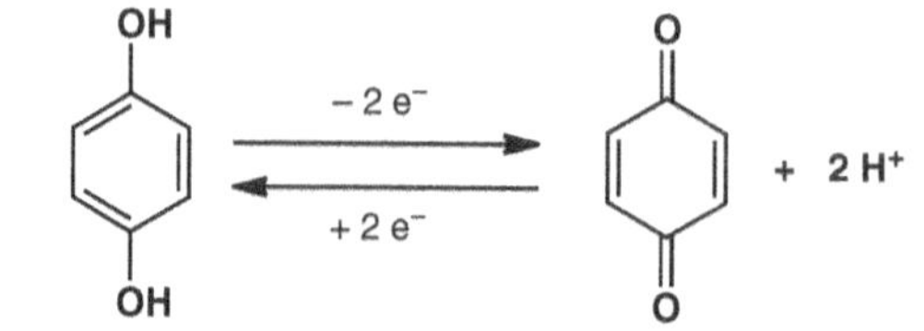

$$\text{benzene-1,4-diol (hydroquinone)} \quad \xrightarrow{-2e^-} \quad \xleftarrow{+2e^-} \quad \text{1,4-benzoquinone (}p\text{-benzoquinone)} \quad + \; 2\,H^+$$

benzene-1,4-diol
(hydroquinone)

1,4-benzoquinone
(*p*-benzoquinone)

The reaction is reversible and the quinone is easily reduced back to hydroquinone by mild reducing agents. Benzene-1,2-diol (catechol) and 1,2-benzoquinone (*o*-benzoquinone) undergo similar reversible reactions.

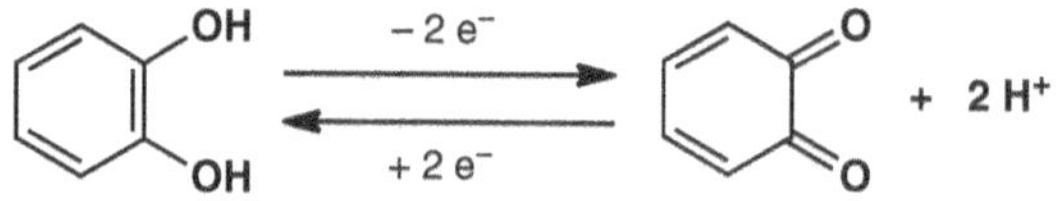

$$\text{benzene-1,2-diol} \quad \xrightarrow{-2e^-} \quad \xleftarrow{+2e^-} \quad \text{1,2-benzoquinone} \quad + \; 2\,H^+$$

benzene-1,2-diol
(catechol)

1,2-benzoquinone
(*o*-benzoquinone)

Panel 16.3 Naturally occurring phenols

Many phenolic compounds are found in nature and representative examples are shown below. Vanillin is one of the simplest; it is the compound principally responsible for the distinctive flavour of natural vanilla, which is obtained from the pods of the vanilla orchid. These days, most of the vanillin used in the catering and food industries is synthetic because the pure natural material is very expensive. Eugenol (which has several uses, e.g. to relieve pain in dentistry) is the main component of clove oil, and gives cloves their characteristic flavour; it is also present in several other culinary herbs and spices.

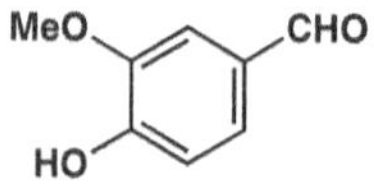

vanillin
(vanilla pods)

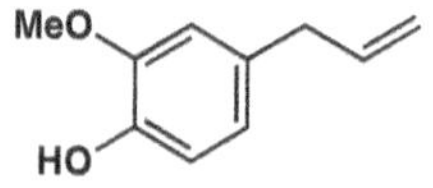

eugenol
(cloves)

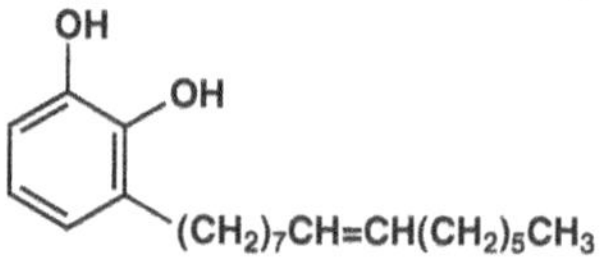

a urushiol
(poison ivy)

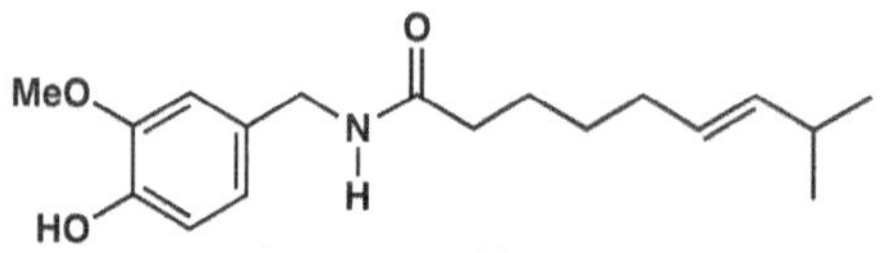

capsaicin
(chilli peppers)

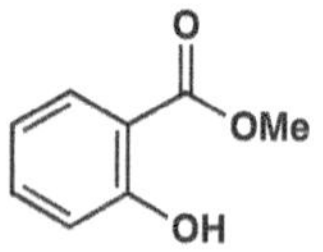

methyl salicylate
(wintergreen)

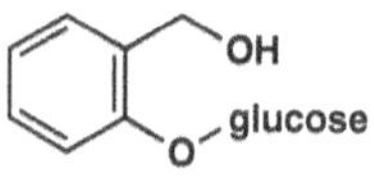

salicin
(willow trees)

A piece of japanned craftwork
(early 18th century)
Original source: Masterpieces of Japan Arts Vol 5th,
TOHTO BUNKA KOEKI Co.,Ltd. Tokyo, 1953-09-30

Urushiol is the generic name for a group of more complex phenols with different alkenyl side-chains; they are responsible for an allergic skin rash caused by poison ivy and are found in the sap of lacquer trees (*urushi* in Japanese: an item of craftwork finished with *urushi* lacquer is said to have been 'japanned'). Capsaicin is the main component of chilli peppers which gives them their hot/burning taste. Whilst it causes pain upon contact with sensitive tissue, e.g. in the mouth, capsaicin is also used in analgesic ointments for relief of muscular aches and pains, and arthritic conditions. Oil of wintergreen is also used in embrocations/liniments to relieve muscular aches and pains. See the Prologue for the development of aspirin, an analgesic and antifebrile drug, from salicin, a phenol derivative which is found in willow bark.

These reversible reduction–oxidation (redox) reactions between quinone and diol can be achieved electrochemically at specific electrode potentials (reduction potentials) and the one for *p*-benzoquinone/hydroquinone is used as an electrochemical standard. (*p*-Benzoquinone and hydroquinone form a greenish black crystalline complex which is known as *quinhydrone*.) This type of redox reaction is used in biological systems to transport electrons in certain kinds of enzyme catalysis.

Coenzymes Q, which are also called *ubiquinones* (ubiquitous quinones found within the inner mitochondrial membrane of every living cell), play an important role in the respiratory chain to produce ATP (adenosine triphosphate, Chapter 24), which acts as a 'biological fuel'. Pyrroloquinoline quinone (PQQ) is a 1,2-quinone of a heteroaromatic compound and acts as a coenzyme in the dehydrogenation of ethanol. Vitamin K_1 is a derivative of 1,4-naphthoquinone and is required for blood coagulation.

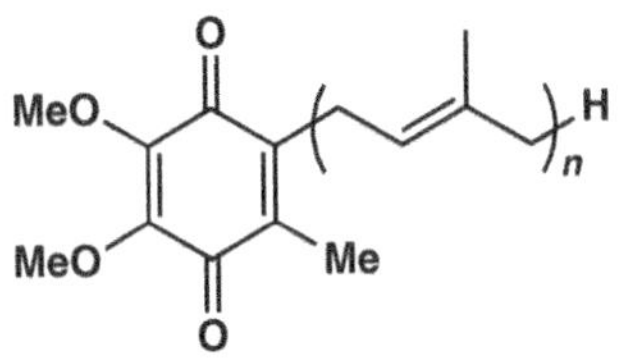

coenzymes Q ($n = 6, 8, 10$)
(ubiquinones)

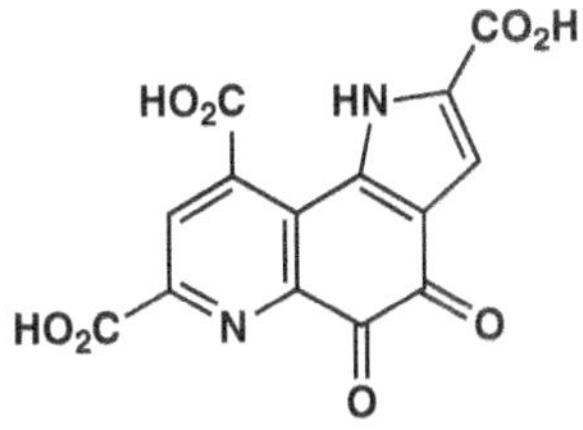

pyrroloquinoline quinone (PQQ)
(a coenzyme for
alcohol dehydrogenase)

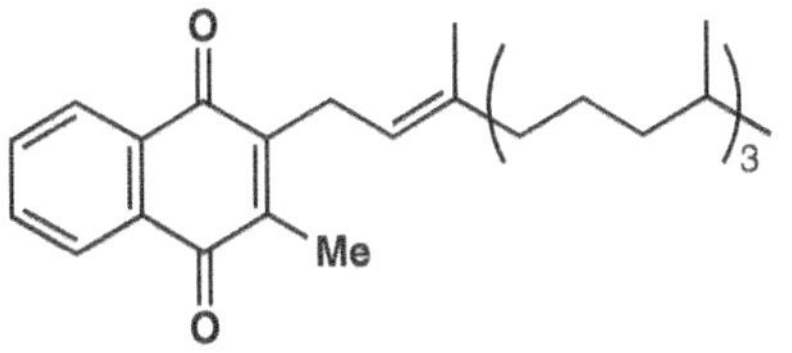

vitamin K_1

Some phenol and polyphenol derivatives are antioxidants which scavenge oxygen radicals in the body and are reputed to suppress ageing (see Chapter 20 for radical reactions—some polyphenols are also related to quinones as covered in Panel 16.2). They include tocopherol (vitamin E), resveratrol found in the skins of red grapes (and, consequently, in red wine), catechins in tea, and anthocyanins which give many flowers and fruit their colours (see Panel 6.2).

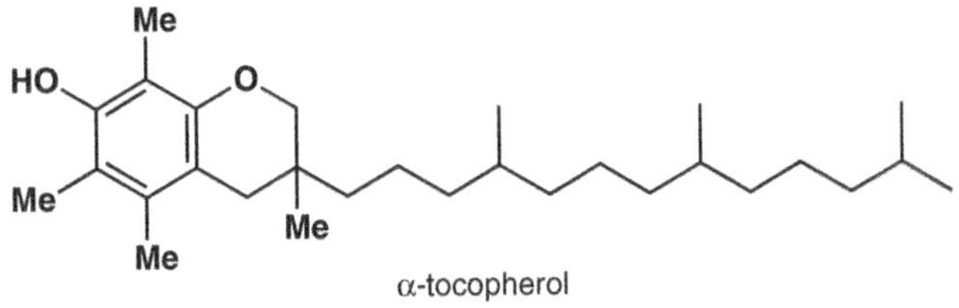

α-tocopherol
(vitamin E)

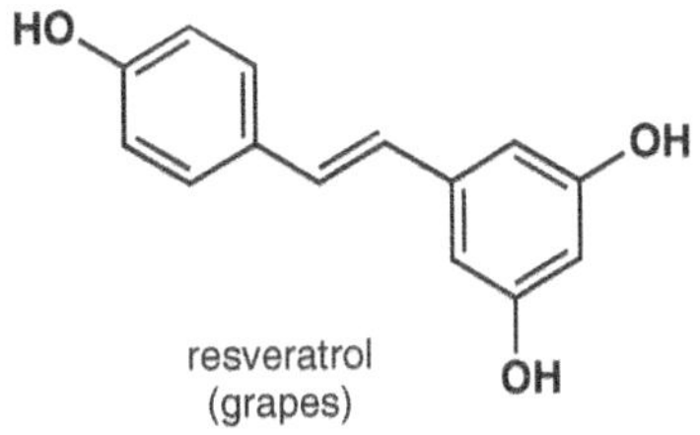

resveratrol
(grapes)

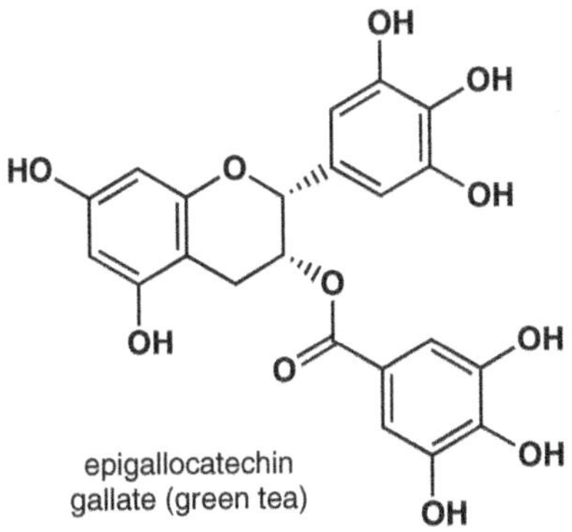

epigallocatechin
gallate (green tea)

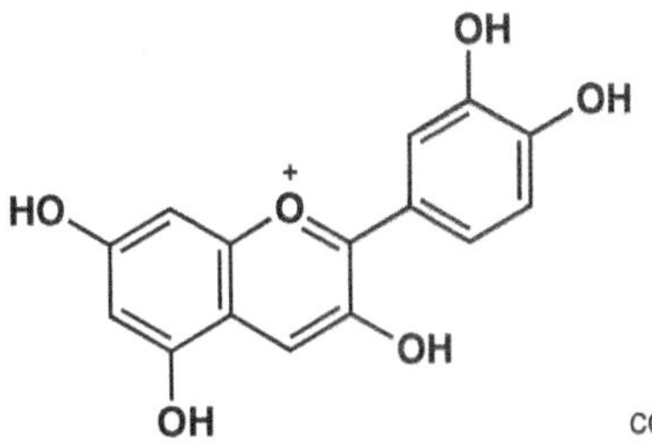

cyanidin (an anthocyanin)
(colours in flowers and fruit)

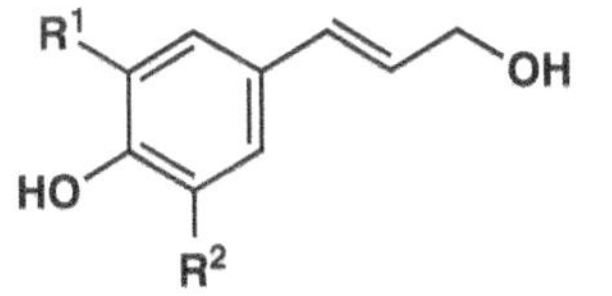

components of lignin (wood)
coniferyl alcohol (R^1 = OMe, R^2 = H)
sinapyl alcohol (R^1 = OMe, R^2 = OMe)
p-coumaryl alcohol (R^1 = H, R^2 = H)

Wood is a complicated and variable material, and about 20–30% of it is lignin, a complex polymeric substance derived from phenolic compounds containing allylic alcohol groups and linked to cellulose.

For example, the product distribution in the nitration of aniline in aqueous HNO_3/H_2SO_4 mixtures depends upon the concentration of the H_2SO_4 (Scheme 16.11).

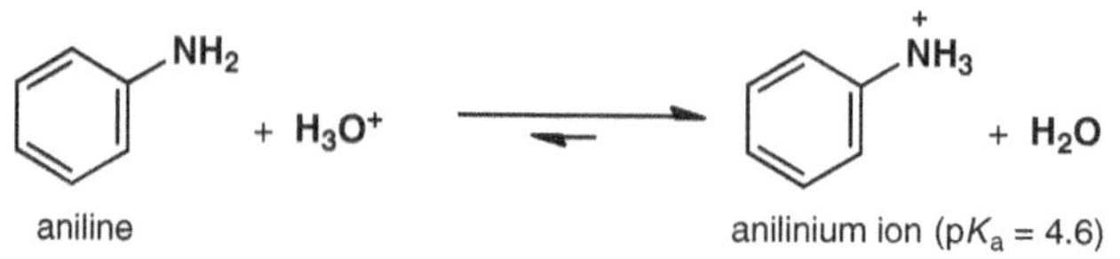

aniline   anilinium ion ($pK_a = 4.6$)

Scheme 16.11
Regioselectivity in the reaction of equimolar aniline and HNO_3 at different concentrations of aqueous H_2SO_4.

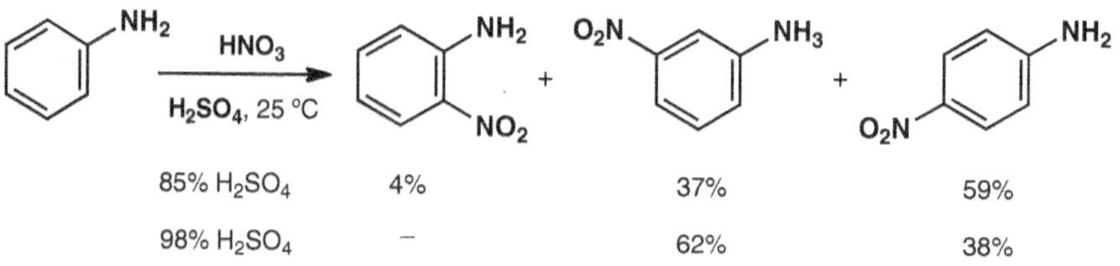

85% H_2SO_4	4%	37%	59%
98% H_2SO_4	–	62%	38%

*Controlled oxidation of $ArNH_2$ with a peracid gives $ArNO_2$.

$$ArNH_2 \xrightarrow{\text{CF}_3\text{CO}_3\text{H}} ArNO_2$$

Additionally, NH_2 is easily oxidized under strongly acidic/electrophilic reaction conditions.*

To avoid these problems with aniline, the basicity and activating effect of the NH_2 are reduced by *N*-acetylation: NHAc is much less easily protonated than NH_2, only modestly electron-donating due to the electron-attracting resonance effect of the acetyl group, and not easily oxidized. The acetyl group can be removed by hydrolysis after the electrophilic substitution, as illustrated in Scheme 16.12 for the preparation of *p*-bromoaniline via *p*-bromoacetanilide (*p*-Br-C$_6$H$_4$NHAc).

Scheme 16.12
Acetylation of aniline to reduce the basicity and activating effect of the amino group in electrophilic substitution.

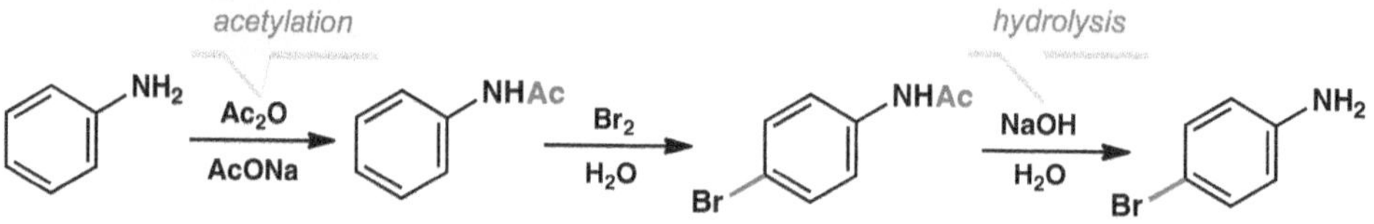

Similarly, nitration of acetanilide gives mainly the *para* product (eqn 16.24) since it is protonated to a much smaller extent than $PhNH_2$ in the nitrating mixture.

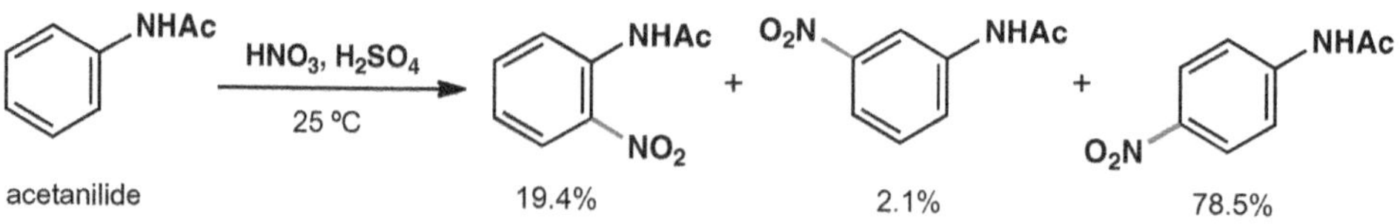

acetanilide   19.4%   2.1%   78.5%

(16.24)

16.6.2 Diazotization

We saw in Section 14.7 that primary, secondary, and tertiary alkylamines undergo characteristically different reactions with nitrous acid. Aniline (and other primary arylamines) are diazotized by nitrous acid just as primary alkylamines are, and secondary arylamines (ArNHR) give N-nitrosoamines just as secondary alkylamines do. However, arenediazonium salts are relatively stable compared with alkanediazonium salts, and can be prepared from primary arylamines in aqueous acidic sodium nitrite with cooling by ice (eqn 16.25). Aqueous tetrafluoroboric acid is the preferred acid if a crystalline arenediazonium salt ($ArN_2^+\ BF_4^-$) is required.

$$Ar\text{-}NH_2 \xrightarrow[\text{H}_2\text{O, 0--5 °C}]{\text{NaNO}_2,\ \text{HX}} \underset{\text{arenediazonium salt}}{Ar\text{-}N_2^+\ X^-} \tag{16.25}$$

The N_2 residue of the benzenediazonium ion is an excellent nucleofuge but the sp^2 C of the phenyl group to which it is bonded cannot allow an S_N2 mechanism (see Section 12.2 for a description of the strict stereoelectronic requirements of the S_N2 mechanism). The S_N1 mechanism is possible, however, and when an aqueous solution of an arenediazonium salt is heated, a phenol is obtained (eqn 16.26).

$$Ar\text{-}N_2^+ \xrightarrow[\Delta]{-\text{N}_2} [\,Ar^+\,] \xrightarrow[-\text{H}^+]{\text{H}_2\text{O}} Ar\text{-}OH \tag{16.26}$$

The diazonio group (N_2^+) can also be replaced by H, i.e. reduction of the arenediazonium ion, and hypophosphorous acid is the usual reagent. This provides a useful technique for removing a substituent altogether (eqn 16.27).

$$Ar\text{-}NH_2 \xrightarrow[\text{H}_2\text{O}]{\text{NaNO}_2,\ \text{HCl}} Ar\text{-}N_2^+\ Cl^- \xrightarrow[\text{H}_2\text{O}]{\text{H}_3\text{PO}_2} Ar\text{-}H \tag{16.27}$$

Arenediazonium ions can act as electrophiles in substitution reactions with reactive aromatic compounds such as phenol and substituted anilines, e.g. eqn 16.28. This so-called *diazo coupling* reaction gives azobenzene derivatives some of which are used as pigments, dyestuffs, and pH indicators; examples are given below.

4-(dimethylamino)azobenzene
(butter yellow)

$$\tag{16.28}$$

methyl orange (a pH indicator:
red at pH < 3.1, yellow at pH > 4.4)

sunset yellow
(a colouring used in food and soft drinks)

Other reactions of arenediazonium salts are very useful in preparations of various substituted benzene derivatives (see Section 18.8).

Tertiary arylamines react differently from their aliphatic analogues with acidified sodium nitrite in aqueous solution; PhNMe$_2$, for example, undergoes electrophilic substitution by nitrous acid to give the *p*-nitroso derivative, *p*-ON-C$_6$H$_4$NMe$_2$.

In aqueous solution, arenediazonium ions react as electrophiles with HO$^-$ to give unstable arenediazohydroxides as the acidity is reduced; in more alkaline conditions, arenediazotate salts are formed.

$$ArN_2^+ \underset{}{\overset{\text{HO}^-}{\rightleftharpoons}} ArN{=}NOH$$
arenediazohydroxide

$$\overset{\text{HO}^-}{\rightleftharpoons} ArN{=}NO^- + H_2O$$
arenediazotate

The same conversion of Ar-N$_2^+$ to Ar-H can be achieved by reductive solvolysis in ethanol or aqueous methanol without an additional reducing agent such as H$_3$PO$_2$ (see Section 18.8).

Butter yellow was used as a colouring in food but has been withdrawn due to its carcinogenicity. It is still used as a pH indicator (red at pH < 2.9, yellow at pH > 4.0) when it is better known as methyl yellow. See Panel 6.2 for pH indicators.

Exercise 16.11

Show using curly arrows how the following diazo coupling reaction with phenol takes place.

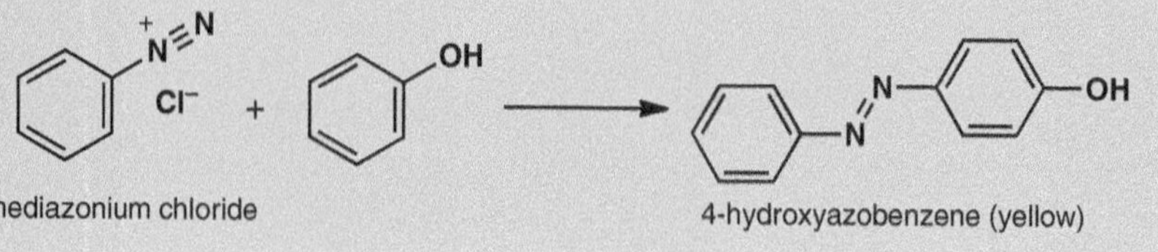

benzenediazonium chloride

4-hydroxyazobenzene (yellow)

16.7 Synthesis of Substituted Benzenes

Because aromatic compounds are so important as industrial products, we need well-developed flexible strategies for the synthesis of multi-substituted benzene derivatives. In the previous section, we saw that $PhNH_2$ is conveniently prepared by reduction of $PhNO_2$. This is an example of a general strategy: when there is no efficient or convenient method of introducing the required substituent, we identify a group which can be readily introduced and then converted into the one we want. Additionally, we need to think about the order in which substituents are introduced since this may affect the substitution pattern in the final product. These matters are best illustrated by examples, and several of the challenges commonly encountered are included in this section.

16.7.1 Limitations to Friedel–Crafts alkylation

Under the normal conditions of Friedel–Crafts alkylation of benzene, the product is more reactive than the starting material because the alkyl introduced has an activating effect. Consequently, further alkylation normally takes place fairly readily. However, it can be minimized easily by using a large excess of benzene (often without an additional solvent), e.g. eqn 16.29, since the difference in reactivity between benzene and a mono-alkylbenzene is usually small (a factor of 1.5–3).

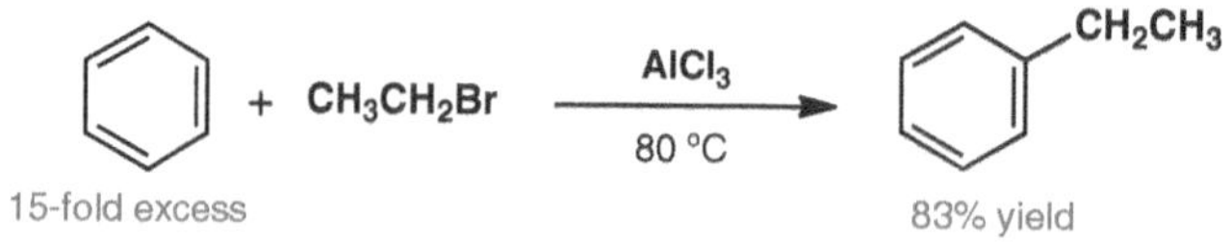

(16.29)

15-fold excess

83% yield

Prior rearrangement of an alkyl group under the conditions of the reaction is another possible complication of Friedel–Crafts alkylation (e.g. eqn 16.30). It arises during generation of the electrophile from the haloalkane and a Lewis acid catalyst by a 1,2-H or 1,2-alkyl shift, as in some S_N1 reactions (Section 14.2).

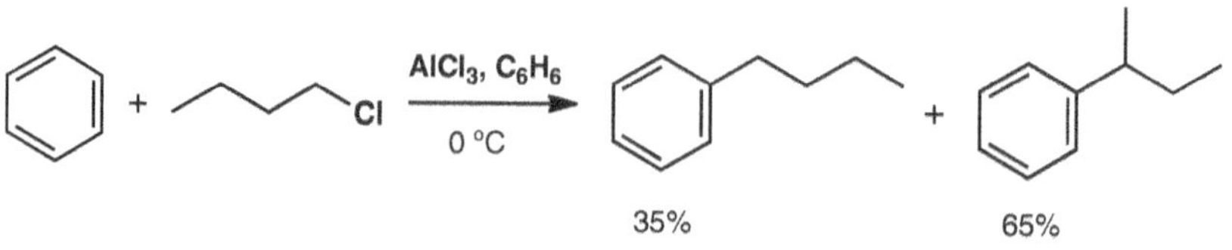

(16.30)

35%

65%

Exercise 16.12

Show using curly arrows how the rearranged product is formed in the reaction of eqn 16.30.

Exercise 16.13

What products would you expect from the Friedel–Crafts alkylation of benzene with 1-chloro-propane and $AlCl_3$?

Yet another restriction to the use of Friedel–Crafts alkylations is that they (like the corresponding acylations; see Sub-section 16.7.5) are not usually successful with strongly deactivated arenes, e.g. eqn 16.31.

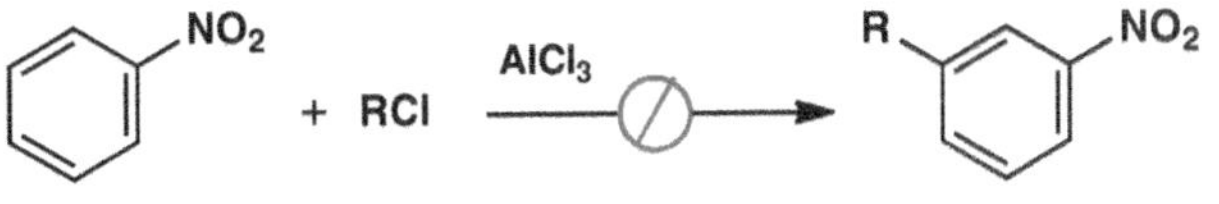

(16.31)

16.7.2 Indirect introduction of a primary alkyl group

An indirect method for preparing a *primary* alkyl-substituted benzene avoids the complications identified in the above sub-section. Friedel–Crafts *acylation* of benzene yields a phenyl ketone which is resistant towards further substitution because acyl is a deactivating substituent. The acyl substituent is then converted into the required primary alkyl group by reduction of the carbonyl to CH$_2$ (Scheme 16.13).

For the reduction of C=O to CH$_2$, we could use the Wolff–Kishner or Clemmensen method (Sub-section 10.2.2).

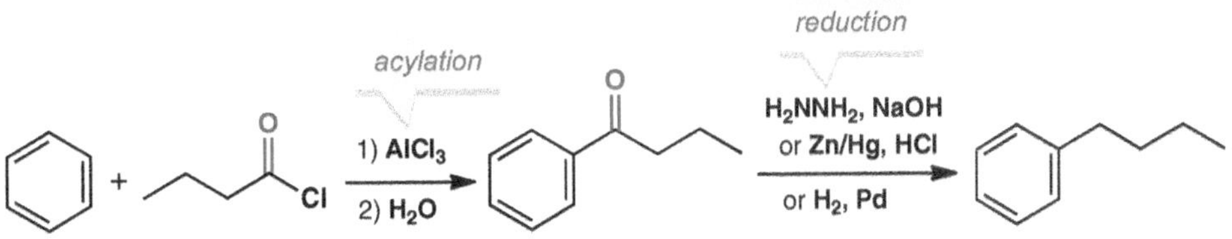

Scheme 16.13
Preparation of a primary alkylbenzene by acylation and reduction.

Propose a method for the preparation of propylbenzene from benzene.

Exercise 16.14

16.7.3 Oxidation of alkyl side-chains

Toluene is oxidized to benzoic acid under vigorous oxidizing conditions such as KMnO$_4$ in an alkaline aqueous solution (eqn 16.32), or Na$_2$Cr$_2$O$_7$ in aqueous acid (eqn 16.33). Other alkyl side chains, regardless of length, are also converted into the carboxylic acid group if they have an α hydrogen (tertiary alkyl groups are not oxidized). This is a useful reaction because the carboxy group can be further transformed into a wide range of other functional groups (see Chapters 9 and 10).

Ar–R $\longrightarrow$ Ar–CO$_2$H

$$\text{Ph–R} \xrightarrow[\text{2) H}_3\text{O}^+]{\text{1) KMnO}_4, \text{ OH}^-, \text{H}_2\text{O}} \text{Ph–CO}_2\text{H}$$ (16.32)

(R = CH$_3$, CH$_2$R', CHR'$_2$)

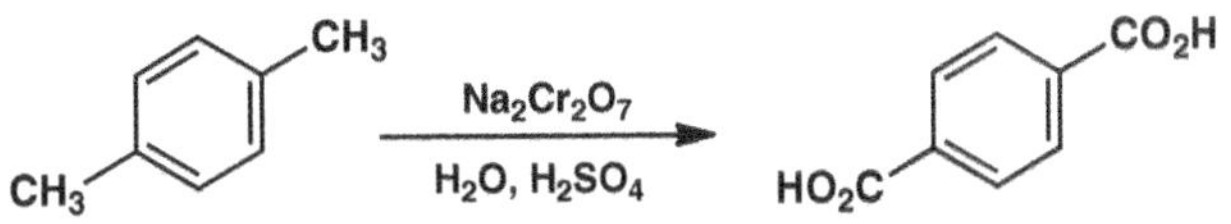

(16.33)

Industrially, toluene and the xylenes are converted to benzoic and phthalic acids, respectively, by atmospheric oxidation with, for example, Co(OAc)$_3$ as catalyst.

Benzylic alcohols are oxidized under very mild conditions which will not affect a simple alkyl group, e.g. dilute nitric acid (eqn 16.34).

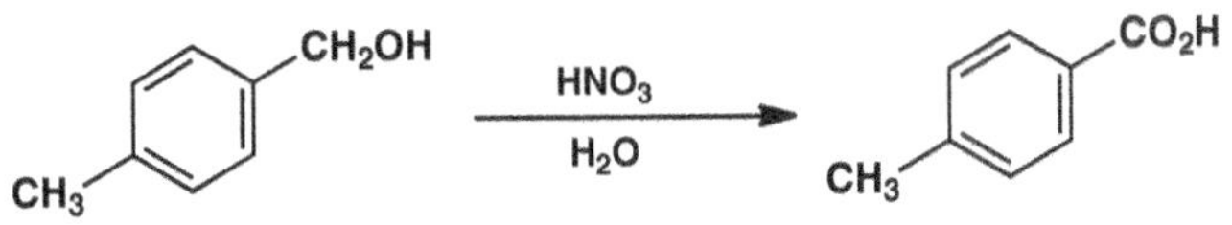

(16.34)

Exercise 16.15

What is the product of each of the following reactions?

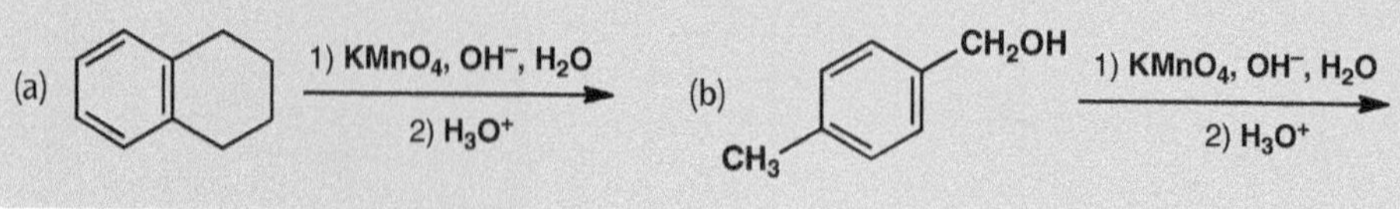

(a)

1) KMnO₄, OH⁻, H₂O
————————————→
2) H₃O⁺

(b)

1) KMnO₄, OH⁻, H₂O
————————————→
2) H₃O⁺

16.7.4 Transformations of haloarenes via Grignard reagents

Because of the versatility of Grignard reagents in organic synthesis (see Chapter 10), this is a widely used reaction. For example, bromobenzene is readily converted into a Grignard reagent which will react with aldehydes and ketones to give alcohols, or with carbon dioxide to give benzoic acid (eqn 16.35).

> The reaction of eqn 16.35 with isotopically labelled carbon dioxide is a procedure for synthesizing ^{13}C- or ^{14}C-labelled benzoic acid.

$$\text{Ph–Br} \xrightarrow[\text{Et}_2\text{O}]{\text{Mg}} \text{Ph–MgBr} \xrightarrow[\text{2) H}_3\text{O}^+]{\substack{\text{1) R}_2\text{C=O} \\ (\text{CO}_2)}} \underset{\overset{|}{\text{OH}}}{\text{Ph–CR}_2} \; (\text{Ph–CO}_2\text{H}) \qquad (16.35)$$

16.7.5 Control of reactivity and regioselectivity in syntheses of substituted benzenes

When we prepare a disubstituted benzene by consecutive electrophilic substitutions, we need to consider the order in which the reactions are carried out since the substituent introduced first affects the rate constant and the regioselectivity of the second reaction.

One restriction to Friedel–Crafts reactions is that a strongly deactivated benzene (one already with a *meta*-directing substituent) is usually an ineffective substrate, as mentioned in Sub-section 16.7.1. *m*-Nitroacetophenone, for example, can be prepared by a combination of nitration and Friedel–Crafts acylation but, although both nitro and acetyl groups are *meta* directing, we have to carry out the acylation first (Scheme 16.14). If nitration is performed first, the subsequent acylation of nitrobenzene is impracticably slow.

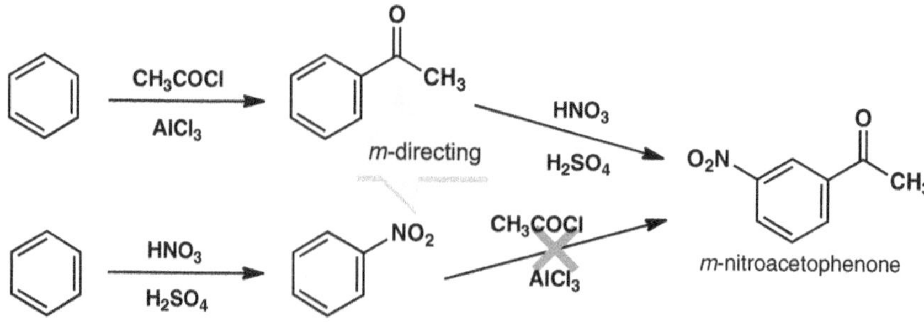

Scheme 16.14
Preparation of
m-nitroacetophenone.

Consider the sequential chlorination and nitration of benzene (Scheme 16.15). If chlorination is carried out first and followed by nitration, the products are the *ortho* and *para* isomers because Cl is *ortho* and *para* directing (reaction 16.36). If the *meta*-directing

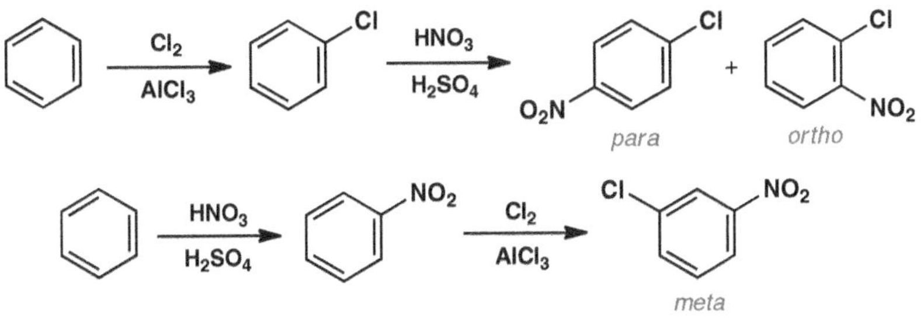

Scheme 16.15 Reactions which give isomeric chloronitrobenzenes.

(16.36)

(16.37)

nitro group is introduced first, however, the subsequent chlorination yields the single *meta* isomer (reaction 16.37).

Sulfonation is exceptional in being a *reversible* electrophilic substitution reaction (see Sub-section 16.3.3); vigorous heating of an arenesulfonic acid in aqueous acid, for example, leads back to the arene (eqn 16.38). This reaction provides an alternative to the sequence in eqn 16.27 for removing a substituent (or for introducing a deuterium atom (^{2}H or D) at a specific location in an arene if D_3O^+/D_2O is used in the desulfonation).

$$Ar\text{-}SO_3H + H_2O \xrightarrow[\Delta]{H_3O^+, H_2O} Ar\text{-}H + H_2SO_4 \qquad (16.38)$$

Consequently, either NH_2 or SO_3H can be introduced temporarily to block a position in an arene, or to control regioselectivity and reactivity. Examples are shown in Schemes 16.16 and 16.17. Initial sulfonation in Scheme 16.16 blocks the *para* and one of the *ortho* positions of the phenol, which allows bromination only at the other *ortho* position. Subsequent removal of the sulfonic acid residues gives the initially required product, *o*-bromophenol.

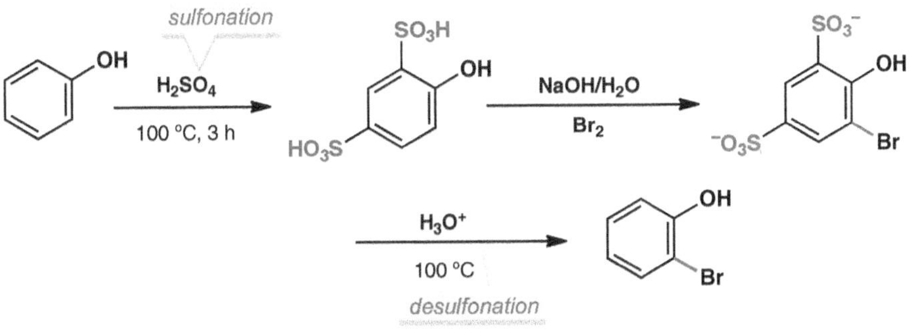

Scheme 16.16 Synthesis of *o*-bromophenol: control of regioselectivity and reactivity by reversible sulfonation.

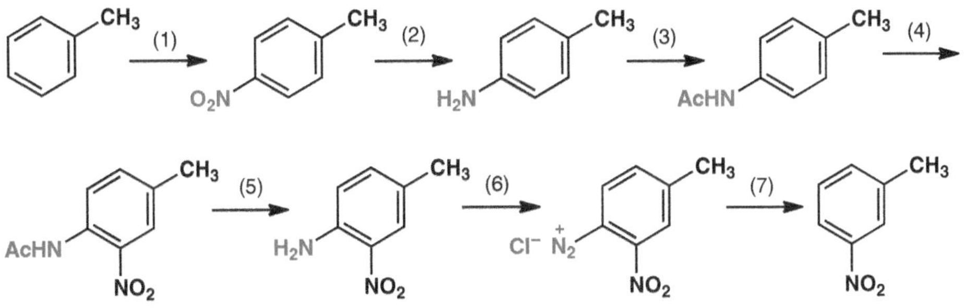

Scheme 16.17 Synthesis of *m*-nitrotoluene: control of regioselectivity by an acetylamino group and its subsequent removal.

Although *o*- and *p*-nitrotoluene can be obtained readily by nitration of toluene, the *meta* isomer is more difficult to make. Friedel–Crafts methylation of nitrobenzene is impracticably slow because the *meta*-directing NO_2 is so powerfully electron-withdrawing. The extended sequence in Scheme 16.17 allows the indirect *m*-nitration of toluene.

Panel 16.4 2-Arylethylamines which have psychological effects

Mescaline has a psychedelic effect and the peyote cactus which contains it has been used in religious ceremonies by native Americans in Mexico since before the arrival there of Europeans. This compound contains a 2-arylethylamine unit (highlighted in the structures below). Ephedrine, which is the active ingredient in a traditional Chinese medicine (*má huáng*) used in the treatment of asthma and bronchitis, contains the same structural feature, as do morphine and codeine (the powerful analgesics mentioned in Panel 6.3). These compounds, along with the others shown, participate in the response of the sympathetic nervous system and have powerful physiological and psychological effects.

Dopamine functions in the brain as a neurotransmitter; it regulates and controls movement, motivation, and cognition, cooperating with other neurotransmitters such as acetylcholine ($AcOCH_2CH_2N^+Me_3$). A dopamine deficiency in the brain is associated with degenerative disorders such as Parkinson's disease.

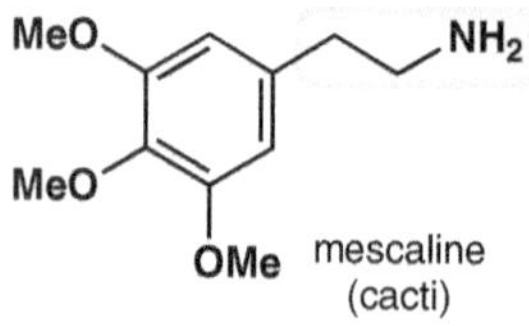

mescaline
(cacti)

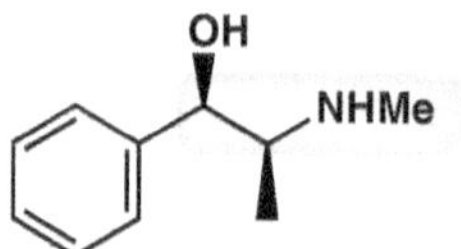

ephedrine
(Chinese herb, *má huáng*
or *Ephedra sinica*)

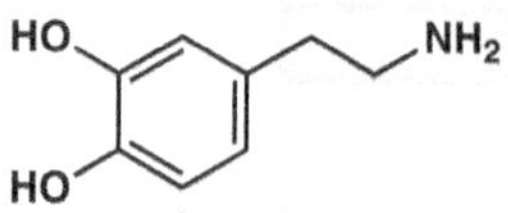

dopamine
(neurotrasmitter)

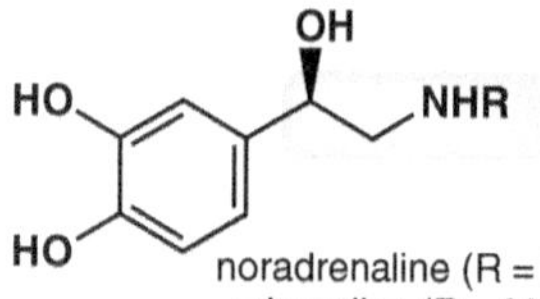

noradrenaline (R = H)
adrenaline (R = Me)
(hormone, neurotransmitter)

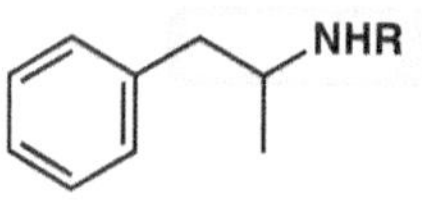

amphetamine (R = H)
methamphetamine (R = Me)
(psychostimulant drugs)

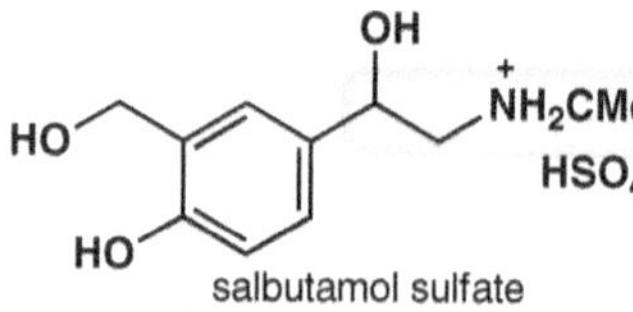

salbutamol sulfate
(a bronchodilator drug)

Physiological oxidation of dopamine leads to noradrenaline (also called norepinephrine) and subsequent methylation gives adrenaline (epinephrine). Adrenaline and noradrenaline are hormones and neurotransmitters, and have similar functions. Both are released when an animal is under stress: they increase blood pressure and heart rate, and dilate air passages to prepare the animal to fight or flee.

Amphetamine and its methylated derivative, methamphetamine, are synthetic drugs which lack the hydroxy group of ephedrine; both are strong psychostimulants. The former has been widely used as a performance-enhancer and the latter (which has various slang names including 'meth') is addictive; both are illegal drugs. Salbutamol (also called albuterol) is a short-acting β_2 agonist used for the treatment of bronchospasms caused by asthma. It is usually administered as sulbutamol sulfate by an inhaler or nebulizer.

Serotonin (5-hydroxytryptamine) is a neurotransmitter in the central nervous system which is sometimes called a 'happiness hormone' but it is not a hormone. It is involved in many central and peripheral physiological functions including contraction of smooth muscle, vasoconstriction, appetite, sleep, pain perception, and memory. Melatonin is a simple derivative of serotonin which is a hormone secreted by the pineal gland in the brain during the night. It is involved in adjusting the internal body clock and controlling the day–night rhythm.

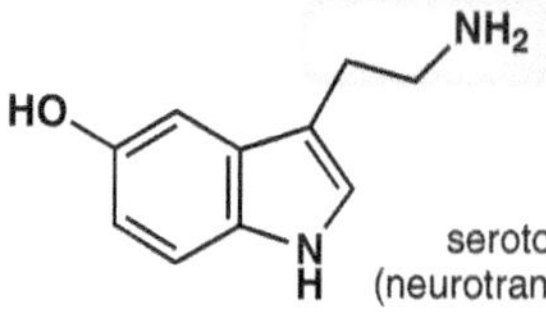

serotonin
(neurotransmitter)

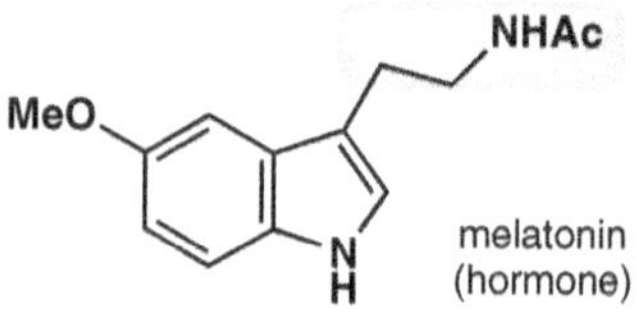

melatonin
(hormone)

Example 16.4

Give the reagents required for each step in Scheme 16.17.

Solution
(1) HNO_3/H_2SO_4 (2) Sn/HCl   (3) AcCl/pyridine   (4) HNO_3/H_2SO_4 (5) $NaOH/H_2O$
(6) $NaNO_2/H_3O^+Cl^-$ (7) H_3PO_2

Exercise 16.16

Propose reactions for the synthesis of each of the following from benzene.

(a) (b) (c) (d)

Summary

- Benzene, with a cyclic 6π electron system, has special properties (including stability) attributable to **aromaticity**.

- The main reactions of benzene are **electrophilic substitutions**: initial electrophilic addition to give a **benzenium ion** intermediate is followed by elimination of a proton to give the substitution product and recovery of aromaticity. Reactions include halogenation, nitration, sulfonation (which is reversible), and Friedel–Crafts alkylation and acylation.

- The reactivity relative to benzene and regioselectivity (orientation) in electrophilic substitution of substituted benzenes are predictable from the relative stabilities of the benzenium ion intermediates involved.

- Substituents are classified as being ***ortho,para* directing** (electron-donating, e.g. OH by resonance, or alkyl groups) or ***meta* directing** (electron-withdrawing, e.g. NO_2 by resonance, or CF_3); the latter are invariably deactivating and the former (except the halogens) are activating. The effects of substituents upon reactivity and regioselectivity in electrophilic substitutions must be taken into account when syntheses of multiply substituted benzenes are being planned.

- Diverse reactions of aniline and phenol, which are very reactive in electrophilic substitution, are also discussed. The reactivity of aniline due to the *ortho,para*-directing NH_2 may be reduced by acylation; acidic media convert NH_2 into the deactivating *meta* directing NH_3^+ substituent. Basic conditions convert the highly reactive phenol into the even more reactive phenoxide.

Problems

16.1 Predict the reactivity order for each set of compounds in electrophilic substitution reactions, and explain the reasons for your predictions.

(a) (b) (c) (d)

16.2 What would be the main mono-nitration product(s) from each of the following.

(a) 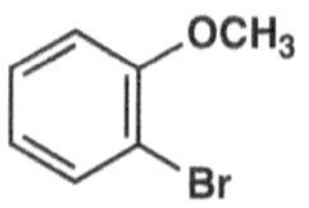 [structure: benzene with OCH₃ and Br]

(b) 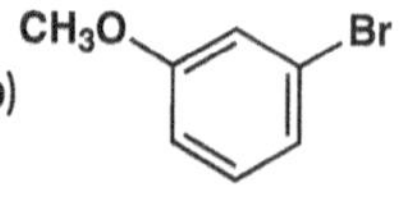 [structure: CH₃O and Br on benzene]

(c) 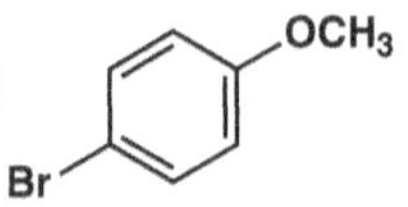 [structure: OCH₃ and Br on benzene]

(d) 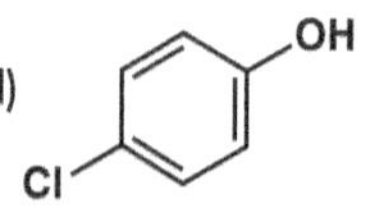 [structure: OH and Cl on benzene]

(e) 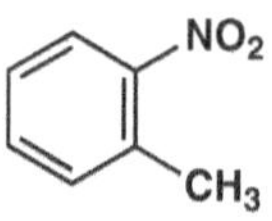 [structure: NO₂ and CH₃ on benzene]

(f) 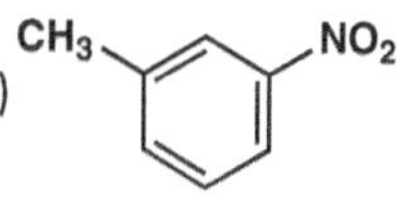 [structure: CH₃ and NO₂ on benzene]

(g) 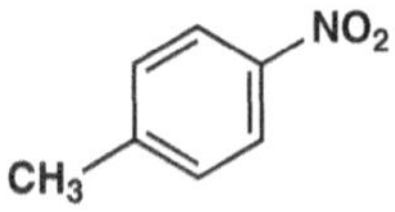 [structure: NO₂ and CH₃ on benzene]

(h) 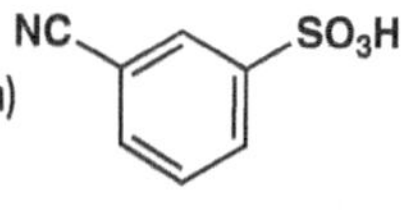 [structure: NC and SO₃H on benzene]

16.3 What would be the main mono-bromination product(s) using Br₂ and FeBr₃ with each of the following?

(a) 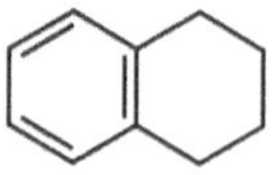 [tetralin structure]

(b) 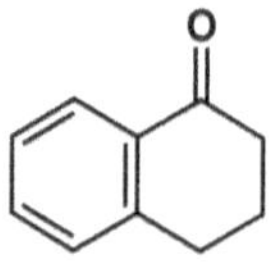 [tetralone structure]

(c) 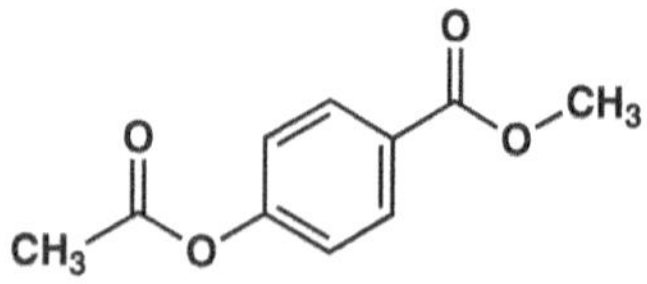 [CH₃-C(=O)-O-aryl-C(=O)-O-CH₃ structure]

(d) 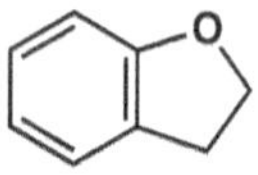 [dihydrobenzofuran structure]

(e) 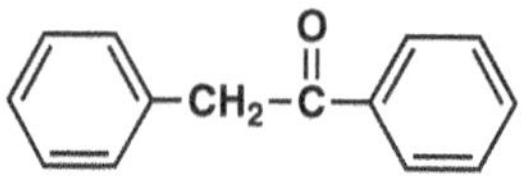 [aryl-CH₂-C(=O)-aryl structure]

(f) 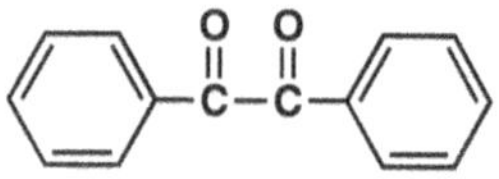 [aryl-C(=O)-C(=O)-aryl structure]

16.4 What would be the main mono-bromination product using Br₂ in ethanoic acid with each of the following?

(a) 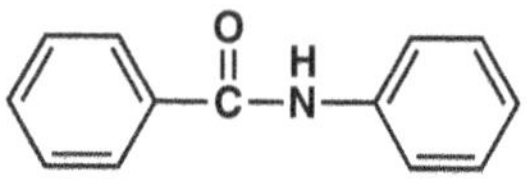 [aryl-C(=O)-N(H)-aryl structure]

(b) 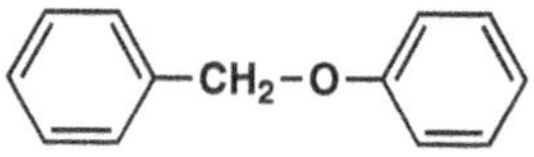 [aryl-CH₂-O-aryl structure]

16.5 N,N-Dimethylaniline gives mainly the p-bromination product in a rapid reaction with Br₂ in methanol but mainly the m-nitration product in a slow reaction with nitric/sulfuric acid mixtures. Explain the difference in regioselectivity and reactivity in the two reactions.

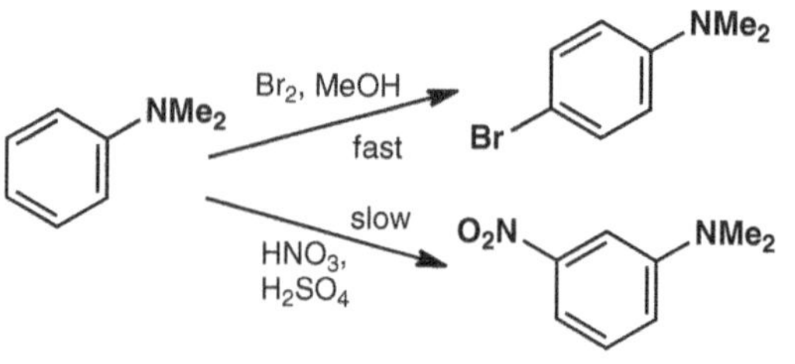

[reaction scheme: NMe₂-benzene → Br₂, MeOH (fast) → p-Br-NMe₂-benzene; → HNO₃, H₂SO₄ (slow) → m-O₂N-NMe₂-benzene]

16.6 Predict the major mono-substitution product in the reaction of benzene with 2,2-dimethyl-1-chloropropane and AlCl₃, and propose a reaction mechanism for its formation.

16.7 Compare the reactions of phenol with Br₂ carried out under acidic and weakly alkaline conditions; explain which reaction is faster and predict the major product under the different reaction conditions.

16.8 Propose a reaction sequence for the preparation of each of the following from benzene.

(a) 1-phenylpentane

(b) p-isopropylacetophenone

(c) m-chloroethylbenzene

(d) 1,4-dinitrobenzene

16.9 Propose a reaction sequence for the preparation of each of the following from toluene.

(a) p-bromobenzoic acid

(b) m-chlorobenzoic acid

(c) m-bromotoluene

(d) o-nitrotoluene

(e) p-methylbenzoic acid

16.10 Propose a reaction sequence for the preparation of each of the following from aniline.

(a) p-aminoacetophenone

(b) 2,6-dichloroaniline

(c) 4-bromo-2-nitroaniline

(d) 1,3,5-tribromobenzene

16.11 Propose a reaction sequence for the preparation of 2,4,6-tribromobenzoic acid from benzoic acid.

16.12 The antioxidant, BHT (2,6-di-*t*-butyl-4-methylphenol), is widely used as a preservative in food as well as in cosmetics and pharmaceuticals, and propofol (2,6-diisopropylphenol), a quick-acting sedative/hypnotic agent, is used to induce and maintain general anaesthesia. Propose a reaction sequence for the synthesis of each of these important compounds from *p*-cresol (*p*-methylphenol) and phenol, respectively, using an alkene as alkylating agent.

(a)

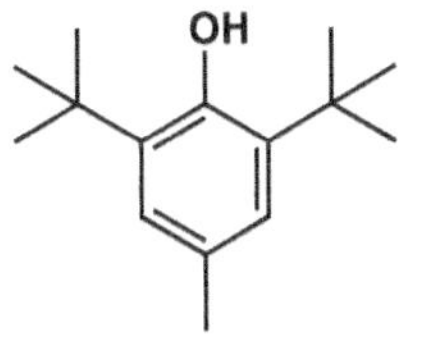

BHT (butylated hydroxytoluene)
2,6-di-*t*-butyl-4-methylphenol

(b)

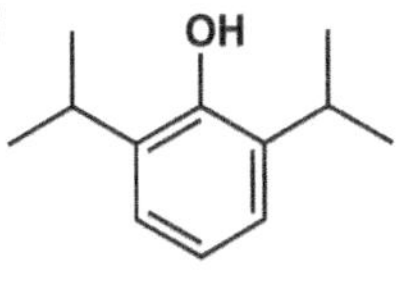

propofol
2,6-diisopropylphenol

16.13 Suggest how the following overall reaction may be achieved.

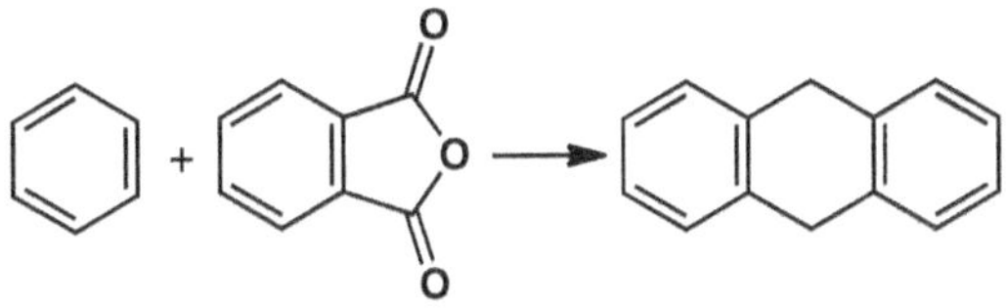

16.14 Give the structures of intermediate products, **A** and **B**, in the following reaction sequence, and suggest appropriate reagents, (a)–(c).

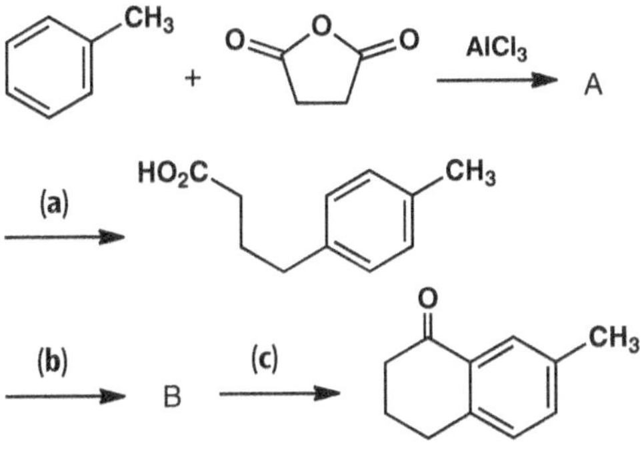

16.15 Explain why and how alkylation of *m*-xylene with 2-methylpropene and H_2SO_4 gives 1-*t*-butyl-3,5-dimethylbenzene upon prolonged reaction.

16.16 When benzene reacts with 3-chloro-2-methylpropene in the presence of H_2SO_4, the main product has the formula $C_{10}H_{13}Cl$. Give the structure of this product and show how it is formed. What would you expect the main product to be if $AlCl_3$ were used as catalyst in an aprotic solvent?

16.17 Explain why alkylation of toluene with (*R*)-1-bromo-1-phenylpropane and $AlCl_3$ would not be a sensible way to try to make chiral 1-(*p*-methylphenyl)-1-phenylpropane.

16.18 Upon reaction of 1,3,5-trideuteriobenzene with Br_2 and $AlBr_3$, a 1:1 mixture of two monobromo products is obtained.

(a) Show structures of the two products.

(b) What does this result imply about the rate-determining step of the electrophilic bromination of benzene? Note that C–D bond cleavage is normally slower than C–H bond cleavage (a kinetic isotope effect).

16.19 Benzene reacts with methanal in the presence of HCl to give benzyl chloride *via* an intermediate **A**. Propose a mechanism for this chloromethylation reaction and identify **A**.

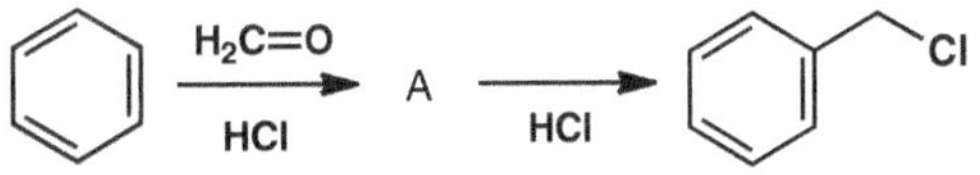

Supplementary Problems

16.20 Acetylsalicylic acid (*o*-acetoxybenzoic acid, aspirin) rearranges to 3- and 5-acetylsalicylic acid (3- and 5-acetyl-2-hydroxybenzoic acid) upon treatment with $AlCl_3$. Propose a mechanism for this rearrangement.

16.21 Propose a method for the preparation of triphenylmethanol from benzene; except for benzene

and solvents, use only C_1 compounds as starting materials.

16.22 When an *N,N*-dimethylamide is treated with phosphorus oxychloride, $POCl_3$, a carbon electrophile is generated which reacts with reactive aromatic nucleophiles to give iminium salts. When

N,N-dimethylaniline is used as the reactive aromatic nucleophile, the product is readily hydrolysed to give a carbonyl compound—the same type of product that is obtained by Friedel–Crafts acylation.

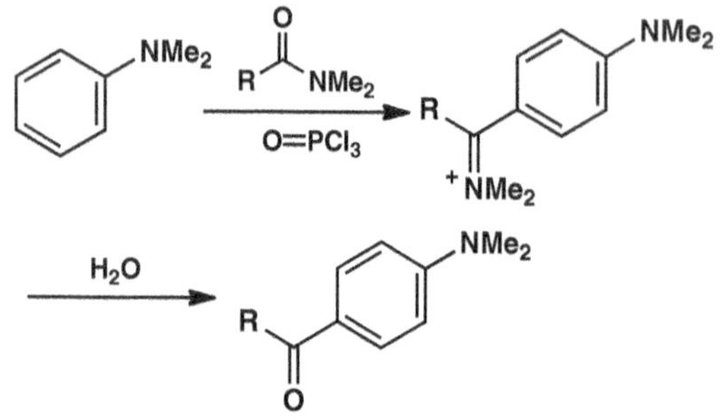

Propose a mechanism for the reaction which gives the iminium ion, the so-called Vilsmeier reaction. (The reaction with *N,N*-dimethylformamide as the reagent is especially useful for the preparation of aromatic aldehydes since a formyl group cannot be introduced by Friedel–Crafts acylation.)

Enolate Ions, their Equivalents, and Reactions

17

Related topics we have already covered:

- **Carbon acids and carbanions** (Section 6.4)
- **Curly arrow representation of reaction mechanisms** (Section 7.2)
- **Reactions of carbonyl compounds** (Chapters 8–10)
- **The S_N2 mechanism** (Section 12.2)
- **Electrophilic addition to alkenes** (Chapter 15)

Topics we shall cover in this chapter:

- Acidity of a C–H α to a carbonyl group
- Keto–enol tautomerism
- Acid–base catalysis of enolization
- Nucleophilicity of enols and enolates
- α-Halogenation and the haloform reaction
- C–C bond formation with enolates: aldol and Claisen reactions
- Enolate ions of 1,3-dicarbonyl compounds
- Enolate equivalents
- Alkylation of enolates and enolate equivalents

Carbonyl compounds have a C=O bond which is reactive toward nucleophiles, as covered in Chapters 8–10, but they often undergo another characteristic type of reaction at the carbon α to the carbonyl group. A hydrogen on the α carbon is weakly acidic so carbonyl compounds are relatively easily deprotonated by strong bases to give their conjugate bases, enolate ions, which have a delocalized electron system and two reaction sites. Typically, an enolate ion is a reactive intermediate which participates in various reactions as a nucleophile with electrophiles including halogens, carbonyl compounds, and alkyl halides. Reactions between enolates and carbonyl compounds provide important methods for making new C–C bonds, and similar reactions are involved in biological systems for the biochemical formation of carbohydrates and fatty acids from smaller molecules (see Panels 17.2 and 17.3). Enols (enolates protonated on oxygen), enamines, and enol silyl ethers are electronically equivalent to enolates, and we shall see that they react similarly.

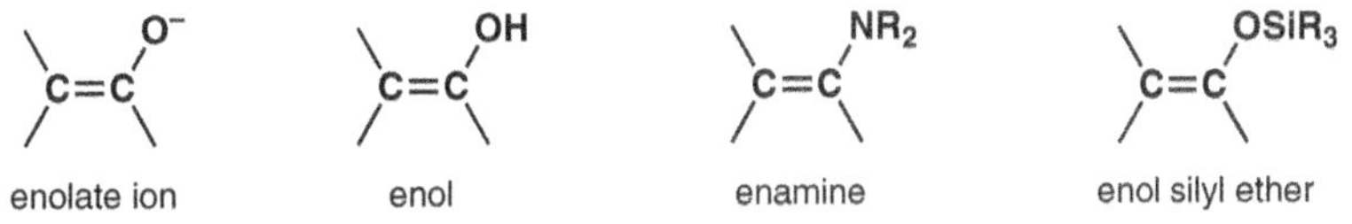

enolate ion   enol   enamine   enol silyl ether

17.1 Keto–Enol Tautomerism

17.1.1 Allylic anions and enolate ions

We have already seen in Chapter 5 that the allyl anion is a carbanion stabilized by electron delocalization, and that **enolate ions**, which are electronically equivalent to the allyl anion, are allyl anion analogues (Section 5.3).

The conjugate acid of the parent allyl anion is propene, which is considerably more acidic ($pK_a \sim 43$) than simple alkanes ($pK_a \sim 50$); see eqn 17.1 with R=H in Scheme 17.1 (see also Section 6.4). A methyl ketone, whose conjugate base is an enolate ion, is even more acidic with $pK_a \sim 20$ (eqn 17.2 in Scheme 17.1). In general, a hydrogen on a C next to a carbonyl group (an α *hydrogen*) is much more acidic than the hydrogen next to a C=C bond (an *allylic hydrogen*), reflecting the relative ease of formation of the corresponding anions.

$$\text{(17.1)}$$

$$\text{(17.2)}$$

Scheme 17.1 Comparison of formations of an allylic anion and an enolate ion.

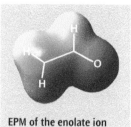

EPM of the enolate ion from ethanal.

An enolate ion, like an allylic anion, is best represented as a resonance hybrid (Scheme 17.1). Although allylic anion resonance contributors, **1a** and **1b**, are (depending on R) almost equivalent, those of the enolate ion, **2a** and **2b**, are not, and **2b** with the negative charge on the more electronegative oxygen atom is the more important of the two. In other words, the electronegative oxygen accommodates the negative charge more effectively than carbon, which strongly stabilizes the enolate ion, and the real electronic configuration of the enolate ion in Scheme 17.1 is closer to **2b** than to **2a**. This is the reason why this anion is better called an *enolate*, i.e. the deprotonated form of an alcohol with a double bond (en+ol = **enol**), than a *carbanion*.

17.1.2 Equilibria involving enols

If the enolate ion in eqn 17.2 is reprotonated on the oxygen instead of the α carbon, a conjugate acid other than the original ketone is formed—the *enol* shown in eqn 17.3. It is a constitutional isomer of the carbonyl compound, often called a **tautomeric isomer** (or **tautomer**), and interconversion between the two occurs by proton transfer, as will be discussed in the next section. An enol is considerably more acidic ($pK_a \sim 11$) than its keto tautomer (eqn 17.2) and even more acidic than a simple alcohol ($pK_a \sim 15$); it is similar in acidity to phenol ($pK_a \sim 10$).

$$\text{(17.3)}$$

The enol and its **keto** tautomer exist in equilibrium in solution (eqn 17.4), but the keto form is generally much more stable. Equilibrium constants for **enolization**, K_E=[enol form]/[keto form], of simple aldehydes and ketones are small ranging from $K_E = 10^{-4}$ to 10^{-9}, as listed in Table 17.1.

Table 17.1

Equilibrium constants for keto–enol tautomerism[a]

Keto form	Enol form[b]	K_E	% enol
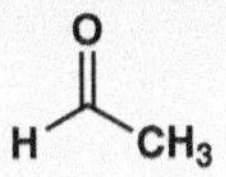 (H–CO–CH₃)	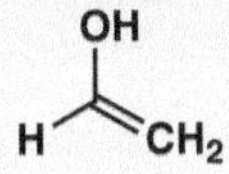 (H–C(OH)=CH₂)	5.9×10^{-7}	6×10^{-5}
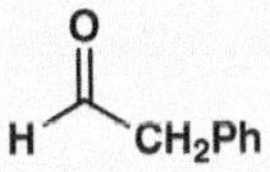 (H–CO–CH₂Ph)	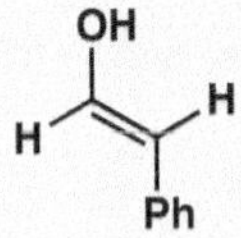 (H–C(OH)=CH–Ph)	8.5×10^{-4}	8.5×10^{-2}
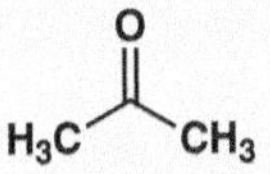 (H₃C–CO–CH₃)	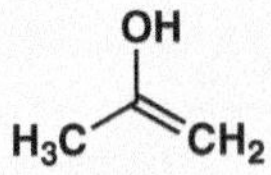 (H₃C–C(OH)=CH₂)	5×10^{-9}	5×10^{-7}
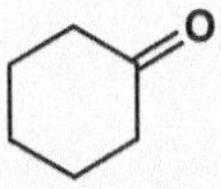 (cyclohexanone)	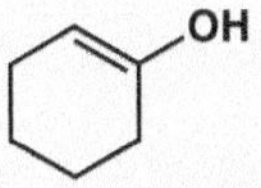 (1-cyclohexenol)	4.1×10^{-7}	4×10^{-5}
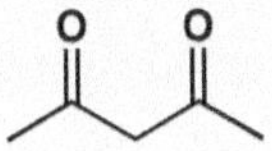 (pentane-2,4-dione)	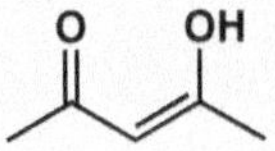 (enol form)	0.15	13

a. At 25 °C in aqueous solution. b. Stereoisomeric enols are possible; $K_E = 4.5 \times 10^{-4}$ for the *cis* enol, the *trans/cis* ratio being about 1.9 at equilibrium for the enol from phenylethanal.

The large K_E value for the 1,3-diketone, pentan-2,4-dione (Table 17.1), is attributable to the stabilization of the enol by intramolecular hydrogen bonding and conjugation of the C=C with the remaining C=O bond.

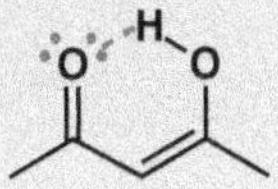

Keto–enol tautomerism:

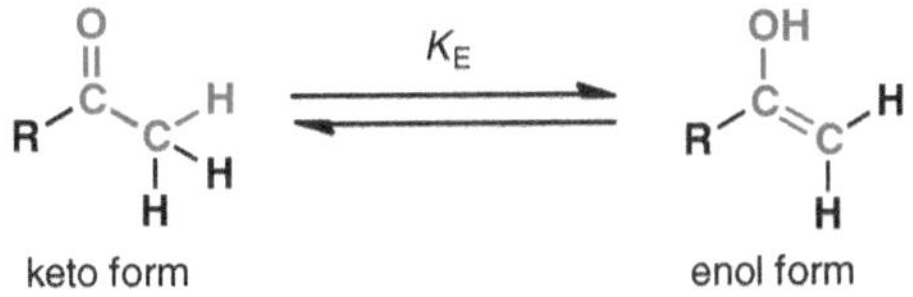

$$\text{keto form} \underset{}{\overset{K_E}{\rightleftharpoons}} \text{enol form} \qquad (17.4)$$

Example 17.1

Draw structures of enol forms of the following carbonyl compounds.

(a) 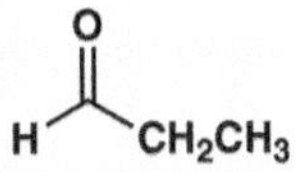 H–CO–CH₂CH₃   (b) 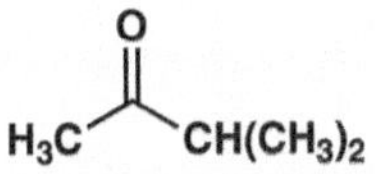 H₃C–CO–CH(CH₃)₂   (c) 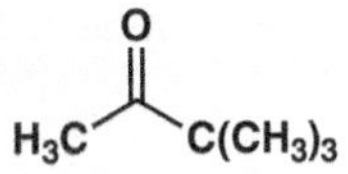 H₃C–CO–C(CH₃)₃

Solution
For all three compounds, find an α C–H, transfer the hydrogen to the O, and redraw double and single bonds to give an enol. In (b) where α and α′ C–H groups are different, two constitutionally isomeric enols are possible. The more alkyl-substituted enol is usually the more stable just as for simple alkenes. Note that the enol in (a) has *cis–trans* isomers, and that there is no α hydrogen in the *t*-butyl group in (c).

(a) 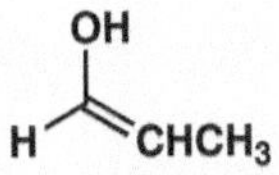 H–C(OH)=CHCH₃   (b) H₃C–C(OH)=C(CH₃)₂ and H₂C=C(OH)–CH(CH₃)₂   (c) H₂C=C(OH)–C(CH₃)₃

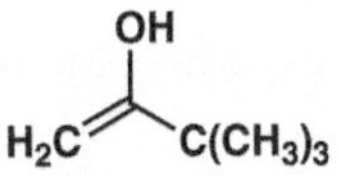

Exercise 17.1

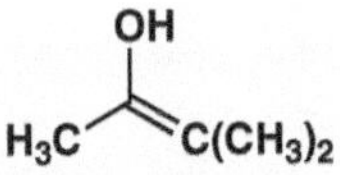

Draw structures of enol tautomers of the following ketones. When more than one enol is possible, explain which is the most stable.

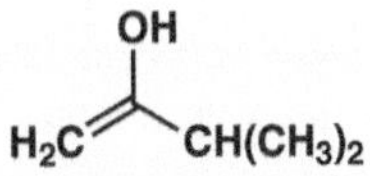

(a) 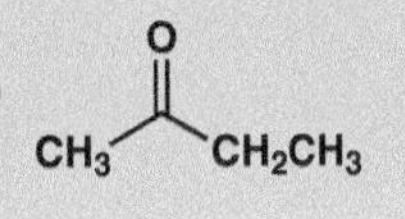 CH₃–CO–CH₂CH₃   (b) 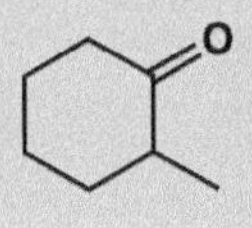 (2-methylcyclohexanone)   (c) 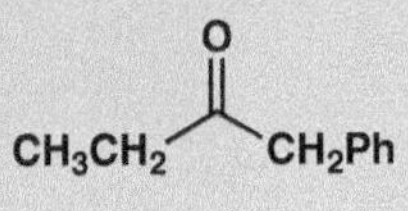 CH₃CH₂–CO–CH₂Ph

Example 17.2

The pK_a values of the keto and enol tautomers of propanone in aqueous solution are about 19 and 11, respectively. Calculate the approximate equilibrium constant for the enolization of propanone.

Solution

Tautomerization and acid dissociations of keto and enol forms are described as a cycle of equilibria below (an example of what is sometimes called a *thermodynamic cycle*). In this, $K_a^K = [\text{enolate}][H_3O^+]/[\text{keto}]$ and $K_a^E = [\text{enolate}][H_3O^+]/[\text{enol}]$, where K_a^K and K_a^E are the acid dissociation constants for the keto and enol forms (see Chapter 6) involving a common conjugate base, the enolate. Since $K_E = [\text{enol}]/[\text{keto}]$, it follows that the equilibrium constants are related by $K_E = K_a^K/K_a^E$. Consequently, $K_E = 10^{-19}/10^{-11} = 10^{-8}$ (the exact value is 5×10^{-9}, as shown in Table 17.1).

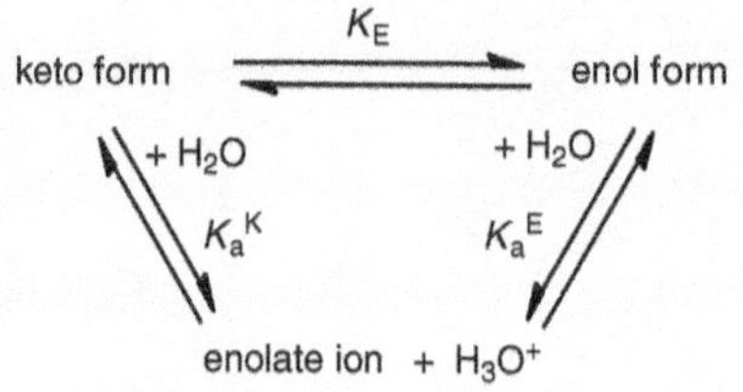

17.2 Mechanisms of Enolization

Enolization is reversible and the reaction from the enol to the keto form is called **ketonization**. Except in the presence of an acid or a base, both forward and reverse reactions are relatively slow in aqueous solutions.

17.2.1 Acid-catalysed enolization

Acid-catalysed enolization of an aldehyde or ketone starts with rapidly reversible protonation of the carbonyl oxygen, as shown in Scheme 17.2 for the oxonium ion and a methyl ketone. The strong electron pull by the protonated carbonyl group enables abstraction of an α proton by even a weakly basic water molecule; this regenerates the H_3O^+, i.e. the enolization has been catalytic.

Note that proton transfer between two O atoms does not involve any rehybridization and is rapid; in contrast, proton transfer to or from a C atom involves rehybridization and is generally not as easy. Consequently, this is normally the rate-determining step in enolization/ketonization (see below).

Scheme 17.2 A mechanism for the acid-catalysed enolization of a methyl ketone.

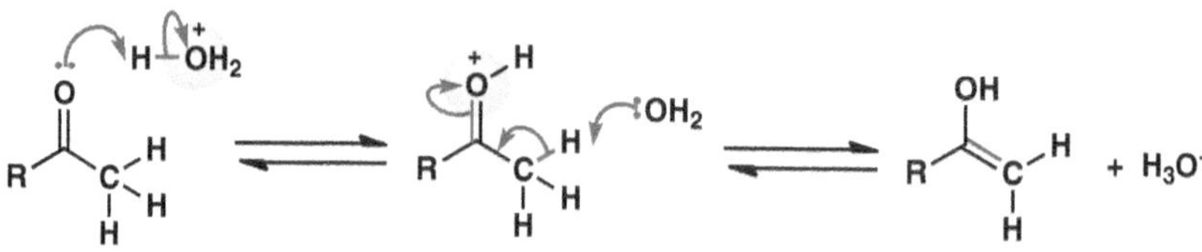

Exercise 17.2

Write a curly arrow mechanism for the acid-catalysed ketonization of an enol.

17.2.2 Base-catalysed enolization

In the first step of the base-catalysed enolization of an aldehyde or ketone in aqueous solution, the base (B with a methyl ketone in Scheme 17.3) abstracts an α proton to give an enolate. This step is reversible but the enolate may also be reprotonated on the oxygen atom to form the enol and regenerate the base.

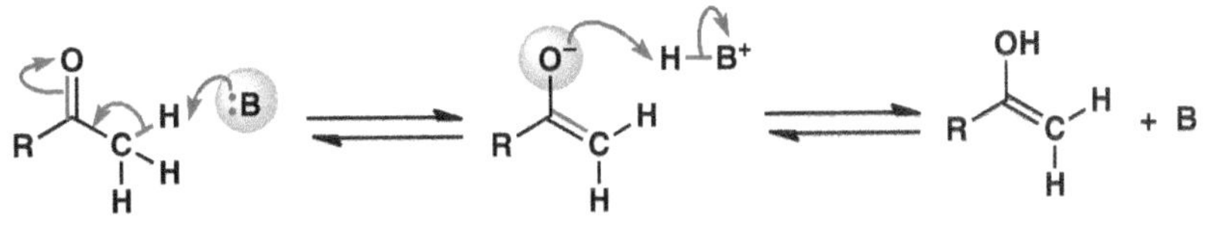

Scheme 17.3 A mechanism for the base-catalysed enolization of a methyl ketone.

Note that in aqueous solution at normal temperatures, the acid- and base-catalysed enolization mechanisms lead to equilibrium mixtures of carbonyl compound and enol (as well as of the protonated carbonyl compound and the enolate). In both mechanisms, the rate-determining step is deprotonation of an α C–H. In the acid-catalysed mechanism, this is by a weak base (water in Scheme 17.2) from the substrate protonated by the catalytic acid; in the base-catalysed mechanism, it is by the stronger catalytic base (B in Scheme 17.3) from the neutral substrate.

If $B=HO^-$ ($pK_{BH^+}=15.7$) in Scheme 17.3, the enolate ion ($pK_{BH^+}\sim 11$) will not deprotonate water to an appreciable extent, i.e. HO^- is not fully regenerated in the second step of Scheme 17.3 and only a minute proportion of enol exists at equilibrium under alkaline conditions. This seldom affects catalysis of the equilibration by HO^- since its depletion by the equilibrium is so small.

Give a curly arrow mechanism for the ketonization of an enol catalysed by a base B.

Exercise 17.3

Since the equilibrium constants for enolization of simple aldehydes and ketones are very small and the equilibrium concentrations of the enol forms are very low, as listed in Table 17.1, enols are seldom observed directly. How, therefore, can we investigate the rate of formation of an enol when its concentration is always too low to measure? The answer is that we do not analyse for the enol directly: we exploit a measurable property of the system which occurs upon enolization (see next section) and this gives us the rate constant for enolization indirectly. But what about the rate constant for ketonization? Modern instrumental techniques enable us to generate the enol form in greater than equilibrium concentrations and to measure the rate of return to equilibrium which, with the rate constant for enolization, leads to the rate constant for ketonization. The experimental rate constants for enolization and ketonization then allow us to calculate accurate keto–enol equilibrium constants, as listed in Table 17.1. Previously, K_Es were determined by analysis of the minute amounts of the enol tautomers in equilibrium mixtures, but such data are not precise enough to give accurate results; old K_E values are not reliable.

17.3 Reactions via Reversible Enolization

17.3.1 Deuterium isotope exchange

When a ketone or aldehyde is dissolved in an acidic or basic deuterated aqueous solution (one containing D_2O in place of H_2O), we observe that deuterium is incorporated at the α position of the carbonyl compound. Correspondingly, the deuterium is lost from an α-deuterated ketone or aldehyde in a non-deuterated aqueous solution. These acid- and base-catalysed **deuterium exchange** processes occur by the reversible enolization/ketonization reactions shown in Scheme 17.4 for a methyl ketone with catalysis (a) by oxonium ion and (b) by hydroxide. In both cases, the rate of the deuterium exchange (which can be measured relatively easily) corresponds to the rate of enolization (with minor adjustments for kinetic isotope effects).

We sometimes include conversion of the keto form to the enolate in the term *enolization* since the enolate invariably gives the enol very rapidly.

Which hydrogens in the following compounds are replaced by deuterium atoms when the compounds in solution in D_2O are treated with a base?

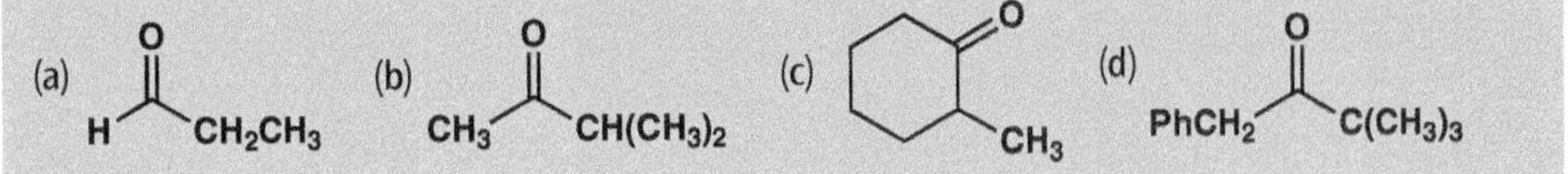

Exercise 17.4

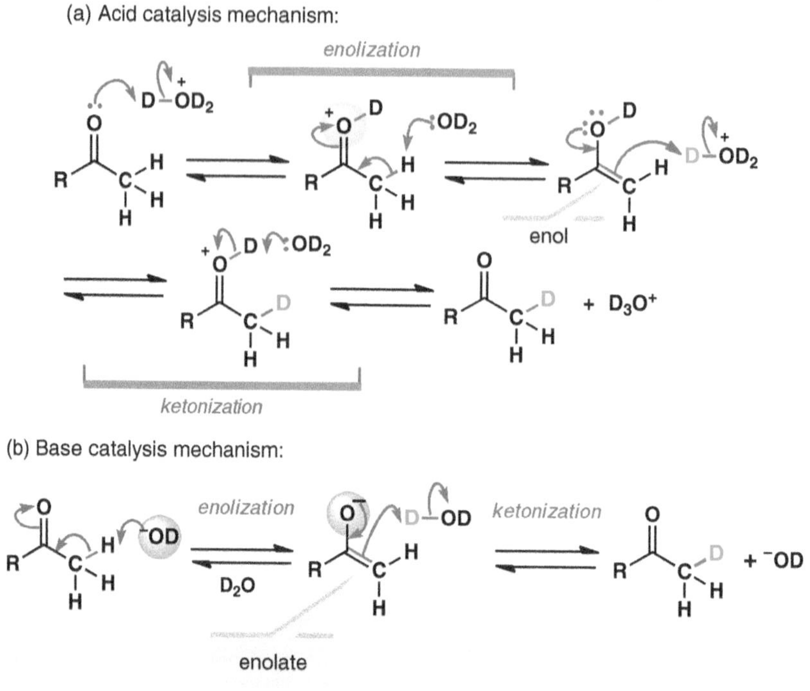

(a) Acid catalysis mechanism:

(b) Base catalysis mechanism:

The enolate in the base catalysis mechanism will give the (deuterated) enol as well as the deuterated ketone but this is rapidly reversible under the conditions of the reaction (see Scheme 17.3).

Scheme 17.4 Acid- and base-catalysed deuterium exchange reactions of a methyl ketone (subsequent steps lead ultimately to $RCOCD_3$).

17.3.2 Racemization

Note that the enol intermediate in Scheme 17.5 is actually a mixture of E and Z isomers. This will be true of many other reactions involving enol (and enolate) intermediates where we indicate a single diastereoisomer. However, when the E or Z configuration of the intermediate does not affect the outcome of the reaction, it does not matter which isomer we include.

An optically active carbonyl compound whose chiral centre is a hydrogen-bearing carbon at the α position gradually loses its optical activity when it is dissolved in an acidic or basic solution—it undergoes **racemization** and this occurs by enolization/ketonization. The chirality is lost when the sp³-hybridized α carbon of the carbonyl compound becomes a planar sp²-hybridized carbon of the enol (or enolate). On re-formation of the ketone, protonation of the sp² carbon occurs on either side with equal probability to give the racemic carbonyl compound, as illustrated in Scheme 17.5. As the rate of racemization is equal to the rate of the initial enolization towards C3, the rate constant for this enolization can be measured by monitoring the optical activity in a polarimeter.

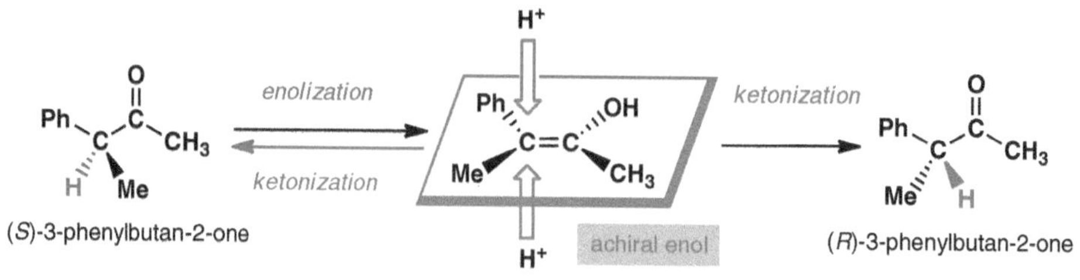

Scheme 17.5 Racemization of a chiral ketone by enolization/ketonization.

Exercise 17.5

When optically active 2-methylcyclohexanone is dissolved in acidic or basic alcoholic aqueous solution, the optical activity is gradually lost. Write curly arrow mechanisms for the racemization under (a) acidic and (b) basic conditions.

17.3.3 Isomerization

β,γ-Unsaturated carbonyl compounds readily isomerize to the α,β-unsaturated compounds when acid or base is added. These isomerizations occur via enol or enolate intermediates and Scheme 17.6 illustrates the hydroxide-catalysed mechanism. The enolate intermediate has a diene structure and the isomerization occurs upon protonation at the γ carbon of the original ketone in the ketonization process. If protonation occurs at the α carbon, the original β,γ-unsaturated carbonyl compound is obtained (reverse reaction). Isomerization is thermodynamically favourable because the C=C and C=O bonds of the α,β-unsaturated compound are conjugated.

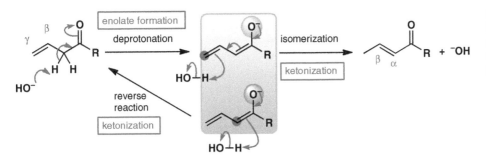

Scheme 17.6
Isomerization of a β,γ- to an α,β-unsaturated carbonyl compound catalysed by hydroxide.

Write a curly arrow mechanism for the isomerization of cyclohex-3-enone to cyclohex-2-enone catalysed by hydroxide.

Exercise 17.6

The *overall* π electron density in an enolate ion is higher at the O than at the terminal C, which is why electrophiles bond at the O when the molecular orbital interactions in the TS are weak; such reactions are sometimes said to be *charge-controlled*. However, the electron pair in the HOMO of the enolate is polarized towards the C as the molecular orbital picture below shows. Consequently, reactions occur at the C terminal of the enolate when overlap between the HOMO of the enolate and the LUMO of the electrophile is the dominant consideration.

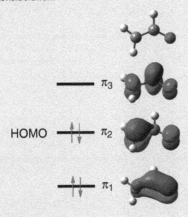

Molecular orbitals of the enolate, $H_2C=CHO^-$.

17.4 α-Halogenation

Enols and enolate ions react as highly nucleophilic alkenes. The ketonization of an enol or an enolate ion is initiated by protonation of the double bond and this may be regarded as an electrophilic addition to an alkene (see Chapter 15) with a proton from a Brønsted acid as the electrophile. In this section, we examine the involvement of enols and enolates in reactions of aldehydes and ketones which have α C–H groups with halogens (X_2, X=Cl, Br, or I) as electrophiles. In these reactions in acidic or basic solutions, the α H of the aldehyde or ketone is replaced by X (**α-halogenation**), as shown in eqn 17.5.

If a base B in eqn 17.5 is appreciably stronger than X^- (as will usually be the case), it will react with HX to give $BH^+ + X^-$; consequently, at least a stoichiometric amount of base will be required and the reaction is not catalytic.

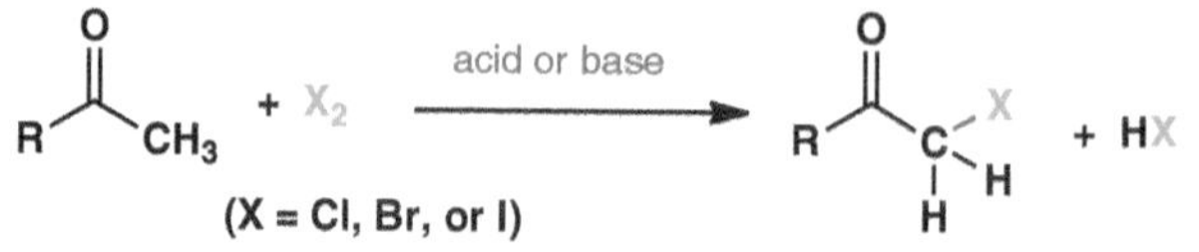

$$(X = Cl, Br, or I) \tag{17.5}$$

17.4.1 Acid-catalysed halogenation

A reaction in which a product functions as a catalyst is sometimes called an *autocatalytic* reaction. Unlike most reactions whose rates decrease continuously from their initial values (see Chapter 7), the rate of an unbuffered autocatalytic reaction first increases as the concentration of the catalytic product increases (but then decreases as the reaction approaches completion).

When an aldehyde or ketone is treated with bromine in ethanoic acid, α-bromination takes place as illustrated in Scheme 17.7 for a methyl ketone. In the initial acid-catalysed enolization, proton abstraction from the α C is generally the rate-determining step of the overall reaction. The enol then rapidly undergoes an electrophilic addition of Br^+ followed by loss of H^+ from the oxygen. So although the overall reaction is a substitution, it proceeds by an addition–elimination mechanism. Note that the HBr which is also produced is a stronger acid and a more effective catalyst than ethanoic acid for further reaction; the build-up of an appreciable concentration of HBr could be prevented by using sodium ethanoate, for example, as a buffer.

The electron-withdrawing α-halo substituent reduces the base strength of the carbonyl group, which causes further enolization and subsequent bromination to be slower than the first. Consequently, the monobromo ketone is easily isolated from the reaction under acidic conditions even if more than one equivalent of Br_2 is used.

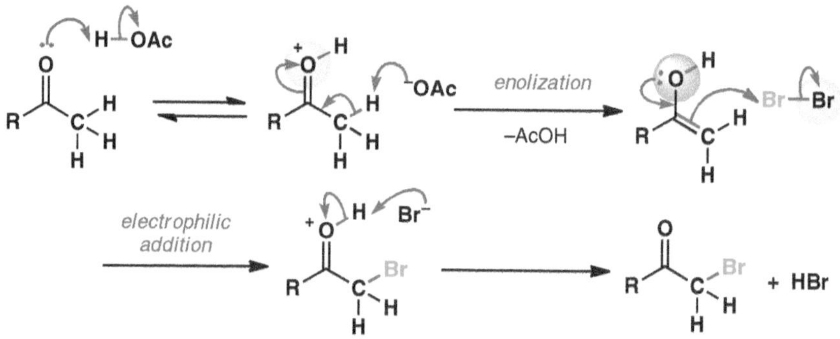

Scheme 17.7 A mechanism for the acid-catalysed α-bromination of a methyl ketone.

Exercise 17.7

Explain why the rate constant for the α-halogenation of a carbonyl compound catalysed by an acid or a base is usually independent of the identity of the halogen and its concentration.

Exercise 17.8

Propose reactions for the preparation of the following starting from cyclohexanone.

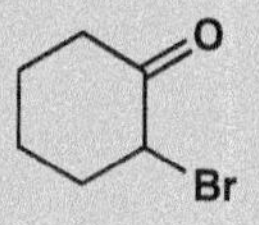

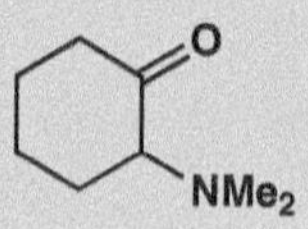

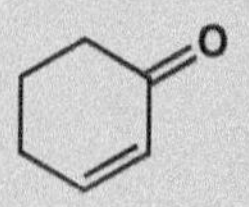

17.4.2 Base-induced halogenation and the haloform reaction

α-Halogenation of a ketone or aldehyde occurs more readily under basic conditions than acidic and, if it has more than one α hydrogen, all will be substituted (one after the other) in the same reaction if sufficient halogen is used. As shown in Scheme 17.8 for ethanal or a methyl ketone under aqueous alkaline conditions, the initially formed enolate ion reacts so quickly with halogen that reverse protonation on C is prevented (except when $[X_2]$ is very low).

A bimolecular reaction, either in the gas phase or in solution, which occurs every time reactant molecules encounter each other is known as a *diffusion* (or *encounter*) *controlled reaction*. The calculated value for a second-order, diffusion controlled reaction in aqueous solution is about 10^{10} dm^3 mol^{-1} s^{-1} at 25 °C, which is similar to experimental values for reactions of enols and enolates with halogens.

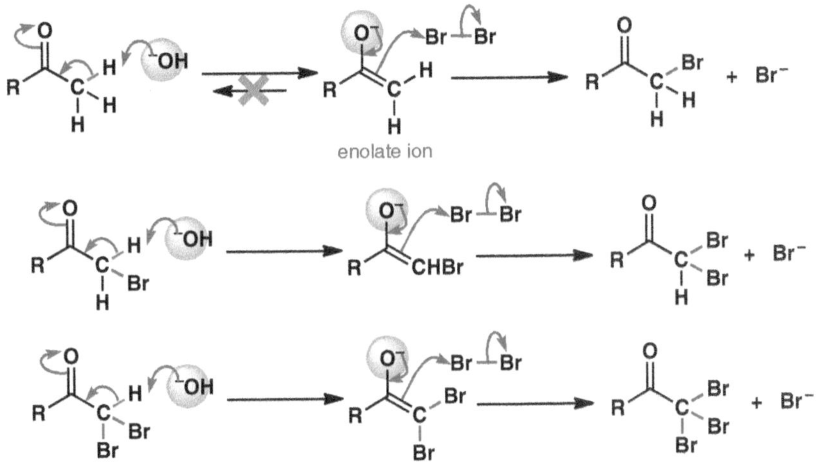

Scheme 17.8 Consecutive α-brominations of ethanal or a methyl ketone under aqueous alkaline conditions with an excess of Br$_2$.

If any of the enolates in Scheme 17.8 are protonated on O before they encounter a Br$_2$ molecule, the resulting enols will also react rapidly with Br$_2$ just as the one does in the acid-catalysed reaction shown in Scheme 17.7.)

The acidity of the two remaining α H atoms of the monobromo compound formed in reaction 17.6 is enhanced by the electron-withdrawing effect of the newly introduced α Br; consequently, the second enolate ion forms more readily than the first, which leads to the second α-bromination (eqn 17.7). The dibromo compound is then even more reactive towards further bromination to give the tribromo compound (eqn 17.8). This increasing ease of enolate formation and, therefore, of successive brominations under basic conditions is in accord with the increasing stabilization of the enolate ions by successive bromine substituents; it is in contrast to what is observed for acid-catalysed bromination (see above).

Even when only one equivalent of Br$_2$ is used, products of multiple bromination may form in base-induced halogenations of carbonyl compounds.

Stabilization of enolate ions by bromine:

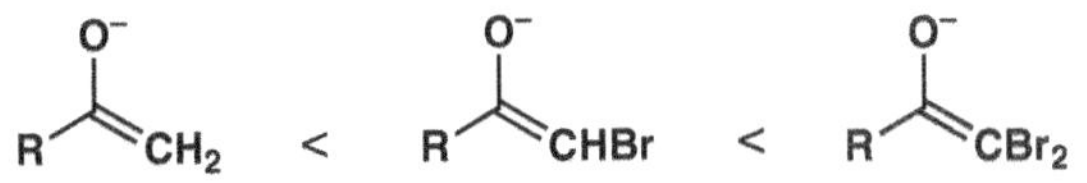

Explain why cyclohexanone plus two equivalents of Br$_2$ give 2,2-dibromocyclohexanone under basic conditions but the 2,6-dibromo-product under acidic conditions.

Exercise 17.9

Depending upon the reaction conditions, the tribromomethyl compound, RCOCBr$_3$, in Scheme 17.8 may react even further as the three electron-withdrawing bromine atoms activate the carbonyl group towards nucleophilic attack by hydroxide. More generally, the tetrahedral intermediate which is obtained in the first step of Scheme 17.9 is reminiscent of ones obtained in reactions of carboxylic acid derivatives (Chapter 9); the

The pK$_a$ value for CHCl$_3$ is about 24 while CHBr$_3$ and CHI$_3$ are even more acidic (pK$_a$ values less than about 15).

The final steps of the haloform reaction shown in Scheme 17.9 may be seen as a mechanistic link between nucleophilic addition reactions of aldehydes and ketones (Chapter 8), and nucleophilic substitution reactions of carboxylic acid derivatives (Chapter 9)— the intermediate trihalomethyl carbonyl compound with R = alkyl is a ketone which reacts as an acid derivative.

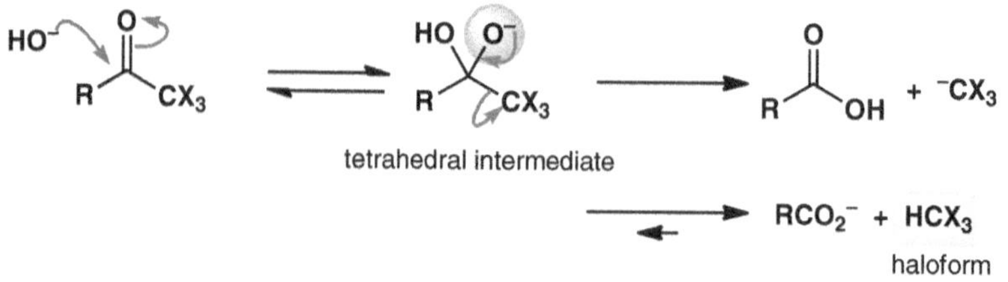

Compounds which contain the secondary alcohol function, $CH_3CH(OH)–$, also undergo the haloform reaction because the reaction conditions will oxidize >CH(OH) to >C=O.

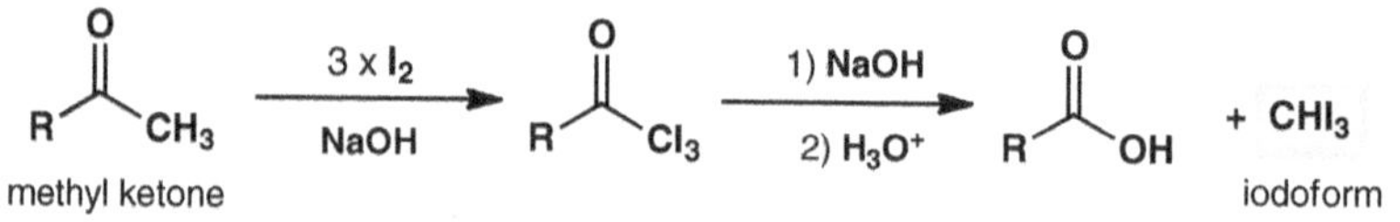

Scheme 17.9 The final steps of a haloform reaction.

trihalomethyl group is a potential nucleofuge because of the carbanion-stabilizing effect of the three electron-withdrawing halogen substituents.

The overall process culminating in Scheme 17.9 is characteristic of methyl ketones and ethanal, and is called the **haloform reaction** since the common names of the trihalomethanes $CHCl_3$, $CHBr_3$, and CHI_3 are chloroform, bromoform, and iodoform. The iodoform reaction shown in eqn 17.9 produces sparingly soluble fine yellow crystals of iodoform (which has a characteristic odour) and was once used as an analytical test for methyl ketones. It is still a useful method for the selective oxidative cleavage of $RCOCH_3$ as shown.

$$(17.9)$$

Contamination of public water supplies by chloroform. This is of some concern because chloroform is carcinogenic. It is produced from organic impurities by the haloform reaction when the water is chlorinated to destroy harmful bacteria. However, because chlorination of water has unquestionably had such a hugely beneficial effect upon public health, and no alternative method of water purification (e.g. use of ozone) has been found to be as effective, it is likely to remain in use for the foreseeable future. Ozone is used in some countries but, even then, some chlorine is still added to prevent subsequent bacterial contamination.

Exercise 17.10

Give a stepwise reaction sequence showing all necessary reagents for the following transformation.

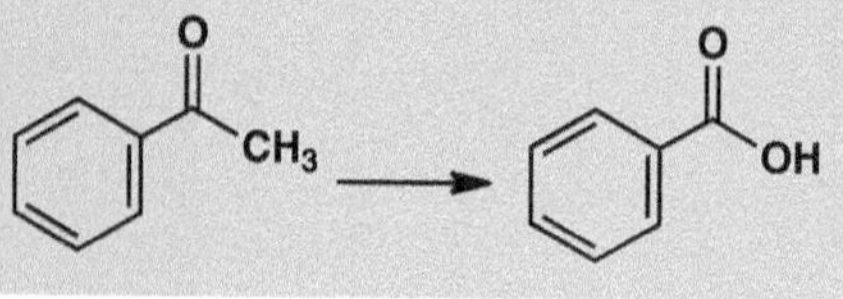

17.5 The Aldol Reaction

17.5.1 Base-catalysed dimerization of simple aldehydes and ketones

We have already seen above that enolate ions react as nucleophiles with halogens, and as bases. Enolates can also add to carbonyl compounds in reactions similar to those covered in Chapter 8. For example, when a small amount of NaOH is added to aqueous ethanal, a low equilibrium concentration of enolate ions form which add to molecules of the remaining aldehyde; this leads to the formation of a β-hydroxy aldehyde, the dimer of ethanal (Scheme 17.10). The product is an aldol (ald+ol) and the reaction is known as the aldol addition or **aldol reaction**.

Scheme 17.10 showing the aldol reaction of ethanal catalysed by hydroxide, with structures:

ethanal (acetaldehyde) → (NaOH) → enolate → ... → 3-hydroxybutanal an aldol + HO⁻

Scheme 17.10 Aldol reaction of ethanal catalysed by hydroxide.

Originally, *aldol* was the trivial name of the dimer of ethanal, but nowadays it is used mainly as the name of the reaction exemplified in Scheme 17.10 and as the common generic name for products of all aldol reactions.

Note that the nucleophile which adds to the electrophilic aldehyde (or ketone) in an aldol reaction is the deprotonated form of another aldehyde (or ketone) molecule. In this reaction, *the new C–C bond is formed between the α carbon of the nucleophilic aldehyde and the electrophilic carbonyl carbon of the other*. This is a very useful reaction in organic synthesis for making a larger molecule with versatile functional groups from two smaller molecules by formation of a just a single new C–C bond.

Hydroxide is consumed in the first step of Scheme 17.10 and regenerated in the final step, so the reaction is catalysed by HO⁻. This reaction should be contrasted with the α-halogenation of a ketone or aldehyde under basic conditions when base is consumed.

Example 17.3

Give the stepwise reaction sequence for the aldol reaction of propanal, and show the structure of the product.

Solution

The enolate formed by deprotonation at the α position of one propanal molecule adds to the carbonyl of another. Note the position of the new C–C bond.

propanal → (NaOH) → enolate → aldehyde → (from enolate / from aldehyde) → (H₂O) → 2-methyl-3-hydroxypentanal

---

Panel 17.1 Borodin: a composer and a chemist

Borodin was born in St Petersburg as the illegitimate son of a Georgian nobleman and a Russian mother. He was a professor of chemistry and an accomplished amateur cellist, but is best known as a member of a group of Russian Romantic composers known as The Five; their aim was to produce a specifically Russian style of music. Borodin's most popular works are his opera 'Prince Igor' and two string quartets. As a chemist, he discovered the aldol reaction in 1872 independently of C.A. Wurtz (1817–1884) who is usually credited with its discovery. He also discovered a method of preparing haloalkanes which is now usually known as the Hunsdiecker reaction ($RCO_2Ag + Br_2 \rightarrow RBr + CO_2 + AgBr$) as early as in 1861 before Hunsdiecker reported the reaction in 1941.

Alexander P. Borodin (1833–1887)

Exercise 17.11

Give the structures of the aldols formed from (a) butanal and (b) phenylethanal following treatment with aqueous NaOH solution.

The reverse of an aldol addition is sometimes called a *retro-aldol reaction*.

*Technical details of how this can be achieved are beyond the scope of this book.

As indicated in the above reaction schemes, the aldol reaction is reversible just like many other nucleophilic addition reactions of carbonyl compounds (Chapter 8). As expected (see Section 8.3), formation of the dimer at equilibrium is more favourable for aldehydes than for ketones. For example, the proportion of dimer from propanone at equilibrium is very modest (eqn 17.10) but a high yield of dimer can be obtained if it is removed from the reaction as it is formed, i.e. when equilibration is prevented.*

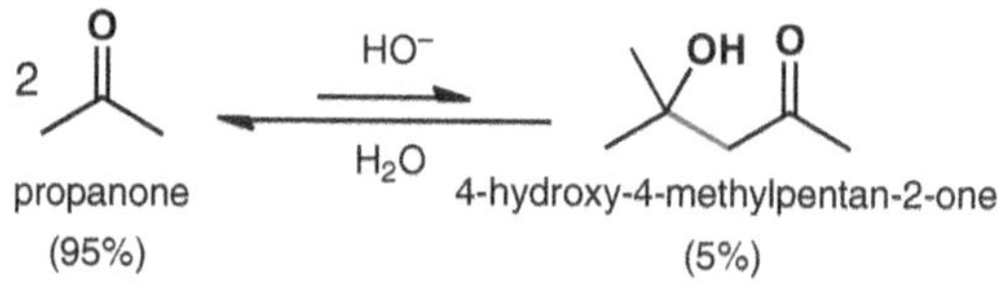

(17.10)

propanone (95%)   4-hydroxy-4-methylpentan-2-one (5%)

Exercise 17.12

Give the mechanism of the hydroxide-catalysed retro-aldol reaction of the aldol dimer of propanone.

Panel 17.2 A biological aldol reaction

An aldol is a β-hydroxy aldehyde or ketone. The ring-opened form of a sugar has a carbonyl and hydroxy groups (one of which is invariably β to the carbonyl) and can be regarded as the product of an aldol reaction which, in principle, is reversible. In fact, a reverse aldol (retro-aldol) reaction is one step in the metabolism of glucose in our body; the reaction is catalysed by an enzyme called aldolase. Glucose can easily be transformed to fructose via enolization (see Problem 17.12), and fructose is reversibly broken down into two three-carbon sugars, glyceraldehyde and dihydroxyacetone (as their phosphate derivatives) in our body.

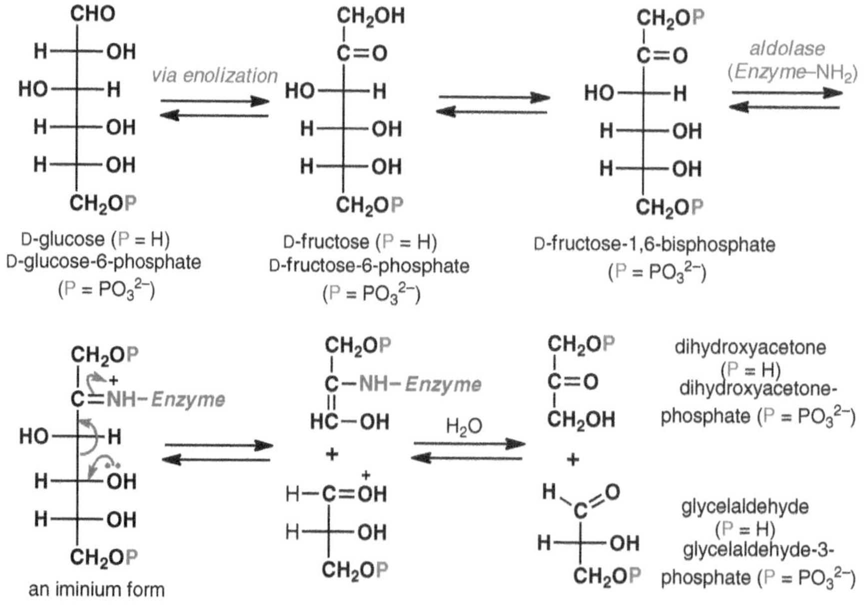

Note that this biological aldol-type reaction actually occurs with an iminium intermediate formed from the substrate and the enzyme, and proceeds under mild physiological conditions; furthermore, the overall scheme is reversible and leads to carbohydrate biosynthesis or degradation according to the requirements of the cell.

17.5.2 Dehydration of aldols

A low concentration of hydroxide is usually sufficient to catalyse the aldol addition. If high concentrations are used and the reaction is heated, the aldol, a β-hydroxy aldehyde, undergoes a further hydroxide-catalysed reaction—dehydration. The product is an α,β-*unsaturated carbonyl compound*, i.e. the C=C bond is conjugated with the carbonyl, e.g. eqn 17.11.

Formation of the stabilized conjugated unsaturated carbonyl product prevents the reverse of the aldol reaction in eqn 17.11.

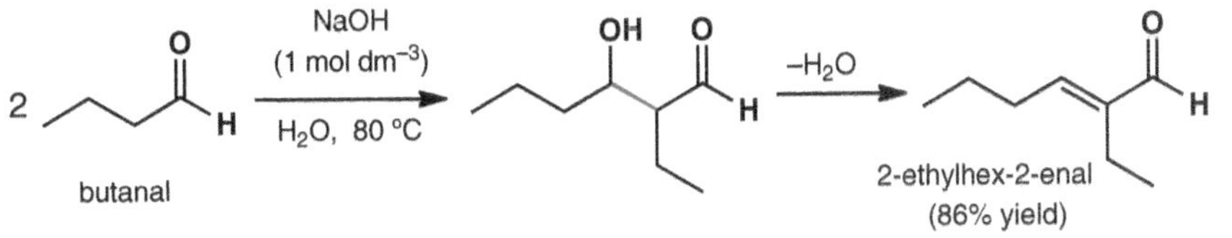

Although dehydration of a simple alcohol to give an alkene under basic conditions is not a thermodynamically favourable process, the dehydration of a β-hydroxy carbonyl compound becomes favourable because of the extra stability of the product, an α,β-unsaturated carbonyl compound (Scheme 17.11). Nevertheless, a favourable reaction still requires a mechanism, and the poor leaving group ability of HO⁻ prevents an E2 mechanism. The enhanced acidity of the α C–H, however, facilitates the E1cB mechanism of elimination (see Section 13.3) via the resonance stabilized intermediate, enolate ion. The combination of the aldol addition and subsequent dehydration is sometimes called the **aldol condensation**.

Reminder (see Sub-section 8.6.1): a *condensation* is a reaction in which two molecules combine with the accompanying elimination of a small molecule such as water or an alcohol.

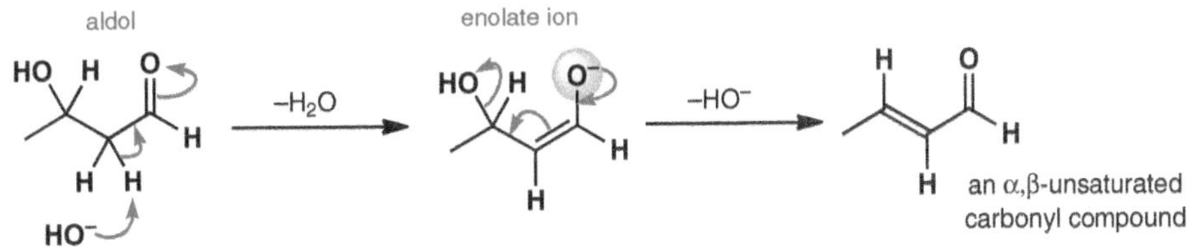

Scheme 17.11 Hydroxide-catalysed dehydration of an aldol: an example of the E1cB mechanism.

The elimination reaction of a β-hydroxy aldehyde occurs more readily under acidic conditions. The acid-catalysed dehydration proceeds by protonation of the HO group followed by an E1 or a solvent-induced E2 elimination mechanism (Sections 13.1 and 13.2), as shown in Scheme 17.12. In fact, the aldol reaction itself can also be catalysed by H_3O^+ and dehydration of the aldol product then usually follows on directly.

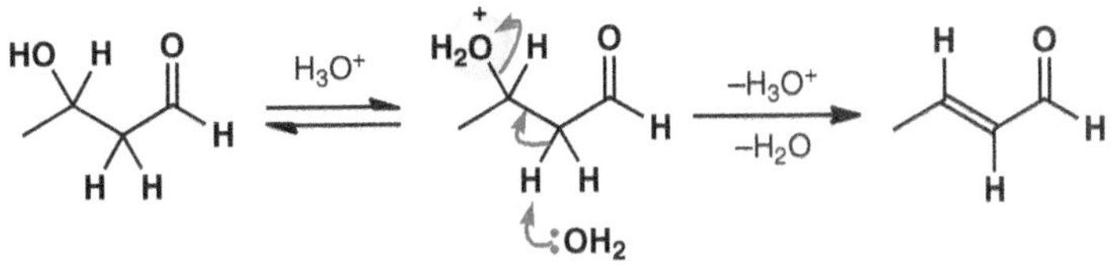

Scheme 17.12 Dehydration of an aldol catalysed by oxonium ion.

Exercise 17.13

Draw structures of the products of aldol condensations of (a) butanal, (b) phenylethanal, and (c) cyclohexanone.

Example 17.4

Give the mechanism of the aldol condensation of propanone catalysed by oxonium ion.

Solution

The reaction begins with the acid-catalysed enolization of propanone. The enol is less nucleo-philic than the enolate but can add to a carbonyl group activated by protonation. The aldol product undergoes dehydration by a mechanism similar to that shown in Scheme 17.12, and the departure of the nucleofuge and proton loss will be stepwise (the E1 mechanism) in this case. The product is 4-methylpent-3-en-2-one.

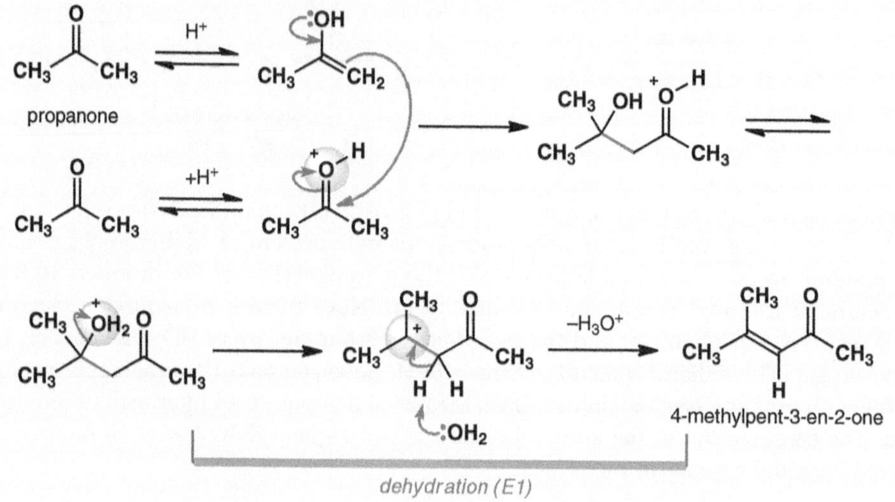

dehydration (E1)

Note that in the acid-catalysed aldol dimerization, the nucleophile is a neutral enol and the electrophile is the protonated carbonyl compound, i.e. electrophilicity of the carbonyl compound is enhanced. In a base-catalysed aldol reaction (e.g. Scheme 17.10), it is the nucleophilicity of the carbonyl compound which is enhanced by deprotonation, i.e. conversion to its enolate. We also encountered examples of this relationship between acid and base catalysis in other reactions of carbonyl compounds in Chapter 8.

17.5.3 Intramolecular aldol condensations

Although simple ketones do not give aldol dimers in high yields, 1,4- and 1,5-diketones (and sometimes 1,6-diketones) undergo intramolecular aldol reactions under basic con-ditions followed by dehydration to give satisfactory yields of five- and six-membered (and sometimes seven-membered) cyclic conjugated enones, e.g. eqn 17.12 with hydrox-ide as the base.

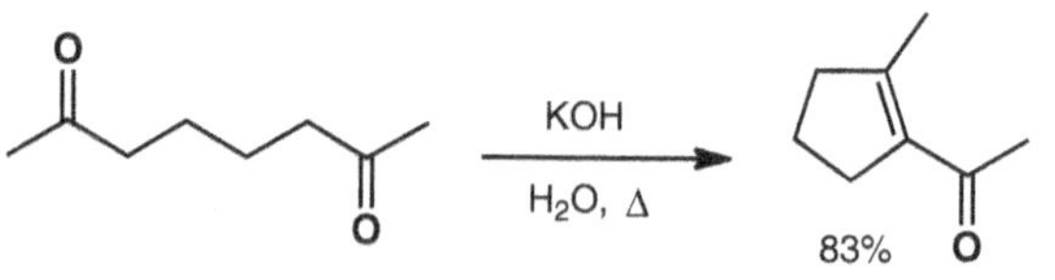

(17.12)

This type of cyclization also occurs, of course, with di-aldehydes and keto-aldehydes.

Exercise 17.14

Give a mechanism for reaction 17.12.

17.5.4 Crossed aldol reactions

An aldol reaction between two different carbonyl compounds is called a **crossed aldol reaction**. If both reactants have α hydrogens, a mixture of four products is possible

under normal basic conditions, and such aldol reactions are of little synthetic importance. For example, when a mixture of ethanal and propanal react in aqueous NaOH, the
product mixture contains two crossed aldol products as well as the two dimeric aldols
(Scheme 17.13).

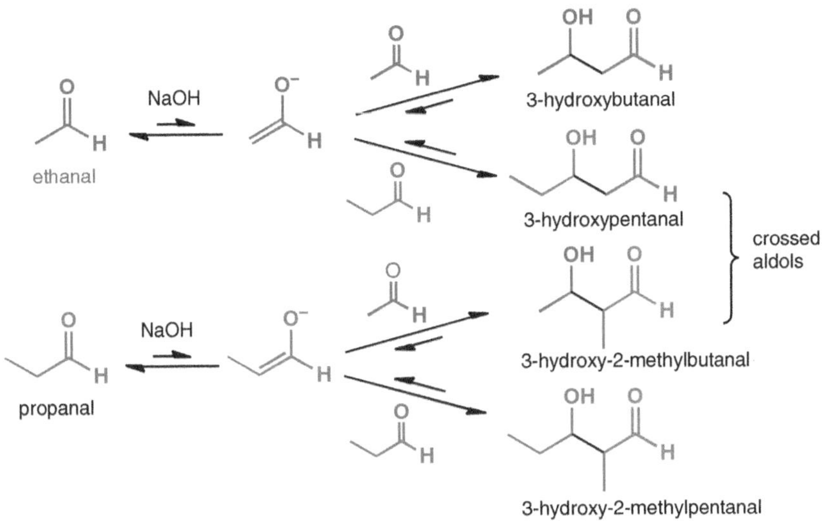

Scheme 17.13 Products of the aldol reactions of a mixture of ethanal and propanal.

Stereochemical issues
of aldol reactions are
important in some
syntheses although we
shall not consider this
aspect of the reactions in
this chapter. As mentioned
earlier, enolates may
be mixtures of *cis–trans*
isomers, and new chiral
centres may be generated
in aldol products (two
in the dimerization of
propanal in Example 17.3).

> **Exercise 17.15**
>
> Identify chirality centres in each of the four aldol products in Scheme 17.13, and specify the
> number of possible stereoisomers of each product.

However, if one of the reactants does not have an α hydrogen (and consequently cannot give an enolate and undergo a normal aldol dimerization), crossed aldol reactions
to give a single product can be of practical value, and an example is given in Scheme
17.14. When aqueous sodium hydroxide is added to a stirred mixture of propanone
and benzaldehyde, the enolate of propanone is generated which adds preferentially
to the carbonyl of the aldehyde (this is more reactive than the carbonyl of undeprotonated propanone). A small proportion of the possible dimer of the ketone is not problematical, however, since it is unstable and readily undergoes a retro-aldol reaction
(see Sub-section 17.5.1), which is also shown in Scheme 17.14. The main aldol from
benzaldehyde then undergoes dehydration in the final step to give the condensation
product which is stabilized by the double bond conjugatively linking the carbonyl and
the phenyl group.

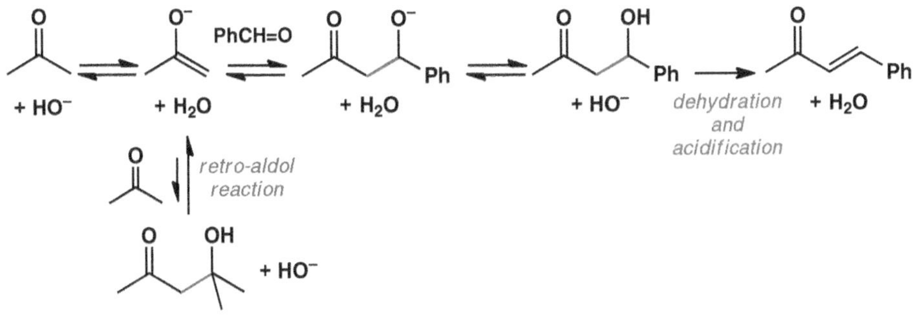

Scheme 17.14 A crossed aldol condensation of propanone with benzaldehyde.

The terminal CH_3 of the
product will be more
acidic than propanone and
could undergo a further
base-induced condensation
with benzaldehyde before
the acidification to give
$(PhCH=CH)_2C=O$. To
minimize this possibility,
an appreciable excess of
propanone is normally
used.

The following are other aldehydes without α hydrogens which are useful for crossed aldol reactions:

methanal
(formaldehyde)

2,2-dimethylpropanal
(pivalaldehyde)

furan-2-carbaldehyde
(furfural)

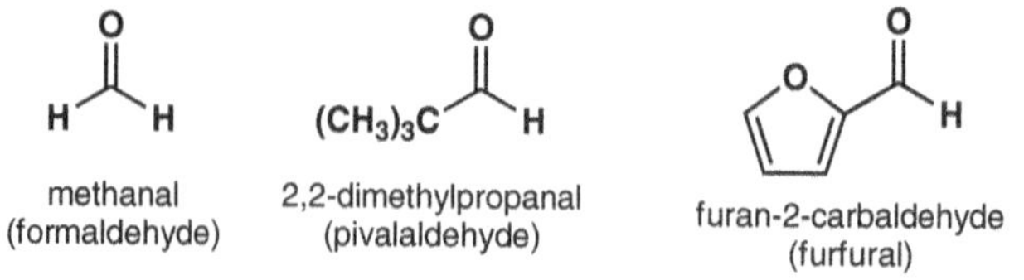

As we shall see in Section 17.9, lithium enolates (generated separately using a strongly basic lithium amide in an aprotic solvent) are used in a more effective general method for crossed aldol reactions.

Example 17.5

An iminium group ($\overset{+}{C}=\overset{/}{N}$), like a carbonyl, is electrophilic and an iminium ion formed from methanal and a secondary amine is especially reactive; it readily gives an amino ketone with even a weakly nucleophilic enol.

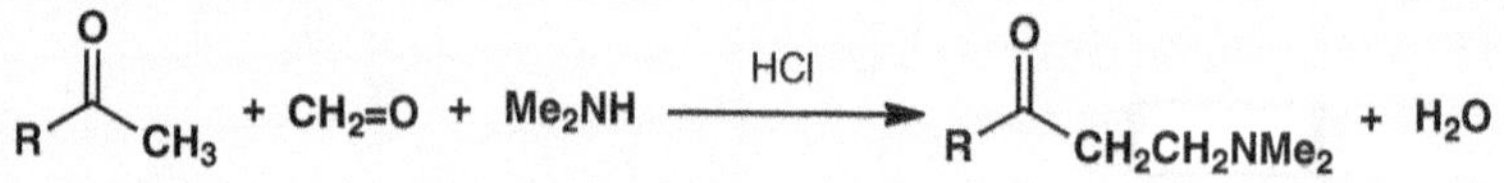

Propose a mechanism for this so-called Mannich reaction.

Solution

Formation of an iminium ion was covered in Section 8.6.

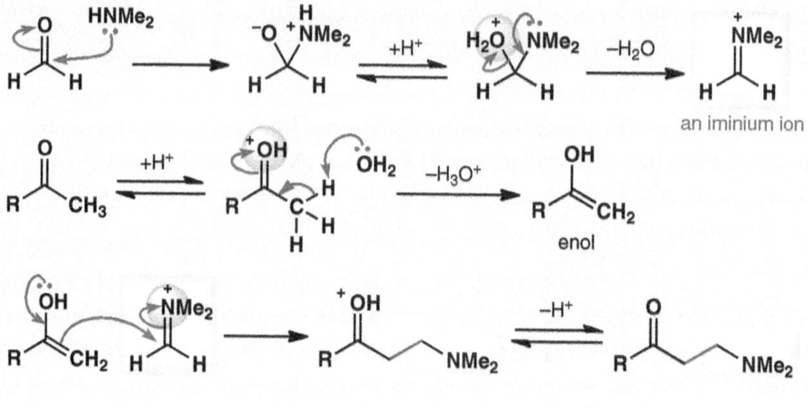

Crossed aldol reactions of carbonyl compounds with methanal lead to hydroxymethylation at the α positions of the carbonyl compounds. If methanal reacts with a carbonyl compound which has two or more α hydrogen atoms, the usual outcome is poly-hydroxymethylation due to the high reactivity of methanal. In contrast, the Mannich reaction very usefully gives solely the mono-aminomethyl product.

17.6 Claisen Condensation

The acidity of an α C–H of a simple ester such as ethyl ethanoate is enhanced by the adjacent alkoxycarbonyl ($pK_a \sim 25.6$), but not to the same extent as that of a simple aldehyde or ketone. However, in alkaline aqueous solutions, nucleophilic attack by HO⁻ at the carbonyl is faster than α hydrogen abstraction and leads on to ester hydrolysis, as described in Chapter 9. But if ethyl ethanoate is dissolved in ethanol containing sodium ethoxide, the product of nucleophilic attack by ethoxide at the carbonyl can only regenerate the original ester. Consequently, these conditions do not inhibit formation and reaction of the low equilibrium concentration of the carbanion formed by deprotonation of an α C–H. The product from ethyl ethanoate following aqueous acidic workup in this so-called **Claisen condensation** is a β-*keto ester*, ethyl 3-oxobutanoate (acetoacetate), as shown in eqn 17.13.

Ludwig Claisen (1851–1930).
(Courtesy of the Royal Society of Chemistry.)

As well as the ester condensation, the rearrangement of allyl phenyl ether (Chapter 21) is also named after this German chemist.

β-keto ester

$$2 \quad CH_3\text{-}C(=O)\text{-}OEt \quad \xrightarrow[\text{2) } H_3O^+]{\text{1) NaOEt/EtOH}} \quad CH_3\text{-}C(=O)\text{-}CH_2\text{-}C(=O)\text{-}OEt \quad + \quad EtOH \qquad (17.13)$$

ethyl ethanoate
(ethyl acetate)

ethyl 3-oxobutanoate
(ethyl acetoacetate)

17.6.1 Mechanism of the Claisen condensation

Ethyl ethanoate dissolved in ethanol containing sodium ethoxide first generates an equilibrium proportion of its conjugate base, the enolate anion. This then adds to another molecule of the ester to give a tetrahedral intermediate which, by expulsion of the ethoxide ion (Scheme 17.15), leads to ethyl 3-oxobutanoate.

Panel 17.3 A biological Claisen condensation

A key step for the biosynthesis of fatty acids (higher carboxylic acids) is a crossed Claisen condensation of thioesters. Malonyl coenzyme A is a thioester formed from acetyl coenzyme A which, in turn, is formed from coenzyme A (CoA-SH: see Section 14.6 for its structure) and ethanoic acid. Malonyl coenzyme A and acetyl coenzyme A then both undergo ester exchange with another thiol called *acyl carrier protein* (ACP-SH) to give acetyl-ACP and malonyl-ACP, which undergo repeated Claisen condensations to produce fatty acids. The enolate in this biochemical system is formed by decarboxylation without using a strong base, and all the reactions are catalysed by enzymes. Inevitably, all naturally occurring fatty acids are composed of even numbers of carbon atoms.

$$CH_3COH \quad + \quad HS\text{-}CoA \quad \xrightarrow{-H_2O} \quad CH_3CSCoA \quad \xrightarrow[carboxylase]{CO_2} \quad HOCCH_2CSCoA$$

ethanoic acid
(acetic acid)

acetyl-CoA

malonyl-CoA

Claisen-type condensation

$$CH_3\text{-}C\text{-}S\text{-}ACP \quad \xrightarrow{-CO_2} \quad CH_3\text{-}C\text{-}S\text{-}ACP \mid CH_2CS\text{-}ACP \quad \xrightarrow{-ACP\text{-}S^-} \quad CH_3CCH_2CS\text{-}ACP$$

$$^-O\text{-}C\text{-}CH_2CS\text{-}ACP$$

reduction | NADPH

$$CH_3(CH_2)_2CS\text{-}ACP \quad \xleftarrow[reduction]{NADPH} \quad CH_3CH=CHCS\text{-}ACP \quad \xleftarrow[dehydration]{-H_2O} \quad CH_3CHCH_2CS\text{-}ACP \mid OH$$

$$CH_3(CH_2)_{2n}COH$$

fatty acid

The Claisen condensation is a nucleophilic substitution reaction of the ester by the enolate derived from another ester molecule; see Chapter 9 for other reactions of esters with nucleophiles. Note that an aldehyde (or ketone) gives a *dimer* in the base-catalysed aldol reaction whereas an ester RCO$_2$R' in R'OH with a base gives a *condensation* product and, although there is an acid-catalysed aldol reaction, there is no acid-catalysed Claisen condensation.

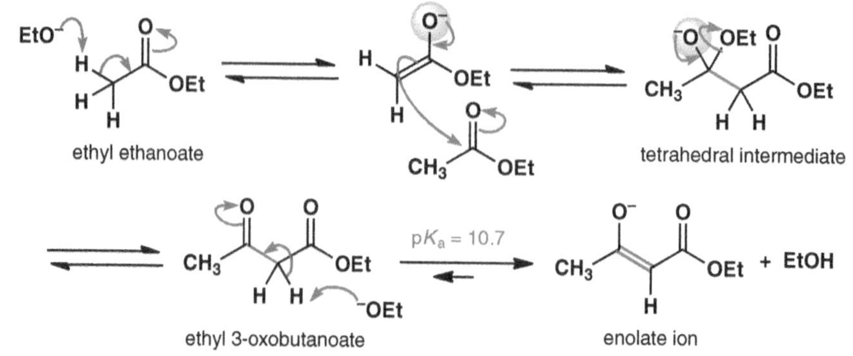

Scheme 17.15 Mechanism of the Claisen condensation of ethyl ethanoate.

The central CH$_2$ of this β-keto ester, however, is flanked by two carbonyl groups, so the compound is more acidic (pK_a = 10.7) than the original ester and the solvent, ethanol (pK_a ~ 16). Consequently, the ethyl 3-oxobutanoate is deprotonated by ethoxide as shown in the final step of Scheme 17.15, and it follows that one equivalent of EtO$^-$ is required, i.e. the reaction is not catalytic. At completion under these basic conditions, the β-keto ester exists predominantly as its conjugate base (the enolate anion, Scheme 17.15) and a subsequent acidic workup is required to liberate the β-keto ester product.

Exercise 17.16

Give a mechanism for the reaction in an acidic aqueous workup of the product mixture in Scheme 17.15 which yields the β-keto ester in equilibrium with small proportions of isomeric enol forms.

Example 17.6

Upon dissolution of ethyl 3-oxo-2,2-dimethylbutanoate, CH$_3$COC(CH$_3$)$_2$COOEt, in ethanol containing sodium ethoxide, C–C bond cleavage occurs to give molecules of two esters. (a) Write a reaction mechanism for this reverse Claisen condensation. (b) Does an analogous reaction occur with ethyl 3-oxobutanoate?

Solution

(a) The reverse Claisen condensation takes place because there is no deprotonation of the 2,2-dimethyl derivative to compete effectively with the nucleophilic attack at the keto carbonyl.

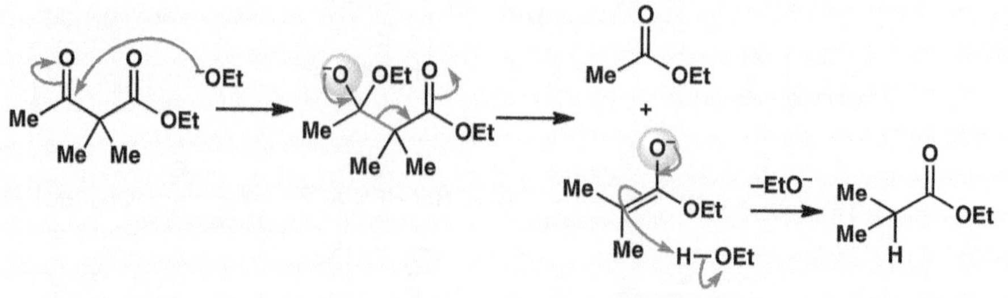

(b) With a CH$_2$ between its two carbonyl groups, ethyl 3-oxobutanoate undergoes ready deprotonation (the last step in Scheme 17.15), which effectively prevents the reverse Claisen condensation.

Show the structure of the product obtained when ethyl propanoate is treated with one equivalent of sodium ethoxide in ethanol followed by an acidic workup.

17.6.2 Intramolecular Claisen condensation

Under the reaction conditions described above, di-esters give cyclized products efficiently when five- or six-membered ring formation is possible. This intramolecular reaction is sometimes known as the Dieckmann condensation, e.g. eqn 17.14.

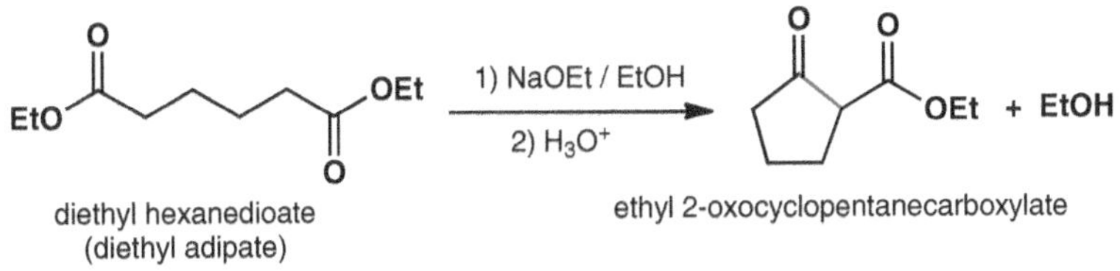

diethyl hexanedioate
(diethyl adipate)

ethyl 2-oxocyclopentanecarboxylate

$$(17.14)$$

Give a mechanism for reaction 17.14.

This reaction is useful for obtaining cycloalkanones because β-keto carboxylic acids easily undergo thermal **decarboxylation** (eqn 17.15).

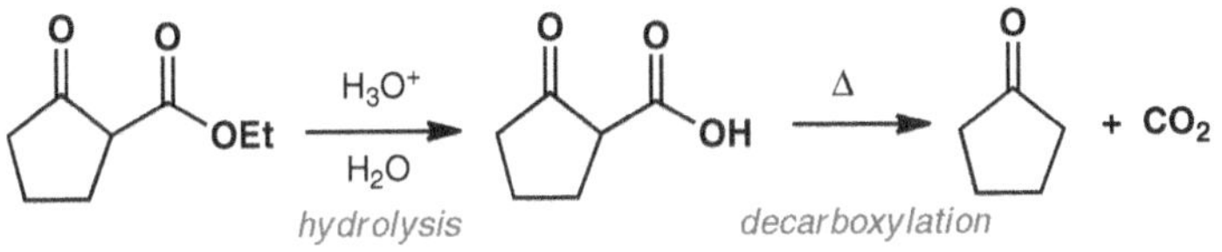

$$(17.15)$$

Intramolecular hydrogen transfer to give an enol intermediate via a six-membered cyclic transition structure facilitates the decarboxylation of β-keto carboxylic acids (Scheme 17.16); the enol then rapidly isomerizes to the more stable ketone.

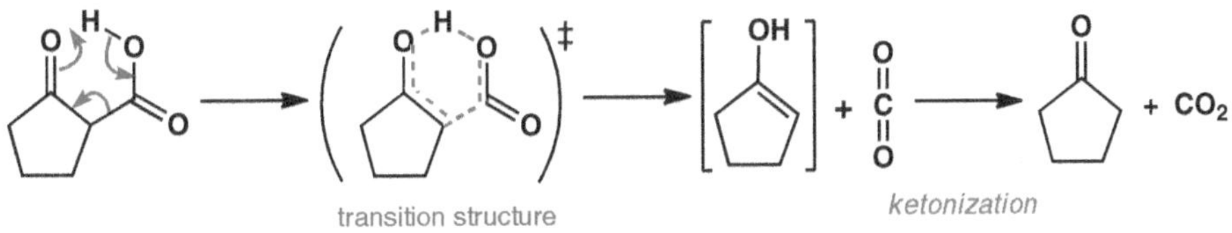

Scheme 17.16
Decarboxylation of a β-keto carboxylic acid via a six-membered cyclic transition structure and an enol intermediate.

17.6.3 Crossed Claisen condensations

Usually, a mixture of two different esters under the conditions for a Claisen condensation will give four products in reactions which are analogous to crossed aldol reactions. In order to obtain a high yield of a single condensation product, only one of the two esters selected should have an acidic α C–H in order that only a single enolate can be generated in the initial deprotonation step. Generally (in a procedure reminiscent of the one used for efficient crossed aldol reactions; see Sub-section 17.5.4), the ester with the acidic α C–H (e.g. ethyl propanoate) is added slowly to an excess of the component without an acidic α C–H (e.g. ethyl benzoate) in the basic solution. The rapidly formed enolate then reacts mainly with the much higher concentration of the undeprotonated ester to give essentially the single crossed condensation product, e.g. eqn 17.16.

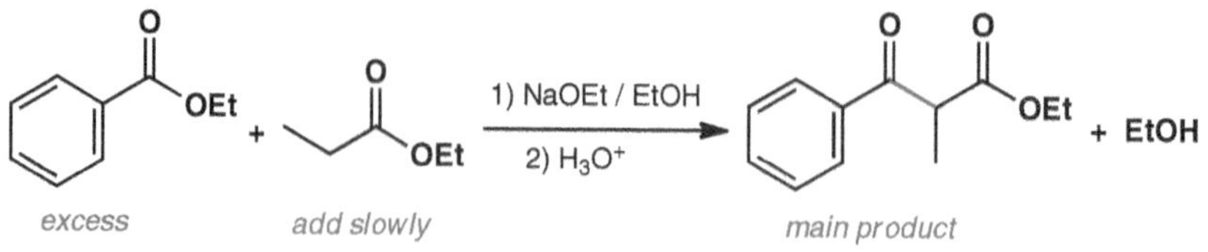

Esters lacking an acidic α hydrogen suitable for use in crossed Claisen condensations include the following:

ethyl methanoate (ethyl formate) ethyl benzoate diethyl carbonate diethyl ethanedioate (diethyl oxalate)

If the ester which does not have the acidic α hydrogen is also the one with the more electrophilic carbonyl group (e.g. ethyl formate and diethyl oxalate), so much the better.

If a solution of a ketone, for example, plus an ester is made basic, aldol dimerization of the ketone will initially take place most readily because the ketone is more acidic than the ester, and more electrophilic. However, the aldol reaction is fully reversible. If the initially predominant enolate (from the ketone) adds to an ester molecule, the tetrahedral intermediate gives a β-diketone which, under the basic conditions gives its conjugate base, and this will be the predominant species at equilibrium if one equivalent of base is used. Subsequent acidification, therefore, in this mixed aldol-ester condensation gives a β-diketone, e.g. eqn 17.17.

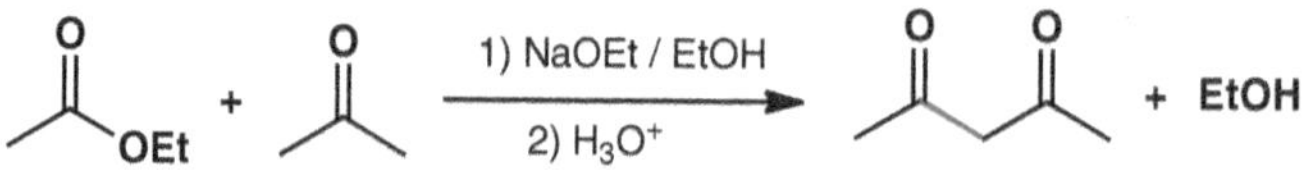

Exercise 17.19

Give a mechanism for reaction 17.17.

17.7 Enolate Ions of 1,3-Dicarbonyl Compounds

The pKₐ values for EtOH and H₂O are 15.9 and 15.7, respectively.

When considering the mechanism of the Claisen condensation, we saw that a hydrogen atom on a carbon between two carbonyl groups, i.e. a C2–H of a 1,3-dicarbonyl compound, is much more acidic than an α C–H of a simple aldehyde, ketone, or ester. Typically, **1,3-dicarbonyl compounds** have pK_a values below 14 so they are almost completely in their enolate forms in aqueous and alcoholic solutions containing at least one equivalent of a strong base.

1,3-Dicarbonyl and related compounds are sometimes called *active methylene compounds* and they include the following.

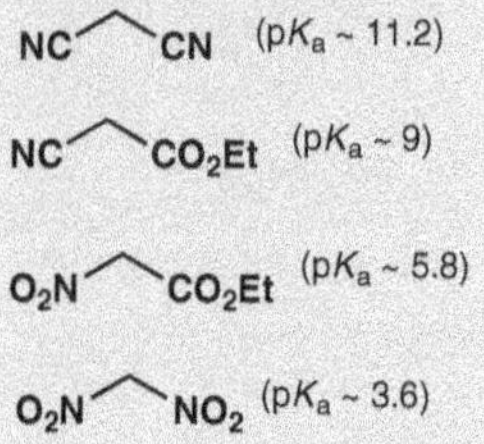

NC⌃CN $(pK_a \sim 11.2)$

NC⌃CO₂Et $(pK_a \sim 9)$

O₂N⌃CO₂Et $(pK_a \sim 5.8)$

O₂N⌃NO₂ $(pK_a \sim 3.6)$

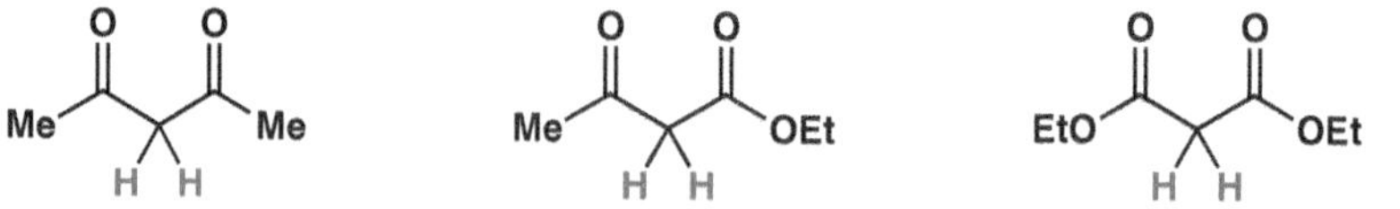

pentane-2,4-dione (acetylacetone) ethyl 3-oxobutanoate (ethyl acetoacetate) diethyl propanedioate (diethyl malonate)

pK_a 8.84   10.7   13.3

The relatively high acidity of a 1,3-dicarbonyl compound is adequately explained by the resonance stabilization of its enolate ion:

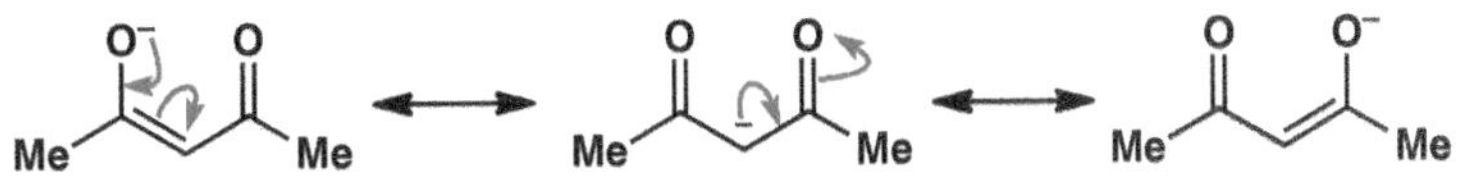

Not surprisingly, therefore, other substituents with an electron-accepting resonance capability have similar effects, and any compound with a CH_2 flanked by two such groups gives a relatively stable enolate ion.

Illustrate the stabilization of the conjugate bases of (a) ethyl cyanoethanoate and (b) ethyl nitro-ethanoate by drawing their resonance contributors.

Exercise 17.20

Explain why the following bicyclic diketone is less acidic than acyclic 1,3-diketones.

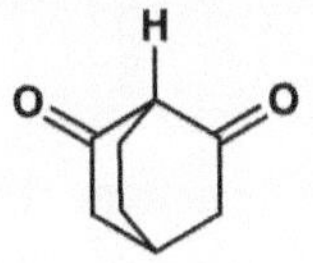

Example 17.7

Solution
The charge generated by proton loss from the bridgehead C between the two carbonyl groups cannot be delocalized into either of the two C=O groups without severe molecular distortion to allow proper alignment of the relevant orbitals (see Panel 13.1 for the instability of bridgehead bicyclic alkenes). Consequently, the most acidic H atoms of this compound are those of the two CH_2 groups adjacent to the C=O groups (as in a simple ketone).

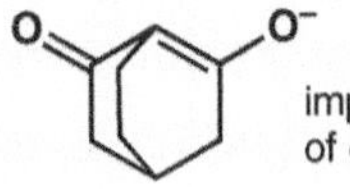

impossible structure of enolate ion

17.8 Alkylation of Enolate Ions

In Sections 17.5 and 17.6, we encountered enolate ions as nucleophilic intermediates in addition reactions with the electrophilic carbonyl compounds from which the enolates were initially derived. If an enolate can be generated which does not add to a carbonyl group (or adds only very slowly), reactions with other types of electrophiles become possible. For example, the enolate could participate as a carbon nucleophile in S_N2 reactions with electrophilic haloalkanes (RX); this results in alkylation at the α C of the original carbonyl compound (eqn 17.18).

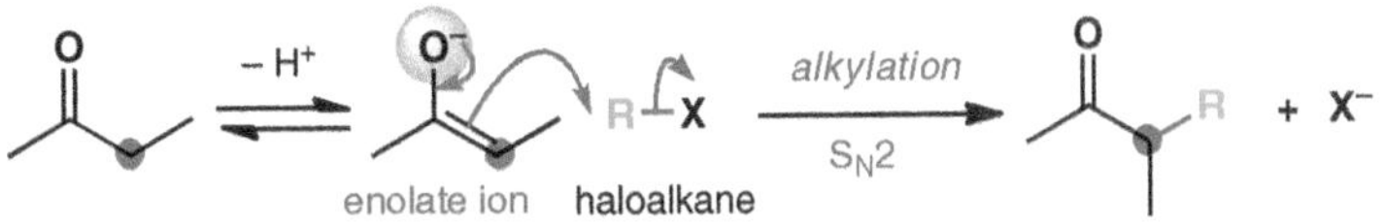

$$\text{(17.18)}$$

17.8.1 Alkylation of 1,3-dicarbonyl compounds

The resonance-stabilized enolate ion of a 1,3-dicarbonyl compound is not sufficiently reactive to add to a carbonyl group in an aldol- or Claisen-type reaction. It is sufficiently nucleophilic, however, to react with haloalkanes by the S_N2 mechanism to give alkylation at C2 of the 1,3-dicarbonyl compound (eqn 17.19 and Scheme 17.17). Similar alkylations are also possible with other comparably stabilized enolates.

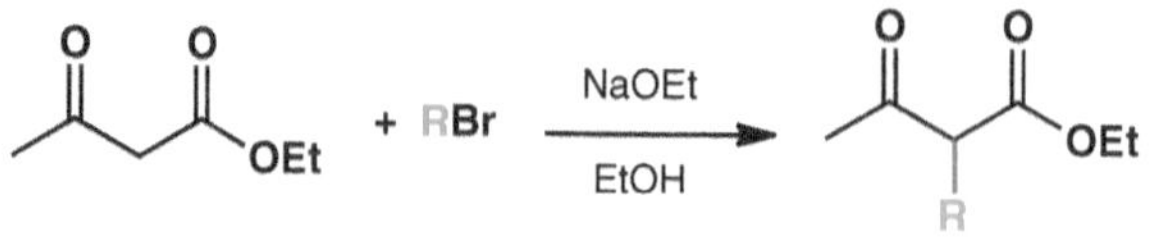

$$(17.19)$$

When there are two hydrogen atoms on C2 of the 1,3-dicarbonyl or related compound, two alkyl groups can be introduced (Scheme 17.17). However, because of the greater steric congestion in the second alkylation, it is considerably slower than the first. Consequently, it is straightforward to limit the extent of dialkylation by controlling the amount of haloalkane used. The relative reactivities of different haloalkanes in these alkylations of 1,3-dicarbonyl compounds are the same as in their other S_N2 reactions (see Sub-section 12.2.1).

Scheme 17.17
Sequential alkylation of a
1,3-dicarbonyl compound.

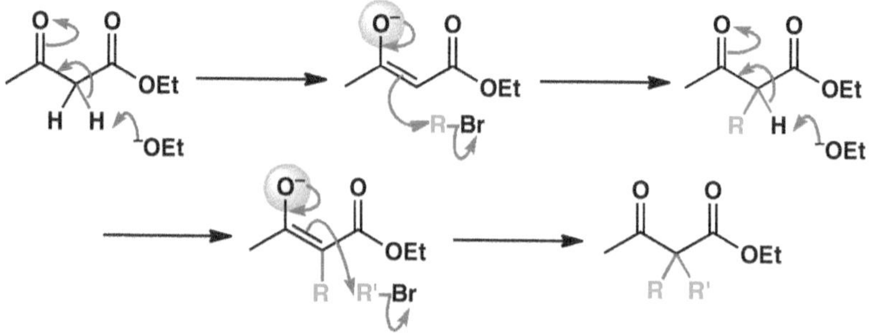

17.8.2 Synthesis of ketones and carboxylic acids via enolates of 1,3-dicarbonyl compounds

Ethyl 3-oxobutanoate (ethyl acetoacetate) and its 2-alkyl derivatives are usually easy to make and readily hydrolysed to the corresponding carboxylic acids which can then be decarboxylated, as we saw in Sub-section 17.6.2. This sequence leads to a generally useful strategy for the preparation of ketones which is sometimes known as the *acetoacetic ester synthesis* for ketones, as exemplified in Scheme 17.18.

Scheme 17.18 An
acetoacetic ester synthesis of
a ketone.

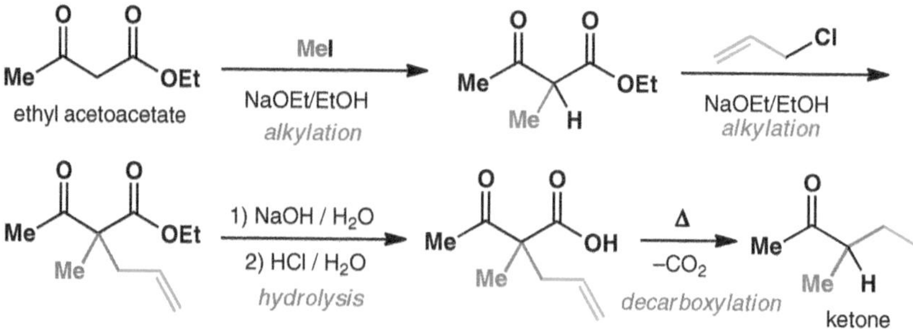

In closely related reactions, the enolate ion of the readily available diethyl propane-1,3-dioate (diethyl malonate) can be alkylated at C2, and hydrolysis followed by mono-decarboxylation of the resulting dicarboxylic acid gives a carboxylic acid in a sequence sometimes known as the *malonic ester synthesis* for carboxylic acids, as illustrated in Scheme 17.19.

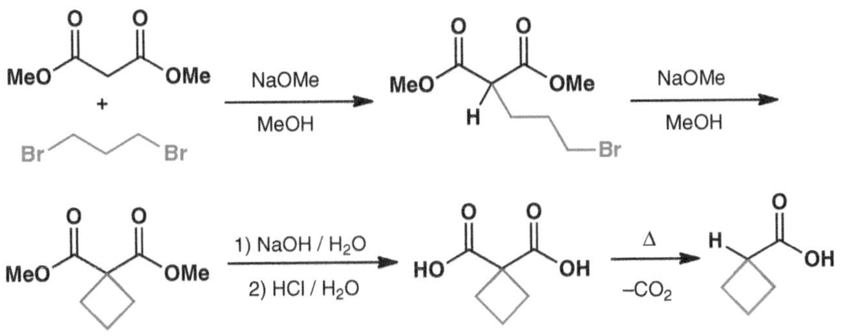

Scheme 17.19 A malonic ester synthesis of a carboxylic acid.

Propose an acetoacetic or malonic ester synthesis for each of the following.

(a) heptan-2-one   (b) 3-benzylhexan-2-one   (c) pentanoic acid   (d) pent-4-enoic acid

Exercise 17.21

17.9 Lithium Enolates

In normal aldol reactions of simple ketones, enolates are generated by the base as transient nucleophilic intermediates which are immediately trapped by undeprotonated ketone. Alternatively, *lithium enolates* of simple ketones can be generated quantitatively in an aprotic solvent at low temperatures by a strong non-nucleophilic base such as lithium diisopropylamide (LDA, $pK_{BH^+} \approx 35$), as shown in eqn 17.20. Under these conditions, the ketone does not trap the enolate as it is formed in a normal aldol reaction, and the sterically hindered LDA does not react as a nucleophile with the carbonyl group of the ketone.

Nucleophilic capture of the undeprotonated ketone by enolate or LDA are both appreciably slower than deprotonation of an α C–H by LDA, so complete formation of the lithium enolate of the ketone is possible.

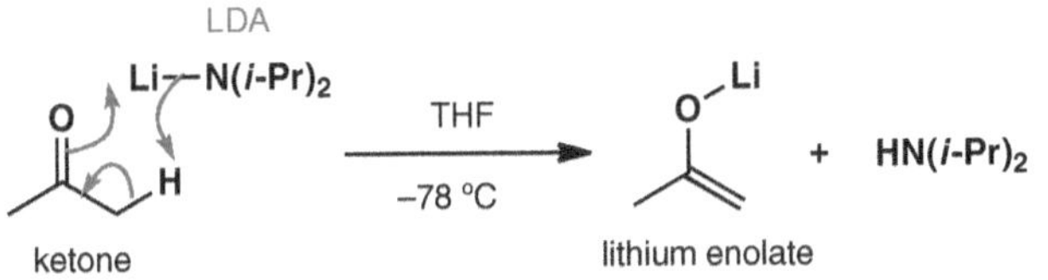

(17.20)

i-Pr$=$(CH$_3$)$_2$CH : isopropyl
LiN(i-Pr)$_2$: lithium diisopropylamide (LDA)

The reaction is completed by addition of any electrophile to the solution of the lithium enolate. Addition of a different carbonyl compound, for example, leads to a crossed aldol reaction. By this method, we purposely obtain a desired single crossed aldol product from a ketone with an α C–H and another carbonyl compound (e.g. eqn 17.21), so the reaction is sometimes called a *directed aldol reaction*.

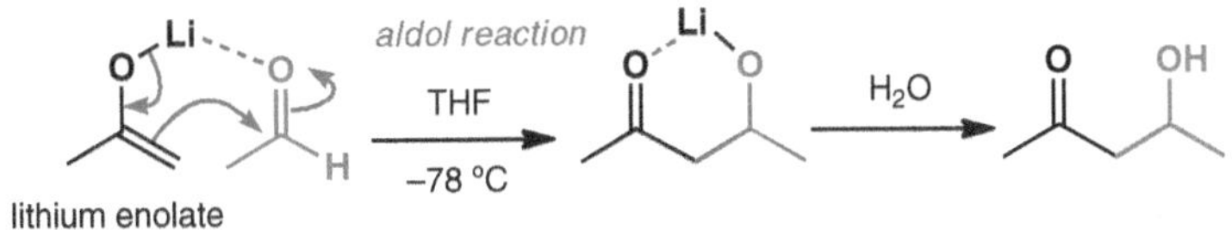

(17.21)

When a haloalkane is added to the solution of the lithium enolate, an S_N2 reaction leads to alkylation (eqn 17.22).

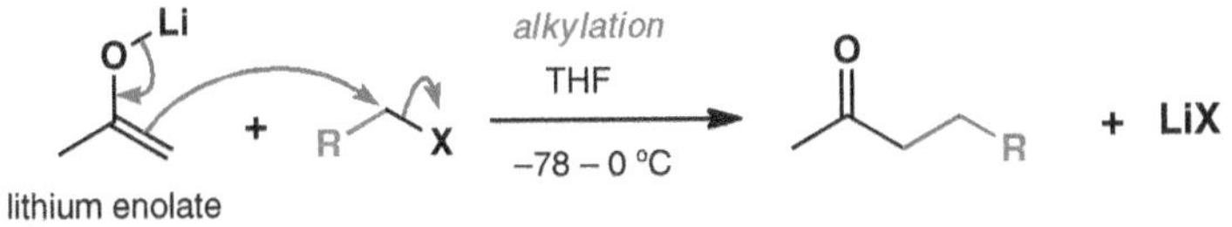

(17.22)

An ester can also give a lithium enolate which can be used in the same way, but an aldehyde is generally too electrophilic. Even at low temperatures, the lithium enolate of an aldehyde is trapped as soon as it is formed by unreacted aldehyde to give the dimer in a normal aldol reaction.

17.9.1 Kinetic and thermodynamic enolates of ketones

When a ketone is unsymmetrical in the sense that the hydrogens on the α and α' carbons are different, two isomeric enolates are possible, e.g. eqn 17.23.

$$R_2CHCCH_2R' \xrightarrow{\text{base}} R_2CHC{=}CHR' \quad + \quad R_2C{=}CCH_2R' \qquad (17.23)$$

kinetic enolate   thermodynamic enolate

Note that a ketone may give a stereoselective mixture of *cis* and *trans* diastereoisomers of the enolate which is formed regioselectively.

The ratio of the two enolates, i.e. the regioselectivity of the proton abstraction, depends on the base used and on the reaction conditions. The less substituted enolate is usually formed faster because the less hindered proton is more readily abstracted by the base. Consequently, this enolate (often called the **kinetic enolate**) is the major product under kinetically controlled conditions. In contrast, if the reaction with the base is allowed to come to equilibrium, i.e. the deprotonation is carried out under thermodynamically controlled conditions, the more stable (more highly substituted) enolate (sometimes called the **thermodynamic enolate**) is the major product. (See Sub-section 15.8.2 for a discussion of thermodynamic versus kinetic control.)

Kinetic enolates are formed when *a strong, sterically hindered base such as LDA is used in an aprotic solvent at lower temperatures for only a short time*; also, the reaction is usually carried out with slow addition of slightly less than one equivalent of the ketone to the basic solution—conditions which minimize the reverse reaction, e.g. eqn 17.24 in Scheme 17.20.

Scheme 17.20
Generation of isomeric enolates under kinetic and thermodynamic control.

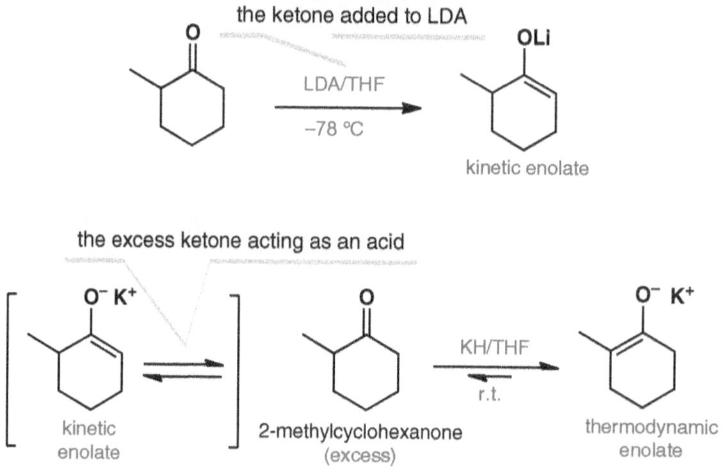

the ketone added to LDA

$$\text{LDA/THF}, -78\ ^\circ C \qquad (17.24)$$

kinetic enolate

the excess ketone acting as an acid

$$\text{KH/THF, r.t.} \qquad (17.25)$$

kinetic enolate   2-methylcyclohexanone (excess)   thermodynamic enolate

In contrast, thermodynamic enolates are generally formed *at higher temperatures for longer reaction times in the presence of a proton donor*. If the solvent is protic (*t*-BuOH with *t*-BuOK as the base is used for this purpose), it will allow the reverse reaction which is required for equilibration; on the other hand, if an aprotic solvent is used (e.g. THF with KH as the base), an excess of ketone acts as proton donor as in eqn 17.25 (Scheme 17.20).

Exercise 17.22

Starting from 2-methylcyclohexanone, propose reactions for the synthesis of (a) 2,2-dimethylcyclohexanone and (b) 2-allyl-6-methylcyclohexanone.

Exercise 17.22

Propose a method involving a directed aldol reaction with a lithium enolate for the preparation of each of the following crossed aldol products.

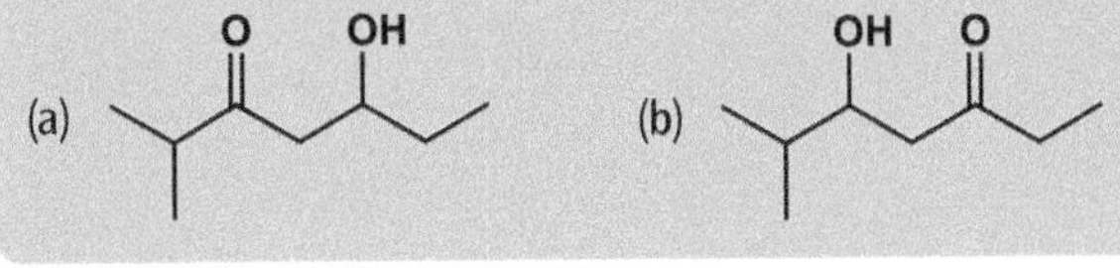

Exercise 17.23

(a)

(b)

17.10 Enolate Equivalents

Nucleophilic alkenes which are electronically equivalent to enolates (or enols) are called **enolate** (or enol) **equivalents**, and are derived from the corresponding carbonyl compounds. Their reactions are similar to those of enolates but under neutral conditions, and the products can be transformed back into carbonyl compounds by hydrolysis.

17.10.1 Enamines

Enamines are formed by the elimination of water in acid-catalysed reactions of aldehydes or ketones with secondary amines, as described in Sub-section 8.6.2; they are nucleophilic alkenes which participate in S_N2 reactions, but only with fairly reactive alkylating agents (Scheme 17.21). Typical electrophiles include allylic and benzylic halides, and α-halo ketones; the alkylated aldehyde or ketone products are isolated after hydrolysis of the intermediate iminium cations (Scheme 17.21).

For the formation of enamines and hydrolysis of iminium ions, see Section 8.6.

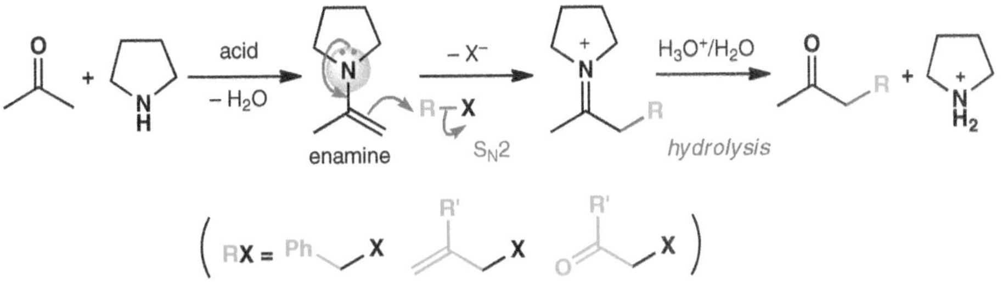

Scheme 17.21
Alkylation of a ketone via an enamine.

Simple alkylating agents such as primary alkyl halides tend to alkylate the enamine at the N atom, which reduces the yield of the required *C*-alkylation (eqn 17.26).

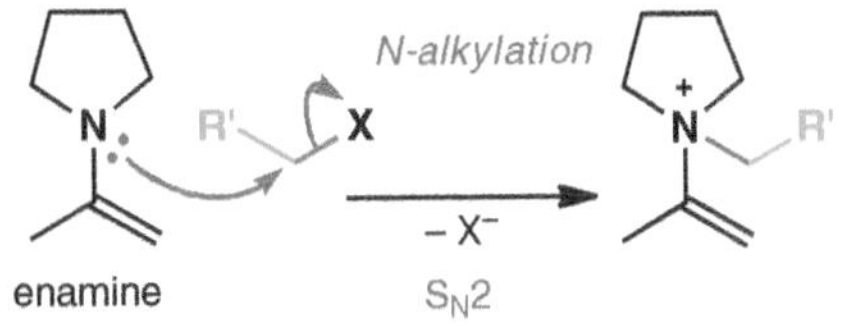

(17.26)

17.10.2 Enol silyl ethers

Enol silyl ethers are best regarded as enolates protected by a silyl group, and are prepared by treatment of a carbonyl compound with a silyl chloride and a tertiary amine. They are less nucleophilic than enamines, so are alkylated only by strongly electrophilic reagents. For example, the carbenium ion generated from a tertiary alkyl halide by a Lewis acid such as $TiCl_4$ or $SnCl_4$ will alkylate an enol silyl ether in a kind of S_N1 reaction (Scheme 17.22).

Scheme 17.22 Lewis acid-promoted alkylation of an enol silyl ether by a carbenium ion.

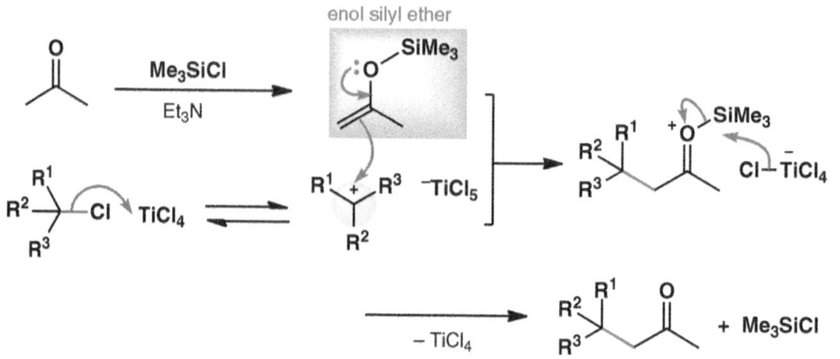

An enol silyl ether will also react with an aldehyde or ketone but only with a Lewis acid to enhance the electrophilicity of the carbonyl group; the product after an aqueous workup is a (crossed) aldol (Scheme 17.23). This method is an alternative to a directed aldol reaction.

Scheme 17.23 Lewis acid-promoted crossed aldol formation from an enol silyl ether and an aldehyde.

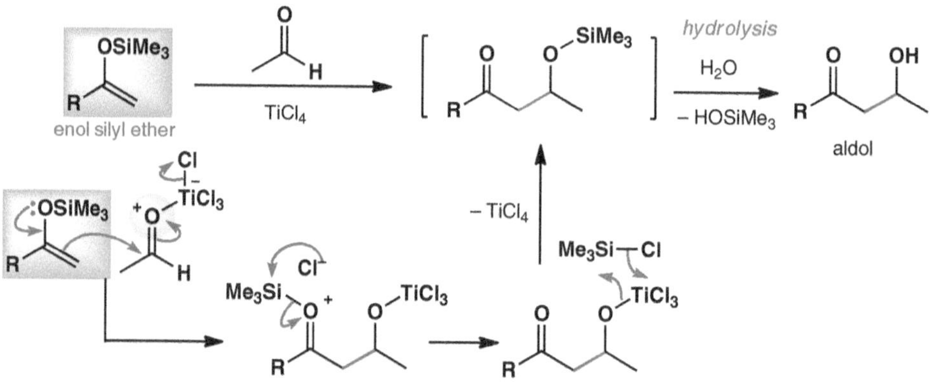

Summary

- The α hydrogens of carbonyl compounds are relatively acidic and readily deprotonated by strong bases to give **enolate ions** which are resonance-stabilized allyl anion analogues.

- An enol and its keto isomer differ only in the location of a hydrogen; this type of isomerism is called **tautomerism**, and the enol and keto forms (which are called **tautomers**) interconvert reversibly. These transformations, **enolization** and **ketonization**, are catalysed by both acids and bases.

- Enolate ions and enols act as nucleophiles and are intermediates in the α-**halogenation** of ketones and aldehydes. Reactions of enolates with carbon electrophiles result in C–C bond formation, and provide useful methods for organic synthesis. Such reactions include **aldol reactions** (with aldehydes and ketones), **Claisen condensations** (with esters), and **alkylations** (with haloalkanes).

- The groups $CH_3C(O)$- and $CH_3CH(OH)$- undergo oxidative cleavage from the rest of the molecule in the **haloform** reaction.

- **1,3-Dicarbonyl compounds** form especially stable enolates; examples are involved in C–C bond-forming reactions in the acetoacetic ester synthesis of ketones and the malonic ester synthesis of carboxylic acids.

Lithium enolates of ketones and esters can be generated quantitatively with strong bases such as LDA in aprotic solvents at low temperatures, and are used for α-alkylation of the carbonyl compounds and directed (crossed) aldol reactions.

Enamines and enol silyl ethers, which are **enolate (enol) equivalents**, participate in reactions similar to those of enolates but under mild non-basic conditions and only with more reactive electrophiles.

Problems

17.1 Draw the structure of the keto form corresponding to each of the following enols.

(a)

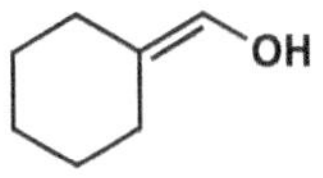

(b)

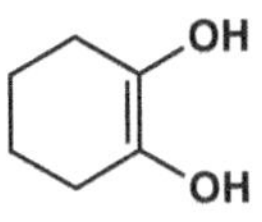

(c)

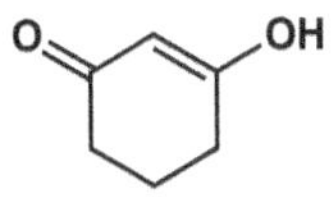

(d)

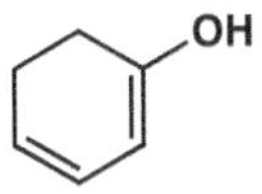

17.2 Draw structures of all possible enolate ions formed by treatment of each of the following ketones with a strong base, and indicate the main enolate obtained at equilibrium in each case.

(a)

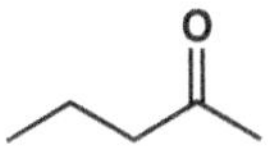

(b)

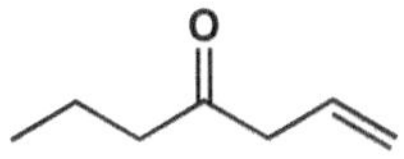

(c)

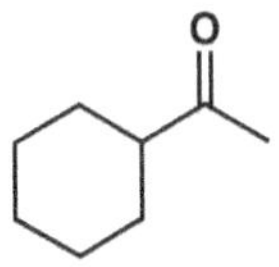

(d)

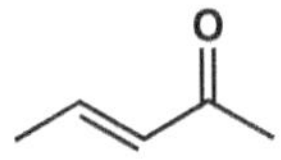

(e)

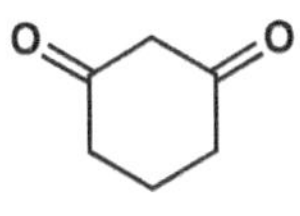

(f)

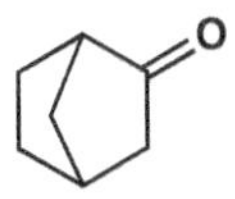

17.3 Explain why the equilibrium proportion of the enol tautomer is greater for phenylethanal than for ethanal (Table 17.1).

17.4 Give the structure of the main product in each of the following reactions.

(a) 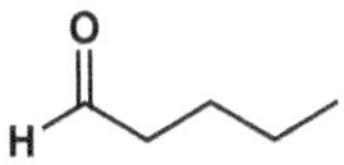 $\xrightarrow[\text{H}_2\text{O}]{\text{NaOH}}$

(b) 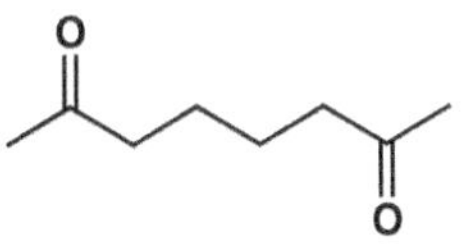 $\xrightarrow[\text{H}_2\text{O}]{\text{NaOH}}$

(c) PhCHO + PhCH$_2$CHO $\xrightarrow[\text{H}_2\text{O}]{\text{NaOH}}$

17.5 Give the structures of carbonyl compounds which give the following in aldol condensation reactions.

(a)

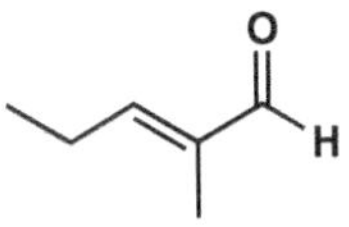

(b)

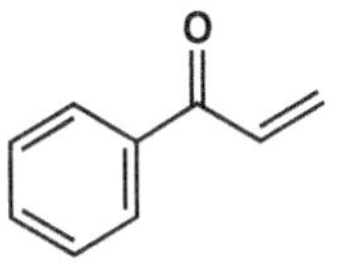

(c)

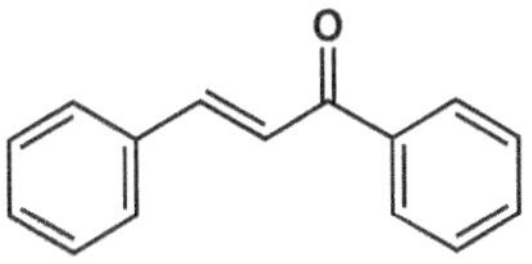

(d)

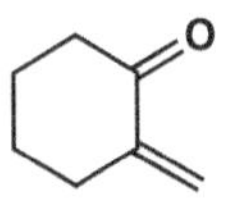

17.6 Propose mechanisms for ketonization of the enol of cyclohexanone by (a) acid catalysis, and (b) base catalysis.

17.7 Isomerization of cyclohex-3-enone to cyclohex-2-enone is catalysed not only by bases (see Exercise 17.6) but also by acids. Propose a mechanism for the isomerization catalysed by oxonium ion.

17.8 Acetophenone undergoes acid-catalysed deuteration in acidic D$_2$O-EtOD to give a product with the molecular formula C$_8$H$_5$D$_3$O. Identify the product and propose a mechanism for its formation.

17.9 When acetophenone is treated with concentrated aqueous sodium hydroxide and an excess of Br$_2$, and the reaction is worked up by an aqueous acidic quench, benzoic acid is obtained. Propose a mechanism for this oxidative cleavage.

17.10 What is the main product from the reaction of acetophenone with more than one equivalent of Br$_2$ in ethanoic acid? Explain the result with a reaction mechanism.

17.11 Propose reactions for the synthesis of each of the following from cyclohexanone.

(a)

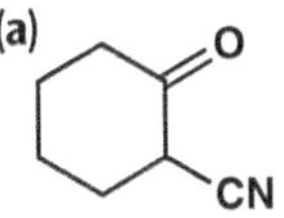

(b)

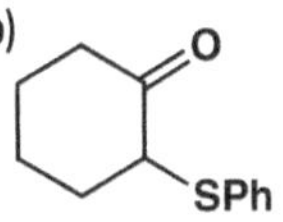

(c)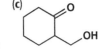

17.12 A carbohydrate with a ketone group (a ketose) can readily isomerize to one with an aldehyde group (an aldose). Propose a mechanism for the base-catalysed isomerization of D-fructose to D-glucose.

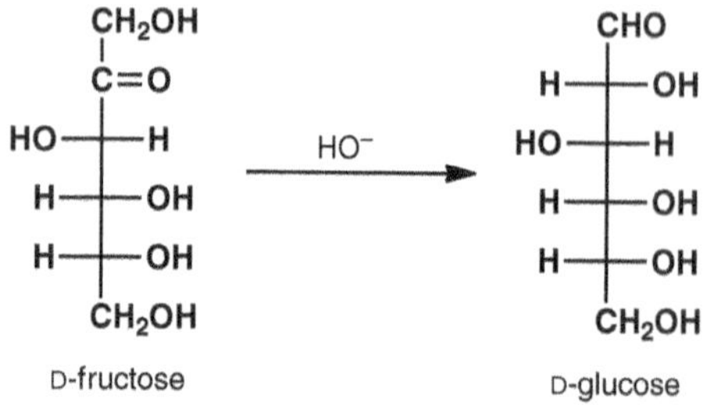

D-fructose → D-glucose

17.13 If two equivalents of benzaldehyde are used, the product shown in Scheme 17.14 reacts further. Give the structure of the final product and show how it is formed.

17.14 Draw the structure of the aldol and its dehydration product obtained in the crossed aldol reaction from each of the following equimolar pairs of compounds: (a) propanone+methanal, (b) butanal+furan-2-carbaldehyde (furfural), and (c) cyclohexanone+benzaldehyde.

17.15 The following keto-aldehyde can lead to three different enolates but the major product in an intramolecular aldol condensation is the one given. Show the other possible products and explain the selectivity observed.

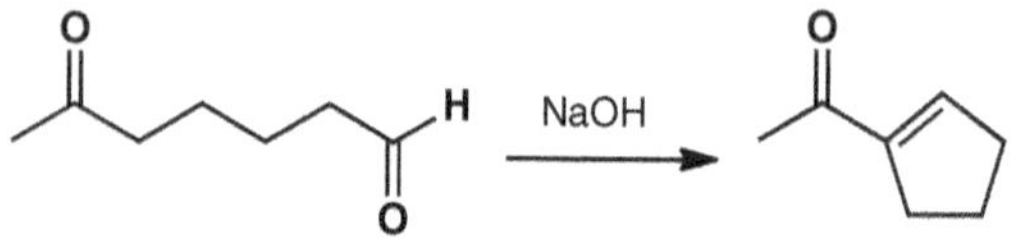

17.16 Give a mechanism for the crossed Claisen condensation in eqn 17.16.

17.17 Give the structure(s) of the crossed condensation product(s) when each of the following equimolar pairs is treated with NaOEt in EtOH and the reaction is quenched with aqueous acid.

(a)

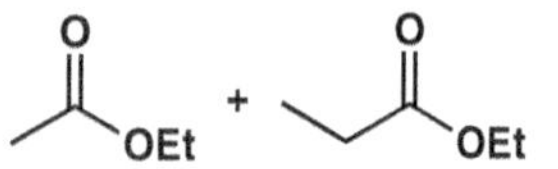

(b)

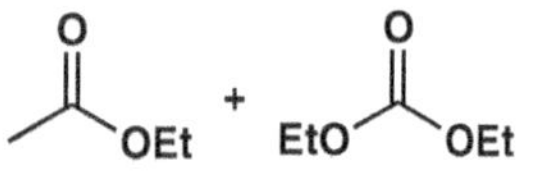

(c)

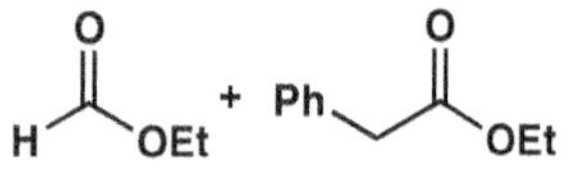

(d)

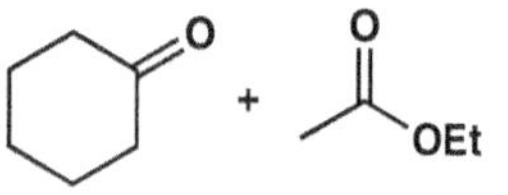

17.18 Draw the structure of the main product in each of the following reactions.

(a) 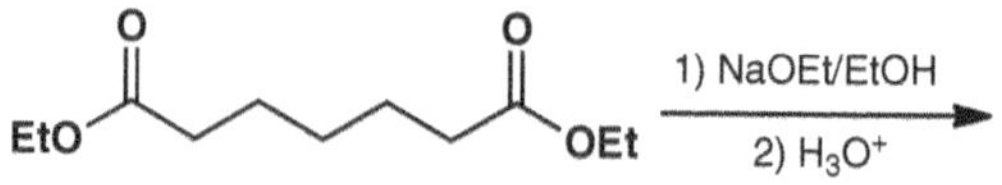   1) NaOEt/EtOH   2) H_3O^+

(b) 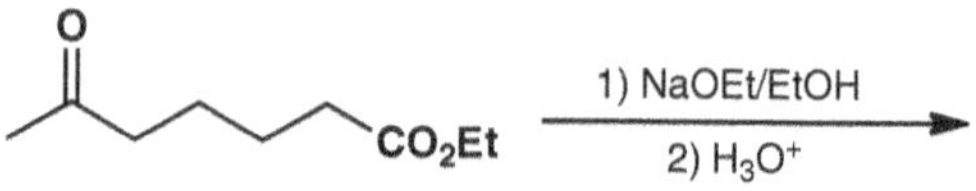   1) NaOEt/EtOH   2) H_3O^+

(c) 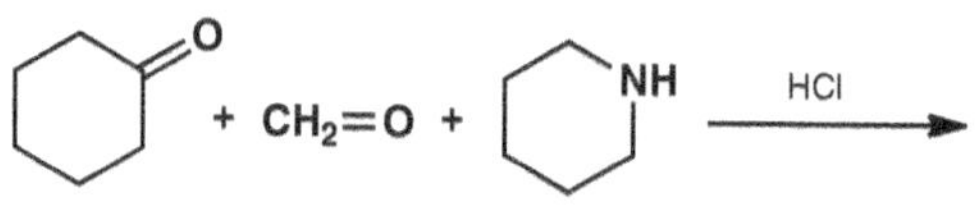   + CH_2=O +   HCl

(d) 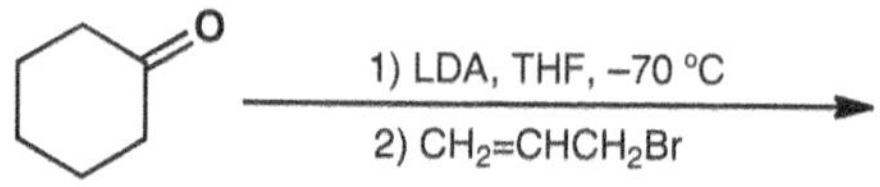   1) LDA, THF, –70 °C   2) CH_2=CHCH$_2$Br

(e) NCCH$_2$CO$_2$Me + PhCH$_2$Br   1) NaOMe/MeOH   2) H_3O^+

17.19 Complete the following reactions by showing the structures of **A–E**.

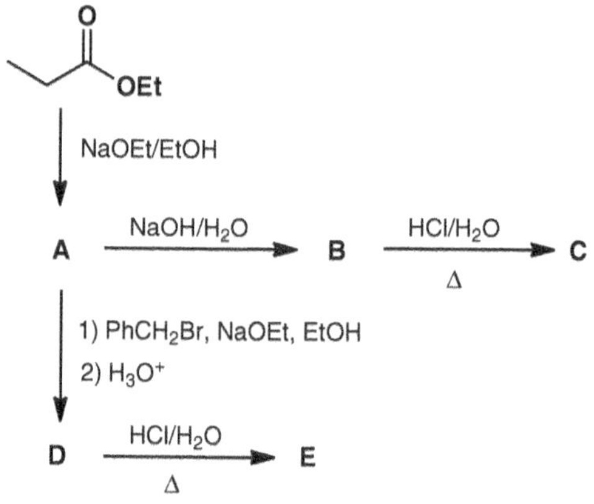

A → NaOH/H_2O → **B** → HCl/H_2O, Δ → **C**

D → HCl/H_2O, Δ → **E**

17.20 Complete the following sequence by identifying compounds **A–C**, and propose a mechanism for each step of the overall reaction.

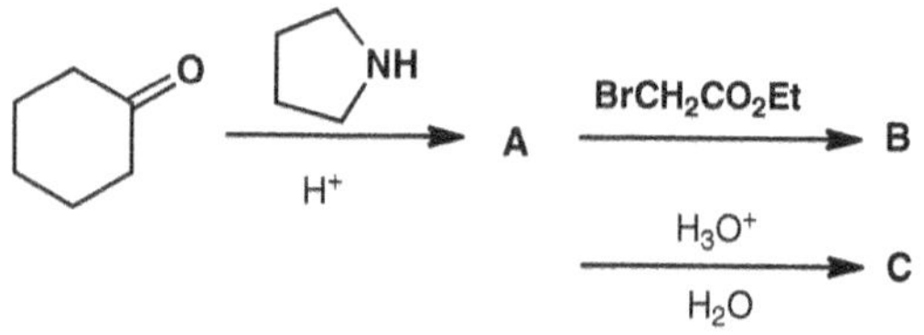

→ H^+ → **A** → BrCH$_2$CO$_2$Et → **B** → H_3O^+, H_2O → **C**

17.21 The lithium enolate of an ester can react with an aldehyde in an aldol-type reaction. Give the mechanism of the reaction of the lithium enolate of ethyl ethanoate with ethanal to give a crossed aldol-type product.

17.22 Propose reactions for the synthesis of the following ketones starting from ketones with no more than five carbon atoms but any alkylating reagent.

(a)

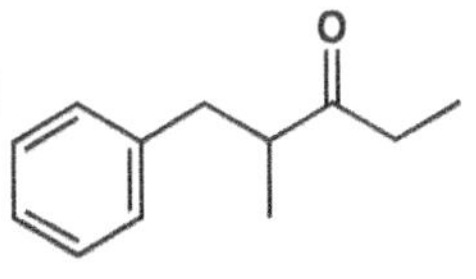

(b)

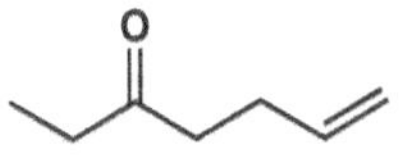

Supplementary Problems

17.23 Propose reactions for syntheses from butanal of (a) 2-ethylhexane-1,3-diol, (b) 2-ethylhex-2-en-1-ol, and (c) 2-ethylhexan-1-ol.

17.24 When ethanal reacts in an alkaline solution of formalin (an aqueous solution of methanal), three equivalents of methanal first react with ethanal and the product reacts further with methanal to give a tetraol called pentaerythritol in an example of the so-called Cannizzaro reaction (Section 10.3). Give reaction mechanisms for all steps of this overall transformation.

17.25 Can the following carboxylic acids be prepared by alkylation of diethyl propanedioate (the malonic ester synthesis)? Write a reaction sequence for each synthesis when it is possible, or explain when the synthesis by this method is not possible. (By retro-synthetic analysis (see Section 10.5), try to identify the haloalkane(s) which, together with diethyl propanedioate, will lead to the desired product most efficiently in each case.)

(a) 4-methylpentanoic acid

(b) 2-methylpentanoic acid

(c) 3,3-dimethylpentanoic acid

(d) 2,2-dimethylpentanoic acid

(e) cyclopentanecarboxylic acid

17.26 Write reactions for the synthesis of each of the following ketones by alkylation of ethyl 3-oxobutanoate (the acetoacetic ester synthesis).

(a) hexan-2-one

(b) 3-methyl-5-phenylpentan-2-one

(c) cyclopentyl methyl ketone

17.27 An ant pheromone (manicone) was prepared via an aldol formed using the enol silyl ether of pentan-3-one. Propose reactions for this synthesis.

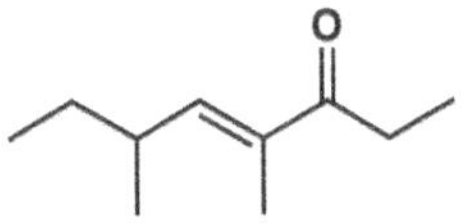

manicone

17.28 1,3-Dicarbonyl and related compounds often undergo an amine-catalysed condensation reaction with aldehydes, i.e. using only mildly basic conditions. Give a mechanism for the following example of this so-called Knoevenagel reaction.

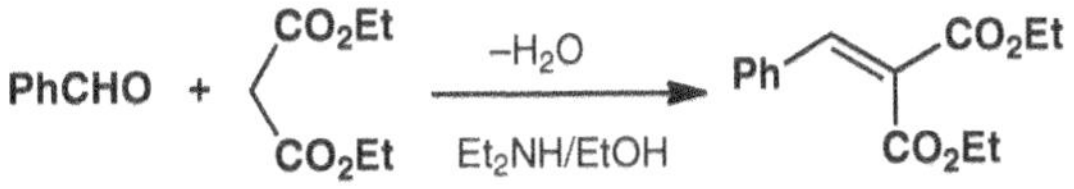

17.29 By giving a reaction mechanism, identify the cyclic product which is obtained in high yield (>90%) in the following reaction.

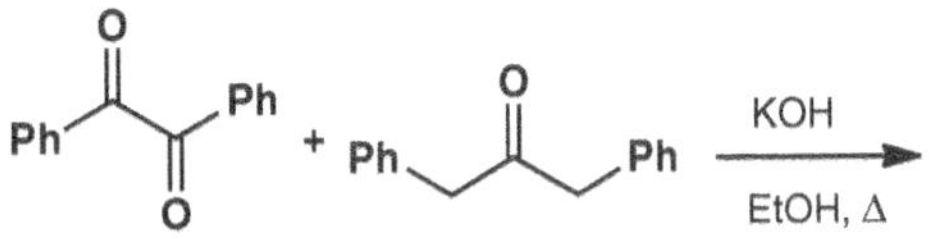

17.30 Outline a mechanism for the following transformation.

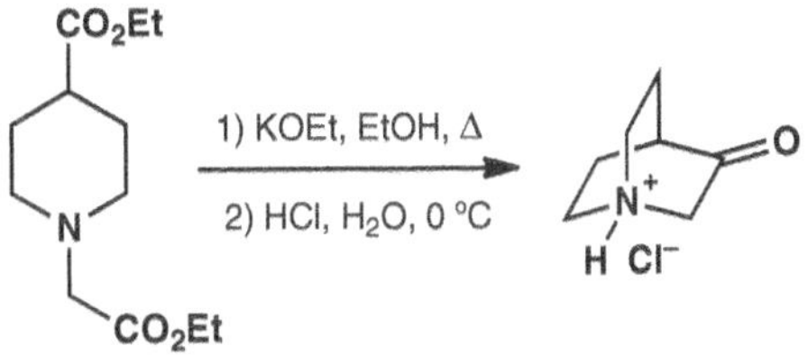

18 Reactions of Nucleophiles with Alkenes and Aromatic Compounds

Related topics we have already covered:

- **Conjugation** (Chapter 5)
- **Curly arrow representation of reaction mechanisms** (Section 7.2)
- **Nucleophilic addition and substitution at the carbonyl group** (Chapters 8–10)
- **Nucleophilic substitution at saturated carbon** (Chapter 12)
- **Electrophilic addition reactions of alkenes** (Chapter 15)
- **Conjugate addition (1,4-addition) to 1,3-dienes** (Chapter 15)
- **Electrophilic aromatic substitution** (Chapter 16)
- **Reactions of enolate ions** (Chapter 17)

Topics we shall cover in this chapter:

- **Conjugate addition to α,β-unsaturated carbonyl compounds**
- **Kinetic and thermodynamic control of carbonyl and conjugate additions**
- **Nucleophilic addition to other electrophilic alkenes**
- **Michael addition**
- **Robinson annulation**
- **Nucleophilic aromatic substitution by addition–elimination and elimination–addition mechanisms**
- **Reactions of arenediazonium salts**

An isolated C=O double bond is electrophilic because of its polarity, and its characteristic reactions are nucleophilic additions, as discussed in Chapter 8. In contrast, the π electrons of the C=C double bond of a simple alkene impart nucleophilic properties and alkenes generally undergo electrophilic addition reactions, as described in Chapter 15. However, a C=C bond becomes *electrophilic* when it is conjugated with a carbonyl group in α,β-unsaturated carbonyl compounds, and undergoes nucleophilic attack at the β carbon in what is called *conjugate addition*; this occurs in competition with direct nucleophilic addition at the carbonyl carbon. The initial enolate product of conjugate addition may be protonated on the oxygen to give an enol (when the reaction may be called *1,4-addition*) or on the α carbon to give the more stable keto product. Consequently, the outcome of conjugate addition to an α,β-unsaturated carbonyl compound under normal conditions is an equilibrium mixture of enol and keto tautomers. Protonation of the initial product of the competing direct addition to the carbonyl group leads only to an unsaturated hydroxy compound by overall 1,2-addition.

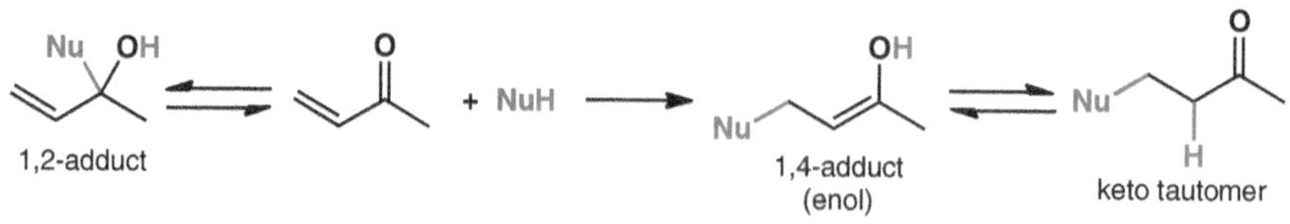

Nucleophilic addition also occurs at the β carbon of alkenes which are conjugated with other electron-withdrawing groups such as CN and NO_2.

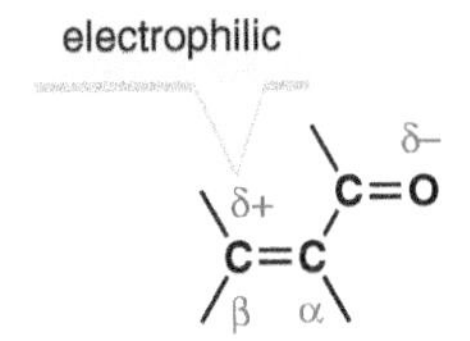

electrophilic

α,β-unsaturated carbonyl compound.

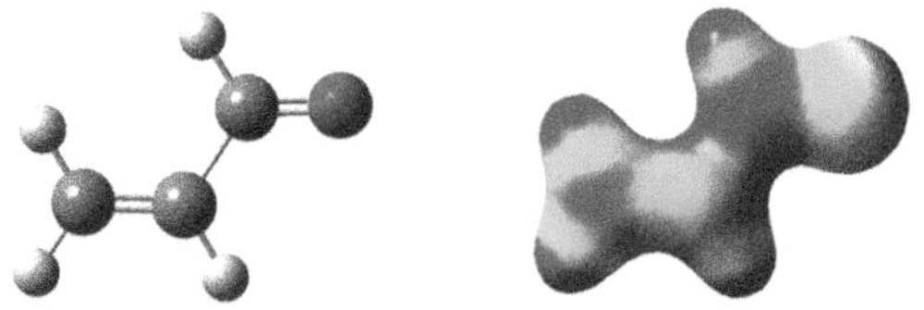

a molecular model and EPM of propenal.

α,β-Unsaturated carbonyl compounds also undergo electrophilic addition of halogens to give the dihalo carbonyl compounds, and are susceptible to hydrogen halide addition although the initial protonation occurs at the carbonyl oxygen (see later).

Benzene and other aromatic compounds are also nucleophilic and their characteristic reactions are electrophilic substitutions with a proton as the electrofuge as described in Chapter 16. However, halobenzenes carrying suitably positioned, strongly electron-withdrawing substituent(s) are activated towards nucleophilic addition and the ensuing elimination of the halide nucleofuge leads to *nucleophilic substitution*.

nucleofuge

electron-withdrawing group

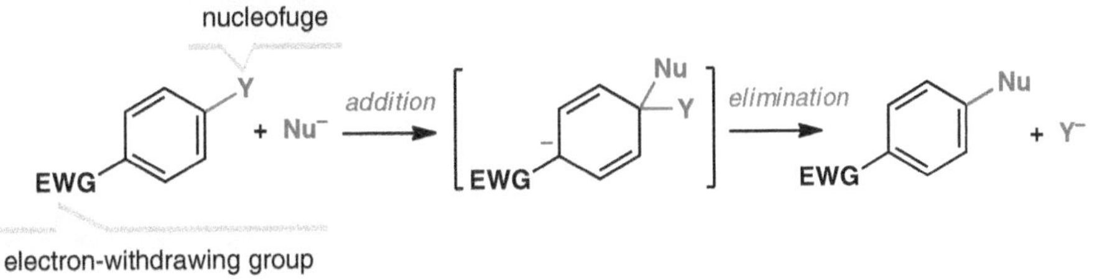

Un-activated haloarenes undergo 1,2-elimination under extremely basic conditions, e.g. with sodium amide in liquid ammonia: the triple-bonded intermediate (called a benzyne) is unstable and immediately undergoes nucleophilic addition. The overall consequence of the elimination and subsequent addition is another type of nucleophilic substitution initiated by proton abstraction. In this chapter, we shall cover reactions of nucleophiles with α,β-unsaturated carbonyl compounds and related electrophilic alkenes as well as with aromatic compounds carrying a nucleofuge—usually haloarenes. We shall also look again at substitution reactions of highly electrophilic arenediazonium ions (which have an excellent nucleofuge, N_2) by the S_N1 and other mechanisms.

18.1 Nucleophilic Addition to α,β-Unsaturated Carbonyl Compounds

α,β-Unsaturated carbonyl compounds contain a C=C double bond conjugated with the carbonyl groups of aldehydes, ketones, and carboxylic acid derivatives.* The resonance representation which describes the electronic structure shows that the β carbon, like the carbonyl carbon, bears a partial positive charge. Consequently, this is often the site of nucleophilic attack.

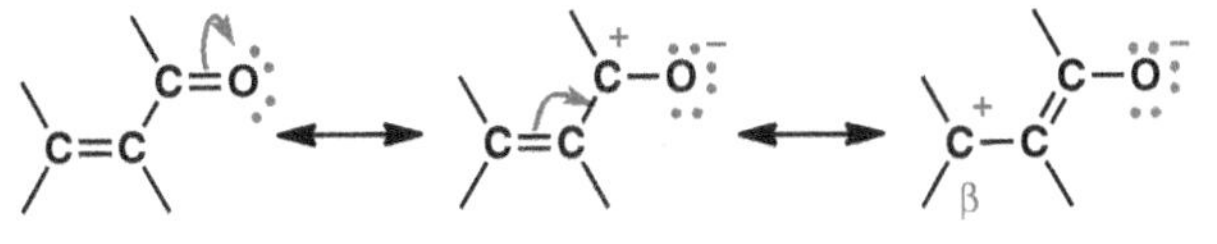

A simple unsaturated ketone is an *alkenone* but this is usually abbreviated to **enone**. The mechanism of conjugate addition of H–Nu to all α,β-unsaturated carbonyl compounds is normally the same so, for the sake of simplicity, we shall usually illustrate them all with enones.

*Reminder: nucleophilic attack directly at the carbonyl of carboxylic acid derivatives usually results in overall substitution (see Chapter 9).

an α,β-unsaturated carbonyl compound (an enone).

18.1.1 Conjugate addition and carbonyl addition

We encountered 1,4-addition to butadiene in which the double bond migrates in Chapter 15. Conjugate addition of H–Nu to an enone to give the enol form of the product also involves migration of a double bond.

α,β-Unsaturated carbonyl compounds are normally regarded as electrophilic alkenes, and (except when the reaction is acid catalysed) addition of H–Nu across the double bond is initiated by nucleophilic attack by a lone pair on the H–Nu (or Nu⁻) at the β carbon, hence the name **conjugate addition**. This generates an enolate intermediate which may be protonated on the oxygen to give an enol by **1,4-addition**. However, the keto tautomer is the more stable product formed by protonation on the α C of the enolate (or by tautomerization of the enol); this corresponds to addition of H–Nu across the C=C bond (Scheme 18.1).

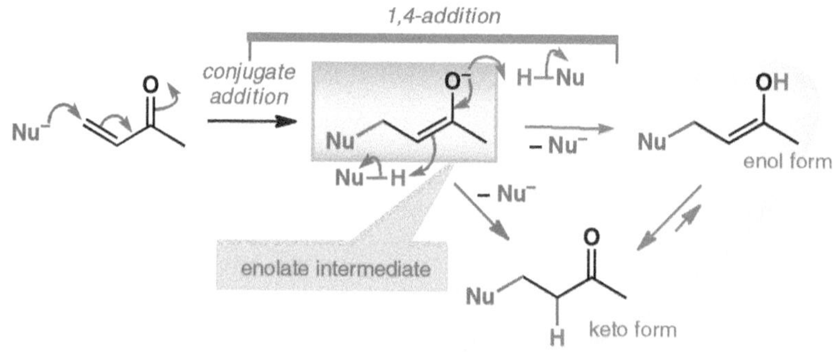

Scheme 18.1 Nucleophilic addition of H–Nu to an enone.

Nucleophiles for these reactions are the same as those involved in the carbonyl addition reactions covered in Chapters 8 and 10: alkoxides, amines, thiols, and cyanide as well as some carbon nucleophiles. A few typical reactions are given below.*

*The product of the conjugate addition of a primary amine to the enone in the first example is a secondary amine. If sufficient enone is used, this initial product can react as the nucleophile in a further conjugate addition to give the following.

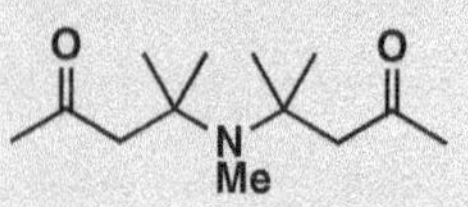

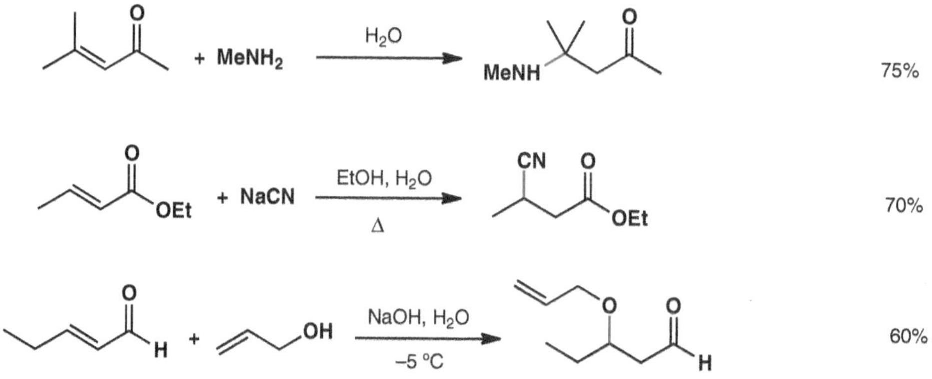

75%

70%

60%

Exercise 18.1

Give the reaction mechanism for the conjugate addition of dimethylamine to pent-3-en-2-one.

Exercise 18.2

What is the main product in each of the following reactions?

(a) + Et₂NH $\xrightarrow{\text{EtOH}}$ (b) + PhSH $\xrightarrow{\text{EtOH}}$

Under basic or neutral conditions, additions of H–Nu are usually initiated by nucleo-philic attack as we have seen. However, if H–Nu is acidic, its conjugate addition to an α,β-unsaturated carbonyl compound may be initiated by proton transfer to the carbonyl oxygen, the most basic centre of the substrate. For example, the addition of a hydrogen halide to give the product of addition of HX across the C=C bond begins with protonation at the carbonyl oxygen (Scheme 18.2).

Although the conjugate addition of HX to an enone with an initial proton transfer is similar to the electrophilic addition of HX to butadiene, we consider it a *nucleophilic addition* because the initial protonation at the carbonyl oxygen is a rapid pre-equilibrium and the rate-determining step is the addition of the nucleophile at the β carbon whose hybridization changes.

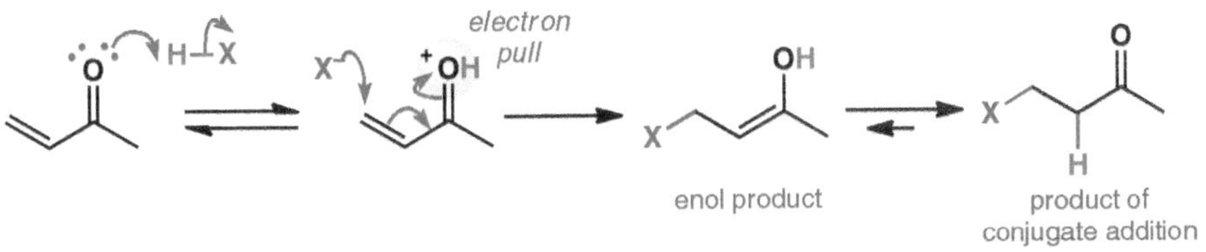

enol product

product of conjugate addition

Scheme 18.2 Conjugate addition of a hydrogen halide to an enone initiated by proton transfer.

Exercise 18.3

Give a reaction mechanism for hydrobromination of methyl propenoate (acrylate) in diethyl ether.

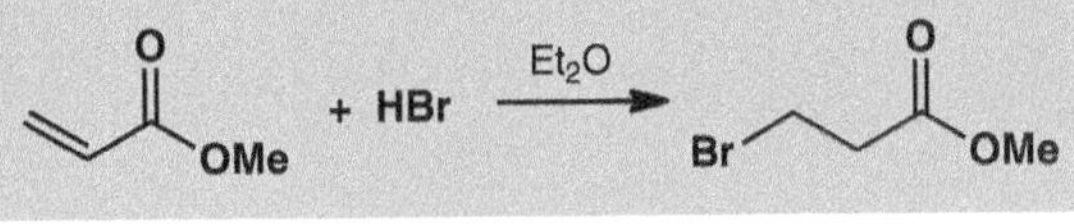

Example 18.1

Additions of weakly nucleophilic alcohols to enones usually require either acid or base catalysis. Give reaction mechanisms for the addition of ethanol to pent-3-en-2-one catalysed by HCl and by EtONa.

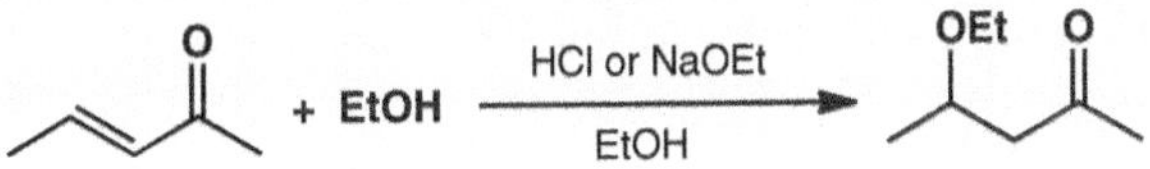

Solution
Acid-catalysed addition:

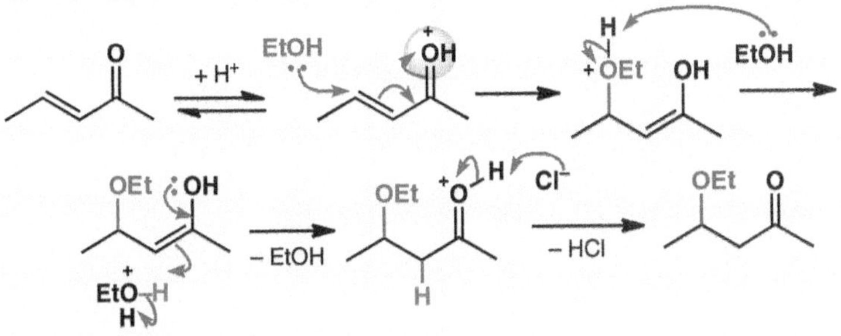

Note that HCl supplies the H+ in the first step then Cl− abstracts H+ to regenerate HCl in the final step, so the reaction is catalytic.

Base-catalysed addition:

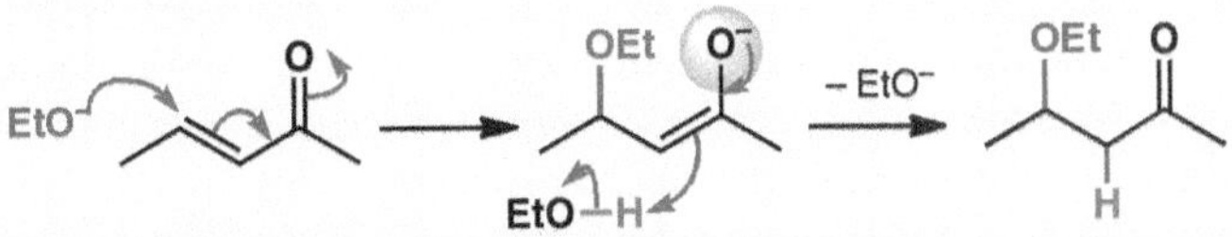

Here, ethoxide ion is consumed in the first step then regenerated in the second step, so this reaction is also catalytic.

*In particular, the product of reversible 1,2-addition of a hydrogen halide to a carbonyl, an α-halo alcohol, would be very unstable and not observable even if it were formed. In contrast, reactions with very strongly basic nucleophiles which bond through carbon, e.g. RLi and RMgBr, are irreversible (see Sub-section 18.1.3).

As we saw in Chapters 8 and 10, the carbonyl carbon is also an electrophilic centre and, in competition with conjugate addition, some nucleophilic attack will occur at this position leading to **1,2-addition** (or **carbonyl addition**), as shown in Scheme 18.3. Addition to the carbonyl is usually reversible when the nucleophile is also a potential nucleofuge,* a process facilitated by the strong electron push from the anionic oxygen.

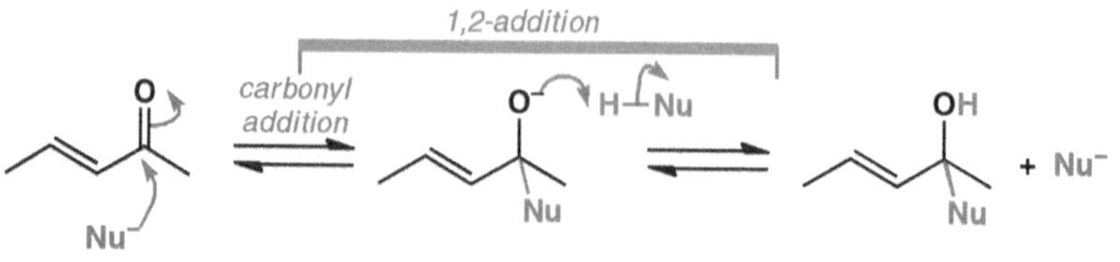

Scheme 18.3 Carbonyl or 1,2-addition of H–Nu to an enone.

18.1.2 Kinetic and thermodynamic control of carbonyl and conjugate additions

Conjugate addition competes with carbonyl addition, and the final product via the former is more stable than that of the latter. Since some carbonyl additions are kinetically favourable and reversible, the kinetic products may be converted to the thermodynamic ones following extended reaction times.

When an enone is treated with cyanide in the presence of an acid at low temperatures, carbonyl addition to give the cyanohydrin is the main reaction (Scheme 18.4). However, in an example of thermodynamic control versus kinetic control, the cyanohydrin is transformed into the product of conjugate addition upon heating.

Conjugate addition leads to retention of the C=O bond, while the carbonyl addition leads to retention of the C=C bond. A C=O bond is considerably stronger than a C=C bond, so the former reaction gives the more stable product.

See Sub-section 15.8.2 for examples of kinetic and thermodynamic control of 1,2- and 1,4-addition to conjugated dienes.

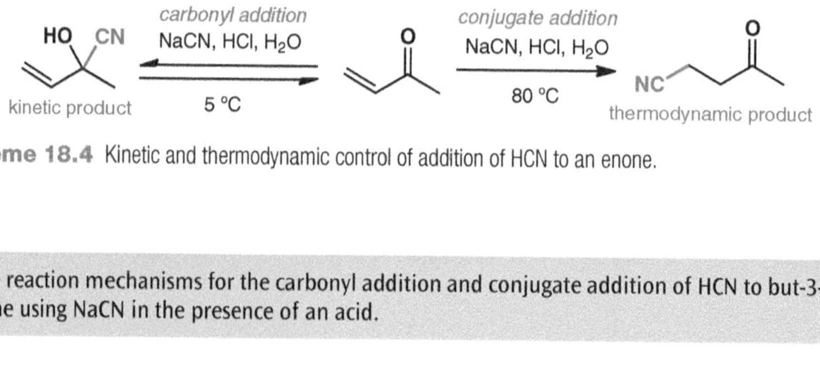

Scheme 18.4 Kinetic and thermodynamic control of addition of HCN to an enone.

Exercise 18.4

Give reaction mechanisms for the carbonyl addition and conjugate addition of HCN to but-3-en-2-one using NaCN in the presence of an acid.

18.1.3 Addition of organometallic reagents and metal hydrides to α,β-unsaturated carbonyl compounds

Additions of organometallic compounds to the carbonyl group are an important method for the formation of new C–C bonds (see Chapter 10).

Organometallic compounds and metal hydrides are strongly basic nucleophiles, and their addition reactions to carbonyl groups are not reversible. Consequently, only kinetic products are obtained in their additions to α,β-unsaturated carbonyl compounds.

The Grignard reagents usually undergo both 1,2- and 1,4-additions to enones, and the regioselectivity depends largely on steric factors. Pent-3-en-2-one, for example, gives carbonyl and conjugate additions in the ratio of about 3 : 1 (eqn 18.1), while the related α,β-unsaturated aldehyde predominantly gives the carbonyl adduct (eqn 18.2).

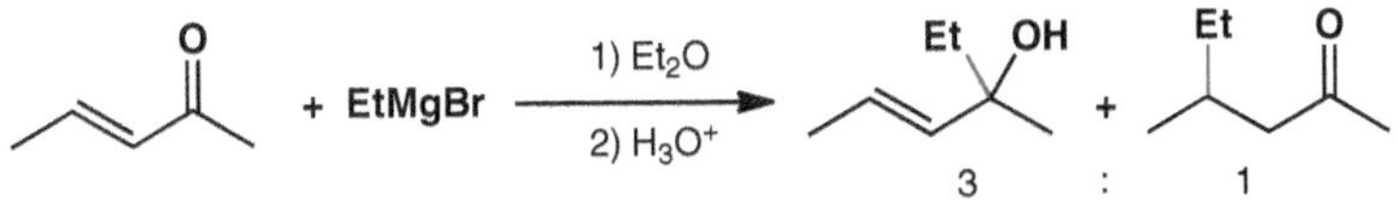

$$\text{(18.1)}$$

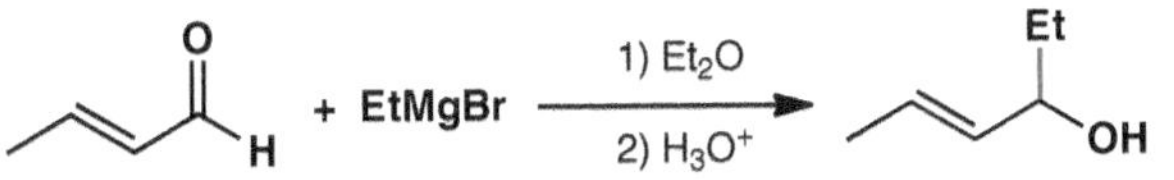

$$\text{(18.2)}$$

In contrast, pent-3-en-2-one selectively undergoes direct carbonyl addition with a simple organolithium reagent, e.g. eqn 18.3, but gives the product of conjugate addition exclusively with the corresponding lithium dialkylcuprate (eqn 18.4).

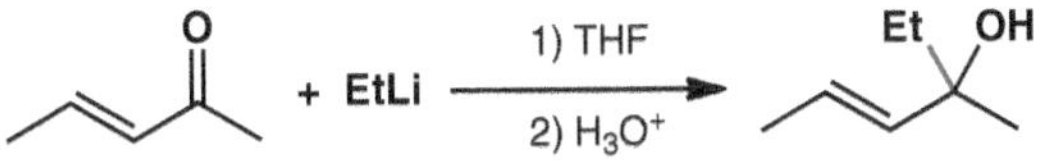

$$\text{(18.3)}$$

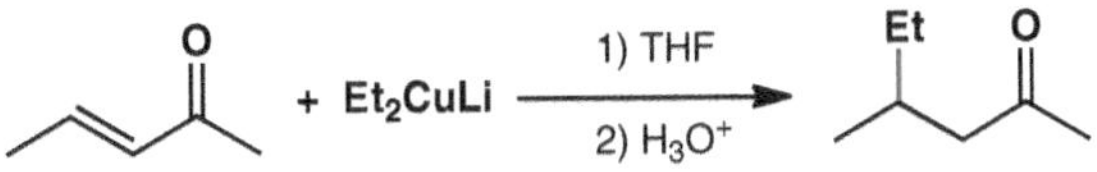

$$\text{(18.4)}$$

The dialkylcuprate reagent R_2CuLi can be prepared by reaction of the alkyllithium with a cuprous compound,

$$2\,RLi + CuX \longrightarrow R_2CuLi + LiX$$

Alternatively, conjugate addition can sometimes be achieved by using a Grignard reagent with just a catalytic amount of a Cu(I) compound.

Hydride reductions of α,β-unsaturated carbonyl compounds by metal hydrides including $LiAlH_4$ and $NaBH_4$ generally give products of both conjugate and carbonyl addition (Scheme 18.5), and the regioselectivity depends very much on the hydride, the substrate, and the reaction conditions.

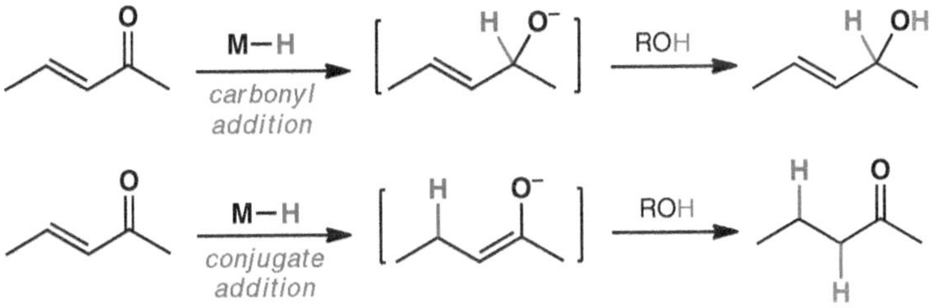

Scheme 18.5 Hydride reductions of an enone.

Regioselective carbonyl reduction is possible by using $NaBH_4$ in the presence of the Lewis acid $CeCl_3$ (eqn 18.5), while reaction with a bulky hydride donor such as KBH(s-Bu)$_3$ is regioselective according to the degree of steric hindrance (Scheme 18.6).

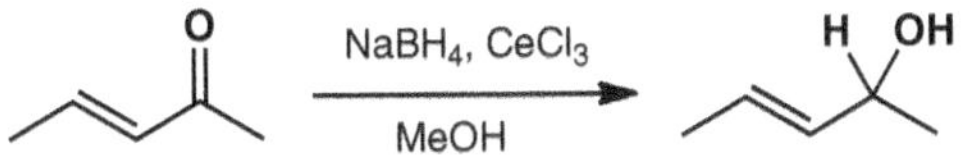

$$\text{(18.5)}$$

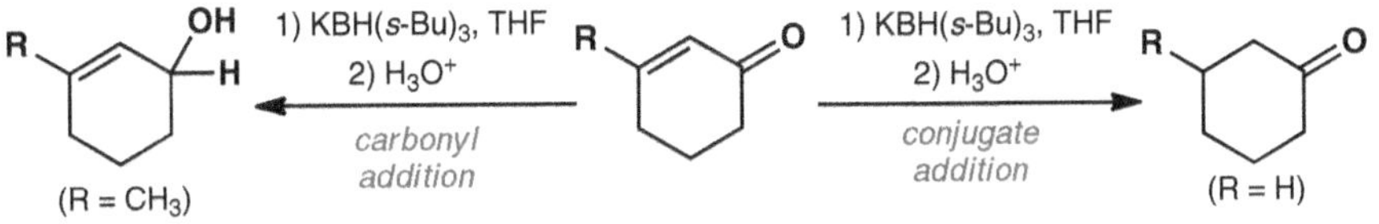

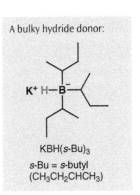

A bulky hydride donor:

KBH(s-Bu)$_3$
s-Bu = s-butyl
(CH$_3$CH$_2$CHCH$_3$)

(R = CH₃) carbonyl addition   1) KBH(s-Bu)₃, THF 2) H₃O⁺   1) KBH(s-Bu)₃, THF 2) H₃O⁺ conjugate addition   (R = H)

Scheme 18.6 Regioselectivity in the hydride reduction of an enone with a bulky hydride donor.

Catalytic hydrogenation can be used to reduce the C=C bond of an α,β-unsaturated carbonyl compound selectively, e.g. eqn 18.6.

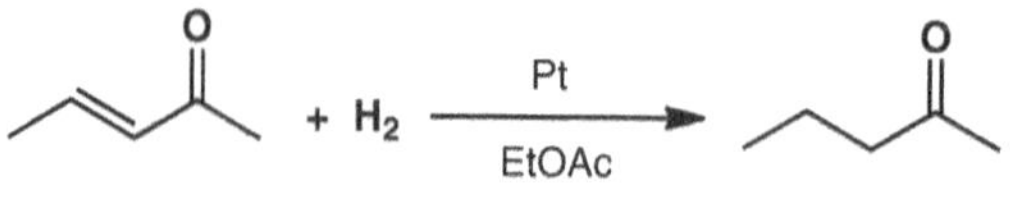

$$(18.6)$$

Example 18.2

What is the product of the following reaction of an alkenoate ester with an excess of butyllithium?

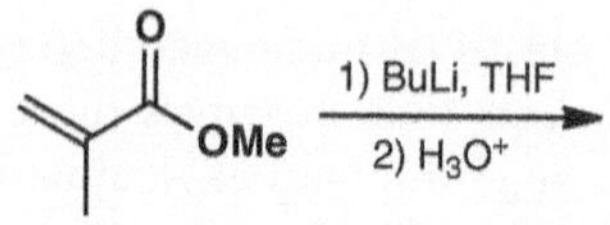

Solution

Butyllithium reacts at the carbonyl group of the ester preferentially to give a tetrahedral intermediate which eliminates the MeO⁻ nucleofuge; the resultant enone then reacts with a second equivalent of BuLi to give a tertiary alcohol, as described in Sub-section 10.4.2.

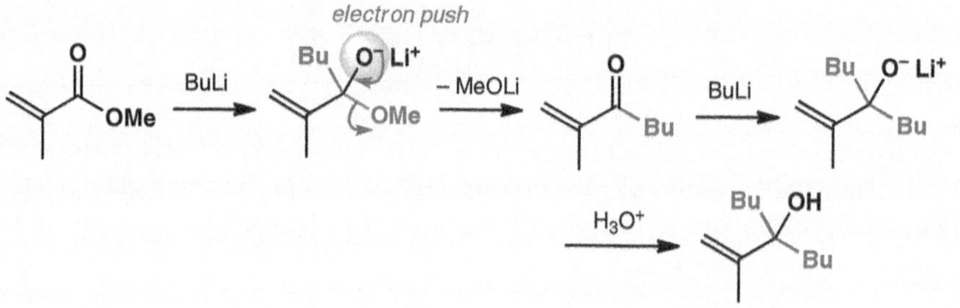

Exercise 18.5

Reduction of ethyl propenoate occurs preferentially at the carbonyl using (a) LiAlH₄, but reduction is mainly by conjugate addition using (b) NaBH₄. Show the reaction sequence to give the main product in each case.

18.2 Nucleophilic Addition to other Electrophilic Alkenes

Conjugate additions to alkenes which are activated by conjugation with electron-withdrawing groups other than carbonyl (e.g. cyano and nitro) are also well known. An example is the addition of an amine to propenenitrile (acrylonitrile) to give the cyano-ethylated amine (Scheme 18.7); this reaction is sometimes called *cyano-ethylation* of the nucleophile.

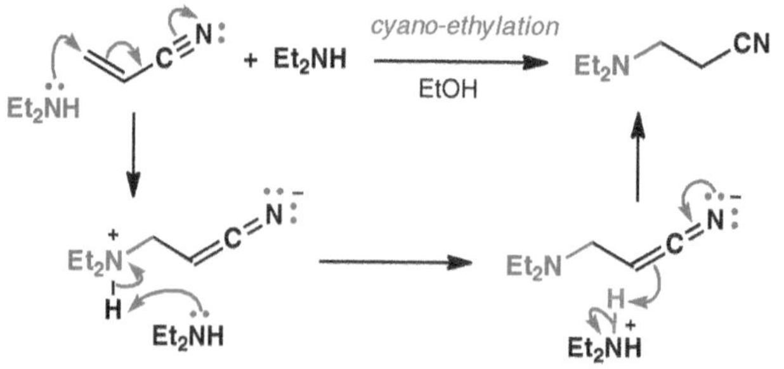

Scheme 18.7 Conjugate addition of an amine to propenenitrile (cyano-ethylation of the amine).

Exercise 18.6

Draw resonance forms which illustrate the electrophilicity of (a) propenenitrile and (b) nitroethene.

18.3 Anionic Polymerization

The conjugate addition of an anionic nucleophile to an electrophilic alkene is completed by the intermediate anion being trapped by an electrophile, usually a proton from a protic solvent. In the absence of an electrophile, the anion may react further with another molecule of the alkene. If this further addition to alkenes is repeated many times, the product is a polymer and we have an **anionic polymerization**; the chain terminates only when the anionic polymer scavenges a proton. The polymerization proceeds as illustrated for propenenitrile (acrylonitrile) in Scheme 18.8.

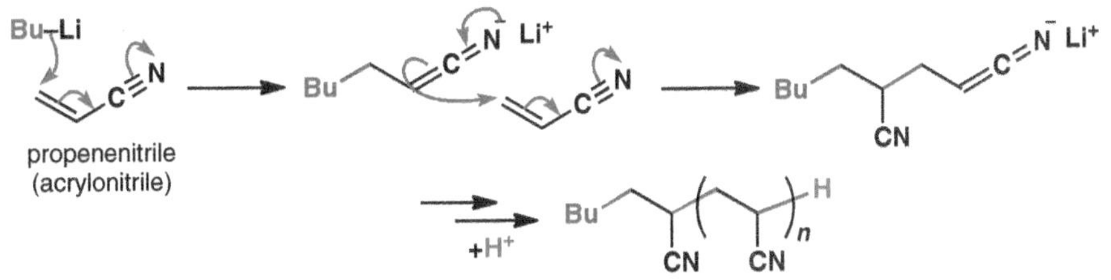

Scheme 18.8 Anionic polymerization of acrylonitrile initiated by butyllithium.

Polyacrylonitrile is used to manufacture synthetic fibres. Poly(methyl methacrylate) is a transparent 'plastic' used as a substitute for glass to make lenses and in the construction of aquariums.

Monomers for anionic polymerization include the following:

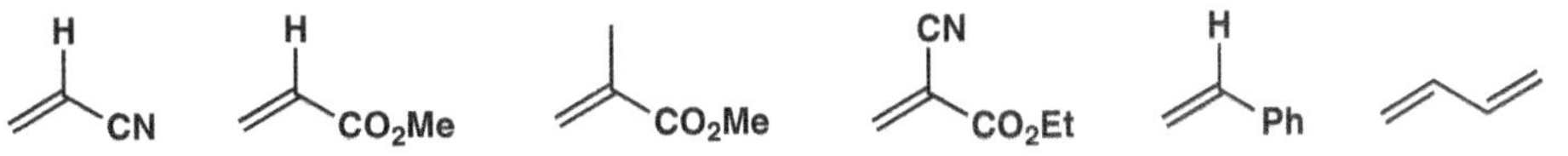

propenenitrile (acrylonitrile)   methyl propenoate (methyl acrylate)   methyl 2-methyl-propenoate (methyl methacrylate: MMA)   ethyl 2-cyano-propenoate (ethyl 2-cyanoacrylate)   phenylethene (styrene)   buta-1,3-diene

Styrene and butadiene are included as monomers for anionic polymerization although the intermediate carbanions from them are stabilized by conjugation with phenyl and vinyl, respectively, rather than a strongly electron-withdrawing group. Their polymerizations are initiated only by a strong nucleophile such as an alkyllithium.

The most reactive of the common anionic polymerization monomers are alkyl 2-cyanoacrylates, and their polymerization is initiated even by weak nucleophiles; they are used in fast-acting adhesives (see Panel 18.1) when polymerization is initiated by moisture on the surfaces of the items to be stuck together.

Exercise 18.7

Show the resonance forms of the anion formed by addition of HO⁻ to ethyl 2-cyanoacrylate.

Exercise 18.8

Give a reaction scheme for the anionic polymerization of styrene initiated by butyllithium.

Panel 18.1 Cyanoacrylate esters in instant glues, for the detection of fingerprints, and in medicine

Alkyl 2-cyanoacrylates are widely used as fast-acting adhesives (sometimes known as instant glues or super glues); their polymerization is very readily induced by nucleophiles including just traces of moisture. The methyl, ethyl, and butyl esters are normally used, but additives are sometimes added to enhance their viscosity or strength.

alkyl 2-cyanoacrylate
(R = Me, Et, or Bu)

poly-2-cyanoacrylate

Cyanoacrylates are also used for detection of fingerprints which are unique to individuals and have been used by the police and forensic scientists for identification purposes. The search for fingerprints at crime scenes, for example, is a well known method of investigation and several methods for their detection have been developed. One general method is to use a very fine powder or chemical reagent to 'develop' the fingerprints on a surface. One reagent is a 2-cyanoacrylate ester whose polymerization is readily induced by the moisture of fingerprint, and leaves a deposit. The reagent may be used as a spray in dry weather, or the sample carrying fingerprints may be left in a vapour of cyanoacrylate in a closed vessel.

Cyanoacrylates are also used as medical adhesives to treat wounds and in surgery. For these purposes, the less harmful 1-methylheptyl (or 2-octyl to use its common name) 2-cyanoacrylate (Dermabond®) is used.

18.4 Conjugate Addition of Enolate Ions to α,β-Unsaturated Carbonyl Compounds

18.4.1 The Michael reaction

Arthur Michael
(1855–1942).
(Kindly supplied by
S. Kounaves.)

Michael was born in Buffalo, New York, and studied in Europe under Bunsen, Hofmann, Wurtz, and Mendeleev; he established the Department of Chemistry at Tufts University in Massachusetts.

Enolate ions, especially those derived from 1,3-dicarbonyl compounds, undergo conjugate additions to α,β-unsaturated carbonyl compounds and other electrophilic alkenes, as illustrated in eqn 18.7.

prop-3-en-2-one   +   diethyl malonate $\xrightarrow[\text{2) H}_3\text{O}^+]{\text{1) NaOEt, EtOH}}$ (18.7)

The reactions of enolate ions with α,β-unsaturated carbonyl compounds are known as *Michael reactions* (or *Michael additions*).

Exercise 18.9

Give a mechanism for the Michael addition shown in eqn 18.7.

Exercise 18.10

When the product of reaction 18.7 is hydrolysed in aqueous NaOH and heated after acidification, a 1,5-dicarbonyl compound is obtained. Show the reactions and give the structure of the final product.

By giving the mechanism for each step in the following reaction sequence, show the structures of compounds A, B, and C.

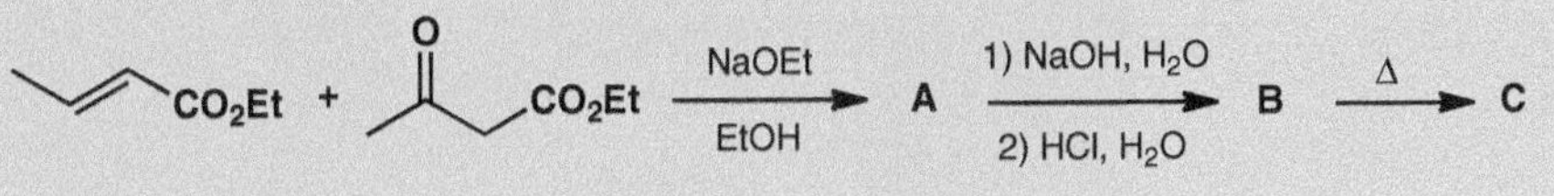

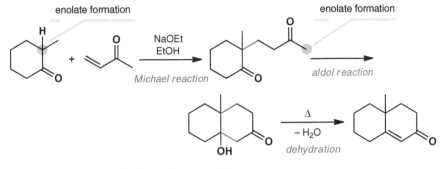

Exercise 18.11

Enamines (see Sub-section 8.6.2) act as enolate equivalents and undergo conjugate addition to enones (Scheme 18.9).

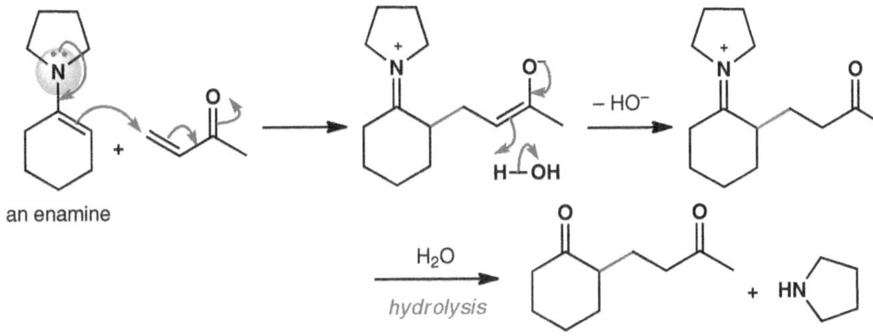

Scheme 18.9 Conjugate addition of an enamine to an enone.

18.4.2 The Robinson annulation

A Michael addition followed by an intramolecular aldol reaction leads to cyclization by formation of two new C–C bonds. This sequence, exemplified in Scheme 18.10, is known as the *Robinson annulation,* and is a useful method for the addition of a cycloalkenone residue in the synthesis of polycyclic compounds such as steroids (Sub-section 24.4.4).

enolate formation   enolate formation

NaOEt
EtOH

Michael reaction   *aldol reaction*

Δ
– H₂O

dehydration

Sir Robert Robinson
(1886–1975)
(Courtesy of the Royal Society of Chemistry.)
Robinson was born in Derbyshire, England, and, following a series of other professorial appointments between 1912 and 1930, became professor of organic chemistry at Oxford University. He contributed in many ways to the development of the subject and received the Nobel Prize in Chemistry in 1947 for his work on alkaloids (see Panel 19.3).

Scheme 18.10 An example of the Robinson annulation.

Give a reaction mechanism for the Robinson annulation in Scheme 18.10.

Exercise 18.12

What is the product of the Robinson annulation using the following reactants?

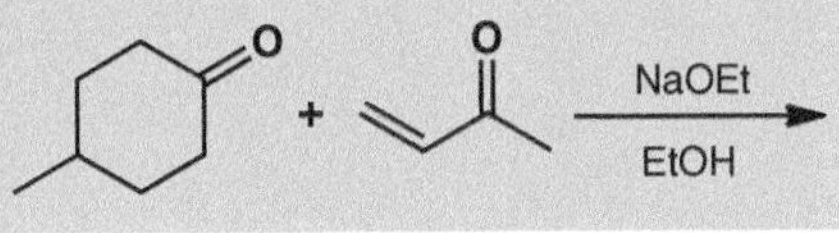

NaOEt
EtOH

Exercise 18.13

18.5 Substitution by a Conjugate Addition–Elimination Mechanism

We encountered nucleophilic substitution reactions of esters and other carboxylic acid derivatives by the nucleophilic addition–elimination mechanism in Chapter 9. If an electrophilic alkene has a nucleofuge at the β position, the enolate formed by addition of a nucleophile at this position may undergo elimination of the nucleofuge (rather than addition of an electrophile). As seen in Scheme 18.11, this leads to overall substitution at the β position rather than 1,4-addition.

Scheme 18.11
A conjugate addition–elimination mechanism for substitution.

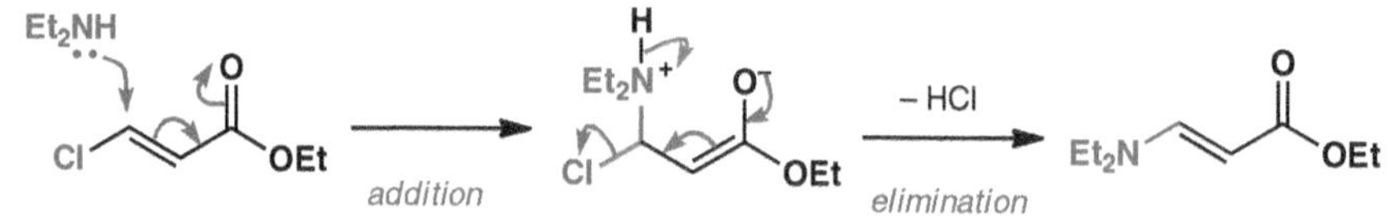

*A similar reaction is also possible with aromatic compounds carrying a halogen nucleofuge and at least one suitably positioned electron-withdrawing substituent, as we shall discuss in the next section.

Departure of the nucleofuge in this reaction is facilitated by recovery of the stabilization through regeneration of the C=C bond conjugated with the carbonyl group.*

The substitution mechanism in Scheme 18.11 should be contrasted with the S_N2 and S_N1 mechanisms for nucleophilic substitutions at saturated carbons. Direct rear-side attack at the sp² hybridized β carbon of an α,β-unsaturated enone with concerted departure of the nucleofuge is highly unfavourable, addition being the preferred reaction. An S_N1-type mechanism with the initial formation of a vinylic cation intermediate destabilized by the conjugated carbonyl is virtually impossible.

The addition–elimination mechanism shown in Scheme 18.11 occurs by attack of the nucleophile from above the molecular plane of the substrate whereas a vinylic S_N2 requires the more sterically hindered in-plane attack; the latter is normally difficult but not totally impossible. A vinylic S_N1 reaction becomes possible only when the intermediate vinylic cation is stabilized by an electron-donating group.

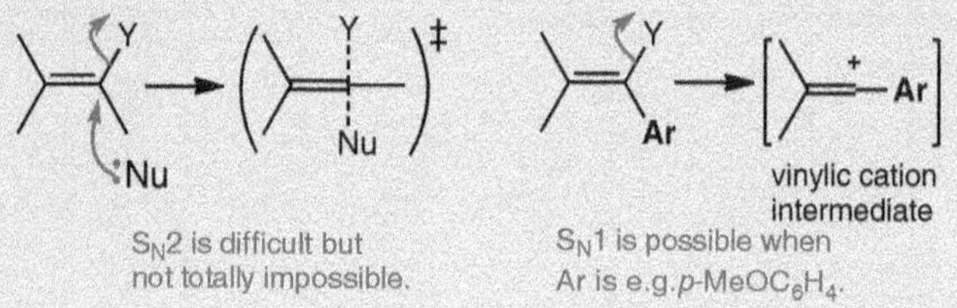

S_N2 is difficult but not totally impossible.

S_N1 is possible when Ar is e.g. $p\text{-MeOC}_6\text{H}_4$.

18.6 Nucleophilic Aromatic Substitution by the Addition–Elimination Mechanism

When a halobenzene has one or more conjugatively electron-withdrawing groups such as NO_2, CN, or C(O)R on the *ortho* and/or *para* position(s), nucleophilic substitution of the halogen becomes possible, e.g. eqn 18.8.

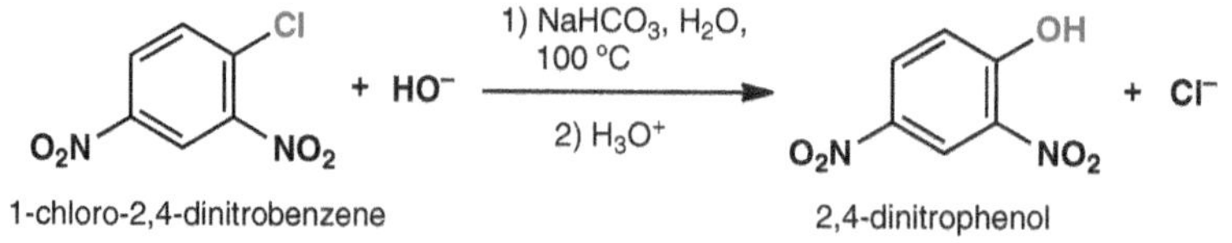

1-chloro-2,4-dinitrobenzene

2,4-dinitrophenol

(18.8)

In the **addition–elimination mechanism** for this reaction, initial nucleophilic addition at the C bearing the halogen gives an intermediate anion (sometimes called a Meisenheimer complex) which is stabilized by conjugative electron delocalization into the *para* and/or *ortho* substituent(s) (Scheme 18.12). The intermediate anion yields the product by loss of the halide ion and recovery of aromaticity.

Examples are known in which haloarenes are activated towards nucleophilic substitution by *inductively* electron-withdrawing groups such as CF_3.

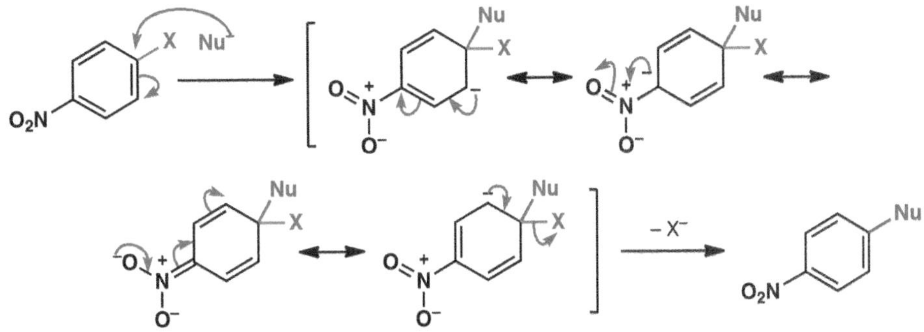

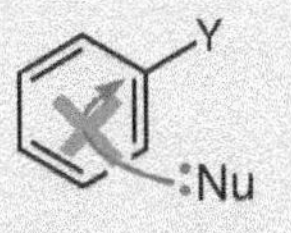

An aromatic S_N2 is impossible.

Scheme 18.12 Addition–elimination mechanism of nucleophilic aromatic substitution.

Exercise 18.14

Explain the order of reactivity of the nitrochlorobenzenes shown below in nucleophilic substitution reactions with sodium methoxide in methanol.

$$Ar\text{--}Cl + MeONa \xrightarrow[\text{MeOH, 80 °C}]{} Ar\text{--}OMe + NaCl$$

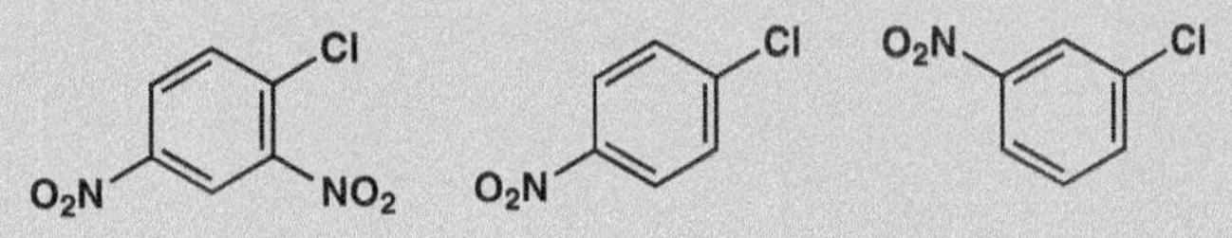

| Relative rate constant: | 34 000 | 1.0 | $< 10^{-4}$ |

Besides hydroxide and alkoxide, amines and thiolate anions also participate in nucleophilic aromatic substitution reactions. Amongst haloarenes, fluoroarenes are much more reactive than the others (Ar–F >> Ar–Cl, Ar–Br, Ar–I) in spite of the generally poor leaving ability of fluoride as a nucleofuge (C–X bond strengths are in the order C–F >> C–Cl>C–Br>C–I). This is because the *rate-determining step of this reaction is the initial nucleophilic addition* and fluorine, being the most electronegative, is best able to stabilize the intermediate anion and the transition structure on the way to its formation (eqn 18.9).

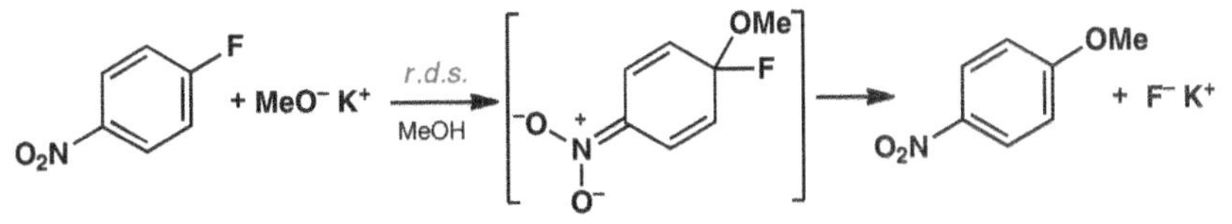

*The reactivities of primary alkyl halides in S_N2 reactions increase in the order: RCH_2F << RCH_2Cl < RCH_2Br < RCH_2I (see Chapter 12).

This order of reactivity is in contrast to what is observed in nucleophilic substitution reactions (S_N2) at saturated carbon when fluoro compounds are notoriously unreactive.*

Exercise 18.15

What is the main product in each of the following reactions?

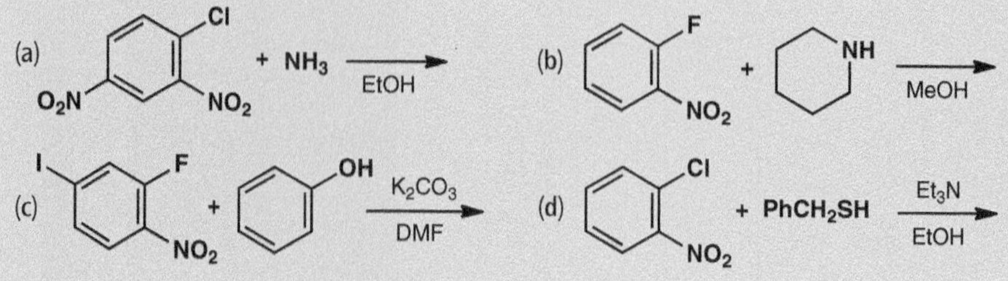

(a), (b), (c), (d) reactions

18.7 Nucleophilic Aromatic Substitution by the Elimination–Addition Mechanism

The nucleophilic addition–elimination mechanism described in the previous section is observed only with halobenzenes activated by suitably positioned electron-withdrawing groups. In order to induce substitution reactions of un-activated halobenzenes, we require very strongly basic reaction conditions whereupon substitution will take place, but by a different mechanism.

Although chlorobenzene does not react when heated in concentrated aqueous NaOH solution, it undergoes substitution to give phenol when fused with NaOH at a high temperature (340 °C) prior to acidification (Scheme 18.13). Only by using even more strongly basic conditions can reaction be induced at low temperatures; for example, chlorobenzene can be converted into aniline by reaction with sodium amide in liquid ammonia (−33 °C, Scheme 18.13).

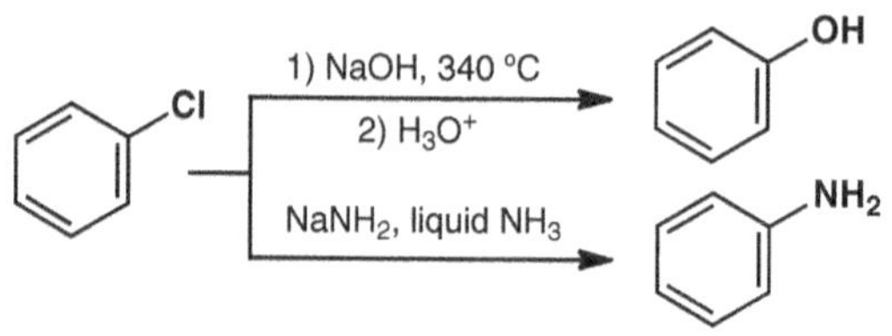

Scheme 18.13
Nucleophilic aromatic substitution under strongly basic conditions.

> The position (or the C) from which the nucleofuge departs is called the *ipso* position (or C) and the one next to it is called the *cine* position. We can also refer to the *ipso* and *cine* substitution products.

A clue about the reaction mechanism of these substitutions is found in the reaction of *p*-bromotoluene with sodium amide in liquid ammonia, which gives equal yields of *p*- and *m*-substituted products (eqn 18.10).

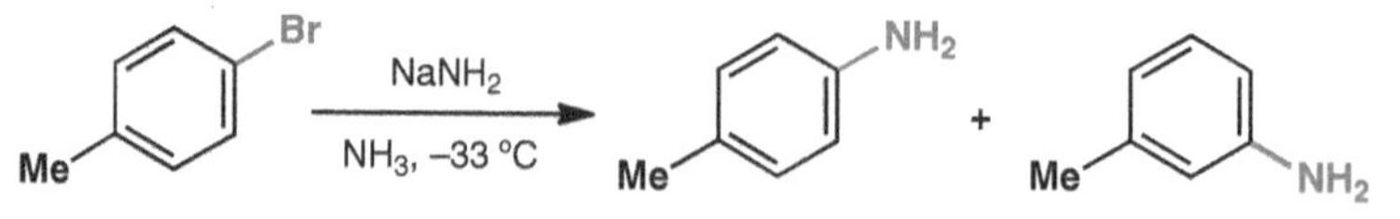

p-bromotoluene   *p*-methylaniline (*ipso* product)   50   :   *m*-methylaniline (*cine* product)   50   (18.10)

This result is best accommodated by an electrophilic intermediate which may be captured equally well at positions *m* and *p* to the methyl group; such an intermediate can be generated by the elimination of HBr with a very strong base, as shown in Scheme 18.14.

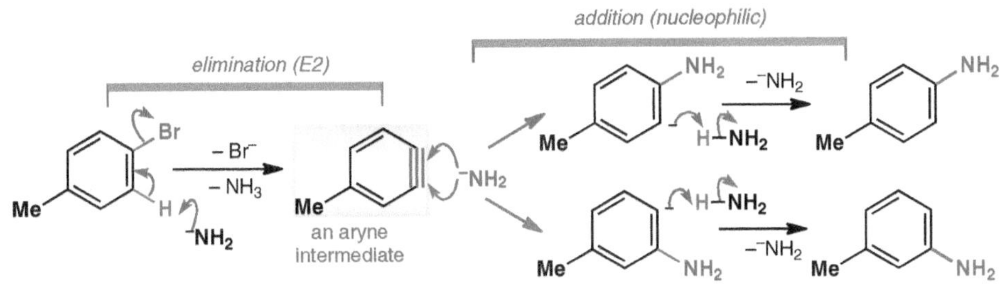

Scheme 18.14 The elimination–addition (E2 benzyne) mechanism for the nucleophilic substitution reaction of p-bromotoluene.

The intermediate is an **aryne** (a substituted **benzyne**) because it has a triple bond, and the **elimination–addition mechanism** is also known as the benzyne mechanism.

> Benzyne is also called dehydrobenzene.

The result shown in eqn 18.11 using chlorobenzene labelled with ^{14}C convincingly shows that the methyl group of p-bromotoluene in the reaction of eqn 18.10 had not significantly affected the outcome (see below).

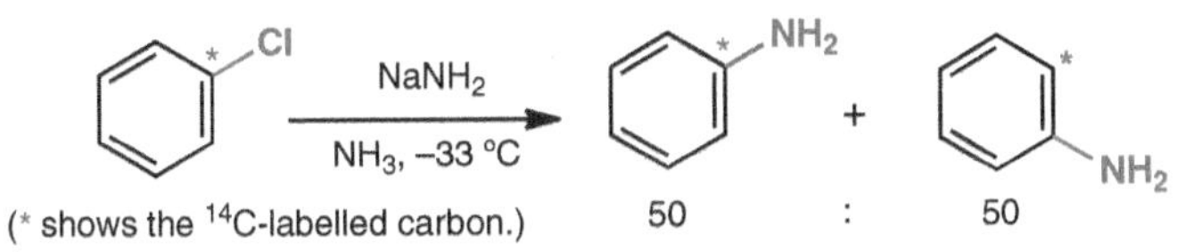

(* shows the ^{14}C-labelled carbon.) 50 : 50 (18.11)

The electrophilic benzyne intermediate may be formed either by the E2 mechanism (as shown in Scheme 18.14) or by the E1cB (as shown in Scheme 18.15) prior to its immediate nucleophilic capture by the amide ion.[1]

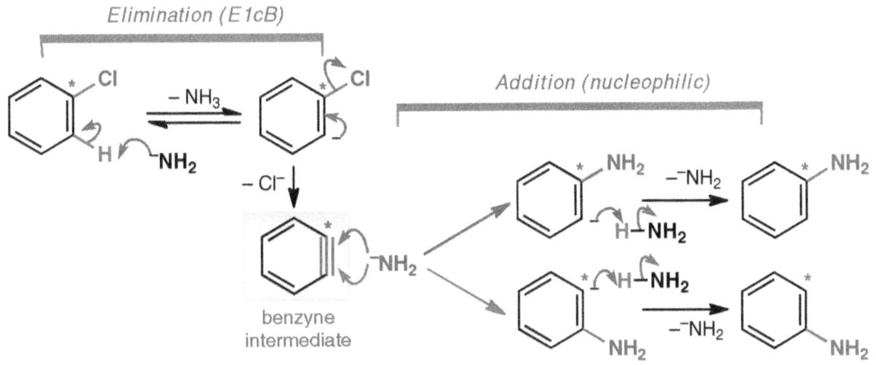

Scheme 18.15 The elimination–addition (E1cB benzyne) mechanism for the nucleophilic substitution reaction of labelled chlorobenzene; * shows the position of the ^{14}C label.

The intermediate benzyne has an unusual structure with a highly strained triple bond within the six-membered ring. One of the π bonds of this triple bond is normal and contributes to the aromatic π system, while the other is perpendicular to the aromatic π

---

[1] The reactions of bromoarenes and iodoarenes occur by the E2 mechanism, while those of chloroarenes are by the E1cB (the leaving abilities decrease in the order I > Br > Cl >> F). The reaction of 2,6-dideuteriobromobenzene is slower than that of normal bromobenzene (kinetic isotope effect), which shows that deprotonation is involved in the rate-determining step (E2). In contrast, isotope scrambling was observed during the reaction of 2,6-dideuteriochlorobenzene, indicating that the deprotonation step is reversible (E1cB). See Chapter 13 for details of elimination mechanisms.

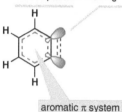

poor overlap of sp² AOs in the plane of the ring

aromatic π system

Figure 18.1 The electronic structure of benzyne.

system and formed by poor side-to-side overlap of the approximately sp² hybrid atomic orbitals which lie in the plane of the six-membered ring (Figure 18.1). This second π bond is very weak and the origin of the high reactivity of benzyne.

Although the methyl substituent in bromotoluene did not affect the regioselectivity in the addition to the aryne implicated in the reaction of eqn 18.10, this is not generally true and o-bromoanisole gives essentially the *cine* product, *m*-methoxyaniline (Scheme 18.16). The observed regioselectivity here is explained by the relative stabilities of the alternative intermediate carbanions formed by nucleophilic addition to the benzyne intermediate, and the transition structures leading to them. The intermediate with the negative charge localized between the inductively electron-withdrawing methoxy and amino groups will be more stable than the one in which the negative charge is more remote from the MeO substituent. Note that the lone pair of the anion is accommodated in a localized sp² orbital which cannot interact conjugatively with the aromatic π system because it is in the plane of the six-membered ring, i.e. perpendicular to the aromatic π system.

Scheme 18.16
Regioselectivity in the benzyne mechanism for a nucleophilic substitution reaction of o-bromoanisole.

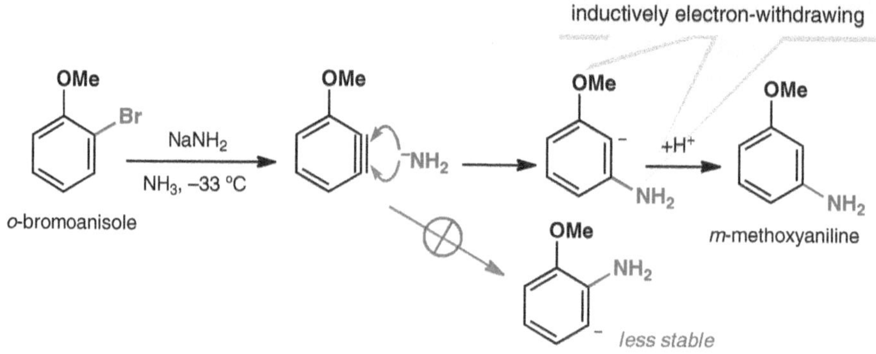

inductively electron-withdrawing

OMe
Br
NaNH₂
NH₃, −33 °C
o-bromoanisole

OMe
⁻NH₂

OMe
⁻
NH₂
+H⁺

OMe
NH₂
m-methoxyaniline

OMe
NH₂
⁻
less stable

Exercise 18.16

Explain why 1-bromo-2,6-dimethylbenzene does not react in liquid ammonia containing sodium amide.

Exercise 18.17

Predict the main product(s) in each of the following reactions.

(a)
Me
Cl
NaNH₂
NH₃, −33 °C

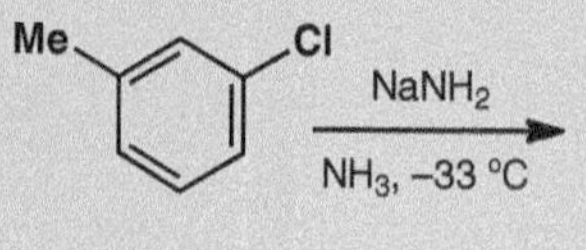

(b)
Br
NaNH₂
CF₃
NH₃, −33 °C

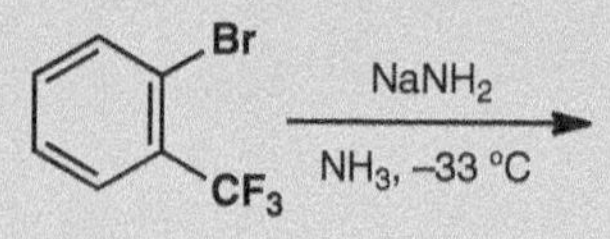

18.8 Reactions of Arenediazonium Salts

As described in Sub-section 16.6.2, arenediazonium salts can be prepared from anilines with cooling in ice-water (eqn 18.12), and are relatively stable in acidic media compared with their aliphatic analogues (see Sub-section 14.7.2).

$$Ar{-}NH_2 \xrightarrow[\text{H}_2\text{O, 0–5°C}]{\text{NaNO}_2,\ \text{HX}} Ar{-}N_2{}^+\ X^-$$

an aniline   an arenediazonium salt

(18.12)

─────── **Panel 18.2** Benzyne intermediates ───────

Benzyne is an unstable intermediate and its formation is rate determining in the elimination–addition mechanism of aromatic nucleophilic substitutions under the strongly basic conditions described in Section 18.7, so its isolation or even detection was impossible. However, it is possible to generate benzyne under milder conditions in the absence of nucleophiles by using a precursor which has a good nucleofuge and a good electrofuge at adjacent positions of benzene. Benzyne can then be generated, characterized by spectroscopic methods, and trapped, e.g. in a Diels–Alder reaction by a diene (see Sections 15.9 and 21.2). Two benzyne precursors are given in the following scheme: diazotization of o-aminobenzoic acid provides a diazonium carboxylate in which N_2 and CO_2 are an excellent nucleofuge and electrofuge; the good nucleofuge and electrofuge in the silylphenyliodonium cation are PhI and Me_3Si^+. Both precursors were used to generate benzyne under mild conditions, and it could be trapped by furan in a Diels–Alder reaction, for example.

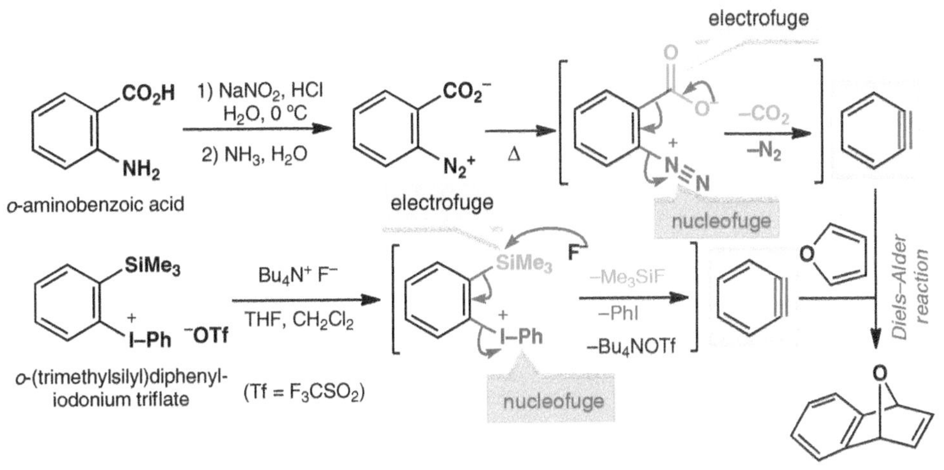

Arenediazonium ions react as electrophiles with various nucleophiles (including HO^- in non-acidic aqueous solution); in particular, they undergo diazo coupling reactions with nucleophilic aromatic compounds such as phenols and amines (see Sub-section 16.6.2).

Because the nitrogen molecule is such a good nucleofuge, arenediazonium ions can undergo substitution reactions by the S_N1 *mechanism*. In practice, however, the reaction is only useful for solvolytic reactions (i.e. when the solvent is the nucleophile which traps the transient aryl cation) because Ar^+ is too short-lived to be intercepted by dilute nucleophilic solutes. For example, when an aqueous acidic solution of an arenediazonium salt is heated, the corresponding phenol is obtained (eqn 18.13) and, if the water is replaced by an alcohol ROH, aryl ethers ArOR are obtained.

$$Ar\text{–}N_2^+ \xrightarrow[\Delta]{-N_2} [\,Ar^+\,] \xrightarrow[-H^+]{H_2O} Ar\text{–}OH \qquad (18.13)$$

an aryl cation

Introduction of iodine is achieved using an iodide salt in aqueous solution (eqn 18.14), but the mechanism remains obscure.

$$Ar\text{–}N_2^+ \, Cl^- \xrightarrow[H_2O]{KI} Ar\text{–}I \qquad (18.14)$$

Depending on the nature of Ar, high yields of the reduction product Ar-H are obtained if ethanol or acidic aqueous methanol is used as the solvent. These reactions involve radical rather than cationic intermediates and are often superior to the older method using hypophosphorous acid (H_3PO_2) as the reducing agent in the conversion of $Ar\text{-}NH_2$ to Ar-H (see Sub-section 16.6.2).

CuCl, CuBr, and CuCN are insoluble in water but may be used as suspensions in the Sandmeyer reaction; alternatively, they can be used as salts of the soluble anionic complexes $CuCl_2^-$ in hydrochloric acid, $CuBr_2^-$ in hydrobromic acid, or $Cu(CN)_2^-$ in aqueous KCN, respectively.

Reactions with chloride, bromide, and cyanide are effectively carried out in the presence of Cu(I) compounds (eqn 18.15). These reactions (often called *Sandmeyer reactions*) are mechanistically complex and involve radicals formed in electron transfer processes.

$$Ar-N_2^+ \; HSO_4^- \xrightarrow[\text{H}_2\text{O}]{\text{CuZ}} Ar-Z \qquad (18.15)$$

$$(Z = Cl, Br, CN)$$

As implied in eqn 18.15, fluoroarenes cannot be made by the Sandmeyer reaction (CuF is unknown). If a crystalline arenediazonium tetrafluoroborate is cautiously heated, however, the fluoroarene is produced (eqn 18.16), but yields are variable.

$$Ar-NH_2 \xrightarrow[\text{0 °C}]{\text{NaNO}_2, \text{HBF}_4} \underset{\text{a crystalline salt}}{Ar-N_2^+ \; BF_4^-} \xrightarrow{\Delta} Ar-F \qquad (18.16)$$

In the so-called Balz–Schiemann reaction, the salt is heated mixed with an inert solid (e.g. sand) to prevent localized overheating; however, it must only be carried out on a small scale with stringent safety precautions as some arenediazonium salts decompose explosively.

Exercise 18.18

What is the main product in each of the following reactions?

(a) [structure: 2-methylaniline] 1) $NaNO_2$, dil. HCl, 0 °C 2) CuCl

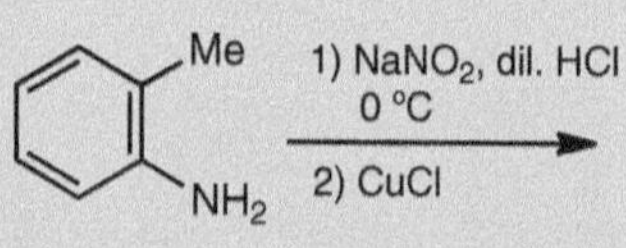

(b) [structure: 3-nitroaniline] 1) $NaNO_2$, dil. H_2SO_4, 0 °C 2) Δ

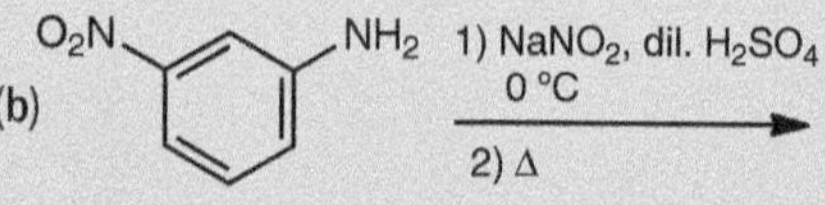

(c) [structure: 2-bromoaniline] 1) $NaNO_2$, dil. HBF_4, −5 °C 2) isolate salt, Δ

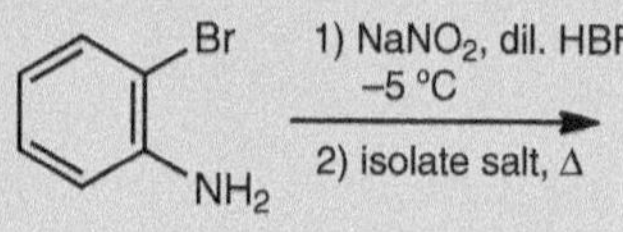

(d) [structure: 4-methoxyaniline] 1) $NaNO_2$, dil. HCl, 0 °C 2) CuCN

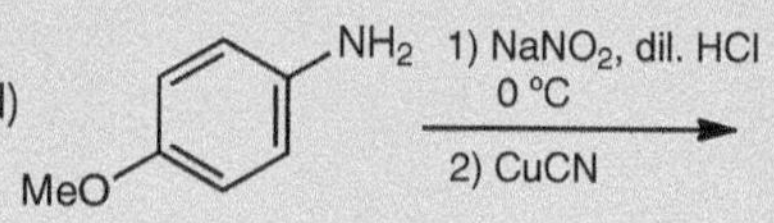

Summary

- A C=C bond conjugated with an electron-withdrawing group such as C=O, CN, or NO_2 is electrophilic and susceptible to nucleophilic attack at the β carbon (**conjugate addition**) to give an intermediate enolate ion (or its equivalent). Subsequent protonation at the α carbon corresponds to overall addition to the C=C bond.

- In the case of α,β-unsaturated carbonyl compounds, conjugate addition followed by protonation of the enolate on oxygen yields the enol (**1,4-addition**) which isomerizes to the more stable keto form. Conjugate addition competes with **carbonyl addition** (also known as 1,2-addition) and the regioselectivity depends on the structures of both the substrate and the nucleophile.

- Carbonyl addition is often reversible and the product of conjugate addition is usually more stable, so the thermodynamic product is generally that of conjugate addition.

- Nucleophiles for these reactions include amines, CN⁻, RO⁻, HO⁻, RSH, and RS⁻, as well as metal hydrides, organometallic reagents, and enolates and their equivalents (conjugate additions of enolates to α,β-unsaturated carbonyl compounds are known as **Michael reactions**).

- α,β-Unsaturated carbonyl compounds also undergo acid-catalysed conjugate additions of hydrogen halides, and acid- and base-catalysed conjugate additions of alcohols.

- Halobenzenes carrying a strongly electron-withdrawing group in the *o*- and/or *p*- position(s) undergo nucleophilic addition followed by departure of halide ion resulting in overall **nucleophilic substitution**.

- Un-activated halobenzenes undergo nucleophilic substitution only under strongly basic conditions, e.g. sodium amide in liquid ammonia, by the elimination–addition mechanism via a **benzyne** intermediate.

- Arenediazonium salts undergo solvolytic substitution reactions by the S_N1 mechanism (N_2 is an excellent nucleofuge) and other substitutions in the so-called **Sandmeyer reactions**.

Problems

18.1 What is the main product in each of the following reactions?

(a) 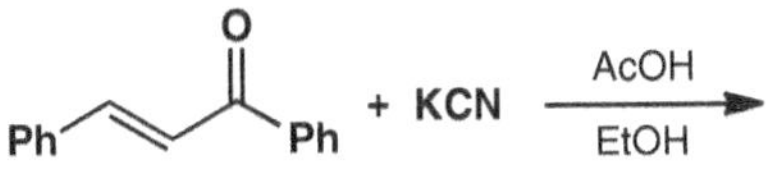 Ph—CH=CH—C(=O)—Ph + KCN $\xrightarrow[\text{EtOH}]{\text{AcOH}}$

(b) Ph—CH=C(CN)—Ph + KCN $\xrightarrow[\text{H}_2\text{O, MeOH}]{}$

(c) 2 CH$_2$=CH—CO$_2$Et + MeNH$_2$ $\xrightarrow{\text{EtOH}}$

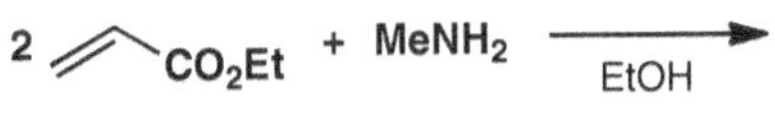

(d) 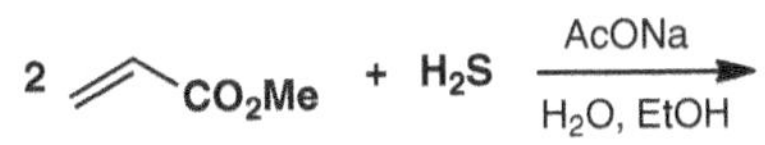 2 CH$_2$=CH—CO$_2$Me + H$_2$S $\xrightarrow[\text{H}_2\text{O, EtOH}]{\text{AcONa}}$

(e) 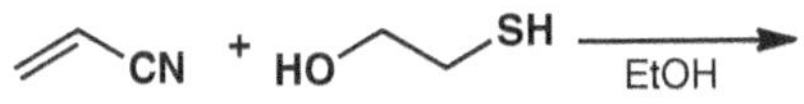 CH$_2$=CH—CN + HO—CH$_2$CH$_2$—SH $\xrightarrow{\text{EtOH}}$

18.2 What is the main product in each of the following reactions?

(a) 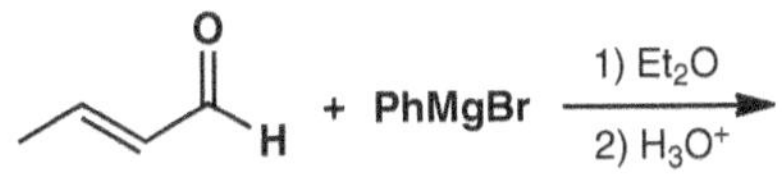 CH$_3$CH=CH—CHO + PhMgBr $\xrightarrow[\text{2) H}_3\text{O}^+]{\text{1) Et}_2\text{O}}$

(b) 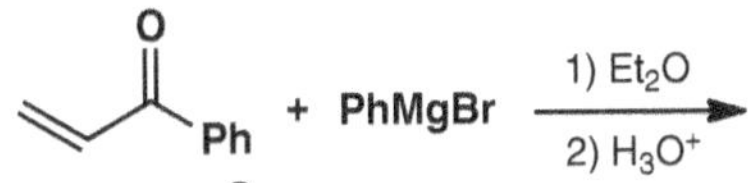 CH$_2$=CH—C(=O)—Ph + PhMgBr $\xrightarrow[\text{2) H}_3\text{O}^+]{\text{1) Et}_2\text{O}}$

(c) Ph—CH=CH—C(=O)—Ph + PhLi $\xrightarrow[\text{2) H}_3\text{O}^+]{\text{1) C}_6\text{H}_6}$

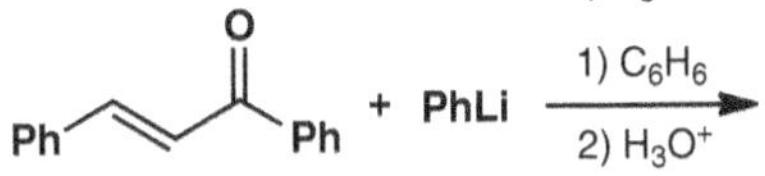

(d) 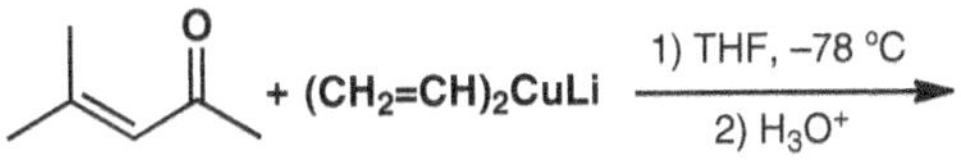 (CH$_3$)$_2$C=CH—C(=O)—CH$_3$ + (CH$_2$=CH)$_2$CuLi $\xrightarrow[\text{2) H}_3\text{O}^+]{\text{1) THF, −78 °C}}$

18.3 What is the main product in each of the following reactions?

(a) 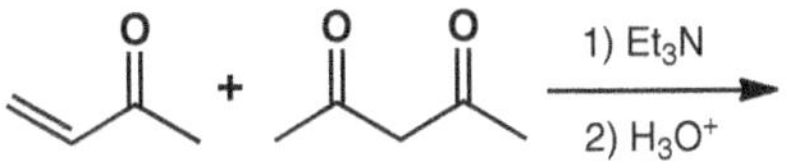 CH$_2$=CH—C(=O)—CH$_3$ + CH$_3$—C(=O)—CH$_2$—C(=O)—CH$_3$ $\xrightarrow[\text{2) H}_3\text{O}^+]{\text{1) Et}_3\text{N}}$

(b) 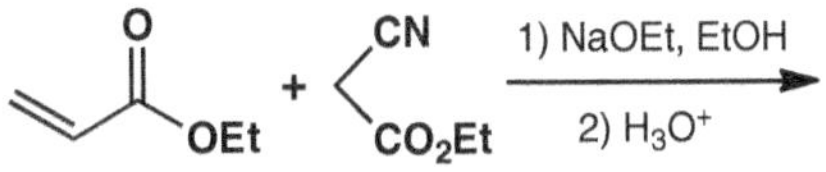 CH$_2$=CH—C(=O)—OEt + NC—CH$_2$—CO$_2$Et $\xrightarrow[\text{2) H}_3\text{O}^+]{\text{1) NaOEt, EtOH}}$

(c) CH$_2$=CH—CN + CH$_2$(CO$_2$Me)$_2$ $\xrightarrow[\text{2) H}_3\text{O}^+]{\text{1) NaOMe, MeOH}}$

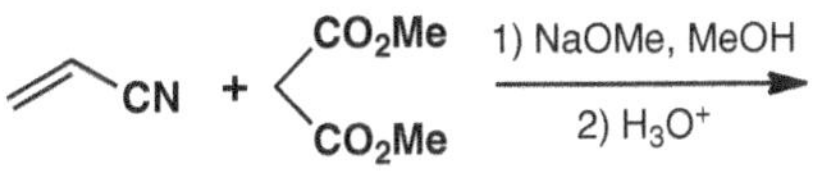

(d) 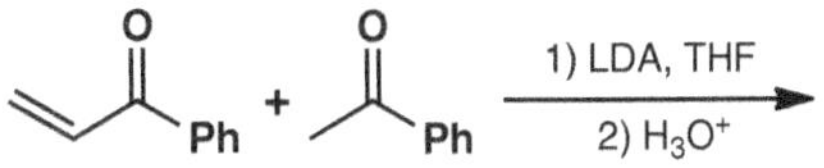 CH$_2$=CH—C(=O)—Ph + CH$_3$—C(=O)—Ph $\xrightarrow[\text{2) H}_3\text{O}^+]{\text{1) LDA, THF}}$

(LDA = *i*-Pr$_2$NLi)

18.4 What are the main products in the following reactions?

(a) 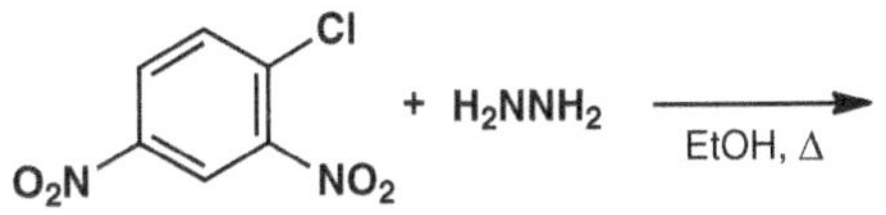 (chlorobenzene with O$_2$N and NO$_2$, Cl) + H$_2$NNH$_2$ $\xrightarrow{\text{EtOH, }\Delta}$

(b) 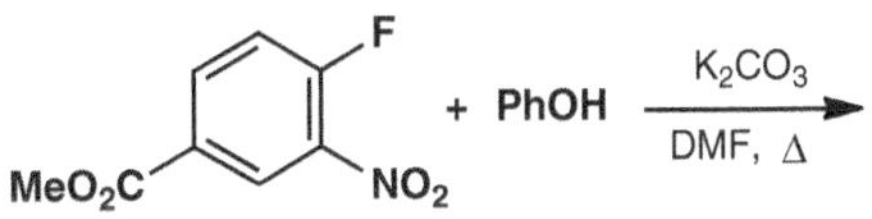 (fluorobenzene with MeO$_2$C and NO$_2$, F) + PhOH $\xrightarrow[\text{DMF, }\Delta]{\text{K}_2\text{CO}_3}$

(c) (dichlorobenzene with O$_2$N and NO$_2$, Cl, Cl) + NH$_3$ $\xrightarrow[\text{HO}\frown\text{OH}]{\Delta}$

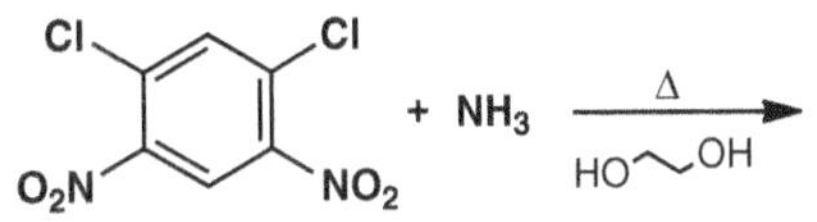

18.5 Give a mechanism for the following transformation.

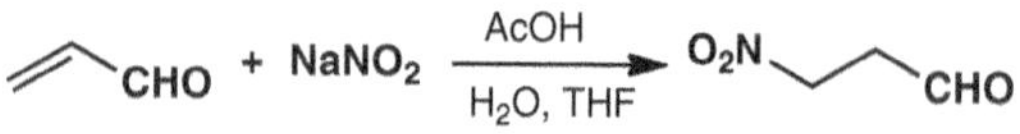

18.6 Indicate the reagent(s) required for the following transformations, and show reaction mechanisms for the first and last steps.

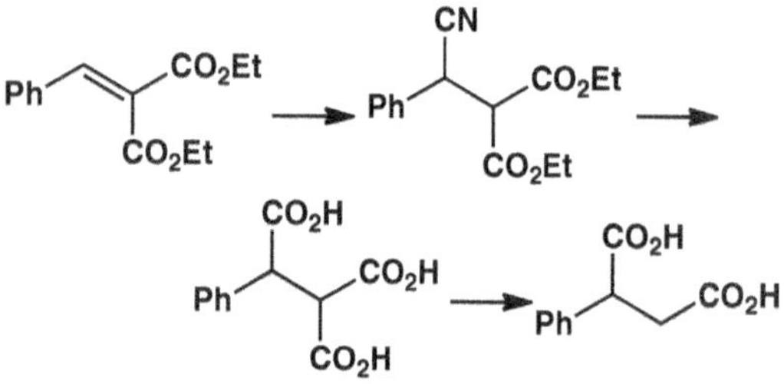

18.7 Propose a method using a conjugate addition for the preparation of each of the following compounds.

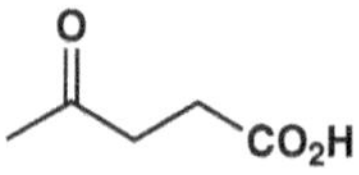

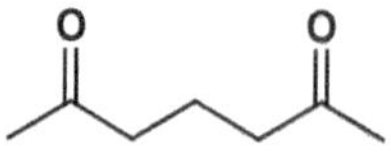

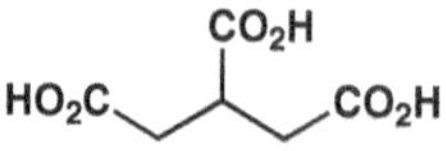

18.8 Show the structure of the product of each of the following Robinson annulations, and give a curly-arrow mechanism for reaction (a).

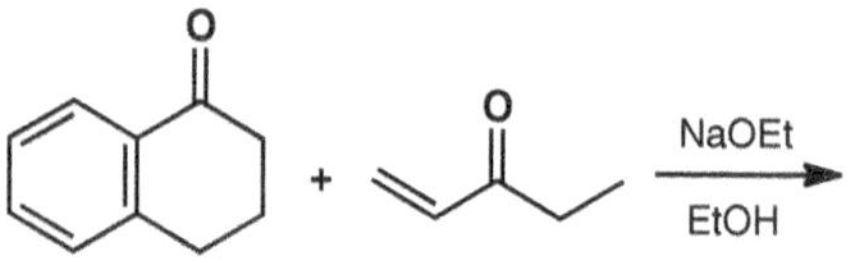

18.9 Give a mechanism for the following transformation.

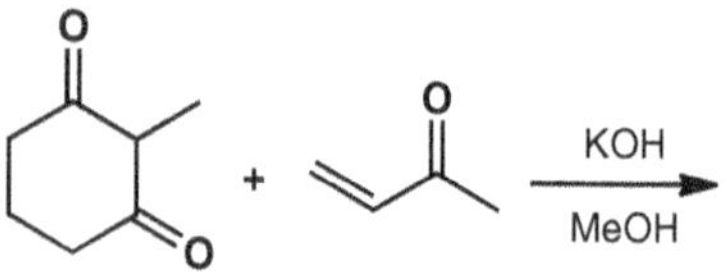

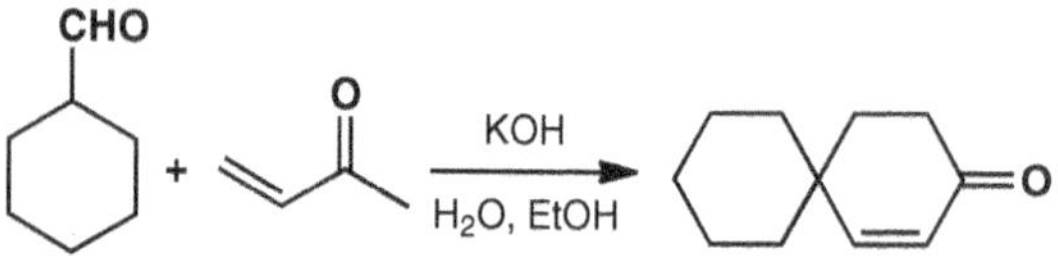

18.10 Propose reactions to obtain each of the isomeric products, **A** and **B**, from pent-3-en-2-one.

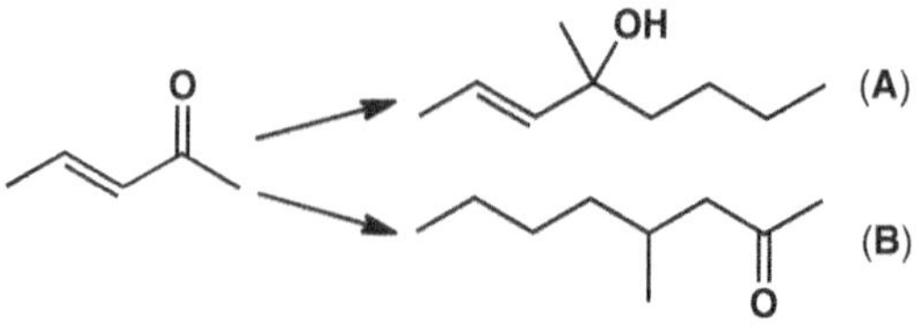

18.11 Give mechanisms to account for the formation of the two main products shown in the following reaction. Note that the tetra-alkylammonium cation facilitates the reaction of hydroxide as a base in this reaction by phase-transfer catalysis (see Panel 12.2).

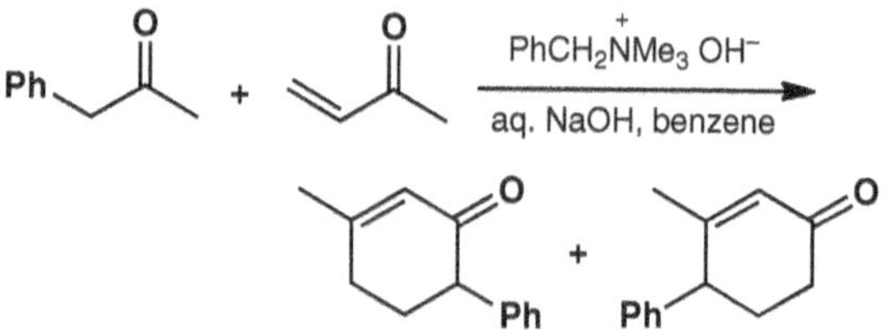

18.12 Indicate the reagents necessary to carry out the following transformation, and explain the reactions involved in each step.

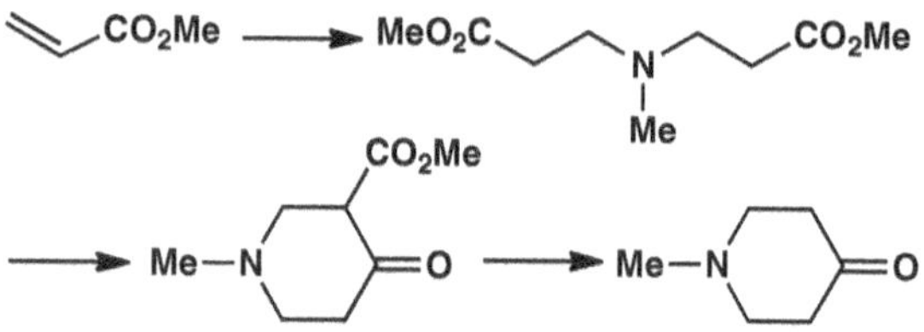

18.13 Propose reactions to obtain the product shown from ethyl propenoate.

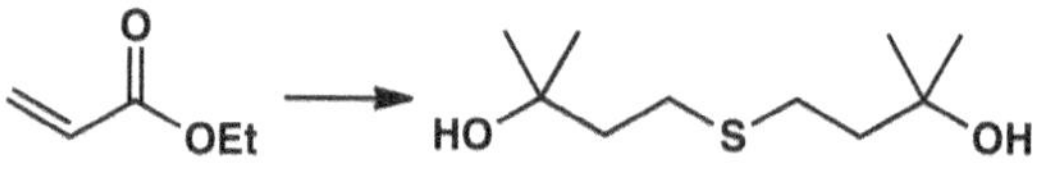

18.14 Explain the relative reactivities in the following sets of compounds in nucleophilic substitution reactions in methanol containing sodium methoxide.

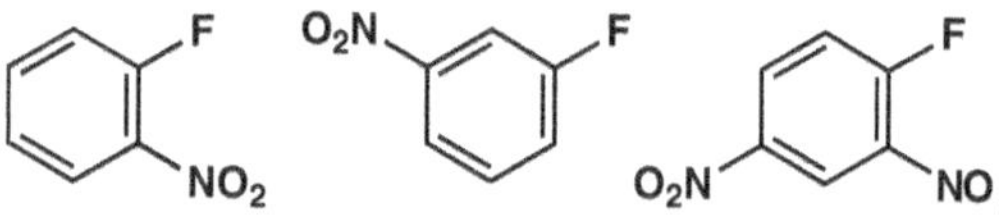

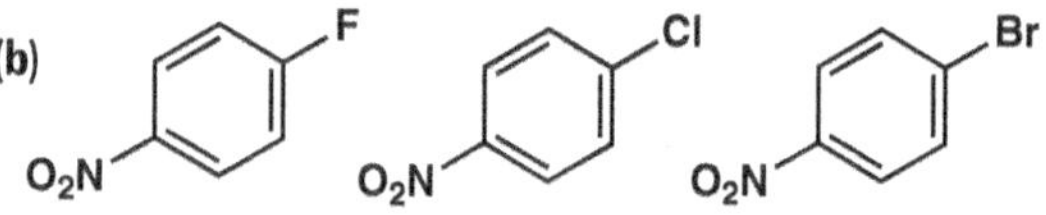

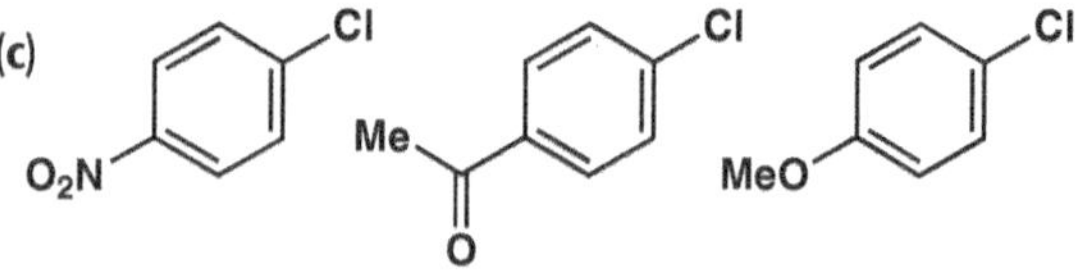

18.15 Predict the main products of the following reactions, and explain the regioselectivity.

(a)

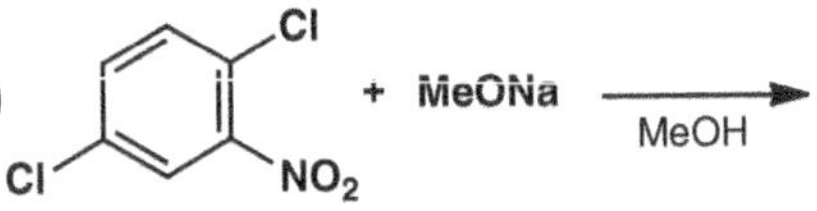

+ MeONa $\xrightarrow{\text{MeOH}}$

(b)

+ PhSH $\xrightarrow[\text{EtOH}]{\text{Et}_3\text{N}}$

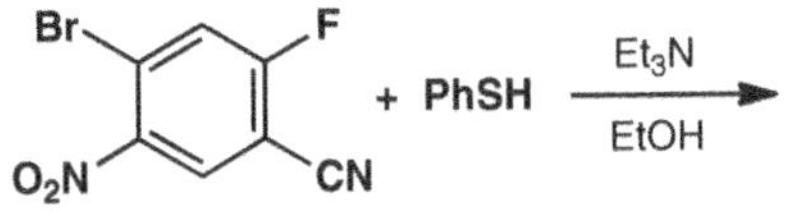

18.16 Give a mechanism for the following transformation.

+ t-BuOK $\xrightarrow[\text{130 °C}]{\text{t-BuOH, DMSO}}$

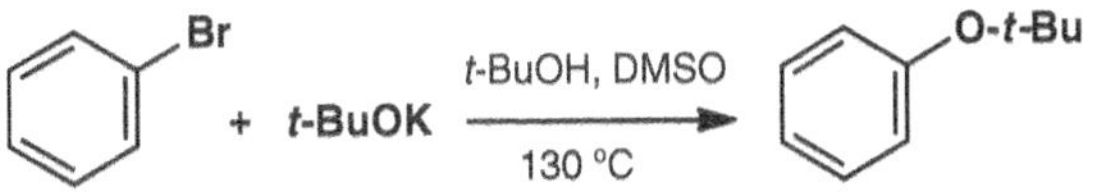

18.17 Predict and account for the main products of the following reaction.

$\xrightarrow[\text{NH}_3, -33\ °C]{\text{NaNH}_2}$

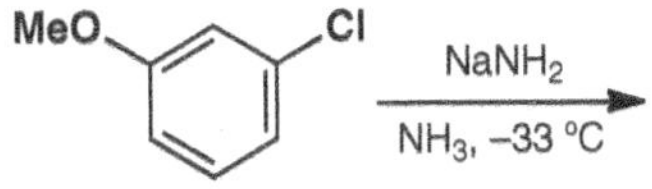

Supplementary Problems

18.18 α,β-Unsaturated carbonyl compounds can be prepared by dehydration of aldols or by elimination reactions of α-halogenated carbonyl compounds. Propose reactions for the preparation of the following.

(a)

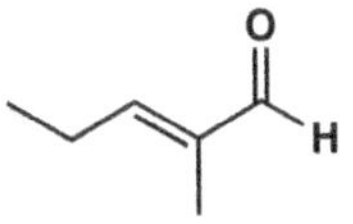

(b)

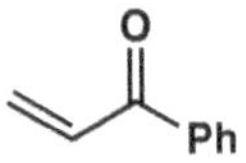

(c)

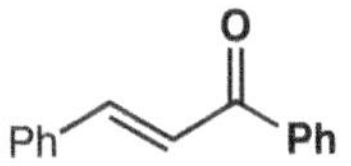

(d)

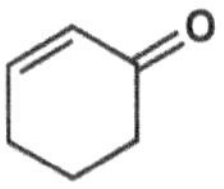

18.19 Give a mechanism for the following transformation.

$\xrightarrow[\text{0 °C}]{\text{HCl, EtOH}}$

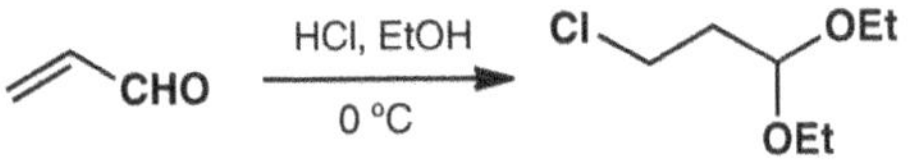

18.20 Identify two possible products of the following reaction. Give a mechanism which describes the formation of the main product, and explain the regioselectivity.

$\xrightarrow[\text{2) H}_2\text{O}]{\text{1) NaNH}_2, \text{NH}_3}$

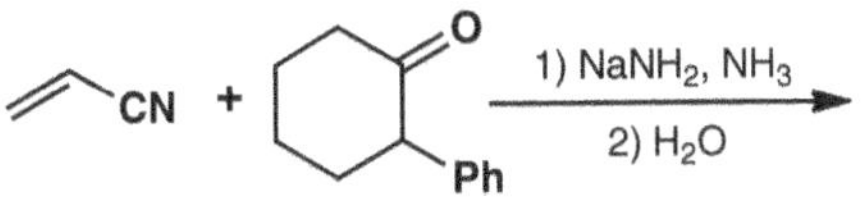

18.21 Although benzene cannot easily be converted directly into phenol, this conversion can be achieved by a sequence of reactions via aniline. Give reagents and essential reaction conditions for such a sequence of reactions.

18.22 Propose a method for the preparation of each of the following from chlorobenzene.

(a) p-nitroaniline   (b) o-nitrophenol

18.23 Propose a method for the preparation of each of the following from aniline.

(a) 1-cyano-4-nitrobenzene

(b) 1-bromo-2-fluorobenzene

18.24 Propose a method for the preparation of each of the following from benzene.

(a) p-t-butylphenol   (b) m-t-butylphenol

(c) p-bromoanisole   (d) m-bromoanisole

18.25 Some esters of p-hydroxybenzoic acid known as parabens are used as preservatives in cosmetic and pharmaceutical formulations. Propose a method for the preparation of methyl p-hydroxybenzoate (methylparaben, E218) from toluene.

18.26 Give a mechanism for each of the following transformations.

(a)

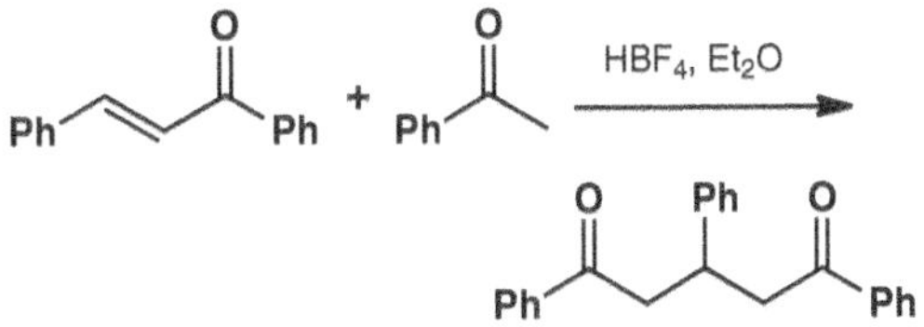

$\xrightarrow{\text{HBF}_4, \text{Et}_2\text{O}}$

(b)

$\xrightarrow[\text{2) H}_3\text{O}^+]{\text{1) AlCl}_3}$

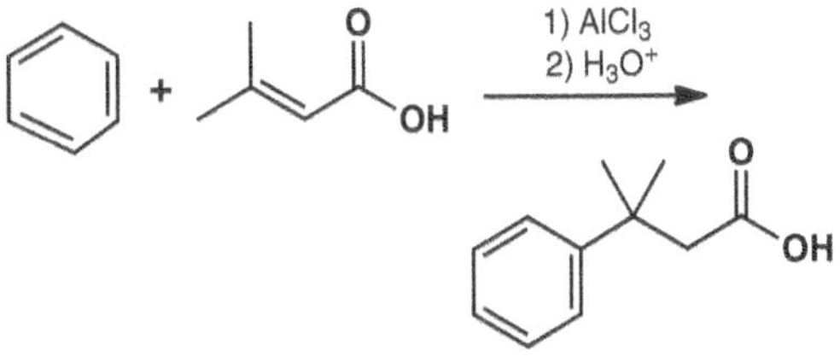

18.27 When o-fluoronitrobenzene reacts with 2-amino-2-phenylethanol in the presence of a base in

DMF (*N,N*-dimethylformamide), the product depends on the nature of the base. Give the structure of the product obtained with (a) K₂CO₃, and (b) NaH, and explain the result in each case.

18.28 The following scheme summarizes transformations of benzene to give various derivatives. Add reagents and any essential conditions required for the reactions by the sides of the arrows.

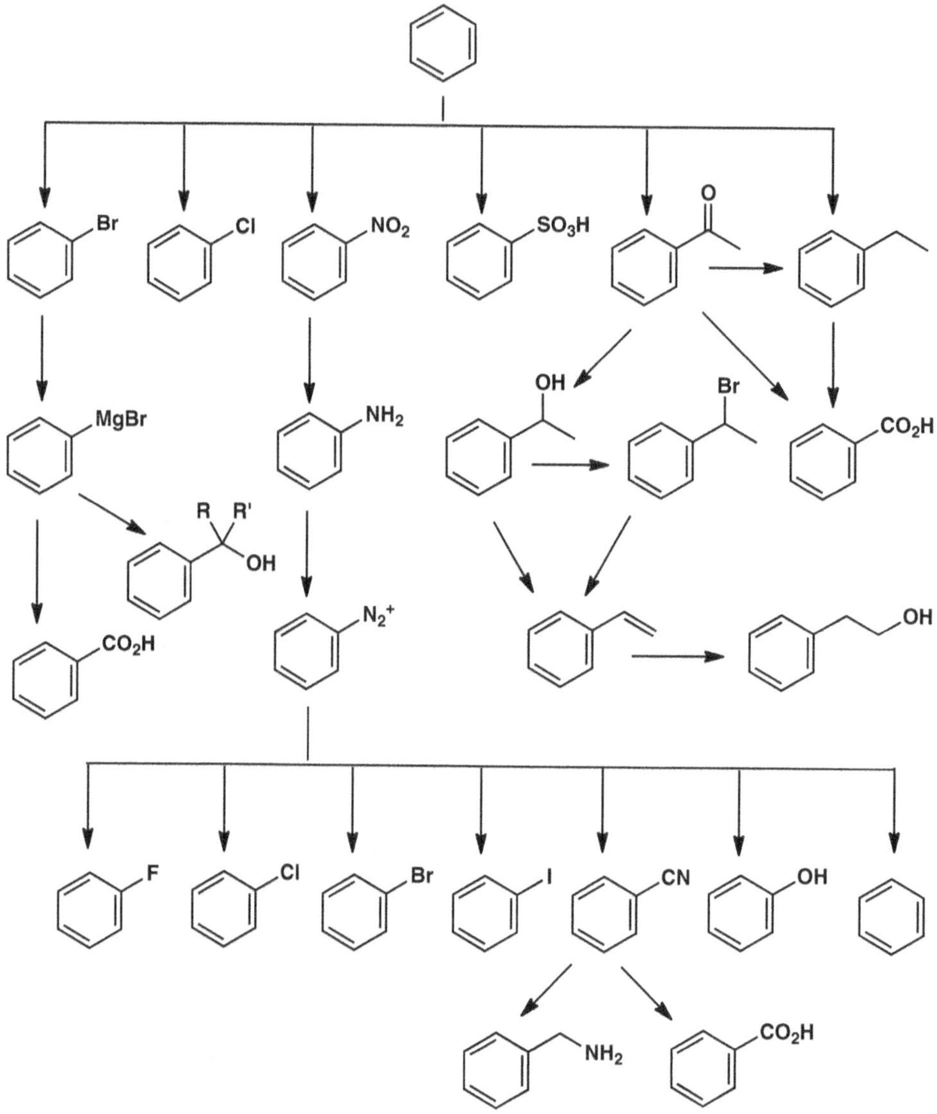

Polycyclic and Heterocyclic Aromatic Compounds

19

Related topics we have already covered:

- **Conjugation and aromaticity** (Chapter 5)
- **Acids and bases** (Chapter 6)
- **Curly arrow representation of reaction mechanisms** (Section 7.2)
- **Reactions of ethers, sulfides, and amines** (Chapter 14)
- **Electrophilic aromatic substitution** (Chapter 16)
- **Nucleophilic aromatic substitution** (Chapter 18)

Topics we shall cover in this chapter:

- **Structures of polycyclic aromatic hydrocarbons**
- **Reactions of polycyclic aromatic compounds**
- **Structures of aromatic heterocycles**
- **Acid-base properties of aromatic heterocycles**
- **Reactions of five-membered aromatic heterocycles: pyrrole, furan, and thiophene**
- **Reactions of pyridine and its derivatives**
- **Synthesis of aromatic heterocycles**

Many aromatic compounds are polycyclic and/or contain one or more heteroatoms in the ring(s). Most **polycyclic aromatic compounds** consist of two or more **benzenoid** rings fused together but some contain **non-benzenoid** rings, e.g. acenaphthylene and azulene. Some polycyclic aromatic compounds including benzopyrenes are carcinogenic.

The most common **heterocyclic compounds** contain one or more of three heteroatoms (N, O, and S) in the ring, and may be saturated or unsaturated. The chemistry of saturated heterocycles is generally similar to that of their acyclic counterparts, and we have already encountered some cyclic ethers, acetals, and amines in previous chapters. In particular, three-membered cyclic oxiranes (epoxides) have ring strain and are highly reactive, as discussed in Section 14.5.

Unsaturated heterocyclic compounds are numerous and many are based on five- and six-membered aromatic rings containing a heteroatom, the following being the most common.

(Z = NH, O, S)

pyridine

pyrimidine

The six-membered aromatic heterocyclic compounds shown have a lone pair (in blue) on each N which is not part of the aromatic 6π electron system whereas the lone pairs (in red) on the heteroatoms of the five-membered heterocycles shown are parts of the 6π electron systems. Aromatic heterocyclic compounds are also called **aromatic heterocycles** or **heteroaromatic compounds**, and some have fused ring structures. Many naturally occurring and biologically important compounds, as well as medicinal drugs, contain aromatic heterocyclic features as illustrated below. In particular, some amino acids contain a heteroaromatic group, and the base components in nucleic acids are heteroaromatic, as we shall see in Chapter 24. Also, an important family of naturally occurring bases found in plants, the alkaloids, are invariably nitrogen-containing heterocyclic compounds, and a selection is given in Panel 19.3.

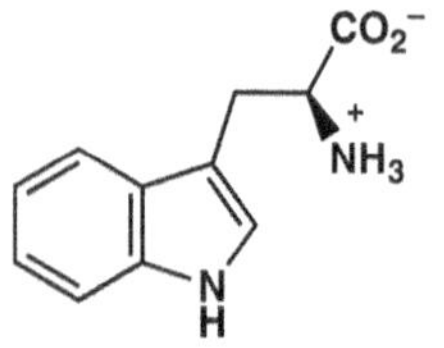

tryptophan
(amino acid)

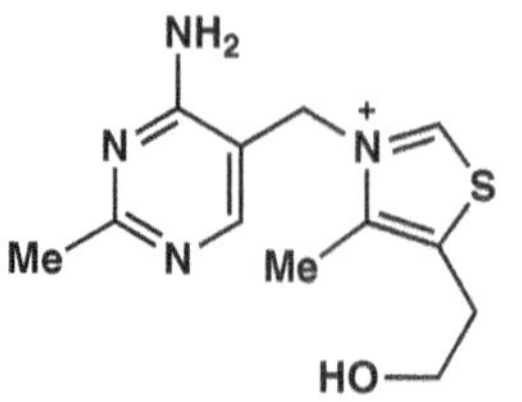

thiamine
(vitamine B₁)

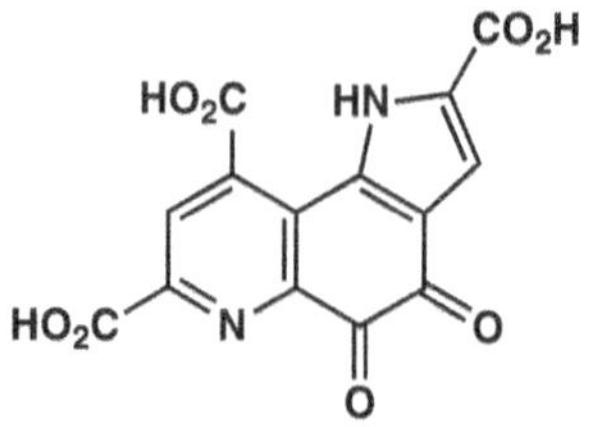

pyrroloquinoline quinone (PQQ)
(a coenzyme for
alcohol dehydrogenase)

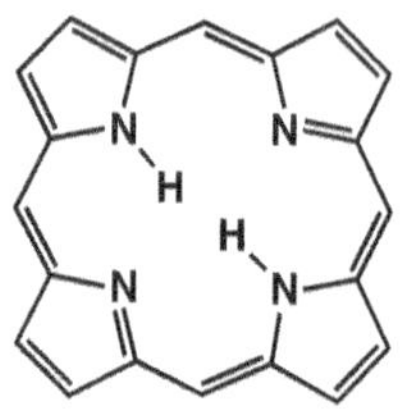

porphine
(the parent skeleton
of porphyrins)

19.1 Polycyclic Aromatic Compounds

19.1.1 Structures of polycyclic aromatic hydrocarbons

Structures of some representative wholly benzenoid **polycyclic aromatic compounds** are given below with position numbers in red.

Aromatic hydrocarbons containing non-benzenoid rings:

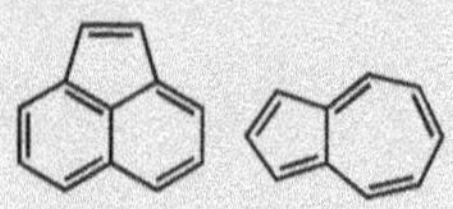

acenaphthylene   azulene

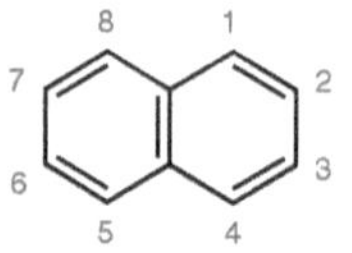

naphthalene

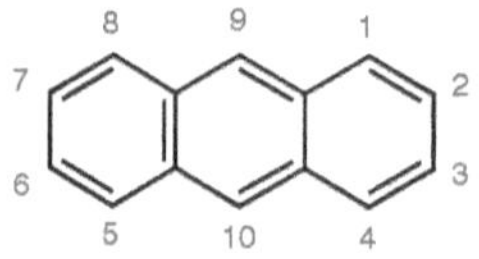

anthracene

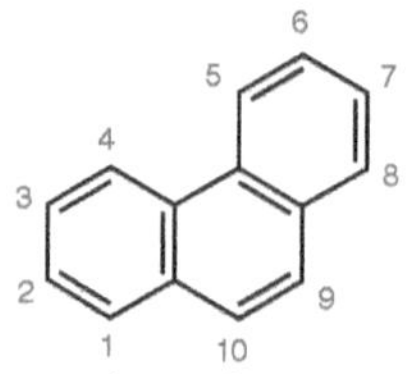

phenanthrene

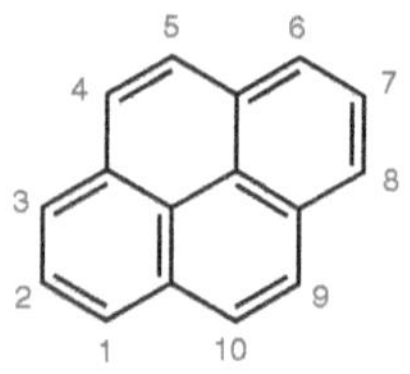

pyrene

The symmetry of any particular molecule indicates which of its C–C bonds are equivalent, and non-equivalent bonds are often of different lengths. For example, the 1,2-bond of naphthalene is shorter than the 2,3-bond. This difference can be rationalized by the resonance description of naphthalene; the 1,2-bond is double in two of the three resonance contributors, while the 2,3-bond is double in only one of them.

The resonance energy of naphthalene is about 250 kJ mol⁻¹, which is less than twice that of benzene, $152 \times 2 = 304$ kJ mol⁻¹ (Sub-section 5.5.3).

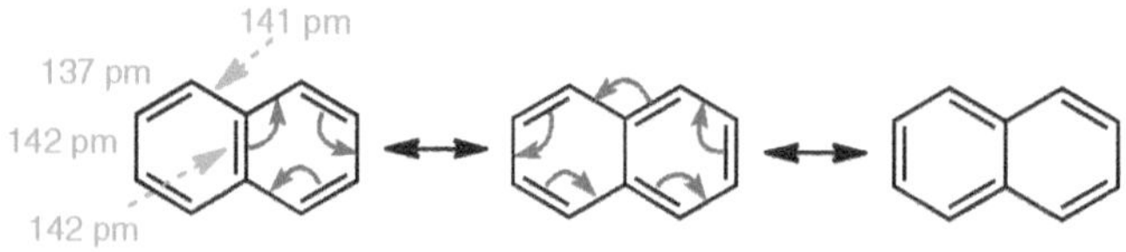

141 pm
137 pm
142 pm
142 pm

Resonance representation of naphthalene.

Exercise 19.1

Draw the four resonance contributors of anthracene and the five of phenanthrene.

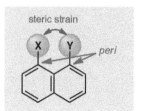

steric strain

X   Y
                peri

Positions 1 and 8 of naphthalene are sometimes called the *peri* positions, and the bonds at these positions are parallel to each other; groups bonded to C1 and C8 are close together which leads to steric strain.

Panel 19.1 Graphene, nanotubes, and fullerenes

Benzene rings fused together in two dimensions (as a film or sheet) give a structure known as **graphene**. It is a very extended condensed aromatic molecule, but a single sheet just one-atom thick is very difficult to isolate because of the strong attractive stacking interactions between the sheets. After searches carried out for several decades, graphene was successfully isolated in 2004 as a single sheet on a silicon wafer. Graphene sheets stack together in layers to give **graphite**, the familiar allotrope of carbon.[1] If the graphene structure forms a cylinder rather than a flat plane sheet, it is called a **carbon nanotube**.

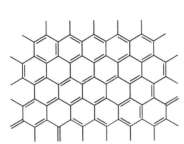

Graphene.
(nobeastsofierce/istockphoto.)

A carbon nanotube.
(nobeastsofierce/istockphoto.)

As indicated, graphene is composed of regular hexagonal rings comprising sp^2 hybridized carbons which, like hexagonal tiles, form a flat plane. If pentagonal ring units are incorporated, the structure becomes concave and the end of a carbon nanotube can be closed with a hemi-spherical cap in this way. If appropriate numbers of hexagonal and pentagonal carbon rings are fused, a spherical or ellipsoidal molecular structure can be made entirely with sp^2 hybridized carbon atoms. These hollow (caged) molecules are called **fullerenes**. When 20 hexagons and 12 pentagons are joined as in a soccer ball, the result is a spherical C_{60} molecule called **buckminsterfullerene** (named after Richard Buckminster Fuller, the designer of geodesic domes which the C_{60} molecule resembles) and is sometimes called buckyball.[2] The C_{70} fullerene is a typical ellipsoidal molecule.

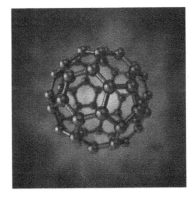

Buckminsterfullerene.
(nobeastsofierce/istockphoto.)

These sp^2 carbon allotropes have semi-metallic properties and are electrical conductors or semiconductors because each carbon has a p orbital containing a single electron. They have been the subject of intensive research because of their structural novelty and possible application in materials science, electronics, optics, and in nanotechnology to develop nanodevices (nanoscale semiconductors and circuits).

---

[1] Allotropes are different structural modifications of an element, and diamond (sp^3 hybridized C atoms) and graphite (sp^2 hybridized C atoms) have long been known to be allotropes of carbon; for the structure of diamond, see Sub-section 4.3.3. *Fullerenes* were recognized as a new allotrope of carbon (sp^2 hybridized C atoms) in the 1980s.

[2] The C_{60} fullerene was first discovered by Sir Harry Kroto (University of Sussex) and Robert Curl and Richard Smalley (Rice University) in 1985, and they were awarded the 1996 Nobel Prize in Chemistry. Carbon nanotubes were discovered by Sumio Iijima (NEC Corporation) in 1991.

19.1.2 Reactions of polycyclic aromatic hydrocarbons

a. Electrophilic substitution reactions

Polycyclic aromatic hydrocarbons are generally more reactive than benzene in electrophilic substitution reactions because the loss of aromatic stabilization energy associated with the formation of the transition structures leading to the arenium ion intermediates is less than in the corresponding reaction of benzene itself. As an example, the bromination of naphthalene takes place without a Lewis acid catalyst (eqn 19.1).

Positions 1 and 2 of naphthalene are sometimes designated α and β, respectively, so 1-bromonaphthalene may also be called α-bromonaphthalene.

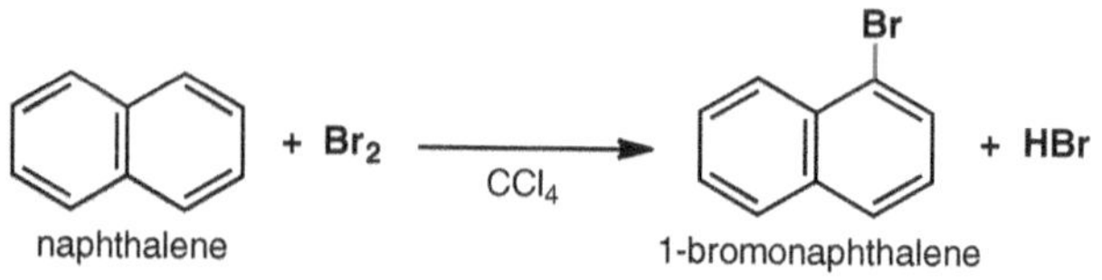

naphthalene + Br$_2$ $\xrightarrow{CCl_4}$ 1-bromonaphthalene + HBr (19.1)

Electrophilic attack at naphthalene occurs preferentially at C1 to form an intermediate naphthalenium ion which is more stable than the one formed by attack at C2, as anticipated from their resonance representations. The C1 intermediate can be described by seven contributors, four of which retain a complete benzene ring, while the C2 intermediate has six contributors only two of which have a benzene ring intact.

- Naphthalenium ion formed by electrophilic addition to C1:

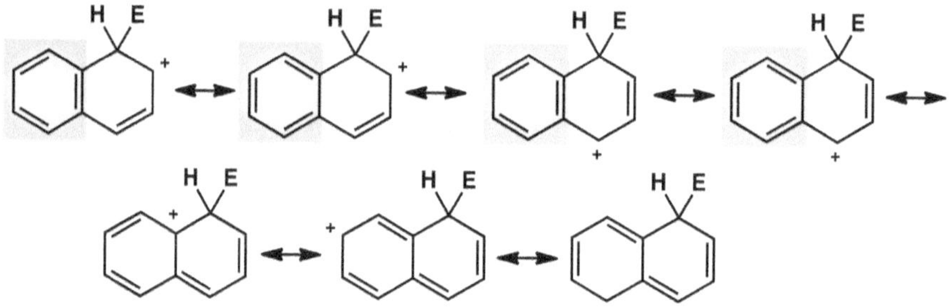

- Naphthalenium ion formed by electrophilic addition to C2:

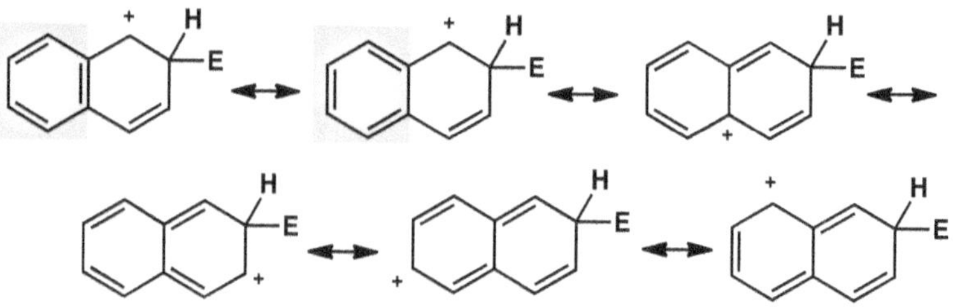

Sulfonation of naphthalene gives naphthalene-1-sulfonic acid at lower temperatures, but it isomerizes to the 2-sulfonic acid at higher temperatures, as shown in Scheme 19.1 (remember—aromatic electrophilic sulfonation is reversible, Sub-section 16.3.3). This is an example of kinetic and thermodynamic control: the 1-isomer is the kinetic product, but it is less stable than the 2-isomer owing to the 1,8-steric interaction.

Scheme 19.1 Sulfonation of naphthalene: kinetic versus thermodynamic control.

naphthalene-1-sulfonic acid (kinetic product) — naphthalene — naphthalene-2-sulfonic acid (thermodynamic product)

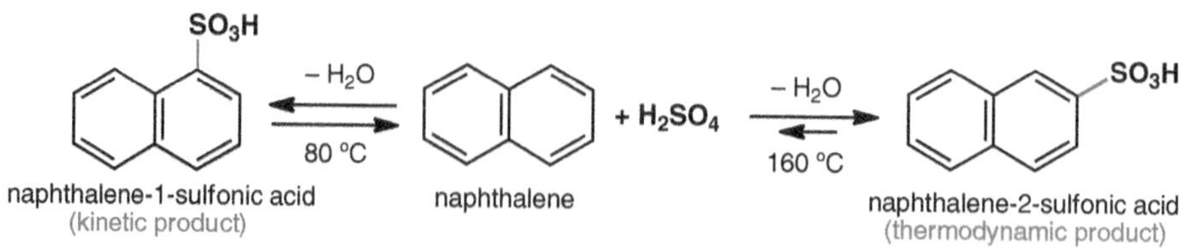

Electrophilic substitution reactions of both anthracene and phenanthrene generally take place at C9 (or C10), as illustrated in reactions 19.2 and 19.3.

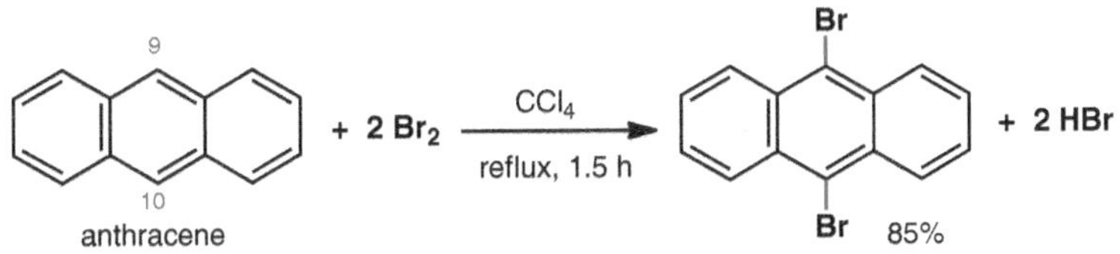

anthracene + 2 Br$_2$ $\xrightarrow[\text{reflux, 1.5 h}]{\text{CCl}_4}$ (product) + 2 HBr   (19.2)

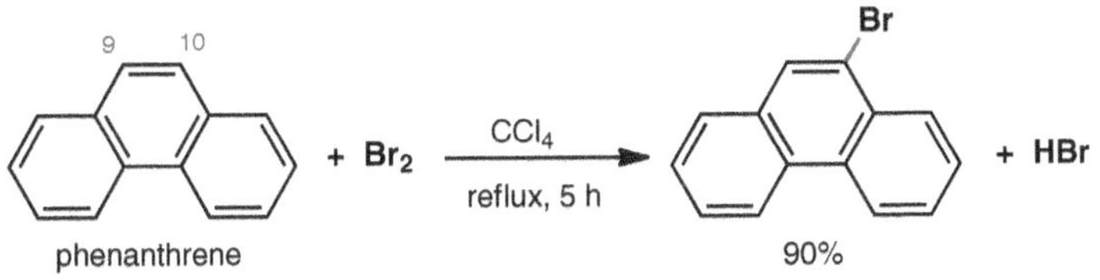

85%

phenanthrene + Br$_2$ $\xrightarrow[\text{reflux, 5 h}]{\text{CCl}_4}$ (product) + HBr   (19.3)

90%

The regioselectivities in eqns 19.2 and 19.3 reflect the relative stabilities of the C9 (or C10) intermediates compared with those following electrophilic attack at any other position, i.e. the loss of resonance energy is smaller upon formation of the C9 (or C10) intermediates. This is because the total resonance energy of the two separate benzene rings in the C9 intermediates from anthracene and phenanthrene is larger than the resonance energy of the naphthalene or biphenyl residues in the intermediates following attack at C1 (or any other position of the outer rings).

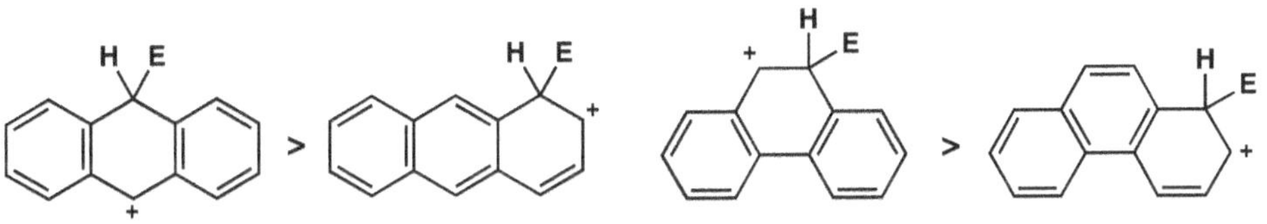

See Sections 15.9 and 21.2 for Diels–Alder reactions.

Example 19.1

The tendency of anthracene to undergo substitution rather than addition can be finely balanced, and depends upon the nature of the reagent and the reaction conditions. Give a reaction scheme to describe the formation of an addition product with bromine which gives 9-bromoanthracene upon dehydrobromination.

Solution

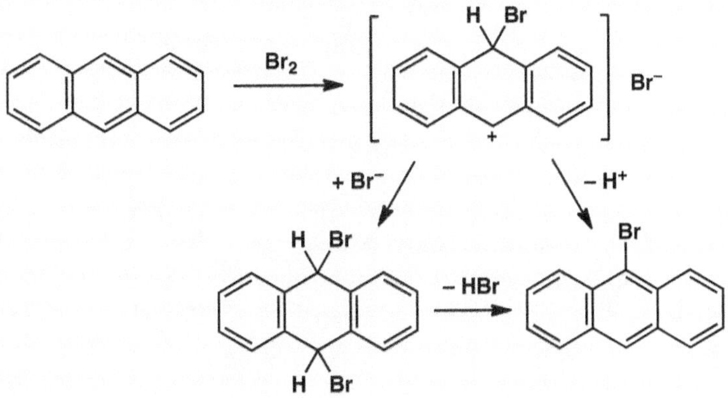

b. Other reactions

Owing to the relatively small loss of aromatic stability upon reaction across its central ring, anthracene can react as a diene in Diels–Alder reactions, e.g. with maleic anhydride (eqn 19.4).

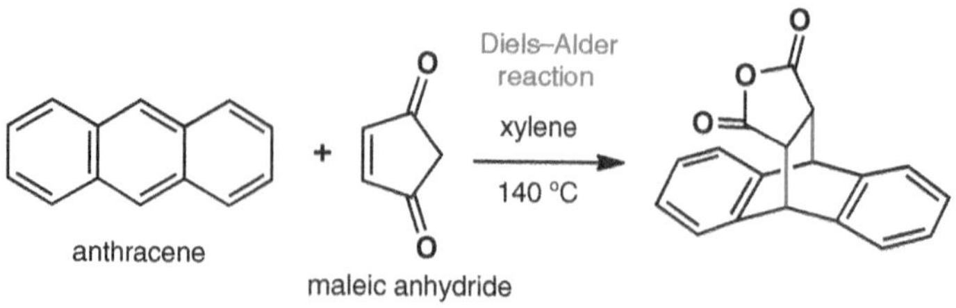

anthracene + maleic anhydride

Diels–Alder reaction

xylene

140 °C

(19.4)

For the same reason, the central rings of anthracene and phenanthrene are readily oxidized to give 9,10-quinones, e.g. eqn 19.5 (see Panel 16.2 for quinones).

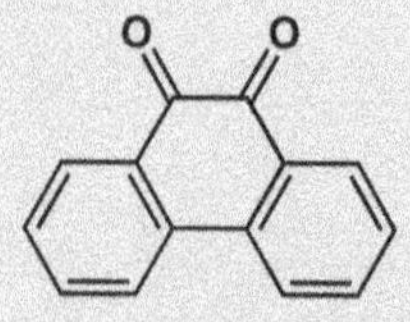

9,10-phenanthraquinone

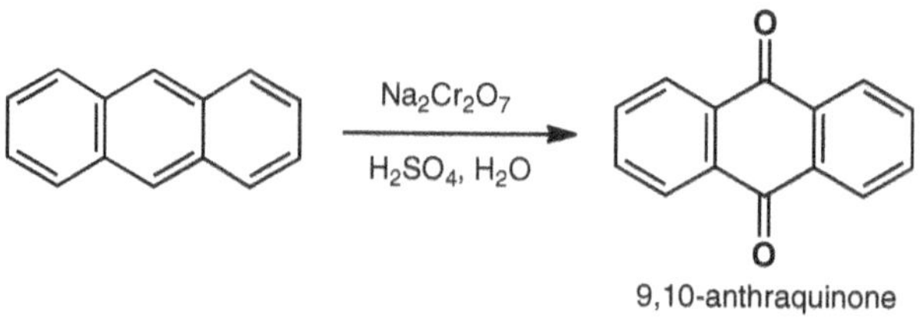

$Na_2Cr_2O_7$

H_2SO_4, H_2O

9,10-anthraquinone

(19.5)

Exercise 19.2

Both anthracene and phenanthrene are easily reduced to give dihydro compounds. Predict the structure of the dihydro product formed from each of them.

Panel 19.2 Carcinogenicity of polycyclic aromatic compounds: epoxide intermediates and detoxification

Benzopyrene (formed by the incomplete combustion of many organic materials and found in cigarette smoke) and aflatoxin B_1 (a fungal toxin found as a contaminant on nuts and grains) are two of the most carcinogenic/mutagenic compounds known. Their toxicity is initiated when they are oxidized within the body by cytochrome P-450, an oxidative metabolic process. Epoxides formed in this process are highly electrophilic and normally are trapped by biological nucleophiles such as glutathione (G-SH), whereupon they are detoxified and excreted.

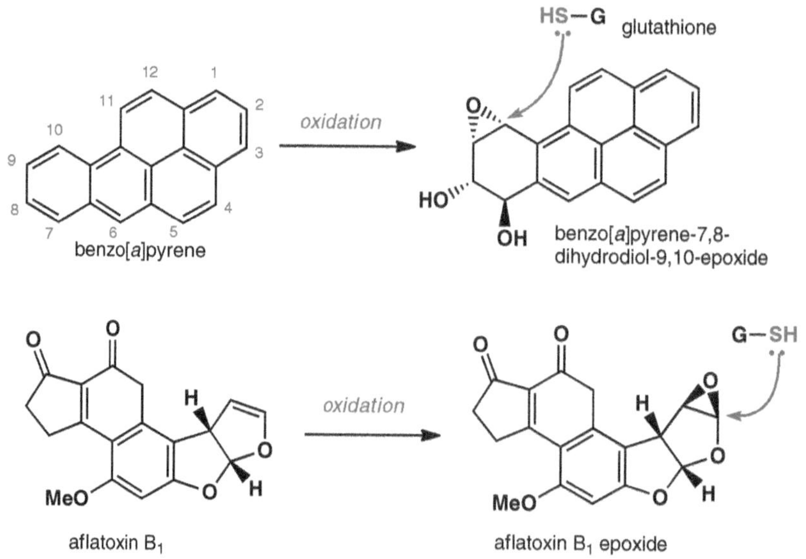

benzo[a]pyrene

oxidation

HS—G glutathione

benzo[a]pyrene-7,8-dihydrodiol-9,10-epoxide

aflatoxin B_1

oxidation

G—SH

aflatoxin B_1 epoxide

However, if the bases of DNA trap the epoxides, the DNA is modified and the integrity of its replication is compromised, which can lead to genetic damage and cancerous growth.

19.2 Structures of Aromatic Heterocyclic Compounds

We saw in Section 5.6 that compounds containing cyclic delocalized $4n + 2$ π electron systems are aromatic, and representative examples with six π electrons ($n = 1$) include benzene and cyclopentadienide ion. The corresponding heterocycles obtained by replacing one or more of the CHs in benzene or cyclopentadienide with one or more heteroatoms are aromatic six- and five-membered heterocyclic compounds. In other words, six- and five-membered heteroaromatics are isoelectronic with benzene and cyclopentadienide, respectively (Figure 19.1). Benzo-fused heteroaromatic compounds are also common; like naphthalene, they have 10π electron systems.

Note that only one lone pair of electrons (red in Figure 19.1) on an N, O, or S of a five-membered heterocycle is involved in its 6π electron system, as shown by the resonance contributors of pyrrole below; other lone pairs (in blue) remain localized.

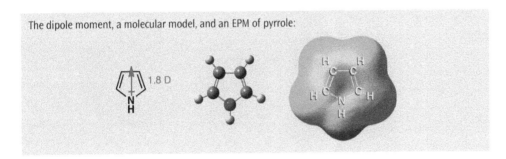

Resonance representation of pyrrole.

The dipole moment, a molecular model, and an EPM of pyrrole:

1.8 D

The contrast between the lone pair on the N of pyrrole (representative of aromatic five-membered heterocycles) and the one on the N of pyridine (representative of aromatic six-membered heterocycles) is illustrated in Figure 19.2. The pyrrole lone pair is in the

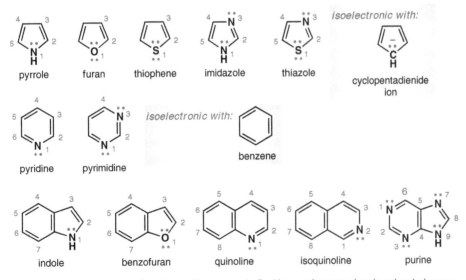

isoelectronic with:

pyrrole   furan   thiophene   imidazole   thiazole   cyclopentadienide ion

isoelectronic with:

pyridine   pyrimidine   benzene

indole   benzofuran   quinoline   isoquinoline   purine

Six-membered aromatic heterocycles containing a heteroatom other than N are possible but not common. Examples include 2-pyrone and the pyrilium ion.

2-pyrone

pyrilium ion

Figure 19.1 Common aromatic heterocyclic compounds. Position numbers are given in red and when we refer to C3, for example, we mean the C at position 3.

Figure 19.2 The 2p orbitals and the lone pairs on N in pyrrole and pyridine.

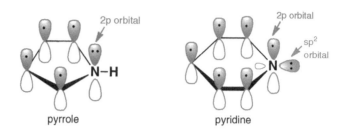

pyrrole   pyridine

unhybridized 2p orbital of the N and part of the aromatic 6π system whereas the pyridine lone pair is in an sp^2 hybrid orbital orthogonal to the π system.

The conjugative electron donation from the N of pyrrole, for example, into the five-membered ring implied by the resonance structures shown above opposes the electron-attracting inductive effect of N; this leads to a dipole moment (1.8 D)[3] with the N the *positive* end, and facilitates electrophilic substitution (Sub-section 19.4.1). There is no such effect to oppose the high electronegativity of the heteroatom in pyridine which is reflected in its high dipole moment (2.26 D with N the negative end) and low reactivity with electrophiles (Sub-section 19.4.2). These different charge distributions are apparent in the EPMs of pyrrole and pyridine.

The dipole moment, a molecular model, and an EPM of pyridine:

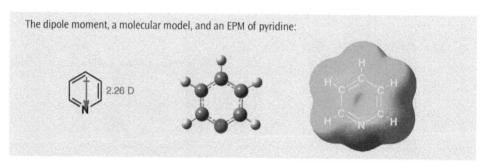

2.26 D

Exercise 19.3

Represent the structure of furan by resonance, and explain why the dipole moment of furan is smaller than that of its saturated analogue, tetrahydrofuran.

Exercise 19.4

Sketch the nitrogen atomic orbital in imidazole which contributes a pair of electrons to the π system, and the one which accommodates the localized lone pair.

19.3 Acid–Base Properties of Heteroaromatic Compounds containing Nitrogen Atoms

19.3.1 Basicity of nitrogen-containing heteroaromatic compounds

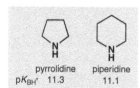

pyrrolidine   piperidine
pK_{BH^+} 11.3   11.1

The nitrogen atoms in heterocyclic compounds are usually basic sites, and saturated nitrogen heterocycles are similar in base strength to normal amines with pK_{BH^+} values of about 11. However, the basicity of nitrogen aromatic heterocycles depends on the

---

3   The unit of dipole moment is D (debye): $1\ D \approx 3.34 \times 10^{-30}\ Cm$.

electronic state of the N involved. As we saw in Example 5.3 of Chapter 5, for example, pyrrole is hardly basic at all because protonation on N would localize the electron pair and prevent aromatic stabilization (Scheme 19.2).

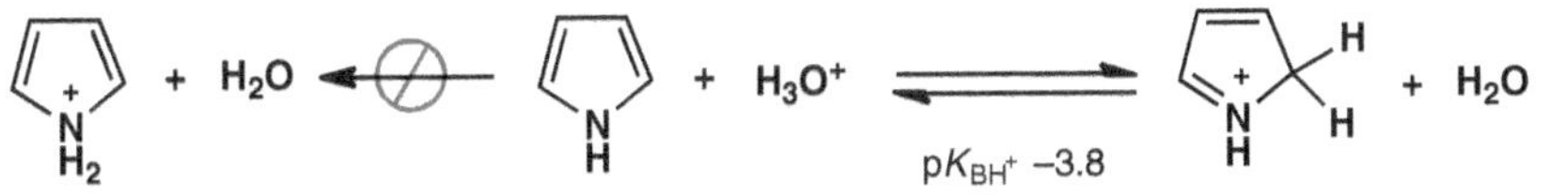

pK_{BH^+} −3.8

Scheme 19.2 Protonation of pyrrole.

Under strongly acidic conditions, however, pyrrole can be protonated ($pK_{BH^+}=-3.8$), but it occurs at C2 rather than at the N (Scheme 19.2). This is because the C2-protonated form, unlike the *N*-protonated form, has significant residual resonance stabilization; even so, this conjugate acid of pyrrole is very reactive and readily polymerizes.

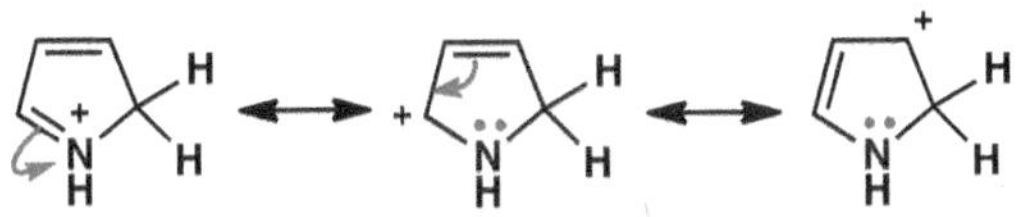

Resonance representation of the C2-protonated conjugate acid of pyrrole.

In contrast to pyrrole, imidazole is moderately basic ($pK_{BH^+}=7.0$) because the lone pair in the sp^2 orbital of the N which accepts the proton is not involved in the aromatic system (eqn 19.6).

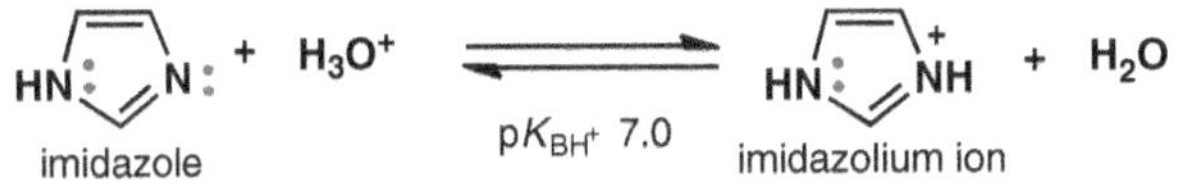

imidazole pK_{BH^+} 7.0   imidazolium ion

(19.6)

Imidazole has two nitrogen atoms. Explain why the basic site is the one indicated in eqn 19.6.

Exercise 19.5

Pyridine is a tertiary amine and the nitrogen lone pair is not part of the aromatic system, i.e. it is localized in an sp^2 orbital in the plane of the six-membered ring, as described in the previous section. Consequently, pyridine is moderately basic (eqn 19.7, $pK_{BH^+}=5.23$) but considerably less so than typical saturated tertiary amines ($pK_{BH^+}\sim 10$). Part of this large difference is attributable to the hybridization of the orbitals accommodating the lone pairs: the lone pair of an alkylamine is in an sp^3 orbital while the pyridine lone pair is in an sp^2 orbital (see Sub-section 6.3.1). The lone pair is more tightly bound (so less available for protonation) in the orbital of greater s character, which contributes to the basicity of pyridine being lower than that of an alkylamine.

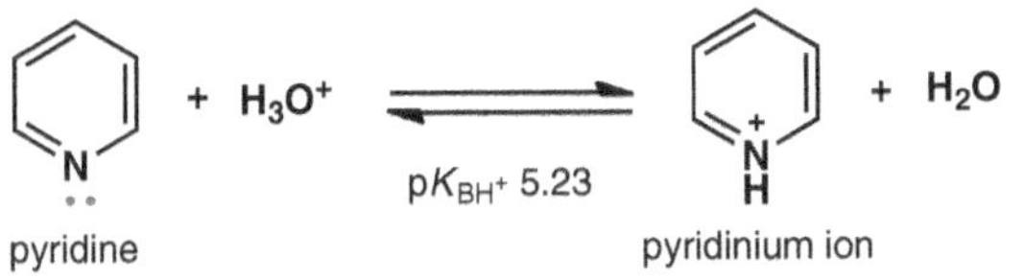

pyridine pK_{BH^+} 5.23   pyridinium ion

(19.7)

The basicities of imidazole and pyridine are both due to a nitrogen lone pair in an sp^2 orbital, but the former is more basic than the latter. The structures of the respective conjugate acids indicate the reason. All the atoms of the two principal equivalent resonance contributors to the overall hybrid of imidazole have full valence shells, and the positive charge is symmetrically delocalized. In contrast, any attempt to delocalize the charge on the pyridinium ion requires resonance forms with atoms having less than full valence shells.

imidazolium ion   pyridinium ion

19.3.2 Acidity of pyrrole and imidazole

Although amines do not normally show acidic properties, pyrrole is considerably more acidic (pK_a=16.5, eqn 19.8) than its saturated analogue, pyrrolidine (pK_a~35). This is partly attributable to the sp^2 hybridized N of pyrrole, which is more electronegative than the sp^3 hybridized N of a saturated amine. The principal reason, however, is that the aromatic stabilization of the conjugate base of pyrrole is greater than that of pyrrole itself—remember that the acidity of cyclopentadiene (pK_a=16) comes from the aromaticity of its conjugate base, cyclopentadienide; see Sub-sections 5.6.1 and 6.4.1.

$$+ \ H_2O \ \rightleftharpoons \ + \ H_3O^+ \qquad (19.8)$$
$$pK_a \ 16.5$$

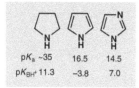

pK_a ~35	16.5	14.5
pK_{BH^+} 11.3	−3.8	7.0

Similarly, the moderately basic imidazole (pK_{BH}^+ = 7.0, discussed above) is also relatively acidic—more so even than pyrrole with pK_a = 14.5 (eqn 19.9). As for pyrrole, this acidity may be explained by the resonance stabilization of the conjugate base being higher than that of imidazole.

$$+ \ H_2O \ \rightleftharpoons \ + \ H_3O^+ \qquad (19.9)$$
$$pK_a \ 14.5$$

Exercise 19.6

Explain using resonance representations why imidazole is more acidic than pyrrole.

19.4 Reactions of Heteroaromatic Compounds

19.4.1 Reactions of pyrrole, furan, and thiophene

a. Electrophilic substitution

These compounds are very reactive towards electrophiles and their main reaction is electrophilic substitution which occurs selectively at C2 (eqns 19.10–19.12).

$$\text{pyrrole} \ + \ (CH_3\overset{O}{\overset{\|}{C}})_2O \ \longrightarrow \ \text{COCH}_3 \qquad (19.10)$$
60%

$$\text{furan} \ + \ (CH_3\overset{O}{\overset{\|}{C}})_2O \ \xrightarrow[\text{AcOH}]{BF_3} \ \text{COCH}_3 \qquad (19.11)$$
75–90%

$$\text{thoiphene} \ + \ AcONO_2 \ \xrightarrow[\text{10 °C}]{Ac_2O} \ \text{NO}_2 \ + \ \text{NO}_2 \qquad (19.12)$$
70%   5%

Panel 19.3 Alkaloids: amines in nature

Alkaloids are nitrogen-containing bases obtained by extraction of plant sources with acid; they are 'alkali-like' organic compounds, hence the origin of the name 'alkaloid'. Most of them are derivatives of heterocyclic amines but some are open-chain amines. Most alkaloids have striking physiological effects on animals: the effects are very different according to the nature of the alkaloid. We have already encountered some of them: **quinine** (an anti-malaria agent) and **coniine** (the powerful toxin which killed Socrates) were mentioned in the Prologue; the opium alkaloids **morphine** and **codeine** are powerful analgesics and were included in Panel 6.3.

Alkaloids have attracted the interest of chemists for centuries and thousands have been isolated; a representative few are given below. The stimulants **caffeine** and **theobromine** are purine derivatives (see Sub-section 24.2.1); the former is found in tea and coffee, the latter in chocolate (cocoa). **Nicotine** is a main alkaloid of the tobacco plant and acts both as a stimulant and a relaxant (but is addictive). Its binding to the nicotinic receptors in the central nervous system increases levels of dopamine in the brain (see Panel 16.4), which may be responsible for its relaxing properties. **Cocaine**, obtained from leaves of the coca plant, is an anæsthetic and a stimulant of the central nervous system. **Atropine**, extracted from the flowering plant commonly called deadly nightshade, is a deadly poison and an antispasmodic agent; it is used as a muscle relaxant in surgery, to dilate pupils for eye examinations, and as an antidote for nerve gas and related poisoning. **Strychnine** and **brucine** are closely related highly toxic bitter compounds which occur in the seeds of the *Strychnos nux vomica* tree (found in South-East Asia). Notice that the last four contain bicyclic pyrrolidine/piperidine structures. **Reserpine** found in the roots of the semitropical plant *Rauwolfia serpentina* is a tranquilizer and antihypertensive agent. **Ephedrine** (which may be extracted from *Ephedra* species) is a stimulant used as a bronchodilator and decongestant in many medicines taken for the common cold (see Panel 16.4).

caffeine (R = Me)
theobromine (R = H)

nicotine

cocaine

atropine

strychnine (R = H)
brucine (R = MeO)

reserpine

ephedrine

This regioselectivity is consistent with the relative stabilities of the alternative intermediate cations formed by addition at C2 and C3, as deduced from their resonance descriptions shown generically in eqns 19.13 and 19.14.

- Electrophilic attack at C2:

$$(19.13)$$

• Electrophilic attack at C3:

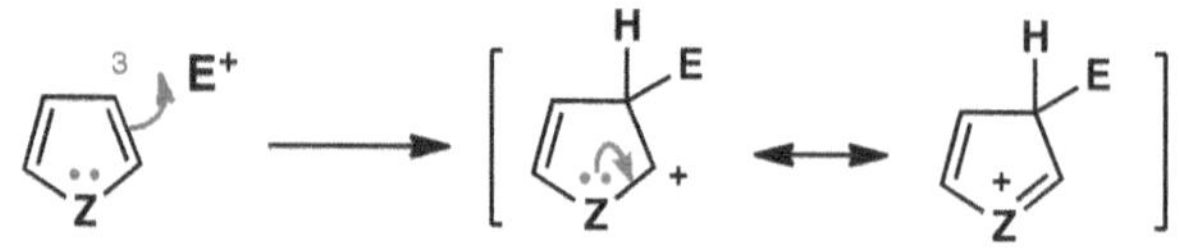

$$(19.14)$$

Approximate relative rate
constants in bromination:

Approximate relative rate
constants in bromination:

3×10^{18}

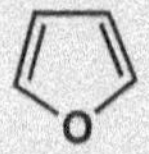

6×10^{11}

5×10^{9}

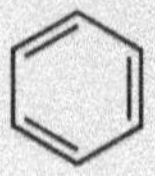

1.0

Pyrrole, furan, and thiophene are hugely more reactive than benzene—their relative reactivities (estimated from the rate constants for bromination) are given in the margin.

The high reactivity of the heterocycles is attributable to (i) their relatively low aromatic stabilization (compared with benzene), and (ii) appreciable stabilization of the rate-determining transition structures in their substitution reactions by the lone pairs on the heteroatoms. We can see this by looking at the electronic structure of the intermediate for substitution at C2 shown in eqn 19.13, for example (reactivity at C3, though lower than at C2, is also hugely greater than of benzene). Very significantly, all five ring atoms in one of the resonance contributors shown have full valence shells; this is not possible for the intermediate from benzene. The degree of this second effect depends on the electronegativity of the heteroatom (NH < O), and on the extent of the orbital overlap of the lone pair orbital of the heteroatom with the 2p orbitals of C (3p of S poorer than 2p of O and N).

The benzo-fused derivatives, indole, benzofuran, and benzothiophene, all react at the heterocyclic ring, but their regioselectivities are different. Electrophilic substitution occurs at C3 of indole, but at C2 of benzofuran (eqns 19.15 and 19.16). Benzothiophene reacts at both C2 and C3 but normally with a slight preference for C3.

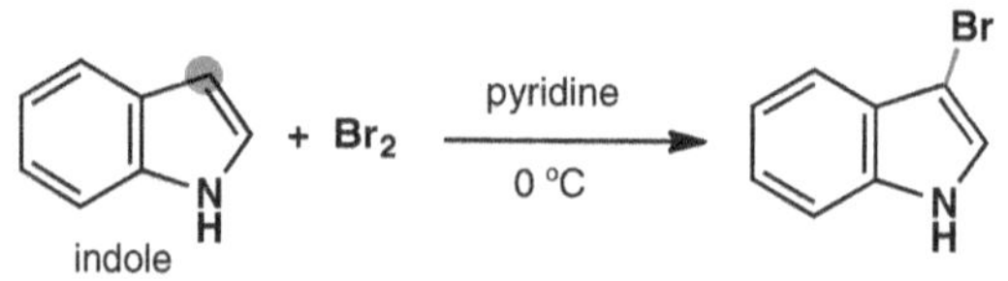

$$(19.15)$$

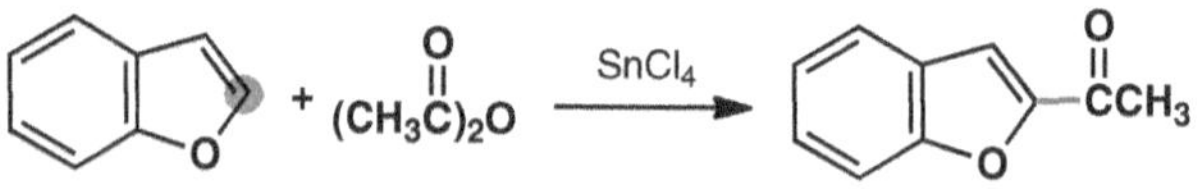

$$(19.16)$$

Exercise 19.7

The difference in regioselectivity of indole and benzofuran can be explained by the different relative contributions from the benzene ring and the heteroatom towards stabilization of the alternative cationic intermediates. Draw resonance contributors of the cationic intermediates formed by electrophilic attack of E$^+$ at C2 and at C3 of the generic benzo-fused heterocycle (**1**), and discuss the relative stabilities of the two intermediates according to whether Z is NH or O.

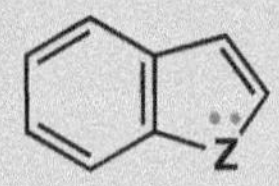

(**1**)

b. Other reactions

The aromatic stabilization of furan is rather small compared with that of benzene, for example, and furan often behaves as a reactive diene. A typical reaction is its Diels–Alder cycloaddition with the dienophile dimethyl butynedioate (Scheme 19.3).

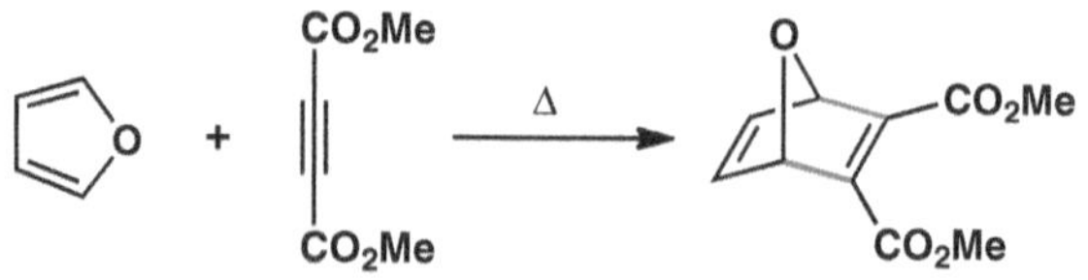

Scheme 19.3 Diels–Alder reaction of furan with dimethyl butynedioate.

Some properties of furans are typical of enol ethers, a class of highly reactive alkenes; for example, their acid-catalysed hydrolysis provides the corresponding 1,4-dicarbonyl compounds, e.g. eqn 19.17.

$$\text{(structure)} + H_2O \xrightarrow[\text{H}_2\text{O, AcOH}]{\text{cat. H}_2\text{SO}_4} \text{(structure)}$$ (19.17)

We previously encountered the Diels–Alder reaction of furan when it was used as a trap for the reactive intermediate, benzyne (which is a dienophile), in Panel 18.2. See Sections 15.9 and 21.2 for further coverage of Diels–Alder reactions.

Give a mechanism for reaction 19.17.

Exercise 19.8

19.4.2 Reactions of pyridine and its derivatives

a. Electrophilic substitution

Pyridine is isoelectronic with benzene and undergoes electrophilic substitution reactions, but only with great difficulty. This is because the high electronegativity of the nitrogen in the pyridine ring causes the π electrons to be more tightly bound than in benzene, which lowers the energy of the HOMO of the π system (see Sub-section 7.3.2 for coverage of HOMO-LUMO interactions in reactions between nucleophiles and electrophiles). Consequently, the electrons of the HOMO are less available and pyridine is less reactive than benzene towards electrophiles, and even less reactive than nitrobenzene.

$$\text{(structure NO}_2\text{)} > \text{(pyridine)} > \text{(structure O}_2\text{N...NO}_2\text{)}$$

Low relative reactivities in electrophilic substitution.

The relative energies of the HOMOs of benzene and pyridine are sketched below: the larger energy gap between the LUMO of an electrophile and the π HOMO of pyridine indicates a poorer HOMO–LUMO interaction and, consequently, lower reactivity than for benzene.

LUMO
E⁺
HOMO
benzene
pyridine

The low reactivity of pyridine in electrophilic substitution is reduced even further under acidic conditions by protonation or by coordination by a Lewis acid at the nitrogen. For example, pyridine undergoes neither nitration by HNO_3/H_2SO_4 nor Friedel–Crafts acylation (Scheme 19.4).

$$\text{(scheme 19.4 structures)}$$

HNO₃ / H₂SO₄ → (pyridinium) → no reaction

RCOCl / AlCl₃ → (pyridine-AlCl₃) → no reaction

Scheme 19.4 Inhibition of electrophilic substitution of pyridine by Brønsted and Lewis acids.

Pyridine reacts as a nitrogen nucleophile in S_N2 reactions, but this involves the nitrogen lone pair in the sp² orbital which is not part of the π system.

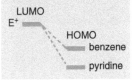

PhCH₂Br → Ph ... Br⁻

Only under strongly electrophilic conditions with potent reagents will pyridine undergo electrophilic substitution, and then the C3-substitution product is obtained, e.g. eqn 19.18.

$$\text{(pyridine)} \xrightarrow[\text{H}_2\text{SO}_4, \text{SO}_3]{\text{Br}_2} \text{(3-bromopyridine)}$$ (19.18)

86%

Exercise 19.9

Explain the regioselectivity of pyridine in reaction 19.18 by comparing the alternative intermedi-ates formed by electrophilic addition to C2, C3, and C4.

As expected, electron-donating substituents enhance the reactivity of pyridine in elec-trophilic substitution reactions, e.g. eqn 19.19.

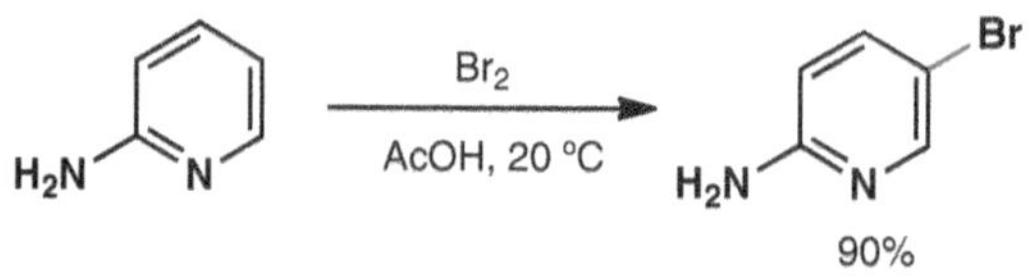

$$\text{H}_2\text{N} \quad \xrightarrow[\text{AcOH, 20 °C}]{\text{Br}_2} \quad \text{H}_2\text{N} \quad \text{Br} \quad 90\% \tag{19.19}$$

b. Nucleophilic substitution

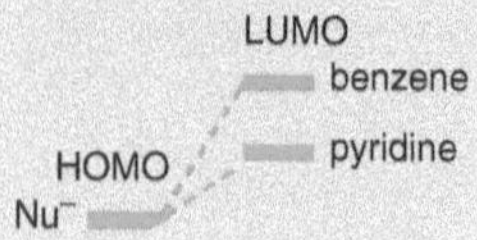

The high electronegativity of the N in pyridine lowers the energy of its LUMO just as it did for the HOMO (see above). This *enhances* the reactivity of pyridine *when it acts as an electrophile*. Consequently, pyridine derivatives bearing a nucleofuge undergo nucleo-philic substitution, and this occurs most readily when the nucleofuge is on C2 or C4 (eqns 19.20 and 19.21).

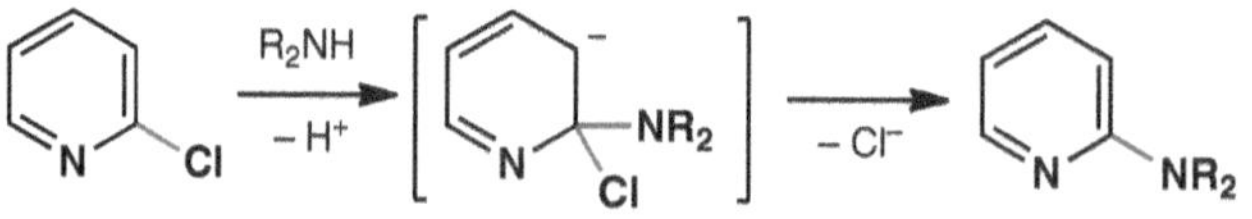

HOMO–LUMO interactions between a nucleophile and a pyridine derivative:

LUMO — benzene
— pyridine
HOMO
Nu⁻

$$\tag{19.20}$$

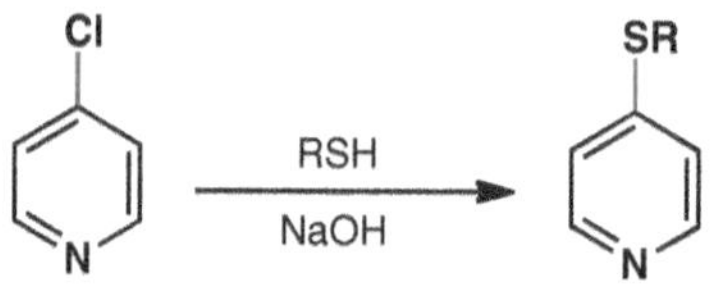

$$\tag{19.21}$$

We saw in Section 18.6 that halobenzenes require an electron-withdrawing substituent at the *ortho* or *para* positions to facilitate the addition-elimination mechanism of nucleophilic substitution, which here proceeds readily for 2- and 4-chloropyridines. In broad terms, the electronic effect of the N in a pyridine derivative compares with the effect of an NO_2 in a substituted benzene.

Exercise 19.10

By drawing resonance contributors of the intermediates involved, compare the relative reac-tivities of 2-chloropyridine and 3-chloropyridine in substitution reactions with nucleophile Nu⁻.

In the strongly basic conditions of $NaNH_2$ in liquid ammonia, pyridine itself gives 2-aminopyridine (eqn 19.22).

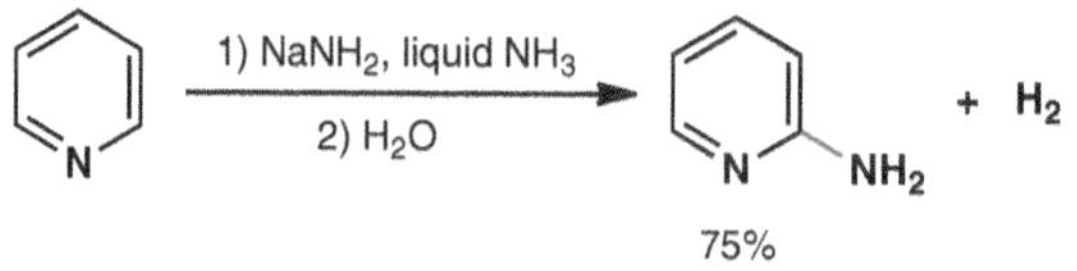

$$\xrightarrow[\text{2) H}_2\text{O}]{\text{1) NaNH}_2, \text{ liquid NH}_3} \quad \text{NH}_2 \quad + \text{ H}_2 \quad 75\% \tag{19.22}$$

This so-called Chichibabin reaction proceeds by the addition–elimination mechanism (see Section 18.6) with hydride (H⁻) as the unlikely nucleofuge (Scheme 19.5).

Diazotization of 2-aminopyridine in aqueous solution provides 2-pyridone (the more stable tautomer of 2-hydroxypyridine) directly (eqn 19.23). Presumably, the

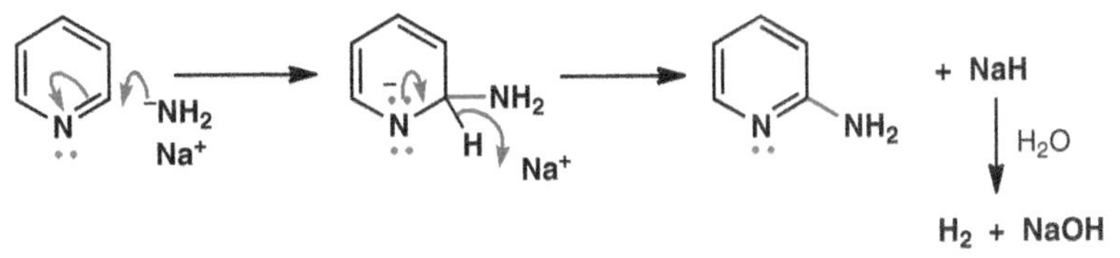

Scheme 19.5 Mechanism of the Chichibabin reaction.

2-pyridinediazonium ion, which has N_2 as a very effective nucleofuge, intervenes as a highly reactive intermediate.

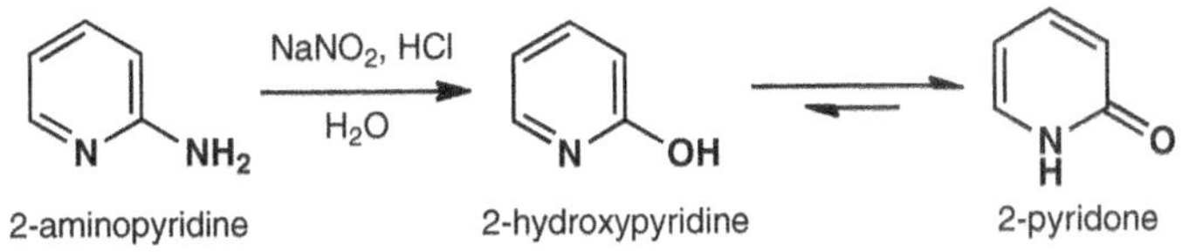

2-aminopyridine   2-hydroxypyridine   2-pyridone   (19.23)

Example 19.2

Explain why 2-pyridone is more stable than 2-hydroxypyridine whereas the keto tautomer of phenol is unstable.

Solution
2-Pyridone has the same number of cyclic 6π electron resonance forms as 2-hydroxypyridine, plus a form with a strong C=O bond. In contrast, the keto form of phenol has no cyclic array of π electrons and, consequently, is not aromatic.

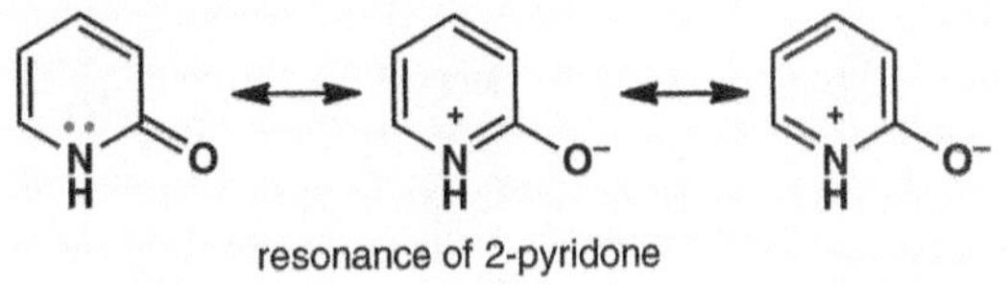

resonance of 2-pyridone

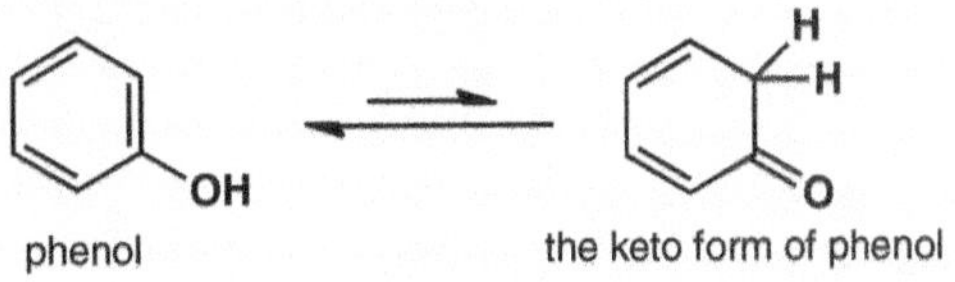

phenol   the keto form of phenol

Exercise 19.11

Draw the structure of the tautomer of 2-aminopyridine and comment on how the relative stabilities of these two isomers compare with the relative stabilities of the pyridone and 2-hydroxypyridine tautomers.

Exercise 19.12

Give a mechanism for the reaction of 2-pyridone with iodomethane.

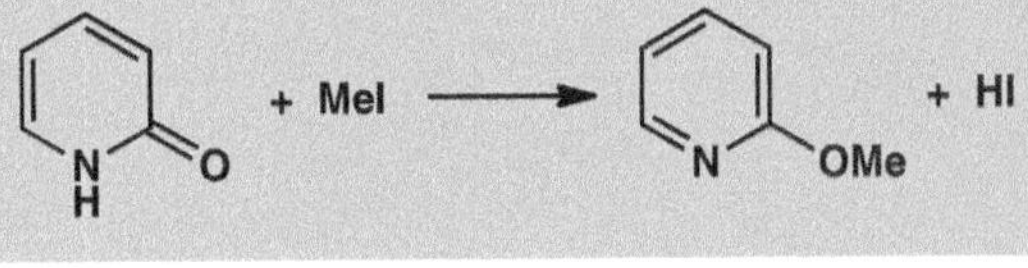

+ MeI   + HI

c. Pyridine *N*-oxide

Pyridine reacts with hydrogen peroxide or a peracid to give pyridine *N*-oxide. This reaction is a nucleophilic substitution at oxygen with pyridine as the nucleophile (e.g. eqn 19.24).

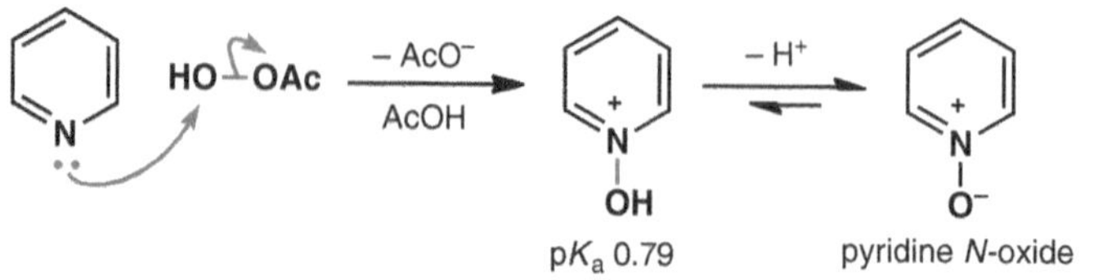

(19.24)

Like the phenoxide ion, pyridine *N*-oxide is activated towards electrophilic substitu-tion by the anionic oxygen. However, although pyridine *N*-oxide is much more reactive than pyridine, the positive end of the zwitterionic *N*-oxide group causes it to be less reac-tive than anionic phenoxide (and even less reactive than neutral phenol). Since the oxy-gen of the *N*-oxide can be removed by PCl_3 or $P(OMe)_3$ after an electrophilic substitution reaction, the overall result corresponds to a substitution on pyridine itself (eqn 19.25).

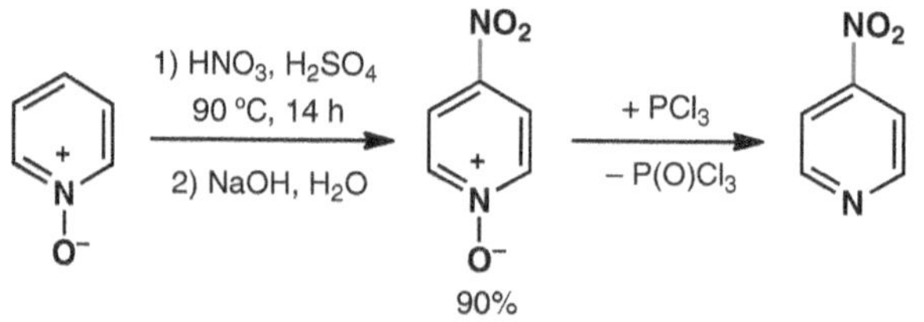

(19.25)

As indicated, substitution occurs mainly *para* to the N, and the overall reaction proceeds as shown in Scheme 19.6.

Scheme 19.6 Nitration of pyridine *N*-oxide and removal of the oxygen.

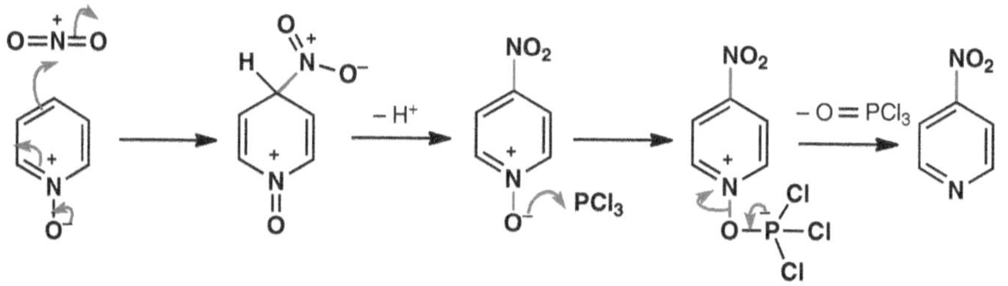

d. Reactions of substituents at C2 and C4

An α hydrogen of a 2- or 4-alkyl group on a pyridine ring is more acidic than an ordinary benzylic hydrogen owing to the conjugatively stabilizing electron-withdrawing effect of the nitrogen upon the conjugate base. This acidity is comparable with that of the α hydrogen of a ketone, and 2- and 4-methylpyridines undergo aldol-type reactions with aldehydes under basic conditions, e.g. eqn 19.26.

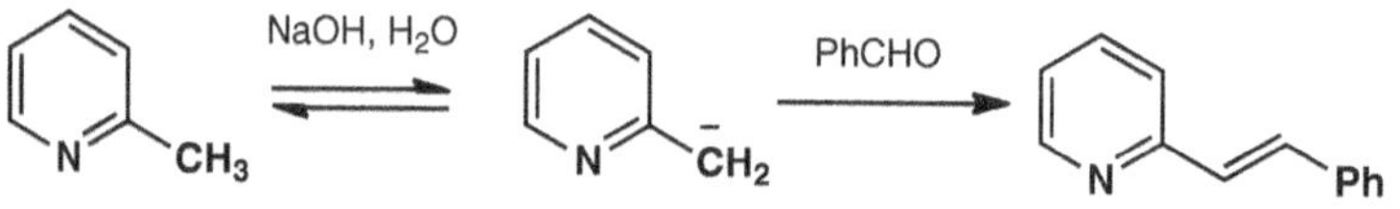

(19.26)

Exercise 19.13

Draw resonance contributors to describe the electronic structure of the carbanion intermediate formed by the proton abstraction from 2-methylpyridine in reaction 19.26.

Exercise 19.14

Give a mechanism for the second step of reaction 19.26.

2-Methylpyridine N-oxide undergoes an interesting reaction with acetic anhydride. When the –O⁻ group is acetylated, the positive N becomes more effective at withdrawing electron density and this facilitates deprotonation from the methyl group. The subsequent concerted rearrangement transfers the acetoxy group from the N to the methylene, as shown in Scheme 19.7.

See Section 21.4 for other rearrangements with concerted bond making and breaking (sigmatropic rearrangements).

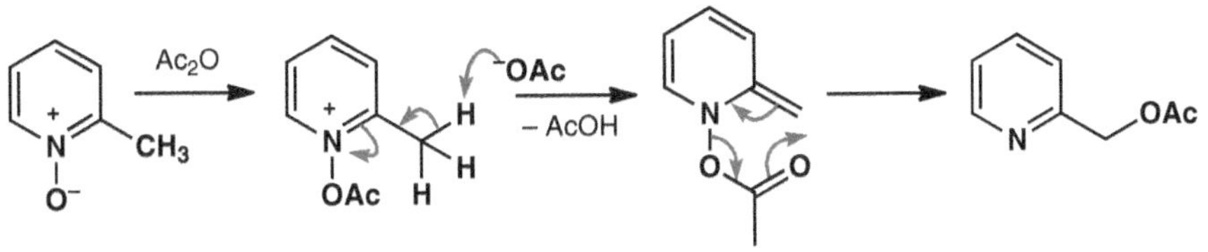

Scheme 19.7 Reaction of 2-methylpyridine N-oxide with acetic anhydride.

e. Reactions of quinoline and isoquinoline

Both of these benzopyridines can be seen as a benzene fused to a pyridine so it is no surprise that (i) they are both more reactive than pyridine but less reactive than benzene in electrophilic substitution, and (ii) electrophilic substitution occurs essentially wholly at the benzene ring, e.g. eqns 19.27 and 19.28.

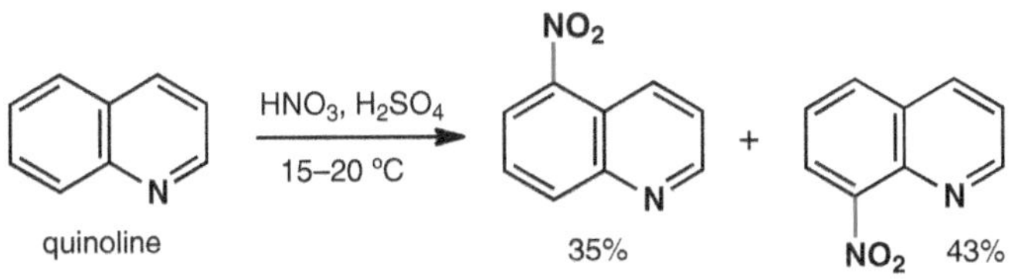

(19.27)

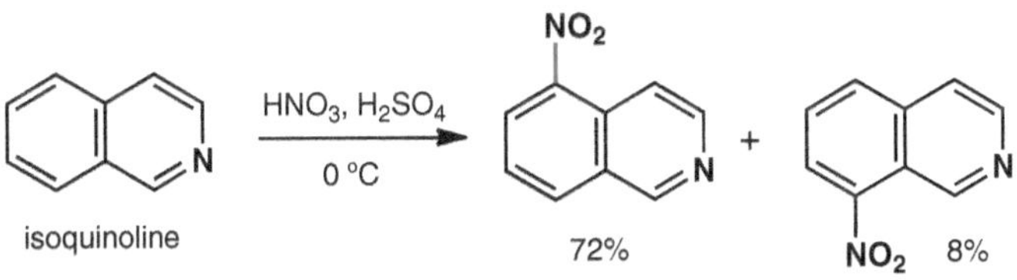

(19.28)

19.5 Synthesis of Aromatic Heterocyclic Compounds

Representative reactions for syntheses of five- and six-membered aromatic heterocycles involving interesting cyclization steps are presented in this section. The so-called Paal–Knorr synthesis is used for the preparation of some pyrroles, furans, and thiophenes from 1,4-diketones. In these, a primary amine or ammonia, or an acid (protic or P_2O_5), or P_2S_5 is used, as appropriate, as the condensation reagent (Scheme 19.8). Reactions proceed via enamine, enol, or enethiol intermediates.

Pyridines can be synthesized by cyclization of 1,5-diketones in the same way, but the initial products are insufficiently unsaturated.* Consequently, dehydrogenation is necessary and is carried out with an oxidizing agent such as nitric acid or a quinone, e.g. Scheme 19.9.

The Hantzsch pyridine synthesis is a well-known method which incorporates reactions analogous to the one in Scheme 19.9 above. In this procedure, a 1,5-diketone is first obtained as an intermediate by the condensation of two molecules of a β-ketoester and an aldehyde in the presence of ammonia (Scheme 19.10). This intermediate then undergoes a cyclization with the ammonia, and an oxidative aromatization gives the substituted pyridine shown.

*Redox properties of pyridine–dihydropyridine groups are the basis of the redox coenzyme, NAD⁺–NADH (see Panel 10.3).

Scheme 19.8 Paal–Knorr syntheses of pyrrole, furan, and thiophene derivatives.

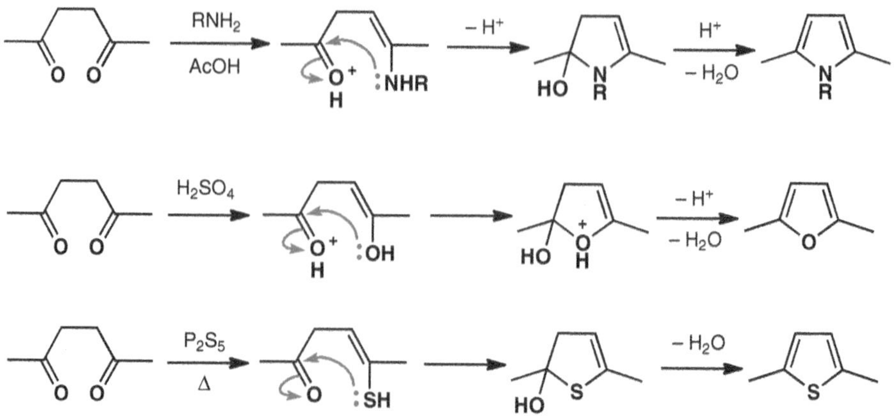

Scheme 19.9 Formation of a pyridine by cyclization of a 1,5-diketone with ammonia followed by oxidative aromatization.

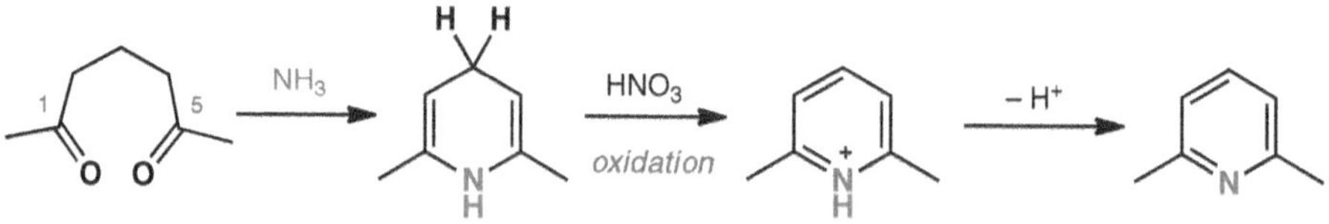

Scheme 19.10 An example of the Hantzsch pyridine synthesis.

The first step of this reaction is reminiscent of the Knoevenagel reaction (see Chapter 17, problem 17.28), and the second step is a Michael reaction (Subsection 18.4.1).

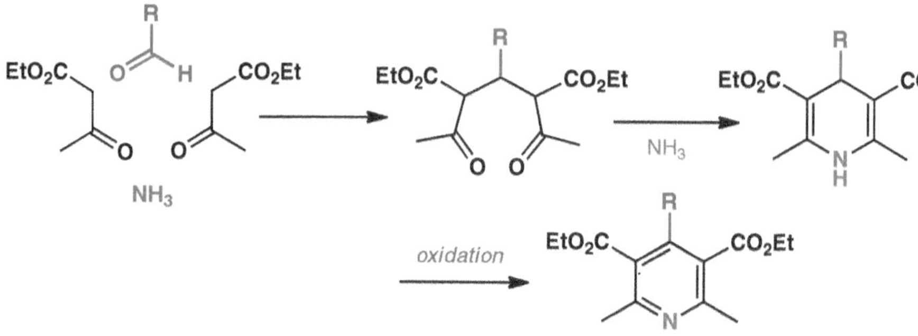

Exercise 19.15

Give mechanisms for the formation of (a) the 1,5-diketone, and (b) the dihydropyridine in the Hantzsch pyridine synthesis in Scheme 19.10.

The reaction of a 1,5-diketone with hydroxylamine (eqn 19.29) is an easier method for the preparation of some pyridines. In this reaction, the dehydration leads directly to the pyridine without the need for an oxidative step.

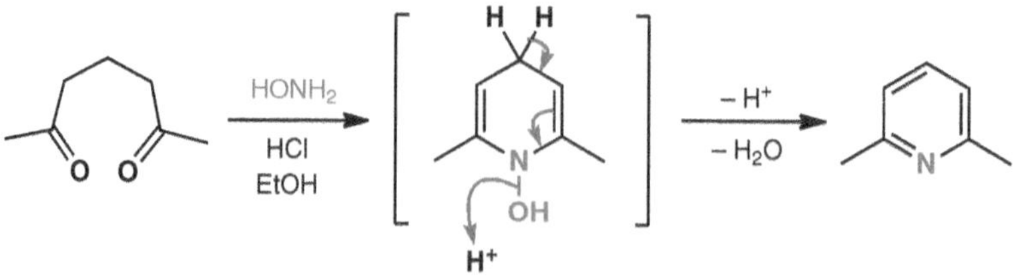

(19.29)

Exercise 19.16

Propose a mechanism for the formation of the dihydropyridine in reaction 19.29.

Summary

- **Polycyclic aromatic compounds** consist of two or more benzenoid or non-benzenoid rings fused together, and are generally more reactive than benzene itself in electrophilic substitution reactions.

- The most common **aromatic heterocyclic compounds** contain N, O, or S and are analogues of benzene and the cyclopentadienide ion.

- The lone pair on the N in **pyrrole** is part of the aromatic 6π electron system and not a basic site; in contrast, the lone pair on the N of **pyridine** is in an sp² orbital, which is not part of the aromatic system, so pyridine is basic. Imidazole with two N atoms in the ring is both basic and weakly acidic.

- The five-membered heterocycles are reactive in electrophilic substitution and furan also acts as a diene in Diels–Alder cycloaddition reactions. Pyridine is very unreactive in electrophilic substitution (although pyridine N-oxide is usefully reactive) but its derivatives containing a suitably positioned nucleofuge undergo nucleophilic substitution reactions.

- Benzo-fused analogues of the common five- and six-membered aromatic heterocycles have properties predictable from knowledge of the properties of benzene and the simple heterocycles.

- A range of aromatic heterocyclic compounds may be prepared by cyclization reactions of 1,4- and 1,5-diketones.

Problems

19.1 Predict the main product formed by mono-nitration of the following.

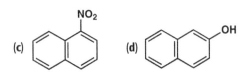

(a)

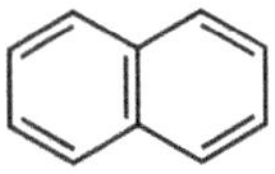

(b) 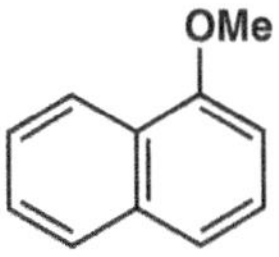 OMe

(c) NO₂

(d) OH

19.2 What is the main product in each of the following reactions?

(a) 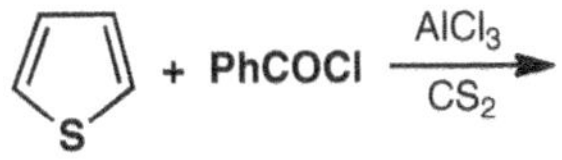 [thiophene] + PhCOCl →(AlCl₃ / CS₂)→

(b) 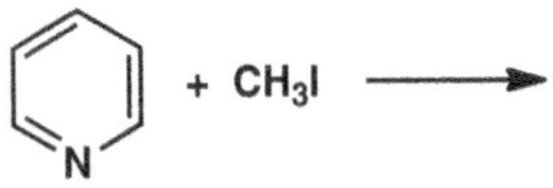 [pyridine] + CH₃I →

(c) 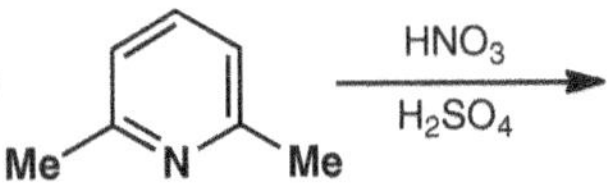 [2,6-dimethylpyridine, Me—N—Me] →(HNO₃ / H₂SO₄)→

(d) 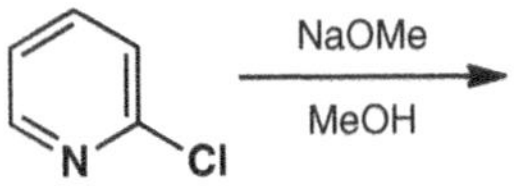 [2-chloropyridine, N—Cl] →(NaOMe / MeOH)→

19.3 The nitrogen atom is the positive end of the dipole of pyrrole whereas it is the negative end of its saturated analogue, pyrrolidine. (a) Explain this difference in the direction of the dipoles. (b) Predict the effect of 3,4-dichloro substituents upon the dipole moment of pyrrole.

19.4 Of pyridine and pyrrole, pyridine is the more soluble in water. Explain the reason for this difference and predict how imidazole will compare in solubility in water with these two compounds.

19.5 4-Dimethylaminopyridine (DMAP), which has two basic sites, is much more basic (pK_{BH^+}=9.9) than pyridine (pK_{BH^+}=5.25) or dimethylaniline (pK_{BH^+}=5.07). Identify the most basic site of the aminopyridine by considering the electronic structures of its isomeric conjugate acids, and explain the reason for its high basicity.

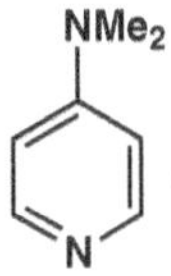

[structure: NMe₂ on pyridine ring]
4-dimethylaminopyridine
(pK_{BH^+} = 9.9)

19.6 Predict the main product of the following reaction, and explain the regioselectivity.

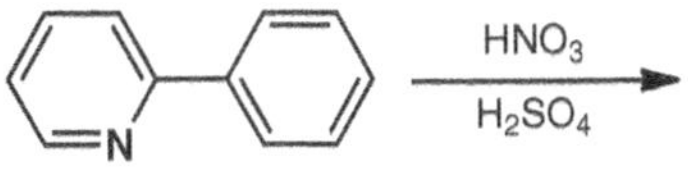

[structure: 2-phenylpyridine] →(HNO₃ / H₂SO₄)→

19.7 Pyrrole-2-carboxylic acid undergoes decarboxylation upon heating. One possibility is that the reaction proceeds via the unfavourable zwitterionic form as shown below. Show the electronic reorganization upon decarboxylation of the zwitterion intermediate with curly arrows.

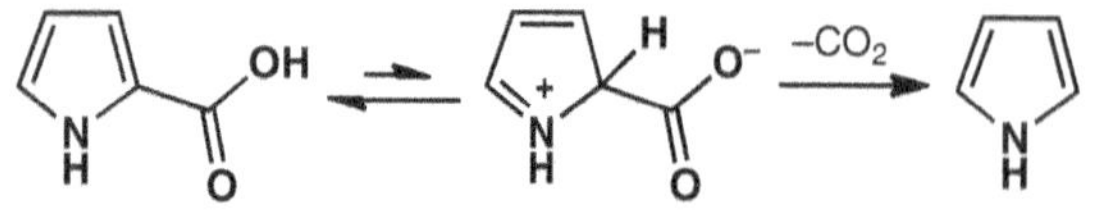

19.8 Pyrrole undergoes polymerization in the presence of a strong acid such as H_2SO_4 to give polypyrrole, a polymer which conducts electricity. Give a mechanism for the acid-induced polymerization.

19.9 Chloropyridines undergo nucleophilic substitutions with sodium amide in liquid ammonia.

However, the reactions of 2-chloro and 3-chloro isomers proceed by different mechanisms. Give mechanisms for the two reactions and explain the difference.

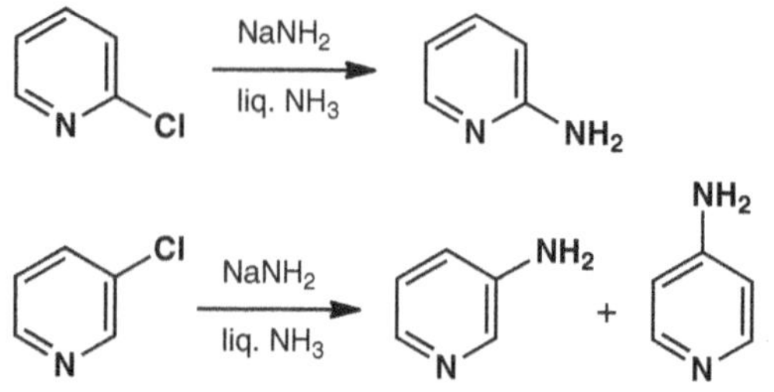

19.10 Reaction of pyrrole with an acid anhydride $(RCO)_2O$ gives different products under neutral and basic conditions. Predict the products obtained (a) by the direct reaction and (b) in the presence of triethylamine, and explain the results.

Supplementary Problems

19.11 One of the methods to generate benzyne is the reaction of 1-bromo-2-fluorobenzene with metallic magnesium in THF. When this reaction is carried out in the presence of anthracene, an interesting molecule called *triptycene* is obtained by a Diels–Alder reaction. Give a reaction scheme showing the formation of triptycene.

19.12 Furan sometimes undergoes electrophilic addition owing to its low aromaticity. Propose a mechanism for the following reaction with bromine in methanol to give the product shown.

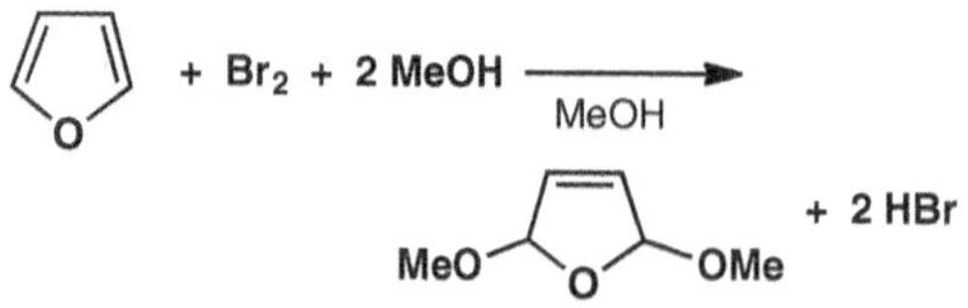

19.13 Indole undergoes the so-called Mannich reaction leading to 3-aminomethylation (see Chapter 17 for the Mannich reaction of a ketone in Example 17.5). Propose a reaction mechanism for the following transformation.

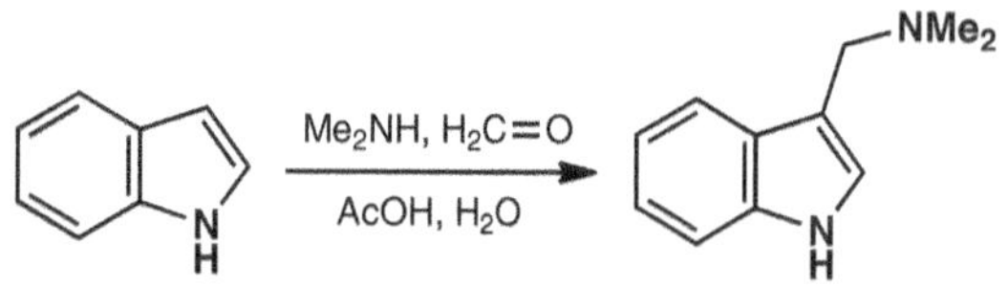

19.14 4-Dimethylaminopyridine (DMAP: Problem 19.5) is a good nucleophilic catalyst. Propose a

mechanism for the reaction by which DMAP catalyses the hydrolysis of phenyl ethanoate in aqueous solution.

19.15 Give a mechanism for each step of the following transformation.

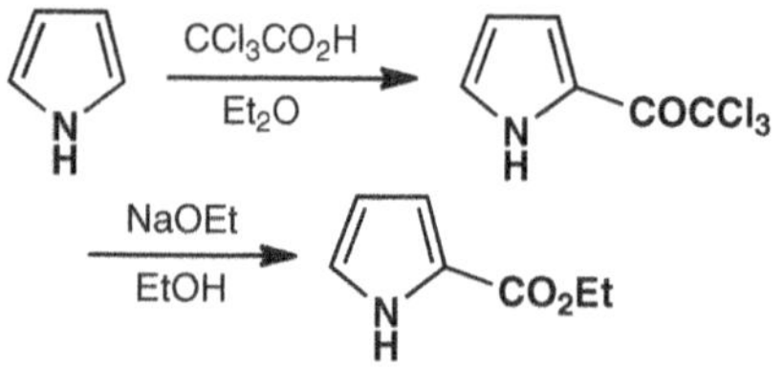

19.16 Phenanthrene can be obtained from naphthalene by the following reaction sequence. Write in the structures of the intermediate compounds **A–J**. Note that **A** and **B** are isomers (formed in the ratio of 56 : 44) and the ensuing pairs of intermediates are also isomeric.

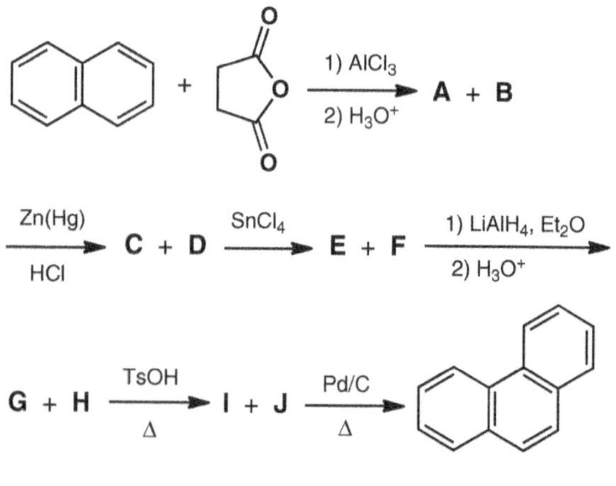

19.17 Propose mechanisms for the steps up to and including the cyclization in the following outline synthesis of quinoline.

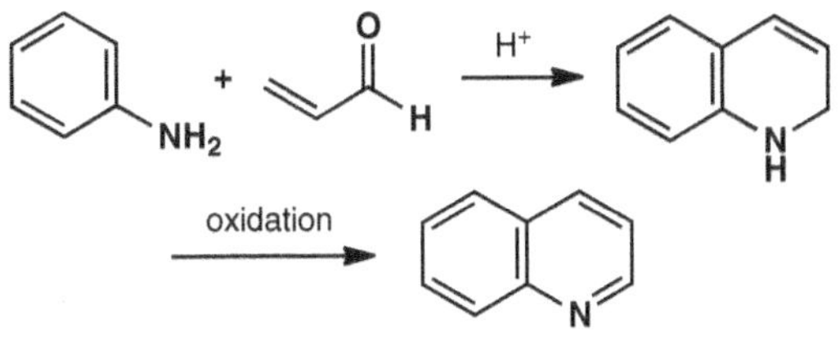

19.18 The following reaction is an outline synthesis of benzimidazole. Give a reaction mechanism for this transformation.

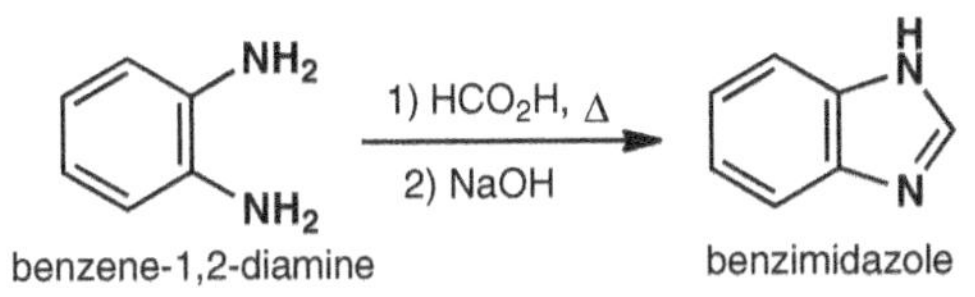

benzene-1,2-diamine   benzimidazole

19.19 In the Seveso disaster in Northern Italy in 1976, a large amount of highly toxic dioxin was released when the reaction converting 1,2,4,5-tetrachlorobenzene into 2,4,5-trichlorophenol was overheated.

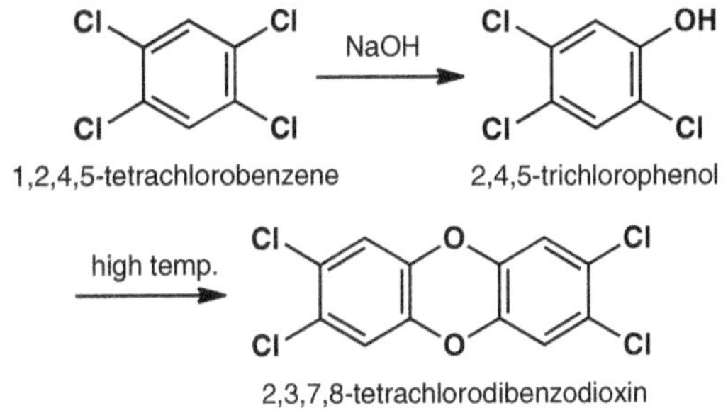

1,2,4,5-tetrachlorobenzene   2,4,5-trichlorophenol

2,3,7,8-tetrachlorodibenzodioxin

Give a mechanism for the formation of the notoriously dangerous dioxin from the trichlorophenol under the strongly basic conditions.

20 Reactions involving Radicals

Related topics we have already covered:

- **Chemical bonding** (Chapter 1)
- **Allyl radical** (Section 5.3)
- **Elementary steps in a chemical reaction and curly arrow representation of reaction mechanisms** (Section 7.2)
- **Nucleophilic substitution at saturated carbon** (Chapter 12)
- **Addition reactions of alkenes** (Chapter 15)

Topics we shall cover in this chapter:

- **Formation of radicals by homolysis**
- **Stability of radicals**
- **Radical chain reactions**
- **Radical polymerization**
- **Formation of radical ions by single electron transfer, and their reactions**
- **Electrolytic reactions**

So far, we have mostly encountered polar reactions which occur by the redistribution of valence electrons in pairs. These reactions involve bonding between nucleophiles and electrophiles or, in reverse, heterolysis of bonds; they proceed via polar transition structures and are often stepwise with cationic or anionic intermediates. In contrast, some organic reactions involve intermediates with unpaired electrons which are called **radicals**; they are formed by homolysis of covalent bonds as described in Sub-section 7.2.1. The mechanisms of radical reactions are quite different from those of polar reactions: most occur by **radical chain mechanisms**.

Radicals are in fact all around us: a normal oxygen molecule ($\cdot$O–O$\cdot$) is a diradical, and radical reactions are important in biology and medicine. Reactive oxygen species such as superoxide anion ($O_2^{\cdot-}$), hydroperoxyl radical (HOO$\cdot$), and hydroxyl radical (HO$\cdot$)[1] are believed to be implicated in the ageing process. Nitric oxide (N=O$\cdot$) is another common radical; it is involved in our physiological control mechanisms including regulation of blood pressure and blood clotting, neurotransmission, and the immune response against tumour cells. Drugs such as sildenafil (Viagra®) and minoxidil (Rogaine®) prolong the effect of nitric oxide in the body and are used principally to treat two conditions associated with ageing males: male erectile dysfunction and hair loss.

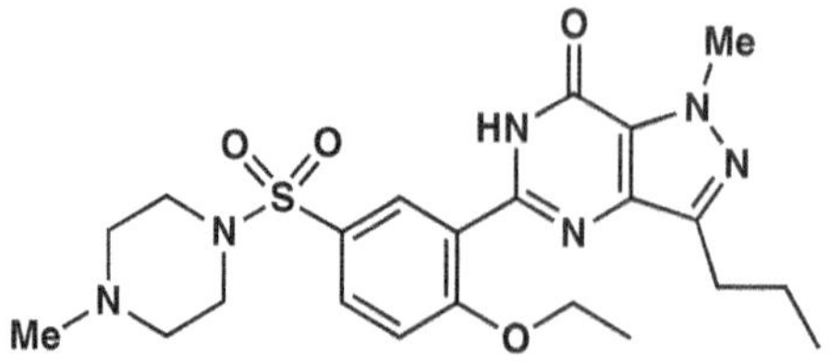

sildenafil (Viagra®) 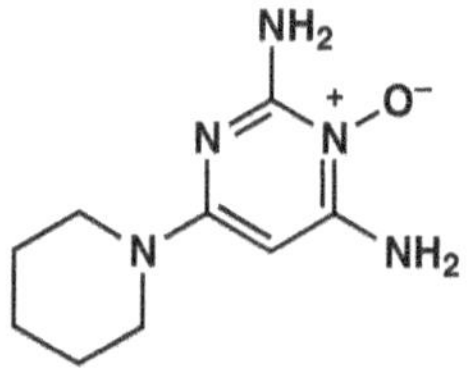   minoxidil (Rogaine®)

---

[1] According to the IUPAC rules on nomenclature, radicals of groups whose names end with –y are named by adding *l*; e.g. alkoxyl (RO·) from alkoxy (RO) and alkylperoxyl (ROO·) from alkylperoxy (ROO).

Radical polymerization is an important industrial process for the production of useful plastics such as polyethylene, polystyrene, poly(vinyl chloride), and Teflon. The cracking process in the petroleum industry (by which useful low molecular weight hydrocarbons are produced from high molecular weight fractions of crude oil) involves thermal decomposition of alkanes via radical intermediates. Radical reactions are also involved in the combustion of fuels which, being strongly exothermic, liberates energy as heat.

20.1 Homolysis

There are two types of bond cleavage, homolysis and heterolysis, as discussed in Section 7.2. In homolysis, the electrons of a σ bond between two atoms become unpaired to give two radicals, each with an *unpaired electron*.

The minimum energy needed for the homolysis of a bond (converted to the molar scale) is called the **bond (dissociation) energy** (strictly, it is an enthalpy term—see later). As summarized on p. 617, C–H and C–C bonds usually have bond energies greater than 350 kJ mol⁻¹, e.g. eqn 20.1, and alkanes do not undergo homolysis unless heated to high temperatures (the industrial thermal cracking of petroleum is carried out at about 500 °C).

The term 'radical' was once used for what we now call a *group*, e.g. an ethyl *group* was called an ethyl *radical* (see Prologue); consequently, the term *'free* radical' was used for what we now simply call a 'radical' (and sometimes still is if we wish to emphasize that the radical is not paired with another).

$$H_3C–CH_3 \longrightarrow 2 \cdot CH_3 \qquad \Delta H = 375 \text{ kJ mol}^{-1} \qquad (20.1)$$

A molecule containing weaker bonds will obviously undergo homolysis more easily. **Peroxides** are representative of compounds which have an O–O bond with a low bond dissociation energy, i.e. < 200 kJ mol⁻¹. In particular, diacyl peroxides, e.g. dibenzoyl peroxide (BPO) in eqn 20.2, readily undergo homolysis to give acyloxyl radicals, which then easily decarboxylate to form alkyl or aryl radicals, e.g. eqn 20.3.

$$\Delta H = 139 \text{ kJ mol}^{-1} \qquad (20.2)$$

dibenzoyl peroxide (BPO)

benzoyloxyl radical

Note that singly barbed (fish-hook) curly arrows are used to represent the movement of single electrons in radical mechanisms, as explained in Section 7.2.

decarboxylation

$$Ph \cdot + CO_2 \qquad (20.3)$$

The double homolysis of azo compounds such as azobisisobutyronitrile (AIBN) with the liberation of nitrogen occurs even more readily (eqn 20.4).

$$2 Me_2C \cdot + N_2 \qquad \Delta H = 133 \text{ kJ mol}^{-1} \qquad (20.4)$$

azobisisobutyronitrile (AIBN)

> **Exercise 20.1**
>
> Draw resonance forms of the benzoyloxyl and cyanoalkyl radicals formed in reactions 20.2 and 20.4.

Compounds such as BPO and AIBN which easily provide radicals are used as **radical initiators**, i.e. to initiate reactions which proceed by radical chain mechanisms, as will be discussed in Section 20.3.

The symbol '*hv*' above a reaction arrow indicates irradiation; the term *photolysis* is also used for bond cleavage caused by irradiation.

When a compound has an absorption band in the ultraviolet and/or visible (UV-vis) regions of the electromagnetic spectrum, molecules interact with the radiation to form transient photoexcited molecules which may undergo a range of reactions including homolysis (see Section 5.7). Bromine and chlorine are elements whose diatomic molecules undergo *photo-dissociation* (e.g. eqn 20.5), i.e. dissociation induced by irradiation.

$$\text{Cl} \overset{\frown\frown}{} \text{Cl} \xrightarrow{hv} 2\ \text{Cl}\bullet \qquad \Delta H = 243\ \text{kJ mol}^{-1} \qquad (20.5)$$

The bond dissociation energies of Cl_2 and Br_2 (243 and 193 kJ mol^{-1}, respectively) are relatively low and dissociation can also be achieved thermally.

20.2 Structure and Stability of Radicals

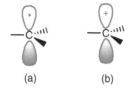

(a)   (b)

Figure 20.1 Time-averaged structures of (a) a radical (electrically neutral with seven valence electrons) and (b) a carbenium ion (positively charged with six valence electrons).

The time-averaged structure of the carbon centre of a radical is usually close to planar and the unpaired electron is accommodated in a 2p orbital (Figure 20.1). This structure is similar to that of a carbenium ion whose central carbon is planar and sp^2 hybridized.

The relative stabilities of alkyl radicals resemble those of carbenium ions, and alkyl radicals, like alkyl cations (Sub-section 12.4.3), are stabilized by hyperconjugation. The stability of a radical R$\bullet$, notionally formed by homolysis of the bond in R–H, is quantified by the R–H bond dissociation energy (*DH*). Some values are given in Table 20.1 (see also the Additional Resources, p. 617); we see that the smaller the bond dissociation energy for its formation, the more stable the radical. Allylic and benzylic radicals are strongly stabilized by conjugation, so the corresponding bond dissociation energies are small. Acyloxyl and cyanoalkyl radicals (eqns 20.2 and 20.4) are other examples with conjugative stabilization.

Panel 20.1 The first radical observed by Gomberg

In 1900, Moses Gomberg, an instructor at the University of Michigan, was trying to prepare hexaphenylethane from triphenylmethyl chloride by treating it with finely divided silver (or zinc) with the exclusion of air (oxygen); he obtained a yellow solution which faded reversibly on heating. From this solution, a colourless crystalline compound, $C_{38}H_{30}$, was isolated which he thought was hexaphenylethane. However, in the 1950s, the product was identified as the dimer formed by the *para* coupling of two triphenylmethyl radicals. The formation of the dimer is reversible owing to the stability of the triphenylmethyl radical.

$$2\ Ph_3C\text{–Cl} \xrightarrow[\substack{2\ AgCl}]{2\ Ag,\ benzene} 2\ Ph_3C\bullet$$

triphenylmethyl chloride — triphenylmethyl radical

$$Ph_3C\text{–CPh}_3 \quad \text{hexaphenylethane}$$

the dimer isolated

$$Ph_3C\text{–O–O–CPh}_3 \quad \text{a peroxide}$$

Moses Gomberg (1866–1947).

If the yellow solution was exposed to air (oxygen), the peroxide $C_{38}H_{30}O_2$ was obtained, which could be reduced to the known alcohol, triphenylmethanol. This was taken as evidence for the first observation of a trivalent carbon species, a radical, and Gomberg is considered to be the father of radical chemistry.

Gomberg was born in the Ukraine and moved to the USA in 1886. He graduated from the University of Michigan, where he was appointed an instructor, and stayed there all his working life.

Table 20.1

Bond dissociation energies (*DH* in kJ mol⁻¹) of the C–H bonds in some hydrocarbons and relative stabilities of the derived radicals

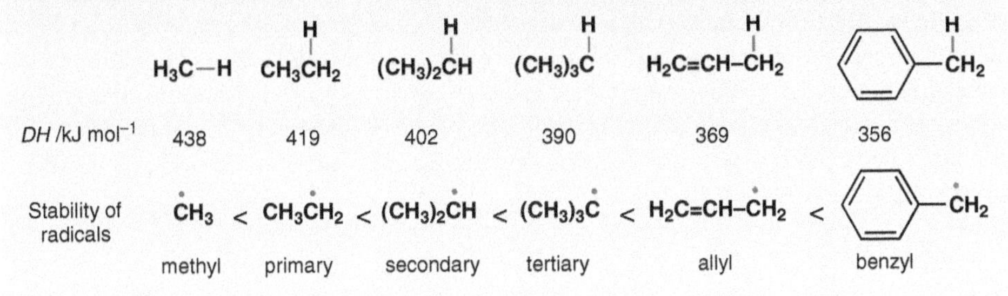

	H₃C—H	CH₃CH₂	(CH₃)₂CH	(CH₃)₃C	H₂C=CH–CH₂	–CH₂ (benzyl)
DH /kJ mol⁻¹	438	419	402	390	369	356

Stability of radicals: $\dot{C}H_3 < CH_3\dot{C}H_2 < (CH_3)_2\dot{C}H < (CH_3)_3\dot{C} < H_2C=CH–\dot{C}H_2 < \text{benzyl}$

methyl   primary   secondary   tertiary   allyl   benzyl

Example 20.1

Explain why the ethyl radical is more stable than the methyl radical by drawing structures with orbitals accommodating their unpaired electrons.

Solution

The radical centre of both is essentially a planar, sp²-hybridized carbon with the unpaired electron in the 2p orbital (SOMO). A C–H bonding orbital of the methyl group of the ethyl radical can interact with the 2p SOMO of the sp²-hybridized C (hyperconjugation), which stabilizes the ethyl radical.

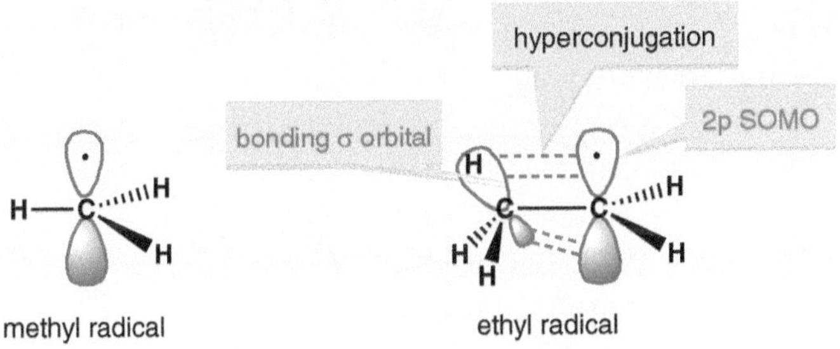

methyl radical   ethyl radical

Arrange the following radicals in order of stability.

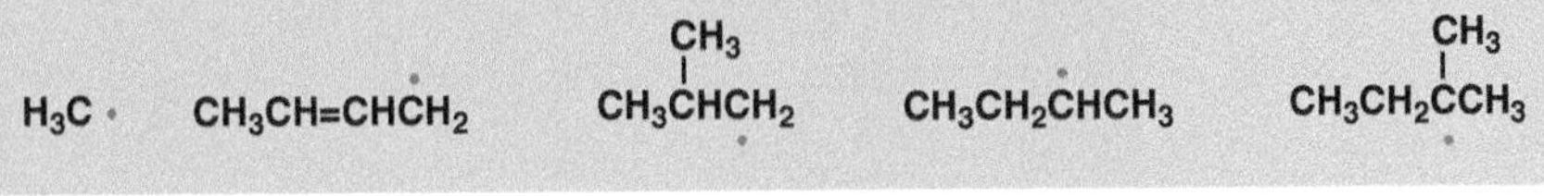

$$H_3C\cdot \qquad CH_3CH=CH\dot{C}H_2 \qquad CH_3\overset{\displaystyle CH_3}{\underset{|}{C}}H\dot{C}H_2 \qquad CH_3CH_2\dot{C}HCH_3 \qquad CH_3CH_2\overset{\displaystyle CH_3}{\underset{|}{C}}CH_3$$

20.3 Halogenation of Alkyl Groups

20.3.1 Chlorination of methane

Radicals are reactive owing to the unpaired electron, and can react even with generally unreactive alkanes at a C–H bond by **hydrogen abstraction**. For example, when a mixture of Cl_2 and methane is irradiated, chloromethane is produced by a **radical chain mechanism**, as shown in Scheme 20.1. Once a radical (Cl·) is formed in the **initiation** step (photolysis of Cl_2, eqn 20.5), it abstracts a hydrogen atom from CH_4 to give HCl and a methyl radical (eqn 20.6); the $CH_3\cdot$ then abstracts a chlorine atom from another Cl_2 molecule to give chloromethane, and generates another $Cl\cdot$ at the same time (eqn 20.7). This $Cl\cdot$ repeats the hydrogen abstraction (eqn 20.6) and generation of another $Cl\cdot$ (eqn 20.7); these are called **chain propagation** steps and are the central parts of the radical chain mechanism which are repeated until one of the reactants is all consumed. The mechanism can also break down when radicals involved in the chain are removed by other steps known as **termination reactions**, e.g. when two methyl radicals combine as in eqn 20.8. The overall reaction is the replacement of an H of methane by a Cl, i.e. a **radical substitution** reaction (eqn 20.9).

Scheme 20.1 A radical chain mechanism for the chlorination of methane.

Initiation reaction:

$$Cl_2 \xrightarrow{\ h\nu\ } 2\ Cl\cdot \qquad (20.5)$$

Chain propagation steps:

$$Cl\cdot \quad H{-}CH_3 \xrightarrow{\ H\ abstraction\ } Cl{-}H + \cdot CH_3 \qquad (20.6)$$

$$H_3C\cdot \quad Cl{-}Cl \longrightarrow H_3C{-}Cl + Cl\cdot \qquad (20.7)$$

Termination reaction:

Disproportionation is another chain termination possibility in reactions involving larger, more complex intermediate radicals (see later).

$$H_3C\cdot \quad \cdot CH_3 \xrightarrow{\ combination\ } H_3C{-}CH_3 \qquad (20.8)$$

Overall radical substitution reaction:

$$CH_4 + Cl_2 \longrightarrow CH_3Cl + HCl \qquad (20.9)$$

When an excess of Cl_2 is present, further chlorination leads to dichloromethane, trichloromethane, and even tetrachloromethane.

$$CH_3Cl \xrightarrow[-HCl]{Cl_2} CH_2Cl_2 \xrightarrow[-HCl]{Cl_2} CHCl_3 \xrightarrow[-HCl]{Cl_2} CCl_4$$

| chloromethane (methyl chloride) | dichloromethane (methylene chloride) | trichloromethane (chloroform) | tetrachloromethane (carbon tetrachloride) |

Exercise 20.3

Explain why the combination of a methyl radical and a chlorine atom is not an important reaction in the formation of chloromethane in Scheme 20.1.

Exercise 20.4

Give chain propagation steps in the radical chain mechanism for the formation of dichloromethane in the photo-induced reaction of chloromethane with Cl_2.

20.3.2 Selectivity in the halogenation of alkanes

Alkanes other than methane undergo chlorination and bromination by radical chain mechanisms in a similar way. An alkane with three or more carbon atoms usually has two or more different kinds of C–H bond, so it yields isomeric haloalkanes depending on which hydrogen is abstracted in the first propagation step. For example, butane gives 2- and 1-halobutanes in the ratios shown in Scheme 20.2 in photo-induced chlorination and bromination.

$$CH_3CH_2CH_2CH_3 \xrightarrow[- HX]{X_2 \quad hv} CH_3CH_2\overset{\overset{\displaystyle X}{|}}{C}HCH_3 + CH_3CH_2CH_2CH_2X$$

butane   2-halobutane   1-halobutane

Cl$_2$, 35 °C   72%   28%

Br$_2$, 127 °C   98%   2%

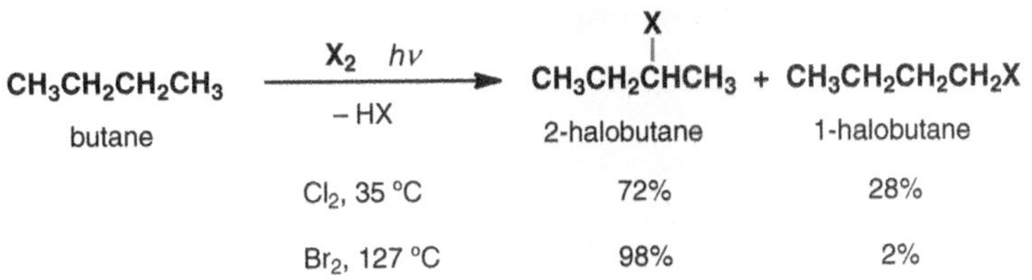

Scheme 20.2 Selectivity in photo-induced chlorination and bromination of butane.

The secondary alkyl halides are formed in higher yields than the primary isomers in both reactions, but the selectivity (even at the higher temperature) is greater in bromination. Preferential formation of the secondary products can be explained by the relative stabilities of the intermediate alkyl radicals formed in the hydrogen abstraction steps— the secondary alkyl radical which leads to the 2-halobutane is more stable than the isomeric primary radical.

Write propagation steps in the radical chain mechanism for the formation of 2-bromobutane from butane and bromine following initiation.

Exercise 20.5

In the radical halogenation of butane, termination steps could involve *disproportionation* in which two alkyl radicals give an alkane and an alkene. One possibility is given below—show how this reaction proceeds using fish-hook curly arrows.

Example 20.2

$$2 \; CH_3CH_2CH_2CH_2{}^{\bullet} \longrightarrow CH_3CH_2CH_2CH_3 + CH_3CH_2CH{=}CH_2$$

Solution
One of the radicals abstracts a hydrogen atom from the other so the one receiving the H becomes an alkane and the one losing the H becomes an alkene.

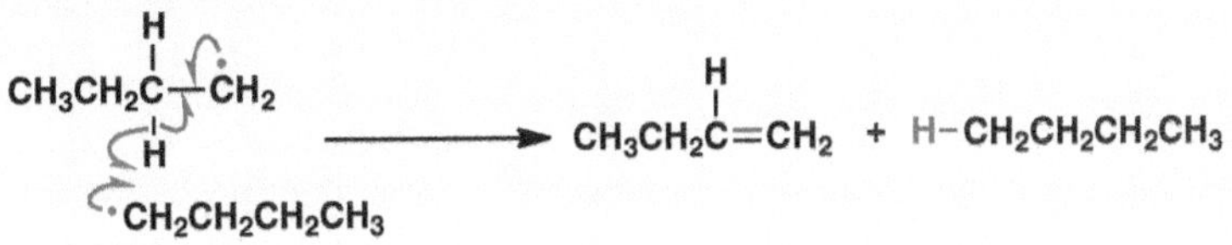

$$CH_3CH_2C{=}CH_2 + H{-}CH_2CH_2CH_2CH_3$$

The greater selectivity in bromination shown in Scheme 20.2 indicates that Br$^{\bullet}$ (being less reactive) is more selective than Cl$^{\bullet}$ in the hydrogen abstractions, and the difference can be understood from the enthalpies of these steps. Bond dissociation energies (*DH*) for H–Cl and H–Br are 432 and 366 kJ mol^{-1}, respectively, while those for the C–H bonds at the primary and secondary carbons are 419 and 402 kJ mol^{-1} (Table 20.1). From eqn 20.10, therefore, we can calculate that the hydrogen abstraction reaction from an alkane is slightly exothermic for chlorine while it is endothermic for bromine, as represented in Figure 20.2.

Figure 20.2 Reaction profiles for hydrogen abstraction by Cl and Br radicals.

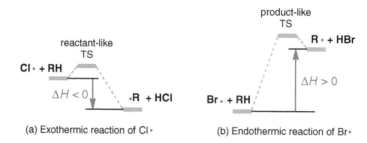

(a) Exothermic reaction of Cl•   (b) Endothermic reaction of Br•

Hydrogen abstraction:

$$X• \ + \ H–R \longrightarrow R• \ + \ H–X \qquad \Delta H = DH(H–R) – DH(H–X) \qquad (20.10)$$

> Although *bond energy* is the commonly used term, the better name is *bond dissociation enthalpy*, which is why eqn 20.10 can be used here to calculate the enthalpy of the reaction shown. It is a particular case of the useful general equation
>
> $$\Delta H = \Sigma DH(\text{bonds broken}) – \Sigma DH(\text{bonds formed})$$
>
> where the symbol Σ means 'the sum of'.

According to the Hammond postulate (Panel 7.2), the transition structure for an exothermic reaction is closer in energy and configuration to the reactants than to the products: the TS is said to be reactant-like. For the chlorination reaction, therefore, the relative stabilities of the isomeric radicals in the competing paths are not reflected in appreciable selectivity. In contrast, the TS for the endothermic bromination reaction is more product-like than reactant-like. Consequently, the significant difference in the stabilities of the isomeric radicals in the competing paths is reflected in an appreciable selectivity in favour of the formation of the secondary alkyl radical leading to the secondary alkyl bromide.

Exercise 20.6

What is the main product when a mixture of 2-methylpropane and Br_2 is irradiated?

20.3.3 Halogenation at allylic and benzylic positions

Allylic and benzylic positions are the alkyl carbons next to the double bonds of alkenes and the aryl groups of arylalkanes, respectively.

As the bond dissociation energies in Table 20.1 suggest, formation of allyl and benzyl radicals is relatively easy. This is attributable to the stabilization of these radicals by delocalization of the unpaired electron (Section 5.4), so allylic and benzylic hydrogen abstraction occurs preferentially when they are available.

$$H_2C{=}CH{-}CH_2 \longleftrightarrow H_2\overset{\bullet}{C}{-}CH{=}CH_2$$

Resonance forms of the allyl radical.

Resonance forms of the benzyl radical.

Halogenation of an alkene, therefore, takes place mainly at the allylic position, e.g. eqn 20.11, and (even more selectively) halogenation of an arylalkane occurs at the benzylic position, e.g. eqn 20.12. The selectivity for allylic and benzylic bromination in particular is generally excellent.

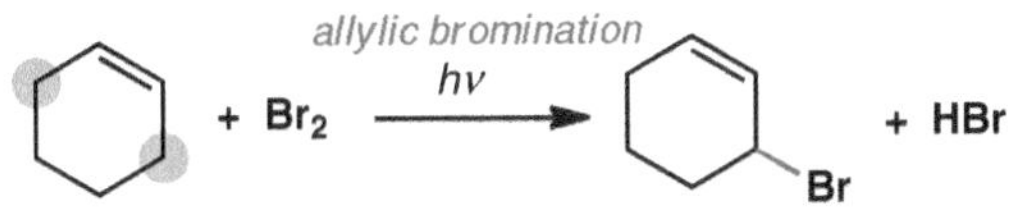

$$\text{(20.11)}$$

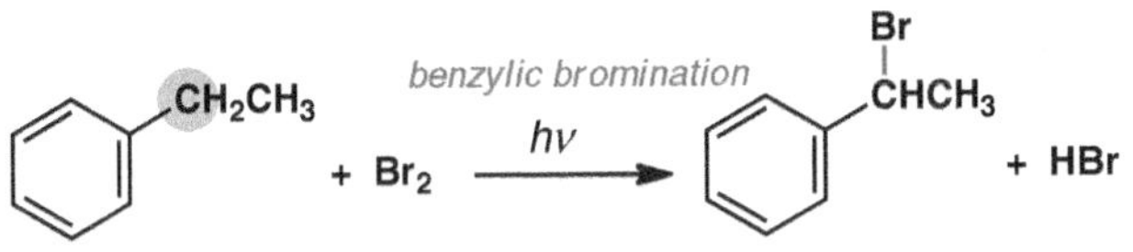

$$\text{(20.12)}$$

> **Give chain propagation steps for the reaction of eqn 20.12.**   **Exercise 20.7**

However, allylic bromination has a readily occurring competitive side reaction: addition of Br_2 to the C=C bond. Since the addition reaction is normally mainly by a polar mechanism, it can be disadvantaged by the use of nonpolar solvents such as CH_2Cl_2 or CCl_4, and/or by keeping the concentration of Br_2 low. For example, the reaction can be carried out using CCl_4 with the very slow addition of the Br_2 in order to keep $[Br_2]$ sufficiently low. However, the slow and controlled addition of bromine so that it remains at a very low concentration is difficult and experimentally very inconvenient.

In a polar solvent, the formation of the bromonium ion and its bromide counter-ion is rate determining, and the reaction is first order in $[Br_2]$. In a less polar solvent such as CH_2Cl_2, however, two molecules of Br_2 are involved leading to Br_3^- as the counter-ion. The order in $[Br_2]$ in the rate law for this polar addition in a less polar solvent, therefore, is greater than in the rate law for radical allylic bromination.

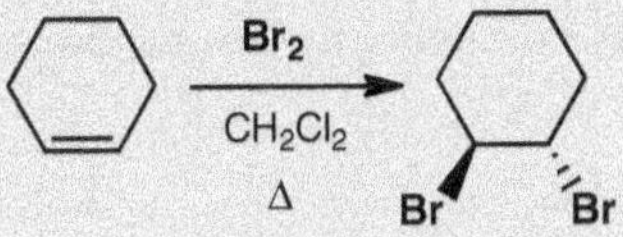

Rate of polar addition = k[alkene]$[Br_2]^2$

Consequently, the polar reaction in a less polar solvent is more disadvantaged compared with the radical reaction when $[Br_2]$ is very low. A radical mechanism for addition of Br_2 to the double bond is also possible, but it is reversible and not a significant reaction at low concentrations of Br_2.

Fortunately, a very useful reagent, *N*-**bromosuccinimide** (**NBS**), is available which provides a steady low concentration of Br_2 for the reaction shown in eqn 20.13, for example.

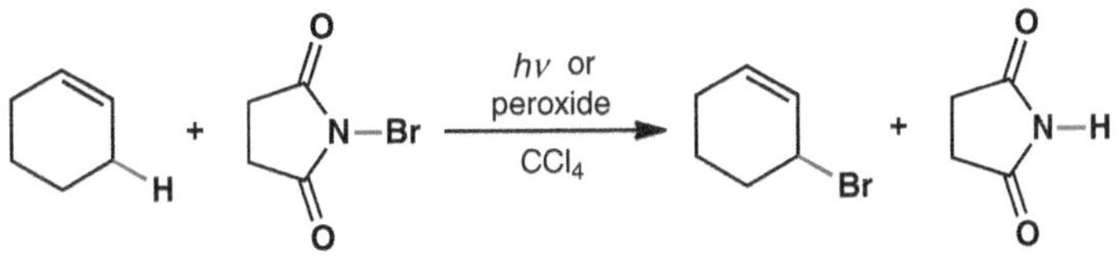

$$\text{(20.13)}$$

N-bromosuccinimide
(NBS)

succinimide

A bromine atom produced by photolysis of NBS, for example, abstracts an allylic H from the alkene to give an allylic radical in the first propagation step of Scheme 20.3. Hydrogen bromide, which is also formed in this step, generates Br_2 in the next (polar) reaction with NBS; the subsequent reaction of the Br_2 with allylic radicals follows immediately to form the allylic bromide and Br·, which propagates the chain.

Although the chain propagation steps in Scheme 20.3 with Br_2 as the main brominating agent largely account for the experimental observations, the complete mechanism may be more complicated. For example, the succinimidyl radical could also abstract an H from the alkene to give the allylic radical, but this is not a main reaction. Curiously, the reaction works well in CCl_4 even though NBS is hardly soluble in CCl_4.

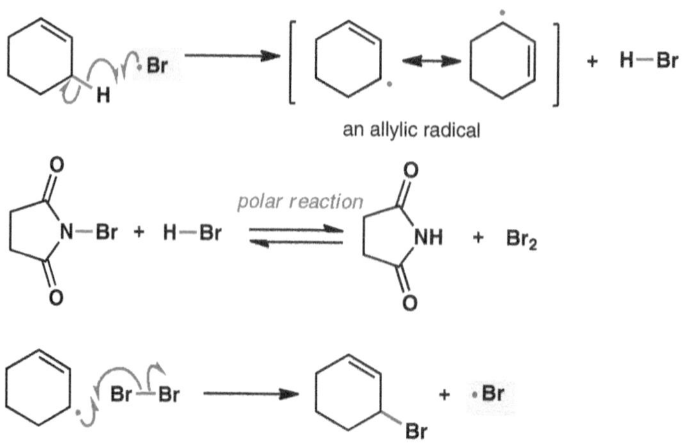

an allylic radical

polar reaction

Scheme 20.3 Chain propagation steps in an allylic bromination by NBS.

Example 20.3

What are the main products of the following reaction?

$$CH_3CH_2CH_2CH=CH_2 + NBS \xrightarrow[CCl_4]{h\nu}$$

Solution

The allylic radical intermediate formed by hydrogen abstraction is a resonance hybrid and gives two regioisomeric products by reaction at the alternative allylic positions. The main product is the internal alkene, which is more stable than the terminal alkene, and has E and Z isomers.

$$\left[CH_3CH_2\overset{\bullet}{C}H-CH=CH_2 \longleftrightarrow CH_3CH_2CH=CH-\overset{\bullet}{C}H_2 \right] \longrightarrow$$

E and Z isomers

Exercise 20.8

What are the main products in the reactions of the following with NBS in the presence of BPO?

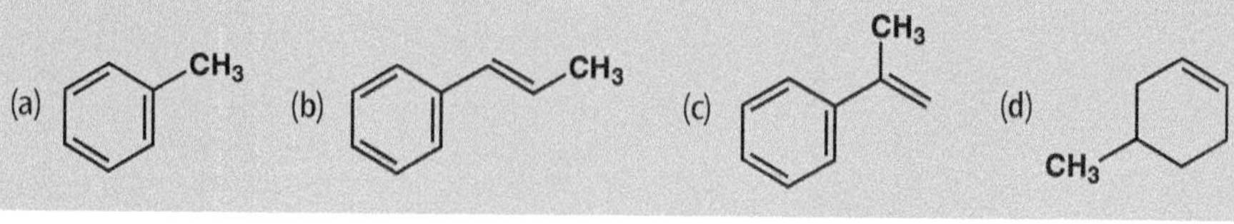

(a)   (b)   (c)   (d)

20.4 Dehalogenation and Related Reductions

We have seen in the preceding section that halogenation of alkyl groups occurs by a radical mechanism. The reverse reaction, replacement of the halogen of a haloalkane by a hydrogen atom, can also be achieved by a radical mechanism using tributyltin hydride (Bu_3SnH) as the reagent, e.g. eqn 20.14. Since the Sn–H bond is weak ($DH = 308$ kJ mol^{-1}), the hydrogen is easily abstracted from the tin hydride and the tin radical can then abstract a halogen to give a chain reaction (Scheme 20.4).

Note that the Sn–Br bond is strong, typically $DH = 552$ kJ mol^{-1}.

$$PhCH_2CH_2Br + Bu_3SnH \xrightarrow{AIBN} PhCH_2CH_3 + Bu_3SnBr \qquad (20.14)$$

Initiation:

$$Me_2C-N=N-CMe_2 \ (CN)(CN) \xrightarrow{\Delta} 2\ Me_2\overset{\bullet}{C}(CN) + N_2 \qquad (20.4)$$

AIBN

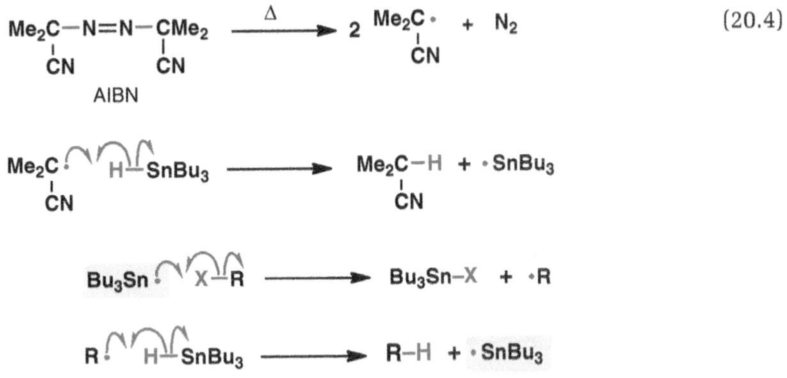

Scheme 20.4 A radical chain mechanism for the dehalogenation of a haloalkane with Bu$_3$SnH initiated by thermolysis of AIBN.

$$Me_2\overset{\bullet}{C}(CN)\ \ H-SnBu_3 \longrightarrow Me_2C-H\ (CN) + \bullet SnBu_3$$

Propagation:

$$Bu_3Sn\bullet\ \ X-R \longrightarrow Bu_3Sn-X + \bullet R$$

$$R\bullet\ \ H-SnBu_3 \longrightarrow R-H + \bullet SnBu_3$$

Alcohols can be deoxygenated indirectly also using Bu$_3$SnH (eqn 20.15). The alcohol is first converted to an alkyl xanthate (dithiocarbonate) ester, for example, which is then reduced by Bu$_3$SnH using AIBN as initiator; this step is mechanistically similar to dehalogenation using Bu$_3$SnH.

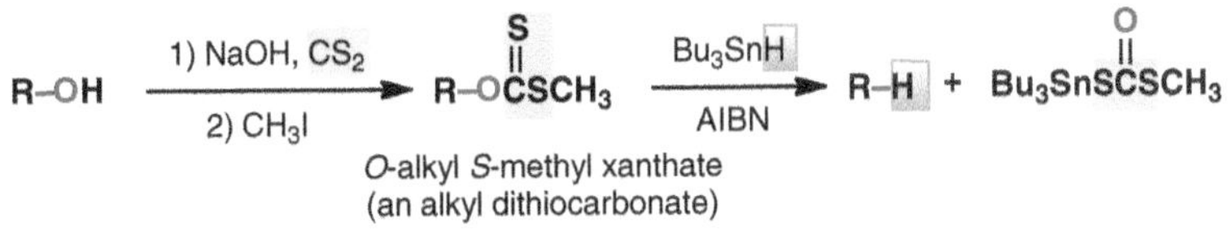

$$R-OH \xrightarrow[2)\ CH_3I]{1)\ NaOH,\ CS_2} R-O\overset{S}{\overset{\|}{C}}SCH_3 \xrightarrow[AIBN]{Bu_3SnH} R-H + Bu_3Sn\overset{O}{\overset{\|}{S}}CSCH_3 \qquad (20.15)$$

O-alkyl *S*-methyl xanthate (an alkyl dithiocarbonate)

This is sometimes called the Barton–McCombie reaction.

Closely similar radical reactions can be used for the removal of hydroxy groups from complex polyfunctional compounds such as carbohydrate and nucleic acid derivatives.

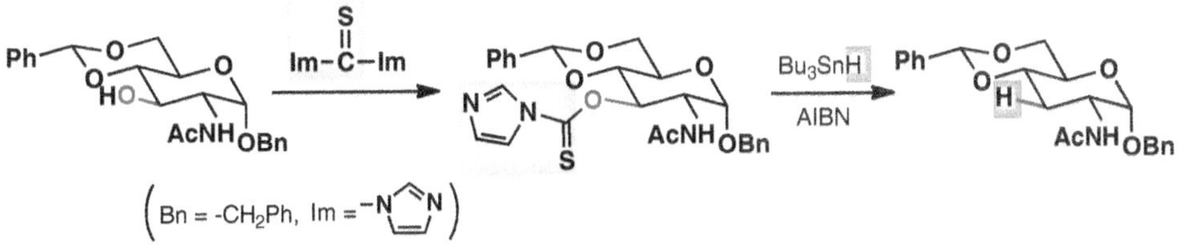

(Bn = -CH$_2$Ph, Im = imidazolyl)

20.5 Radical Addition Reactions

20.5.1 Radical addition of HBr to alkenes

Hydrogen halides usually undergo electrophilic additions to unsymmetrical alkenes to give haloalkanes with 'Markovnikov' regioselectivity (e.g. eqn 20.16) via an initial proton transfer step, as covered in Section 15.2. Unlike HCl and HI, however, HBr sometimes yields an *anti-Markovnikov* product (e.g. eqn 20.17), in an apparently anomalous reaction. The seemingly inconsistent results with HBr puzzled chemists in the early days of mechanistic organic chemistry until they were found to be caused by different reaction conditions and, in particular, the purities of the materials used. The different regioselectivities are now known to be attributable to different reaction mechanisms.

When hydrobromination of an unsymmetrical alkene is carried out in a polar medium, rigorously free of any peroxide contaminants, the electrophilic addition mechanism leads to the expected Markovnikov regioselectivity as we discussed in Section 15.2.

454 . . . 20 Reactions involving Radicals

Electrophilic addition:

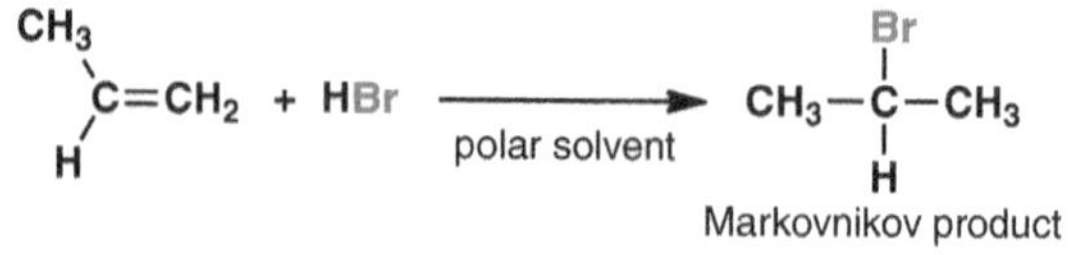

$$(20.16)$$

In contrast, anti-Markovnikov regioselectivity is observed using HBr (especially in a nonpolar solvent) in the presence of a compound such as a peroxide which can initiate a radical chain mechanism (Scheme 20.5).

Radical addition:

Scheme 20.5 A radical chain mechanism for the addition of HBr to an unsymmetrical alkene.

$$(20.17)$$

Initiation:

Propagation:

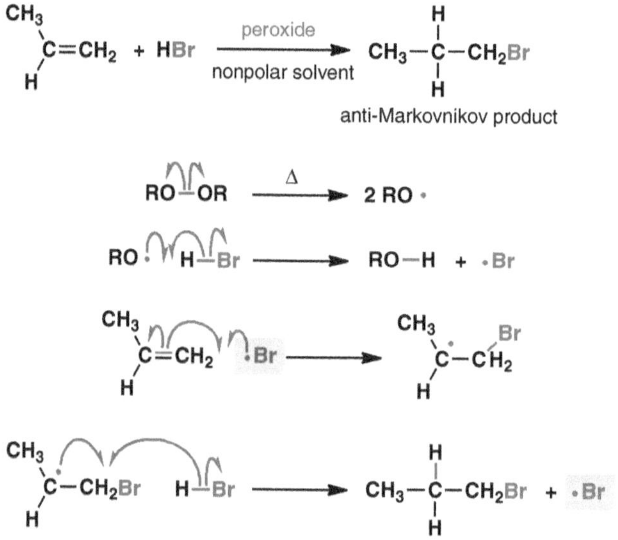

The radical chain mechanism involves Br˙ which is initially generated by hydrogen abstraction from HBr by the alkoxyl radical formed by homolysis of the peroxide. The bromine atom adds to the alkene to give a carbon radical which then reacts with HBr to generate another Br˙; these two reactions constitute the propagation.

The regioselectivity of the hydrobromination of an unsymmetrical alkene is determined by the Br˙ addition step: to which end of the C=C does the bromine atom add? As we have already seen, the relative stabilities of radicals are similar to those of carbenium ions, so a secondary radical is generally more stable than a primary radical.

Stability of the radicals following addition of Br˙ to propene:

Stability of the carbenium ions following addition of H^+ to propene:

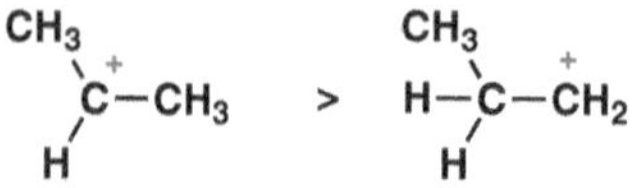

Consequently, although the overall regiochemistry of the radical reaction is indeed opposite to that of the electrophilic addition, this is only because the bromine adds first by the radical mechanism whereas the hydrogen adds first in the electrophilic addition.

Exercise 20.9

What are the main products when the following compounds react with HBr in the presence of a peroxide?

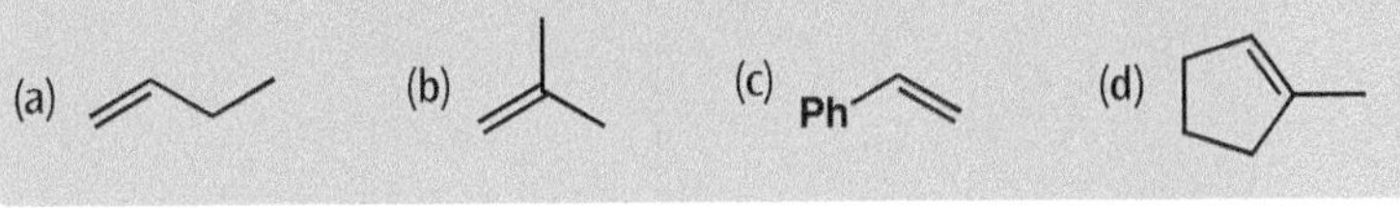

(a) (b) (c) Ph (d)

Addition of HCl and HI to alkenes by radical mechanisms never appear to compete with addition by polar mechanisms. Thiols, however, add to alkenes by the radical mechanism with the expected anti-Markovnikov regioselectivity but it is usually necessary to use a large excess of the thiol to prevent polymerization of the alkene (see Section 20.7).

Exercise 20.10

Give a mechanism and the main product of the following reaction.

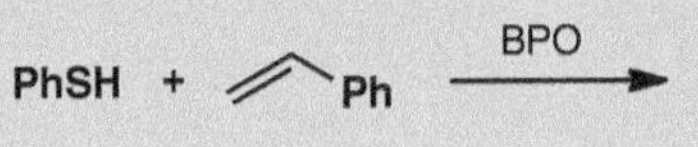

$$PhSH \ + \ \text{[alkene]}Ph \xrightarrow{\text{BPO}}$$

20.5.2 Radical additions to alkenes involving Bu$_3$SnH

In the dehalogenation of a haloalkane RX in Scheme 20.4, the intermediate alkyl radical abstracts H from the tin hydride to give the reduction product, RH. However, if the reaction is carried out in the presence of an alkene, CH$_2$=CHY, the intermediate alkyl radical can be intercepted by the alkene to give another radical intermediate; this can then abstract H from the tin hydride to give the *addition* product, RCH$_2$CH$_2$Y as illustrated in Scheme 20.6 and exemplified in eqn 20.18.

Note that there is one more propagation step in the mechanism of Scheme 20.6 than in the reduction mechanism of Scheme 20.4. Furthermore, radical R• must add to the alkene faster than it abstracts H from Bu$_3$SnH, and radical RCH$_2$CHY must abstract H from Bu$_3$SnH faster than it adds to another alkene molecule (otherwise, the result would be polymerization; see Section 20.7).

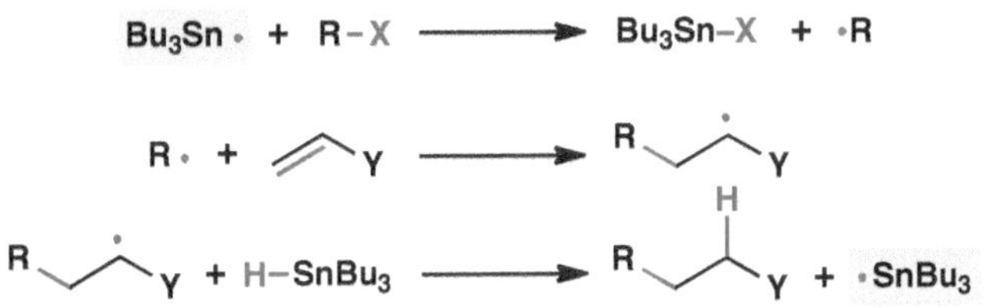

$$Bu_3Sn• \ + \ R{-}X \longrightarrow Bu_3Sn{-}X \ + \ •R$$

$$R• \ + \ \text{[alkene]}Y \longrightarrow R{\sim}Y$$

$$R{\sim}Y \ + \ H{-}SnBu_3 \longrightarrow R{\sim}Y \ + \ •SnBu_3$$

Scheme 20.6 Radical chain propagation steps for the indirect addition of R–H to an alkene involving Bu$_3$SnH. Initiation steps are the same as in Scheme 20.4.

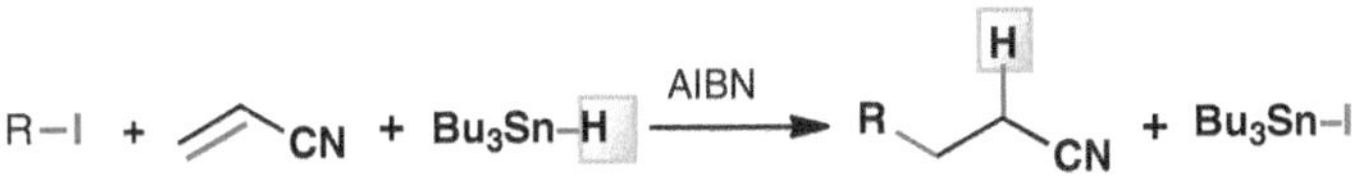

$$R{-}I \ + \ \text{[alkene]}CN \ + \ Bu_3Sn{-}H \xrightarrow{\text{AIBN}} R{\sim}CN \ + \ Bu_3Sn{-}I \qquad (20.18)$$

20.6 Intramolecular Reactions of Radicals

We shall see in Section 22.6, just as we saw in Section 12.5, that a nucleophilic group can engage with an electrophilic centre in the same molecule if the two can come into close proximity, and the outcome is an S$_N$2-like intramolecular nucleophilic substitution reaction. In a similar way, a radical can undergo an intramolecular reaction if there is a centre susceptible to radical attack elsewhere in the molecule, and the two can come sufficiently close.

20.6.1 Cyclization of alkenyl radicals

As with polar intramolecular reactions, cyclization is a possible outcome of an intramolecular radical reaction. We saw in Sub-section 20.5.2 just above that a carbon radical generated from a haloalkane can add to an alkene (Scheme 20.6); the intramolecular version of this reaction occurs when 6-bromohex-1-ene, for example, reacts with Bu_3SnH in the presence of AIBN (eqn 20.19).

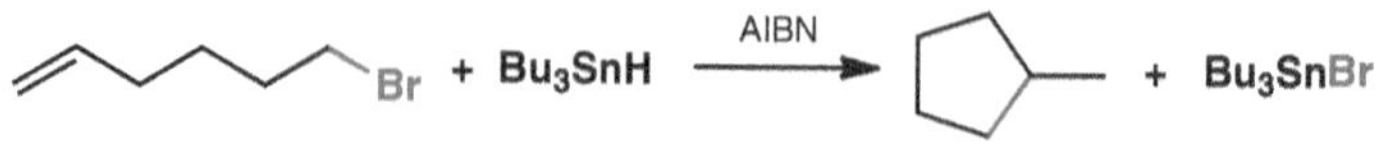

$$\text{(20.19)}$$

The cyanoalkyl radical generated from AIBN abstracts H from Bu_3SnH to form the tin radical (as in Scheme 20.4) which then propagates the cyclization sequence shown in Scheme 20.7.

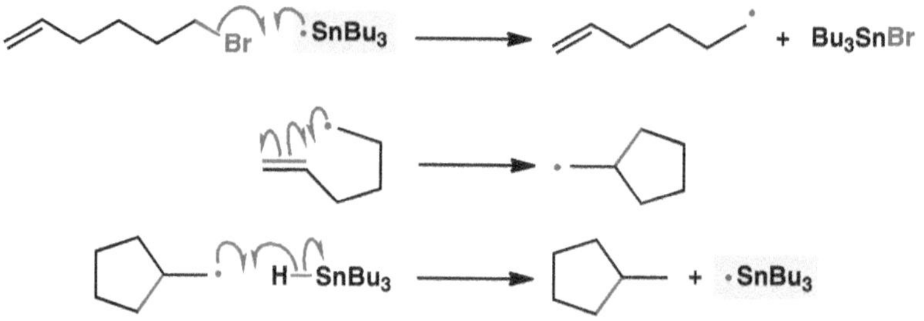

$$\text{(20.20)}$$

Scheme 20.7 Chain propagation steps in the radical cyclization of an alkene.

The cyclization in eqn 20.20 to give the five-membered ring is called a 5-*exo*-cyclization (*exo* because the other end of the double bond to which the radical centre becomes attached points out of the newly formed ring), and the one in eqn 20.21 to give a six-membered ring is a 6-*endo*-cyclization (*endo* because the other end of the double bond to which the radical centre becomes attached becomes part of the newly formed ring).

Note that cyclization to form the cyclohexyl radical is also possible (eqn 20.21), and (for electronic reasons) this secondary radical is more stable than the primary radical with the five-membered ring in eqn 20.20.

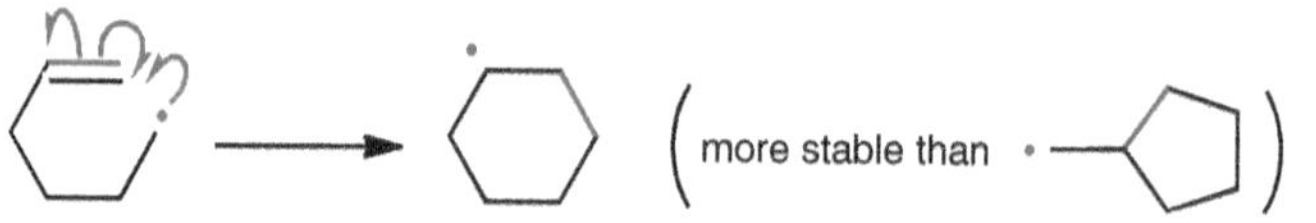

$$\text{(20.21)}$$

It follows that the TS for formation of the five-membered ring (eqn 20.20) must be lower in energy than the one for formation of the six-membered ring (eqn 20.21). This could be because the $-CH_2$ bearing the unpaired electron needs to interact from above the plane of the double bond, and the necessary conformation is less strained for the TS leading to the five-membered ring.

20.6.2 1,5-Hydrogen transfer

Intermolecular hydrogen atom transfers (abstractions), as in dehalogenations using Bu_3SnH (Scheme 20.4) or disproportionation termination steps in radical chain reactions (Example 20.2), are characteristic reactions of radicals.

An intramolecular 1,5-H transfer was observed when 1-(o-bromophenyl)-5-phenylpentane was heated with Bu_3SnD and AIBN, as shown in Scheme 20.8. Deuteration occurs both at the original aromatic position of the Br to give **1** and at the saturated carbon C3 in **2** in the ratio of 14 : 86. The aromatic radical is formed first by the abstraction of Br by the tin radical, then the 1,5-shift of the H from C3 to the aromatic position of the initial radical results in the C3 radical; this rearranged radical abstracts deuterium from Bu_3SnD to give the product of the intramolecular 1,5-H shift, **2**.

A 1,5-H transfer from carbon to carbon in a radical involves a C–H bond only changing its position, so the enthalpy of the isomerization will generally be very small and there is little reason for the reaction to proceed. In the example of Scheme 20.8, however, the 1,5-H transfer involves replacement of a weaker sp^3 C–H in the initial radical being replaced by a stronger sp^2 C–H in the rearranged radical, i.e. the rearranged radical is more stable than the initial radical.

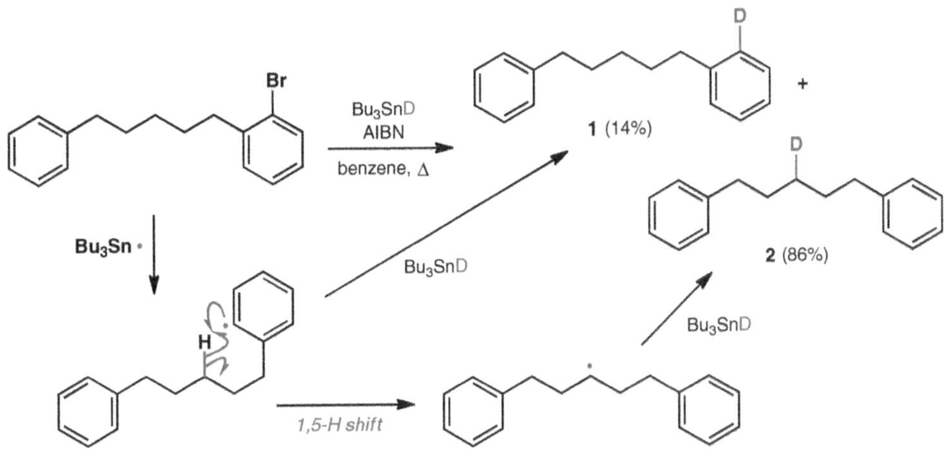

Scheme 20.8
A 1,5-hydrogen transfer detected using tributyltin deuteride.

In the so-called Barton reaction, which was originally introduced as a method for functionalizing methyl groups in steroids (eqn 20.22), a photolytically generated oxygen radical undergoes a 1,5-H transfer from C to O.

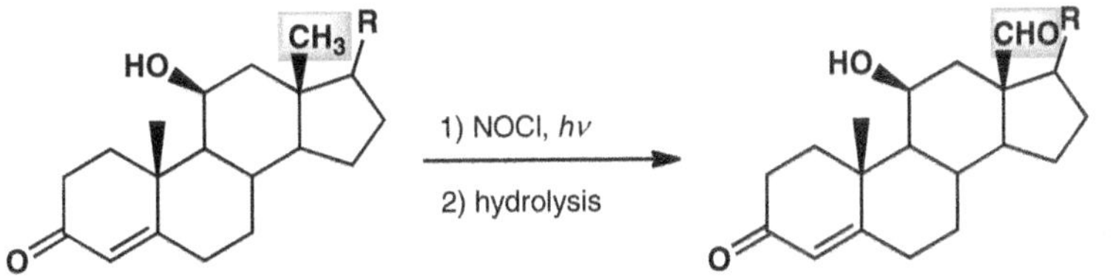

(20.22)

The British chemist Sir Derek Barton (1918–1998) contributed widely to organic synthesis and also shared the 1969 Nobel Prize in Chemistry with the Norwegian chemist, O. Hassel (1897–1981), for their recognition of the relationship between molecular shape and chemical reactivity (conformational analysis) of organic compounds.

As shown in Scheme 20.9 with partial structures, nitrosyl chloride converts the hydroxy group into a nitrite ester which, upon photolysis, gives an oxygen radical and nitric oxide. Following the 1,5-H shift and combination of the carbon radical with the nitric oxide, a nitroso-methyl compound is obtained which isomerizes to an oxime; hydrolysis of the oxime gives the aldehyde product.

Two important features of this reaction are (a) it is otherwise difficult specifically to functionalize a C-bonded methyl group, and (b) the aldehyde group can be readily converted into a range of other functional groups. Note also that the reaction is possible only when the atoms between which the 1,5-H transfer takes place are held together in the proper spatial relationship by the relative rigidity of the reactant—a steroid in the above example.

Note the difference between a nitrite ester, $RONO$, and a nitroalkane, RNO_2.

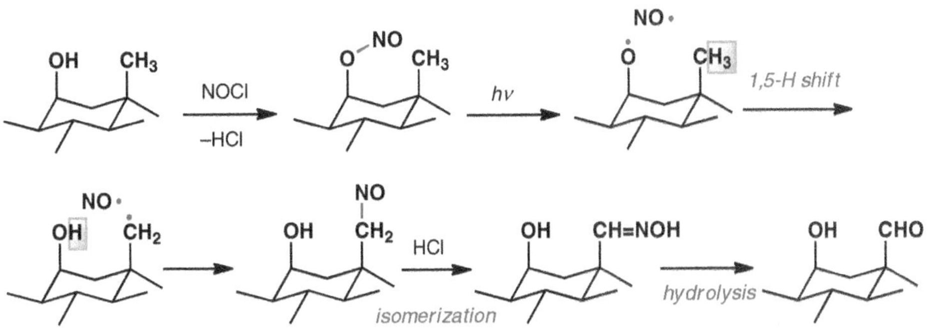

Scheme 20.9 1,5-H transfer within an oxygen radical in the Barton reaction.

20.6.3 Fragmentation of radicals

We saw earlier (eqn 20.3) that the carboxyl radical formed in the homolysis of dibenzoyl peroxide undergoes decarboxylation, a reaction which is generalized in eqn 20.23. This is a specific example of a range of unimolecular reactions in which a larger radical fragments to give a smaller radical and (usually, but not invariably) a stable small molecule.

$$R \overset{\curvearrowleft}{-} C \overset{O}{\underset{\underset{\cdot}{O}}{\big\langle}} \qquad \longrightarrow \qquad R\cdot \; + \; CO_2 \qquad\qquad (20.23)$$

*These reactions are
sometimes called
β-fragmentations.

Fragmentations of tertiary alkoxyl radicals are similar examples, e.g. eqn 20.24. In both cases, homolysis of the bond β to the radical centre and redistribution of valence electrons gives a new radical and liberates a stable molecule.* The driving force for reaction 20.24 is the formation of a stable ketone, just as the formation of carbon dioxide is the driving force for reaction 20.23.

$$R \overset{\curvearrowleft}{-} \overset{\overset{\displaystyle CH_3}{|}}{\underset{\underset{\displaystyle CH_3}{|}}{C}} \overset{\curvearrowleft}{-} O\cdot \qquad \longrightarrow \qquad R\cdot \; + \; \overset{\displaystyle CH_3}{\underset{\displaystyle CH_3}{C}}{=}O \qquad\qquad (20.24)$$

The synthesis of a medium ring keto ester in eqn 20.25 involves a reversible intramolecular radical addition to a ketone carbonyl bond, and a C–C bond homolysis within the intermediate oxy radical prior to H-abstraction from Bu_3SnH. The driving force for the homolytic cleavage is the regeneration of the carbonyl group but, because the molecule is cyclic, the carbonyl group is not part of a separate new molecule.

Note that the main driving
force of the homolyses
in the intramolecular
reaction of eqn 20.25
is the same as that for
the β-fragmentation in
eqn 20.24.

(20.25)

Exercise 20.11

Give a radical chain mechanism for the reduction of the intermediate xanthate to the alkane in eqn 20.15 (Section 20.4).

If there is sufficient driving force, cleavage of a C–C bond β to a *carbon* radical is possible. The rearrangement of the radical intermediate in eqn 20.26 is driven by the relief of strain following opening of the three-membered ring.

(20.26)

The bond cleavage of eqn 20.26 is one of the fastest measured first-order reactions ($k_1 = 1.3 \times 10^8$ s^{-1} at 25 °C) and is used to evaluate the rate constants of bimolecular radical reactions by a competition method: this is known as the 'radical clock' method. Another radical clock reaction is the cyclization of eqn 20.20 in Scheme 20.7 with $k_1 = 1.0 \times 10^5$ s^{-1} at 25 °C.

Give a mechanism and the main product of the following reaction.

$$\triangleright\!\!\!\!\diagdown\!\!\!\diagup \quad + \quad PhSH \xrightarrow{\text{AIBN}}$$

20.7 Radical Polymerization of Alkenes

As we saw in Sub-section 20.5.2, addition of a radical to an alkene gives another carbon radical which may abstract an H atom from a donor if one is present. If the carbon radical adds to another molecule of the alkene, however, and this is repeated many times without any appreciable side reactions such as hydrogen abstraction, a polymer is obtained. Polymerization of alkenes by this radical chain mechanism is one of the most important industrial chemical reactions, and millions of tons of polymers are produced every year by this process, including polyethylene (polythene), poly(vinyl chloride) (PVC), and polystyrene. Alkene monomers which undergo **radical polymerization** most readily are those with a substituent which is conjugatively electron-donating or withdrawing.

Vinyl is the common name of the CH_2=CH– group and vinyl compounds used in polymerizations are sometimes referred to as *vinyl monomers.*

Radicals are stabilized by conjugation regardless of whether the substituent is electron-donating or -withdrawing.

Alkene monomers for radical polymerization:

$$H_2C=C\overset{H}{\underset{Ph}{\big\langle}} \qquad H_2C=\overset{H}{\underset{H}{C}}-C=CH_2 \qquad H_2C=C\overset{H}{\underset{Cl}{\big\langle}} \qquad H_2C=C\overset{H}{\underset{OAc}{\big\langle}}$$
styrene   buta-1,3-diene   vinyl chloride   vinyl acetate

$$H_2C=C\overset{H}{\underset{CN}{\big\langle}} \qquad H_2C=C\overset{H}{\underset{CO_2Me}{\big\langle}} \qquad H_2C=C\overset{Me}{\underset{CO_2Me}{\big\langle}} \qquad F_2C=CF_2$$
acrylonitrile   methyl acrylate   methyl methacrylate   tetrafluoroethene

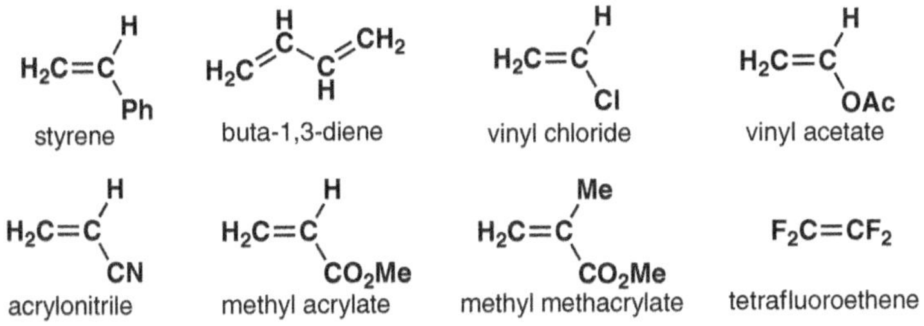

A radical chain mechanism leading to polymerization when an alkene monomer is heated with a small amount of a radical initiator such as a peroxide or AIBN is shown in Scheme 20.10.

Initiation:

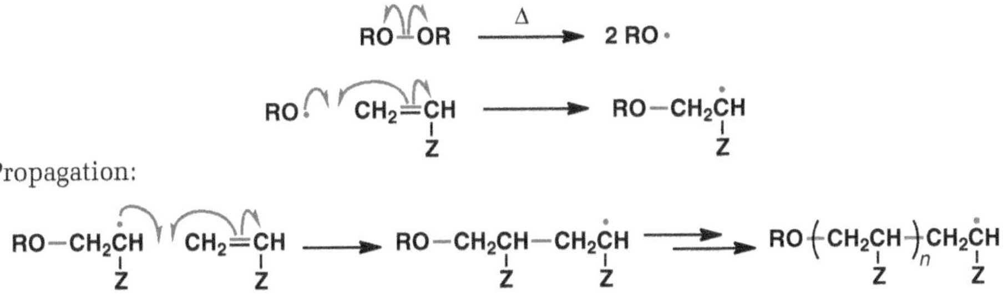

$$RO\!-\!OR \xrightarrow{\ \Delta\ } 2\ RO\cdot$$

$$RO\cdot \quad CH_2\!=\!\overset{}{\underset{Z}{CH}} \longrightarrow RO-CH_2\overset{\bullet}{\underset{Z}{CH}}$$

Propagation:

$$RO-CH_2\overset{}{\underset{Z}{CH}} \quad CH_2\!=\!\overset{}{\underset{Z}{CH}} \longrightarrow RO-CH_2\overset{}{\underset{Z}{CH}}-CH_2\overset{\bullet}{\underset{Z}{CH}} \longrightarrow RO\!\left(\!CH_2\overset{}{\underset{Z}{CH}}\!\right)_{\!n}\!CH_2\overset{\bullet}{\underset{Z}{CH}}$$

Scheme 20.10 Initiation and propagation of the chain reaction for radical polymerization of a vinyl monomer.

Chain termination in radical polymerization occurs when two polymer radicals either combine or disproportionate; these two termination processes are just as in the radical chain mechanism for the halogenation of alkanes discussed in Section 20.3. How often they occur determines the length (molecular weight) of the polymer. Hydrogen abstraction from other molecules in the reaction mixture can also terminate the main propagation chain, but this produces another radical which then may start another chain reaction: this process is sometimes called a *chain transfer.*

Draw structures of the products of the disproportionation termination steps in the mechanism in Scheme 20.10.

Ethene is not a particularly good monomer for radical polymerization but polyethylene is produced industrially on a huge scale by the radical process at high temperatures (100–350 °C) and pressures (1000–4000 atm).[2] Branching occurs when a polymeric radical abstracts a hydrogen from another chain, and the new radical then adds to further ethene units (Scheme 20.11).

Scheme 20.11
Branching during the radical polymerization of ethene.

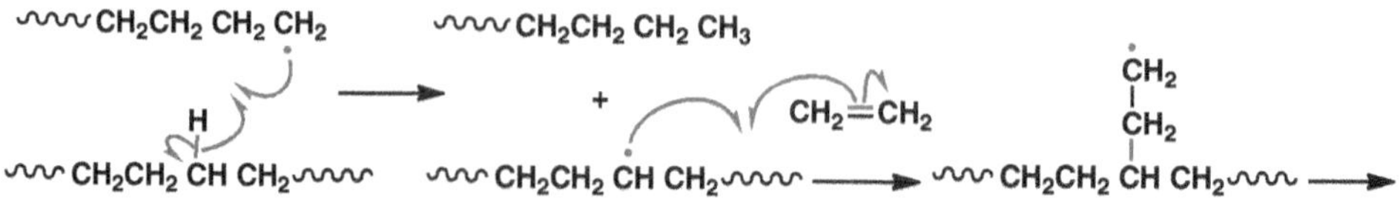

Exercise 20.14

Intramolecular 1,5-hydrogen transfer sometimes occurs within a polymer radical during radical polymerization of ethene. This reaction is sometimes called 'back-biting', and butyl branches are formed on the polyethylene chain. Give a mechanism to show how this happens.

20.8 Autoxidation

C–H groups next to the oxygen of an ether are particularly susceptible to autoxidation. Consequently, hydroperoxides are formed by autoxidation of ether solvents which are left exposed to air in the laboratory. Explosions may occur when ether solutions contaminated in this way are concentrated and heated to dryness.

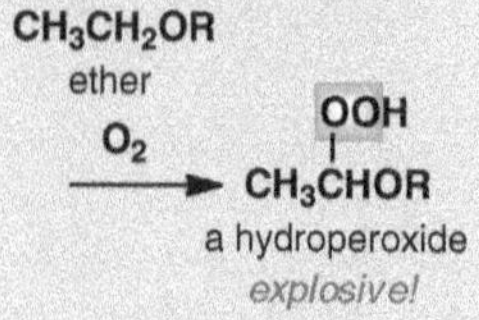

CH_3CH_2OR
ether

$$\xrightarrow[O_2]{} \quad \overset{\displaystyle OOH}{\underset{\displaystyle |}{CH_3CHOR}}$$

a hydroperoxide
explosive!

The slow oxidation of organic materials by atmospheric oxygen is called **autoxidation** and it occurs by a radical chain mechanism. Molecular oxygen is a diradical (·O–O·) which can abstract hydrogen atoms, but is not very reactive. However, when radicals which can initiate a chain reaction are generated by irradiation (or by other means) oxygen can participate in propagation steps. Unsaturated compounds in food are susceptible at their allylic (or benzylic) positions, and polyunsaturated fatty acid residues of lipids (Chapter 24) are readily oxidized, which leads to the food becoming rancid. Scheme 20.12

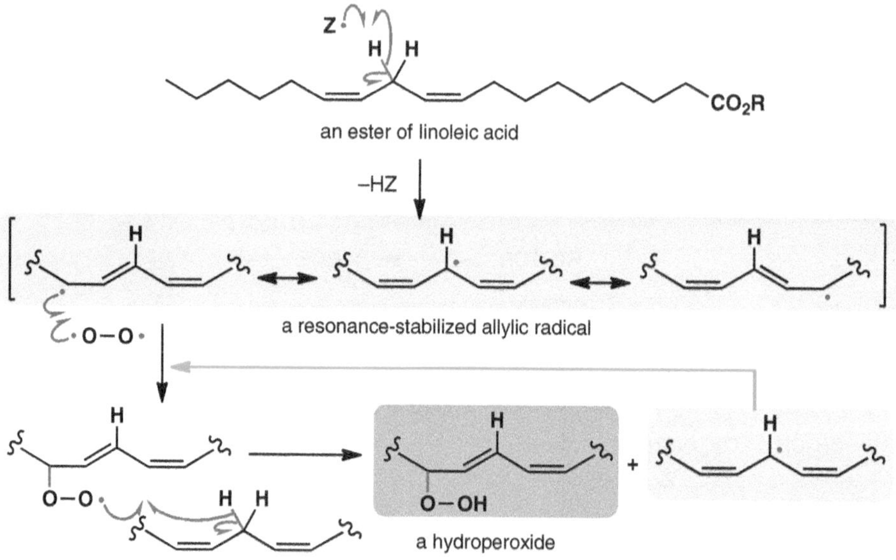

Scheme 20.12 Initiation and propagation in the autoxidation of an ester of linoleic acid.

---

[2] Polyethylene produced by this high pressure method has many branches and is known as low-density polyethylene (LDPE), its density being 0.91–0.93 g cm^{-3}. Polyethylene is also manufactured under lower pressures using a so-called Ziegler–Natta catalyst (e.g. AlEt$_3$–TiCl$_4$). By this process, which does not involve radicals, the polymer is largely unbranched and known as high-density polyethylene (HDPE), the density being 0.93–0.97 g cm^{-3}. LDPE is flexible and used for plastic bags, for example, while HDPE is rigid and used for plastic containers, for example. Ziegler–Natta catalysts are also used to control the stereoselectivity as each monomer is added in the polymerization of propene and other vinyl monomers; the outcome affects the physical properties of the polymer. K. Ziegler (1898–1973, Germany) and G. Natta (1903–1979, Italy) were jointly awarded the Nobel Prize in Chemistry in 1963.

represents a radical chain reaction of a linoleic acid ester initiated by radical Z•, and the doubly allylic radical propagates the chain mechanism.

The intermediate hydroperoxide may fragment to give hydroxyl (HO•) and an alkoxyl radical which also fragments to give an alkyl radical and an aldehyde; the latter then undergoes further autoxidation to give to a carboxylic acid (Scheme 20.13).

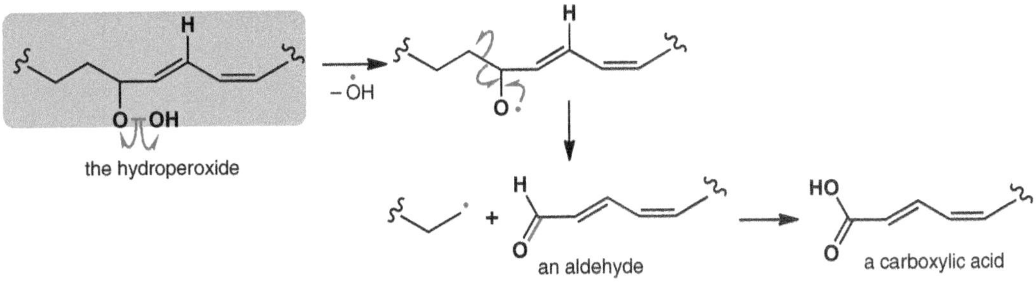

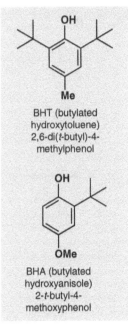

Scheme 20.13 Stepwise fragmentation of a hydroperoxide.

Autoxidation reactions are usually difficult to control, and are not generally useful. However, it is a ubiquitous natural process: the slow deterioration in air of organic materials such as plastics, rubber, paint, and oil is caused by autoxidation.

Antioxidants, including some phenolic compounds known by the abbreviations BHT and BHA, are added to some foods as preservatives; they act by inhibiting the autoxidation reactions by which food becomes rancid. A radical intermediate (R•) in the autoxidation mechanism abstracts a hydrogen from an antioxidant to form a resonance-stabilized and (therefore) unreactive radical which disrupts the chain mechanism.

$$R• + Ar–OH \longrightarrow Ar–O• + R–H$$
a stabilized radical

We have in our body natural antioxidants such as vitamins C (Sub-section 24.1.3) and E (Sub-section 24.4.4), and the coenzyme, glutathione (Sub-section 14.6.2).

Autoxidation is not always an unwanted chemical process, however. Isopropylbenzene (cumene) is used in the industrial manufacture of phenol (and acetone), and the first step is its autoxidation (Scheme 20.14). For the subsequent acid-catalysed rearrangement, see Chapter 22.

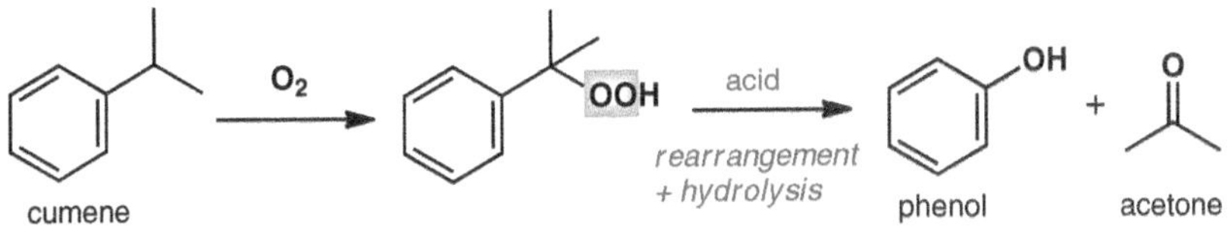

OH

Me

BHT (butylated hydroxytoluene)
2,6-di(*t*-butyl)-4-methylphenol

OH

OMe

BHA (butylated hydroxyanisole)
2-*t*-butyl-4-methoxyphenol

O_2 — acid rearrangement + hydrolysis

cumene — OOH — phenol + acetone

Scheme 20.14 Autoxidation in the manufacture of phenol and acetone from cumene.

Cumene is produced by the Friedel–Crafts alkylation of benzene with propene (see Example 16.2 in Chapter 16).

20.9 Formation of Radical Ions by Single Electron Transfer and their Reactions

A radical can be generated by adding or removing one electron from a normal molecule or ion with all electrons paired. In this way, a radical or a radical ion bearing an unpaired electron is formed by a **single electron transfer** (SET), i.e. in an oxidation–reduction process.

Metals (e.g. Li, Na, Mg, Al, Ti) of low ionization energy or salts of metals in low oxidation states [e.g. Fe(II), Cu(I), and Ti(III)] are used to provide single electrons. Conversely, metal ions in high oxidation states [e.g. Fe(III), Cu(II), Mn(III), and Ag(I)] as well as oxidizing agents such as quinones and I_2 act as electron acceptors. Electron transfers at electrodes in electrochemical cells (*electrode reactions*, see Sub-section 20.9.4) are also possible.

> Electron donors have a relatively high HOMO, while electron acceptors have a relatively low LUMO.

20.9.1 Dissolving metal reduction

Group 1 metals (usually Na or Li) become cations by releasing electrons in mixtures of liquid ammonia and an alcohol, e.g. eqn 20.27 in Scheme 20.15. The electron may be accepted into the LUMO (π^* orbital) of an arene, as in eqn 20.28, to form a **radical anion**. The radical anion then abstracts a proton from a solvent molecule, e.g. ethanol in eqn 20.29, to give a neutral radical. Another electron transfer to the radical gives an anion which then abstracts another proton to form the reduction product (eqn 20.30). The overall reaction in Scheme 20.15 is the **dissolving metal reduction** of benzene to give a non-conjugated cyclohexadiene (eqn 20.31). This method of reduction of aromatic compounds (for example) is known as **Birch reduction** after the Australian chemist (A.J. Birch) who developed it.

> The intense blue solution obtained when Na or Li is added to liquid ammonia is due to *solvated electrons*, but the colour gradually fades as the electrons reduce NH_3 to NH_2^- and H_2.
>
> Na $\xrightarrow[fast]{liq.\ NH_3}$ Na$^+$ + e$^-$(NH$_3$)$_n$
> blue
>
> $\xrightarrow{slow}$ NaNH$_2$ + (1/2)H$_2$
> colourless

> The intermediate anion in eqn 20.30 in the Birch reduction of benzene is extensively delocalized but its protonation gives the non-conjugated cyclohexa-1,4-diene rather than its more stable conjugated isomer, cyclohexa-1,3-diene. This is not easily explained but protonation at C3 of the cyclohexadienyl anion must be faster than at C1/C5 (otherwise the more stable isomer would form). Birch reduction of a mono-substituted benzene, C_6H_5Z, generally gives a single main product but, depending on the nature of Z, it may be 1-Z- or 3-Z-cyclohexa-1,4-diene.

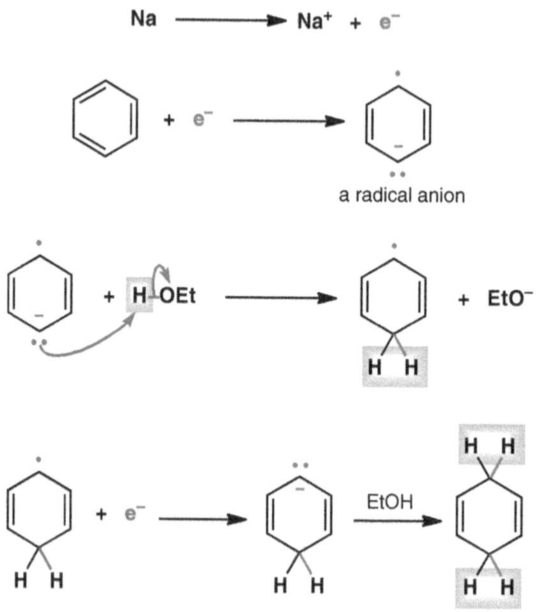

$$\text{Na} \longrightarrow \text{Na}^+ + \text{e}^- \tag{20.27}$$

(20.28)

a radical anion

(20.29)

(20.30)

Overall reaction (Birch reduction):

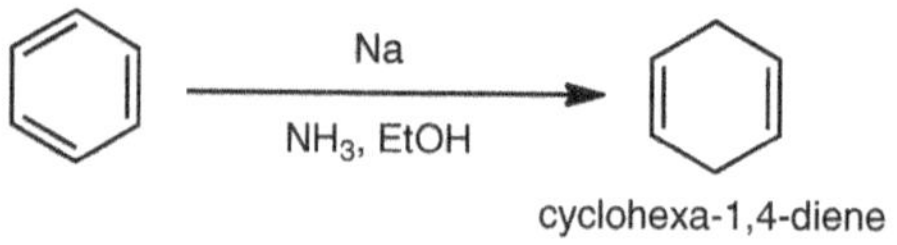

(20.31)

cyclohexa-1,4-diene

Scheme 20.15 Dissolving metal (Birch) reduction of benzene.

Exercise 20.15

The product of Birch reduction of 1,2-dimethylbenzene is 1,2-dimethylcyclohexa-1,4-diene. Give a reaction sequence for this transformation.

Dissolving metal reduction of a ketone in ethanol (i.e. without ammonia) proceeds in a similar way via a radical anion called a *ketyl* which yields an alcohol by sequential protonation, another electron transfer, then another protonation upon acidification (Scheme 20.16). This transformation could also be carried out with a metal hydride; note that two SETs + a proton (H⁺) transfer are equivalent to a hydride (H⁻) transfer.

This procedure, sometimes called the Bouveault–Blanc method, will also reduce an ester RCO_2R' to $RCH_2OH + R'OH$, but hydride reduction (discussed in Chapter 10) is generally more convenient for carbonyl compounds.

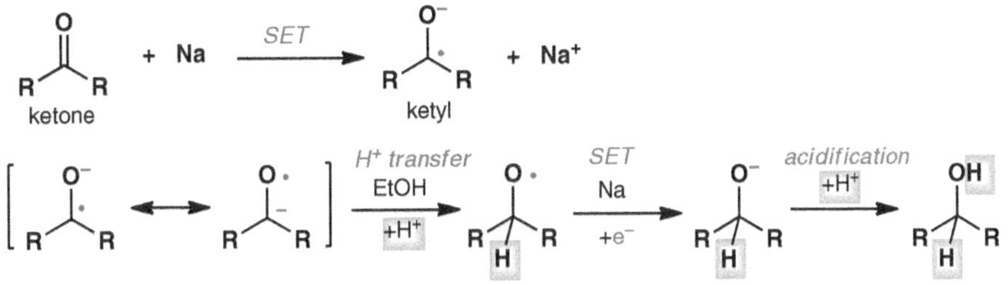

Scheme 20.16 Dissolving metal reduction of a ketone with ethanol as the proton donor.

20.9.2 One-electron reduction of carbonyl compounds and radical coupling

As seen above, a ketyl gives a reduction product if a proton donor is immediately available (e.g. as the solvent or a co-solvent), but dimerization occurs by radical coupling in aprotic solvents. Although a ketyl is anionic and two anionic species will repel each other electrostatically, the dimerization proceeds smoothly with metals such as Mg or Al because coordination of the metal ion by the anionic oxygens holds the reactants together; the reaction is completed by addition of an aqueous acid (Scheme 20.17). This reaction is known as the **pinacol reaction** because the first known example was the preparation of pinacol (2,3-dimethylbutan-2,3-diol) from propanone (acetone).

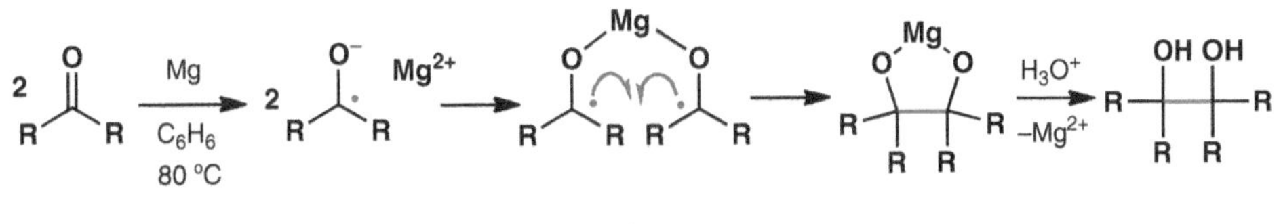

Scheme 20.17 The pinacol reaction.

The radical anion from an ester will dimerize in the same way in the absence of a proton donor, but the first-formed dimeric species has the structure of a double tetrahedral intermediate from which two alkoxide ions are eliminated to give a diketone (Scheme 20.18). The diketone is then further reduced to another dianion by two more SETs. When the reaction is quenched with aqueous acid, proton transfers give an enediol which tautomerizes to an α-hydroxy ketone (an acyloin) as the final product. The overall reaction is called the **acyloin condensation** of an ester.

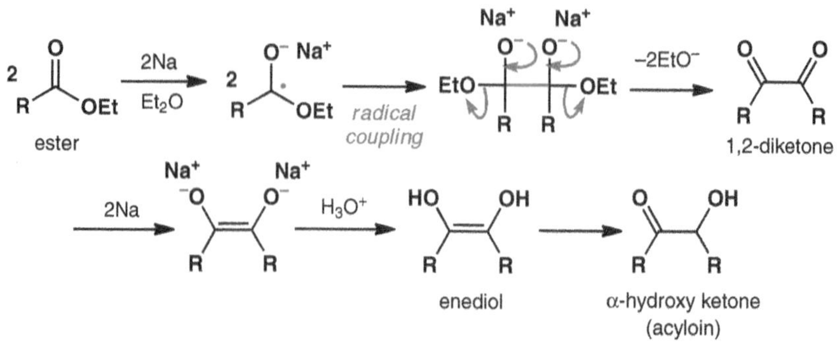

Scheme 20.18 The acyloin condensation of an ester.

Exercise 20.16

What are the main products of the following reactions?

(a) PhCOCH$_3$ $\xrightarrow[\text{NH}_3, \text{ EtOH}]{\text{Na}}$

(b) PhCOCH$_3$ $\xrightarrow[\text{2) H}_3\text{O}^+]{\text{1) Mg, C}_6\text{H}_6}$

(c) PhCO$_2$Et $\xrightarrow[\text{2) H}_3\text{O}^+]{\text{1) Na, Et}_2\text{O}}$

20.9.3 The radical mechanism of nucleophilic substitution

Nucleophiles and electrophiles in polar reactions are electron-pair donors and acceptors: a nucleophile donates an electron pair to an electrophile which leads to the sharing of an electron pair and formation of a new bond. There are special cases, however, in which a nucleophile (acting as a single electron donor) completely transfers just one electron to the electrophile to form a radical anion as an intermediate in an overall nucleophilic substitution. This electron transfer initiates the S$_{RN}$1 **mechanism**, which involves the unimolecular dissociation of the radical anion intermediate, and proceeds by a radical chain mechanism, as shown in Scheme 20.19. The initiating SET is sometimes promoted by UV-vis irradiation.

Initiation:

Scheme 20.19 A radical chain mechanism for the S$_{RN}$1 nucleophilic substitution reaction.

$$R{-}Y + Nu^- \xrightarrow{SET} R{-}Y^{\cdot-} + Nu\cdot$$

Propagation:

$$R{-}Y^{\cdot-} \longrightarrow R\cdot + Y^-$$

$$R\cdot + Nu^- \longrightarrow R{-}Nu^{\cdot-}$$

$$R{-}Nu^{\cdot-} + R{-}Y \xrightarrow{SET} R{-}Nu + R{-}Y^{\cdot-}$$

Overall reaction:

$$R{-}Y + Nu^- \longrightarrow R{-}Nu + Y^-$$

Both aliphatic and aromatic substitutions by the S$_{RN}$1 mechanism are known, e.g. eqns 20.32 and 20.33. The electron-withdrawing *p*-nitro substituent in the tertiary alkyl halide in eqn 20.32 inhibits an S$_N$1 mechanism, and an S$_N$2 mechanism is prevented by the steric bulk of the reactants.

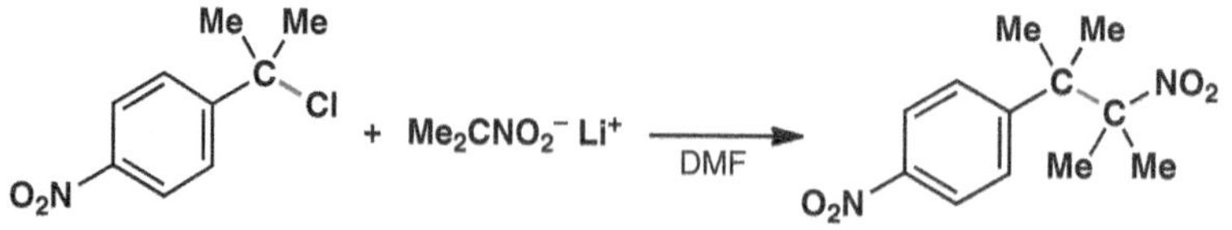

The nucleophile in eqn 20.32 can be obtained by the reaction:

Me$_2$CHNO$_2$ + BuLi $\longrightarrow$

Me$_2$CNO$_2^-$ Li$^+$ + BuH

$$\text{(dibromobenzene)} + 2 \text{ PhS}^- \xrightarrow[\text{liq. NH}_3]{h\nu} \text{(bis(phenylthio)benzene)} + 2 \text{ Br}^- \quad (20.33)$$

The substrate in the aromatic nucleophilic substitution of eqn 20.33 is not activated by an electron-withdrawing group for the normal addition–elimination aromatic nucleophilic substitution mechanism, and the reaction conditions are not sufficiently basic for the elimination-addition mechanism via an aryne intermediate (see Chapter 18). Under these circumstances, photoexcitation of an electron into a higher orbital of the thiophenolate facilitates its transfer to the dibromobenzene and the $S_{RN}1$ mechanism becomes possible.

> Give a radical chain mechanism for the reaction in eqn 20.32.

Exercise 20.17

20.9.4 Electrode reactions

Oxidation of a carboxylate ion by removal of an electron at the anode of an electrolytic cell gives a carboxyl radical which decarboxylates to yield an alkyl radical; the alkyl

---

Panel 20.2 Chlorofluorocarbons and the ozone layer

Chlorofluorocarbons (CFCs, also called freons) such as $CFCl_3$, CF_2Cl_2, CF_3Cl, and CF_3CF_2Cl were used as aerosol propellants and refrigerants until recently. They are stable, nontoxic, nonflammable gaseous compounds, and seemed to be ideal for the purposes. However, their stability (lack of reactivity) led to environmental problems. They survive the conditions of the lower atmosphere of the Earth and diffuse unchanged into the stratosphere where the more intense UV radiation causes homolysis of the C–Cl bonds.

$$Cl{-}CF_3 \xrightarrow{h\nu} Cl \cdot + \cdot CF_3$$

In the stratosphere, there is a dynamic equilibrium involving O, O_2, and O_3. However, the intrusion of chlorine atoms disturbs this equilibrium and leads to the depletion of the ozone by the following chain mechanism.

$$Cl \cdot + O_3 \longrightarrow ClO \cdot + O_2$$
$$ClO \cdot + O \longrightarrow Cl \cdot + O_2$$
$$\overline{}$$
$$O_3 + O \longrightarrow 2\, O_2$$

Ozone is only a minor component (2–8 ppm) of the stratosphere but it acts as a filter by absorbing the high energy UV radiation (wavelength $< 320\,\text{nm}$) which is so harmful to living organisms. Only a relatively small increase in the level of this radiation at the surface of the earth would cause a large increase in the incidence of skin cancer.

The role of Cl atoms from CFCs in depleting the ozone concentration in the stratosphere was originally a controversial proposal by F.S. Rowland and M.J. Molina in 1974, but was subsequently confirmed. Following the Montreal Protocol on *Substances that Deplete the Ozone Layer* in 1987, the use of CFCs was banned and substitutes such as hydrochlorofluorocarbons (HCFCs, e.g. CF_3CHCl_2) and hydrofluorocarbons (HFCs, e.g. CF_3CH_2F) are now used. These compounds are less stable and consequently do not survive long in the lower atmosphere but, like CO_2, they still have a greenhouse effect. Furthermore, while being less damaging than CFCs, they still have an adverse effect on the ozone layer, so they also are now being phased out. (Rowland and Molina were awarded the 1995 Nobel Prize in Chemistry jointly with P.J. Crutzen who carried out related work on the effect of increased levels of nitric oxide, $NO \cdot$, on ozone depletion.)

Bromine-containing halomethanes analogous to the CFCs, e.g. $CBrF_3$ and $CBrClF_2$, are used as clean fire-extinguishing agents; when used, they leave no dirty or corrosive residues. However, because they are similar to CFCs in their effect upon the ozone layer, their application is allowed only where alternatives cannot be used, e.g. in aircraft and electrical facilities.

See the Prologue for biographical information about Hermann Kolbe (1818–1884).

radicals then couple, i.e. dimerize, to form an alkane. This reaction is known as *Kolbe electrolysis* and has been known for many years.

$$RCO_2^- \xrightarrow[\text{anodic oxidation}]{-e^-} RCO_2^{\bullet} \xrightarrow[\text{decarboxylation}]{-CO_2} R\bullet \xrightarrow[\text{coupling}]{\text{radical}} (1/2)\,R\text{–}R$$

Propenenitrile (acrylonitrile) is reduced at the cathode of a different type of electrolytic cell to form a radical anion which adds to another molecule of acrylonitrile on the way to hexanedinitrile (adiponitrile) as the final product. This process, followed by reduction of the dinitrile, is applied in the manufacture of hexane-1,6-diamine, one of the monomers for nylon 66 (the other being hexane-1,6-dioic acid, adipic acid).

$$H_2C{=}CH{-}CN \xrightarrow[H_2O]{\substack{\text{cathodic reduction} \\ +\,e^-}} \left[\, H_2C{=}CH{-}CN \,\right]^{\overset{\bullet}{-}} \longrightarrow \quad NC \diagup\diagdown\diagup CN$$

acrylonitrile   adiponitrile

Exercise 20.18

Propose a mechanism for the formation of adiponitrile from acrylonitrile by cathodic reduction.

Summary

- Radicals are formed by **homolysis** of a covalent bond. Compounds with a weak bond such as peroxides are used as radical initiators in **radical chain mechanisms**.

- Typical radical chain reactions include halogenation of alkanes (substitution), addition to alkenes, alkene polymerization, and autoxidation.

- Halogenation at allylic and benzylic positions, and dehalogenation using Bu_3SnH, are particularly useful radical reactions.

- Radical addition of HBr to unsymmetrical alkenes follows anti-Markovnikov regioselectivity.

- Intramolecular reactions of radicals include cyclization of alkenyl radicals, 1,5-hydrogen transfer, and β-fragmentation.

- Antioxidant food preservatives act by inhibiting autoxidation.

- Radical ions are formed by **single electron transfer** (SET) mechanisms: alkali metals, as single electron donors, induce dissolving metal reduction and radical coupling reactions.

- The electrodes in electrochemical reactions involving radical intermediates either supply electrons to, or accept electrons from, organic compounds in reduction and oxidation steps, respectively.

Problems

20.1 Predict the order of homolytic bond dissociation energies of the C–H or C–halogen bonds indicated in each set of compounds in (a)–(c) and give reasons for your predictions.

(a) $H{-}CH_3$ $H{-}CH_2CH_3$ $H{-}CH(CH_3)_2$ $H{-}C(CH_3)_3$

(b) $(CH_3)_2C\overset{H}{\underset{|}{C}}CH_2CH_3$ $(CH_3)_2CH\overset{H}{\underset{|}{C}}HCH_3$ $(CH_3)_2CHCH_2\overset{H}{\underset{|}{C}}H_2$

(c) $H_3C{-}F$ $H_3C{-}Cl$ $H_3C{-}Br$ $H_3C{-}I$

20.2 Give the main steps in the photo-induced reaction of ethane with Br_2 to form bromoethane.

20.3 Photochemical chlorination of propane gives two monochloropropane products in the ratio shown. Calculate the relative reactivity of the primary and secondary hydrogens of propane in this reaction.

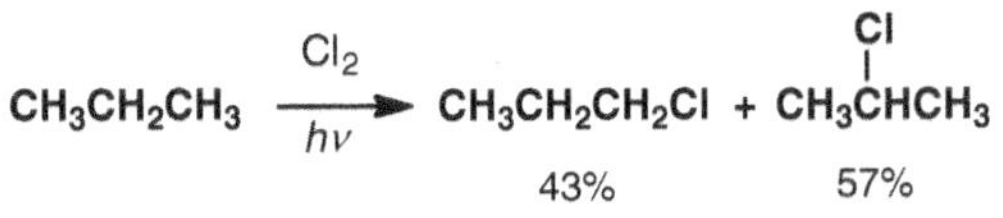

$$CH_3CH_2CH_3 \xrightarrow[hv]{Cl_2} CH_3CH_2CH_2Cl + CH_3\overset{Cl}{\underset{|}{C}HCH_3}$$

43%   57%

20.4 When propane reacts photochemically with an equimolar mixture of Br_2 and Cl_2, a different ratio of monobromination products is observed compared with in the reaction with Br_2 alone. Explain this difference in selectivity.

20.5 Give all the possible monobromo products in radical bromination of the following, and predict which is the main product in each reaction.

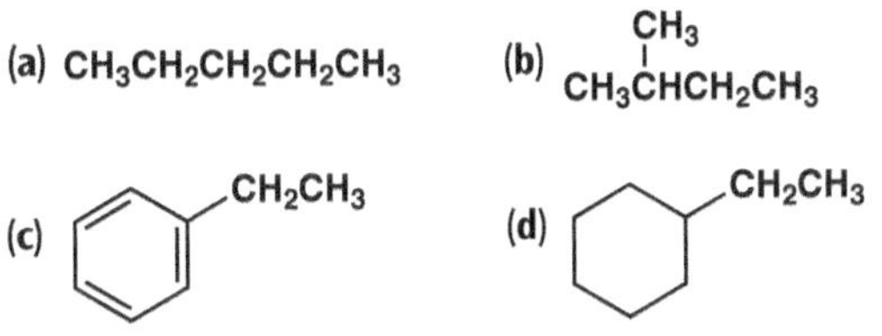

(a) $CH_3CH_2CH_2CH_2CH_3$

(b) $CH_3\overset{CH_3}{\underset{|}{C}H}CH_2CH_3$

(c) [benzene ring with CH₂CH₃ substituent]

(d) [cyclohexane ring with CH₂CH₃ substituent]

20.6 What are the main products in the following reactions?

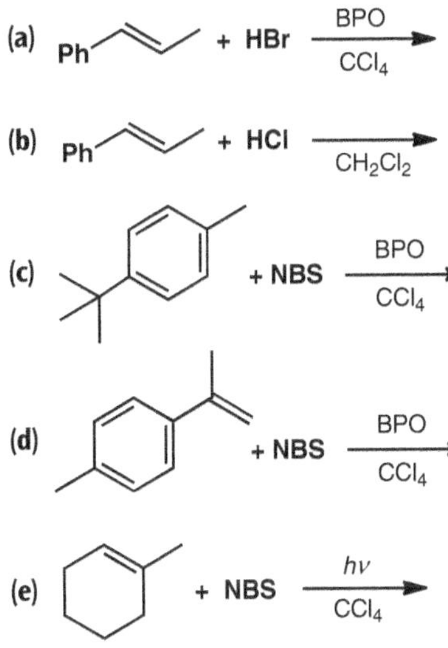

(a) Ph ═══ + HBr $\xrightarrow[CCl_4]{BPO}$

(b) Ph ═══ + HCl $\xrightarrow{CH_2Cl_2}$

(c) [benzene ring with tert-butyl and methyl substituents] + NBS $\xrightarrow[CCl_4]{BPO}$

(d) [benzene ring with isopropenyl and methyl substituents] + NBS $\xrightarrow[CCl_4]{BPO}$

(e) [cyclohexene with methyl substituent] + NBS $\xrightarrow[CCl_4]{hv}$

20.7 Give a mechanism and the main product for the following reaction.

[cyclohexene with methyl substituent] + EtSH $\xrightarrow[\Delta]{BPO}$

20.8 Give radical chain mechanisms for the following transformations.

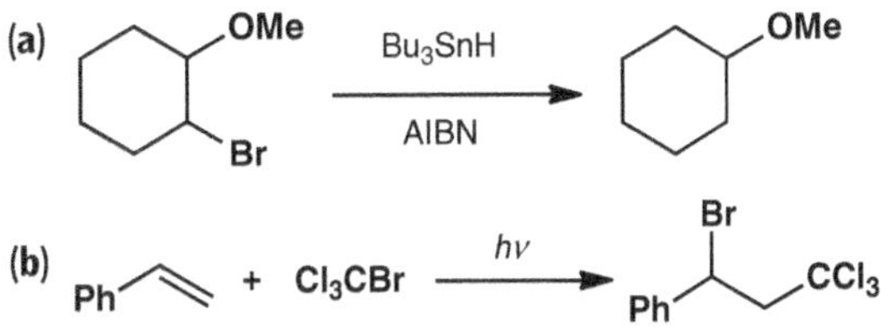

(a) [cyclohexane ring with OMe and Br substituents] $\xrightarrow[AIBN]{Bu_3SnH}$ [cyclohexane ring with OMe substituent]

(b) Ph ═══ + Cl_3CBr $\xrightarrow{hv}$ Ph [with Br and CCl₃ substituents]

20.9 Give a radical chain mechanism for the reaction of an alk-1-yne (RC≡CH) with Bu_3SnH in the presence of AIBN which gives an alk-1-enyltin product.

20.10 Give a radical chain mechanism for the polymerization of vinyl chloride initiated by dibenzoyl peroxide.

20.11 Photochemical chlorination of 2,2-dimethylhexane with Cl_2 gives a mixture of monochloro compounds with 2-chloro-5,5-dimethylhexane as the main product. Explain why this secondary alkyl chloride is more readily formed than the other secondary products. Note that chlorination is not generally very selective and that a radical rearrangement is possible.

20.12 Hydroquinone is a good radical inhibitor because each molecule intercepts two alkyl radicals to form a quinone and two alkane molecules. Give a fish-hook curly arrow mechanism for this reaction.

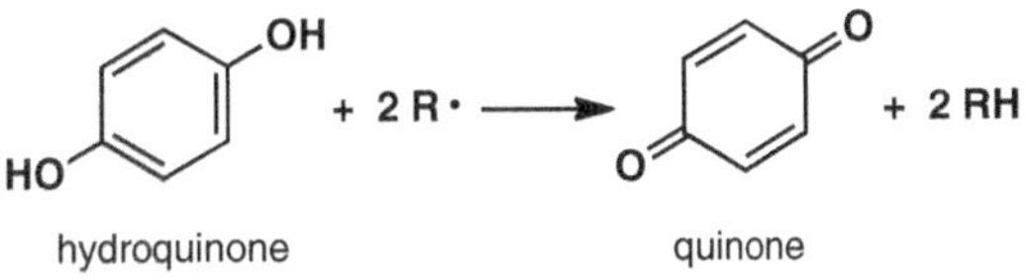

[hydroquinone structure with OH and HO groups] + 2 R· ⟶ [quinone structure with O groups] + 2 RH

hydroquinone   quinone

20.13 By giving a reaction sequence, deduce the structure of the product from the reaction of cyclopentanone with magnesium in benzene followed by treatment with an aqueous acid.

Supplementary Problems

20.14 Reaction of anisole (methoxybenzene) with lithium in liquid ammonia containing ethanol gives 1-methoxycyclohexa-1,4-diene. Hydrolysis of this product in aqueous acid provides cyclohex-2-en-1-one. Write a reaction sequence for the overall transformation. Note that cyclohex-3-en-1-one can be isolated under mild hydrolytic conditions.

20.15 Reaction of enantiomerically pure (R)-2,2,3-trimethylpentane with Cl_2 under UV-vis irradiation gives several monochloro products.

(a) Draw structures of all the monochloro products which (in principle) can be separated by distillation.

(b) Identify the monochloro product which would not show any optical activity in spite of the optical activity of the starting material, and explain your answer.

20.16 The industrial photochemical reaction of cyclohexane with nitrosyl chloride gives cyclohexanone oxime via nitrosocyclohexane, and the oxime is then converted to ε-caprolactam (the monomer for the production of nylon-6) using sulfuric acid; this is known as the *Toray process*.

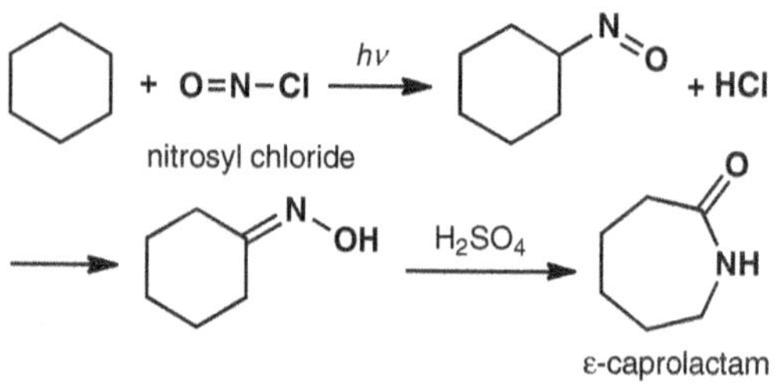

nitrosyl chloride

ε-caprolactam

Give mechanisms for (a) the nitrosation, and (b) the acid-catalysed rearrangements of nitrosocyclohexane to the oxime and the oxime to the lactam.

20.17 Give radical chain propagation steps for the following reaction.

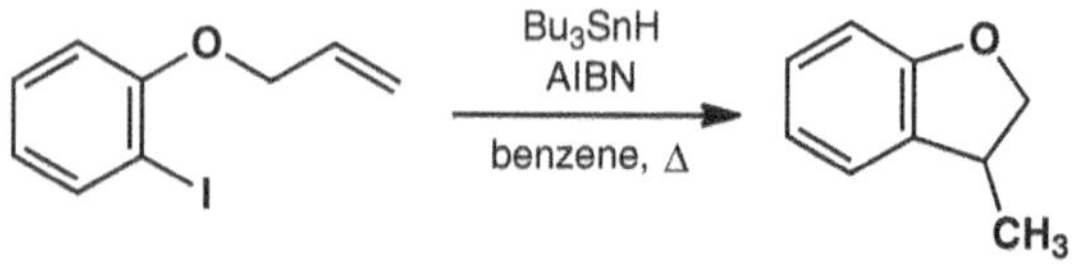

20.18 Give radical chain propagation steps for the following reaction.

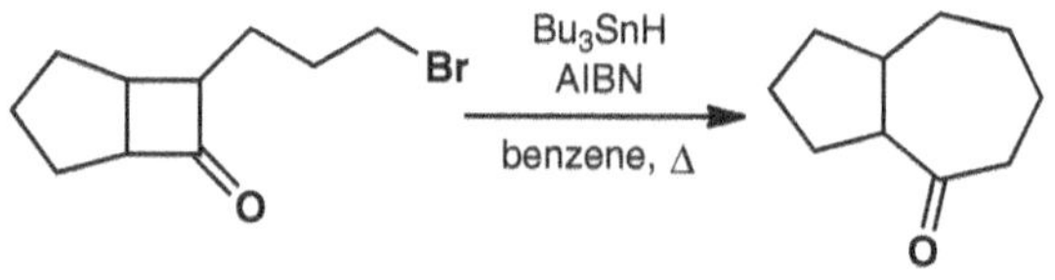

20.19 Birch reduction of 3-methylbenzoic acid with addition of iodomethane immediately before an acidic quench gives the product shown below. Give a mechanism for this reaction.

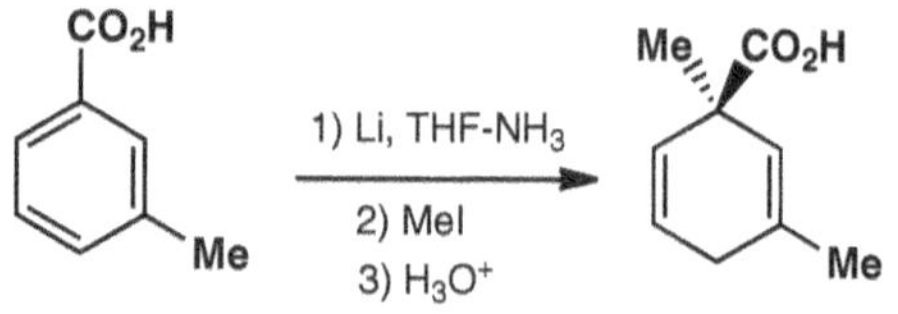

20.20 Birch reduction of 6-methylcyclohex-2-enone followed by addition of iodomethane gives a compound $C_8H_{14}O$. Give a reaction mechanism and identify the organic product(s).

Pericyclic Reactions: Cycloadditions, Electrocyclic Reactions, and Sigmatropic Rearrangements

21

Related topics we have already covered:

- **Molecular orbitals of conjugated systems** (Chapter 5)
- **Aromaticity** (Section 5.6)
- **Photochemical reactions** (Sub-section 5.7.3)
- **Curly arrow representation of reaction mechanisms** (Section 7.2)
- **Molecular orbital interactions in reactions** (Section 7.3)
- **Diels–Alder reactions** (Section 15.9)

Topics we shall cover in this chapter:

- **Types of pericyclic reactions**
- **Molecular orbital considerations of pericyclic reactions**
- **Cycloadditions: Diels–Alder reactions and 1,3-dipolar cycloadditions**
- **Other cycloadditions and related reactions**
- **Electrocyclic reactions**
- **Sigmatropic rearrangements**

Most of the organic reactions we have covered so far have been polar reactions which proceed with the pairwise redistribution of valence electrons involving nucleophilic and electrophilic reaction sites; they are often stepwise with cations or anions as intermediates. We have also discussed radical reactions in Chapter 20; these involve unpaired electrons, (usually) in multistep (chain) homolytic mechanisms with radicals as intermediates. There is a third major class of reaction which we introduced in Chapter 7 and briefly encountered in Chapter 15, and which involves neither heterolysis nor homolysis. In these so-called **pericyclic reactions**, all the bond forming and bond breaking occur together with a nonpolar, concerted *cyclic redistribution of valence electrons via aromatic transition structures.*

Polar (heterolytic) reactions are often facilitated by catalysis and reaction rates depend on the solvent polarity. Investigations of these phenomena provide evidence about the mechanisms of polar reactions, as we saw in Chapter 7. The involvement of intermediates in stepwise reactions can be investigated, for example, and the natures of rate-determining transition structures. In contrast, what are now called pericyclic reactions are little influenced by changes in the nature of the solvent and they are not generally subject to catalysis; with few means of probing their mechanisms, mechanistic investigations were slow and uncertain (they were once called 'no-mechanism' reactions!). Then, elucidation of their mechanisms rapidly followed the development of molecular orbital theory and, in particular, the recognition of the importance of orbital symmetry by Woodward and Hoffmann in 1965 (see the Prologue).

21.1 Three Main Types of Pericyclic Reactions

Pericyclic reactions fall into three main groups, cycloadditions, electrocyclic reactions, and sigmatropic rearrangements: the last two are both isomerization reactions. In **cycloadditions**, two unsaturated molecules combine to form a cyclic compound with two new σ bonds and the loss of two π bonds. An **electrocyclic reaction** involves ring formation of a conjugated polyene with the gain of one σ bond and loss of one π bond, while a **sigmatropic rearrangement** occurs with a shift of bonds within a molecule but no change in the numbers of σ and π bonds. All these reaction types are in principle *reversible*, as indicated in Scheme 21.1.

Cycloaddition:

Scheme 21.1 Three main types of pericyclic reactions illustrated.

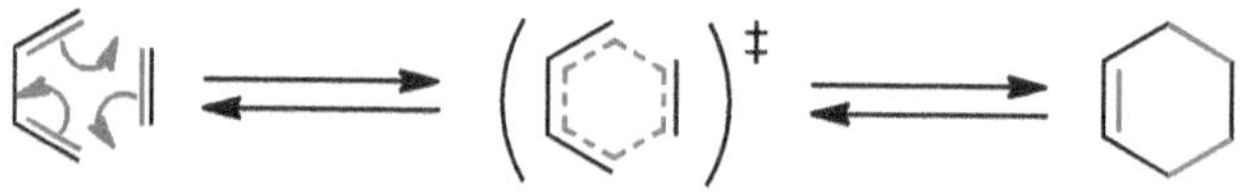

$$(21.1)$$

Electrocyclic reaction:

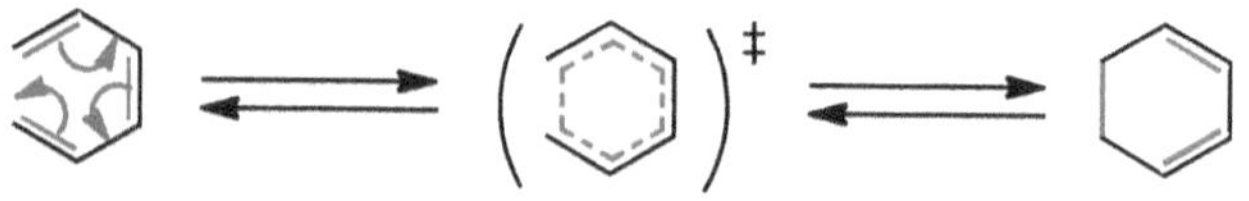

$$(21.2)$$

Sigmatropic rearrangement:

$$(21.3)$$

Note that the curly arrows used to describe the concerted pericyclic reactions via cyclic transition structures in Scheme 21.1 do not have the same significance as curly arrows in heterolytic reaction mechanisms such as the S_N2 or E2, as noted in Section 15.9. Here, they simply identify the valence electrons of the reactant(s) which are involved, and could circulate in the opposite direction:

The reaction of eqn 21.1 is the prototype [4+2] cycloaddition between a diene and an alkene (a *dienophile*) forming a cyclohexene ring. This type of cycloaddition is also known as the Diels–Alder reaction and was introduced in Section 15.9. In an electrocyclic reaction (eqn 21.2), π bonds of a linear conjugated polyene shift to form a new σ bond between the terminal carbons and generate a cyclic structure. In the sigmatropic rearrangement of eqn 21.3, reorganization of the bonding in hexa-1,5-diene occurs via a cyclic transition structure (TS) resulting in a different atom connectivity. This particular reaction is known as the Cope rearrangement. We encountered the same type of rearrangement of a similar unsaturated system containing an oxygen atom in Chapter 7 (Scheme 7.2).

All the three reactions of Scheme 21.1 have a six-membered cyclic TS with six electrons delocalized in a configuration similar to the one in benzene (Section 5.5) and sometimes referred to as an *aromatic TS*. The reactions are *concerted* and *stereospecific* when stereochemical issues are involved. The stereospecificity and the ease of reaction are related to the number of electrons involved in the TS, and can be predicted from the symmetries of the participating molecular orbitals, i.e. the interactions require in-phase overlap between orbitals of compatible symmetries for the reaction to proceed smoothly.

21.2 Cycloadditions

Cycloadditions are concerted reactions in which two components form two new σ bonds at the ends of the π systems involved to form a new ring. These concerted reactions, like other pericyclic reactions, are controlled by the symmetries of the molecular orbitals involved.

21.2.1 Diels–Alder reactions

As mentioned above, cycloadditions are reversible in principle and the reverse processes are called *cycloreversions* or *retro-cycloadditions*.

Diels–Alder reactions, discussed briefly in Section 15.9, are the most important type of cycloaddition and a simple example is given in eqn 21.4.

buta-1,3-diene   ethyl propenoate
(a diene)   (a dienophile)

(21.4)

In general, a conjugated **diene** reacts at positions 1 and 4 with another unsaturated group, usually a C=C or C≡C bond in a compound known as a **dienophile**, to form a new cyclohexene ring (or a cyclohexa-1,4-diene if the dienophile is an alkyne). The conjugated diene provides *four* π electrons and the dienophile provides *two*, hence the name **[4+2] cycloaddition**. The TS of the concerted stereospecific reaction (see Sub-section 15.9.1) includes a cyclic array of six π electrons similar to the aromatic sextet in benzene.

Draw the structure of the Diels–Alder product of each of the following combinations of diene and dienophile.

Exercise 21.1

a. Molecular orbital interactions in [4+2] cycloadditions

The HOMO–LUMO (frontier orbital) interactions in the prototype [4+2] cycloaddition between butadiene and ethene (eqn 21.1) are illustrated in Figure 21.1. We see the four π MOs of the diene on the left with the HOMO and LUMO labelled in red and blue, respectively (see also Figure 5.2 in Chapter 5). In the second column, the two π MOs of the dienophile (ethene) are represented; again, the HOMO is labelled in red, the LUMO in blue. The arrows (a) and (b) between the two columns identify two in-phase HOMO–LUMO interactions. These two favourable pairwise interactions between orbitals of compatible symmetry allow this reaction to occur, and the reaction is said to be *symmetry allowed*.

In some Diels–Alder reactions, the diene has an electron-withdrawing group and the dienophile has an electron-donating group; these are called *inverse electron demand* Diels–Alder reactions.

In most known Diels–Alder reactions, dienophiles have at least one electron-withdrawing substituent, e.g. eqn 21.4. This lowers the energy of the LUMO of the dienophile which improves the HOMO-LUMO interaction (a) in Fig. 21.1.

Dienes carrying an electron-donating group are more reactive in normal Diels–Alder reactions. Explain the reason for this enhanced reactivity.

Exercise 21.2

b. Stereochemistry of Diels–Alder reactions

As discussed in Sub-section 15.9.1, a concerted [4+2] cycloaddition occurs stereospecifically by *syn* addition at both the diene and the dienophile. The *cis* isomer of a dienophile

Figure 21.1 Molecular orbital interactions between buta-1,3-diene and ethene in the Diels–Alder reaction of eqn 21.1; symmetries of the HOMO–LUMO interactions identified as (a) and (b) on the left are represented on the right.

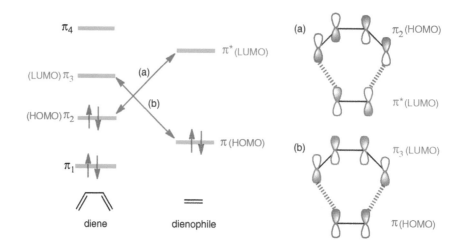

gives the *cis* cyclohexene product, while the *trans* isomer gives the *trans* product. Correspondingly, the stereochemistry of the diene is conserved in a [4+2] cycloaddition.

Exercise 21.3

Draw the structure of the Diels–Alder product of each of the following combinations of diene and dienophile.

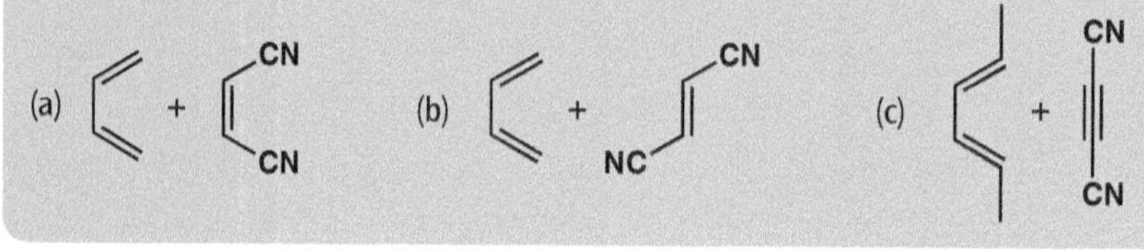

(a) | (b) | (c)

The terms *endo* and *exo* can describe groups within a bicyclic molecule, or be parts of the names of the diastereoisomers themselves. A substituent on one bridging group of a bicyclic molecule is *endo* if it is closer to (pointing towards) the longer of the other two bridging groups. Correspondingly, the *exo* diastereoisomer has the substituent further (pointing away) from the longer of the other two bridging groups.

Reaction between a cyclic 1,3-diene and vinylic dienophiles gives *bicyclic* products, which introduces a new stereochemical issue. Consider, for example, the cycloaddition of cyclopentadiene and methyl propenoate (acrylate). In principle, two diastereoisomeric bicyclic products can form–the so-called *endo* and *exo* isomers (Scheme 21.2) which differ in the configuration of the C bearing the methoxycarbonyl substituent. These diastereoisomeric products are formed in unequal yields from non-stereoisomeric reactants, so this cycloaddition is *stereoselective* (see Section 11.8 in Chapter 11).

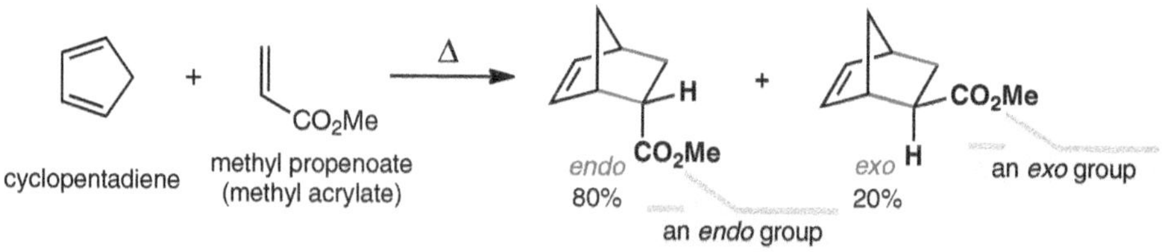

cyclopentadiene   methyl propenoate (methyl acrylate)   *endo* CO_2Me 80% an *endo* group   *exo* H 20% an *exo* group

Scheme 21.2 A stereoselective [4+2] cycloaddition of cyclopentadiene to form *endo* and *exo* products.

Perhaps surprisingly, the kinetic product is generally the more congested *endo* isomer, and this preference over the *exo* diastereoisomer is greater when the dienophile has a *cis*-disubstituted double bond as in dimethyl *cis*-butenedioate (maleate) or maleic anhydride. This faster formation of the *endo* product is ascribed to an attractive interaction

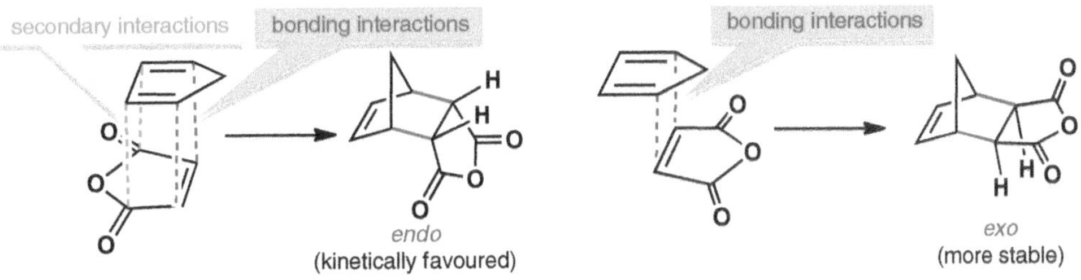

Scheme 21.3 Formation of the *endo* and *exo* products in a Diels–Alder reaction: transition state secondary orbital interactions leading to a kinetic advantage are shown in green (bonding interactions are indicated in red).

within the TS between π orbitals of the diene and ones *of substituents* on the double bond of the dienophile, as illustrated for the [4+2] cycloaddition of cyclopentadiene and maleic anhydride in Scheme 21.3. These are called *secondary orbital interactions* because they are between orbitals not *directly* involved in the bonding process. However, the sterically less congested *exo* diastereoisomer is usually more stable than the *endo* and, since the Diels–Alder reaction is reversible, it predominates at equilibrium, i.e. the *exo* is the thermodynamic product (see Sub-section 15.8.2 for further information about kinetic versus thermodynamic control).

Draw a composite reaction profile for the Diels–Alder reaction of cyclopentadiene and maleic anhydride, the *endo* product to the left and the *exo* product to the right from the reactants in the centre, to illustrate kinetic and thermodynamic control of the stereoselectivity.

Exercise 21.4

Show the products of (i) kinetic and (ii) thermodynamic control in each of the following Diels–Alder reactions.

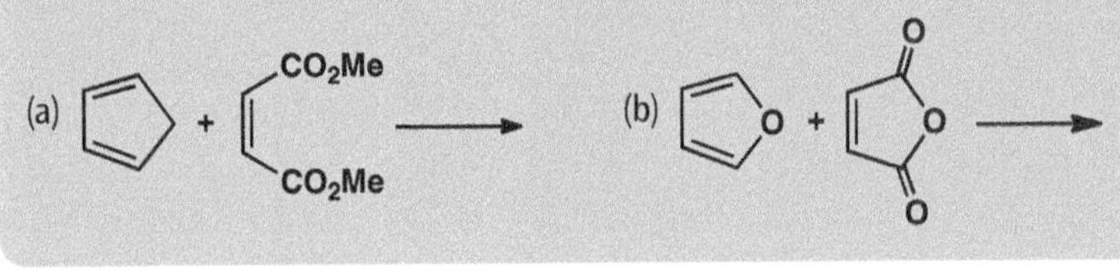

Exercise 21.5

Stereochemical relationships between reactants and products are not so straightforward when *both* diene *and* dienophile are stereoisomeric. To illustrate the issues, consider eqn 21.5 where we designate the substituents at the terminal carbons of the 1,3-diene within (2E,4E)-hexa-2,4-diene as *outer* and *inner* groups (this diene may be compared with cyclopentadiene shown in the margin).

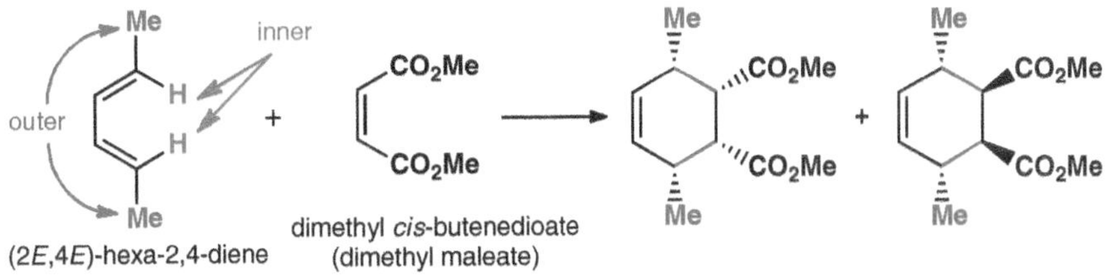

(21.5)

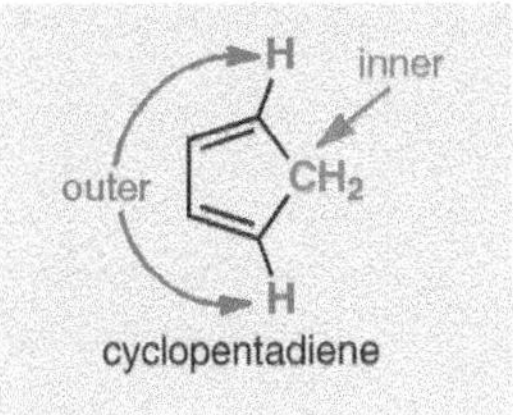

(2E,4E)-hexa-2,4-diene
dimethyl *cis*-butenedioate
(dimethyl maleate)

Reaction 21.5 can now be represented in Scheme 21.4 in the way we considered the *endo–exo* selectivity in the reactions of cyclopentadiene in Scheme 21.3.

Note that the two outer groups of the diene become *cis* to each other (as do the inner groups) in *both* products of *syn* addition. In other words, two diastereoisomeric products are formed in parallel concerted stereospecific *syn* cycloadditions according to the

Scheme 21.4 Formation of the diastereoisomeric products **1** and **2** by stereospecific *syn* additions in reaction 21.5.

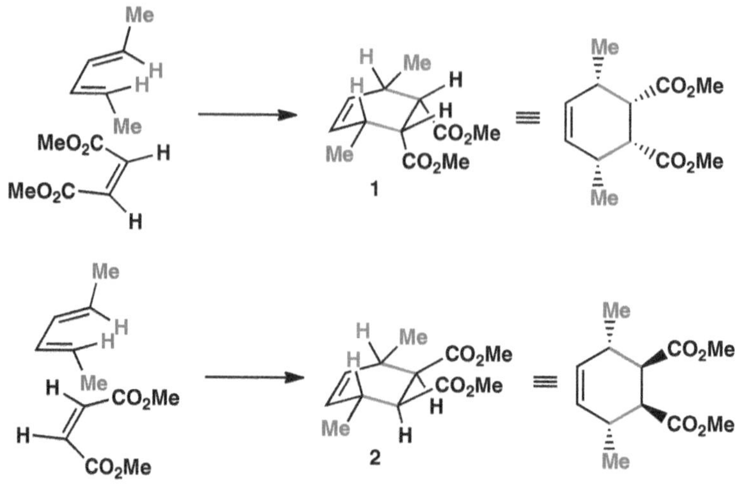

relative configurations of the molecules as they come together (Scheme 21.4). The outer groups of the diene end up in either a *cis* or a *trans* relationship with the substituents of the dienophile, and the all-*cis* product **1** is kinetically favoured in eqn 21.5; the reason is the same as for the *endo* selectivity in the Diels–Alder reaction of cyclopentadiene with maleic anhydride in Scheme 21.3—the TS leading to **1** involves stabilizing secondary orbital interactions.

Exercise 21.6 Draw structures of the products of cycloadditions of (2E,4E)-hexa-2,4-diene with dimethyl *trans*-butenedioate (dimethyl fumarate).

Exercise 21.7 Draw the structures of the products of cycloadditions of (2E,4Z)-hexa-2,4-diene with (a) dimethyl maleate and (b) dimethyl fumarate.

c. Regioselectivity in Diels–Alder reactions

When both the diene and dienophile are unsymmetrical, the question of the regioselectivity of the cycloaddition arises. Of two important types of substituted 1,3-diene, one has an electron-donating group at C1, e.g. 1-methoxybuta-1,3-diene, and the other has such a group at C2, e.g. 2-methoxybuta-1,3-diene. When these two dienes react with propenenitrile (acrylonitrile, an example of a dienophile bearing a single electron-withdrawing substituent), just one main product is formed in each case, as shown in Scheme 21.5.

The curly arrows in Scheme 21.5 show the cause of the regioselectivity—the conjugation in each case between the electron-donating MeO group and the electron-withdrawing C≡N in the developing TS; such conjugation is not possible in the formation of the transition structures leading to the isomeric '*meta* product'.[1]

---

[1] The regioselectivity of the Diels–Alder reactions between an electrophilic dienophile and the two types of substituted nucleophilic dienes exemplified in eqns 21.6 and 21.7 may be summarized as follows.

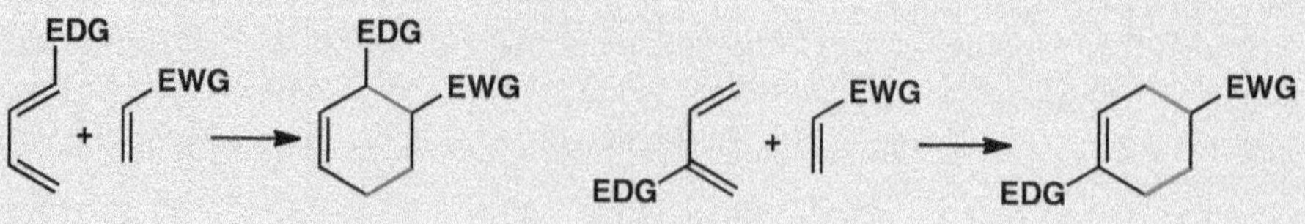

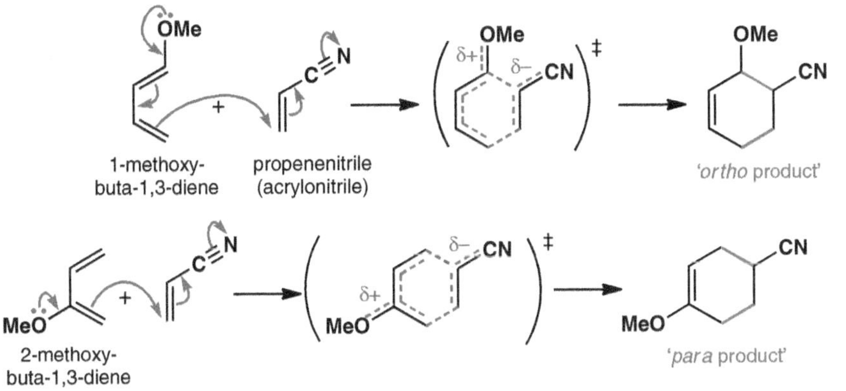

$$(21.6)$$

$$(21.7)$$

A cyclohexene product with substituents at C1 and C2, as in eqn 21.6, is sometimes called the 'ortho product', and one with substituents at C1 and C4, as in eqn 21.7, is the 'para product'. Note that the product in reaction 21.6 is a mixture of cis and trans isomers.

Scheme 21.5 Examples of regioselective Diels–Alder reactions.

What are the regiochemically favoured products of Diels–Alder reactions of the following pairs of reactants? Predict the stereoselectivity if relevant.

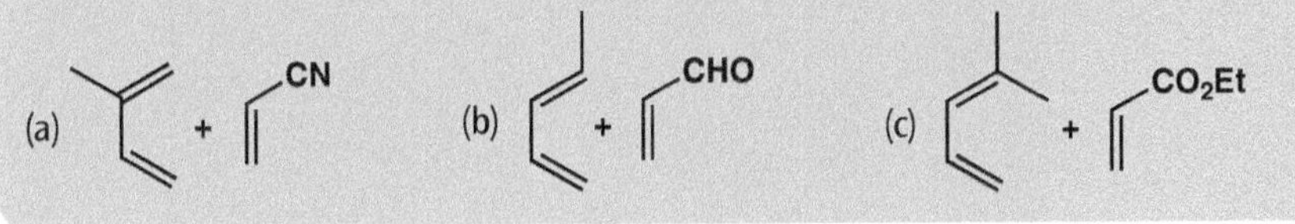

Exercise 21.8

Lewis acids such as SnCl$_4$, AlCl$_3$, and ZnCl$_2$ accelerate some Diels–Alder reactions and enhance the regioselectivity by coordinating with the dienophile at the electron-withdrawing group, which enhances its electrophilicity.

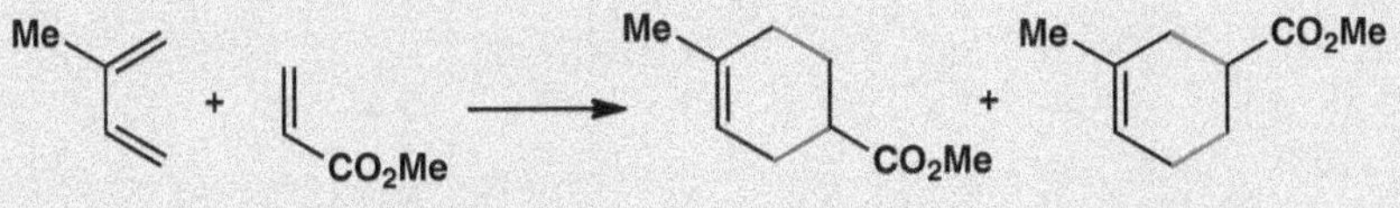

Uncatalysed reaction: 120 °C, 6 h	70%	30%
AlCl$_3$-promoted reaction: 20 °C, 3 h	95%	5%

21.2.2 1,3-Dipolar cycloaddition

Remember that the [4+2] cycloaddition occurs between a 4π electron system and a 2π electron system. The allyl anion and its analogues are a different type of 4π electron system which we encountered previously in Section 5.3. The HOMO and LUMO of the allyl anion are shown in Figure 21.2 (also see Figure 5.4): their symmetries are compatible with the LUMO and HOMO of an alkene, respectively, so orbital interactions for cycloaddition between an allyl anion analogue and an alkene are possible. This type of [4+2] cycloaddition is in fact known, as we shall see in Scheme 21.6 below. It is called a **1,3-dipolar cycloaddition** because the allyl anion analogue is also known as a **1,3-dipole**; the counterpart two-electron system is correspondingly called a **dipolarophile**.

1,3-Dipolar cycloadditions are sometimes called [3+2] cycloadditions by counting the numbers of atoms across the bonding sites of the two components rather than the numbers of π electrons which are involved.

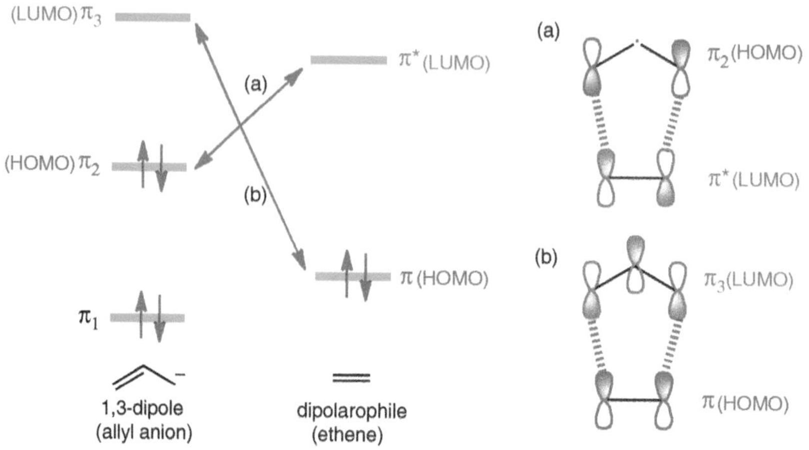

Figure 21.2 HOMO–LUMO interactions between a 1,3-dipole (allyl anion) and a dipolarophile (ethene).

Note that the dot in representation (a) of the orbital interactions represents a nodal plane in π_2 which contains the central atom of the 1,3-dipole.

(LUMO) π_3

π^* (LUMO)

(a)

(HOMO) π_2

(a)

(b)

π_2 (HOMO)

π^* (LUMO)

π (HOMO)

(b)

π_3 (LUMO)

π_1

π (HOMO)

1,3-dipole
(allyl anion)

dipolarophile
(ethene)

1,3-Dipoles which can participate in 1,3-dipolar cycloadditions include the following, and examples are given in Scheme 21.6. These reactions provide excellent methods for the synthesis of some heterocyclic compounds.

1,3-Dipoles:

Note that formal charges on 1,3-dipoles are on adjacent atoms (regardless of whether we write alternative resonance forms), but the dipoles react at positions 1 and 3. Attempts to write a mechanism for a cycloaddition across these adjacent positions lead to structures with impossible electronic configurations.

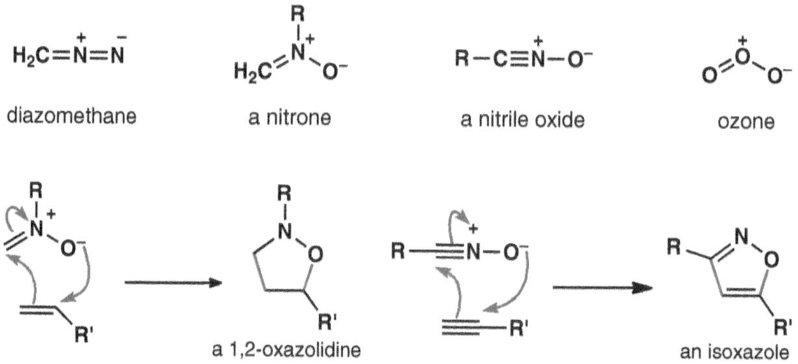

$H_2C=\overset{+}{N}=\overset{-}{N}$

diazomethane

$H_2C=\overset{\underset{R}{|}+}{N}-O^-$

a nitrone

$R-C\equiv\overset{+}{N}-O^-$

a nitrile oxide

$\overset{+}{O}\overset{}{O}\overset{}{O}^-$

ozone

a 1,2-oxazolidine

an isoxazole

Scheme 21.6 Examples of 1,3-dipolar cycloadditions.

1,3-Dipolar cycloaddition reactions of ozone with alkenes are covered in the next sub-section.

Exercise 21.9

Draw other resonance forms of the 1,3-dipoles given above.

21.2.3 Ozonolysis of alkenes

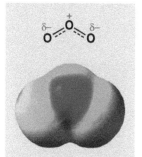

EPM of ozone.

The 1,3-dipolar cycloaddition of ozone[2] to an alkene occurs readily to give a transient cyclic adduct, a *molozonide*, which rearranges with C–C bond cleavage by dissociation/recombination to a second unstable cyclic compound, an *ozonide* (Scheme 21.7). The

---

[2]   Ozone, an allotrope of oxygen, is harmful even at low concentrations in the atmosphere at low altitudes. It is unstable and normally used in the laboratory as a dilute mixture in normal oxygen. Because of the instability of ozone and the initial products, ozonolyses of alkenes are normally carried out at low temperatures (typically, –70 °C); the reaction mixture is then allowed to warm up during the workup.

ozonide may be decomposed reductively in the workup by treatment with Zn, Me$_2$S, or H$_2$/Pd to give aldehyde and/or ketone fragments; alternatively, an oxidative workup with H$_2$O$_2$ gives carboxylic acid and/or ketone fragments.

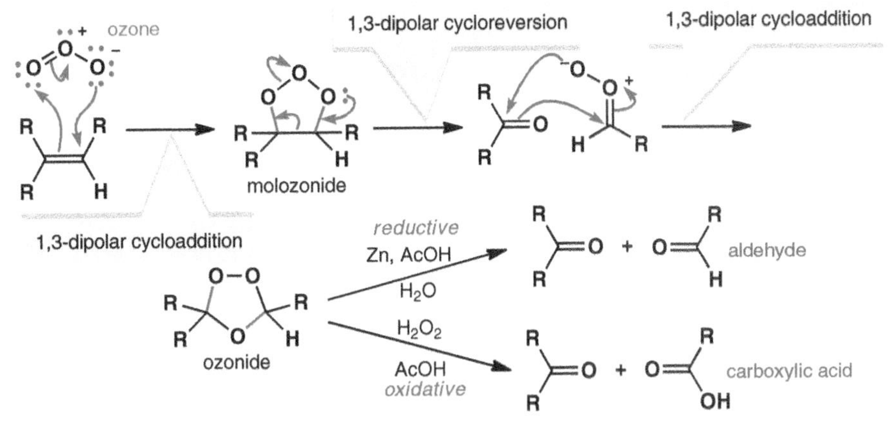

Scheme 21.7 Ozonolysis of an alkene.

The cleavage of an alkene by ozonolysis followed by identification of the fragments produced can be used as method of structural elucidation; in particular, the procedure may allow location of the double bond of an alkene. For example, characterization of butanone and methanoic acid as the products from an alkene C$_5$H$_{10}$ uniquely identifies it as 2-methylbut-1-ene (eqn 21.8).

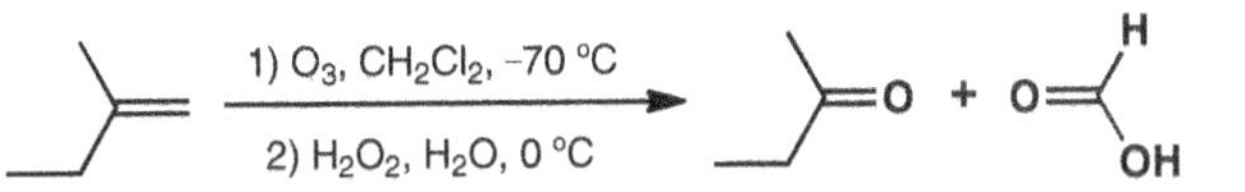

(21.8)

The products of ozonolysis of four alkenes followed by workup with aqueous H$_2$O$_2$ are shown below in (a)–(d). What are the structures of the four alkenes?

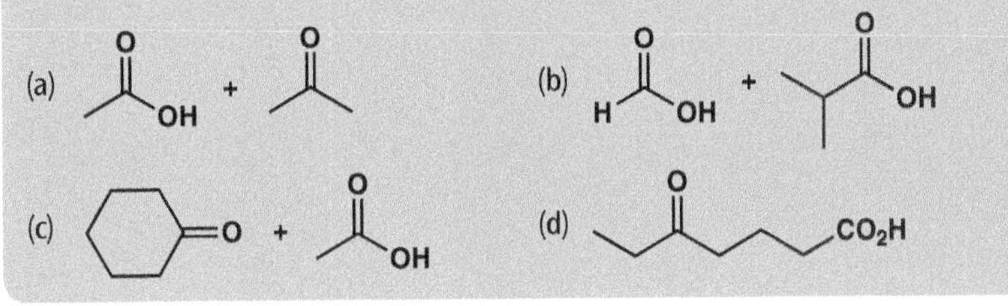

Exercise 21.10

21.2.4 Reaction of osmium tetroxide with alkenes

The reaction of osmium tetroxide with an alkene can be formally represented as a 1,3-dipolar cycloaddition although the electronic structure of OsO$_4$ is not straightforward owing to the electronic configuration of an atom of the heavy metal, Os. The initial product, an osmate ester, yields a diol in a reductive workup with Na$_2$SO$_3$, NaHSO$_3$, or Na$_2$S (e.g. Scheme 21.8). The overall reaction is useful as a stereospecific **dihydroxylation** of alkenes via a *syn cycloaddition*.

Osmium tetroxide is volatile, expensive, and extremely poisonous, however. To minimize these disadvantages, a catalytic method has been developed which requires only

The rearrangement of the molozonide to ozonide occurs by consecutive reactions: a reverse cycloaddition (cycloreversion) to give a carbonyl compound and a 1,3-dipole (a carbonyl oxide) which then undergo a second 1,3-dipolar cycloaddition.

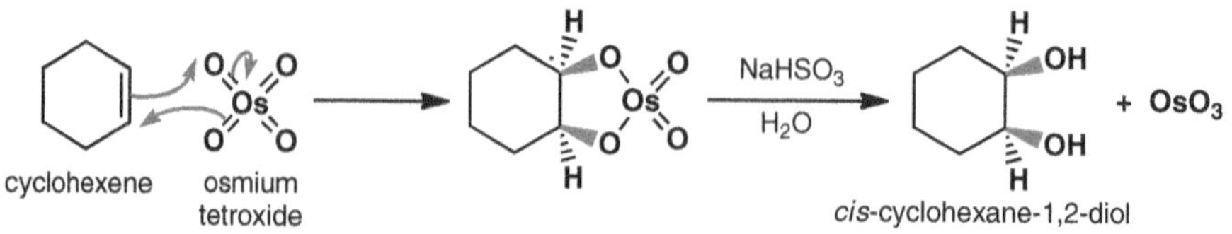

Scheme 21.8 *syn*-Dihydroxylation of an alkene via cycloaddition of osmium tetroxide.

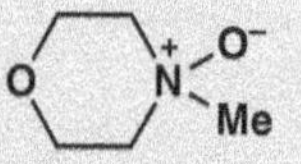

N-methylmorpholine *N*-oxide (NMO)

a catalytic amount of OsO_4 in conjunction with a co-oxidant, e.g. *N*-methylmorpholine *N*-oxide (NMO) or $KFe(CN)_6$, which converts the Os(VI) back to Os(VIII).[3]

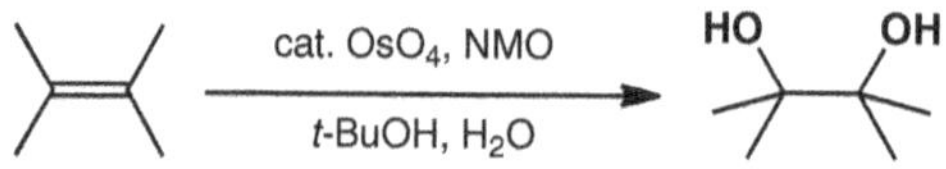

Exercise 21.11

Draw the structures of the products obtained by the reactions of (*E*)- and (*Z*)-but-2-ene with OsO_4 followed by treatment with $NaHSO_3$, and explain the stereochemical relationship between the products.

Alkaline potassium permanganate has been used for many years in organic chemistry as a cheap reagent for the oxidation of alkenes to diols via *syn* dihydroxylation; the mechanism is similar to the one with OsO_4. However, it is a powerful indiscriminate oxidant and results are not always reproducible.

21.2.5 Other cycloadditions and related reactions

Cycloadditions are classified according to the number of electrons involved. The [4+2] cycloadditions so far examined are the most common, but other combinations of reacting components are possible.

a. [2+2] Cycloaddition

Is cycloaddition possible between two alkene molecules (eqn 21.9) when both components of the reaction contain two π electrons? According to the $4n+2$ rule (see Sub-section 5.6.1), there will be no aromatic stabilization of the planar delocalized 4π electron transition structure in a [2+2] cycloaddition reaction.

$$\| + \| \xrightarrow{?} \square \qquad (21.9)$$

Further insight is possible, however, if we consider the orbital interactions of a [2+2] cycloaddition. There cannot be a productive interaction between the HOMO of one ethene and the LUMO of another because their symmetries are not compatible when both molecules are in their electronic ground states, as shown in Figure 21.3: if the interaction at one end is in-phase, the one at the other end is out-of-phase (Figure 21.3(a)). This (thermal) reaction is said to be *orbital symmetry forbidden*. The situation is quite different, however, when one electron in the π orbital of one ethene molecule is transferred into the LUMO of the molecule. This electronic excitation can be achieved by

[3] If sodium periodate, $NaIO_4$, is used as the co-oxidant in the catalytic process with OsO_4, carbonyl compounds will be obtained in a 'one pot' overall reaction because periodate will also oxidatively cleave *cis*-1,2-diols.

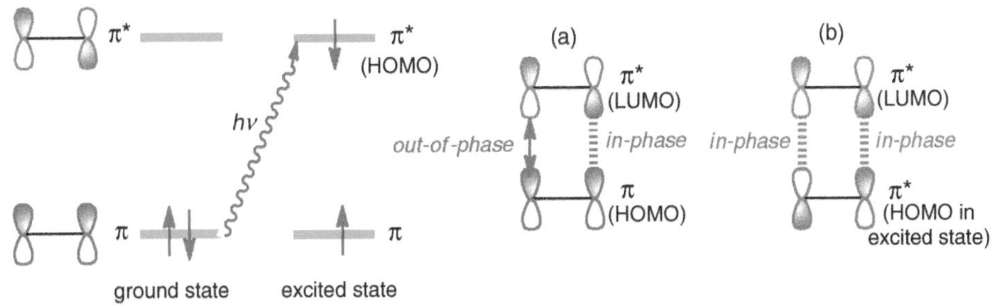

Figure 21.3 Molecular orbitals of two ethene molecules, and HOMO–LUMO interactions (a) between two ground state molecules in a thermal [2+2] cycloaddition, and (b) between a ground state molecule and an excited molecule in a photochemical [2+2] cycloaddition.

photo-irradiation, and the original LUMO (π^*) becomes a SOMO (and a HOMO at the same time) of the electronically excited state. Now, the symmetries of the SOMO/HOMO of the photoexcited ethene and the LUMO of a ground state ethene are compatible and their interaction is in-phase at the both ends (Figure 21.3(b)). The [2+2] cycloaddition is *photochemically allowed*.

In general, cycloadditions involving $4n+2$ π electrons are thermally allowed but photochemically forbidden, and those involving $4n$ π electrons are thermally forbidden but photochemically allowed.

Identify each of the following as thermally or photochemically allowed.

Exercise 21.12

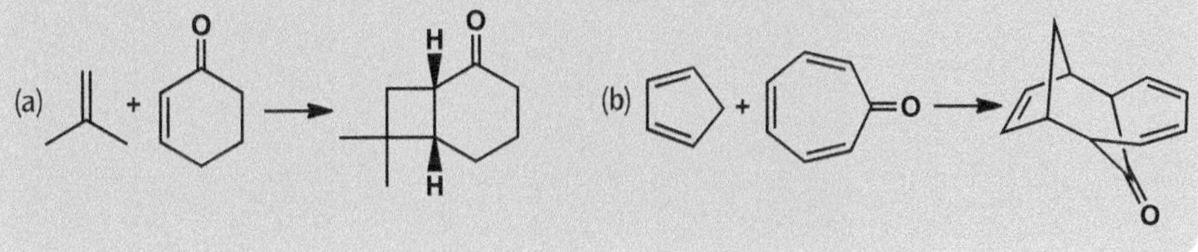

b. Cheletropic reactions

Cycloadditions in which two new σ bonds are made to the same atom (or cycloreversions in which they are broken) are sometimes called cheletropic reactions, and a well-known example is the stereospecific addition of a carbene to an alkene (Section 15.6). At first sight, this might seem to be a symmetry-forbidden thermal reaction because a singlet carbene has two unshared electrons so, with the two of the alkene, we have a four-electron system. However, we know that the reaction takes place so there seems to be an inconsistency between experiment and theory. The paradox is resolved if we look at the symmetries of the orbitals involved.

Across the top of Figure 21.4, we see methylene approaching an alkene from above with the plane of the H–C–H perpendicular to the plane of the alkene. On the left, we see the LUMO of the carbene has the wrong symmetry to interact with the HOMO of the alkene, and, on the right, the HOMO of the carbene has the wrong symmetry to interact with the LUMO of the alkene.

Across the lower half of Figure 21.4, we see methylene approaching the alkene, again from above, but this time with the plane of the H–C–H parallel with the plane of the alkene. This time on the left, the LUMO of the methylene can interact with the HOMO of the alkene, and, on the right, the HOMO of the methylene can interact with the LUMO of the alkene. By mutual approach of the carbene and alkene with their molecular planes parallel as shown in Figure 21.4(b), therefore, the thermal cycloaddition becomes symmetry

For the electronic configuration of a singlet carbene, see Figure 15.3 (Chapter 15).

Figure 21.4 HOMO–
LUMO interactions in the
approach of methylene to
an alkene; (a) and (b) show
alternative orientations of
the methylene relative to the
alkene.

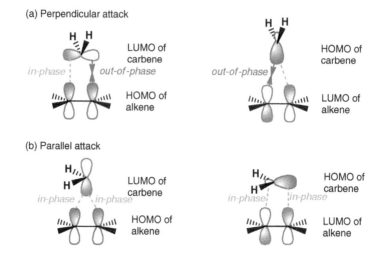

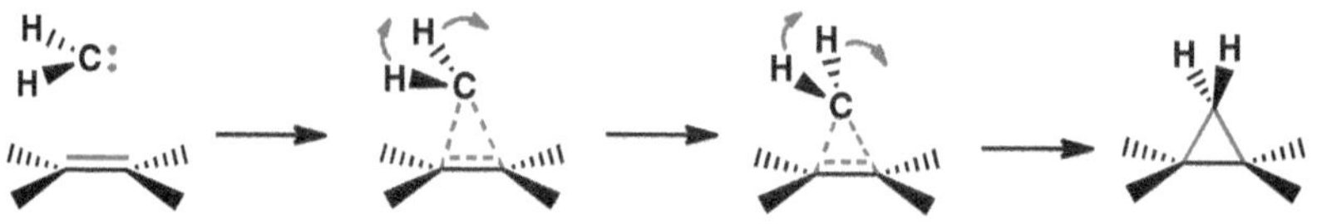

(a) Perpendicular attack

LUMO of
carbene

HOMO of
carbene

HOMO of
alkene

LUMO of
alkene

(b) Parallel attack

LUMO of
carbene

HOMO of
carbene

HOMO of
alkene

LUMO of
alkene

Figure 21.5 Schematic
representation of the close
approach and cheletropic
addition of methylene to an
alkene.

allowed. As the carbene approaches the alkene more closely, it rotates through 90° to
allow the bonds to develop fully, as indicated in Figure 21.5.

The extrusion of carbon monoxide in the reaction of eqn 21.10 is a [4+2] cycloreversion
which may also be regarded as a cheletropic reaction.

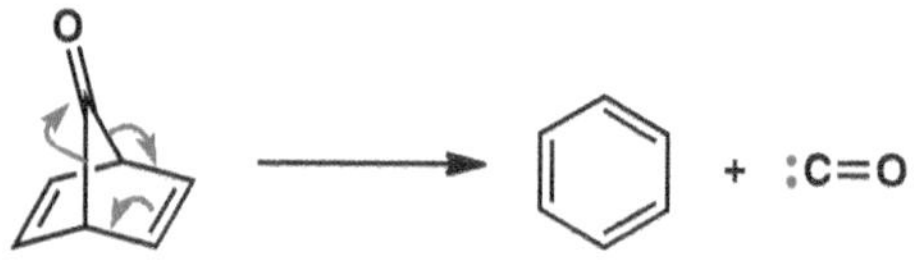

$$\qquad\qquad\longrightarrow\qquad + \quad :C{=}O \qquad\qquad (21.10)$$

c. The ene reaction

The so-called **ene reaction** (sometimes known as the Alder ene reaction) closely resembles a Diels–Alder reaction but with a C–H of the ene residue in place of one of the double
bonds of the diene, as illustrated in eqn 21.11. In other words, the four-electron system
(the ene of the ene reaction) comprises an alkene π bond and an allylic C–H σ bond rather
than a diene as in the Diels–Alder reaction, and the dienophile becomes an **enophile**.
Overall, the reaction involves transfer of an allylic H to one end of the enophile, and an
allyl group to the other end via a cyclic six-electron TS.

$$\qquad\qquad\longrightarrow\qquad\qquad\qquad (21.11)$$

ene

The reaction takes place more readily if the enophile bears at least one conjugated electron-withdrawing group, as in a Diels–Alder reaction. When the enophile is unsymmetrical with a single conjugated electron-withdrawing group, the reaction is regioselective:

in the major product, the allylic hydrogen transfers to the α carbon of the enophile and the other end of the allyl group bonds to the β carbon.

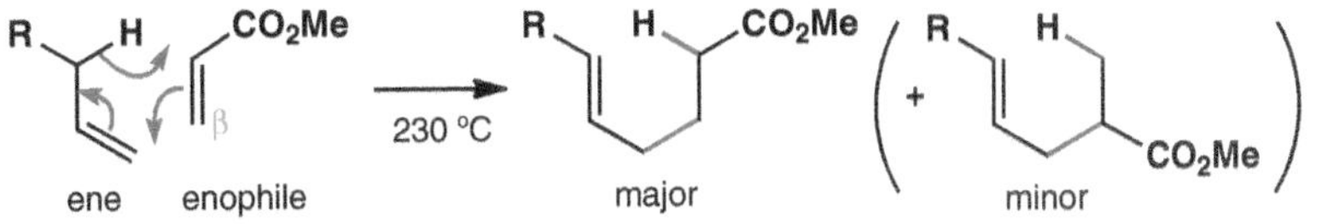

Decarboxylation of a β-keto acid (see Sub-section 17.6.2) may be seen as a heteroatom analogue of a retro-ene reaction.

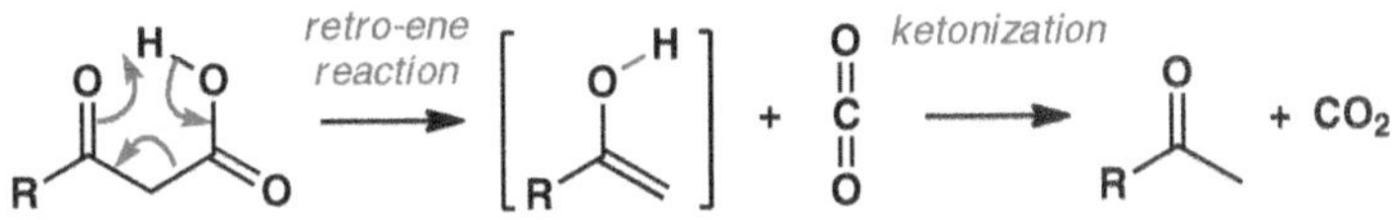

21.3 Electrocyclic Reactions

An electrocyclic reaction, e.g. eqn 21.2 given earlier, is typically an isomerization of a conjugated polyene in which a new ring is formed and the molecule loses a π bond; both terminal sp^2 carbons of the π system must rotate to allow formation of the new σ bond between them. When the participating double bonds of the reactant have substituents, the stereochemistry of the cycloalkene product is determined by the configurations of the original double bonds, and the relationship between the rotations of the ends of the π system. For example, (2E,4Z,6E)-octa-2,4,6-triene (**1** in Scheme 21.9) gives the *cis* product in a thermal reaction, but the *trans* product in a photochemical reaction. Other stereo-isomeric octatrienes give different, but still contrasting thermal and photochemical ring closures; this shows that electrocyclic reactions are *stereospecific* in two senses—different stereoisomers give different products under the same reaction conditions, and thermal and photochemical reactions of a single stereoisomer give stereochemically different products.

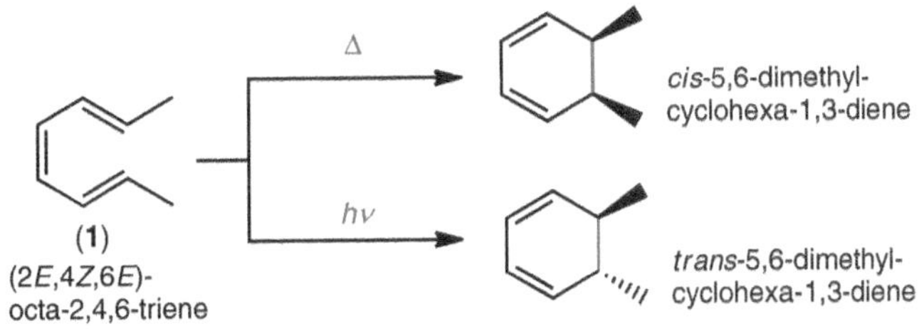

Scheme 21.9 Thermal and photochemical electrocyclic reactions of the octatriene, **1**.

The intriguing stereochemical outcomes of these thermal and photochemical cycliza-tions can be rationalized in terms of the symmetries of the HOMOs of the ground and photoexcited states, respectively, of the polyene. Figure 21.6 shows the π MOs of hexa-1,3,5-triene (i.e. the conjugated π electron system of **1** in Scheme 21.9) and π_3 is the HOMO of the electronic ground state of the molecule.

We see in Figure 21.7(a) that the ends of the HOMO of the hexatriene system of **1** are in phase when C2 and C7 come close together in a particular conformation. To achieve the compatible overlap between the ends of the orbital required for the formation of a new σ bond, the terminal carbons of the hexatriene system (C2 and C7) need to rotate in opposite directions. This mutual rotational mode for bonding is called **disrotatory**, and the thermal disrotatory electrocyclic ring closure of **1** inevitably yields the *cis*-dimethyl product shown.

Note that the number of nodes of the MOs in Figure 21.6 shown by red dotted lines increases as the energy of the MOs increases.

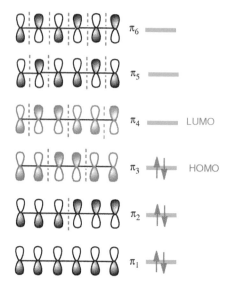

Figure 21.6 Molecular orbitals of hexa-1,3,5-triene.

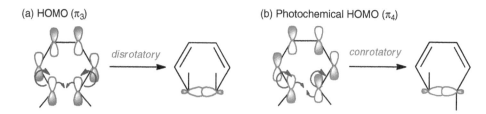

(a) HOMO (π_3)   disrotatory   (b) Photochemical HOMO (π_4)   conrotatory

Figure 21.7 Symmetries of the HOMOs of ground and photoexcited states of the hexatriene system of **1** in (a) thermal disrotatory and (b) photochemical conrotatory electrocyclic ring closures.

In contrast, photochemical reactions take place from the first electronically excited state and, for a hexa-1,3,5-triene, π_4 becomes the HOMO. When a photoexcited molecule of (2E,4Z,6E)-octa-2,4,6-triene (**1**) assumes the conformation required for ring closure, carbons C2 and C7 must rotate in the *same* direction to achieve the in-phase interaction between the ends of the HOMO appropriate for σ bond formation (Figure 21.7(b)). Consequently, the two methyls inevitably become *trans* in the product of the photochemical **conrotatory** electrocyclic ring closure of **1**.

By similar reasoning, it can be shown that, in general, electrocyclic ring closure of a conjugated π system containing 4n+2 electrons must be thermal disrotatory or photochemical conrotatory processes. In contrast, the symmetry-allowed electrocyclic ring closures for 4n π electron systems are thermal conrotatory or photochemical disrotatory reactions.

Exercise 21.13

Draw structures of products of thermal and photochemical electrocyclic reactions of the following.

(a) (2E,4Z,6Z)-octa-2,4,6-triene   (b) (2E,4Z)-hexa-2,4-diene

21.4 Sigmatropic Rearrangements of Nonpolar Molecules

A sigmatropic rearrangement is an isomerization of a π electron system in which one σ bond breaks and a new one forms, i.e. a σ bond migrates within an unsaturated molecule; whether the reaction is allowed is determined principally by the symmetries of the MOs involved. The reactions are characterized by two numbers (shown in brackets) associated, respectively, with the two parts of the molecule connected by the σ bond which migrates. One number corresponds to the number of atoms from the initial to the final location of the migrating σ-bond in one part of the molecule; the other corresponds to the number of atoms from the initial to the final location of the migrating σ-bond in the other part of the molecule. Reaction 21.3 (the Cope rearrangement given earlier) and its oxygen analogue given in Scheme 7.2 of Chapter 7 (a Claisen rearrangement) are both [3,3] sigmatropic rearrangements, as illustrated here in the margin. The migration of a hydrogen atom from one C to the fifth along the diene system also shown in the margin is another common type of sigmatropic rearrangement, this time a [1,5] sigmatropic rearrangement; sigmatropic hydrogen migrations are often labelled [1,n]-H shifts where n indicates the number of atoms in the chain from the initial to the final location of the H.

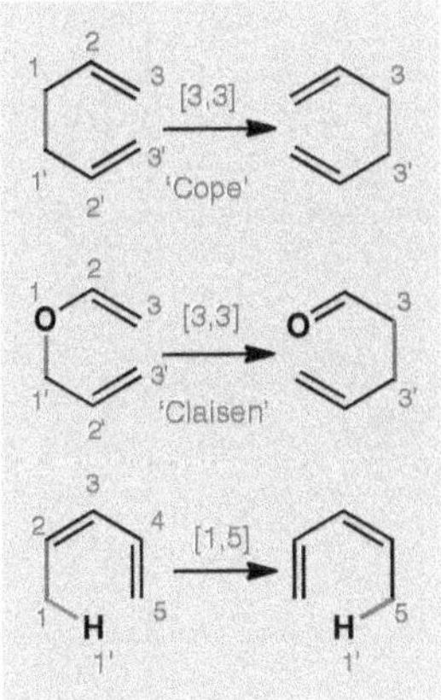

21.4.1 [3,3] Sigmatropic rearrangements

The MO interactions in the TS for the [3,3] sigmatropic rearrangement of eqn 21.3 at the beginning of the chapter can be represented as the overlap between MOs of two allyl radicals. As each allyl radical has three π electrons (i.e. we have a 6π electron system), the HOMO of each is a SOMO (the π_2 orbital in Fig. 5.6 in Chapter 5). We see in Figure 21.8(a) that the symmetries of the SOMOs of the allyl radicals allow in-phase interactions between their positions 1 and 3,* so the formation of this TS is thermally allowed. Moreover, the paired allyl components will be in the form of a chair cyclohexane (Figure 21.8(b)) because this is the lowest energy conformation of a six-membered carbocyclic ring, as we saw in Chapter 4.

We can now show in a specific example how the interactions between the MO lobes in the TS of the thermally-allowed [3,3] sigmatropic rearrangement determine the stereochemical outcome. The stereoisomer of 3,4-dimethylhexa-1,5-diene shown in Scheme 21.10, or its enantiomer (or the racemate), is converted stereospecifically into achiral (2E,6E)-octa-2,6-diene as illustrated.

*When the interactions at one component are both at the same face, they are called *suprafacial* so the symmetries of the SOMOs of the two allyl radicals require *suprafacial–suprafacial* interactions, as shown in Figure 21.8(a).

(3S,4S)-3,4-dimethylhexa-1,5-diene   (2E,6E)-octa-2,6-diene

Scheme 21.10
Stereochemistry of the [3,3] sigmatropic rearrangement of 3,4-dimethylhexa-1,5-diene.

> Predict the stereochemistry of the product of the thermal rearrangement of *meso*-3,4-dimethylhexa-1,5-diene.

Exercise 21.14

The Claisen rearrangement of alk-1-enyl alk-2-enyl ethers (allyl vinyl ethers) to give carbonyl compounds illustrated in the margin at the beginning of this section is closely similar to the Cope rearrangement discussed above. The driving force is the formation

(a) SOMO of allyl radical

SOMO of allyl radical

(b)

Figure 21.8 Frontier orbitals of two allyl components in the TS for the [3,3] sigmatropic rearrangement of eqn 21.3; (a) shows suprafacial–suprafacial interactions, and (b) shows the favourable chair cyclohexane-like conformation of the TS.

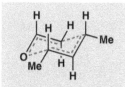

TS from one enantiomer of the reactant in eqn 21.12.

of the strong carbonyl bond, and a further example is given in eqn 21.12. Note that the cyclohexane-like TS with the two methyl groups in *pseudo*-equatorial positions of the chair-like conformation leads to the *trans* product shown.

$$170\ ^\circ C,\ 15\ min \quad [3,3] \quad 80\%$$

(21.12)

A Claisen rearrangement is also involved in the following synthesis of γ,δ-unsaturated carbonyl compounds.

acetal exchange ... *elimination*

an acetal   an allylic alcohol   an allyl vinyl ether

$$\xrightarrow[{[3,3]}]{\Delta}$$

a γ,δ-unsaturated ketone

The aromatic Claisen rearrangement was discovered before the aliphatic version or the Cope rearrangement.

Scheme 21.11 shows the prototype of the well-known Claisen rearrangement of an allyl aryl ether—the isomerization of allyl phenyl ether to *o*-allylphenol. As shown, the thermal [3,3] sigmatropic rearrangement of allyl phenyl ether initially gives a keto product which enolizes to the more stable phenol in a subsequent step. When one end of the allyl group was isotopically labelled with ^{14}C (shown with * in Scheme 21.11), it was confirmed that the allyl group had bonded to the benzene ring from its opposite end. The intramolecular nature of this reaction is also confirmed by this labelling experiment.

Scheme 21.11 An example of the aromatic Claisen rearrangement.

$$200\ ^\circ C \quad [3,3] \quad enolization \quad 85\%$$

Exercise 21.15

Draw the structure of the product of the Claisen rearrangement of but-2-enyl phenyl ether and point out the implication of the result regarding the reaction mechanism.

21.4.2 [1,5] Sigmatropic rearrangements

The migration of a hydrogen atom from one end of a penta-1,3-diene system to the other end with an associated reorganization of the intervening double bonds is a [1,5] sigmatropic rearrangement (a 1,5-H shift). The transition structure for this may be regarded as a penta-1,3-dienyl radical paired with a hydrogen atom, as shown in Figure 21.9. Since the penta-1,3-dienyl radical has five electrons in its π system and the H atom contributes one, we have a six-electron system.

Two stereochemical courses are conceivable: a *suprafacial* migration (see the previous sub-section) and an *antarafacial* migration. In the former process, the hydrogen atom stays on the same face of the penta-1,3-dienyl π system, while in the latter, the H moves from one face of the π system to the opposite face. As seen in Figure 21.9, the symmetry

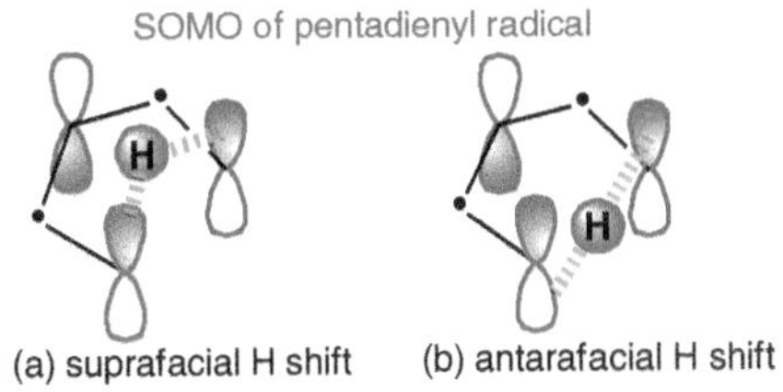

SOMO of pentadienyl radical

(a) suprafacial H shift   (b) antarafacial H shift

Figure 21.9 Frontier orbital interactions in the TS for a [1,5] hydrogen shift in penta-1,3-diene. The SOMO of the penta-1,3-dienyl radical and the 1s AO of a hydrogen atom are shown for (a) the suprafacial migration and (b) the antarafacial migration.

of the SOMO of the pentadienyl radical allows the overlap of orbitals required for the suprafacial hydrogen migration (a), but not for the antarafacial process (b).

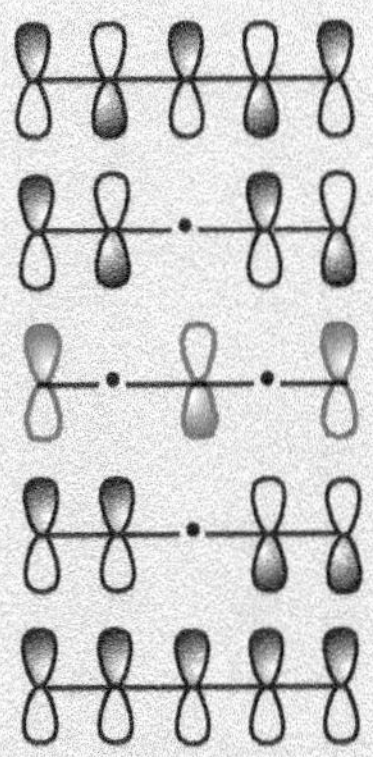

For comparison, the TS in a [1,3] hydrogen shift of an allyl system may be regarded as an allyl radical (three electrons) associated with an H atom (one electron). The symmetry of the SOMO (π_2) of the allyl radical (see Figure 5.4 in Chapter 5 for the MOs of the allyl radical) does not allow a suprafacial [1,3] hydrogen shift. Although the symmetry of the SOMO of the allyl radical allows a thermal antarafacial H migration (in a four-electron system, Figure 21.10), the process is actually difficult. This is because the small 1s atomic orbital of the H atom cannot achieve adequate overlap with the appropriate lobes of the two C atoms involved: the distance between them is too great as they are on opposite faces of the allyl system.

The stereospecificity of the thermally allowed [1,5] sigmatropic hydrogen shift can be seen in the example given in Scheme 21.12. The suprafacial H migration in the stereochemically pure isotopically labelled (2E,4Z)-diene results in both the E,Z and Z,Z isomers (but only these) depending on the conformation of the reactant through which the reaction takes place. Notably, the configuration of the carbon carrying the deuterium in the product is predicted to be R in the E,Z isomer and S in the Z,Z isomer, respectively, and this was the result observed.

MOs of penta-1,3-dienyl radical with five π electrons. The HOMO in the electronic ground state (we are dealing with a thermal reaction) is the SOMO shown in red.

SOMO of allyl radical

Figure 21.10 Frontier orbital interaction in an antarafacial [1,3] hydrogen shift. Only the SOMO of the allyl radical and the 1s AO of the hydrogen atom are shown.

rotation about C5–C6

250 °C
[1,5]

250 °C
[1,5]

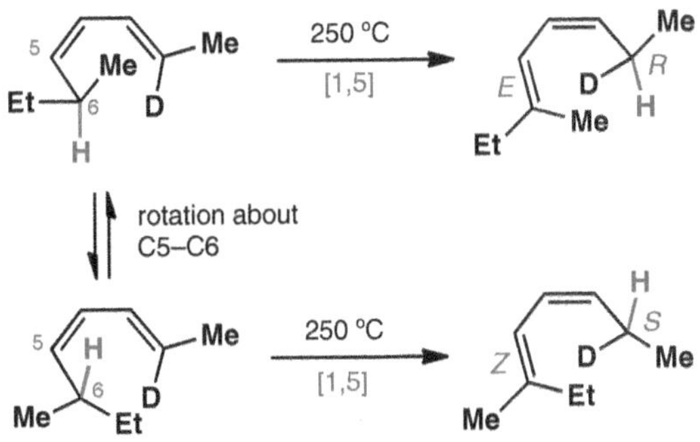

Scheme 21.12 Stereospecific suprafacial [1,5] sigmatropic rearrangement of (2E,4Z,6S)-2-deuterio-6-methylocta-2,4-diene.

A hydrogen atom migrates around the ring with an associated redistribution of double bonds in the thermal isomerizations of 5-methylcyclopentadiene, as shown in eqn 21.13. These transformations are the outcome of sequential [1,5]-H shifts which lead to an equilibrium mixture of 1-, 2-, and 5-methylcyclopentadienes.

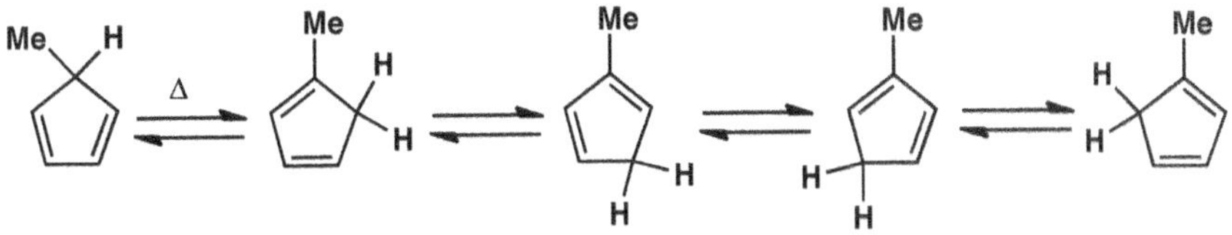

(21.13)

Exercise 21.16

Give alternative curly arrow mechanisms for just one single-step rearrangement of (1-^{13}C,5-^{2}H)-5-methylcyclopentadiene into labelled 1-methylcyclopentadiene.

Panel 21.1 Biological pericyclic reactions in vitamin D formation

Rickets is a childhood disease characterized by poor bone growth due to lack of vitamin D and it is known that sunlight helps to prevent rickets. A similar disease in adults called osteomalacia is prevalent among Arab women who remain completely covered out of doors. How does sunlight lead to the formation of vitamin D?

One of the active forms of vitamin D is vitamin D$_3$ and this is produced photochemically in the skin from a precusor called 7-dehydrocholesterol (see Chapter 24 for cholesterol).

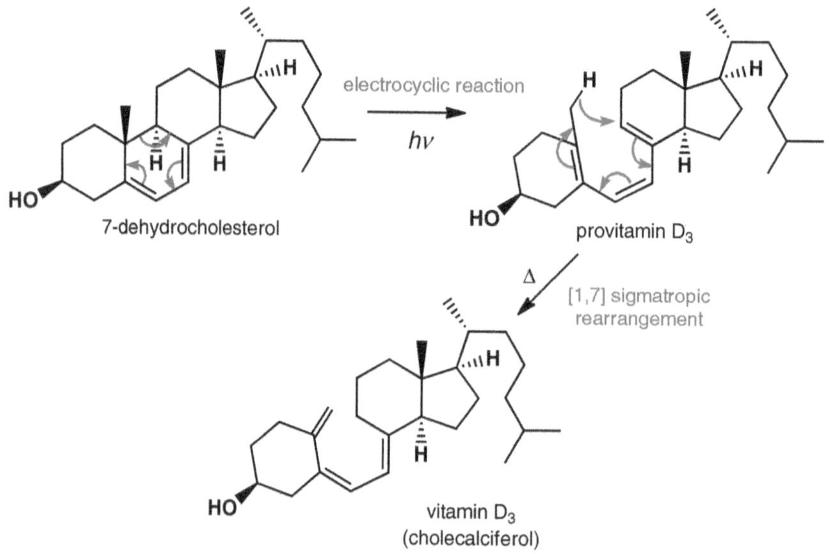

Vitamin D$_3$ is further hydroxylated in the liver and kidney, and is involved in calcium deposition in bone.

Summary

- Pericyclic reactions are concerted and stereospecific when stereochemical issues are involved; thermal variants proceed via cyclic transition structures involving redistribution of $4n+2$ electrons. Cyclic transition structures in photochemical pericyclic reactions involve $4n$ electrons. The kinetics and stereochemistry of the reactions are controlled by the symmetries of the molecular orbitals involved.

- Pericyclic reactions are classified into three main types, cycloadditions, electrocyclic reactions, and sigmatropic rearrangements.

- In cycloadditions, two unsaturated molecules combine to give a cyclic product; [4+2] cycloadditions (Diels–Alder reactions and 1,3-dipolar cycloadditions) are the most common. Ozonolysis and dihydroxylation of alkenes with osmium tetroxide are useful 1,3-dipolar cycloadditions.

- [2+2] Cycloadditions are symmetry forbidden thermally but allowed photochemically.

- Electrocyclic reactions of conjugated polyenes are cyclizations by formation of a new σ bond between the terminal sp^2 carbons of the π system and the loss of one π bond; they are a type of isomerization.

In a sigmatropic rearrangement of a π electron system, one σ bond breaks and a new σ bond forms. The Cope and Claisen rearrangements are representative thermal [3,3] sigmatropic rearrangements; [1,5] hydrogen shifts are another well-known type.

Problems

21.1 What are the main kinetic products of the Diels–Alder reactions of the following combinations of diene and dienophile? Show the stereochemistry where appropriate.

(a)

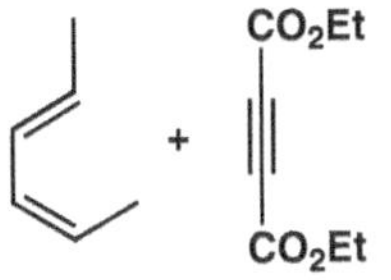

(b)

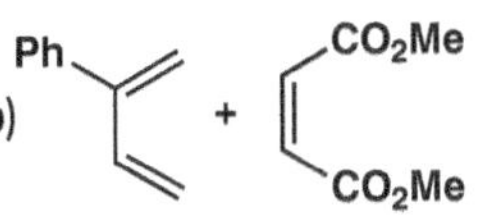

(c)

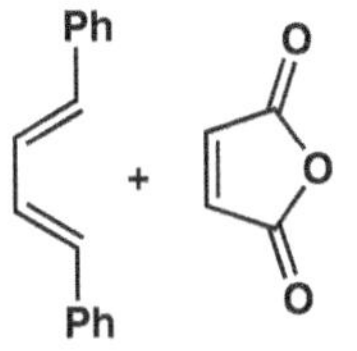

(d)

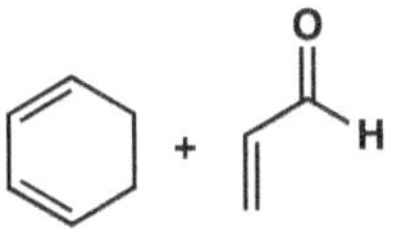

21.2 What are the main products of the Diels–Alder reactions of the following combinations of diene and dienophile?

(a)

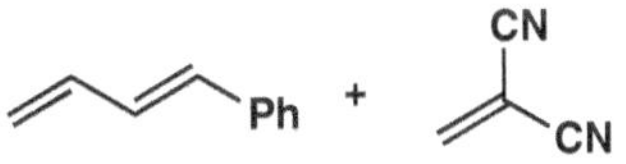

(b)

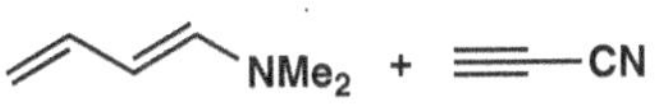

(c)

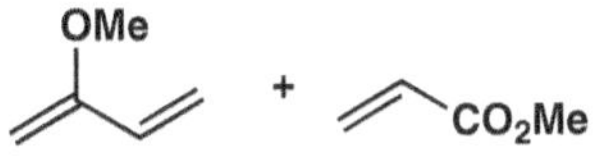

(d)

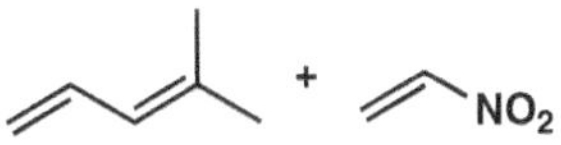

21.3 Explain the relative reactivities of the (2E,4E), (2E,4Z), and (2Z,4Z) stereoisomers of hexa-2,4-diene in reactions with dimethyl maleate. Note that we saw reactions of two isomers of the hexadiene in eqn 21.5 and Exercise 21.7.

21.4 Liquid cyclopentadiene dimerizes readily at room temperature, while the dimer dissociates upon distillation at 140 °C. Show how the reversible reaction takes place and give the stereochemical structure of the dimer.

21.5 When buta-1,3-diene is heated, it slowly dimerizes in a cycloaddition reaction. By giving the mechanism, predict the structure of the dimer of butadiene.

21.6 By giving the steps of the following reaction, identify the product.

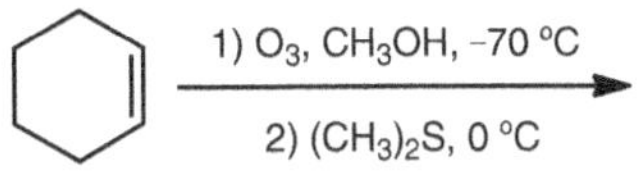

21.7 The following all-*cis* cyclic polyenes readily give bicyclic products upon heating. Explain the reactions and give the structures of the products.

(a) cyclonona-1,3,5-triene

(b) cyclodeca-1,3,5,7,9-pentaene

21.8 (2E,4Z,6Z,8E)-Deca-2,4-6-8-tetraene readily undergoes two consecutive electrocyclic reactions. Explain the reactions and give the structures of the intermediate and final products.

21.9 Give structures of the rearrangement products obtained by heating the following, and identify each reaction type.

(a)

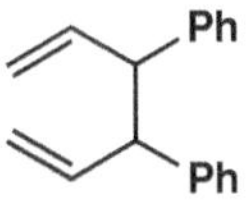

(b)

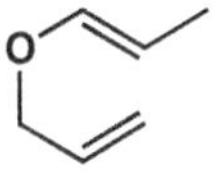

(c)

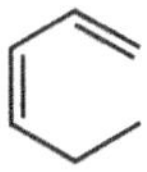

(d)

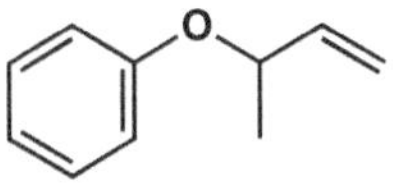

21.10 Although *trans*-1,2-divinylcyclopropane is stable, the *cis* isomer readily rearranges upon heating. What is the product?

21.11 5-Methoxycyclohexa-1,3-diene gives an equilibrium mixture with 1- and 2-methoxycyclohexa-1,3-dienes upon being heated.

(a) Give a mechanism for this equilibration.

(b) Which is the most stable isomer?

21.12 The following allyl phenyl ether substituted with two methyl groups in the *ortho* positions gives the *p*-allyl phenol upon heating. Give a mechanism for this rearrangement.

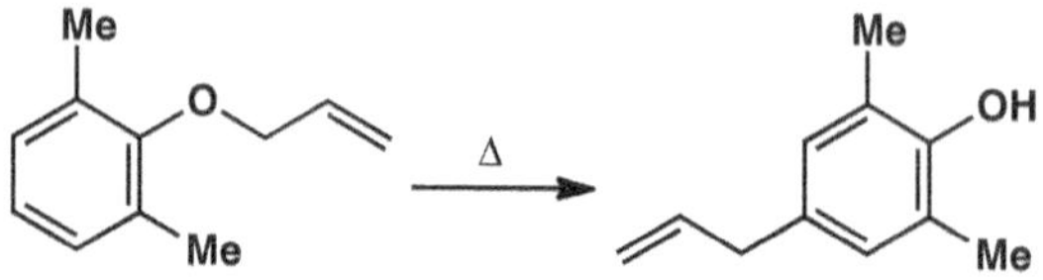

21.13 Give a mechanism for the following transformation (an example of what is sometimes called the *oxy-Cope rearrangement*).

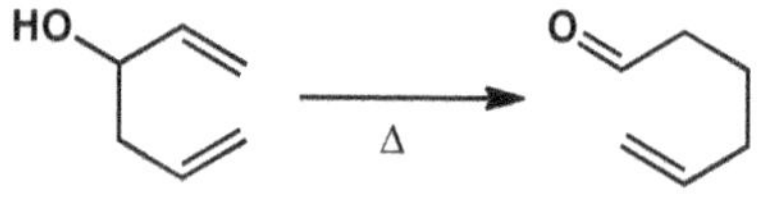

21.14 Explain whether the higher analogue of a Claisen rearrangement of the pentadienyl phenyl ether shown is thermally allowed or not.

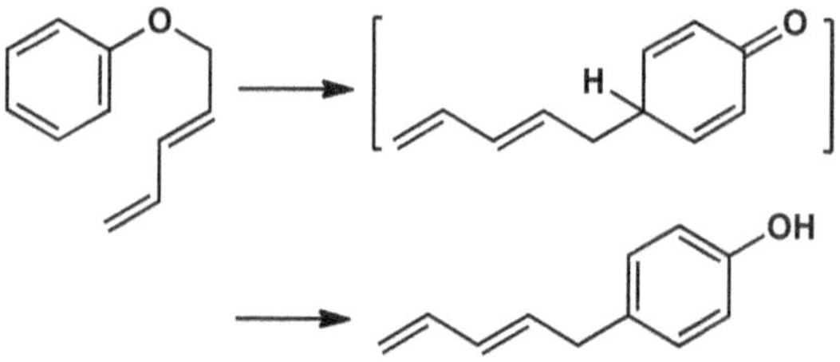

21.15 So-called Dewar benzene is a highly strained unstable isomer of benzene, but it does not readily isomerize to benzene. Explain this kinetic stability of Dewar benzene.

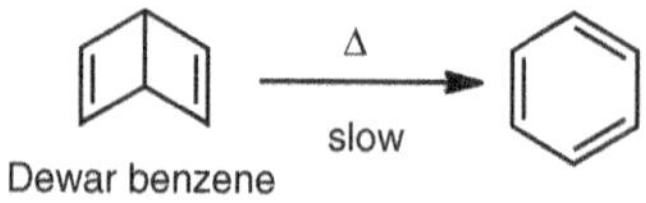

21.16 Enolization could be represented as a concerted [1,3] sigmatropic rearrangement. Explain why this reaction does not occur thermally.

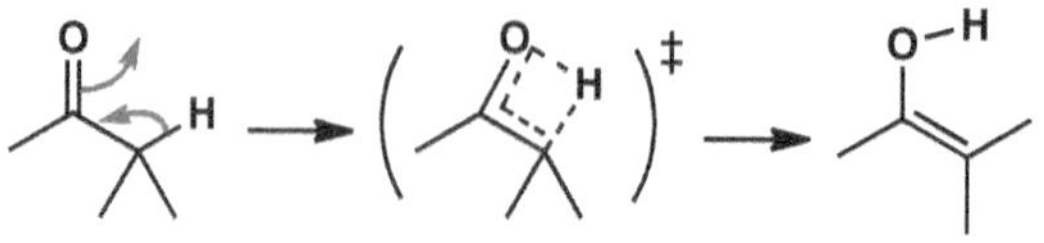

21.17 The following reaction of a benzocyclobutene to give a steroidal skeleton proceeds by an initial electrocyclic ring opening followed by an intramolecular cycloaddition. Show the processes and explain how they account for the stereochemical relationships between the ring hydrogens shown.

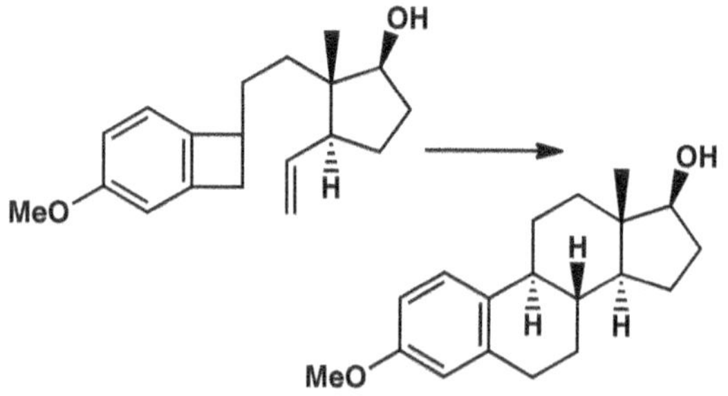

21.18 In rare cases, reverse electron demand Diels–Alder reactions are possible in which the cycloaddition occurs between a diene carrying electron-withdrawing group(s) and a dienophile with electron-donating group(s). Explain the molecular orbital interactions in this type of Diels–Alder reaction.

Supplementary Problems

21.19 Show how the following sequence of reactions proceed using curly arrows.

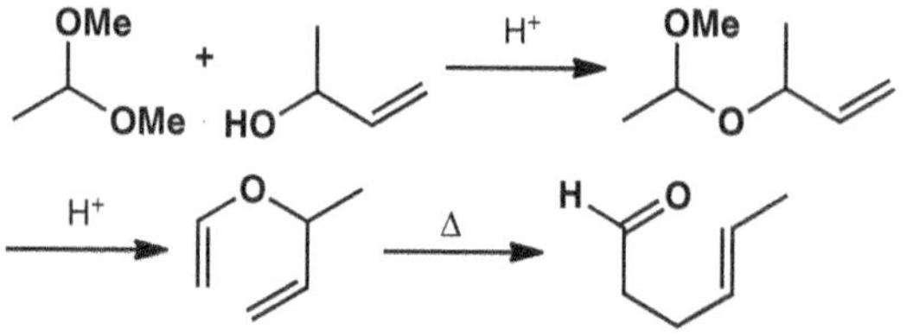

21.20 An industrial synthesis of geranial, a terpenoid contributing to the flavour of lemons, incorporates sigmatropic rearrangements as outlined below. Give mechanisms for the last two steps from the allylic enol ether intermediate.

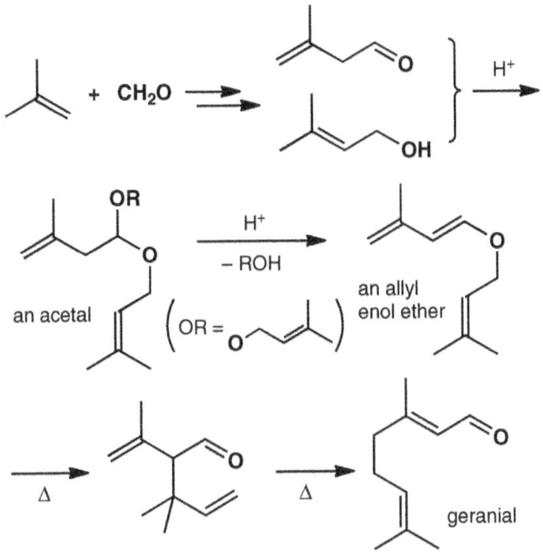

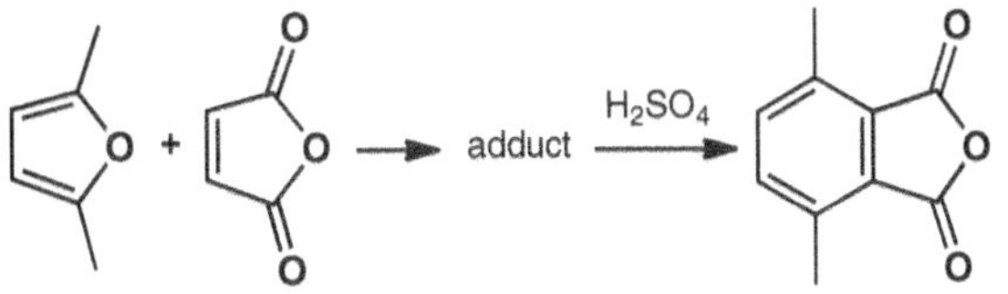

21.22 The apparently abnormal Claisen rearrangement given below occurs thermally via a normal Claisen rearrangement with an enolization followed by an intramolecular ene-type of reaction to give a cyclopropane intermediate which then undergoes a retro-ene-type of reaction. Give a mechanism for each step of the overall transformation.

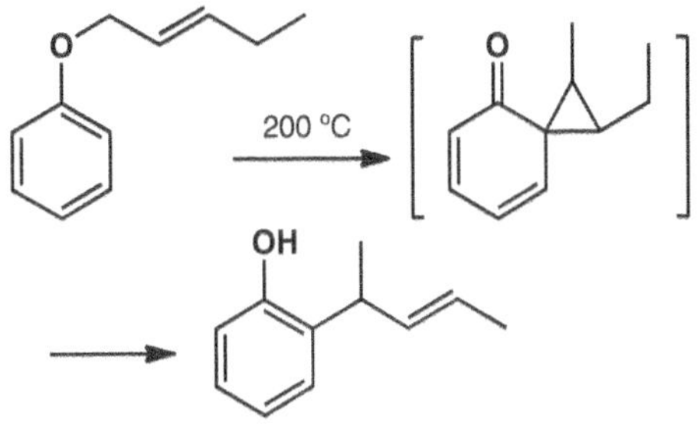

21.21 2,5-Dimethylfuran undergoes a Diels–Alder reaction with maleic anhydride to give an adduct which can be transformed into 3,6-dimethylphthalic anhydride by an acid-catalysed dehydration. Give the structure of the Diels–Alder adduct and a mechanism for its acid-catalysed dehydration.

22 Rearrangement Reactions involving Polar Molecules and Ions

Related topics we have already covered:

- Curly arrow representation of reaction mechanisms (Section 7.2)
- Molecular orbital interactions in reactions (Section 7.3)
- Rearrangements of carbenium ions (Section 14.2)
- Pericyclic reactions (Chapter 21)
- Sigmatropic rearrangements of nonpolar compounds (Chapter 21)

Topics we shall cover in this chapter:

- [1,2] Sigmatropic rearrangements
- Concerted and stepwise 1,2-shifts in carbenium ion reactions
- Catalysed rearrangement of carbonyl compounds involving 1,2-shifts
- 1,2-Shifts from carbon to oxygen and nitrogen
- Rearrangements involving carbenes and nitrenes or their precursors
- Rearrangements involving neighbouring group participation

The sigmatropic rearrangements covered in the previous chapter were concerted single-step reactions of nonpolar molecules which could be carried out in nonpolar solvents or in the gas phase. However, some multistep reactions of polar organic compounds or ions also involve rearrangements and take place only in polar solvents. These have been investigated experimentally since well before the application of MO theory to organic chemistry but we now know that they share some of the characteristics of the sigmatropic rearrangements covered in the previous chapter. In this chapter, we shall show briefly how 1,2-shifts in carbenium ion intermediates in solvolytic reactions, for example, can be considered as [1,2] sigmatropic rearrangements, and (consequently) related to concerted sigmatropic rearrangements of nonpolar compounds. The MO description of these simple 1,2-shifts in carbenium ions can be extended to describe the rearrangement steps in more complicated multistep heterolytic reactions. Stepwise rearrangements via cyclic intermediates formed by neighbouring group participation are also covered.

22.1 1,2-Shifts in Carbenium Ions

We encountered carbenium ion rearrangements in Section 14.2 where the migrating group was hydrogen or alkyl. The frontier orbital interactions in the TS of the simplest example (eqn 22.1, which is normally thought of as a 1,2-hydride shift) may be regarded as between the empty 1s AO (LUMO) of a proton and the HOMO of ethene (Figure 22.1).

The compatible orbital symmetries shown in Figure 22.1 indicate that this suprafacial 1,2-shift involving a three-membered cyclic two-electron TS is thermally allowed.

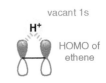

vacant 1s

HOMO of ethene

Figure 22.1 Frontier orbital interactions in the TS for a 1,2-H shift in a carbenium ion.

$$(22.1)$$

This type of rearrangement is commonly observed during reactions involving carbenium ion intermediates such as S_N1 and E1 reactions (Chapters 12–14), electrophilic additions to alkenes (Chapter 15), and Friedel–Crafts alkylation (Chapter 16). For example, Scheme 22.1 shows the electrophilic addition of HCl to an alkene in which a methyl shift competes very effectively with nucleophilic capture of the first-formed carbenium ion intermediate by Cl⁻.

Rearrangements of carbenium ions are sometimes called Wagner–Meerwein rearrangements.

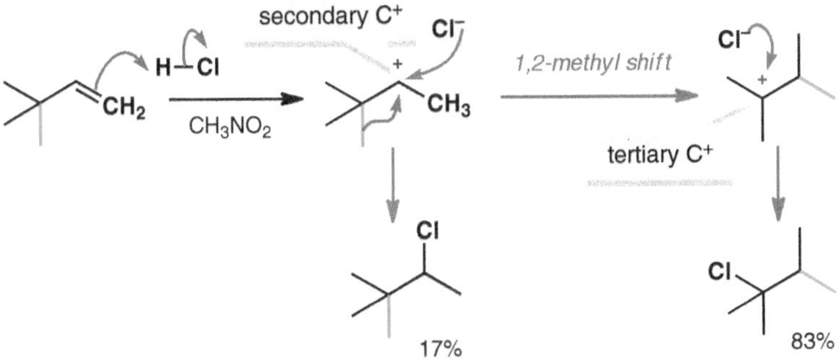

secondary C⁺

1,2-methyl shift

tertiary C⁺

17%

83%

Scheme 22.1 Carbenium ion rearrangement during the electrophilic addition of HCl to an alkene.

In Scheme 22.1, the three methyl groups on the C next to the electron-deficient centre of the first-formed carbenium ion are equivalent, so there is no question of which will migrate. If there are different groups which, in principle, could migrate, we need to consider their *migratory aptitudes*. Although the outcome ultimately depends on the relative stabilities of the potential product carbenium ions, *the group which is more stable as a cation itself generally has the greater migratory aptitude* (although this may be affected by steric and geometric factors). Hydrogen, however, is generally a good migrator and this is probably because its migration will always lead to the most stable of the possible alternative product carbenium ions.

Migratory aptitudes : H > 3°R > 2°R > 1°R > Me

When an alkyl group migrates, the configuration of the migrating carbon centre is retained if it is chiral. This is because the alkyl group never becomes free in the suprafacial migration as illustrated by the MO representation of the TS in Figure 22.2.

In a related reaction known as the pinacol rearrangement, a 1,2-diol rearranges to a ketone (or sometimes an aldehyde) with acid catalysis (Scheme 22.2).[1] The 1,2-shift of

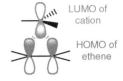

LUMO of cation

HOMO of ethene

Figure 22.2 Frontier MO interactions for a 1,2-alkyl shift in a carbenium ion.

---

[1] The name of the pinacol reaction is derived from the old name (pinacol) of the prototype reactant (2,3-dimethylbut-2,3-diol) which rearranges to 3,3-dimethylbutan-2-one (known trivially as pinacolone), a reaction which was discovered as long ago as 1860 in Germany. It is now the generic name of analogous reactions of any 1,2-diol. The name of this *rearrangement* should not be confused with the *pinacol reaction*, which is the common name for the formation of pinacol from propanone (acetone); see Sub-section 20.9.2.

Scheme 22.2 The pinacol rearrangement.

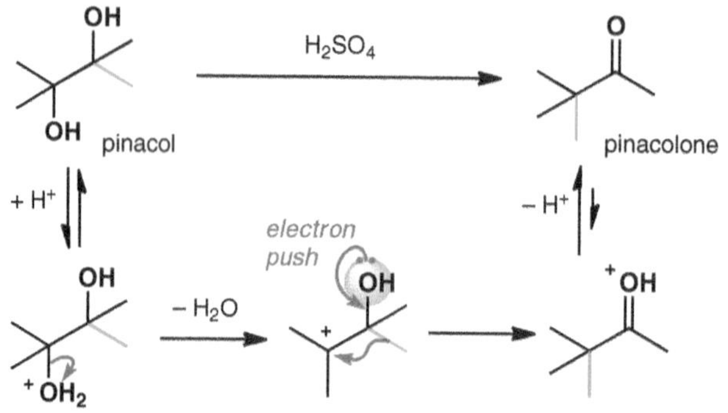

the methyl group with its bonding pair to the electron-deficient centre of the intermediate carbenium ion is facilitated by an electron-push from the OH group.

In the case of an unsymmetrical diol, we need to address the question of which protonated OH will depart as the nucleofuge in the pinacol rearrangement. We do this by considering the relative stabilities of the alternative carbenium ions which would be obtained. For example, 2-methylpropane-1,2-diol in eqn 22.2 mostly gives an aldehyde following the initial formation of a *tertiary* carbenium ion intermediate and the ensuing 1,2-hydride shift; departure of the other protonated OH would give a *primary* carbenium ion which could not compete with the reaction shown.

Note that it is sometimes difficult to distinguish between *stepwise* departure of the nucleofuge with a *subsequent* 1,2-shift in a pinacol rearrangement, and the alternative in which the 1,2-shift is *concerted with* the departure of the nucleofuge.

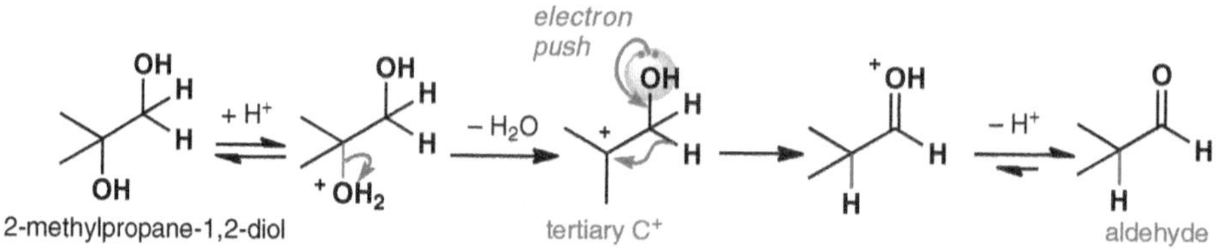

$$(22.2)$$

When the same diol is first tosylated with tosyl (*p*-toluenesulfonyl) chloride and pyridine, and the tosylate is hydrolysed under neutral conditions, the main product is a ketone (eqn 22.3).

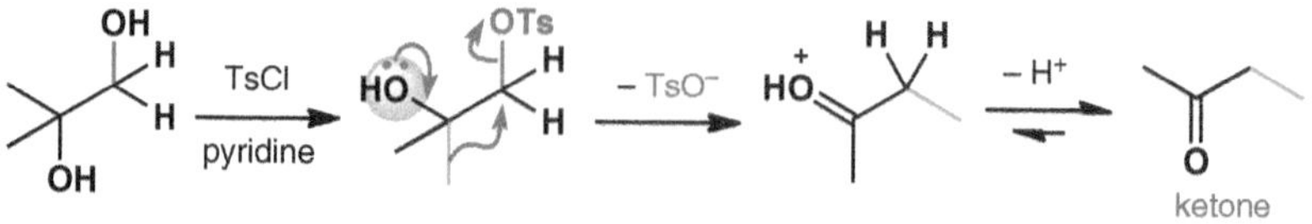

$$(22.3)$$

Here, the tosyl chloride and pyridine react mainly at the less hindered primary OH group to give the product shown and this site-selectivity based upon steric effects determines the outcome of the subsequent reaction. The departure of the tosylate assisted by participation of a lone pair on the OH (electron push) and the concerted 1,2-migration of a methyl group with its bonding pair of electrons lead to the formation of the ketone. A simple S_N1 mechanism would not be competitive since the intermediate primary alkyl carbenium ion would be too unstable, and a solvent induced S_N2 mechanism would be sterically hindered. Such pinacol-type rearrangements in which there is no ambiguity about the site from which the nucleofuge departs are now generally called *semi-pinacol rearrangements*.

Exercise 22.1

Give a mechanism for the following rearrangement.

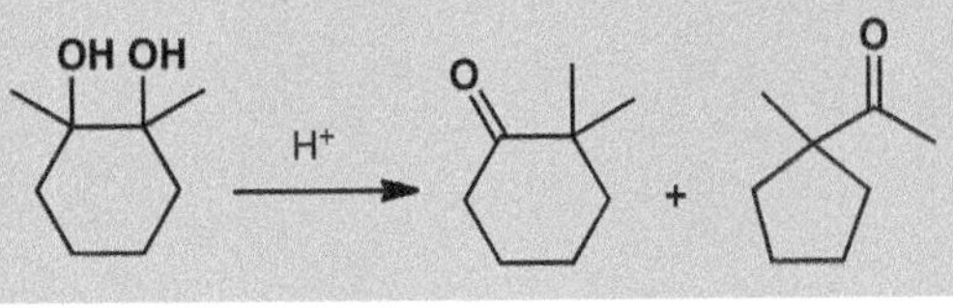

Exercise 22.2

Propose mechanisms for the following reactions and identify the main product in each case.

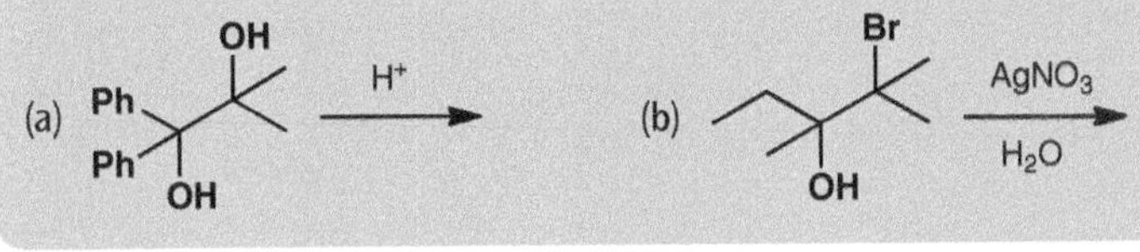

22.2 Concerted 1,2-Shifts bypassing the Formation of Unstable Carbenium Ions

In some dissociative reactions, a 1,2-shift provides a route avoiding the formation of a highly unstable carbenium ion intermediate, as illustrated in the semi-pinacol rearrangement of eqn 22.3 above. A simple primary alkyl substrate bearing a good nucleofuge, but which cannot undergo an S_N2 or an E2 reaction, may still ionize with a concerted 1,2-H or alkyl shift to the carbon from which the nucleofuge departs even without the assistance of a β OH. The solvolytic reaction of 2,2-dimethylpropyl (neopentyl) tosylate in eqn 22.4 is an example. The reactant is a primary alkyl derivative but it cannot undergo an S_N2 reaction owing to the severe steric hindrance, and there is no β H for a solvent-induced E2.

We encountered a similar (but acid-catalysed) reaction in Chapter 14 (Scheme 14.7).

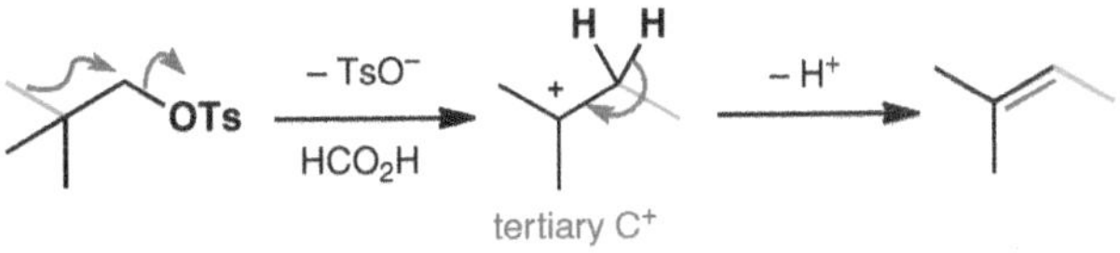

$$(22.4)$$

Example 22.1

When a primary amine is diazotized with nitrous acid in aqueous solution, the amino group is converted into N_2, an excellent nucleofuge (see Sub-section 14.7.3). Following diazotization, the optically active amine below gives a rearranged alkene; deduce the structure of the product by giving a mechanism for the reaction including the rearrangement.

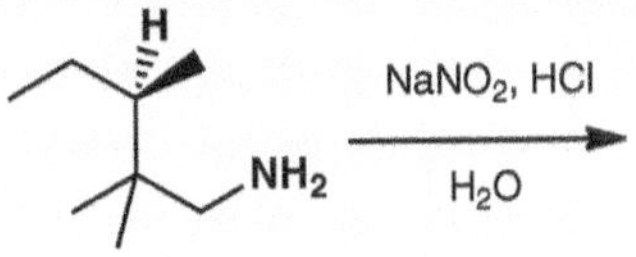

Solution
The main reaction involves the 1,2-shift of the chiral secondary alkyl group upon departure of the N_2 from the initially formed alkanediazonium ion. The configuration of the migrating alkyl group is retained since it never becomes free during its 1,2-shift (concerted with departure of the N_2) via the three-membered cyclic TS in the suprafacial process.

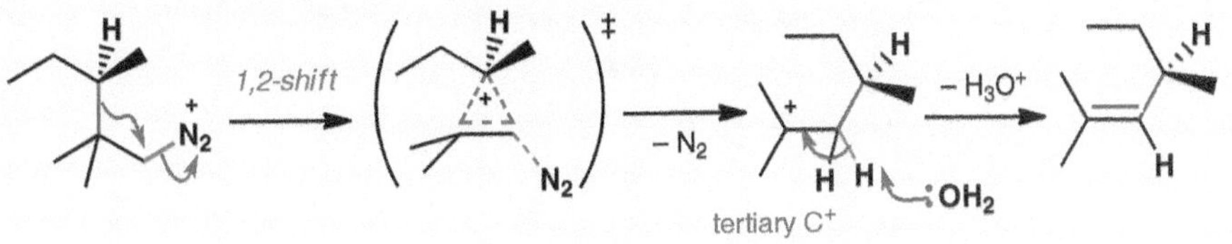

Exercise 22.3

Give a mechanism for the following transformation.

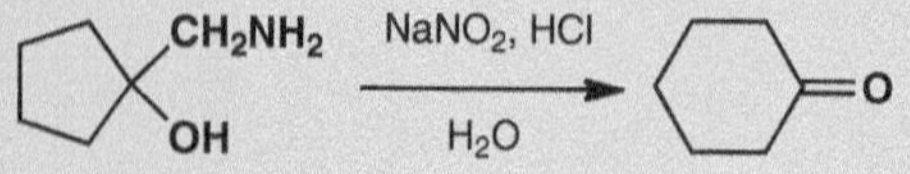

22.3 Catalysed Rearrangement of Carbonyl Compounds involving 1,2-Shifts

Some α-hydroxy ketones undergo an acid-catalysed rearrangement to give isomeric hydroxy ketones, e.g. Scheme 22.3. The mechanism of this reaction can be understood easily if we think of the carbenium ion resonance form of the protonated carbonyl which provides an electron pull. This, together with a push of electron density by a lone pair on the α-hydroxy group, facilitates the 1,2-alkyl shift, just as in the concerted semi-pinacol rearrangement in eqn 22.3.

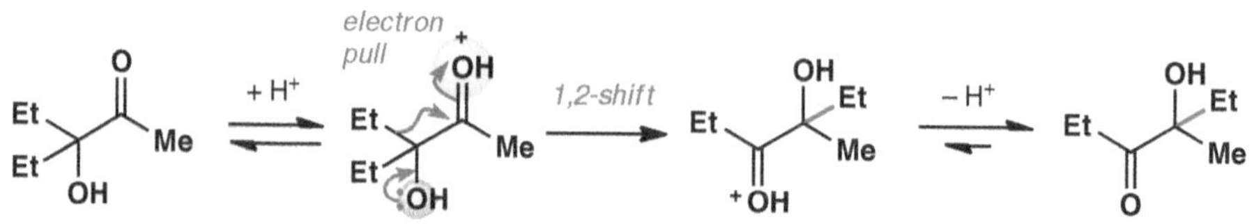

Scheme 22.3 Acid-catalysed rearrangement of an α-hydroxy ketone.

The same transformation can also be achieved by base catalysis (Scheme 22.4); whereas acid catalysis involves a protonated intermediate, a deprotonated intermediate is involved in the base-catalysis mechanism. In both mechanisms, the 1,2-shift within the respective intermediates is induced by an electron push at one end and an electron pull at the other.

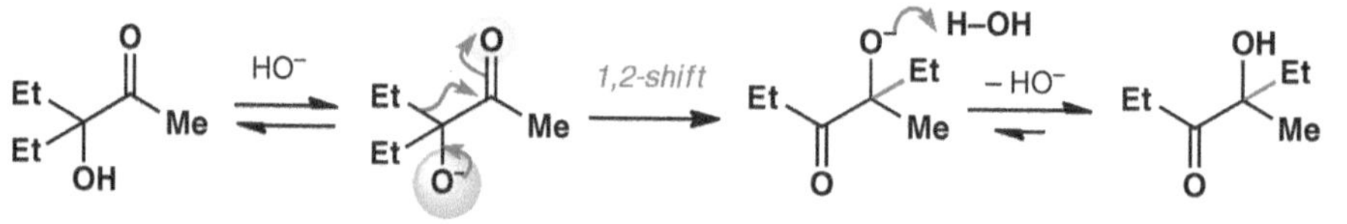

Scheme 22.4 Base-catalysed rearrangement of an α-hydroxy ketone.

Exercise 22.4

Benzil is an α-diketone which can be converted into the salt of an α-hydroxycarboxylic acid known as benzilic acid by being heated in a strongly alkaline solution. Propose a mechanism for this so-called benzilic acid rearrangement.

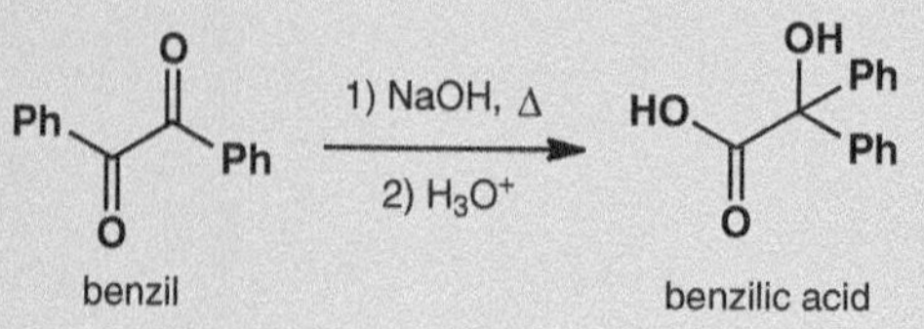

22.4 Concerted 1,2-Shifts from Carbon to Oxygen and Nitrogen

22.4.1. The Baeyer–Villiger oxidation

Ketones may be oxidized to esters by insertion of an O next to the carbonyl using per-acids in the so-called Baeyer–Villiger oxidation (Scheme 22.5). The reaction is initiated by the nucleophilic addition of the peracid to the carbonyl group (with an associated proton transfer) to give a tetrahedral intermediate. Upon departure of the excellent nucle-ofuge ($CF_3CO_2^-$) from one of the two oxygen atoms bonded to the central C, a concerted 1,2-alkyl shift is facilitated by an electron push and proton loss from the other oxygen.

Note that the peracid acts as a nucleophile in the Baeyer–Villiger reaction but as an electrophile in the epoxidation of alkenes encountered in Section 15.5.

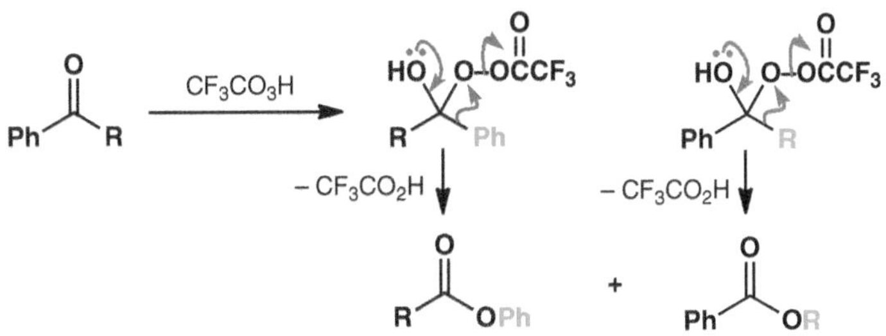

Scheme 22.5 Baeyer–Villiger oxidation of phenyl alkyl ketones.

R =	$R\text{-}CO\text{-}OPh$	$Ph\text{-}CO\text{-}OR$
Me	100%	0%
Et	94%	6%
i-Pr	34%	66%
t-Bu	3%	97%

Examples of the Baeyer–Villiger oxidation in Scheme 22.5 demonstrate the relative ease of migration to a potentially electron-deficient site of alkyl groups compared with phenyl: the ease of migration (migratory aptitude) increases in the order, methyl < pri-mary R < phenyl < secondary R < tertiary R. Not surprisingly, this order is similar to that observed in carbenium ion rearrangements.

Phenyl and H usually have greater migratory aptitudes than alkyl groups in carbenium ion rearrangements. Phenyl migrations usually occur in the context of *neighbouring group participation* which is covered in Section 22.6.

> **Exercise 22.5**
>
> Deduce the product of the reaction of cyclohexanone with peracetic acid by giving a mechanism for the reaction.

Example 22.2

Isopropylbenzene (cumene) is the starting material in an industrial process for the production of phenol and propanone (acetone). In this process, the hydroperoxide formed by atmospheric oxi-dation of cumene is treated with an acid which induces the 1,2-shift of the phenyl group. Give a mechanism for the rearrangement of the hydroperoxide which leads to phenol and propanone.

Solution

Protonation of the OH of the hydroperoxide provides a very good leaving group, H_2O, and a concerted 1,2-phenyl shift accompanies its departure. This 1,2-phenyl shift bypasses the forma-tion of a highly unstable electron-deficient oxygen species, and gives a cation reminiscent of a

(continues …)

We encountered a similar rearrangement during the oxidation of the alkylborane product of hydroboration in Sub-section 15.3.3.

(... continued)

protonated carbonyl group. Nucleophilic capture of the cation by water leads to a hemiacetal which decomposes to phenol and propanone.

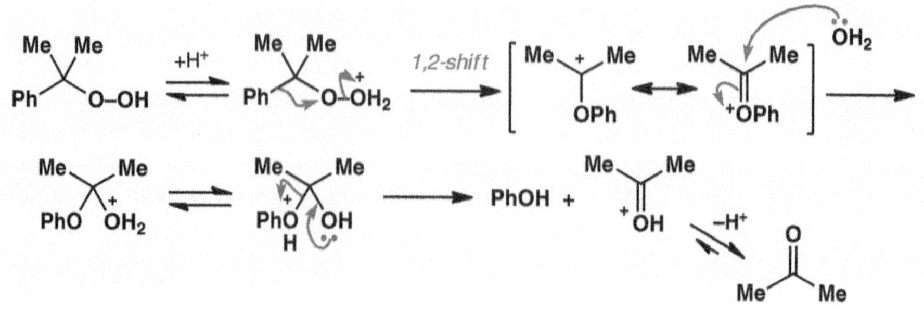

22.4.2 The Beckmann rearrangement

Oximes (formed from ketones or aldehydes with hydroxylamine, Sub-section 8.6.1) undergo acid-catalysed isomerizations to give amides. This reaction (the Beckmann rearrangement) involves departure of H_2O from the protonated oxime accompanied by the concerted 1,2-shift of an alkyl group to the nitrogen (Scheme 22.6). This 1,2-alkyl shift avoids formation of a highly unstable intermediate containing a doubly bonded nitrogen with only six valence electrons.

Scheme 22.6 A mechanism for the Beckmann rearrangement.

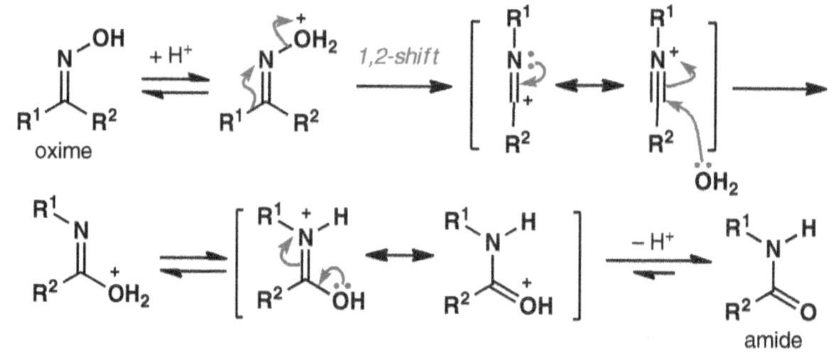

Since there are *cis–trans* isomers of an unsymmetrical oxime, the possibility of stereo-specificity arises; experiments have established that the group *trans* to the departing nucleofuge migrates stereospecifically. However, *cis–trans* isomerization of an oxime can be rapid in the presence of a protic acid, so we usually observe that the alkyl group with the greater migratory aptitude moves irrespective of the configuration of the original oxime.

Exercise 22.6

Cyclohexanone oxime undergoes an acid-catalysed rearrangement to a cyclic amide, ε-caprolactam, which is the monomer for the production of nylon 6. Give a mechanism for this rearrangement.

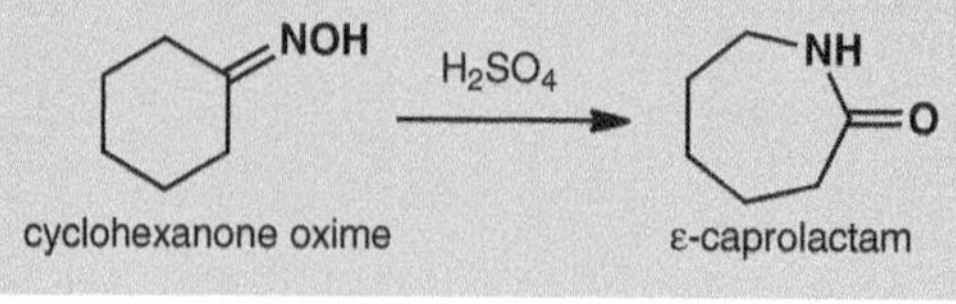

22.5 Rearrangements involving Carbenes and Nitrenes or their Precursors

22.5.1 Carbenes

Carbenes (R_2C:) are neutral divalent carbon compounds encountered in Chapter 15. They can be generated by the α-elimination of HX from a haloalkane, and may isomerize to an alkene by a 1,2-shift of alkyl or aryl to the divalent carbon (e.g. eqn 22.5) which has only six valence electrons and is, therefore, electron-deficient.

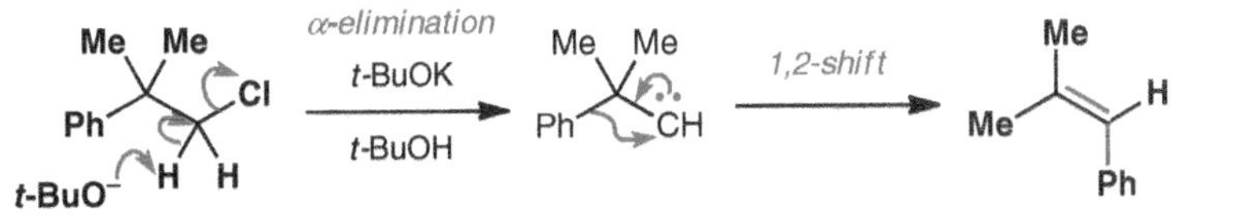

(22.5)

The reaction of eqn 22.5 might seem unlikely at first sight—bases/nucleophiles with alkyl halides normally give substitution or β-elimination reactions. Such reactions are not possible in this case, however, because of the nature of the reactants and solvent: steric hindrance inhibits an S_N2 reaction, the primary alkyl halide and a relatively nonpolar solvent prevent an $S_N1/E1$ mechanism, and there is no β H for an E2 mechanism.

An α-ketocarbene can be generated by the decomposition of a diazoketone, and its isomerization is sometimes called the Wolff rearrangement (Scheme 22.7).* The immediate product of the isomerization of the ketocarbene is a *ketene* which is normally rapidly trapped by a nucleophilic solvent to yield a carboxylic acid in water or an ester in an alcohol.

*The decomposition of the diazoketone in this sequence is usually catalysed by Ag_2O.

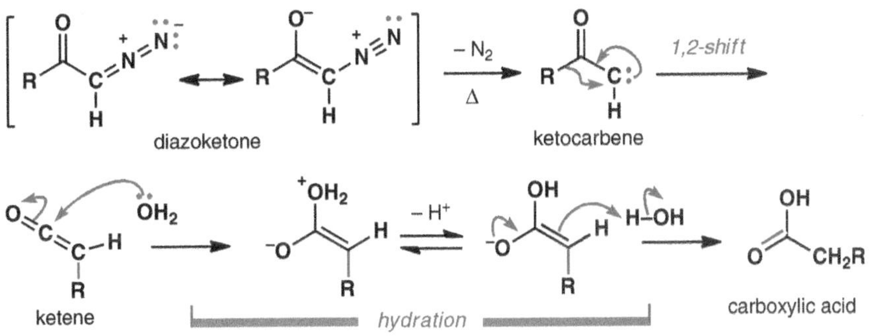

Scheme 22.7 The Wolff rearrangement in the conversion of a diazoketone to a carboxylic acid.

The departure of the N_2 and the 1,2-alkyl shift in Scheme 22.7 may be stepwise as shown, or concerted (i.e. bypassing the ketocarbene).

The prior preparation of the diazoketone is normally by reaction of an acyl halide with diazomethane, and application of this reaction starting with a carboxylic acid provides a method for the conversion of RCO_2H into RCH_2CO_2H with the addition of one carbon unit (Scheme 22.8).

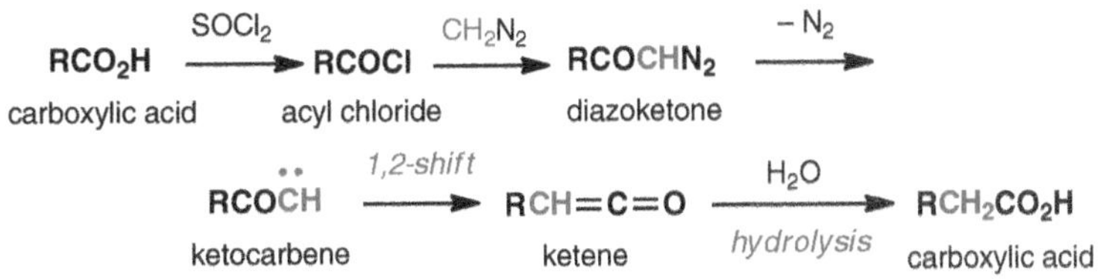

Scheme 22.8 The Arndt–Eistert synthesis.

This sequence is sometimes known as the Arndt–Eistert synthesis. The insertion of the CH_2 next to the carbonyl in this reaction may be compared to the corresponding insertions of an NH in the Beckmann rearrangement and an O in the Baeyer–Villiger reaction.

22.5.2 Nitrenes

The neutral singly bonded nitrogen species which is isoelectronic with a carbene is called a **nitrene** (the parent being HN); like carbenes, nitrenes are usually very short-lived reactive intermediates. One method of generating an acylnitrene, $RC(O)N$, begins

The overall transformation of the amide to the amine with the loss of the carbonyl group of the starting material as in Scheme 22.9 is sometimes called the Hofmann rearrangement. However, Hofmann's name is also associated with elimination reactions (see Chapter 13) so there is scope for confusion.

with the halogenation of a primary amide under basic conditions (the first line of Scheme 22.9). The initially formed *N*-haloamide undergoes an α-elimination to give an acylnitrene, rearrangement of which affords an isocyanate. Under the usual aqueous alkaline conditions, the isocyanate is immediately hydrolysed to give an amine, so the overall transformation is $RC(O)NH_2 \rightarrow RNH_2$ (Scheme 22.9).

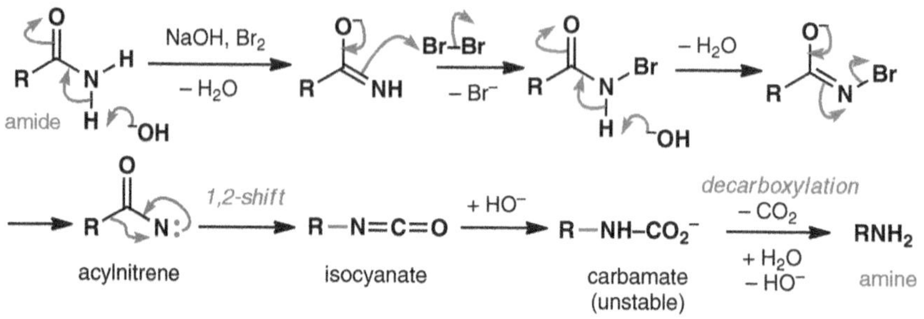

Scheme 22.9 Hofmann rearrangement.

The reaction of Scheme 22.10 is sometimes known as the Curtius reaction (or rearrangement); sodium azide was originally used but modified procedures using either tetrabutylammonium azide or diphenylphosphoryl azide are generally more effective. Note that acyl azides are potentially explosive.

The acylnitrene implicated in the Hofmann rearrangement can also be generated by thermal decomposition of an acyl azide which, in turn, may be prepared from an acyl chloride and an azide (Scheme 22.10). The rearrangement of the acylnitrene to an isocyanate in Schemes 22.9 and 22.10 is closely similar to the rearrangement of an acylcarbene to a ketene in the decomposition of a diazoketone (Scheme 22.8). Correspondingly, the departure of the nucleofuge and the 1,2-alkyl shift in Schemes 22.9 and 22.10 may be stepwise as shown, or concerted (i.e. bypassing the acylnitrene).

The isocyanate may be isolated in the Curtius reaction and subsequently hydrolysed via the unstable carbamic acid, $RNHCO_2H$, to give the amine RNH_2 plus CO_2. Alternatively, it can be converted into a stable urethane, e.g. $RNHCO_2Et$ using ethanol.

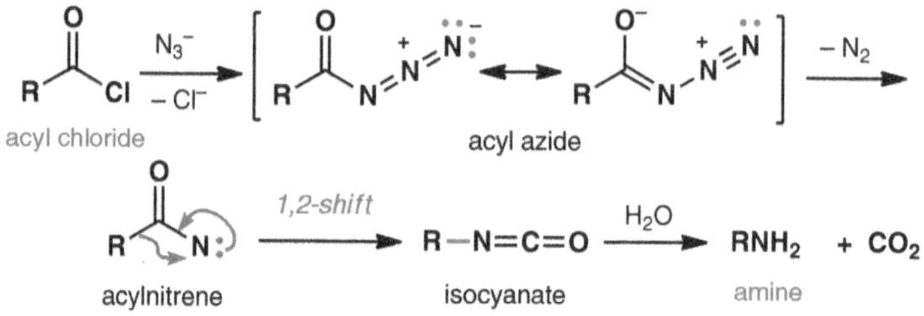

Scheme 22.10 The Curtius reaction.

Exercise 22.7

What is the product of the reaction of butanamide with Br_2 and NaOH in aqueous solution? Why does *N*-methylbutanamide not undergo a similar reaction?

22.6 Rearrangements involving Neighbouring Group Participation

22.6.1 Participation by groups with lone pairs

We saw in Chapter 12, Section 12.5, that some electrophilic substrates bearing a group with unshared electron pairs may undergo substitution reactions with enhanced rates attributable to neighbouring group participation (n.g.p.). The mechanisms of these reactions are heterolytic and stepwise through cyclic intermediates and give unrearranged substitution product with retention of configuration; frequently, rearranged substitution products are also obtained. The ethanolysis of 4-methoxypentyl *p*-bromobenzenesulfonate (brosylate, **1**),* for example, leads to considerable amounts of rearranged material, which implicates the mechanism shown in Scheme 22.11 involving a cyclic intermediate.

*The brosylate nucleofuge is like tosylate but with Br in place of the CH₃.

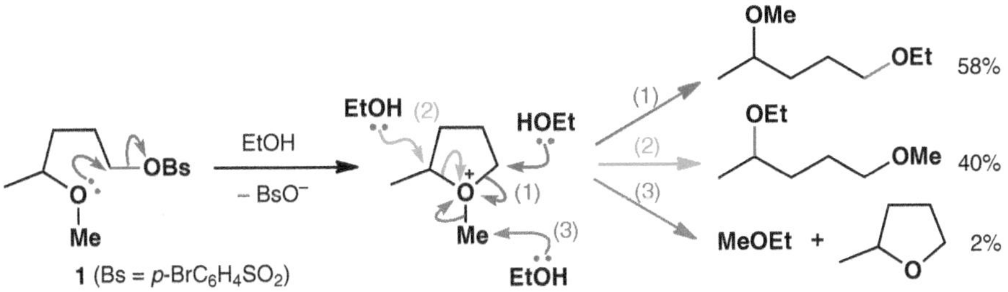

Scheme 22.11 An example of neighbouring group participation in a solvolytic reaction leading to rearrangement products.

The first-order rate law for the ethanolysis of **1** indicates a unimolecular rate-determining step and the rate constant is about 4000 times greater than the one for the ethanolysis of the pentyl analogue without the methoxy group. The mechanism proposed has S_N2 and S_N1 characteristics: concerted displacement of a nucleofuge by a nucleophile, and capture of a cationic intermediate by a (weakly) nucleophilic solvent in a stepwise mechanism.

The methoxy oxygen of **1** first displaces the nucleofuge from the electrophilic carbon to give a five-membered cyclic oxonium ion as an intermediate (an S_N2-type reaction with *intramolecular* rear-side displacement of the nucleofuge). In subsequent parallel steps, the cyclic oxonium ion gives a separate product upon capture by the solvent at each of the three electrophilic carbons next to the positive oxygen; two of the three products are rearranged.

The rearrangement of the five-membered cyclic amine to a six-membered cyclic isomer shown in eqn 22.6 is a non-solvolytic example. This mechanism is also like a double S_N2 reaction, the first one being intramolecular.

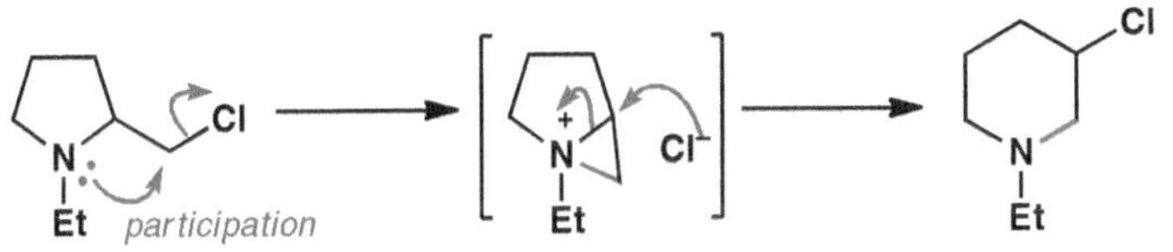

(22.6)

The reactant of eqn 22.6 has the molecular features which would allow a [1,2] sigmatropic rearrangement facilitated by the concerted participation of the lone pair on the N—compare the ionization step in eqn 22.6 with the ionization of the β-hydroxy-alkyl tosylate in eqn 22.3. However, this can be ruled out as a different product would have been formed, as shown.

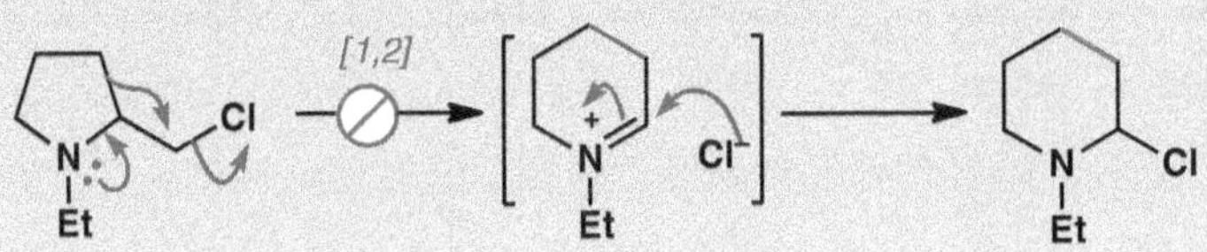

As we shall now see, intramolecular groups other than ones with lone pairs may also act as internal nucleophiles by assisting the departure of a nucleofuge (which leads to enhanced rates) and giving a cyclic intermediate (which leads to some rearranged products).

Exercise 22.8

Reactions of the following isomeric alcohols with hydrochloric acid give the same product. Explain the reactions and identify the product.

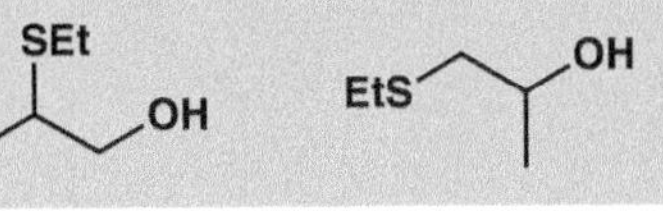

22.6.2 Participation by aryl groups

Both the isomeric tosylates bearing a β-phenyl group in Scheme 22.12 undergo solvolysis in ethanoic acid and give the same pair of products. In each case, one is the product of a simple substitution but with stereochemical retention of configuration, and the other is a rearranged product in which the carbons at the origin and the terminus of the migration are both of inverted configuration. These results are best accommodated by a mechanism involving a common benzenium ion intermediate formed by participation of the phenyl group in the ionization. A π electron pair of the benzene ring assists departure of the nucleofuge and forms a new σ bond in the three-membered ring of the benzenium ion intermediate; this common intermediate leads to the same product mixture regardless of its precursor. Because of the conjugative stabilization of the rate-determining TS leading to the benzenium ion intermediate, the 1,2-shift of the phenyl group is very easy compared with [1,2] sigmatropic rearrangements of hydrogen and alkyl groups. Participation by the Ph in these solvolyses leads to enhanced rates of reaction.

Formation of the rearranged products in Scheme 22.12 may be regarded as intramolecular *ipso* electrophilic substitutions (see Section 16.2 for electrophilic substitution, and Section 18.7 for the meaning of *ipso*).

The benzenium ion intermediate in Scheme 22.12 may be represented as

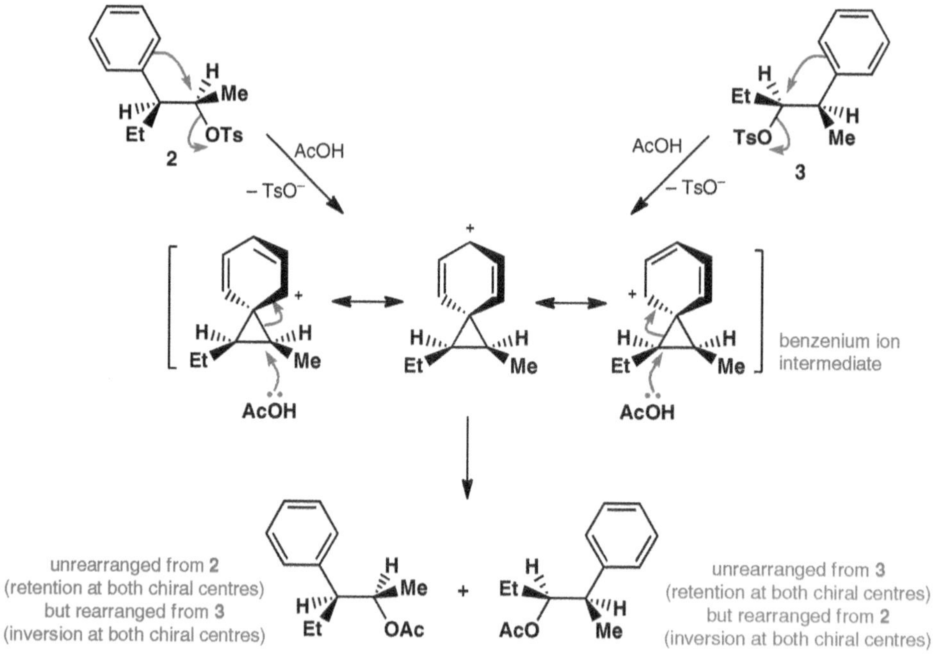

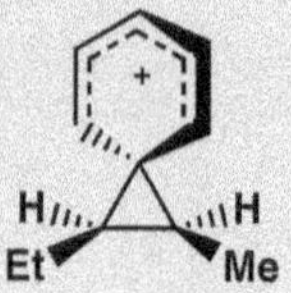

Scheme 22.12 Phenyl migration in the solvolysis of β-phenyl-substituted tosylates.

Exercise 22.9

Explain the following results for the solvolysis of stereoisomeric tosylates of 3-phenylbutan-2-ol in ethanoic acid.

(a) Both the (2R,3S) and (2S,3R) isomers give the same racemic ethanoate.
(b) The (2R,3R) isomer gives an optically active product.

22.6.3 Participation by carbon–carbon double bonds

Carbon–carbon double bonds are well known to participate in solvolytic reactions and some rate enhancements are huge, as shown by the solvolytic rate ratio for the tosylates in Figure 22.3.

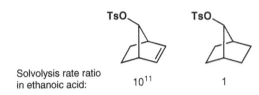

Solvolysis rate ratio in ethanoic acid: 10^{11} 1

Figure 22.3 Rate enhancement by intramolecular participation by a C=C bond in 7-norbornenyl tosylate.

However, only few known examples show all the three features we have identified as characteristic of n.g.p.—rate enhancement, unrearranged substitution with retention of configuration, and rearrangement product(s), although the steroidal tosylate (**4**) in Scheme 22.13 is one. In buffered ethanoic acid, it is about 100 times more reactive than cyclohexyl tosylate, a reasonable model compound. In MeOH, **4** gives either unrearranged substitution product (**6**) with retention of configuration, or the less stable rearranged substitution product (**7**) in MeOH buffered with potassium ethanoate. These results are an early example of how the C=C bond in **4** intervenes as a neighbouring group to facilitate departure of the nucleofuge and give a cationic intermediate (**5**) stabilized by interaction between the π electrons of the C=C bond and the 2p orbital of the C to which the tosylate had been bonded.* The intermediate is then trapped by the solvent at one of two sites with high regioselectivity according to the reaction conditions.

*The correct alignment for overlap of the interacting orbitals is easily possible because of the relatively rigid steroidal molecular framework.

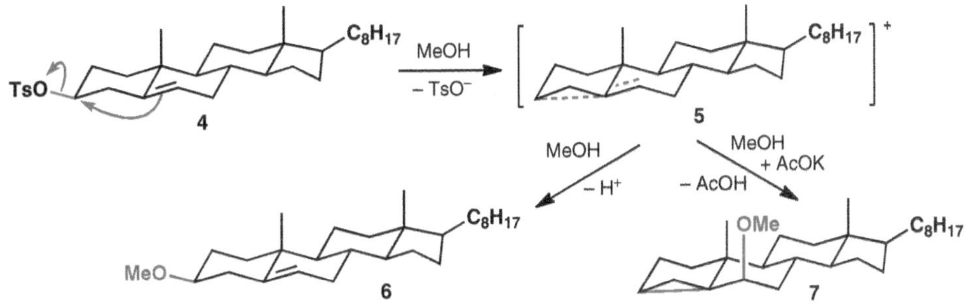

Scheme 22.13 Neighbouring group participation by a C=C bond leading to rearrangement or unrearranged substitution with retention of configuration in a solvolytic reaction.

In a separate experiment in ethanoic acid, the rearranged ether (**7**) gave the ethanoate corresponding to the unrearranged ether (**6**) of retained configuration via the same stabilized intermediate (**5**).

22.6.4 Participation by carbon–carbon σ bonds

For many years during the twentieth century, this was a controversial issue; immense effort was deployed to settle the matter which led to major instrumental developments in chemistry, e.g. in high resolution NMR spectroscopy (see Chapter 25). Computational chemistry was particularly influential in helping to resolve the controversy by supporting carbenium ion structural studies by NMR under highly non-nucleophilic conditions at very low temperatures, and deuterium kinetic isotope effect measurements.

Investigations of solvolytic reactions of some *exo*-norbornyl** compounds identified rate enhancements, unrearranged substitution with retention of configuration, and rearrangement products—all three characteristics of neighbouring group participation. The main product of the solvolysis of enantiomerically enriched *exo*-2-norbornyl brosylate (**8**) in ethanol is completely racemic *exo*-2-norbornyl ethyl ether (**9a** and **9b**), as shown in Scheme 22.14. We see that compound **9a** corresponds to unrearranged substitution with

**For the systematic names of bicyclic alkanes, see Panel 4.2.

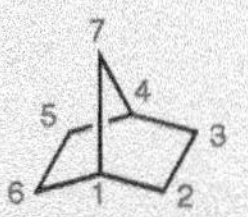

bicyclo[2.2.1]heptane (norbornane)

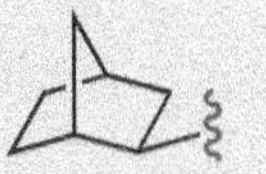

the *exo*-bicyclo[2.2.1]hept-2-yl (*exo*-2-norbornyl) group

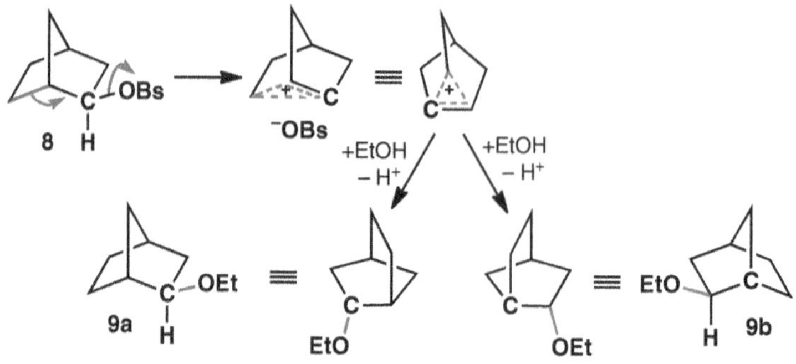

Scheme 22.14 Solvolysis of *exo*-2-norbornyl brosylate (**8**) in ethanol. (To help track the atom connectivity, the C originally bearing the nucleofuge is identified throughout.)

retention of configuration, and **9b** is its enantiomer, i.e. of inverted configuration but with a rearranged carbon framework.

The σ bond of **8** marked red in Scheme 22.14 provides rear-side nucleophilic assistance to the departure of the nucleofuge in the initial ionization (hence the rate enhancement); the symmetry of the intermediate carbenium ion leads to unrearranged substitution with retention of configuration and rearranged substitution of inverted configuration in equal amounts, i.e. racemic substitution product.

Example 22.3

Show how n.g.p. mechanisms account for the products obtained in the reactions of the diastereoisomeric tosylates shown below, and comment on the *anti/syn* rate ratio = 10^7.

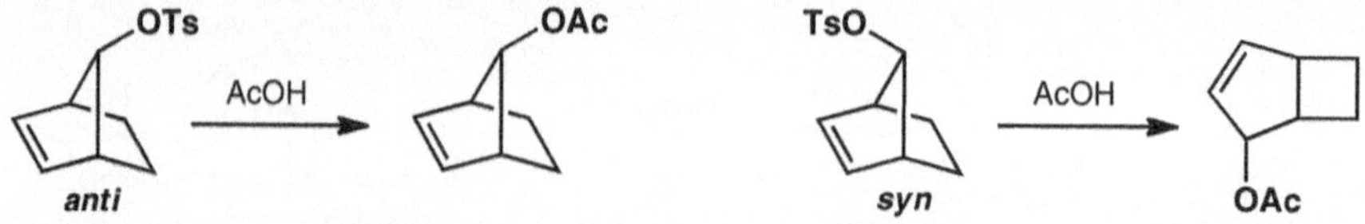

Solution
In the ionization of the *anti* isomer, the π electron pair from the C=C bond becomes the three-centre two-electron bond of the intermediate carbenium ion. The two electrons then re-form the C=C bond in the product when the nucleophile adds to the electrophilic carbon from which the nucleofuge had departed.

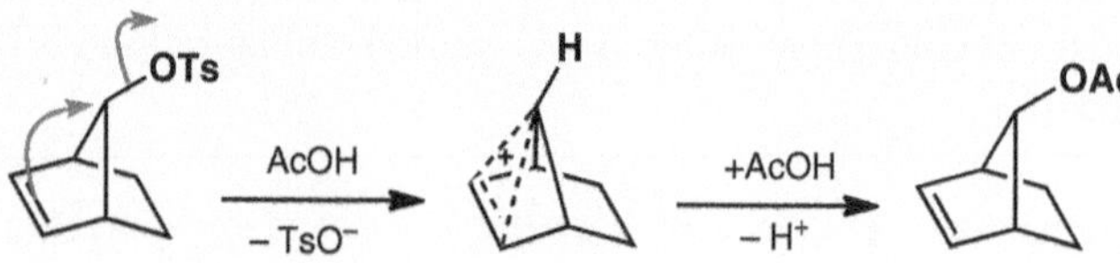

The intramolecular nucleophilic assistance in the *syn* isomer is provided by either of two equivalent rear-side carbon–carbon σ bonds (only one curly arrow is shown) and gives an allylic cation, capture of which by the solvent yields the (racemic) product. However, a carbon–carbon σ bond in the *syn* isomer is much less effective as a neighbouring group than the π bond in the *anti* isomer, hence the large rate ratio.

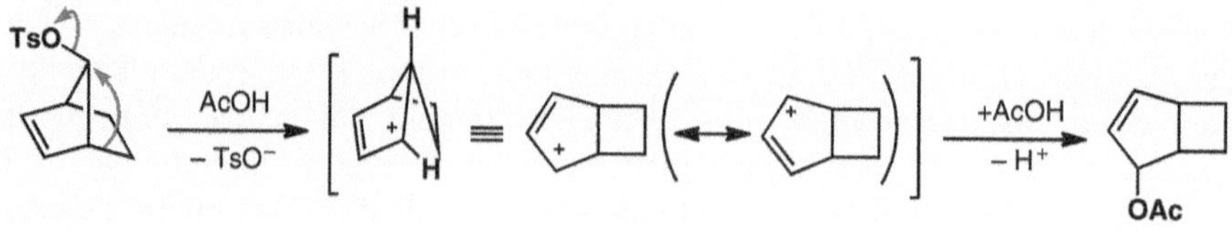

Exercise 22.10

Give a mechanism involving n.g.p. to account for the following reaction.

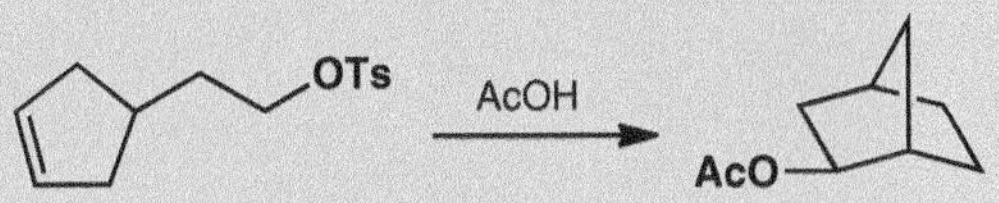

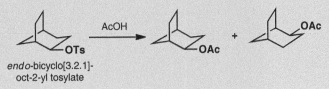

Exercise 22.11

Give a mechanism to explain why the main product of the reaction of enantiomerically enriched *endo*-bicyclo[3.2.1]oct-2-yl tosylate in ethanoic acid is racemic *endo*-bicyclo[3.2.1]oct-2-yl ethanoate.

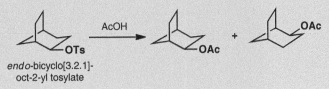

endo-bicyclo[3.2.1]-
oct-2-yl tosylate

Summary

- Rearrangements of carbenium ions and related intermediates by 1,2-H shifts, for example, may be regarded as [1,2] sigmatropic rearrangements via substituted cyclic 3-centre 2-electron transition structures.

- The pinacol rearrangement of a 1,2-diol to give an aldehyde or a ketone involves a 1,2-shift of H, alkyl, or aryl.

- Some 1,2-shifts are concerted with the departure of a nucleofuge and bypass the formation of unstable carbenium ion intermediates.

- Some α-hydroxy ketones rearrange with acid or base catalysis to give isomeric hydroxy ketones.

- The Baeyer–Villiger oxidation and Beckmann rearrangement can be used to insert O or NH, respectively, next to the carbonyl of a ketone; the Arndt–Eistert synthesis inserts CH_2 next to the carbonyl of a carboxylic acid.

- The Wolff rearrangement involves a 1,2-shift in a carbene (or its precursor); the Hofmann and Curtius rearrangements involve 1,2-shifts in nitrenes (or their precursors).

- Solvolytic reactions involving rear-side participation by groups with lone pairs, aryl groups, C=C bonds, and even C–C σ bonds are characterized by the intervention of cyclic intermediates and some or all of the following: rate enhancements, the formation of unrearranged substitution products with retention of configuration, and rearrangement products.

Problems

22.1 Give a mechanism for the formation of the *t*-butyl cation from 1-fluorobutane under extremely acidic non-nucleophilic conditions (SbF_5–SO_2ClF) at low temperatures.

22.2 Give a mechanism for the following hydrolytic rearrangement reaction induced by Ag^+.

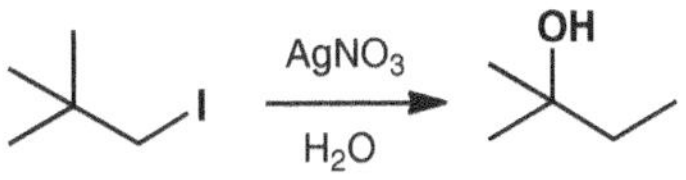

22.3 By giving reaction mechanisms, predict the structures of the rearrangement products when the following diols are treated with an acid.

(a)

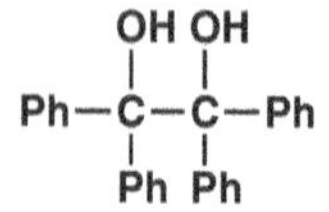

(b)

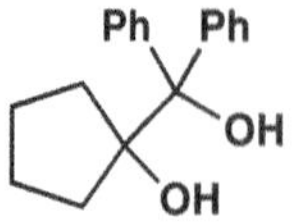

22.4 Give a mechanism for the following dehydration reaction.

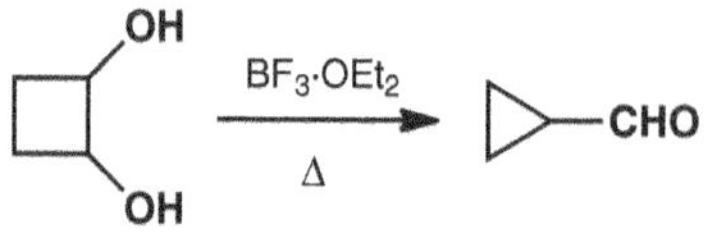

22.5 Give a mechanism for the following rearrangement.

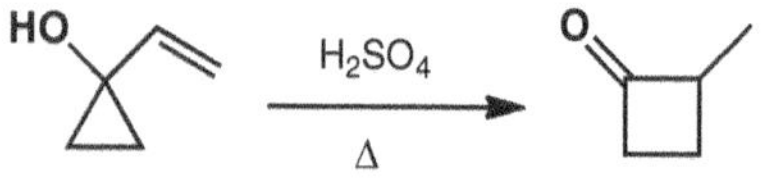

22.6 Give a mechanism for the following reaction.

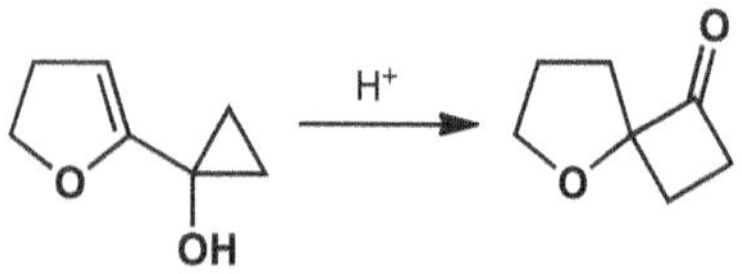

22.7 Give a mechanism for the following reaction.

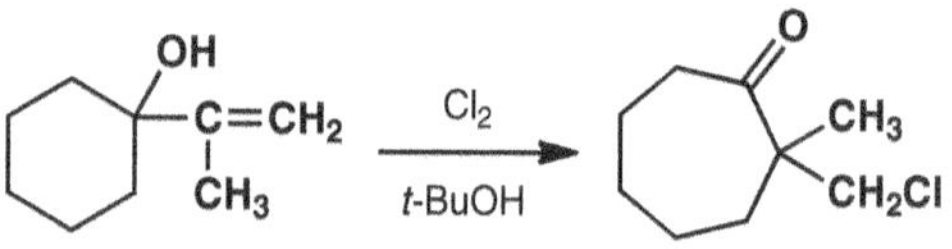

22.8 3-Deuterated 3-methylbutan-2-ol is transformed into 3-deuterated 2-methylbutan-2-ol without loss of deuterium upon treatment with dilute sulfuric acid. Give a mechanism for this transformation and explain how the deuterium label is fully retained.

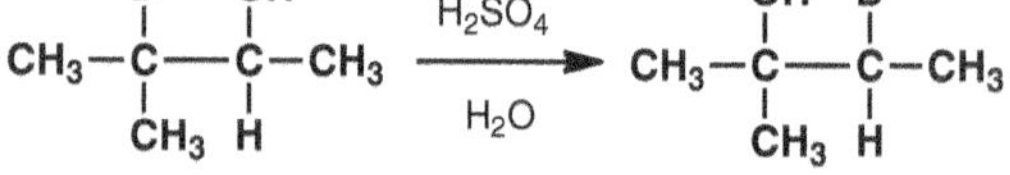

22.9 What are the main products of the following reactions?

(a)

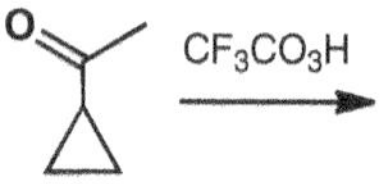

(b)

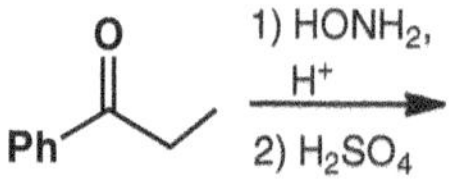

(c)

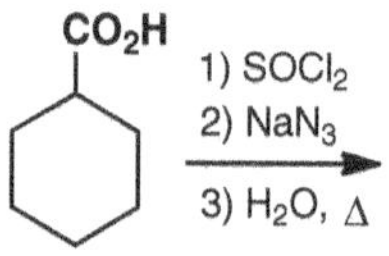

(d)

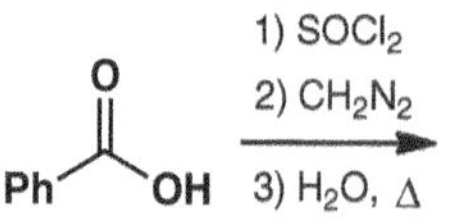

22.10 Give mechanisms for the following rearrangements of epoxides to aldehydes with Lewis acid catalysis.

(a)

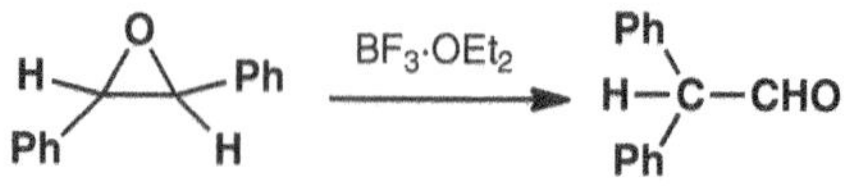

(b)

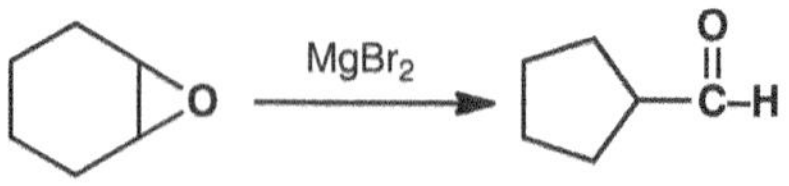

22.11 Give a mechanism for the following reaction.

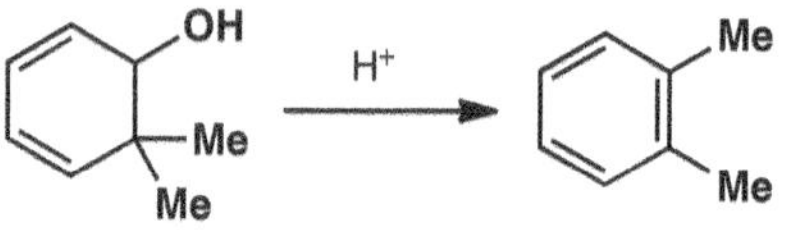

22.12 Propose a mechanism for the following reaction (sometimes called the *dienone-phenol rearrangement*) which involves a 1,2-shift to an electron-deficient carbon.

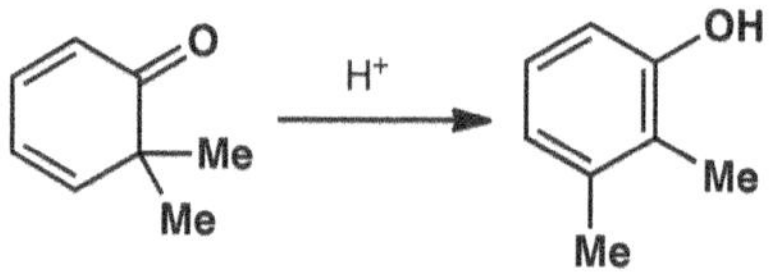

22.13 Compound **1** in which Y is a carboxylate nucleofuge is about 5×10^4 times more reactive than the cyclooctyl analogue **2** under hydrolytic conditions. In buffered aqueous propanone, **1** gives a rearrangement product (**3**) and the unrearranged alcohol (**4**), as shown. Give a mechanism involving n.g.p. which accounts for these results.

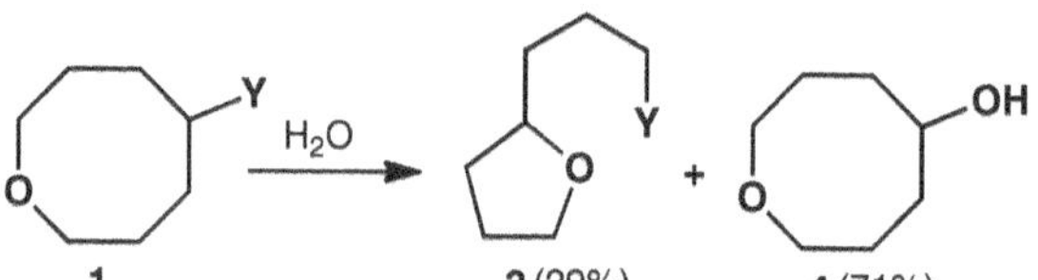

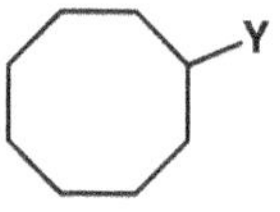

22.14 Bromination of (*E*)-6-benzylthiohex-2-ene gives benzyl bromide plus a five-membered cyclic product which rearranges to a six-membered cyclic isomer by heating.

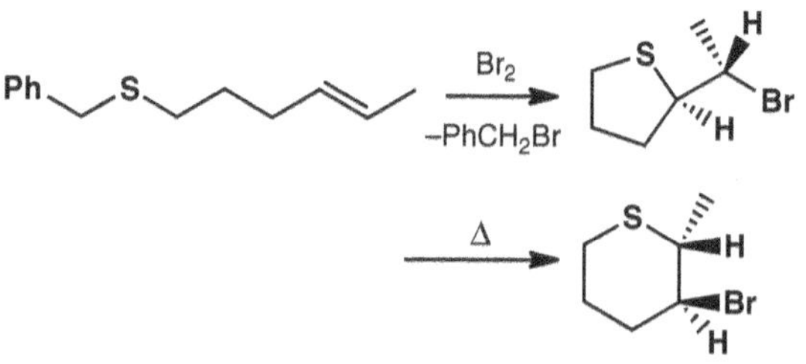

Give mechanisms for (a) the formation of the five-membered cyclic intermediate, and (b) the ensuing rearrangement.

22.15 The main products of the hydrolysis of bicyclo[2.2.2]oct-2-yl tosylate in aqueous propanone are bicyclo[2.2.2]octan-2-ol and *exo*-bicyclo[3.2.1]octan-2-ol, as shown below. Give a reaction mechanism and explain the formation of these two products in unequal amounts. On the basis of your mechanism, comment on the enantiomeric purity of each product if the reactant had been enantiomerically pure.

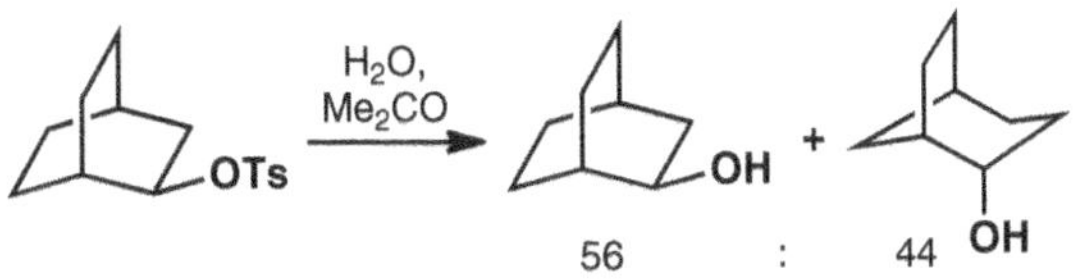

22.16 Give a mechanism which explains (i) why the following bicyclic alcohol is the product of the hydrolysis of the 4-nitrobenzoate ester of (*E*)-6-hydroxycyclodecene in aqueous propanone, and (ii) why the first-order rate constant is much (× 1500) greater than that for the analogous cyclodecyl ester.

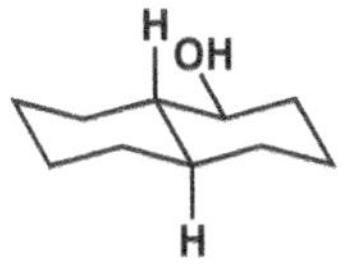

Supplementary Problems

22.17 When the following diastereoisomeric 2-aminocyclohexanols are diazotized with nitrous acid, the immediate decomposition of the resulting diazonium ions gives different carbonyl compounds as main products. Propose mechanisms for the decomposition of the diastereoisomeric diazonium ions and identify the carbonyl compounds formed.

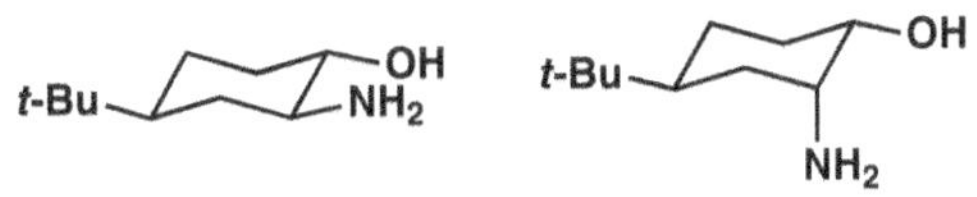

22.18 The isomeric α-chloro derivatives of phenylpropanone (phenylacetone) give the same methyl ester by reactions in basic methanol. Propose mechanisms for these examples of the so-called Favorskii rearrangement which proceed through a common cyclopropanone intermediate.

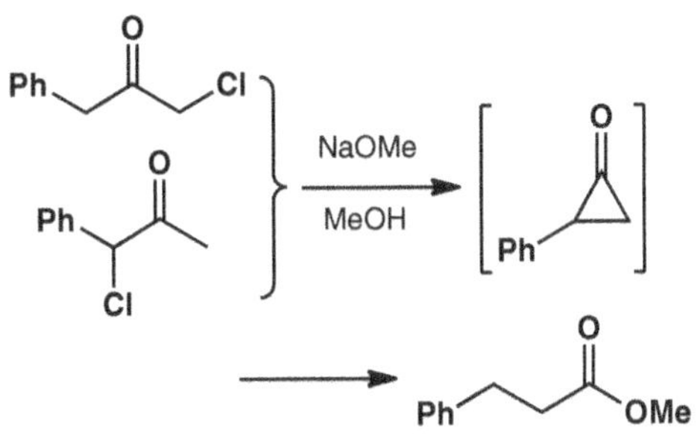

22.19 Give a mechanism for the following reaction which is an example of what is sometimes called the Schmidt rearrangement.

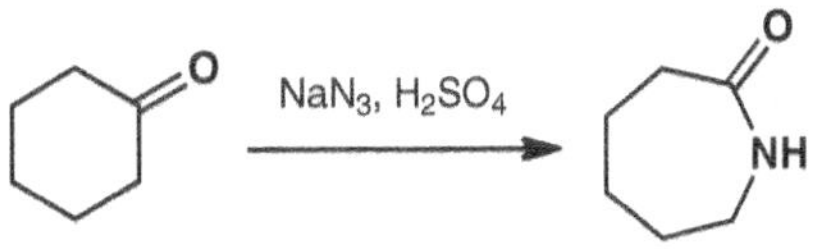

22.20 When a mixture of two oximes, each with a tertiary alkyl group, is treated with an acid, a mixture of amides including alkyl-crossed products is obtained. The reaction is sometimes called the Beckmann fragmentation. What makes these oximes different from those which undergo the usual (intramolecular) Beckmann rearrangement?

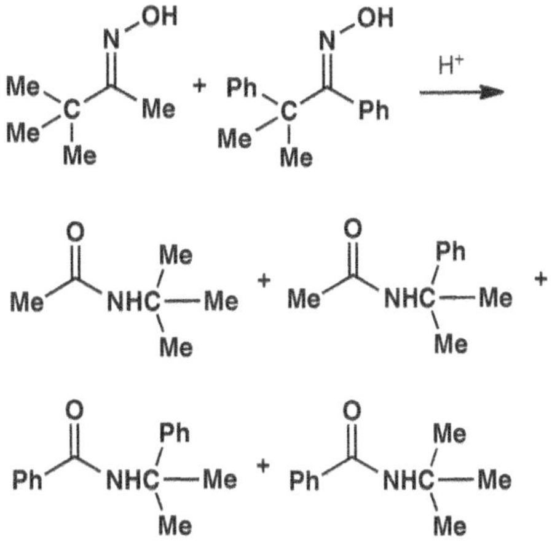

23 Organic Synthesis

Related topics we have already covered:

- **Functional group interconversions** (see Appendices 2 and 3 for a summary)
- **Carbon–carbon bond forming reactions** (see Appendix 4 for a summary)
- **Planning syntheses of alcohols using Grignard reactions** (Section 10.5)
- **Syntheses of substituted benzenes** (Section 16.7)

Topics we shall cover in this chapter:

- **Reactions used in organic synthesis**
- **Retrosynthetic analysis**
- **Synthons and the corresponding reagents**
- **Reaction selectivity**
- **Protection–deprotection**
- **Linear and convergent strategies**
- **Examples of organic synthesis**

Organic synthesis is the directed preparation of a sought compound usually, but not invariably, with some structural complexity from simple, readily available compounds. A typical synthetic target might be a novel organic compound of theoretical importance, or a useful compound which can be obtained only in minute amounts from natural sources, or is not found in nature at all.

When the synthesis of an organic compound is achieved by a sequence of reactions, the overall method should be economical and environmentally benign. To design an efficient method for the synthesis of a complex target compound, we work backwards from its molecular structure step by step: this strategy was introduced briefly in Section 10.5 for the synthesis of alcohols by Grignard reactions. In this chapter, we shall develop the concept of **retrosynthesis** further and see how to identify routes for syntheses of target compounds by applying organic reactions we have encountered in previous chapters.

Organic synthesis is one of the ultimate goals of organic chemistry and provides limitless opportunities to challenge the scientific imagination. In the past, particular challenges have led to the discovery of new classes of compounds, new chemical reactions, and the development of new strategies for the construction of complex molecules. There is occasionally an aesthetic aspect to the work of the best organic chemists as they strive to match the efficiency (and elegance) of nature in the synthesis of natural products.

23.1 Reactions used in Organic Synthesis

An organic synthesis starts with a plan of the route as a sequence of organic reactions. For the synthesis of a complex molecule, they include **carbon–carbon bond formations** and **functional group interconversions** (FGIs). Reactions we have covered in previous chapters are summarized at the end of the book for reference under *Principal reactions of functional groups* (Appendix 2), *Syntheses of classes of compounds* (Appendix 3), and *Reactions for the formation of carbon–carbon bonds* (Appendix 4).

In planning a synthesis, the construction of the carbon framework is particularly important because it has a major influence on the choice of starting materials. The main reactions used for C–C bond formation fall into two broad categories. First, we have reactions between carbon nucleophiles and carbon electrophiles; the former include carbanions generated at the carbon α to a carbonyl group, organometallic compounds, and aromatic compounds; the latter include carbonyl compounds, haloalkanes, electrophilic alkenes, and carbenium ions generated from haloalkanes, alcohols, or alkenes. Secondly, there are concerted cycloadditions and sigmatropic rearrangements.

In order to identify the best reaction for a particular C–C bond formation, it is important that we appreciate the patterns which are structurally characteristic of each C–C bond-forming reaction: these patterns are found in Appendix 4. Remember also that functional group interconversions sometimes help us to find alternative methods for C–C bond formation.

23.2 Planning Organic Syntheses: Retrosynthetic Analysis

In designing an organic synthesis, we work backwards step by step starting from the **target molecule** (the molecule to be made) to appropriate starting materials. This analytical

---

Panel 23.1 Recent C–C bond-forming reactions using catalytic organometallic complexes

Complementing conventional methodologies (but not included in our examples), recent procedures for C–C bond formation using transition metal complexes have opened up new strategies for molecular construction. In particular, palladium-catalysed **cross-coupling reactions** allow C–C bond formation involving sp² hybridized C (the development of these reactions led to the award of the Nobel Prize in Chemistry in 2010 to R.F. Heck, E. Negishi, and A. Suzuki). The Heck reaction of a haloarene (or haloalkene) with an alkene catalysed by a Pd(0) complex gives an alkenylarene (or diene); this is also called the Mizoroki–Heck reaction to recognize the Japanese chemist who first discovered it, T. Mizoroki (1933–1980).

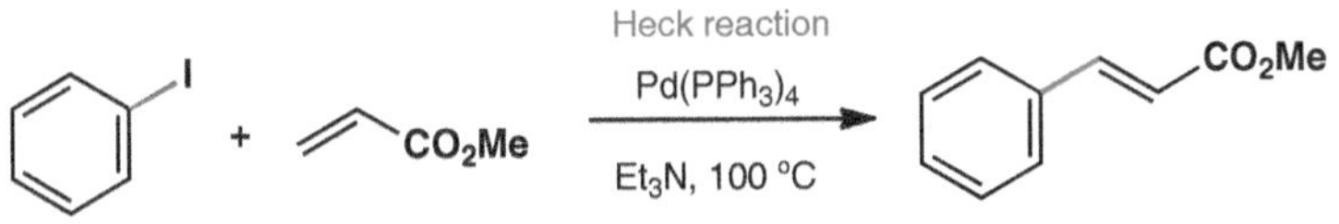

Cross-coupling reactions between organic halides and organometallic compounds of various metals (or metalloids) such as Zn, Sn, and B also proceed with catalysis by a Pd complex. An example is the Suzuki–Miyaura coupling between an alkenylboronate and a haloalkene illustrated below. Both reactants are readily available and stable, and their configurations are maintained in the coupling. The reaction is also compatible with most functional groups.

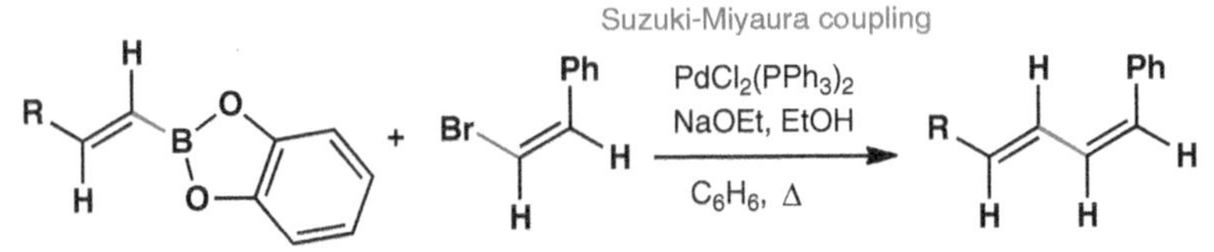

Alkene metathesis is a reaction catalysed by a metal-carbene complex (a so-called Grubbs complex) which allows ring closure of an α,ω-diene with the formation of a new C=C bond and loss of ethene. This reaction is also compatible with many functional groups, and its development led to the award of the 2005 Nobel Prize in Chemistry to Y. Chauvin, R.H. Grubbs, and R.R Schrock.

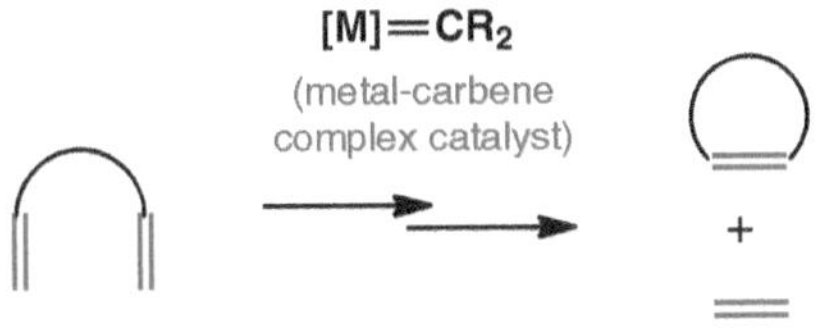

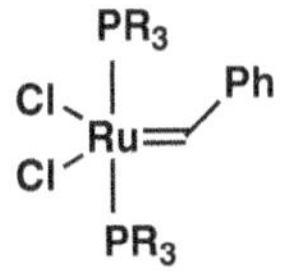

(Kindly supplied by E.J. Corey.)

E.J. Corey (1928–), Professor of Chemistry, Harvard University, USA, coined the term 'retrosynthetic analysis' and developed the concept; he was awarded the Nobel Prize in Chemistry in 1990 for his contributions to organic synthesis.

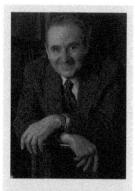

approach is the opposite of describing the conversion of the starting materials into the final product through the intermediate compounds one by one in the sequence of the organic reactions deployed. Starting with the target molecule, we identify a (potential) precursor by disconnecting a bond or identifying a transformation leading to a functional group of the target molecule. From the immediate precursor of the target molecule, we then work back another step, and so on until we arrive at readily available starting materials. This process (introduced in Section 10.5) is called **retrosynthetic analysis**, and each step in a retrosynthesis is represented by a double-shafted arrow (⇒) as exemplified for the Robinson annulation in Scheme 23.1, a two-step synthesis encountered in Chapter 18.

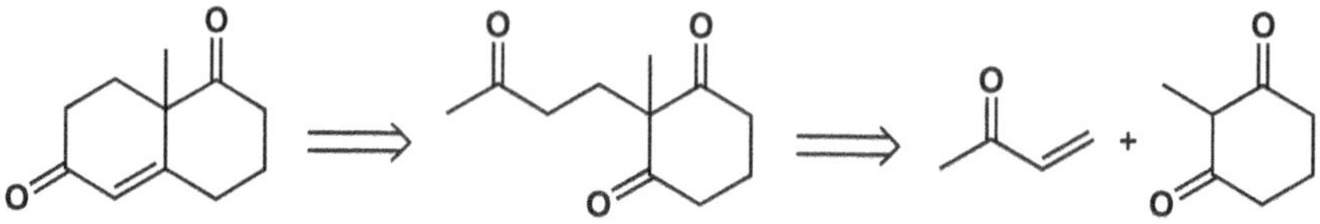

Scheme 23.1 Retrosynthetic steps for the Robinson annulation.

23.2.1 Disconnections: synthons and the corresponding reagents

Choices of **disconnections** of C–C bonds to give immediate precursors in all steps of a multistep synthesis are important in the retrosynthetic analysis of any large molecule. For both steps in Scheme 23.1 (and examples in Chapter 10), a retrosynthetic arrow connects a product molecule with its immediate precursors—real chemical compounds. More generally, however, the disconnection arrows lead backwards to structures of *hypothetical* ions (or radicals) derived *directly* by the heterolytic (or homolytic) disconnection, and these are called **synthons**; eqn 23.1 in Scheme 23.2 is an example of a disconnection

Disconnection:

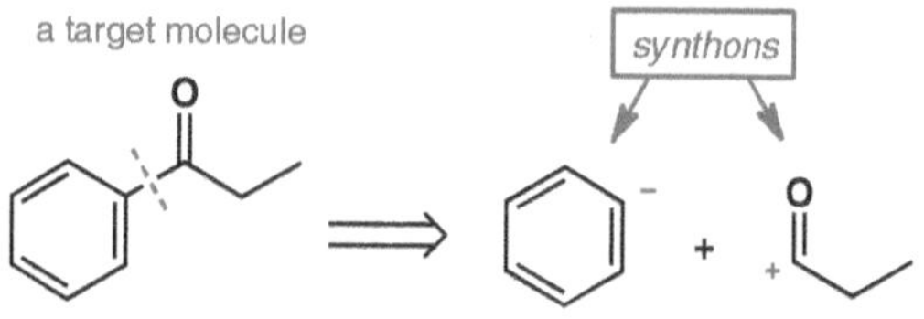

(23.1)

Synthetic reaction 1:

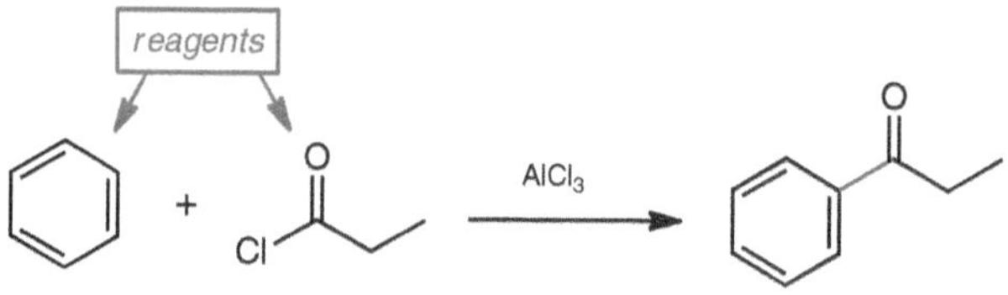

(23.2)

Synthetic reaction 2:

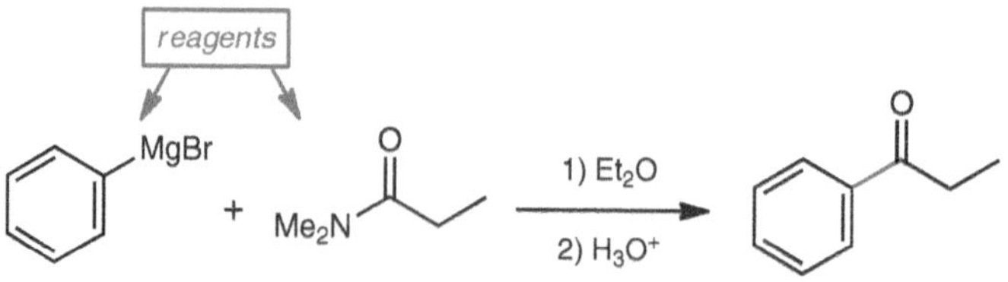

(23.3)

Scheme 23.2 Synthons following a disconnection and alternative possible corresponding reagents and synthetic reactions.

to a pair of synthons. From the synthons, we identify possible corresponding reagents (sometimes called *synthetic equivalents*), and the number of alternative reagents identifiable as corresponding to a single synthon depends upon one's knowledge of organic chemistry. In the example in Scheme 23.2, the heterolytic disconnection at the phenyl group of the ketone gives the phenyl anion and an acyl cation as synthons. This single disconnection (an exercise in scientific imagination) provides a pair of synthons which lead to alternative pairs of reagents for practical synthetic reactions (eqns 23.2 and 23.3), based upon the content of earlier chapters.

> Give a synthesis corresponding to the retrosynthesis in Scheme 23.1.

Exercise 23.1

The disconnection in eqn 23.1 provides the phenyl anion and acyl cation synthons but it could be in the opposite heterolytic sense to give a phenyl cation and an acyl anion, or homolytic (to give two radicals). Furthermore, the disconnection could have been on the alkyl side of the carbonyl; in this case also, we can identify alternative pairs of synthons and, for each, various synthetic reactions will be possible. It follows that, for a single target molecule of only modest complexity, there will be more than a few possible disconnections, and each of them may be heterolytic (with either polarity) or homolytic; in all, there could be many possible retrosynthetic steps with each corresponding to several synthetic reactions. As a result, the range of possibilities could be represented as a multi-branched tree of alternative synthetic routes, and the job of the aspiring synthetic chemist is to develop the skills needed to eliminate the impracticable ones and try to identify the most efficient. Knowledge of common synthons and their corresponding reagents will help to minimize the bewilderment arising from multiple disconnection patterns, and Table 23.1 summarizes some of them.

Table 23.1

Representative synthons and possible corresponding reagents

Synthon	Reagent	Synthon	Reagent
R^+	R–Y (Y = halogens, OSO_2R', OH)	R^-	R-MgX, R-Li
		Ar^-	Ar-MgX, Ar-Li, Ar-H

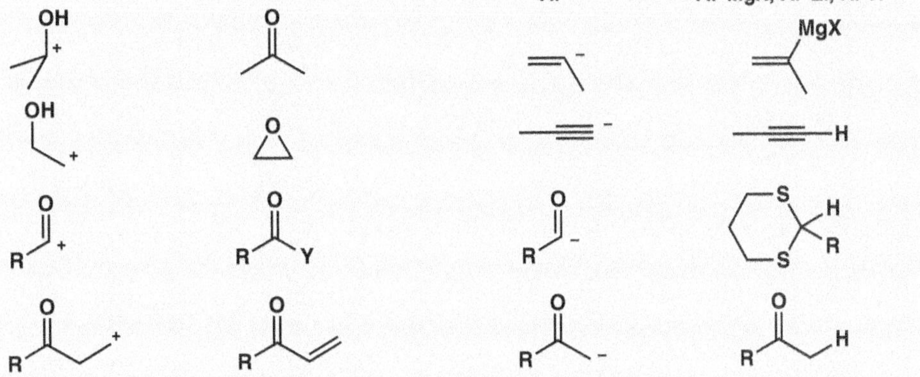

> Give an alternative disconnection on the alkyl side of the carbonyl group in the target molecule in Scheme 23.2, and propose synthetic reactions corresponding to the synthons identified.

Exercise 23.2

23.2.2 Exploiting functional group interconversions: synthesis of a representative secondary alcohol

Let us consider the synthesis of 1-phenylpentan-2-ol as another example. The three dis-connections *a–c* are readily identified (Scheme 23.3), and Grignard reactions in eqns 23.4–23.6 are possible corresponding preparations. The choice of the best (however that may be defined) from these possibilities depends on the scope and versatility of the reactions, and known peculiarities of particular reagents. For example, the Grignard reaction with the benzyl reagent in eqn 23.5 tends to give side reactions and would not be a good choice.

Scheme 23.3
Disconnections of
1-phenylpentan-2-ol.

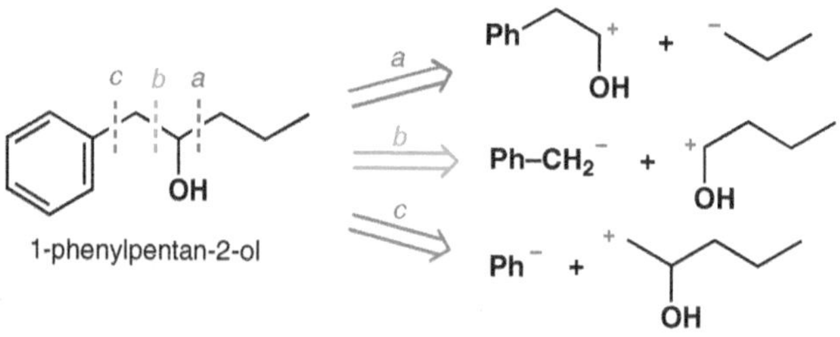

Disconnection *a*:

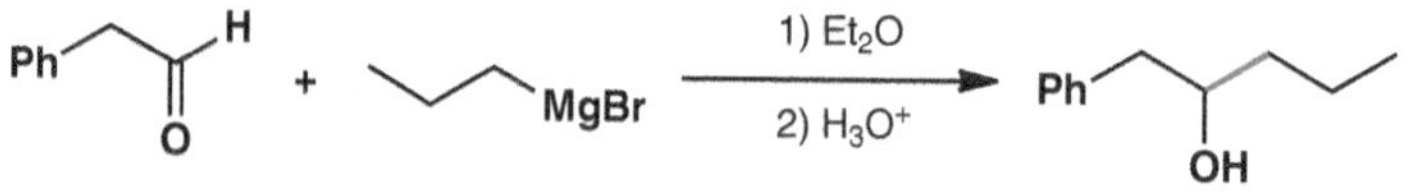

(23.4)

Disconnection *b*:

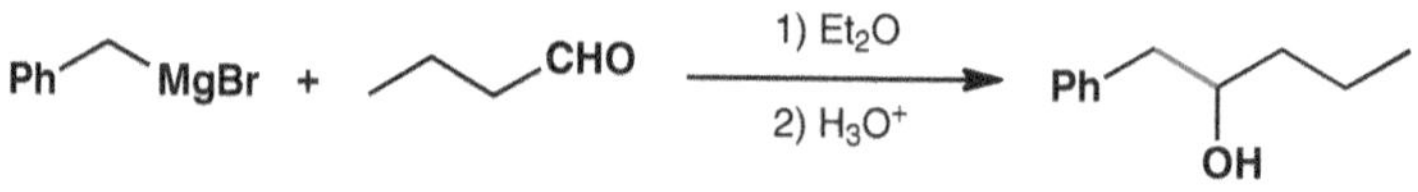

(23.5)

Disconnection *c*:

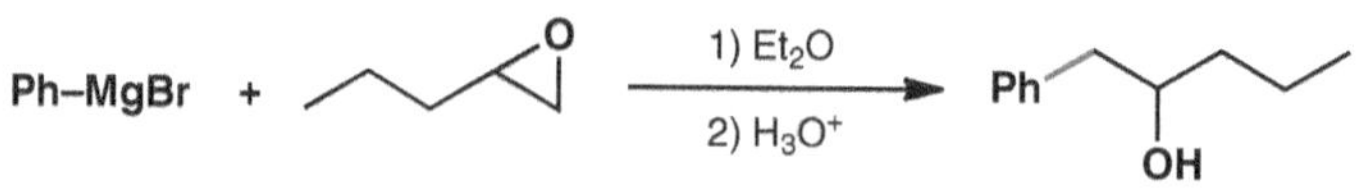

(23.6)

Secondary alcohols are readily obtained from ketones by reduction, so the initial tar-get may be replaced by a ketone as a secondary target molecule. In other words, the functional group interconversion (FGI) from ketone to secondary alcohol is easy, so the synthetic challenge becomes the ketone, 1-phenylpentan-2-one (Scheme 23.4).

Scheme 23.4
A functional group
interconversion to modify
the target molecule and new
disconnections of bonds *d* and
e of the secondary target.

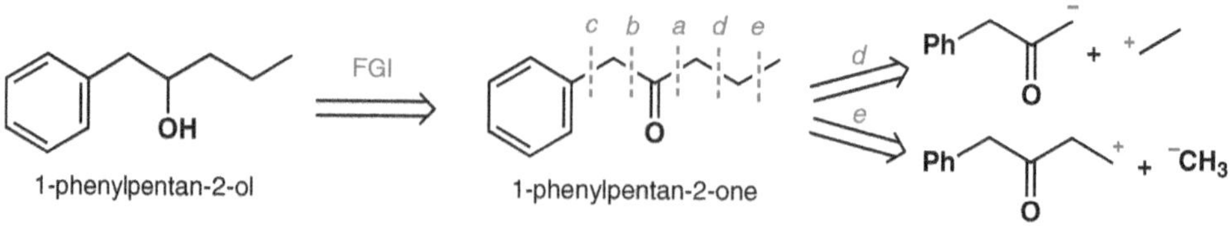

For 1-phenylpentan-2-one, disconnections *d* and *e* become feasible in addition to disconnections *a–c*, and possible synthetic reactions for them are given in eqns 23.7 and 23.8. In reaction 23.7, a hydrolysis and a decarboxylation are required after the Et group has been introduced in an S_N2 reaction between an enolate anion and bromoethane. Conjugate addition of the Grignard reagent in reaction 23.8 is promoted by using a Cu(I) compound as catalyst (see Sub-section 18.1.3).

Disconnection *d*:

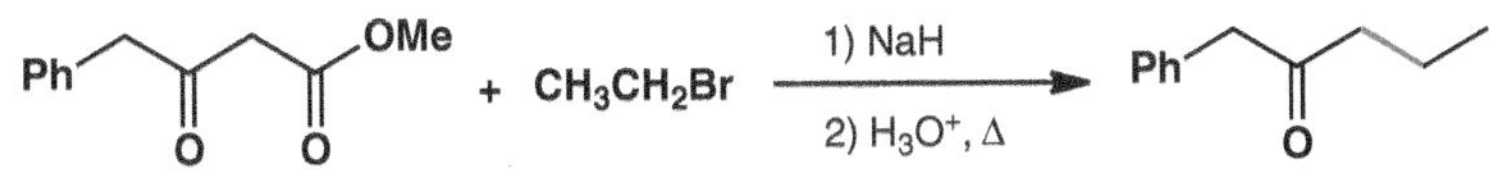

$$(23.7)$$

Disconnection *e*:

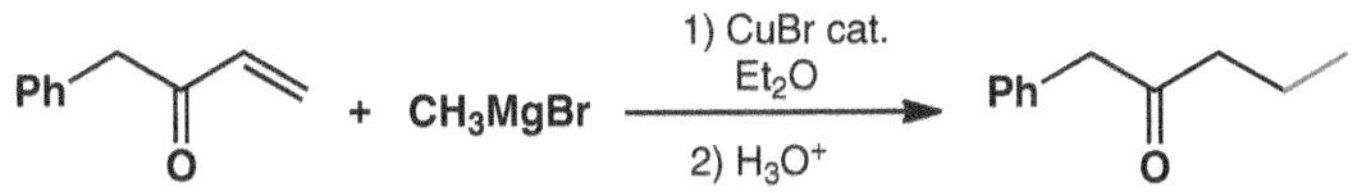

$$(23.8)$$

(a) Give a mechanism for reaction 23.7. (b) Compared with the β-keto ester in reaction 23.7, why is the simpler phenylpropanone not an efficient reagent corresponding to the nucleophilic synthon of disconnection *d*?

Exercise 23.3

Pairs of synthons of opposite polarities are given in Scheme 23.5 for disconnection *a* of 1-phenylpentan2-one.

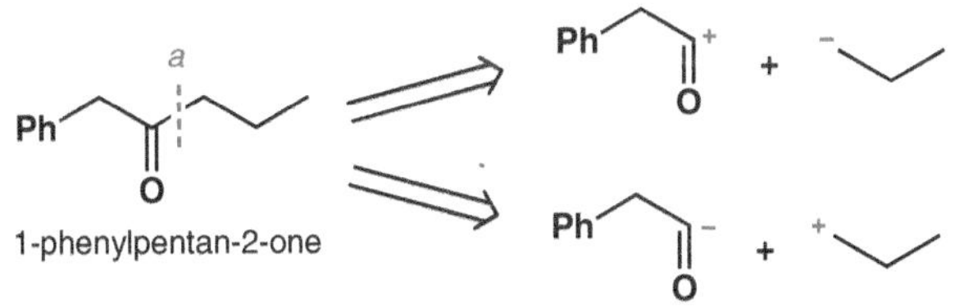

Scheme 23.5
Disconnection *a* of
1-phenylpentan-2-one.

The first pair of synthons lead to the Grignard reaction of eqn 23.9, while the second pair indicate the reliable reaction in eqn 23.10.

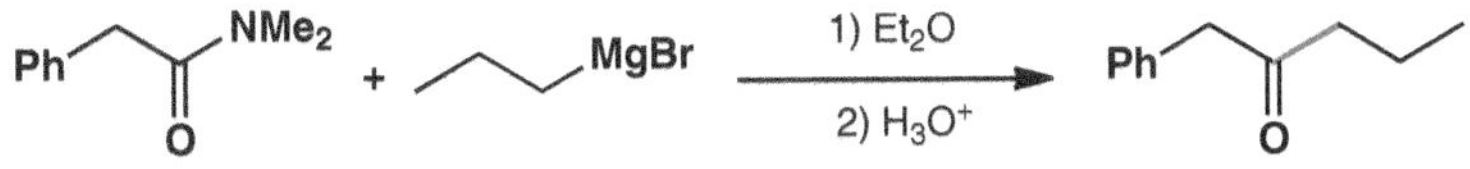

$$(23.9)$$

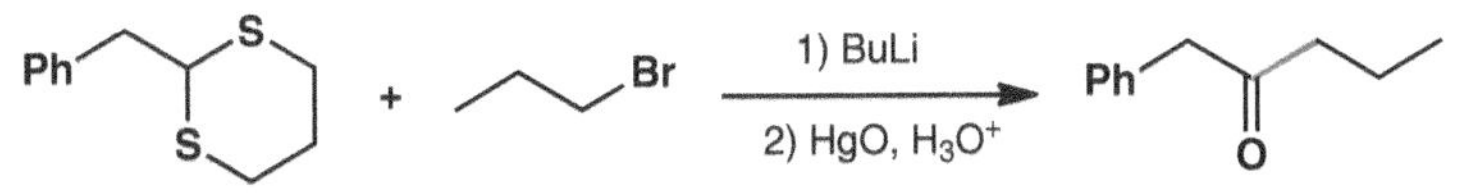

$$(23.10)$$

The dithioacetal (dithiane, $pK_a \sim 31$) in eqn 23.10 can be transformed easily into a carbanion which is stabilized by the sulfur atoms (Sub-section 14.6.3, p. 307); this provides a reagent for an anionic carbonyl synthon because a dithioacetal can be readily obtained from an aldehyde (Sub-section 8.4.3, p.180), and easily hydrolysed back to the carbonyl compound (Sub-section 10.5.3, p.222). Note that, while a carbonyl compound acts as a typical electrophile, the dithioacetal gives a nucleophile upon deprotonation. This reversal of polarity is sometimes described by the German word '**Umpolung**', and is illustrated by a comparison of reactions 23.10 and 23.4 (formation of a Grignard reagent from a haloalkane is also an Umpolung).

Exercise 23.4

Show a reaction sequence with mechanisms for a preparation from phenylethanal of the dithio-acetal which is used in eqn 23.10, and give a mechanism for the reaction of eqn 23.10.

Exercise 23.5

Give the reactions of the reagents corresponding to the alternative pairs of synthons of opposite polarities following disconnection *b* of 1-phenylpentan-2-one in Scheme 23.4.

23.2.3 Disconnections at heteroatoms

When a molecule consists of two carbon residues connected by a heteroatom, the first disconnection is usually made at the heteroatom. The decision to make the disconnection on one side or the other is important and usually depends on the availabilities of good corresponding synthetic reactions. In Scheme 23.6, for example, disconnection *b* of the ether does not lead to a good convenient synthetic reaction.

Scheme 23.6 Alternative disconnections of an ether but only one feasible retrosynthetic analysis.

(no reliable reagents)

reagents for S_N2 reaction

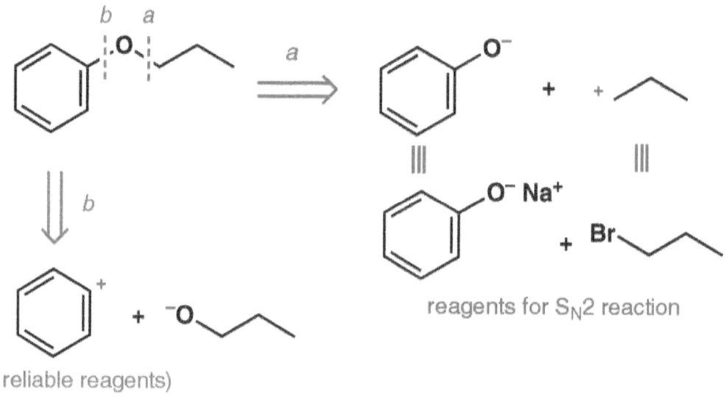

For the same reason, the disconnection of the sulfide in Scheme 23.7 can be identified as the better alternative and leads to an S_N2 reaction for the synthetic process. For this thioether, there are no convenient synthetic reagents corresponding to the synthons obtained by disconnection on the other side of the S, regardless of the polarity of the disconnection.

Scheme 23.7 The better disconnection of a sulfide leading to a viable retrosynthetic analysis.

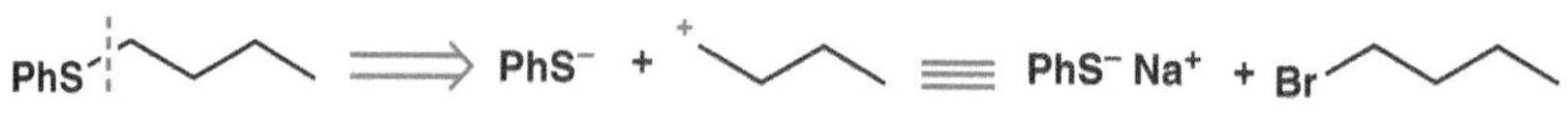

A second functional group near the principal one often influences the choice of a reagent. The sulfide in Scheme 23.8 contains a carbonyl group (the compound is a β-thio

Scheme 23.8 Retrosynthetic analysis of a β-thio ketone.

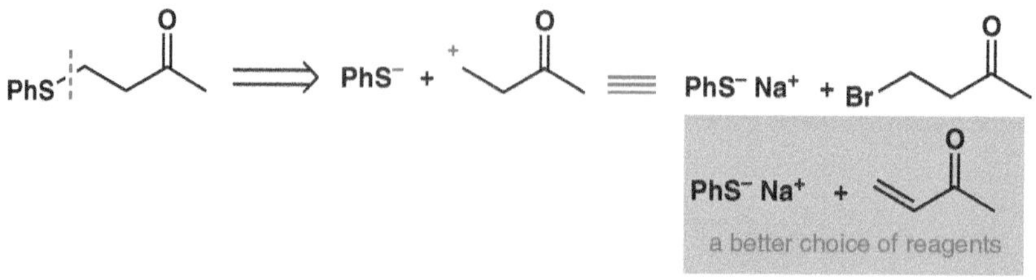

a better choice of reagents

ketone) and could be disconnected as shown. However, the carbonyl in the electrophilic synthon allows a better alternative electrophilic reagent than a halide (see Table 23.1); the conjugate addition of the thiolate to an α,β-unsaturated carbonyl compound is a superior synthetic reaction corresponding to the disconnection shown.

The 1,2-difunctional component of the structure in Scheme 23.9 is another fairly common feature: an OH group with a nucleophilic residue (an amino group in this case) on the β carbon. A useful reagent for the electrophilic synthon bearing the hydroxy group is an epoxide (Table 23.1) which yields the 1,2-difunctional compound upon ring opening by a nucleophilic reagent—here, the amine.

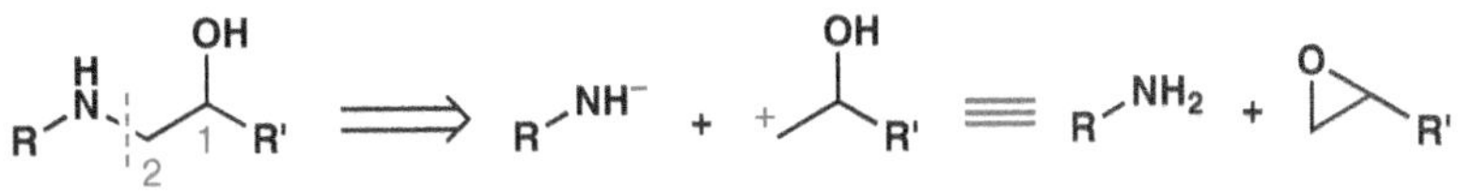

Scheme 23.9
Retrosynthesis of a
1,2-difunctional compound.

Propose a retrosynthetic analysis and the corresponding synthetic reaction for each of the following compounds.

Exercise 23.6

(a) 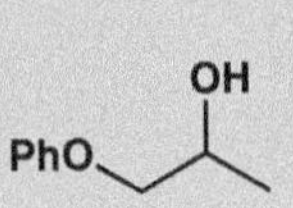 PhO

(b)

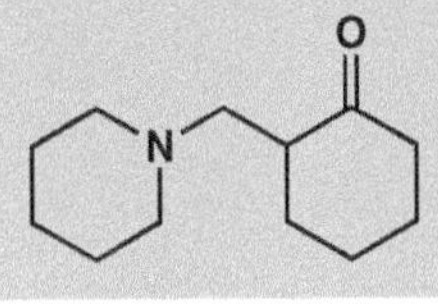

23.2.4 Multiple functionalities which lead to standard disconnections

The examples in Sub-sections 23.2.1 and 23.2.2 included simple heterolytic C–C disconnections next to a single isolated functional group. We then saw in the preceding Sub-section 23.2.3 that the choice of a reagent can be influenced by a second functional group when the disconnection is made at a heteroatom. We shall now see how some multifunctional (but fairly common) molecular features lead to standard disconnection strategies and familiar C–C bond forming reactions.

a. 1,3-Dioxygenated compounds

In Chapter 17, we saw reactions of enolates which give products containing 1,3-dioxygenated functionalities, e.g. the aldol reaction and Claisen condensation (Scheme 23.10). Consequently, by recognizing such a pattern, we should be led to a disconnection which indicates an aldol reaction or Claisen condensation.

reagents for aldol reaction

reagents for Claisen condensation

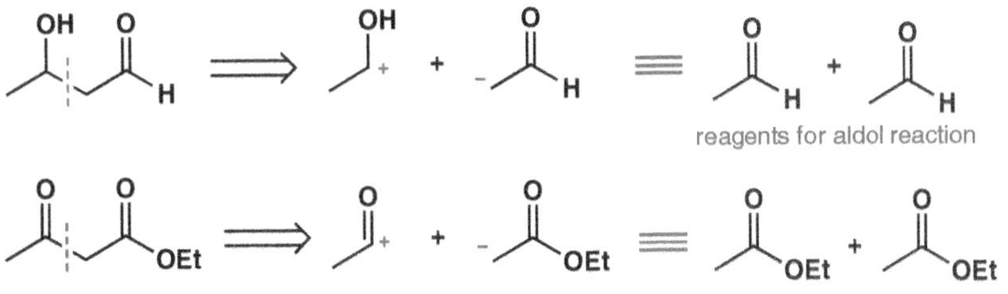

Scheme 23.10
1,3-Dioxygenated
functionalities with
disconnections leading to the
aldol reaction and Claisen
condensation.

Examples of disconnections indicating initial aldol reactions are given in Schemes 23.11 and 23.12. The first one is intramolecular and straightforward as the target molecule is a β-hydroxy ketone.

Scheme 23.11 An intramolecular example of the retrosynthetic analysis of a 1,3-dioxygenated compound.

Disconnection:

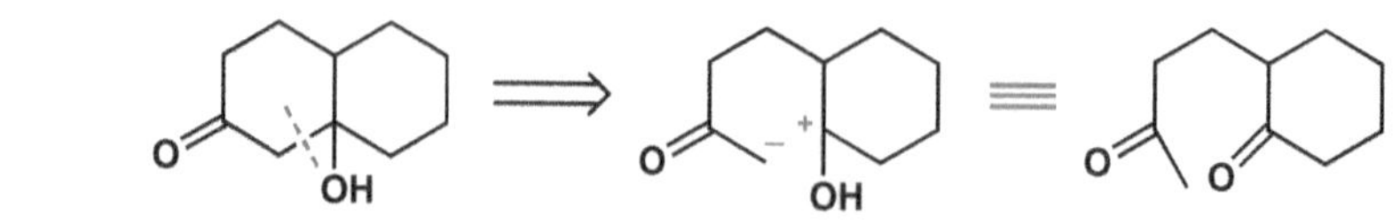

Synthetic reaction:

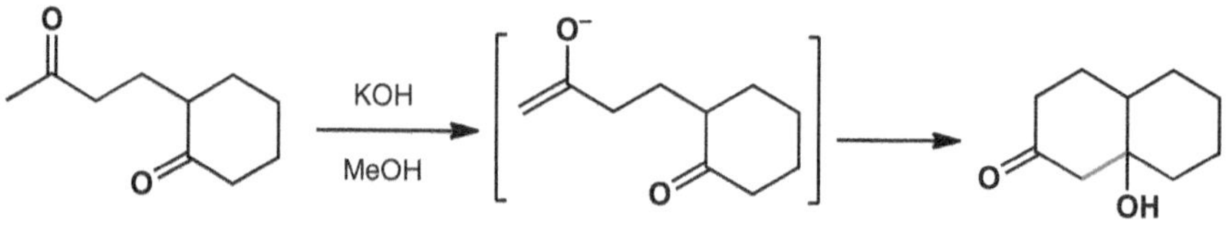

The second involves an FGI to relate the initial target to the product of a crossed aldol reaction.

Disconnection:

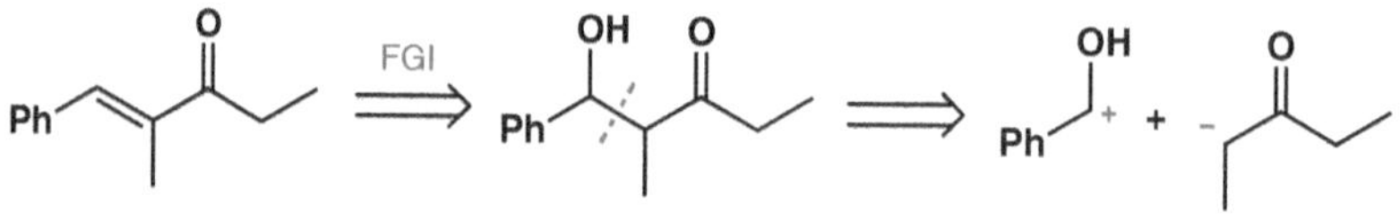

A simple aldol condensation via (reversible) dimerization of the ketone reactant in Scheme 23.12 is thermodynamically less favourable than formation of the product shown with the extended conjugated system.

Synthetic reaction:

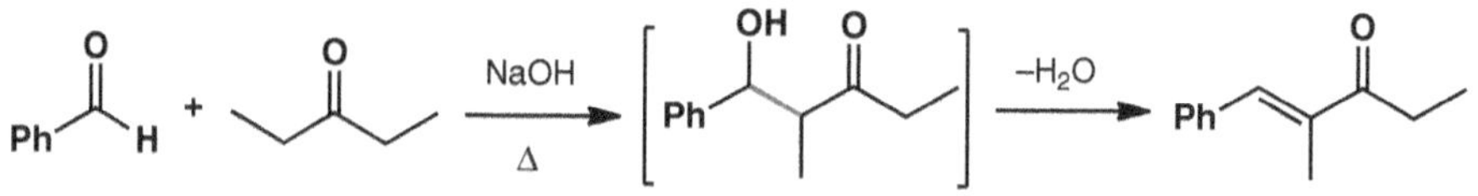

Scheme 23.12 Retrosynthetic analysis of a compound which can be converted into a 1,3-dioxygenated compound by an FGI.

Exercise 23.7

Propose a retrosynthesis and corresponding synthetic reaction(s) for each of the following target molecules.

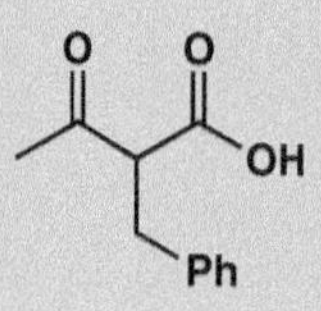

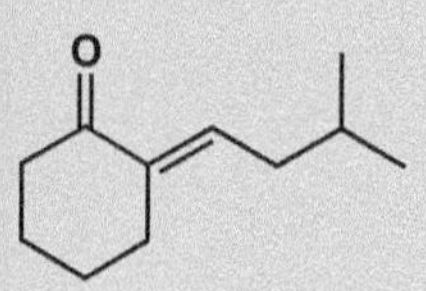

b. 1,5-Dioxygenated compounds

In principle, a 1,5-dicarbonyl compound may be obtained by the Michael addition of an enolate to an α,β-unsaturated carbonyl compound (Sub-section 18.4.1, p. 410). The disconnection of the 1,5-diketone in Scheme 23.13 gives an enolate and a C$_4$ synthon with an electrophilic C$_3$ residue, indicating a Michael addition to an α,β-unsaturated ketone. Since a simple enolate is not generally a good nucleophile for the Michael addition, an enamine is used as an enolate equivalent (Sub-section 18.4.1, p. 411) in the corresponding synthetic reaction.

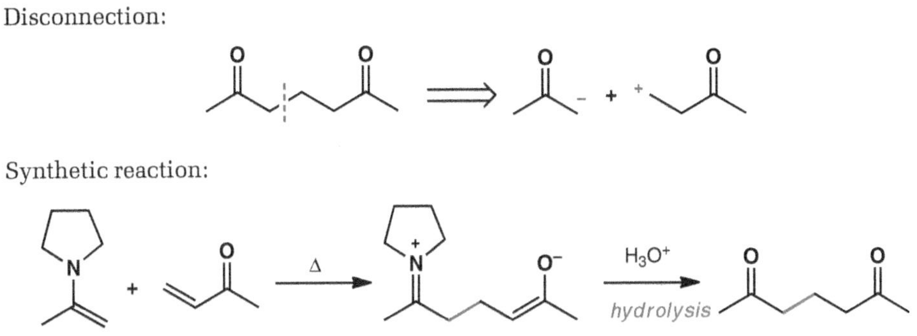

Disconnection:

Synthetic reaction:

Scheme 23.13
Retrosynthetic analysis and synthesis of a 1,5-dicarbonyl compound.

Alternatively, we could use a β-keto ester to generate a stabilized enolate for use as the nucleophile in the Michael addition corresponding to the disconnection of Scheme 23.13 with subsequent removal of the ester group by hydrolysis and decarboxylation (Sub-section 17.6.2, p. 391).

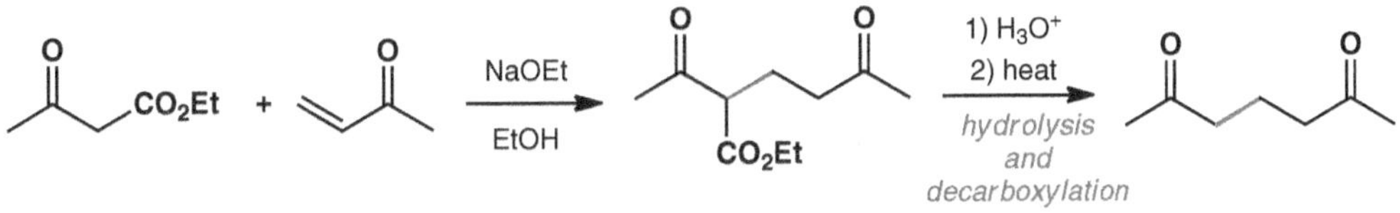

c. Functionalized cyclohexane rings

A functionalized six-membered carbocyclic ring is another easily recognised structural feature which points to a particular disconnection and synthetic strategy. When the target is a derivative of cyclohexene (or can be transformed into one by FGIs), a disconnection can often be related to a Diels–Alder reaction; a concerted disconnection then gives neutral molecules—the actual reactants instead of synthons. If the cyclohexene bears an electron-donating group (EDG) on the double bond, so much the better as we know that Diels–Alder reactions between nucleophilic dienes and alkenes (especially alkenes with electron-withdrawing groups (EWGs)) are particularly effective.

Note that it is sometimes necessary to give real molecules (rather than synthons), even following a non-concerted disconnection in a multistep retrosynthesis of a complex molecule.

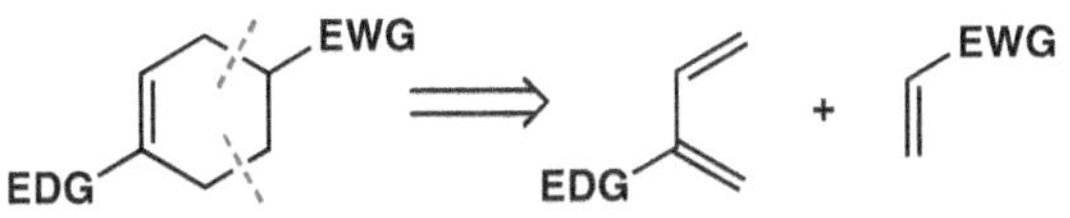

Scheme 23.14 shows how a retrosynthesis including FGIs of the target, 4-(1-hydroxy-ethyl)cyclohexanone, indicates a synthesis beginning with a Diels–Alder reaction. As ketones are readily formed by acid-catalysed hydrolysis of enol ethers, an initial FGI of the carbonyl provides us with a 1-alkoxycyclohexene. An oxidative FGI of the 1-hydroxy-ethyl group then allows disconnections which provide a 2-alkoxybutadiene as one reactant and butenone, an electrophilic dienophile, as the other.

Scheme 23.14
A retrosynthetic analysis involving FGIs and a concerted disconnection indicating a Diels–Alder reaction.

Retrosynthetic analysis:

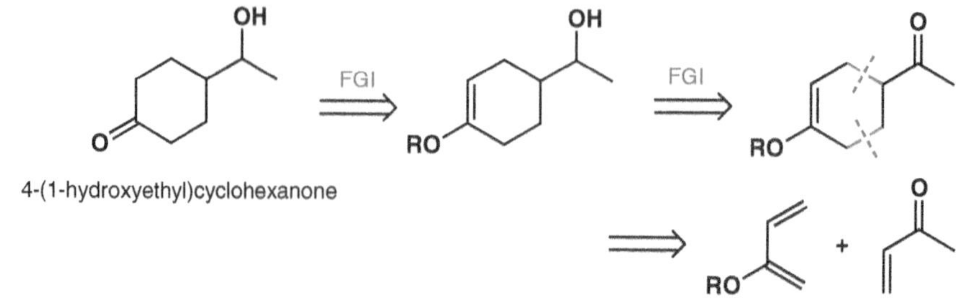

4-(1-hydroxyethyl)cyclohexanone

Synthesis:

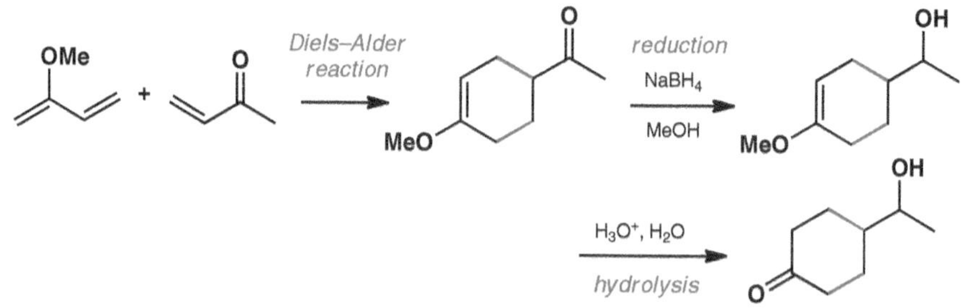

Give a retrosynthesis and corresponding synthetic reactions for each of the following target molecules.

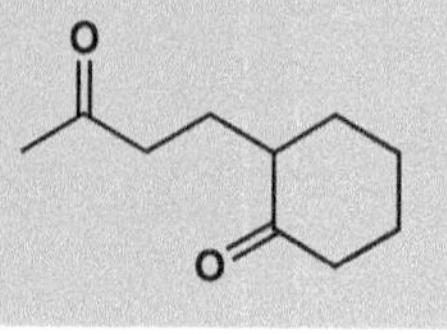

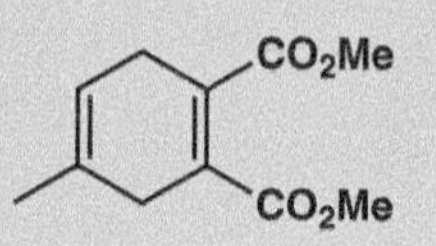

23.3 Chemoselectivity and Functional Group Protection

23.3.1 Selectivity in chemical reactions

Selectivity is an important factor in choosing synthetic reactions. On the one hand, many organic compounds have two or more functional groups so we need to know how they react and their relative reactivities. On the other hand, a single functional group may be able to react in more than one way even with a single reagent. In both cases, we need to be able to anticipate which reaction will occur under the specified reaction conditions, and (preferably) be able to control (rather than simply predict) the outcome.

We can remind ourselves briefly of three broad types of selectivity: **chemoselectivity** is about which of two or more potential reaction sites (e.g. functional groups) in a molecule will react faster, or which of alternative reagents is better for a particular transformation; **regioselectivity** is about the direction (or orientation) in which a single functional group will react when there are alternatives; **stereoselectivity** is about which of alternative stereoisomeric products will be obtained from a non-stereoisomeric reactant.

Investigations of relative reactivity have been a main concern of those studying organic reaction mechanisms, and have provided a huge volume of data to assist in the quest for chemoselectivity. We have already seen that the choice of reagent is important to achieve selectivity in the oxidation of alcohols in Section 14.4, and in the selective reduction of carbonyl groups in Section 10.1. However, a reagent which will bring about the desired transformation of one functional group may be incompatible with other functional groups in the molecule, as will be discussed in the next section. When we want to keep a more reactive group intact during the transformation of the less reactive group, we need to protect the former.

Regioselectivity has been a major aspect of a number of reactions encountered in previous chapters: Zaitsev vs. Hofmann direction in elimination reactions (Chapter 13), Markovnikov vs. anti-Markovnikov orientation in additions to alkenes (Chapter 15), 1,4- or 1,2-addition to conjugated unsaturated compounds (Sections 15.8 and 18.1), orientation in aromatic electrophilic substitution reactions (Chapter 16), and kinetic or thermodynamic enolate formation (Sub-section 17.9.1).[1]

The stereochemical course of a reaction,* like regioselectivity, is often dependent upon the stereoelectronic requirements of its mechanism; the following are examples we have already encountered: inversion of configuration in substitution by the S_N2 mechanism (Section 12.2), *anti* elimination by the E2 mechanism (Section 13.2), *anti* and *syn* addition to alkenes (Chapter 15), and the retention of stereochemical integrity in concerted reactions (Chapter 21). In other reactions, stereoselectivity may be controlled simply by steric factors or by prior coordination of the reactants to a catalyst (see below).

*This may refer to stereoselectivity or stereospecificity (see Section 11.8), as we shall see in particular examples.

23.3.2 Protection and deprotection

We saw cases in Sub-section 10.5.3 where protection of a carbonyl group is necessary in Grignard reactions and hydride reductions. For example, the carbonyl group of an aldehyde or ketone will need to be protected by acetalization (or thio-acetalization) if we wish to reduce an ester group elsewhere in the molecule: both protection and deprotection usually involve acid catalysis (Scheme 23.15). Groups which are used for protection are called **protecting groups**, e.g. acetals, as in Scheme 23.15, or the similar thioacetals.

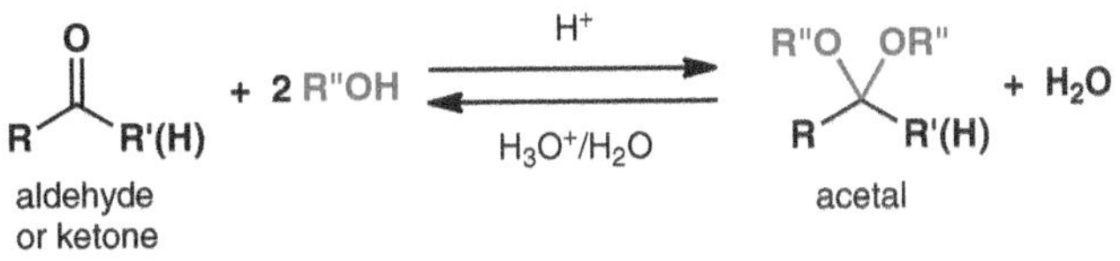

Scheme 23.15
Protection and deprotection of an aldehyde or ketone.

Protic groups such as OH and NH are highly reactive as proton donors towards organometallic reagents and some metal hydrides, and as nucleophiles towards electrophilic centres; they may also be fairly easily oxidized. Consequently, their protection is often essential and Schemes 23.16 and 23.17 summarize the principal protection/deprotection methods for OH and NH groups, respectively.

---

[1] Some authors have extended the original meaning of the term *regioselectivity* (the direction or orientation of a reaction at a *single* reaction site or functional group) to include selectivity between *unrelated* reaction sites (what we have occasionally called *site selectivity*). In doing so, they have lost some of the precision of the original meaning. Whether propene undergoes Markovnikov or anti-Markovnikov addition with HBr, for example, is quite different from whether HBr adds to one double bond of a non-conjugated diene, or the other, when the two double bonds are independent and unrelated.

The ease of removal of the triphenylmethyl (trityl) protecting group under mild acidic conditions (e.g. aqueous ethanoic acid) can be increased by the introduction of *p*-methoxy substituents in one, two, or even three of the phenyl groups. Other protecting groups can also be modified by substituents or different R groups to control their reactivity.

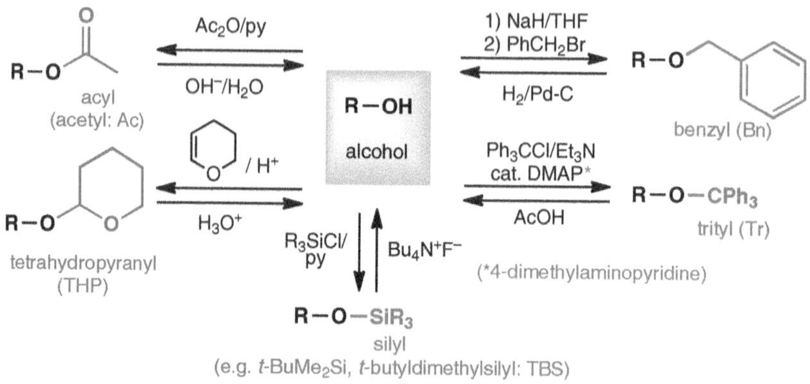

Scheme 23.16 Protection of alcohols and deprotection. Protecting groups are given in red with abbreviations in parentheses.

Exercise 23.9

Explain the increasing ease of deprotection of 4-methoxytrityl, 4,4′-dimethoxytrityl, and 4,4′,4″-trimethoxytrityl ethers under acidic solvolytic conditions.

Note how reaction conditions for deprotection depend on the protecting groups, even when they are similar. In particular, amines RR'NH are often protected as carbamate esters, RR'N(CO)OR″, but different deprotection reaction conditions are employed according to the nature of the alkyl group, R″.

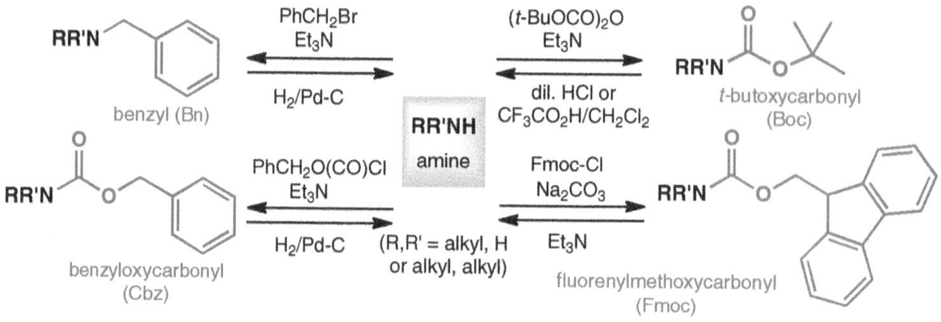

Scheme 23.17 Protection of primary and secondary amines, and deprotection. Protecting groups are given in red with abbreviations in parentheses.

Exercise 23.10

Give mechanisms for the deprotection of (a) R_2NBoc with H_3O^+ and (b) R_2NFmoc with Et_3N.

The selection of a protecting group depends on the reaction conditions used for the synthetic reaction. For example, the OH of prop-2-yn-1-ol (propargyl alcohol) is more acidic (pK_a 13.1) than the acetylenic hydrogen ($pK_a \sim 25$) so a strong base will deprotonate the OH first. In order to deprotonate the acetylenic function with a strong base, the OH must be masked, and the tetrahydropyranyl (THP) group (unaffected by bases) is the normal choice of protecting group (Scheme 23.18). The THP group is introduced with acid catalysis under anhydrous conditions, and removed in aqueous solution also with acidic catalysis.

Scheme 23.18 An example of the protection of OH and deprotection.

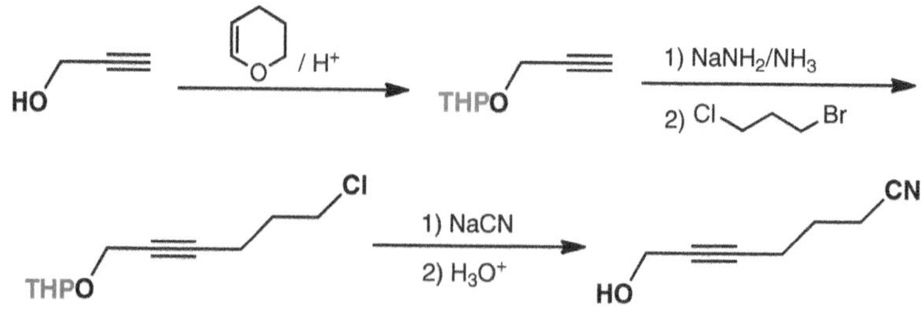

Give reaction mechanisms for the THP protection of an alcohol, ROH, and the deprotection.

Protecting groups are especially important in the syntheses of polyfunctional compounds such as sugars and peptides (see Chapter 24). Reaction 23.11 shows an example of a reaction for connecting two multiply-protected monosaccharides (see Sub-section 24.1.2), galactose and rhamnose.

protected galactose (Tr = Ph₃C) protected rhamnose the diol is protected as an acetal with propanone

(23.11)

L-Rhamnose is a naturally occurring deoxy sugar. It is unusual in being one of the few L sugars found in nature (see Section 24.1).

The OH groups in reaction 23.11 are protected in four ways: using acetyl (Ac), benzyl (Bn), trityl (Tr), and acetal groups. All these protecting groups are stable under the conditions of the coupling reaction, and do not retard it. The trityl group may be removed by a weak acid such as ethanoic acid, while the acetal can be hydrolysed by a dilute aqueous strong acid. Only the acetyl groups can be removed under basic conditions (ester hydrolysis). The benzyl groups in the product can be removed selectively by hydrogenolysis with Pd–C to regenerate the OH. It follows that the different protecting groups used in reaction 23.11 can be removed one at a time in any order using appropriate deprotection reactions.

23.4 Efficiency in Organic Synthesis

The main factors which determine the efficiency of an organic synthesis are (a) the number of steps, (b) the availability of starting materials and reagents, (c) the ease of carrying out each step, (d) the yield of each step, (e) the ease of isolation and purification of products, and (f) the minimization and responsible disposal of waste. However, in some situations, industrial and social factors have to be taken into account as well, so the selection of a 'best' synthetic route may be quite complex. And even when a good synthetic route is established, a single new reaction for one step could improve the overall synthesis quite dramatically; consequently, established procedures are usually under constant review.

In addition to the efficiencies of individual steps, there are two strategies for a synthesis which could very much affect the overall result. These are the *linear* synthesis and the *convergent*. Consider a five-step synthesis of a hypothetical target molecule ABCDEF, and assume (optimistically) that the yield of each step is 90%. The overall yield of the target is 59% ($=0.9^5 \times 100$) if the reactions are carried out consecutively, as illustrated in a **linear synthesis** in Figure 23.1(a). Alternatively, two components of the target molecule could be synthesized separately and then combined in the final step of a **convergent synthesis** (Figure 23.1(b)). Although the total number of reaction steps is the same, and the yield of each step is the same, the number of *consecutive* steps is decreased to three in the convergent strategy for this example. Consequently, the overall yield is much improved at 73% ($=0.9^3 \times 100$).

We see these alternative strategies illustrated in the historical context by syntheses of prostaglandin E₂ (PGE₂). The first total synthesis was achieved by Corey in 1969 and improved methods followed. An outline of the Corey synthesis, which comprises 21 consecutive steps (a linear synthesis) starting from cyclopentadiene, is given in Scheme 23.19.

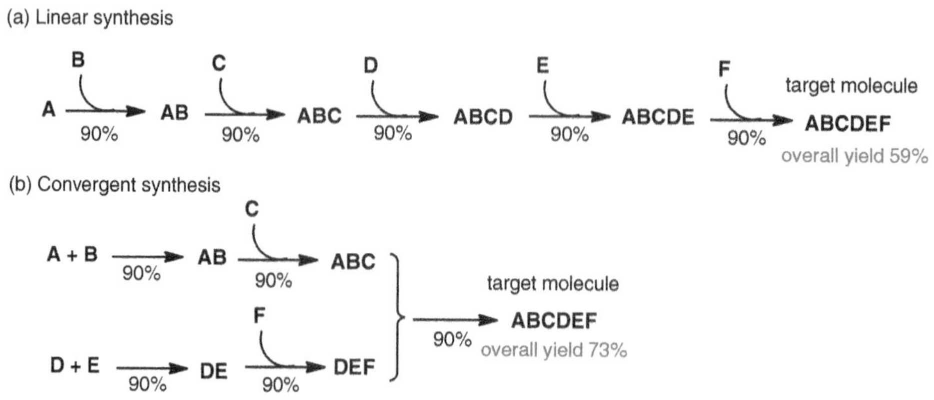

Figure 23.1 Overall yields in linear and convergent syntheses involving five steps.

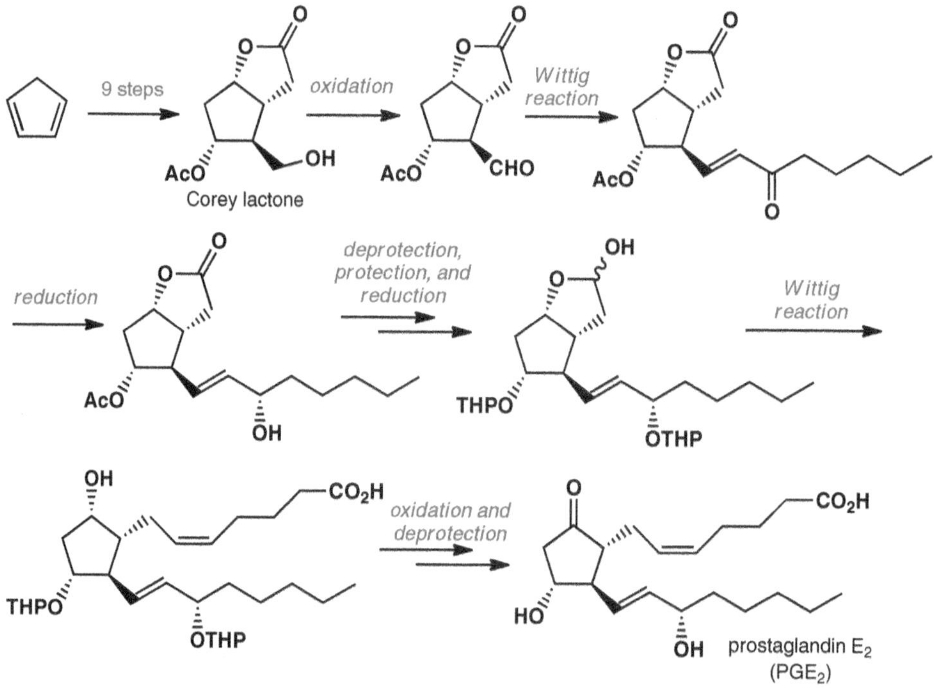

(a) Linear synthesis

(b) Convergent synthesis

Scheme 23.19 The first total synthesis of PGE$_2$ by Corey (1969).

Note that protecting groups, Ac, THP, and TBS, which we introduced in Sub-section 23.3.2, are used in Schemes 23.19 and 23.20.

About 20 years later, Noyori presented an improved method by a convergent strategy, combining three components in one vessel (Scheme 23.20). Conjugate addition of an organo-lithium reagent (**A**) to a protected cyclopentenone (facilitated by CuI/PBu$_3$) gives a lithium enolate (**B**); replacement of the lithium by SnPh$_3$ using Ph$_3$SnCl provides a tin enolate (**C**) which is then trapped by the electrophilic allylic iodide (**D**) to give **E** on the way to PGE$_2$. The longest sequence of steps, therefore, is down to eleven.

In addition to the smaller number of consecutive steps in the convergent synthesis, there are other merits of this strategy. The operations are usually easier because reactions can be carried out on a smaller scale to obtain the same amounts of products. Furthermore, the loss following failure in one step in a planned convergent synthesis would be limited compared with a failure in one step of the linear alternative. The material from all the preceding steps would be lost in the linear approach and the whole strategy might be at risk. However, the loss of material would be limited to that from the preceding steps in just one strand of the convergent strategy, and only that one strand might need to be reconsidered.

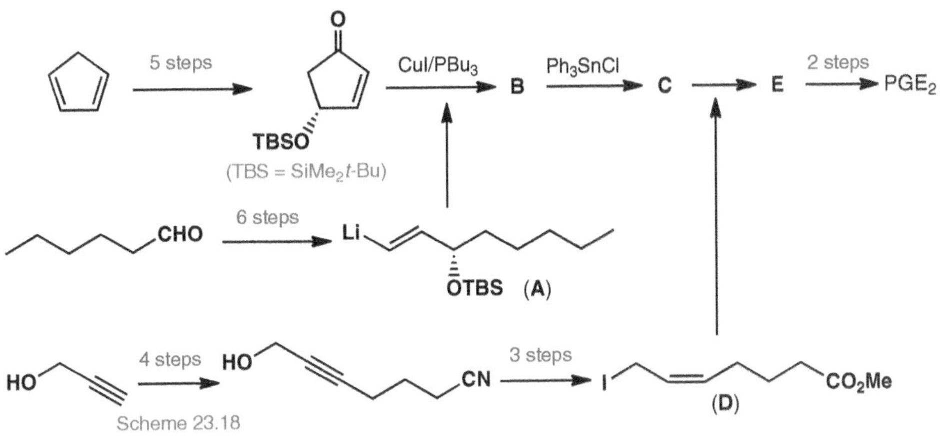

Scheme 23.20 Noyori's convergent total synthesis of PGE$_2$ (1988).

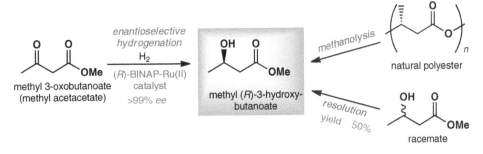

Scheme 23.18

23.5 Stereoselectivity and Asymmetric Synthesis

Many biologically active compounds, such as the prostaglandin encountered in the preceding section, are chiral, and it is important to be able to synthesize them in stereochemically pure forms. The synthesis of a chiral compound as a single stereoisomer from achiral precursors is called an **asymmetric synthesis** and requires a *chiral source*, as discussed in Section 11.8; there are two general strategies—use of a **chiral catalyst** or a **chiral auxiliary**.

We can illustrate the first (the use of an enantiomerically pure catalyst as the chiral source) by an industrial example, the enantioselective synthesis of methyl (R)-3-hydroxybutanoate. This ester can be obtained by methanolysis of a natural polyester produced by bacterial fermentation, or by resolution of the synthetic racemate, but neither of these methods is satisfactory for large-scale production. In contrast, enantioselective reduction of methyl 3-oxobutanoate (acetoacetate) is quite effective using a chiral BINAP-Ru(II) complex as a catalyst (Scheme 23.21).*

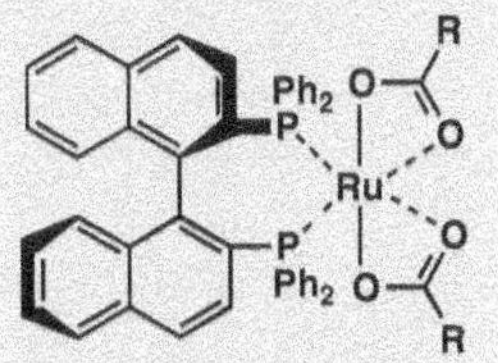

(R)-(+)-BINAP-Ru(II) catalyst

*This chiral catalyst was developed by the Japanese chemist, R. Noyori, who shared the Nobel Prize in Chemistry in 2001 with the Americans, W.S. Knowles and B. Sharpless (see below).

For the definition of the enantiomeric excess $(ee = |R\% - S\%|)$, see Sub-section 11.5.3, p. 241.

Scheme 23.21 Methods for producing methyl (R)-3-hydroxybutanoate.

The enantioselectivity of the catalytic hydrogenation in Scheme 23.21 was achieved by differentiation of the *enantiotopic* faces of the *prochiral* substrate (see Section 11.8) by the chiral catalyst. In the hydrogenation of the ketone represented in Figure 23.2, addition of hydrogen from above the face of the carbonyl bond leads to the R enantiomer, while addition from below gives the S isomer

In the well-known Sharpless asymmetric epoxidation procedure for allylic alcohols, illustrated in eqn 23.12, the catalytic system comprises (*i*-PrO)$_4$Ti and an optically active tartrate ester.

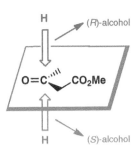

Figure 23.2 Enantiotopic faces in the hydrogenation of the prochiral methyl 3-oxobutanoate.

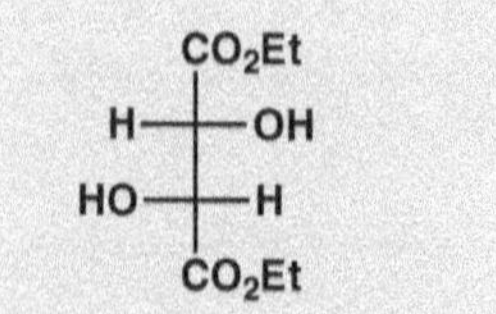

$$\text{(23.12)}$$

*This auxiliary is easily prepared from the amino acid, valine (Section 24.3).

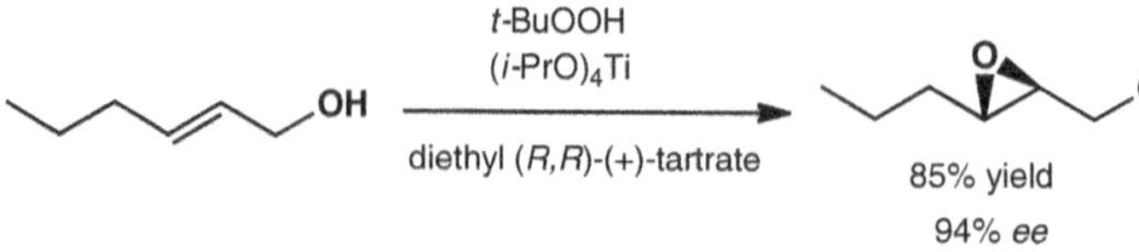

The second general strategy in asymmetric synthesis is to use a *chiral auxiliary* as the chiral source. In the example shown in Scheme 23.22, the chiral auxiliary (shown in red) is an enantiomerically pure oxazolidinone derivative attached directly to the carbonyl group of the substrate.* The electrophilic boron reagent Bu_2BOTf reacts at this carbonyl group with loss of triflate, and proton abstraction from the adjacent C by Et_3N, to give the boron enolate, enantiomer **A**. The double bond has the Me *trans* to the oxazolidinone residue and the Lewis acid/base interaction between the electrophilic boron atom and the oxygen of the ring carbonyl holds **A** in the shape indicated.

The boron atom of the reagent Bu_2BOTf is electrophilic (Lewis acidic) and triflate ($OTf=OSO_2CF_3$) is an excellent nucleofuge.

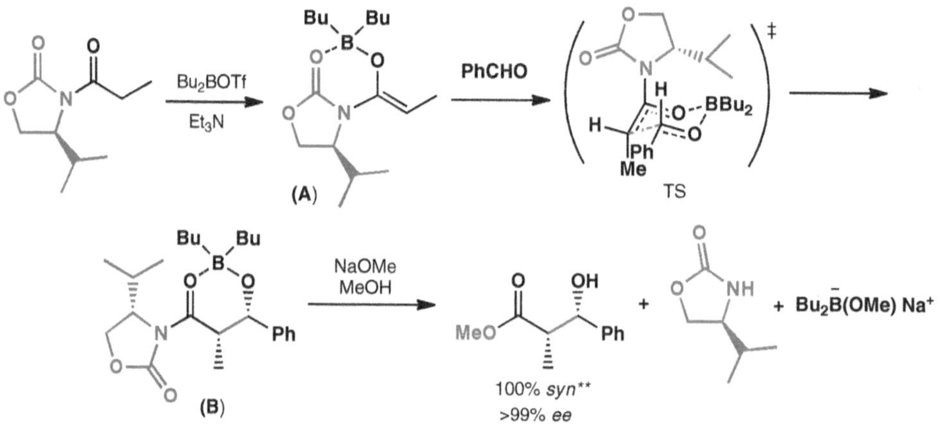

Scheme 23.22 An example of an asymmetric synthesis using a covalently bonded chiral auxiliary.

**The relative configuration of pairs of substituents in linear open-chain compounds is based on a zigzag representation of the chain, and uses the terms *syn* and *anti*. The zigzag chain is taken to be in the plane of the paper and substituents at chirality centres above or below the plane of the zigzag are indicated by solid or broken wedges, respectively. If groups on different Cs are on the same side of the zigzag plane, they are said to be *syn*; if they are on opposite sides, they are *anti*.

Upon reaction of the boron enolate of **A** with benzaldehyde via the TS shown, the covalent bond between the boron and the enolate residue becomes a new Lewis acid/base interaction. At the same time, the 'old' Lewis acid/base interaction between boron and the oxazolidinone carbonyl is replaced by a covalent bond from the boron to the O which was the O of the benzaldehyde; the result is the single stereoisomer, **B**. Treatment of the reaction mixture with methoxide in methanol gives the β hydroxy ester with excellent overall stereoselectivity, and liberates the chiral auxiliary.

A recent development of the above strategy is to avoid isolation of the substrate bonded to the chiral auxiliary (and hence the need for a stoichiometric amount of the chiral auxiliary), and use instead only a catalytic amount in a 'one pot' reaction. L-Proline, the only naturally occurring α-amino acid containing a secondary amino group (see Section 24.3), and related enantiomerically pure secondary amines can be used in this way, as exemplified by the aldol reaction in Scheme 23.23.

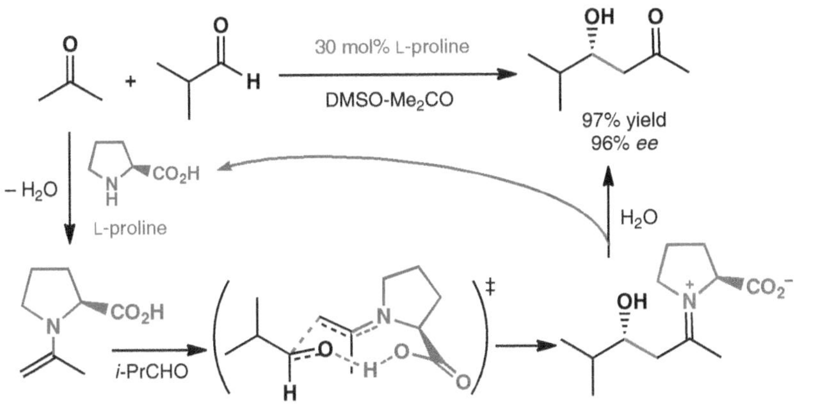

Scheme 23.23 A catalytic asymmetric synthesis of an aldol via a chiral enamine intermediate. The reaction is carried out in one vessel.

Term *organocatalysis* has been introduced to distinguish what we are describing here, i.e. catalysis by a (relatively) small organic molecule, from catalysis by H_3O^+, HO^-, metals or metal-containing compounds, and enzymes; the mechanistic details, however, are very similar to those observed in biochemical reactions catalysed by coenzymes and enzymes.

The enantiomerically pure enamine formed from propanone (which is used in excess) and L-proline (which, as an acid, may also catalyse this first step) reacts with the aldehyde to generate an iminium zwitterion with a new chirality centre. Its spontaneous hydrolysis liberates the aldol shown and the proline (which can recycle within the reaction mixture, hence the need for only a catalytic amount). The high enantioselectivity of the overall reaction indicates preferential formation of the single diastereoisomer of the iminium zwitterion shown most likely via a TS involving H-bonding between the carboxy group of the proline residue and the oxygen of the aldehyde.

See Section 8.6 for information about the formation and hydrolysis of imines and enamines, and Panel 17.2 for a biological aldol reaction via an enamine derivative of a catalytic enzyme. Imine/enamine intermediates are also involved in reactions catalysed by the coenzyme, pyridoxal (Panel 8.2).

Exercise 23.12

L-Proline was used as a catalytic chiral auxiliary for the following stereoselective reaction. Give reactions for this transformation as well as the one to provide the starting material.

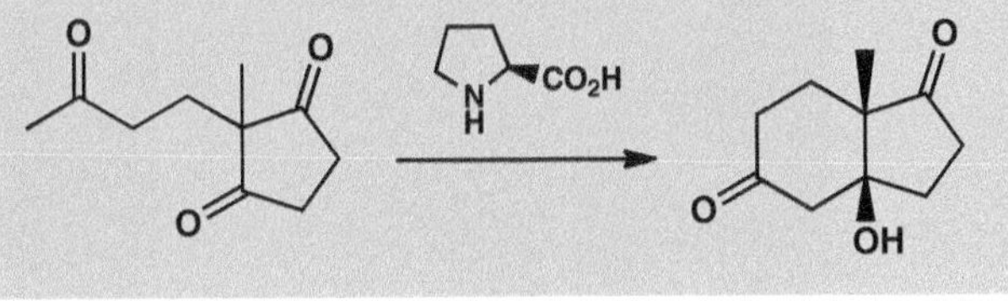

23.6 An Example of a Multistep Synthesis

We shall now examine the synthesis of an anti-neoplastic agent, bexarotene (or Targretin®), which is used for the treatment of lung cancer, breast cancer, and Kaposi's sarcoma. The target molecule has two aromatic groups and the retrosynthesis in Scheme 23.24 includes an FGI to give the ketone **A** which allows a reasonable disconnection into components, **B** and **C**.

A synthesis is outlined in Scheme 23.25 with yields of the main steps. Component **B** is constructed by the Friedel–Crafts alkylation of toluene using a doubly tertiary dichloride which is readily obtained by the reaction of HCl with the corresponding diol. The Friedel–Crafts acylation of **B** is then carried out with acid chloride **C** (prepared from the monomethyl ester of benzene-1,4-dicarboxylic acid) to provide **A**. The ketone carbonyl group of **A** is converted into methylene by a Wittig reaction, then ester hydrolysis of the product **D** gives the target molecule, bexarotene.

Scheme 23.24 A retrosynthesis of bexarotene.

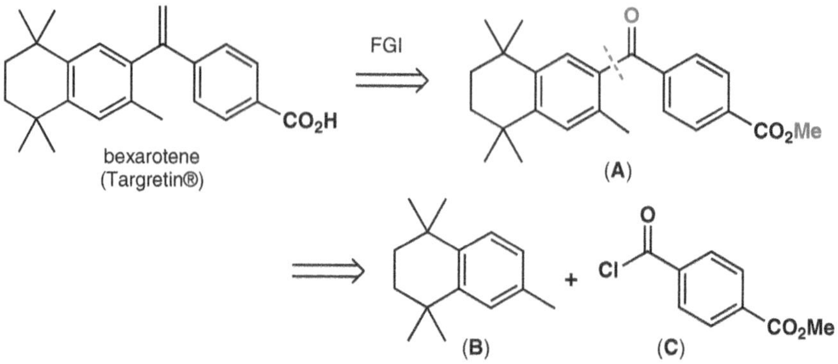

Scheme 23.25 Synthesis of bexarotene.

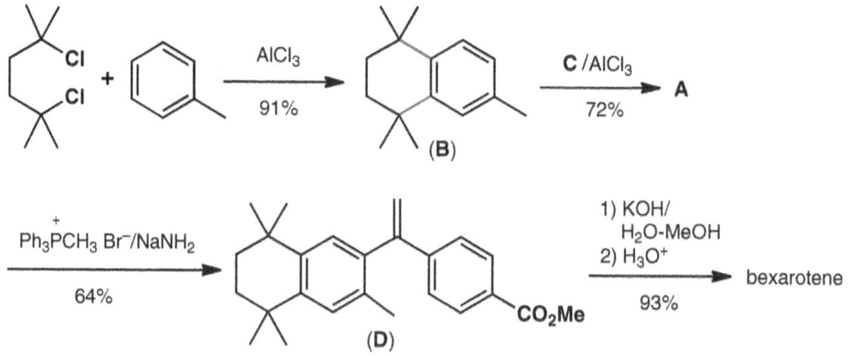

Exercise 23.13

Bexarotene is represented as shown below and analogues **E–G** are given in correspondingly abbreviated forms. Give reaction schemes for syntheses of **E–G** which were required for investigation of their biological properties.

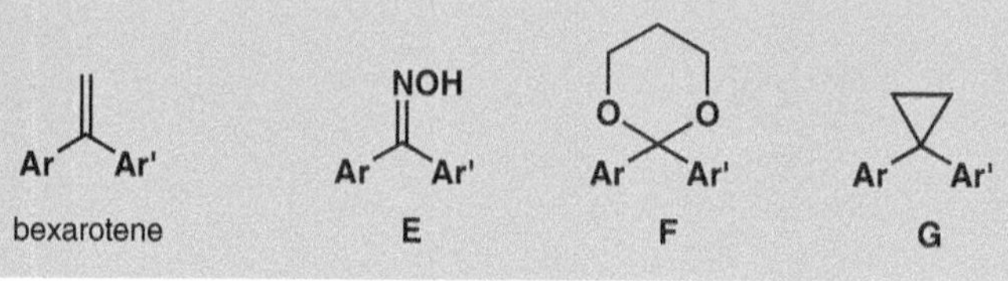

Summary

- Organic synthesis is the design and execution of an efficient combination of organic reactions to make a target compound.

- **Retrosynthetic analysis**, which involves working backwards from the target to precursors (and ultimately to readily available starting materials) via **disconnections** and **synthons**, is an effective approach to the planning of a synthesis.

- Selectivity in each reaction step is important for an efficient synthesis, especially stereoselectivity in syntheses of chiral compounds.

- Reactive functional groups usually need to be protected when a less reactive group in a multifunctional compound is to be modified.

- A *convergent* synthesis is usually superior to a *linear* approach.

Problems

23.1 Give reasonable synthons and a synthetic reaction using corresponding reagents for each of the disconnections indicated in the following target molecules.

(a)

(b)

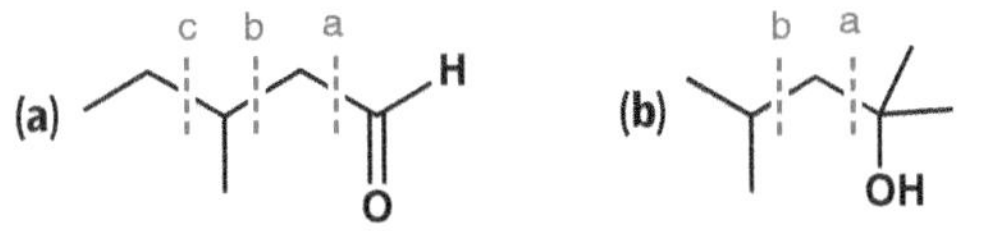

23.2 Give a retrosynthesis and the corresponding synthesis for each of the following target molecules.

(a) PhCH₂OPh

(b) PhNH ... OEt

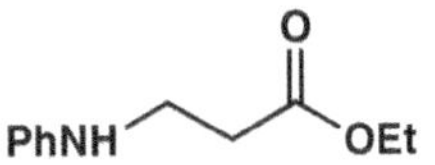

(c)

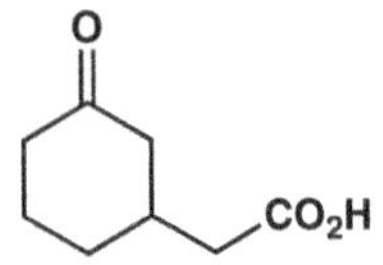

(d)

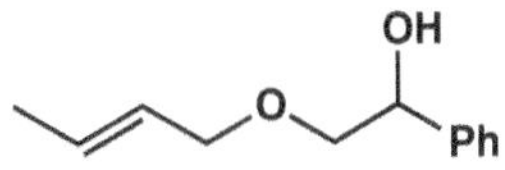

23.3 Identify synthons of opposite polarities for the following disconnection of 4-oxopentanoic acid, and give corresponding synthetic reactions for both possibilities.

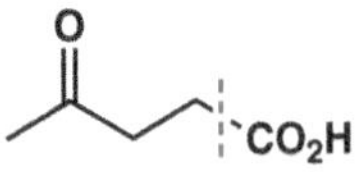

CO₂H 4-oxopentanoic acid

23.4 Propose reactions for a multistep synthesis corresponding to the following retrosynthesis using appropriate protecting groups.

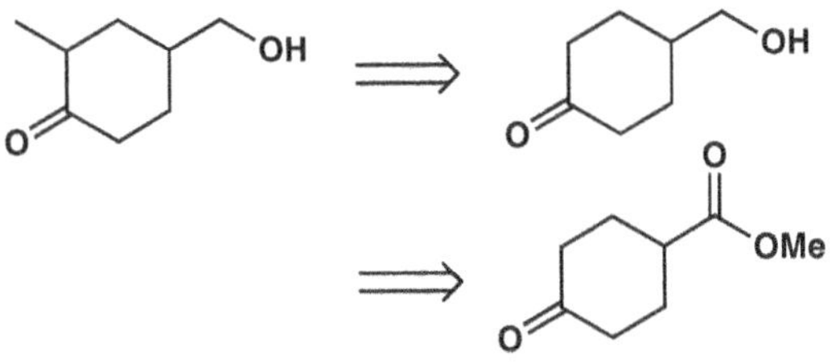

23.5 Propose a synthesis for propranolol, a widely used pharmaceutical (a β-blocker) for reducing blood pressure and the symptoms of stress and anxiety. Begin by giving a retrosynthesis with disconnections next to the heteroatoms, and then a corresponding synthesis.

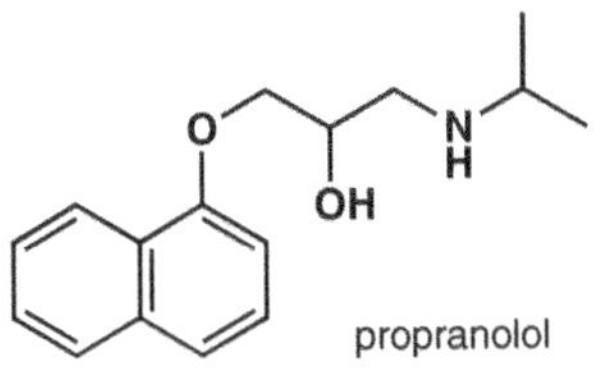

propranolol

23.6 Propose a brief retrosynthesis and synthetic reactions starting from the compounds indicated for each of (a) procaine (a local anaesthetic) and (b) salbutamol (also called albuterol: used in asthma inhalers).

(a) procaine

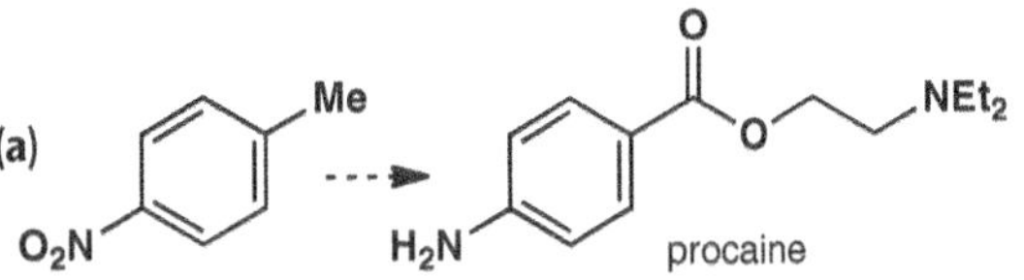

(b) salbutamol (albuterol)

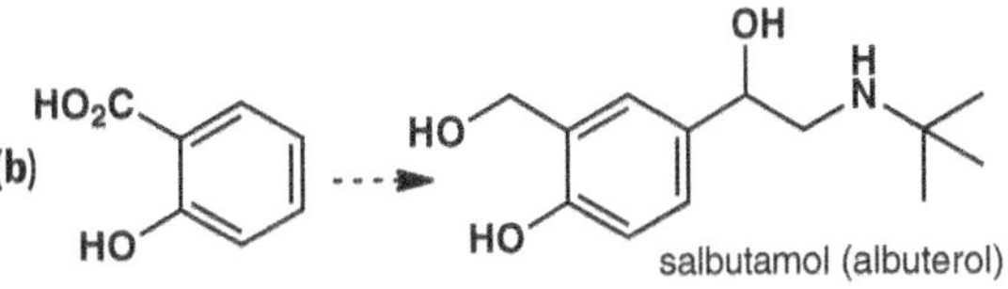

23.7 Geraniol may be transformed into a higher homologue by the following three-step reaction sequence. Give appropriate reagents for steps (a)–(c).

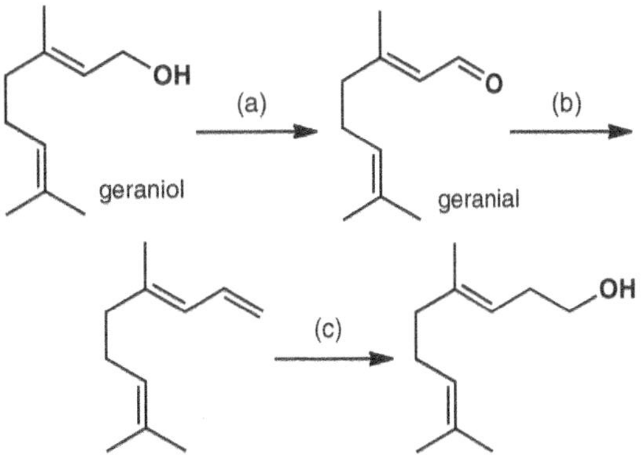

geraniol   (a)   geranial   (b)

(c)

23.8 The following is part of a synthetic sequence for vitamin A.

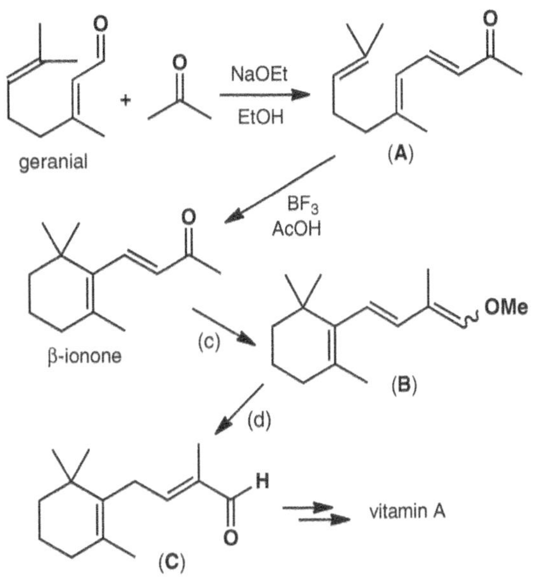

geranial   NaOEt / EtOH   (A)

BF₃ AcOH

β-ionone   (c)   (B)

(d)

(C)   vitamin A

(a) Propose a mechanism for the reaction which yields intermediate **A**.

(b) Propose a mechanism for the cyclization of **A** into β-ionone.

(c) Give a reagent for the conversion of β-ionone into the intermediate enol ether **B**.

(d) Propose reaction conditions and a mechanism for the conversion of **B** into the aldehyde **C**.

23.9 Rebamipide is a drug used for mucosal protection, and the treatment of gastritis and gastroduodenal ulcers. A retrosynthesis is given below showing only the main compounds.

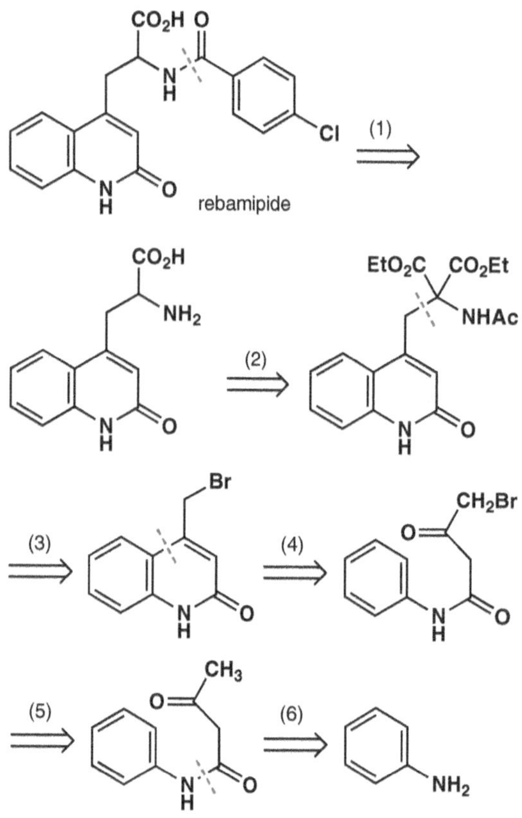

rebamipide

(a) Give synthetic reactions indicated by the retrosynthesis starting from the materials indicated.

(b) Propose a mechanism for the synthetic reaction corresponding to step 4.

23.10 The following is a retrosynthesis of nuciferal, a terpenoid found in the wood oil of the Japanese *kaya* tree (*Torreya nucifera*).

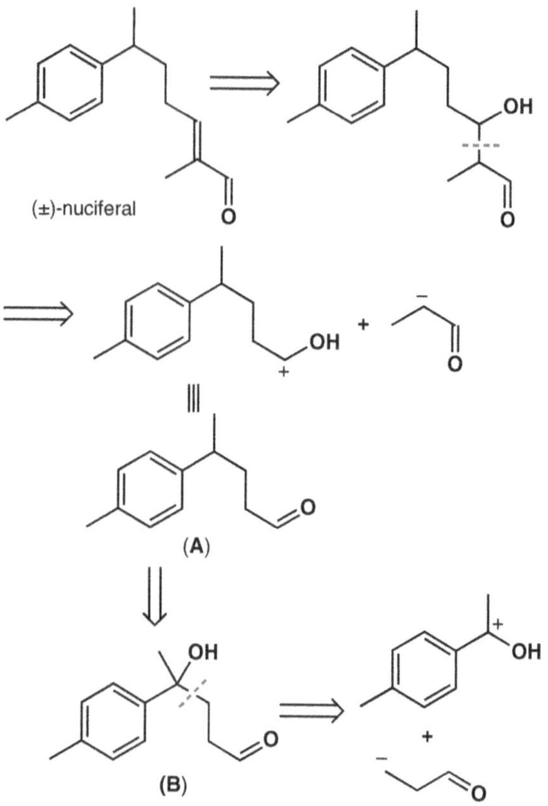

(±)-nuciferal

(a) Give reactions for a synthesis of **B** starting from toluene.

(b) Give reactions for the conversion of **B** to **A**.

(c) Give reactions for the synthesis of nuciferal from **A**.

23.11 Scheme 23.19 in Section 23.4 shows an outline of Corey's total synthesis of PGE$_2$. The initial part to make the so-called Corey lactone from cyclopentadiene, and its oxidation to the aldehyde, is shown below.

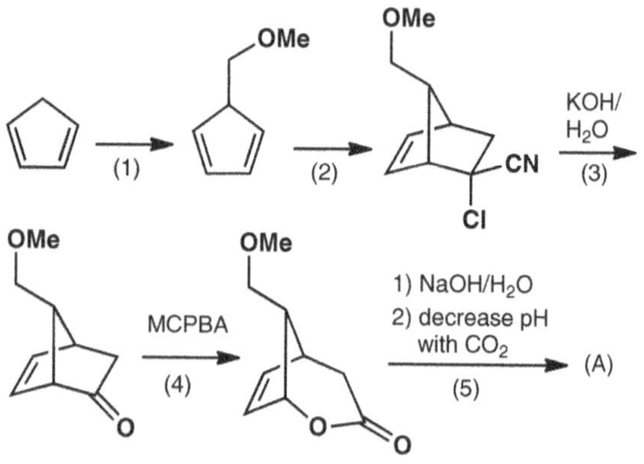

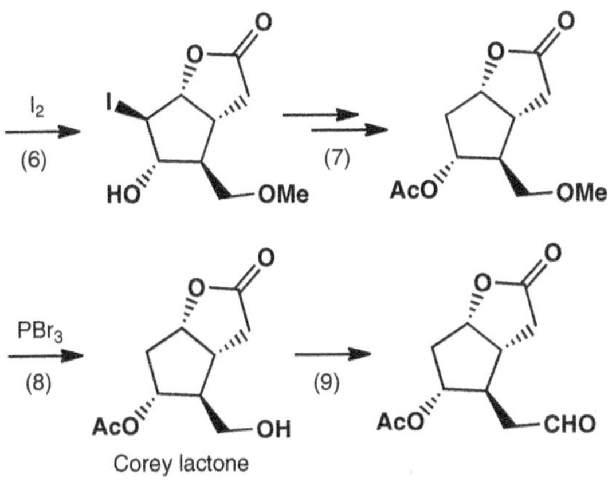

I₂
(6)

PBr₃
(8)

(7)

(9)

Corey lactone

(a) Give reactions for steps (1) and (2).

(b) Give a reaction mechanism for step (3).

(c) Give a reaction mechanism for step (4).

(d) Deduce the structure of intermediate **A** by giving a mechanism for step (5).

(e) Give a mechanism for step (6) which is sometimes called iodo-lactonization.

(f) Transformation (7) involves two reactions: protection of the OH and de-iodination; give suitable reagents for each.

(g) The product of transformation (7) is demethylated in step (8) to give the Corey lactone which is then oxidized to the aldehyde in step (9). What is a suitable reagent for the oxidation?

(h) The subsequent reactions (given in Scheme 23.19) involve introduction of two side chains, both using the Wittig reaction. Give the steps involved in the transformation below, which is the second of these two Wittig reactions.

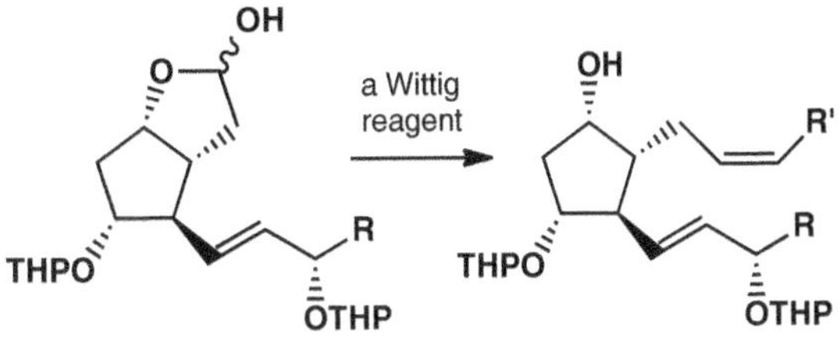

a Wittig reagent

23.12 The following highly stereoselective cyclization combines three components using an enantiomerically pure secondary amine as a chiral catalyst. Give a sequence of individual steps for this overall transformation, but without stereochemical details.

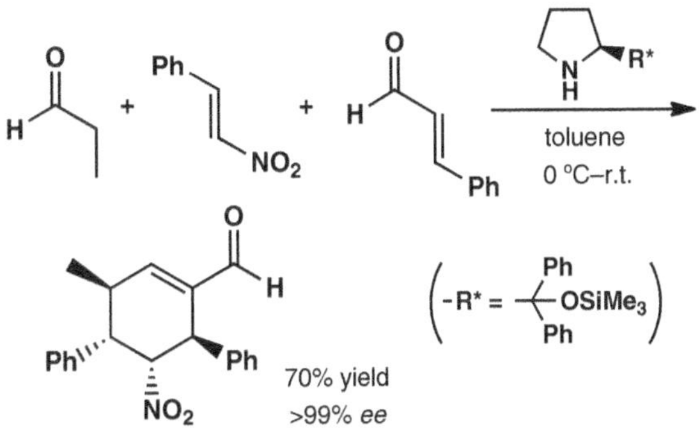

toluene
0 °C–r.t.

70% yield
>99% *ee*

$\left(-R^* = -\!\!\!\begin{array}{c} Ph \\ OSiMe_3 \\ Ph \end{array} \right)$

24 Chemistry of Biomolecules

Related topics we have already covered:

- **Stereochemistry and molecular conformations** (Chapter 4)
- **Acids and bases** (Chapter 6)
- **Chemistry of carbonyl compounds** (Chapters 8 and 9)
- **Chirality** (Chapter 11)
- **Chemistry of alcohols, thiols, and amines** (Chapter 14)
- **Aromatic heterocyclic compounds** (Chapter 19)

Topics we shall cover in this chapter:

- **Carbohydrates**
- **Nucleosides, nucleotides, and nucleic acids**
- **Amino acids, peptides, and proteins**
- **Fats and oils, and phospholipids**
- **Terpenes, steroids, and eicosanoids**

Organic chemistry was once the chemistry of compounds produced by living organisms (see the Prologue), and we have already mentioned in previous chapters some biomolecules (mainly small ones) found in nature which are biologically active as medicines, poisons, or nutrition, and ones used as flavouring and colouring materials. In this chapter, we shall cover some of the major constituents of our body: carbohydrates, nucleic acids, proteins, and lipids, and related compounds. Although structures of many biomolecules may be complicated, their chemistry is not especially difficult. Their reactions are governed by the principles we have already encountered in our coverage of the chemistry of simple organic compounds. In this chapter, therefore, we shall introduce the kinds of compounds which play major roles in living systems, and identify the molecular features which allow them to function in their biological roles. We shall not, however, dwell on biochemical details or revisit basic chemical aspects already covered.

24.1 Carbohydrates

Animals and most microorganisms are dependent on carbohydrates which are produced by photosynthesis in plants; they function as important food reserves and as structural components of cell walls.

The name **carbohydrate** (meaning hydrates of carbon) comes from the general formula of many of this class of compounds, $C_m(H_2O)_n$. **Sugars** and **saccharides** are alternative widely used names for carbohydrates, especially for ones with sweetening properties. A **monosaccharide** is a carbohydrate monomer (usually a C_3–C_9 compound); when a few monosaccharides are linked together in a particular way (as we shall see), the product is called an **oligosaccharide** (typically 2–10 monosaccharide units), and **polysaccharides** are polymers comprising many monosaccharide units.

24.1.1 Monosaccharides

A monosaccharide is a polyhydroxyaldehyde (an **aldose**) or a polyhydroxyketone (a **ketose**). The simplest are C3 compounds: glyceraldehyde (2,3-dihydroxypropanal) and 1,3-dihydroxypropanone. The former is chiral and (for reasons discussed in Panel 11.3) its *R* and *S* enantiomers are designated D- and L-glyceraldehyde, respectively; the latter is achiral.

The systematic names of sugars usually end in –ose.

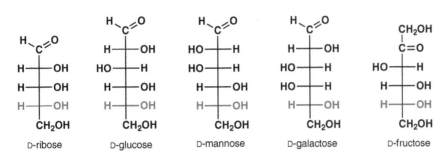

(R)-glyceraldehyde or D-glyceraldehyde   (S)-glyceraldehyde or L-glyceraldehyde   1,3-dihydroxypropanone

The most abundant monosaccharide units in nature are five- and six-carbon sugars (**pentoses** and **hexoses**) which have up to four and five chirality centres, respectively; their structures are conveniently represented by Fischer projections (see Section 11.3), as shown in Figure 24.1.

*We encountered one exception, L-rhamnose, in Sub-section 23.3.2; L-ascorbic acid, vitamin C, see later, is another.

Because of the number of its chirality centres, any typical sugar will generally have many stereoisomers, each one being a different compound. For example, an aldohexose (a C_6 aldose) could have four chiral carbons leading to 16 possible stereoisomers (2^4) comprising eight pairs of enantiomers. They are divided into a D series whose chiral carbon farthest from their carbonyl group has the same configuration as the central C of D-glyceraldehyde, and an L series which have the opposite configuration at this carbon. Naturally occurring sugars are almost all D compounds.*

**The names, *furanose* and *pyranose*, come from furan and pyran.

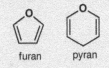

furan   pyran

Of the 16 aldohexoses, therefore, only half (the D sugars) are present in nature. One of these eight, D-glucose (the best-known hexose) is included in Fig. 24.1 along with two diastereoisomers: D-mannose has the opposite configuration at C2, and D-galactose has the opposite configuration at C4. The other two representative sugars in Figure 24.1 are D-ribose (a C_5 aldose) and D-fructose (a C_6 ketose); the former, being the monosaccharide unit of ribonucleic acids (see later), is a very important pentose.

These monosaccharides exist mainly in their **cyclic hemiacetal** forms as five- or six-membered rings (**furanoses** or **pyranoses**, respectively).** The open-chain form of D-glucose, for example, is in equilibrium with two pyranose forms designated α and β (Scheme 24.1). These six-membered cyclic diastereoisomers are called α-D-glucopyranose and β-D-glucopyranose, and the OH groups at their C1 atoms are axial and equatorial, respectively, with the other OHs on ring carbons being equatorial in the chair forms of the six-membered rings. Carbon 1 is the carbonyl of the open-chain form and becomes a new chirality centre upon the cyclization which creates the two new stereoisomers: they are called **anomers** and C1 is termed an **anomeric carbon**. The α and β anomers can readily interconvert via the open-chain form.***

***If we need to establish in general whether an anomer is α or β, we compare the *R/S* configuration of the anomeric carbon with that of its highest numbered chirality centre (e.g. C5 in an aldohexose). If they are the same, the anomer is β; if they are different, it is α. For example,

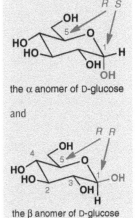

the α anomer of D-glucose

and

the β anomer of D-glucose

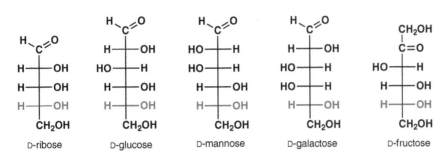

H C=O	H C=O	H C=O	H C=O	CH₂OH
H—OH	H—OH	HO—H	H—OH	C=O
H—OH	HO—H	HO—H	HO—H	HO—H
H—OH	H—OH	H—OH	HO—H	H—OH
CH₂OH	H—OH	H—OH	H—OH	H—OH
	CH₂OH	CH₂OH	CH₂OH	CH₂OH
D-ribose	D-glucose	D-mannose	D-galactose	D-fructose

Figure 24.1 Fischer projections of representative monosaccharides found in nature.

Scheme 24.1 Equilibrium of open-chain and pyranose forms of D-glucose.

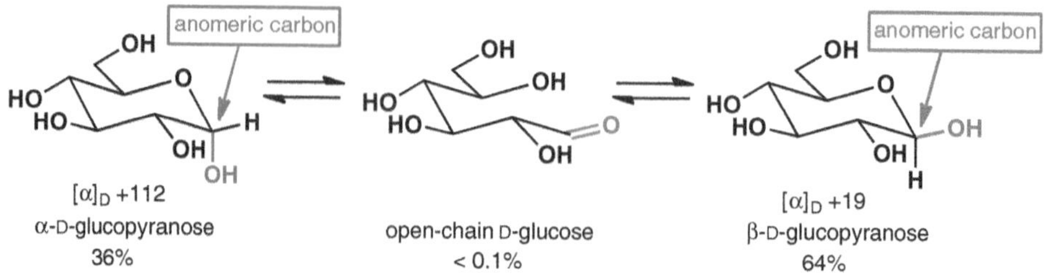

$[\alpha]_D$ +112
α-D-glucopyranose
36%

open-chain D-glucose
< 0.1%

$[\alpha]_D$ +19
β-D-glucopyranose
64%

The interconversion in Scheme 24.1, like the normal reversible formation of hemiacetals, is catalysed by acids and bases (Sub-section 8.4.1). Both glucose anomers can be obtained pure, each with its own specific rotation (Sub-section 11.5.2), and a solution of either gradually becomes the equilibrium mixture. This process is sometimes called a *mutarotation* since the optical rotation of the solution of one anomer changes to become that of the equilibrium mixture.

Exercise 24.1

We encountered the C_4 monosaccharides, erythrose and threose (2,3,4-trihydroxybutanals), in Chapter 11, Sub-section 11.4.1. Give the names of the four diastereoisomers using D/L designations.

Exercise 24.2

Give a mechanism for the anomerization in Scheme 24.1 catalysed by hydroxide.

Exercise 24.3

Represent a six-membered cyclic structure of D-mannose in a chair conformation.

24.1.2 Glycosides

The name of a glycoside is derived from the name of the sugar by replacing the suffix *–ose* with *–oside*, e.g. ethyl β-D-glucopyranoside in eqn 24.1. Note that *glycoside* is a general name, and *glucoside* is a glycoside of glucose. Glycosides are found widely in nature: one example is salicin (*o*-hydroxymethylphenyl β-D-glucopyranoside) which is found in the bark of willow trees and led to the discovery of aspirin (see the Prologue).

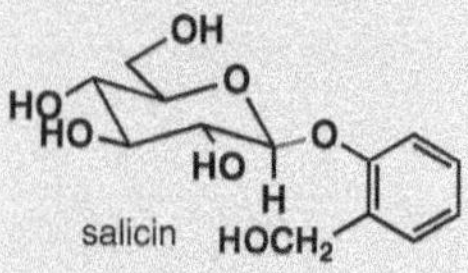

The cyclic hemiacetal of a monosaccharide can form an acetal by reaction with an alcohol just as a simple hemiacetal does (Section 8.4). Such an acetal of a sugar is called a **glycoside**, and the non-sugar portion is referred to as an **aglycone**; the C–O bond connecting the sugar residue to the aglycone is called a **glycosidic bond** (or link). Both α- and β-glycosides are possible (eqn 24.1).

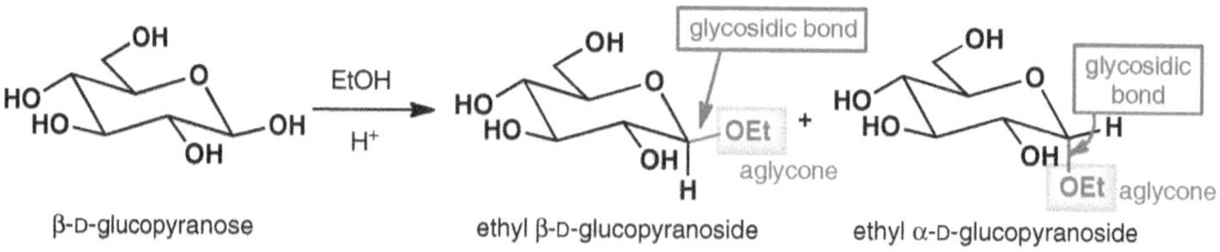

β-D-glucopyranose

ethyl β-D-glucopyranoside

ethyl α-D-glucopyranoside

(24.1)

salicin

Reaction of a monosaccharide and an amine in the presence of a catalytic acid gives an *N*-**glycoside**, and nucleosides (Section 24.2) are the β-*N*-glycosides of D-ribose or deoxy-D-ribose (ribose without the OH at C2, see later) with certain heterocyclic bases (see Sub-section 24.2.1).

Exercise 24.4

Give a mechanism for the acid-catalysed reaction of hemiacetal forms of D-glucose in methanol to form methyl D-glucosides.

Panel 24.1 The anomeric effect

When an alcoholic solution of D-glucose is treated with an acid catalyst, two isomeric glycosides are obtained, the α and β anomers, usually with the α anomer in excess. The higher equilibrium proportion of the α anomer with the axial aglycone was originally surprising and regarded as anomalous—bulky substituents on the chair form of a six-membered ring generally prefer to be equatorial. This tendency for the glycosidic bond to be axial became known as the **anomeric effect**.

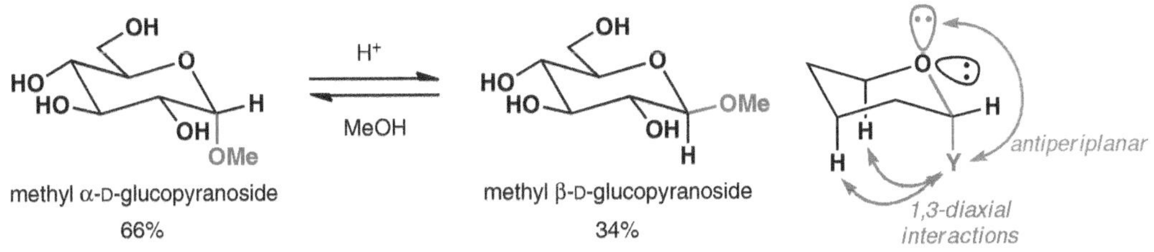

methyl α-D-glucopyranoside
66%

methyl β-D-glucopyranoside
34%

antiperiplanar

1,3-diaxial interactions

In the α anomer shown to the right of the equation, one lone pair of the endocyclic O and an axial electronegative group Y are antiperiplanar. This allows a stabilizing hyperconjugative interaction between the axial lone pair and the vacant antibonding σ* orbital of the C–Y bond which is of relatively low energy (see Sub-section 14.6.3 for a similar hyperconjugative effect). Note, however, that α-D-glucopyranose itself is less stable than the β anomer; this is probably because of the steric bulk of the OH solvated by water molecules through hydrogen bonding leading to greater adverse 1,3-diaxial effects for the α anomer.

24.1.3 Reduction and oxidation of monosaccharides

The carbonyl group of the open-chain form of a monosaccharide can be reduced to a hydroxy group by $NaBH_4$ or other reducing agents; the product is an **alditol**. For example, D-glucose gives D-glucitol (also called D-sorbitol), as shown in eqn 24.2. Many alditols (and derivatives of them) are found in nature including D-glucitol and xylitol which are used as sweeteners.

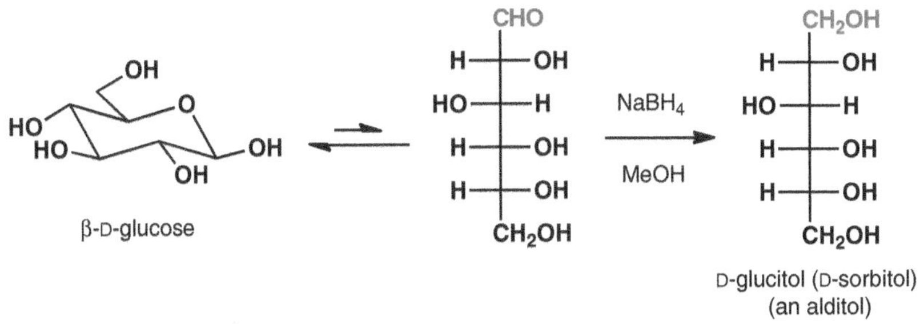

β-D-glucose

CHO	
H—	—OH
HO—	—H
H—	—OH
H—	—OH
	CH_2OH

$NaBH_4$

MeOH

CH_2OH	
H—	—OH
HO—	—H
H—	—OH
H—	—OH
	CH_2OH

D-glucitol (D-sorbitol) (an alditol)

(24.2)

Exercise 24.5

Xylitol is used as a sweetener in 'sugar-free' chewing gum, which does not promote tooth decay. Explain why xylitol is not called D-xylitol when it is obtained by reduction of D-xylose.

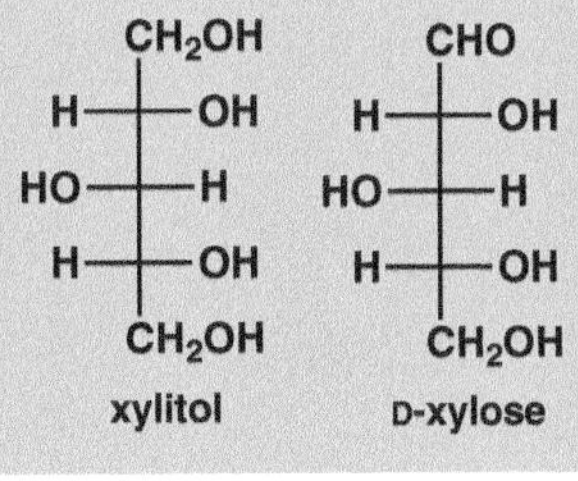

CH_2OH	
H—	—OH
HO—	—H
H—	—OH
	CH_2OH

xylitol

CHO	
H—	—OH
HO—	—H
H—	—OH
	CH_2OH

D-xylose

Products of the oxidation of monosaccharides depend on the oxidizing agents and reaction conditions. Nitric acid oxidizes both the aldehyde group and the primary alcohol group of an aldose to give a dicarboxylic acid called an **aldaric acid** (eqn 24.3).

D-glucose → glucaric acid (an aldaric acid), HNO$_3$, H$_2$O, 0 °C (24.3)

Historically, sugars which can be oxidized easily are called *reducing sugars*. Cyclic hemiacetals are in equilibrium with open-chain aldose (or ketose) forms, so they are reducing sugars. Glycosides (acetals) of the same monosaccharide units, however, are not easily oxidized and so are non-reducing sugars.

Milder oxidizing agents will selectively oxidize the aldehyde group of an open-chain aldose to give an **aldonic acid**; the one from D-glucose, for example, is D-gluconic acid (eqn 24.4). Some Ag(I) and Cu(II) salts are suitably mild oxidants which were developed as visual tests for the aldehyde group. Generally used reagents include Tollens' reagent (Ag$_2$O in aqueous ammonia which forms a mirror of Ag(0)) and Benedict's and Fehling's reagents (blue alkaline solutions of Cu^{2+} complexed by tartrate and citrate anions, respectively, are reduced to give a brick-red precipitate of Cu$_2$O), as shown in eqn 24.4. Under the basic conditions of these tests, however, a ketose readily isomerizes to an aldose (see Chapter 17, Problem 17.12), so ketoses also give positive results with the Ag(I) or Cu(II) reagents.

D-glucose → D-gluconic acid (an aldonic acid) (or the carboxylate salt); Ag$_2$O, NH$_3$, H$_2$O or Cu^{2+}, HO$^-$, H$_2$O or Br$_2$, H$_2$O; + Ag(0) mirror or Cu$_2$O (s), brick-red or Br$^-$ (24.4)

Bromine, however, can also selectively oxidize aldoses to aldonic acids in aqueous solution, but not ketoses because the isomerization is too slow in neutral solutions. Consequently, decoloration of aqueous bromine will distinguish an aldose from a ketose, and bromine is a suitable oxidant for the preparation of an aldonic acid.

Exercise 24.6

Which of the following aldohexoses gives an optically inactive aldaric acid by oxidation with HNO$_3$?

(a) D-glucose   (b) D-mannose   (c) D-galactose

The structure of L-ascorbic acid (vitamin C) resembles that of a monosaccharide. Vitamin C is readily oxidized to the dehydro form, so it acts as an antioxidant, and both forms are present in our body. Vitamin C is synthesized from D-glucose both biochemically (in plants and many animals) and industrially. However, humans cannot synthesize vitamin C in our bodies, so we have to take it in our diet—principally in fresh fruit and vegetables.

L-ascorbic acid (vitamin C) ⇌ L-dehydroascorbic acid; oxidation / reduction

(© Die Post)
A molecular model of vitamin C is shown on this postage stamp, which commemorates the International Year of Chemistry, 2011.

24.1.4 Disaccharides and polysaccharides

A disaccharide comprises two monosaccharide units connected by a glycosidic bond at the anomeric carbon of (at least) one unit; in other words, it is derived from two mono-saccharides with the elimination of a water molecule. **Maltose** and **cellobiose** are stereoisomeric disaccharides, each containing two units of D-glucose. Maltose is a hydrolysis product of **starch** (see below) and has an α (1→4)-glycosidic link. In contrast, cellobiose is a hydrolysis product of **cellulose** and has a β (1→4)-glycosidic link.

maltose

cellobiose

Note that the configuration of the OH at C1 of the second sugar unit may be α or β and the two diastereoisomers (being hemiacetals) can interconvert just as monosaccharide hemiacetal anomers do. The systematic name of maltose is α-D-glucopyranosyl-(1→4)-α(or β)-D-glucopyranose and that of cellobiose is β-D-glucopyranosyl-(1→4)-α(or β)-D-glucopyranose.

Lactose, another disaccharide, is a main sugar constituent of the milk of mammals and is composed of D-galactose and D-glucose units linked by a β (1→4) bond.[1] Its structure is similar to that of cellobiose, but the configuration at C4 of its galactose unit is opposite to that at C4 of the glucose unit of cellobiose. Another disaccharide, **sucrose**, is familiar to us as table sugar, and comprises D-glucose and D-fructose components linked at the anomeric carbons of both sugar units.

lactose

sucrose

The systematic name of lactose is β-D-galactopyranosyl-(1→4)-α(or β)-D-glucopyranose and that of sucrose is α-D-glucopyranosyl β-D-fructofuranoside.

---

[1] The intestinal enzyme *lactase* in infants who are breast-fed catalyses lactose hydrolysis to give D-galactose and D-glucose. Those who are 'lactose intolerant' lack this enzyme and may suffer health problems if their diet contains lactose (e.g. from dairy products). Sucrose is hydrolysed enzymatically in the digestive system to give D-glucose and D-fructose.

Exercise 24.7	Explain why lactose is a reducing sugar but sucrose is not.

Two important polysaccharides, **starch** and **cellulose**, are both (condensation) polymers of D-glucose, and one important difference between them is the stereochemistry of their glycosidic linkages. Starch has α (1→4)-glycosidic links whereas they are β (1→4) in cellulose. This difference causes the two polysaccharides to have quite different chemical and physical properties: starch, for example, is soluble in water but cellulose is not. When we consume starch in our diet, it is digested mainly through the action of our enzymes, α-*amylase* and *maltase*, which convert starch into glucose. The glucose is then metabolized by glycolysis in cells to produce energy. However, we cannot digest cellulose (the most abundant organic material on Earth)—we do not have the necessary enzymes.

> Enzymes which catalyse the hydrolysis of cellulose are called *cellulases* and are produced mainly by fungi and bacteria. Although most insects and mammals lack cellulases, some (e.g. termites and cattle) can digest cellulose with the help of symbiotic microorganisms in their digestive systems which produce cellulases.

Example 24.1	Two types of polymers, each with up to about 1000 glucose units, are found in starch: *amylose* and *amylopectin*, and most starch contains both. Amylose is a linear polysaccharide joined by α (1→4)-glycosidic links as discussed above. The polymer backbone of amylopectin is also joined up by α (1→4)-glycosidic links but α (1→6)-glycosidic links lead to branching as well. Show a partial structure of amylopectin.

Solution

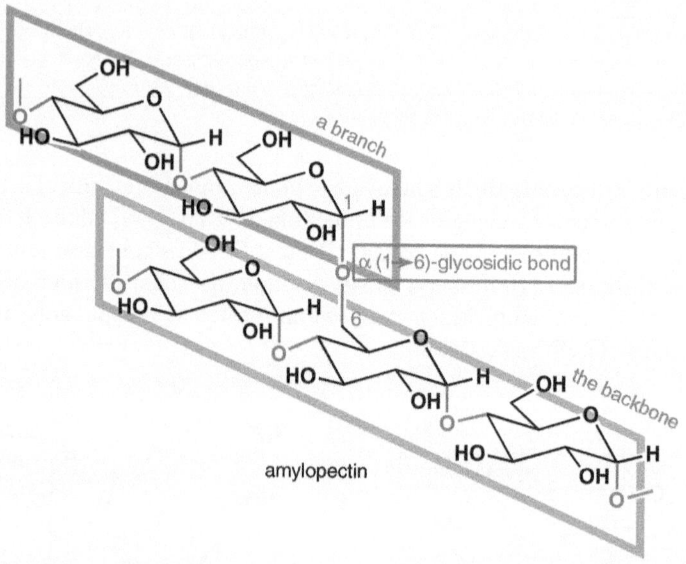

amylopectin

24.2 Nucleic Acids

Nucleic acids, **RNA (ribonucleic acid)** and **DNA (deoxyribonucleic acid)**, play a fundamental role in the storage and transfer of genetic information. They are polymers of **nucleotides** which consist of three parts: a heterocyclic base, a sugar, and phosphate

(Figure 24.2). The sugar in RNA is D-ribose and it is 2-deoxy-D-ribose (which lacks an OH at C2) in DNA. The combination of the sugar and the heterocyclic base is called a **nucleoside**.

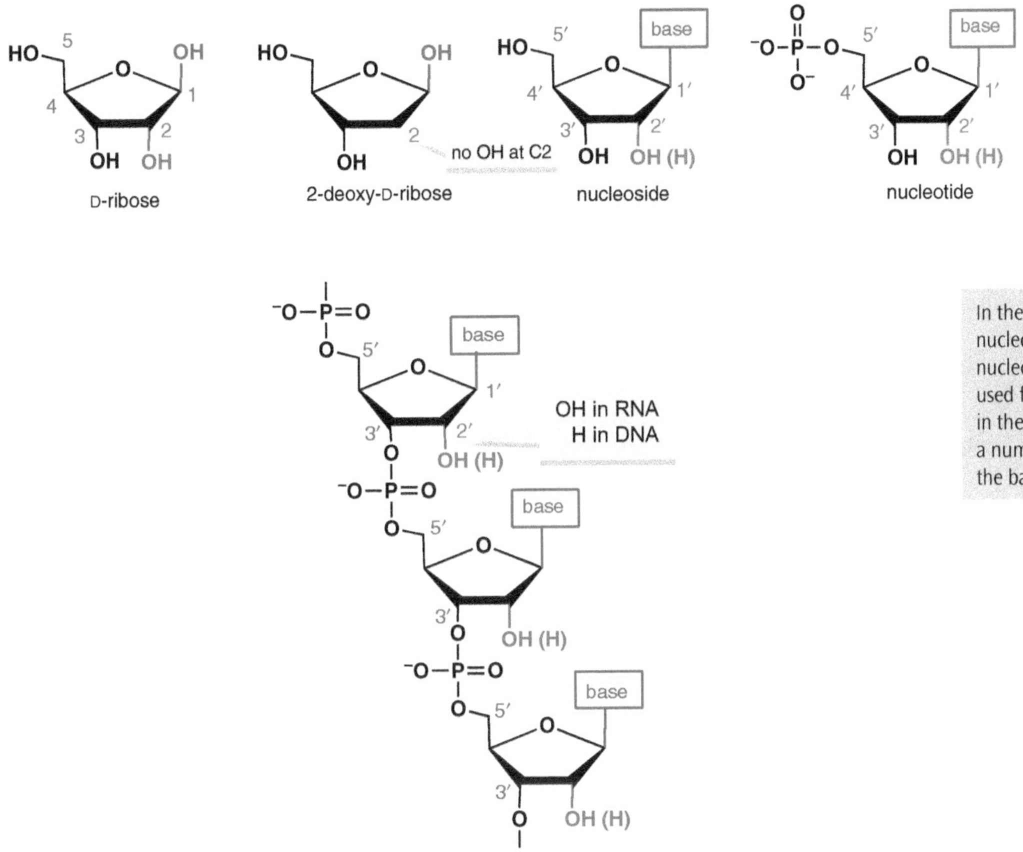

Figure 24.2 The structure of nucleic acids.

In the names of nucleosides and nucleotides, a prime (') is used for position numbers in the sugar but not (if a number is needed) in the base.

24.2.1 Nucleosides and nucleotides

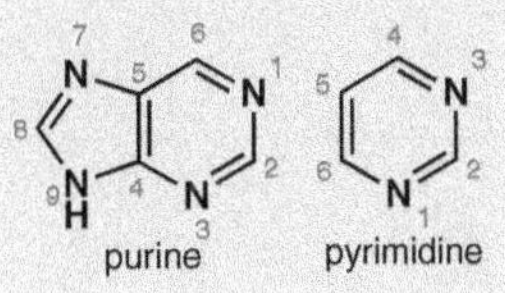

Nucleosides are *N*-glycosides of ribose or deoxyribose in which heterocyclic bases are β-linked to the anomeric carbon. As mentioned above, nucleotides (the monomers of the two types of nucleic acids) are the phosphate derivatives of nucleosides. There are five different bases involved in nucleosides and nucleotides, and they are all heteroaromatic compounds, **purines** and **pyrimidines**, as listed in Table 24.1. The two purine bases, **adenine** (**A**) and **guanine** (**G**), are common to both DNA and RNA, as is the pyrimidine, **cytosine** (**C**). Of the other two pyrimidines, **uracil** (**U**) is found only in RNA and **thymine** (**T**) only in DNA. So, there is a set of four ribonucleosides (containing A, G, C, and U) for RNA, and four deoxyribonucleosides (containing A, G, C, and T) for DNA.

Nucleosides and nucleotides are important not only as components of nucleic acids; they also play crucial roles in various biochemical reactions and some examples have already been mentioned. *S*-**Adenosylmethionine** (**SAM**) is a biochemical methylating agent (Panel 12.1) while **coenzyme A** (**CoA**) is a thiol which acts as an acylating agent via a thioester (Sub-section 14.6.2 and Panel 17.3). **Nicotinamide adenine dinucleotide** (**NAD+**, Panel 10.3) is one of Nature's oxidizing agents, and **adenine triphosphate** (**ATP**) is another important molecule whose exothermic hydrolysis to **adenine diphosphate** (**ADP**) and phosphate facilitates various biochemical processes.

Table 24.1

Structures of bases and nucleosides found in nucleic acids

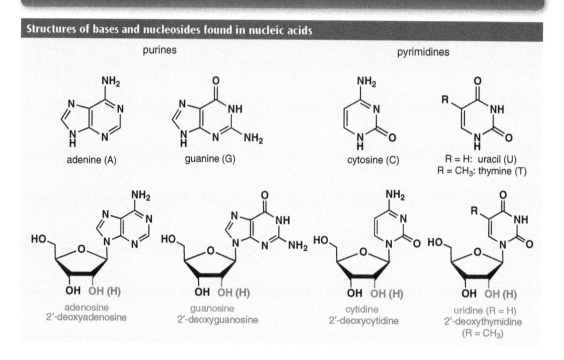

Some biologically important derivatives of a nuleoside, adenosine

The degree of protonation of phosphate groups in nucleotides will depend upon the pH of the aqueous environment.

Exercise 24.8

Pyrimidines have tautomeric forms. Draw structures of four tautomers of each of (a) cytosine and (b) thymine.

24.2.2 DNA and RNA

Nucleotides are joined together through phosphate ester links between the 5′-hydroxy group of one sugar and the 3′-hydroxy of another to form a nucleic acid molecule. In other words, the **nucleic acid** chain is a long unbranched (condensation) polymer of nucleotide monomers, with alternating sugar and phosphate units, and the bases attached to the anomeric carbons of the sugars. By convention, the sugar–phosphate backbone is written from the 5′-end of the sugar residues on the left to the 3′-end on the right, as shown in Figure 24.2 (p. 535), and with representative bases included in Figure 24.3 for DNA. The **base sequence** along the sugar–phosphate backbone (again, with the 5′-ends of the sugars to the left and the 3′-ends to the right) is the critical structural feature of a nucleic acid by which it retains genetic information; the sequence is often shown in an abbreviated form using capital letters to represent the bases. In this way, the partial structure in Figure 24.3 of a DNA molecule is fully described by the base sequence –ACGT–.

Note that two of the OH groups of a phosphoric acid molecule, O=P(OH)$_3$, are involved in each (di-ester) link between two sugar units in nucleic acids, and the other is dissociated at the physiological pH (7.3).

24.2.3 Base pairing in nucleic acids

Nucleic acids include intermolecular **hydrogen bonds** between specific pairs of bases; in particular, the hydrogen bonds are exclusively between purine–pyrimidine pairs, i.e. A–T (or A–U) and G–C, as shown in Figure 24.4.

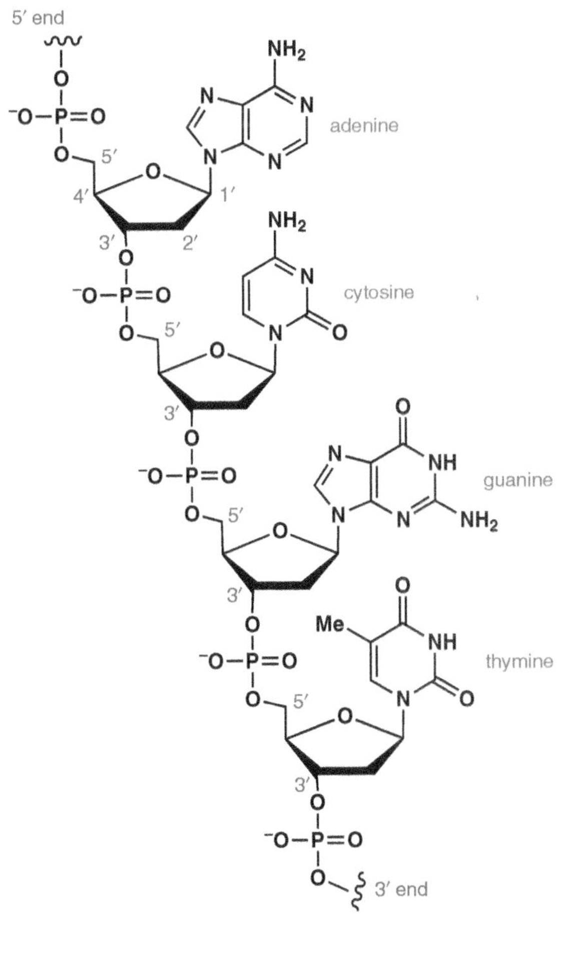

Figure 24.3 Four nucleotide units in part of a DNA molecule with the base sequence –ACGT–.

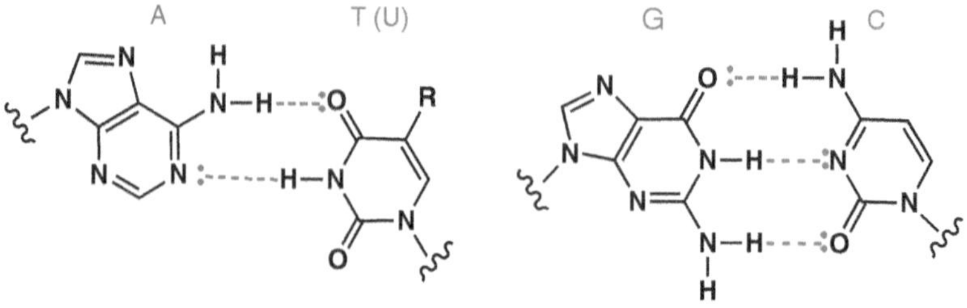

Figure 24.4 Hydrogen bonds formed between specific pairs of bases.

These specific hydrogen bonds (G will only H-bond to C, and A will only H-bond to T or U) are fundamental to the **double helix** structure of DNA, and the replication and transmission of genetic information. Because of the specificity of the base pairing, the two strands of a DNA double helix are **complementary**, i.e. a given sequence of one strand requires a unique sequence in its complementary strand which runs in the opposite direction (one strand is said to be *antiparallel* to the other).

In the **replication** of DNA, the two strands of a double helix first unwind, then each strand acts as a template for the synthesis of a new strand. Each new strand is complementary to an original owing to the specificity of the base pairings (Figure 24.5). In this way, two new double helices are formed, and both are the same as the original, so the genetic information is passed on to a new DNA chain. This simplistic description is of quite a complex process catalysed by enzymes—DNA polymerases.

How is the genetic information stored? The information is for the synthesis of proteins, and the base sequence in a section of a DNA specifies the amino acid sequence of a protein to be biosynthesized. A sequence of three bases, called a **codon** (or a **triplet code**) specifies a single amino acid. With four different bases, $4^3=64$ different sequences of three bases (codons) are available, which are more than enough for the 20 different amino acids found in proteins (see Section 24.3). Most amino acids are specified by two or more different codons, and some other codons are used to terminate the synthesis.

An RNA molecule differs structurally from a DNA molecule as described above, and usually exists as a single strand. Although DNA stores the genetic information, RNA participates in the processes for using this information, and RNA is classified according to its function or cellular location.

Figure 24.5 Replication of DNA.

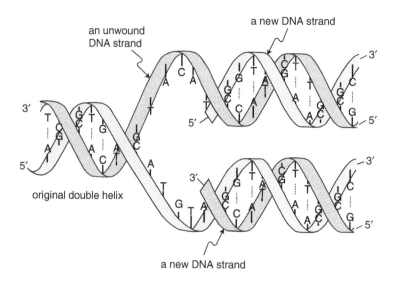

an unwound DNA strand

a new DNA strand

original double helix

a new DNA strand

Draw the structure of a fragment of DNA which is complementary to –ACGT–.

Exercise 24.9

24.3 Amino Acids, Peptides, and Proteins

The amino group of one α-**amino acid** can form an **amide bond** with the carboxy of another (the same or different) and the product is called a **dipeptide**; if another adds by another amide bond, we have a **tripeptide**, and so on.[2] Up to a dozen or so α-**amino acid** molecules joined in this way give an **oligopeptide** and, when many are joined, we have a **polypeptide**. If we do not wish to differentiate between these different types, we can simply say **peptide**.

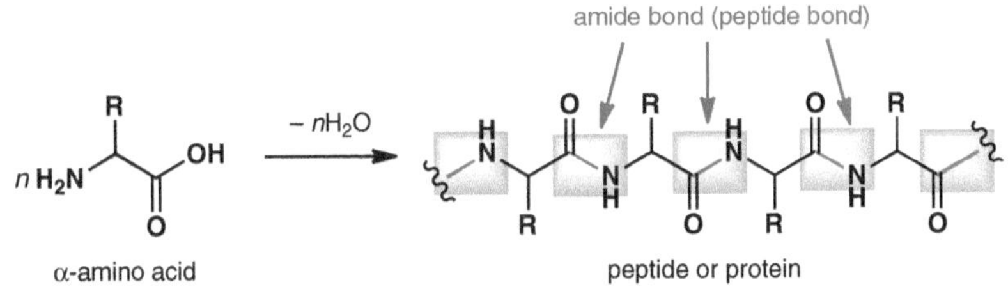

Proteins are naturally occurring polypeptides and are hugely important biological molecules. They are found in structural components of the bodies of animals and have diverse functions within and between cells. For example, they facilitate the transport of ions and molecules through cell membranes and, as enzymes, proteins catalyse biochemical reactions; some peptides act as hormones.

24.3.1 α-Amino acids

There are 20 α-amino acids listed in Table 24.2 which occur in natural proteins. Common names are given together with three-letter and one-letter abbreviations. These α-amino acids are all chiral except glycine and their configuration (related to glyceraldehyde) is L at the α carbon, so they are L-α-amino acids: the amino group appears to the left of the α carbon in the conventional Fischer projection formula.

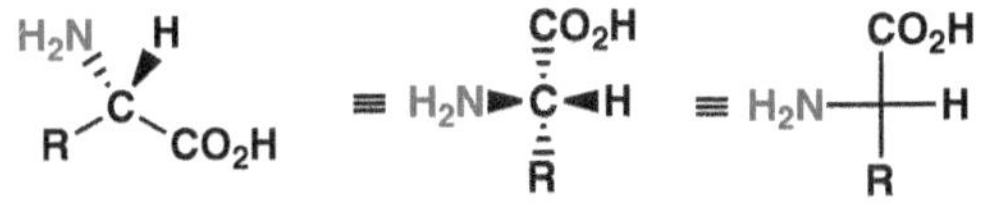

Structure of an L-α-amino acid.

We humans have to take in our diet the nine so-called *essential* amino acids which we need but cannot synthesize in our body: they are indicated by superscript b) in Table 24.2.

With the exception of cysteine, the configuration of L-α-amino acids at the α carbon is *S* according to the Cahn–Ingold–Prelog convention (see Section 11.2).

---

[2] Formation of an amide bond (in the present context, also called a **peptide bond**) between two α-amino acid molecules is formally a condensation reaction (see Chapter 9), but (in practice) it has to be achieved indirectly.

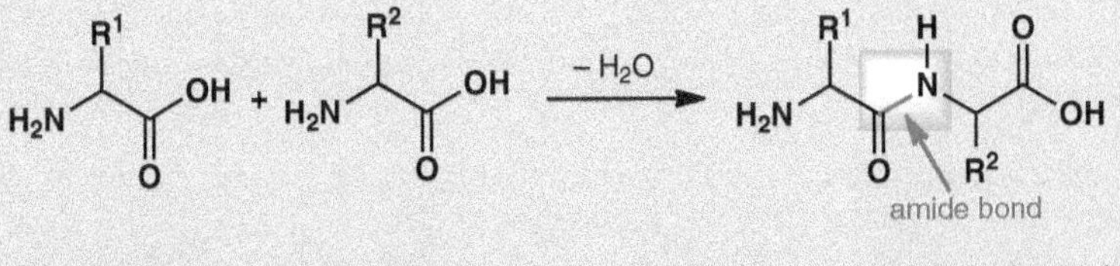

In other words, polypeptides, like polysaccharides and nucleic acids, are condensation polymers (Section 9.6), each connection between one monomer unit and the next (which can be the same or different) being formed with the elimination of a water molecule. The links between monosaccharide units are glycosidic and the ones between nucleotides are ester bonds.

Table 24.2

α-Amino acids[a]

(a) Amino acids with a nonpolar side chain

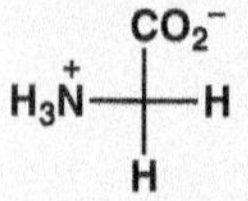

glycine
Gly, G
2.34, 9.60

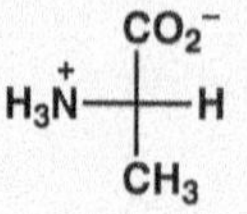

alanine
Ala, A
2.34, 9.69

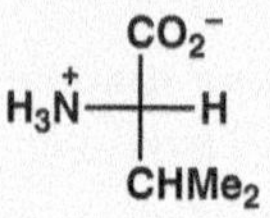

valine[b]
Val, V
2.32, 9.62

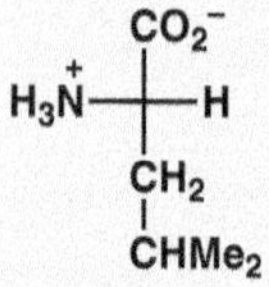

leucine[b]
Leu, L
2.36, 9.60

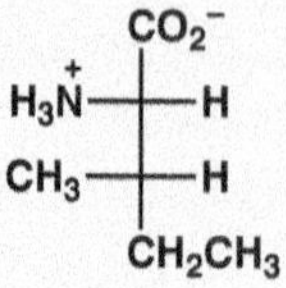

isoleucine[b]
Ile, I
2.36, 9.68

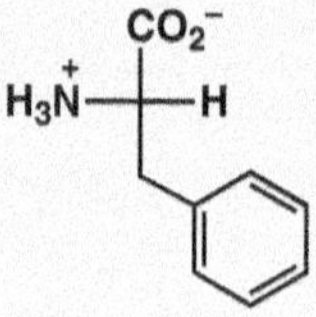

phenylalanine[b]
Phe, F
2.16, 9.18

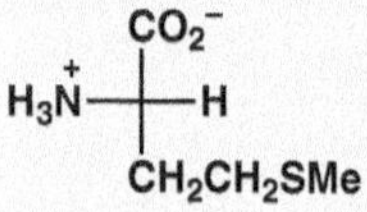

methionine[b]
Met, M
2.34, 9.69

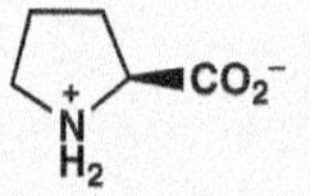

proline
Pro, P
1.99, 10.60

(b) Amino acids with a polar side chain

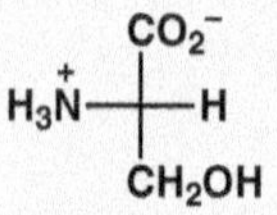

serine
Ser, S
2.21, 9.15

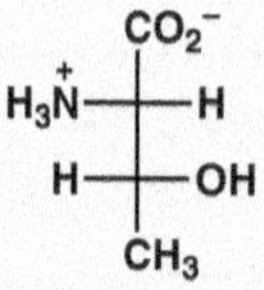

threonine[b]
Thr, T
2.63, 9.10

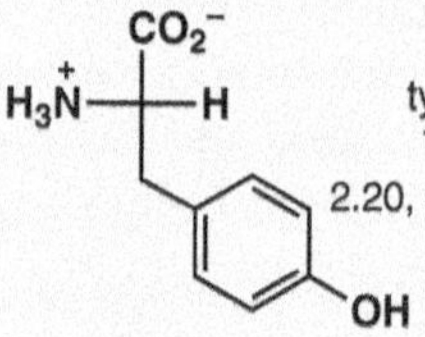

tyrosine
Tyr, Y
2.20, 9.11, 10.07

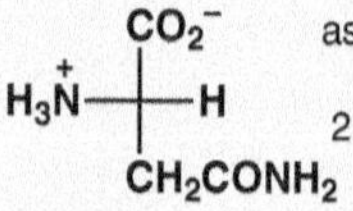

asparagine
Asn, N
2.02, 8.84

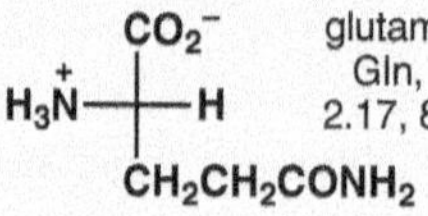

glutamine
Gln, Q
2.17, 8.84

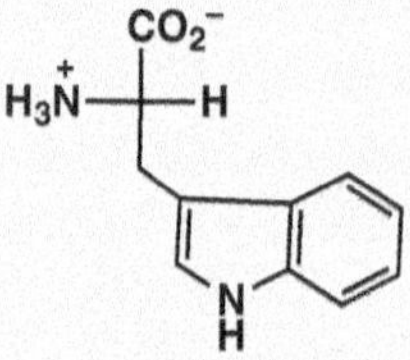

tryptophan[b]
Trp, W
2.38, 9.39

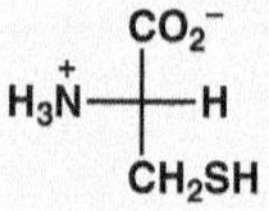

cysteine
Cys, C
1.92, 10.46, 8.35

(c) Amino acids with an acid/base side chain

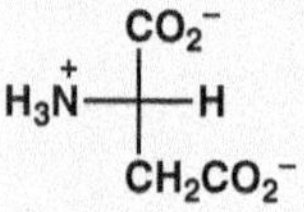

aspartic acid
Asp, D
2.09, 9.82, 3.86

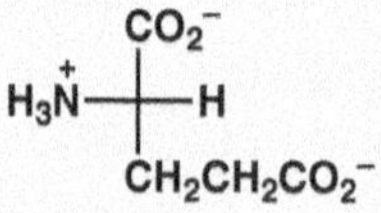

glutamic acid
Glu, E
2.19, 9.67, 4.25

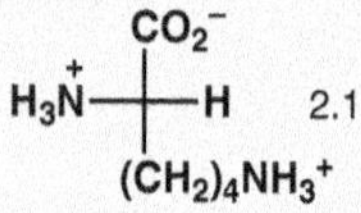

lysine[b]
Lys, K
2.18, 8.95, 10.79

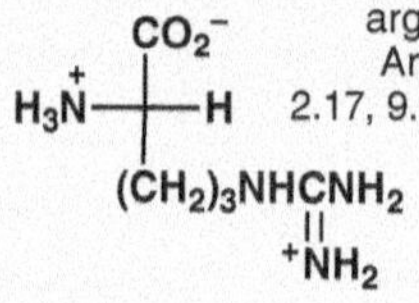

arginine
Arg, R
2.17, 9.04, 12.48

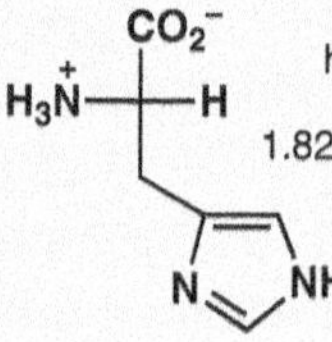

histidine[b]
His, H
1.82, 9.17, 6.04

a. Structures are those in aqueous solution buffered at pH 7. Standard abbreviations are given together with pK_a values in the order: $-CO_2H$, α-ammonio, and side-chain groups (in their conjugate acid forms).
b. Essential amino acids.

Explain why the configuration of the α C of cysteine is *R* when its Fischer projection appears analogous to the others in Table 24.2 which are all *S*.

Isoleucine and threonine each have a second chirality centre so diastereoisomers are possible. However, the natural form of each is the single diastereoisomer shown (as a Fischer projection) in Table 24.2. Assign the *R/S* configuration of the chirality centres of these amino acids, and draw their structures by line-angle drawings with wedged bonds.

The protonated forms of α-amino acids (in acidic solution) have pK_a values of about 2 (carboxy group) and 9 (ammonio group), so α-amino acids exist as zwitterions in a buffered neutral aqueous solution, as shown in Scheme 24.2 (see Section 6.6).

Note that the pK_a value of the carboxy group of α-amino acids is relatively low owing to the electron-withdrawing effect of the adjacent positive ammonio group.

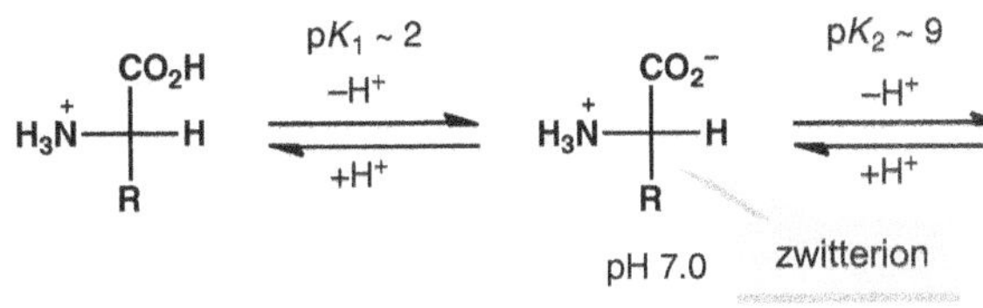

Scheme 24.2 Dissociation of an α-amino acid.

Some amino acids have another ionizable group so they have a third dissociation constant (Table 24.2). These side-chain functional groups are usually carried over when the amino acid is part of a protein, and their properties influence the behaviour of the protein—typically the catalytic property of enzymes.

The pH at which the concentration of the zwitterionic (neutral) form is a maximum (i.e. when the concentrations of cationic and anionic forms are equally small) is called the **isoelectric point**, p*I*. For a simple amino acid, this is numerically equal to the mean of its two pK_a values:

$$pI = (pK_1 + pK_2)/2$$

The concept of the isoelectric point is useful in the practical separation and purification of amino acids and peptides using *electrophoresis* in which compounds in solution migrate in an electric field according to their overall charge.

When there is a third ionizing group, we need to think more carefully about the relative concentrations of the different possible ionic forms under different pH conditions. Explain why p*I*=(pK_1+pK_2)/2 for α-amino acids with an extra acidic group using aspartic acid as an example.

Solution

For α-amino acids with an extra acidic group, the cationic form is first converted to the neutral, then to the singly negative anionic form, by dissociations corresponding to pK_1 and pK_2, as illustrated for aspartic acid, so p*I* is the mean of pK_1 and pK_2, i.e. p*I*=(pK_1+pK_2)/2.

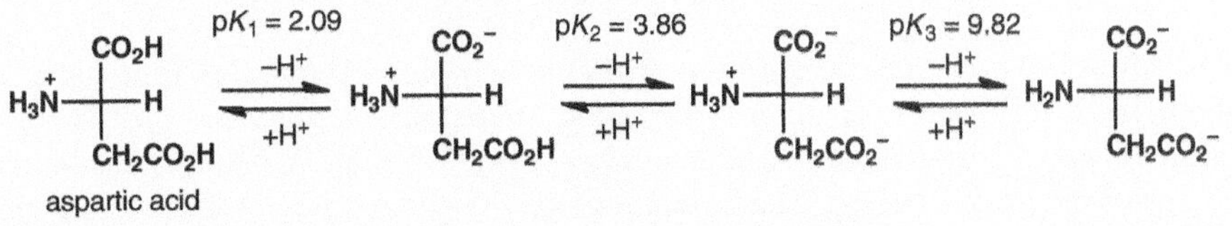

aspartic acid

Consequently, the p*I* for aspartic acid=2.98. The third dissociation corresponding to pK_3 generates the doubly negative anionic form and is not relevant to the calculation of p*I*.

Exercise 24.12

Using lysine as an example, explain why $pI = (pK_2 + pK_3)/2$ for α-amino acids with an extra basic group.

Exercise 24.13

Calculate isoelectric points of (a) alanine, (b) glutamic acid, and (c) histidine from the data in Table 24.2.

24.3.2 Structures of peptides

Peptides, like α-amino acid monomers, usually exist as zwitterionic forms at neutral pH but we shall give non-ionized neutral forms because of possible complications caused by ionization of some side chain groups.

When two different amino acids are joined by a peptide bond, two isomeric dipeptides are possible. For example, phenylalanine (Phe) and aspartic acid (Asp) give Phe-Asp and Asp-Phe, depending on which amino group connects to which carboxy group in the formation of the peptide bond.

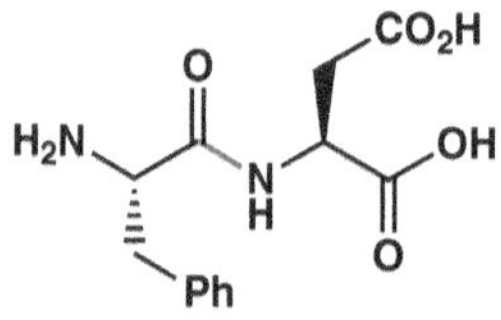

Phe-Asp (F-D) 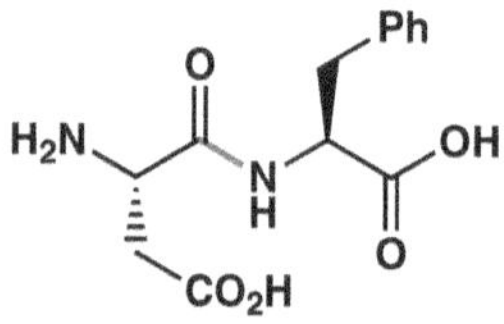   Asp-Phe (D-F)

Aspartame, an artificial sweetener, is the monomethyl ester of Asp-Phe.

aspartame
(sweetener)

By convention, peptides and proteins are represented from the amino end (**N-terminus**) on the left to the carboxy end (**C-terminus**) on the right, so that the left-most amino acid residue has a free NH_3^+ group and the right-most amino acid residue has a free CO_2^- group. We shall now look at a few representative biochemically important peptides.

Two pentapeptides found mainly in the brain and nerve tissue, and which have analgesic and opiate properties, are called enkephalins. They differ in the amino acid at the *C*-terminus: [met]enkephalin is Tyr-Gly-Gly-Phe-Met and [leu]enkephalin is Tyr-Gly-Gly-Phe-Leu.

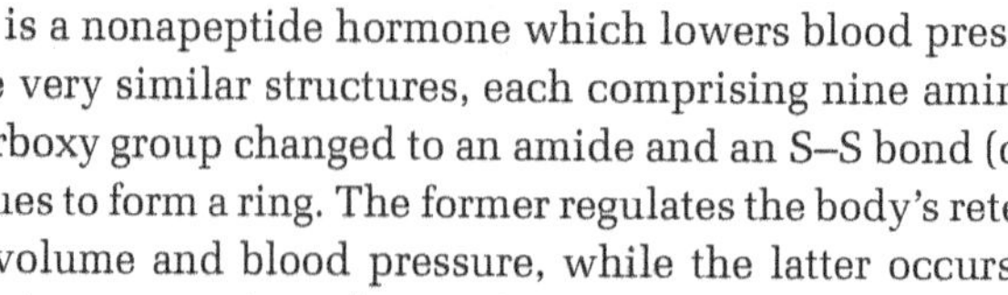

enkephalin $\left(\begin{array}{l} R = CH_2SMe: \text{[met]enkephalin} \\ R = CHMe_2: \text{[leu]enkephalin} \end{array} \right)$

Bradykinin is a nonapeptide hormone which lowers blood pressure. Vasopressin and oxytocin have very similar structures, each comprising nine amino acid units with the *C*-terminal carboxy group changed to an amide and an S–S bond (or bridge) between the two Cys residues to form a ring. The former regulates the body's retention of water, which affects urine volume and blood pressure, while the latter occurs only in females and stimulates uterine contractions during childbirth and facilitates breastfeeding.

Bradykinin:

Arg-Pro-Pro-Gly-Phe-Ser-Pro-Phe-Arg

Vasopressin:

$$\text{Cys-Tyr-Phe-Gln-Asn-Cys-Pro-Arg-Gly-NH}_2$$
$$\text{S}\rule{2.5cm}{0.4pt}\text{S}$$

Oxytocin:

$$\text{Cys-Tyr-Ile-Gln-Asn-Cys-Pro-Leu-Gly-NH}_2$$
$$\text{S}\rule{2.5cm}{0.4pt}\text{S}$$

Insulin, a hormone secreted by the pancreas, regulates the metabolism of glucose and consists of a peptide chain (A) of 21 amino acid residues linked by two disulfide bridges to another chain (B) of 30 amino acid residues, with a third disulfide bridge connecting cysteine units within chain A.

> **Exercise 24.14**
>
> Glutathione is a tripeptide coenzyme represented by γ-Glu-Cys-Gly, in which one of the peptide bonds involves the side-chain carboxy group of the glutamic acid residue. Draw the structure of glutathione.

24.3.3 Synthesis of peptides

The synthesis of a specific peptide is not straightforward because of the difunctional nature of amino acids. Even from just two different amino acids, four isomeric dipeptides are possible; e.g. from alanine and glycine, Ala-Ala, Gly-Gly, Ala-Gly, and Gly-Ala may be formed.

To solve this problem, the amino group of the amino acid required at the N-terminus is protected, and its carboxy group is activated before the second amino acid is added. Widely used amine protecting groups are t-butyloxycarbonyl (**Boc**) and 9-fluorenylmethoxycarbonyl (**Fmoc**), which we encountered in Sub-section 23.3.2. The former can be introduced using di-t-butyl dicarbonate, and the latter with 9-fluorenylmethyl chloroformate. The carboxy group of the second amino acid is usually protected as an ester prior to the coupling.

To illustrate the strategy using Boc protection, we can look at a synthesis of the dipeptide Gly-Ala in Scheme 24.3. The first reaction is the protection of the amino group of glycine (eqn 24.5), and the protected glycine is then treated with dicyclohexylcarbodiimide (DCC) to activate the carboxy group (eqn 24.6). With its amino group protected and its carboxy group converted into a more electrophilic centre, the methyl ester of the second amino acid, alanine, is then added. There is now no alternative to the desired reaction: the electrophilically enhanced carbonyl group at the C-terminus of glycine suffers nucleophilic attack by the amino group of the alanine ester, which leads to the formation of a peptide bond (eqn 24.7). Finally, the protecting groups are removed: the Boc group is removed with acid while the alkyl ester is hydrolysed under basic conditions (eqns 24.8 and 24.9).

(**Boc$_2$O**)
di-t-butyl dicarbonate

9-fluorenylmethyl chloroformate

Other more reactive forms of a carboxylic acid such as an acyl chloride are too reactive towards side-chain groups of some amino acids. DCC yields a mildly activated derivative.

Protection of the amino group:

$$\text{H}_2\text{N}\sim\text{CO}_2^- + t\text{-BuO-(Boc}_2\text{O)-O-}t\text{-Bu} \xrightarrow[\substack{-t\text{-BuOH} \\ -\text{CO}_2}]{\text{Et}_3\text{N}} t\text{-BuO-C(O)-NH}\sim\text{CO}_2^- \quad (24.5)$$

from glycine

$$\left(\text{Boc}\underset{\text{H}}{\text{N}}\sim\text{CO}_2^- \right)$$

Activation of the carboxy group:

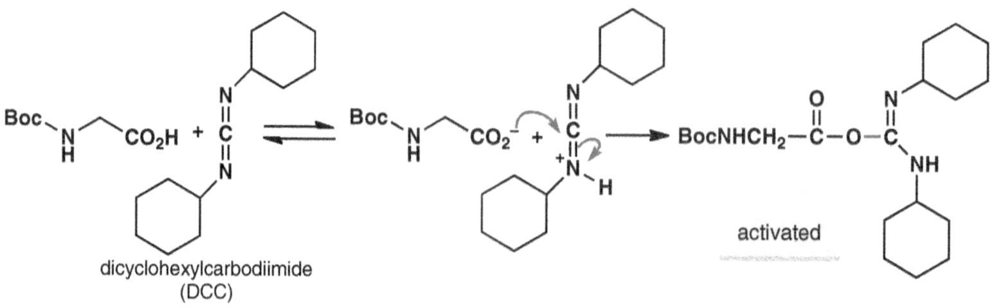

(24.6)

Coupling:

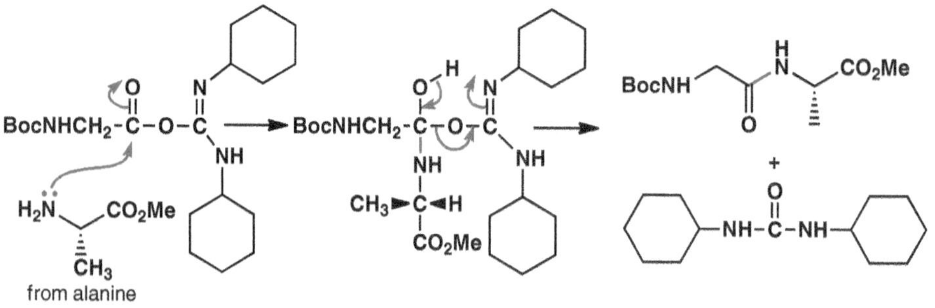

(24.7)

Deprotection:

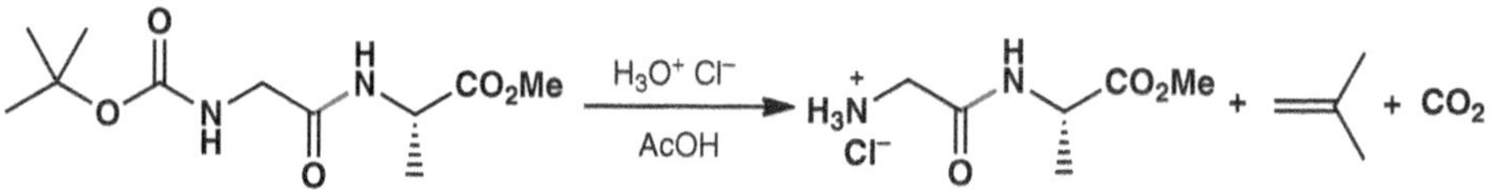

(24.8)

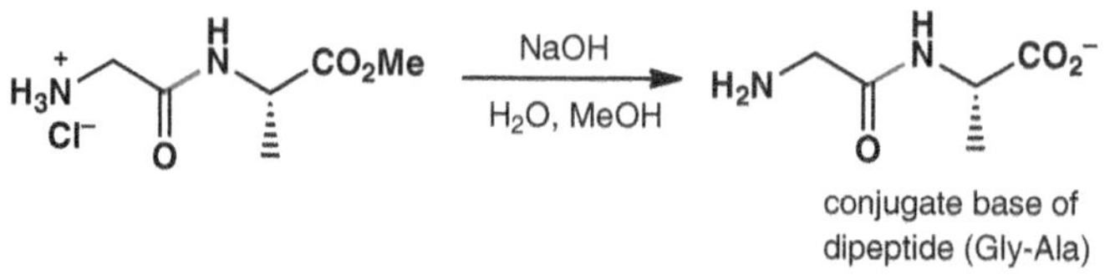

(24.9)

Scheme 24.3 An illustrative synthesis of a dipeptide.

In order to attach additional amino acid residues in the synthesis of longer chain peptides, the activation and coupling reactions must be repeated. Inevitably, the overall yield becomes lower as the number of steps increases in this multiple step, linear synthetic strategy, as discussed in Section 23.4. Moreover, the intermediate products must be isolated and purified after each step, and these manipulations are extremely laborious and time-consuming. A much more efficient method was developed, however, in the early 1960s—**solid-phase peptide synthesis**.

The Merrifield method allows a peptide synthesis to be carried out with very simple manipulations, in a short time, and in high yield. In this method, the first (*C*-terminal) amino acid is covalently bonded to an insoluble polymer, and other amino acid residues are added, one by one towards the *N*-terminus, to give the sequence of the required peptide. Impurities and by-products are easily washed away with a solvent at each stage because they, unlike the developing peptide, are not attached to the polymer. Because there is no loss of the peptide intermediate along the way, yields are high. Furthermore, the Merrifield method can be completely automated using a 'peptide synthesizer'.

Solid-phase peptide synthesis was devised by the American chemist, R.B. Merrifield, who was awarded the 1984 Nobel Prize in Chemistry.

Polystyrene modified by chloromethylation of some pendant phenyl groups is a commonly used polymer, and a representative solid-phase synthesis of a tripeptide is illustrated in Scheme 24.4.

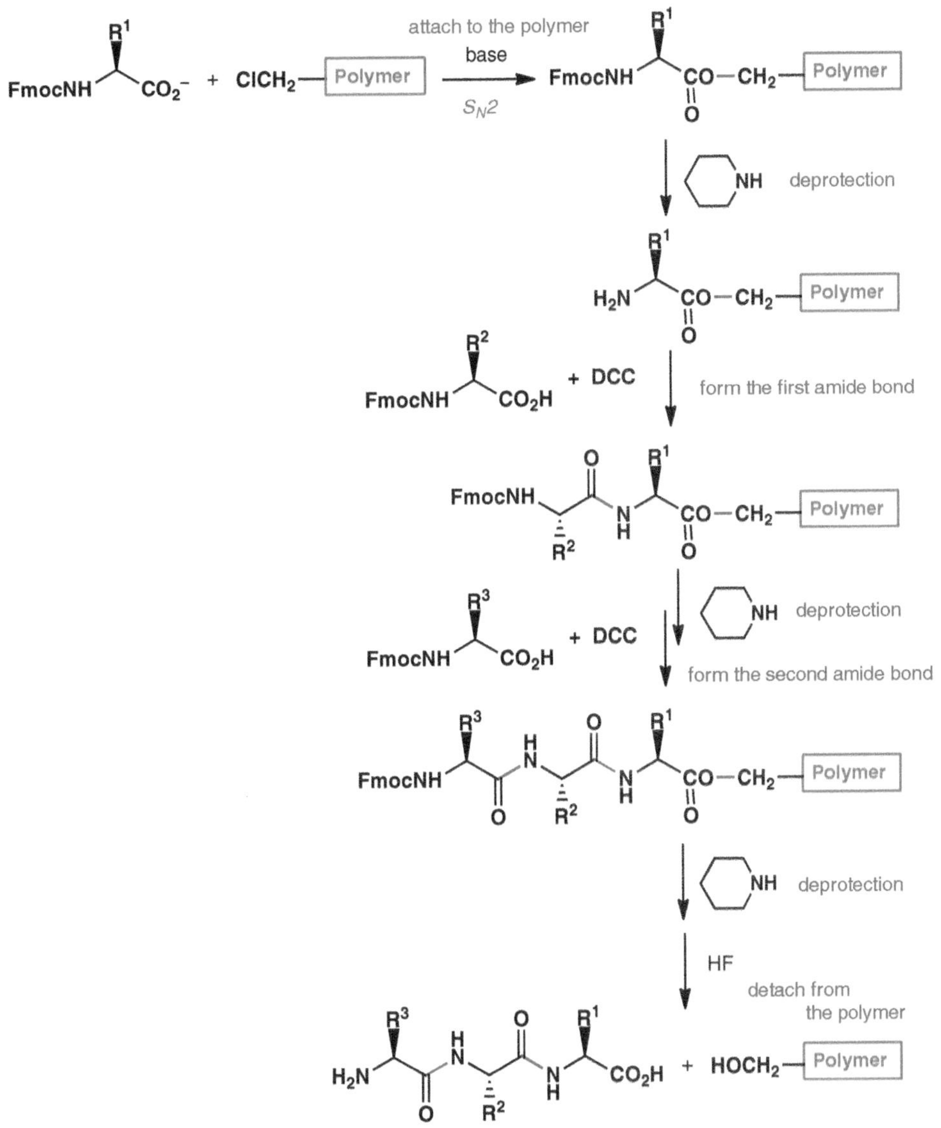

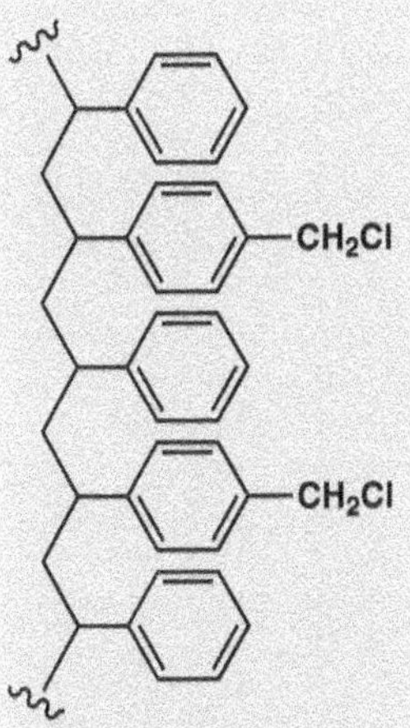

a modified polystyrene used as a solid-phase support

The amino group of each amino acid to be attached is normally protected with Fmoc in an automated peptide synthesis because only mild conditions are required for deprotection (a mechanism for which was considered in Chapter 23, Exercise 23.10).

Scheme 24.4 A schematic representation of a solid-state synthesis of a tripeptide.

The initial bonding of a protected amino acid to the polymer occurs by an S_N2 reaction of the carboxylate end of the amino acid with an electrophilic side chain of the polymer (chloromethyl in the one shown). The amino end of the polymer-bound initial amino acid is then deprotected by piperidine prior to the attachment of the next NH_2-protected amino acid by a condensation reaction of its carboxy group promoted by DCC. After deprotection of the attached dipeptide, the next condensation reaction follows using a third NH_2-protected amino acid using DCC in the same way, and so on, until the required peptide is detached from the polymer using liquid HF at 0 °C in the final stage of the synthesis.

Note that HF (bp 19.5 °C) is a highly corrosive, dangerous material and should be used carefully in a special vessel under a hood. It is able to penetrate the skin and the vapour can cause blindness.

24.3.4 Determination of peptide and protein sequences

Chemical methods for the determination of the amino acid sequence of a peptide start with identification of the *N*-terminal amino acid residue. One method, outlined in Scheme 24.5, is based on the nucleophilic aromatic substitution reaction of *2,4-dinitrofluorobenzene* (sometimes called the **Sanger reagent**) with amines. After treatment of the peptide with the Sanger reagent, the modified peptide is hydrolysed to give a mixture of its constituent amino acids; the *N*-terminal amino acid is identified by its *N*-dinitrophenyl group.

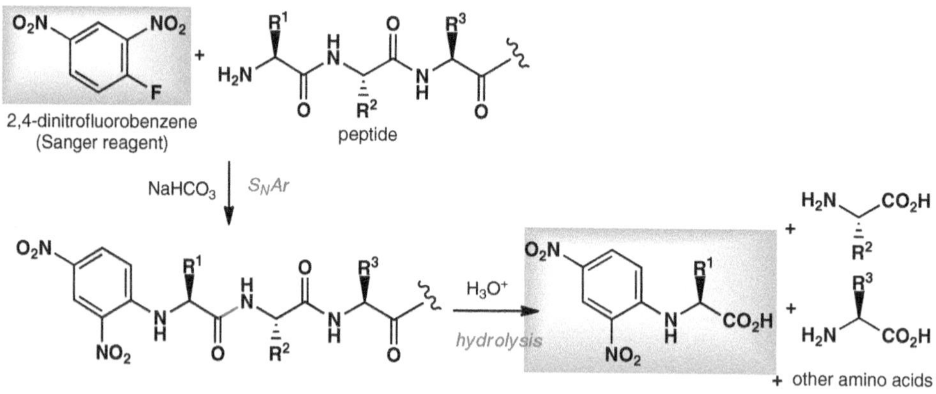

For nucleophilic aromatic substitution, see Section 18.6.

The British biochemist F. Sanger (1918–) was awarded the Nobel Prize in Chemistry in 1958 for his work on the structure of proteins, and then shared the 1980 Nobel Prize in Chemistry with others for their work on the biochemistry of nucleic acids.

Scheme 24.5 Identification of the *N*-terminal amino acid of a peptide using the Sanger reagent.

The **Edman degradation** is a more versatile method which allows determination of the sequence of amino acids; *phenyl isothiocyanate* is the key reagent in this procedure (Scheme 24.6). It reacts with the *N*-terminal amino acid unit to give a thiourea intermediate.

Scheme 24.6 Peptide sequencing by the Edman degradation.

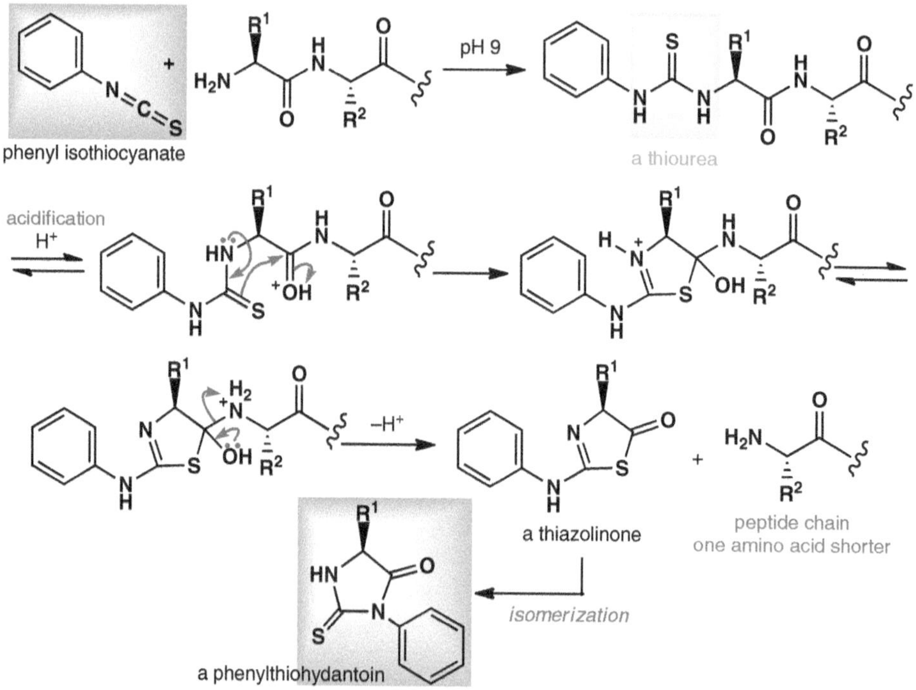

Acidification removes the *N*-terminal unit as a substituted thiazolinone, which isomerizes to a more stable phenylthiohydantoin, but leaves the rest of the chain intact. A second reaction is then carried out on the remaining peptide (which is one unit shorter) to give the thiazolinone corresponding to its *N*-terminal amino acid, and so on. The phenylthiohydantoin derivatives are characterized to determine the *N*-terminal amino acids of the ever shortening peptides, one by one. The repeated cycles to identify a sequence of *N*-terminal amino acids have been automated in a *peptide sequenator (sequencer)*.

Larger peptides and proteins are dissected into smaller peptides by hydrolysis catalysed by enzymes. Depending on the enzyme used, the cleavages are at specific positions, which provides additional information about the amino acid sequence. For example, **chymotrypsin** cleaves peptides on the *C*-terminal side of aromatic amino acid residues (Phe, Tyr, and Trp) and (more slowly) some other residues with bulky side-chains, e.g. Leu, Met, Asn, and Glu. **Trypsin**, on the other hand, cleaves peptides on the *C*-terminal side of the basic residues, Arg and Lys. Shortened peptide chains are then sequenced and their original connectivity is deduced, which leads to the entire amino acid sequence.

We can illustrate this process with a simple hypothetical example of a peptide with 12 amino acid residues (Figure 24.6), although (in this case) the peptide is short enough to be sequenced by an automatic sequencer. The sequence could be confirmed by synthesis of the original peptide from the fragments.

A polypeptide can now be sequenced by mass spectrometry (Chapter 25). The protonated (charged) peptide is energized to undergo fragmentation in the gas phase. Analysis of the fragments allows their identification and deduction of the sequence of the peptide, but details of the technique are beyond the scope of this book.

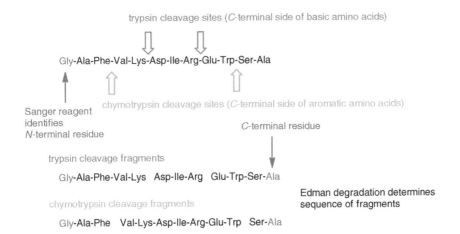

Figure 24.6 Determination of the amino acid sequence of a peptide with twelve amino acid units.

Which of the following tripeptides can be hydrolysed with chymotrypsin or trypsin?

(a) Phe-Ser-Arg   (b) Gly-Lys-Leu   (c) Ala-Trp-Glu

Exercise 24.15

24.3.5 Structures of proteins

The amino acid sequence of a protein which is defined by the covalent bonds is called the **primary structure**; it tells us the atom connectivity of the molecule but provides no information about its shape, for example. The three dimensional structure of a protein with a particular sequence of amino acids is controlled principally by specific non-covalent interactions between different parts of the polypeptide chain. The interacting groups are sometimes quite far apart along the chain, but held close together by the interactions. Although any single interaction may be quite weak, the effect of several acting together is sufficient to impose a **secondary structural *motif***, i.e. *a local conformation of the polypeptide chain*; major local features of this sort are the **α helix** and the **β-pleated sheet**.

The planarity of the amide bond is due to the partial double bond character of the C–N bond (~40%); rotation about this bond would severely reduce the quality of the overlap between the unhybridized 2p orbitals on the carbonyl C and the N.

Due to the partial positive charge on the N and the partial negative on the carbonyl O, hydrogen bonding between amide groups is strong.

The main stabilizing influences upon the secondary structure are (i) *the planarity of the amide bond*, (ii) *hydrogen bonding between the N–H of one amide and the C=O oxygen of another*, and (iii) *attractive electrostatic interactions between side chain groups.* Adverse steric interactions between side chain groups also have an effect, as we shall see.

Although the amide bond is relatively rigid, rotation about the bond between an α carbon and the nitrogen is relatively free, which allows peptide chains some conformational freedom. An extended polypeptide chain can fold back through 180° to give *antiparallel* portions of the chain held in place by hydrogen bonding between adjacent segments, as shown in Figure 24.7. Several such *antiparallel* alignments could conceivably form a flat sheet-like structure, but it would not be stable because of the steric congestion between the side chains, R.

Slight coupled rotations of some bonds allow the R groups to point above or below the notional sheet, which reduces the steric repulsion between them and causes the sheet to becomes pleated; the resulting structural feature is called a β-**pleated sheet** (or simply a β sheet), as shown in Figure 24.8. The β sheet regions are usually 5–10 residues long, and the widths depend upon the number of additional turns or loops: each turn gives an additional strand of an aligned *antiparallel* portion but long loops allow portions with *parallel* alignment. The pleated sheet partial structure can only accommodate small and medium-sized R groups—larger R groups may partially disrupt a pleated sheet or even prevent its formation.

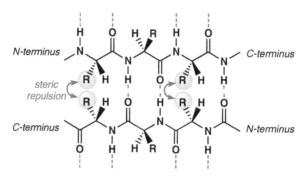

Figure 24.7 Part of a hypothetical flat sheet-like partial structure of antiparallel linear portions of a polypeptide chain.

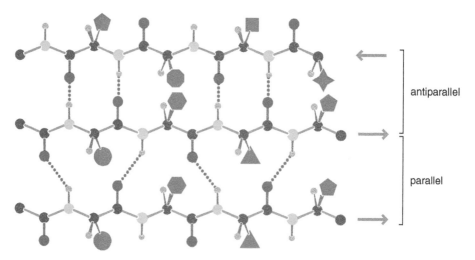

Figure 24.8 A β-pleated sheet partial structure of a protein. (Reproduced from Craig et al, *Molecular Biology: Principles of Genome Function*, Oxford University Press, 2010.)

The far more important structural feature of naturally occurring proteins is the α **helix** which twists in a right-handed manner. The helical structure of a single chain is held together by hydrogen bonding between pairs of amide groups four units apart in the chain, as illustrated in Figure 24.9. Each full turn of the helix contains 3.6 amino acid residues.

In a natural protein, α helices and β-pleated sheets are connected by loops and turns of the chain to form the overall three-dimensional shape, which is called the **tertiary structure** of the whole protein. Various additional features are involved in stabilizing tertiary structures, including disulfide bonds between cysteine residues. In a protein's natural aqueous environment, the folding of the chain exposes the maximum number of polar groups on the outer surface and encloses the maximum number of nonpolar groups within the interior.

Some complex proteins function as ordered non-covalently associated aggregates of more than one polypeptide chain. The overall structure of a protein comprising (perhaps) several subunits is called the **quaternary structure**. For example, hemoglobin consists of two pairs of polypeptide chains called the α and β subunits, each of which carries a heme unit (a porphyrin complex of an Fe^{2+} ion).

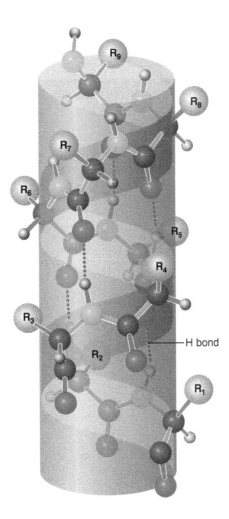

Figure 24.9 An α helix portion of a protein. (Reproduced from Craig et al, *Molecular Biology: Principles of Genome Function*, Oxford University Press, 2010.)

24.4 Lipids

Lipids are nonpolar natural products which can be extracted with organic solvents such as diethyl ether or chloroform. Since they are classified by a physical property, i.e. solubility, they do not have a common functional group. They have a variety of structures and many different functions in cells. The main types of compounds classified as lipids include fats and oils, phospholipids, terpenoids, steroids, and eicosanoids, which we shall look at in turn.

24.4.1 Fats and oils

Saturated and unsaturated fatty acids:

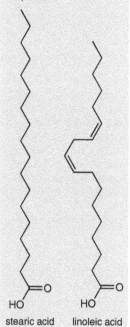

stearic acid   linoleic acid

Fats and oils are *fatty acid* triesters of glycerol and are also called **triacylglycerols** or **triglycerides**. Fatty acids are naturally occurring unbranched carboxylic acids with an even number of carbon atoms,* usually 12–22, and some common examples are included in Table 24.3. They may be saturated or unsaturated with double bond(s) of Z configuration. If they are solid at room temperature, they are called fats and are generally found in animals; if they are liquid, they are called oils and are mainly obtained from plants. Fats are mostly derivatives of saturated fatty acids while oils are derived more from unsaturated fatty acids, the lower melting point being attributable to their irregular structures caused by kinks in the chains at Z double bonds.

Hydrolysis of a triacylglycerol yields glycerol and three fatty acids which can be the same or different (eqn 24.10). Sodium and potassium salts of fatty acids obtained by alkaline hydrolysis are **soaps** (Sub-section 9.2.2) which act as surfactants. Soaps form micelles in water (which leads to their cleaning properties) as summarized in Panel 24.2.

$$
\begin{array}{c}
CH_2OC{-}R \\
CHOC{-}R \ + \ 3\ H_2O \\
CH_2OC{-}R
\end{array}
\xrightarrow{\ H_3O^+ \text{ or } HO^-\ }
\begin{array}{c}
CH_2OH \\
CHOH \ + \ 3\ RCO_2H \\
CH_2OH \quad (RCO_2^-)
\end{array}
\qquad (24.10)
$$

triacylglycerol   glycerol   fatty acid
(triglyceride)   (carbon number, 12–22)

Natural waxes found in plants and animals are typically esters formed from a high molecular weight alcohol and a fatty acid. Their functions include providing a protective coating on leaves and fruit to prevent evaporation of water, and acting as a water repellant on the feathers of birds. Paraffin waxes from petroleum are hydrocarbons of high molecular weight.

Table 24.3

Some common fatty acids			
Name	Number of C atoms (C=C bonds)	Structure	mp/°C
Saturated fatty acids			
lauric acid	12 (0)	$CH_3(CH_2)_{10}CO_2H$	44
myristic acid	14 (0)	$CH_3(CH_2)_{12}CO_2H$	58
palmitic acid	16 (0)	$CH_3(CH_2)_{14}CO_2H$	63
stearic acid	18 (0)	$CH_3(CH_2)_{16}CO_2H$	70
Unsaturated fatty acids			
palmitoleic acid	16 (1)	$CH_3(CH_2)_5CH{=}CH(CH_2)_7CO_2H$	1
oleic acid	18 (1)	$CH_3(CH_2)_7CH{=}CH(CH_2)_7CO_2H$	4
linoleic acid	18 (2)	$CH_3(CH_2)_4(CH{=}CHCH_2)_2(CH_2)_6CO_2H$	−5
linolenic acid	18 (3)	$CH_3CH_2(CH{=}CHCH_2)_3(CH_2)_6CO_2H$	−11
arachidonic acid	20 (4)	$CH_3(CH_2)_4(CH{=}CHCH_2)_4(CH_2)_2CO_2H$	−49

Soaps are sodium or potassium salts of fatty acids obtained by alkaline hydrolysis of fats or oils. The anion of a soap has a lipophilic (hydrophobic) nonpolar hydrocarbon chain and a hydrophilic carboxylate ion end group. Soaps are miscible with water, but they do not dissolve as individual dissociated ions. The soap anions form clusters in water with their lipophilic tails on the inside to avoid contact with the water and the ionic hydrophilic head groups on the outside in contact with the aqueous phase. The spherical clusters formed in this way and surrounded by the soap cations are called **micelles**; one is represented schematically below. Molecules of grease, fat, or oil (dirt) become solubilized in an aqueous soap by being taken into the lipophilic interiors of the micelles.

A schematic representation of a soap micelle.

Soaps are surfactants and reduce the surface tension of water. Consequently, soap ion pairs which coexist with micelles easily penetrate fabrics, and their lipophilic tails dislodge the oily layers of dirt from the fibres; the separated lipophilic molecules are then absorbed into the soap micelles. However, soaps have a drawback—they form precipitates in hard water. Soap anions form insoluble salts with Ca^{2+} and Mg^{2+} in hard water which float as scum and are minimally effective as cleansing agents. This drawback is avoided by using synthetic detergents, typically sulfonate salts.

Branched alkylbenzenesulfonates were once widely used as detergents, but they were not biodegradable and caused environmental problems by polluting rivers and lakes. Linear alkylbenzenesulfonate (LAS) salts are now the most commonly used synthetic detergents, with cationic and non-ionic detergents being used for specific purposes.

sodium p-dodecylbenzenesulfonate
a linear alkylbenzenesulfonate (LAS)

a cationic detergent

$CH_3(CH_2)_{11}O(CH_2CH_2O)_7CH_2CH_2OH$

a polyether
(a non-ionic detergent)

Bile salts, which are produced in the liver to help the digestion of fats (see Sub-section 24.4.4), may be regarded as physiological detergents.

24.4.2 Phospholipids

Phospholipids have a characteristic structure with two long hydrocarbon chains connected to a polar phosphate group, and there are two types: phosphoacylglycerols (or phosphoglycerides) and sphingolipids.

If one end acyl group of a triglyceride is replaced by a phosphono group ($-PO(OH)_2$), we have a **phosphatidic acid**, which is chiral (see below), and **phosphoacylglycerols**

HOCH$_2$CH$_2$NH$_2$

2-aminoethanol
(ethanolamine)

HOCH$_2$CH$_2$NMe$_3^+$

N-(2-hydroxyethyl)trimethyl-
ammonium ion (choline)

are phosphatidyl esters (or phosphodiesters). Phosphatidylethanolamines (or *cephalins*) and phosphatidylcholines (or *lecithins*), which are the phosphodiesters of 2-aminoetha-nol (ethanolamine) and N-(2-hydroxyethyl)trimethylammonium ion (choline), respec-tively, are two representative examples.

phosphatidic acid
(R is saturated
R' is unsaturated)

phosphatidylethanolamine
(cephalin)

phosphatidylcholine
(lecithin)

Exercise 24.16

Draw the structure of a phosphatidylserine, an important phosphoacylglycerol derived from L-serine.

Sphingolipids are derived from *sphingosine*, and *sphingomyelin* is a representative example which has two nonpolar groups and one polar group.

sphingosine

sphingomyelin

Phospholipids resemble soaps in that they both have polar and nonpolar groups, but an important difference is that a phospholipid molecule has *two* nonpolar (lipophilic) tails with one ionic (hydrophilic) head, while a soap molecule has only one nonpolar tail. Phospholipids, like soaps, form micelles in aqueous media but, more importantly, they also associate to give **lipid bilayers** (Figure 24.10). The outer faces of the bilayer are hydrophilic and the area sandwiched between them is lipophilic.

Phosphoacylglycerol bilayers are the main component of **cell membranes**, which form a hydrophobic (lipophilic) barrier between the aqueous extracellular environment and

Figure 24.10
A representation of a section through a lipid bilayer.

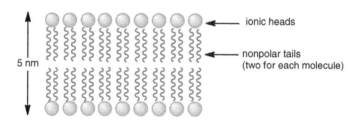

ionic heads

nonpolar tails
(two for each molecule)

5 nm

the intracellular liquid (the cytosol). On the other hand, sphingolipids make up the protective coating (called a myelin sheath) which encloses nerve fibres and insulates the electrical signals from the extracellular environment.

24.4.3 Terpenes

Terpenes are types of alkenes which, like their oxygenated derivatives (often called *terpenoids*), are found in plants; we have encountered some examples previously (e.g. in Chapters 8 and 15). Some of them are fragrant compounds used in scents and perfumes, others occur in fruit and fragrant culinary herbs. Most terpenes have skeletons of 10, 15, 20, or 30 carbon atoms and characteristic structures comprising so-called **isoprene units**; they are classified according to the number of carbon atoms (i.e. the number of isoprene units) as follows:

monoterpenes, C_{10}; sesquiterpenes, C_{15}; diterpenes, C_{20}; triterpenes, C_{30};

and so on.

Representative terpenes and terpenoids are given bellow with the isoprene units (separated by red bonds) linked mostly head-to-tail (their main sources are given in parentheses):

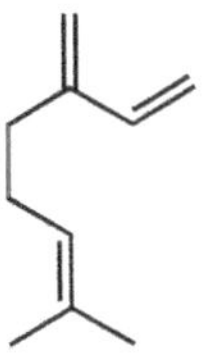

myrcene (bay laurel) 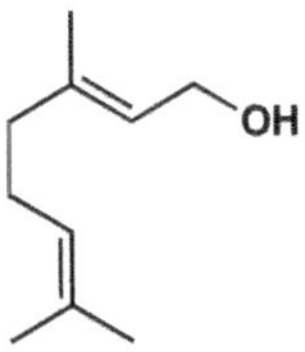 · geraniol (roses) 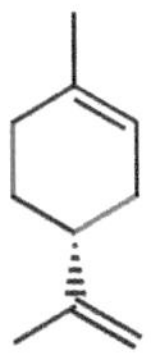 · (R)-(+)-limonene (oranges and lemons) 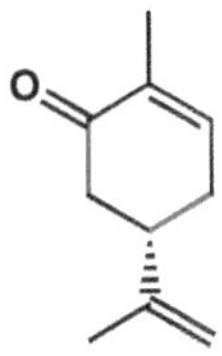 · (S)-(+)-carvone (caraway) 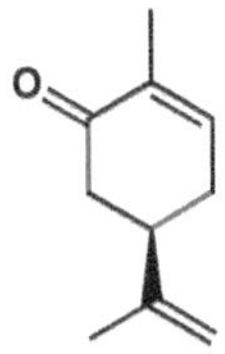 · (R)-(–)-carvone (spearmint) 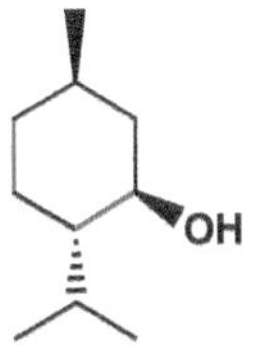 · menthol (peppermint)

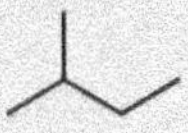

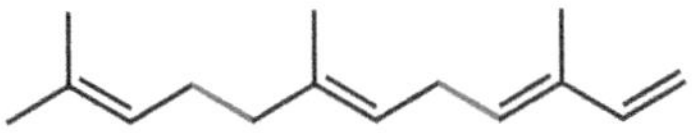

α-farnesene (natural coating of apples)

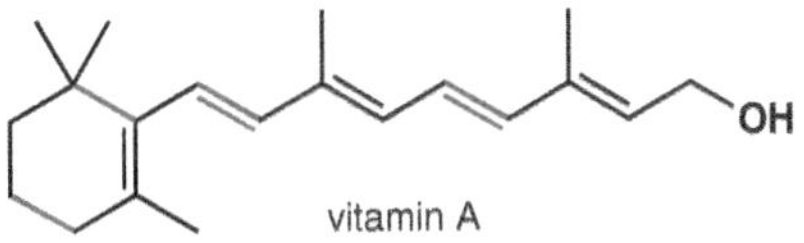

vitamin A

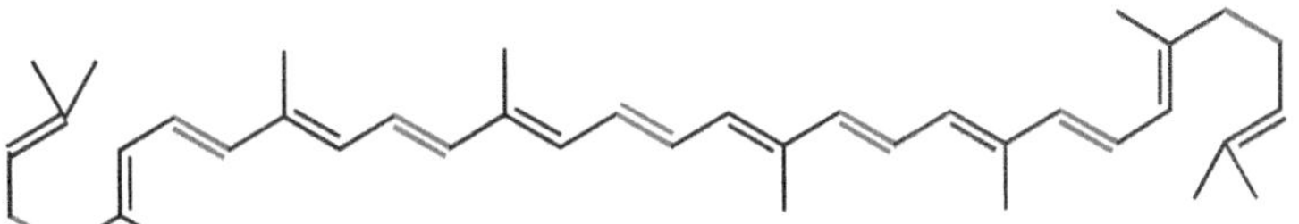

lycopene (tomatoes)

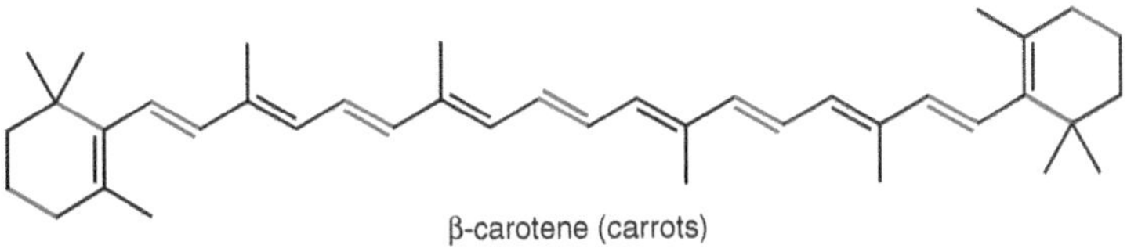

β-carotene (carrots)

Isoprene is the trivial name of the C_5 diene, 2-methylbuta-1,3-diene.

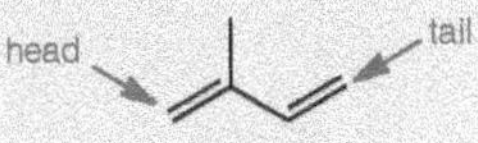

The prefix 'sesqui' means *one and a half*.

Terpenes are biosynthesized in nature from isoprene units by S_N1 reactions (see Panel 12.3).

The menthol structure given is of the naturally occurring form: it is one of eight stereoisomers resulting from three chirality centres.

Natural rubber is a polymer formed by 1,4-additions of isoprene units, and its double bonds are all *Z*.

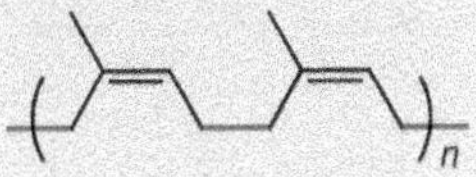

natural rubber

Classify the terpenes and terpenoids given above as monoterpenes, sesquiterpenes, diterpenes, or others.

Exercise 24.17

Panel 24.3 Origin of the isoprene unit for terpene biosynthesis

Acetoacetyl-CoA (formed by the Claisen condensation of acetyl-CoA, Panel 17.3) undergoes an aldol-type reaction at the keto group with acetyl-CoA; enantioselective hydrolysis at one end of the achiral product leads to formation of β-hydroxy-β-methylglutaryl-CoA (HMG-CoA). HMG-CoA is partially reduced to (R)-mevalonic acid by NADPH (the 2'-phosphate of NADH, Panel 10.3) which undergoes phosphorylation by ATP and elimination with loss of carbon dioxide to give isopentenyl diphosphate. This isopentenyl compound (whose IUPAC name is 3-methylbut-3-enyl diphosphate) is the isoprene unit involved in the biosynthesis of terpenes, which was covered in Panel 12.3; the triterpene squalene is the biological precursor of cholesterol and, hence, all other steroids (Panel 24.4). All steps in the sequence shown below are catalysed by enzymes.

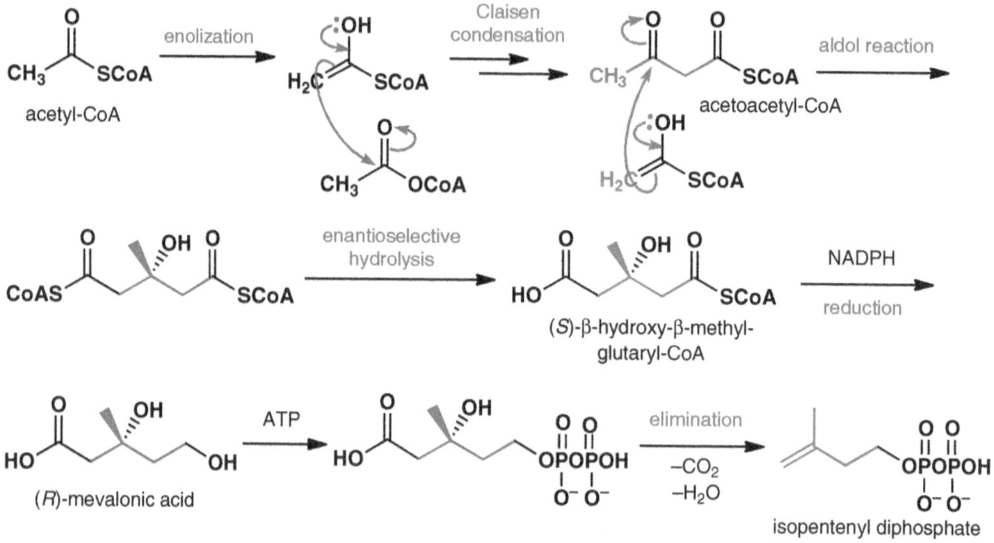

One common cause of health problems in people from middle age onwards is excessive cholesterol in the blood. One strategy to combat this problem is to take a medication which inhibits the enzyme HMG-CoA reductase, and *statins* are a family of compounds which act in this way. Atorvastatin (Lipitor®) is one of the best-selling statins; it inhibits the synthesis of mevalonic acid in the liver and thereby reduces the formation of cholesterol in the body.

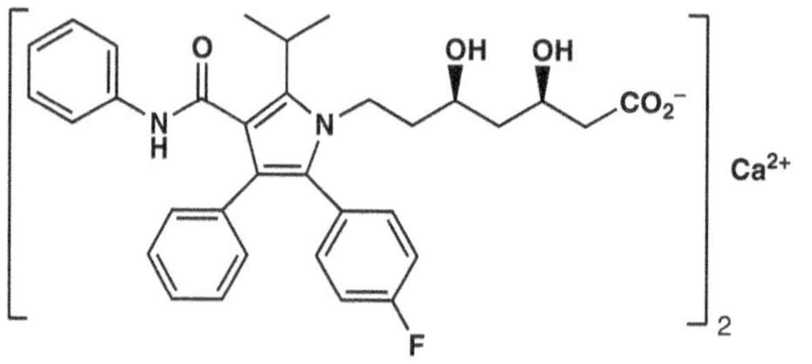

atorvastatin (Lipitor®)

24.4.4 Steroids

Steroids are a group of tetracyclic lipids which are physiologically active. They have three six-membered rings and one five-membered ring designated A, B, C, and D. When saturated, all the ring junctures are usually *trans*; consequently, *the ring system is approximately planar*. Most steroids contain two *angular methyl groups* (numbered 18 and 19) which are axial to chair cyclohexane partial structures, and point upwards in the usual representation of the steroid. The side chain R at C17 is generally *cis* to the angular Me groups.

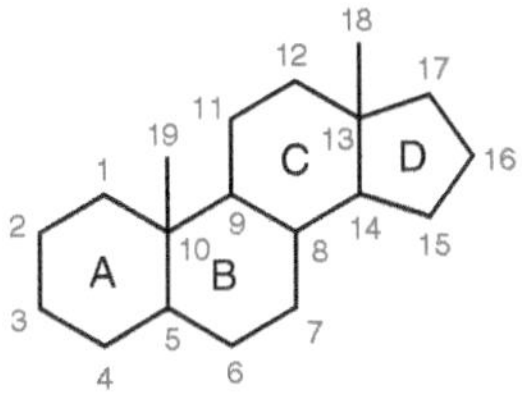

general steroid skeleton

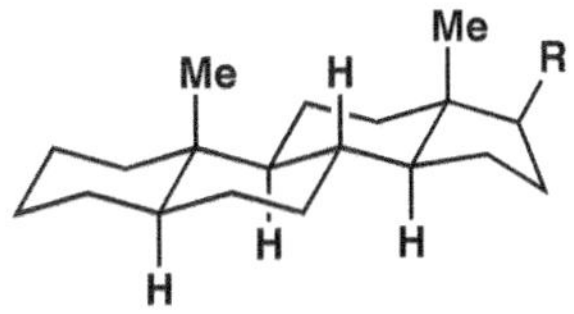

general structure of steroids

In some steroids, the A,B ring juncture is *cis*, and the two chair cyclohexane rings are almost perpendicular.

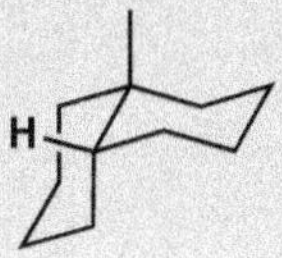

The best known steroid, cholesterol, is synthesized in the body from a triterpene called squalene, as summarized in Panel 24.4, and is present in all animal cells as a component of their membranes, for example. Consequently, it is a significant component of the diet of non-vegetarians. Cholesterol is also a precursor of all other steroids.

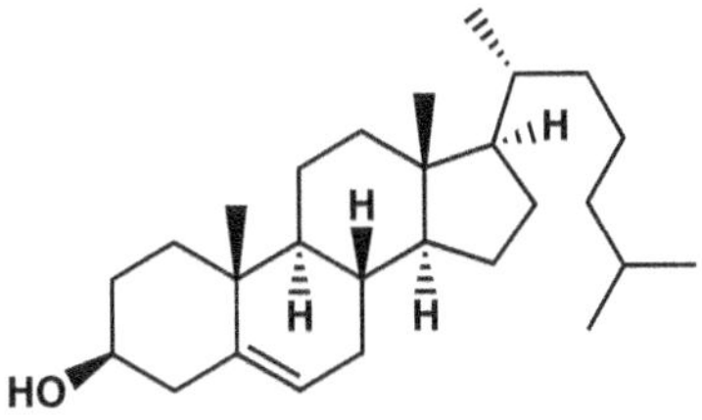

cholesterol

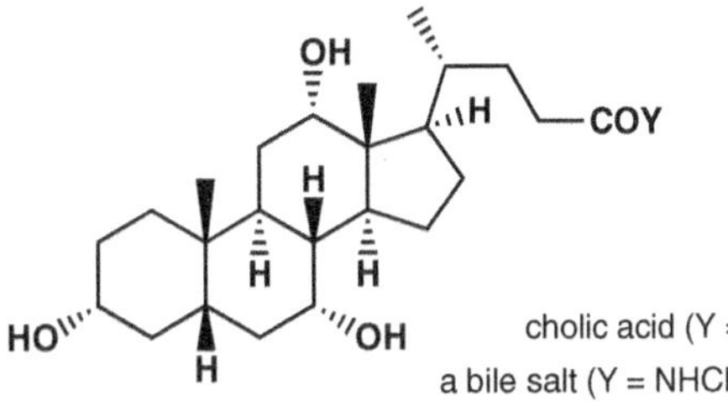

cholic acid (Y = OH)
a bile salt (Y = NHCH$_2$CH$_2$SO$_3^-$ Na$^+$)

Panel 24.4 Biosynthesis of cholesterol from squalene

Cholesterol is synthesized in the body in a series of enzyme-catalysed reactions from the triterpene, squalene. Squalene is first oxidized by oxygen to give an epoxide. Protonation of the epoxide then initiates a cyclization by a sequence of carbenium ion additions to double bonds. The result is the formation of four new C–C bonds and the generation of the tetracyclic steroid skeleton in the form of the protosterol cation. This cation rearranges by a series of 1,2-shifts of two hydrogens and two methyl groups to form a tertiary carbenium ion, deprotonation of which leads to the dienol, lanosterol. Lanosterol is then converted to cholesterol in a multistep process which involves loss of three methyl groups. All the other steroids are derived from cholesterol.

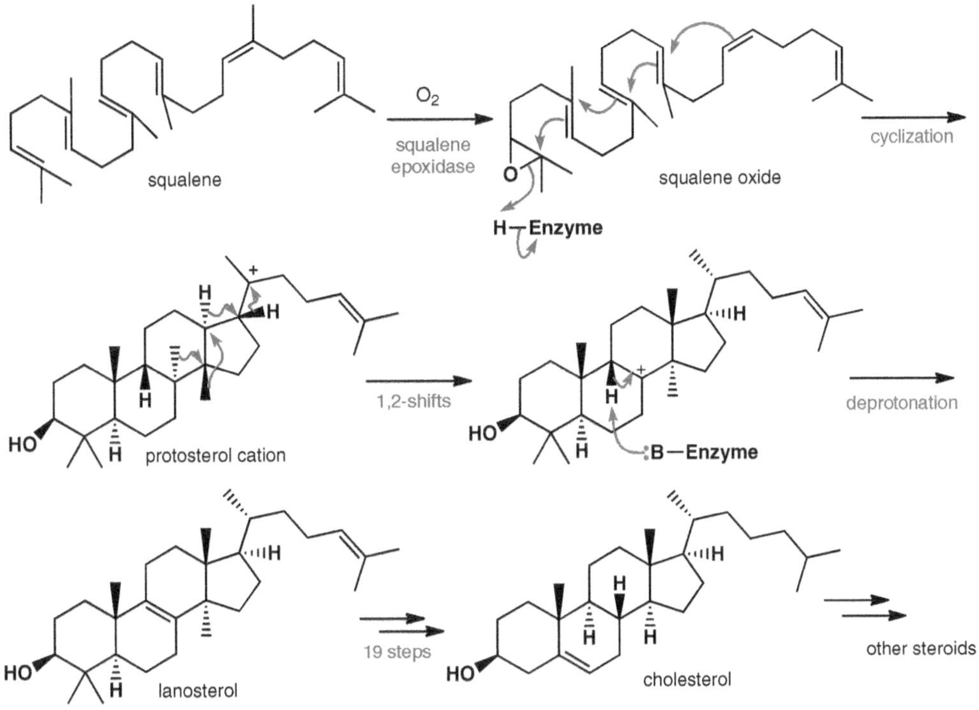

Vitamin D is a group of compounds, and vitamin D₃ is derived from 7-dehydrocholesterol by photochemical ring opening as covered in Panel 21.1.

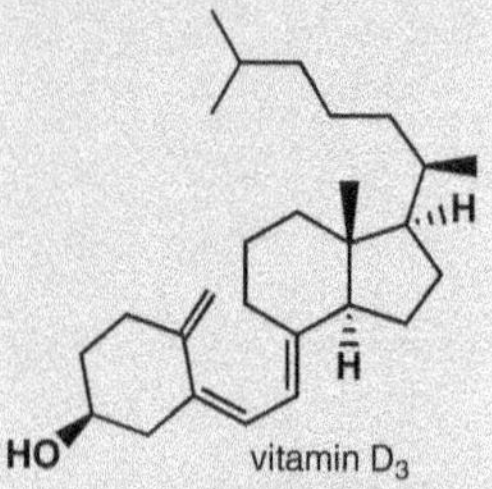

vitamin D₃
(cholecalciferol)

Cholic acid is one of the main bile acids and forms salts in the bile produced by the liver. A bile salt, with a lipophilic steroidal skeleton and an ionic 'tail', is a kind of detergent which emulsifies fats in food (as micelles) to help their digestion, and (when excreted) removes cholesterol from the body.

Many other important steroids are hormones secreted by the endocrine glands; they include **sex hormones** and adrenal cortical steroids. Sex hormones are classified into three major groups: the female hormones called estrogens, the male hormones called androgens, and the pregnancy hormones called progestins; the following are representative examples.

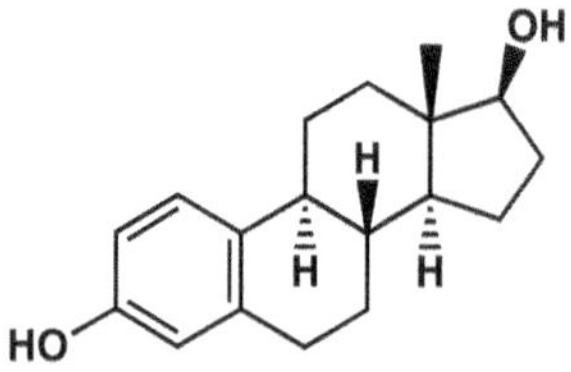

estradiol
(a female hormone)

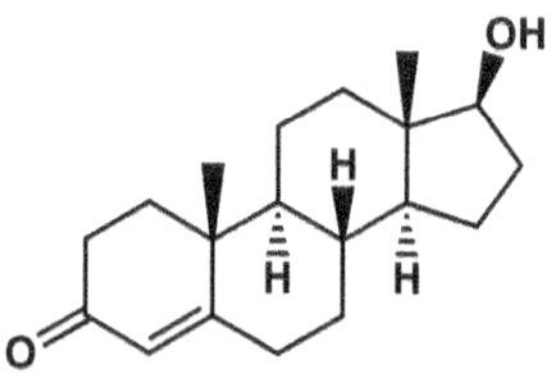

testosterone
(a male hormone)

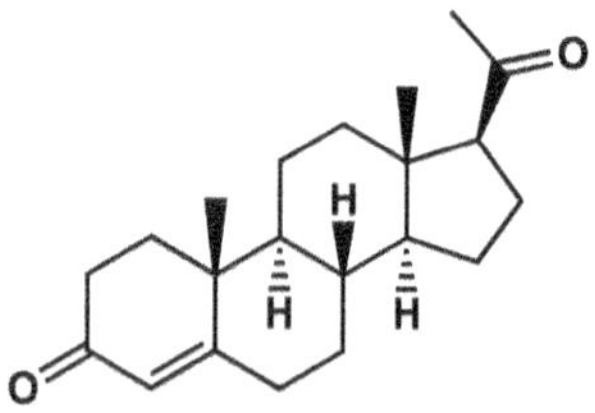

progesterone
(a progestin)

The adrenal gland is one of the endocrine glands which produce hormones; its two parts sit on top of the kidneys and their outer part is called the adrenal cortex.

Three examples of **adrenal cortical hormones** are:

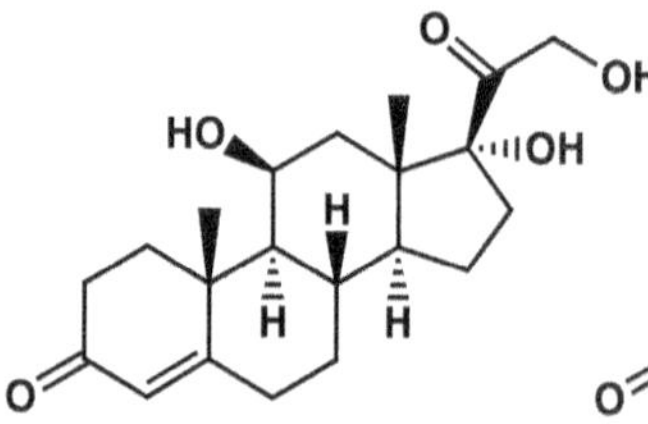

cortisol

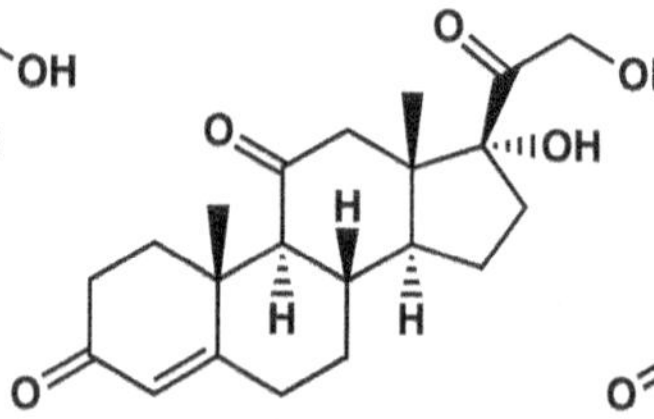

cortisone

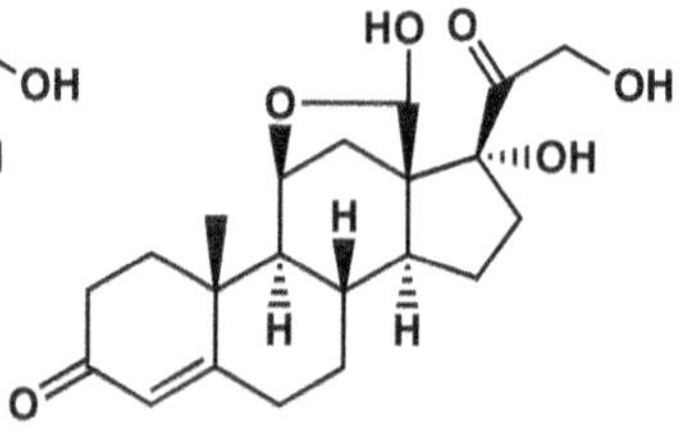

aldosterone

Cortisol and cortisone serve as anti-inflammatory agents and also regulate carbohydrate metabolism, while aldosterone regulates blood pressure by controlling the concentrations of Na^+ and K^+ in body fluids.

Exercise 24.18

How many chirality centres are there in (a) cholesterol and (b) cortisol?

Lipophilic vitamins We encountered vitamins A and D in this section; the former is classified as a terpene and the latter is a group of steroid derivatives. Both are sometimes called lipophilic vitamins (as opposed to water soluble ones such as vitamin C), a group which also includes vitamins E and K. Vitamin E is a group of phenolic compounds which act as biochemical antioxidants (Section 20.8), while vitamin K is a group of related compounds which contain a quinone structure and are involved in blood coagulation.

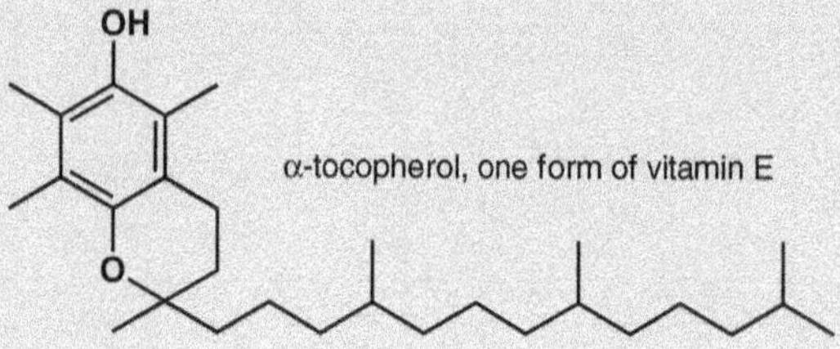

α-tocopherol, one form of vitamin E

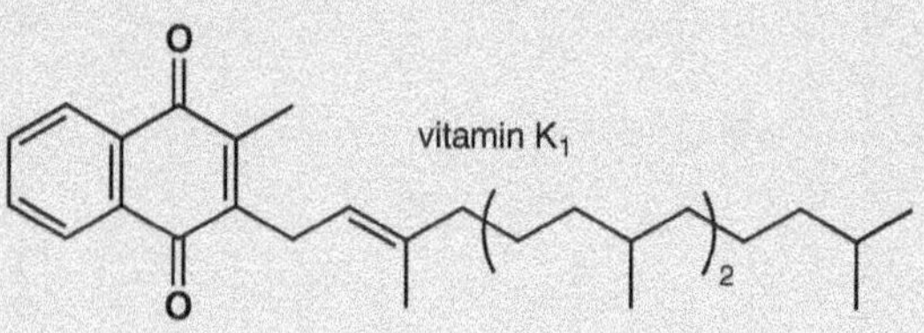

vitamin K₁

24.4.5 Eicosanoids

Eicosanoids are a group of biologically active compounds derived from a C_{20} unsaturated acid, arachidonic acid (Section 24.4.1). There are four classes of them: **prostaglandins**, **thromboxanes**, **prostacyclins**, and **leukotrienes**, and the following are examples.

The name 'eicosanoid' comes from the old name of a C_{20} alkane, *eicosane*, which we now spell icosane. The IUPAC rules recommend the name 'icosanoid' in accord with the new name of the C_{20} alkanes.

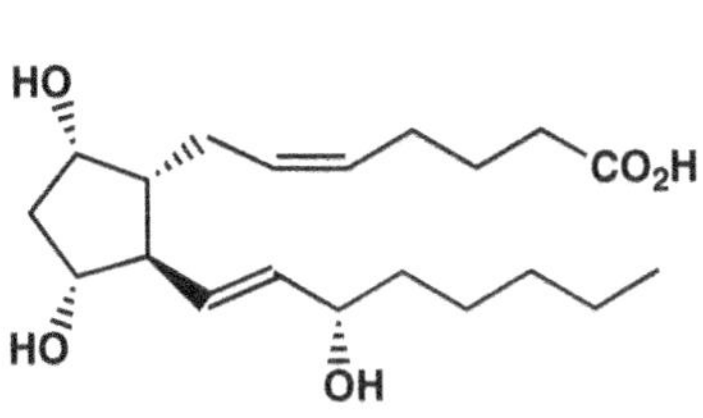

prostaglandin $F_{2\alpha}$

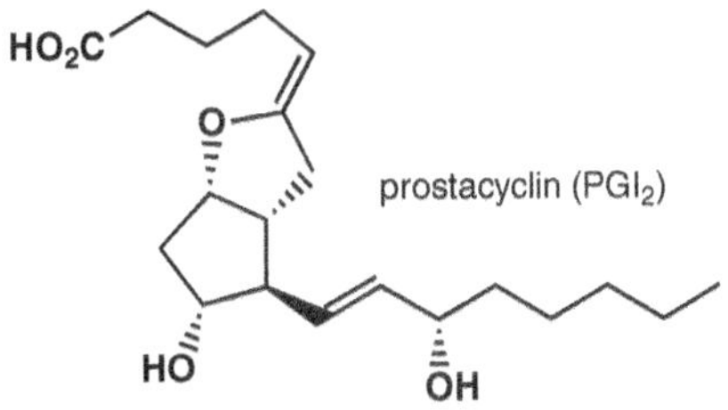

prostacyclin (PGI$_2$)

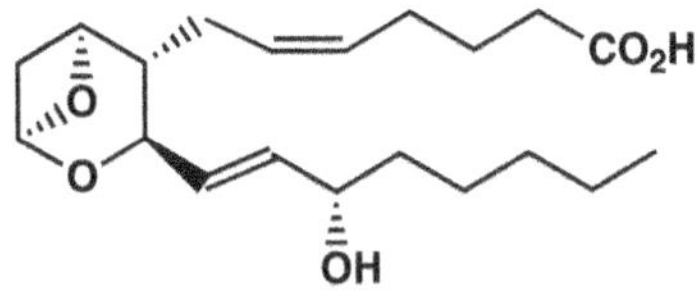

thromboxane A_2

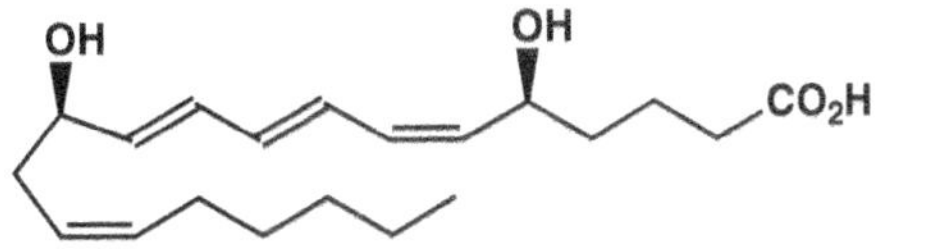

leukotriene B_4

They are very potent but relatively unstable compounds present in low concentrations in the cells where they are synthesized and carry out their functions, which distinguishes them from hormones which are produced in glands and secreted into the bloodstream. Their main functions relate to heart rate, blood pressure, blood clotting, conception, and allergic and inflammatory responses.

They are synthesized in the body from arachidonic acid in response to an external stimulus and act where they are formed; the main synthetic route begins with enzyme-catalysed oxidation of arachidonic acid by O_2 to give PGG$_2$ which is then converted into other prostaglandins, thromboxanes, and prostacyclins. Laboratory syntheses of a prostaglandin were included in Chapter 23.

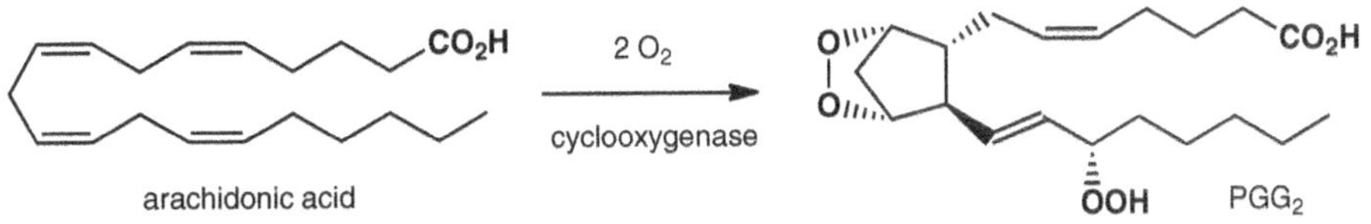

arachidonic acid

$2\ O_2$

cyclooxygenase

PGG$_2$

How aspirin works By acetylating the enzyme cyclooxygenase, aspirin (acetylsalicilic acid: see the Prologue) blocks the synthesis of prostaglandins thereby reducing inflammation and fever, and inhibiting the transmission of signals to the brain which lead to pain. In the same way, aspirin also blocks the synthesis of thromboxanes which cause the aggregation of platelets following damage to a blood vessel. Low daily doses of aspirin are now widely taken to prevent development of blood clots which cause heart attacks, but bleeding in the stomach is a possible side effect.

Summary

● **Carbohydrates** include monosaccharides, disaccharides, trisaccharides, etc., up to polysaccharides, which include starch and cellulose.

● **Monosaccharides** are aldehydes (aldoses) or ketones (ketoses) typically containing six or five carbon atoms with several OH groups, and exist mainly as cyclic hemiacetals. They generally have several chiral carbons, and the

C atom most remote from the carbonyl group in the open-chain form of naturally occurring sugars has the ᴅ configuration, i.e. *R* absolute configuration.

○ Reaction of another hydroxy compound at C1 of a sugar gives an acetal called a **glycoside**; amino compounds may also react at C1 to give *N*-glycosides.

○ **Nucleic acids** (DNA and RNA) are polymers of nucleotides which are phosphate esters of nucleosides. Nucleosides are *N*-glycosides of 2-deoxy-ᴅ-ribose or ᴅ-ribose with four different heterocyclic bases (purines or pyrimidines) in each case.

○ The **double helix** form of DNA molecules is held together by hydrogen bonds between specific pairs of purine and pyrimidine bases (G with C, and A with T in DNA or U in RNA) in the complementary strands. The sequence of the bases of DNA contains the code for genetic information which is translated into the amino acid sequence of proteins.

○ α-**Amino acids** link together through amide bonds to give **peptides**; **proteins** (including enzymes) are naturally occurring **polypeptides** composed of 20 ʟ-α-amino acids.

○ Functional group protection and deprotection are important features of the laboratory synthesis of peptides. Chemical determination of peptide sequences uses reagents which react specifically at the *N*-terminal amino acids.

○ The amino acid sequence is called the primary structure of a **protein** and non-covalent interactions lead to α **helix** and β-**pleated sheet** local features of the secondary structure. Additionally, polypeptide chains form loops and turns which are held by disulfide bridges, for example, to give the polypeptide an overall three-dimensional tertiary structure. Some complex proteins function as ordered aggregates of several subunits, which leads to a quaternary structure.

○ **Lipids** include triacylglycerols, phospholipids, terpenes, steroids, and eicosanoids.

○ **Triacylglycerols**, the triesters of C_{12}–C_{22} fatty acids with glycerol, are fats and oils.

○ **Phospholipids** resemble triacylglycerols structurally, but one terminal ester group is replaced by a phosphate ester with an ionic group. The two lipophilic chains and hydrophilic head group are the characteristic structural features responsible for the formation of phospholipid bilayers, the main component of cell membranes.

○ **Terpenes** are alkenes composed of isoprene units and, along with terpenoids (their oxygenated derivatives) are found in the 'essential oils' of a wide range of plants.

○ **Steroids** are physiologically active compounds with a tetracyclic carbon structure.

○ **Eicosanoids** are unstable, physiologically active compounds derived from arachidonic acid, and are classified into prostaglandins, thromboxanes, prostacyclins, and leukotrienes.

Problems

24.1 Draw structures of the furanose and pyranose forms of ᴅ-fructose which are present in equilibrium in solution.

24.2 Which of the alditols obtained by reduction of the monosaccharides in Figure 24.1 is optically inactive?

24.3 The primary alcohol residue of a hexose can be enzymatically oxidized to a carboxy group, and the product is called a uronic acid; ᴅ-glucose gives ᴅ-glucuronic acid in this way. Give the structure of ᴅ-glucuronic acid in its open-chain form as a Fischer projection and in its cyclic pyranose form.

(D-Glucuronic acid acts as a detoxifying agent in the liver by solubilizing a wide range of hydroxy compounds as glycosides to facilitate their excretion.)

24.4 D-Gluconic acid easily gives cyclic products; draw structures of two of them.

24.5 L-Ascorbic acid (vitamin C) is a diprotic acid with pK_a values of 4.1 and 11.8. Explain the acidity of vitamin C by describing the electronic structures of its conjugate base forms.

24.6 The aldehyde group of an aldose forms a cyanohydrin by reaction with NaCN in the presence of an acid. The partial hydrogenation of the CN group of the cyanohydrin followed by hydrolysis provides two isomeric aldoses, each with one additional carbon atom. Give the reactions in this so-called Kiliani–Fischer synthesis starting with a pentose, D-arabinose, and identify the two hexoses obtained.

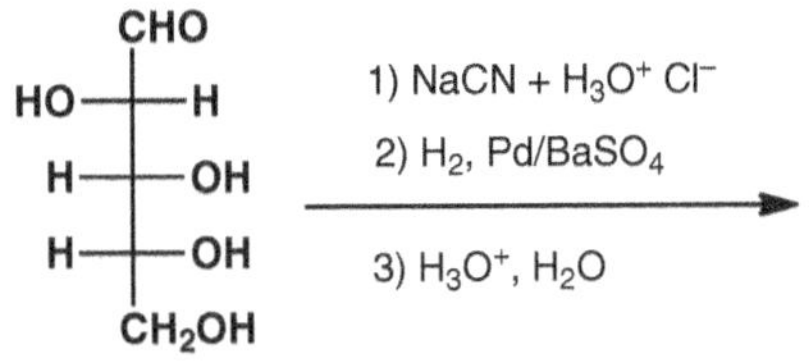

D-arabinose

1) NaCN + H_3O^+ Cl^-
2) H_2, Pd/BaSO$_4$
3) H_3O^+, H_2O

24.7 Chitin, the structural component of the exoskeleton of invertebrates such as crabs, lobsters, and insects, is a polymer of N-acetylglucosamine with β (1→4)-glycosidic bonds. N-Acetylglucosamine is a derivative of D-glucose with an acetylamino group in place of the OH on C2. Draw a partial structure of chitin.

24.8 Trehalose, which is the blood sugar of some insects and is found in mushrooms, is a disaccharide with two D-glucose units linked through their anomeric carbons by an α,α(1↔1)-glycosidic bond. Draw a structure of this disaccharide and explain whether it is a reducing sugar.

24.9 One of the two strands of a DNA double helix has a partial sequence 5'-ACCTGAATCG-3'. What is the sequence of the associated complementary strand?

24.10 Nucleosides are stable in a dilute aqueous base but they undergoes hydrolysis in dilute aqueous strong acids to give heterocyclic bases (which become protonated) and a pentose. Give a mechanism for the hydrolysis of 2'-deoxyadenosine in aqueous acid.

24.11 The 6-thio analogues of the purine bases, 6-mercaptopurine and 6-thioguanine, are anticancer drugs used to treat leukemia. Draw structures of their tautomers.

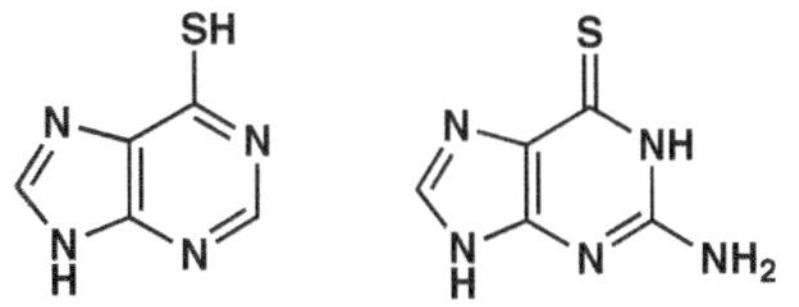

6-mercaptopurine   6-thioguanine

24.12 RNAs are less stable than DNAs and rapidly give cyclic phosphates in alkaline solution. This has been attributed to the *cis* relationship between the 2'-OH and the 3'-phosphate. Give a mechanism for the following reaction.

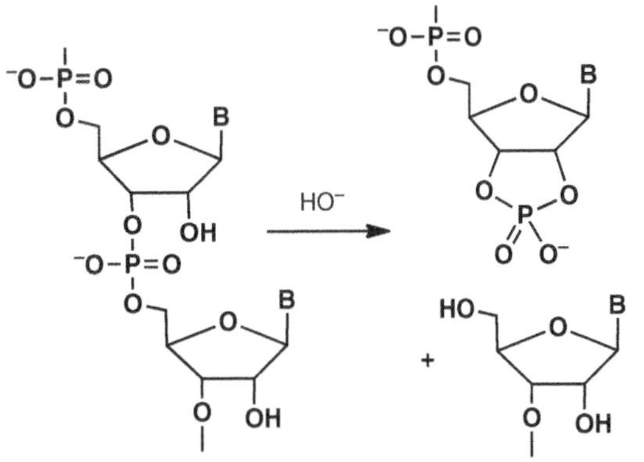

24.13 One of the methods for the synthesis of α-amino acids includes hydrolysis of the amino nitrile formed by the reaction of an aldehyde with an acidified aqueous cyanide plus ammonia. Outline this so-called Strecker synthesis by a reaction sequence.

24.14 Primary amines are not usually prepared by the reaction of haloalkanes with ammonia as follows.

$$RBr + 2\ NH_3 \xrightarrow{H_2O} RNH_2 + NH_4^+\ Br^-$$
low yield

However, α-amino acids can be obtained in a fairly good yields by a similar reaction:

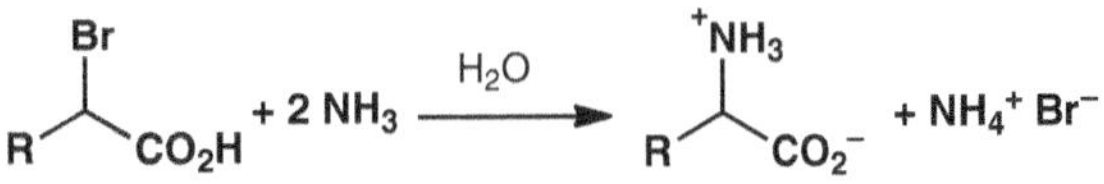

Discuss why the reactions are different.

24.15 Ninhydrin is used as a colour-developing reagent for the detection of α-amino acids. The same purple pigment is formed regardless of the structure of the amino acid. Give mechanisms to show how the following colour-developing reaction proceeds.

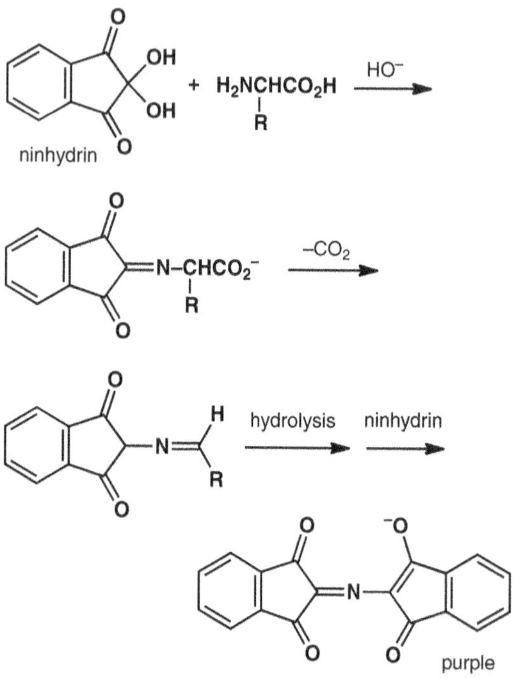

ninhydrin + H₂NCHCO₂H (R) →(HO⁻)

=N—CHCO₂⁻ (R) →(−CO₂)

—N=CH—H (R) →(hydrolysis) →(ninhydrin)

purple

24.16 Determine the amino acid sequence of a nonapeptide from the following results. When the peptide is treated with phenyl isothiocyanate under Edman degradation conditions, the structure of the first phenylthiohydantoin is as given below. Hydrolysis of the nonapeptide with chymotrypsin gives three fragments: Arg, Ser-Pro-Phe, and Arg-Pro-Pro-Gly-Phe.

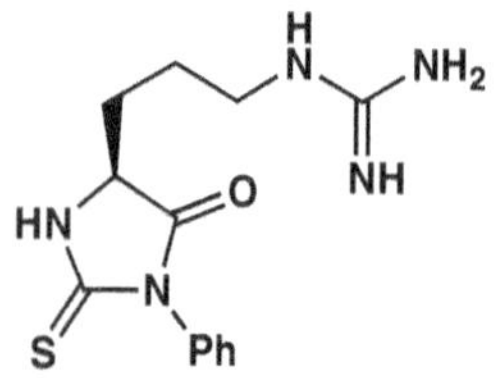

24.17 What are the products of ozonolysis of linolenic acid (see Table 24.3) followed by reductive treatment with dimethyl sulfide?

24.18 Indicate the isoprene units of each of the following terpenes and terpenoids.

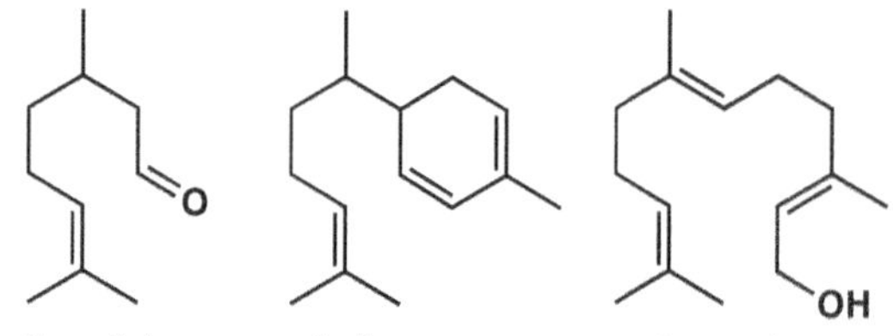

citronellal (lemon)   zingiberene (ginger)   farnesol (lemon grass)

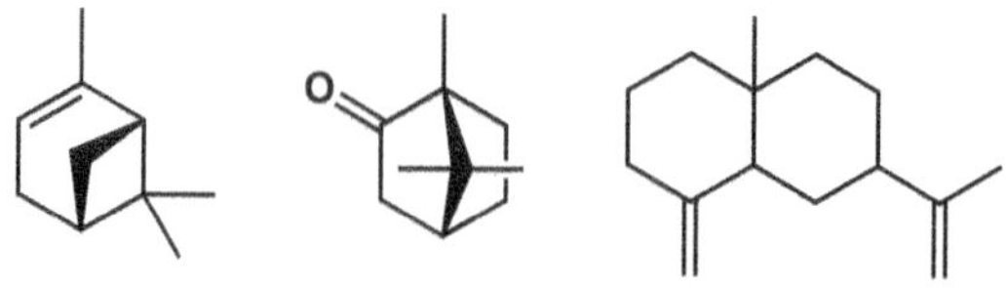

α-pinene (pine trees)   camphor (camphor trees)   β-selinene (celery seed oil)

24.19 The lecithin of egg yolk (see Sub-section 24.4.2) acts as an emulsifying agent in the preparation of mayonnaise. Which parts of the lecithin molecule interact with the oils and which with the aqueous components, respectively?

24.20 Benzaldehyde undergoes an acid-catalysed reaction with D-glucose to give the bicyclic acetal, **A**.

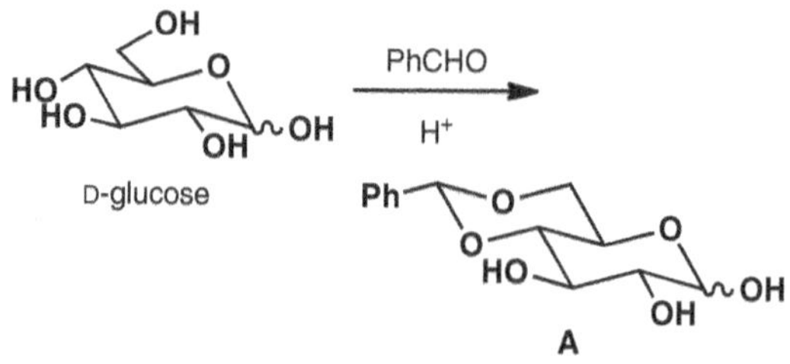

D-glucose →(PhCHO)(H⁺)

A

A similar reaction of propanone with D-glucose does not give the expected bicyclic acetal, **B**; the main product is the diacetal, **C**.

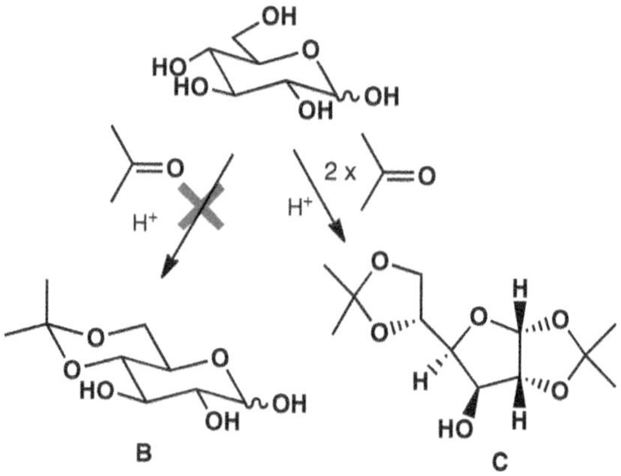

B   C

(a) Give a reaction mechanism for the formation of **A**.

(b) Explain why the acetal **B** is not stable.

(c) Propose a reaction scheme to explain the formation of **C**.

24.21 Give a mechanism for the isomerization of the thiazolinone to the phenylthiohydantoin in Scheme 24.6.

Structural Determination of Organic Compounds

25

Related topics we have already covered:

- **Chemical bonding** (Chapter 1)
- **Molecular structure** (Chapter 2)
- **Functional groups** (Chapter 3)
- **Stereochemistry** (Chapters 4 and 11)
- **Molecular vibrations** (Section 4.1)
- **Electron delocalization and aromaticity** (Chapter 5)
- **Photoexcited organic molecules** (Section 5.7)
- **Perception of colours** (Panel 5.2)

Topics we shall cover in this chapter:

- **Interactions of electromagnetic radiation with molecules**
- **Ultraviolet and visible spectroscopy**
- **Infrared spectroscopy**
- **Proton and carbon nuclear magnetic resonance spectroscopy**
- **Mass spectrometry**

As we have seen in previous chapters, the physical properties, chemical reactivity, and biological activity of a compound are determined by its molecular structure, and organic chemists talk about chemistry on the basis of the structures of molecules. In spite of the importance of molecular structure, we have not so far in this book concerned ourselves much with how it is determined. This is mainly because structural determination nowadays is principally a technical instrumental issue and seldom impinges on reactions and reactivity of organic compounds. Even so, it is important that organic chemists understand the *basis* of each principal method they are likely to use for determining the molecular structures of organic compounds, especially ones they have prepared themselves. In this last chapter, we shall also indicate the *scope* of each method we discuss so that, when faced with a structural problem, the most appropriate method for its solution can be identified.

IR and mass spectral data in this chapter are taken from the NIST Chemistry WebBook (http://webbook.nist.gov/chemistry/).

25.1 Electromagnetic Radiation and Spectroscopy

25.1.1 The electromagnetic spectrum and types of spectroscopy

Most modern techniques for determining molecular structures are based on **spectroscopy**, which involves *interactions of electromagnetic radiation (light) with molecules*. The spectral range which is used in the determination of molecular structure is broadly classified into X-ray, ultraviolet (UV), visible (vis), and infrared (IR) radiations, and

Figure 25.1 The electromagnetic spectrum.

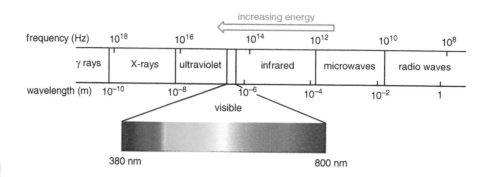

1 nm (nanometre) = 10^{-9} m

microwaves and radio waves, in order of decreasing frequency (i.e. energy) or increasing wavelength (Figure 25.1).

Techniques involving radiation in different ranges of the spectrum provide different kinds of information about atoms and molecules. The high energy of X-rays corresponds to excitation of electrons in the inner shells of atoms; spectroscopic methods using X-rays do not give information about structures of organic molecules. However, the wavelengths of X-rays match bond lengths, so X-rays are diffracted by crystals to give diffraction patterns which allow determination of the three-dimensional structures of molecules (**X-ray crystallography**). Interactions between molecules and ultraviolet and visible light lead to the excitation of valence electrons, as covered in Section 5.7: **UV-vis spectra** provide information about molecular orbitals (especially those involved in bonding) when a molecule contains unsaturated groups (e.g. a carbonyl) or a conjugated π electron system. The energy of IR radiation corresponds to transitions between molecular vibrational energy levels: **IR spectra** provide information about functional groups in organic molecules. Microwaves can be related to molecular motion, in particular low-energy vibrations and rotations, and (as a rather specialized technique which we shall not discuss) microwave spectroscopy provides precise information about bond angles and bond lengths in small molecules. Radio (or radiofrequency, rf) waves induce transitions between energy levels of atomic nuclei in a magnetic field (sometimes called spin inversions) leading to **nuclear magnetic resonance (NMR) spectra**; ^{1}H and ^{13}C NMR spectra provide information about the carbon–hydrogen framework of a molecule. **Mass spectrometry (MS)**[1] is a useful technique for determining the molecular formula of an organic compound as well as structures of parts of a molecule if not its whole structure. In this method, the masses of ionized molecules and their ionic fragments are measured, so the technique is fundamentally different from the spectroscopic methods mentioned above.

In order to determine the structure of an organic compound, we generally need a pure sample and purification of compounds depends on intermolecular interactions (see Chapter 3). Chromatography is the most useful technique for separation and purification of organic compounds (Panel 3.3).

The absorption of microwave energy by food (mainly the polar bonds of water) in domestic microwave ovens leads to increased molecular motion, i.e. the food becomes hot.

25.1.2 Interactions of electromagnetic radiation with molecules

Electromagnetic radiation (light) is characterized by its wavelength (λ), which is directly related to its frequency (ν) by the equation $c = \lambda\nu$, where c is the velocity of light (3.00×10^8

---

[1]   The term *spectroscopy* is used for techniques involving interactions between molecules and electromagnetic radiation. In MS, a beam of electrons is normally used to energize and ionize molecules, and the term *spectrometry* is used.

ms^{-1}). The unit for frequency is s^{-1} (cycles per second) or Hz (hertz). The energy (E) of a photon is directly proportional to the frequency of the radiation and inversely proportional to its wavelength, as shown in eqn 25.1.

$$E = h\nu = hc/\lambda, \quad \text{where} \quad h = \text{the Planck constant } (6.626 \times 10^{-34} \, \text{J s}) \qquad (25.1)$$

When a beam of electromagnetic radiation passes through a compound, molecules interact with light of only some wavelengths. Since the energy of a molecule is *quantized* (i.e. the molecule has only discrete energy levels), radiation of only the wavelengths matching the differences between pairs of energy levels (ΔE) can be absorbed, as outlined in Section 5.7 for electronic excitation. A graph of the absorption of radiation versus its wavelength (or frequency) is a **spectrum** (plural, spectra) which provides information about differences between electronic or vibrational energy levels (for example) of the molecule. This information can then be used to elucidate the structure of the molecule since electronic or vibrational energy levels (for example) are directly related to how the atoms of a molecule are joined together.

Note that eqn 25.1 gives the energy of a single quantum of light (a photon) which interacts with a single molecule. If we want the energy on a molar scale, the energy of one photon is multiplied by the Avogadro constant ($6.022 \times 10^{23} \, \text{mol}^{-1}$).

25.2 Ultraviolet and Visible Spectroscopy

Absorption of electromagnetic radiation in the UV and visible region corresponds to the excitation of an electron from a lower energy orbital in a molecule to a higher energy orbital, as illustrated in Figure 25.2. Since there are technical difficulties in measurements below about 200 nm, the conventional UV-vis range is limited to about 200–800 nm; this corresponds to $\pi \rightarrow \pi^*$ or $n \rightarrow \pi^*$ electronic transitions.

A typical UV-vis spectrum, that of buta-1,3-diene, is shown in Figure 25.3; the absorption is due to a $\pi \rightarrow \pi^*$ transition and the absorption maximum is at 217 nm (we say $\lambda_{max} = 217$ nm). The λ_{max} becomes longer as the extent of the conjugation increases, as exemplified by the data in Table 25.1. The observation of UV-vis absorption at $\lambda_{max} > 200$ nm, therefore, generally indicates a conjugated π electron system in a molecule; compounds with highly conjugated π systems (e.g. β-carotene, see Sub-section 24.4.3) absorb radiation in the visible region, so they are coloured. Carbonyl compounds also have absorption bands due to $\pi \rightarrow \pi^*$ transitions, but at shorter wavelengths than ones of dienes; as expected, conjugation of a carbonyl group pushes λ_{max} to longer wavelengths. In addition, carbonyl compounds have absorption bands corresponding to $n \rightarrow \pi^*$ transitions

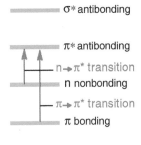

Figure 25.2 Energy levels of molecular orbitals in an organic molecule.

Measurements of UV spectra below 200 nm (sometimes called the far UV region) are experimentally difficult because the O_2 of air and quartz glass absorb radiation in this range.

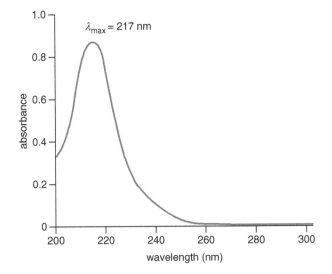

Figure 25.3 The UV spectrum of buta-1,3-diene.

Electronic transitions involving σ orbitals require absorption of radiation of higher energy, i.e. $\lambda < 200$ nm.

UV-vis absorption bands of some unsaturated compounds	
	λ_{max} /nm (ε)
ethene	165 (15 000)
buta-1,3-diene	217 (21 000)
hexa-1,3,5-triene	256 (50 000)
deca-1,3,5,7,9-pentaene	334 (125 000)
β-carotene	453, 483
benzene	204 (7 900), 254 (200)
propanone	195 (9 000), 274 (13.6)
but-3-en-2-one	203 (9 600), 331 (25)

Absorption bands in UV-vis spectra are broad to a greater or lesser degree because transitions between the two electronic energy levels involve numerous vibrational and rotational sub-levels.

at longer wavelengths (λ_{max} > 250 nm), but of lower intensity (which is measured by the *molar absorptivity* of the compound, ε, as defined below).

In UV-vis spectra, the **absorbance** (*A*) of the radiation by a compound is plotted against the wavelength, where *A* defined by eqn 25.2.

$$A = \log(I_0/I) \qquad (25.2)$$

In eqn 25.2, I_0 is the intensity of the light entering the sample and *I* is the intensity of light emerging from the sample. The absorbance is directly proportional to the molar concentration of the substance (c_S) and the path length (*l*) through the sample:

$$A = \varepsilon c_s l.$$

The Lambert–Beer equation is the basis of a routine analytical method to measure concentrations in solution (*A* is measured so c_S can be calculated when ε and *l* are known). The method is very sensitive when ε values are high, and values of 10^4–10^5 dm³ mol⁻¹ cm⁻¹ are common for compounds with conjugated π systems.

This equation is an expression of the **Lambert–Beer law**, and the proportionality constant, ε, is the *molar absorptivity* (also called the molar absorption coefficient and, formerly, the extinction coefficient); since *A* is dimensionless, ε has the units dm³ mol⁻¹ cm⁻¹ when the units of c_S and *l* are mol dm⁻³ and cm.

Exercise 25.1

Give the energy in kJ mol⁻¹ corresponding to radiation of the following wavelengths.

(a) 217 nm   (b) 254 nm   (c) 500 nm

Exercise 25.2

Determine the concentration of a solution of anthracene which has absorbance A = 0.850 at its λ_{max} (256 nm where ε = 1.80×10^5 dm³ mol⁻¹ cm⁻¹) when the UV spectrum is measured in a cell of path length 1.00 cm.

UV-vis spectroscopy was the first spectroscopic technique to be developed as a method of structural determination, especially for the characterization of natural products in the mid-twentieth century. Although it is still widely used for analytical purposes, it is much less used nowadays for the determination of structure following the development of other methods, especially NMR (see later).

25.3 Infrared Spectroscopy

25.3.1 Introduction to IR spectroscopy

Covalent bonds in a molecule are not rigid—they are constantly stretching and bending (Section 4.1). Different kinds of bonds vibrate with different frequencies and radiative excitation of a particular vibration corresponds to absorption of IR radiation of a particular frequency (energy). The absorption pattern plotted against the radiation frequency is the IR spectrum. Since functional groups have characteristic bonds and hence characteristic vibrational frequencies, *IR spectroscopy may be used analytically to identify functional groups* in a substance.

Whereas wavelength is given for UV-vis radiation in a spectrum, IR radiation is normally described by its *wavenumber* ($\bar{v}$ in cm^{-1} where $\bar{v}=1/\lambda$) and the range is typically 4000–400 cm^{-1} (corresponding to λ of 2.5–25 µm). Another difference is that **transmittance** ($T=I/I_0$) is recorded in IR spectra (although we still usually refer to 'absorption bands' or 'peaks') rather than absorbance (A, eqn 25.2) as in UV-vis spectra. Characteristic absorption bands (peaks) which are useful for identifying particular molecular structural features are summarized in Tables 25.2 and 25.3.

Table 25.2 lists the characteristic IR absorption bands due mainly to bond stretching vibrations of the main functional groups of organic compounds. Most of the characteristic bands are above 1500 cm^{-1}; the region below 1500 cm^{-1} is called the **fingerprint region** where many peaks correspond to various C–O, C–C, and C–N single bond stretching and bending vibrations. Some characteristic out-of-plane bending vibrations for C–H in planar groups and molecules such as alkenes and benzene derivatives are listed in Table 25.3.

1 µm (micrometre) = 10^{-6} m

The complex spectral pattern below about 1500 cm^{-1} is unique to each organic compound and acts as its 'fingerprint'.

Table 25.2

Characteristic IR absorptions of some functional groups[a]				
alkane	C–H	2850–2960 (m-s)		
alkene (arene)	C(sp^2)–H	3020–3100 (m)	C=C	1620–1680 (m-w)[b]
alkyne	C(sp)–H	~3300 (s)	C≡C	2100–2260 (m-w)
alcohol (ether)	O–H[c]	3200–3600 (s, br)	C–O	1050–1250 (s)
	free O–H	3590–3650 (m)		
aldehyde (ketone)	C(O)–H	~2720 and 2820 (w)	C=O	1630–1750 (s)[d]
carboxylic acid	O–H	2500–3300 (s, br)	C=O	1710–1780 (s)[d]
amine	N–H	3300–3500 (m, br)	C–N	1020–1230 (m)

a. Absorption bands are given in cm^{-1} with indications of their intensities in parentheses: s, strong; m, medium; w, weak; br, broad. b. The skeletal vibrations of the benzene ring are typically four bands of variable intensities in the region 1450–1600 cm^{-1}. c. H-bonded OH. d. The wavenumber of the stretching band for carbonyl compounds RC(O)X is characteristically dependent on the nature of the X group.

Table 25.3

Out-of-plane bending vibrations for C–H in planar groups and molecules

25.3.2 Examples of IR spectra

Stronger bonds have larger force constants resulting in higher vibrational frequencies, as indicated by Hooke's law: $v = (1/2\pi)(\kappa/\mu)^{1/2}$ where κ is the *force constant* and μ is the *reduced mass.* Hooke's law also shows why C–H bonds have higher frequency stretching bands (around 3000 cm^{-1}) than similarly strong C–C bonds (around 900 cm^{-1}), for example: the reduced mass is smaller.

We shall now examine IR spectra of some representative compounds, and see how to correlate characteristic absorption bands in a spectrum with particular functional groups of the compound. Figures 25.4–25.6 show the IR spectra of hydrocarbons, heptane, hept-1-ene, and hept-1-yne.

All three spectra have peaks in the region 2850–3000 cm^{-1} which are associated with C(sp^3)–H stretching vibrations; in addition, Figures 25.5 and 25.6 have sharp peaks at 3080 and 3320 cm^{-1} due to the C(sp^2)–H and C(sp)–H bonds of a terminal alkene and alkyne, respectively. The frequency (which is proportional to the wavenumber) of a C–H stretching vibration reflects the strength of the bond: C(sp^3)–H < C(sp^2)–H < C(sp)–H, and the stronger the bond, the higher the frequency (wavenumber). Absorption bands for C=C and C≡C bond stretching vibrations are seen at 1645 and 2120 cm^{-1}, respectively, in Figures 25.5 and 25.6 while those for stretching vibrations of C–C single bonds appear in the range 800–1000 cm^{-1} in Figure 25.4; these differences arise from differences in

Figure 25.4 The IR spectrum of heptane (vapour).

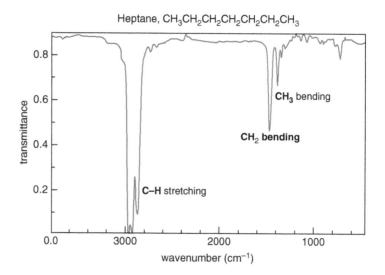

Heptane, CH$_3$CH$_2$CH$_2$CH$_2$CH$_2$CH$_2$CH$_3$

Figure 25.5 The IR spectrum of hept-1-ene (10% CCl$_4$ solution).

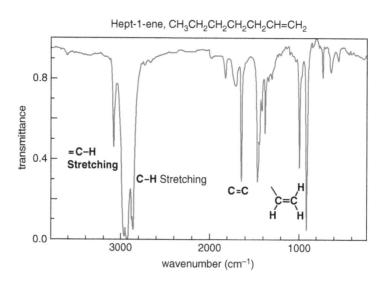

Hept-1-ene, CH$_3$CH$_2$CH$_2$CH$_2$CH$_2$CH=CH$_2$

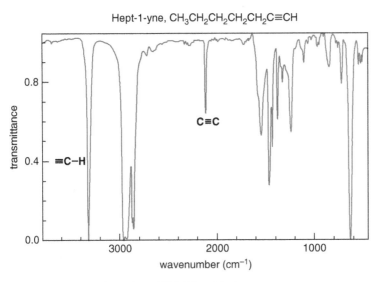

Figure 25.6 The IR spectrum of hept-1-yne (10% CCl₄ solution).

the strengths or stiffnesses of the carbon-carbon bonds, C–C<C=C<C≡C. Two peaks in the range 900–1000 cm⁻¹ in Figure 25.5 are attributable to the out-of-plane C–H bending vibrations characteristic of mono-substituted alkenes.

The absorption peak corresponding to the O–H stretching vibration of an alcohol usually appears in the region 3200–3600 cm⁻¹ as a band which is broad owing to inter-molecular hydrogen bonding. The presence of a free OH group can be recognized by a sharp peak at the high end of this range in the vapour phase or in a dilute solution in an aprotic solvent. Both features can be seen in the IR spectrum of butan-2-ol in Figure 25.7.

Carbonyl compounds have a strong absorption band around 1700 cm⁻¹, as seen in the spectra of an aldehyde in Figure 25.8 and a ketone in Figure 25.9. Aldehydes also have a characteristic pair of peaks due to C–H stretching vibrations at about 2720 and 2820 cm⁻¹ (Figure 25.8).

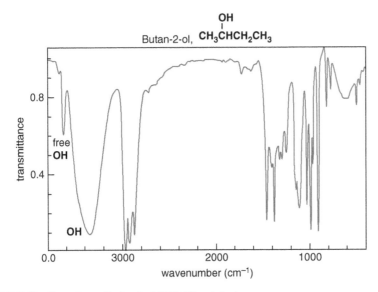

Figure 25.7 The IR spectrum of butan-2-ol (10% CCl₄ solution).

Figure 25.8 The IR spectrum of butanal (vapour).

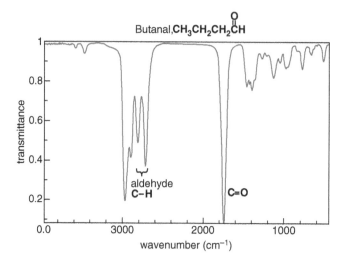

Figure 25.9 The IR spectrum of butanone (10% CCl_4 solution).

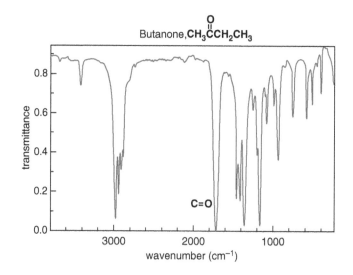

The weak band at about 3400 cm^{-1} in the spectrum of butanone is an *overtone* of the strong carbonyl band at about 1700 cm^{-1}. It corresponds to an excitation from the ground vibrational state to the *second* excited vibrational state.

Example 25.1

The following is the IR spectrum of an aromatic compound with the molecular formula C_8H_{10}. Give the structure of this compound and explain your reasoning.

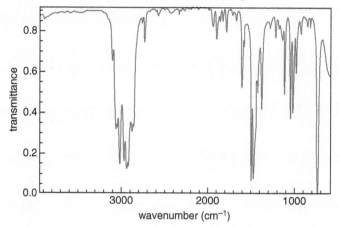

Solution

The molecular formula corresponds to ethylbenzene and isomers of dimethylbenzene, but the strong single peak at 742 cm^{-1} indicates an *ortho* disubstituted benzene (Table 25.3). Other peaks may be attributed as follows: 1606, 1583, 1495, and 1467 cm^{-1} (the benzene ring); 2800–3000 cm^{-1} (C(sp^3)–H stretching vibrations); 3000–3200 cm^{-1} (C(sp^2)–H stretching vibrations).

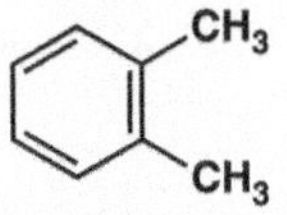

o-dimethylbenzene (*o*-xylene)

Exercise 25.3

IR spectra of two constitutional isomers C$_4$H$_{10}$O, butan-1-ol and diethyl ether, are given. Explain which is which.

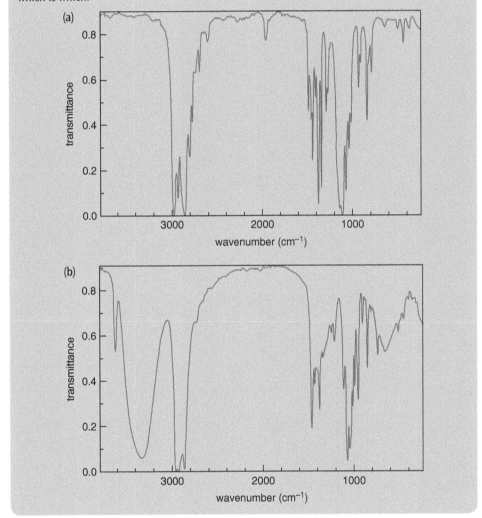

(a)

(b)

Exercise 25.4

Match IR spectra (a) and (b) with two of the compounds given and explain your reasons.

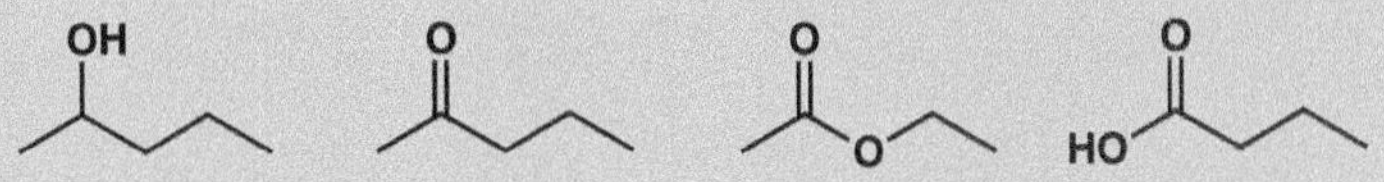

(continues ...)

(... continued)

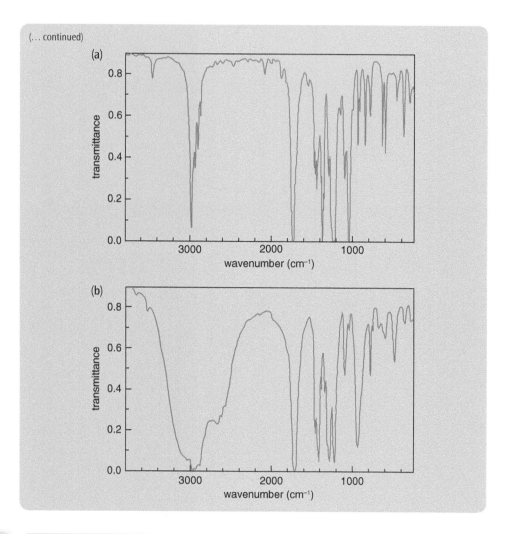

Exercise 25.5

Match the following spectrum to one of the compounds given and explain your reasons.

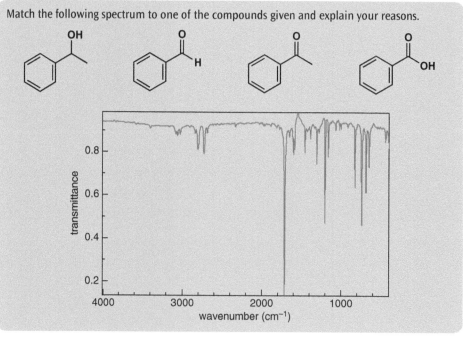

25.4 Nuclear Magnetic Resonance Spectroscopy: Proton NMR Spectra

Nuclear magnetic resonance (NMR) is a phenomenon involving the atomic nuclei of certain isotopes of some elements in a magnetic field and, within organic chemistry, the two most important nuclei are of ^{1}H (the most abundant isotope of hydrogen) and ^{13}C (a minor isotope of carbon). NMR techniques applied to these nuclei are the most important spectroscopic methods available today for the determination of molecular structures of organic compounds. Together, they provide detailed information about the environments of hydrogen and carbon atoms in organic molecules, and the carbon-hydrogen frameworks of molecules. Furthermore, modern NMR techniques enable the elucidation of stereochemical details.

> NMR spectroscopy of ^{19}F (the only stable isotope of fluorine) and ^{31}P nuclei is also fairly routine.

> In this section, we deal only with solution-state NMR spectroscopy; applications of solid-state NMR spectroscopy are beyond the scope of this book.

25.4.1 Physical basis of NMR

The nuclei of some isotopes have magnetic properties which are a function of a nuclear property called *spin* (but this term should not be taken too literally) and an associated spin quantum number, S (for ^{1}H and ^{13}C, $S=\pm\frac{1}{2}$). In the absence of an external magnetic field, the small magnetic moments of these nuclei are orientated randomly (as represented by the arrows in Figure 25.10(a)) and the nuclei have a single magnetic energy level. However, the small magnetic moments of these nuclei with $S=\pm\frac{1}{2}$ can interact with an external magnetic field. They orientate so that their fields become aligned either with the external field ($S=+\frac{1}{2}$ or with $\boldsymbol{\alpha}$ **spin**) or against it ($S=-\frac{1}{2}$ or with $\boldsymbol{\beta}$ **spin**), and two energy levels become available—the lower α spin state and the higher β spin state, as illustrated in Figure 25.10(b).

The energy difference between the two spin states, ΔE, is proportional to the external magnetic field ($\boldsymbol{B}_0$), and the proportionality constant is called the *magnetogyric ratio* (γ) which is a characteristic parameter of the nucleus: $\gamma=2.675\times10^{8}$ T^{-1} s^{-1} for ^{1}H and $\gamma=6.688\times10^{7}$ T^{-1} s^{-1} for ^{13}C (T or *tesla* is a unit of the strength of a magnetic field):

$$\Delta E=\gamma(h/2\pi)\boldsymbol{B}_0$$

where h is the Planck constant (6.626×10^{-34} J s).

In an external magnetic field, electromagnetic radiation whose energy corresponds to the energy difference between the two energy states available to the nuclei will induce transitions between them with $\Delta E=h\nu$. The energy difference between the two states is extremely small (so the equilibrium excess of nuclei in the lower state is extremely small) and the radiation required to induce an excitation is in the radiofrequency (rf, i.e. MHz) range. The transition between the two energy levels induced by the rf radiation is

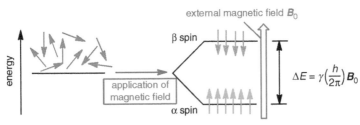

external magnetic field $\boldsymbol{B}_0$

β spin

application of magnetic field

α spin

$\Delta E=\gamma\left(\dfrac{h}{2\pi}\right)\boldsymbol{B}_0$

(a) Nuclear magnetic moments are random in the absence of an external magnetic field.

(b) Nuclear magnetic moments are with or against the external field.

Figure 25.10 Nuclear magnetic moments (a) in the absence of an external magnetic field and (b) in an external magnetic field.

sometimes called **spin-flip**, and the term *resonance* is used to describe the equilibration of nuclei between the two nuclear magnetic states induced by rf radiation of precisely the right frequency, hence the name **nuclear magnetic resonance** (NMR).

Since the nucleus of each isotope has its own magnetogyric ratio value, all ^1H nuclei (for example) might be expected to come into resonance at the same frequency in the same applied magnetic field. In this event, we would observe a single rf resonance signal for ^1H, which would be of little use. Fortunately, however, this is not the case. All the nuclei in a molecule are surrounded by electrons and, when an external magnetic field, B_0, is applied to the molecule, the electrons around the nuclei induce *local magnetic fields* (B_{local}) which oppose the applied field. Consequently, the *effective field* ($B_{effective}$) felt by the nucleus at its particular location within the molecule is somewhat smaller than the applied external field:

$$B_{effective} = B_0 - B_{local}.$$

We describe the effect of the local induced field by saying that (generally) *the electron density around the nucleus shields it from the external magnetic field*, i.e. it has a **shielding effect**. Consequently, if two ^1H atoms, for example, within the same molecule are in different environments, the shielding at their nuclei by their different electronic surroundings will be different.

Because the shielding for different types of ^1H in a molecule at a fixed value of B_0 is not the same, different ^1H nuclei will achieve resonance at different radiofrequencies, i.e. the rf has to be varied (scanned) so that the unique resonance conditions are attained for each type of ^1H in turn. The small rf differences required for resonance of the ^1H nuclei in different magnetic environments within a molecule at constant B_0 are measured by a Fourier transform (FT) method using a modern high-resolution NMR spectrometer, and the output is the ^1H NMR spectrum of the compound.

> Modern high-resolution FT NMR spectrometers use a superconducting magnet (cooled to 4.2 K with liquid helium) to produce enormously strong stable fields of up to 21.6 T which require radiofrequencies at about 920 MHz to bring ^1H nuclei into resonance; more common instruments use field strengths in the 4.7–19 T range corresponding to radiofrequencies in the 200–800 MHz range for ^1H NMR and 50–200 MHz for ^{13}C. Together, these very high magnetic field strengths and the high value of γ for ^1H lead to high sensitivity for ^1H NMR measurements. Consequently, proton NMR spectra can be obtained on relatively small amounts of material (typically a few micromoles for 400–600 MHz instruments).

25.4.2 Proton chemical shifts

> TMS was chosen as a standard in the early days of NMR spectroscopy because its ^1H nuclei are strongly shielded by surrounding electron density (due to the low electronegativity of Si) and its signal is away from those of the less shielded (more *deshielded*) hydrogens affected by organic functional groups. Furthermore, TMS has 12 equivalent Hs which give a single signal; also, it is inert and volatile (bp 27 °C). So, a very small amount of TMS gives a relatively strong signal which seldom interferes with the spectra of organic compounds, and TMS can be removed from the solution by evaporation after the measurement if necessary.

Virtually all organic compounds contain hydrogen atoms and, as mentioned, the effective magnetic field experienced by the nucleus of a particular ^1H in a molecule depends on its electronic environment. In broad terms, the higher the electron density around a nucleus, the greater the shielding, and the lower the local effective magnetic field sensed by the nucleus. As a result, the frequency (energy) of the electromagnetic radiation required by a nucleus to achieve resonance in a constant applied field becomes lower the more strongly it is shielded.

Whereas UV-vis and IR instruments record absorbance (or transmittance) at absolute values of wavelength (for UV-vis) or wavenumber (for IR), NMR spectrometers record the rf required for resonance of a particular type of ^1H *relative to the rf required for resonance of the ^1H of a standard compound* under identical conditions (applied field strength, temperature, and solvent). In the case of ^1H (and ^{13}C) spectra, the standard compound is **tetramethylsilane** (TMS, $Si(CH_3)_4$), which is usually added to the solution of the sample; the spectral parameter which is recorded (delta, δ) is called the **chemical shift**, and is defined by eqn 25.3.

$$\delta = \frac{\text{(resonance frequency of the sample)} - \text{(resonance frequency of TMS)}}{\text{(resonance frequency of TMS)}} \times 10^6 \qquad (25.3)$$

The huge advantage of using eqn 25.3 is that δ is independent of the applied field so results from NMR spectrometers operating at different magnetic field strengths can be legitimately compared. Note, however, that the difference between the absolute value of the resonance frequency of the signal for a typical H in an organic compound and the one for the TMS signal is very small, so the actual ratio is multiplied by 10^6 in eqn 25.3 to provide manageable numbers, and the units of δ are ppm (parts per million). The chemical shift of TMS is 0 by definition with the proton chemical shift scale increasing from right to left across a conventional spectrum.

a. The chemical shift range

The chemical shift scale defined by eqn 25.3 is in terms of resonance frequencies at a constant applied field strength but, in principle, these could be translated into applied field strengths at a constant radiofrequency (which is how early NMR spectrometers operated). This is why we refer to the left hand side of the chemical shift scale as low-field (or **downfield**) and the right hand side as high-field (or **upfield**), as indicated below the spectrum of 2,2-dimethylpropan-1-ol, for example, in Figure 25.11. The ^{1}H signals of most organic compounds ($\delta>0$), therefore, are downfield of the TMS signal and in the range 0–10 ppm with the exception of acidic hydrogens which have $\delta>10$. The range of proton chemical shifts for typical types of H in organic compounds is summarized in Figure 25.12, which indicates how extremely useful ^{1}H NMR spectroscopy is for establishing the different types of hydrogens in organic molecules.

Proton chemical shifts in alkanes generally increase in the order, methyl $CH_3<$ methylene $CH_2<$ methine CH, but the differences are modest, as shown in Figure 25.12; polar substituents on the C bearing the H have a larger effect. Since the shielding of a nucleus depends on the electron density around it, the chemical shift of a given type of H will increase (be moved downfield) if its surrounding electron density is reduced by an electron-withdrawing group. This is illustrated in the spectrum of 2,2-dimethylpropan-1-ol in Figure 25.11: the CH_2 has a higher chemical shift (larger δ value) than the methyls mainly because of the electron-withdrawing effect of the OH.

In the ^{1}H NMR spectrum of $(CH_3)_3CCH_2OH$ of Figure 25.11, the methyl groups give an intense signal at the highest field ($\delta=0.91$) and the OH and CH_2 peaks are seen at 1.48 and 3.29 ppm, respectively. (The very small signal at 7.26 ppm is due to $CHCl_3$ which contaminates the solvent, $CDCl_3$.)

Chemical shifts of Hs directly bonded to heteroatoms are variable, and depend on the solvent and concentration; although related to the acidity of the H, they are difficult to understand.

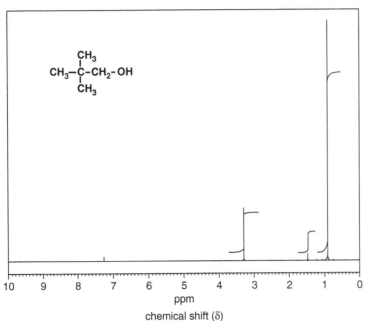

Figure 25.11 The ^{1}H NMR spectrum of 2,2-dimethylpropan-1-ol (400 MHz, $CDCl_3$ solution).

chemical shift (δ)
higher frequency $\leftarrow \rightarrow$ lower frequency
lower magnetic field (downfield) $\leftarrow \rightarrow$ higher magnetic field (upfield)
lower $\leftarrow$ shielding effect $\rightarrow$ higher

O‖—C—H	benzene H	C=C H	Z —C—H (Z = O, N, X)	—C—H

10   9.0   6.5   4.5   2.5   0

O‖
—C—OH 10–12

$R_2C=CH_2$ 4.5–5.0
$R_2C=CH$ 5–6

ArCH₃ ~2.5
—C≡C-H ~2.5
ROH) 1–5
RNH₂) (variable)

RCH₃ ~0.9
R₂CH₂ ~1.3
R₃CH ~1.6
Z=C-CH₂ 1.5–2.5
(Z = C,N,O)

Figure 25.12 Approximate proton chemical shifts (δ, ppm).

b. Diamagnetic anisotropy

An induced ring current in benzene and other aromatic compounds in a magnetic field is distinctive, and its occurrence is regarded as one of the criteria of aromaticity (Section 5.6). If a proton is located above (or below) the centre of the aromatic ring, it comes into resonance at a lower radiofrequency because the local induced field opposes the constant applied field at these locations; this corresponds to an upfield chemical shift (see Panel 25.1).

Proton chemical shifts are also appreciably influenced by the hybridization of the carbon to which the H is bonded and by its stereochemical location. The chemical shift of an H bonded to an sp² carbon is larger than expected purely from electron density considerations. This comes from the so-called local **diamagnetic anisotropy** due to the π electrons.

We saw in Sub-section 25.4.1 that electrons around a nucleus induce a local magnetic field when a molecule is in an external magnetic field; in a similar way, an applied external magnetic field causes a π electron system to induce a local magnetic field. It is as though a **ring current** is generated in the π electrons in response to the external magnetic field which, in turn, induces a local magnetic field, as illustrated in Figure 25.13. The direction of this induced field is in line with the applied field in the vicinity of the hydrogen atoms of an alkene and benzene (Figures 25.13(a) and (b)), so the (constant) applied field is reinforced by this additional induced field; consequently, resonance occurs at a higher radiofrequency. In an instrument operating at constant radiofrequency, a lower applied field is sufficient to achieve resonance for an H where the induced field reinforces the applied field. Regardless of the spectrometer details, a greater δ is observed for an H where the induced field reinforces the applied field, and the effect is described as a downfield shift.

The diamagnetic anisotropy is slightly more complicated for terminal alkynes. The ring current of the triple bond analogous to that of an alkene illustrated in Figure 25.13(a) should lead to a chemical shift similar to that for an alkene ($\delta \sim 5$ ppm) rather than the appreciably different observed value of about 2.5 ppm. The difference is caused by an additional effect, superimposed upon the one mentioned above, which leads to reduced shielding. When the molecule is aligned with the applied magnetic field, the induced π

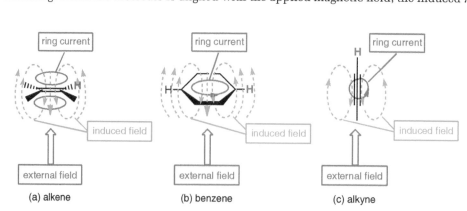

Figure 25.13 The effects of diamagnetic anisotropy due to the ring current of π electrons in (a) an alkene, (b) benzene, and (c) an alkyne.

electron ring current of the triple bond is *around* the axis of the bond (Figure 25.13(c)), and this induces a local field which *opposes* the applied field at the acetylenic H. The time-averaged result of these two opposed induced effects leads to a chemical shift of $\delta \sim 2.5$ ppm for terminal alkynes.

How many signals are observed in the ¹H NMR spectra of the following?

(a) ethane   (b) methyl ethanoate   (c) 1,4-dimethylbenzene

(d) 1,3-dimethylbenzene   (e) 1,2-dimethylbenzene

Exercise 25.6

Estimate approximate chemical shifts of the signals in a ¹H NMR spectrum of methyl methanoate.

Exercise 25.7

25.4.3 Integration: proton counting

The signals in the ¹H NMR spectrum of 2,2-dimethylpropan-1-ol in Figure 25.11 are not all the same size. The area under each signal peak, which can be obtained by integration, is proportional to the number of equivalent protons giving rise to the peak. The integrated peak areas are usually given with the spectrum either as printed numbers or by superimposed integration ('stair-step') curves in which the heights of the steps are proportional to the areas of the associated peaks. In the spectrum of Figure 25.11, the relative integral values are 9 : 1 : 2 from the upfield side, as expected for the three methyls, one OH, and one CH_2 group of $(CH_3)_3CCH_2OH$.

> Being able to integrate signals allows NMR spectra to be used analytically to measure the purity of a compound, and the ratio of components in a mixture. The method is not particularly sensitive, however.

Panel 25.1 Aromaticity and ring currents

As discussed in Sub-section 25.4.2, an applied magnetic field induces a ring current in the delocalized π electron system of benzene (and other aromatic compounds). In turn, the ring current then induces a local magnetic field which is responsible for the diamagnetic anisotropy. Hydrogen atoms in the plane of the benzene ring, but outside it, experience a deshielding local magnetic field; this results in ¹H NMR chemical shifts well downfield of the values for hydrogens bonded to simple sp²-hybridized carbons and are typically at $\delta \sim 7$ ppm; see Figures 25.12 and. 25.13(b). In contrast, any protons inside the ring current are shielded and their ¹H NMR signals are appreciably upfield (smaller δ values). Examples of this kind upfield shift are shown in the compounds below.

[7]paracyclophane   [18]annulene   1,6-methano[10]annulene

Paracyclophanes have a methylene bridge between the two *para* positions of benzene, and the signals of the methylene protons in the middle of the chain above the centre of the ring are at high field ($\delta = -0.6$ ppm in [7]paracyclophane); this compares with the chemical shift of $\delta = 7.07$ ppm for the four hydrogens bonded directly to the benzene ring. [18]Annulene is predicted to be aromatic by the Hückel $4n+2$ rule (see Section 5.7), and in fact the six protons inside the ring are subject to a very high upfield shift at $\delta = -2.9$ ppm. [10]Annulene (see Figure 5.10 in Chapter 5) cannot be planar with two hydrogens inside the ring because of the non-bonded strain which would be involved. However, the 1,6-methano derivative shown is almost planar with the bridging methylene protons just above the centre of the 10π aromatic system and strongly shielded ($\delta = -0.5$ ppm). These large differences in chemical shifts between the inside and outside protons, or the abnormally high upfield shifts of the inside protons, are taken as good evidence for aromaticity.

25.4.4 Spin–spin splitting

In the ^{1}H NMR spectrum of Figure 25.11, the signal for each distinct type of proton is seen as a single peak (a **singlet**). This is not always the case: we often observe a signal as multiple peaks (a **multiplet**). For example, the ^{1}H NMR spectrum of 2-bromo-1,1-dimethoxyethane in Figure 25.14 shows the BrCH$_2$– protons appearing as two peaks (a **doublet**) centred at 3.42 ppm, a tall singlet for the methoxy groups at 3.47 ppm, and the methine proton at C1 as a **triplet** at 4.61 ppm; the relative integrated areas are 2 : 6 : 1 from high field to low.

a. The multiplicity of spin–spin splitting patterns

The phenomenon of peak multiplicity, called **spin–spin splitting**, is caused by the interaction (or **spin–spin coupling**) between the nuclear spins of neighbouring atoms. In other words, the local effective magnetic field felt by one nucleus is affected by the tiny magnetic fields of adjacent nuclei as well as by surrounding electrons as discussed in Sub-section 25.4.2. Each ^{1}H nucleus is either in the α or β spin state (aligned with or against the applied field), and these different spin states affect a nearby nucleus differently. If a proton is affected by the spin of only one adjacent H in a molecule, its signal becomes a doublet—one peak due to the half of the molecules with the interacting H of α spin, and one due to the half with the interacting H of β spin. If there are two equivalent adjacent protons, a triplet signal is observed; more generally, *when the adjacent carbon has n equivalent protons, the signal splits into n+1 peaks*, but how does this arise?

As shown in Figure 25.15, coupling of a single proton H$_A$ with one H$_X$ gives a doublet for H$_A$ due to the magnetic influence of α and β spin possibilities of H$_X$. If there are two equivalent H$_X$ protons, both of which can be α or β, each of the two peaks of the doublet for H$_A$ splits into two again with the same separation (the coupling constant), resulting in *three* lines (a triplet) as the central two of the four are superimposed. The relative intensities of the three peaks of the H$_A$ triplet (in principle) are 1 : 2 : 1, which reflects the effect of the three ways the α and β spins of the two H$_X$ nuclei can combine, αα, (αβ, βα), and ββ. In the same way, three equivalent H$_X$ nuclei can couple with the neighbouring H$_A$ to give a quartet signal for H$_A$, and we can predict the relative intensities of the four to be 1 : 3 : 3 : 1 owing to the four ways the α and β spins of the three H$_X$ nuclei can combine: ααα, (ααβ, αβα, βαα), (αββ, βαβ, ββα), and βββ.

> More generally, a signal splits into $2nS+1$ peaks where S is the spin quantum number of the isotope causing the splitting ($S=\frac{1}{2}$ for ^{1}H).

> It is important to appreciate that the *multiplicity* (number of peaks) of the signal of H$_A$ is related to the number of nuclei with which it couples, not to the number of H$_A$ nuclei. The signal of the H$_X$ nuclei which is not shown in Figure 25.15 (regardless of how many H$_X$ nuclei there are) will be a doublet due to coupling with the single H$_A$ nucleus.

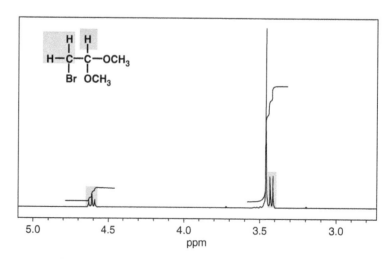

Figure 25.14 The ^{1}H NMR spectrum of 2-bromo-1,1-dimethoxyethane (270 MHz, CDCl$_3$).

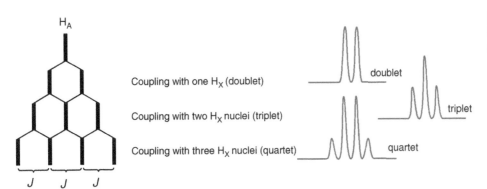

Figure 25.15 Relative intensities of spin–spin coupling patterns.

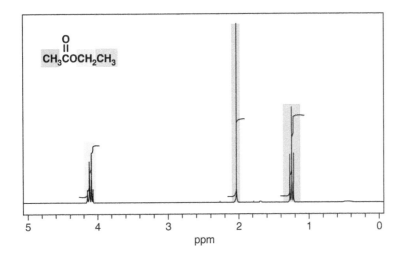

Figure 25.16 The ¹H NMR spectrum of ethyl ethanoate (270 MHz, CDCl₃).

b. Coupling constants

In the ¹H NMR spectrum of ethyl ethanoate (Figure 25.16), the acetyl protons appear at 2.02 ppm as a singlet which integrates for 3H, while the protons of the ethyl group give two signals—a triplet (3H) at 1.25 ppm and a quartet (2H) at 4.12 ppm. The former is the signal of the methyl which is a triplet due to coupling with the two protons of the CH_2, and the latter is the signal of the methylene which is a quartet due to coupling with the three protons of the CH_3. The equal gaps (in Hz) between the components of the two multiplets are the **coupling constant** (J) which is a measure of the strength of the coupling between the two types of nuclei. They are equal in the two multiplets ($J=7.0$ Hz for this ethyl group) because they reflect a single mutual magnetic interaction. Typical values of J are summarized in Table 25.4.

c. The coupling constant range

The *geminal* spin–spin coupling (^2J) between two protons bonded to the same carbon, i.e. separated by two bonds, is influenced by the bond angle and by the other groups, X and Y, bonded to the same carbon (the first entry in Table 25.4). The *vicinal* coupling (^3J) between two protons separated by three bonds is influenced by the dihedral angle (ϕ) between the two C–H bonds about the C–C bond (illustrated by the Newman projection in the margin) and follows the Karplus equation, $^3J=A\cos^2\phi+B\cos\phi+C$, where A, B, and C are constants.

This relationship arises from the way the interaction between the orbitals of the two C–H bonds depends upon the dihedral angle; the interaction and hence the ^3J coupling

Table 25.4	Spin–spin coupling constants (Hz)[a]

| 6–18 | 6–8 (free rotation) | 8–12 | 2–3 | 2–3 |

| 0–2 | 6–12 | 12–18 | 7–10 | 2–3 | 0–1 |

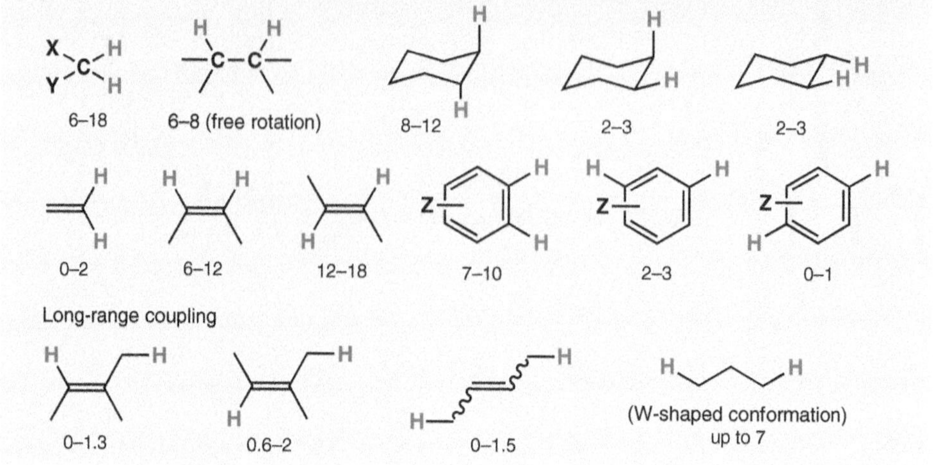

Long-range coupling

| 0–1.3 | 0.6–2 | 0–1.5 | (W-shaped conformation) up to 7 |

a. Magnitudes of coupling constants are independent of the applied magnetic field.

is minimal at $\phi=90°$ and maximal at $\phi=0°$ ($^3J=9$–12 Hz) and 180° ($^3J=10$–15 Hz). The Karplus relationship is illustrated by the ^3J coupling constants between Hs on adjacent carbons of a cyclohexane ring; average values are observed, however, when bond rotation about the adjacent carbons is sufficiently rapid as in a simple acyclic molecule. Coupling constants generally decrease with the number of the intervening bonds between the interacting nuclei, and coupling between protons separated by four bonds (or more) are usually negligible ($^4J<1$ Hz), an exception being when the C–H bonds are held in a planar W-shaped conformation (last entry in Table 25.4).

Exercise 25.8	How many peaks due to spin–spin coupling are expected for the signal of each of the protons (a)–(i)?

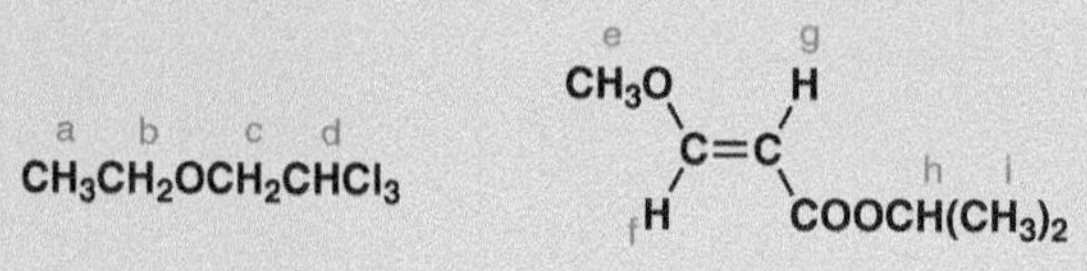

$$\underset{a}{CH_3}\underset{b}{CH_2}\underset{c}{OCH_2}\underset{d}{CHCl_3}$$

d. Roof effect in spin splitting

We now examine relative intensities of signals with spin splitting in more detail. When we observe a pair of doublets corresponding to two protons (H_A and H_B) which couple with each other (coupling constant $=J$ in Hz), the two peaks of each doublet have the same intensity if the chemical shifts (ν expressed in Hz) are appreciably different compared with the coupling constant ($\Delta\nu/J>10$), as shown in Figure 25.17(a). However, the shapes of the signals are different if their chemical shifts are more similar: the relative intensity of the inner two peaks compared with the outer two depends on the value of $\Delta\nu/J$, as shown in Figures 25.17(b)–(d) for $\Delta\nu/J$ values of 5, 2, and 1. This phenomenon is sometimes called a *roof effect*, and it is useful when looking for partners with mutual coupling. (When the chemical shifts ultimately become the same [$\Delta\nu=0$ in Figure 25.17(e)], we see only one signal for the two Hs.)

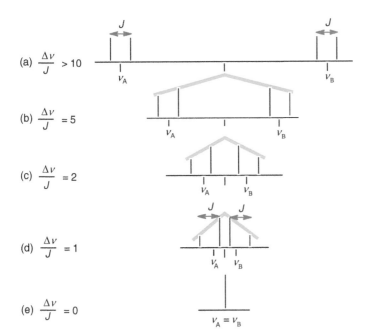

Figure 25.17 Schematic representation of how a pair of doublets due to different protons with mutual coupling depend upon the ratio $\Delta v / J$.

(a) $\dfrac{\Delta v}{J} > 10$

(b) $\dfrac{\Delta v}{J} = 5$

(c) $\dfrac{\Delta v}{J} = 2$

(d) $\dfrac{\Delta v}{J} = 1$

(e) $\dfrac{\Delta v}{J} = 0$

The chemical shift of each 'roofed' doublet is the weighted average of the two peaks. The aromatic signals at about 7.1 and 6.8 ppm in the spectrum of p-methoxytoluene is an example with $\Delta v / J = 8.2$ (Figure 25.18).

We saw at the beginning of our discussion of NMR that chemical shifts defined by eqn 25.3 are independent of the external magnetic field. However, unlike a coupling constant, the *absolute value* of the radiofrequency for resonance of a particular ^{1}H atom does depend upon the field strength. Consequently, Δv and hence $\Delta v / J$ in Figure 25.17 depend on the external magnetic field. It follows that the shape of a splitting pattern in a spectrum will depend upon the field strength of the NMR spectrometer.

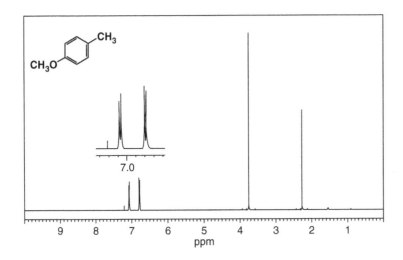

Figure 25.18 The ^{1}H NMR spectrum of p-methoxytoluene (400 MHz, CDCl$_3$).

Interpret the ^{1}H NMR spectrum of p-methoxytoluene given in Figure 25.18.

Exercise 25.9

25.4.5 Interpretation of ¹H NMR spectra

We shall illustrate in this sub-section how ¹H NMR spectra of some organic compounds are interpreted to obtain information about molecular structures.

a. An isopropyl ketone

The first example is the spectrum of 3-methylbutanone (isopropyl methyl ketone) in Figure 25.19. We see three signals in this spectrum—a doublet at 1.11 ppm, a singlet at 2.15 ppm, and a small multiplet at 2.60 ppm; the first two belong to the two kinds of methyl groups, and the small multiplet to the methine of the isopropyl group. The methine proton is coupled with the six equivalent methyl protons, so its signal is a septet (seven peaks) as we can see in the enlarged inset.

Ideally, the relative intensities of the seven peaks of a septet are 1:6:15:20:15:6:1. Note also that the six adjacent protons to which the methine H of the isopropyl group is coupled (J=7.2 Hz) are equivalent even though they are on different carbons.

Exercise 25.10

Assign (with reasoning) signals in the spectrum of Figure 25.19 to the two kinds of methyl groups of 3-methylbutanone.

b. A vinylic compound

The second example is the spectrum of 2-chloroethyl vinyl ether in Figure 25.20. We can see more clearly in the enlarged inset a pair of triplets at 3.70 and 3.95 ppm (J=5.8 Hz) corresponding to the two kinds of CH_2 of the chloroethyl group. Consequently, the two multiplets at δ=4.08 and 4.23 ppm, and the more remote one at 6.48, must be the signals of the three alkene protons. The former two signals near 4 ppm are at quite a high field for hydrogens on sp² carbon. They correspond to the terminal methylene protons, and their upfield δ values are attributable to the high electron density on the β carbon because of the conjugative electron donation from the lone pair of the alkoxy oxygen. The other multiplet at 6.48 ppm (which is rather downfield owing to the inductive effect of the adjacent O) is the signal of the other vinylic hydrogen. Each of these three non-equivalent vinylic protons is coupled to the other two with different coupling constants (J_{trans}=14.4, J_{cis}=6.8, and J_{gem}=2.4 Hz), so each multiplet is a doublet-of-doublets.

The four peaks of a doublet-of-doublets are of similar intensities and must not be confused with a quartet due to coupling with three equivalent protons which are equally spaced and, in principle, in the ratio 1:3:3:1 (see above).

The two triplets of the coupled protons of the chloroethyl group in Figure 25.20 show the roof effect described in Sub-section 25.4.4d.

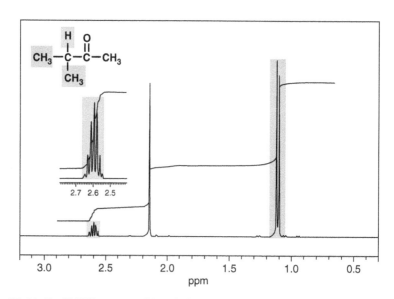

Figure 25.19 The ¹H NMR spectrum of 3-methylbutanone (400 MHz, CDCl₃).

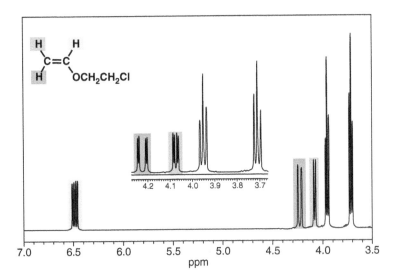

Figure 25.20 The ^{1}H NMR spectrum of 2-chloroethyl vinyl ether (400 MHz, CDCl$_3$).

The ^{1}H NMR spectrum of 4-chlorobutanoic acid shows the following signals. Assign the signals and explain the multiplicity of each.

(a) 11.8 ppm (broad singlet)   (b) 3.62 ppm (triplet)   (c) 2.56 ppm (triplet)
(d) 2.11 ppm (five peaks)

c. A compound with diastereotopic hydrogens

In examples so far, hydrogens on the same saturated carbons (methyl and methylene) have appeared to be equivalent. This is usually the case when bond rotations are rapid compared with the time scale of NMR measurements which otherwise would allow such protons to be differentiated (see Sub-section 25.4.6).

The two hydrogens of a methylene group in a molecule with a chiral centre, however, are not completely equivalent even if free rotation of the bonds is possible. Such non-equivalent hydrogens are examples of *diastereotopic* atoms, and the methylene hydro-gens of 1,2-dibromopropane which is chiral at C2, are examples, as seen in its spectrum (Figure 25.21).

The hydrogens of a terminal methylene as in the vinyl group in Figure 25.20 are not equivalent even though they are bonded to the same carbon because there is no rotation about the double bond.

Atoms or groups are *diastereotopic* if replacement of one of them gives one of a pair of diastereoisomers. See Panel 25.2, and Section 11.8, for the related concept of *enantiotopicity*.

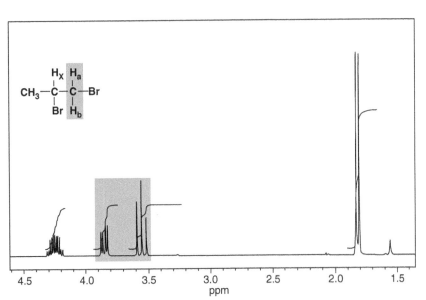

Figure 25.21 The ^{1}H NMR spectrum of 1,2-dibromopropane (270 MHz, CDCl$_3$).

Panel 25.2 Topicity

Atoms or groups in a molecule which are chemically identical in all respects are said to be *homotopic*, e.g. the hydrogens in CH_4, in CH_3Br, or in CH_2Br_2 (and the Br atoms in CH_2Br_2). The methylene hydrogens in XCH_2Y are equivalent in many respects but, if one is replaced by a third group (Z), the product (HCXYZ) is chiral. Consequently, the methylene hydrogens differ in the sense that replacement of one of them gives one enantiomer of HCXYZ and replacement of the other H gives the other enantiomer. Such methylene hydrogens are described as *enantiotopic* and ethanol provides an example. Normally, enantiotopic protons are not distinguishable by NMR.

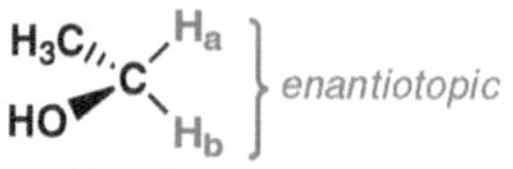

If one of the groups X or Y in XCH_2Y contains a chiral centre, and we have a single enantiomer exemplified by (*R*)-1,2-dibromopropane, replacement of, say, H_a by Z will generate a new chiral centre and yield a single *diastereoisomer*; replacement of H_b will give the alternative diastereoisomer. The original methylene hydrogens H_a and H_b are said to be *diastereotopic* and give separate ^1H NMR signals.

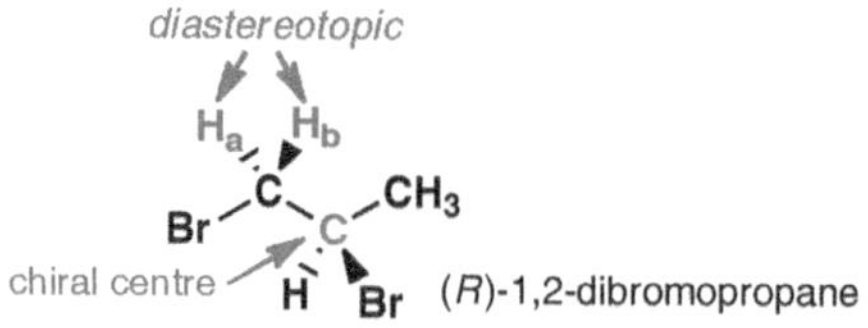

We have already referred to the enantiotopic *faces* of an unsymmetrical ketone $R^1R^2C{=}O$ (**1**) in Section 11.8 in the context of nucleophilic addition to the prochiral trigonal C of its carbonyl group. Enantiotopic *atoms* (or *groups*) arise from a prochiral tetrahedral centre and are exemplified by the methylene hydrogens in generic structure **2** which may be compared with **1**.

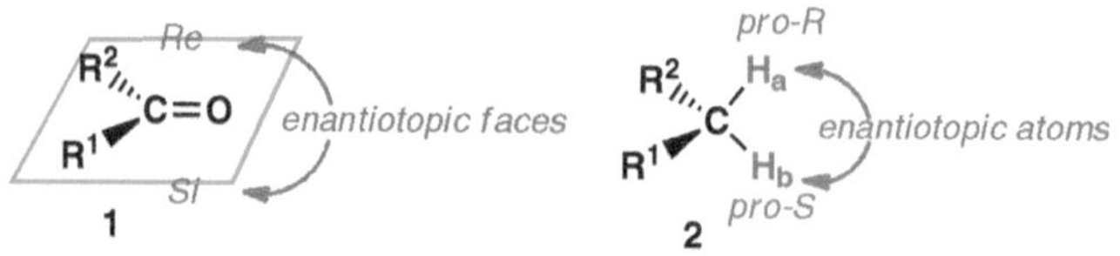

(Priority: $O>R^1>R^2$)

Just as we need to be able to specify one of a pair of enantiomers, we may also need to be able to specify one of two enantiotopic faces (e.g. of **1**) or atoms (or groups, e.g. of **2**). Enantiotopic faces are designated using the identifiers *Re* and *Si* just as we label chiral centres *R* or *S*. If we assume that the priority of the groups attached to the prochiral C of **1** is $O>R^1>R^2$ according to the Cahn–Ingold–Prelog (CIP) sequence rules, their decreasing order of priorities viewed from above is clockwise, and this defines the upper face as *Re*. It follows that the decreasing order of priorities of the same groups viewed from below appears anticlockwise, so the lower face of the ketone is *Si*.

Correspondingly, an enantiotopic atom (or group) is designated *pro-R* (as shown above in **2**) if, when it is arbitrarily assigned CIP priority over the other enantiotopic atom, the configuration of the newly generated chiral centre is *R*; the other enantiotopic atom (or group) is *pro-S*.

The following are some examples of molecules featuring different topicities.

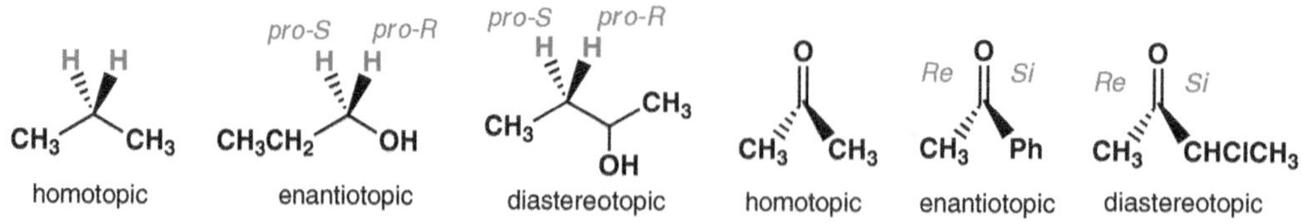

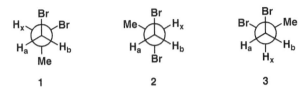

Figure 25.22 Staggered conformations of 1,2-dibromopropane. (These conformers are of one enantiomer, but the other enantiomer is equivalent in this context.)

Three possible staggered conformations of this compound are shown as **1–3** in Figure 25.22. We see that, regardless of conformation, the relationships of H_a and H_b with the rest of the molecule are different, i.e. the two protons on C1 (H_a and H_b) are not equivalent—they are diastereotopic. As a result, their signals appear at 3.55 ($J_{ax}=10.1$ and $J_{ab}=10.0$ Hz) and 3.85 ppm ($J_{bx}=4.3$ and $J_{ab}=10.0$ Hz), respectively, as doublets-of-doublets (the former looks like a triplet because $J_{ab}\sim J_{ax}$). (The methyl signal at 1.83 ppm is a doublet due to coupling ($J=6.5$ Hz) with the methine whose signal at 4.25 ppm is a complicated multiplet owing to different couplings with the methyl, H_a, and H_b.)

Exercise 25.12

^{1}H NMR spectra of two isomeric compounds with the molecular formula $C_8H_{11}N$ are given. Show structures of the compounds and assign the signals.

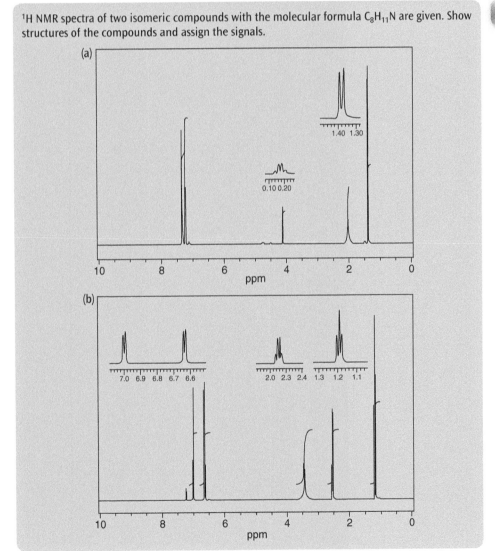

25.4.6 Disappearance of spin–spin coupling

a. Proton exchange

In the ¹H NMR spectrum of 2,2-dimethylpropan-1-ol in Figure 25.11, the OH and CH protons both appear as single peaks. Since these different protons are next to each other, they should couple. The observed lack of coupling is due to the rapid intermolecular exchange of hydroxy protons (they do not stay long enough at one site to couple with other nuclei), accelerated by adventitious water and/or acid in the sample solution. If pure alcohol is used in a pure aprotic solvent, the proton exchange becomes slower and spin–spin coupling can be observed (J=6.2 Hz) as a triplet corresponding to the OH (1.44 ppm) and a doublet (3.29 ppm) for the CH_2 (Figure 25.23).

The time scale of NMR excitation/relaxation is usually 10^{-4}–5 s (but can be much longer) depending on the coupling constant of the signal and the *relaxation times* of the excited nuclei (see Sub-section 25.4.7). When an exchange (or another dynamic process) is faster than the excitation/relaxation, separate signals cannot be observed—they coalesce into a single time-averaged peak. When the half-life of a dynamic process is comparable with the NMR time scale, its rate constant can be determined by analysis of the shape of the signal (line-shape analysis). This technique is known as the dynamic NMR method of chemical kinetics.

If a drop or two of deuterium oxide (^2H_2O) is added to the sample of 2,2-dimethylpropan-1-ol in $CDCl_3$ and the spectrum is re-recorded, the signal of the OH in Figure 25.23 will have disappeared. This is because the ¹H atoms of the hydroxy group of the alcohol will have exchanged with the ²H atoms of the heavy water which do not show an NMR signal in the same range as ¹H. This is a very useful general procedure for identifying which peaks in a spectrum are due to exchangeable hydrogens.

b. Spin decoupling

Spin–spin splitting can be removed by additional radiation of the sample at the radiofrequency of one particular signal of the coupled system. This will cause very rapid spin inversions of the H whose resonance corresponds to the additional rf which will average its magnetic influence upon neighbouring nuclei including the one(s) to which

Figure 25.23 The ¹H NMR spectrum of pure 2,2-dmethylpropan-1-ol in $CDCl_3$ (400 MHz).

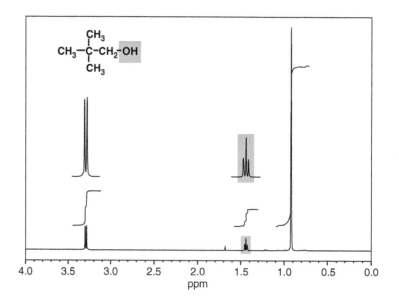

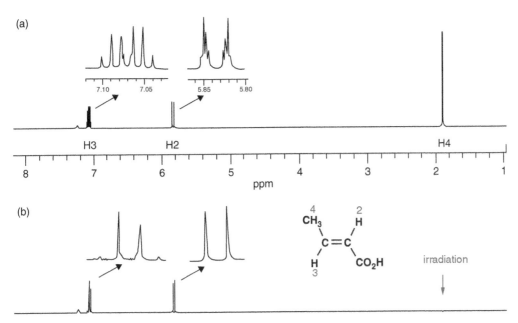

Figure 25.24 ^{1}H NMR spectra of (E)-but-2-enoic acid (crotonic acid) in CDCl$_3$ at 400 MHz. (a) The normal spectrum and (b) the spin-decoupled spectrum obtained by irradiation at the methyl resonance (H4 at 1.87 ppm).

it is coupled. In other words, the irradiated proton does not stay in one spin state long enough to interact with neighbouring protons. As a result, the spin–spin coupling disappears just as when decoupling is induced by rapid proton exchange. A spectrum with a complicated coupling pattern is usually simplified by this technique, which makes the identification of coupling partners and analysis of the whole spectrum easier. It is sometimes called the *double resonance method* and is illustrated in Figure 25.24.

The 'normal' ^{1}H NMR spectrum of (E)-but-2-enoic acid (crotonic acid) is shown in Figure 25.24(a) and we see a complicated pattern due to coupling with the methyl superimposed on the mutual coupling between the two alkene H atoms. The effect of spin decoupling by irradiation at the signal of the Me (H4 at 1.87 ppm) is shown in Figure 25.24(b): the alkene signals at 5.83 (H2) and 7.07 ppm (H3) have both become simple doublets (*J*=15.7 Hz) due only to their mutual coupling.

25.4.7 The nuclear Overhauser effect

Relationships between C–H groups close across space in a molecule can be investigated by the so-called **nuclear Overhauser effect** (or nuclear Overhauser enhancement, **NOE**). This phenomenon is related to the *relaxation* of spin-excited nuclei, i.e. the mechanisms by which the spin-excited nuclei return to their lower energy state. By one mechanism, excited nuclei relax by emission of rf energy (the same frequency by which they were initially excited); a second mechanism of dissipating their excess energy is by energy transfer to other nearby nuclei. This latter relaxation process (sometimes called spin–lattice relaxation) induces spin inversion of neighbouring nuclei; consequently, stronger NMR signals of these neighbouring nuclei can be observed during the irradiation of the first signal. This enhancement occurs by a through-space mechanism, not by a through-bond interaction, and is inversely proportional to the sixth power of the distance (*r*) between the interacting nuclei (NOE $\propto 1/r^6$).

In practice, an NOE spectrum is obtained as a *difference spectrum*, so enhanced signals can be identified easily. This technique is very useful for investigating the

The observed sensitivity of ^{13}C NMR using samples which are not isotopically enriched is only about 1/6000th that of ^{1}H NMR. The value of its magnetogyric ratio is about 1/4th that of ^{1}H and the NMR signal intensity is proportional to γ to the power three (γ^3).

*This method also intensifies ^{13}C NMR signals by the NOE.

There are solvent and diamagnetic anisotropy effects upon ^{13}C chemical shifts but, because ^{13}C chemical shifts are so large, these effects are much less significant than the ones upon proton chemical shifts, so assignment of signals is easier. The ^{13}C signal of CDCl$_3$ which is commonly used as a solvent usually appears as a triplet at 77.16 ppm (remember from the margin note in Sub-section 25.4.4a that the number of lines in a multiplet is $2nS+1$, and $S=1$ for ^{2}H).

stereochemistry of a compound. For example, in the semi-synthetic penicillin shown in the margin of the previous page, the orientation of the six-membered sulfur-containing ring was established by NOE. Irradiation at the resonance frequency of the protons of one of the methyl groups leads to enhancements of both H3 and H10 signals, so the sulfur of the six-membered ring is *trans* to the other S as shown.

25.5 Carbon-13 NMR Spectra

25.5.1 Introduction to ^{13}C NMR spectra

Although the most abundant carbon isotope (^{12}C) has no nuclear spin, ^{13}C (1.1% natural abundance) does have a nuclear magnetic moment which gives rise to an NMR signal. The sensitivity of ^{13}C NMR is much smaller than that of ^{1}H NMR because of the lower occurrence of the isotope and the smaller magnetogyric ratio (γ) of the ^{13}C nucleus. However, modern computer-assisted techniques (Fourier transform methods with pulsed irradiation and accumulation of multiple scans) have enabled the measurement of ^{13}C NMR spectra of organic compounds to have become a routine method for the elucidation of organic molecular structures.

The physical basis of ^{13}C NMR is the same as that of ^{1}H NMR and carbon chemical shifts are also given in ppm using the signal of the carbons of TMS as a reference. However, because of the low natural abundance of ^{13}C, the probability of two ^{13}C atoms being close enough in a molecule to couple with each other is very small. Consequently, ^{13}C–^{13}C splitting is not observed in routine spectra. Furthermore, all couplings with protons can be removed by additional irradiation in the rf region of proton resonance. By this technique of *broad-band proton decoupling*,* ^{13}C signals are normally observed as singlets instead of being complicated by couplings with protons.

Chemical shifts in ^{13}C spectra extend widely over about 200 ppm and, because signals are normally singlets, the likelihood of signal overlap is small, so we can easily establish the number of nonequivalent carbons in a molecule. Different types of carbons can be identified from their chemical shifts (a carbon bonded to an electronegative atom shows a downfield shift), and Table 25.5 gives a summary. ^{13}C chemical shifts are very dependent on the hybridization: an sp^2 carbon appears at lower fields than sp and sp^3 carbons, and carbonyl carbons are at particularly low fields.

Table 25.5

^{13}C chemical shifts (ppm)

(CH$_3$)$_4$Si	0	C–I	0–40	R–C(=O)–Y	165–185
R–CH$_3$	8–35	C–Br	25–65		
R–CH$_2$–R	15–50	C–Cl	35–80	R–C(=O)–R	195–205
R–CH–R (R)	20–60	C–N	40–60		
R–C–R (R, R)	30–40	C–O	50–80	R–C(=O)–H	195–215
≡C	65–85	=C	100–150	(benzene)C	110–170

25.5.2 Interpretation of ¹³C NMR spectra

As a first example, we see the ¹³C NMR spectrum of ethyl ethanoate in Figure 25.25. There are four signals corresponding to the four different carbon atoms of the molecule (plus the CDCl₃ signal). The signal at the lowest field (~171 ppm) is due to the strongly deshielded carbonyl carbon and the two high field signals at ~21 and ~14 ppm correspond to the two methyl groups; the more downfield of these two signals belongs to the methyl next to the carbonyl. The signal of the methylene group at ~60 ppm is downfield of those of the methyls owing to the deshielding influence of the oxygen to which it is bonded. All the signals are within the expected regions of chemical shifts listed in Table 25.5.

Figure 25.26 shows a ¹³C NMR spectrum of 4-methylpent-3-en-2-one. The assignment of three signals to the three sp² carbons is straightforward. The signal at the lowest field (~199 ppm) is of the carbonyl and the next lowest (~155 ppm) corresponds to the

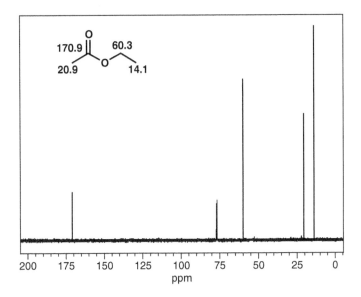

Figure 25.25 The ¹³C NMR spectrum of ethyl ethanoate (150 MHz, CDCl₃).

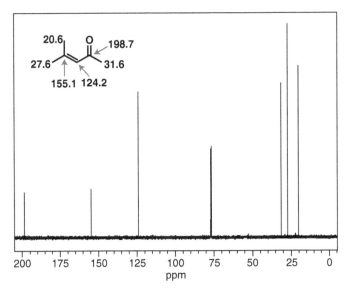

Figure 25.26 The ¹³C NMR spectrum of 4-methylpent-3-en-2-one (150 MHz, CDCl₃).

Figure 25.27 The ^{13}C NMR spectrum of a solution of ethyl 3-oxobutanoate (150 MHz, CDCl$_3$).

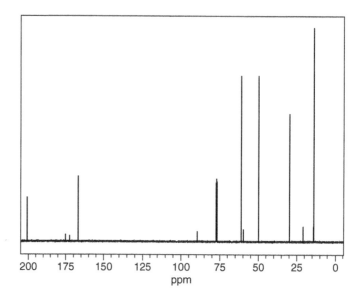

β alkenyl carbon which is deshielded by the conjugative electron-withdrawing effect of the carbonyl; this leaves the signal at (~124 ppm) belonging to the α alkenyl carbon. The assignment of the three methyl carbons is not as simple and based largely upon analogies.

A ^{13}C NMR spectrum of ethyl 3-oxobutanoate (acetoacetate) in CDCl$_3$ solution is shown in Figure 25.27 and is somewhat complicated owing to enolization (see Section 17.1). This β-keto ester is in equilibrium with about 10% of an enol form and chemical shifts are given in the equation. As seen, the signals of the carbonyl carbons of the ester (167.1 and 172.6 pm) are upfield of the ketone signal (200.7 ppm).

> The proportions of the tautomers were measured by integration of the ^{1}H NMR spectrum.

167.1	172.6
200.7	175.4
61.3	59.8
30.1 50.0 14.0	21.1 89.7 14.2
90%	10%

Example 25.2

Compare and comment on the chemical shifts of the alkene carbons of the enol of ethyl 3-oxo-butanoate (Figure 25.27) and those of 4-methylpent-3-en-2-one (Figure 25.26).

Solution

The difference between the chemical shifts of the β C and α C in the enol of ethyl 3-oxobutanoate (175.4 and 89.7 ppm) is appreciably greater than the difference between the corresponding Cs of the enone (155.1 and 124.2 ppm). The resonance forms for the enol illustrated below are not available to the enone and have the effect of deshielding the β C of the enol and shielding its α C compared with the corresponding carbons of the enone.

One shortcoming of ^{13}C NMR spectroscopy illustrated in the spectra above is that the signal intensity is not normally proportional to the number of equivalent carbons in a molecule. This is because the relaxation times of the excited ^{13}C nuclei are relatively long and dependent on the environment of the nucleus in a molecule. When ^{13}C NMR spectra are obtained, the data are collected at short intervals and accumulated. However, the relaxation of some carbons after one pulse might not be complete and hence unable to respond to the next pulse, i.e. incompletely relaxed nuclei give smaller signals. In particular, quaternary carbons bearing no hydrogens (and hence with weak spin-lattice relaxation mechanisms) generally give less intense signals.

Furthermore, although the broad band irradiation of the proton resonances for ^{13}C–^{1}H spin decoupling slightly enhances ^{13}C signals by the NOE, the effect is variable and depends on the nearness of hydrogens to the ^{13}C (and their number), as discussed in Subsection 25.4.7. This signal enhancement mechanism is especially weak for quaternary carbons as they bear no H atoms.

The ^{13}C NMR spectra of the two isomers of cyclohexadiene have the following signals. Explain which signals belong to which isomer.

(a) 26.0 and 124.5 ppm   (b) 22.2, 124.4, and 126.2 ppm

Exercise 25.13

The ^{13}C NMR spectra of the three isomers of dichlorobenzene have the following signals. Explain which signals belong to which isomer.

(a) 129.8 and 132.6 ppm   (b) 127.7, 130.5, and 132.6 ppm
(c) 126.9, 128.7, 130.4, and 135.1 ppm

Exercise 25.14

25.6 Mass Spectrometry

Mass spectrometry (MS) can determine the molecular formula of a molecule ionized by loss of an electron following collision with a high-energy electron in an electron beam. In addition, the ionic fragments formed by subsequent decomposition of the molecular ion provide important information about the molecular structure. Molecular formulas were once deduced from elemental analyses combined with molecular weights usually determined by the effect of the compound upon the boiling point (an elevation) or freezing point (a depression) of a solvent; these measurements were time- and labour-consuming tasks, and required appreciable amounts of pure compounds. Mass spectrometry now allows us to obtain the same information (and much more) in less time with only minute amounts of material.

25.6.1 The mass spectrometric method

In a mass spectrometer, a sample of the compound under investigation is introduced as a molecular beam into a high vacuum chamber and ionized (usually) by bombardment with a beam of high-energy electrons (typically 70 eV), as illustrated in Figure 25.28.[2]

---

[2] The electron volt, eV, is an energy unit applicable to electrons, or processes involving electrons, and $1 \text{ eV} = 1.602 \times 10^{-19}$ J. Since the eV is defined for a single electron process, we need to take account of the Avogadro constant (6.022×10^{23} mol^{-1}) if we want to convert electron volts into the more usual molar energy units used in chemistry; the conversion factor is $1 \text{ eV} \rightarrow 96.5$ kJmol^{-1}. Clearly, the amount of energy imparted to molecules upon impact with 70 eV electrons is hugely in excess of the energy required to break bonds.

Figure 25.28 Schematic representation of a mass spectrometer.

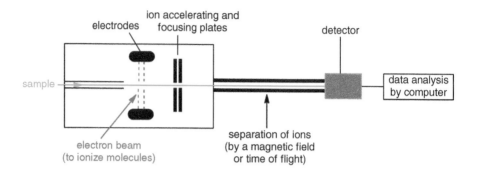

In this **electron impact** (EI) ionization mass spectrometric method, the molecular ion (a radical cation), formed by ejection of an electron following collision with one of the high-energy electrons, mainly decomposes into fragment radicals and cations.

$$\begin{array}{ccccccc} M & + & e^- & \xrightarrow{\text{ionization}} & M^{\bullet+} & + & 2\,e^- \\ \text{molecule} & & \text{high-energy} & & \text{molecular ion} & & \\ & & \text{electron} & & & & \end{array}$$

$$\downarrow \text{fragmentation}$$

$$\text{radicals} + \text{cations}$$

> Because the charge of the cation (z) is almost always +1, the numerical values of its m/z ratio and its mass (m) are normally the same.

The cationic species (the molecular ion and its charged fragments) are separated by their mass (more exactly, their **mass-to-charge ratio**, m/z) to provide a mass spectrum by plotting the intensity of each ion against its mass (or m/z ratio). The separation is achieved by diverting the flow of gaseous ions with a magnetic field (the ions are deflected according to their m/z) or by their time of flight (the velocity of an ion depends on its m/z).

A conventional low resolution mass spectrometer records intensities of whole-number m/z ratios, and the highest (most intense) peak in the spectrum is called the *base peak*; intensities of all other peaks are given relative to this. The *molecular ion* (identified as $M^{\bullet+}$, M^+, or sometimes simply as M in spectra) generated in the ionization chamber by loss of an electron will appear at the highest m/z value in the mass spectrum if it is sufficiently stable to reach the detector before fragmenting.

25.6.2 The mass spectrum and fragment ions

a. Mass spectra of alkanes

> The symbol $[M]^{\bullet+}$ used for a molecular ion means that the radical cation is the molecule inside the square brackets *less* an electron. From where the electron has been lost may be obvious if, for example, the molecule is unsaturated or contains a heteroatom with one or more lone pairs (it will be from the HOMO).

The mass spectrum of pentane is shown in Figure 25.29; we see the base peak which corresponds to the propyl cation ($C_3H_7^+$) at $m/z=43$, and the molecular ion of C_5H_{12} is at $m/z=72$. The relatively low intensity of the molecular ion indicates that an appreciable proportion of them decompose (fragment) before they reach the detector. As indicated above, the main fragmentation process of a radical cation is to split into a radical (which will not be detected as it is uncharged) and a cation. Four possibilities are shown for pentane in Scheme 25.1 and these account for the peaks at $m/z=15$, 29, 43 (the base peak), and 57.

The peak for a fragment ion is generally accompanied by peaks at $m/z-1$ and/or $m/z-2$ corresponding to loss of one or two hydrogen atoms from the carbenium ion. Fragmentation of the propyl cation in Scheme 25.1, for example, proceeds as follows, which accounts for the strong signals at $m/z=41$ and 42 observed in Figure 25.29.

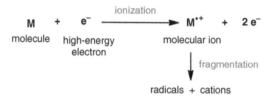

$$\underset{m/z=43}{CH_3CH_2\overset{+}{C}H_2} \xrightarrow{-H\bullet} \underset{m/z=42}{[\,CH_3CHCH_2\,]^{\bullet+}} \xrightarrow{-H\bullet} \underset{m/z=41}{CH_2{=}CH\overset{+}{C}H_2}$$

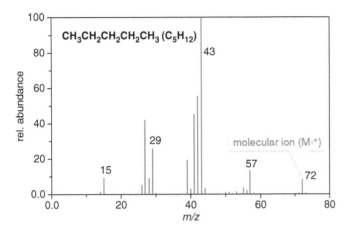

Figure 25.29 Mass spectrum of pentane.

Scheme 25.1 Some fragmentations of the pentane molecular ion.

$CH_3\dot{C}H_2$ + $\overset{+}{C}H_2CH_2CH_3$
$m/z = 43$

$CH_3\overset{+}{C}H_2$ + $\dot{C}H_2CH_2CH_3$
$m/z = 29$

$[CH_3CH_2CH_2CH_2CH_3]^{\cdot +}$
$m/z = 72$

$\dot{C}H_3$ + $\overset{+}{C}H_2CH_2CH_2CH_3$
$m/z = 57$

$\overset{+}{C}H_3$ + $\dot{C}H_2CH_2CH_2CH_3$
$m/z = 15$

Mass spectra of 2-methylbutane and 2,2-dimethylpropane (isomers of pentane) are shown in Figures 25.30 and 25.31, respectively. The mass spectrum of the former (Figure 25.30) resembles that of pentane (Figure 25.29), but the peak at $m/z=57$ is larger relative to the base peak at $m/z=43$ (which this time corresponds to formation of the isopropyl cation, Me_2CH^+, by loss of Et·) and the molecular ion peak ($M^{\cdot +}$) is smaller. This is because loss of a methyl radical from the 2-methylbutane molecular ion gives a secondary cation which is more stable than the primary cation formed by loss of a methyl from the pentane molecular ion.

$CH_3\overset{+}{C}HCH_3$ + $\dot{C}H_2CH_3$
$m/z = 43$

$\begin{bmatrix} CH_3 \\ | \\ CH_3CHCH_2CH_3 \end{bmatrix}^{\cdot +}$
$m/z = 72$

$CH_3\overset{+}{C}HCH_2CH_3$ + $\dot{C}H_3$
$m/z = 57$

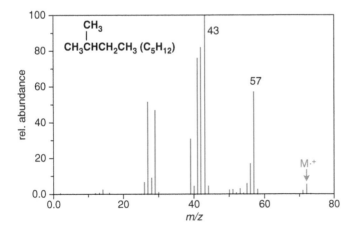

Figure 25.30 Mass spectrum of 2-methylbutane.

Figure 25.31 Mass spectrum of 2,2-dimethylpropane.

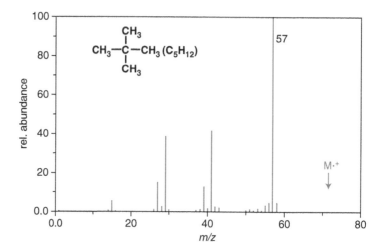

This tendency to give the more stable cation of the possible alternatives is more apparent in the mass spectrum of 2,2-dimethylpropane (Figure 25.31): a methyl radical is lost so readily from the 2,2-dimethylpropane molecular ion to give the stable *t*-butyl cation (which is now the base peak at $m/z=57$) that the molecular ion peak is hardly detectable.

$$\left[\begin{array}{c} CH_3 \\ CH_3-C-CH_3 \\ CH_3 \end{array} \right]^{\cdot+} \longrightarrow (CH_3)_3C^+ + \dot{C}H_3$$

$m/z=72 \qquad\qquad m/z=57$

The other two major peaks at $m/z=41$ and 29 (and smaller ones) are results of complicated cation rearrangements, such as those covered in Chapter 22, and subsequent fragmentations.

Exercise 25.15

Predict the m/z value of the base peak in the mass spectrum of 3-methylpentane.

Another mass spectrum (although not of an alkane) in which fragmentation of the molecular ion is determined by the stability of the cation generated is that of toluene (Figure 25.32). A strong molecular ion is seen at $m/z=92$ (so it is not so unstable), and the base peak is at $m/z=91$ corresponding to M−1. The loss of a hydrogen atom from the toluene molecular ion gives the (relatively) stable benzyl cation which rearranges to the even more stable tropylium ion.

> Note that loss of a hydrogen atom to give a tertiary carbenium ion from the molecular ion of 2-methylbutane is not a significant fragmentation, i.e. it does not compete with loss of methyl or ethyl to form *less stable* secondary carbenium ions. A likely cause is that the C–H bond energy is higher than the C–C bond energy in the molecular ion. Loss of H· to give the benzyl cation from the toluene molecular ion is the main route only because alternative fragmentations involve disruption of the aromatic system and give hugely less stable cations.

Exercise 25.16

Predict the m/z value of the base peak in the mass spectrum of ethylbenzene.

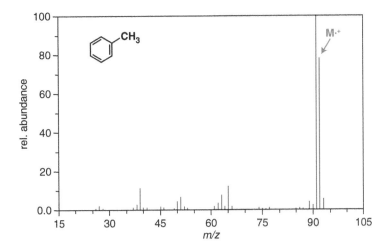

Figure 25.32 Mass spectrum of toluene.

b. Mass spectra of haloalkanes and isotopic peaks

Most of the common elements have naturally occurring heavier isotopes as listed in Table 25.6. For three common elements in organic compounds, C, H, and N, the principal heavier isotope is one mass unit greater than the most abundant. These elements in a compound give rise to a small isotopic peak at M+1. For four other elements, O, S, Cl, and Br, the principal heavier isotope is two mass units greater than the most abundant one. These elements are responsible for an isotopic peak at M+2.

The relative abundances of the heavier isotopes given in Table 25.6 allow calculation of the intensities of the small peaks to be expected at m/z values greater than the principal one. For example, the molecular ion peak of C_5H_{12} should have an M+1 peak with the intensity $5 \times 1.11 + 12 \times 0.016 = 5.74\%$ due to the ^{13}C and ^2H isotopes. Chlorine and bromine have high proportions of isotopes with masses two units greater than the most abundant, so a compound containing one of these elements has a large M+2 peak. Natural abundances of bromine isotopes, ^{79}Br and ^{81}Br, are about 1 : 1 while those of

Natural abundances of stable isotopes of some elements							Table 25.6
Element	Relative natural abundance of main isotopes						
carbon	^{12}C	100	^{13}C	1.11			
hydrogen	^1H	100	^2H	0.016			
nitrogen	^{14}N	100	^{15}N	0.38			
oxygen	^{16}O	100	^{17}O	0.04	^{18}O	0.20	
sulfur	^{32}S	100	^{33}S	0.78	^{34}S	4.40	
fluorine	^{19}F	100					
chlorine	^{35}Cl	100			^{37}Cl	32.5	
bromine	^{79}Br	100			^{81}Br	98.0	
iodine	^{127}I	100					

Figure 25.33 Mass
spectrum of 1-bromopropane.

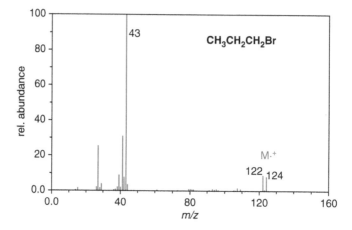

Figure 25.34 Mass
spectrum of 2-chloropropane.

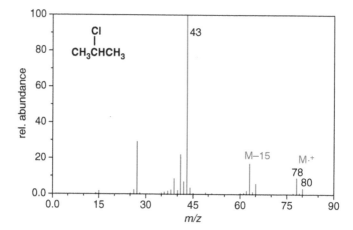

chlorine, ^{35}Cl and ^{37}Cl, are about 3 : 1. As a result, a compound containing these elements shows characteristic pairs of peaks in the ratio 1 : 1 or 3 : 1 for ions containing one bromine or one chlorine, respectively, as exemplified in Figures 25.33 and 25.34.

In the mass spectrum of 1-bromopropane (molecular weight 123.0) in Figure 25.33, we see two peaks with nearly equal intensities at $m/z=122$ and 124 corresponding to molecular ions containing either ^{79}Br or ^{81}Br. The appearance of this pair of parent peaks of nearly equal intensity is characteristic of a compound containing one bromine. Loss of the bromine atom gives the base peak (the propyl cation) at $m/z=43$.

The base peak at $m/z=43$ in the mass spectrum of 2-chloropropane (Figure 25.34) is due to the isopropyl cation following loss of a chlorine atom from the molecular ion. The two molecular ion peaks (at $m/z=78$ and 80) in the ratio ~3 : 1 (the molecular weight of 2-chloropropane is 78.5) are a very distinctive feature of a mono-chloro compound. A corresponding pair of peaks is also evident for the (M−15) fragments at $m/z=63$ and 65 by loss of the methyl radical. The observation of this pair of peaks indicates that loss of methyl competes with loss of chlorine, *i.e.* that the C–Cl and C–C bonds in the molecular ion are comparably strong.

c. Mass spectra of alcohols

Relatively stable hydroxycarbenium ions (or oxonium ions, according to their dominant resonance forms) are very readily generated from the molecular ions of alcohols. Consequently, the molecular ion peak ($m/z=74$) is hardly perceptible in the mass spectrum of butan-2-ol (Figure 25.35). Peaks are seen at $m/z=45$ (the base peak) and 59 by loss of ethyl and methyl radicals, respectively, from the molecular ion (eqn 25.4).

$$\left[\begin{array}{c} \overset{OH}{\underset{CH_3\overset{+}{C}H}{|}} \longleftrightarrow \overset{+OH}{\underset{CH_3\overset{||}{C}H}{}} \end{array}\right] \xleftarrow{-\overset{\cdot}{C}H_2CH_3} \overset{\overset{\cdot+}{O}H}{\underset{CH_3\overset{|}{C}HCH_2CH_3}{}} \xrightarrow{-\overset{\cdot}{C}H_3} \left[\begin{array}{c} \overset{OH}{\underset{H\overset{||}{C}CH_2CH_3}{}} \longleftrightarrow \overset{OH}{\underset{H\overset{+}{C}CH_2CH_3}{|}} \end{array}\right]$$

$m/z = 45 \qquad\qquad\qquad m/z = 74 \qquad\qquad\qquad m/z = 59$

$$(25.4)$$

The isomeric primary alcohol, butan-1-ol, gives a mass spectrum appreciably different from that of butan-2-ol, as seen in Figure 25.36. The base peak at $m/z=56$ (M−18) corresponds to the radical cation fragment formed by expulsion of H_2O from the molecular ion (which is too weak to be observed). The M−18 fragment ion is characteristic of primary alcohols with a suitably placed C–H, and a credible mechanism for its formation in this case is shown in eqn 25.5.

$$\text{(structures)} \qquad \xrightarrow{\quad} \qquad \text{(structures)} \xrightarrow{-H_2O} \qquad H_2\overset{\cdot}{C} \qquad \overset{+}{C}H_2 \qquad (25.5)$$

$m/z = 74 \qquad\qquad\qquad\qquad\qquad\qquad m/z = 56$

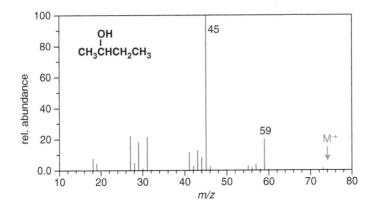

Figure 25.35 Mass spectrum of butan-2-ol.

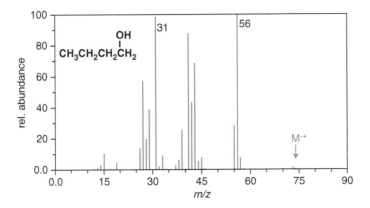

Figure 25.36 Mass spectrum of butan-1-ol.

A peak of comparable intensity is also seen at $m/z=31$ which belongs to the hydroxymethyl cation (protonated methanal) but there is only a very small peak at M − 1 due to formation of protonated butanal by loss of a hydrogen atom from C1 (see the note near the end of Sub-section 25.6.2a).

$$CH_3CH_2CH_2\overset{\overset{\displaystyle\overset{+}{\ddot{O}}H}{|}}{CH_2} \longrightarrow CH_3CH_2\overset{\displaystyle\cdot}{C}H_2 + \left[\overset{\overset{\displaystyle\overset{+}{O}H}{||}}{H\overset{\displaystyle\cdot}{C}H} \longleftrightarrow \overset{\overset{\displaystyle OH}{\overset{+|}{}}}{H\overset{\displaystyle\cdot}{C}H} \right]$$

$m/z = 74$ $m/z = 31$

The next two major peaks at $m/z=41$ and 43 in Figure 25.36 are attributable to the allyl and propyl cations.

Exercise 25.17

Predict the m/z value of the base peak in the mass spectrum of propan-2-ol.

d. Mass spectra of carbonyl compounds

The mass spectra of butanone and butanal in Figures 25.37 and 25.38, respectively, are representative of ketones and aldehydes generally. The mass spectrum of butanone is particularly simple with the two prominent peaks corresponding to acylium ion fragments at $m/z=43$ and 57 (eqn 25.6). Dominant peaks attributable to acylium ion fragments are characteristic features of mass spectra of carbonyl compounds; in the case of an aldehyde, formation of an acylium ion (M−1) fragment is by loss of a hydrogen atom.

$$CH_3-\overset{+}{C}\equiv\overset{\cdot\cdot}{O} \xleftarrow{-\ \overset{\displaystyle\cdot}{C}H_2CH_3} CH_3\overset{\overset{\displaystyle\overset{\cdot+}{\ddot{O}}}{||}}{C}CH_2CH_3 \xrightarrow{-\ \overset{\displaystyle\cdot}{C}H_3} CH_3CH_2-\overset{+}{C}\equiv\overset{\cdot\cdot}{O} \qquad (25.6)$$

$m/z = 43$ $m/z = 72$ $m/z = 57$

> This reaction may be regarded as a retro-ene reaction as covered in Sub-section 21.2.5c.

However, the base peak for butanal is not due to loss of hydrogen and formation of an acylium cation. As seen in Figure 25.38, it is at $m/z=44$ (M−28) corresponding to an enol radical cation following loss of ethene, C_2H_4, as shown in Scheme 25.2. This mass spectral elimination reaction (known as the **McLafferty rearrangement**) occurs via a cyclic six-electron transition structure, and is characteristic of carbonyl compounds with a γ C−H bond.

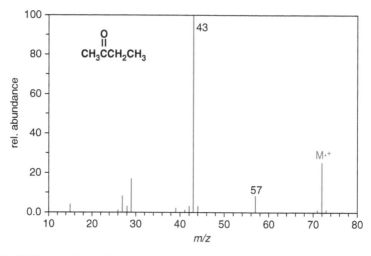

Figure 25.37 Mass spectrum of butanone.

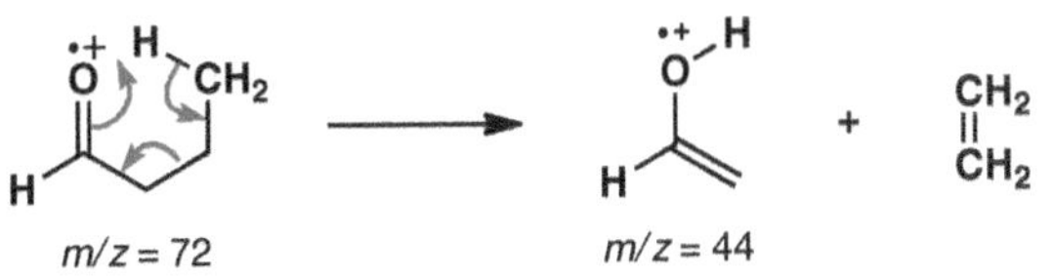

Scheme 25.2
A McLafferty rearrangement.

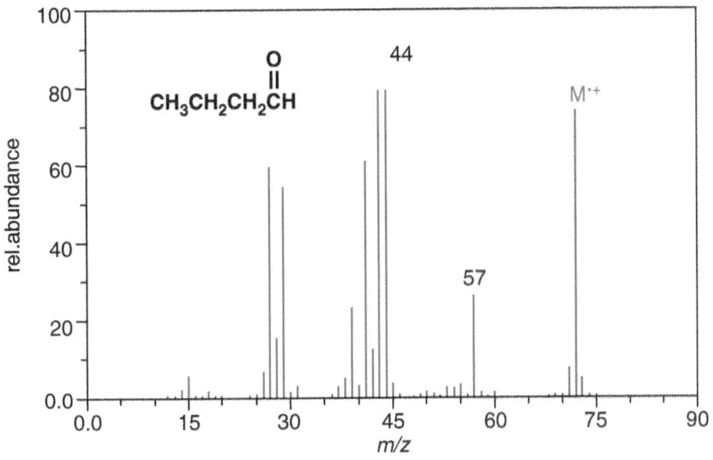

$m/z = 72$

$m/z = 44$

$+$ CH_2
      $\|$
      CH_2

Figure 25.38 Mass
spectrum of butanal.

100

80

$CH_3CH_2CH_2\overset{\overset{\textstyle O}{\|}}{C}H$

44

M$^{\cdot+}$

60

40

57

20

0.0

rel.abundance

0.0   15   30   45   60   75   90
                    m/z

Explain the main peaks in the mass spectrum of benzaldehyde shown below.

Example 25.3

100

80

77

105 106

60

40

20

0.0

rel. abundance

20   40   60   80   100   120
                m/z

Solution

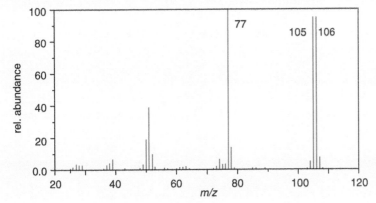

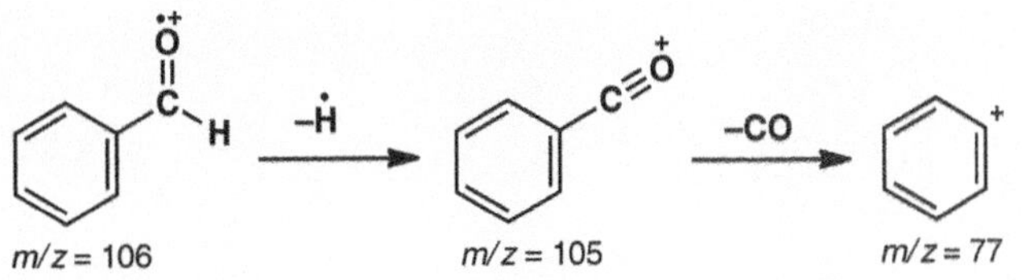

$m/z = 106$ $-\overset{\cdot}{H}$ $m/z = 105$ $-CO$ $m/z = 77$

Exercise 25.18

Explain the base peak in the mass spectrum of butanoic acid shown below.

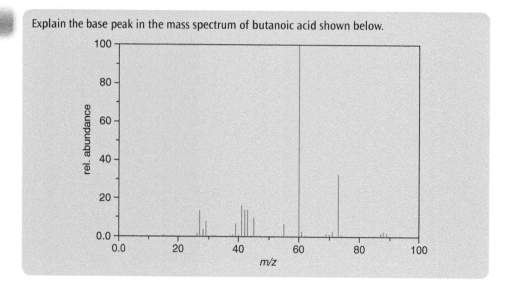

25.6.3 High-resolution mass spectrometry: determination of molecular formulas

The mass spectra we have seen so far have been low-resolution spectra which report m/z ratios to the nearest whole number. However, high-resolution mass spectrometers can measure m/z values with a precision of ± 0.0001 amu (atomic mass unit). Consequently, although compounds with the following molecular formulas all have a molecular mass of 72 to the nearest whole number, they have different *exact* molecular masses, as shown.

molecular formula	C_5H_{12}	C_4H_8O	C_4H_5F	$C_3H_8N_2$	$C_3H_4O_2$
molecular mass	72.0939	72.0575	72.0375	72.0687	72.0211

This is because exact atomic masses of individual isotopes are not whole-numbers, as seen in the list of some values in Table 25.7. Using these values, we can usually identify a unique molecular formula for a molecular ion (or a fragment) from its precisely measured molecular mass.

Table 25.7

Exact masses of common isotopes			
Isotope	Mass	Isotope	Mass
^1H	1.007825	^{32}S	31.97207
^{12}C	12.00000	^{35}Cl	34.96886
^{14}N	14.00307	^{37}Cl	36.96590
^{16}O	15.99492	^{79}Br	78.91835
^{19}F	18.99840	^{81}Br	80.91635

Exercise 25.19

Compounds with molecular formulas $C_{10}H_{17}N$, $C_9H_{13}NO$, $C_8H_9NO_2$, and $C_7H_5NO_3$ all have the whole-number molecular mass of 151. Calculate exact molecular masses of these compounds using the most abundant isotopes in each case.

25.6.4 Advanced types of mass spectrometry

a. Gas chromatography–mass spectrometry (GC–MS)

Gas chromatography (GC) and mass spectrometry (MS) can be combined to analyse mixtures of compounds. The gas chromatograph separates a mixture, and the individual components pass directly into the mass spectrometer so that individual mass spectra can be recorded. The quantitative analysis of the components of the mixture is carried out by the GC and, at the same time, identification of the components is achieved by MS.

b. Mass spectra of high molecular weight biomolecules

Until the 1980s, MS was limited to compounds (usually of <800 amu) which could be vaporized thermally under very low pressure. Since then, new techniques such as electrospray ionization (ESI) and matrix-assisted laser desorption-ionization (MALDI) have been developed to generate gas-phase ions of non-volatile large molecules (with molecular weights >100000 amu). This allows mass spectrometric analysis of large biomolecules such as proteins, nucleic acids, and carbohydrates, but details are beyond the scope of this book.

> The 2002 Nobel Prize in Chemistry was shared between J.B. Fenn (USA) and K. Tanaka (Japan) who developed these MS techniques, and K. Wüthrich (Switzerland) who developed an NMR method for investigating three-dimensional structures of biological macromolecules in solution.

Summary

- Spectroscopic methods are the principal accessible instrumental techniques for the elucidation of molecular structures. The **spectrum** is the output of an instrument which measures interactions of electromagnetic radiation with molecules, and is classified by the frequency (or wavelength) range of the radiation involved.

- A **UV-vis** (ultraviolet and visible) spectrum relates to transitions between electronic energy levels, and provides information about the electronic structure of a molecule, particularly its π electron system.

- An **IR** (infrared) spectrum is associated with vibrational energy levels of covalent bonds (bond stretching and bending) in a molecule and usually allows identification of the functional groups present.

- **NMR** (nuclear magnetic resonance) spectra are concerned with energy levels of ^1H and ^{13}C (and some other nuclei) in a magnetic field and, in particular, give information about the carbon–hydrogen framework of a molecule.

- **Mass spectrometry** (**MS**) is based on the ionization (loss of an electron) of molecules (mainly) by electron impact and the fragmentation of molecular ions; a mass spectrum displays the mass/charge ratios of the molecular ion and cationic species formed by its fragmentation. The molecular formula of a compound (and of cationic fragments) can be deduced from a high-resolution mass spectrum.

Problems

25.1 How many signals in principle will be observed in the ^1H and ^{13}C NMR spectra of each of the following? Assume that temperatures are high enough for conformational changes to be fast compared with the NMR time-scale.

(a) cyclohexanol

(b) cyclohexanone

(c) cyclohexane-1,4-dione

(d) 4-hydroxycyclohexanone

(e) *trans*-cyclohexane-1,4-diol

(f) *cis*-cyclohexane-1,4-diol

(g) *trans*-cyclohexane-1,3-diol

(h) *cis*-cyclohexane-1,3-diol

(i) *trans*-cyclohexane-1,2-diol

(j) *cis*-cyclohexane-1,2-diol

25.2 Arrange the following sets of compounds in order of the chemical shifts (δ values) of the coloured hydrogens.

(a) CH_3OH, CH_3NH_2, CH_3F

(b) CH_3CH_2Ph, $(CH_3CH_2)_2O$, $CH_3CH_2NO_2$

(c) $CH_3CH_2CH_3$, $CH_2=C(CH_3)_2$, $CH_3CH_2CH=CHCH_3$

25.3 Arrange the hydrogens identified in the following compounds in order of their increasing chemical shifts (δ values).

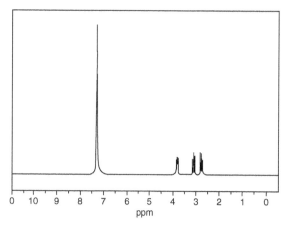

(a)

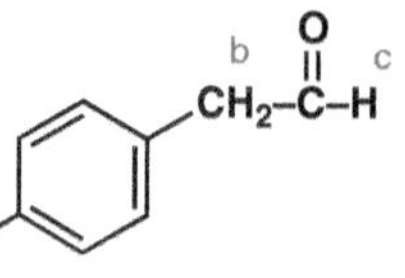

(b) a b c d e
$CH_3CH_2CH=CH-C-OCH_3$

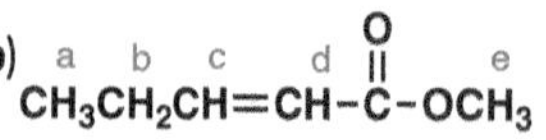

25.4 Predict the multiplicity of the NMR signal of each type of hydrogen in the following compounds.

(a) $(CH_3)_3CCH_2OCH_3$ **(b)** $CH_3OCH_2CHBr_2$

(c) $ClCH_2CH_2CH_2Cl$ **(d)** $p\text{-}ClC_6H_4COCH_3$

25.5 The ^1H NMR spectrum of a compound with the molecular formula C_8H_8O is shown below. Give a structure for this compound and explain your reasoning.

(Each of the three signals seen in the 2.5–4.0 ppm range comprises four peaks.)

25.6 Predict the ^1H NMR spectrum of each of the isomeric carboxylic acids with the molecular formula $C_3H_5ClO_2$.

25.7 The following is the ^{13}C NMR spectrum of methyl 2-methylpropenoate (methyl methacryrate). Assign the five signals to the carbon atoms of the molecule.

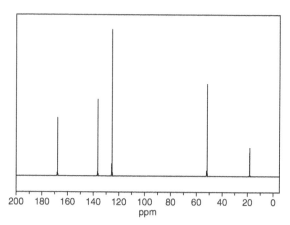

25.8 The following are mass spectra of 1-methoxybutane and 2-methoxybutane. Explain which is which.

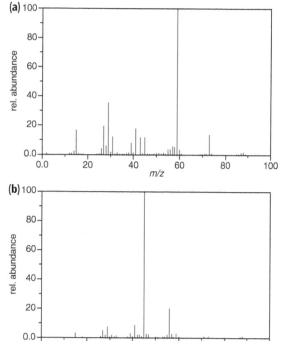

25.9 Deduce the structure of the compound with the molecular formula C_8H_8O which gave the following mass spectrum.

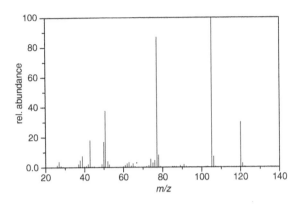

25.10 Explain how the following pairs of compounds could be differentiated using UV-vis, IR, ¹H NMR, and/or ¹³C NMR spectroscopy, and/or mass spectrometry.

(a)

(b)

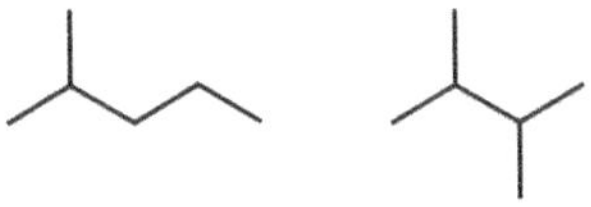

(c)

(d)

(e)

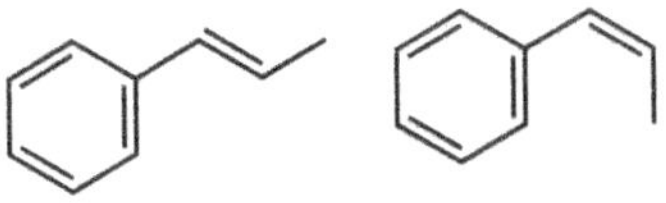

(f)

(g)

(h)

(i)

(j)

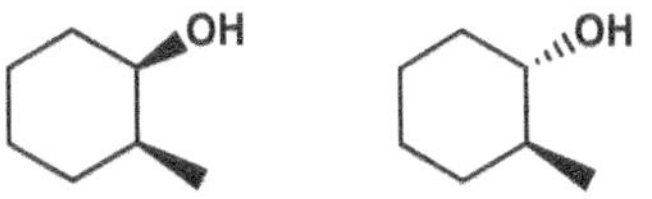

(k)

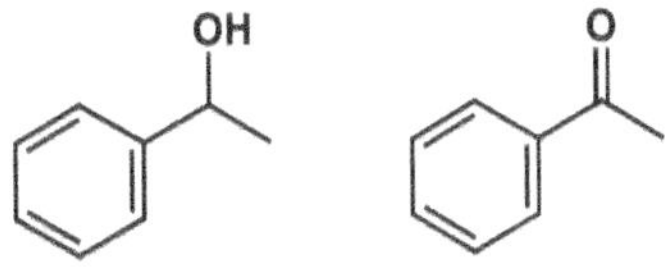

(l)

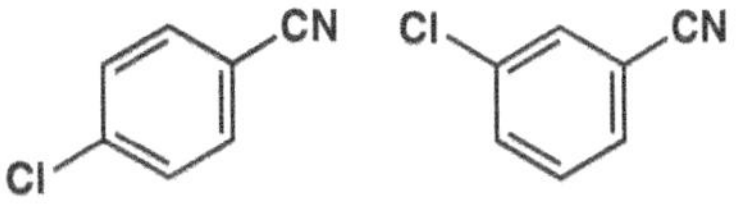

(m)

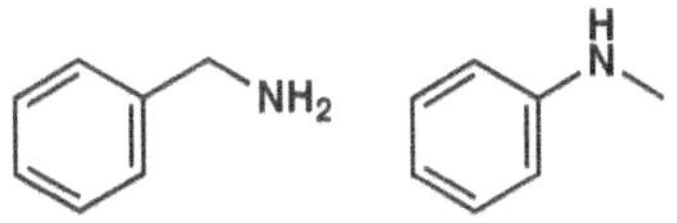

(n)

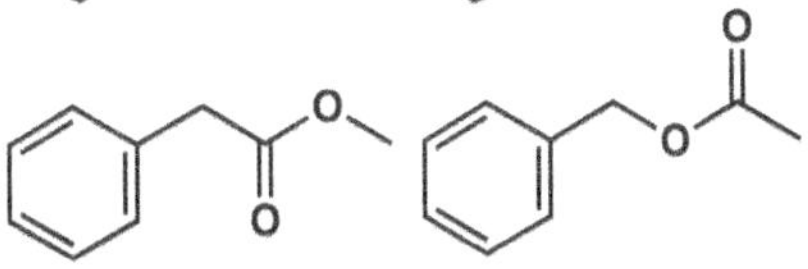

(o)

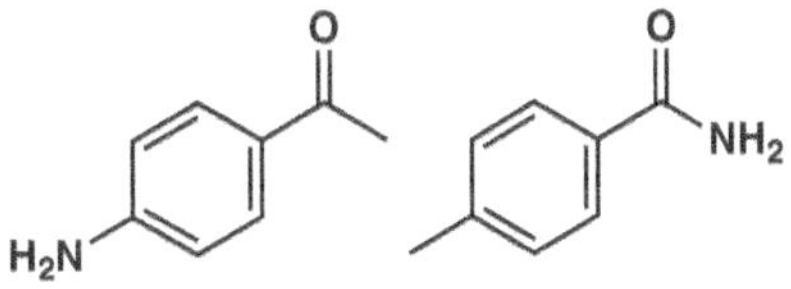

(p)

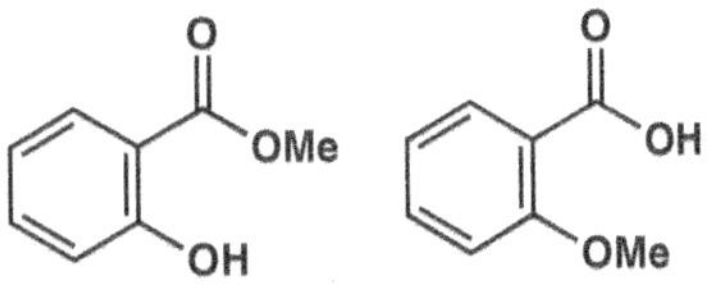

Appendix 1
pK$_a$ Values of Representative Compounds[1]

Inorganic Compounds

H_2O	15.74
H_3O^+	−1.74
HI	−10*
HBr	−9*
HCl	−7*
HF	3.17
$HClO_4$	−10*
H_2SO_4	−3*
HSO_4^-	1.99
HNO_3	−1.64
HNO_2	3.29
H_3PO_4	1.97
$H_2PO_4^-$	6.82
HPO_4^{2-}	12.3
H_2CO_3	6.37
HCO_3^-	10.33
HOOH	11.6
H_2S	7.0
NH_4^+	9.24
$HONH_3^+$	6.0
$H_2NNH_3^+$	8.07
NH_3	35*

Organic Compounds
Alcohols

CH_3OH	15.5
CH_3CH_2OH	15.9
$(CH_3)_2CHOH$	17.1
$(CH_3)_3COH$	19.2
$HOCH_2CH_2OH$	15.4
CF_3CH_2OH	12.4
$(CF_3)_2CHOH$	9.3
$(CF_3)_3COH$	5.1

Phenols

C_6H_5OH	9.99
p-$NO_2C_6H_4OH$	7.14
$2,4$-$(NO_2)_2C_6H_3OH$	4.1
$2,4,6$-$(NO_2)_3C_6H_2OH$	0.3

Carboxylic acids

HCO_2H	3.75
CH_3CO_2H	4.76
$(CH_3)_3CCO_2H$	5.03
$HOCH_2CO_2H$	3.46
CF_3CO_2H	−0.6
$H_3\overset{+}{N}CH_2CO_2H$	2.35
$CH_2{=}CHCO_2H$	4.25
$C_6H_5CO_2H$	4.20
p-$NO_2C_6H_4CO_2H$	3.44
$HO_2CCH_2CO_2H$	2.85
$^-O_2CCH_2CO_2H$	5.70

Sulfonic acids

$C_6H_5SO_3H$	−2.8
CH_3SO_3H	−1.9
CF_3SO_3H	−5.5

Hydroperoxy compounds

$CH_3C(O)O_2H$	8.2
CH_3OOH	11.5
$(CH_3)_3COOH$	12.8

Thiols and thioacids

CH_3SH	10.33
C_6H_5SH	6.61
$CH_3C(O)SH$	3.43
$CH_3C(S)SH$	2.57

[1] Values with an asterisk are approximate.

Amines and amides

$(Me_2CH)_2NH$	38[b]
$C_6H_5NH_2$	27.7
CH_3CONH_2	15.1

Ammonium ions

$CH_3NH_3^+$	10.64
$(CH_3)_2NH_2^+$	10.73
$(C_2H_5)_3NH^+$	10.75
$(CH_3)_3NH^+$	9.75
$C_6H_5NH_3^+$	4.60
p-$NO_2C_6H_4NH_3^+$	0.99
2,4-$(NO_2)_2C_6H_3NH_3^+$	−4.31
2,4,6-$(NO_2)_3C_6H_2NH_3^+$	−10.04
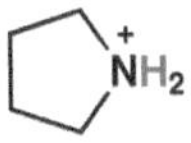	11.30
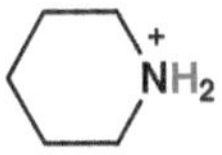	11.12
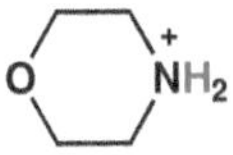	8.4
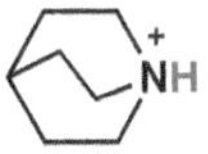	11.0
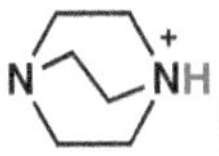 (DABCO)	8.4
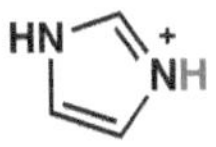	6.99
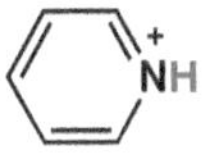	5.25
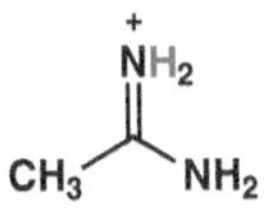	12.4
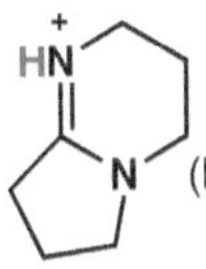 (DBN)	13.5*
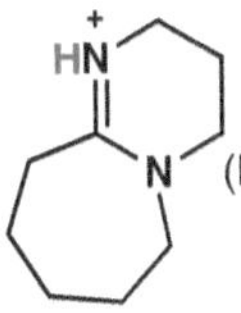 (DBU)	12.5*

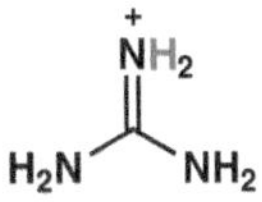	13.6
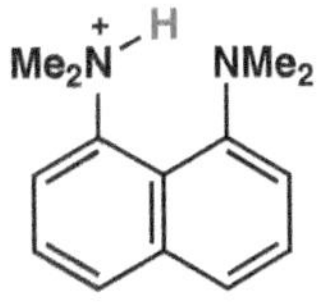	12.1
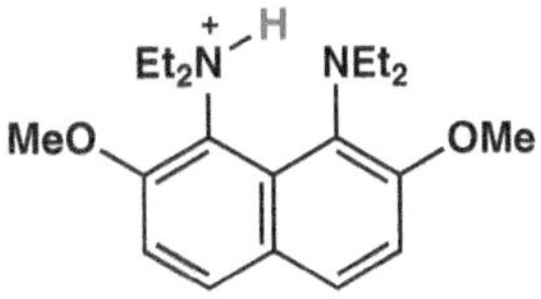	16.3

Carbon acids

$HC\equiv N$	9.1
$HC\equiv CH$	25*
$H_2C{=}CH_2$	44*
$H_3C{-}CH_3$	50*
$H_2C{=}CHCH_3$	43*
$C_6H_5CH_3$	41*
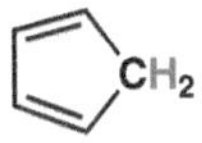	16
$CH_3\overset{O}{\overset{\parallel}{C}}CH_3$	19.3
$CH_3\overset{O}{\overset{\parallel}{C}}OEt$	25.6
$CH_3\overset{O}{\overset{\parallel}{C}}CH_2\overset{O}{\overset{\parallel}{C}}CH_3$	8.84
$CH_3\overset{O}{\overset{\parallel}{C}}CH_2\overset{O}{\overset{\parallel}{C}}OEt$	10.7
$EtO\overset{O}{\overset{\parallel}{C}}CH_2\overset{O}{\overset{\parallel}{C}}OEt$	13.3
CH_3CN	28.9
CH_3NO_2	10.2
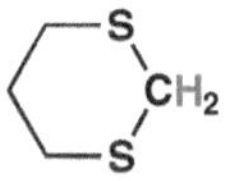	31*
$CH_3\overset{O}{\overset{\parallel}{S}}CH_3$	33*
$CHCl_3$	24*

Conjugate acids of miscellaneous organic compounds

$CH_3\overset{+}{O}H_2$ −2.05

$(CH_3)_2\overset{+}{O}H$ −2.48

$(CH_3)_2\overset{+}{S}H$ −6.99

$\overset{+}{O}H$
$CH_3\overset{\|}{C}CH_3$ −3.06

$\overset{+}{O}H$
$CH_3\overset{\|}{C}OMe$ −3.90

$\overset{+}{O}H$
$CH_3\overset{\|}{C}NMe_2$ −0.21

$\overset{+}{O}H$
$CH_3\overset{\|}{S}CH_3$ −1.54

$CH_3-\overset{+}{N}(=O)OH$ −12*

$Ph_2C=\overset{+}{N}HCH_2CH_2CH_3$ 7*

$CH_3-C\equiv\overset{+}{N}-H$ −10*

Appendix 2
Principal Reactions of Functional Groups[1]

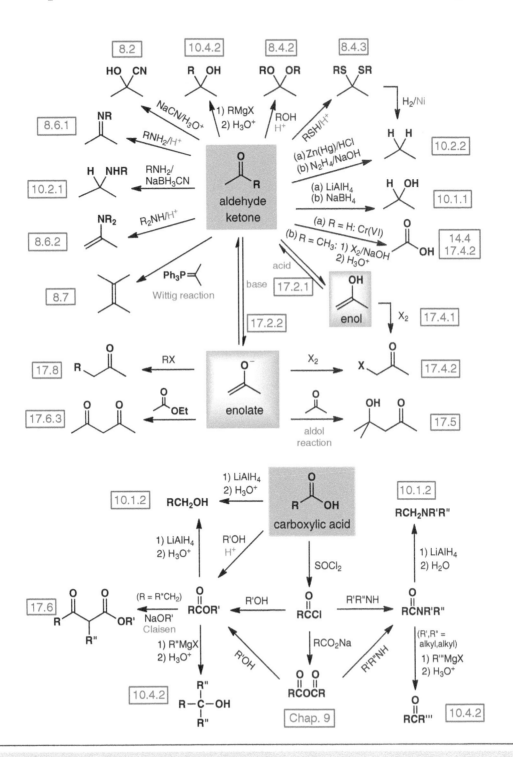

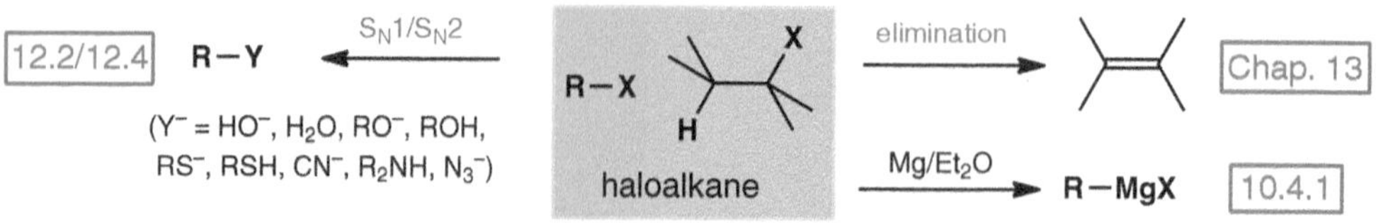

Haloalkane reactions

12.2/12.4	**R—Y** $\xleftarrow{S_N1/S_N2}$

$(Y^- = HO^-, H_2O, RO^-, ROH,$
$RS^-, RSH, CN^-, R_2NH, N_3^-)$

R—X haloalkane

$\xrightarrow{\text{elimination}}$ | Chap. 13 |

$\xrightarrow{Mg/Et_2O}$ **R—MgX** | 10.4.1 |

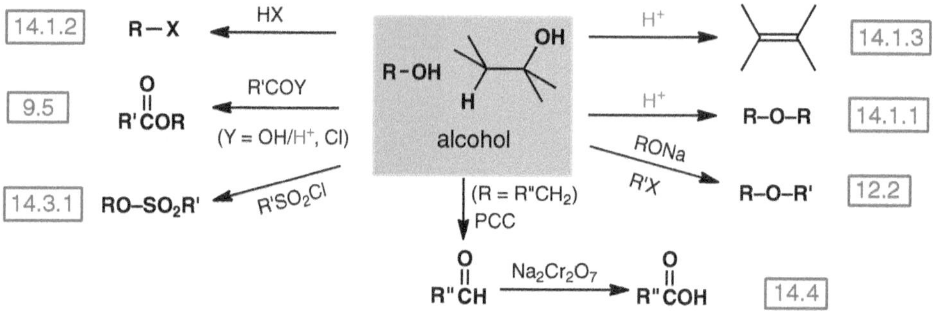

Alcohol reactions

14.1.2	**R—X** $\xleftarrow{HX}$

9.5	$\underset{\overset{\|}{R'COR}}{\overset{O}{\|}}$ $\xleftarrow{R'COY}$

$(Y = OH/H^+, Cl)$

14.3.1	**RO—SO$_2$R'** $\xleftarrow{R'SO_2Cl}$

R—OH alcohol

$\xrightarrow{H^+}$ | 14.1.3 |

$\xrightarrow{H^+}$ **R—O—R** | 14.1.1 |

$\xrightarrow[R'X]{RONa}$ **R—O—R'** | 12.2 |

$\xrightarrow[PCC]{(R = R''CH_2)}$ $\underset{R''CH}{\overset{O}{\|}}$ $\xrightarrow{Na_2Cr_2O_7}$ $\underset{R''COH}{\overset{O}{\|}}$ | 14.4 |

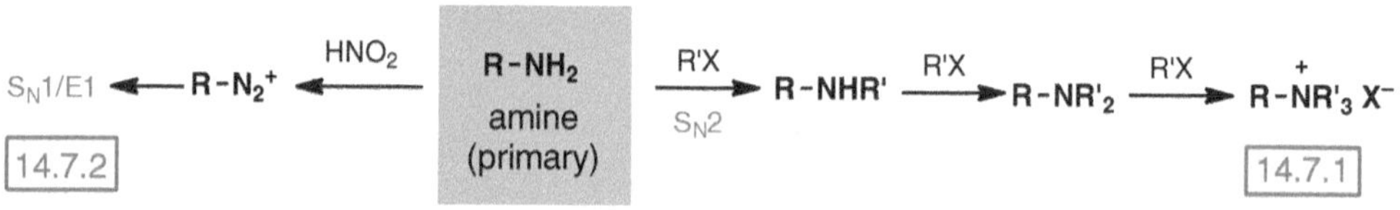

Amine reactions

$S_N1/E1 \xleftarrow{} $ **R—N$_2^+$** $\xleftarrow{HNO_2}$ **R—NH$_2$** amine (primary) $\xrightarrow[S_N2]{R'X}$ **R—NHR'** $\xrightarrow{R'X}$ **R—NR'$_2$** $\xrightarrow{R'X}$ **R—$\overset{+}{N}$R'$_3$ X$^-$**

| 14.7.2 | | | | 14.7.1 |

Alkene reactions

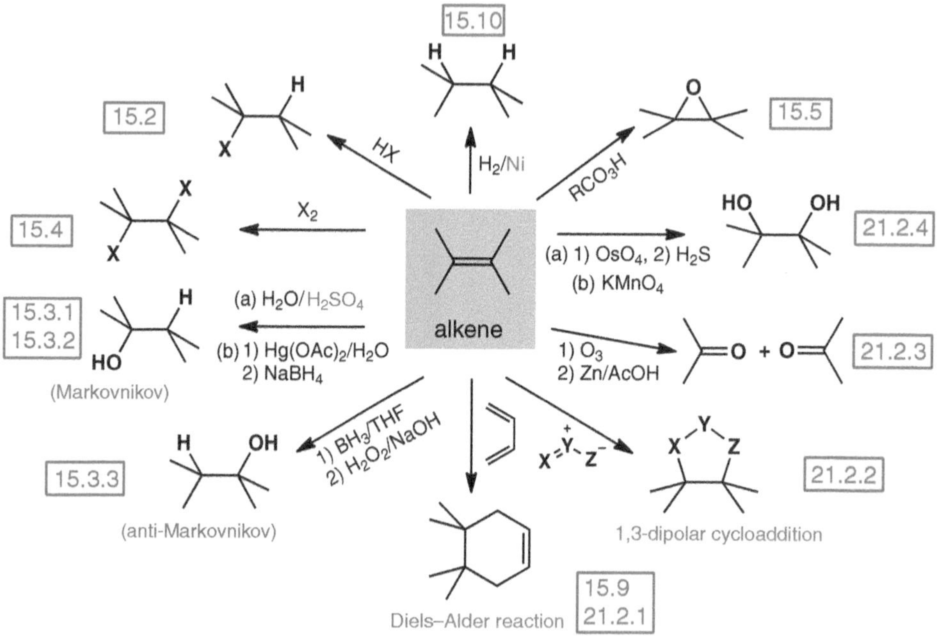 15.10	$\xrightarrow{H_2/Ni}$

15.2	$\xleftarrow{HX}$

15.5	$\xrightarrow{RCO_3H}$

15.4	$\xleftarrow{X_2}$

21.2.4	(a) 1) OsO$_4$, 2) H$_2$S (b) KMnO$_4$

15.3.1 15.3.2	(a) H$_2$O/H$_2$SO$_4$ (b) 1) Hg(OAc)$_2$/H$_2$O 2) NaBH$_4$ (Markovnikov)

alkene

21.2.3	1) O$_3$ 2) Zn/AcOH

15.3.3	1) BH$_3$/THF 2) H$_2$O$_2$/NaOH (anti-Markovnikov)

21.2.2	X=Y—Z 1,3-dipolar cycloaddition

15.9 21.2.1	Diels–Alder reaction

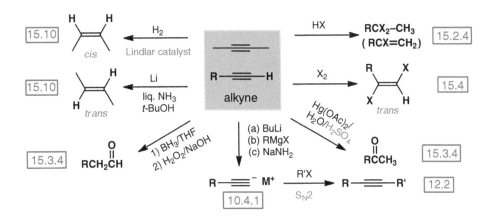

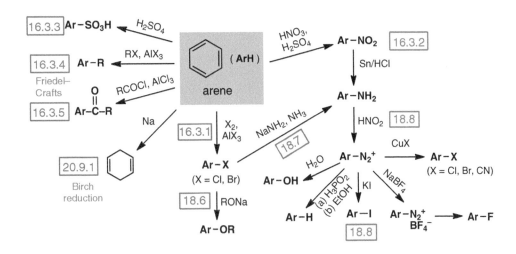

Appendix 3

Syntheses of Classes of Compounds[1]

X = Cl, Br, I

(1) Alkanes (R–H)

	Type of reaction	Relevant entries
R–X →(Bu₃SnH) R–H	Reduction	20.4

R–X →(Mg/Et₂O) R–MgX →(H₃O⁺) R–H — Reduction *via* Grignard reagent — 10.4.2

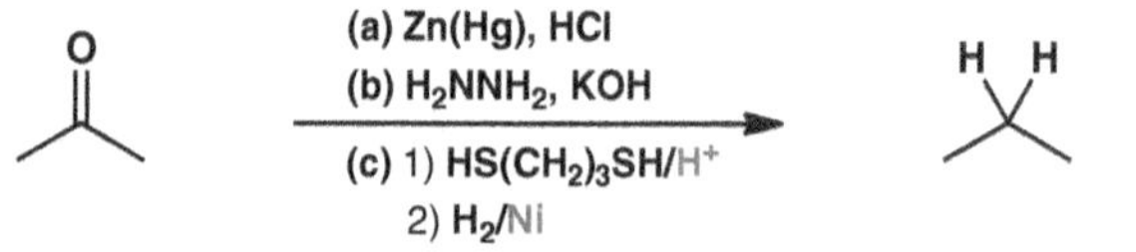

(ketone) →(a) Zn(Hg), HCl / (b) H₂NNH₂, KOH / (c) 1) HS(CH₂)₃SH/H⁺ 2) H₂/Ni → (alkane, H H) — (a) Clemmensen reaction / (b) Wolff–Kishner reaction / (c) Thioacetal desulfurization — 10.2.2

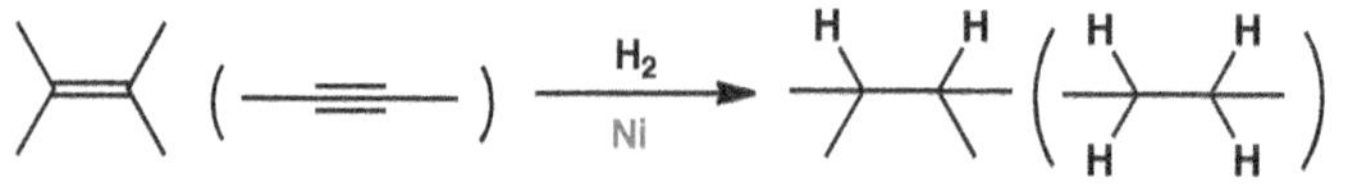

(alkene) (═) →(H₂ / Ni) (alkane with H's) — Hydrogenation — 15.10

2 R–CO₂⁻ →(– 2e⁻, – 2CO₂) R–R — Kolbe reaction (anodic oxidation) — 20.9.4

(2) Alkenes (C=C)

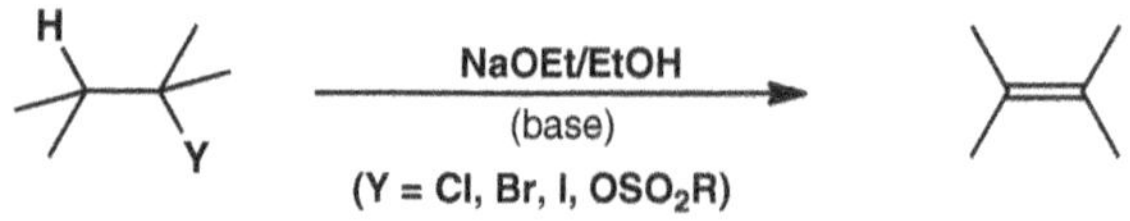

(H, Y substituted) →(NaOEt/EtOH (base)) (Y = Cl, Br, I, OSO₂R) → (alkene) — Elimination — 13.2

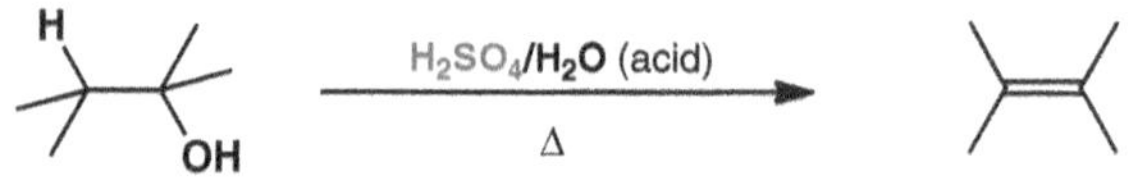

(H, OH substituted) →(H₂SO₄/H₂O (acid), Δ) → (alkene) — Dehydration — 14.1.3

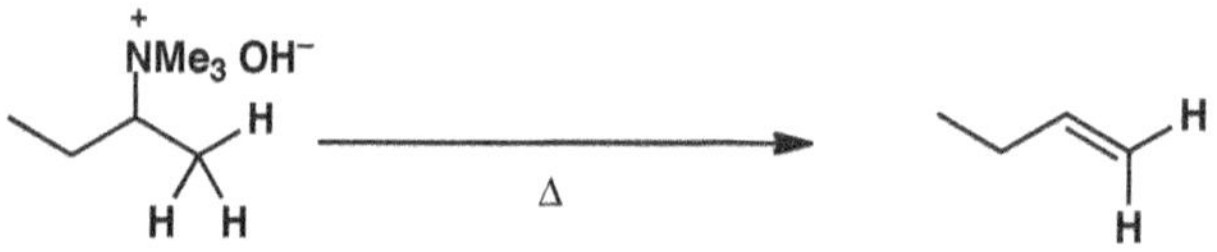

(⁺NMe₃ OH⁻, H, H H) →(Δ) → (cis alkene, H, H) — Hofmann elimination — 14.7.1

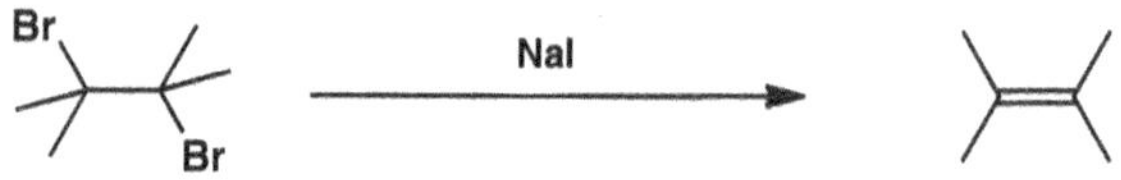

(Br, Br substituted) →(NaI) → (alkene) — Nucleophile-induced elimination — Problem 13.16

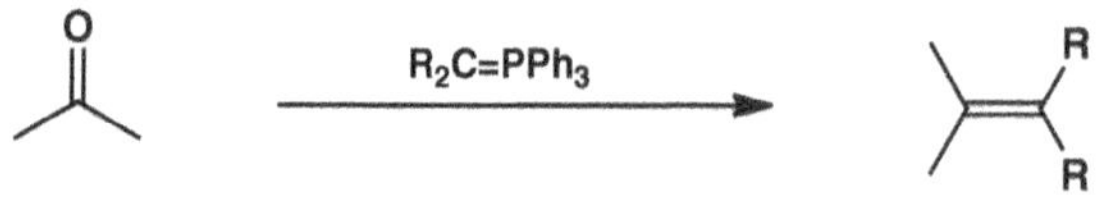

(ketone) →(R₂C=PPh₃) → (alkene with R, R) — Wittig reaction — 8.7

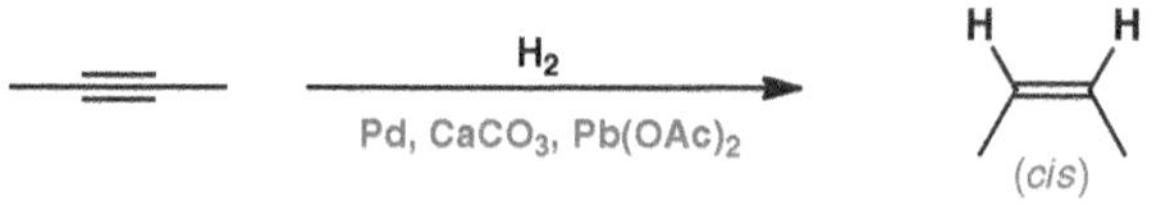

(alkyne) →(H₂ / Pd, CaCO₃, Pb(OAc)₂) → (*cis* alkene, H, H) — Hydrogenation (Lindlar catalyst) — 15.10

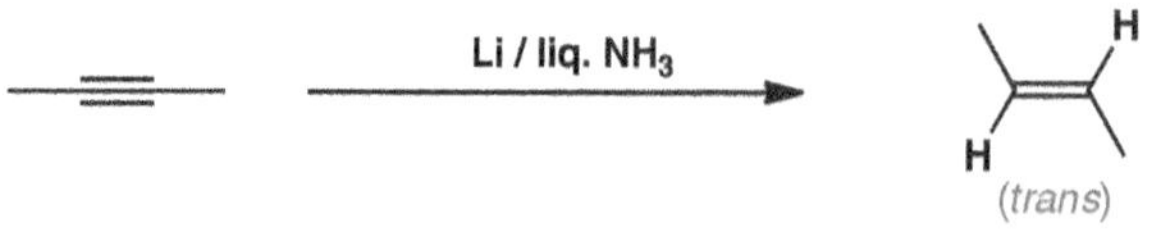

(alkyne) →(Li / liq. NH₃) → (*trans* alkene, H, H) — Dissolving metal reduction — 15.10

---

(3) Alkynes (C≡C)

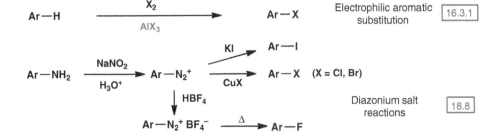

$$\text{(structures)} \xrightarrow[\text{(strong base)}]{\text{NaNH}_2} \quad \equiv \quad \text{E2 elimination} \quad \boxed{\text{Chap. 13}}$$

$$\text{RCH}_2\text{—X} \xrightarrow[\text{(M = Li, MgX, Na)}]{\equiv^-\text{M}^+} \text{RCH}_2\equiv \quad \text{S}_\text{N}2 \text{ reaction} \quad \boxed{12.2}$$

(4) Haloalkanes (R–X)

$$\text{R—H} \xrightarrow[hv \text{ or } \Delta]{\text{X}_2} \text{R—X} \quad \begin{array}{c}\text{Radical substitution}\\ \text{(low selectivity)}\end{array} \quad \boxed{20.3}$$

$$\text{R—OH} \xrightarrow{\substack{\text{(a) HX}\\ \text{(b) PX}_3 \text{ (or PX}_5)\\ \text{(c) SOX}_2}} \text{R—X} \quad \text{Halogenation} \quad \boxed{\substack{14.1\\14.3}}$$

$$\text{R—OSO}_2\text{R}' \xrightarrow{\text{NaX}} \text{R—X} \quad \begin{array}{c}\text{Sulfonate}\\ \text{S}_\text{N}2 \text{ reaction}\end{array} \quad \boxed{14.3}$$

$$\xrightarrow{\text{HX}} \text{ (H, X product)} \quad \begin{array}{c}\text{Electrophilic}\\ \text{addition}\end{array} \quad \boxed{15.2}$$

$$\xrightarrow{\text{X}_2} \text{ (X, X product)} \quad \begin{array}{c}\text{Electrophilic}\\ \text{addition}\end{array} \quad \boxed{15.4}$$

Haloarenes (Ar–X)

$$\text{Ar—H} \xrightarrow[\text{AlX}_3]{\text{X}_2} \text{Ar—X} \quad \begin{array}{c}\text{Electrophilic aromatic}\\ \text{substitution}\end{array} \quad \boxed{16.3.1}$$

$$\text{Ar—NH}_2 \xrightarrow[\text{H}_3\text{O}^+]{\text{NaNO}_2} \text{Ar—N}_2{}^+ \begin{array}{c}\xrightarrow{\text{KI}} \text{Ar—I}\\ \xrightarrow{\text{CuX}} \text{Ar—X} \quad (\text{X = Cl, Br})\end{array}$$

$$\downarrow \text{HBF}_4$$

$$\text{Ar—N}_2{}^+ \text{ BF}_4{}^- \xrightarrow{\Delta} \text{Ar—F} \quad \begin{array}{c}\text{Diazonium salt}\\ \text{reactions}\end{array} \quad \boxed{18.8}$$

(5) Alcohols (R–OH)

R–X	$\xrightarrow{\text{NaOH}}$	R–OH	S_N2	12.2

$\xrightarrow[\text{(b) 1) Hg(OAc)}_2/\text{H}_2\text{O} \\ \text{2) NaBH}_4]{\text{(a) H}_2\text{O, H}_2\text{SO}_4}$ Hydration (Markovnikov)   14.3

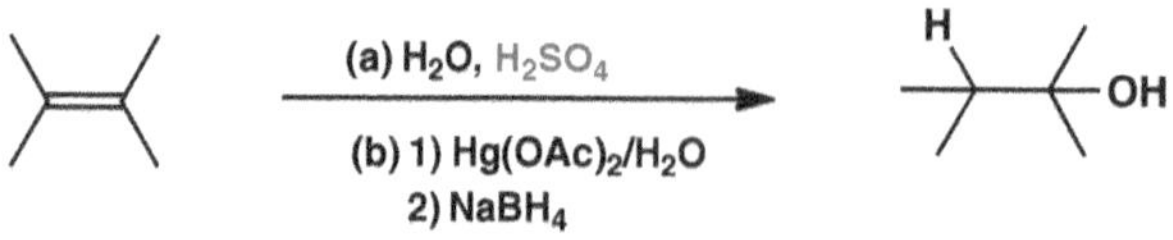

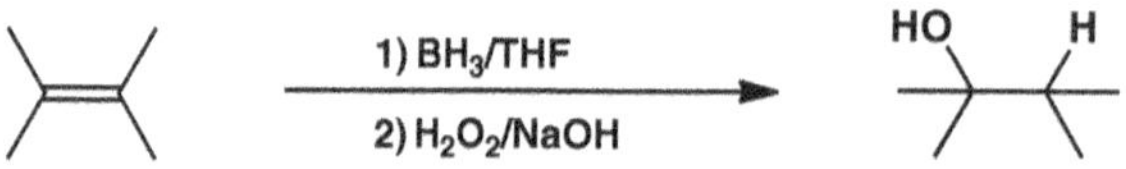

$\xrightarrow[\text{2) H}_2\text{O}_2/\text{NaOH}]{\text{1) BH}_3/\text{THF}}$ Hydroboration–oxidation (anti-Markovnikov)   14.3.3

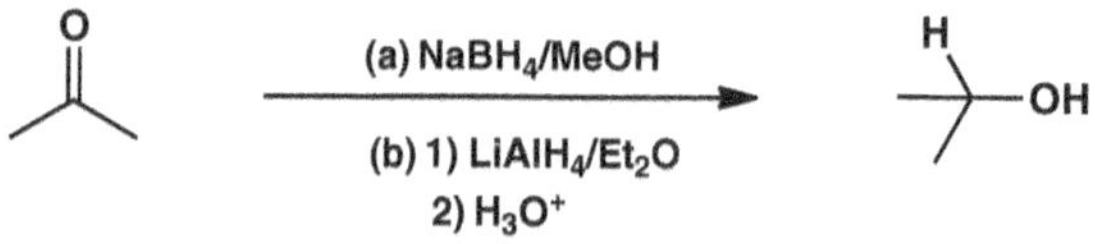

$\xrightarrow[\text{(b) 1) LiAlH}_4/\text{Et}_2\text{O} \\ \text{2) H}_3\text{O}^+]{\text{(a) NaBH}_4/\text{MeOH}}$ Hydride reduction   10.1.1

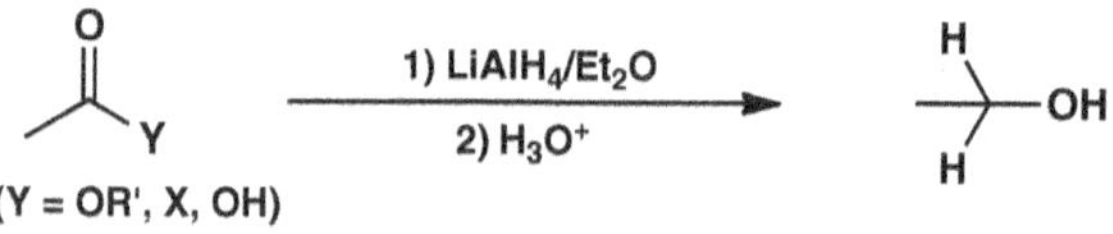

$\xrightarrow[\text{2) H}_3\text{O}^+]{\text{1) LiAlH}_4/\text{Et}_2\text{O}}$ (Y = OR', X, OH)   Hydride reduction   10.1.2

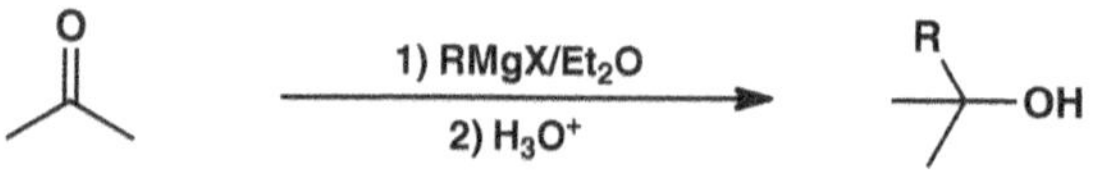

$\xrightarrow[\text{2) H}_3\text{O}^+]{\text{1) RMgX/Et}_2\text{O}}$ Grignard reaction   10.4.2

$\xrightarrow[\text{2) H}_3\text{O}^+]{\text{1) RMgX/Et}_2\text{O}}$ 3° alcohol   Grignard reaction   10.4.2

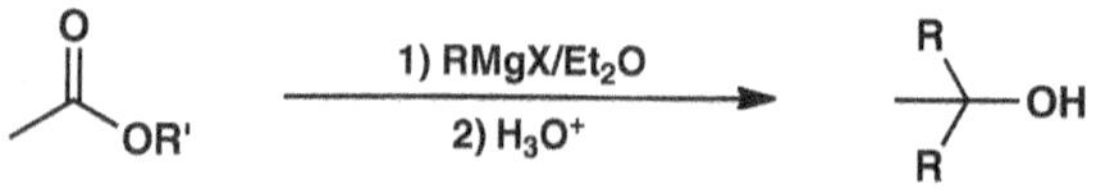

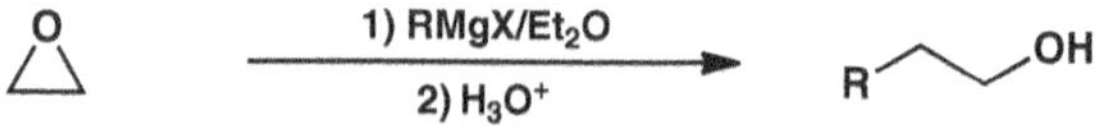

$\xrightarrow[\text{2) H}_3\text{O}^+]{\text{1) RMgX/Et}_2\text{O}}$ Grignard reaction   10.4.2

$\xrightarrow[\text{(b) 1) OsO}_4 \text{ 2) H}_2\text{S}]{\text{(a) KMnO}_4/\text{NaOH}}$ *syn*-diol   Hydroxylation   21.2.4

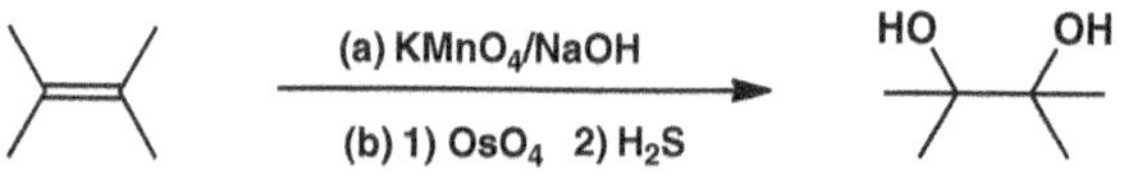

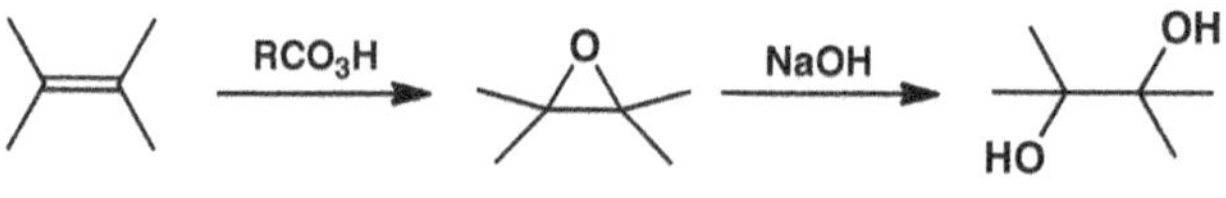

$\xrightarrow{\text{RCO}_3\text{H}}$ $\xrightarrow{\text{NaOH}}$ *anti*-diol   Epoxidation– ring opening   15.5 / 14.5.2

Phenols (Ar–OH)

Ar–X	$\xrightarrow[\Delta]{\text{NaOH}}$	Ar–OH	With EWGs in Ar or benzyne mechanism	18.6 / 18.7

Ar–NH$_2$ $\xrightarrow[\text{H}_3\text{O}^+]{\text{NaNO}_2}$ Ar–N$_2^+$ $\xrightarrow{\text{H}_2\text{O}}$ Ar–OH   Diazotization– substitution   16.6.2 / 18.8

(6) Ethers (**R–O–R**)

$$R-X \xrightarrow{\text{NaOR'}} R-OR' \qquad \text{Williamson synthesis} \quad \boxed{12.2}$$
$$(S_N2)$$

$$2\ R-OH \xrightarrow[\Delta]{\text{H}_2\text{SO}_4} R-O-R \qquad \text{Acid-catalysed substitution} \quad \boxed{14.1.1}$$

Aromatic ethers

$$Ar-X \xrightarrow[\Delta]{\text{NaOR}} Ar-OR \qquad \text{With EWGs in Ar} \quad \boxed{18.6}$$

Epoxides

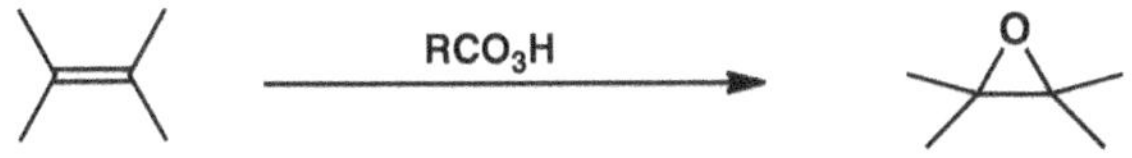

$\xrightarrow{\text{RCO}_3\text{H}}$ Epoxidation $\boxed{15.5}$

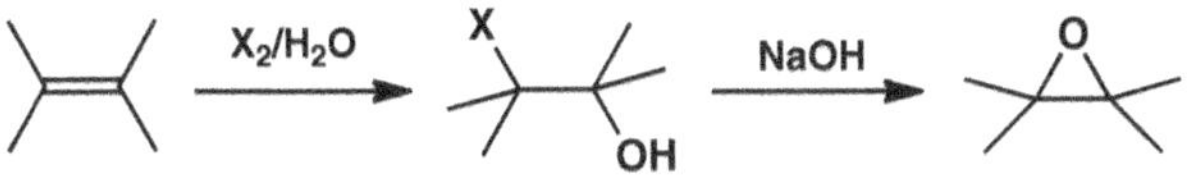

$\xrightarrow{X_2/\text{H}_2\text{O}}$ $\xrightarrow{\text{NaOH}}$ Electrophilic addition intramolecular S_N2 $\boxed{\text{Problem } 15.19}$

(7) Amines (**R–NR₂**)

$$R-X \xrightarrow{\text{NH}_3} \begin{array}{l} R\text{-NH}_2,\ R_2\text{NH}, \\ R_3\text{N},\ R_4\text{N}^+\ X^- \end{array} \qquad S_N2 \quad \boxed{14.7}$$

$$R-X \xrightarrow[\text{3) H}_2\text{O}]{\text{1) NaN}_3 \quad \text{2) LiAlH}_4} R-NH_2 \qquad S_N2,\ \text{reduction} \quad \boxed{\begin{array}{c}12.2\\ \text{Chapt. }10\end{array}}$$

$$R-X \xrightarrow[\text{3) H}_2\text{O}]{\text{1) NaCN \quad 2) LiAlH}_4} R-CH_2-NH_2 \qquad S_N2,\ \text{reduction} \quad \boxed{\begin{array}{c}12.2\\ \text{Chapt. }10\end{array}}$$

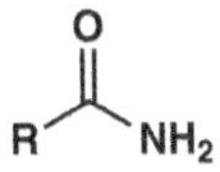

$\xrightarrow{\text{1) LiAlH}_4 \quad \text{2) H}_2\text{O}}$ $R-CH_2-NH_2$ Hydride reduction $\boxed{10.1.2}$

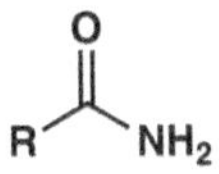

$\xrightarrow{\text{1) Br}_2/\text{NaOH \quad 2) H}_2\text{O}}$ $R-NH_2$ Hofmann rearrangement $\boxed{22.5.2}$

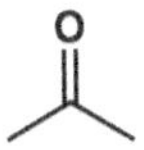

$\xrightarrow{\text{RNH}_2/\text{NaBH}_3\text{CN}}$ Reductive amination $\boxed{10.2.1}$

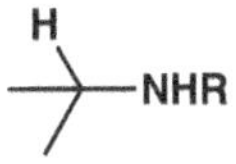

Arylamines (**Ar–NH₂**)

$$Ar-NO_2 \xrightarrow{\text{Sn/HCl}} Ar-NH_2 \qquad \text{Reduction} \quad \boxed{16.6}$$

$$Ar-X \xrightarrow{\text{NaNH}_2\ /\ \text{liq. NH}_3} Ar-NH_2 \qquad \text{Benzyne mechanism} \quad \boxed{18.7}$$

(8) Nitriles (R–CN)

R—X	$\xrightarrow{\text{NaCN}}$	R—CN	S$_N$2 reaction	12.2

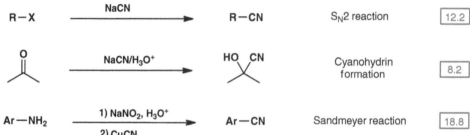

acetone $\xrightarrow{\text{NaCN/H}_3\text{O}^+}$ cyanohydrin (HO, CN) — Cyanohydrin formation — 8.2

Ar—NH$_2$ $\xrightarrow[\text{2) CuCN}]{\text{1) NaNO}_2,\ \text{H}_3\text{O}^+}$ Ar—CN — Sandmeyer reaction — 18.8

(9) Aldehydes (R–C–H) with O double bond on C

R–CH$_2$OH $\xrightarrow[\text{(b) (COCl)}_2/\text{DMSO/Et}_3\text{N}]{\text{(a) pyridinium} \overset{+}{\text{N}}\text{H CrO}_3\text{Cl}^-}$ R–CHO
(a) PCC oxidation
(b) Swern oxidation — 14.4

R–C(=O)–Y $\xrightarrow{\text{1) DIBAL \quad 2) H}_3\text{O}^+}$ R–CHO — Reduction — 10.1.2
(Y = OEt, NMe$_2$)   [DIBAL = Al(CH$_2$CHMe$_2$)$_2$H]

H–C(=O)–NMe$_2$ $\xrightarrow[\text{2) H}_3\text{O}^+]{\text{1) RMgX/Et}_2\text{O}}$ R–CHO — Grignard reaction — 10.4.2

alkene $\xrightarrow{\text{1) O}_3 \quad \text{2) Zn/AcOH}}$ 2 CH$_3$CHO — Ozonolysis–reductive workup — 21.2.3

R—≡—H $\xrightarrow[\text{2) H}_2\text{O}_2/\text{NaOH}]{\text{1) BH}_3/\text{THF}}$ RCH$_2$–CHO — Hydroboration–oxidation — 15.3.4

(10) Ketones (R–C–R)

H, OH on R–CH(R')	$\xrightarrow[\text{H}_2\text{SO}_4/\text{H}_2\text{O}]{\text{Na}_2\text{Cr}_2\text{O}_7}$	R–C(=O)–R'	Oxidation	14.4
R–C(=O)–NMe$_2$	$\xrightarrow[\text{2) H}_3\text{O}^+]{\text{1) R'MgX/Et}_2\text{O}}$	R–C(=O)–R'	Grignard reaction	10.4.2
(CH$_3$)$_2$C=C(CH$_3$)$_2$	$\xrightarrow{\text{1) O}_3 \quad \text{2) Zn/AcOH}}$	2 CH$_3$–C(=O)–CH$_3$	Ozonolysis–reductive workup	21.2.3
R–C≡C–H	$\xrightarrow{\text{Hg(OAc)}_2/\text{H}_2\text{SO}_4/\text{H}_2\text{O}}$	R–C(=O)–CH$_3$	Oxymercuration	15.3.4
R–X + CH$_3$–C(=O)–CH$_2$–C(=O)–OEt	$\xrightarrow[\text{2) H}_3\text{O}^+ \quad \text{3) }\Delta]{\text{1) NaOEt/EtOH}}$	CH$_3$–C(=O)–CH$_2$R	Acetoacetate synthesis	17.8.2
Ar–H	$\xrightarrow{\text{RCCl / AlCl}_3}$	Ar–C(=O)–R	Friedel–Crafts acylation	16.3.5

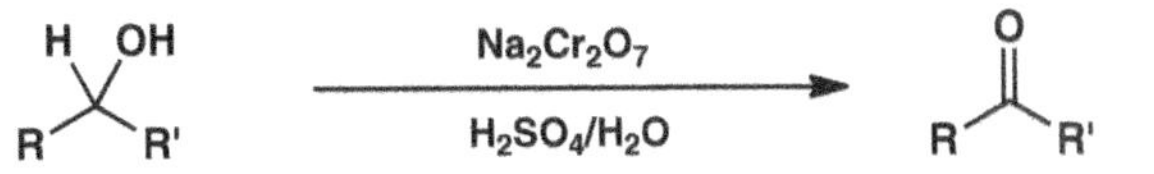

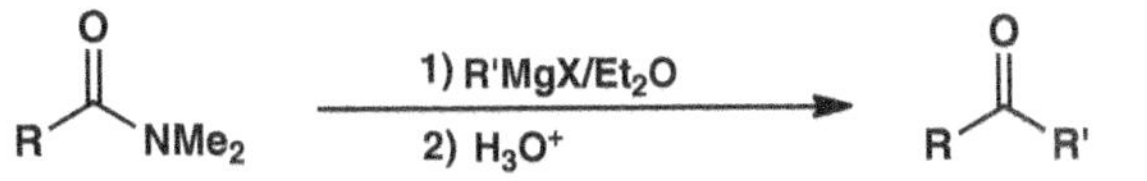

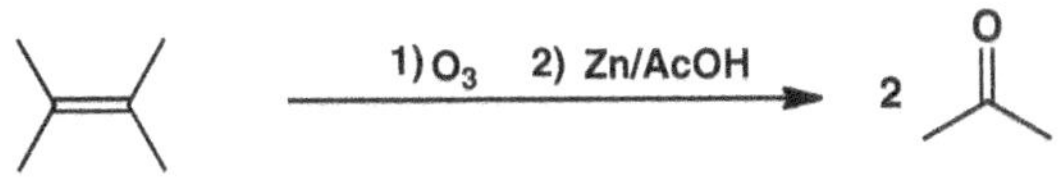

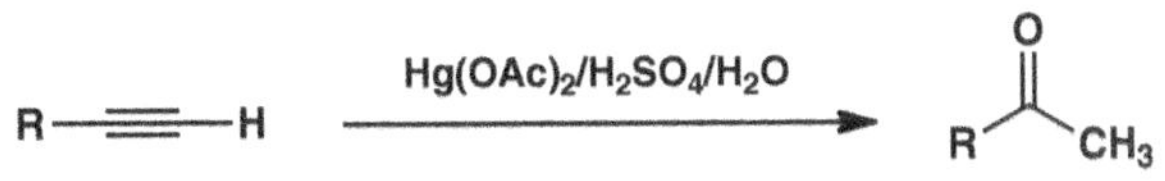

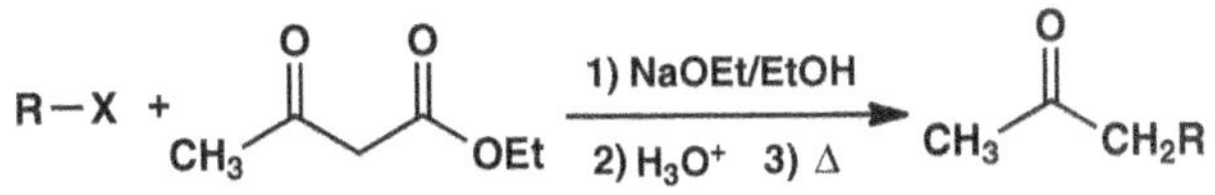

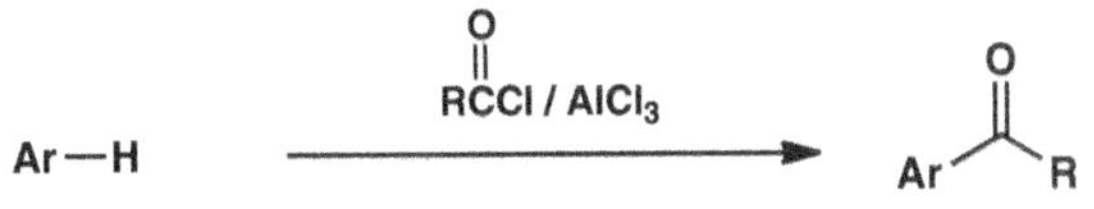

(11) Carboxylic acids (R–COH)

R–CH$_2$–OH (R–C(=O)–H)	$\xrightarrow[\text{H}_2\text{SO}_4/\text{H}_2\text{O}]{\text{Na}_2\text{Cr}_2\text{O}_7}$	R–C(=O)–OH	Oxidation	14.4
R–X $\xrightarrow{\text{Mg/Et}_2\text{O}}$ R–MgX	$\xrightarrow[\text{2) H}_3\text{O}^+]{\text{1) CO}_2}$	R–CO$_2$H	Grignard reaction	10.4.2
R–CN	$\xrightarrow[\text{(b) H}_2\text{SO}_4/\text{H}_2\text{O}]{\text{(a) NaOH/H}_2\text{O}}$	R–CO$_2$H	Hydrolysis	Chapt. 9
R–C(=O)–CH$_3$	$\xrightarrow{\text{1) X}_2/\text{NaOH} \quad \text{2) H}_3\text{O}^+}$	R–C(=O)–OH	Haloform reaction	17.4.2
R–X + EtO–C(=O)–CH$_2$–C(=O)–OEt	$\xrightarrow[\text{2) H}_3\text{O}^+ \quad \text{3) }\Delta]{\text{1) NaOEt/EtOH}}$	R–CH$_2$CO$_2$H	Malonate synthesis	17.8.2
Ar–CHR$_2$	$\xrightarrow[\text{(b) Na}_2\text{Cr}_2\text{O}_7/\text{H}_2\text{SO}_4/\text{H}_2\text{O}]{\text{(a) 1) KMnO}_4/\text{OH}^-/\text{H}_2\text{O} \quad \text{2) H}_3\text{O}^+}$	Ar–CO$_2$H	Oxidation	16.7.3

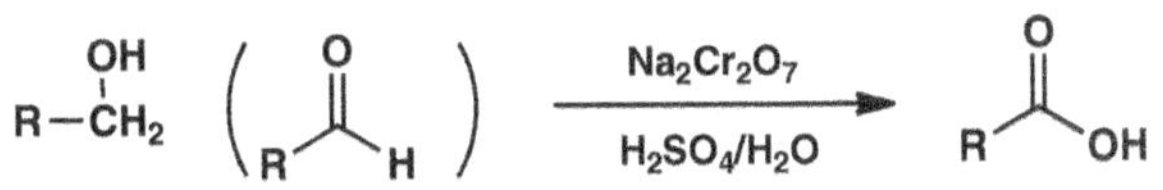

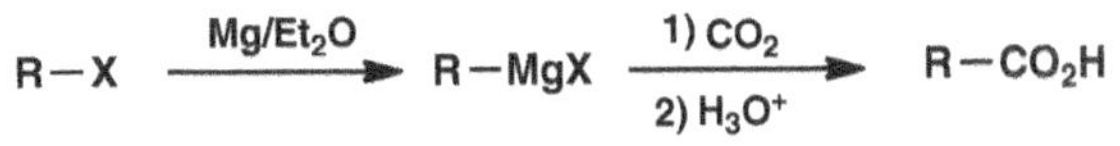

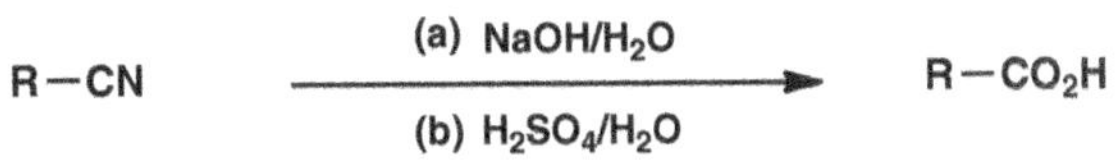

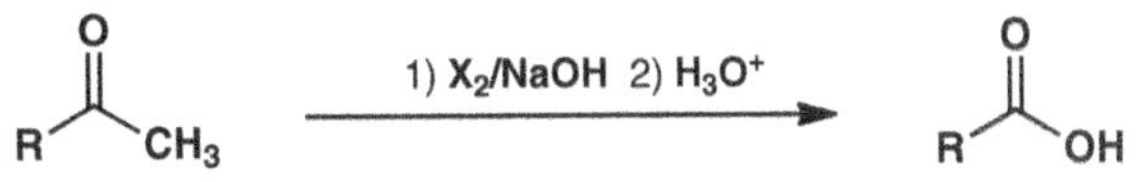

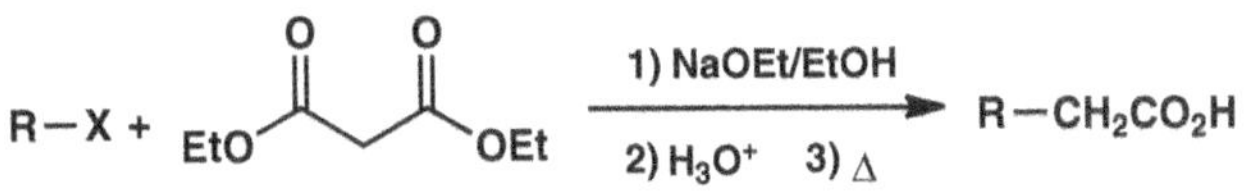

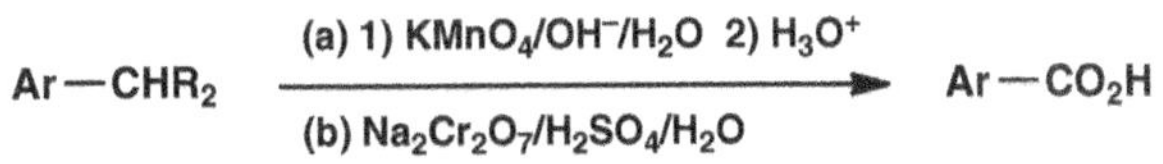

(12) Carboxylic acid derivatives ($R-\overset{\overset{\displaystyle O}{\|}}{C}-Y$)

$R-\overset{\overset{\displaystyle O}{\|}}{C}-OH$ $\xrightarrow{\text{SOCl}_2}$ $R-\overset{\overset{\displaystyle O}{\|}}{C}-Cl$		Halogenation	9.5.1
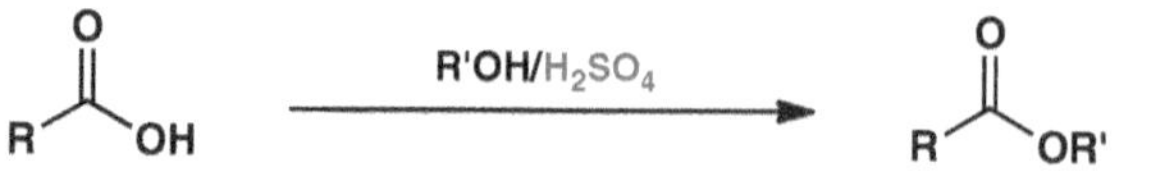 $R-\overset{\overset{\displaystyle O}{\|}}{C}-OH$ $\xrightarrow{\text{R'OH/H}_2\text{SO}_4}$ $R-\overset{\overset{\displaystyle O}{\|}}{C}-OR'$		Fischer esterification	9.2.3
$R-\overset{\overset{\displaystyle O}{\|}}{C}-Cl$ $\xrightarrow{\text{YH}}$ $R-\overset{\overset{\displaystyle O}{\|}}{C}-Y$ (Y = R'COO, R'O, R'$_2$N)		Substitution	9.5.1
$R-\overset{\overset{\displaystyle O}{\|}}{C}-R'$ $\xrightarrow{\text{RCO}_3\text{H}}$ $R-\overset{\overset{\displaystyle O}{\|}}{C}-OR'$		Baeyer–Villiger oxidation	22.4.1
$R-\overset{\overset{\displaystyle NOH}{\|}}{C}-R'$ $\xrightarrow{\text{H}_2\text{SO}_4}$ $R'-\overset{\overset{\displaystyle O}{\|}}{C}-NHR$		Beckmann rearrangement	22.4.2

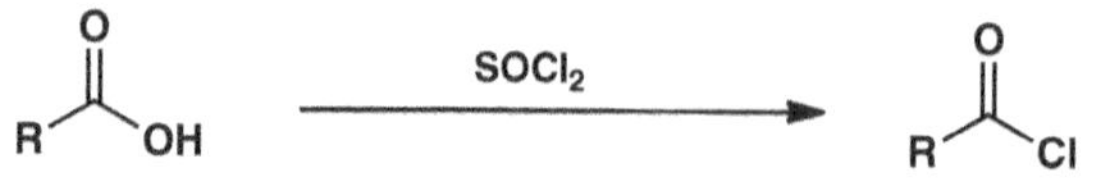

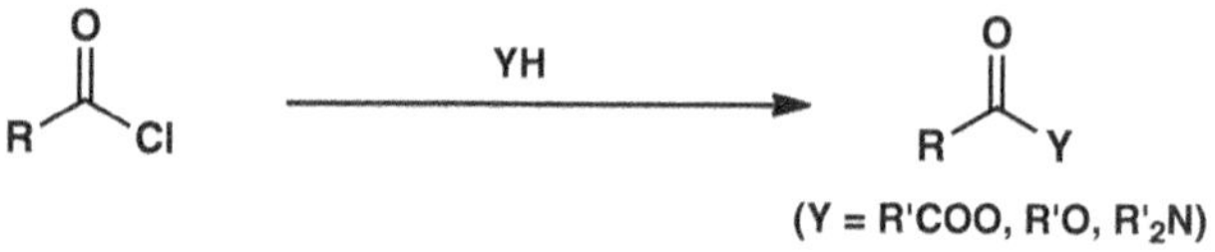

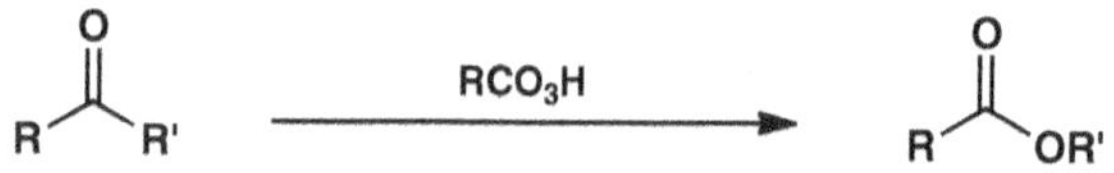

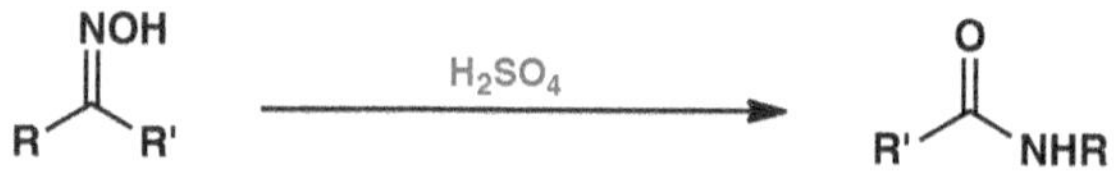

Appendix 4
Reactions for the Formation of Carbon–Carbon Bonds

	Retro-synthesis	Electrophile	Nucleophile	Type of reaction	Relevant section
organometallic compounds as nucleophiles			RMgX	Grignard reaction	10.4.2
			RMgX	Grignard reaction	10.4.2
			RMgX	Grignard reaction	10.4.2
		(—CN)	RMgX	Grignard reaction	10.4.2
		CO_2	RMgX	Grignard reaction	10.4.2
			R_2CuLi	Conjugate addition of organocuprates	18.1.3
enolates and enolate equivalents as nucleophiles			O^-	Aldol reaction	17.5
			O^-	Claisen condensation	17.6
		RX	OLi	Alkylation of lithium enolates	17.9
		RX	N	Alkylation of enamines	17.10.1
		RX	$OSiR_3$ (TiCl$_4$)	Alkylation of enol silyl ethers	17.10.2
		RX	(NaOEt)	Acetoacetate synthesis	17.8.2
		RX	(NaOEt)	Malonate synthesis	17.8.2
			(NaOEt)	Michael reaction	18.4.1

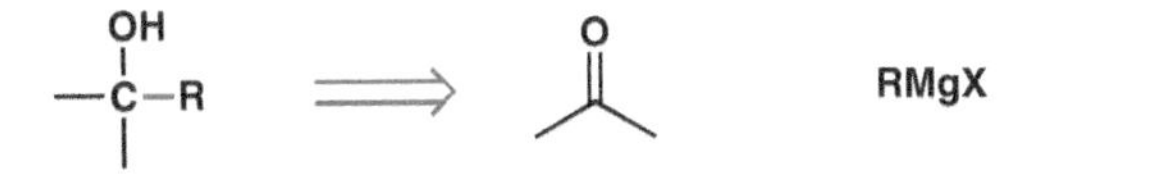

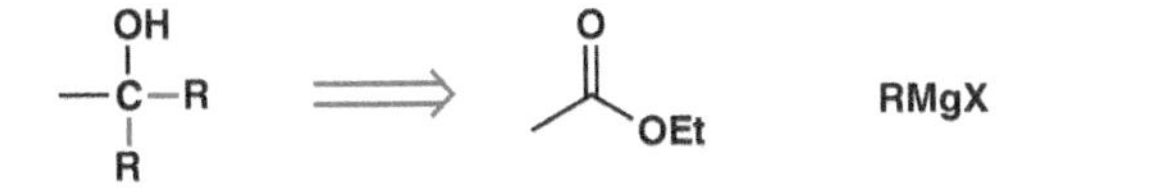

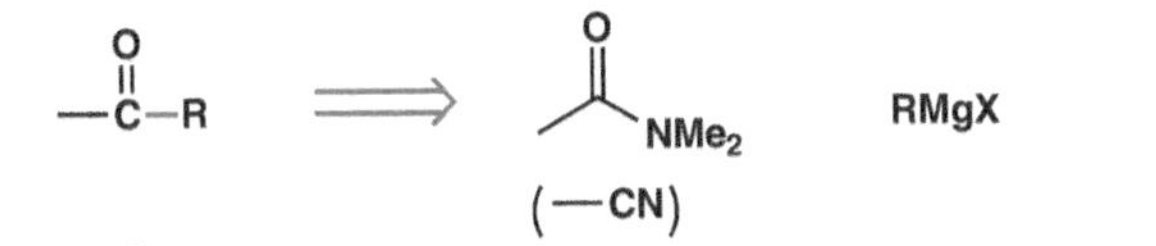

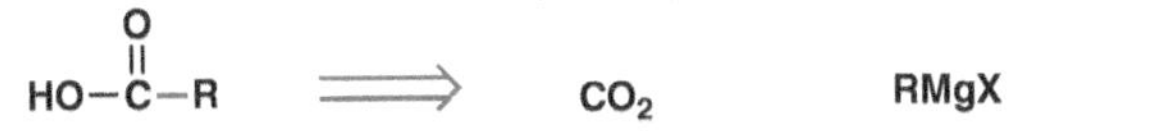

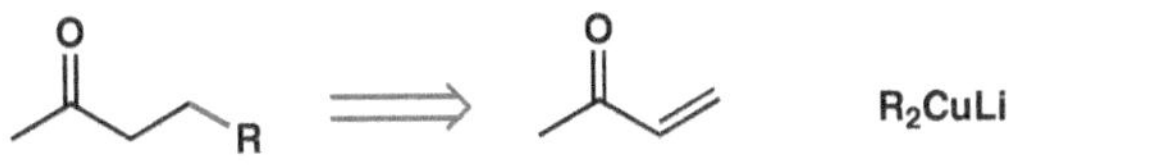

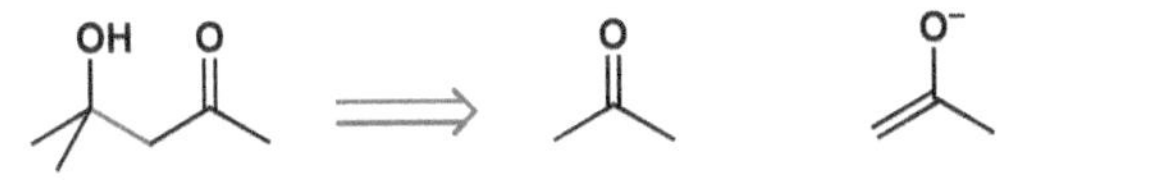

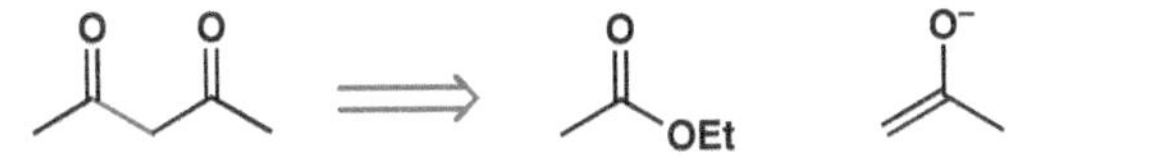

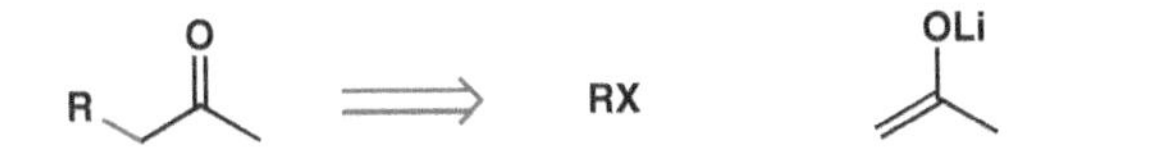

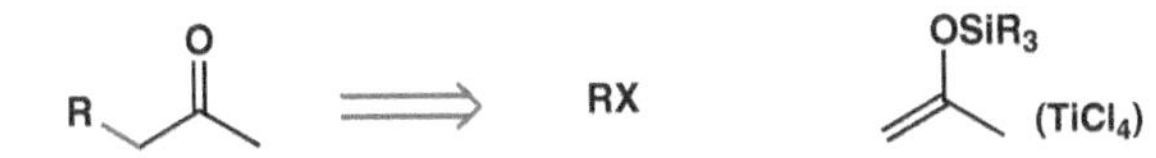

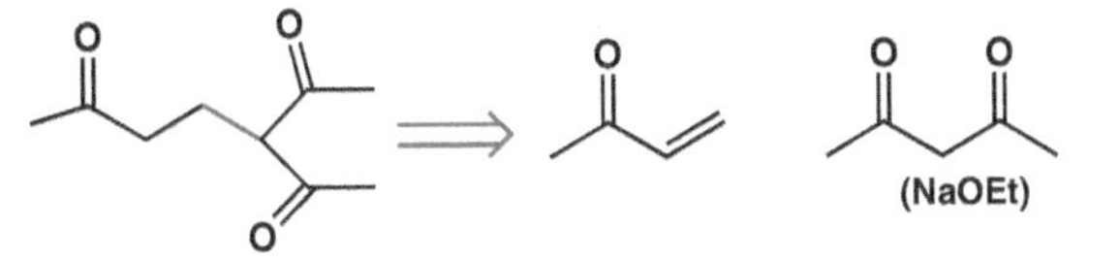

other types of polar reactions

R–C(=O)–	⟹	RX	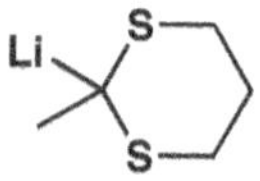 Li, dithiane	Umpolung S_N2 — 23.2.2
RCH_2–≡	⟹	RCH_2X	≡–M^+ (M = Li, MgX, or Na)	S_N2 — 12.2
R–C≡N	⟹	RX	NaCN	S_N2 — 12.2
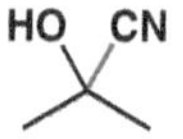 HO, CN	⟹	 ketone	NaCN, HCl	Cyanohydrin — 8.2
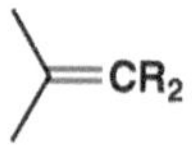 =CR_2	⟹	 ketone	$R_2C=PPh_3$	Wittig reaction — 8.7
R–Ar	⟹	RX (AlX_3)	ArH	Friedel–Crafts reaction — 16.3.4
R–C(=O)–Ar	⟹	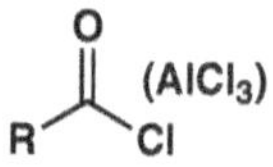 R–C(=O)–Cl ($AlCl_3$)	ArH	Friedel–Crafts reaction — 16.3.5

homolytic and metal-catalysed reactions

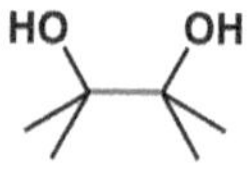 HO, OH (pinacol)	⟹	 acetone	 acetone (Mg)	Pinacol reaction via SET (Mg) — 20.9.2
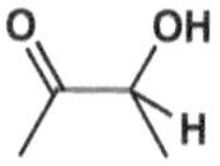 O, OH	⟹	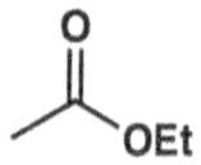 EtO ester	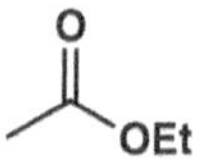 ester (Na) OEt	Acyloin condensation via SET (Na) — 20.9.2
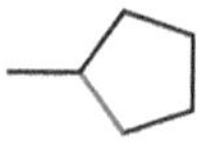 cyclopentane	⟹	 hexenyl bromide Br (Bu_3SnH, AIBN)		Radical cyclization — 20.6.1
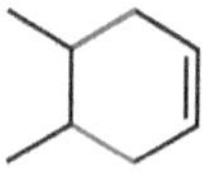 cyclohexene	⟹	 alkene	 diene	Diels–Alder reaction ([4+2] cycloaddition) — 15.9 / 21.2.1
 cyclohexadiene	⟹	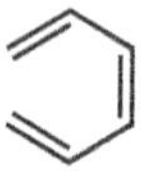 hexatriene		Electrocyclic reaction — 21.3
 cyclopropane	⟹	 propene	$R_2C:$	Addition of carbenes — 15.6
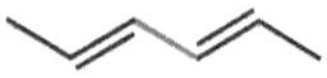 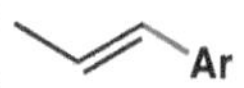 triene (Ar)	⟹	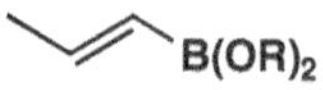 B(OR)$_2$ Br, (ArBr) ($L_2Pd(0)$)	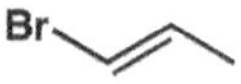	Cross-coupling reactions — Panel 23.1
 cycloalkene	 ⟹	alkenes ([M]=CR_2)		Alkene metathesis — Panel 23.1

Additional Resources

Typical bond lengths of some covalent bonds (pm)							
H–H	74	Csp³–H	109	Csp³–Csp³	154	C=C	133
H–N	101	Csp²–H	108	Csp³–Csp²	150	C=N	135
H–O	96	Csp–H	106	Csp²–Csp²	146	C=O	121
H–F	92	C–F	140	Csp–Csp	138	C≡C	120
H–Cl	127	C–Cl	181	Csp³–N	147	C≡N	116
H–Br	141	C–Br	196	Csp³–O	143		
H–I	161	C–I	217				

Bond dissociation enthalpies (kJ mol⁻¹)[a]				
X	H–X	CH$_3$–X	Ph–X	X–X
H	436	438	463	436
CH$_3$	438	375	432	375
NH$_2$	450	354	426	275
OH	499	386	464	210
F	568	(472)[b]	523	158
Cl	432	349	398	243
Br	366	293	335	193
I	298	238	270	151

$H_2C=CH_2$	654[c]	$H_2C=CH-H$	442[c]
$H_2C=O$	750[c]	$HC≡C-H$	555[c]
$HC≡CH$	965[c]		

a. Calculated from reliable thermochemical results. b. Value at 0 K. c. Values for the coloured bonds.

Symbols and Recommended Values of Some Physical Constants

Avogadro constant	N_A	6.022141×10^{23} mol⁻¹
Gas constant	R	8.31446 J K⁻¹ mol⁻¹
Planck constant	h	6.626070×10^{-34} J s
Boltzmann constant	k_B (R/N_A)	1.380649×10^{-23} J K⁻¹
Velocity of light in a vacuum	c	299 792 458 m s⁻¹

Units Conversions

temperature	0 °C	273.15 K
pressure	1 atm	101 325 Pa
energy	1 cal	4.184 J
	1 eV	96.485 kJ mol⁻¹
length	1Å	10^{-10} m (100 pm)

Symbols of SI Prefixes for Multiplicities and Fractions

tera	T	10^{12}	milli	m	10^{-3}
giga	G	10^{9}	micro	μ	10^{-6}
mega	M	10^{6}	nano	n	10^{-9}
kilo	k	10^{3}	pico	p	10^{-12}
deci	d	10^{-1}	femto	f	10^{-15}
centi	c	10^{-2}			

Fundamental Classes of Reactions and Guidelines for Writing Curly Arrows

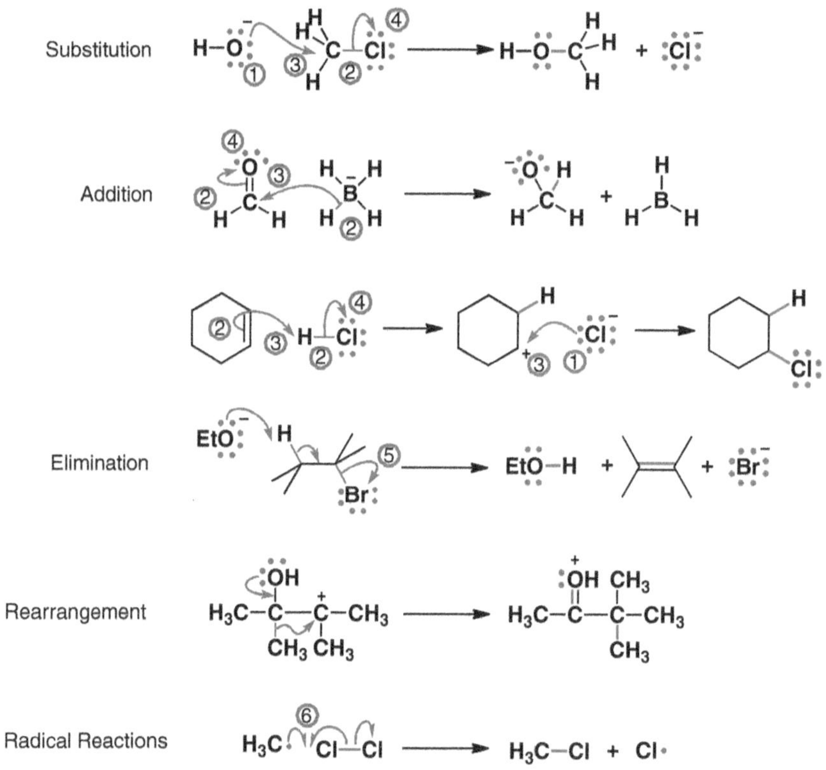

1. Start from a lone pair ① or a bonding pair of electrons ②.
2. Point the arrow *either* towards the atom to which a new bond will form ③, *or* onto the atom which will accept a new lone pair ④.
3. Do not exceed an octet of valence electrons.
4. The total charges (or electrons) must remain unchanged.
5. Electron pairs flow in one direction ⑤.
6. Use curly arrows with a singly barbed head ⑥ for the movement of an unpaired electron.

Index

S